ARCTIC ENERGY RESOURCES

ENERGY RESEARCH

Vol. 1 **Geophyscial Aspects of the Energy Problem** (Rapolla, Keller and Moore, Editors)
Vol. 2 **Arctic Energy Resources** (Rey, Editor)

ENERGY RESEARCH 2

# ARCTIC ENERGY RESOURCES

Proceedings of the Comité Arctique International Conference on Arctic Energy Resources, held at the Veritas Centre, Oslo, Norway, September 22–24, 1982

Edited by

LOUIS REY

*Comité Arctique International, Monte Carlo, Monaco*

with the assistance of

CHRISTIAN BEHRENS

*Det norske Veritas, Oslo, Norway*

This set of papers has been published as a special issue of *Cold Regions Science and Technology,* Volume 7

ELSEVIER
Amsterdam – Oxford – New York 1983

ELSEVIER SCIENCE PUBLISHERS B.V.
Molenwerf 1
P.O. Box 211, 1000 AE Amsterdam, The Netherlands

*Distributors for the United States and Canada:*

ELSEVIER SCIENCE PUBLISHING COMPANY INC.
52, Vanderbilt Avenue
New York, N.Y. 10017

Library of Congress Cataloging in Publication Data

Conference on Arctic Energy Resources (4th : 1982 :
Veritas Centre)
Arctic energy resources.

(Energy research ; 2)
"Fourth of a series of meetings organised by Comité
Arctique International"--
Bibliography: p.
Includes index.
1. Power resources--Arctic regions--Congresses.
I. Rey, Louis René, 1931- . II. Behrens, Christian.
III. Comité Arctique International. IV. Title.
V. Series.
TJ163.15.C653 1982 333.8'2'09113 83-11629
ISBN 0-444-42218-8 (U.S.)

ISBN 0-444-42218-8 (Vol. 2)
ISBN 0-444-41685-4 (Series)



Printed in The Netherlands

COMITÉ ARCTIQUE

INTERNATIONAL

This conference was the fourth of a series of meetings organised by *Comité Arctique International* created in Monaco in 1979 with the active support of H.S.H. Prince Rainier III. The objectives of the Committee are to improve knowledge and understanding of Arctic areas and to that end to promote research in different fields on an international and multi-disciplinary basis. Composed of experts from many countries concerned with polar activities and acting in a private capacity, *Comité Arctique International* works as a non-political group with a view to enhancing and sustaining a harmonious, well-balanced development of Arctic regions in close cooperation with existing relevant national and international authorities and organisations.

It is formally established in the Principality of Monaco.

DET NORSKE
**VERITAS**

*P.O.Box 300, N-1322 Høvik, Norway*

Veritas is an independent foundation working for safety of life, environment and property. Veritas' main activities are classification of ships and mobile offshore units, certification of offshore structures and systems, and quality assurance services towards most high risk industries in today's society. All activities are supported by an extensive and comprehensive research and development effort.

# Contents

## SECTION I
### Opening Session



## SECTION II
### Occurrence of Energy Resources



## SECTION III
### Technological/Economic Aspects of Exploration/ Exploitation of Arctic Energy Resources

**REVIEW NOTICE**

The papers have been reviewed by the Editors and approved for publication. Approval does not necessarily signify that the contents reflect the views and policies of the Comité Arctique International.

# The Contributors


Terence Armstrong, Dr., Acting Director,
Scott Polar Research Institute
Univ. of Cambridge
Cambridge CB2 1ER
England

Richard Bailey, Manager of Planning
Petro Canada
P.O.Box 6530 Stn.D.
Calgary, Alberta
Canada T2P 1H4

Egil Bergsager, Deputy Director
Norwegian Petroleum Directorate
P.O.Box 600
4001 Stavanger
Norge

Torkild Carstens, Head of Research
The Norwegian Hydrodynamic Laboratories
P.O.Box 4118
Valentinlyst
7001 Trondheim
Norge

Svein Fjeld, Dr. Senior Principal Surveyor
Det norske Veritas
P.O.Box 300
1322-Høvik
Norge

Lindsay J. Franklin, Vice President, Operation
Panarctic Oils Ltd.
P.O.Box 190
Calgary, Alberta
Canada T2P 2H6

Tore Gjelsvik, Dr., Managing Dir.,
Norwegian Polar Research Institute
Rolfstangveien 12
1330 Oslo Lufthavn
Norge

Franklyn Griffiths, Dr., Professor
Dept. of Political Science
University of Toronto
100 St. George Street
Toronto, Ontario
Canada M5S 1A1

Donn K. Haglund, Dr., Professor
Dept. of Geography
University of Wisconsin
Milwaukee Wisconsin 53201
USA

John Hnatiuk, Manager Frontier
Environmental Engineering
Gulf Canada Resources Inc.
Box 130
Calgary, Alberta
Canada T2P 2H7

Richard Hoos, Director, Environmental
Assessment
Dome Petroleum Ltd.,
P.O.Box 200 Calgary, Alberta
Canada T2P 2H8

Howard Hume, Manager, Environmental
Protection Group
Petro Canada
P.O.Box 2844 Calgary, Alberta
Canada T2P 3E3

Lawrence R. Jacobsen, Program Manager
General Dynamics
97 East Howard Street
Quincy, Massachusetts 02169
USA

Olavi M. Kaustinen, Vice President, Engineering
Polar Gas Project
P.O.Box 90 Commerce Court West
Toronto, Ontario
Canada M5L 1H3

Jeffery Larminie, General Manager,
Environmental Control Centre
British Petroleum Company
Britannic House — Moore Lane
London, EC2Y 9BU
UK

Arthur A. Meyerhoff, Dr., Vice President,
Associated Resource Consultants Inc.
P.O.Box 4602
Tulsa, Oklahoma 74159
USA

James J. Murphy, Vice President, Marketing
General Dynamics
97 East Howard Street
Quincy, Massachusetts 02169
USA

Walter W. Nassichuk, Director
Institute of Sedimentary & Petroleum Geology
Geological Survey of Canada
3303-33rd St. N.W.Calgary, Alberta
Canada. T2L 2A7

Alv Orheim, Chief Geologist
Store Norske Spitsbergen Kulkompani A/S
9170 Longyearbyen
Svalbard

Louis Rey, Dr.
Professor, President
Comité Arctique International
16, Boulevard de Suisse
MC-98000 Monte Carlo
Principauté de Monaco

Robert E. Smith, Civil Eng., Director
ARCO Oil and Gas Co.
P.O.Box 2819
Dallas, Texas 75221
USA

Finn Sollie, Dr.
Fridtjof Nansens Foundation
Fridtjof Nansensvei 17
1324 Lysaker
Norge

Erik Sprunk-Jansen, Managing Director
Greenex A/S
Landmarket 8
DK-1119 København
Danmark

Patrick Taylor, Dr.
Geophysicist,
NASA
Goddard Space Flight Centre
Code 922 NASA/GSFC
Greenbelt MD 20771
USA

Hellmut Wilckens, Manager Offshore
Projects & Production Development
Thyssen Nordseewerke Gmbh
P.O.Box 2351
2970 Emden
West-Germany

# Section I

## *Opening Session*

# Opening of the conference

*H.R.H. Crown Prins Harald of Norway*

I am very pleased to be invited to open this conference on Arctic Energy Resources. The conference has been initiated by Comité Arctique International, which has chosen Norway as the host country for this event.

We certainly appreciate this choice. Norway has a long tradition in exploring the Arctic region and in utilizing its resources. For hundreds of years, Norwegians have sailed north to fish and to hunt, or have endeavoured to discover shorter maritime passages to other continents. Fridtjof Nansen's «Fram» expedition in 1893 and Roald Amundsen's voyage on «Gjøa» through the North-West Passage in 1903 are well known examples of early Norwegian achievements in arctic research.

In recent times, we have exploited the energy resources of Svalbard and our offshore oil and gas exploration and development are approaching Arctic areas. Accordingly, it is of great importance that we develop our nation's skill in arctic technology and gain a broader understanding of the processes that take place in these areas that are characterized by extreme vulnerability to human interference and by tough and often unpredictable physical conditions.

When developing the resources of the Arctic, we should keep in mind the somewhat unique life conditions which prevail there and adapt both technology and human activity to this very sensitive environment. Any activity should be carried out in an environmentally and socially responsible manner. I note with great pleasure that the programme of this conference reflects such issues.

The arctic areas are gaining increased international importance. The challenge of mastering the harsh environment should therefore call for a broad international cooperation and a common understanding of the problems involved. I think that a conference like this will contribute hereto.

In conclusion, I wish the Comité Arctique International, Det norske Veritas and all participants good luck with the conference.

I hereby declare the conference opened.

*Cold Regions Science and Technology*, 7 (1983) 5—10
Elsevier Science Publishers B.V., Amsterdam — Printed in The Netherlands

# The Role of the Arctic Regions in the Energy Challenge*

*Louis REY,*
Comité Arctique International
Monaco

Energy harnessing is as old as mankind and, already, in the Sinanthrop'us diggings, ashes were found mixed with stone implements. The magic and powerful fire, provided from time to time by Nature, was, then, carefully secured and maintained by wandering tribes who did not know, yet, how to strike flint or develop burning embers at the end of a rotating rod. For many millennia, and until very recent historical times, wood and peat warmed huts, houses and castles and the flickering flame of oil lamps dissolved the gloomy nights and chased away the devils. Winds and rivers were used to power mills and sail ships in Persia and in the Egypt of Pharaoh, and the tides drove the grinding wheels in Brittany.

Coal, mined since Antiquity in ancient China, came into use in Europe in the early Middle Ages. Petroleum, the dark blood of Modern Society, collected from natural seepages in the Far East and in Syria, allowed the Byzantine scientists in the 7th century to develop the «Greek Fire» which burnt so many ships and saved Constantinople several times.

## A Difficult Issue: The Energy Challenge

Human societies at that time were evolving at a slow pace and the described situation could have lasted were it not for the striking outburst of sciences which, from the end of the 18th century till the present times, boosted forward the Western World into the Technological Age.

In 1769 *Watt* designed the steam-engine. The production of electricity, reported by *Volta* in 1800 and demonstrated by *Gramm* in 1873, was first achieved by the California Electric Co. (1879) using steam powered dynamos, and quickly followed by hydro-electricity pioneered in the 1880's by *Berges*. Oil discovered in Pennsylvania by *Drake*, in 1859, was soon extracted from numerous fields in Baku, Rumania, Burma and Indonesia. In 1903, William *Knox D'Arcy*, sponsored the drilling of the first production wells in Persia and created, six years later, the Anglo-Persian Oil Company, the ancestor of British Petroleum. We had then entered into the 20th century and the following decades demonstrated an unbelievable explosion of technology with its derived industrial achievements based upon striking advances in fundamental and applied sciences. Numerous landmarks spot this evolution and it would be impossible to quote them all; however, a special mention should be given to the glamorous, but questionable newcomer in the energy field: Nuclear Power. From classical fission to fast breeder reactors it has known a major development in the last twenty years, but it still aims at achieving the more advanced goal of nuclear fusion — somewhere in the beginning of the years 2000. Despite its obvious success, nuclear power has also triggered a deep environmental concern, though often based upon purely emotional and political issues. Curiously, this has resulted in a marked interest in the so called «soft, natural, renewable energies», such as: wind, waves, solar, geothermy, and biomass.

* Oslo Conference Introductory lecture.

0165-232X/83/$03.00 

In view of this spectacular evolution and until very recently, (roughly twenty years ago) the world energy situation looked fine: energy was plentiful and cheap and, in the wake of scientific discoveries, industry enjoyed a fast growth, whilst more and more new markets were eager to join the «western consumption club» and share the wealth of what was deemed to be an advanced form of civilization.

Then, quite suddenly, the whole pattern changed: Petroleum prices shot up abruptly in successive waves; political unstability resulted in a substantial decrease in world trade; the international monetary system, placed under high pressure in this changing economic order, had to face major disorders whilst the rising energy prices provoked a sharp increase in manufacturing, transportation and distribution costs for most industrial products and processed goods. As a consequence, the consumption rates dropped: the post-war Golden Age was now over and the gloomy spectrum of recession, unemployment and social unrest emerged in most places.

At the same time, a growing disparity developed between technologically advanced countries and the Third World, and it became an element of global unstability that neither the U.N. System nor the North-South Conferences were able to exert adequately control. As an example, the Swedish population of eight millions consumes more energy than half a billion Indians; and the United States, with seven tons per capita per annum, absorb one third of the world's available energy for less than six per cent of its population. Even worse, should the past average growth rate continue within the next 25 years, world energy consumption would no longer be covered by production and a large deficit would be met for petroleum products. For the first time in history, mankind discovered that the world resources were limited and that irresponsible plunder would deplete them irretrievably. Finally, it became clear that the ever-growing consumption of «dirty fuels» polluted the environment, seriously damaged the ecosystems and endangered human health. The «Quality of Life» appeared, then, as more important for the world citizen than the «Standard of Living» and serious doubts were cast on the adequacy of advanced technology, resulting into a marked loss of confidence in policy makers and scientists.

The world, though reacting as one single great organism, was, nevertheless, unable to maintain the cohesion between its different elements, which were drifting apart in an uncontrolled way. And, on the verge of a new millenium, it entered into what is probably one of the most severe crisis in human history. Here we are today, and, if the situation has not yet reached a cataclysmic point, it is serious enough to request an immediate merging of all available forces to challenge one of its key components: Energy.

## Harsh and Sensitive Arctic

This is precisely where the Arctic Regions come in, though, for many reasons, they should have been kept far away from the turbulences of mid-latitude policies. Indeed, as we will see, the Boreal Lands constitute a world of its own with its distinctive geographical, meteorological and environmental characters and its highly specialized cultures going back into prehistoric times. Placed far away from the «omphalos» of the western civilizations, and despite very early post-glacial human settlements up to the very rim of the polar sea, the Arctic Regions have been essentially ignored for centuries and our knowledge on their exact structure is a fairly recent acquisition, dating back less than fifty years. The essential feature of that structure is the presence, on the world axis, of the Arctic Ocean, the «Polar Mediterranean». Essentially connected to the North Atlantic through the deep trench of the Fram Strait, it can only be trans-navigated by nuclear propelled submarines or ice-breakers.

Despite the fertile nature of the adjacent ice-free northern seas, the Arctic Regions are fairly sensitive because of the severe constraints imposed by the climate and it is quite easy to understand that any additional load on the every-day polar stress might inbalance highly delicate equilibria and damage ecosystems. This is why the accidental or intentional release of foreign organics or chemicals in the Arctic Regions, as well as any other operation which might disturb the «normal» biological and human rhythms, might very well provoke large, unforecastable, permanent scars. In arctic conditions there are no such things as sub-toxic levels or unsignificant interactions, since the limits of detection and action are definitely bound to be lowered because of the synergistic effect of the climate.

This equally applies to the human societies

settled in the circumarctic world, and no one should ever underrate or neglect the potential negative impacts of newly introduced technologies on the elaborate socio-cultural patterns that they have painfully developed over thousands of years.

## The Arctic Eldorado

It is in this context only that we can think upon the contribution of Arctic Regions to the Energy Challenge, and the difficult issue is precisely that: they are susceptible in rendering a major contribution. Indeed all around the Pole lie enormous deposits of fossil fuels, rare metals, uranium ... and many fields extend even further northwards, offshore, under the continental platforms of the coastal seas. At the same time, large northern rivers, powerful coastal currents, deep bays and fjords, wind beaten barren lands and huge circumpolar forests offer a wide range of replenishable resources; whilst, in some selected areas, additional possibilities exist to recover geothermal energy, thanks to local volcanism and associated high thermal gradients.

Everyone is familiar with the coal mines in Svalbard and aware of the importance of the Prudhoe Bay oil and gas fields in Alaska as well as of the striking development of the lower Ob' basin in northern USSR. The Canadian arctic archipelago has been found extremely rich in natural gas near Melville Island, and large fertile structures have been recently identified in the deep formations which extend offshore eastwards of Labrador and New Foundland. The Davis Strait, Ellesmere Island, Northern and Eastern Greenland, the Barents, Kara, Laptev, East-Siberia and Chukchi Seas, alike the Barrow Sea, hold huge potential hydrocarbon deposits and some specialists claim that more than 50 per cent of the world's petroleum reserves lie north of the Arctic Circle.

Undoubtedly the Arctic Regions are, in the Energy Challenge, the modern Eldorado that could balance and even surpass the traditional Middle-East fields, prevent progressive throttling of Western economies and still preserve the European and American reservoirs from rapid depletion.

The same situation is found with mineral ores, and to quote a few examples: on the lower course of River Yenisey, Dudinka-Noril'sk has become the Nickel City; and in Alaska, the molybdenum deposit found near Ketchikan is the second largest in the world. The fabulous Siberian lands conceal many other mineral resources, and a systematic prospection and inventory of the millions of square kilometres of Canadian and Alaskan northern wastes have only started recently.

It is, thus, quite logical that both the Western and Eastern Worlds demonstrate high interest in their northern lands, irrespective of obvious strategic considerations, and keep a vigilant and careful eye on the entire Arctic basin. Moreover it is no wonder that latecomers and not the least the circumpolar populations themselves wish to keep a close eye on all arctic developments and that the latter are quite determined to fight for what they consider as their aboriginal rights to monitor and control the main resources of their home lands. Tens of thousands of people are involved — millions in the USSR. Billions of dollars are at stake, and the whole game is played thousands of kilometres from the main expertise centres, far away from the supporting facilities and out of reach of most decision makers.

Remote, difficult, hostile to man, still wild and poorly understood, but holding formidable resources and potential to bring long-lasting wealth to all partners, the Arctic, today, is both the «Last Frontier» for wilderness and the largest world theatre to deploy advanced modern technology, a Land of Paradox, Misery and Promise, which can become a wonderland or a grave.

## The Industrial Development of the North

A harmonious, balanced development of the North in light of its industrialization is certainly, nowadays, our major challenge and, because of its multifocal character, it is very uneasy to control.

Operations in arctic areas are extremely difficult and often one order of magnitude more expensive than in the temperate zone. Erecting large structures or mining on delicate and unstable permafrost, drilling in the offshore pack-ice, supporting active sites in completely isolated areas, in spite of the polar night, the blizzard, whirling snow and radio black out, require a tremendous amount of effort, exceptional skill and long arctic experience.

Transportation is even more difficult to achieve should large amounts of raw materials be shipped over long distances the whole year round. Freighters can operate only for a short period unless they are ice-worthy and their route is opened by ice-breakers, as is currently the case along Northern Siberia. Laying pipelines is often the best solution; but, here again, the costs involved rocket to unbelievable levels should the line cross mountains, rivers, stand on the permafrost, resist earthquakes and not bar the migration routes of the wandering caribous. Developing mammoth ice-breaking tankers or LNG carriers could offer an interesting alternative; but, unlike the Northern Siberian Route, the North American archipelago channels offer solid pack-ice, the year round, and its continuous dislocation by regular tankers' routes in this special location raise environmental problems which are still unsolved, and can only be overcome by submarine tankers.

The building of support facilities, power plants, accommodation blocks, radio stations, airfields, repair and maintenance units on shipping lanes, road and railway networks, all present delicate environmental issues which, even when they are under strict control, still induce many emotional reactions within ecologists and politicians. Environmental concern is thus becoming the major element in any project discussion and, knowing the particular sensitivity of the Arctic, this looks logical. Accordingly, long preliminary studies are compulsory so that appropriate environmental base-lines and time-zero situations are well established in order to be able to assess, later on, through a continuous monitoring programme, the potential impact on the local ecosystems of the newly introduced technologies.

All this takes time, costs vast amounts of money and dampers somehow the enthusiasm of entrepreneurs for arctic resources. The result is, often, a mere postponement or, even, a cancellation of the original projects, as was the case, in May 1982, for the Trans-Alaska gasline.

Arctic development has definitely become a big business, and it can only be implemented if it is properly rated against conventional operations in classical sites.

However, technical, environmental and financial considerations are not the only ones and arctic development faces also major political and sociocultural issues. Actually, many local populations are, indeed, strongly hostile to massive investments in the North because they fear, with some good reasons, that this will accelerate the decay of their traditional culture. Indeed, they are afraid to see their small groups being drowned into the uncontrolled flux of southern newcomers and they are quite certain that the abrupt wealth that royalties, and job opportunities can give to them, would not be enough to balance the potential loss of their socio-cultural patterns. In those conditions, some of them are even asking for an immediate moratorium as they see no point in dilapidating, today, highly precious non renewable resources which could bring them an optimum wealth, later on, when they reach a more advanced stage of their progressive and cautious «opening» to the western mode of civilization. As a matter of compensation, they place a main emphasis on the replenishable resources provided by tidal energy, hydroelectricity, wood, bio-mass, fish and game and geothermy.

Conversely, western investors are not too keen to pursue their ventures in those difficult conditions and are often reluctant to carry on large operations in an unsteady political climate. They also feel dubious, in the present state, about the validity of the replenishable resources approach, as they do not always have the appropriate solutions to match local situations, despite the existence of some well documented propositions. It took, for instance, a long time to project and implement Churchill Falls in Labrador, which delivers more than 5,200 MW of electric power, and the Bay of Fundy Complex is still under hard discussion. Very elegant plans have been proposed for hydroelectric plants in Greenland where close to 10.000 GWh could be produced annually using the quick fall of the ice-cap melt-waters into the adjacent fjords. In Alaska alone, more than 70 sites have been identified on the main rivers which could support hydroelectric plants of 2,500 MW or greater. And, altogether, they would be susceptible to deliver a clean, replenishable energy of some 175,000 MW, equivalent, annually, to 180 million tons of oil, roughly 1.5 times the yearly production of the Prudhoe Bay Fields.

A recent study has even found larger resources in the magma heated grounds and a rough estimate of Alaska geothermal energy goes as far as 31 quads already identified, plus an additional potential 1000 quads which correspond to seven to eight times the annual production of Prudhoe Bay.

Are these projects wishful thinking, expert's dreams, unrealistic approaches or long-sighted views of gifted pioneers? Nobody really knows, but it is quite certain that, in order to develop these local resources, it is absolutely necessary to resort to another «Energy System». Actually, a «green-field-off-market» production of electricity is of no use, unless it allows the manufacturing of an appropriate industrial fuel which can be stored, shipped and distributed on a world wide basis, like Iceland does already with that concentrated energy block, which is aluminium. Then, the potential energy production of arctic areas would be almost infinite and this is why, before additional major investments are made in the extraction and transportation of fossil fuels, all interested parties should review in great detail the present status and future prospects of the «Hydrogen Energy System». Should it become operational within the next decades, then, the Arctic would be able to ship, by Liquid $H_2$ tankers, or even thanks to the then existing oil and gas lines, millions of tons of hydrogen fuel produced by water electrolysis and liquefaction thanks to its fantastic hydro-electric and geothermal powers, or, even, with OTEC (Ocean Thermal Energy Conversion) plants sitting on the polar seas and making use, in winter, of the large temperature difference between surface air and deep water layers.

This would also mean the coming of a new era: the era of a clean abundant replenishable energy — electricity generated hydrogen — which, together with nuclear fusion, would, then, be able to sustain the world development, almost indefinitely, preserving the precious fossil fuels for appropriate chemical synthesis instead of burning them into primitive engines.

## The Human Price of Energy

This development would also suppress completely, the deleterious pollution plumes stemming out of the coal — and fuel — fired power stations which, not only pour acid rains on Southern Canada and Northern Europe but also, by long distance air-transport through the eurasiatic pathway, spray heavy contamination on the northern rims of the circumarctic lands — demonstrating that even the remote boreal territories are not protected from the ill effects of mid-latitudes industrial activities.

Indeed, this would be the obvious sign of new relationships between Man and Nature where, instead of considering the earth as a vast colony which can be conquered and plundered without restriction, an advanced Society would develop a humanistic approach to the surrounding world based upon mutual respect and partnership. This would, then, clearly demonstrate that the 20th century man has finally understood that he has not absolute ownership on the natural resources but, that, single element in a long chain of world citizens, he was just fortunate enough to receive from his fore-fore-fathers a unique heritage which he ought to manage in a responsible way, in order to be able to transfer it unstained to the rising generations. This is neither a problem of knowledge nor one which can be solved by state or international regulations, it is a matter of mere education, and it is too late to deal with it in the Universities since the respect of life, of environment, of natural resources has to be taught as early as the kindergarden so that the «ethics of the commons» become an intrinsic part of everyone's natural behaviour.

The attitude of man towards Energy falls in the same pattern and is derived directly from his attitude in front of the modern Society. It is quite evident that Technology is no longer «neutral» and that any technological choice means a political choice and implies the selection of a given type of Society. It is equally clear that the rather primitive virtues of efficiency and productivity have less and less appeal for an evolving world which enters the «post-industrial» era and dreams of a more human, less hierarchized and more fraternal Society. It has been amply demonstrated that if, unfortunately happiness does not come along with under-development, it is definitely not the compulsory consequence of wealth. In an advanced, balanced Society, where the old ghosts of cold, hunger and epidemics have faded away, human and individual oriented values become the essential assets. More than being overhelmed by electronics, informatics, robotics of all kinds, the citizens are eager to live their lives in the plenitude of their human contents, they wish to have time to exist, to love, to be loved, they request individual consideration and hope to stay long enough to be able to thrive in peace.

Accordingly they are not ready to pay any price for any products or services, and they are quite aware of the fact that every Energy System has its own limits and bears its own ecological cost. They no longer want to be instructed by would-be technical prophets and are suspicious in front of the glittering new scientific ventures. After having indulged to exhaustion in the charms of Modern Times they are coming back, slowly, to a serene appraisal of the old traditional values.

When we deal with the Arctic Regions, let us bear in mind that they might very well be one of the last Sanctuaries of our human enterprise and let us hope that we can bring, there, prosperity, in confidence, so that, for many many years ahead, in the crystalline arctic night, where billions of stars twinkle over the frosted fjells, we hear the hearty joik of the lonely Saami, paying the traditional tribute to the long, mobile, phosphorescent veils of the «Northern Lights».

*Cold Regions Science and Technology*, 7 (1983) 11—25
Elsevier Science Publishers B.V., Amsterdam — Printed in The Netherlands

# The Arctic Regions in the Light of Industrial Development: Basic Facts and Environmental Issues *

*Louis REY***
Comité Arctique International
Monaco

## Mysterious and Challenging Arctic

Lost at the rim of the Earth, sometimes shining white and blue under the permanent sun, sometimes frightening and elusive in the deadly whiteout of threatening mists, the Arctic Regions have been considered for millenniums as forbidding areas of mystery and desolation.

For Classical Antiquity these regions lay on the boreal side of the ocean beyond any known landmarks, since «ita est rerum natura, post omnia oceanus, post oceanum nihil». There in the midst of «mare cronium» (the dead sea), *Hippocrates* placed the happy *Hyperboreans* enjoying a permanent felicity among flowering gardens. But there, also, the most reliable scholars described nothing but impressive wastes, shrouded in drifting ice where land and sea fused into the ill-defined, misty horizon of that «sea lung» discovered by *Pytheas*, the only man who dared to challenge the quiet kingdom of gods in the northern waters.

Many centuries would have to elapse before explorers and cosmographers, at much human cost, could unveil the mysteries of the Septentrio and before the famous charts of *Eratosthene, Adam of Bremen, Edrisi, Olaus Magnus, Mercator, Linschoten, Nansen* and *Nobile* spread more than 2000 years, could stand as the milestones of our evolving arctic geographical knowledge.

Today, regular airliners criss-cross the North Pole area, numerous research stations drift on the polar pack, powerful nuclear propelled icebreakers circumnavigate the arctic ocean and, in the silent depths of northern waters, strategic submarines swim, undetected under the protective shield of the permanent ice, carrying their formidable load of transcontinental ballistic missiles. At the rim of the Polar Sea, new airports and research bases hum with activity and, over the vast expanse of the frozen, barren lands, the yellow flares of the test-wells roar into the arctic night. Caribou herds still migrate in the tundra and wander within the oil sites, derricks and huge radar and transmission antennae dwell side by side with traditional native huts whilst, in the dark-blue waters of the loose summer pack, the mechanical groan of the semi-submersible drilling rigs has replaced the occasional flap of sails in the wind. These exciting ventures, however impressive locally, remain dispersed and, despite this sudden uprise of multifarious activity, there lie hundreds, thousands of miles of arctic desert between these tiny human settlements.

Under open blue skies of crystal transparency or dense clouds of deadly ice-fogs, lie areas of huge potential mineral resources, but they are scattered about in one of the most hostile environments on earth: unlimited areas for human expansion, but with communities fenced in by cultural, historical or political constraints: hundreds and thousands of square miles of fishing grounds over the fertile banks of arctic waters, but made dangerous by violent gales or drifting ice and subject to fierce competition between home fleets and interlopers: vast taigas of in-

* Derived from the original key note address given at the Alaska Alpha Health Congress, Anchorage, May 1982

** The author is also Distinguished Visiting Professor of Arctic Science and History, The University of Alaska-Fairbanks.

0165-232X/83/$03.00 

calculable biological resources, susceptible to being wiped out for decades or even centuries by accidentally or irresponsibly started fires: fascinating wild life, last remnant, perhaps, of the world's original schemes, incredibly resistant to the harsh northern climate, but liable to extinction should the delicate ecological balance be disturbed. It is a region of contrast: of power and weakness: of dramatic and destructive beauty: a wonderland which can become a grave. Therein lies the paradox of the Arctic.

## Glimpses into the Geological Evolution of the Arctic and its Paleoclimatology

The geological history of the Arctic Regions is also, to some extent, paradoxical since very old formations happen to have appeared around the North Pole at relatively recent geological times.

Indeed, circumpolar spaces were occupied originally by a vast ocean and the present arctic lands were fused together into a large continent lying far south in the temperate to sub-tropical zones. Then, some 200 million years ago, the expansion of the mid-oceanic ridges started to push the plates towards higher latitudes whilst the original Pangaea was split into separate continents.At the beginning of the Cretaceous Period, Africa drifted away from South America and, at the end of the same period, 80 million years ago, the North Atlantic Ocean opened and severed Europe from North America. Meanwhile, the plates moved slowly northward, still pushed apart by the expansion of the mid-ocean cordillera which opened the Labrador Sea, 50 million years b.p. At that time the *Lomonossov* ridge got separated from Northern Siberia and the Eurasian basin appeared in the Arctic Sea. However, near the end of the Cretaceous Period, the two oceanic ridges *Ran* and *Alpha* came to a standstill and, at the same time, world temperature dropped substantially. After a period of stability of nearly 20 million years, temperature dropped again at the onset of the Pliocene and the winter cooling of the large continental masses, positioned north, increased the meridionality of the mean surface winds. Then cold land surfaces quickly became covered with ice and, 3 million years b.p. there was an extended continental glaciation in the northern hemisphere which lasted until the end of the *Cenozoic* period. From then on, northern regions have been submitted to wide climatic fluctuations where alternation of cold and warm periods provoked the increase and decrease of the ice caps, thus influencing the circulation patterns within the Arctic Ocean.

Meanwhile, near 1.5 million b.p., the same forces of expansion widened the Norwegian Sea and, pushing Greenland westward, opened the Greenland Sea. The geographical frame of the arctic regions, as we know them in the present period, was then established.

This does not mean that the medio-oceanic ridges stopped their action at that point and, indeed, today, they still demonstrate continuous volcanic and seismic activity; both in the Atlantic and Pacific, but this will show up only on the geological scale and, for us, in human, or even human paleontology terms, the Neo-Pleistocene structure can be considered as a stable pattern.

However, despite this apparent «stability» the northern habitat was, thus, substantially different from the present period, for purely climatic reasons. Indeed, in the last 200,000 years, and since the *Mindel-Riss* I period, northern lands have been submitted to major successive glaciations separated by warmer interstadials. The *Mindel-Riss I* and *Riss II* to *Riss III* periods each lasted 30,000 years but the real glacial age came some 90,000 years ago, with the onset of the *Würm-Wisconsin* period and it lasted for nearly 50,000 years with an acute period of bitter cold between 20,000 and 10,000 years b.p. As we can see, this is a fairly recent event and the soils and ecosystems of the arctic regions seriously damaged by the heavy, aggressive ice-caps which extended as far south as mid-latitudes are just starting to recover from this severe stress. This explains why the arctic regions, despite the fact they are constituted in great part by very old shields, dating more than 1 billion years, pushed northward some million years ago, have, nevertheless, extremely young biological ecosystems which have not had time to mature yet on equally young soils.

## An old Dweller of the North: The Arctic Man

Northern lands have been inhabited by Ancient Man for tens of thousands of years, and there is archaeological evidence that, as early as the Neo-Pleistocene period, some 40,000 — 35,000 years ago, Neanderthal man moved north-eastward in the cold arctic tundra to follow the big game, mammoths, bears and bos-biso.

Indeed, as soon as ice receded during an interstadial, wandering tribes took over the subarctic and arctic lands of northern Europe and

Siberia. Reindeer had already appeared, 1 million years b.p. and, since the Riss-Würm period, musk-ox descending from the late Pliocene Chinese ox. Later on, at the turn of the Aurignacian period, the polar bear stemmed out of North-Siberian brown bears. As time went by, the primitive populations moved eastward receding towards southern milder climates or sheltered in the depths of caves during the longer cold episodes.

Near 25,000 years b.p. at the peak of the Wurm IV glaciation, the climate got so cold that most of the run-off of northern lands was frozen into ice-caps. The mean level of the seas went down by nearly 200 meters and the Arctic Ocean turned into an inland frozen sea. The Bering strait became, then, a dry land bridge where cold arctic tundra conditions prevailed and, as such, it offered a safe path-way for the Siberian arcto-mongoloid hunters following the occasional migrations of their game. Thus, ancient man, walking east, took foot in North America and, there, diversified.

From the common Epi-Gravetian nucleus *Athabascan* and *Algonkian* Indians developed inland whilst the *Paleo-Eskimo*, hunting sea-mammals or following the «musk-ox way» established themselves all along the northern rim of the Arctic lands.

Thus, at the time of the *Lascaux* optimum, 25,000 — 15,000 years ago, all the arctic circumpolar lands were inhabited by populations who developed highly specialized cultures and were able to establish and maintain themselves in this difficult environment until the present day.

Severed from Siberia at the end of the Wurm IV period by the final rewarming of the world climate and the subsequent rise of the ocean level flooding the Bering strait again, the Asiatic hunters, who had been locked in Alaska for a while, started to move again and, when the coastal ice-caps disappeared, they reached the very tip of South America within the next 10,000 years.

Considering this history of human penetration it is quite clear that the arctic regions differ widely from their corresponding southern part. Indeed, Antarctica has always been a barren land and has never supported any human development whatsoever until very recent times when white explorers made pioneering expeditions and established scattered temporary bases on the coastline. Conversely, boreal countries have been the motherland of native circumpolar people for more than 30,000 years. *Inupiat, Yup'ik, Aleut, Koniag, Chugach, Eyak, Athabascan, Tlingit, Haida, Tsimshian,* all were born there and their great-great-grandfathers also and it can be easily understood that they have good reason to claim aboriginal rights. They, nevertheless, did not remain the sole inhabitants of the Great North and, in the Middle Ages, they had to withdraw in some regions and give way to the wandering *Vikings* who, in pursuit of their maritime expansion, sailed their drakkars westward and, eventually, reached Iceland, Greenland, and in the year one thousand, North America, close to 500 years before Columbus.

The route was open, then, for the daring fleets, which from the 15th till the 19th centuries, sailed north to fish cod, collect precious furs, sea-mammals ivory and hides, to hunt whales or try to discover, in the north-west and north-east, shorter maritime passages toward the riches of *Cathay* and *China*. Even at the turn of this century, the north became again an attractive magnet for all those, who, at the peak of the gold rush, tried to probe *Job's* maxim «aurum ab aquilone venit». Sixty years later, the same northern lands once more attracted men and money for the challenging conquest of the viscous, brown blood of modern society which springs hot from the depths of Mother Earth, - natural oil.

## Some Basic Features of the Arctic Regions

### a) GEOGRAPHICAL AND PHYSICAL CHARACTERISTICS

**— The Arctic Ocean**

The core of the arctic regions is occupied by the Polar Sea, a large deep central ocean of more than 9.5 million km$^2$, accounting for nearly 2.6 per cent of the total oceanic surface. It is the largest land-locked maritime expanse in the world and its central part is occupied by several deep basins (*Eurasian* — 4300, *Fram* — 5180, *Makarov* — 4030, *Canadian* — 3490 meters), which are surrounded by large epicontinental coastal seas (*Barents, Kara, Laptev, New Siberia, Chukchi*), that cover almost 36 per cent of the surface but hold only 2 per cent of the total water mass.

From a hydrographical point of view it is interesting to note that the Arctic Ocean connects with the world oceans only by restricted, narrow (*Bering Strait, Arctic Canada Channels, Nares Strait*) or shallow ( *Barents Sea*) passages with the exception of a large deep communication pathway, the *Fram Strait*, which lies between Greenland and Spitsbergen. Through these openings there is a continuous double flow. Warm, saline, Atlantic and Pacific waters enter the polar sea and ice covered cold waters bear south along Greenland and Labrador or along the coasts of Iceland and the Feroes. The overall transport is enormous, more than 9.4 Sverdrup (9.4 million $m^3$/sec) of which 7.1 take place in the Fram Strait alone. (To be compared to a little less than 1 million $m^3$/sec in the Gibraltar Strait resulting from the cross exchange between the Mediterranean and the Atlantic Ocean). This formidable thermal and mass transfer plays a major role in the energy balance of the northern hemisphere and it exerts a critical influence on the world climate (hence the major interest given to the research project on Fram Strait hydrography recently initiated by the Comité Arctique International together with SCOR).

To this maritime influx should also be added the important run-offs coming from the mighty North American and Siberian rivers (*Mackenzie, Yenisey, Ob', Lena, Kolyma,* etc.), which, altogether, feed some 90.000 $m^3$/sec of fresh water within the upper layers of the Arctic Ocean (nearly half the Amazon) draining a surrounding continental area of nearly 10 million $km^2$.

Those incoming waters, whether of maritime or continental origin, merge together in the central pool where they whirl for several years, nicely structured in depths with upperlying cold, ice-covered fresh waters, intermediate saline, warmer atlantic layers and deep stable cold polar water masses which, further south, in the abyssal areas, fuse with the corresponding cold bottom sheets issued from Antarctica. Powerful steady currents constantly drive these waters within the arctic basin, east to west in the *Transpolar drift*, offshore Siberia, and clockwise in the huge vortex of the *Pacific gyre*, north of America, thus distributing the incoming waters almost evenly all around the arctic basin.

Another major characteristic of the Polar Sea is its ice-cover which is permanent and can reach, at its annual maximum, nearly 10 million $km^2$ (30 per cent of the world's sea ice) and plays a critical role in surface energy transfers because of its «skin-effect» and its high albedo, reflecting back into the upper atmosphere most of the incident solar radiation. It represents also a major — and until recently, insuperable — obstacle to surface navigation and transport. Indeed, the mean air temperatures over the Arctic Ocean are very low and, in winter, average $-32\,^\circ$ C in January and seldom increase over $+2\,^\circ$C in July. This explains why the sea is permanently frozen. However, pack-ice, though almost continuous, is never very thick and it seldom exceeds 2 to 3 meters except in the pressure ridges (hummocks), where rafting and compression due to regular tides and wind can build up ice structures as high as 20 meters with corresponding deep underwater keels (bummocks). Quite certainly, this layer of ice is sufficient to shield more than 99 per cent of the sunlight, thus spreading little illumination in the upper oceanic layers. It also suppresses gas exchanges between the free surface of the sea and the overlying atmosphere and helps to damp and smooth out temperature changes which, by alternate melting or freezing, are converted essentially into phase-changes. Nevertheless, it is not thick enough to prevent the huge mass of the arctic ocean playing the role of a formidable «heat-vent» in the cold upper latitudes and acting as a thermal buffer. As such, it helps to build over the central polar areas a rather stable anticyclone where air-masses coming from the sub-tropical zones sink before flowing out, radially, as regular north-easterly winds, sweeping the adjacent circumpolar lands.

That particular set of conditions is highly remarkable and it has some very important bearings on the arctic environment. Indeed, because of this special structure, the arctic ocean is the melting pot in which all chemicals, particles, dusts or aerosols coming from lower mid-latitudes gather and get concentrated, whether carried by currents or by drifting air-masses. Once there, they get locked below the ice trap of the frozen pack and then circulate for years all around the arctic basin driven by the currents. Thus, eventually, they get evenly distributed to all adjacent neighbouring coasts and shore-lines. This demonstrates the unique features of the Arctic Ocean and underlines its central role as a «Polar Mediterranean» where the prevailing rule is that everything has to be shared in common between all bordering nations.

### — Circum-Polar Arctic Lands

If the Arctic Ocean is a well defined geographical unit, on the contrary it is much more delicate to draw a precise southern frontier to the surrounding circumpolar arctic lands. Indeed, because of the basic asymmetry of the glacial ocean, the sea remains open to unrestricted maritime traffic the whole year round up to about 75° N in the North Atlantic sector, for instance near *Bear-Island*, at the edge of the *Barents Sea*. Conversely, at lower latitudes (70° N) in the Canadian arctic archipelago most channels are frozen solid 12 months a year; in the same area of northern Canada bays, basins, straits and a great length of the coast lines are covered with thick close pack-ice for several months, as far south as 45°N, more than 2,300 km south of the Arctic Circle. This clearly demonstrates that a mere astronomical definition of the arctic countries, as «those which are lying poleward of the Arctic Circle», does not reflect the climatic truth.

A similar situation applies also to many northern lands where «arctic conditions» are met in many places which are not located within the Arctic Circle and where, on the contrary, some purely «arctic» countries benefit from a much milder climate. To obviate this semantic difficulty, several criteria have been proposed to help delineate the arctic areas:

- Through climatology: The regions where the mean air temperature never exceeds 10° at the warmest period.
- Through ecology: The regions lying north of the ultimate extension of the forest and poleward of the tree-line.

Both are certainly pertinent and closer to the environmental appraisal of the arctic world but they do remain rather «artificial» and should be balanced by numerous exceptions. It is thus advisable to keep the «boreal boundaries» rather «flexible» and give preference to a global environmental approach based upon a certain number of physical, climatic, ecological and human features which, pooled together, give a comprehensive view of the arctic kingdom.

### — Energy Balance and Arctic Climatology

Because of their circumpolar location the arctic regions receive an overall low input from the sun in the continuous day-light season and present substantial energy losses during the dark period.

- *In summer and intermediate seasons*, the sun is always low over the horizon which tends to exaggerate the geographical profiles and delivers a weak average energy because of the slant illumination, which «spreads» the solar light and results in substantial losses in the atmosphere by diffusion or absorption (50 per cent of the incident energy is lost when the sun is 8° above the horizon). For instance Fairbanks gets only 65 W/m$^2$ as compared to 200 W/m$^2$ in tropical areas. Moreover, the presence of a considerable cloud coverage impedes the radiation transfer by reflecting back to the upper atmosphere a substantial amount of the incident energy. This is further increased by the permanent pack-ice or snow which, with its high albedo, sends back most of the solar energy.
- *In winter*, when the sun has disappeared, the balance is negative and from October to April, the lands radiate most of their energy reserves into the infinite heat-sink of outer space, especially in the High Arctic where very dry conditions are met and where dry barren lands do not profit from the blanketing effect of snow. Indeed, the major part of the High Arctic is a real desert and, drier, by far, than most southern desertic areas. Precipitations never occur and Phoenix, Arizona, has nearly three times more waterfalls than Northern Greenland!!

Accordingly, since these conditions have prevailed for thousands of years, during the world glaciations, the soils are permanently frozen to great depths and this «permafrost» layer extends sometimes as deep as 700 meters on emerged lands as well as below the sea-bottom in the shallow epicontinental coastal seas (with the exception of a surface transition layer which adjusts itself to the ambient temperature, both on land and at the bottom of the sea). This deep frozen layer exerts a major action on all surface ecosystems and acts as a barrier to most transfers, especially water transfers, acting thus as an impermeable sheet which turns the surface «active layer» into a fractured brittle irregular zone in winter, or a spongy unstable marsh in summer, which is quite often flooded, at spring time, when the ice breaks on the main rivers.

## b) ARCTIC ECOSYSTEMS

In these conditions, animals and plants have very little time to adjust themselves to this changing environment and are compelled to «expedite» their growth, biosynthesis, reproduction and migrations in a few months, sometimes a few weeks, before their metabolism initiates appropriate resistance mechanisms to help them survive the winter period. The overall productivity is very low on lands (30 kg/ha/y for Salix arctica compared to 2 — 5.000 kg/ha/y in temperate areas) and substantially reduced in northern ice-covered maritime expanses, despite their high nutrient content and active convection, because of the length of the dark period and the «filtering effect» of the pack, which dramatically decreases solar energy and associated primary biosynthetic production, in sharp contrast with the fertile ice-free arctic fishing grounds (*Lofoten, Barents Sea, Icelandic Waters, Davis Strait*).

**— Animals**

Some animals, like birds, migrate towards southern climatic paradises. Others make use of an increased protection against thermal losses by the development of a dense fur and the accumulation of large lipidic and carbohydrate reserves in the body and/or under the skin to help reduce the energy seepage. Musk oxen and reindeers are thus able to stand intense cold and survive, provided the snow cover is thick enough to protect the lichens and small tundra plants from becoming frozen solid, and that no intermediate episode of cycling melting/refreezing of the surface provokes the formation of a hard, sharp, ice «frigid layer», which would injure their legs and prevent them from foraging for their food.

Other smaller animals like shrews, voles, hazel-hens and weasels, manage to survive in the sub-nivean bioclimate, which extends between the surface frozen layers and the deeper permafrost and benefit from the good thermal insulation provided by the loose crystals of the fluffy «api» snow and the constant flow of geothermal energy coming from the soil itself and from the progressive release of the solar energy stored during the short summer months.

Finally, there are also animals, which, instead of fighting against the hostile environment, give way partly to thermal stress by falling into hibernation. Then, in the protected haven of an appropriate den or natural refuge, a polar bear, for instance stops all its activities, reduces its metabolism in a spectacular way and falls into a deep sleep which can last several months. During this period it burns its organic reserves (the brown fats) and its main physiological functions are depressed: respiration is low, the heart rate substantially reduced, and, most often, the central temperature drops. In that state, however, the hibernating animal is just asleep and is able to resume full activity in a very short time in case of danger. Its physiological activity and behaviour are thus completely different from that of a non-hibernating homeotherm, artificially hypothermized which, although it has the same appearance, is nevertheless in full coma and needs to be reanimated.

Many other different interesting adaptation mechanisms exist in that direction and it has been shown that insects, for instance, are able to withstand very low temperatures, either because they can support a high degree of desiccation preventing the freezing of the remaining concentrated body fluids or else that, manufacturing glycerol within their internal media, they can stand partial reversible freezing of their organs. Marine animals, also demonstrate in the cold polar waters very active metabolic rates, 3 to 5 times higher than temperate zone organisms when placed in identical conditions.

**— Vegetation**

Plants and trees are also challenged by the same thermal stress and, here, the effects of the low temperatures are much more acute, since the vegetation does not have a heat release mechanism of any importance, and it stands unprotected in the wind-swept zone and has no possibility of escaping or hiding from outside aggression. It is not surprising, then, that some types of vegetation have developed special characteristics: small size, waxy cuticles, pubescent leaves and stems, settings in cushions or rosettes — unusual growth rate at the onset of the «optimum season» - high resistance to freezing — energy storage in roots etc.

The conditions vary from south to north and they become harsher when switching from the taiga to the forest-tundra, the tundra or even the wind eroded frost heaved fellfields of the High Arctic. Here again, the «active physiological cycle» is very short and summer is followed by a long winter where, on the barren lands, cotton sedges, dward birchs, lichens, mosses, heath and small perennial plants seek protection under the wind hardened «upsik». Despite the fact that this compact snow transfers heat much better than the fluffy, porous, low density api of the taiga (an increase of 4 in heat transfer for a density multiplied by 2), it still constitutes that ecotone which separates two drastically different environments: a harsh, violent, unprotected macro-environment, beaten by gales and burnt by extreme temperatures with wild oscillations and, near the soil surface, under the snow cover, a constant, humid, balanced sub-niveal micro-environment where, fortunately, lie the infrastructure and the mooring system of trees and large shrubs. Indeed, the survival mechanisms are there, in the deep reserves of the roots, but the upper aerial structures, holding the metabolic photosynthetical reproduction capacities, are essential for the perennial life of the plant. They are, however, under much greater stress, because their stems grow out unprotected into the arctic environment and can expect extreme temperature changes, from + 35°C on a warm sticky summer day, down to —60 °C or even lower in a period of bitter cold and high winter wind. They also have to bear a heavy load of snow and ice and resist the destructive effects of blizzards and gusts which often result in twisted tortuous trees: the «drunken forest». This is why, despite its fantastic resistance and remarkable adaptation abilities arctic vegetation remains fragile and has very little regeneration faculty. Curiously enough, it can also function as an open system and become an «energy and nutrients sink» as in the formation of peat and thus deplete the environment by the immobilization of large amounts of essential minerals, which are not further recycled in the biosphere.

**— Photo-Periods and Biological Rhythms**

Thermal stress is not the only factor to act on biological ecosystems in arctic areas and great importance should also be given to another parameter, light.

Because of their upper latitudes, arctic regions experience the most important changes in the photo-period which can be found on earth, switching from complete darkness in the peak of winter, to continuous daylight in mid-summer. This, undoubtedly, exerts a critical effect on all living organisms, man included, as the ever evolving light periodicity interferes constantly with the circadian rhythm and the numerous internal biological clocks. To what extent this solar light acts as a powerful Zeitgeber and maintains the synchronization of internal clocks with the daily pulsations of the environment, is still a matter of controversy and there are many examples, in northern latitudes, of desynchronization of animals like small mammals (voles), birds (Tengmalm's owl), fish (lavaretus, Finland) and even plants (oxalis acetosella) during the periods of maximum disequilibrium like summer or winter.

This, indeed, constitutes a major difference between the arctic and alpine environments which show, in many other areas, remarkable analogies. Like boreal barren lands, mountains in high temperate-zones present huge climatic seasonal variations in temperature. They also have wind-beaten high altitude tundras, but the mean atmospheric pressure is much lower, the sun intensity substantially higher (because of both less atmospheric filtration due to altitude and increased orthogonality of the solar rays as a function of lower latitude) and there is almost no significant switch in the annual photo-period, whilst the magnetic environment is generally undisturbed.

This latter point is also a significant specific characteristic of the circumpolar countries and its role is not yet perfectly understood. The strong magnetic disturbances, induced in the auroral zone by the abrupt invasion of particles from solar flares pumped within the narrow channels of the magnetosphere overlying the geo-magnetic poles, have well-known effects on compass orientation, radio transmissions, navigation systems etc. Their biological action, on the contrary, is almost unknown.

### c) MAN IN THE ARCTIC

As we can easily understand from this brief review, the arctic is very demanding and it exerts on all living organisms a constant environmental pressure, which results in a heavy load on both the physiological and psychological performance of individuals. In the arctic zone animals and plants are under constant stress and, as such, much more vulnerable to outside interference. Man is no exception to this basic rule and, in this difficult area where there is no room for error, he has his full share of physical strain and mental tension. To the obvious thermal constraints should be added the more subtle effects of changing photo-periods, magnetic disturbances and, maybe, still more elaborate effects, like atmospheric pressure waves and other factors connected to the wind, like low-frequency cyclic sounds and pumping effects on habitats.

Numerous medical and occupational health research works have shown that, for instance, in Finmark, a large proportion of the population, in mid-winter, complains of sleeplessness, often followed by neurosis and reduced physical and mental capacity. The so-called «Ruska» reaction, in Finland, where natives who look at the bright colours of the vegetation at the end of August react by falling into an abrupt depressive mood at the beginning of September, shows that, for people living constantly in the area, any prior warning of the irretrievable return of the long dark winter is considered as a sign of death and causes deep anxiety which may be, later on, associated with depression and psychotic disorders.

The same occurs when, in mid-winter, or mid-summer, continuous darkness or daylight fails to synchronize the biological clocks and adjust the circadian rythm with the earth's rotation thus breaking the delicate balance between individual rhythms. The discovery of polar regions is filled with narratives of the effects of such events on isolated explorers. The only thing, then, which can prevent a total dissociation with the environment, is the occurrence of «social signals», which become the most important of the environmental cues and which work as effective new Zeitgebers.

Still unknown is the action of the natural or induced magnetic fields, which seem to play some role on brain activity and may influence sleep patterns. It has equally been demonstrated that circadian rhythms not only concern physiological variables (like temperature, heart rate etc.) but that they have a marked effect on psychological sensibility and performance, the sensitivity to external stimuli and even the reaction to drugs, poisons etc. Chronopharmacology in arctic areas is still a virgin subject but it deserves a lot of attention, since it might lead to important discoveries of high practical value.

In fact, when dealing with man in arctic conditions, we should resist the temptation to make an easy extrapolation using data derived from populations in temperate regions such as New York, Mexico or Paris but not applicable to the human groups living in the arctic environment. What is true for sleep, biological rythms, or drug sensivity is equally true in the nutritional field. Clinical research carried out in Canada's Northwest Territories has shown spectacular regression of the number of days illness in children who are breast-fed in Inuit, Indians and non-natives and has emphasized the absolute need to delay food supplementation, to encourage the use of native meat and disregard sweet and excess carbohydrates. The controversial virtues of pemmican have been discussed many times and its supporters claim openly that its mixture of high fats and proteins «restore strength and ambition and holds mentally stimulatory qualities» whereas the opponents maintain that pemmican can lead to «grim tragedies». Whatever the truth of the matter it has been shown by explorers in the field like *Robert Peary* and *Vilhjalmur Stefansson* that men can work hard and for extended periods «not on bread alone» but on «the fat of the land».

It is equally clear that traditional game and wild life resources are essential, not only to the very existence of some populations like Inuit, Lapps or Indians, but that their protection and development (through appropriate game management) is vital for the maintenance of their cultural identity. This is why, in assessing the potential impact of new technologies and large scale industrialization in any circumarctic land, we should keep in mind the life conditions peculiar to the region and, as has been said quite often by *Fred Roots*, adjust the flux of technology to avoid inadequacy and protect not only a highly stressed and sensitive environment but also a unique cultural heritage, built successfully by trial and error over millenniums.

## The Impact of Industrialization on the Arctic Regions

As we have seen previously, the arctic regions have stood for centuries outside the main field of activity of occidental people and their sole permanent inhabitants were the boreal native populations. However, at the end of the 15th century some of the arctic resources were already sought after, like the arctic cod on the *Newfoundland* banks. In the following centuries the arctic and sub-arctic seas roused sustained interest, since they offered good commercial prospects for whaling and the hunting of sea-mammals; at the same time deep, in the taiga, trappers became actively engaged in the precious fur trade. These enterprises lasted several centuries and they were still important in the course of the 19th century, and it is true that commercial ventures made an important contribution to our knowledge of the arctic areas and even aroused the interest of Monarchs. A good example is *Svalbard*, in the 17th century where Kings and Heads of States became so actively involved that they occasionally sent some «men-of-war» to protect their interests and care interlopers off the whaling grounds.

Unfortunately, the overall result turned out to be a disaster as the main biological stocks of the arctic seas vanished, rapidly exhausted by irresponsible fleets which plundered all sites in the most ruthless way. We remain speechless today in the face of such operations as the one carried out in 1606 by Captain *Bennet* of the *Muscovy Company* in *Bear Island* (*Cherry Island* at that time) where, in less than six hours, nearly 700 walruses were slaughtered. A small part of the tusks and blubber were recovered but the carcasses and blood-spattered hides were left to decay for months on the shore.

We are not quite like that today but our consciousness of the environment and our earnest desire to manage it as an essential part of the human patrimony over the long term, is a fairly recent attitude which, unfortunately, is not yet shared by all interested parties. In fact, we still have a lot of difficulty in developing a reasonable policy which safeguards natural non-renewable resources for future generations.

It is also extremely difficult for most entrepreneurs and political leaders to understand that our planet Earth is unique and that every area is interrelated and affected by others. This is particularly true of the vast desolate barren lands of the arctic and its adjacent frozen seas.

There are certainly many valid reasons which encourage increased prospection and subsequent exploitation of the potentially large resources of arctic areas. The World Energy Crisis, the desire for technologically advanced countries to acquire political freedom by better access to key industrial raw materials within their respective zones of influence, the rise of nationalism and the strong determination of less developed native populations or ethnic groups to control the mineral and wild-life resources of their homeland, all these factors lead to an increased activity in more and more remote areas. The race for sources of energy and rare metals, thus, succeeds to the whalers, the trappers and the gold rush of past centuries.

This continuous trend endangers many sensitive natural sites as we have seen above, but it should be given credit for the wealth which results from these operations which benefits not only the pioneering teams and countries but also all the indigenous people. There is certainly no evil desire for world domination or economic slavery behind all this but a sincere wish to take advantage of the availability of these large modern Eldorados to boost national economies, raise standards of living and provide people with better services at lower cost.

The key problem is how to change this wishful dream into practical reality so that additional wealth does not lead to a lower quality of life, so that increased resources and industrial power do not result in permanent environmental scars, so that poor ecological strategy does not induce technological tragedy as an inevitable result of economic growth. What is really essential is to know how to achieve the greatest economic benefit for the existing communities of the Arctic with the least social, cultural and political disruption possible, together with a reasonable reward to the outside entrepreneurs who take the major technical risks and bear the heaviest financial load. It is absolutely necessary to keep in reasonable balance environmental issues and national developments and needs and to design an appropriate policy to prevent any permanent irreversible damage to major environmental sites and never to weigh present and long term public health against economic premiums however glittering the prizes may seem.

From this aspect, arctic regions show many features in common with other pioneering areas but, as well, they have some intrinsic peculiarities which give them a high sensitivity which need, accordingly, a very careful, comprehensive approach for their industrialization. This is a vast issue which is difficult to discuss in an exhaustive way, but which presents, nevertheless, some remarkable features on which we would next like to focus attention.

### a) SOURCES AND FATE OF INDUSTRIAL POLLUTANTS

Any industrial activity, as any human activity, is bound to end up with wastes. Such materials may be harmless and can be released with care into the environment, provided that their biological degradation is efficient and does not unbalance the natural ecosystems. On the other hand, some wastes are very harmful and should be kept under strict control. This is true for numerous industrially manufactured substances, which require special handling and careful distribution and usage, such as chemicals derived from synthesis plants and metal mills, and for natural products like crude oil and its derivatives. Within this category some substances are present in the environment only by natural seepage, leaks or accidents, like petroleum products; others are released as dust, smoke or aerosols, like the aerial emissions of coal and fuel-fired power-stations and the unburnt oil from ship's furnaces (nearly 2 per cent) and are swept by the winds; some are intentionally spread over large areas for plant protection and fertilization as in the case of agrochemicals. They can, then, concentrate locally or be carried away by air distillation-redeposition processes or be transported by streams. Generally most end in the sea.

Modern technology, together with the stringent rules established in occupational hygiene now prevent any major pollution risk and we can claim that the accidental release of highly toxic compounds in the environment is very rare. However, industry is a human enterprise and, as such, liable to human as well as to technical error.

Obviously, industrial operations in arctic regions follow the same rules and resort to the same type of engineering, but it is, nevertheless, advisable in any northern venture, to reinforce security measures substantially, to take into account the very special conditions which prevail there. For instance, it is quite easy to understand that fire-fighting and rescue procedures are extremely difficult with surrounding low temperatures, high velocity winds, reduced visibility in snow storms, icing of structures, and the like. Telluric movements, occasional land slides, solifluction, irregular permafrost, are potential hazards for land and rail transportation and for civil engineering as well as for the stability of complex installations having huge piping networks and massive reactors. Magnetic disturbances can also shield radio-transmissions, fool aircraft navigation and seriously impede rapid intervention in critical areas. Finally, last but not least, navigation on rivers, seaways or intracontinental straits and bays is permanently challenged by solid pack-ice, and drifting floes whilst, in the coastal sector huge icebergs are dreadful threats to offshore platforms, sub-sea structures and anchored semi-submersibles because of their terrific mass and the ploughing and gouging action they exert with their keels on the sea floor.

Of course, engineering studies, as well as strict field procedures, help to reduce the risks to a strict minimum but it should be accepted that this minimum will never reach nil and that responsible authorities, as well as local populations, should be prepared to cope with emergency situations.

The interference of different chemicals with the environment is of very high complexity and, indeed, forms the basis of a new science, Eco-toxicology that Professor René *Truhaut* has been pioneering for more than 15 years. In this new branch of toxicology, basic standard effects are hard to define, since the action of a potential toxic can vary enormously according to the given environment. The final result will, thus, depend upon the nature and the concentration of the toxic, and upon its state of dispersion and mode of administration. It will also be a function of the state of development and/or physiological activity of the contaminated organisms and might show different effects if incorporated during growth, at the reproduction phase or in a stable hibernating period. Moreover, its action might be specific on selected species and, thus, the result of a given pollution on a mixed population could be the destruction or inhibition of some individuals and the enhancement of others, which, in turn, could modify completely the original balance of the ecosystem. A typical example is the elimination of limpets and barnacles on wave-beaten reefs which, several years after contamination by an oil slick, are almost entirely covered by an odd dense vegetation. Should the pollution be of a complex nature, there might also be crossed interaction between the individual pollutants, and some chemicals might sensitize animals or plants to a given toxic which would not have affected them otherwise.

Finally, in arctic waters, pollution might interact with crucial periods of biosynthesis, like the spring bloom, and then exert devastating effect on bio-coenoses of limited diversity and variable population densities exhibiting narrow resitance and almost no adaptation capacity. Because of the presence of ice, the pollutant, diluted in the surface layers, cannot be eliminated by sun-powered distillation or wind dispersion and it might, furthermore, become entrapped in the highly active ice biota which develop in the brine pockets of the floes.

In fact, it is quite clear that no simple single rule applies and that each precise case has to be studied individually, always bearing in mind that the toxics act on highly sensitive «stressed» populations where reactions might be quite acute and even irreversible. The toxic effects are «potentialized» by the climatic constraints and, conversely, the sensitivity of the organism to environmental stresses is dramatically increased by a prior weakening of the organic reactions induced by the absorption of the toxic. This explains why, in arctic conditions, sub-toxic levels of pollutants might nevertheless have significant actions. An example is the screening effect that some chemicals exert, at the mating period of some crustacea, on the transmission of «sexual messages» in the water, reducing substantially, or even suppressing totally, the attraction between opposite sexes with subsequent disastrous effects on the populations.

Thus, before any major step is taken towards a given industrial development which might release chemicals in the arctic environment or else introduce major changes in its parameters (effects on temperature, on hydrography, soil drainage etc.) indepth assessment of the original environment should be made with the utmost care. This would obviously include such studies as determination of the different types of biota, evolution of the ecosystems, and evaluation of their inter-relationship etc.

Certainly it is a lot of work but it is a real must. For instance, nobody would really like to take up the challenge of sending regular LNG tankers through the North West passage before a better understanding of the ecology of the *Lancaster Sound* is reached or of building large industrial mining or chemical operations in areas critical for wild-life until its basic patterns and requirements are fully understood.

### — Long-Distance Pollution

From this angle, the situation looks rather simple and it might be felt that, thanks to an appropriate reglementation, and careful environmental work, most of the parameters could be kept under control. Unfortunately, for the arctic regions, this is not the case, since, because of their geographical and meteorological peculiarities, they are at the confluence of a massive flux of pollutants coming from industrial activities taking place in the temperate or even sub-tropical zones.

This long-distance pollution has in fact been neglected and poorly understood for many years but, today, thanks to such pioneering research as that of *Ottar, Shaw, Rahn* and others, we have begun to discover that «dust and chemical stable aerosols» can travel over thousands of miles and be deposited far away from their sources of emission provoking heavy contamination in remote isolated lands.

Much publicity has already been given, in this area, to the acid rains, which have progressively destroyed most of the lake district of southern Scandinavia and have been observed as far away as Iceland. It has been shown, without any doubt, that they come, not only from the industrial smoke of Scandinavia itself but, essentially from the long-range transport of aerial pollution from northern Europe and western USSR and, more precisely, from the power-stations and steel mills of the English, French, Belgian, Ruhr and East European districts. The low grade fuels used to fire these plants release large amounts of chemicals and especially $SO_2$ which, by air oxidation, is transformed into air-borne $SO_4$ giving rise to acid rains.

This transport system is not limited to Europe and similar smoke is unfortunately released in many other areas like Japan, Korea, the United States and Canada. However, the general atmospheric circulation is such that most of these other transport routes have little effect on the arctic lands, either because of their natural deflection by mountain ranges like the *Rocky Mountains* or the *Brooks Range* or because their course includes an oceanic leg, with rather high precipitation rates, where most of the pollutants get deposited.

On the contrary, due to the presence of powerful «action centers» like the *Aleutian* and *Icelandic lows* and the *Canadian* and *Asiatic Highs* , there is a strong, persistent, continental atmospheric pathway from mid-latitudes towards the arctic regions, which, being almost free of precipitation, is ideal for long-range aerosol transport from large European and western USSR sources. This route is now well understood and the air masses from the emission zones are driven northward near *Novaya Zemlya* over the Arctic ocean and funnelled further west over northern Scandinavia, Iceland and Greenland and all along northern Canada and arctic Alaska. Precise chemical evaluation in *Barrow* and *Fairbanks* has shown that these aerosols — which might be 3 to 4 kms thick and travel in the lower troposphere — are definitely of continental origin (because of their $^{210}Pb$ content) and pollution derived (by the measurement of the ration V/Al aerosol/V/Al crust). Furthermore, the evaluation of their aging process ($SO_4$/V ratio) demonstrates, on the evidence, that they have been air-borne for long periods extending to more than two weeks.

This unfortunate set of circumstances is thus responsible for the so-called «Arctic haze» which invades northern skies in winter and affects significantly the local climatology by its radiative, depositional and nucleation effects. This also explains why, irrespective of the absence of any major industrial activity close by, cities like *Barrow* display atmospheric concentrations of $SO_4$ as high as $2 \mu g/m^3$, which is nearly one third of the peak levels in the heavy industrial pollution areas.

Driven into the polar Highs, these pollutants get deposited finally on the cold trap of the frozen sea and lands and are eventually collected in the arctic ocean by means of land drainage and river transport. There is thus a formidable reception area of 9.5 million $km^2$ of arctic seas + nearly 10 million $km^2$ of adjacent circumpolar drainage area which is offered, unprotected, through the northern eurasian pathway, to the uncontrolled airpollution generated at mid-latitudes. These are facts which, unfortunately, cannot be challenged any more and which can be easily demonstrated. As one pointed example let us just mention that in *Barrow*, Pacific air-masses only bring clear skies, whilst the northern easterly winds blowing from *Eurasia* drive in a thick, polluted, dark arctic haze!!

This is a matter of serious concern, but unfortunately, this is not the only one, because long-distance pollution is also carried by the seas and there are ample reasons to believe that every day the North Atlantic Drift, derived from the Gulf Stream, pumps thousands of tons of undesirable agrochemicals and chlorinated pesticides into the *Barents Sea* and the Arctic Ocean, which have been collected off eastern North America and western Europe as a result of the run off from farm lands.

To this should also be added a substantial load of volatile mercury and polychlorinated hydrocarbons, which, as demonstrated by *Ottar* are progressively driven towards polar areas from as far as tropical countries, by successive evaporation and redeposition mechanisms. 2000 tons of DDT in *Antarctica*, heavy concentrations in the *Baltic* and northern Norway, substantial amounts in the arctic sea, easily demonstrate the ill effects of this permanent deleterious substance which, when concentrated in the trophic chain and food-webs, poisons the upper level of fish and birds, blinds the seals and still floats around the globe at atmospheric levels in quantities as high as 300,000 tons. No wonder then, why environmentalists, epidemiologists and all those who are responsible for the quality of life and public health share deep concern over this unrestrained evolution of the «Chemical Age».

Unfortunately, it is not always possible to amend and correct wrong procedures in those territories where the ill-effects appear since, as we have just shown, every year, through atmospheric riverine and oceanic pathways hundreds of thousands of tons of harmful, undesirable, semi-permanent «foreign» pollutants are fed into the arctic regions and the arctic sea. Indeed, it is a rather bitter sight to watch a yearly riverine and atmospheric deposit of nearly 70,000 tons of $SO_4$ alone in an area where almost none of it is actually released!!

This, of course, should be further developed in the light of its international legal and political implications. The transborder pollution convention painstakingly set up in Geneva years ago, has not yet been signed by all interested parties and governments. Remote pollution is still unpunished and very limited action, if any, is taken in the critical areas of emission.

Time, unfortunately, has not yet roused the world-conscience for the protection of human health and the environment.

### b) THE ROLE OF SUPPORTING STRUCTURES

Any sizable development in arctic areas requires ad-hoc supporting structures: powerlines, water and gas ducts, roads, occasionally railway-tracks, airports or harbours, together with the appropriate buildings on the operational site and near-by housing facilities with their ancillary services. Thus, every venture into a «green-field» area implies, in fact, a complex planning procedure because of its intimate interference with the original environment.

Quite clearly, along these lines, all these operations are bound to be much more difficult and costly in arctic areas than in standard temperate locations. Low temperatures, frequent occurrence of large snow-falls and severe winds, permafrost layers ... have to be integrated into the numerous parameters for detailed project planning and subsequent design and construction.

We have already mentioned this point and do not feel it appropriate to elaborate any further with the exception, may be, of some basic considerations on transportation.

On land, the adduction of gas, oil, water, or power as well as the development of tracks, roads or railways will be carried out, by necessity, on «structured» lands and soils, presenting unstable surface «active layers» and deep underlying sensitive permafrost. This is, in no way, a simple issue and *Brown* has shown many times how delicate the permafrost structures are and how carefully constructions should be built to avoid undue mechanical and, especially, thermal stresses to the supporting layers. Indeed, that complex, dynamically balanced system reflects the post-climatic and vegetational history, recent climate, present environmental conditions, animal and human activity, and is influenced by local topography, the nature of the substratum, and the tendency to solifluction. Plants and trees could play a major insulating role by the development of thick vegetation mats or they might reduce stability by infiltration into the thermokarst. A deep knowledge of the properties of permafrost has already been acquired but much more is needed to ensure safe planning. It is enough, for instance, to recall the recent erosion and partial collapse of trails in *Ogoturuk Valley*, subsequent to active transport in this area in the years 1959 to 1961, or the development of wide, deep gullies on the arctic slopes as a consequence of the dense traffic of winter tractor trains in the late forties. Human activities do definitely degrade permafrost and any reasonable steps have to be taken to minimize, and hopefully, avoid, these deleterious effects.

The marine environment, especially in the coastal area is equally sensitive and one should avoid undue disturbance of its basic equilibrium. Whales, for instance, appear to be easily disturbed by underwater noises of given frequencies which seem to interfere with their communication system, particularly in the nursing period when they also react strongly to traces of pollutants, mainly hydrocarbons. Even extended cracks in the pack, as those opened by ice-breakers, might impair their usual feeding layers.

Along these lines particular importance should be given to the shore-lead which is an essential part of the coastal pattern because of the critical role played by its extended open surface for light and gas exchanges with the overlying atmosphere and also because dominant winds might induce major upwelling effects there.

However important these different phenomena are, they are less dramatic in the ocean than on land because of the «buffering action» of the seamasses which smooth out the most acute transitions, especially temperature fluctuations, which, as we have already seen, are converted into phase-changes by the freezing or melting of the surface waters. Basically, pollution or environmental disruptive processes act on the arctic seas as they would in any other part of the ocean but the results, there, are more spectacular because of the particularly high productivity of the cold, oxygen-rich, nutritive, polar waters where a constant and active convection maintains an intense biosynthesis. Thus the accidental release of undesirable pollutants or the occurrence of ill-adapted technologies in this area might seriously impair dense plancton layers and large schools of fish or damage great numbers of sea-mammals.

The same, indeed, applies to the continental sea-ways which penetrate the arctic archipelago, which may become in the near future major commercial routes for huge deep draught tankers built for ice conditions shipping North Slope oil or arctic liquefied gas to the large eastern atlantic markets. Here, again, the continuous dislocation of pack-ice might impair its permanent structure and influence the associated wild-life. How this can affect the hunting practices of local native populations is still unknown, highly debatable and might turn into a purely emotional issue. On top of these socio-cultural and behavioural effects, the massive use of frozen arctic sea-lanes will definitely involve the development of large supporting facilities including navigation aids and communication systems, ice-warning observation platforms and patrols, manned maintenance and rescue teams and their associated dry-docks and under-sea inspection and repair units, plus the numerous beacons, harbours, and terminals needed to service the main production areas and the associated processing plants in charge of gas separation, scrubbing, liquefaction and storage, together with the neighbouring loading and unloading piers and moored buoys. If we add, as we should, the necessary administrative offices, intelligence and security ships, medical care, emergency fire or pollution fighting units and, of course, all that is necessary to sustain the daily life of dispersed human communities (education, recreation, religious and social structures), then, we can easily understand that the industrialization of the arctic sector is a huge multiform enterprise which cannot just grow by itself but which does require previous, extensive, detailed studies and subsequent close monitoring and control services.

### c) THE ULTIMATE HUMAN CHALLENGE: DEVELOPMENT AND HARMONY IN THE ARCTIC

Arctic areas have been considered until fairly recent times as strange, remote, difficult territories of minor importance on a world scale. As such, they have remained isolated from the major currents of development although they have supported native populations living in harmony with their surroundings for millenniums.

Then, man moved in and his lack of knowledge of the basic requirements of the rigid environment of boreal lands, as well as his total absence of interest in their future evolution, had a definite upsetting impact on the original balance in most places.

For instance, whilst the East-Greenland eskimos managed to survive for several hundred years on the shores of their ice-covered rocky fjords, the settlement of inexperienced Vikings in Iceland, 1100 years ago, resulted in an accelerated degradation of the environment. As Sturla *Fridriksson* pointed out, men would not have survived otherwise but their expansion, even limited, did alter in depth the existing ecosystems: half the vegetation cover was lost (dropping from 40,000 $km^2$ to 20,000 $km^2$), most of the birch-woods disappeared and many species vanished. The introduction of new agricultural practices and crops progressively altered animal and plant communities and it is more than evident that, even with the utmost good will and associated skill, pre-settlement conditions will never again be reestablished.

The same applies in Northern America where, in recent historical times, human irresponsibility has started fires which have destroyed large shares of the aboriginal taiga. Three fourths of the spruce forests in *Alaska*, the totality in *Saskatchewan* have been burnt and may never recover. Contrary to the claims of the professionals of the timber cartels «Fire is not good» for northern forests!!

We have already mentioned the massive slaughter of sea-mammals and, especially of whales, from the 17th to the 19th centuries : they are now almost extinct in many areas and cannot even support the needs of local populations.

For too long, the Arctic Regions have been plundered and it is now time to turn the profit-orientated selfish exploitation of the North into a balanced development which encompasses physical, biological and ecological requirements as well as the basic intangible needs of the human communities. This, indeed, is a major task since these human communities have diversified and present widely different cultural and economical patterns today.

As has been rightly proposed by Professor V.A. *Kaznacheev* from *Novosibirsk*, the time has now come to carry out anthropo-ecological research in the North and study the behaviour of its permanent dwellers who according to him, fall in two categories: the stayers and the sprinters. In fact, they have been identified, quite long ago, under different names and do represent two widely different human types:

- The low energy, introvert, «stayers» are essentially those for whom the Arctic is a homeland. They have little or no impact on the environment and limited entrepreneurship. Perfectly well adapted to the boreal conditions they live by tradition and, generally, suffer chronic diseases. Most have been for generations in arctic regions and always in rather small numbers.

- The high energy, extrovert, «sprinters», consider the Arctic as a frontier that they have to reach, conquer and tame. They attack the environment and never «fuse» with their surroundings. They are men of challenge and enterprise but are frequently struck by acute diseases. They have made recent massive moves into the arctic, more than 10 million people in ten years in Northern USSR, according to Kaznacheev, and, in many areas, they have completely superseded the local populations.

This is precisely where the acute difficulties lie since this powerful influx of «modern» civilization has had, without any doubt, both a beneficial and disruptive impact upon the natives and old dwellers and may rupture their traditions. This is such a sensitive issue that, in the USSR, the Party Central Committee and the Council of Ministers has made strong recommendations on the ways and means to help the evolution of northern people without impairing their individual socio-cultural inheritages. It can easily be understood that this is an extremely difficult issue, which, unfortunately, in many instances, is purely wishful thinking!

However, every effort should be made to resolve this problem and establish a reasonable working philosophy which will secure a harmonious development of arctic regions with the active support and participation of local populations, and which will keep in balance the socio-economic interests and the ecological requirements while preserving the cultural patrimony.

This has to be achieved, of course, according to each nation and its own traditions, rules and policies but it has, also, a much more general implication. For the numerous reasons that we have already given, the success or failure of Arctic Development will depend, entirely, upon the possibility of developing coherent management schemes of the circumpolar coastal areas, in order to ensure global protection to the northern environment.

Unfortunately, this is a very difficult operation since human and ecological considerations are not the only ones. Major economic interests may conflict with environmental values, which, too often, present also a dubious political profile and, last but not least, military and/or diplomatic considerations will still play an overriding role.

One thing is sure : the development of the arctic regions is one of the biggest economic programmes of the future, as claimed by Academician *Aganbegyan*, and their huge potential resources will require a tremendous effort in very difficult conditions. This implies an increased labour force moving into boreal countries, not necessarily for the exploitation of petroleum products, which require only limited skilled personnel, but essentially for mining, transport and support logistics. Millions of square kilometers of Exclusive Economic Zones are still unexploited, and, despite its spectacular planned development, (more than 40 per cent between 1981 — 1985) the Northern Sea Route is but the first achievement in arctic transportation. The polar route from London to Tokyo is only 6,500 miles versus 11,200 by the conventional lanes and the nuclear-propelled ice-breakers Arktika and Sibir have already demonstrated its feasibility.

Undoubtedly the once neglected northern regions will become within the next twenty years a zone of intense activity and, hopefully, friendly competition.

Just as medieval craftsmen gathered together to use their skills to send vaulted columns soaring up into lofty pinnacles, let us hope that the peoples and nations of today will rise above self interest and, whilst keeping faith with the arctic environment, pull together to harness the resources of this pivotal region of the earth.

*Cold Regions Science and Technology*, 7 (1983) 27—32
Elsevier Science Publishers B.V., Amsterdam — Printed in The Netherlands

# Basic Features of the Arctic

*Tore Gjelsvik*

Norwegian Polar
Research Institute,
Oslo

## Abstract

*The basic features of the Arctic are caused in part by the geomagnetic relationships, in part by the inclination of the axis of the earth in relation to its orbit, as well as by the physiographic and topographic conditions.*

*To start from the very top: the geomagnetic condition allows interaction between solar plasma and the upper atmosphere, causing magnetic storms and aurora which disturb magnetic measurements and the propagation of the radio waves. This gives navigation and communication difficulties. However, it also makes the Arctic a most important scientific window to the outer space.*

*The geography of the Arctic can be described as a centrally located ocean, surrounded by low land and a few high mountain ranges and narrow straits leading to the world oceans. In this way it differs fundamentally from the Antarctic, which is a continent surrounded by a big ocean. The Arctic climate, although polar, is less severe than the Antarctic one.*

*The features of the Arctic, which give most problems to the activities of man, are of two types: the permanent pack ice filling up the Arctic Basin and expanding into adjacent seas during winter, the permafrost and particularly the thawing during the summer of the 1—2 m thick uppermost layers of the soil.*

*In addition to the description of the geophysical characteristics of the various Arctic regions or zones mention is made of biologic features, the indigenous people and of environmental matters.*

## Introduction

What is the Arctic? Being asked this question, most people think about the area around the North Pole as a region with complete winter darkness or midnight sun, polar bears and reindeer, eskimos, tundra and permafrost, and a cold and harsh climate — generally, something inhospitable and unpleasant. Those of us who know the Arctic feel different although at times we admit it can be rough up there. To us the Arctic is an area of unbelievable beauty and solitude, of dreams and challenge.

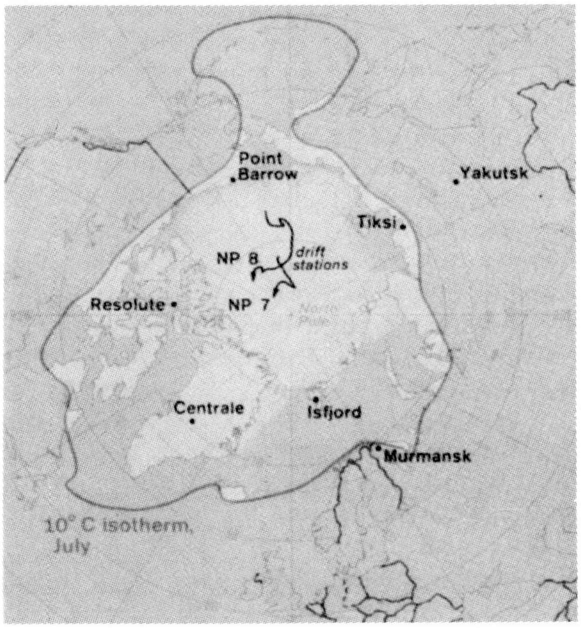


FIGURE 1.

0165-232X/83/$03.00 

These are the feelings, what are the realities? In order to describe them, one has to define the borders of the Arctic. Some people use the latitudes such as the Arctic circle or the 60's degree N. In the U.S. or Canada the latter is often used, which means that even most of southern Norway is included. Anyone having sweated through the tropical temperatures during last summer here in Oslo, will have difficulties in accepting that definition. There are other alternatives such as the northern tree line or the limit of contineous permafrost in the lowlands. None of these gives us a borderline in the sea. I prefer to use the +10°C isotherm for the warmest month, July, which coincides relatively well with the tree line, and which is shown in Fig. 1. According to this definition, the Arctic comprises 14 mill. square kms of sea and 8 of land.

## Cosmic Geophysics

In describing the basic features of the Arctic, I would like to start from the very top, the upper atmosphere and the near-earth space in polar regions. Fig. 2 shows a sketch of the earth's magnetic field envelope and of the flowlines of the solar wind. The energetic particles of the solar wind are picked up by the magnetic field lines and directed towards the polar caps, where they interact with atoms of the upper atmosphere to produce the beautiful polar lights, or

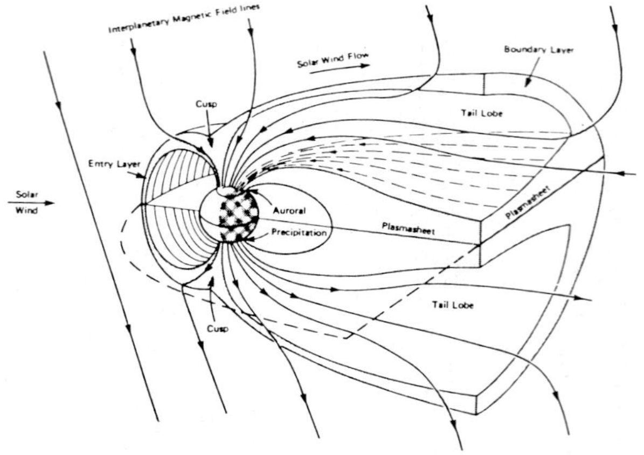


FIGURE 2. *Sketch of the Earth's magnetic-field envelope, or magnetosphere, and its connection to the polar regions (not to scale). Relevant plasma regions and sample magnetic field lines connecting the polar cap to the interplanetary field are shown.*

aurora, but at the same time disturb the magnetic field and radio emissions. Before the satellites were introduced, telecommunication and navigation in the high arctic were rather difficult. North of approximately 78 degrees N, satellite telecommunication is still not possible.

At the same time, these effects have made it possible for cosmic physicists and space scientists to observe directly manifestations of phenomena occurring in the deep space. The polar areas have thus become the Earth's window for remote sensing and other studies of the space and its effects on our environment.

## Physiography

The Arctic may briefly be described as a perennially ice-covered, deep ocean, surrounded by islands, continental coastal areas, and of peripheral , seasonally ice-covered seas. The landmasses mostly consist of lowland. High mountain ranges are found only in Alaska, easternmost Siberia and Canada, and in Greenland. The bottom of the peripheral seas consists largely of extensive and shallow continental shelves.

The three to five thousand m deep polar basin has a rather complicated morphology. It is intersected by two major subsea mountain ridges called the Alpha-Mendeleyev-Ridge and the Lomonosov Ridge, the latter crossing near the North Pole with a crest of 1000 to 1500 m below surface. A median rift valley splits the basin on the Eurasian side of the Lomonosov Ridge in two sub-basins. The 500 km wide and 2500 m deep Fram Strait between Greenland and Svalbard provides the most important channel for exchange of water between the polar basin and the world oceans. The Polar Ocean constantly receives fresh water from the big north-flowing Siberian and Canadian rivers, giving a brackish surface layer that enhances the formation of ice. Along the east side of the Fram Strait warm atlantic water enters the polar basin. To balance these inputs, the excess water and large masses of polar packice flow south along the opposite side of the strait.

## Geology

Only accessible parts of the Arctic continental shelves and the floor of some peripeheral seas have been subjected to systematic geological surveys up to now. In those parts the basement is covered by from 2 to more than 10 km of sediments, with a substantial hydrocarbon potential in some areas. This aspect will be treated by later speakers. From the central, perennially ice-covered basin the data are few, unevenly distributed and preliminary. What we know, is based upon measurements from Soviet, U.S. and canadian scientific stations of floating ice and from airborne geophysical surveys. In general the basement of the Arctic sea floor seems to be covered by sedimentary sequences of 2 km thickness, which is more than in any other ocean basin. In some places, even 5 km or more of sediments have been inferred. In the Canada basin the layers thicken seawards, and structural traps have been observed. The Lomonosov Ridge is by most scientists considered a fragment of the Eurasian continental shelf. The origin of the Alpha-Mendeleyev-Ridge is more questionable. Although some scientists are in favour of a continental crust, it must be studied in more detail before safe conclusions can be drawn.

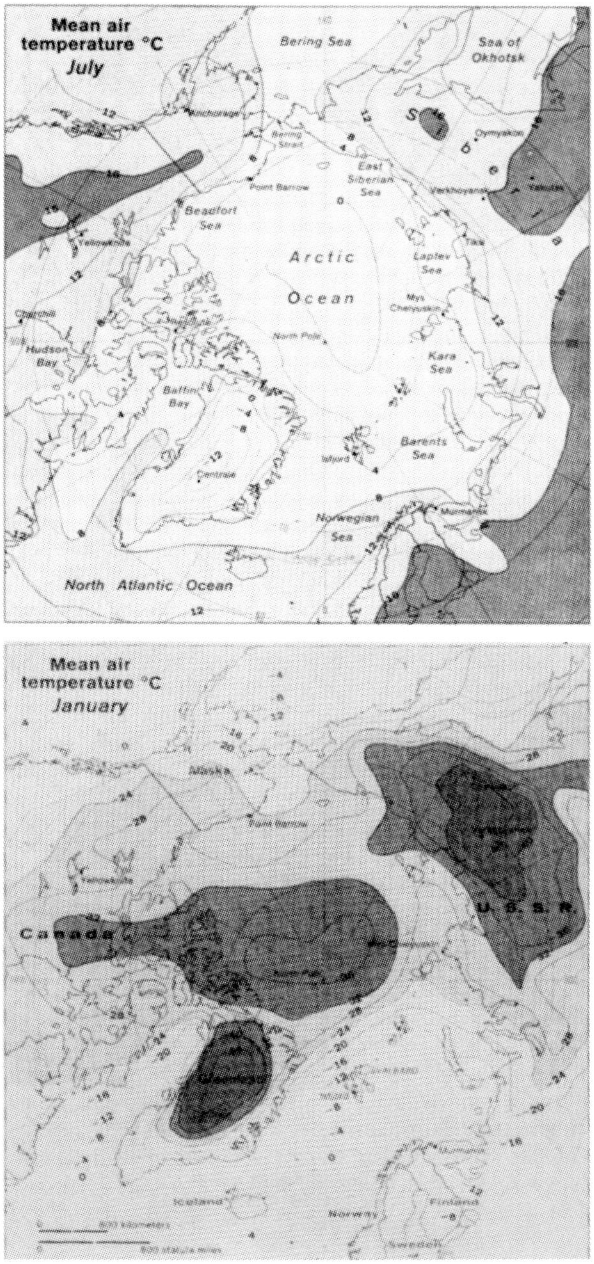


FIGURE 3.

## Biology

Biological conditions in the Arctic are characterized by low primary production (although in some arctic and sub-arctic seas it can be high), and ecosystems with few species and populations ranging from few to very abundant. They often show remarkable behavioral adaptations, high vulnerability to change, and slow recovery if locally disrupted, but are tough over the Arctic as a whole.

## Climate

The popular concept of the Arctic as a region of perpetual bitter cold and deep snow is rather incorrect. The extreme temperatures and snowfalls are found in the arctic borderlands or even in the sub-arctic areas. The mean air temperature over the Arctic basin during winter is $-30$ to $-35\,^{\circ}$C, the lowest readings around $-50\,^{\circ}$C. In the summer, surface air temperatures are near the freezing point and may go up to a few degrees avbove it (Fig. 3). Annual precipitation is very low, 100 to 150 mm, most of it falling in the summer or autumn. Low clouds and fog are a common nuisance, especially during the summer season. Winds are moderate to low.

The climate of the Arctic as a whole is dependent on a number of factors, such as: the latitudes determining solar altitude and the length of the day; the proportion of land and sea and of ice-covered and open waters; the distribution and temperatures of ocean currents; the proportion of snow-covered and barren surfaces on land as well as of the areal extension of the sea ice; the topography of the land areas. Closely related to most of these factors are the frequency, intensity and tracks of the cyclones.

The interaction of all these factors gives rise to a variety of climate regimes in the peripheral arctic regions. Large parts of Greenland have temperatures about 10°colder than the central Arctic, both in the summer and winter, primarily because of the height of the ice cap. Annual precipitation is very low except for the southern and western parts, where it is around 500 mm. South-western Greenland has a moister and milder climate, moderated by the nearby ice-free ocean. The coastal areas of north America and Eurasia have continental climate, i.e. dry and cold winters and mild summers. During summer the great temperature contrast between the cold Arctic Ocean and the warm interior landmasses gives variable weather with average temperatures around +10°C, but hotter spells may occur. Precipitation is rather low, mostly between 100 and 250 mm.

The Svalbard area experiences rather special weather: Thanks to the north-trending, warm ocean current and the cyclonic activity in the Norwegian and Barents seas, much higher temperatures occur during winter than should be expected at these high latitudes. Open water therefore exists outside the west coast of Spitsbergen up to about 80 degrees N all year round. Average winter temperatures on the west coast of Spitsbergen, at 78 degrees N, is about 15°C warmer than at Cape Chelyuskin at the same latitude in Siberia. In the winter and spring, high pressure areas with cold arctic air masses may frequently dominate the weather situation in Svalbard. Thus, great temperature variations are typical , too, and even at high altitudes rain can fall in winter time. Precipitation is higher than in most other arctic areas, especially on glacier slopes on the east side, where it may reach 1000 mm annually. The strongest winds in the Arctic occur in south Greenland and in the Norwegian and Barents seas.

## Ice

The formation of ice at sea and on land are the most important climatic effects for the development of arctic energy resources. The sea ice covers on the average a minimum of 5.2 mill square km in September and a maximum of 11.7 mill square km in March-April (Fig. 4). Only a small part is smooth fast ice, up to a couple of m thickness formed in the winter in near-shore areas, and largely melting in the summer time. Most of it is moving ice — packice — with a thickness of 2 to 6 m depending upon age and location. Due to wind pressure, tide and ocean currents this ice is in constant motion, forming high pressure ridges of ice rubble, and open leads in a confusing pattern. Variations in size and age of the ice floes can be considerable within short distances. Pressure ridges are mostly 1-2 m high, max. 15 m, and may extend to a depth of 50 m under the surface.

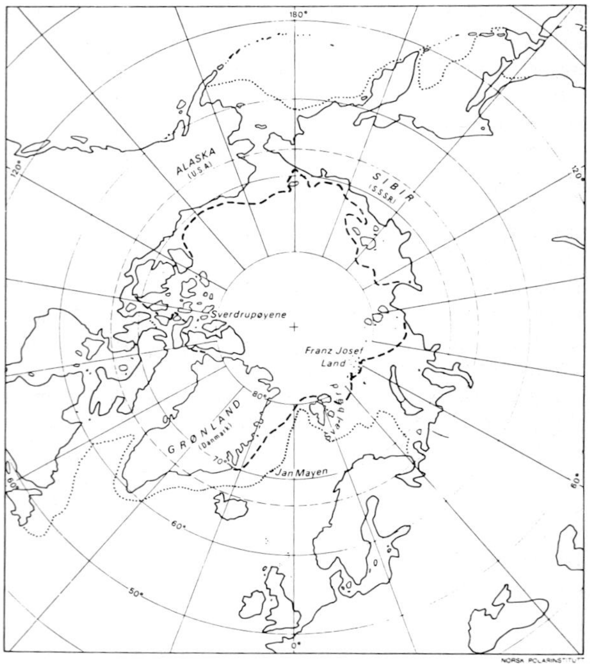


FIGURE 4.

- - - - Average minimum extent of sea ice

........ Average maximum extent of sea ice

In the Beaufort Sea the packice moves in a great gyre with low speeds in the centre, and increasing outwards. In the Eurasian basin, the ice generally moves with an average speed of 3 cm/sec from the siberian Shelf towards the Fram Strait where the speed accelerates to 10-15 cm/sec. Along the borders, smaller gyres are formed. The winter ice is hard and tough, in the summer the surface gets slushy and puddled, making surface transport and airplane landing difficult and hazardous.

In most of the arctic highlands, glaciers are ubiquitous, the Greenland ice cap being by far the most important, covering an area of 1.7 mill square km with up to more than 3 km depth. Glacier surges of high speed, up to 35 m per day, are not uncommon in arctic mountainous areas. Greater danger to human activity stems from icebergs calving from fiord glaciers, ice caps and ice shelves. The largest are the rare tabular icebergs which may be up to 30 km wide and 60 m thick, but also the irregular icebergs can be large and have even deeper drafts, up to 450 m. Most of them are from Greenland and Ellesmere Land. Quite a few are carried southwards into the shipping lanes of the North Atlantic by the swift east Greenland current or along the coast of Labrador. Future petroleum activity on the east Greenland continental shelf will no doubt be greatly disturbed by them, but icebergs on the east side of the Svalbard archipelago should not be disregarded, either.

## Permafrost

Perennially frozen ground — permafrost — occurs where the ground temperature remains constantly below 0°C over a period of years. Its depth depends on the average annual temperature, the topography and the length of time of the negative temperature balance. Ordinarily, it is absent under glaciers and sea, although it may be present under shallow and cold seas, possibly as a relic from the time when the sea level was lower than today. The thickest permafrost — 1500 m — is recorded in Siberia. From Canada 600 m is reported. In central Spitsbergen, 3-400 m is found under the mountains, 100-200 m under the valleys. Ice is concentrated in the pores of unconsolidated sediments, but also in wedges and sheets (ground ice). During summer the surface layers will melt to a depth of 1-2 m, maximum 5. This so-called active layer is biologically very important, since it supports a sparse vegetation. Due to the impermeability of the permafrost below, the surface layer gets very wet and represents a serious problem for ground transportation and construction work in the Arctic. The alternation of thawing and freezing produces very striking structures in arctic soils, called patterned ground. The material is often sorted according to size, stones are concentrated and oriented in polygons or rings, some centimetres to tenth of metres across, with the finer material in the middle. On sloping hills, the whole active layer slides by gravity (solifluction), and the patterned structures are drawn out to parallel stripes or bands. The frequent freezing and thawing lead to an extensive fracturing of the surfaces of exposed rocks. Thus, in the Arctic, mechanical disintegration greatly outweighs chemical processes in the weathering of rocks. In combination with slush flow, solifluctuion, river and glacial erosion this results in a rapid denudation of mountains. The very muddy rivers deposit their load in lower valleys, deltas and other near-shore areas. Tides, currents and ice scour rapidly alters the configuration of coastlines.

## Other resources

This conference is focused upon the Arctic energy resources. However, it should not be forgot that the Arctic also contains other important mineral resources, as well as rich biological resources. The reindeer and sea mammals are no longer important on a global scale, but they continue to be of great interest to the indigenous peoples. The fish resources of the Barents and Bering seas and the Greenland waters are among the greatest in the world. Any petroleum activity up there is going to last for perhaps a century or two. For the peoples settled on the northern coasts, the fishing industry will continue to be a source of living for ever.

The other arctic mineral resources are presently mostly exploited in the Soviet Arctic. In the Kola Peninsula and Siberia, nickel, cobalt, platinum, copper, gold, tin, and iron are produced on a large scale in Greenland, the unique cryolite mine is exhausted; only a lead-zinc mine is in operation. Two rich deposits of the same metals have been put into operation in Arctic Canada, also. More will be found as exploration goes on.

## Indigenous People

The social impact of petroleum development on the indigenous people of the Arctic will be considered by other speakers. But to the basic features of the Arctic also belongs the very existence of these peoples, who have lived and worked under harsh conditions for thousands of years, and who have become part of the ecosystem. They are the eskimos and indians of North America and Greenland (110,000), the Chukchis, Samoyeds, and other arctic groups of Siberia and Northern Russia (7-800,000). These figures also include peoples living in nearby subarctic areas, where the 30-40,000 laps of northern Fenno-Scandia also live.

## Gaps in Knowledge

I should like to end this exposé of the basic features of the Arctic by reminding you that there are still large gaps in our knowledge. The understanding of the processes affecting the sea ice, its distribution and dynamics is most important. For the exploitation of energy resources in the Arctic, including transportation to markets a thorough study of the ecosystem in the ice-infested waters, in particular its marginal zones, is another must. We have some genereal information on the physical and biological features of the Arctic, but we often lack precise data on critical areas, as well as information about the extreme conditions, needed both for industrial standards and governmental monitoring and regulation.

# *Section II*

## *Occurrence of Energy Resources*

*Cold Regions Science and Technology*, 7 (1983) 35—40
Elsevier Science Publishers B.V., Amsterdam — Printed in The Netherlands

# Magnetic Data over the Arctic from Aircraft and Satellites

*Patrick T. Taylor*

NASA Goddard Space Flight Center
Greenbelt

## Abstract

*Extensive low-level aeromagnetic flights have covered almost half the Arctic ocean basin, while satellite missions have encompassed the entire region (except around the North Pole). These data sets reveal the age, nature and tectonic relations between the major physiographic features of the Arctic. The Alpha and Lomonosov ridges are continental, sea-floor spreading has occurred in the Amerasian, Makarov and Canada basins. These magnetic data set the framework for Arctic tectonics.*

## Introduction

Two types of magnetic measurements will be discussed. The first are recorded by a polar-orbiting satellite, Magsat. The second data type is a reprocessing and reinterpretation of the southern Canada Basin east-west aeromagnetic anomaly profiles (Taylor et al., 1981). Mathematical processing of these latter measurements were carried out in an effort to remove diurnal variations of the magnetic field. Satellite magnetic profiles cover the Arctic ocean basin (south of 83°N latitude) far more completely than the aeromagnetic measurements; but at some sacrifice of anomaly resolution. Magsat altitudes are generally 400 km.

Magnetic anomaly data record contrasting magnetization within the crust or upper mantle from which we can infer the geologic structure, and hopefully, the major petrology. These remotely sensed data (satellite and aircraft) are of particular importance since the ice-bound Arctic is denied to surface research ships. The main geological/geophysical problems of the west Arctic or Amerasian Basin are the origin and age of the Canada Basin and Alpha-Mendeleyev Ridge. These two morphological features dominate this region; they are best presented in the recent GEBCO chart of the Arctic (Johnson et al., 1979). To avoid confusion, the Amerasian or west Arctic Basin is defined, in this report, to extend from the north slope of Alaska to the Lomonosov Ridge (Fig. 1). The Canada Basin is approximately a million square kilometers in extent (Fig. 1) and it lies between the Alpha-Mendeleyev Ridge and Alaskan north slope.

Aeromagnetic results have been presented in summary form by Kovacs et al. (1983) while satellite magnetic data were given by Coles et al. (1982). It is the purpose of this report to try and understand the significant problems of the Amerasian Basin from an interpretation of these satellite and aeromagnetic data.

## Satellite Magnetic Anomaly Data

There have been two satellite systems which have been used to measure magnetic anomaly data for the Earth's crustal layer. The first was the POGO (Polar Orbiting Geophysical Observatories) series of satellites (the data used were from three satellites, 2, 4 and 6, orbit altitude 400 to 1500 km; Cain and Langel, 1971). These earlier POGO satellite magnetic anomaly data, for the Arctic, were published by Langel (1979). Recently, however, there was the lower orbiting (352 to 561 km) Magsat (Magnetic Satellite) which recorded both vector and scalar magnetic field data.

0165-232X/83/$03.00  

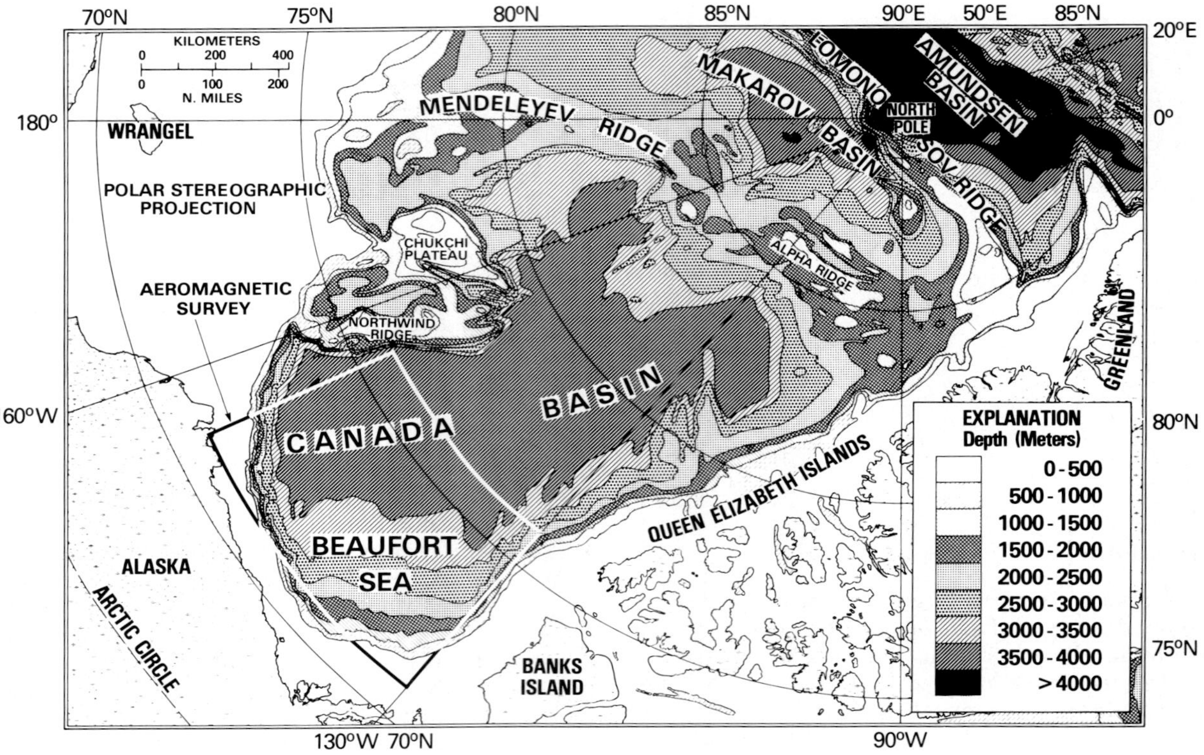


FIGURE 1: *Generalized bathymetric map of the Arctic Ocean adapted from Johnson et al., (1979).*

Before any significant effort was invested in the geological interpretation of these satellite data, it was decided to compare them with aeromagnetic measurements. All available aeromagnetic data in the Arctic region (Kovacs et al., 1983) were analytically continues upwards to the altitude of the satellite measurements over the Amerasian Basin. A good qualitative agreement with anomaly amplitude was found. Since satellite magnetic measurements represent a completely new geological tool it is advisable to confirm or verify them with ground-based observations wherever possible.

As a first step in the interpretation of such long wavelength data it was decided to simulate or model these anomalies by large, homogeneous bodies whose dimensions are controlled by the known bathymetry and crustal thickness. The crustal thickness used (30 km) and bulk effective susceptibility for the model were obtained from the preliminary results from the LOREX (Lomonosov Ridge Expedition; Sweeney et al., 1982 and Sweeney, personal communication, 1982). The upper surface of the body was controlled by the GEBCO (Johnson (Fig. 2). This extremely simplified geometry is justified

F

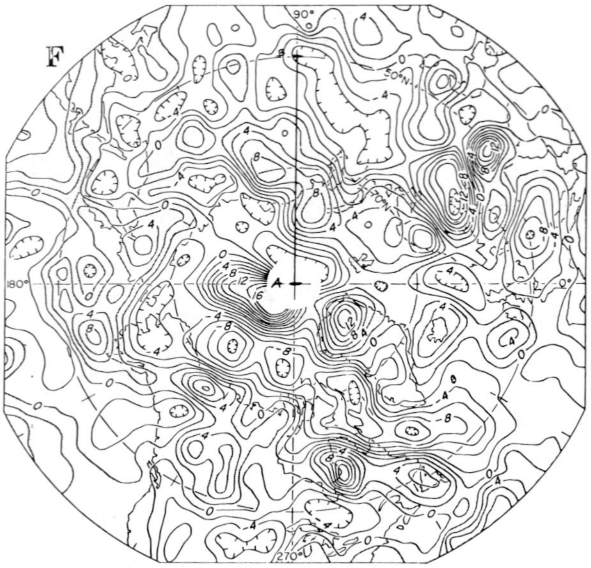


FIGURE 3: *Simplified block models used to simulate or represent the Alpha and Mendeleyev Ridges and produce a synthetic total-magnetic field anomaly at satellite altitude (400 km).*

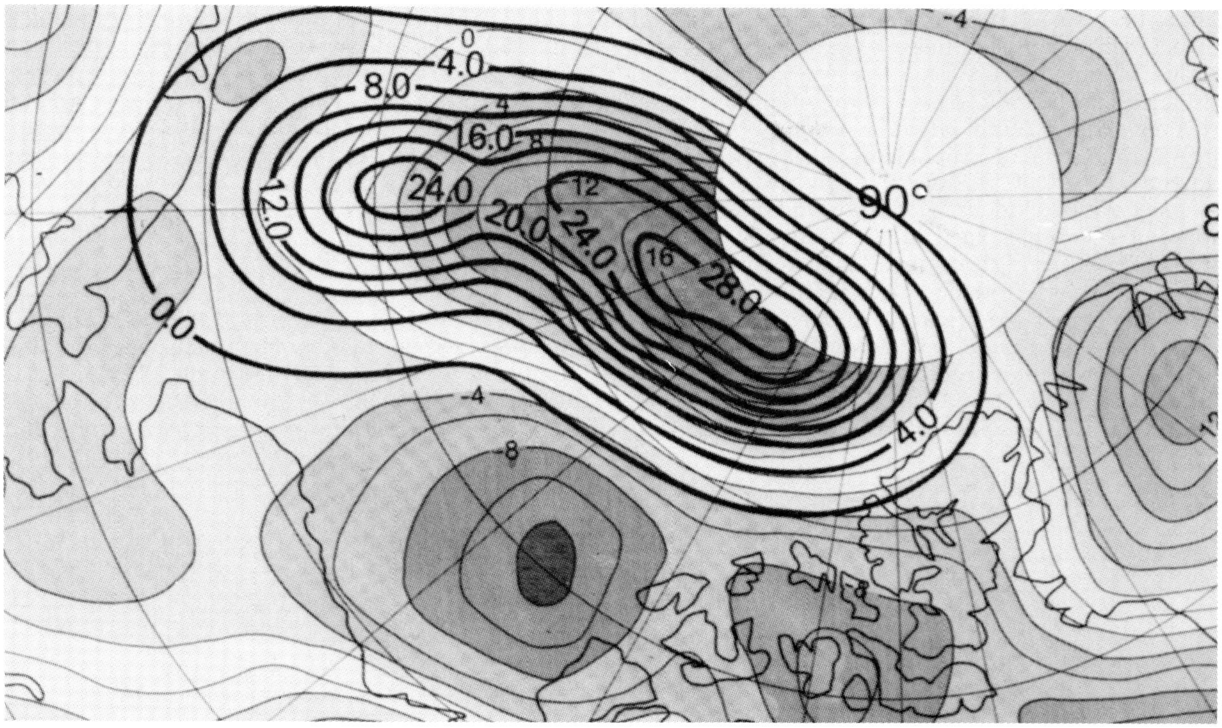


FIGURE 2: *Total-field magnetic anomaly map (F) derived using Magsat data (contours in nanoteslas; nT); from Coles et al. (1982).*

by the effect of modeling at satellite altitudes (about 400 km) which tends to smooth out any small-scale topographic or geologic details. The computer programs used to model these ridges (von Frese et al., 1981) account for the sphericity of the Earth and the variation of magnetic field inclination over the extent of the body.

For the purposes of modeling, the Alpha-Mendeleyev Ridge was simulated by two separate blocks (Fig. 4). This procedure was justified since it is not known if the Alpha-Mendeleyev represents one or two geologically distinct morphologies. After the magnetic anomaly field produced by the model was computed it was compared with the presently available Magsat model derived by Coles et al. (1982). The match (Fig. 2) was good for the Alpha Ridge but poorer for the Mendeleyev section. It was, therefore, thought that these two ridge segments were indeed different geological structures.

From this satellite map (Fig. 3) we can determine that there is a large amplitude high over the Alpha Ridge. This is one of the largest global magnetic anomalies measured at satellite altitude. There are also prominent magnetic highs over the MacKenzie River Delta and along the western half of the Aleutian Island chain. Between the Alpha Ridge and MacKenzie Delta magnetic highs there is a broad region of negative magnetic anomaly values. The amplitude of these negatives are approximately half the flanking positive anomalies. These observed negative magnetic values are consistent, in both

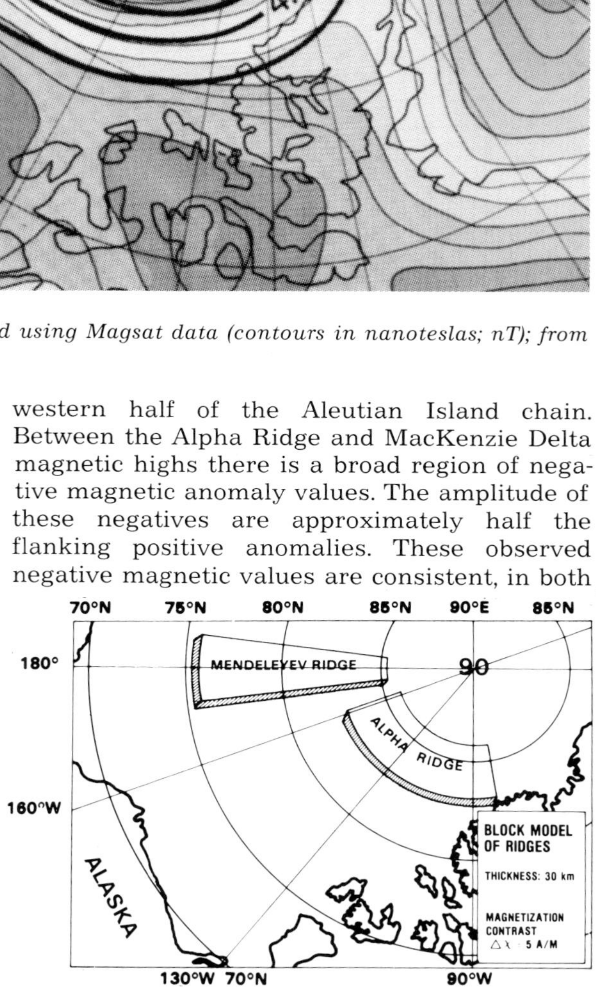


FIGURE 4: *Detail of the Magsat Arctic map (Fig. 3) centered on the northern Canada Basin. Dark contour lines are synthetic anomaly produced from model calculations (Fig. 4). Contour interval is 4 nT.*

sign and size, with a mathematical model of areally larger blocks having positive magnetizations (about 2.5 A/m) for the Alpha Ridge and MacKenzie Delta and zero magnetization in between. That is, the negative magnetic field in the Canada Basin proper can be explained as a geometric effect produced by two distinct and separate large blocks with high positive magnetizations being polarized in a nearly vertical field. Therefore, it is doubtful if negative magnetic anomalies, at these high inclinations, have any geological significance; unless they represent reversely magnetized features.

The above interpretation of the Magsat anomaly maps of the Arctic is, of course, preliminary. Refinements of the proposed models, simulating vector data and fitting the models to individual satellite passes and not equal-area averages are interpretation goals to be pursued in the future. There are obviously large and significant, magnetic anomalies present at satellite altitude in the Arctic and their interpretations should prove to be geologically rewarding.

## Aeromagnetic Data

In 1977 and 1978 twenty-three aeromagnetic profiles across the Canada Basin were obtained between 71° and 76°N latitude (Taylor et al., 1981). Recently, these Canada Basin data were subjected to further processing in an effort to

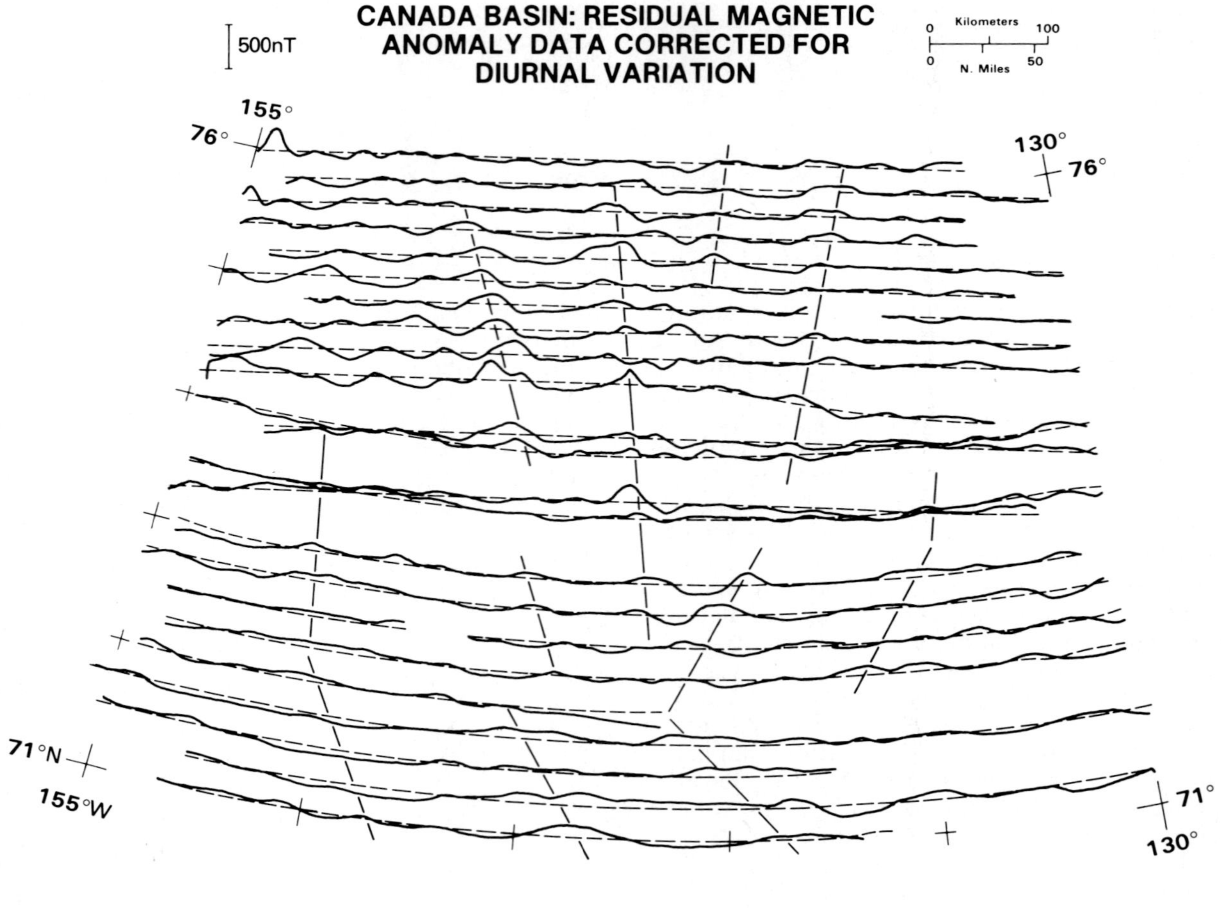


FIGURE 5: *Aeromagnetic profiles across the southern Canada Basin (modified from Taylor et al., 1981; their Figure 3). Short-dashed lines are flight paths. Solid lines represent the anomaly profiles. Long-dashed lines indicated possible magnetic anomaly correlations.*

remove these diurnal variations. This process simulated a high-pass filter, with the longer wavelength having been suppressed (Fig. 5). An alternative interpretation to that given by Taylor et al. (1981; their Figure 2) is also shown in Figure 5. If the anomaly lineations, indicated by the correlation lines, are related to the opening of the Canada Basin (see, for example, Vogt et al., 1982) then, obviously, this tectonic process was complex and not a simple scissors-like motion. It should not be surprising that the small-scale nature of this rifting process would have a jagged or rough edge. Of course, we are not certain that the filter is removing only diurnal variations, however, this is one of the disadvantages of filtering.

## Discussion

A nearly complete magnetic anomaly map of the Arctic region has been acquired by the Magsat Project. While these data are still preliminary, and subject to ambiguities, it is still possible to deduce some geologic significance from these measurements.

Over the Alpha Ridge we have recorded one of the largest (greater than 16 nT) global satellite magnetic anomalies. Other significant magnetic features (independent of sign, since a body magnetized in a positive direction south of the magnetic equator will possess a negative magnetic anomaly) occur over Bangui in Central Africa (Regan and Marsh, 1982); Kentucky (Mayhew et al., 1982); Broken Ridge, Indian Ocean (Frey, 1982); and Antarctica (Ritzwoller and Bentley, 1982). In practically every instance where these extremely large amplitude anomalies occur, the apparent causative body is continental or pseudo-continental in nature. This fact is extremely significant with respect to the Arctic since it strongly suggests that the Alpha Ridge is, most likely, of a continental nature. While all continental structures do not possess extreme (greater than 10 nT) magnetic anomalies, at satellite altitude, every extreme magnetic anomaly is associated with a continental or pseudo-continental feature.

There have been four major theories proposed for the origin of the Alpha Ridge: former center of sea-floor spreasing (Vogt and Ostenso, 1970); remanent subduction zone (Herron et al., 1974); hot-spot trail (Vogt et al., 1979) and to have a continent-like nature (Eardley, 1948 and Soviet workers). These satellite magnetic data support the last theory. This hypothesis results not only from the amplitude of the anomaly but from the dimensions of the block used to model this feature. A crustal thickness of 30 km is consistent with a continental crust. As mentioned previously this value was obtained from the preliminary results from the LOREX expedition (Sweeney et al., 1982) and was used as the model for the Alpha-Mendeleyev Ridge. The 30 km crustal thickness value was obtained by seismic refraction techniques over the Lomonosov Ridge and was used to model or simulate the magnetic anomaly over the Alpha-Mendeleyev Ridge. An effective magnetization of 5 A/m was used to model this anomaly (Fig. 4). But from Figure 2 note that while the shape of the anomaly is reasonably well matched the amplitude is not. The observed value was 16 nT while the model calculation is 28 nT. This means that either the 5 A/m magnetization used was too large and should be about 2.8 A/m or that the absolute level of the reference field has not been established. The lower magnetization value (2.8 A/m) is near the number mentioned previously during the discussion of the negative anomaly in the Canada Basin. The arithmetic manipulation used to arrive at the lower magnetization is justified since once the geometry of the body has been established the anomaly amplitude is directly proportional to the magnetization. In any case, the value of magnetization used for the model is not unique, and the petrology of this feature can not be established from the value.

The Mendeleyev ridge section appears to be essentially non-magnetic (Fig. 2). This negative result would suggest that the Alpha-Mendeleyev Ridge is actually two geologically different features. While the Alpha is proposed to be continental-like the Mendeleyev sector is composed, in bulk, by non-magnetic material; either sediments or rock of low magnetization (for example, gneisses or schists).

Suppression of the diurnal variations on the aeromagnetic profiles over the Canada Basin by data processing have permitted a revised interpretation of the linear anomalies in this region. While the broad line spacing between profiles does not permit a definitive interpretation to be made, these linear anomalies (Fig. 5) suggest that the opening of the Canada Basin was not an abrupt or uniform tectonic event. This rifting was complex, non-uniform, process analogous with the opening of the Bay of Biscay.

Magnetic anomaly maps of most of the Arctic region, derived from both aircraft (Kovacs et al.,

1983) and satellite (Coles et al., 1982) have recently become available. While geological interpretations of these maps are preliminary and developing, there are sufficient information revealed to warrant further analysis, both of the map construction process and the interpretative models. This type of study is an activity which more effort should be invested. The present report should be considered as only a beginning.

A comparison was recently made (Taylor and Langel, 1982) between the Arctic Magsat map of Coles et al. (1982) and a preliminary, but independently derived, Magsat map produced at the Goddard Space Flight Center for the same region. There were disagreements between the two versions. This variance might have resulted from either the different satellite passes chosen to produce each version or to the dissimilar filtering process which was applied to each data set. While it may be justifiable to construct geological models from these satellite data, until reduction procedures produce more consistent results these interpretations must be viewed with caution. The disagreement between the two Magsat anomaly maps was of a second-order, the major features of the field were consistent.

## References


Cain, J.C. and Langel, R.A., 1971, Geomagnetic Survey by the Polar-Orbiting Geophysical Observatories, in, World Magnetic Survey 1957-1969, edited A.J. Zmuda, IAGA Bulletin No. 28, 65-75.

Coles, R.L., Haines, G.V., Jansen van Beak, G. Nandi, A. and Walker, J.K., 1982, Magnetic Anomaly Maps from 40°N to 83°N Derived from Magsat Satellite Data, Geophys. Res. Letts., 9, 281-284.

Eardley, A.J., 1948, Ancient Arctica, Jour. of Geology, 56, 409-436.

Frey, H., 1982, Magsat Scalar Anomaly Distribution: The Global Perspective, Geophys. Res. Letts., 9, 277-280.

Herron, E.M., Dewey, J.F. and Pitman, W.C., III, 1974, Plate Tectonic Model for the Evolution of the Arctic, Geology, 2, 377-380.

Johnson, G.L., Monahan, D., Gronlie, G. and Sobczak, L., 1979, General Bathymetric Chart of the Oceans (GEBCO) Sheet 5.17: Canadian Hydrographic Service, 1:1,000,000.

Kovacs, L.C., Bernero, C., Johnson, G.L., Pilger, R.H., Taylor, P.T. and Vogt, P.R., 1983, Aeromagnetic and Tectonic Chart of the Arctic Region, Geol. Soc. Amer., in press.

Langel, R.A., 1979, Crustal Anomaly Representation, in, Earth Survey Applications Division, Research Report-1979, ed., L. Carpenter, NASA Goddard Space Flight Center, Greenbelt, TM 8064,2/13-21.

Mayhew, M.A., Thomas, H.H. and Wasilewski, P.J., 1982, Satellite and Surface Geophysical Expression of Anomalous Crustal Structure in Kentucky and Tennessee, EPSL, 58, 395-405.

Regan, R.D. and Marsh, B.D., 1982, The Bangui Magnetic Anomaly: Its Geological Origin, Jour. Geophys. Res., 87, 1107-1120.

Ritzwoller, M.H. and Bentley, C.R., 1982, Magsat Magnetic Anomalies over Antarctica and the surrounding Oceans, Geophys. Res. Letts., 9, 285-288.

Sweeney, J.F., Weber, J.R. and Blasco, S.M., Continental Ridges in the Arctic Ocean: LOREX Constraints, Tectonophysics, 88, 1982, Vol 89, p 217-238.

Taylor, P.T., Kovacs, L.C., Vogt, P.R. and Johnson, G.L., 1981, Detailed Aeromagnetic Investigation of the arctic Basin. 2. Jour. Geophys. Res., 86, 6323-6333.

Taylor, P.T. and Langel, R.A., 1982, Interpretation of Satellite-Derived Magnetic-Anomaly Data over the Arctic, EOS Trans. Am. Geophys. Un., 63, 312.

Vogt, P.R. and Ostenso, N.A., 1970, Magnetic and Gravity Profiles Across the Alpha Cordillera and their Relation to Arctic Sea-Floor Spreading, Jour. Geophys. Res., 75, 4925-4937.

Vogt, P.R., Taylor, P.T., Kovacs, L.C. and Johnson, G.L., 1979, Detailed Aeromagnetic Investigation of the Arctic Basin, Jour. Geophys. Res., 84, 1071-1089.

Vogt, P.R., Taylor, P.T., Kovacs, L.C. and Johnson, G.L., 1982, The Canada Basin: Aeromagnetic Constraints in Structure and Evolution, Tectonophysics, 89, 1982, Vol 89, p 295-326.

von Frese, R.R.B., Hinze, W.J. and Braile, L.W., 1981, Spherical Earth Gravity and Magnetic Anomaly Analysis by Equivalent Point Source Inversion, EPSL, 53, 69-83.

*Cold Regions Science and Technology*, 7 (1983) 41—50
Elsevier Science Publishers B.V., Amsterdam — Printed in The Netherlands

# Current Petroleum Exploration of the Norwegian Arctic

*Egil Bergsager*

Norwegian Petroleum Directorate
Stavanger

## Introduction

This paper discusses the exploration in the Norwegian Arctic Area. Exploration drilling onshore Spitsbergen started in 1963, while offshore drilling commenced in 1980, in the Troms 1 Area. Data availability is discussed. An evaluation of the prospectivity of the area is given and it is discussed in the light of development costs in such remote areas. Questions related to jurisdiction of the area are also discussed. The paper is intended for a broad group of technical people rather than a small group of specialists.

## Norwegian Arctic — Definition of the area

60° N is a common borderline for the southern extension of the Arctic Areas. This is reasonable in most areas when the climate is considered. In Norway, however, the climatic situation is different. The influence of the Gulf Stream results in an abnormally mild climate all the way to the northern part of the Norwegian mainland, north of 70° N (see Fig. 1). For the purpose of this paper I have defined the Norwegian Arctic as the area north of approximately 70° N. The major part of the area is offshore. Onshore areas are the islands of Spitsbergen, Hopen, Bjørnøya and Jan Mayen.

## Jurisdictional Questions

The basis for the Norwegian offshore claims are the Norwegian mainland, the islands of Spitsbergen and the Jan Mayen Island (Fig 2).

Some questions related to delimitation and legal status are not solved. They are illustrated in Fig 2.

In the Barents Sea the delimitation between Norway and USSR is unsettled. The disputed area is around 155 000 km$^2$. This is larger than the Norwegian North Sea south of 62° N which is 142 352 km$^2$.

The delimitation between Jan Mayen and Greenland is likewise disputed.

The only borderline which is settled in the Norwegian Arctic Area is that between Jan Mayen and Iceland. An agreement on that was reached in 1981.

In addition to problems related to delimitation there are also questions related to the Svalbard Treaty. The land territory to which the Treaty applies is defined as all islands within an area between 10° E, 81° N, 35° E and 74° N. Norwegian sovereignty is recognized under the Treaty but the signatories have certain rights. One of them is a nondiscriminatory right, based on certain objective criteria, to acquire concessions for the purpose of exploration and exploitation of minerals, with considerable tax benefits.

The Treaty applies only to land and territorial waters (4 n. miles). According to another interpretation, the Treaty would also apply to the continental shelf beyond territorial waters. This is the position of the USSR, while other countries i.e. UK, US and France have reserved their position in this regard. This legal dispute has not lead to any concrete conflict and seems unlikely to do so.

0165-232X/83/$03.00  

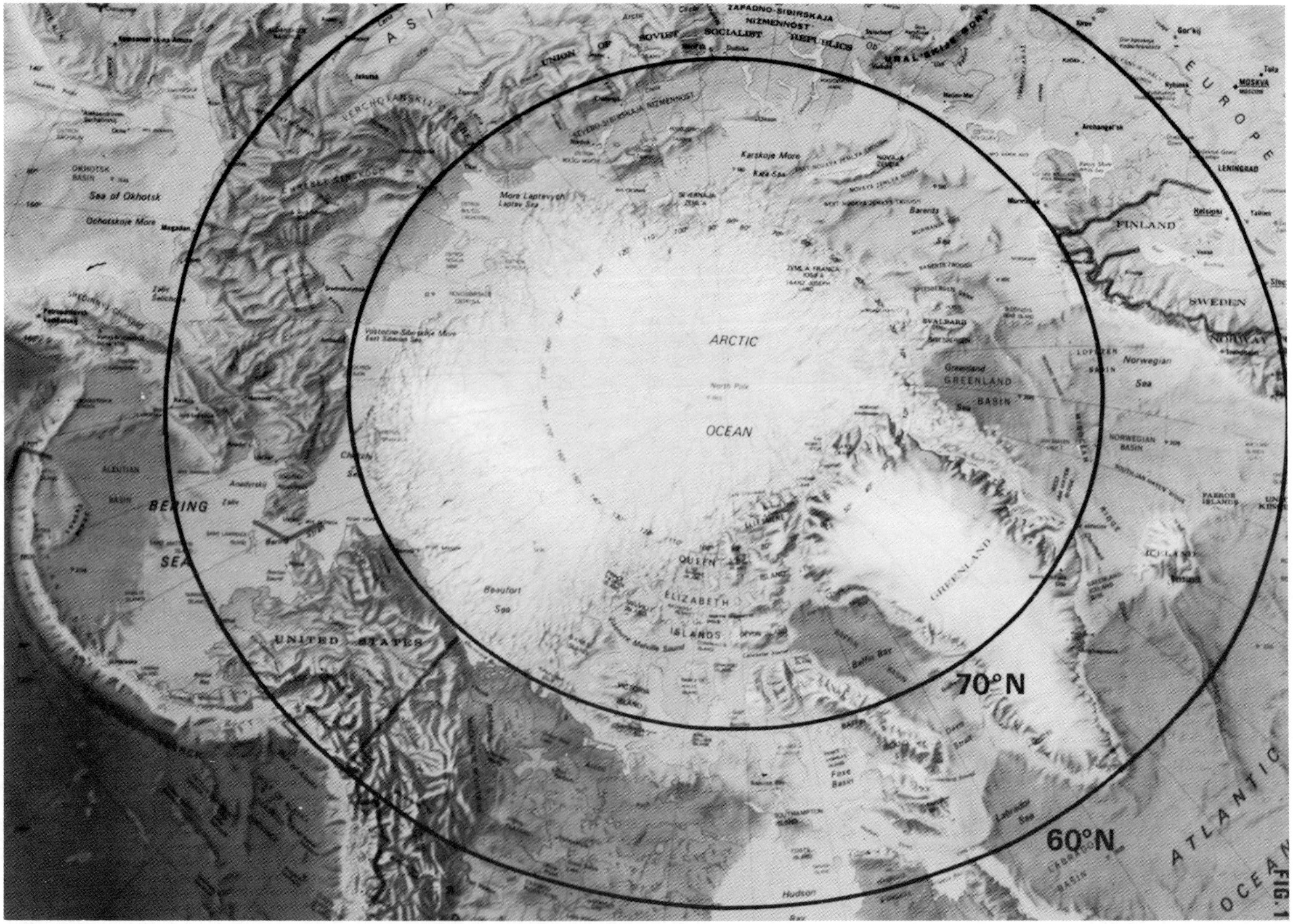
FIG. 1
70°N
60°N
ARCTIC
OCEAN
North Pole
ASIA
EUROPE
UNION OF SOVIET SOCIALIST REPUBLICS
ZAPADNO-SIBIRSKAJA NIZMENNOST'
FINLAND
SWEDEN
NORWAY
LENINGRAD
MOSKVA
Helsinki
Archangel'sk
Murmansk
Karskoje More
Kara Sea
NOVAJA ZEMLJA
Barents
Sea
ZEMLJA FRANCA IOSIFA
SVALBARD
Greenland
GREENLAND
BASIN
Norwegian
Sea
NORWEGIAN
BASIN
ICELAND
FAEROE ISLANDS
GREENLAND
QUEEN
ELIZABETH
ISLANDS
Beaufort
Sea
Baffin Bay
Davis Strait
Foxe Basin
Hudson
Labrador Sea
LABRADOR BASIN
ATLANTIC
OCEAN
UNITED STATES
BERING
SEA
Anadyrskij Zaliv
ALEUTIAN
BASIN
More Laptevych
Laptev Sea
Vostočno-Sibirskoje More
East Siberian Sea
Sea of Okhotsk
Ochotskoje More
OKHOTSK
BASIN
Magadan
VERCHOJANSKIJ CHREBET
Jakutsk

FIGURE 1.

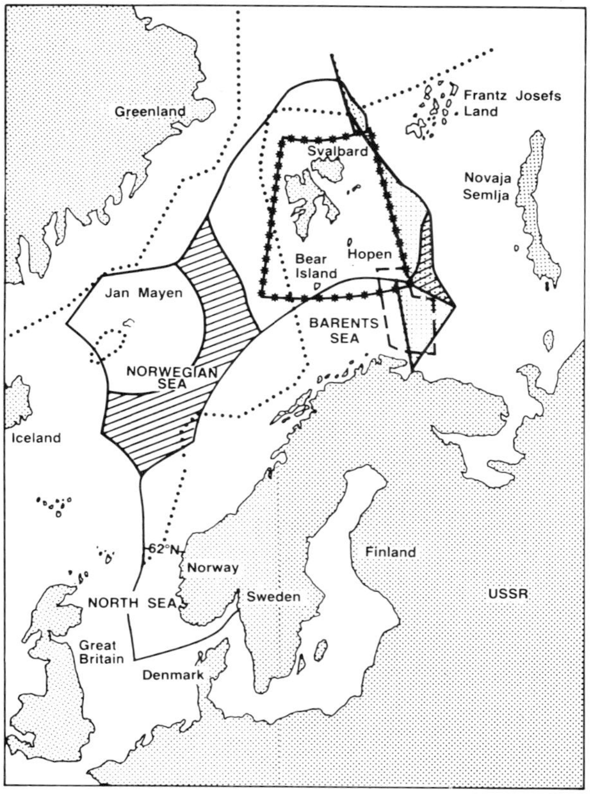


FIGURE 2. **NORWEGIAN ARCTIC Jurisdictional situation**

- - - Grey zone Fishery zone
— Delimitation of economic zone Median lines or 200 n. miles
······ Outer limit of continental shelf
Disputed area between USSR and Norway
Area beyond 200 n.m. from island or land area
Delimitation of area where Svalbard Treaty applies

## Exploration Activities

Evaluation of the area is based on three main types of surveys. Those are accomplished by:

i) Academic institutions
ii) Norwegian Petroleum Directorate
iii) Commercial companies

Each of them are discussed below:

### Academic institutions

Investigations by academic institutions have been carried out for many years, and over large areas. A wide variety of geological and geophysical methods have been used. Seismic surveys are the most significant for the evaluation of petroleum potential offshore. Such surveys have been performed over the last two decades. Around 20 000 km of seismic have been acquired. This includes single channel seismic. Around 8 000 km is multifold reflection seismic and considered to be of industrial standard.

The onshore areas of the Norwegian Arctic are mainly the Spitsbergen Islands. Traditional geological mapping with rock sampling and analysis is a most significant tool for the evaluation of the petroleum potential onshore. Such mapping has been accomplished for many decades, but the larger amount of work has been carried out after the second world war. There are several reasons for that. The Norwegian Polar Research Institute was established in 1948 illustrating the higher priority given to this area by the authorities, the communication system has been significantly improved and money is more easily available for scientific surveys. Numerous maps and publications are publicly available from the onshore area.

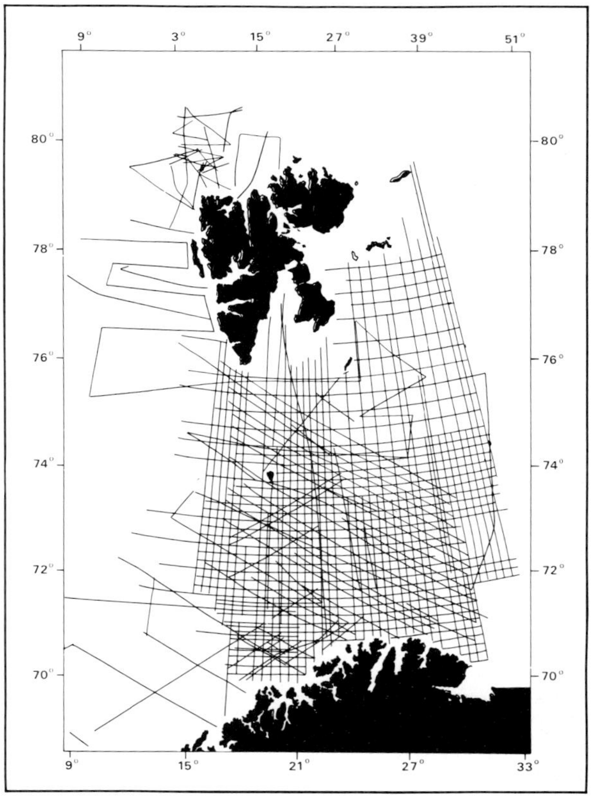


FIGURE 3. **SEISMIC SURVEYS IN THE BARENTS SEA**

## Norwegian Petroleum Directorate (NPD)

One of the functions of the NPD is to evaluate the petroleum potential of the Norwegian Continental Shelf. Part of that is to carry out exploration in areas which are not open to commercial companies.

Seismic surveys are the most important, but bottom sampling, shallow wells, magnetic and gravimetric surveys are also accomplished. NPD's exploration has been carried out since 1973 and 72 500 km of multifold reflection seismic has been acquired. The distribution of the seismic profiles is shown in Fig 3.

The main part of the data are not publicly available. This is due to political and commercial reasons. Several papers are, however, published.

## Commercial companies

Only a limited area has been opened for commercial companies offshore. This area is called Troms 1 and is located at Tromsøflaket. It is shown in Fig 4 and 5. Nine exploration wells have been drilled and extensive seismic surveys have been accomplished by the companies.

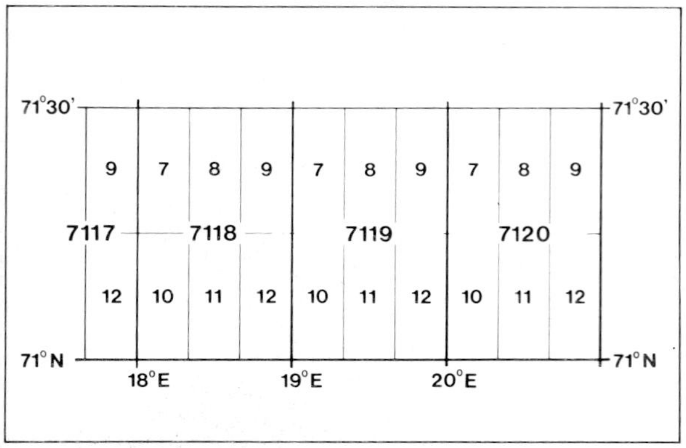


FIGURE 4.

**BLOCK SUBDIVISION OF TROMS 1**

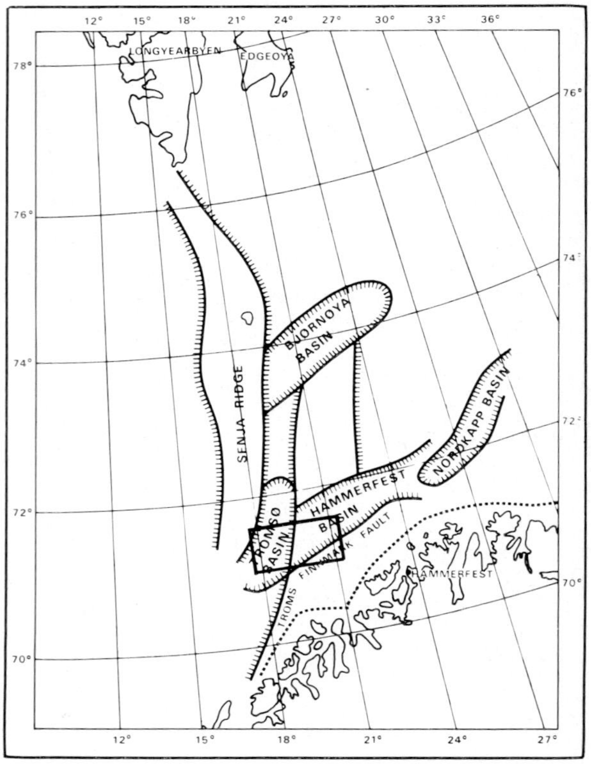


FIGURE 5.

**Sedimentary basins off Northern Norway and the Troms 1 area**

## Basis for the selection of areas to be opened for detailed surveys

The Norwegian Government has decided to go stepwise and open only limited areas for companies to do detailed surveys and drilling in the areas north of the North Sea. There are several reasons for this. Some of the more important are:

— Desire to keep full control of the activity level and its geographic distribution.

— Fishing is very important, especially in the northern part. Conflict of interest between the oil and the fishing activities should be minimized.

— The legal status of parts of the shelf is not settled or the Norwegian view is questioned by other nations. The opening of new areas must therefore be viewed in a broader political context and only limited areas are opened.

If such a policy shall be undertaken it is necessary to have a technical/geological state organization to propose such areas. In Norway this is organized through the Petroleum Directorate, which perform extensive pre-drilling exploration. Several thousand km of seismic profiles have been acquired in areas which have not been opened for commercial activities by companies.

General political guidelines determine the areas which are given priority. The Petroleum Directorate will not normally carry out detailed surveys. The seismic grid is limited to a semi-regional nature with 4-6 km distance between the lines. This is considered sufficient to define the main geological trends.

Once a company has bought the seismic data package it is allowed to do detailed surveys within that area. This is an important principle since it allows various geological prospect evaluations to be tested by different companies.

## Drilling history

The Troms 1 Area was recommended already in 1974 in a white paper to the Norwegian Storting (Parlament). For various political reasons the actual allocation of licenses was delayed for several years. The first production licenses were awarded early 1980. This allocation was followed by two new phases. They are all considered part of the 5th concession round on the Norwegian Continental Shelf. A total of seven licenses have been awarded. The three phases of allocation are shown in table I below:

**Table I:** *The three phases of block allocation in the Troms I Area:*

| Date of allocation | | Number of concessions | Blocks |
|---|---|---|---|
| 18 Jan | 1980 | 2 | 7119/12, 7120/12 |
| 27 March | 1981 | 2 | 7117/9, 7120/8 |
| 23 April | 1982 | 3 | 7120/7, 7120/9, 7119/7 |

The first offshore exploration well was spudded on 1 June 1980. The wells are shown in table II below:

**Table II:** *Exploration wells in the Troms 1 Area:*

| Well | Spudded | Finished | Total depth | Number of days |
|---|---|---|---|---|
| 7120/12-1 | 1.6.80 | 12.10.80 | 2548 m | 134 |
| 7119/12-1 | 14.6.80 | 10.10.80 | 3088 m | 119 |
| 7119/12-2 | 16.4.81 | 26.6.81 | 1877 m | 72 |
| 7120/12-2 | 15.4.81 | 11.9.81 | 4655 m | 150 |
| 7120/8-1 | 28.6.81 | 10.9.81 | 2585 m | 75 |
| 7117/9-1 | 24.4.82 | 16.7.82 | 3177 m | 84 |
| 7120/8-2 | 15.4.82 | 29.7.82 | 2565 m | 106 |
| 7120/9-1 | 25.7.82 | 26.9.82 | 2277 m | 64 |
| 7120/7-1 | 31-7.82 | 8.10.82 | 2817 m | 70 |

**Table III:** *Exploration wells on the Svalbard islands:*

| | | Spudded | Finished | Days on location | Operator |
|---|---|---|---|---|---|
| 1 | Grønnfjorden | 9.6.63 | 8.9.65 | 287 * | Norsk Polar Navigasjon |
| 2 | Ishøgda | 1.8.65 | 15.3.66 | 227 | Caltex |
| 3 | Bellsund | 23.8.67 | 70.8.81 | 299 ** | Norsk Polar Navigasjon |
| 4 | Hopen 1 | 71.8.71 | 29.9.71 | 50 | FINA |
| 5 | Raddedalen | 2.4.72 | 10.7.72 | 100 | Caltex |
| 6 | Plurdalen | 29.6.72 | 12.10.72 | 106 | FINA |
| 7 | Kvadehuken 1 | 21.4.73 | 10.8.73 | 112 | Norsk Polar Navigasjon |
| 8 | Hopen 2 | 20.6.73 | 10.70.73 | 123 | FINA |
| 9 | Kvadehuken 2 | 13.8.73 | 10.3.74 | 186 *** | Norsk Polar Navigasjon |
| 10 | Sarstangen | 15.8.74 | 1.12.74 | 109 | Norsk Polar Navigasjon |
| 11 | Haketangen | 11.9.76 | 19.9.77 | 109 *** | Norsk Polar Navigasjon |
| 12 | Colesbukta | 13.11.74 | 1.12.75 | 373 | Trust Arktikugol |

* Drilled in four phases
** Drilled in seven phases, not yet finished
*** Drilled in two phases

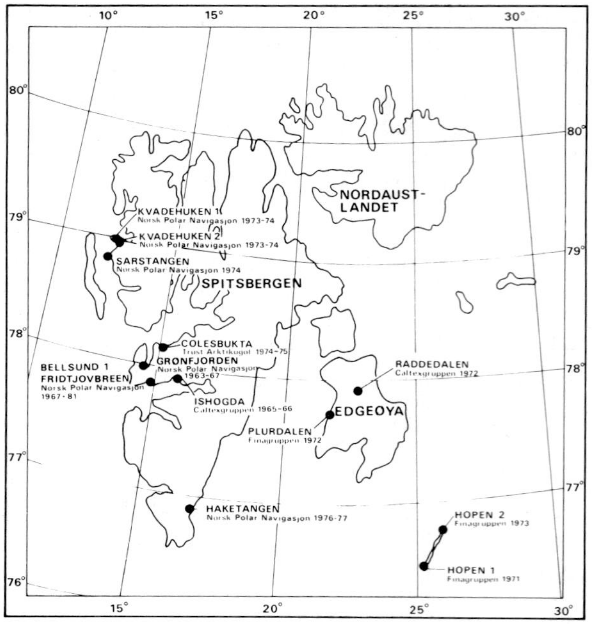


FIGURE 6.
**EXPLORATION WELLS ON SVALBARD**

12 exploration wells have been drilled onshore the Spitsbergen Islands. The locations are shown in Fig 6.

The first well was spudded in 1963 by Norsk Polarnavigasjon. The wells are shown in table III.

## Drilling results

The logs and the detailed drilling results have not yet been released from any exploration wells either offshore or onshore. However, extensive informations are publicly available and it is possible to give a general evaluation of the area.

### Onshore Spitsbergen

No discoveries have been made in the wells drilled on Spitsbergen. Thick sediments with rich source rocks, reservoir rocks, cap rocks and significant structures have been encountered. A very negative aspect is the low porosity in many of the potential reservoir rocks. It causes high rock velocity. Such high velocities have been

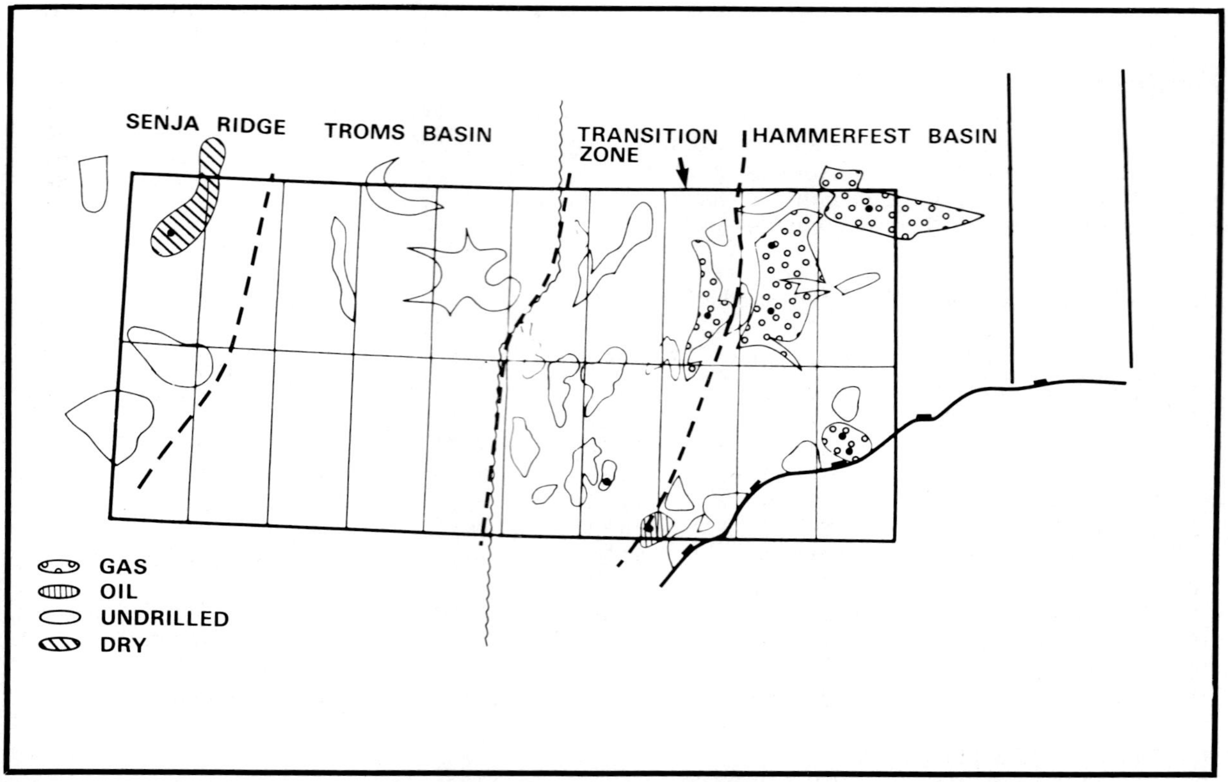


FIGURE 7. **TROMS I — PROSPECTS AT JURASSIC - CRETACEOUS LEVEL**

reported from several seismic investigations of the Spitsbergen islands. The maturity of the source rocks show variations and may represent a problem in some areas.

Because of the regional nature of those negative factors it is necessary to attach a high geologic risk to the possibility of finding commercial amounts of hydrocarbons on the central Spitsbergen islands. The possibility cannot be ruled out and occasional discoveries could be made, but the islands will probably not become a major hydrocarbon province.

### Offshore Troms

The results of the wells in the Troms 1 Area are considered very encouraging. Five gas discoveries have been made with recoverable reserves in the order of 200 x $10^9$ $m^3$. This is about the same amount as that of the major Frigg field in the North Sea. Around half of the reserves are found in the Askeladden Field in block 7120/8 (Fig 7).

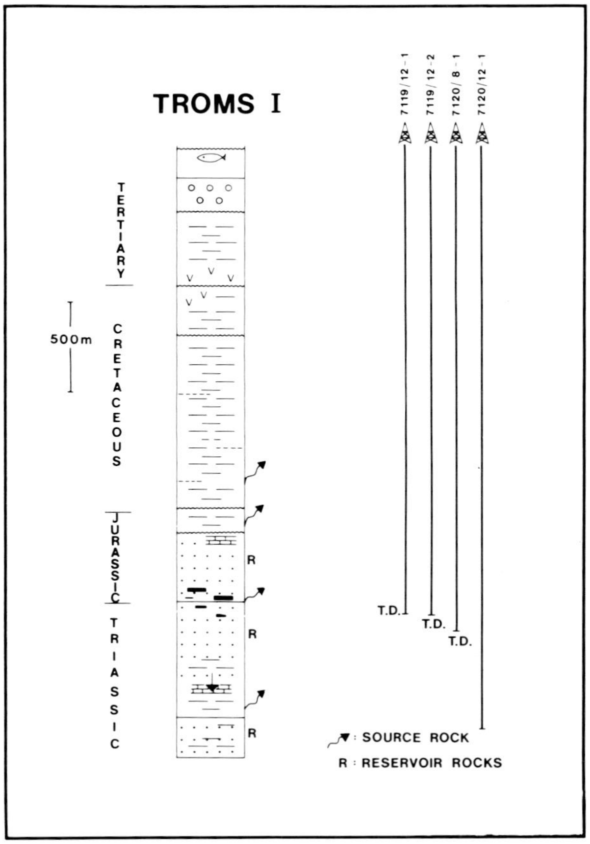


FIGURE 8.

The other discoveries have been made in the blocks 7120/7, /9, /12. In the latter block two small discoveries were made at two different stratigraphic levels in the same well (7120/12-2). They are all found in sandstones. Four of them are of Jurassic age while the deep discovery in 7120/12-2 is of Triassic age.

No oil discoveries have been made, but significant oil shows have been encountered. In 7119/12-2 excellent oil shows were encountered in porous Jurassic sandstone. The cores were bleeding oil, but the logs and subsequent testing showed that the water saturation was high and only water produced. It does, however, demonstrate that oil must have been generated in the area.

A generalized lithostratigraphic column and four of the wells are shown in Fig 8.

The wells have demonstrated potential reservoir rocks and source rocks in several horizons. Porosity is considered to be a critical factor. It seems to deteriorate with depth. A better understanding of porosity distribution within the different basins will therefore be important for the future exploration.

Organically rich shales have been encountered in several horizons. Analyses have shown that they are potential source rocks for oil as well as gas. It is therefore reasonable to assume that oil will eventually be found in the area.

## Further plans

The Norwegian parliament has said that further activity should be based on expansion of the area which has already been opened, rather than opening a new area. In accordance with that, the Ministry plans to enlarge the Troms 1 Area towards north and east. Another nineteen blocks are expected to be opened for the companies in 1983. The timing and number of blocks actually being allocated in the new area will be viewed in relation to the allocation of concession areas further south on the Norwegian Shelf. It will also depend on the allocation within Troms 1 where several of the remaining blocks are considered very prospective.

A white paper addressing these questions is expected to be discussed in the Norwegian Parliament (Stortinget) during the spring of 1983. The first phase of allocation of new blocks could subsequently occur late 1983.

## Exploration strategies

Seen from the point of view of the authorities there are three main objectives for the allocations to come:

i) *Find more gas*

Although large, the gas discoveries made until now are not considered commercial at this stage. If Askeladden had been found in the North Sea, it would probably have been considered commercial, but in this area with the long distance to potential markets, more gas will be needed to make a pipeline viable. It will probably require 2-3 times the amount discovered by now. Statoil as operator for 7120/8 is evaluating the economy of a possible LNG-plant. Such a transportation system may require less reserves to be commercially attractive. Operational costs are, however, expected to be significantly higher than those of a pipeline. The economy of each development could therefore be better and more discoveries developed if sufficient reserves for a pipeline project were discovered. Viewed from a national point it is therefore desireable to allocate concession areas which holds a good chance of finding more gas.

Production of gas is very much a question of finding a market. The development of the gas consumption in the main consumer countries of the world is shown in Fig 9.

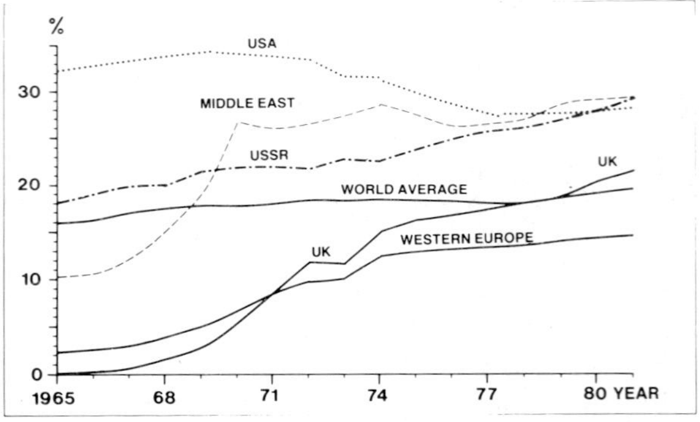


FIGURE 9.
**CONSUMPTION OF NATURAL GAS AS A PERCENTAGE OF T.P.E. FOR SELECTED MARKETS (Source: BP)**

A possible pipeline from the Norwegian Arctic will have to go to Western Europe and/or UK. Prognosis for future demand in Western Europe is shown in Fig 10 and that of UK is shown in Fig 11. They demonstrate that towards the end of the century there is a significant uncovered

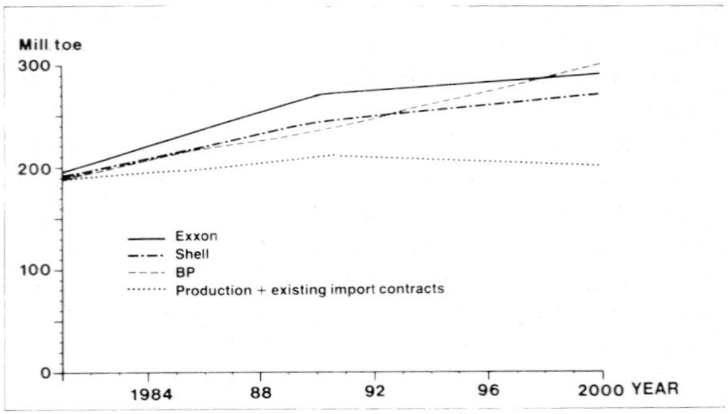


FIGURE 10.
**PROGNOSIS FOR THE DEMAND FOR NATURAL GAS IN EUROPE**

demand. This corresponds to the earliest possible time when gas from the Norwegian Arctic could be on stream.

The conclusion is that potential markets are available at the time when gas from the Norwegian Arctic could be developed.

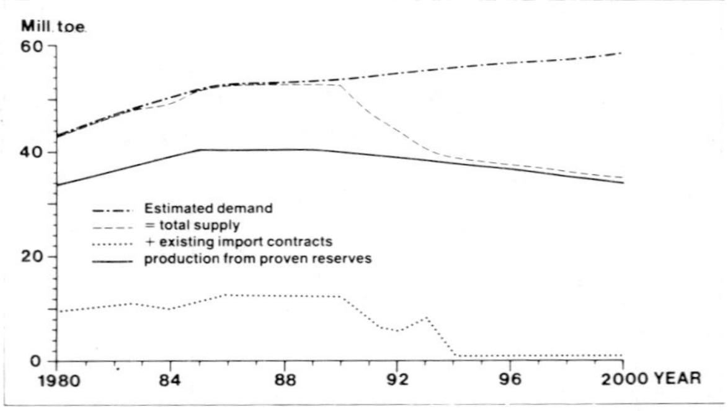


FIGURE 11.
**PROGNOSIS FOR THE DEMAND FOR NATURAL GAS IN GREAT BRITAIN (Source: Financial Times)**

ii) *Find oil*

As described above, no oil discoveries have been made. Significant oil shows have, however, been encountered and there are good reasons to assume that oil has been generated.

Analyses show that rich source rocks occur in several horizons and they are of such a nature that both oil and gas can be generated.

For the generation of oil, it is necessary that the source rock is mature. This requires burial to a depth where it is exposed to a temperature somewhere between 100°C and 200°C. Gas has a wider temperature window. The oil-shows which are encountered, demonstrate that oil has been generated in the Troms Area. Seismic interpretation indicates that the oil window has been reached over large areas. An oil discovery which can be economically developed could probably be on stream faster than a gas discovery. It would leave the operator with a choice between offshore loading and/or pipeline and there are broader market possibilities. Development of production in the Northern Norway has high political priority. For the further block allocations it is therefore a line of action to secure the testing of structures in an optimal position for the discovery of oil.

Every explorationist will know that it is very difficult to find significant amounts of hydrocarbons. To predict whether oil or gas will be found is even more difficult — especially in «new» areas with few wells like the Troms Area. It should therefore be stressed that although the two strategies of finding oil and gas are well defined it is not all that clear how to achieve it.

iii) *Secure the testing of a variety of prospect types*

Experience show that for the thorough mapping of the resources it is necessary to secure the drilling of a broad variety of prospect types. Although agreement on the main geologic trends and the definition of different sediment basins can be arrived at, the views differ considerably when it comes to the detailed prospect evaluations and the decisions on where to drill. This was well demonstrated when the companies applied for blocks in the Troms Area. 20 blocks were announced for allocation and 25 companies applied. There was at least one application for 13 of the blocks and seven blocks were ranked as 1st priority by at least one company.

Thorough mapping of the resources can best be achieved by allowing several companies with different geologists to compete for the concessions within the areas which are opened for drilling at any given time. This has been Norwegian policy from the very beginning and it seems like there is broad political support for such a policy also in the years to come.

It could perhaps be said that two main factors determine how much oil and gas will be found. The one is the actual amount which is preserved in the ground. The other is the application of a concession policy which allows for competition between several companies and thereby secure the input of different geologic evaluations.

## Prospectivity of the Barents Sea

The whole of the Barents Sea contains thick sediments. The major part is classified as an intracratonic basin. The western part borders the oceanic Norwegian Sea with a passive shear margin.

The Barents Sea Shelf is subdivided in several sedimentary basins and ridges. Some of them are shown in Fig 5. They provide a variety of prospects. Changing depositional environments through the different geologic periods give rise to a variety of prospects within each basin. We are in a very early stage of the exploration of the area and it is only in the southern Barents Sea that the main geologic trends have been defined.

The western margin contains the youngest deposits where thick tertiary sediments are found. The eastern part reveals Mesozoic sediments immediately below thin quaternary bottom sediments.

Onshore USSR both oil and gas are produced from the Pechora Basin. It is mainly from Paleozoic rocks (Devonian and Permian sandstones and limestones). Source rocks have been reported in beds as old as Ordovician. Sediments from the same stratigraphic horizons which are expected to be found in the Barents Sea have been studied onshore Spitsbergen, Bjørnøya and Pechora Basin. Eastern Greenland and Andøya in the Lofoten Area are also containing relevant sedimentary rocks. Based on information from those areas, and well results, paleogeographic reconstructions have been made for the main stratigraphic horizons.

Rich source rocks are expected to be widespread and occur in several stratigraphic horizons. Seismic profiles show that they have been buried to such a depth that they are mature over large areas. Good reservoir rocks in several stratigraphic horizons and cap rocks are expected to have a widespread occurrence. Significant structuration is known from the major part of the Barents Sea.

The conclusion is that the Norwegian Arctic offshore areas are considered very prospective. Significant discoveries have already been made and further drilling is expected to reveal more.

## Development costs

The area is considered prospective, but the cost of possible development will be very high. The water depth is considerable (2-500 m), the

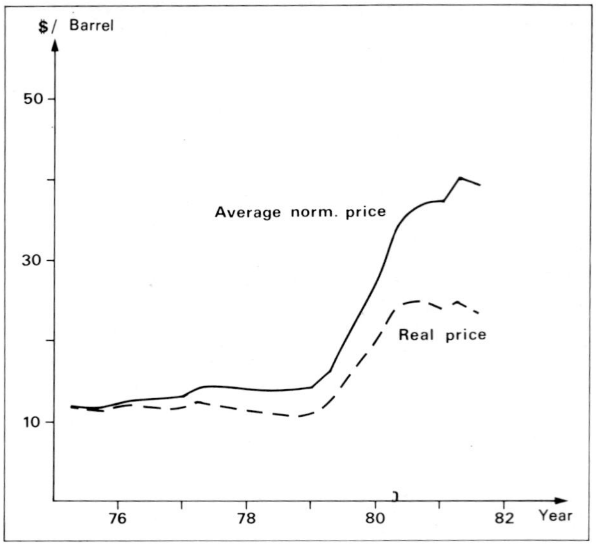


FIGURE 12.

**DEVELOPMENT OF OIL PRICE**
**nominal and real price 1975=100**

distance to possible market is great and ice related problems will also add to the cost. As a result of the very long period of investments before any significant income can be acquired, the capital risk will be much higher than that of projects in less hostile areas.

The discoveries will therefore have to be large if they are going to be commercial. In the absence of existing infrastructure this is particularly true for the first field to be developed.

If such a field is found, however, smaller fields in the area could take advantage of the production installations and thereby become economic. This happened in the North Sea when Ekofisk was discovered.

The price of oil and gas will also be very important for the economy of the fields. After a long period of low and stable prices the last decade has shown rather dramatic changes. Two major price jumps have occurred. The first in 1973 and the latter in 1979.

The last year has shown a decreasing trend. The price development is shown in Fig 12.

I believe the price will increase in the long run, also in real terms, but it will probably be vulnerable and subject to fluctuations. The development of the oil price is dependent on factors outside Norway. But the vulnerability has added another element of economic risk, especially in remote areas like the Norwegian Arctic.

The remarks regarding development costs will probably be the same for any area in the Arctic.

*Cold Regions Science and Technology*, 7 (1983) 51–88
Elsevier Science Publishers B.V., Amsterdam — Printed in The Netherlands

# Petroleum Potential in Arctic North America and Greenland

*Walter W. Nassichuk*

Geological Survey of Canada
Calgary

## Abstract

*Sedimentary deposition in Arctic North America and Greenland, in Phanerozoic time, occurred north of the Precambrian Canadian Shield and extended from the Wandel Sea in the east to the Chukchi Sea in the west. Seven geological provinces considered to be highly prospective for oil and gas exist within the region. Lower Paleozoic sediments may exceed 3000 m (10 000 ft) in the several separate basins in the Arctic Interior Platform of the Central Stable Region. To the north of the Arctic Interior Platform, lower Paleozoic sediments thicken into the Franklinian Geosyncline, the second geological province. Twelve thousand metres of lower and middle Paleozoic rocks are folded and faulted along a sinuous belt extending from the Wandel Sea in northeastern Greenland to Prince Patrick Island in the western Canadian Arctic Islands. Unconformably overlying the Franklinian Geosyncline are two successor basins, the third and fourth geological provinces considered in this report. The Sverdrup Basin in the Canadian Arctic Islands contains over 13 000 m (43 000 ft) of upper Paleozoic, Mesozoic and Tertiary sediments. The Wandel Sea Basin in and adjacent to northeastern Greenland, analogous to the Sverdrup Basin, contains a similar complement of upper Paleozoic, Mesozoic and Tertiary sediments. Onshore in Greenland, a composite thickness of 3000 m (10 000 ft) of sediments has been described and offshore, beneath the continental shelf, sediments in the basin may thicken to 10 000 m (33 000 ft). The Arctic Slope of Alaska, the fifth province, exhibits characteristics of a successor basin that has been rotated, translated or otherwise displaced from the Arctic Ocean Basin. Beneath a thickness of some 10 000 m (33 000 ft) of upper Paleozoic, Mesozoic and Tertiary strata is a folded and faulted lower and middle Paleozoic terrane similar to the Franklinian Geosyncline «Fold Belt». The southern margin of this province, now containing the Brooks Range, was folded and faulted during the Laramide Orogeny. A sixth province, the Arctic Coastal Plain, extends along the continent's northern margin from Alaska, to northern Greenland. Strata ranging in age from Jurassic through Pleistocene thicken abruptly from a featheredge to more than 12 000 m (39 000 ft) as they prograde into the Arctic Ocean Basin. The seventh major geologic province is the Baffin Bay Basin, which contains some 10 000 m (33 000 ft) of Mesozoic and Tertiary sediments.*

*With the exception of the Wandel Sea Basin, each of the geological provinces has had its petroleum potential evaluated from well-data. The Arctic Slope of Alaska has hydrocarbon reserves in the order of 1.6 x $10^9$ $m^3$ (10.0 Bbbl) of oil and 750 x $10^9$ $m^3$ (26.5 Tcf) of natural gas. Recoverable reserves for the Beaufort Sea-Mackenzie Delta region of the Arctic Coastal Plain are tentatively considered to be 0.15 x $10^9$ $m^3$ (943 MMbbl). The potential natural gas reserves for the Beaufort Sea-Mackenzie Delta region are in the range of 0.25 x $10^{12}$ $m^3$ (8.83 Tcf). The Sverdrup Basin has yet to have significant reserves of oil indicated but the potential reserves of natural gas range from a minimum of 0.45 x $10^{12}$ $m^3$ (15.89 Tcf) up to what will probably prove to be more than 565 x $10^9 m^3$ (20 Tcf) with effective delineation drilling. The Franklinian Geosyncline*

0165-232X/83/$03.00  

*contains both oil and gas but no significant reserves have been developed. The Wandel Sea Basin, Baffin Bay Basin and Arctic Stable Platform have yet to produce indications of significant pooled hydrocarbons.*

*The undiscovered resource potentials of the Arctic Slope and contiguous Arctic Coastal Plain of Alaska have been estimated to have median values of 3.0 x $10^9$ $m^3$ (18.9 Bbbl) of oil and 1.39 x $10^{12}m^3$ (49.08 Tcf) of natural gas. The undiscovered resource potentials of the Beaufort Sea-Mackenzie Delta region of the Arctic Coastal Plain have expected values of 1.35 x $10^9$ $m^3$ (8.45 Bbbl) of oil and 2.91 x $10^{12}$ $m^3$ (102.75 Tcf) of natural gas. The combined resource potential of the Arctic Interior Platform, Franklinian Geosyncline, Sverdrup Basin and Baffin Bay Basin have expected values of 0.82 x $10^9$ $m^3$ (5.16 Bbbl) of oil and 2.98 x $10^{12}$ $m^3$ (105 Tcf) of natural gas. The resource potential of the Arctic Islands Coastal Plain has not yet been effectively evaluated because the greater portion of the region lies beneath constantly shifting pack-ice and data required for a meaningful assessment are not available. It is possible however, that the Coastal Plain in the Arctic Islands area may have resource potentials as great as any other previously assessed Arctic province. Formal estimates of the resource potentials of northern Greenland and the Wandel Sea Basin are not available but on the basis of comparable geology, values can be expected to be proportional to those of the Arctic Interior Platform, the Franklinian Geosyncline and the Sverdrup Basin, respectively.*

## Introduction

Sedimentary basins in Arctic regions of the Western Hemisphere are diverse in sedimentary and structural style; at least two of them, the Alaskan Arctic Slope and the Sverdrup Basin, contain giant accumulations of hydrocarbons. The discovery of oil at Prudhoe Bay, Alaska, the first giant oil pool to be found in Arctic North America, was the beginning of a new and exciting period of Arctic research and exploration. An effort requiring the expenditure of many billions of dollars is still necessary to find the other Arctic giants that almost certainly exist.

FIGURE 1. *Prospective Petroleum Provinces in the Western Arctic Hemisphere.*

In this report, the geology and hydrocarbon potential of basins in seven major geological provinces of Arctic North America and Greenland will be reviewed. It is in one, two or perhaps three of these provinces that other «giants» will be found. These petroleum resources must be developed without causing major and irreversible stress to delicate environments or to the way of life of native peoples.

The basins discussed herein are underlain by «basement» rocks, generally non petroliferous, that vary from Precambrian crystalline rocks to orogenic complexes to oceanic crust. Precambrian basement is mainly the immense Canadian Shield (Fig. 2) of Archean and younger Precambrian ages that forms the fundamental continental structure for most of Canada and Greenland. The orogenic complexes are composed of metasedimentary and igneous rocks that were deformed and emplaced during the Ordovician through Lower Devonian Caledonian Orogeny, the Upper Devonian Ellesmerian Orogeny or the Jurassic through Paleogene Columbian and Laramide Orogenies. One orogenic complex, the Cordillera of Alaska and northwestern Canada (Fig. 3), is a complex of accreted, allochthonous terranes, the origins and ages of which remain conjectural.

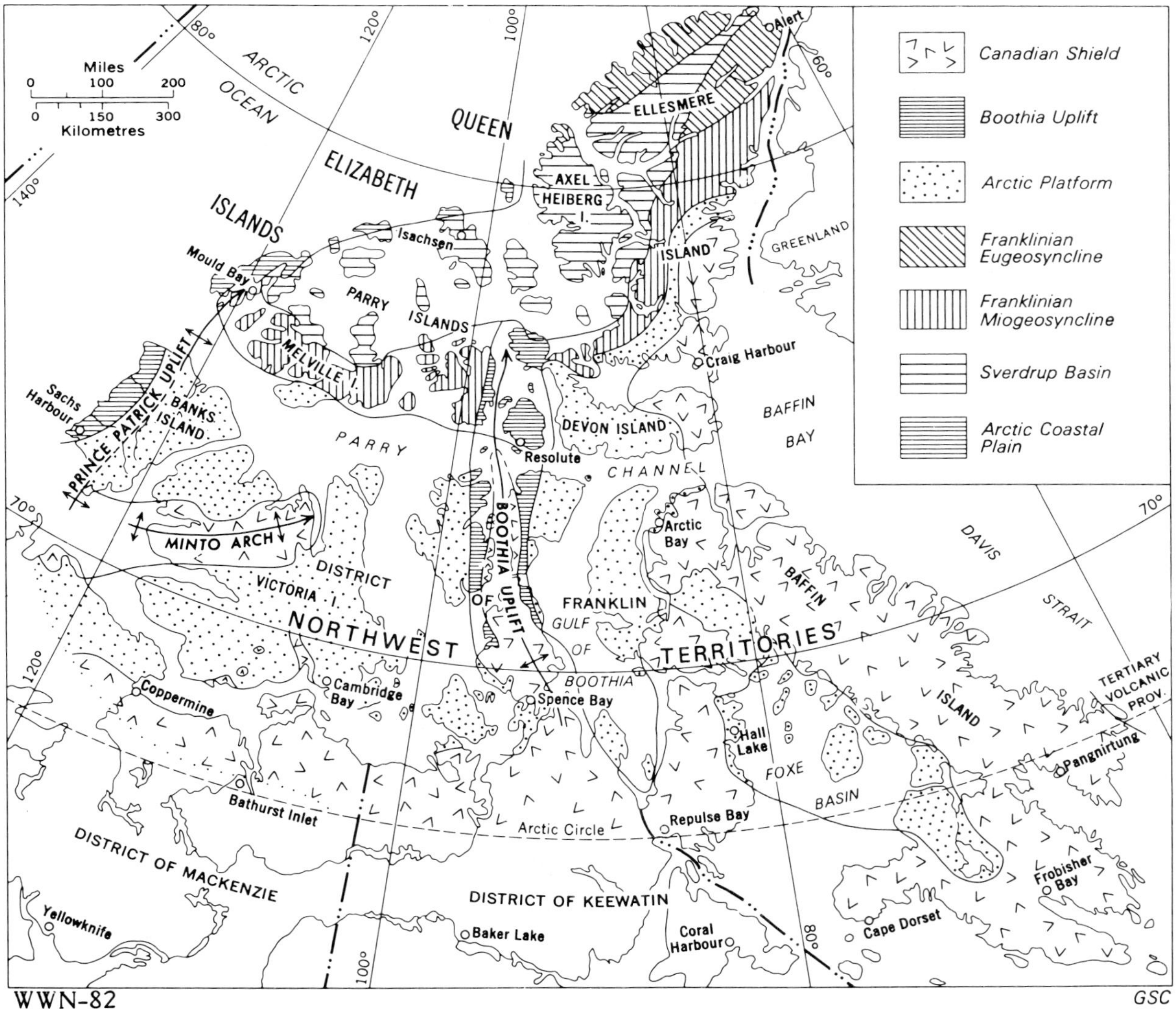


FIGURE 2. *Geological Provinces of the Canadian Arctic Archipelago; from Thorsteinsson and Tozer, 1970.*

The Arctic Coastal Plain, Baffin Bay Basin and the Wandel Sea Basin (Figs. 1, 3) lie in part on oceanic crust or attenuated continental crust proximal to oceanic crust. Although its age is not everywhere known the oceanic crust is thought to be younger than Jurassic and in some instances as young as Tertiary.

Most of the basins discussed here are actually structural basins but the present boundaries probably approximate the limits of the original depositional basins. Some of the basins are relatively shallow and the contained sediments form a thin, discontinuous cover on the Shield or craton; other basins are deeper and linear, and lie parallel with or form the edge of the existing continent.

The sediments within the various Arctic basins represent all geologic systems from Precambrian to Recent. With the exception of the youngest sediments, which have not been deeply buried, and the oldest sediments, which have not yet been adequately tested, all the depositional successions have yielded indications of hydrocarbons. To date, the largest accumulations have been found in rocks of Devonian or younger age. Geochemical data, however, also indicate a possibility of significant hydrocarbons in lower Paleozoic sediments.

The writer is indebted to colleagues in the Geological Survey of Canada, the United States Geological Survey and in industry, who have assisted with preparation of this summary; R.L. Christie, K. Osadetz, A. Embry, J. Dixon, T. Powell, D.K. Norris, R. Procter, G.C. Taylor, I. Tailleur and N. Haimila.

## Geological Provinces and Sedimentary Basins

Four structural-stratigraphic provinces in Arctic Canada and Greenland contain one or more sedimentary basins that are already known to be or are potentially petroleum-bearing: A) the Arctic Interior Platform of the Central Stable Region, B) the continent-marginal Franklinian and Cordilleran geosynclines, C) successor basins of the geosynclinal areas, and D) young basins of the coasts and channels. The sedimentary basins shown in Figure 1 include: the several cratonic basins of the Interior Platform of the Central Stable Region,

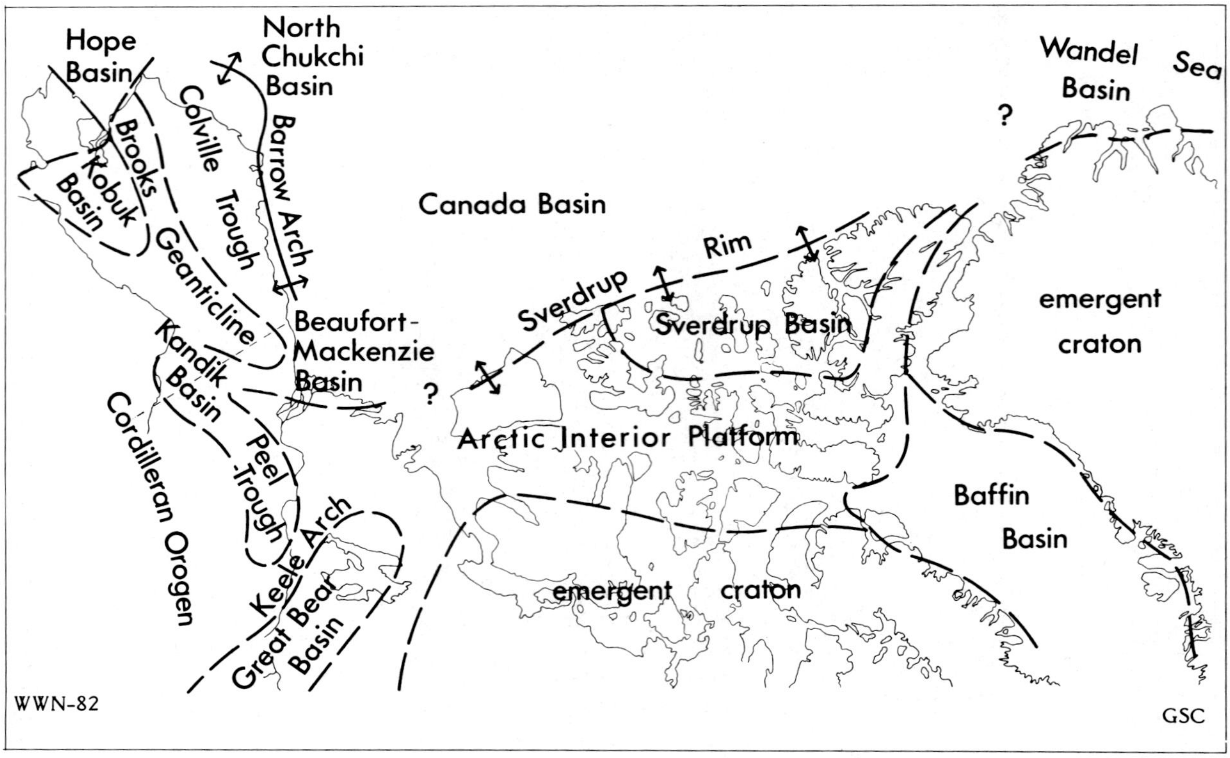


FIGURE 3. *Major tectonic provinces, mid-Early to Late Cretaceous (Brookian) tectonic regime; from Balkwill et al., in press.*

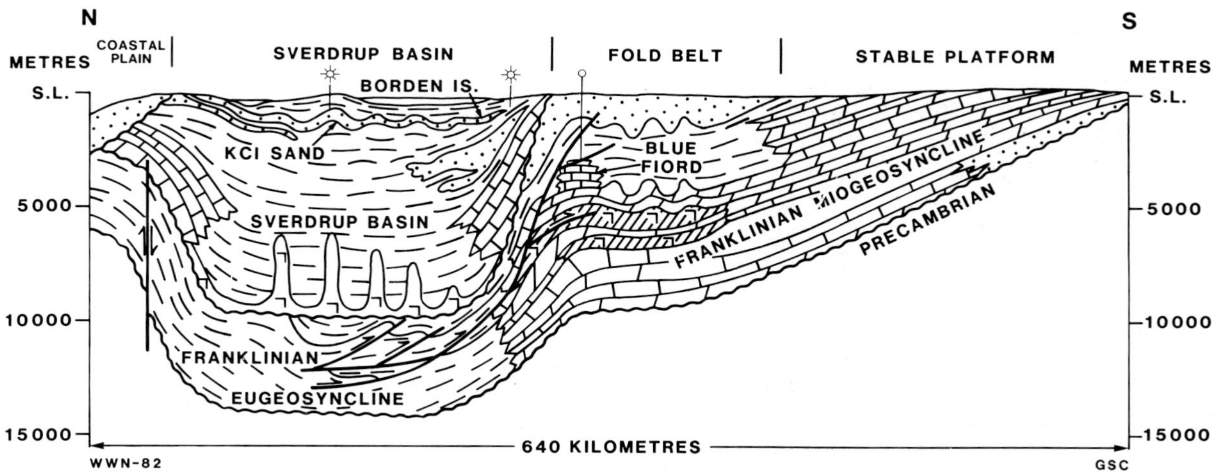


FIGURE 4. *Diagrammatic restored section in Canadian Arctic Islands showing relationships between Stable Platform, Franklinian Geosyncline, and Sverdrup Basin; from Rayer, 1981.*

(A), the Franklinian Geosyncline of Canada and Greenland (B), the Wandel Sea Basin of Greenland (C), the Sverdrup Basin (C), the Arctic Slope of Alaska (C), the Arctic Coastal Plain (D), and the Baffin Bay Basin (D).

The basins of the Arctic Interior Platform province of Greenland and northern and western Canada are shallow and relatively undisturbed; that is, the beds are still nearly flat-lying, as when deposited. The Arctic Interior Platform and the Canadian Shield together make up the Central Stable Region, which has been characterized by relatively stable tectonic conditions since Cambrian time or earlier. The deeper basins — the Franklinian Geosyncline and the Arctic Slope of Alaska — are moderately to severely folded and faulted and now form parts of two zones of deformation, or orogenic belts: 1) the Innuitian orogenic region to the north, which consists of the Franklinian Geosyncline and Sverdrup Basin (Fig. 4); and 2) the Cordilleran orogenic region to the west (Fig. 3). The orogenic belts, in contrast with the Central Stable Region, were characterized by long periods of subsidence and nearly continuous, abundant sedimentation interrupted by major periods of folding, faulting and intrusion of igneous rocks. The successor basins between the Arctic Slope of Alaska and the Wandel Sea Basin were also of considerable depth, were involved in later stages of deformation in the orogenic belts, and, as indicated earlier, are now part of the orogens. Young sediments of the Arctic Coastal Plain and Baffin Bay form wedge-shaped prisms and have undergone negligible tectonic deformation.

## Regional Sedimentation — A Summary

A succession of clastic, carbonate and volcanic rocks, generally unmetamorphosed and with a thickness of nearly 6000 m (20 000 ft) was deposited in several Precambrian basins in the Canadian Shield. These little known strata contain traces of hydrocarbons and clearly call for increased exploration in the future. These strata are overlain by some 3000 m (10 000 ft) of Cambrian to Devonian sandstones, limestones, dolomites, conglomerates and anhydrite, all widely distributed in relatively shallow cratonic basins within the Arctic Interior Platform region. The Cambrian to Devonian cratonic basin successions grade northerly into strata of the Franklinian Geosyncline (Fig. 4) which received some 12 000 m (39 000 ft) of clastic, carbonate, evaporitic and some volcanic rocks between Late Precambrian and Late Devonian time (Fig. 5).

Between Late Devonian and Early Carboniferous time, that is, during the Ellesmerian Orogeny, the Franklinian Geosyncline was uplifted, intruded and metamorphosed. Deformed strata of the geosyncline subsequently subsided and were overlain by thick, 13 000 m (43 000 ft) successions of Carboniferous (Viséan) to Tertiary strata including clastic, carbonate, evaporitic

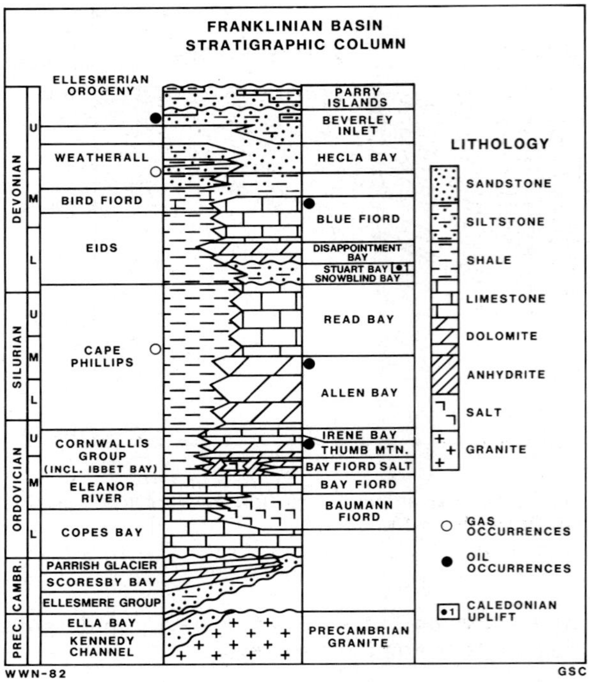


FIGURE 5. *Generalized stratigraphic succession in the Franklinian Geosyncline; from Rayer, 1981.*

and volcanic rocks; these successions form the Sverdrup Basin in the Canadian Arctic (Figs. 6, 9) and the Wandel Sea Basin in Greenland (Figs. 11, 12). Both successor basins were folded and faulted during orogenesis that continued intermittently from Late Cretaceous to Middle Tertiary time.

In the Arctic mainland region of Canada, late Precambrian and early Paleozoic sedimentary rocks of the Arctic Interior Platform (Fig. 3) and Cordilleran Geosyncline evidently represent shelf and miogeosynclinal facies that bordered the northern and western margins of the North American craton. Greater total thicknesses occur in the geosynclinal region and the shelf sequences are marked by thinning and truncation of formations along ancient arches and geanticlines, and by unconformities with locally great angular discordance. Shallow-water conditions of the shelves and miogeosynclines are represented by fine-grained clastic rocks and by carbonates. Sedimentation in the interconnecting troughs of the geosynclinal region is represented by shale, argillaceous limestone, siltstone, and sandstone. Deposits near former geanticlinal regions include feldspathic clastic rocks, conglomerate, limestone, and orthoquartzite.

A great marine transgression occurred during Early, Middle, and early Late Devonian time and carbonate «sheets» of these ages extend from the geosyncline well onto the craton. Thick clastic sequences were deposited in local basins during and after Late Devonian orogenesis, and granites were emplaced in the geosynclinal region. Geanticlines separated local troughs in late Paleozoic and Mesozoic times, and sediments of these ages rest with angular unconformity on rocks folded during late Paleozoic orogenesis. Jurassic and Early Cretaceous clastic sediments were deposited in foredeeps in northern Yukon Territory, Alaska, and south and east of these regions. Late Mesozoic and early Tertiary orogeny folded the younger sediments of the Cordilleran region; equivalent beds, for the most part undisturbed Lower Cretaceous sandstones and shales, underlie large areas in the Mackenzie Delta region east of the Cordilleran fold belt (Fig. 3).

Overlying and transecting eroded structures in the orogenic regions from northern Alaska through to the northern Queen Elizabeth Islands (Fig. 2) is the Arctic Coastal Plain, a narrow belt containing Upper Tertiary and Lower Quaternary sand and gravel at the surface and sediments possibly as old as Jurassic in the subsurface. The Coastal Plain is continuous with the Arctic Continental Shelf, which extends about 160 km into the Arctic Ocean.

The Baffin Bay Basin (Figs. 1, 3), which may have originated during early stages of the rift-opening of the Atlantic Ocean, contains clastic sediments of late Mesozoic and Tertiary age. The Baffin Bay sedimentary domain may extend into the broad, inter-island channels to the north and west.

## Evaluation of Petroleum Resources

Since 1971 the Geological Survey of Canada has been developing techniques for estimating, on a scientific basis, the amount of oil and gas that is known to exist or presumed to exist in the major sedimentary basins of Canada. Estimates of total hydrocarbon resources comprise two parts: discovered resources (reserves), and undiscovered or inferred resources (potential). Current estimates for recoverable oil and gas resources in Arctic regions of Canada and Alaska are shown in Table 1.

Techniques for estimating oil and gas resources have, in the last decade, evolved into sophisticated procedures that use the data base that is acquired in the course of petroleum exploration. This approach contrasts with that used in former times, when resources were estimated from volume of sediments using «petroleum yield factors» for the various basins.

The assessment procedures now followed use probabilistic analyses of 'plays', where a play is a group or family of potential drilling targets or 'prospects' (usually certain stratigraphic intervals) with similar physical attributes. Each attribute, or parameter such as areal extent, reservoir thickness, or porosity may be represented by a range of measurable values, the values distributed in a cumulative frequency form. The several parameters for each prospect occur in different mutual combinations; volumes of hydrocarbons for each combination may be calculated using engineering factors. All possible parameter combinations can be simulated mathematically with the aid of computers, and the result is a complete range of possible sizes of hydrocarbon accumulation that may occur in the prospect, or, by extension, in the play. The 'typical prospect size', furthermore, is represented by the 'expected value' or mean value of the cumulative frequency distribution. The potential size of the play can be obtained as a sum of the potential hydrocarbon accumulations of the prospects, or as a product of typical prospect size and the number of prospects, the calculation modified by a 'risk' factor taking into account appropriate levels of probability or uncertainty. Finally, the sum of the potential sizes of the plays within a basin provides a potential resource for the basin.

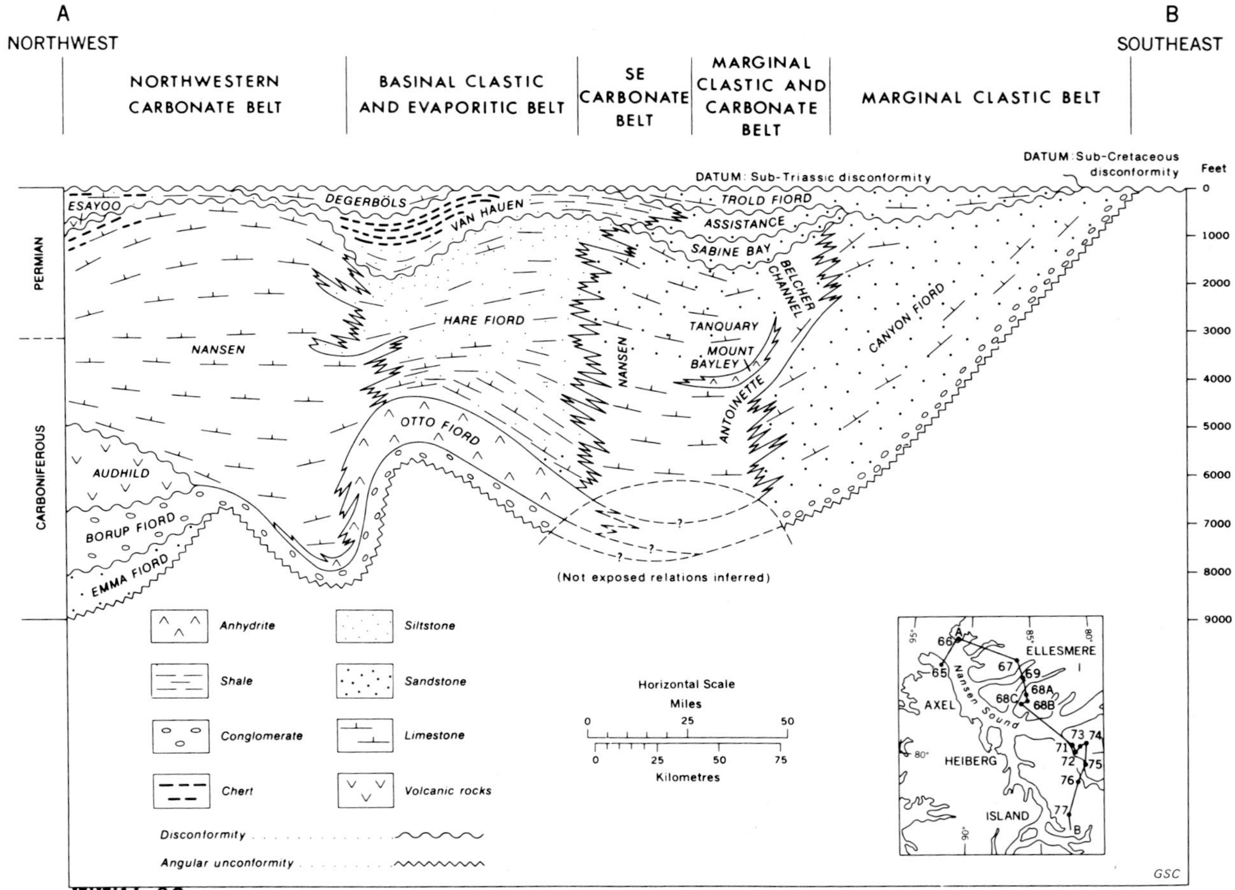


WWN-82

FIGURE 6. *Diagrammatic restored section of Carboniferous and Permian rocks across Sverdrup Basin, from Thorsteinsson, 1974.*

The range of sizes of resource potential is usually expressed as a «greater than» cumulative frequency distribution in which the estimated values are related to appropriate probabilities or levels of certainty.

The statistical technique described above is a simulation process using random numbers, or a 'Monte Carlo technique'. This technique, and certain others (such as a log-probability simulation) have been used successfully for some years now in preparing estimates for exploration planning or competitive bidding for leases.

An important feature of the statistical method described here is that resource estimates can be made even from poor data bases. With few data, however, the range of estimates will be wide and the levels of certainty low. As data accumulate, of course, updated estimates with improved certainties can be made.

All of the accessible sedimentary basins of Canada have been evaluated and estimates of resource potential have been made. The data tested by well penetrations. Substantial reserves of both oil and natural gas have been discovered on the Arctic Slope in northern Alaska, and in the Beaufort Sea-Mackenzie Delta (Fig. 3) and Sverdrup Basin regions of Canada.

## The Petroliferous Basins

### A Summary of Oil and Gas Discovery

The seven potentially to significantly petroliferous geological provinces reviewed in this report were enumerated earlier. Following are remarks on the nature of these basins, their contained sediments and their petroleum potential. Many of the hydrocarbon accumulations discovered in Arctic regions may become economically viable once a threshold of reserves is established in the area. Furthermore, as an infrastructure for exploration, development and transportation is developed, additional pooled hydrocarbons will become available for exploitation.

**Table 1.**

*Oil and Gas Potential (Discovered and Undiscovered Recoverable Resources)*

| Region | Mean Oil $m^3$ (Bbbl) | Mean Gas $m^3$ (Tcf) |
|---|---|---|
| Arctic Slope Alaska . . . . . . . . . . . . . | $2.74 \times 10^9 m^3$ (17.2) | $1.37 \times 10^{12} m^3$ (48.4) |
| Arctic Coastal Plain . . . . . . . . . . . . | $2.0 \times 10^9 m^3$ (12.6) | $0.76 \times 10^{12} m^3$ (26.8) |
| Mackenzie Delta-Beaufort Sea . . . . . | $1.50 \times 10^9 m^3$ (9.4) | $3.16 \times 10^{12} m^3$ (111.6) |
| Arctic Interior Platform . . . . . . . . . . | $0.17 \times 10^9 m^3$ (1.1) | $0.24 \times 10^{12} m^3$ (8.5) |
| Franklinian Geosyncline . . . . . . . . . | $0.08 \times 10^9 m^3$ (0.5) | $0.22 \times 10^{12} m^3$ (7.8) |
| Sverdrup Basin . . . . . . . . . . . . . . . | $0.46 \times 10^9 m^3$ (2.9) | $1.98 \times 10^{12} m^3$ (69.9) |
| Baffin Bay . . . . . . . . . . . . . . . . . . | $0.11 \times 10^9 m^3$ (0.7) | $0.54 \times 10^{12} m^3$ (19.1) |

bases for most areas presently being explored consist of extensive seismic coverage, gravity and magnetic surveys, surface geological maps, well data, and geochemical, reservoir and biostratigraphic data. Such full coverage may be described as an excellent data base.

The data suggest that potential petroleum-bearing sedimentary basins of several ages lie to the north of the ancient, crystalline «basement» or cratonic rocks of North America and Greenland. These basins extend from the Chukchi Sea (*see* Chukchi Basin, Fig. 3) in the west, to the Wandel Sea Basin east of Greenland. With the exception of the Wandel Sea Basin and a portion of the Arctic Coastal Plain, the petroleum potential of the basins has been

The Arctic Interior Platform part of the Central Stable Region, the Wandel Sea Basin, parts of the Arctic Coastal Plain and the Baffin Bay Basin do not as yet have any established reserves. Potential resource estimates are not available for the Wandel Sea Basin or for other areas of northern Greenland, on-land and offshore, but geological similarities with Arctic Canada suggest that significant resources will be identified eventually.

Oil and gas are known to occur in Devonian strata within the Franklinian Geosyncline; oil was encountered a number of years ago in several of the «Bent Horn» wells on Cameron Island and gas was indicated in the first well ever to be drilled in the Canadian Arctic Islands at Winter

Harbour, Melville Island in 1961. Neither deposit is, however, of commercial size. Other wells drilled within the geosyncline in Canada have encountered shows of oil and gas (Fig. 5). Thus far, no exploratory drilling for oil and gas has been carried out within the Greenland portion of the Franklinian Geosyncline.

Several oil discoveries and some fifteen gas discoveries have been made in the Sverdrup Basin. The oil discoveries have not yet been fully delineated but reserves are expected to be rather small. Two of the gas fields in the Sverdrup Basin rank in the «giant» category, with over 90 x $10^9 m^3$ (3 Tcf) of marketable natural gas in each. Several other fields are now known to contain approximately 30 x $10^9 m^3$ (1 Tcf) of marketable natural gas each.

Geochemical analyses and exploration indicate that the Mesozoic rocks of the Sverdrup Basin area are prone to the generation of natural gas when they reach geochemical maturity. The upper Paleozoic strata of the Sverdrup Basin have been only sparsely tested but favorable carbonate-shale facies environments with numerous Carboniferous and Permian «reefs» are known to exist in the basin, and maturation levels appear to be suitable for oil and gas generation.

The Prudhoe Bay Field (Fig. 13) on the Arctic Slope of Alaska was discovered in 1968; earlier discoveries, in the late 1940s and 1950s, had been made in the Naval Petroleum Reserve No. 4 but none were comparable to that at Prudhoe Bay. Permian and Triassic reservoirs in the Prudhoe Bay Field are reported to contain 1.5 x $10^9 m^3$ (9.44 Bbbl) of oil and 0.7 x $10^{12} m^3$ (25 Tcf) of gas. Two million barrels of oil have already been produced from the field. The Kuparuk River Field is estimated to have 0.56 x $10^9 m^3$ (3.5 Bbbl) of oil-in-place. Eight other oil and gas accumulations have also been found on the Arctic Slope of Alaska.

Seven potential source rock levels, ranging from Mississippian to Lower Cretaceous have been identified on the Arctic Slope of Alaska. Oil has also been found in Tertiary reservoirs in that area and therefore potential source rocks probably exist in the Tertiary section also. The contiguous Arctic Coastal Plain in the Beaufort Sea-Mackenzie Delta area also contains hydrocarbon accumulations in the Tertiary section. Exploration of the Arctic Coastal Plain in Alaska has not yet extended far into offshore regions but nearer shore Upper Cretaceous «targets» are being sought. Four hydrocarbon accumulations have been indicated for the Arctic Coastal Plain of Alaska but reserves have not yet been fully evaluated.

The Arctic Coastal Plain of Canada is being explored only in the Beaufort Sea-Mackenzie Delta area. Considerable success has been experienced in the area and more than 25 gas and oil accumulations have been identified to date: eight oil pools, eight oil and gas pools, and nine gas pools. Industry submissions to the National Energy Board of Canada in 1977 indicated that approximately 0.17 x $10^{12} m^3$ (6 Tcf) of marketable natural gas was available at that time. The Parsons Lake and Taglu gas fields, the largest fields found to date, each contain marketable gas in the 56-84 x $10^9 m^3$ (2-3 Tcf) range.

Recent press releases indicate that the Kopanoar, Koakoak, Issungnak and Tarsiut areas in the Beaufort Sea each have appreciable oil accumulations. Although the fields have not been fully delineated by drilling, well-log, sample and seismic data have allowed probable reserve estimates to be made. Estimates of oil in place for the Koakoak area indicate the possibility of between 0.32 and 0.80 x $10^9 m^3$ (2.0 to 5.0 Bbbl) and for the Kopanoar area the possibility of between 0.29 and 0.72 x $10^9 m^3$ (1.8 to 4.5 Bbbl) Issungnak and Tarsiut may also contain accumulations of comparable size.

Hydrocarbons in the Beaufort Sea-Mackenzie Delta area were evidently derived from source rocks of three different ages between the Jurassic and Tertiary: 1) the Jurassic — Early Cretaceous 2) Late Cretaceous to Tertiary and 3) Tertiary. Most of the organic material in the Mackenzie Delta — Beaufort Sea area is terrestrial in origin and is prone to gas generation. The oil in the area is thought to be derived from anomalous source material and will be discussed later.

## Geology of the Arctic Interior Platform, Central Stable Region

The Arctic Interior Platform is made up of large areas of relatively undisturbed Paleozoic and younger sedimentary rocks that lie between the Canadian Shield (Fig. 2) and the Innuitian Region. The contact between the Paleozoic rocks of the platform and the exposed Precambrian Shield is markedly irregular. This is due in part to the considerable topographic relief and the innumerable straits and channels of the North American Arctic archipelago. Also, it is due to exposure, through erosion, of Precambrian topographic highs and to the presence of arches and uplifts that postdate the Paleozoic rocks. The most prominent of the northerly-trending arches is the Boothia Uplift in the south-central Canadian Arctic (Fig. 2).

Five more-or-less discrete Paleozoic basins are known from the Interior Platform of the Canadian Arctic but a relatively regular homocline of Paleozoic beds forms the Arctic Interior Platform of northern Greenland. Some of the basins in the Interior Platform in Canada are clearly ancient features that persistently affected Paleozoic sedimentation, but others appear to have been produced by tectonic events that post-date the youngest rocks in the area.

Numerous outliers of lower Paleozoic sedimentary rocks occur at widely separated localities in the Central Stable Region. Cambrian beds form the base of some outliers, and Middle Ordovician, others; it is evident that there were widespread marine transgressions of the Central Stable Region in Cambrian and in Ordovician-Silurian times. The Paleozoic rocks of the Arctic Interior Platform today are but remnants of a former, more extensive cover.

Paleozoic beds are relatively thin at the southern margin of the Arctic Interior Platform, but thicken northward gradually until, at the former continental margin or «hinge line», they merge with those of the Franklinian Geosyncline (Fig. 4). North of the hinge line, the rate of thickening is considerably greater. The thick sedimentary units along the northern edge of the Canadian Interior Platform now constitute a synform belt of foreland basins (*see* King, 1959; p. 28) conforming to the southern boundary of the Ellesmerian Orogen (the folded part of the Franklinian Geosyncline). The foreland basins are characterized by relatively deep sedimentary columns compared to those of basins of the interior parts of the craton, and by mild deformation due to tectonic activity along the orogenic front. The youngest beds of the foreland basins represent clastic wedges derived from early phases of Ellesmerian Orogeny (late Devonian-Early Carboniferous). The foreland basins of Arctic Canada are broadly synclinal, with axes parallel to the front of Ellesmerian folds. Large areas, or embayments, of Paleozoic rocks south of the foreland basins are evidently basinal in form and more cratonic in nature than the foreland basins. Sedimentary thicknesses in the Paleozoic cratonic basins are also generally uncertain, but aeromagnetic data suggest thicknesses of more than 3000 m (10 000 ft) on either side of the Boothia Uplift (Gregory, et al., 1961, pp. 22-24). Even greater thicknesses probably occur in the foreland basins, marginal to the Franklinian Geosyncline (Fig. 5).

## Petroleum Potential of the Arctic Interior Platform, Central Stable Region

Throughout much of the Arctic Interior Platform a widespread transgression occurred during the Cambrian and Ordovician and a basal sandstone rests directly on crystalline rocks of the Precambrian Shield. These clastic sediments are overlain by a platform sequence dominated by shelf carbonates. Minor eustatic sea level changes or epeirogenic warpings were responsible for depositional thinning against pre-existing topographic highs, for internal unconformities, and for onlap and local erosional truncations as sediments were deposited across the shelf during transgression.

Reservoir rocks are plentiful in the basal Paleozoic sandstones and primary and leached porosity exists within carbonate units. Argillaceous interbeds, shale units, and nonporous carbonates are available as seals for any potential reservoirs.

Geochemical analyses indicate that Cambrian dolomite, although low in total organic carbon, may constitute a medium to good source rock and is of intermediate maturity (T.G. Powell, pers. comm.). Middle and Upper Cambrian carbonates in northern Greenland are bituminous and have favourable source rock characteristics (Henriksen, 1980, p. 13). Where they occur, Middle and Upper Ordovician shale units, the Silurian Cape Phillips Formation and the Devonian Eids, Bird Fiord and Weatherall formations (Fig. 5) may contain fair to excellent source rocks. In the Central Stable Region the transition from mature to overmature sediments may be found at well depths up to 1800 m (5900 ft).

As mentioned earlier, no reserves have been identified on Central Stable Region and resources are entirely undiscovered. Several plays have been evaluated however, and the expected or mean values of potential resources are 0.17 x $10^9 m^3$ (1.1 Bbbl) of oil and 0.24 x $10^{12} m^3$ (8.4 Tcf) of natural gas.

## Geology of the Franklinian Geosyncline

The Franklinian Geosyncline (Figs. 1, 4) is widely exposed on northern Ellesmere Island

and in northern Greenland, but it is partially obscured to the northwest by the younger Sverdrup Basin and the Arctic Coastal Plain. The Wandel Sea Basin forms a partial cover in northeastern Greenland. The stratigraphic and structural characteristics of the geosyncline are more or less typical of mountain belts that border large cratonic masses: a miogeosynclinal belt continuous with platformal successions; deepwater shale with a flyschoid facies; and a volcanic and intrusive belt at the continent-margin. It must be emphasized, however, that the volcanic-intrusive, «outboard» belt of the geosyncline lies along the «inner» margin of a geanticlinal welt, the Pearya Geanticline (Trettin, 1971) and not on oceanic crust. Indeed, both the Pearya Geanticline and the basement of the Franklinian Basin are considered to comprise sialic Precambrian crust (Trettin and Balkwill, 1979)

Sedimentation in the Franklinian Geosyncline, during much of early Paleozoic time, occurred along three depositional belts (Trettin and Balkwill, 1979, p. 752): an unstable southeastern shelf, a central trough, and a complex northwestern belt. The southeastern and central belts are widely exposed in the Canadian Arctic Islands and northern Greenland, but the northwestern belt is known only in northernmost Ellesmere Island. The relationships in northernmost Greenland are obscured by structural complexity and regional metamorphism (*see* Dawes, 1976, p. 276).

The sedimentary belt along the southeastern margin of the Franklinian Geosyncline lies jacent to the Arctic Interior Platform; a series of «hinge lines» forms an obscure boundary. The southeastern belt is characterized by platformal and slope carbonate rocks, craton-derived clastic sediments, and gypsum-anhydrite and other units of lagoonal deposition. The central Hazen Trough received relatively thin, starved-basin sediments from Early Cambrian to Early Silurian time compared to relatively thick carbonates deposited on the shelf. The trough sediments include carbonates and clastic materials derived from both the southeastern and the northwestern shelves. The upper part of the trough succession is marked by radiolarian chert and a flysch phase of deposition. The strata of the northwestern shelf form a complex of volcanic, shallow marine carbonate rocks, and shallow marine to nonmarine clastic sediments with tongues of the starved-basin and flysch facies of the Hazen Trough. The volcanic rocks are felsic to intermediate in composition. Fluvial and shallow-marine, coastal-plain and shelf environments are indicated, and the grade and abundance of clastic sediments changes along the length of the belt, which is exposed only in northern Ellesmere Island.

## Petroleum Potential of the Franklinian Geosyncline

Cambrian and Ordovician platform carbonates in the Franklinian Miogeosyncline (Fig. 2) extend across the Arctic Interior Platform and beneath the Fold Belt, shown in Figure 4. The carbonates grade into shales to the northwest of the Fold Belt beneath the Sverdrup Basin. During Early and Middle Ordovician time the Fold Belt was the site of lagoonal evaporite deposition behind a reef front. During Silurian and Devonian time the carbonate-to-shale transition shifted to the southeast toward the hinge line between the Stable Platform and the Franklinian Geosyncline.

During the Ellesmerian Orogeny in Late Devonian and Early Mississippian time compression from the north and northeast produced discordant folds and thrusts in the Fold Belt, with décollements soled in evaporitic sequences. Upward coarsening clastic cycles within the Weatherall, Hecla Bay and younger Devonian formations terminated deposition in the Franklinian Geosyncline (Fig. 5).

Reservoir rocks exist in the Ordovician platform sequence, in reefal carbonates adjacent to the shale basin and in Upper Devonian clastic rocks. Shale intervals above the reservoir rocks might conceivably serve as seals for any possible reservoired hydrocarbons.

In the Bent Horn «field» on Cameron Island, several wells encountered oil in a carbonate reef-edge within the Devonian Blue Fiord Formation (Fig. 5). The reef rock appears to have been incorporated in a faulted fold and was enveloped in shales that sealed and probably provided a source for the oil. Original expectations for the Bent Horn «field» were optimistic but delineation tests failed to indicate a commercial accumulation.

The Winter Harbour well on Melville Island tested 1.7 x $10^3 m^3/d$ (61.2 mcf/d) of natural gas from within Upper Devonian clastic rocks.

Middle and Upper Ordovician shale units in the Cornwallis Group, the Silurian Cape Phillips Formation and the Devonian Eids, Bird Fiord and Weatherall Formations (Fig. 5) may contain moderate to excellent source rocks. The transi-

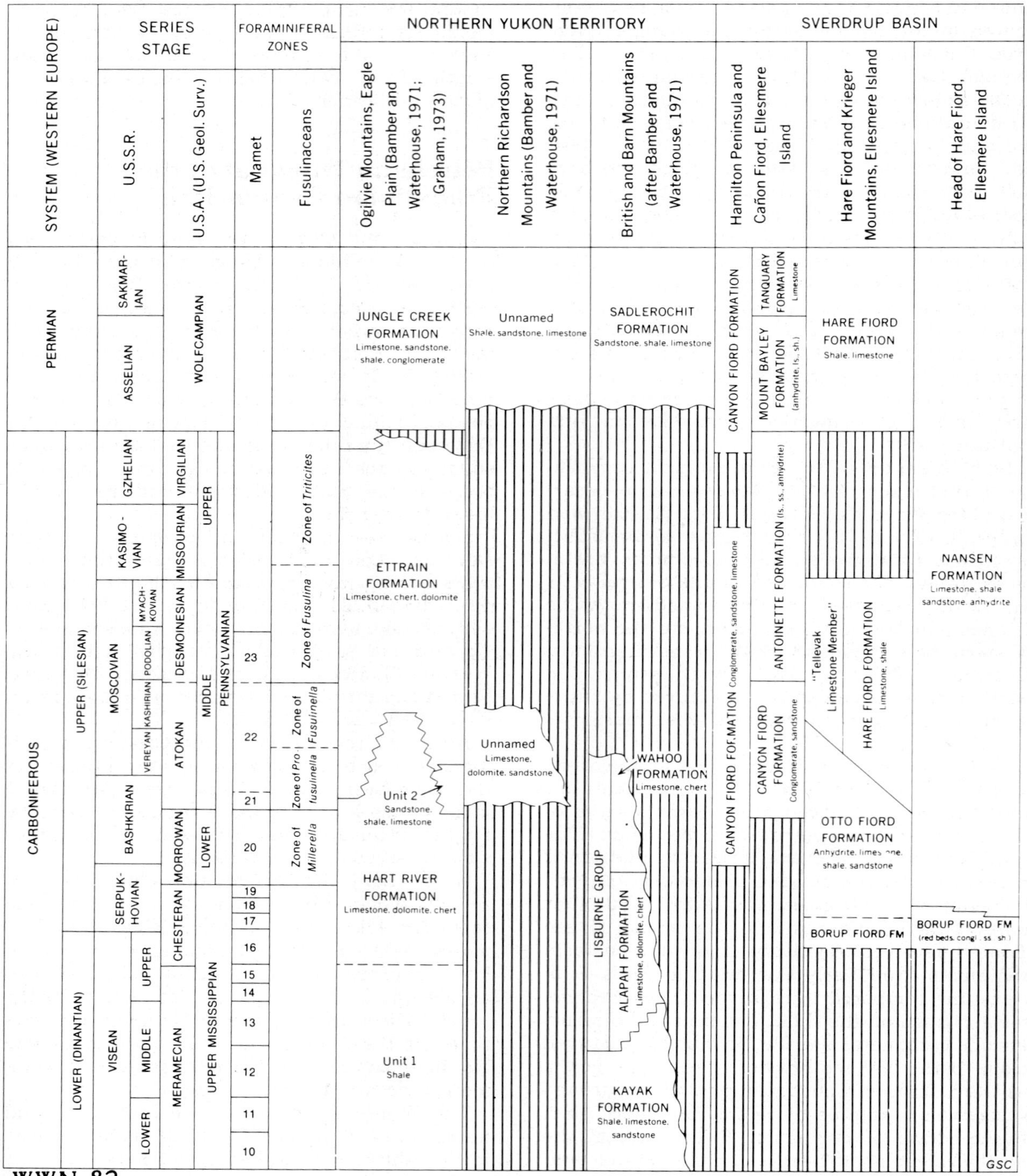


WWN-82

FIGURE 7. *Relationships between Carboniferous and Lower Permian Formations in the Sverdrup Basin and the Yukon Territory, from Nassichuk and Bamber, 1979.*

tion from mature to overmature sediments in the Franklinian Geosyncline probably occurs at depths greater than 3350 m (11 000 ft).

Undeclared oil reserves exist in the Bent Horn «field» on Cameron Island and all other oil shows or oil indications in the Franklinian Geosyncline must be considered undiscovered resources. The Franklinian Geosyncline in northern Greenland has not been evaluated but promising Cambrian and Ordovician prospects are known. Within the Franklinian Geosyncline in Canada, the mean or expected value of potential resources is 0.08 x $10^9 m^3$ (0.5 Bbbl) of oil and 0.22 x $10^{12} m^3$ (7.7 Tcf) of natural gas.

## Geology of the Sverdrup Basin

The Sverdrup Basin is a regional depression that occupies a large part of the northwestern Queen Elizabeth Islands (Figs. 1, 2) and contains an essentially conformable sequence of upper Paleozoic, Mesozoic and Tertiary strata (Figs. 6, 7, 9, 10). In summary, the Sverdrup Basin is an elongate, northeasterly trending pericratonic depression containing a maximum thickness of 13 000 m (43 000 ft) of Lower Carboniferous (Viséan) to lower Tertiary (Eocene) rocks, which rest unconformably on Devonian to Precambrian rocks of the Franklinian Geosyncline

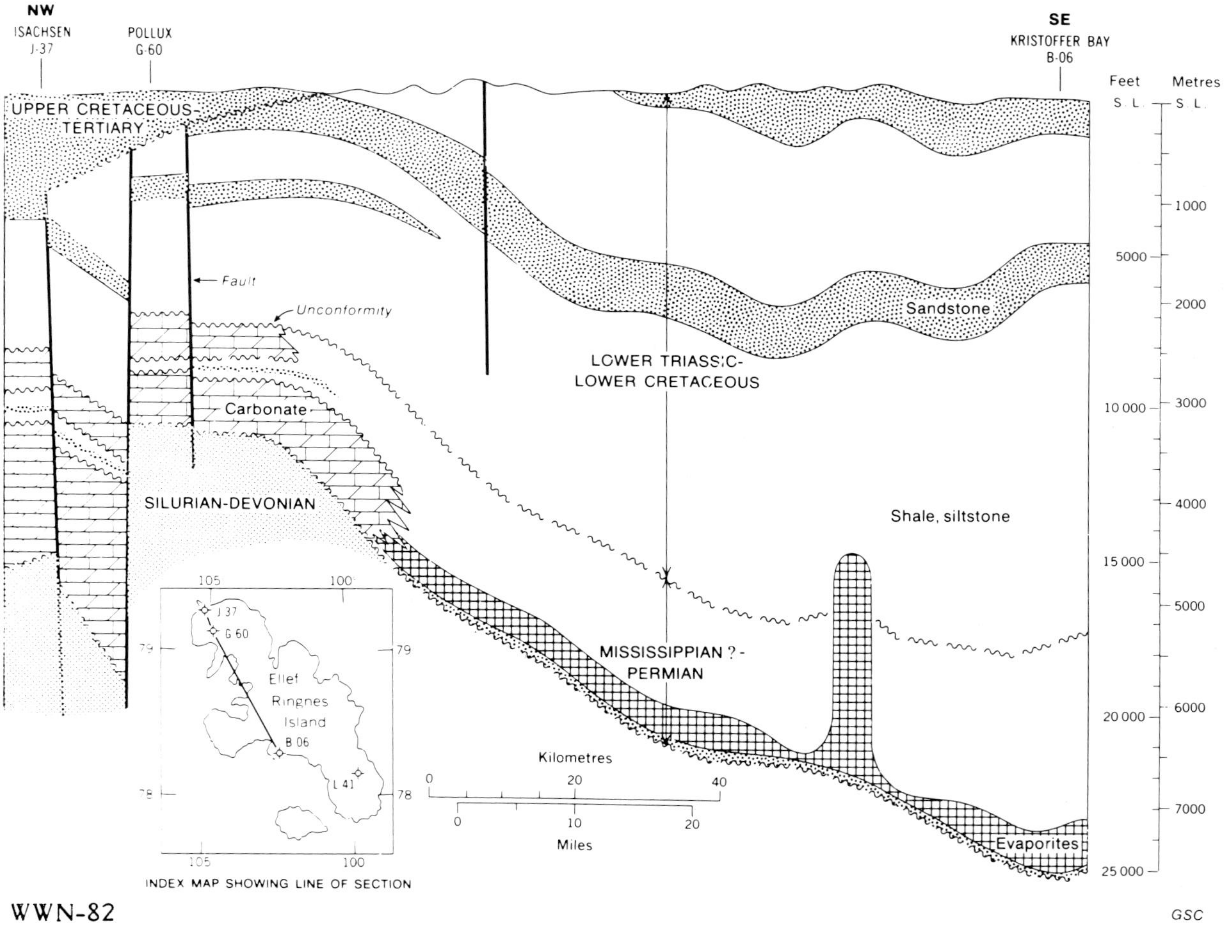


FIGURE 8. *Schematic structural-stratigraphic section across the western half of the Sverdrup Basin at Ellef Ringnes Island (modified from Meneley et al., 1975). The section illustrates the distribution of Carboniferous evaporite facies in the subsurface with local diapirism, and the presence of coeval carbonates over a structural high forming the northwestern rim of the basin.*

(Fig. 4). The latter rocks were deformed by the Ellesmerian Orogeny between Late Devonian (Frasnian) and Early Carboniferous (Tournaisian) time. Along with the Franklinian Geosyncline the Sverdrup Basin constitutes the Innuitian orogenic system embracing the Canadian Arctic Islands and northern Greenland.

The first sediments deposited within the Sverdrup Basin were laterally discontinuous nonmarine coaly shales and siltstones of the Emma Fiord Formation (Viséan). Initial expansion and deepening of the basin coincided with the first marine transgression, represented by a sequence of red-weathering lowermost Upper Carboniferous sandstones and conglomerates throughout the Basin. As deposition continued into the Lower Permian, several distinctive facies belts remained relatively constant in the Basin (Fig. 6).

Recently completed regional mapping, seismic and structural studies in the Sverdrup Basin have provided sufficient data to interpret early depositional history. The geometry and depositional fabric of the basin suggest that it may have evolved through episodes of crustal fracturing, which resulted in periodic subsidence and filling accompanied by volcanic extrusion and intrusion. According to Trettin et al. (1972), the axial trough of the Sverdrup Basin coincides approximately with the position of the lower Paleozoic Hazen Trough in the underlying Franklinian Geosyncline and a structural high bordering the axial trough on the northwest side of the basin coincides with a lower Paleozoic structural high. This northwestern structural high is illustrated in a cross-section of Ellef Ringnes Island (Fig. 8).

The Sverdrup Basin, which now is a structural synclinorium, was subjected to two episodes of deformation. Between Late Pennsylvanian and Early Permian time, the Melvillian Disturbance initiated localized folding and faulting in marginal regions of the basin; folding has been recognized in the southwestern part of the basin in the vicinity of northwestern Melville Island (Tozer and Thorsteinsson, 1964) and faulting on northern Ellef Ringnes Island (Meneley et al., 1975). Structural deformation also is apparent between Pennsylvanian and Permian strata near the eastern margin of the Sverdrup Basin at Trold Fiord (Thorsteinsson, 1970) and at Bjorne Peninsula. Following the Melvillian Disturbance clastic deposition was predominant and, during the upper Paleozoic, sediments were derived principally from the southwest, south, southeast and east. Patterus of deposition established during late Early Permian and early Late Permian time continued throughout Mesozoic time. The southerly margins of the Basin were sites of prograding, deltaic deposition with local depocentres migrating along the basin margin. The upper Paleozoic and lower Mesozoic sandstones extend farthest into the Sverdrup Basin from the southeast as several overstepping and prograding sandstone tongues. The distal ends of sandstone tongues extend into the western portion of the Sverdrup Basin (Fig. 9). Upper Triassic and Jurassic sandstone units now known to contain oil and

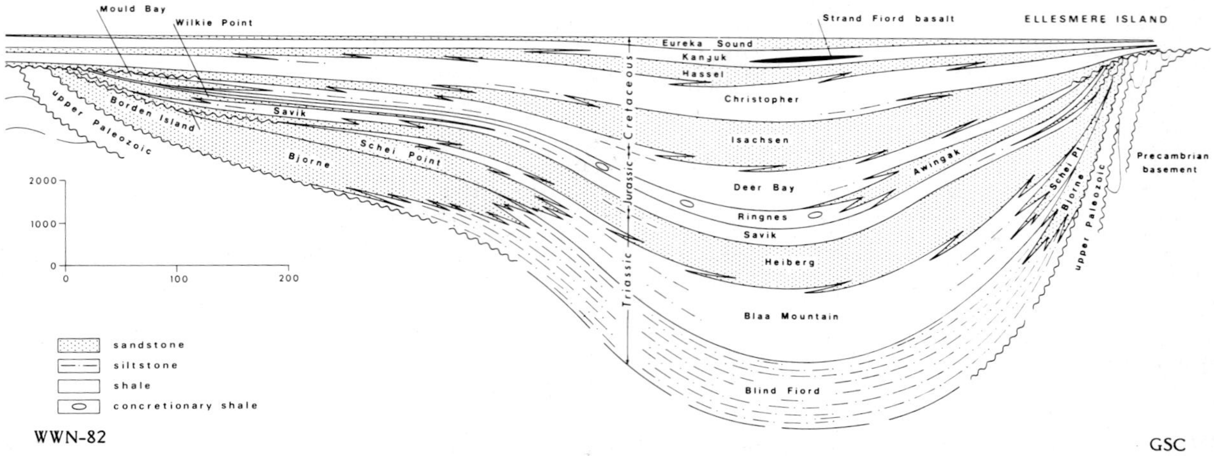


FIGURE 9. *Schematic stratigraphic cross-section, Mesozoic strata from West to East across Sverdrup Basin, Canadian Arctic Archipelago; from Balkwill et al., in press.*

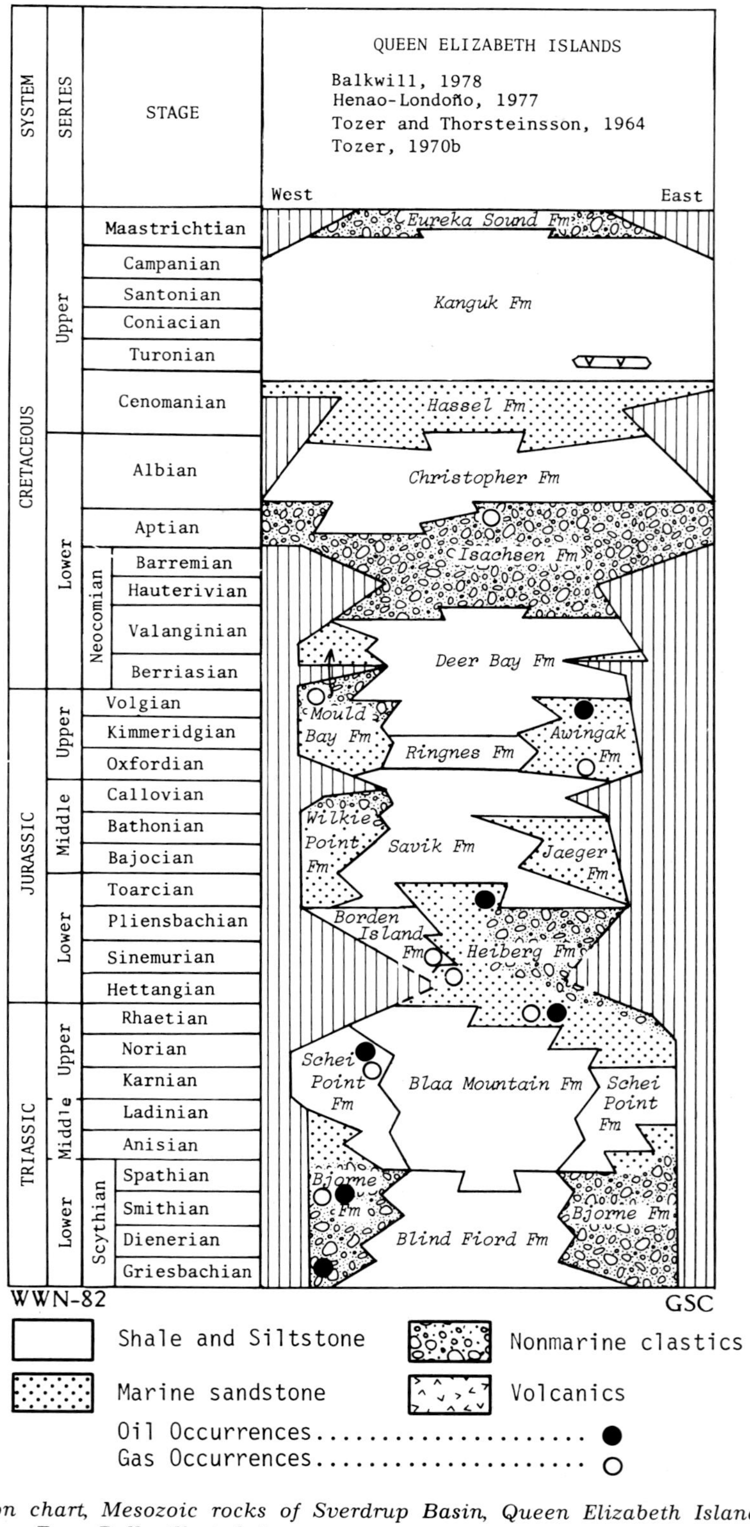


FIGURE 10. *Correlation chart, Mesozoic rocks of Sverdrup Basin, Queen Elizabeth Islands, Canadian Arctic Archipelago. From Balkwill et al., in press.*

gas, particularly in western and southern parts of the Basin were deposited as overlapping and coalescing formations (Fig. 10).

Most of the major structural characteristics presently observed within the basin, such as broad asymmetrical folds, abundant normal faults and some reverse faults, were imprinted during the Eurekan Orogeny, an event dominated by widespread vertical movements between Late Cretaceous (Senonian) and middle Tertiary (Miocene) time. On Axel Heiberg and Ellesmere Islands, where major upper Paleozoic and Mesozoic depocentres were situated, deformation was particularly intense. Elsewhere it was relatively mild, especially in western parts of the basin. In some central parts of the basin, detachment occurred above Carboniferous evaporites; locally, intrusion of evaporites into the cores of folds and along faults affected structural style. However, intrusion of large halite-cored diapirs (Fig. 8) in the central and southern axes of the basin (Fig. 2) probably was not triggered by the Eurekan Orogeny but may have been activated by normal sediment loading as early as Jurassic time. Whereas the Eurekan Orogeny generally is considered to mark the climax of widespread marine deposition within the basin, it is particularly clear that the basin ceased to exist as a separate depositional entity even earlier, in the Early Cretaceous, when transgressions progressed far beyond the basin boundaries that existed in pre-Cretaceous time. Alternating sandstone and shale continued to fill the Basin in local depressions into Late Cretaceous time. Sedimentation in the Sverdrup Basin was terminated with deposition of the synorogenic, Tertiary Eureka Sound Formation.

Finally, although structural and stratigraphic information for the basin is abundant, little is known of its relationships vis-a-vis circumpolar tectonics and deposition. It is evident that upper Paleozoic and lower Mesozoic strata of the Sverdrup Basin, of northeastern Greenland (Wandel Sea Basin) and of some of the islands of Svalbard and Novaya Zemlya are remarkably similar. The last two groups of islands are on the Barents Shelf; Svalbard is some 800 km (500 miles) northeast of northernmost Ellesmere Island and Novaya Zemlya is an equal distance beyond Svalbard. Clearly, all regions were affected by comparable tectonism, and deposition occurred under similar environmental conditions during much of the late Paleozoic.

Outcrop distribution shows that the Sverdrup Basin extended at least 1000 km from northeastern Prince Patrick Island to northeastern Ellesmere Island. The maximum width of the basin, from southwestern Ellesmere Island to northwestern Axel Heiberg Island, is about 290 km (180 miles). These must be considered minimum dimensions; recently recovered subsurface data from the westernmost Queen Elizabeth Islands indicate that deposition extended far beyond present outcrop limits. The present eastern and southern erosional margins of the Sverdrup Basin are bounded by deformed Devonian and older Paleozoic rocks that are, in turn, bounded to the east and southeast by northern extensions of the Canadian Shield.

The northwestern margin of the basin is less clearly defined than other margins since much of it is concealed by Tertiary and younger sediments of the Arctic Coastal Plain (Fig. 2). Erosional edges, which may parallel depositional margins, are delineated in northwesternmost Ellesmere Island and northwestern Axel Heiberg Island by exposures of Devonian and older Paleozoic sedimentary, igneous and metamorphic rocks (Trettin et al., 1972), against which abut younger Paleozoic rocks of the Sverdrup Basin succession. Thorsteinsson and Tozer (1960, 1970) and Tozer and Thorsteinsson (1964) recognized nearshore characteristics in Mesozoic rocks in the western Queen Elizabeth Islands and suggested that a periodically positive ridge separated the Sverdrup Basin from the Arctic Ocean during late Paleozoic and Mesozoic time. Gregory et al. (1960) suggested on the basis of magnetic and radiometric profiles that an arch, possibly originating in the Franklinian Geosyncline, parallels the edge of the basin in this area. Meneley et al. (1975) showed from data assembled from seismic profiles and drillholes that the basin margin from Prince Patrick Island to Axel Heiberg Island was influenced by a tectonically active fault bounded ridge. The trend of this ridge parallels the trend of the mid-Paleozoic Grantland Uplift on Ellesmere Island, and of the outcrop belt of lower Paleozoic metamorphic and other rocks along the northwestern coast of Ellesmere Island (Trettin et al., 1972, Fig. 11). It is important to note that although the intermittently active northwestern marginal ridge may have formed a bathymetric or even topographic (subaerial) barrier at various times during Sverdrup Basin deposition, sedimentation was probably contemporaneous on both sides of the ridge.

Reconstruction of both northwestern and northeastern basin margins on northern Ellesmere Island is made difficult by discontinuous outcrops (outliers) of late Paleozoic to

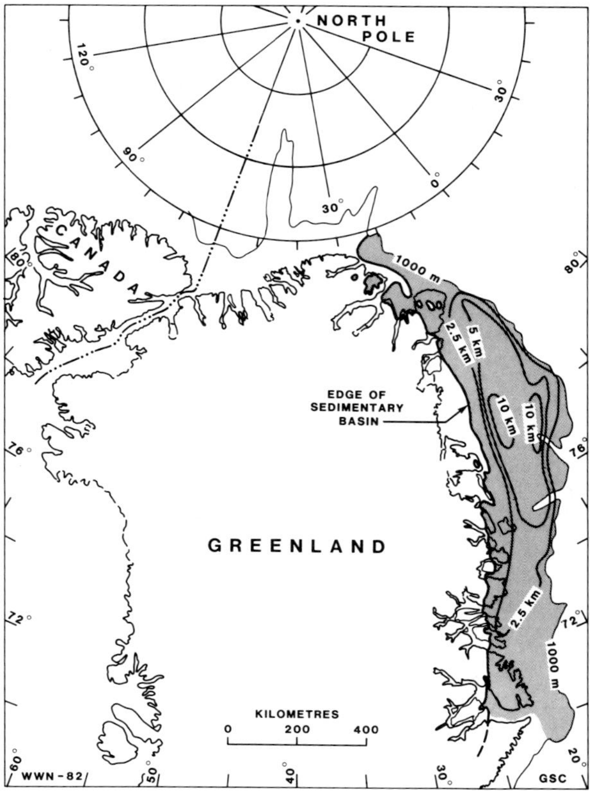


FIGURE 11. *Wandel Sea Basin along the northeastern edge of Greenland (stippled pattern) with depth-to-basement indicated by contours; after Henderson, 1976.*

Cenozoic age. An extensive linear sequence of upper Paleozoic to Cenozic strata on the northeastern side of the basin parallels the axis of the basin at Lake Hazen. These strata are separated from the basin proper by the Lake Hazen Fault Zone in which Paleozoic rocks occur at the surface. Similarly, it is not known whether Carboniferous and Permian outliers on the northwestern side of the basin along the north coast of Ellesmere Island are allochthonous or authochthonous. They may have been isolated from the Sverdrup Basin by Tertiary tectonism or, alternatively, they may represent deposition in satellite depressions separated from the Sverdrup Basin. Another alternative is that the outliers were once continuous with the Sverdrup Basin and, prior to Tertiary plate movements, with sequences now preserved in Svalbard and parts of Greenland.

## Petroleum Potential of the Sverdrup Basin

During the upper Paleozoic, the Sverdrup Basin experienced a relatively rapid subsidence and, given an abundant source of sediments, filled more or less continuously from Early Carboniferous to Late Permian time. Depositional environments changed progressively from continental lacustrine, to continental fluvial, to restricted marine, and finally to open marine.

Carbonates of the Northwest Carbonate Belt (Fig. 6) and the Southeast Carbonate Belt (Fig. 6) intertongue with shales and evaporites in the Basinal Clastic and Evaporitic Belt. In the Marginal Clastic and Carbonate Belt porous carbonates may be in direct contact with shale and coarse clastic facies. Sandstones of the Marginal Clastic Belt thin basinward and prograde over shale units. Locally, in both Carboniferous and Permian sequences carbonates intertongue with sandstone beds.

Carbonate fronts on both the southeast and northwest sides of the basin are ideal reservoir rocks with a well-developed porosity shoreward from the shelf edge. In addition, many prograding sandstone tongues constitute reservoir rocks in both Carboniferous and Permian clastic sequences. Faults and shale intertongues seal many prospective targets; shelf-edge carbonates rendered non-porous by early marine cementation seal many others. Structural closures overlain by shale units form the trapping mechanisms for numerous prospects.

Uppermost Lower Permian, Upper Permian and Mesozoic strata are predominantly clastic in origin and reservoir rocks are mainly sandstones. Entrapment configurations for these strata may be truncated, upturned edges at the eroded basin margin or structural closures capped with shale beds.

Due to lack of maturity or to the type or source of organic matter deposited in the Sverdrup Basin, Mesozoic strata have commonly yielded only gas. The Middle Triassic Schei Point Formation is an exception in containing hydrogen-rich organic matter. It is mature in the central part of the Basin and is the principal source for oil shows and tar sands exposed on Melville Island. The Schei Point is also the source for more recent oil discoveries in the central part of the basin; i.e. the Cisco, McLean, Skate, Balaena and Char wells.

The main reservoirs in the Sverdrup Basin are the Borden Island Formation and the various members of the Heiberg Formation (Fig.

10). The Borden Island is the reservoir for natural gas in the Drake and Hecla Fields in Melville Island. There, the Borden Island is the distal and reworked portion of one of several prograding tongues of Heiberg sandstone that are common in the western portion of the Sverdrup Basin (Fig. 9). Approximately 85 per cent of the natural gas reserves in the western Sverdrup Basin are located within the Borden Island Formation.

In central regions of the Sverdrup Basin 80-85 per cent of the natural gas reserves are located within the King Christian Sand at the top of the Heiberg Formation. The diachronous Heiberg Formation consists of several prograding tongues of sandstone with intervening shale intervals. Progradation is mainly from the east and southeast toward the central and western portions of the Sverdrup Basin. The sandstone units, where they have structural closure, and are sealed by overlying shale units, constitute highly significant petroleum traps and prospects throughout the basin.

Upper Paleozoic rocks in the Sverdrup Basin are also prone to the generation of gas. Most tests in the Upper Paleozoic have been drilled near the southern margin of the Basin where sediment was derived from terrestrial sources. There is a paucity of geochemical data for upper Paleozoic strata in central portion of the basin, but from the definite marine character of most of the upper Paleozoic section of that region it is reasonable to expect eventual discovery of oil source rock. Gaseous hydrocarbons were formed in the mature zone from terrestrial organic matter and in the overmature zone from both terrestrial and marine organic matter by the cracking of liquid hydrocarbons and kerogen. In the Sverdrup Basin, the marginally mature and mature zones occur at approximately 1500 m (5000 ft) and 3000 m (10 000 ft) maximum burial depth respectively. The transition from mature to overmature occurs in the vicinity of 4300 m (14 100 ft) to 4600 m (15 100 ft) maximum burial depth. Isotopic data from all parts of the Basin show that gases were derived from late mature to overmature zones. They migrated to their present position through fracture zones.

Delineation drilling in the Sverdrup Basin has been minimal but recoverable oil reserves have been estimated at 0.05 x $10^9$ $m^3$ (0.31 Bbbl). Marketable natural gas reserves in the Sverdrup Basin are estimated to be 0.45 x $10^{12}m^3$ (15.9 Tcf).

Several exploration plays have been assessed within both upper Paleozoic and Mesozoic strata and undescovered resource potentials for both intervals have been aggregated; 0.46 x $10^9m^3$ (2.9 Bbbl) of oil and 1.98 x $10^{12}m^3$ (69.9 Tcf) of natural gas.

## Geology of the Wandel Sea Basin

Successor basins of northeasternmost Greenland and central east Greenland rest respectively on deformed Upper Devonian and on Ordovician to Middle Devonian beds. The northern basin has been named the Wandel Sea Basin (Dawes and Soper, 1973, p. 131); the central East Greenland basin is as yet unnamed. The two basins are clearly contiguous and, indeed, may prove to be a single, integral basin (*see* Henderson, 1976, and Fig. 11, this paper). For brevity in the present discussion, the Wandel Sea-East Greenland basin will be referred to as the Wandel Sea Basin.

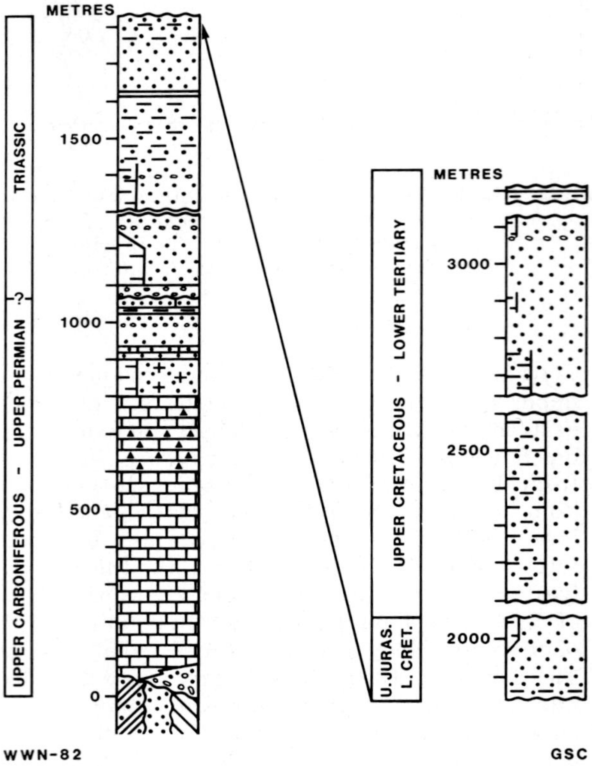


FIGURE 12. *Upper Paleozoic to Tertiary succession in Wandel Sea Basin after Håkansson, 1979.*

The Wandel Sea Basin is exposed in Peary Land, in Kronprins Christian Land, and discontinuously along the coast of eastern Greenland, but the major portion of the basin lies offshore on the Continental Shelf between latitudes 70° N and 82° N. The basin is 300 km (190 miles) wide at its widest point and extends approximately 1450 km (900 miles) in a north-south direction. It is bounded on the north and east by the continental slope and rise, on the south by the East Greenland Tertiary Volcanic Province and to the west by depositional and structural truncations against the Caledonian and Precambrian «basement». Aeromagnetic and refraction profile data indicate that sediments in the Wandel Sea Basin must certainy be more than 5 km (16 000 ft) thick and may attain thicknesses up to 10 km (33 000 ft); (Fig. 11).

A composite stratigraphic section from Kronprins Christian Land (Fig. 12), near the depositional edge of the main basin and now preserved as an outlier, contains more than 3 km (10 000 ft) of Carboniferous to Tertiary strata. Transgressive basal conglomerate, sandstone and sandy limestone are similar to basal units in the Sverdrup Basin. Upper Carboniferous (Upper Pennsylvanian) to Lower Permian carbonates and minor evaporites overlie the basal clastics. Fossils in the Carboniferous and Permian succession are closely comparable to forms from the Sverdrup Basin, Spitsbergen and Eurasia. As in the Sverdrup Basin, Late Permian-Triassic units in the Wandel Sea Basin are dominated by upward-coarsening clastic units which may represent progradation into the basin. Unconformities within Triassic, Jurassic and Cretaceous sequences, reflecting epeirogenic warpings and eustatic sea level changes, have produced incomplete sections near the basin margins. Offshore, one might expect more complete sections.

To the south, in eastern Greenland, the Upper Permian and Triassic succession contains evaporitic beds correlated with the Zechstein and Keuper in the North Sea. Also in the south, the Mesozoic section has been intruded by Cretaceous and Tertiary basaltic rocks. Aeromagnetic data indicate that the west side of the basin in the offshore areas also contains igneous or volcanic rocks.

Early Tertiary sediments in the northern onshore areas are similar to and have been correlated with the Eureka Sound Formation in the Canadian Arctic. These are the youngest sediments preserved onshore, but offshore even younger Tertiary and Quaternary sediments must occur.

The northern Wandel Sea Basin is dominated by northwest-southeast trending faults that parallel the fracture zone separating northern Greenland from Spitsbergen. Southward fault-bounded horsts and grabens lie subparallel to the coastline and are truncated by northwest-southeast faults. Locally, the axes of minor folds trend east-southeast within the fault blocks. These structural trends are related to tectonic activity that was associated with the rifting of the Atlantic area in post-Paleocene-Tertiary times.

Pre-Atlantic rifting paleogeography would place Greenland proximal to Norway and therefore the northern Wandel Sea area would be contiguous with the Barents Sea shelf between Spitsbergen and Norway.

## Petroleum Potential of the Wandel Sea Basin

The Wandel Sea Basin contains as much as 10 km (33 000 ft) of Carboniferous through Tertiary strata that are closely similar to those in the Sverdrup Basin. Clastic, carbonate and evaporitic sequences in facies belts comparable to the Sverdrup Basin indicate that numerous source rocks and traps are probably available. Reservoirs rocks are probable in coarse clastic and carbonate units that might be sealed by evaporitic and shale beds.

Geochemical analyses have been performed with encouraging results on a number of Wandel Sea sediments. One sample of upper Paleozoic limestone contained migrated hydrocarbons (Henderson, *ibid.*, p. 502). Source rocks can be anticipated in Upper Permian shale units, Middle Triassic limestone and in two distinctive shaly units in the Upper Jurassic succession. Additionally, mud volcanoes in East Greenland indicate the former existence of methane gas seeps.

The presence of continuous pack ice in the Wandel Sea has precluded exploration and, therefore, no reserves are known. The data base in the offshore area continues to expand but no formal resource estimates are known to be available.

The Wandel Sea Basin is similar in size to the Sverdrup Basin and it may be transitional in depositional character between the Sverdrup and North Sea Basins. Both of the latter contain appreciable reserves and potential resources in the form of natural gas and oil. By analogy it may be reasonable to assume that large potential resources also exist in the Wandel Sea Basin.

## Geology of the Arctic Slope Alaska

The Arctic Slope of Alaska (Fig. 1) extends from the Chukchi Sea in the West 1050 km (660 miles) to the Canadian Boundary in the east. In the vicinity of Barrow, Alaska, it is over 400 km (250 miles) wide. At its eastern extremity the Arctic Slope is only a few tens of kilometres wide between the Romanzof Uplift of the Brooks Range and the Arctic Ocean. The western extremity of the Arctic Coatal Plain (*see* following section of this paper) is included, for purposes of discussion, in the Alaskan Arctic Slope province.

The Arctic Slope is divided into three physiographic sub-provinces that reflect the main structural elements of the area. The southern province, the Brooks Range, lies within the Columbian (Jurassic-Early Cretaceous) and Laramide (Late Cretaceous-Early Tertiary) Brooks Range Orogen; to the north are the Southern and Northern Foothills, which overlie the Colville Trough; and farther north the Arctic Coastal Plain overlies the deeply buried Barrow Arch (Figs. 3, 13, also Bird, 1981).

Devonian and older rocks are exposed at the surface in the southern Brooks Range; northward, these strata are overlain by a succession of Carboniferous, Permian, Triassic and Lower Cretaceous sediments (Fig. 13). Mainly Lower and Upper Cretaceous rocks outcrop in the Foothills areas, and Upper Cretaceous and Tertiary rocks in the Arctic Coastal Plain.

Both «basement» and younger rocks in the Brooks Range have been uplifted over the adjacent Colville Trough, which contains a succession of Carboniferous through Upper

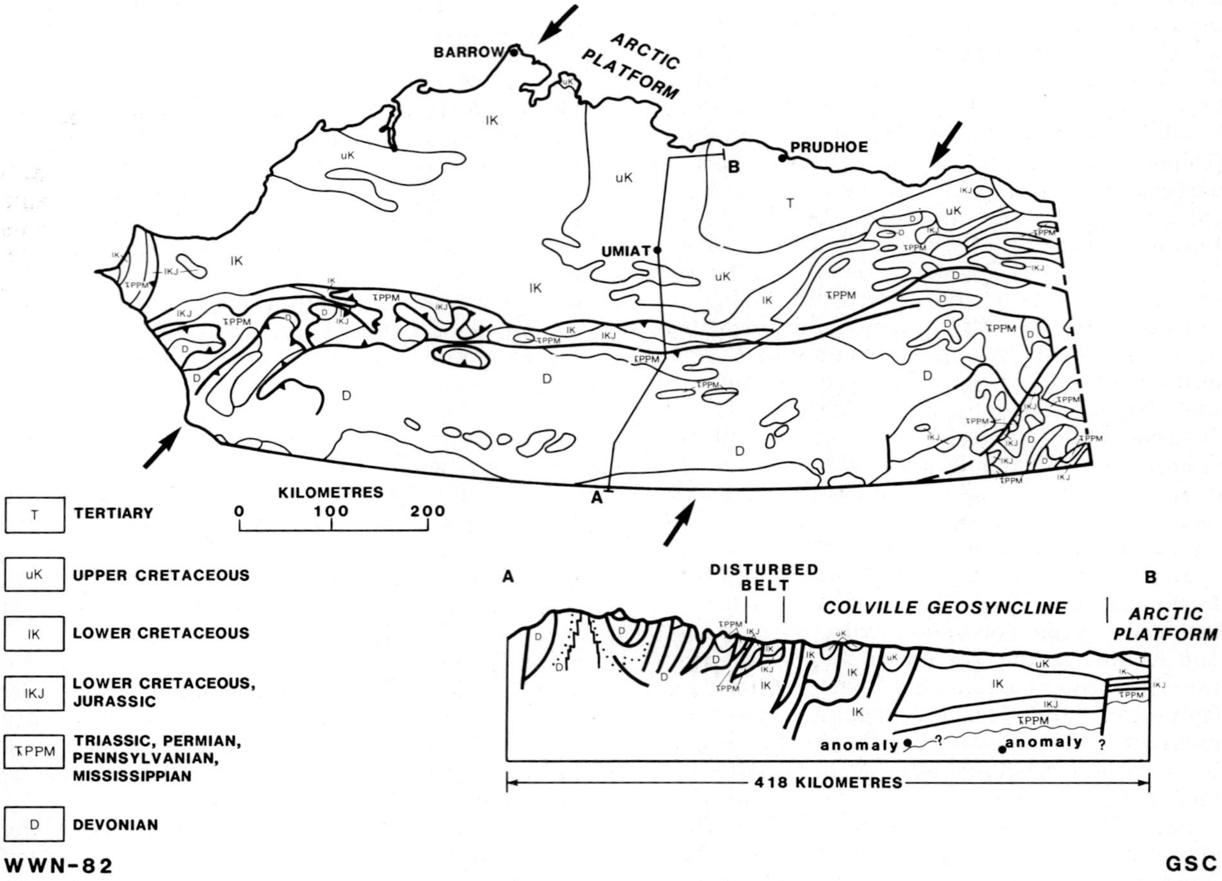


FIGURE 13. *General geological map of northern Alaska showing a north-south schematic structural cross-section (after I. Tailleur, pers. comm.).*

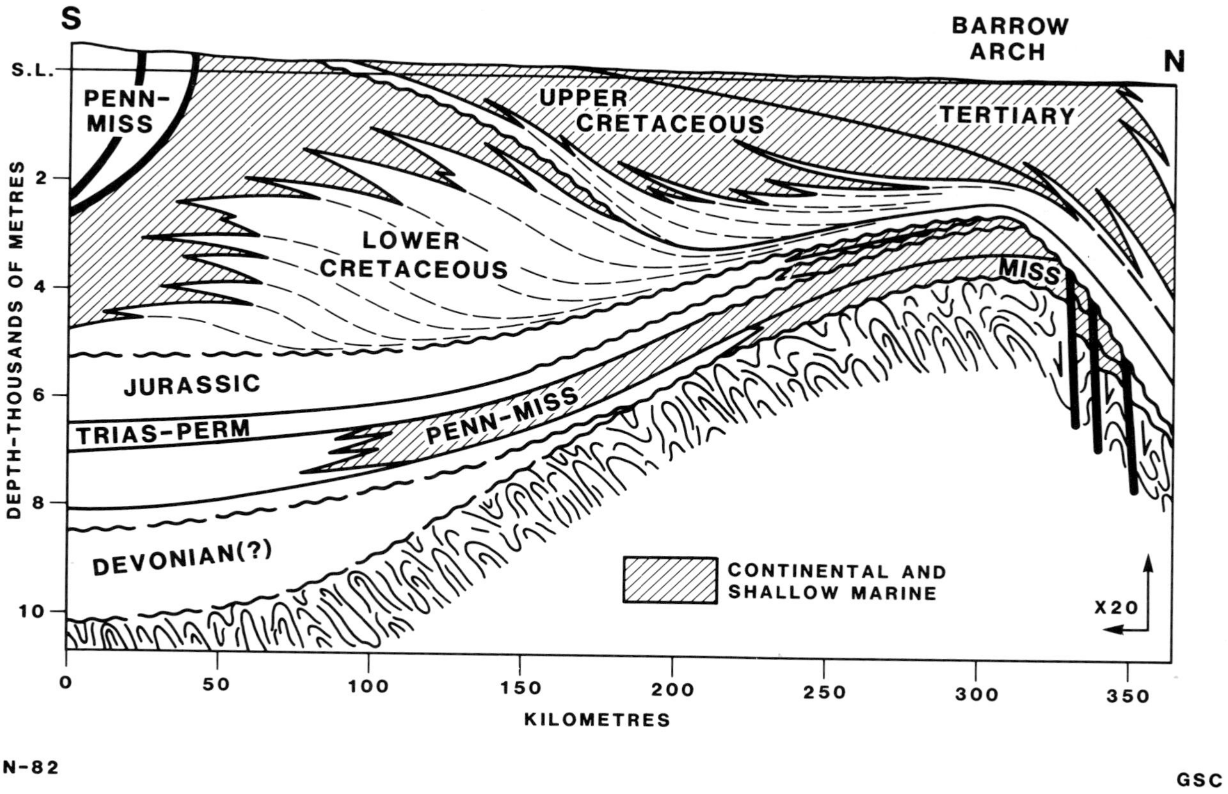


FIGURE 14. *Diagrammatic structural stratigraphic cross-section (NS) from the Foothills to the Arctic Coast; after Bird and Jordan, Oil and Gas Journal, Apr. 18, 1977.*

Cretaceous rocks that may be up to 10 000 m (33 000 ft) thick. The Colville Trough is separated from the Arctic Ocean by the Barrow Arch, a broad, regional basement high that lies subparallel with the present coast line. The Arctic Coastal Plain sequence overlies the Barrow Arch and the Upper Cretaceous and Tertiary sediments thicken northward into the Arctic Ocean (Figs. 13, 14).

«Basement» in the Arctic Slope of Alaska includes all pre-Ellesmerian (pre-Late Devonian-Early Mississippian) rocks. The basement rocks are composed of mildly deformed and metamorphosed Devonian carbonates and Devonian and older clastic sequences. Ordovician and Silurian graptolites have been recovered from argillites in wells that have penetrated this interval. «Basement» strata in the Arctic Slope are similar to more intensely deformed strata within the Fold Belt of the Franklinian Geosyncline (Fig. 4) but in Alaska they are intruded by igneous rocks of granitic and locally, of mafic composition. In the Arctic Slope, pre-Ellesmerian strata are considered non-prospective for petroleum exploration. The truncated and southerly inclined surface of the pre-Mississippian basement provided a ramped platform for the deposition of subsequent sediments.

As in the Sverdrup and Wandel Sea basins, initial deposition in the Alaskan Arctic Slope is characterized by coarse clastic, lacustrine strata associated with Early Carboniferous marine redbeds. The clastic sediments are overlain by lagoonal and shallow marine, shale, siltstone, coal and local evaporitic deposits, the Endicott Group. These lower clastic strata are diachronously overlain by Mississippian and Pennsylvanian platform carbonate rocks of the Lisburne Group (Figs. 14, 15). Grain and crystal size, onlapping relationships, depositional thinning, internal unconformities and truncated surfaces indicate a transgression to the north and northeast onto a landmass that lay north of the present Alaskan coastline.

A mild tectonic event similar to the Melvillian

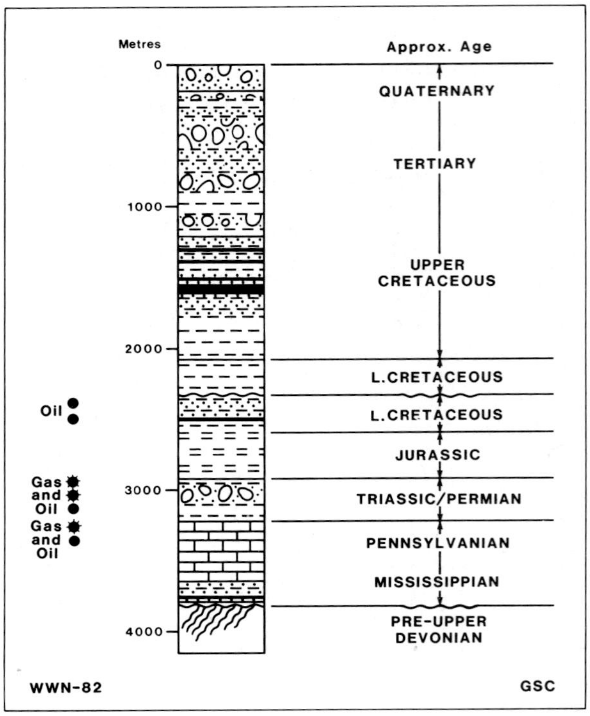


FIGURE 15. *Stratigraphic column in the Prudhoe Bay area showing principal oil and gas reservoir; Oil and Gas Journal, May 24, 1971.*

Disturbance in the Arctic Islands of Canada resulted in marine regression during Permian time; carbonate deposition terminated with this event. Clastic sediments predominated in subsequent deposits, which resulted from uplift in the north and accompanying marine transgression elsewhere. Coarse clastics of the Sadlerochit Group prograded southward into the Colville Trough and laterally equivalent siltstone and shale have been identified as potential source rocks. Clastic deposition from northern sources continued from Permian through Early Cretaceous times. The carbonate and coarse clastic units constitute the main reservoirs in the Prudhoe Bay Area (Fig. 15) and the intervening shale units constitute primary source rocks.

A major unconformity during Early Cretaceous time truncated the succession from the Lower Cretaceous down to the «basement». The developing Brooks Range provided sediments that prograded northward into the foredeep of the starved Colville Trough. The distal, organic rich shale facies of the southerly derived clastics lapped onto the Barrow Arch and sealed the truncated lower sedimentary sequence. The same process placed the Lower Cretaceous shales, which contain the richest source rocks in the region, in juxtaposition with leached reservoir rocks below the truncated surface.

Progradation and infilling of the Colville Trough continued periodically, and, with the basin filled, upward-coarsening clastic material surmounted the Barrow Arch. Subsequently, Upper Cretaceous and Tertiary clastic sediments were deposited on the collapsed northern continental margin and in troughs contiguous with the Arctic Ocean. These latter sequences constitute the sediments of the Arctic Coastal Plain of Alaska.

Deformation in the Brooks Range is characterized by low-angle thrust faults in the west and higher angle reverse faults in the east. In the Southern Foothills, structural style is similar to that in the Brooks Range and deformation diminishes northward in a series of detached concentric folds.

The southern flank of the Barrow Arch dips gently into the Colville Trough and the northern flank is down-stepped in a series of fault blocks and horst-and-graben terranes. The Barrow Arch generally plunges to the southeast, with local culminations along its axis. Hydrocarbon accumulations may be located along the axis of the arch, in down-faulted blocks on the northern flank, and in onlap structures on the southern flank.

## Petroleum Potential of the Arctic Slope of Alaska

The most promising petroleum potential of the Arctic Slope of Alaska lies within the post-Devonian succession that was deposited after the Ellesmerian Orogeny. Onlapping Lower Carboniferous, coarse clastic and carbonate strata underlie most of the Arctic Slope except in northernmost areas, where Early Cretaceous erosion truncated all units down to and including the basement. The Cretaceous event also removed northerly derived, coarse clastic strata of Permian, Triassic, Jurassic and Early Cretaceous age. Permian to Lower Cretaceous coarse clastics grade southward into siltstones and shales that are preserved in the Colville Trough.

Uplift of the Brooks Range during the Columbian and Laramide Orogenies provided coarse clastic material for the Colville Trough foredeep.

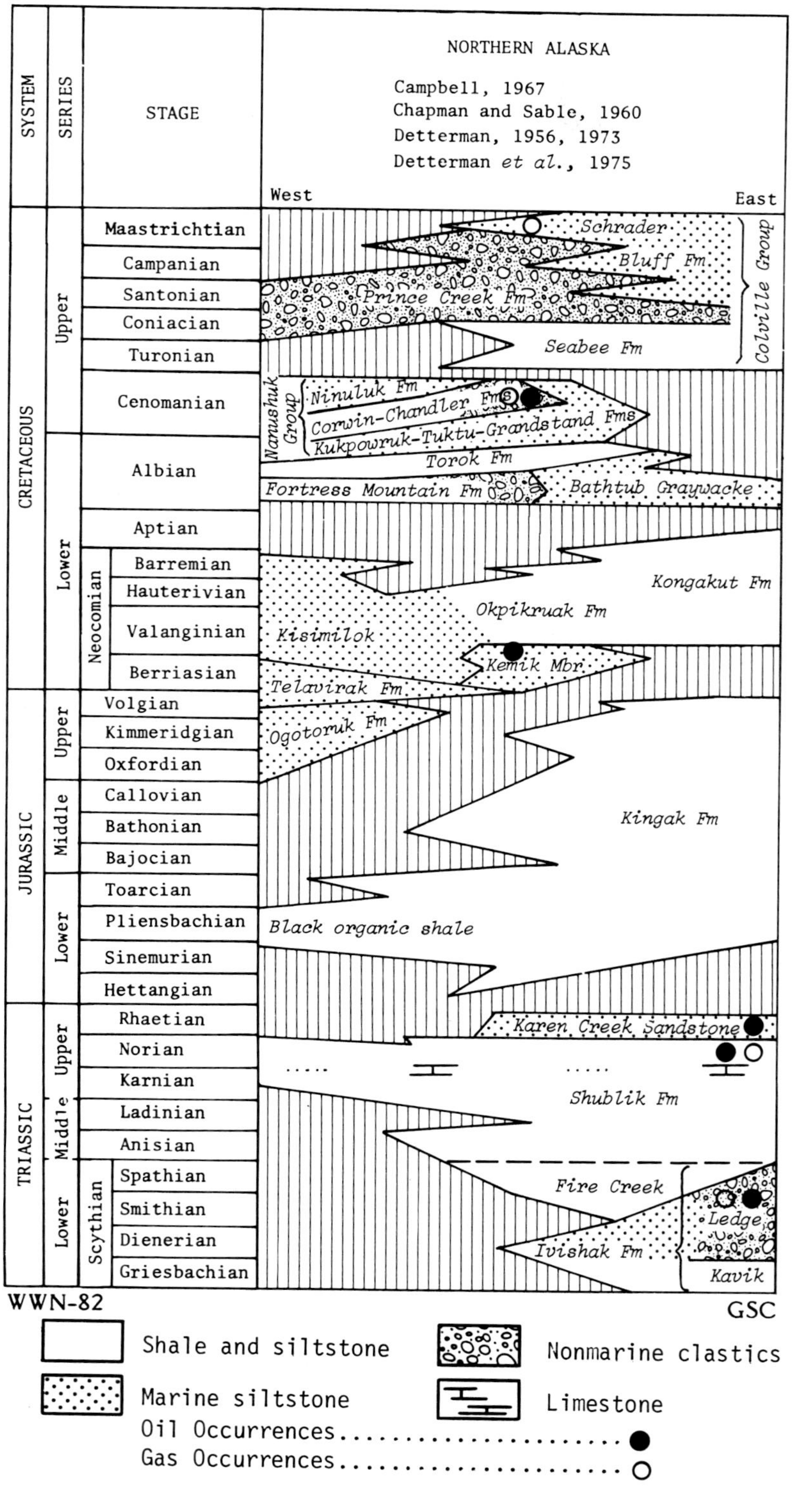


FIGURE 16. *Correlation chart, Mesozoic Formations of northern Alaska. From Balkwill et al., in press.*

Shale of excellent source-rock quality was deposited contemporaneously in distal portions of the starved Colville Trough, especially near the Barrow Arch. Upper Cretaceous and Tertiary sandstone and shale interbeds advanced, prograding northward over the Barrow Arch into the Arctic Ocean Basin.

Reservoir rocks exist in Lower Carboniferous, coarse clastic units, especially where they have been truncated by the Lower Cretaceous unconformity. This same unconformity enhanced porosity in the Carboniferous and Permian carbonates along the Barrow Arch. The northward coarsening and truncated sandstones and conglomerates of Permian, Triassic, Jurassic and Early Cretaceous age also had their porosity enchanced by leaching at the unconformity.

In Upper Cretaceous and Tertiary sequences good reservoir quality was developed in the delta front sandstones of prograding megacycles; the quality of reservoirs tends to improve up-section.

Geochemical research has shown principal petroleum sources to be organic-rich, phosphatic units within the Triassic Shublik Formation, the Jurassic Kingak Shale and in deeply buried Lower Cretaceous, post-Neocomian shale (Fig. 16). Basal Mississippian shales and redbeds, Carboniferous carbonates and Permian shales have also been analyzed geochemically. The Mississippian redbeds and Carboniferous carbonates have organic carbon contents below levels considered to constitute a good source rock but the other units are more likely to have contributed to the generation of hydrocarbons.

Reserves of oil exist in seven known accumulations on the Arctic Slope of Alaska. The Prudhoe Bay Field contains 3.6 x $10^9 m^3$ (22.4 Bbbl) of oil-in-place and 1.5 x $10^9 m^3$ (9.6 Bbbl) of recoverable oil in Permian-Triassic clastic reservoirs. The reserves in the underlying carbonate reservoirs have not been calculated. The Kuparak River Field has been estimated to contain approximately 0.56 x $10^9 m^3$ (3.5 Bbbl) of

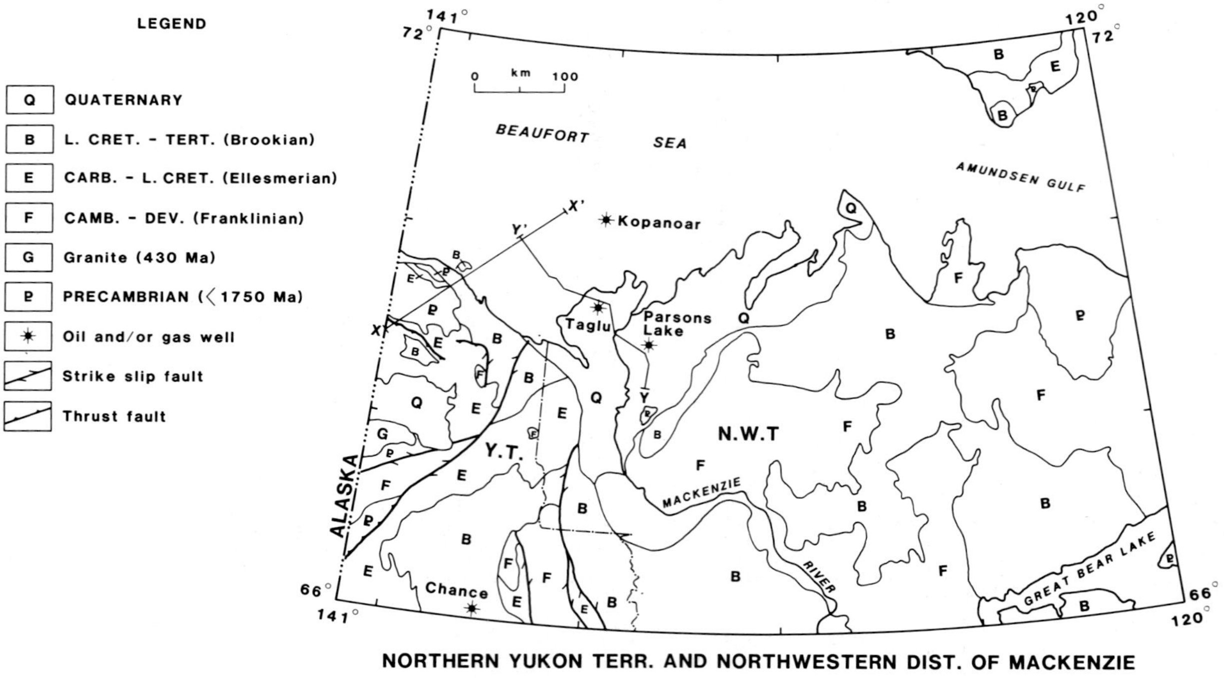


FIGURE 17. *Geological map of the northern Yukon Territory and northwestern District of Mackenzie; courtesy D.K. Norris, Geological Survey of Canada.*

oil-in-place. The Umiat Field may have been between 11-15 x $10^6m^3$ (70-100 MMbbl) and the Simpson Field 0.39 x $10^6m^3$ (2.5 MMbbl) of oil in place. Undeclared reserves of oil are found in Upper Cretaceous-Tertiary pools, in the basal Mississippian clastic rocks and in the West Sak discovery area.

Gas reserves have also been found in Triassic clastic rocks of the Prudhoe Bay Field. Approximately 0.73 x $10^{12}m^3$ (26 Tcf) of natural gas are contained in this field. The Kavik Field contains gas in Jurassic and Permian-Triassic sands but no reserves are available. Jurassic sands in the South Barrow Field contain 0.56 x $10^9m^3$ (20 Bcf) of natural gas. The Kemik Field also contains gas in Jurassic sands but the reserves are not known. Upper Cretaceous sandstone in the Gubik Gas Field contains up to 8.5 x $10^9m^3$ (300 Bcf) of natural gas.

Reserves in exess of 1.74 x $10^9m^3$ (10.9 Bbbl) of recoverable oil have been discovered on the Arctic Slope and reserves of over 0.74 x $10^{12}m^3$ (26.1 Tcf) of gas are available for exploitation if transporation and markets are available.

The National Petroleum Council of the United States, after polling 20 organizations that prepare resource estimates has published averaged resource estimates for several regions in the Arctic (Oil and Gas Journal, Nov. 23, 1981). When aggregated, the onshore portion of the Arctic Slope has an undiscovered oil resource potential of 1.0 x $10^9m^3$ (6.29 Bbbl) of recoverable oil. Offshore, in the Alaskan Arctic Coastal Plain the undiscovered resource potential is 2.0 x $10^9m^3$ (12.6 Bbbl) of recoverable oil. The undiscovered gas resource potential for the onshore areas of the Arctic Slope is 0.63 x $10^{12}m^3$ (22.4 Tcf) and the offshore undiscovered gas resource potential is 0.76 x $10^{12}m^3$ (26.8 Tcf).

## Geology of the Arctic Coastal Plain

The Arctic Coastal Plain extends for 2800 km (1750 miles) from Axel Heiberg Island in the Arctic Islands to the Chukchi Sea northwest of Alaska. Progressing southwesterly from Axel Heiberg Island, sediments of the Arctic Coastal Plain are superimposed upon sediments of the Sverdrup Basin, the Franklinian Geosyncline, portions of the Arctic Interior Platform and the Arctic Slope of Alaska and Northern Yukon. Onshore, the boundary of the province is defined by the landward limit of Tertiary and Upper Cretaceous sediments. Offshore, the province extends to the base of continental slope and rise.

The sediments of the Arctic Coastal Plain attain a thickness of more than 12 000 m (39 000 ft) in the Beaufort Sea-Mackenzie Delta region but elsewhere it may be considerably thinner.

Along the Arctic Slope of Alaska the Coastal Plain is composed primatily of Upper Cre-

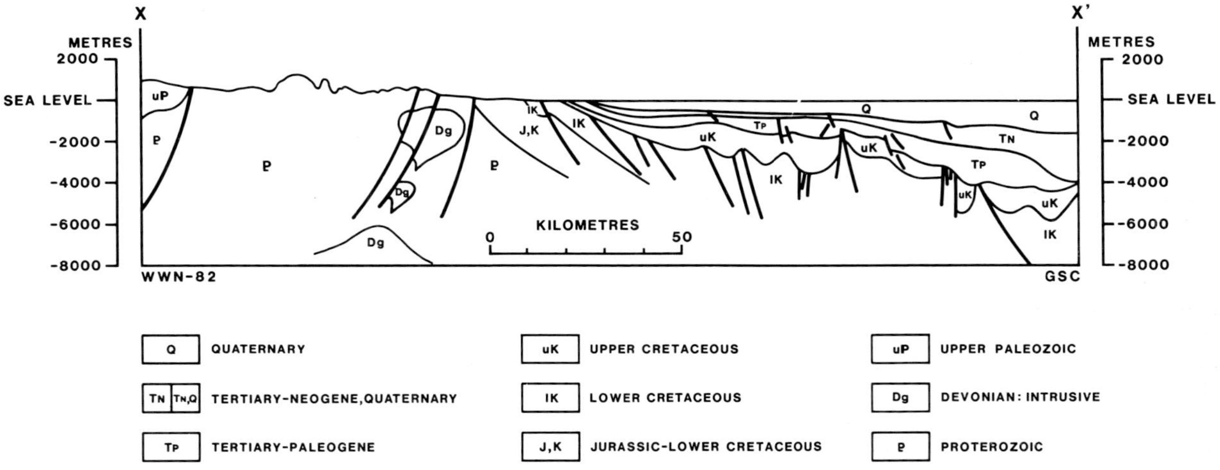


FIGURE 18. *Diagrammatic cross-section along x-x', from Mackenzie River Delta into Beaufort Sea, shown on map in Figure 17; from Yorath, Norris and Young, 1980.*

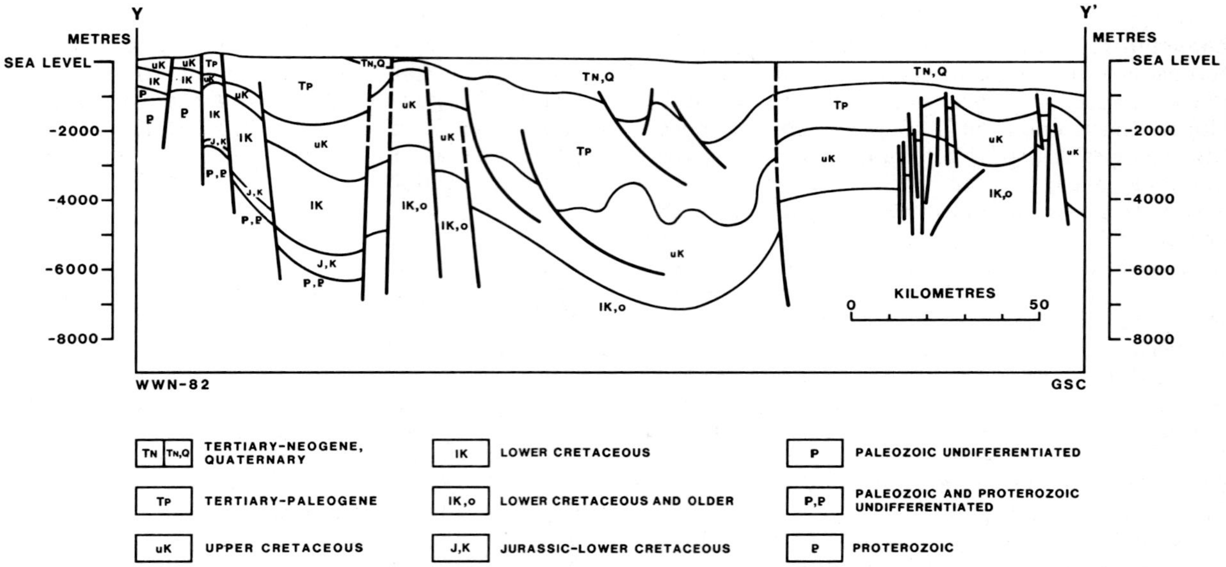


FIGURE 19. *Diagrammatic cross-section along y-y', from Mackenzie River Delta into Beaufort Sea, shown on map in Figure 17; from Yorath, Norris and Young, 1980.*

taceous and Tertiary sediments. Onshore in the Beaufort Sea-Mackenzie Delta area strata may be as old as Early Jurassic. In the offshore areas of the Arctic Islands Jurassic or Early Cretaceous sediments appear to have onlapped onto the Arctic Ocean rim of the Sverdrup Basin; Jurassic sediments are known to occur in grabens on Banks and Prince Patrick Islands.

The Beaufort-Mackenzie Basin (Fig. 3) is situated at the continental margin between the Arctic Interior Platform and the Alaskan and Northern Yukon Arctic Slope. In the Beaufort-Mackenzie area sediments of Cambrian to Devonian age were designated «Franklinian» by Lerand (1973). Similarly, Carboniferous to Jurassic sediments were designated «Ellesmerian» and the Lower Cretaceous to Tertiary sediments «Brookian».

Two sections across the margins of the Beaufort-Mackenzie Basin (Figs. 18, 19) illustrate development of Jurassic-Cretaceous deposits on older basement rocks at the edge of the continent. Section X-X' (Fig. 18) shows Jurassic-Cretaceous and Lower Cretaceous strata overlying Precambrian basement. Offshore, the section is complicated by faulting and diapirism. Section Y-Y', also complicated by faulting and diapirism, illustrates a thickening Jurassic-Cretaceous section deposited on Proterozoic and Paleozoic basement.

The northward-thickening wedge representing Jurassic-Cretaceous deposition in the early Beaufort-Mackenzie Basin was initiated with deposition of coarse clastic units of the Bug Creek Formation (Fig. 21) in a trough between the Arctic Interior Platform and the Brooks Geanticline (Fig. 2). Two additional prograding and onlapping coarse clastic cycles occurred during the Early Cretaceous; both contain hydrocarbons.

As early as Jurassic time, a pattern of deltaic megacycle deposition, was established that included a marine transgressive shale overlain diachronously by delta-front sandstone and siltstone. Delta-plain mixed lithologies of siltstone, sandstone, mudstone and coal usually completed the cycle, after which deposition was interrupted by a minor hiatus. In general, each deltaic megacycle was initiated by marine transgression and prograded successively farther northward into the early Beaufort-Mackenzie Basin.

By Late Cretaceous time the slope-shelf break between the prodelta shale and the delta-front sandstone had prograded beyond the present shoreline. Between Latest Cretaceous and Recent time six major deltaic megacycles advanced the slope-shelf break northward to its present position (Willumsen and Coté, in press). Four of the deltaic megacycles are illustrated in

Fig. 20; the stippled pattern represents coarse clastic, delta-front and delta-plain facies and the underlying unpatterned units represent prodelta and transgressive marine shale units. Turbidites may exist in the unpatterned units also. As the shelf-edge prograded basinward, slumps, traction loads and turbidite flows may have deposited additional sands on the continental slope and in ponded basins in front of the shelf. As the shelf advanced northward over previously deposited sediments, the load of the overburden caused the underlying over-pressured and undercompacted prodelta shales to deform into shale-cored diapiric anticlines and domes. These deformed sediments are also illustrated at the base of the section in Fig. 20.

The generalized depositional events described above were complicated by syndepositional tectonic events. During Jurassic and Early Cretaceous time, sedimentation occurred on the collapsing continental margin. Along the southeast side of the Beaufort-Mackenzie Basin the relief on the Aklavik Arch Complex was enhanced by down-to-basin faults and onlapping sediments along structurally high regions. The Aklavik Arch Complex also served as a local source for sediments. Jurassic and Early Cretaceous sediments are preserved on the downfaulted blocks and these structures may serve as hydrocarbon traps.

Faults that were active during the Laramide Orogeny have been traced into the Beaufort-Mackenzie area. Onshore, considerable dextral displacement is evident in certain of the faults. Juxtaposition of sedimentary facies and antithetic structures probably result from faulting within the southeastern Beaufort Shelf. Dextral motion appears to have been accompanied by compression in a northeast-southwest direction; this resulted in reverse faulting and produced an enhanced relief on the elongate diapiric structures in the offshore areas. Onshore, thrust faults occur in Jurassic Cretaceous strata.

The age of diapiric structures appears to vary with position in the basin. For instance, a particularly prominent shale-cored anticline near the southwest coast appears to have developed in Paleocene time, but succeeding groups of diapirs to the northeast appear to be younger. This relationship is apparent in Figure 20, where structures at the base of the section in

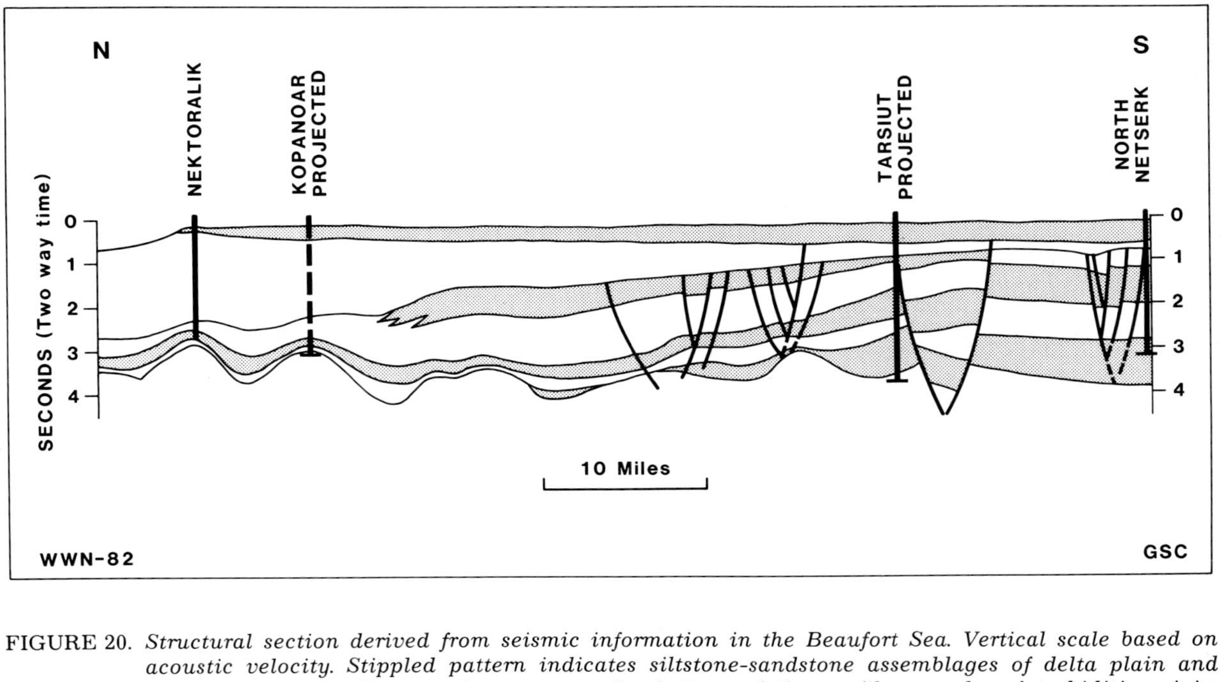


FIGURE 20. *Structural section derived from seismic information in the Beaufort Sea. Vertical scale based on acoustic velocity. Stippled pattern indicates siltstone-sandstone assemblages of delta plain and delta front origin. Som sandstones near the bottom of the profile may be of turbiditic origin. Unpatterned intervals between the stippled pattern are mainly shales and siltstones of prodelta origin; after Willumsen and Coté (in press).*

the south are shown affecting only the oldest deltaic cycle while diapiric structures in the north deform younger, overlying sediments.

Diapirism apparently occurs in the Beaufort-Mackenzie Basin and extends westward along the Coastal Plain into Alaska. Sediments offshore from Banks Island (Fig. 2) onlap and thin shoreward. There, tectonic style is dominated by mainly down-to-basin faults although apparent antithetic faults may be accounted for by host and graben structures.

## Petroleum Potential of the Arctic Coastal Plain

Deltaic sedimentation as old as Early Jurassic occurred in onshore regions of the Beaufort-Mackenzie Basin and Jurassic and Early Cretaceous sediments onlap the continental margin beneath the Arctic Coastal Plain in the Arctic Islands. Prograding deltaic megacycles are apparent above basal sequences in the Beaufort-Mackenzie Basin and they may also exist elsewhere along the northeastern Arctic Coastal Plain.

A nearly complete stratigraphic succession from Lower Jurassic to Recent is preserved in the Beaufort-Mackenzie Basin. Regional erosional or depositional breaks occur within the Upper Cretaceous, Lower Eocene and Upper Miocene intervals.

Potential reservoirs exist in each of the delta-front sandstones within the deltaic megacycles as well as in coarser clastic intervals in the delta-plain environments. Reservoirs also exist in the turbiditic sandstones in prodelta slope environments.

Trapping mechanisms for these reservoir rocks are provided by roll-over and down-to-basin faults, draping over diapiric structures, tilting with truncations against the flanks of diapiric structures, and onlap onto pre-existing structures. Transgressive marine shales of deltaic cycles and prodelta shales provide excellent seals for trapped hydrocarbons.

Geochemically, oils and condensates of the Beaufort-Mackenzie Basin can be divided into three genetic families. Gas condensates found in the Lower Cretaceous «Parsons» sandstone (Fig. 21) probably were derived from Jurassic and Lower Cretaceous sources (Husky and Kingak equivalents). Oil on the southeastern flank of the basin was derived from a conventional source rock, the bituminous zone of the Upper Cretaceous Boundary Creek Formation. The oils and condensates of the Tertiary reservoirs appear to be unique. The Tertiary oils are not high wax oils typical of terrestrial sources, but rather have high contents of napthenes and aromatics and occur in sediments of low geochemical maturity. The present geochemical data suggest that the Tertiary oils were derived from terrestrial organic material enriched in resin that produces gas-condensate and oil at lower than 'normal' maturity levels.

The conventional marginally mature zone occurs in the depth range of 2500-3700 m (8200-12 100 ft) and the main phase of oil generation begins at depths of approximately 5000 m (16 400 ft) in the Beaufort-Mackenzie area. The Tertiary oils and condensate were generated at depths of about 3100 m (10 200 ft), which is much shallower than the main oil generating zone. The transition from mature to overmature sediments occurs at depths in excess of 6000 m (19 700 ft).

Oil reserves may exist in several fields in the Beaufort-Mackenzie Basin but formal declarations of these reserves have not been published. Oil has been reported in Devonian carbonates directly below the pre-Cretaceous unconformity in the Mayogiak well but this oil is thought to be derived from Lower Cretaceous shales. The Lower Cretaceous «Parsons sandstone» (Fig. 21) is the reservoir for oil in the Kugpik, Imnak and North Parsons Fields. Other Lower Cretaceous sandstones are reservoirs for oil in the Kamik, Atkinson and West Atkinson wells. The Boundary Creek Formation is thought to be the source for these oils. Lower Eocene sediments are the reservoirs for oil or gas condensates in the Taglu, Niglintgak, Ya-Ya, Kumak, Adgo, Garry and Nektoralik areas. Upper Eocene sediments are the reservoirs for oil or gas condensate in the Mallik and Ivik areas.

Estimates of the sizes of discoveries in the Beaufort Sea vary widely. At one extreme there are industry-based press releases reporting the ranges that various structures are capable of containing, such as:

| | | |
|---|---|---|
| Kopanoar | 0.29 to 0.72 x $10^9 m^3$ | (1.8 to 4.5 Bbbl) |
| Koakoak | 0.32 to 0.80 x $10^9 m^3$ | (2.0 to 5.0 Bbbl) |
| Issungnak | 0.27 to 0.43 x $10^9 m^3$ | (1.7 to 2.7 Bbbl) |
| Tarsiut | 0.15 to 0.25 x $10^9 m^3$ | (1.0 to 1.6 Bbbl) |

At the other, less optimistic extreme are tenuous estimates of established recoverable reserves that may total less than 0.15 x $10^9 m^3$ (943 MMbbl). Until delineation drilling occurs, the wide spread between in-place values that reflect gross structure capacity versus reserve

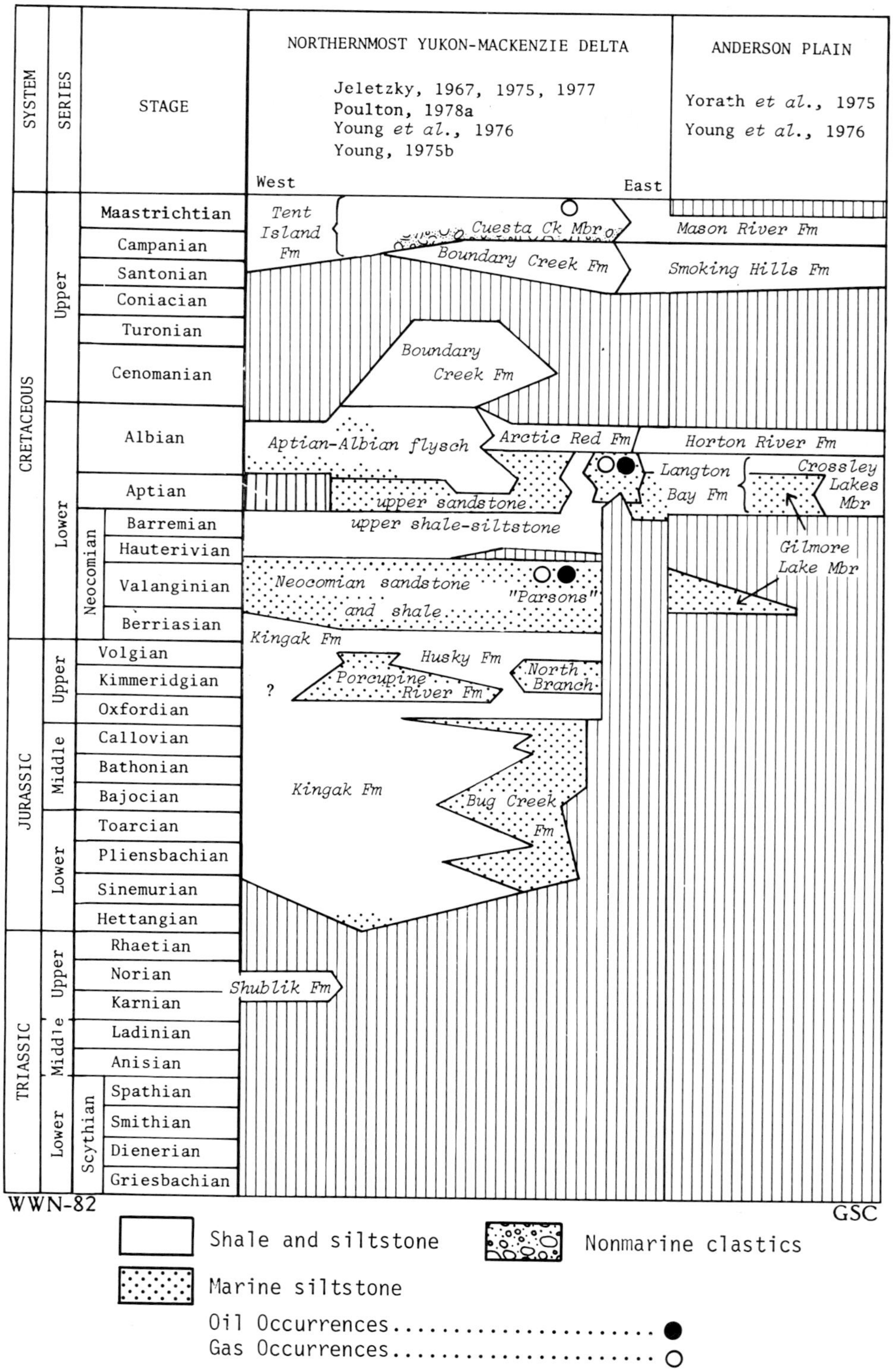


FIGURE 21. *Correlation chart, Mesozoic Formations of northern Yukon and northwestern Northwest Territories. From Balkwill et al., in press.*

estimates based on carefully determined net pay, and pool area, porosity etc. will remain, and such values should be used with caution.

The most published estimates of reserves for the gas fields in the Beaufort-Mackenzie were submitted to the National Energy Board in 1977. The National Energy Board's Reasons for Decision, Northern Pipelines, Volume 1, June, 1977, p. 2-130 released the following Marketable Natural Gas reserves (sanctioned by the National Energy Board) as follows:

| | | | |
|---|---|---|---|
| Adgo | (2.8 x $10^9m^3$) | 100 | billion cubic feet |
| Garry | (5.6 x $10^9m^3$) | 200 | « |
| Kumak | (0.42 x $10^9m^3$) | 15 | « |
| Mallik | (2.5 x $10^9m^3$) | 90 | « |
| Netserk | (1.7 x $10^9m^3$) | 60 | « |
| Niglintgak | (18.4 x $10^9m^3$) | 650 | « |
| Parsons Lake | (42.4 x $10^9m^3$) | 1500 | « |
| Reindeer | (0.14 x $10^9m^3$) | 5 | « |
| Taglu | (70.6 x $10^9m^3$) | 2500 | « |
| Titalik | (0.28 x $10^9m^3$) | 10 | « |
| Ya-Ya N. | (1.4 x $10^9m^3$) | 50 | « |
| Ya-Ya S. | (2.8 x $10^9m^3$) | 100 | « |
| TOTAL | * (149.2 x $10^9m^3$) | * 5280 | billion cubic feet |

* Additional reserves discovered since 1977 may raise total natural gas reserves into the range 170-280 x $10^9m^3$ (6000-10 000 Bcf); 250 x $10^9m^3$ (8.8 Tcf) is employed in Figure 25 of this report.

The undiscovered recoverable oil potential for the Beaufort-Mackenzie area has been estimated to be 1.5 x $10^9m^3$ (9.4 Bbbl) at the expected value. The undiscovered marketable gas potential for the area has been estimated to be 3.16 x $10^{12}m^3$ (112 Tcf) at the expected value.

The resource potential for the Arctic Coastal Plain in Alaska has been discussed in the section dealing with the petroleum potential of the Arctic Slope of Alaska. The resource potential for the Arctic Coastal Plain to the northeast of the Beaufort-Mackenzie area has not been fully assessed using the probabilistic play method. Many plays exist along the Arctic Coastal Plain in the Canadian Arctic Islands. Because of constantly shifting ice, however and, extremely high costs of exploration in the area, the data base for assessment of the Arctic Coastal Plain is incomplete but an excellent resource potential is envisaged for the region.

## Geology of the Baffin Bay Basin

The Baffin Bay Basin is a small but immature ocean basin between the coasts of Greenland and Baffin Island (Fig. 1). The Basin extends northward from the Arctic Circle, a distance of some 1600 km (1000 miles), and is approximately

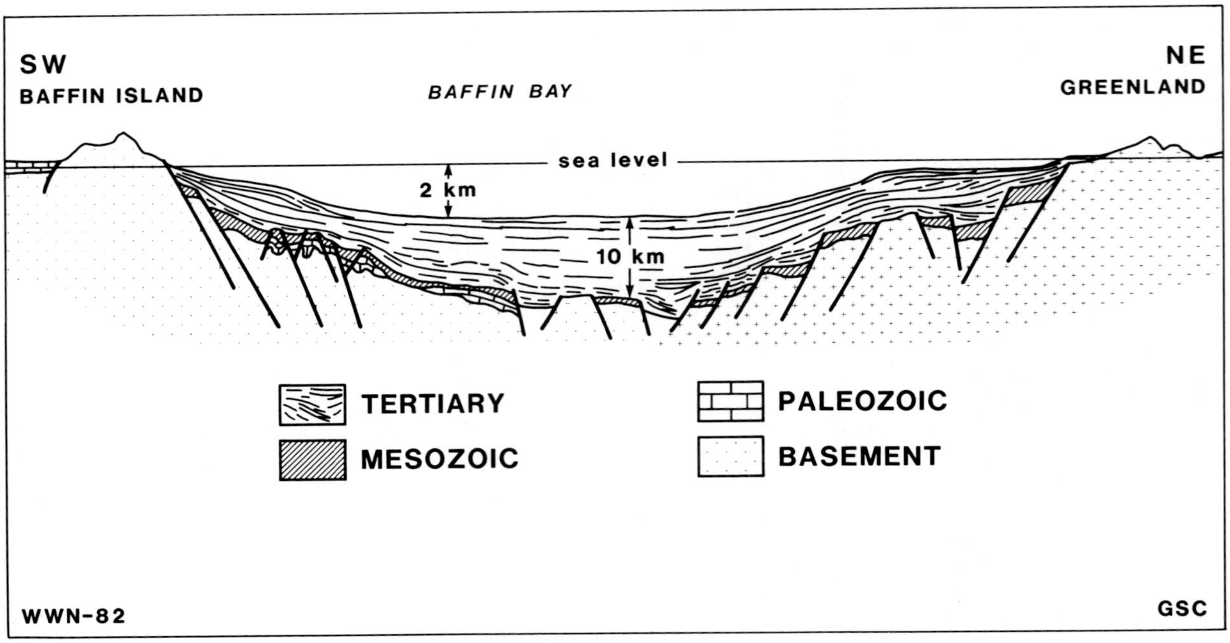


FIGURE 22. *Schematic structural stratigraphic cross-section (NE-SW) across Baffin Bay from Baffin Island to Greenland.*

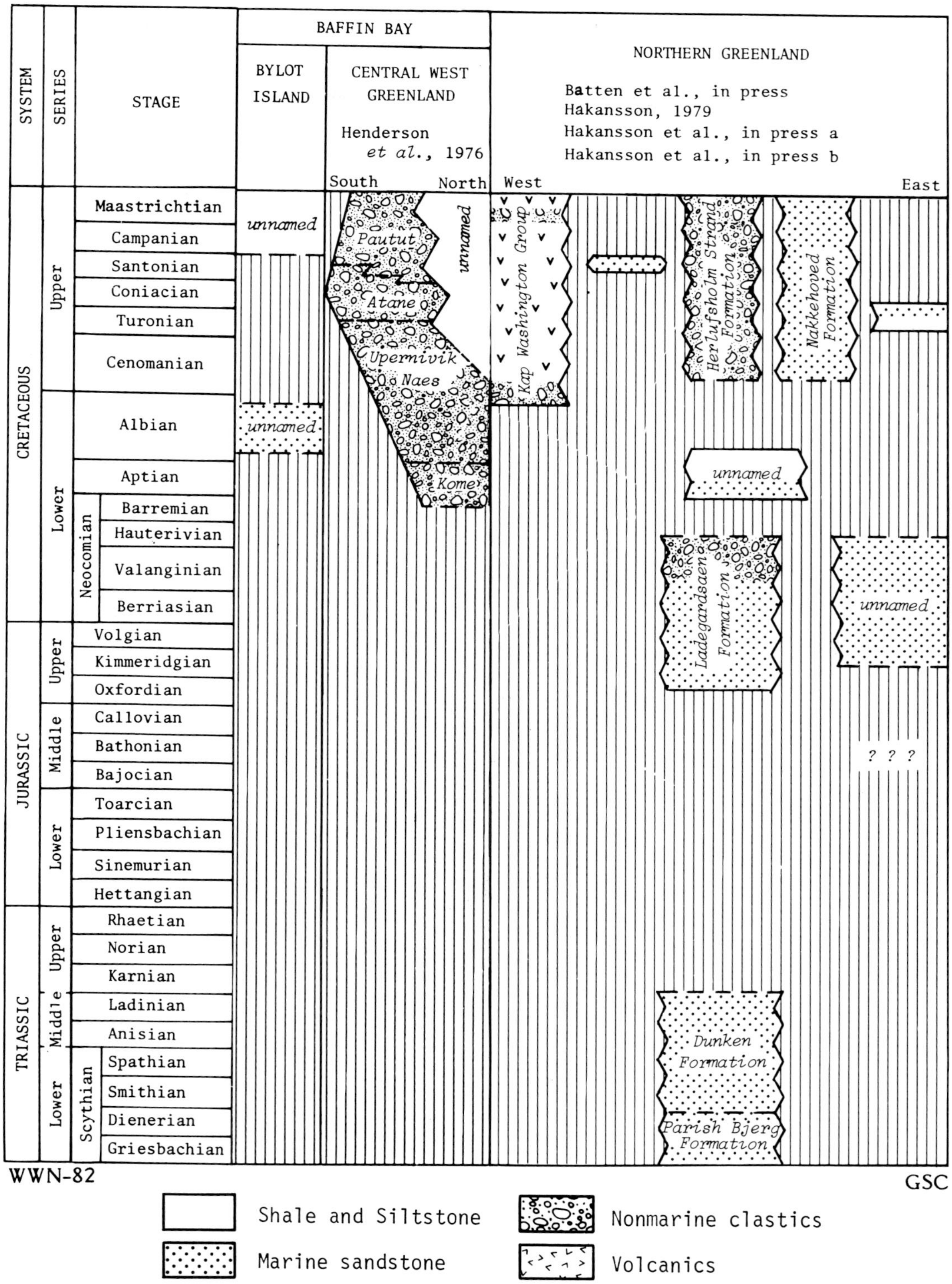


FIGURE 23. *Correlation chart, casozoic strata of northern and western Greenland, and Bylot Island, Canada. From Balkwill et al., in press.*

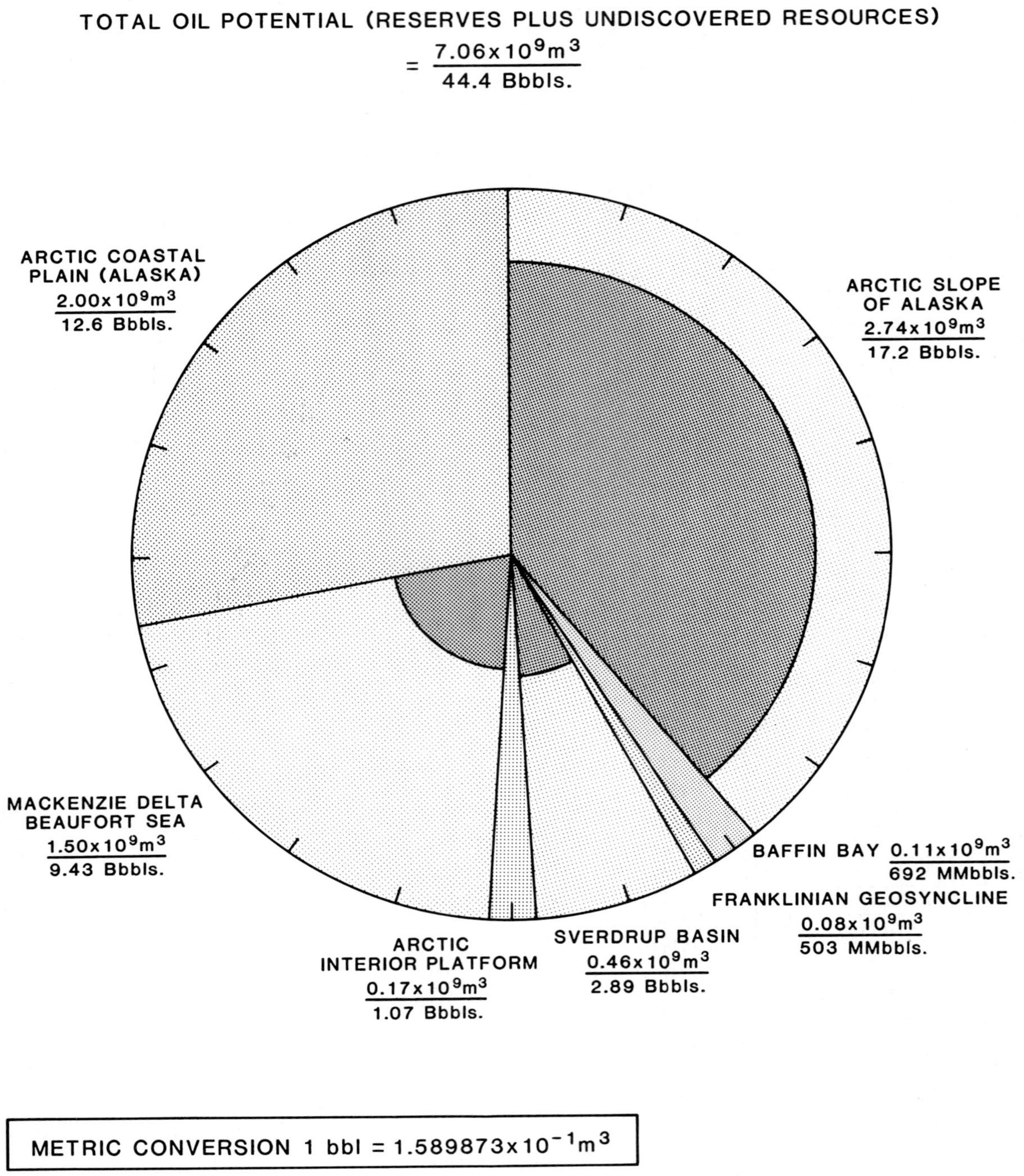


FIGURE 24: *Crude oil potential within the Arctic regions of North America. Each sector represents a fraction of the total recoverable potential (7.06 x $10^9 m^3$) attributed to each geological province. The best current estimates (B.C.E.) of recoverable reserves are indicated by the darkest area of each sector. The lightest area represents the estimated undiscovered recoverable resources.*

480 km (300 miles) wide. Except for outcrops on western Greenland, Baffin Island and Bylot Island, all sediments of the Basin lie offshore. Numerous hypotheses have been presented to explain the nature of material forming the floor of the basin in its deeper parts (*see* McMillan, 1973 and Srivastava et al., 1981). The hypotheses cover a range of possibilities from a floor consisting entirely of continental crust to a floor consisting entirely of oceanic crust. It has also been suggested that the floor may be oceanized, attenuated or transformed continental crust, attenuated continental and oceanic crust, or may contain varying amounts of oceanic crusts.

Aside from uncertainties conceruing the composition of the basement floor, seismic data indicate that much of it is arranged as a series of horsts and grabens (Fig. 22). Greenland and Baffin Island are formed of Precambrian granitic basement rocks; minor Tertiary and Cretaceous sediments overlie those rocks onshore.

Within Baffin Bay proper, regional geological and geophysical data suggest that basement rocks may be overlain by eroded remnants of Paleozoic rocks, possibly Ordovician, Silurian and/or Carboniferous carbonates. Sub-graben within the basin complex may also contain lower Mesozoic clastic units.

From outcrops in western Greenland it is known that a fairly complete succession of Lower Cretaceous (Barremian) to Tertiary (Eocene) marine and nonmarine clastic sediments exists (Henderson et al., 1976). Using seismic data and projecting stratigraphy basinward, it appears that up to 2500 m of these sediments may be present relatively nearshore. The Cretaceous sequence appears to consist of three deltaic megacycles (Fig. 23). Coarse sandstones and conglomerates predominate nearshore, and these coarse clastics grade basinward into sandstones then into siltstones and finally into prodelta shales. Unconformities occur between the Lower and Upper Cretaceous and between the Upper Cretaceous and Tertiary. In western Greenland and at Cape Dyer on Baffin Island, volcanic pillow breccias and basaltic flows are intercalated with Paleocene sediments.

Late Cretaceous and Tertiary sediments in the Baffin Bay Basin become more marine in origin upwards in the succession. Unconformities occur in the Tertiary section; the most pronounced of these appears to be of Middle Tertiary, possibly Earliest Oligocene age.

The central part of the basin may have received up to 10 000 m (33 000 ft) of sediments during the Cretaceous and Tertiary, and the northern part up to 14 000 m (46 000 ft). Nearshore sub-basins may contain sediments up to 2500 m (8200 ft) thick.

The formation of grabens in the Baffin Bay Basin may have taken place during Lower Cretaceous to Middle Tertiary time. After Middle Tertiary time the sediments were affected only by compaction.

## Petroleum Potential of the Baffin Bay Basin

Several facies favourable to petroleum entrapment may occur within Baffin Bay Basin, although their presence is somewhat speculative at this time. The base of the section may contain carbonate rocks; periods of erosion and leaching could produce or enhance porosity in such facies. The initial graben filling cycles of a rift zone usually consist of porous, poorly sorted and loosely consolidated coarse clastic units; if this area is a typical passive rift zone then the basal clastic units should possess excellent porosity. The Early Cretaceous, Late Cretaceous and Early Tertiary sequences are composed of prograding deltaic megacycles; the delta front sandstones would constitute excellent reservoir rocks.

Geochemical analyses of Upper Cretaceous to Tertiary sediments of Western Greenland indicate that several potential source rocks may exist in the section. Oil and gas seeps from Upper Cretaceous and Eocene sediments in areas of the northeast coast of Baffin Island are encouraging as signs of oil generation in the area. Substantial amounts of organic carbon are found in Upper Cretaceous rocks. The organic material in those rocks is of terrestrial origin and only marginally mature in the shallowly buried areas.

Certain Paleocene and Eocene rocks in the vicinity of Baffin Island contain resinite (McMillian, 1982, p. 12; Fig. 16); in the Beaufort-Mackenzie area resinite is thought to generate appreciable quantities of oil at low maturation levels.

No reserves have been found in the Baffin Bay Basin but south of Davis Strait the Hekja well near Baffin Island tested $0.28 \times 10^6 m^3/d$ (10 MMcf/d) of wet gas from a pool that is estimated to contain reserves of over $65.0 \times 10^9 m^3$ (2.3 Tcf) of natural gas.

Five exploration wells have been drilled near the southern margin of the Baffin Bay, in the offshore area of Greenland. Data from one of

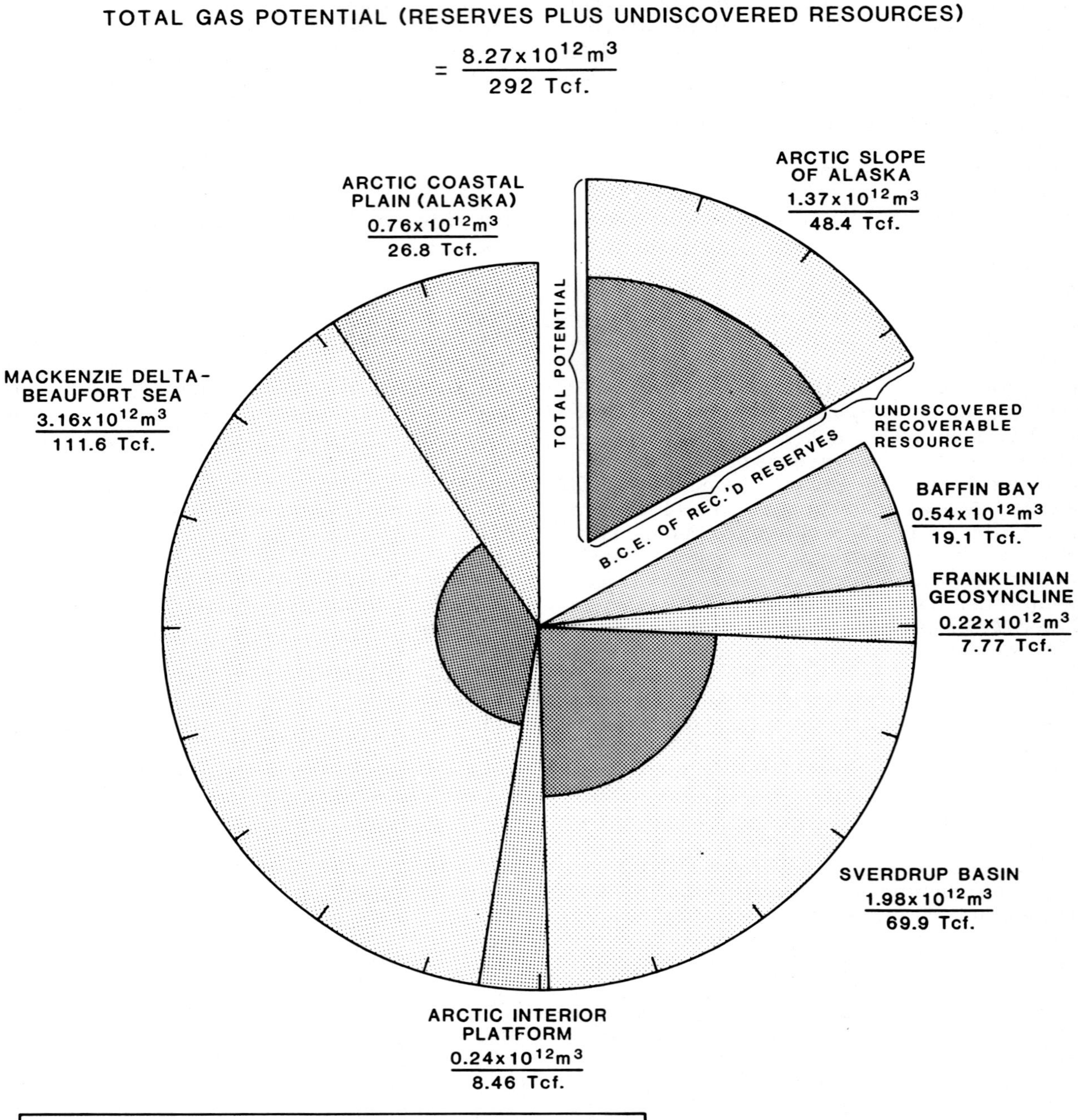


METRIC CONVERSION 1 ft$^3$ = 2.831685x10$^{-2}$ m$^3$

FIGURE 25: *Natural gas potential within the Arctic regions of North America. Each sector represents a fraction of the total recoverable potential (8.27 x 10$^{12}$m$^3$) attributed to each geological province. The best current estimates (B.C.E.) of recoverable reserves are indicated by the darkest area of each sector. The lightest area represents the estimated undiscovered recoverable resources.*

the wells indicate that the absence of petroleum in that well is due to absence of reservoir rocks and to the lack of geochemical maturity at the depths the well penetrated (Manderscheid, 1980).

The undiscovered resource estimates for the Baffin Bay Basin consist of 0.1 x $10^9 m^3$ (0.7 Bbbl) of potential recoverable oil at the expected value and 0.54 x $10^{12} m^3$ (19 Tcf) of potential marketable natural gas at the expected value.

## CONCLUSION

Arctic regions of North America, from the North Slope of Alaska to the Wandel Sea Basin of northeast Greenland, have varying but generally excellent hydrocarbon potentials. The cost of recovering petroleum in these regions may be extraordinatily expensive but exploration and discovery will nevertheless be carried out. Several regions, such as the North Slope of Alaska and the Arctic Coastal Plain, and especially the Beaufort-Mackenzie Basin, contain significant petroleum reserves. The Sverdrup Basin has extensive gas reserves and continues to show promise as a highly prospective petroleum province. Most conditions favouring large pools are indicated and recent relatively small discoveries have significantly improved prospects for more drilling successes. The success ratio for the Sverdrup Basin has been excellent when compared with exploration in other frontier regions such as the North Sea. The Wandel Sea Basin is geologically comparable with the Sverdrup Basin and comparable success might be anticipated when the serious technical problems relevant to drilling in regions of shifting pack-ice have been overcome. The same reasoning can be applied to the Arctic Coastal Plain contiguous to the Arctic Islands, where similar technical problems relative to shifting ice must be overcome before interesting prospects can be drilled. Numerous attractive structures within the Basin, an extremely thick sedimentary succession, and oil seeps adjacent to Baffin Island suggest that this basin too may prove to be a productive oil province.

## References


Balkwill, H.R.
1978: Evolution of Sverdrup Basin, Arctic Canada; American Association of Petroleum Geologists, Bulletin, v. 62, p. 1004-1028.

Balkwill, H.R., Cook, D.G., Detterman, R.L., Embry, A.F., Håkansson, E., Miall, A.D., Poulton, T.P. and Young, F.G.
in press: Mesozoic of Arctic North America and northern Greenland; in The Phanerozoic of the World II, The Mesozoic; Moulland, M. and Nairn, A.E.M., eds.; Elsevier Publishing Company, Amsterdam, New York.

Bird, Kenneth, J.
1981: Petroleum exploration of the North Slope in Alaska, U.S.A.; U.S. Geological Survey, Open File Report 81-227, 43 p.

Bird, Kenneth, J. and Jordan, Clifton, F. Jr.
1977: Petroleum Potential of the Lisburne Group, eastern Arctic Slope, Alaska; Oil and Gas Journal, April 18, p. 90-94.

Campbell, R.H.
1967: Areal geology in the vicinity of the Chariot Site, Lisburne Peninsula, northwestern Alaska; U.S. Geological Survey, Professional Paper 395, 71 p.

Chapman, R.M. and Sable, E.G.
1960: Geology of the Utukok-Corwin region, northwestern Alaska; U.S. Geological Survey, Professional Paper 303-C, p. C47-C167.

Churkin, M.
1969: Paleozoic tectonic history of the Arctic Basin north of Alaska; Science, v. 165, p. 549-555.

Dawes, P.R.
1976: Precambrian to Tertiary of northern Greenland; in Geology of Greenland; Escher, A. and Watt, W.S., eds.; Geological Survey of Greenland, p. 284-303.

Dawes, P.R. and Soper, N.J.
1973: Pre-Quaternary history of North Greenland; American Association of Petroleum Geologists, Memoir 19, p. 112-134.

Detterman, R.L.
1956: New and redefined nomenclature of the Nanuskuk Group; in Mesozoic Sequence in Colville River region, northern Alaska; Gryc, G. et al.; American Association of Petroleum Geologists, Bulletin, v. 40, p. 233-244.

1973: Mesozoic sequence in Arctic Alaska; in Proceedings of 2nd International Arctic Symposium; American Association of Petroleum Geologists, Memoir 19, p. 376-387.

Detterman, R.L., Reiser, H.N., Brosgé, W.P. and Dutro, J.T., Jr.
1975: Post-Carboniferous stratigraphy, northeastern Alaska; U.S. Geological Survey, Professional Paper 886, 46 p.

Gregory, A.F., Bower, M.E. and Morley, L.W.
1961: Geological interpretation of aerial, magnetic and radiometric profiles, Arctic Archipelago, Northwest Territories; Geological Survey of Canada, Bulletin 73.

Håkansson, E.
1979: Carboniferous to Tertiary development of the Wandel Sea Basin, eastern North Greenland; Grønlands Geologiske Undersogelse, Rapport 88, p. 73-83.

Håkansson, E., Konnernp-Madsen, J., Piasecki, S. and Thomsen, E.
in press: A late, thermal event in the Wandel Sea Basin, eastern North Greenland.

Håkansson, E. and Pederson, S.A.S.
in press: Tectonic evolution of the continental margin in North Greenland, Proceedings Third Arctic Symposium, Calgary, 1981.

Henao-Landono, D.
1977: Correlation of producing formations in the Sverdrup Basin; Bulletin of Canadian Petroleum Geology, v. 25, no. 5, p. 969-980.

Henderson, G .
1976: Petroleum Geology; in Geology of Greenland; Escher, A. and Watt, W.S. eds.; Geological Survey of Greenland, p. 488-505.

Henderson, G., Rosenkrantz, A. and Schiener,E.J.
1976: Cretaceous-Tertiary sedimentary rocks of West Greenland; in Geology of Greenland; Escher, A. and Watt, W.S. eds., Geological Survey of Greenland, p. 338-362.

Jeletzky, J.A.
1967: Jurassic and (?)Triassic rocks of the eastern slope of Richardson Mountains, northwestern District of Mackenzie, 106M and 107B (part of); Geological Survey of Canada, Paper 66-50.

1975: Jurassic and Lower Cretaceous Paleogeography and Depositional Tectonics of Porcupine Plateau, adjacent areas of northern Yukon and those of Mackenzie District, N.W.T.; Geological Survey of Canada, Paper 74-16.

1977: Porcupine Formation; a new Upper Jurassic sandstone unit, northwest Yukon Territory; Geological Survey of Canada, Paper 76-27.

Kerr, J.Wm.
1968: Stratigraphy of central and eastern Ellesmere Island, Arctic Canada, Part Il, Ordovician; Geological Survey of Canada, Paper 67-27, Part Il.

King, P.B.
1959: The evolution of North American; Princeton University Press.

Lerand, M.
1973: Beaufort Sea; in The Future Petroleum Provinces of Canada — their geology and potential; McCrossan, R.G. ed., Canadian Society of Petroleum Geologists, Memoir 1, p. 315-386.

Manderscheid, Guy
1980: The geology of the offshore sedimentary basin of West Greenland in Facts and Principles of World Petroleum Occurrence; Miall, Andrew, D. ed.; Canadian Society of Petroleum Geologists, Memoir 6, p. 951-973.

McMillan, N.J.
1973: Labrador Sea and Baffin Bay; in The Future Petroleum Provinces of Canada — their Geology and potential; McCrossan, R.G. ed.; Canadian Society of Petroleum Geologists, Memoir 1, p. 473-518.

1982: Canada's east coast: the new super petroleum province; Journal of Canadian Petroleum Technology, v. 21, no. 2, p. 1- 15.

Meneley, R.A., Henao, D. and Merritt, R.
1975: The Northwest margin of the Sverdrup Basin; in Canada's continental margins and offshore petroleum exploration; Yorath, C.J., Parker, E.R. and Glass, D.J. eds.; Canadian Society of Petroleum Geologists, Memoir 4, p. 531—544.

Nassichuk, W.W. and Bamber, E.W.
1979: Middle Pennsylvanian biostratigraphy, eastern Cordillera and Arctic Islands, a summary; in Western and Arctic Canadian biostratigraphy, P.S. Warren Memorial Volume; Stelck, C.R. and Chatterton, B.D.E., eds.; Geological Association of Canada, Special Paper 18, p. 395-413.

Nassichuk, W.W. and Davies, G.R.
1980: Stratigraphy and sedimentation of the Otto Fiord Formation — a major Mississippian — Pennsylvanian evaporite of subaqueous origin in the Canadian Arctic Archipelago; Geological Survey of Canada, Bulletin 286, p. 1-87.

Plauchut, B.D.
1971: Geology of the Sverdrup Basin; Bulletin of Canadian Petroleum Geology, v. 19, p. 659-679.

Poulton, T.P.
1978: Internal Correlations and Thickness Trends, Jurassic Bug Creek Formation, Northeastern Yukon and Adjacent Northwest Territories; Geological Survey of Canada, Paper 78-1B, p. 27-30.

Rayer, F.G.
1981: Exploration prospects and future petroleum potential of the Canadian Arctic Islands; Journal of Petroleum Geology, p. 367-412.

Srivastava, S.P., Falconer, R.K.H. and MacLean, B.
1981: Labrador Sea, Davis Strait, Baffin Bay: geology and geophysics — a review; in Geology of the North Atlantic Borderlands; K rr, J.Wm., Ferguson, A.J., and Machan, L.C., eds.; Journal of Canadian Society of Petroleum Geologists, Memoir 7, p. 333-398.

Thorsteinsson, R.
1974: Carboniferous and Permian stratigraphy of Axel Heiberg Island and western Ellesmere Island, Canadian Arctic Archipelago; Geological Survey of Canada, Bulletin 224.

Thorsteinsson, R. and Tozer, E.T.
1960: Summary account of structural history of the Canadian Arctic Archipelago since Precambrian time; Geological Survey of Canada, Paper 60-7.
1970: Geology of the Arctic Archipelago; in Geology and Economic Minerals of Canada; Geological Survey of Canada; Economic Geology 1, 5th ed., p. 548-590.

Tozer, E.T.
1970: Mesozoic and Cenozoic; in Geology of the Arctic Archipelago; in Geology and Economic Minerals of Canada; Thorsteinsson, R. and Tozer, E.T.; Geological Survey of Canada, Economic Geology 1, 5th ed., p. 574-588.

Tozer, E.T. and Thorsteinsson, R.
1964: Western Queen Elizabeth Islands, Arctic Archipelago; Geological Survey of Canada, Memoir 332, 242 p.

Trettin, H.P.
1979: Contributions to the tectonic history of the Innuitian Province, Arctic Canada; Canadian Journal of Earth Sciences, v. 16, p 748-769.

Trettin, H.P., Frisch, T.O., Sobczak, L.W., Weber, J.R., Niblett, E.R., Law, L.K., de Laurier, I. and Whitham, K.
1972: The Innuitian Province; in Variations in Tectonic Styles in Canada; Price, R.A. and Douglas, R.J.W. eds.; Geological Association of Canada, Special Paper 11, p. 83-180.

Willumsen, P.S. and Coté, R.P.
in press: Tertiary sedimentation in the southern Beaufort Sea; Proceedings, Third Arctic Symposium, Calgary, 1981.

Yorath, C.J. and Norris, D.K.
1975: The tectonic development of the Southern Beaufort Sea and its relationship to the origin of the Arctic Ocean Basin; in Canada's Continental Margins and Offshore Petroleum Exploration; Yorath, C.J., Parker, E.R. and Glass, D.J. eds.; Canadian Society of Petroleum Geologists, Memoir 4, p.589-611.

Yorath, C.J., Norris, D.K. and Young, F.G.
1980: Geology of the Beaufort-Mackenzie Basin, District of Mackenzie and Yukon Territory; Geological Survey of Canada, Map 1509A (Compiled by B.R. Pelletier).

Young, F.G.
1975: Stratigraphic and Sedimentologic Studies in Northeastern Eagle Plain, Yukon Territory; Geological Survey of Canada, Paper 75- lB, p. 309-323.

Young, F.G., Myhr, D.W. and Yorath, C.J.
1976: Geology of the Beaufort-Mackenzie Basin; Geological Survey of Canada, Paper 76-1 1, 65 p.

*Cold Regions Science and Technology*, 7 (1983) 89—166
Elsevier Science Publishers B.V., Amsterdam — Printed in The Netherlands

# Energy Resources of Soviet Arctic and Subarctic Regions

*Arthur A. Meyerhoff*

Associated Resource Consultants, Inc.
Tulsa

## Abstract

*Broadly defined, the region considered here is north of 60° N lat, with only a few exceptions: the Baltic syneclise on the west; the southern part of the West Siberian basin in the center; and the Sea of Okhotsk massif on the east, in the midst of the Pacific mobile belt. Thus, the region is naturally divisible into three great geologic provinces. From west to east, these are the Russian platform with its ancient, stable, Precambrian floor that is exposed on the west on the Baltic shield; the similarly ancient and stable Siberian platform in the center, separated from the Russian platform by the Paleozoic Urals fold belt; and the Pacific mobile belt on the east, a region that has been deformed repeatedly into mountain ranges since early in the earth's history. Petroleum is reviewed first, followed by coal. The area is discussed from west to east.*

*The westernmost basin, the Baltic syneclise, is a Cambrian through Middle Devonian feature, with mainly Cambrian oil. Reserves are about 100 million barrels. Most future discoveries-the resources-may be 500 million barrels or more. Most will be in Cambrian sedimentary rocks with lesser amounts in Ordovician through Devonian strata. The Timan-Pechora basin in the northwestern USSR has production from the Silurian through the Triassic. Source rocks are mainly in the Devonian through Permian section. This basin bifurcates offshore, with the southwestern part of the basin appearing to join basins north of Norway on the west, and the eastern part of the basin continuing northward along the western side of Novaya Zemlya. Reserves in the Timan-Pechora basin are about 5.1 billion barrels of liquids and 19.4 trillion cubic feet (Tcf) of gas, all of it onshore. Although the onshore has important resource potential (4 billion barrels and 50 Tcf), the offshore has the greater potential (16 billion barrels and 100 Tcf), the offshore has the greater potential (16 billion barrels and 100 Tcf). One discovery has been made in the Norwegian sector of the Barents Sea, where both the Triassic and Jurassic have considerable potential, in addition to the Paleozoic section beneath.*

*East of the Urals and Novaya Zemlya, the West Siberian basin (including the Kara Sea) occupies an area of 2,100,000 sq km. The northern onshore part of the basin contains the world's largest concentration of gas reserves, most of it Cenomanian biogenic gas. Smaller amounts of gas are in the Neocomian and Upper Jurassic. The principal source of the biogenic gas is Aptian-Albian coal. The oil and gas of the remainder of the basin have a mainly Upper Jurassic source, although some source beds are in the Lower Cretaceous. Proved oil reserves are about 18 billion barrels; gas reserves are 545 Tcf. The resource potential is an additional 54 billion barrels and at least 948 Tcf.*

*Several basins are known in East Siberia and the Soviet Far East. The largest are offshore and have major potential. Of the onshore basins, the Vilyuy has 17 Tcf of proved gas reserves in Jurassic, Triassic, and Permian rocks; the resource potential is an additional 19 Tcf or more. The volume of liquids found to date is small. The gas sources appear to be Permian or older strata. West of the Vilyuy basin, the Lena-Tunguska province contains about 225 million barrels of oil and condensate and 18 Tcf of gas in rocks of*

  

*Proterozoic and Early Cambrian ages. Sources are of the same ages. Resource potential is at least 2 billion barrels and 83 Tcf. On Sakhalin Island, along the Pacific coast, oil and gas are present in Miocene and lower Pliocene sandstones. Reserves are 24 million barrels of liquids and 2 Tcf of gas, with resource potentials of about 1 billion barrels and 5 Tcf.*

*The Okhotsk, Bering, Chukchi, East Siberian, and Laptev Seas have huge potential-27 billion barrels and 175 Tcf. Yet these areas are insufficiently explored, some hardly at all. Production from the first two areas should be mainly Tertiary; from the others, Mesozoic and lower Tertiary. Because of the enormous logistics problems, important production from the offshore Arctic USSR will be minuscule until after the year 2000.*

*Coal reserves have been established mainly in the Timan-Pechora basin where they have been mined extensively since 1932. Reserves of 20,000 million metric tons are established; an additional 303,000 million tons is inferred from field studies and test drilling. Most of the coal is Lower Permian, but important reserves and resources are present in the Upper Permian and Triassic. Coal rank increases from hard brown coal in the west to high-rank bituminous hard coals (fat and coking coals) at Vorkuta. In the Polar Urals and the Pay-Khoy, the rank increases to semi-anthracite and, locally, to anthracite.*

*Very little exploration has been conducted for coal east of the Polar Urals. Permian and Triassic coals, however, are present in large volumes in the Taymyr fold belt, an extension of the Urals-Pay Khoy-Novaya Zemlya fold belt. Estimates of the coal resources of the Taymyr differ widely. However, it would be no exaggeration to say that the coal resource potential is much greater than 20,000 million tons.*

*Farther east, in the northern part of the Lena-Tunguska province, huge coal deposits are known to underlie about 1,000,000 sq km of Triassic trap rocks on the Siberian platform. These coals also mainly Carboniferous, Permian, and Jurassic. Several Soviet geologists have stated that the Lena-Tunguska area contains the largest coal resource found anywhere in the USSR-well in excess of 4,000 billion tons. However, it will be many decades before the potential of this area is tapped, not only because of its remoteness, but also because of the thick traprock cover (up to 1,800 meters) and the readily available, more accessible coal fields elsewhere.*

*South of the Vilyuy basin is another huge coal-bearing region, the South Yakutia coal province (500 million tons of high-calorific coking coal proved; more than 40,000 million tons inferred), largely of Jurassic age. Farther east, in the Far East and Maritime provinces, coal resources are widespread, are mainly Jurassic, Cretaceous, and Tertiary, but are used only locally.*

*In all, the energy resource potential of Arctic regions is great. The principal problems are economic ones-the very high cost of exploration and development and the remoteness of many regions, especially in areas covered by ice and permafrost.*

## Introduction

For the purposes of this paper, I define the arctic and subarctic as the region which generally is north of lat 60° N (Fig. 1). I have included the Baltic syneclise because it borders the USSR's coastline, along which the USSR has few ice-free ports in northern waters. Moreover, little has been written in English about the geology and petroleum prospects of the Baltic syneclise. The entire West Siberian basin is included because it is a single geologic unit, extending from 54° N to 76° N (including the offshore southern half of the Kara Sea). Sakhalin Island is included because it lies in the Arctic's «iceberg alley» of the northern Pacific. The Lena-Tunguska province is included because it is an area where major gas and oil reserves are being discovered in Proterozoic rocks, and to a lesser extent, in Lower Cambrian strata.

The area under consideration is huge-at least 17,000,000 sq km, including the offshore. The USSR has one of the world's largest offshore shelves, and its Arctic Ocean frontage occupies 52 percent of the total land area which faces the Arctic Ocean.

The region is divided naturally into three geologic provinces. From west to east these are first, the Russian, or East European platform, rimmed on the west by an ancient Paleozoic fold belt, the Caledonides, and the exposed Archean Baltic shield. This platform has remained tectonically stable since the middle or latter part of Precambrian time-more than 1 billion years. East of the Russian platform, and separated from it by the Paleozoic fold belt of the Ural Mountains, is the Siberian platform, possibly one of the most stable areas on earth. Both the

Russian and Siberian platforms include extensive areas of the Arctic shelf. The third, and last geologic province is the Pacific mobile belt, a broad, mountainous region east of the Lena River and extending to the shores of the Pacific Ocean. This region is a part of the so-called «ring of fire», a complex of deep-sea trenches, island arcs, youthful mountains, and volcanoes which completely encircle the Pacific Ocean basin. The Pacific mobile belt within the USSR has been actively deformed since at least the last half of Precambrian time, and therefore, unlike the Russian and Siberian platforms, is geologically and tectonically unstable.

The volume of energy reserves and resources of these three conterminous regions is related directly to the degree of tectonic stability. Thus the size of individual petroleum basins, oil and gas fields, and coal fields is far greater on the Russian and Siberian platforms than in the Pacific mobile belt, which has been folded and faulted repeatedly by tectogenesis, with the result that basins and fields have been broken up repeatedly into smaller and smaller entities.

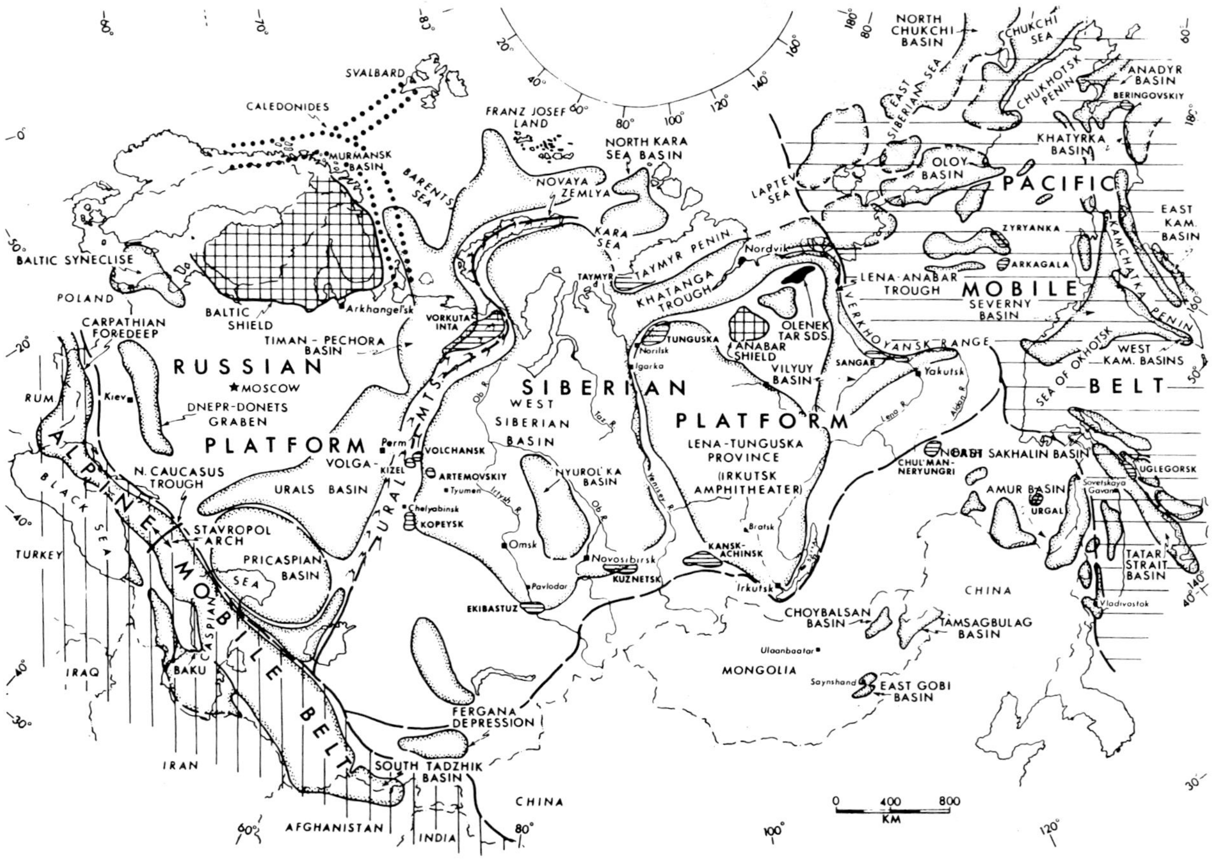


*Fig. 1* — Generalized index map to the petroleum- and coal-bearing and potentially petroleum- and coal-bearing basins of the USSR.

## Introduction

The USSR is a petroleum-oriented country, with 64 percent of its energy needs supplied by petroleum and only 30 percent supplied by coal. Of the 64 percent that is supplied by petroleum (equal to 12,220,000 barrels of oil and condensate and 45 billion cubic feet of gas per day), fully 58 percent of the oil and 32 per cent of the gas come from the West Siberian basin (Fig. 1) alone. Within a few years, more than 50 percent of the gas also will come from the West Siberian

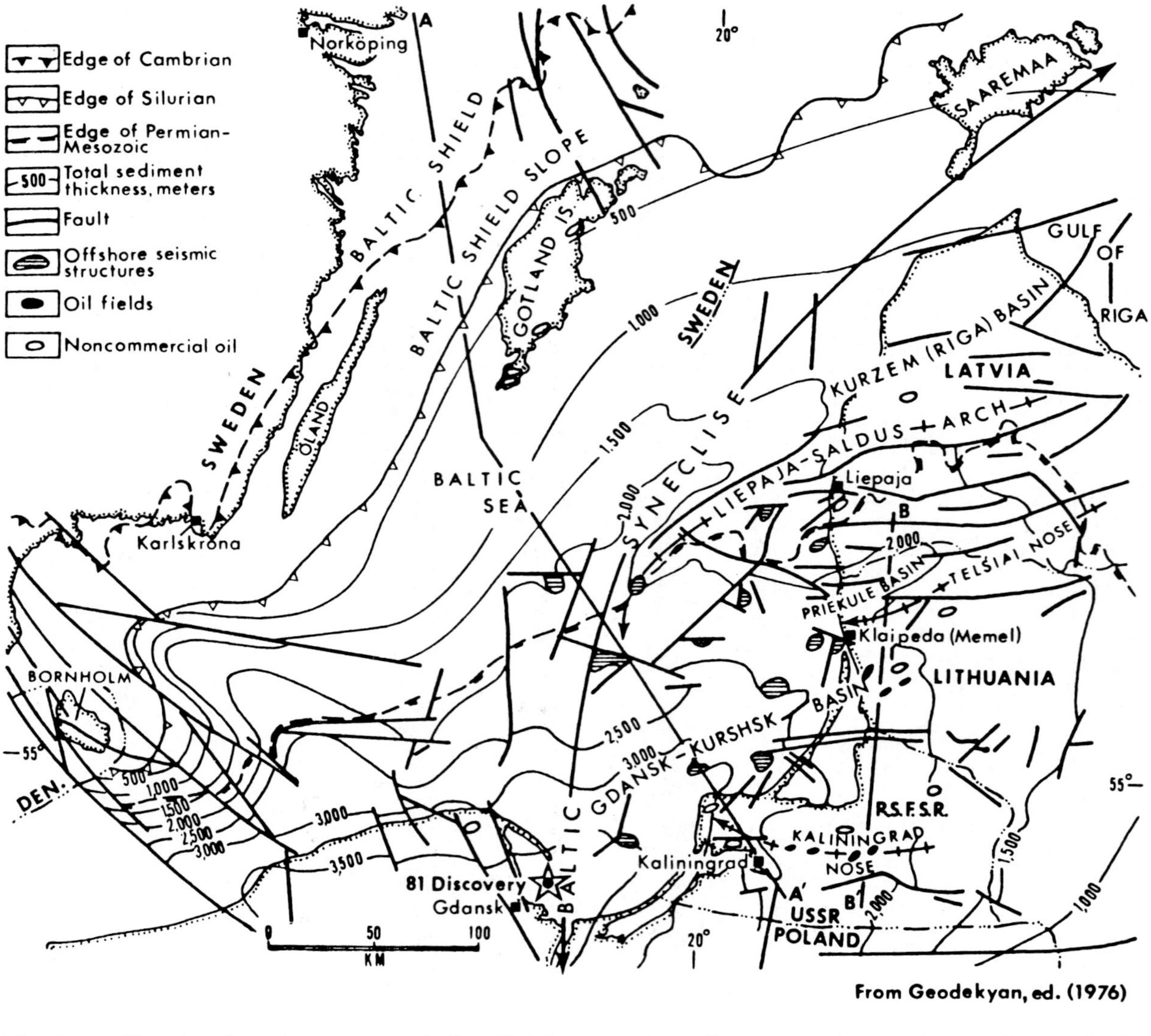


*Fig. 2* — Structural-contour map of the Baltic syneclise. Structural datum is the base of the Cambrian. The map shows various subbasins, some of the onshore oil fields, and offshore structures as yet undrilled. Not all oil fields are shown. Depths are in meters. Locations of Figs. 3 and 4 are shown. From Geodekyan (1976).

basin. Thus the Soviet arctic and subarctic regions are of vital importance to the country.

In contrast, only 34 percent of the nation's coal comes from the Soviet arctic and subarctic, if one includes the giant Kuznetsk and Ekibastuz coal producing areas (Fig. 1). Neither of these two basins, technically speaking, is in the arctic, but they are at the rim of the West Siberian basin. Ultimately should coal continue to be used as a major energy source-most of the nation's coal will come from the northern Urals, the Taymyr Range, the Tunguska basin, the Vilyuy basin, and the South Yakutia basin at Neryungri (Fig. 1), because these areas hold fully 60 percent of the country's coal resources.

I shall begin with the Russian platform, and will move eastward. I deal first with petroleum.

## Baltic Syneclise

### Introduction

The Baltic syneclise (basin) at the western end of the Russian platform, borders the ancient Baltic shield. It occupies at least 180,000 sq km in Latvia, Lithuania, the Kaliningrad district of the RSFSR, northern Poland, the central Baltic Sea, and the Øland-Gotland Islands area of coastal Sweden (Geodekyan, 1976; Dikenshteyn, 1977; Geodekyan et al., 1978; Fig. 2). Of this amount, 100,000 sq km belongs to the USSR. Exploratory work began here in 1949. In 1960 and 1963, two small noncommercial oil discoveries were made in the USSR. Swedish exploration began with stratigraphic tests in 1967, and the first discovery was made on Gotland Island in 1974 (Dahl, 1982). In the USSR, the first two commercial fields were found in 1968 and, by 1980, 16 commercial fields (38-45° API) had been discovered in USSR territory, in addition to 6 noncommercial fields. Five noncommercial finds had been made on Gotland Island, Sweden, and one in coastal Poland. Sixteen offshore wells had been drilled by the end of 1981-all dry, and 150 shallow wells had been drilled on Gotland Island, where small Ordovician reefs produce 2,000 bbl/d of 40° API oil from 200 to 300 m depths; 25 wells still were productive on January 1, 1982 (Anonymous, 1981). In Polish waters, the first offshore discovery was made near Gdansk (Fig. 2) during 1981; four

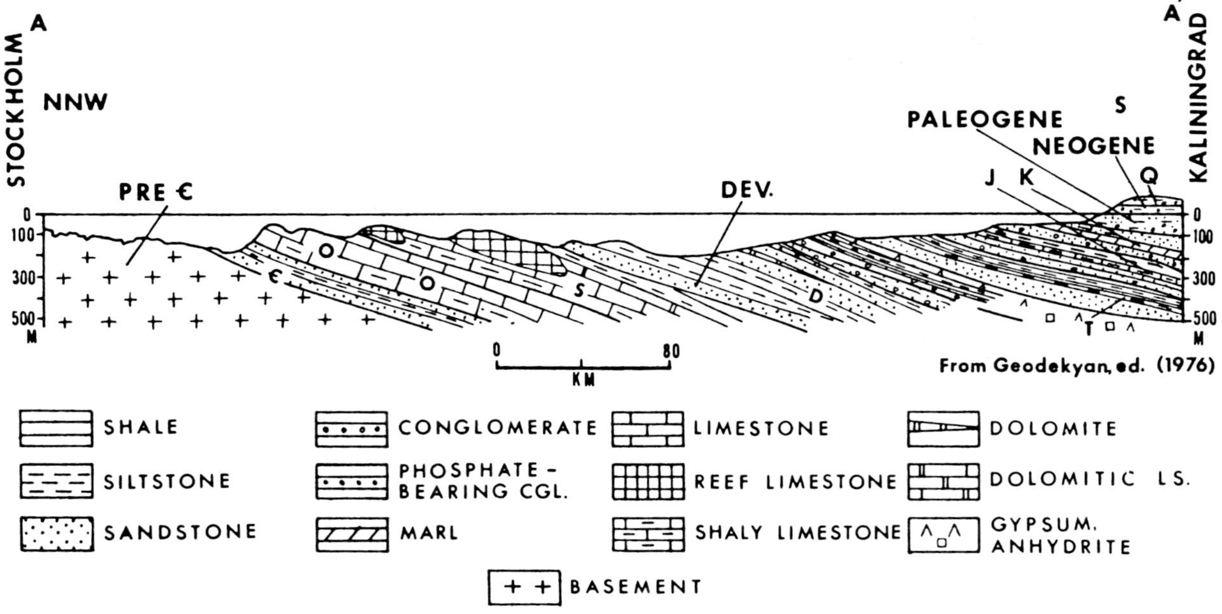


*Fig. 3* — NNW-SSE cross section A-A' of Baltic syneclise, Stockholm (Sweden) to Kaliningrad (RFSFR). Depths are in meters. Location is on Fig. 2.

C = Cambrian; T = Triassic;
O = Ordovician; J = Jurassic;
S = Silurian; K = Cretaceous;
D = Devonian; Q = Quaternary.

From Geodekyan (1976).

later wells were dry. Therefore, the discovery may be noncommercial. These five wells were a joint venture of the Petro-Baltik Trest, the members of which are Poland, East Germany, and the USSR.

In late 1981, a fixed platform was built just offshore from Kaliningrad in the USSR, and the Soviet Union's first offshore drilling in the Baltic Sea commenced during 1982. The mobile rig that found oil offshore from Gdansk will move to the USSR to drill one of the several structures known there (Fig. 2). Until 1982, the deepest well drilled in the basin was 2,984 m (onshore, near Kaliningrad) near the basin center (Fig. 2). The offshore test drilled in 1981 near Gdansk reached 2,520 m.

## Regional Geologic Framework

The total sedimentary rock thickness is shown on Figure 2, which is a structural-contour map of the base of the Cambrian. The stratigraphic sequence appears on Figures 3 and 4. The three figures illustrate that the Baltic syneclise is a supracratonic depression of Early Cambrian through Middle Devonian ages. The Devonian overlies the Silurian unconformably. After Middle Devonian time, the basin was overlapped from the south by Carboniferous and younger sediments of the Northwest German basin.

## Stratigraphy

The Cambrian consists of up to 478 m (the thickest section known, northwest of Gdansk) of arkosic conglomerate which grades upward into increasingly finer grained terrigenous rocks. Some argillaceous and organogenic phosphatic limestone is present in the Middle and Upper Cambrian, together with quartzose sandstone, siltstone, and black shale. Marine fossils are abundant. The principal production is from quartz sandstone in the Middle Cambrian. The Cambrian becomes very fine grained offshore and siliceous in a southerly direction.

The Ordovician is 40 to 200 m of marine shallow-water and biostromal carbonates, conformable above the Cambrian. Argillaceous limestone, minor sandstone, and some glauconite are present. Approximately 100 km west of Gdansk near Slupsk, the Ordovician thickens abruptly to more than 1,400 m of deep-water black shale (Geodekyan, 1976). The Silurian ranges in thickness from 500 to 2,200 m, and is complete in the central part of the syneclise. It is conformable on the Ordovician. Around the flanks of the basin, the Silurian was eroded before deposition of the Devonian. It is a sequence of organic-rich bituminous and organic shales with beds of siltstone, sandstone, and carbonate rocks, including bioherms and biostromal limestone. Deep-water conditions prevailed in the central part of the basin and in

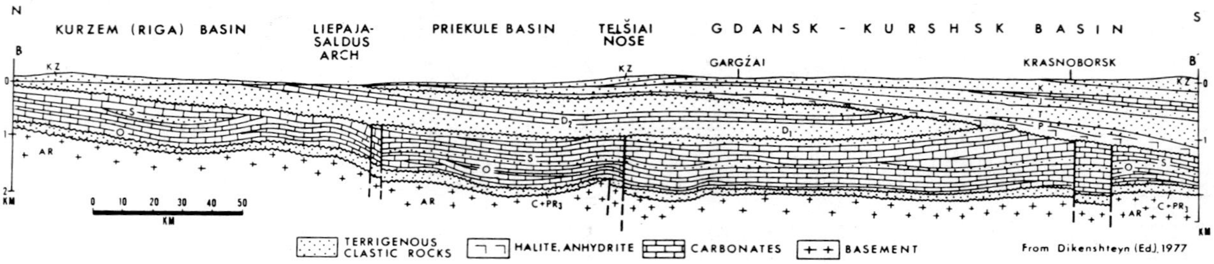


*Fig. 4* — N-S cross section of the eastern part of the Baltic syneclise, RSFSR. Depths are in meters. Location is on Fig. 2.

| | | | | | |
|---|---|---|---|---|---|
| AR | = | Archean; | $D_2$ | = | Middle Devonian; |
| $PR_3$ | = | Upper Proterozoic; | P | = | Permian; |
| C | = | Cambrian; | T | = | Triassic; |
| O | = | Ordovician; | J | = | Jurassic; |
| S | = | Silurian; | K | = | Cretaceous; |
| $D_1$ | = | Lower Devonian; | KZ | = | Cenozoic. |

From Dikenshteyn (1977).

the west. Devonian time records a shallowing of the sea and its eventual retreat from the basin. The Lower Devonian, 50 to 400 m thick, overlies the Silurian unconformably, particularly along the basin flanks. The basin center shifted from near Gdansk to near Klaipeda (Fig. 2). The strata are lagoonal, marginal marine, and nonmarine sandstone, siltstone, shale, and marl. Colors range from gray to red and red violet. The lithology and depositional environment indicate the onset of the Caledonian tectogenesis.

The Middle Devonian is thickest near Klaipeda, and consists of sandstone, siltstone, shale, marl, dolomite, and gypsum. The thickest known section is 760 m. Much of the section is nonmarine, and fish remains abound. Only in the carbonates are marine faunas prevalent.

The Caledonian and Hercynian tectogeneses occupied much of the time interval from Middle Devonian until Early Permian time. During the Permian, an arm of the Northwest German (Zechstein) basin transgressed the southern third of the syneclise, and up to 350 m of terrigenous-clastic marine and nonmarine strata was deposited (Figs. 3, 4). Mesozoic and some Cenozoic strata complete the section-zero in the north but up to 600 m in the southern third of the basin.

## Petroleum Geology

Of the 23 commercial and noncommercial discoveries in the USSR, and of the two in Poland, 20 are in Middle Cambrian sandstones and 5 in Ordovician carbonates. Only minor oil shows have been found in Silurian rocks. Oil and gas shows are known from the Permian. Small flows of oil have been recorded in Devonian strata. Sources of the oil appear to be in Cambrian and Silurian strata, with some source materials present in the Ordovician. The noncommercial flows of oil found in five localities on Gotland Island may have been generated in rocks of the same ages.

In the 16 commercial fields, all of them in the USSR, the Soviets have publically stated that

**Table 1.** Petroleum Resources of the USSR — Produced, Proved, and Predicted through 1979.

| Basin | Produced oil and condensate (million bbl) | Estimate of Proved liquids (million bbl) | Resource Potential (million bbl) | Produced Gas (Bcf) | Proved Gas (Bcf) | Resource Potential (Bcf) |
|---|---|---|---|---|---|---|
| Baltic syneclise | 195 | 100 | 500 | — | 100 | 500 |
| Carpathian trough | 639 | 100 | 300 | 8,600 | 100 | 500 |
| Dnepr-Donets graben | 1,004 | 200 | 1,000 | 30,460 | 19,600 | 10,000 |
| Black Sea-Crimea (Sivash trough) | 1 | 1 | 10 | 200 | 1,000 | 10,000 |
| North Caucasus-Mangyshlak trough | 4,825 | 1,950 | 5,000 | 22,715 | 3,500 | 30,000 |
| Baku-Cheleken district | 8,090 | 2,100 | 10,000 | 4,780 | 1,000 | 2,000 |
| Kopet Dag trough | — | — | 1,000 | 37,307 | 91,800 | 50,000 |
| Volga-Urals basin | 30,037 | 15,400 | 7,000 | 14,460 | 79,200 | 50,000 |
| Timan-Pechora basin | 548 | 5,150 | 4,000 | 4,200 | 19,420 | 50,000 |
| North Caspian basin | 492 | 1,000 | 2,000 | 2,500 | 700 | 20,000 |
| Fergana depression | 375 | 90 | 100 | 200 | 200 | 500 |
| Chu-Sarysu basin | — | 5 | 50 | — | 300 | 1,000 |
| Ustyurt platform | — | 5 | 100 | — | 250 | 1,000 |
| West Siberian basin | 10,990 | 16,350 | 54,000 | 26,427 | 545,050 | 948,000 |
| Vilyuy syneclise | — | — | 500 | 60 | 17,000 | 19,000 |
| Lena-Tunguska province | — | 225 | 2,000 | — | 18,100 | 83,000 |
| East Sakhalin basin | 486 | 24 | 1,000 | 496 | 2,000 | 5,000 |
| **SUBTOTAL (ONSHORE)** | **57,682** | **42,700** | **78,560** | **152,405** | **808,320** | **1,280,500** |
| Barents Sea | — | — | 16,000 | — | — | 100,000 |
| Kara Sea | — | — | 2,000 | — | — | 500,000 |
| Laptev Sea | — | — | 1,000 | — | — | 50,000 |
| East Siberian Sea | — | — | 10,000 | — | — | 50,000 |
| Chukchi Sea | — | — | 10,000 | — | — | 50,000 |
| Bering Sea | — | — | 100 | — | — | 10,000 |
| Sea of Okhotsk | — | — | 6,000 | — | 200 | 12,000 |
| Lost (estimate) | — | — | — | 35,000 | — | — |
| **TOTALS** | **57,682** | **42,700** | **123,660** | **187,405** | **808,520** | **2,052,500** |

recoverable A+B reserves initially were 219,000,000 bbl (30 million mt). This is one of the few oil-reserve estimates ever announced officially in the USSR. About 195 million bbl had been produced by the end of 1979 (Table 1), by which time estimated A+B reserves were 100 million bbl, and a potential for 500 million additional barrels existed, perhaps half of which is offshore. Gas potential is considered to be very low. Production early in 1981 averaged 29,000 bbl/d from 80 wells in 14 fields; two fields were to be put on stream during 1981.

The principal objective is the sandstone section of the Middle Cambrian. Dry porosity values range from 10 to 30 percent; permeabilities of 1,000 to 3,000 md are common (Geodekyan et al., 1978). The traps are small (2 x 3 km or less), although a few are large (22 x 25 km) (Bakirov, 1979). Reservoirs become tighter toward the west and are silicified in the southwest, particularly in Poland. For this reason only two flows of oil have ever been recorded in the Middle Cambrian rocks of Poland, one onshore and one offshore (Fig. 2). The other 15 structures drilled in Poland were dry.

A few wells in the USSR and Gotland Island have found Ordovician oil flows, from a few barrels a day to as much as 20 to 50 bbl/d. However, most of the flows have been short-lived. Of the 150 wells drilled on Gotland, only 25 produce oil. Sixteen offshore tests and six others drilled on Øland were dry (Dahl, 1982). Organic carbon content in a few limestone units ranges from 1 to 12 percent, so that rich carbonate source rocks are present in some areas. Vug and fracture porosity is prerequisite, however, for accumulations to form in Ordovician rocks, and in general, little porosity or permeability has been found in this carbonate section.

Silurian (especially Ludlovian and Downtonian) dolomites exhibit porosity values up to 26 percent and permeabilities of 20 md. Secondary dolomites with leached organisms show porosity values of 17 percent and permeabilities of 24-105 md. However, no flows of oil have been found, despite the presence of both reservoirs and rich source beds.

The Devonian has good porosity and permeability values, but a problem, except in the southern third of the basin, is the absence of seals.

Overall, the basin is poorly explored and possibilities abound for many small fields to be discovered. Larger fields eventually may be found in stratigraphic traps. The greatest unknown is the offshore. Most traps to be found will be stratigraphic-reefs, bioherms, or pinchouts. However, seismic technology capable of detecting such traps has only begun to be used in the USSR where, until two years or so ago, moderately sophisticated CDP methods (12-fold or more) were not used. Today 12-fold stacking techniques are available, but whether they are being used in this area, I do not know.

## Timan-Pechora Basin

### Introduction

The Timan-Pechora basin is another primarily Paleozoic feature, which occupies an onshore area of 350,000 sq km, of which 200,000 to 250,000 sq km is prospective (Dikenshteyn, 1977). It lies east of the Baltic shield within the Russian platform, and is bordered on the east by the Paleozoic fold belt of the Ural Mountains. The onshore part is in the Nenets Autonomous Okrug (Arkhangelsk Oblast) and Komi ASSR. Of the offshore area (Barents Sea) is about 1,147,000 sq km, of which 800,000 sq km is prospective (Meyerhoff, 1980a). Only the onshore area has been explored and drilled. Seismic work has been conducted locally since 1960 (single-fold shooting); multifold studies began about 1978. The first offshore drilling began in early 1982.

Oil seeps attracted attention to the basin during the 1500s, possibly earlier. Exploration began in 1869 and, between that year and 1917, 30 shallow but noncommercial wells were drilled. Exploration recommenced in 1929, and the Chib'yu oil field was discovered during 1930 in the southwestern part of the basin (Fig. 5). A total of 78 fields has been discovered (through 1979). of which four, and possibly five, are giants.

### Regional Geologic Framework

The Timan-Pechora basin occupies the northeastern corner of the Russian platform. It is bounded on the southwest by the Timan Ridge (a Baykalian to early Caledonian fold belt which extends from Norway-and probably Ireland to the Pacific); on the east by the Hercynian Urals fold belt, including Novaya Zemlya; and on the northwest by the Svalbard platform. Offshore the basin bifurcates, with an eastern branch extending northward along the western side of the Urals (Novaya Zemlya) and a western branch striking toward the Varanger and Nordkapp basins north of Norway.

The Timan-Pechora basin actually is a northern extension of the Volga-Urals basin, but is partly separated from the latter where the Timan Ridge intersects the Urals. The oldest rocks dated by fossils in the Timan-Pechora basin are Ordovician (Kushnareva and Rasska-

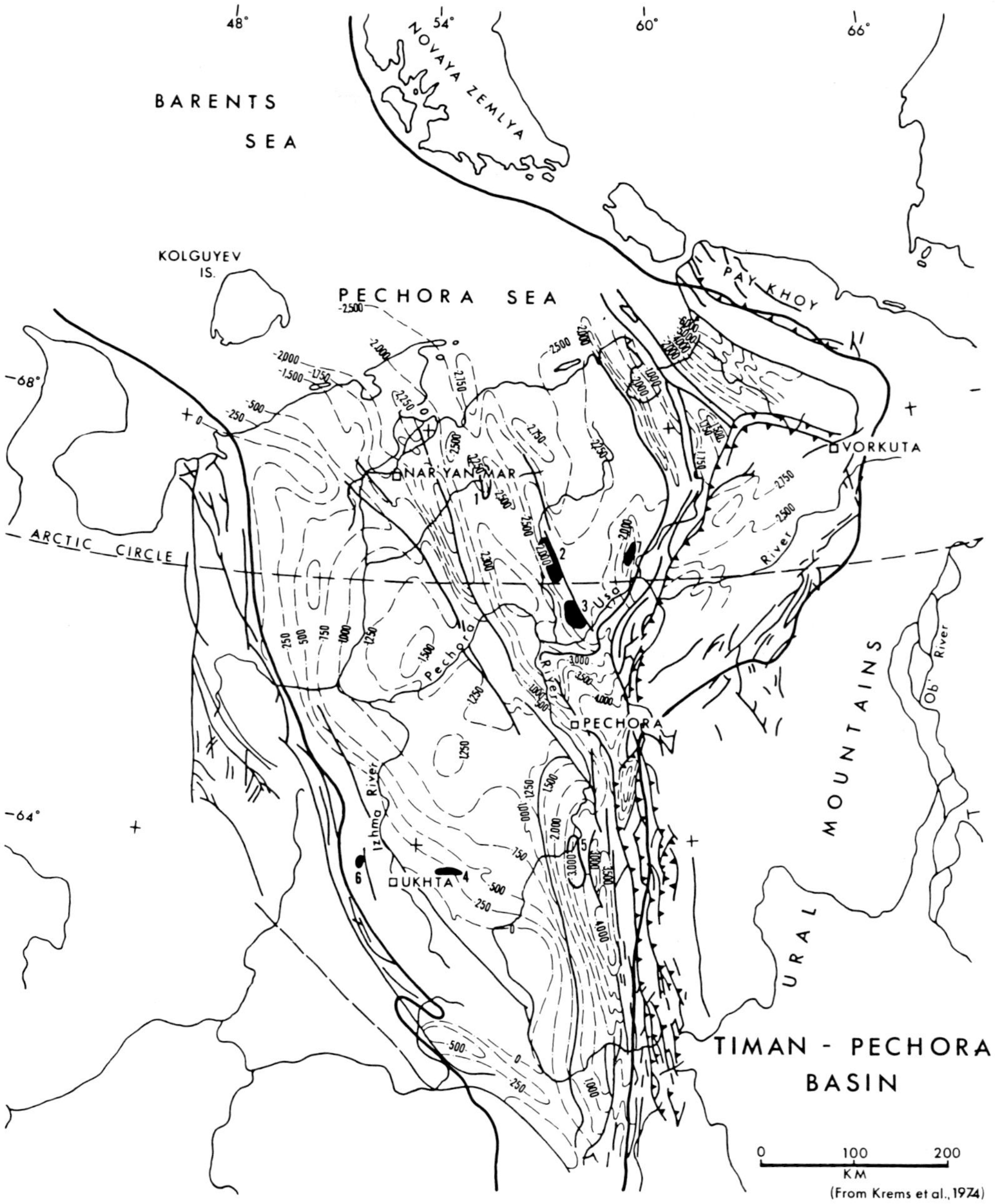


*Fig. 5* — Structural-contour map of Timan-Pechora basin. Structural datum is the top of the Lower Permian (Artinskian) carbonates (base of Kungurian anhydrite). Depths are in meters. Giant fields and other fields mentioned in text are shown.

1 = Layavozh;
2 = Vozey;
3 = Usa;
4 = Zapadno-Tebuk (Figs. 7, 8);
5 = Vuktyl' (Figs. 8, 9);
6 = Chib'yu (1930 discovery field for the basin);
7 = Salyukhin.

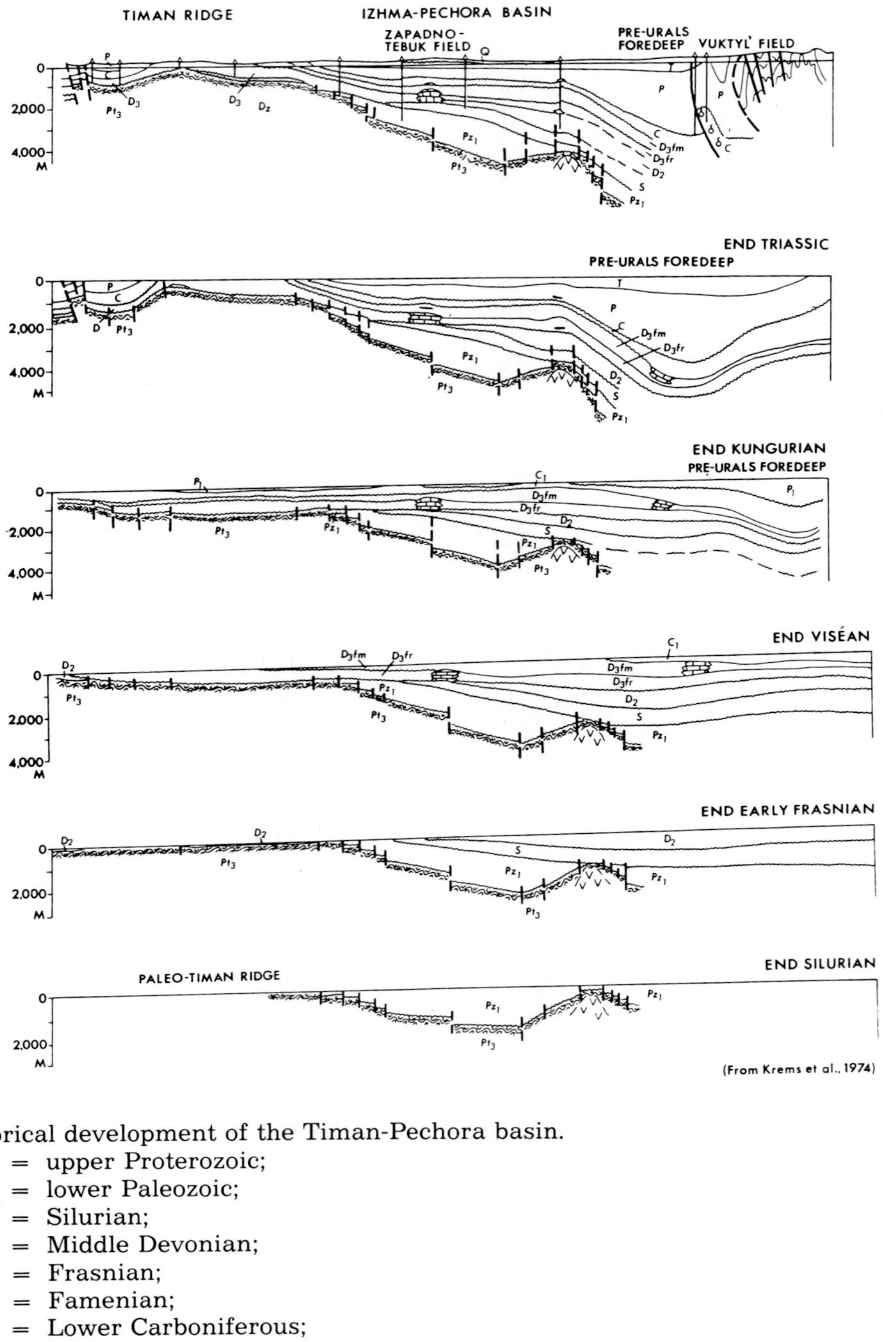


*Fig. 6* — Historical development of the Timan-Pechora basin.

$Pt_3$ = upper Proterozoic;
$Pz_1$ = lower Paleozoic;
S = Silurian;
$D_2$ = Middle Devonian;
$D_3fr$ = Frasnian;
Dfm = Famenian;
$C_1$ = Lower Carboniferous;
$P_1$ = Lower Permian;
P = Permian
T = Triassic;
Q = Quaternary.

From Krems et al. (1974). Republished with permission of *Geological Magazine* zova, 1977) and the youngest with production potential are Triassic. In the offshore, younger rocks may be found to be prospective. The onshore productive section extends from the Silurian through the Triassic. The geologic development of the basin is shown on Figure 6.

### Stratigraphy

Pre-Ordovician rocks include Precambrian metaquartzite, metasiltstone, metaconglomerate, slate, marble, etc. Basalt flows of unknown age overlie the metamorphic rocks (Krems et al., 1974).

Ordovician strata, 200 to 1,090 m thick, form a succession of marine sandstone, siltstone, and shale. Tuff is interbedded in the northwest. Gas shows were found in one field. The Ordovician grades upward into the Silurian, which is mainly limestone and dolomite with some shaly layers. Both oil and gas are produced from the Silurian.

The Lower Devonian is absent in much of the basin, particularly in the west. The base of the Devonian is an unconformity, presumably a re-

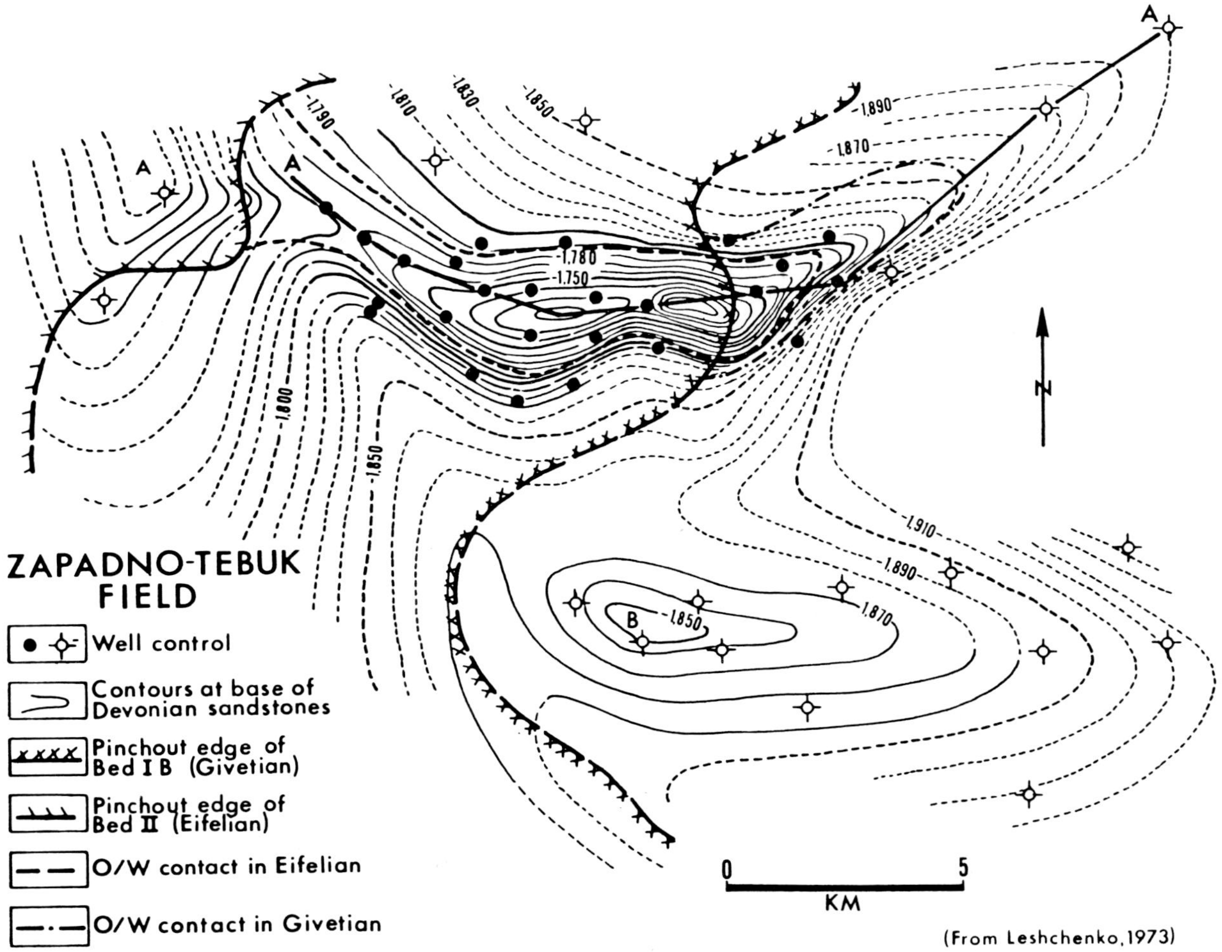


*Fig. 7* — Structural-contour map, Zapadno-Tebuk oil field, Timan-Pechora basin. Field location is on Fig. 5. The field was discovered in 1959. Ultimate recoverable reserves are about 100 million bbl. Structural datum is the base of the Middle Devonian. Depths are in meters.
A = Ban'yusk structure;
B = Tebuk structure.
Ban'yusk is dry because productive sandstones have pinched out. In the Tebuk structure, the Bed 1B reservoir is structurally low. Shows location of Fig. 8. From Leshchenko (1973). Republished with permission of *Geological Magazine*

flection of the Caledonian orogeny. In the north, up to 800 m of marine carbonates is present. Near the Urals, the Lower Devonian is several hundred meters thick and includes some oil-productive deltaic sandstones.

The Middle Devonian is 0 to 700 m thick, and consists of deltaic to marine conglomerate, sandstone, siltstone, and shale, with increasing amounts of marine marl and limestone in the west. Some chert is in the section. The Upper Devonian, 330 to 1,400 m thick, is very similar. Several Middle and Upper Devonian sandstones and limestones produce oil and gas, and anhydrite beds are developed locally along the western margin of the basin, in lagoonal sediments behind bioherms (e.g., Fig. 8).

The Lower Carboniferous, 40 to 950 m thick, consists of dark-gray organic shale, mudstone, sandstone, and limestone. Coal is present in some areas, as is anhydrite (anhydrite is present only in the Viséan). The Upper Carboniferous, 70 to 310 m thick, is mainly limestone

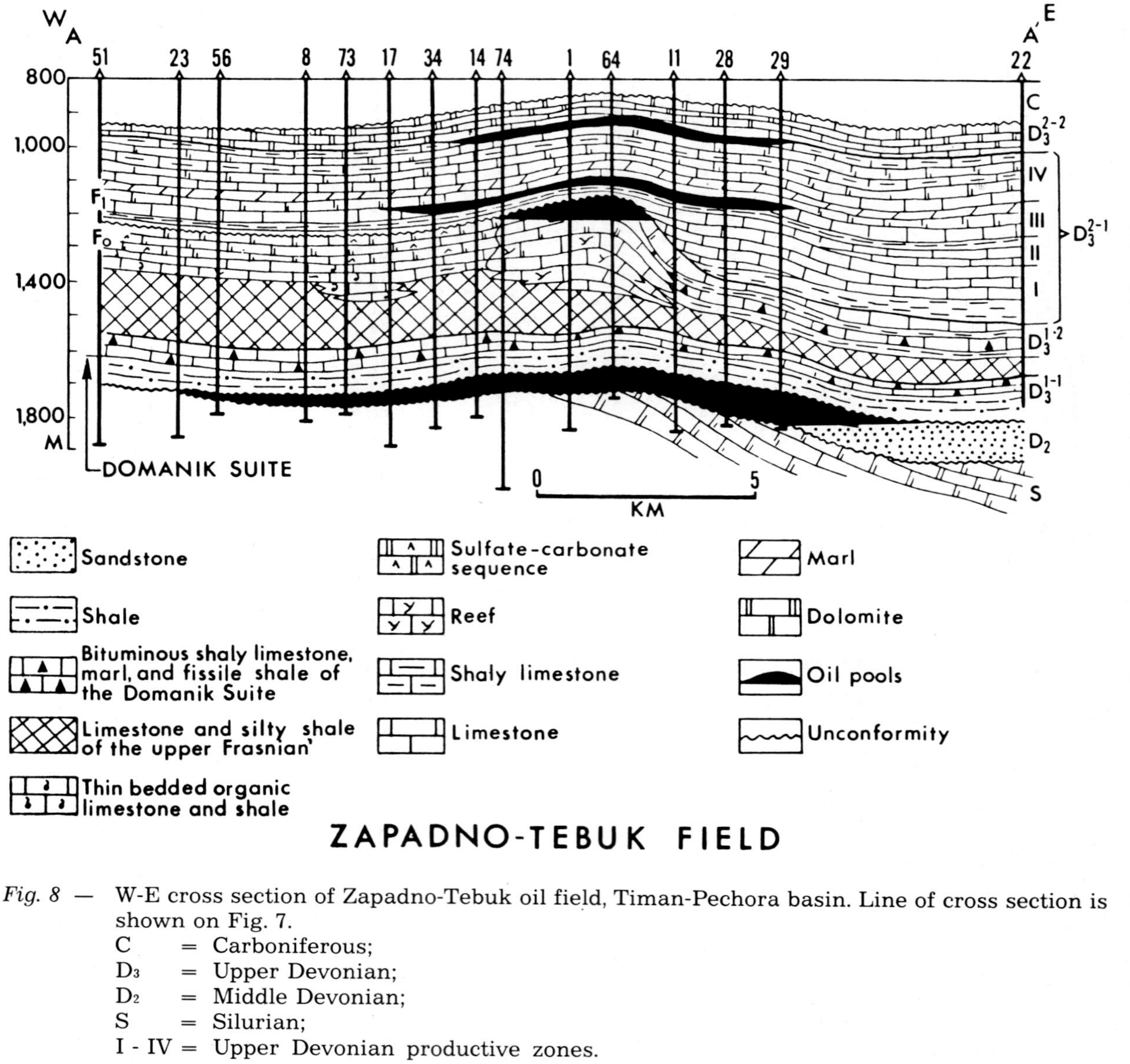


*Fig. 8* — W-E cross section of Zapadno-Tebuk oil field, Timan-Pechora basin. Line of cross section is shown on Fig. 7.
C = Carboniferous;
$D_3$ = Upper Devonian;
$D_2$ = Middle Devonian;
S = Silurian;
I - IV = Upper Devonian productive zones.

From Krems et al. (1974). Republished with permission of *Geological Magazine*

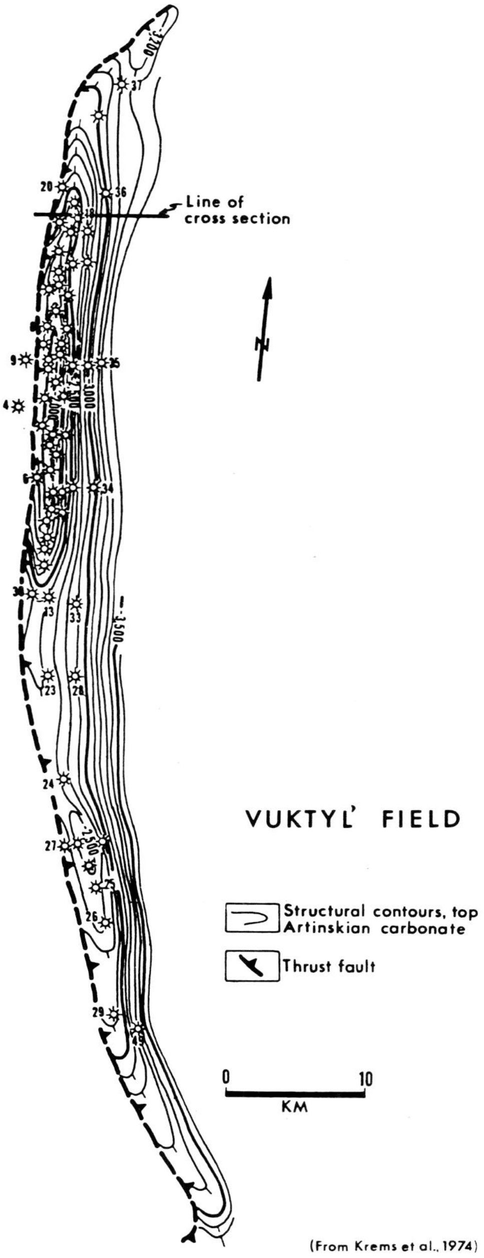


*Fig. 9* — Structural-contour map of Vuktyl' gas field, Timan-Pechora basin. Field location is on Fig. 5. Structural datum is the top of the Artinskian carbonates (base of Kungurian anhydrite). Depths are in meters. Location on Fig. 10 is shown. From Krems et al. (1974). Republished with permission of *Geological Magazine*

and dolomite. Chert, marl, and shale are present in small amounts. Several zones are oil- and gas-productive in both the Lower and Upper Carboniferous.

The Permian also is productive, especially below a widespread Kungurian anhydrite-halite sequence which forms an important seal in the basin. Most of the pre-evaporite section is shallow-water dolomite and limestone, and is 60 to 1,550 m thick. Some shale, siltstone, and sandstone are present in the south and west. The post-evaporite Permian section ranges in thickness from 400 to 2,900 m and consists of sandstone, siltstone, shale, marl, limestone, and coal. The carbonate content increases westward. Several zones are productive.

A Mesozoic section covers the basin, and ranges in thickness from 170 m to 1,400 m, becoming thicker toward the north. Two fields produce gas from the Lower Triassic.

### Petroleum Geology

A total of 78 fields had been found by the end of 1979 in the onshore part of the basin, but not all are proved to be commercial. At least 100 surface structures near the Urals have not been drilled. About 50 additional dry structures have been drilled. Many offshore structures are known to exist.

There is an abundance of structural traps-compressional mountain-front folds and gently dipping drape-type or compaction folds characteristic of the platform environment. The latter result mainly from vertical movements of old basement structures or compaction on reefs, mainly of Devonian age. An example of the latter is Zapadno-Tebuk field (Figs. 7, 8). An example of a compressional mountain front fold is the giant Vuktyl' gas field (Figs. 9, 10). The principal production in the basin is oil from the Devonian and gas from the Permian.

A major, but still untested play is a subthrust zone along the western flank of the Urals geosyncline, from the southern end of the basin to the northern tip of Novaya Zemlya. This play has been discussed by Borisov et al. (1976). The offshore extensions of the basin are shown on Figure 11. Figures 12 and 13 are seismic cross sections of the offshore-one in the South Barents basin and one in the Pre-Novaya Zemlya foredeep.

Proved and probable (A + B) oil reserves at the end of 1979 were about 5.15 billion bbl (Table 1), 3 of which were in the Usa field alone (Toporskaya, 1970). Another 0.5 to 1.0 billion bbl is in the nearby Vozey field (Fig. 5). Proved gas reserves were 19.42 Tcf. The Vuktyl' field alone originally

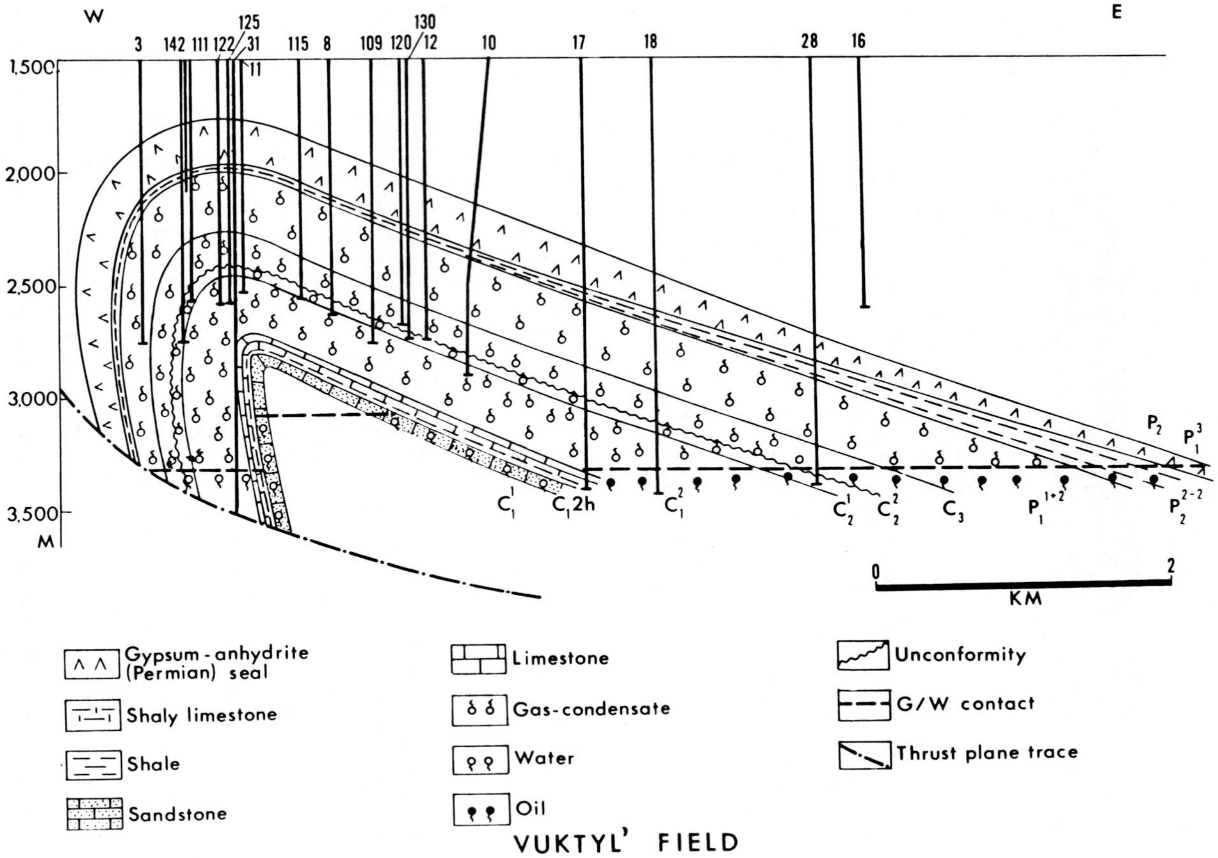


*Fig. 10* — W-E cross section, Vuktyl' field, Timan-Pechora basin. Location is on Fig. 9.

The C letters refer to various levels in the Carboniferous

$C_1$ = Lower;
$C_2$ = Middle;
$C_3$ = Upper

and the P letters are Permian
$P_1$ = Lower;
$P_2$ = Upper

The caprock and the décollement plane are the Kungurian anhydrite. From Krems et al. (1974). Republished with permission of *Geological Magazine*

held a recoverable reserve of 17.5 Tcf. The estimated resource potential (still to be discovered) is 20 billion bbl (4 onshore; 16 offshore) and 150 Tcf (50 onshore; 100 offshore). The offshore is discussed briefly under the general heading of the Barents Sea. Details of individual fields in the basin have been published by Vasil'yev (1968a, 1968b), Krems et al. (1974), Vasil'yev and Zhabrev (1975), Dikenshteyn (1977), Bakirov (1979), Meyerhoff (1980a), and many others. Production during 1980 was 370,000 bbl/day and 1.7 Bcf/d.

## West Siberian Basin

### Introduction

The West Siberian basin is the depressed western half of the Siberian Precambrian platform. The onshore area of the West Siberian basin (Figs. 1, 14), geographically the largest in the world, is 1,600,000 sq km (some authors assign to this basin an area of 1,750,000 sq km). The southern part of the Kara Sea, an offshore continuation of the basin, contains an additional

350,000 sq km, or a total minimum prospective area of 1,950,000 sq km. The basin is bounded on the west and northwest by the Uralian fold belt; on the north and northeast by the Taymyr fold belt; on the east by the Siberian platform; and on the south by the Kazakhstan fold belt. Thus all boundaries are early and/or late Paleozoic fold systems which probably originated in Precambrian time.

Most of the basin is low and flat-lying (less than 100 m elevation) and is in the tundra and taiga regimes, with normal temperate-boreal floras in the southern third. The rivers drain northward, so that the spring thaw which begins in the south causes the rivers to overflow their banks early in the year. Vast areas thus are flooded in the central and northern parts of the basin, which also are underlain by permafrost. The permafrost, 1,500 m thick near the Arctic coast, thins to zero just north of the Middle Ob' oil district (Fig. 12), and the flooding converts the entire region from south of the Middle Ob' to the Arctic coast into a quagmire of mud and water in which field operations become incredibly difficult and expensive. The total population is less than nine million people, most of them in the southern part of the basin near the Trans-Siberian Railway. Huge areas are

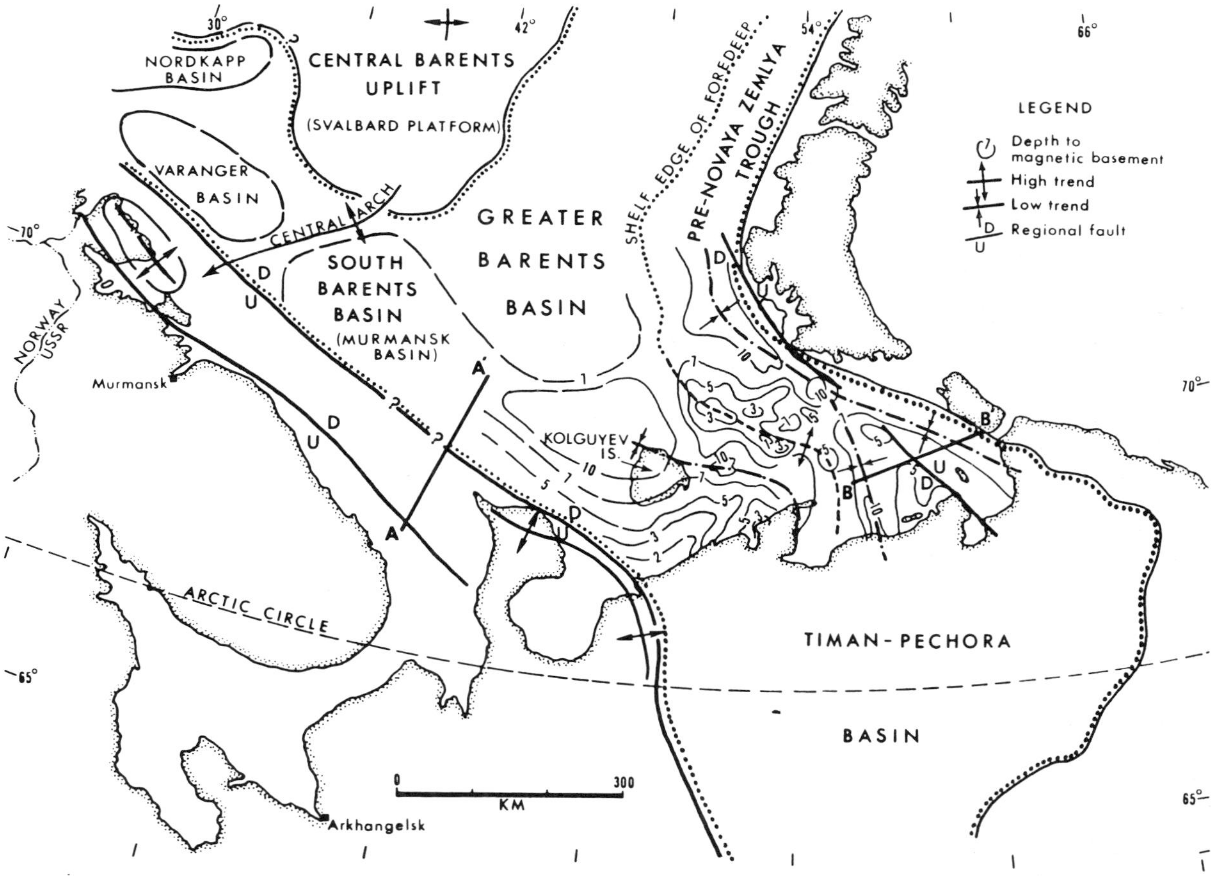


*Fig. 11* — Offshore structural interpretation of Timan-Pechora basin and possible connection with Norwegian basins. Based on aeromagnetic data and seismic data taken from several sources, especially Emelyanov et al. (1971). Shimarayev et al. (1971), Alekhin and Verba (1979), and Rønnevik (1981). The connection shown between the deep which passes through Kolguyev Island (South Barents basin) and the Varanger basin is pure speculation. Depths in km. Shows locations of Figs. 12 and 13.

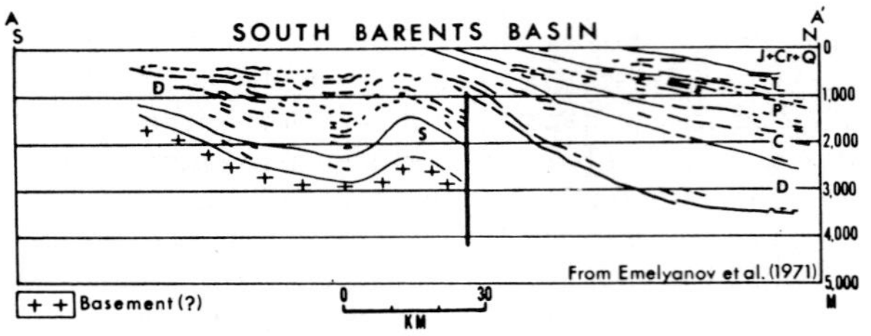


*Fig. 12* — S-N single-fold reflection-seismic line from mainland northward across the South Barents basin.
S = Silurian;
D = Devonian;
C = Carboniferous;
P = Permian;
T = Triassic;
J = Jurassic;
Cr = Cretaceous;
Q = Quaternary (including all Cenozoic).
Location is on Fig. 11. From Emelyanov et al. (1971).

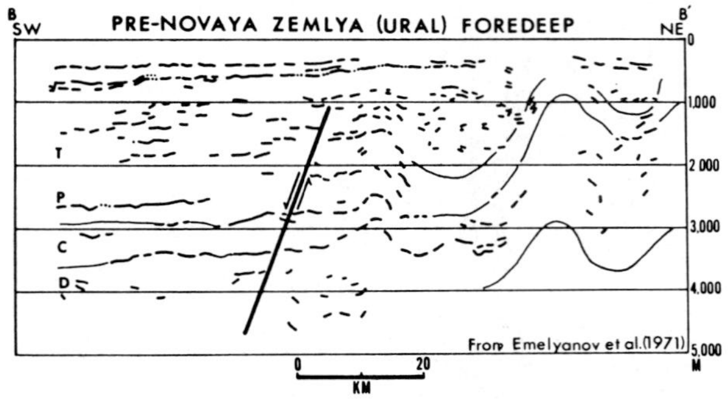


*Fig. 13* — SW-NE single-fold reflection-seismic line across the Prenovozemlya foredeep (Pre-Novaya Zemlya foredeep). Letter symbols are the same as on Fig. 12. Location is on Fig. 11. From Emelyanov et al. (1971).

uninhabited, although numerous petroleum-related communities are springing up.

The first petroleum activity began in 1912, when the Tsarist government permitted several small tracts to be leased for drilling. Nothing came of these ventures, and serious exploration did not commence until 1948, mostly around the basin flanks where flooding is less of a problem (Salmanov, 1974). The Berezovo gas field on the eastern flank of the Urals (Fig. 14) was discovered in 1953, a small (280 Bcf) discovery at 1,266-1,350 m in sandstone and detrital organogenic limestone of Callovian, Oxfordian, and Kimmeridgian ages. (The field now is nearly depleted.) This gas discovery was a great disappointment to the Soviet petroleum industry, whose geologists were searching for a new, major, oil-bearing basin to replace the Volga-Urals basin, which from 1953 to 1975 was the country's major source of oil. Intensive geophysical surveys across the basin were carried out after 1953. In 1960, near Shaim and south of the Berezovo gas discovery, the Trekhozero oil field (70 million bbl) was found at 1,439-1,530 m, also in Callovian-Kimmeridgian marine reservoirs (Fig. 14). The discovery of this oil field abruptly stimulated widespread drilling in the basin. In 1961, the first two giant oil fields (Ust'-Balyk, Megion) were found in Neocomian paralic, fluviatile, and nearshore sandstones along the Middle Ob' (the east-to-west flowing part of the river) near the basin center (Fig. 14). Through 1978. a total of 308 fields had been discovered (Trebin and Rubtsova, 1978), including at least 21 giant oil fields and more than 31 giant gas fields. The latter contain the largest proved plus probable volumes of gas (mainly Cenomanian; some is Neocomian) known in the world-about 1,000 Tcf being in these two categories. These discoveries and their geology were described by Nesterov et al. (1971), Salmanov (1974), and Kontorovich et al. (1975). An English summary was published by Meyerhoff (1980a).

**Regional Geologic Framework**

The West Siberian basin is a Mesozoic feature superimposed on Paleozoic and Precambrian trends. Thus it is a successor basin. It began to form in Triassic time, when taphrogenesis created major north-south trending graben systems (Fig.15). The largest of these is in the basin center, and extends the full length of the basin; this is the Koltugor-Urengoy graben (Fig. 15). A second major graben system parallels the Koltugor-Urengoy graben on the east, about midway between it and the Siberian platform. Parts of at least five additional north-south trending grabens have been discovered by seismic work and deep drilling.

The shape of the West Siberian basin is related directly to structural trends in the Precambrian and Paleozoic fold belts surrounding

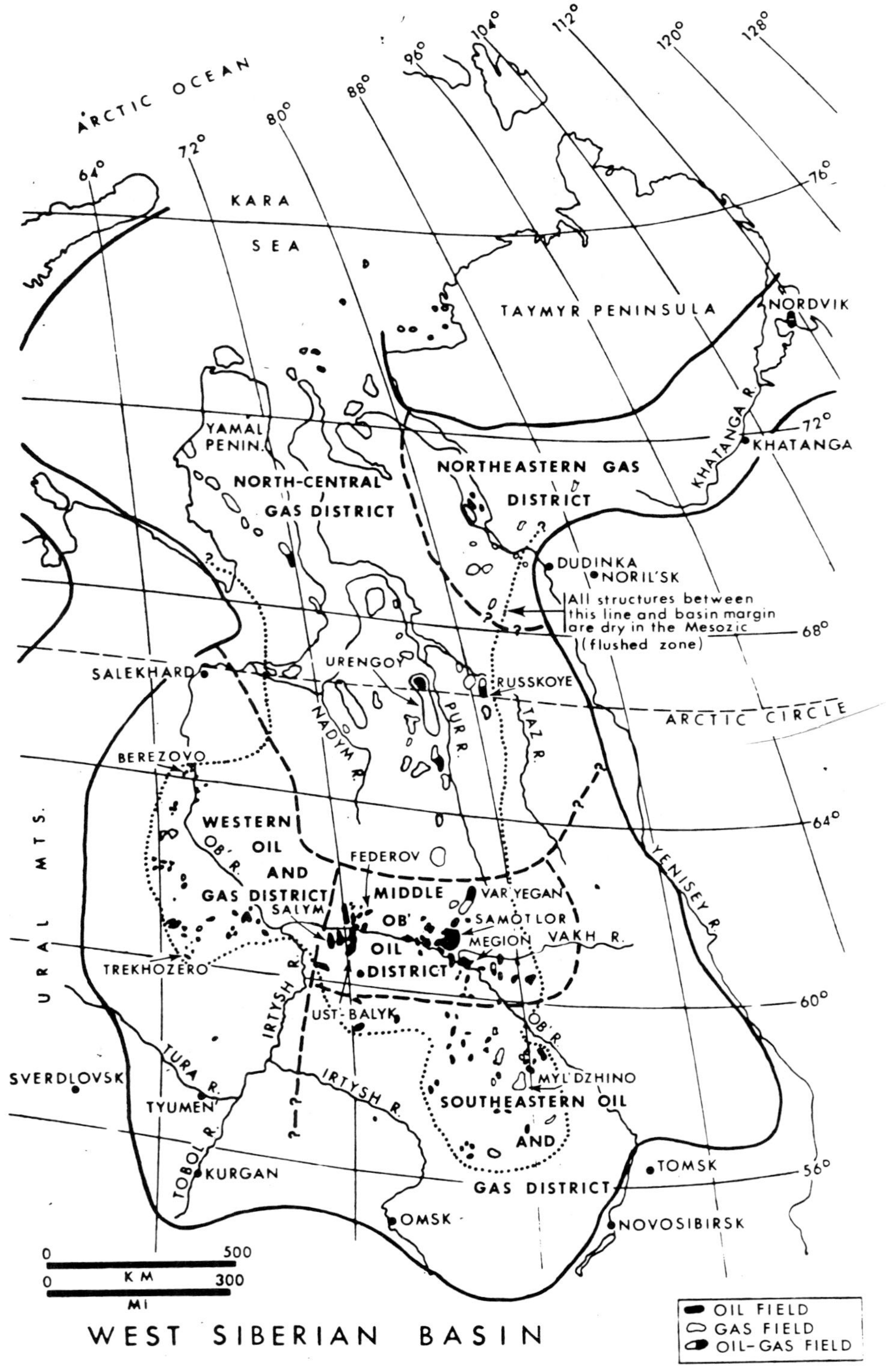


*Fig. 14* — Major oil and gas provinces, and oil and gas fields, of the West Siberian basin. From Meyerhoff (1980a). Republished with permission of *Geological Magazine*

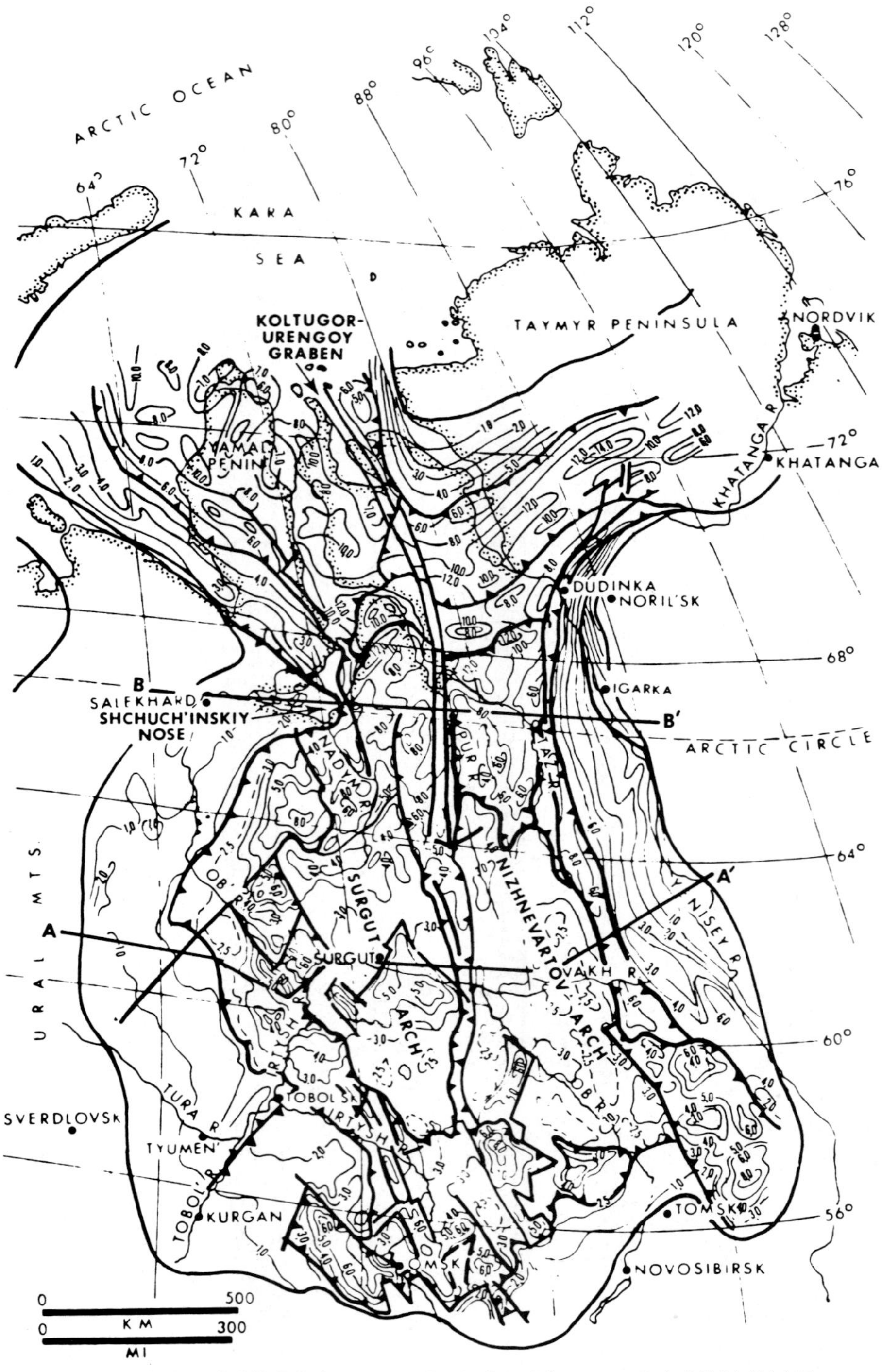


*Fig. 15* — Structural-contour map of the West Siberian basin. Structural datum is the basement (6.0-6.4 km/sec). Shows locations of Figs. 16 and 17 (A—A', B—B'). From Rudkevich (1980).

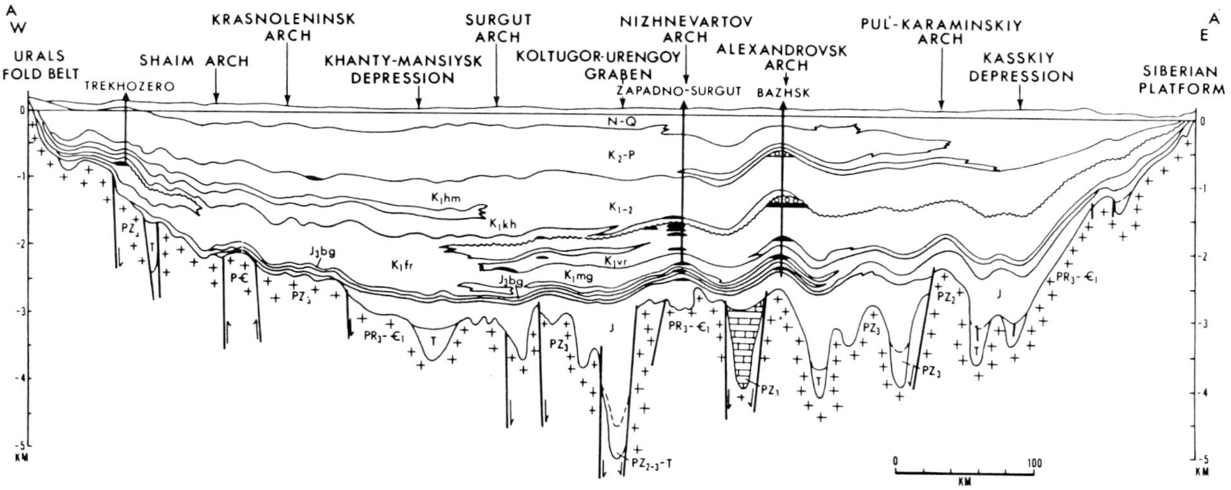


*Fig. 16* — W-E schematic cross section A—A' across the southern part of the West Siberian basin. The shallow accumulations (Cenomanian) are mainly gas; deeper accumulations generally are oil and are in Neocomian strata; both gas and oil are present in Upper Jurassic rocks.

PC = Precambrian;
$PR_3$ = Vendian (late Proterozoic);
$PZ_2$ = middle Paleozoic;
$PZ_3$ = upper Paleozoic;
$C_1$ = Lower Carboniferous;
T = Triassic (may include some Permian);
J = Jurassic;
$J_3bg$ = Upper Jurassic Bazhenov and Georgiyev Formations, important source rocks in the basin;
Kfr = Federov Formation;
Kkl = Kulomzin Formation;
Kmg = Megion Formation;
Kvr = Nizhnevartov Formation;
Kkh = Kharosoimsk Formation;
Khm = Khantimansinsk Formation;
$K_1$ = Lower Cretaceous;
$K_2$ = Upper Cretaceous;
P = Paleogene;
N = Neogene;
Q = Quaternary.

Location is on Fig. 15. Modified from Dikenshteyn (1977).

and beneath the basin. Subsidence of the entire basin began in Early Jurassic time (Korzh, 1978), and subsidence continued through Tertiary time, although subsidence was greater in the north than in the south. This is shown clearly on Figure 15 (Rudkevich, 1980). Basement is 2 to 4 km deep in the south, and 5 to 12 km deep in the north.

In the southeastern quadrant of the basin, between the Middle Ob' and the city of Novosibirsk, a thick platform sequence of large areal extent is present, consisting mainly of carbonates of Silurian, Devonian, and Early Carboniferous ages (Zapivalov et al., 1975, 1978). The depth to these deposits ranges from about 2,000 m to more than 4,000 m. This platform sequence forms the so-called Nyurol'ka basin and, because it has oil-bearing formations unrelated to those in the overlying Mesozoic cover, is treated under a separate heading.

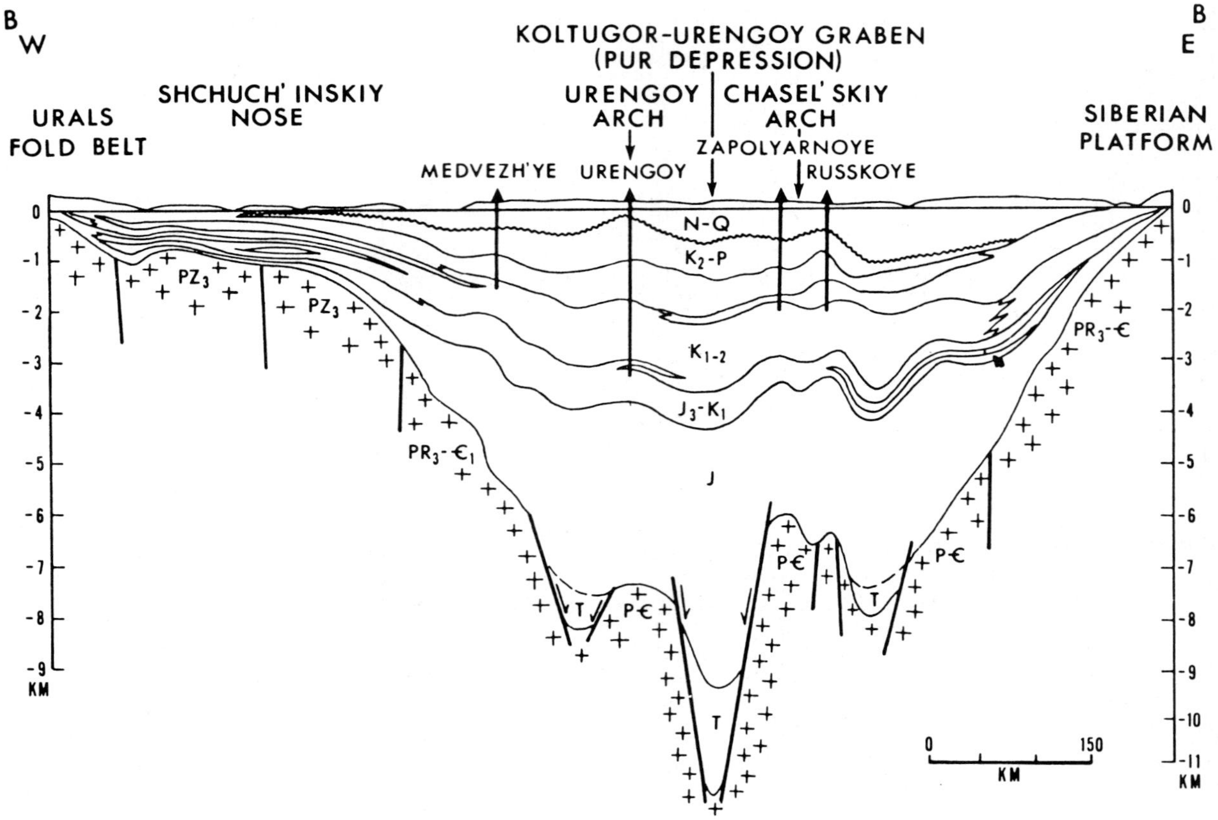


*Fig. 17* — W-E schematic cross section B-B' across the northern part of the West Siberian basin. Symbols are the same as for Fig. 16. Location is on Fig. 15. Modified from Dikenshteyn (1977).

**Stratigraphy**

The Koltugor-Urengoy and related north-south trending graben systems formed on a heterogeneous basement of Paleozoic and Precambrian eugeosynclinal, miogeosynclinal, and platform deposits. Sections up to 6,000 m thick may have accumulated in the Triassic grabens. The strata consist of continental dark-gray, greenish, and reddish-brown (redbeds), arkosic sandstone, siltstone, and shale, with some conglomerate, tuff, and volcanic rocks (flows or sills). A distinctive Triassic flora is present.

The Lower and Middle Jurassic-the oldest objective section within the Mesozoic West Siberian basin-covers the entire basin. Its thickness is up to 2,300 m. The depositional environment is mainly continental, but fossiliferous marine tongues are found in the far north near the present coast. The Jurassic oversteps and overlaps all older rocks-Triassic, Paleozoic, and Precambrian. Figures 16 and 17 are schematic cross sections across the West Siberian basin, and show that the Jurassic deposits filled in and smoothed the pre-Jurassic topography.

The Lower and Middle Jurassic strata are of fluviatile, swamp, and lacustrine origin, except where marine facies is present in the far north near the Kara Sea. The principal source of the conglomerates, sandstones, and arkoses in the section was highlands on the site of the present Siberian platform. Rivers generally drained from the south-southeast to the north-northwest (Korzh, 1978). Sediment sources on the present site of the Ural Mountains were few and localized. Large areas of the basin, particularly in the southwest, were the sites of shale deposition with few potential reservoirs. Marine bars probably are present near the present Kara Sea coast.

The Lower and Middle Jurassic generally has a basal conglomeratic zone, with fairly coarse-grained sandstone above it. Light-, medium-, and dark-gray to green-gray sandstone, siltstone, and shale comprise the remainder of the section. Coal is abundant and, in total volume, may make up one of the most extensive coal fields in the world. Grain size generally decreases upward. Most of the sandstone is polymictic, although arkose is developed locally. Sideritic marl and siliceous shale are well developed in some places. The overall color of the section becomes more variegated toward the south. Where sandstone channels are developed, porosity values range from 10 to 25 percent. The older sandy units have permeability values of 0.1 to 73 md, but locally are much higher. In the Middle Ob' oil district, values as high as 740 md are common.

Late Jurassic time (Callovian through Tithonian) was the beginning of a major marine transgression in the basin. Upper Jurassic deposits are about 450 to 600 m thick. The section is coarsest and most continental near the base, and becomes finer upward in the section. Most sandstone units in the upper formations are fine grained. Most of the Kimmeridgian and Tithonian beds are marine, although marginal deltas were developed, especially in the southeast, south, and west. The highest unit, the Tithonian (Volgian) Bazhenov Formation, was deposited in a silled basin and is extremely organic-rich. It probably was deposited in a reducing environment.

The lithologies include conglomerate (especially in the Callovian), gray sandstone (arenite and graywacke[1]), light- to dark-gray siltstone, glauconitic siltstone and sandstone, and medium- to dark-gray and dark-gray-brown shale. Some organic-rich shale is black. Small reefs are present around the fringes of basement highs along the eastern flank of the Urals. These reefs commonly are productive.

Although organic-rich shale is present at several levels, most of it is in the two highest formations-the Kimmeridgian Georgiyev Formation and the Tithonian Bazhenov Formation. These are believed to be the major source rocks for the Jurassic and Neocomian oils in the basin; coal in the Lower to Middle Jurassic probably generated much of the gas in the southern part of the basin. The Bazhenov contains up to 15-20 percent organic matter by weight (Khabarov et al., 1980), with values generally in the 6.75 to 11.7 percent range (Skorobogatov, 1980). The Georgiyev also is rich, with values up to 7.12 percent (Skorobogatov, 1980).

Porosity values in the fields adjacent to the Urals commonly are in the 23 to 35 percent range, but are as low as 11-14 percent. Permeability values in some fields range from 500 to 6,700 md, with known low values in the 0.1 to 145 md range (average is 20-40) and a high value of 33,800 md recorded in one reservoir (Kontorovich et al., 1975). This range of values of porosity and permeability applies to both the limestone detritus (calcarenite and calcirudite) reservoirs and the associated sandstone reservoirs. Farther east, in the areas of the Surgut and Nizhnevartov arches, porosity values in the south are 15-17 percent; permeabilities are 25-30 md, with some values of 100 md. North of the Ob' River, the reservoir quality is better, with porosity values of 15-23 percent (average, 19) and permeability values of 6 to 835 md, averaging 200 md. Porosity and permeability values appear to be even better at Var'yegan field (Tikhomirov et al., 1978; see Fig. 14).

The oldest commercial production found is in the Callovian-Kimmeridgian along the western side of the basin (e.g., Berezovo and Trekhozero, Fig. 14); in the Callovian marine section of the Surgut arch (Fig. 15); and in the Late Jurassic continental to marginal-marine strata of the Nizhnevartov arch (e.g., Myl'dzhino, Var'yegan, Fig. 15; see Nesterov et al., 1971; Kontorovich et al., 1975; Tikhomirov et al., 1978). Noncommercial to marginally commercial oil production has been found at Salym field in fractures within silicified shales of the Bazhenov Formation (Tithonian; see Abdullin, 1980; Skorobogatov, 1980; Fig. 21).

The Lower Cretaceous (Neocomian through Cenomanian) is up to 1,400 m thick in the Middle Ob' region. It is predominantly sandstone and siltstone, continental to deltaic in the far east and south, becoming more marine and shaly toward the west and north (Barentsov et al., 1968). The sandstone becomes more quartzose upward, reaching values of 65-70 percent. At least 30 deltaic and shelf-bar sandstones produce oil on both the Nizhnevartov and Surgut arches (Fig. 15). In the west, along the eastern flanks of the Urals, sandstone is less

[1] West Siberian sandstones, according to Salmanov (1974, p. 32), are actually arkosic and lithic arenites and feldspathic and lithic graywackes (Pettijohn et al., 1972, p. 158). Rock fragments comprise 20-40 percent; 25-30 percent is quartz; up to 5 percent is mica; the rest is feldspar. Clay matrix and calcite cement may be 30 to 40 percent of the rock.

common and is developed only locally. A few beds (three to five) are productive. The ages of the productive reservoirs are Neocomian to early Aptian. Farther north, in the north-central and northeastern gas districts (Fig. 14), gas condensate and gas are present in the Neocomian to lower Aptian zones, but the main production is dry gas in the Cenomanian.

Porosity and permeability values in the Neocomian-lower Aptian range from 14 to 29 percent; porosity decreases southward and westward. Permeabilities are 3 to 1,780 md, most commonly in the 80-200 md range in productive reservoirs. Cenomanian porosities are 10-44 percent, normally about 22-26 percent. Permeabilities are 1-5,000 md, commonly in the 200-500 md range.

The Lower Cretaceous is predominantly marine in the central, western, and northern parts of the basin, and appears to be sandstone-poor in the area between the Surgut arch and the eastern flank of the present-day Urals. Colors are predominantly gray, with a full range from light to dark gray. Greenish colors are common. Southward, the section becomes more variegated and reddish in the continental facies. Along the eastern margin of the basin, deltas developed with coal swamps-particularly during Aptian through Cenomanian times.

The Upper Cretaceous (800 m thick), Paleogene (800 m). and Neogene (250 m) rocks lie above the Cenomanian. The sea became greatly constricted during Tertiary time; Neogene strata were deposited only in the northern and northwestern parts. The section is increasingly nonmarine upward.

Few pools of hydrocarbons have been found in post-Cenomanian strata. Those pools that are known are small and probably represent accumulations that migrated upward along faults and fractures. The most important unit in the entire Upper Cretaceous-Neogene section is the Turonian Kuznetsov Formation — a dark-gray to black marine shale unit which covers the entire basin, except around the edges. This shale is 20 to 200 m thick and forms an effective seal for the underlying deposits of gas and oil.

**Petroleum Geology**

A total of 308 fields had been found by the end of 1978 (Trebin and Rubtsova, 1978); by 1974, 174 structures had been found to be dry (Kontorovich et al., 1975). Most of these appear to be in a flushed (?) zone around the basin flanks. Figure 14 shows the boundary of the flushed(?) zone. However, flushing may not be the only reason for the presence of these dry structures around the basin margins. The Kuznetsov shale seal is absent in many areas near the basin margin; porous beds extend nearly to the surface; and source beds may not reach to the basin flanks. Of the approximately 1,100 structures mapped with the seismograph by the end of 1974, about 600 remained to be drilled, but had not been drilled because of remote location, small size, or location in a gas-prone area. (Oil was the principal exploration target.) A major problem in the USSR is that only four giant oil fields have been found in West Siberia since the discovery of Samotlor in 1965. Two are Var'yegan and Federov. However, production from the two fields is only a third of Samotlor's, so that no replacement for Samotlor has been discovered. Today, as the oil shortage in the USSR becomes more critical, increasing numbers of smaller structures are being drilled. Drilling for stratigraphic traps along the flanks of the major arches has barely begun.

Serious consideration has been given to searching for fractured siliceous shale reservoirs of the type found in the Bazhenov Formation at Salym field (Fig. 14), but little has been done to develop — or even to search for — such reservoirs (Abdullin, 1980). Published opinions on the likelihood of finding major reserves in such reservoirs generally have been negative. Therefore, mainly structural traps have been drilled, although some stratigraphic pinchouts have been found along the Urals flanks on the west and along the flanks of the Surgut and Nizhnevartov arches. Salym is the only large fractured shale reservoir described in the literature. Patch reefs have been found only along the eastern margin of the present day Urals.

A major problem which is the topic of a large Soviet literature is why the northern parts of the basin are gas-prone and the central part is oil-prone. When exploration first began, it was expected that major oil reserves would be found below the Cenomanian gas in the northern part of the basin. This has not happened. Instead, gas condensate has been found. The Cenomanian to Aptian dry gas seems to be mainly biogenic (Rice and Claypool, 1981); the deeper Neocomian gas condensate seems to be mainly thermal. However, drilling has not been sufficiently deep to determine whether the Aptian and Neocomian gas condensate is just above the oil window, or below it. It is possible that the condensate reservoirs are below the oil window, and that the oil window is mainly in the Aptian and Albian where few reservoirs are present. Only a few Albian reservoirs produce and they also are gas condensate. Another hypothesis which has been proposed is that oil is found only where the Bazhenov and Georgiyev source beds were deposited. Only one large oil deposit has been found — Russkoye (Fig. 14), which has 7 billion bbl of heavy oil. Its origin is unknown. It appears to be biodegraded. In my opinion, the possibility that the oil window has not yet been reached in the northern gas districts still is a viable hypothesis.

In the central part of the basin (Middle Ob' oil district; Fig. 14), there is some shallow gas in the Cenomanian, but it is judged to be noncommercial, at least for the present. Oil is the major hydrocarbon found, and is present in the lower Aptian and in the Neocomian. Very little gas (other than associated gas) has been found in the lower Aptian and Neocomian. The largest gas cap is in lower Aptian sandstones at Samotlor, where at least 25 Tcf was present when the field was discovered. Several Tcfs have been flared.

The Jurassic is rather unpredictable. It contains both oil and gas. The giant Myl'dzhino gas field is entirely in strata of Jurassic age (Fig. 14), and the gas seems to have been generated in Jurassic coals.

The oil in the basin comes from several sources, the principal of which are the Georgiyev and Bazhenov Formations of Late Jurassic (Kimmeridgian and Tithonian) age. It is possible that other organic-rich shales in the Upper Jurassic and Lower Cretaceous sections are oil and gas sources. It also is possible that the Bazhenov and Georgiyev Formations *must* be present for commercial oil deposits to form.

The Jurassic oil presumably comes from the same sources, and the Jurassic gas probably comes in part from Jurassic coal measures. The giant gas accumulations of the northern parts of the basin were generated in Aptian, Albian, and Cenomanian coals deposited during a retreat of the sea from the eastern part of the basin during the latter part of Early Cretaceous time.

However, an area the size of the West Siberian basin should contain about 5,000 to 6,000 separate fields, ranging in size from only a million barrels to giants, such as Samotlor. The hostile environment has prevented the detailed exploration necessary to find large numbers of fields, and Soviet drilling has concentrated understandably on the major structures. Small structures, except near the Urals and in the Paleozoic of the southern part of the basin (Nyurol'ka basin; see next section), have been ignored and undoubtedly hundreds of small closures remain undetected because the seismic grids are so broad. Drilling for stratigraphic traps along the flanks of the major arches and the basin flanks has has begun only recently, so that there are distinct possibilities for finding such traps. Many parts of the arches still are undrilled, although these are few; those that are the least drilled appear to be sandstone-poor regions. Drilling in the centers of the synclines and grabens of the basin has been nil, because Soviet drilling equipment does not drill economically or efficiently below 2,600 to 3,100 m, except under unusual circumstances. Moreover, Soviet seismic investigations have been concentrated on the basin arches and flanks, not in the lows, despite the likelihood of finding major structures within them. The Jurassic is almost untested in most regions. Finally, the huge area of the southern Kara Sea never has been drilled. Thus the prospects for finding (1) numerous small structural traps, (2) stratigraphic traps, (3) traps in the lows between arches, (4) traps at 3,200 m + , and (5) fields offshore are excellent. It is almost certain that large discoveries will be made in deep reservoirs, particularly in the northern part of the basin, both on- and offshore. The end of West Siberia is nowhere in sight. However, there will have to be a vast improvement in the area's infrastructure and the quality of seismic and drilling equipment before the West Siberian basin can be explored thoroughly.

To date, 21 giant oil fields and 31 giant gas fields have been discovered among the 308 fields. It is safe to predict that several additional large to giant fields will be found, both at greater depths and offshore. By the end of 1979, 10.99 billion bbl (now about 15 billion bbl) had been produced; an estimated 16.35 billion bbl-possibly 18 had been discovered; and 54 billion remains to be found. (Table 1). Also at the end of 1979, 26.5 Tcf of gas had been produced; 545 Tcf of gas had been discovered (most of it in the Cenomanian); and 948 Tcf remained to be discovered. Production during 1981 was 6,460,000 bbl/d (nearly 2,800,000 bbl/d from Samotlor alone) and 16.9 Bcf of gas per day. Samotlor production began to decline in 1981.

Two fields are shown here for the purpose of illustration. Figure 18 is a structural-contour map of the largest oil field, Samotlor, discovered in 1965. It is the USSR's largest find to date.

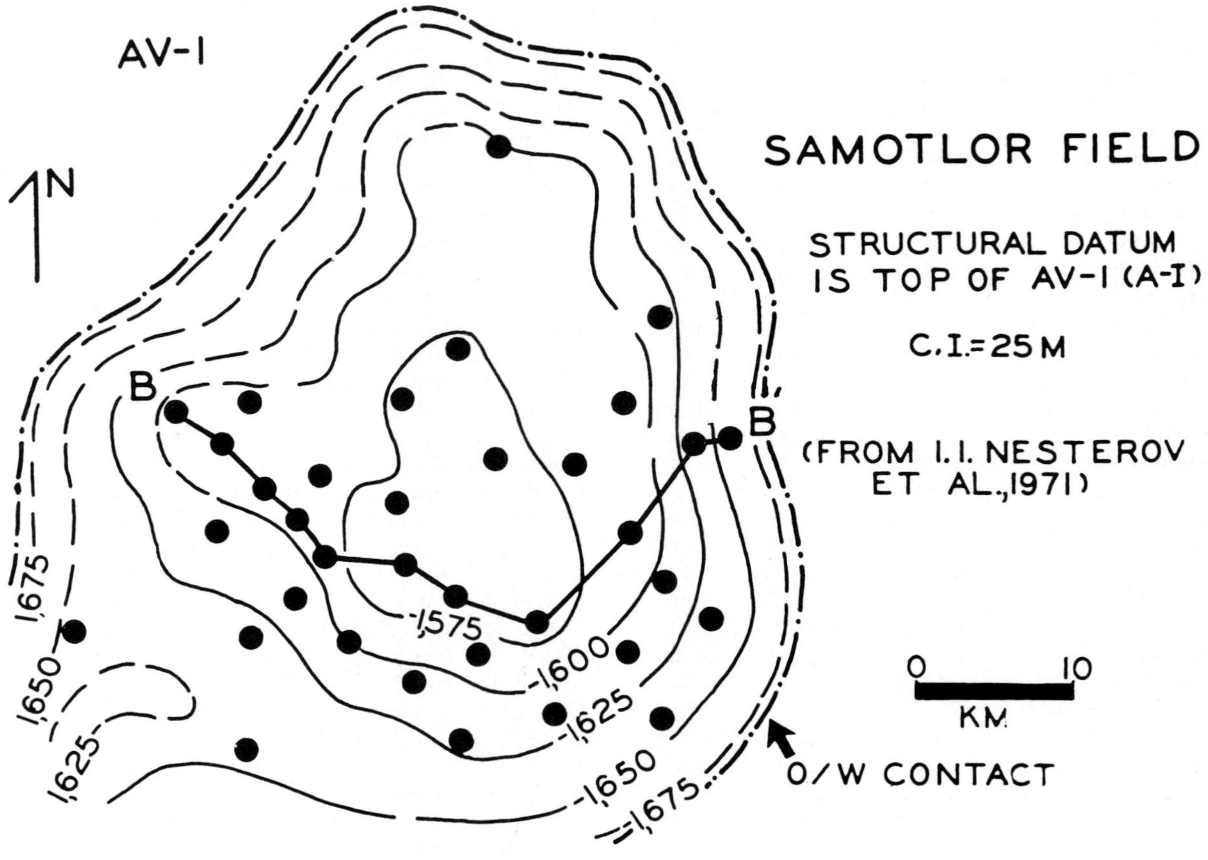


*Fig. 18* — Structural-contour map of the Samotlor oil field, West Siberian basin, the largest oil field in the USSR. Structural datum is the top of the Aptian $AV_1$ reservoir. Depths are in meters. Location of Fig. 19 is shown. From Nesterov et al. (1971)

Ultimately recoverable reserves, originally stated to be 14.6 to 15.1 billion bbl, are now known to be 18 billion bbl. In July 1981, Samotlor produced its billionth metric ton (7.3 billion bbl). A cross section of the field is shown on Figure 19. The largest volume (55 percent) of reserves is in the lower Aptian $AV_1$ through $AV_5$ sandstones; the Valanginian (middle Neocomian) $BV_8$ sandstone originally held 41 percent of the reserves. Dry gas is present in the Cenomanian, and a fair amount of oil (4 percent) is in deeper sandstones of the lower Neocomian ($BV_{10}$).

Urengoy is the world's largest gas field, and contains more unproduced, but proved gas, than is left in the entire United States (including Alaska). Ultimate recovery recently has been upgraded to 262.5 Tcf, most of it in the Cenomanian. Figure 20 is a structural-contour map of the field. Figure 21 shows cross sections of Urengoy, including the main Cenomanian dry gas reservoir and five of the 10 lower Albian, Aptian, and Neocomian reservoirs. The scale shows the immense size of this basement drape structure — at least 209 km long. Urengoy soon will supply much of the nation's gas, as well as a sizable part of the gas consumed in Europe. In 1980, production was 7.67 Bcf/d.

## Nyurol'ka Basin

### Introduction

The Paleozoic Nyurol'ka basin is in Tyumen and Novosibirsk oblasts (provinces), in the southeastern quadrant of the West Siberian basin. The size of the basin is not known with certainty, but drilling to date indicates that it covers an area beneath the Mesozoic cover of at least 180,000 sq km (Fig. 1). The basin was discovered during the 1960s as wells penetrated the overlying Jurassic in search of oil and gas. The first commercial oil discovery was made during 1971 in Devonian carbonates (Shnip et al., 1981). Until 12-fold CDP seismic techniques were introduced during the very late 1970s, however, it was nearly impossible to map the Paleozoic strata which comprise the basin. The Mesozoic cover ranges in thickness from about 2,600 to 4,800 m.

## Geologic Framework

The objective section known thus far consists of 3,280 m+ of Late Silurian, Early, Middle, and Late Devonian, and Early to Middle Carboniferous shallow water marine sediments, largely shelf limestone, dolomite, and shale. (The Carboniferous was largely eroded before the Triassic grabens formed.) As a result, only a few occurrences of Middle Carboniferous (Bashkirian) are known, and the number of Lower Carboniferous occurrences also is limited (Viséan, Tournaisian). The depositional area perhaps should not be regarded as a true basin, but the remnant of a middle Paleozoic platform or stable zone between two Caledonian (early to middle Paleozoic) fold belts (Kontorovich et al., 1975). Maksimov and Dikenshteyn (1979) showed that the Nyurol'ka platform deposits in fact are a post-Baykalian to post-Caledonian stable basin which coexisted with late Caledonian to early Hercynian tectogenesis in the Urals (i.e., it overlies the Baykalian fold belt, which extends possibly from the Timan Ridge to Lake Baykal (Fig. 22)).

## Stratigraphy

Zapivalov et al. (1975, 1978, 1980) and Shnip et al. (1981) have described the section in general terms. Very detailed lithologic and paleontologic

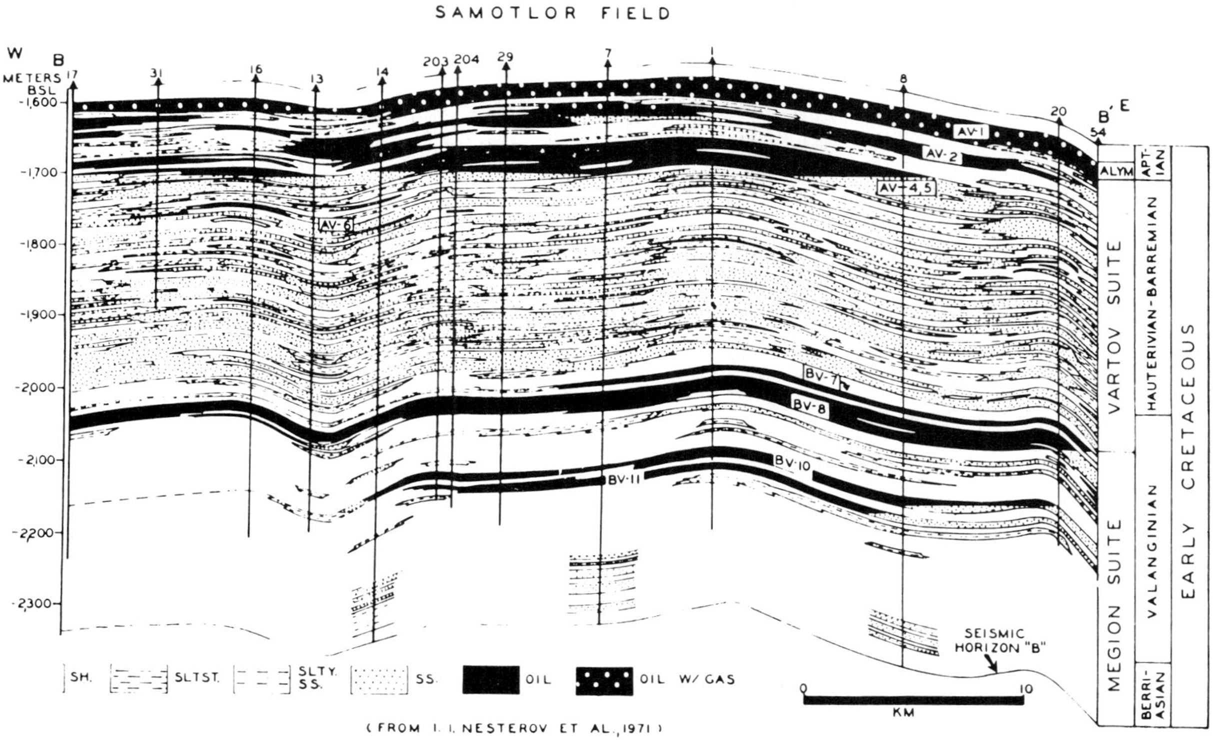


*Fig. 19* — W-E cross section of Samotlor, West Siberian basin. Location is shown on Fig. 18. From Nesterov et al. (1971).

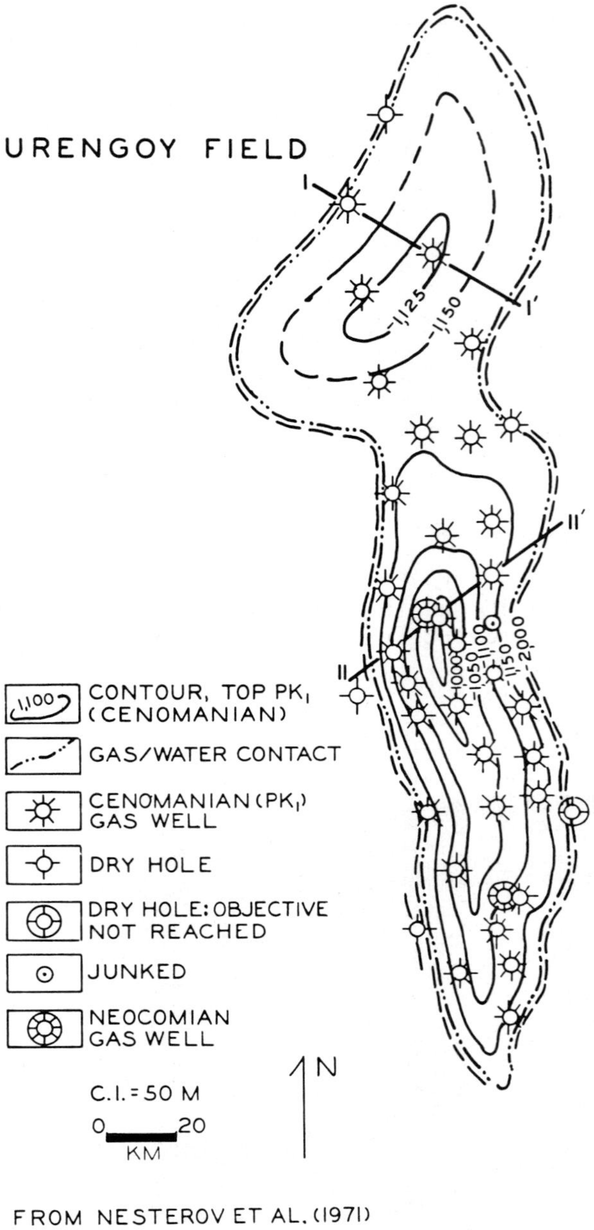


*Fig. 20* — Structural-contour map of Urengoy gas field, West Siberian basin. This is the world's largest gas field (262.5 Tcf). Structural datum is the top of the Cenomanian. Depths in meters. Shows location of Fig. 21. From Nesterov et al. (1971).

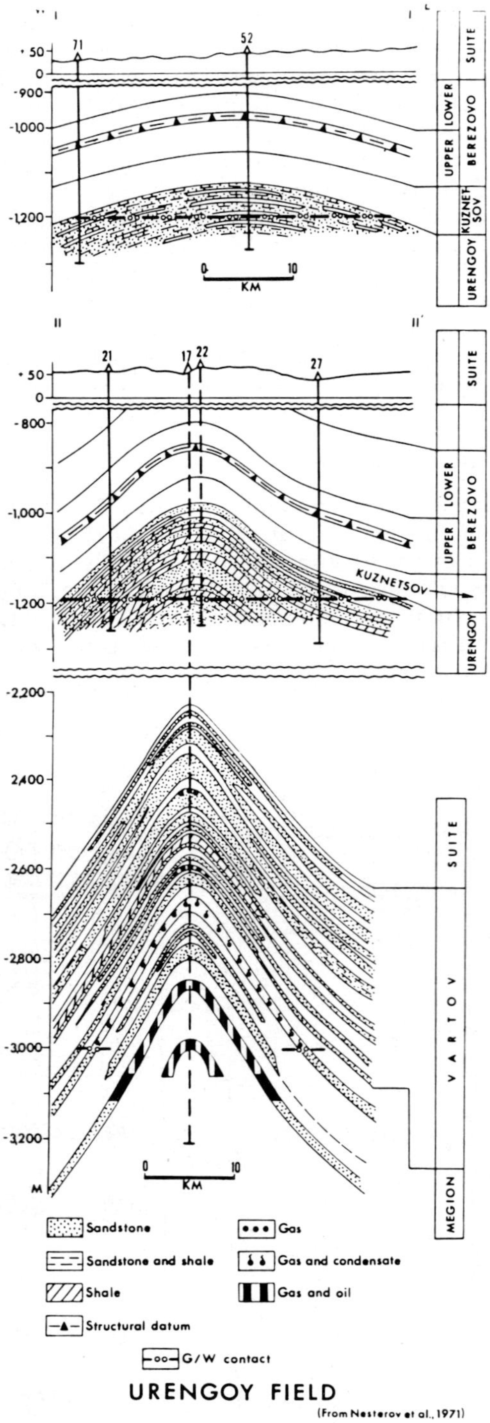


*Fig. 21* — W-E cross sections, Urengoy field, West Siberian basin. Locations are shown on Fig. 20. From Nesterov et al. (1971).

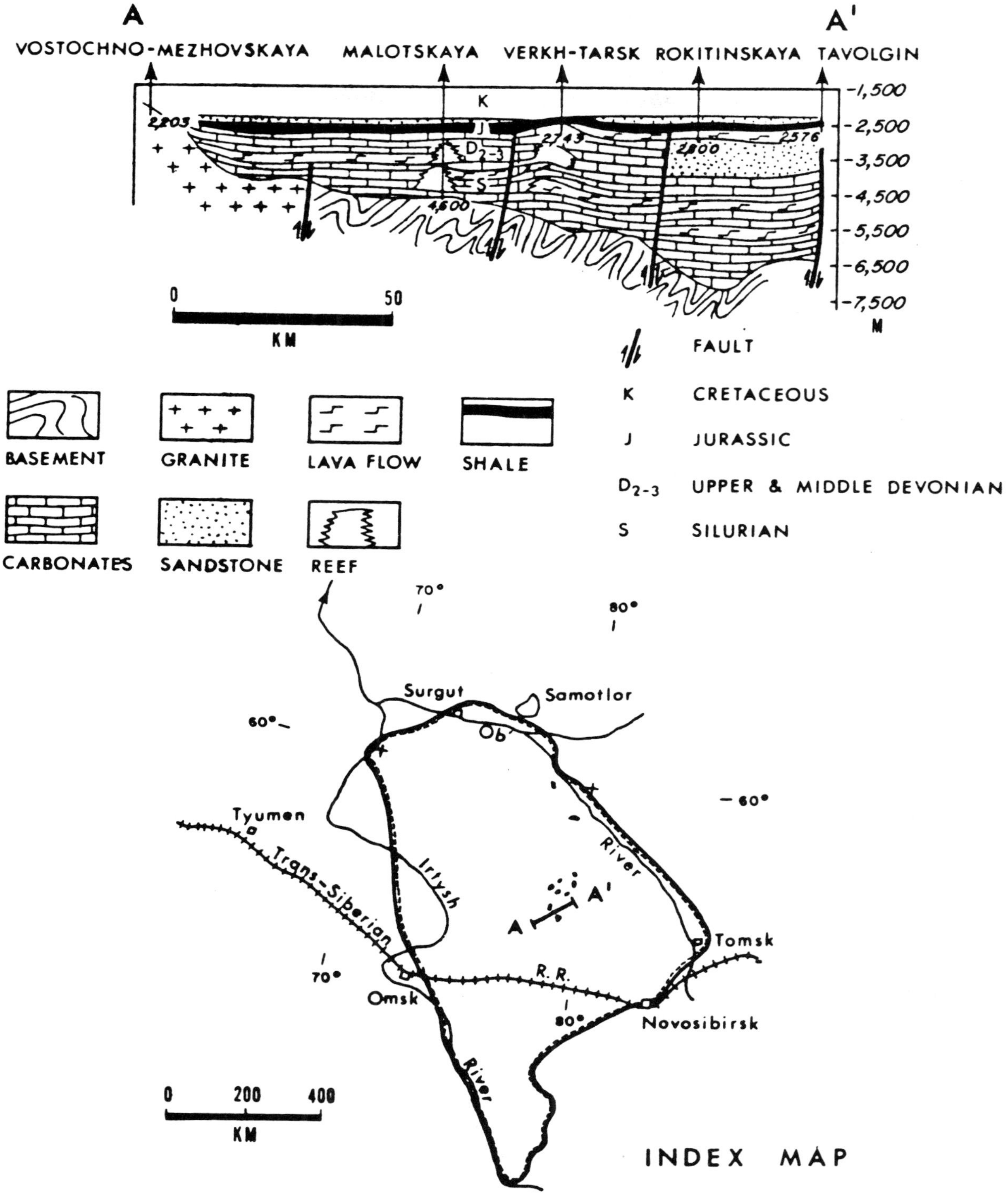


*Fig. 22* — SW-NE cross section of central part of Nyurol'ka basin, West Siberia. Index map shows outline of the basin and location of the cross section. From Zapivalov (1979) and Maksimov and Dikenshteyn (1979).

descriptions were published by Trofimuk and Vyshemirskiy (1976), Balabanova (1981), and Dubatolov (1981). Zapivalov et al. (1978, 1980) proved that the source of the oil is Paleozoic, not Mesozoic-probably Silurian and Devonian. Kanareykin and Yashkov (1980) and Vedernikov et al. (1980) showed that, before deposition of the continental Triassic graben deposits and associated intrusive-extrusive activity, the Silurian through Carboniferous section was folded gently and block faulted. Dips average less than 10°, but in the south, near their outcrops along the northern margin of the Altay, approach 30° Many of the seismic sections give the impression that the objective sequence is preserved only in grabens, but more detailed surveys demonstrate the essential continuity of the Silurian and Devonian in the basin. There are places where intrusives (Triassic ?) and uplifts of Proterozoic rocks interrupt the platform continuity, but these occupy a rather small area. No Ordovician or Cambrian has been reported. This may not be the result of the absence of rocks of these ages, but of the inability of Soviet rigs to penetrate to sufficient depth to reach Ordovician and Cambrian rocks. Strata of these ages should be below 5,000 m in most areas, below the economic depth of most rigs being used in the basin.

The Upper Silurian consists of 699 m+ (base not reached) of fossiliferous dolomitized limestone, organogenic limestone, and dolomite, much of it fractured. Some is bituminous and organic, so may be source rock. The Lower Devonian (906 m in one well), Middle Devonian (706 m in one well), and Upper Devonian (819 m in one well) are composed of organogenic shelf limestone, dolomitized limestone, dolomite, red silty marl, gray argillaceous siltstone, and light- to dark-gray organic shale. Fractures are common. The beds are very fossiliferous. Pelletoid limestone also is present. The Viséan-Bashkirian section is similar (Balabanova, 1981; Shnip et al., 1981). Gabbro and diabase porphyry sills are common, but are not very thick. Locally a wedge of Permian (?) continental strata is present, up to 1,000 m thick (Kanareykin and Yashkov, 1980).

### Petroleum Geology

Most of the oil is in vuggy and fractured Devonian carbonates. A leached or fossil soil zone at the top of the section and the overlying Jurassic shales form the seal. In some areas, the oil is in Lower Carboniferous or Silurian carbonates. Porosity is low, except where pellets and fossil remains are abundant. In the more fossiliferous layers, porosity values can reach 6 to 19 percent. Permeability is low except where abundant fractures are present. Field sizes are small, 30 million bbl or less. Total reserves proved to date are about 90 million bbl in the first three fields (Meyerhoff, 1982b). By 1979, ten fields had been found (Maksimov and Dikenshteyn, 1979). Despite the small field sizes, the basin will be developed because most of the structures are shallow (less than 3,200 m), the fields are close to existing rail and pipeline facilities, and the oil is of good quality. Reserves and resources of the Nyurol'ka are included in the West Siberian basin figures of Table 1. No gas reserves are included, inasmuch as only oil with associated gas has been found. Considered separately, the Nyurol'ka could produce more than one billion bbl, and probably more.

## Lena-Tunguska Petroleum Province

### Introduction

The Lena-Tunguska petroleum province (which includes the Irkutsk amphitheater, the Kansk basin, and the Tunguska basin) occupies an area of 2,827,000 sq km, the eastern half of the Siberian platform. This province includes the exposed Archean Anabar shield (Figs. 1, 23). Of this amount, 1,737,000 sq km is prospective (Dikenshteyn, 1977). Some authors (e.g., Dikenshteyn, 1977; Bazhenova et al., 1980) include the Vilyuy basin within this province. In this paper, I exclude the Vilyuy basin because its main prospective stratigraphy is related to that of the Verkhoyansk fold belt on the east. However, if one wishes to be absolutely correct, the Anabar shield, the Tunguska, Kansk, Lena, *and* Vilyuy basins, and the arches which separate the basins, together constitute the Siberian platform.

A systematic investigation of the region's hydrocarbon and other mineral potential began in 1932. However, it was not until 1938 that intensive investigations began. The first stratigraphic test was drilled in 1938 near the mouth of the Tuolba River (200 km east of Olekminsk) and flowed 86 liters/day from Cambrian dolomite at 373 m (Fig. 23). A dry hole was drilled in 1940 along the western margin of the Tunguska basin at Turukhansk (Fig. 23). Additional tests

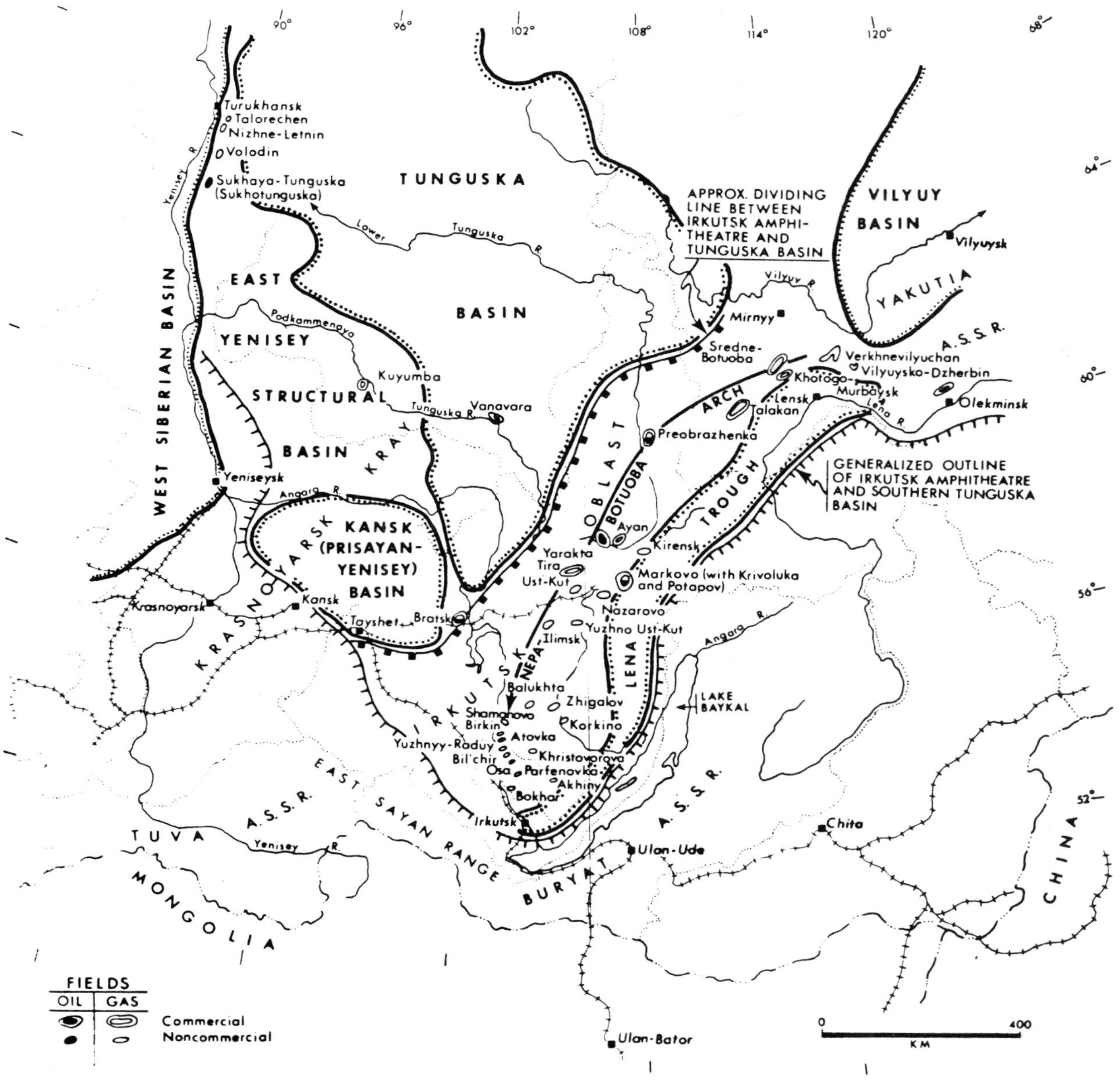


*Fig. 23* — Major basins of Lena-Tunguska petroleum province, Siberian platform. Commercial and noncommercial oil and gas fields shown.

were drilled during 1941 and 1942. Surface and subsurface work led to the delineation of the huge Nepa-Botuoba arch which strikes north-northeast through the area (Fig.23). The narrow, elongate Lena trough is on the eastern flank of the arch next to Lake Baykal; the Kansk and Tunguska basins were found on the western flank of the arch. Between 1948 and 1967, at least 160 shallow and intermediate-depth tests were drilled (Vasil'yev, 1968a). In 1954, a noncommercial discovery was made in Lower Cambrian carbonates at Osa (Fig. 23), on the southern plunge of the Nepa-Botuoba arch. On March 18, 1962, the Markovo-1 well on the Lena River (Fig. 23) tested 2,040 bbl/d of 42° API oil on a 25-mm choke in the Lower

Cambrian fractured carbonates. However it was not until the Krivoluka-1 well was drilled, just northeast of Markovo, that production was found in the late Proterozoic (Vendian) sandstones below the Lower Cambrian carbonates (Weaver, 1962; Trofimuk et al., 1964; Drobot and Isayev, 1966; Vasil'yev, 1968a, 1968b; Samsonov, 1975; Dikenshteyn, 1977; Meyerhoff, 1980b).

By 1980, two giant gas fields had been discovered, the first Proterozoic giants in the world. More than 700 wells of different types had been drilled in exploration of the region — these in addition to about 120 stratigraphic tests. By 1981, 15 commercial gas and oil fields and 18 noncommercial discoveries had been made (Dikenshteyn, 1977; Ovcharenko and Bakin, 1979; Vorob'yev, 1980; Bodunov et al., 1981; see Fig. 23). A 16th discovery — probably commercial — was made during late 1981.

## Regional Geologic Framework

The eastern part of the Siberian platform (Siberian platform *sensu stricto*) separates the West Siberian basin on the west from the Pacific fold belt on the east. On the north are the Yenisey-Khatanga and Lena-Anabar troughs separating the platform from the Hercynian Taymyr fold belt. On the southwest, south, and southeast is the Baykal fold belt (Baykalides), which may extend from Finnmark (northern Norway) on the west to the Pacific coast at the Sea of Okhotsk. The Baykalides may reach the northwestern part of Ireland. Thus the deformed geosyncline could have a length of not less than 6,000 km, and possibly as much as 8,500 km.

The Siberian platform is underlain by metamorphosed and igneous rocks of Archean to middle Proterozoic age. The Proterozoic rocks show three periods of metamorphism and intrusion, according to K/Ar dates: 1,350-1,450 ma, 1,050-1,100 ma, and 700-800 ma. The Baykal fold belt, which may extend from Finnmark to the Pacific, rims the platform on the southeast, south, and west.

The stratigraphic section within the Baykal fold belt (miogeosyncline, according to Bakirov, 1979) ranges in thickness from 12 to 20 km. In contrast, that of the platform ranges in thickness from 0 (where the Anabar shield is exposed) to more than 12 km in the Tunguska basin, 8 km in the Kansk basin, and 3-4 km in the Lena basin, east and southeast of the Nepa-Botuoba arch (Fig. 23). The miogeosynclinal section exhibits a great «sameness,» thus having a monotonous appearance. The section includes argillaceous polymictic conglomerate and sandstone which are poorly sorted; flysch-like strata with graded bedding; carbonate clastic rocks and argillaceous limestone; and — especially in the upper part of the section — increasing amounts of redbeds. The section is most marine in the Riphean through Silurian; it is least marine in the upper part, becoming nonmarine in the Devonian. The basal part of the upper Riphean (middle-late Proterozoic) is 2,500 to 4,000 m thick and has no time equivalents on most of the platform where drilling has reached basement. However, time equivalents of these upper Riphean rocks probably are present in the deepest parts of the Kansk and Tunguska basins, and very possibly in the Lena basin close to the present Lake Baykal shoreline.

The miogeosynclinal strata are severely folded. At the platform margin, these strata grade laterally and abruptly into platform deposits which are much thinner than their miogeosynclinal equivalents. The abrupt platformward facies change is accompanied by an almost as abrupt dying out of the folds. On the platform, almost no deformation has taken place and most movements have been vertical.

The platform includes the exposed Anabar shield in the northeast, and the north-northeast-striking Nepa-Botuoba arch, which extends from the northeastern part of the platform, southward to near the city of Irkutsk at the western end of Lake Baykal (Fig. 23). Lake Baykal itself is a Cenozoic graben whose margins are old fault lines, some as old as early to middle Proterozoic. On the eastern side of the platform, east of the exposed shield and the Nepa-Botuoba arch, three basins are developed. From north to south, these are the small Sukhana basin of Vendian and Cambrian ages, just northeast of the Anabar shield (see Fig. 36); the Vilyuy syneclise of Carboniferous through Jurassic age (it probably had an early Paleozoic or late Proterozoic precursor; and the narrow Lena basin or trough between Lake Baykal and the Nepa- Botuoba arch. On the western side of the platform, two large basins are present (Fig.23). These are the huge Riphean through Devonian Tunguska basin in the northwest, and the small Kansk basin of the same age farther south.

The Vilyuy syneclise or basin is considered separately, because its productive strata are Permian and younger, and inseparably related to the history of the Verkhoyansk Range (Verkhoyansk fold belt) east of the Siberian platform. The Verkhoyansk Range belongs to the Circum-Pacific fold belt. In contrast, west of the Siberian platform is the great West Siberian basin which overlies or is marginal to Baykalian, Caledonian, and Hercynian fold belts.

**Stratigraphy**

The Proterozoic section consists of «tight» polymictic conglomerate (basal conglomerates), polymictic sandstone (subgraywacke), and light-, medium-, dark and greenish-gray shale. The section is mainly marine. A few anhydrite beds are known in the Kansk and Tunguska basins (Samsonov, 1975; Danilkin, 1980). The oldest beds are latest Riphean and Vendian, as determined from microfossils and K/Ar dates on glauconites (summarized in Meyerhoff, 1980b). These dates are consistently 590 to 680 ma, except for one date which is 925 ma (Anatol'yeva et al., 1966; Samsonov, 1975). The last date is inconsistent with the 700-800 ma dates that come from the youngest metamorphosed and intrusive sequence of the underlying basement (Samsonov, 1975). The sandstones are light and medium gray, fine to medium grained, feldspathic, and porous, with porosity values up to 26 percent (but usually 11-19 percent) on the Nepa-Botuoba arch. The excellent state of preservation of the porosity and permeability for a period of about 680 ma attests to the great stability of the Siberian platform. The thickness of the unit ranges from 0 to several hundred meters, with values of 100 m to 400 m in wells drilled on and along the flanks of the Nepa-Botuoba arch. The section is much thicker in the Kansk and Tunguska basins. In the Kansk basin and near the platform margins, redbeds are present in the gradational zone between the platform and geosyncline. These redbeds are present at all levels from Proterozoic to Devonian. At least five reservoirs — all sandstone — are present in the latest Riphean and Vendian (Samsonov, 1975). Three of these reservoirs contain commercial quantities of gas, condensate, and oil along the flanks of and on the Nepa Botuoba arch from just north of Irkutsk to Olekminsk, and also is in the Tunguska basin.

On the Nepa-Botuoba arch and its eastern flank, the section above the Vendian is Cambrian and Ordovician. The whole section contains evaporites, but evaporitic strata are especially abundant in the Lower Cambrian. The Lower Cambrian is 1,200 to perhaps 6,000 m of limestone, dolomite, anhydrite, and halite, with some sylvite and heavier salts (Yanshin, 1962; Zharkov, 1969). Individual halite beds are known to be as thick as 370 m. Zharkov (1969) calculated the minimum volume of halite and heavier salts to be 558,000 cu km, but a more likely value, according to his calculations, is 800,000 cu km. Thus this is one of the most important halite (and heavier salts) basins in the world. This part of the section contains eight carbonate vuggy and/or fractured dolomite reservoirs, the youngest reservoirs in the area. A regional unconformity is at the top of the Lower Cambrian.

The Middle and Upper Cambrian are less marine and more continental, with sandstone, siltstone, and shale, commonly with a reddish color. Some anhydrite and salt are present. The same description applies to the Ordovician.

In the Kansk and Tunguska basins, marine Silurian carbonates and Devonian redbeds, with some marine tongues, are present (Bakirov, 1979). Equivalent strata are not present in the Lena basin. The total thickness of the post-Early Cambrian section ranges from a feather edge (0 m on and east of the Nepa-Botuoba arch) to more than 2,500 m in the Tunguska basin.

Continental Lower or Middle Carboniferous through Triassic beds are widespread, especially in the central and northern parts of the platform. These strata are associated with Carboniferous -Permian coal measures, which Bakirov (1979) thinks form one of the largest coal basins in the world. Bakirov wrote (p. 325) that, at the very least, the coal measures of the Tunguska basin are the largest in the USSR. Diabase, gabbro, basalt, and intermediate intrusives of Permian and Triassic ages comprise the Siberian traps, the largest sill and flood basalt province in the world (1,000,000 sq km). These trap rocks are found mainly north of 50° N lat, and are the host for numerous rare earth and metallic deposits, among the richest of their kinds in the world. The thickness of this section ranges from a feather edge to about 2,000 m. At Yarakta (Yaraktin in Meyerhoff, 1980b) field, the oil- and condensate-bearing Proterozoic sandstones underlie 1,750 m of gabbro, dolerite, diabase, and basalt (Bazanov, 1973). This field was discovered serendipitously, and not by any scientific method (Ministry of Oil Industry personnel, oral commun., 1979).

**Petroleum Geology**

Thirty-four oil, gas, and gas-condensate fields — 15 of them commercial — have been found in vuggy and fractured dolomite reservoirs in the

Lower Cambrian (8 potentially productive zones, at least four of which have established reserves of commercial value); and in arkosic sandstones of the Vendian and uppermost Riphean. Four of the Vendian zones have commercial reserves. Most of the hydrocarbons are gas and gas-condensate, but some oil has been found, particularly at Markovo (oil ring), Preobrazhenka, and Yarakta (Fig. 23). All discoveries made (the first was Markovo in 1962) are shut in, and few fields have been fully delineated. Of the four Riphean-early Paleozoic basins, the Lena and the Tunguska are the most prospective. The Tunguska in particular offers very large potential. The Kansk basin has unfavorable structure (tight folds, numerous faults) and facies (more geosynclinal and more redbeds) (Gol'dberg et al., 1981). The Sukhana basin, northeast of the Anabar shield, is an unexplored area, but has low potential (Gol'dberg et al., 1981). The arches between the various basins have good potential.

Porosity values in the Vendian sandstones range from 8 to 26 percent; permeability values range from a few millidarcys to 4,300 md, with the average in the 50 to 150 range. Effective thickness values range from a few meters to 90 m, with the average 15 to 30 m. The only Riphean reservoir found to date has porosity and permeability values that are too low for commercial production (Ovcharenko and Bakin, 1979; Meyerhoff, 1980b).

Seals for the Proterozoic reservoirs are dark marine shales with organic matter up to 3 or 4 percent by weight. Many reservoirs are offshore bars and stream channels that are closed reservoir systems. The Riphean and Vendian hydrocarbons are very different from the Lower Cambrian hydrocarbons, and have been proved to be indigenous to the Proterozoic (Drobot and Isayev, 1966; Samsonov and Tyshchenko, 1970; Gavrilov et al., 1971; Antsiferov, 1980; Bodunov et al., 1981).

Traps in the Riphean and Vendian beds are both stratigraphic and structural. Figures 24 and 25 show the stratigraphic trap developed within the Parfenovo Horizon (Parfenovka) in the Markovo field. The trap is a northeast-southwest-striking offshore or beach bar, on the northwest side of which is part of a north to northwest-trending stream distributary system (Zolotov et al., 1968). Danilkin (1980) showed one of these stream-offshore bar-sandstone distributary systems; their reservoir properties were described by Zheleznova and Kul'kova (1981). Porosity values, as would be expected, are greatest in the main parts of the stream channels and bars, as are the permeability values. The Yarakta field (Fig. 23), beneath 1,750 m of Permian and Triassic diabase traps, is another northeast-striking offshore or beach-bar system. The updip limit of production is the sandstone pinchout, except at the eastern end of the field where a partial fault seal is present (Bazanov, 1973; Meyerhoff, 1980b). In contrast, fields at the northeastern end of the Nepa-Botuoba arch are

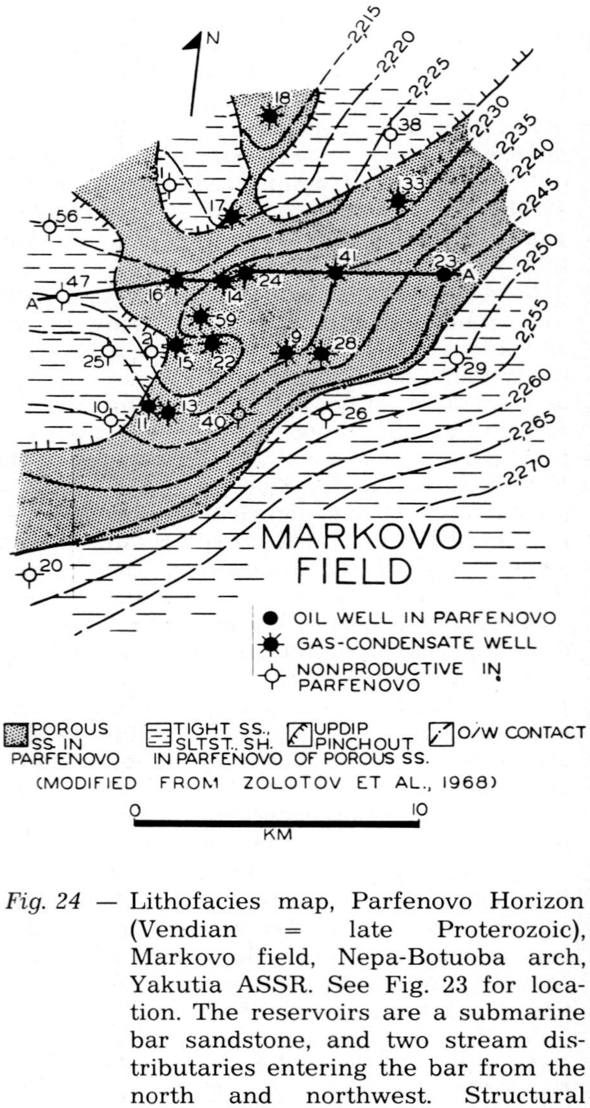


*Fig. 24* — Lithofacies map, Parfenovo Horizon (Vendian = late Proterozoic), Markovo field, Nepa-Botuoba arch, Yakutia ASSR. See Fig. 23 for location. The reservoirs are a submarine bar sandstone, and two stream distributaries entering the bar from the north and northwest. Structural datum is the top of the Parfenovo. Line of cross section is Fig. 25. From Zolotov et al. (1968). Republished with permission of the *American Association of Petroleum Geologists.*

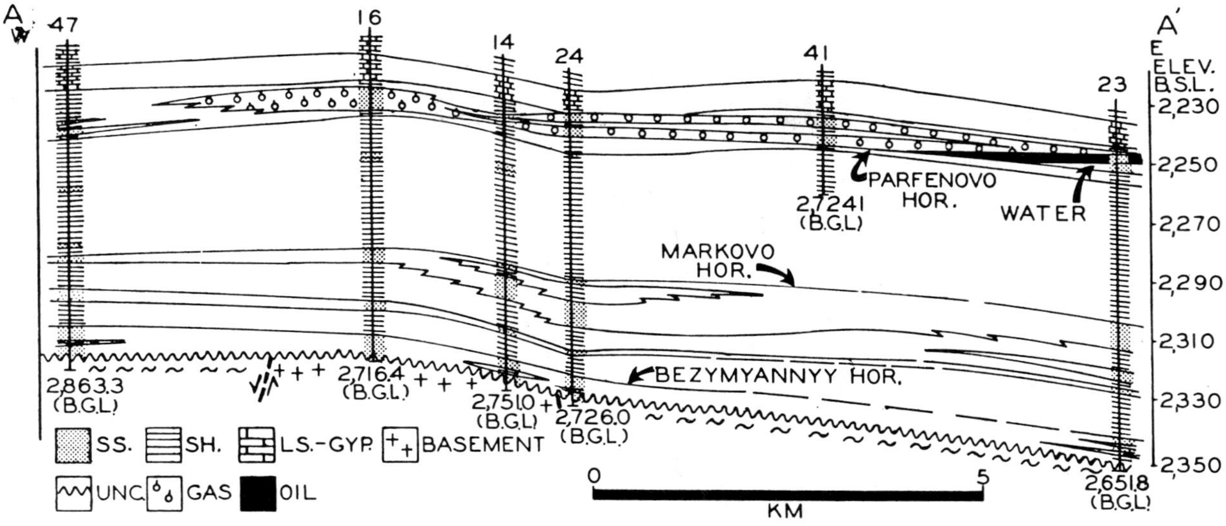


*Fig. 25* — W-E structural cross section, Parfenovo (Vendian) zone, Markovo field. Depths in meters. Location is on Fig. 24. From Zolotov et al. (1968). Republished with permission of the *American Association of Petroleum Geologists.*

large basement drape anticlines. Two are the world's first giant Proterozoic fields (e.g., Sredne botuoba, Verkhnevilyuchan, erroneously called Verkhnevilyuy by Meyerhoff, 1980b). Others include Talakan, and Vilyuysko-Dzherbin (Figs. 23, 26). Figure 26 shows a map of the structure, a columnar section, and a cross section of the Srednebotuoba field (from Belen'kiy and Kunin, 1978). Figure 27 is an electric log cross section of the same field, as interpreted by Meyerhoff (1980b).

Drilling depths are in the 1,200- to 2,700 m range on the Nepa-Botuoba arch. They will become deeper as exploration shifts into the deep basins.

Porosity values in the Lower Cambrian fractured and vuggy dolomite zones range from 0.5 to 25 percent and generally are 4—11 percent; permeabilities range from 0 to 350 md. Much of the permeability is produced by fractures. The thickness of the productive zones ranges from 2 to 100 m, but averages about 30 to 45 m (Ovcharenko and Bakin, 1979; Meyerhoff, 1980b).

Seals for the dolomite reservoirs are anhydrite and halite beds (Fig. 28). Source rocks are the associated limestones with organic percentages ranging from 0.5 to 2.4 by weight. The hydrocarbons in the Lower Cambrian are distinctive chemically from those in the Proterozoic (see Meyerhoff, 1980b, for summary of properties; see also Bodunov et al., 1981).

Traps found to date in the carbonate section depend on the development of salt swells and the degree of salt deformation. The oil and gas are trapped in fractured dolomites within and beneath the salt swells. Kalinkina (1981) recently noted the presence of Early Cambrian bioherms in the Nepa-Botuoba arch region (Fig. 29). Such traps could be numerous.

Historically, the first and only traps to be found with the seismograph in this region were the salt swells at depths of 1,000 to 1,400 m (Fig. 29). Drilling beneath these swells led eventually to the discovery of hydrocarbons in Vendian stratigraphic traps, and eventually to exploration for structural traps in the Vendian.

There has been no production from these fields except for local use. Proved reserves are 225 million bbl of condensate and oil, and 18.1 Tcf of gas (Table 1). Predicted resources are 2 billion bbl of liquids and 83 Tcf of gas. Prospecting in this area will be extremely difficult because of the thick igneous trap cover, the thick carbonate section, and the stratigraphic nature of many of the traps. Eventually the basin should become very important in providing fuel for Siberia and perhaps even the European

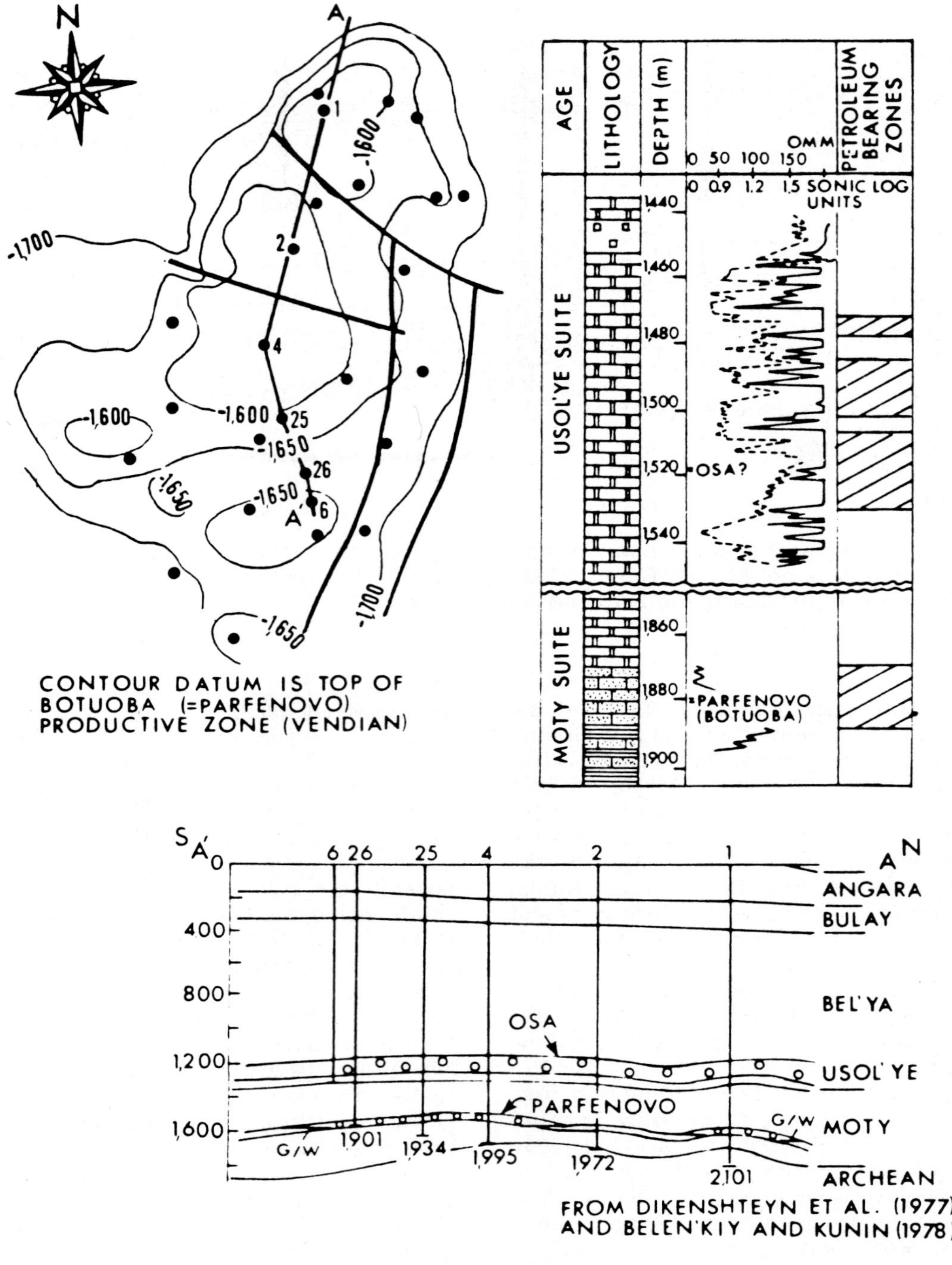


*Fig. 26* — Srednebotuoba gas-condensate field, Nepa-Botuoba arch, Yakutia ASSR. See Fig. 23 for location. Structural-contour map in meters subsea. Structural datum is top of Parfenovo Horizon. Lines of cross sections of this figure and of Fig. 27 are shown. Cross section strikes S-N, and shows both the basal Cambrian Osa producing zone and the Vendian Parfenovo producing zone. Columnar section shows lithologic, electric-log, and producing-interval details. From Dikenshteyn (1977) and Belen'kiy and Kunin (1978). Republished with permission of the *American Association of Petroleum Geologists.*

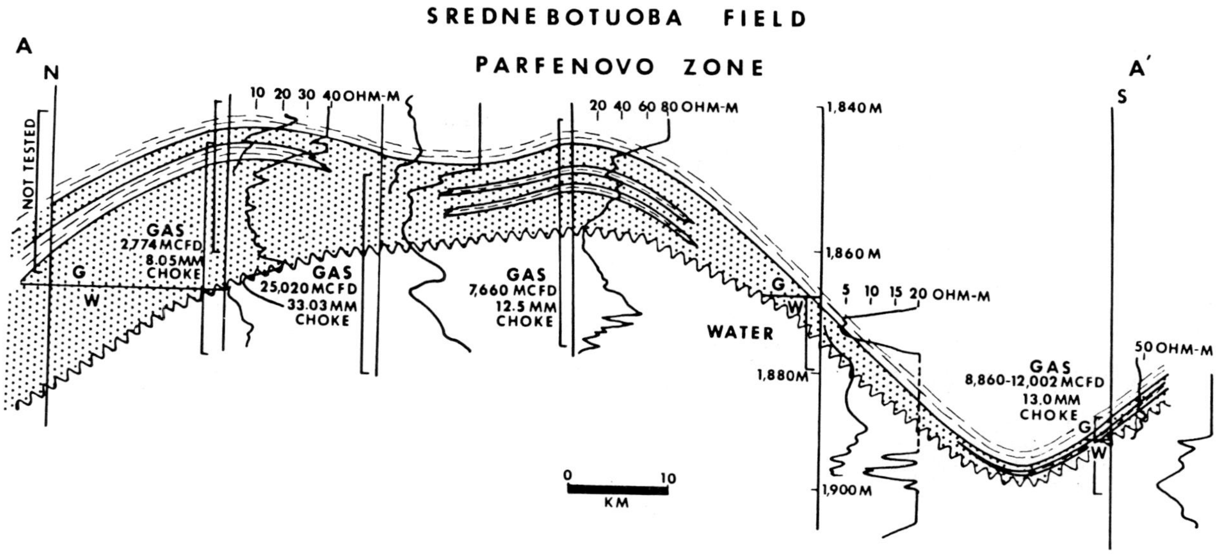


*Fig. 27* — West-east structural and electric-log cross section, Srednebotuoba field, Nepa-Botuoba arch, Yakutia ASSR. Shows tested intervals. Faults are omitted. Approximate location is on Fig. 33. From Meyerhoff (1980a). Reproduced with permission of the *American Association of Petroleum Geologists.*

USSR. Two of the gas fields discovered to date are giants. These are Srednebotuoba with 5.9 Tcf of gas and 149 million bbl of liquids; and Verkhnevilyuchan with 10.5 Tcf of gas and approximately 300 million bbl of liquids (Vasil'yev and Zhabrev, 1979; Bakirov, 1979). Yarakta, the principal oil field (with some gas and condensate) found to date, could have as much as 200 million bbl of oil, but it is not certain how much of this is recoverable.

## Vilyuy Basin (Syneclise)

### Introduction

The Vilyuy basin, or syneclise, is a V-shaped, northeast-striking re-entrant into, and on, the Siberian platform (Fig. 30). It is the easternmost basin on the platform. Bazhenova et al. (1980) showed that the basin began to form late in Ordovician time, but its present shape and orientation are Devonian and younger phenomena (Gusev, 1979; Bazhenova et al., 1980). This is true because Devonian carbonates and evaporites everywhere underlie the Carboniferous and/or Permian of the present Vilyuy basin, Lena-Anabar trough, and Yenisey-Khatanga trough. Even though the Vilyuy basin overlies the eastern margin of the Siberian platform, the productive section within it is of Permian through Jurassic age and correlates, lithologically and paleontologically, with the western Verkhoyansk Range section east of the Siberian platform (Fig. 31). Because of this, the Vilyuy basin is treated apart from the Siberian platform.

The prospective section within the Vilyuy basin ranges in age from at least Carboniferous through Jurassic (Fig. 31). Eventual production from Devonian and older units — and even Early Cretaceous — cannot be ruled out. The section is a thick monotonous sequence of grayish deltaic, continental, paralic, and marginal marine sandstones and shales with abundant amounts of woody organic debris. The units are nearly identical to and correlate with the Verkhoyansk Range section the so-called «Verkhoyansk complex» of Soviet literature (Nalivkin, 1973; Gusev, 1979). The section is at least 11 to 12 km thick beneath the Lena River, between the Verkhoyansk Range and the eastern part of the Vilyuy basin (Fig. 30).

The basin occupies an area of 200,000 sq km west of the Verkhoyansk Range and the Lena River. The basin's stratigraphy continues northward along the Verkhoyansk front to the head

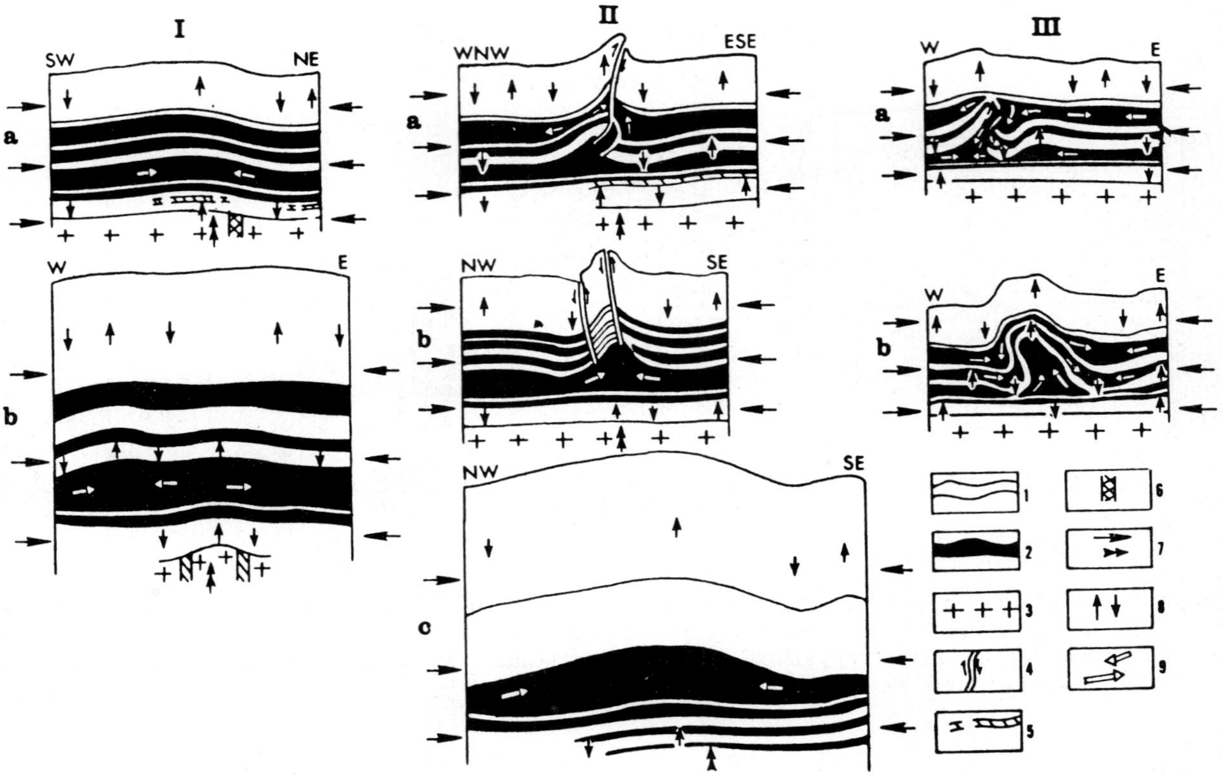


*Fig. 28* — Various types of traps in Lower Cambrian carbonate rocks of the Nepa-Botuoba arch. Some are synsedimentary, and some are post-sedimentary.

Column I = structures in Cambrian salt that involve slight amounts of . basement flexuring (e.g., Bratsk, Atovka, Birkin);

Column II = salt movements accompanied by faulting (e.g., Zhigalovo, Akhino);

Column III = intrabed deformation of the Markovo and Kirensk type.

*Legend*:

1 = bedding planes;
2 = salt;
3 = crystalline basement;
4 = faults;
5 = traps;
6 = deep basement fault;
7 = compression directions;
8 = sense of vertical movements;
9 = directions of salt flow.

From Dubronin (1976). Republished with permission of the *American Association of Petroleum Geologists.*

of the Lena delta (Lena-Anabar trough), where it swings sharply westward around the northeast corner of the Siberian platform (Fig. 1). The section continues beneath the western, east-west-striking arm of the Lena-Anabar trough into the Nordvik area, whence it continues west ward into the Yenisey-Khatanga trough (Yaskevich et al., 1980). Even the Devonian salt which forms diapirs in the western end of the Vilyuy basin, continues around the northeast end of the Siberian platform through the Kharaulakhskiy Range to Nordvik and farther west, well into the Yenisey part of the Yenisey-Khatanga trough (Vasil'yev, 1968a; Nalivkin, 1973; Gusev, 1979; Yaskevich et al., 1980).

The Vilyuy basin is primarily a gas province

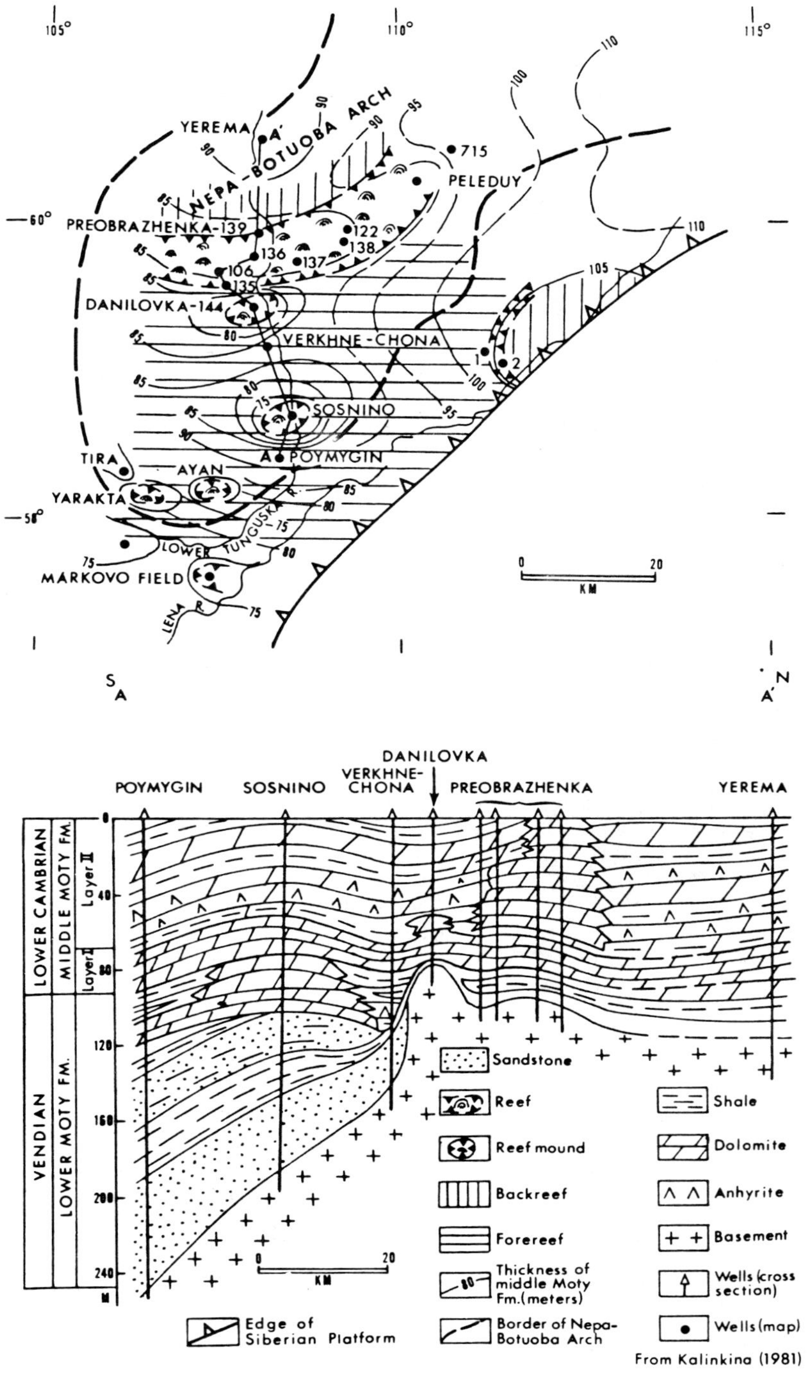


*Fig. 29* — S-N stratigraphic cross section (with index map) of known bioherms in middle Moty Formation (basal Cambrian), just above the Vendian part of the Moty (sandstone, shale) on Nepa-Botuoba arch. From Kalinkina (1981).

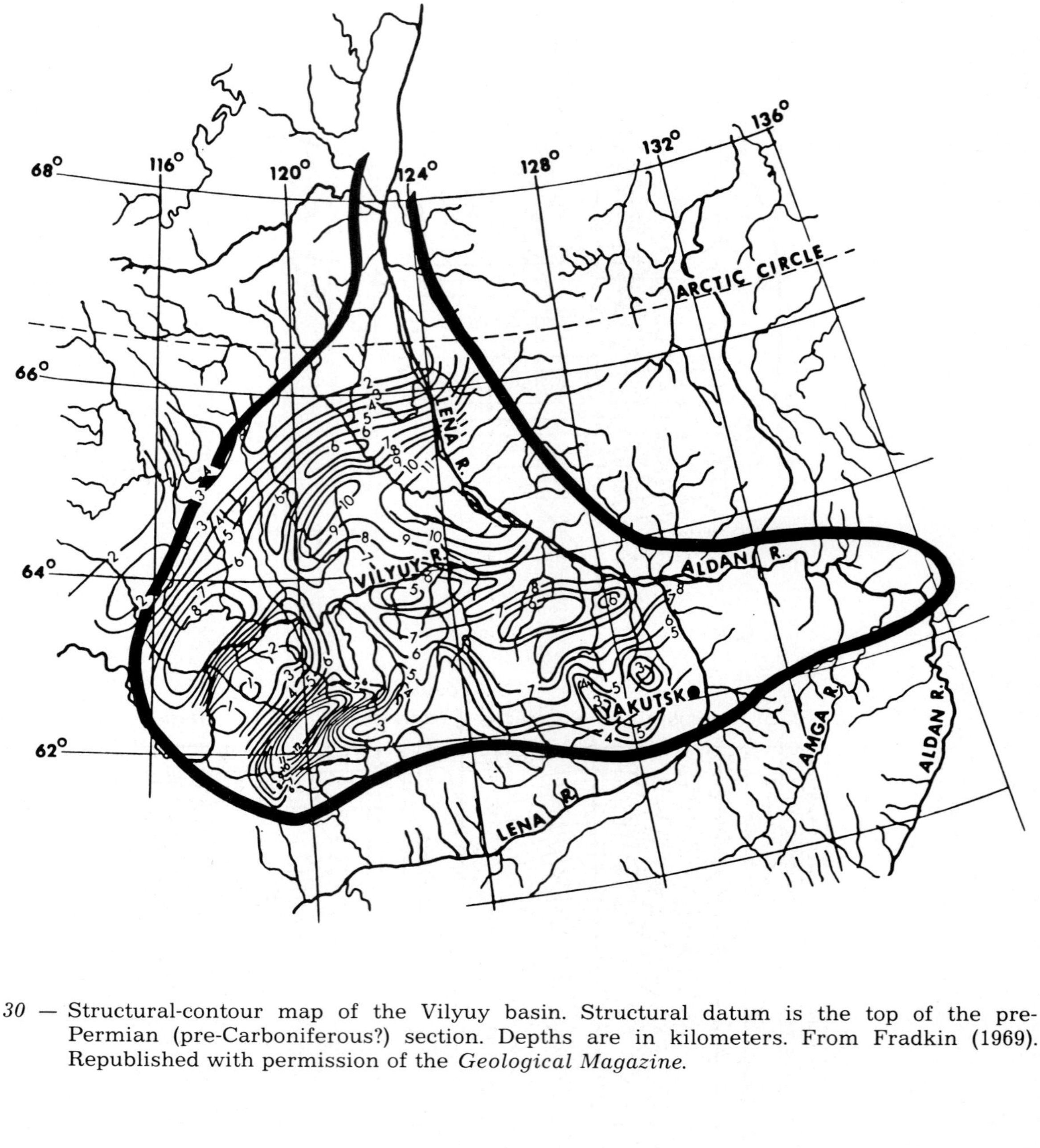


*Fig. 30* — Structural-contour map of the Vilyuy basin. Structural datum is the top of the pre-Permian (pre-Carboniferous?) section. Depths are in kilometers. From Fradkin (1969). Republished with permission of the *Geological Magazine.*

with producing zones from the Upper Permian through Lower to Middle Jurassic. Eight gas fields have been found, and eventually the basin should be the site of many more gas fields. Two major trends are found, the northeast-trending structures of the Siberian platform that overlie the Siberian platform; and the northwest-striking compressional folds of the Verkhoyansk Range. Total reserves found to date are about 17 Tcf (and possibly 250 million bbl of condensate — not shown on Table 1). The northeast-trending basement structures are large and hold the greatest share of the known reserves. These structures and those related to them, presumably, will hold most of the resources still to be discovered.

The first petroleum investigations within the basin were conducted from 1931 to 1932. Some

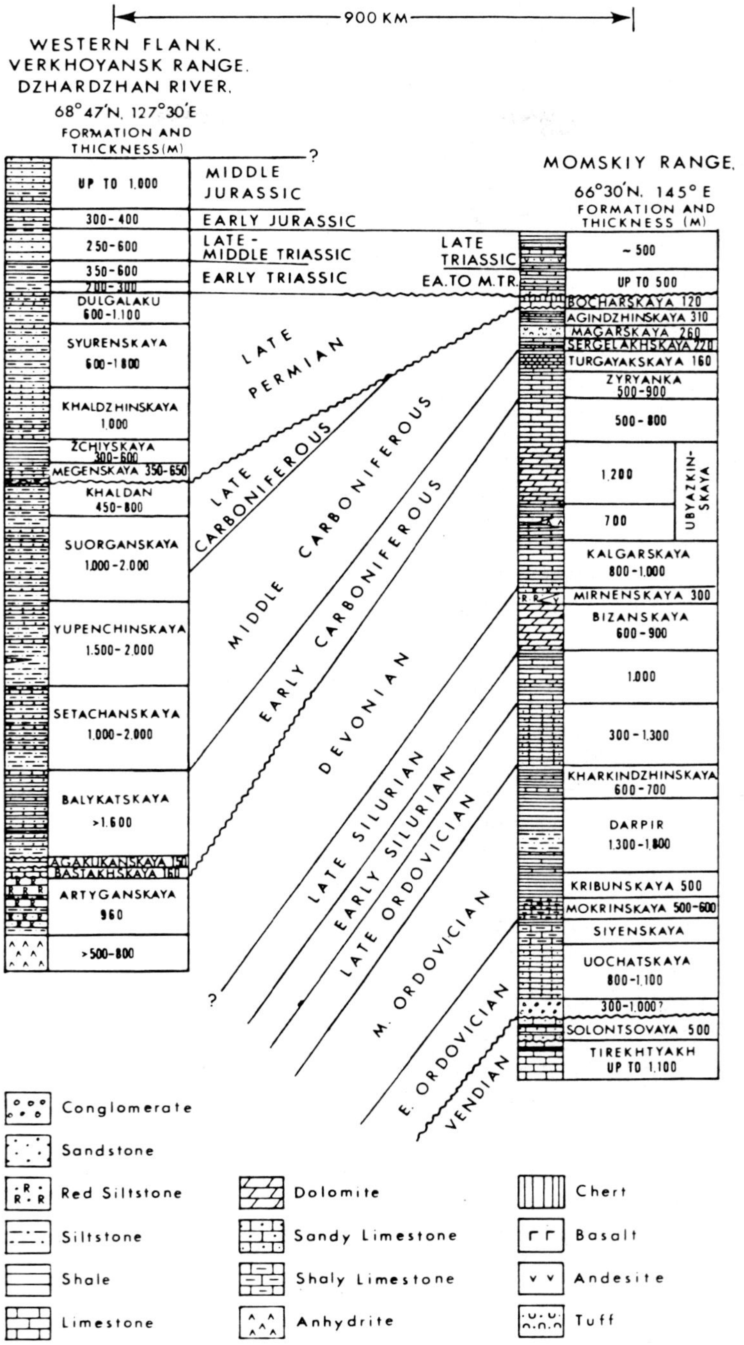


*Fig. 31* — Columnar sections from the western Verkhoyansk Range (northern end of Lena-Anabar trough at 68° 47'N, 127°30'E, Dzhardzhan Valley) and eastern Momskiy Range (66°30'N, 145°E). The near identity of the Tournaisian and Devonian sections of the two regions, with nearly identical lithologies and faunas, makes it very unlikely that the fault zone separating the Verkhoyansk and Cherskiy Ranges is a «collision zone» between the North American and Eurasian «plates,» as some have alleged (e.g., Churkin, 1972). From Gusev (1979).

shallow structural drilling was carried out during 1936, but serious investigations were not made until the early 1940s.

Because the first discovery in the West Siberian basin was a gas discovery (which the government did not want), exploration for oil began in the Vilyuy basin during 1956. This exploration in an extremely remote region was undertaken by the first Minister of Geology and Conservation of Natural Resources (MGION), Dr. P. Ya. Antropov, who also was responsible for the bold, far-seeing exploration of the West Siberian basin. The first discovery was made in 1957 — Ust'-Vilyuy, a tiny mountain-front fold parallel with the Verkhoyansk trend. This was a gas discovery, and all subsequent discoveries were gas — eight fields to date. The failure to find oil was the final excuse needed to dismiss P. Ya. Antropov. Antropov, who was a controversial person in his own right, one who had made many enemies by his aggressive and unconventional methods, was fired in 1961, and was attacked almost universally for his «errors.» Ironically, one of these «errors,» the West Siberian basin, proved that some of his judgments were correct. It is too early to assess his prognosis of the Vilyuy basin. The last gas field found in the basin was in 1972.

During the 1970-1971 period, emphasis in the Vilyuy basin was switched to the Srednebotuoba area of the Siberian platform, where permafrost problems were nil, transport conditions better, and drilling depths only 50 to 70 percent of those in the Vilyuy basin proper (1,200-2,700 m vs. 2,300-4,800 m). Effective exploration of the Jurassic through Permian sediments has stopped because of the excessive depths (too deep for Soviet equipment), terrain logistics problems, and the almost fanatical emphasis on the West Siberian basin where the success ratio has dropped dramatically (Filanovskiy, 1980; Lalayants, 1980).

### Regional Geologic Framework

The Vilyuy basin, or its precursor, evidently was already in existence as the Verkhoyansk foredeep (geosyncline), extending from the Arctic to the Sea of Okhotsk, began to subside during Carboniferous time and be filled with sediments. The general trends are parallel with the axis of the basin, east-west to northeast-southwest, at right angles to the margin of the Siberian platform. A major central arch, the Khapchagay arch, dominates the axial part of the basin (Fig. 32). The biggest fields of the basin, six of them are on the crest of this arch. Two other fields are mountain-front fields in compressional folds which strike at right angles to the basin trends.

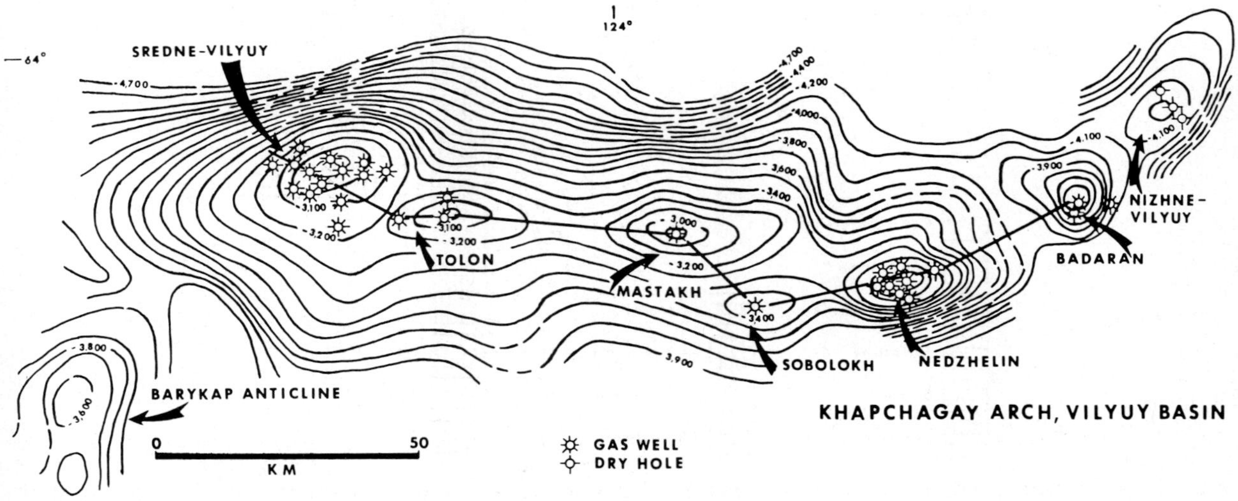


*Fig. 32* — Structural-contour map of the Khapchagay arch, central Vilyuy basin. The arch (striking west to east) is plainly visible on Fig. 30. Structural datum is the top of the Permian. Location of Fig. 33 is shown. Modified from Fradkin (1969). Reproduced with permission of the *Geological Magazine*

## Stratigraphy

The objective section in the basin center and within reach of Soviet drilling equipment consists almost entirely of Upper Permian and younger strata. As newer equipment with a capability for deeper drilling comes into use, the Lower Permian, Carboniferous, and even deeper formations may become objectives, including a Proterozoic-lower Paleozoic section similar in age and lithology to that of the Lena-Tunguska petroleum province. Because of the practical limits of Soviet drilling rigs, only the Carboniferous and younger section equivalent to the section of the Verkhoyansk fold belt (i.e., the Vilyuy basin sedimentary section *sensu-stricto)* is considered here.

The Carboniferous, in the deepest part of the basin beneath the present Lena River, may be 5,200 to 5,300 m thick; farther east in the Verkhoyansk Range proper, it is estimated to be close to 8,700 m thick (Gusev, 1979). The basal part is the typical Tournaisian limestone that is found from the upper Kolyma region of northeastern USSR to the British Isles. In the Verkhoyansk Range, it is about 150 m thick. The Tournaisian is unconformable on the Devonian. Above it, the Lower, Middle, and Upper Carboniferous consist almost entirely of coarse- to fine-grained, poorly to moderately sorted lithic arenite, some arkosic arenite, siltstone, silty shale, and coaly beds. The section is carbonaceous in many places. Marine strata are most abundant on the eastern flanks of the Verkhoyansk Range, whereas coal seams are found along the western flank adjacent to the Siberian platform and the Vilyuy basin (Nalivkin, 1973).

About 1,600 m of Permian beds have been penetrated in wells drilled along the Khapchagay arch; a much thicker section is present in the Verkhoyansk Range. A maximum of 5,200 m of Permian is believed to be present along the eastern end of the Khapchagay arch (Fradkin, 1969; Meyerhoff, 1980a). The section consists of thick-bedded paralic and deltaic quartzose arenite and arkosic arenite, which alternate with siltstone, black shale, carbonaceous and coaly shale, and coal. The coal and the carbonaceous shale presumably are the sources of most of the gas in the basin (Maksimov et al., 1979). The sandstones have porosity values in the 9—18 percent range. The section is characterized by a rather monotonous and repetitive sameness.

The Triassic, 1,165 to 1,360 m thick, consists of gray sandstone, shaly sandstone, siltstone, and dark shale. Conglomerate appears in the Upper Triassic. Plant fossils predominate, with some coaly beds but, locally, marine strata are known. Porosity values of the sandstones are typically 18 to 21 percent (Vasil'yev, 1968b).

The Lower Jurassic consists of two sequences separated by an unconformity which increases in angularity eastward (Afanas'yev et al., 1966; Mikulenko and Gaydeburova, 1966). The whole Lower Jurassic section is 300 to 680 m thick, and includes a monotonous sequence of thin-bedded dark shale, siltstone, and arkosic to lithic arenite. The Middle Jurassic, 140 to 175 m thick, is very similar. In contrast, the Upper Jurassic is a post-tectogenic molasse, consisting of 535 to 1,120 m of more variegated thin-bedded sandstone, siltstone, carbonaceous shale, and coal measures.

The Lower and Upper Cretaceous are another post-orogenic molasse sequence, up to 2,780 m thick, with sandstone, sand, siltstone, silt, shale, clay, and coal. On the whole the Cretaceous strata are poorly consolidated. They appear to have no oil or gas potential.

## Petroleum Geology

More than 100 structures have been mapped in the Vilyuy basin. Many have been drilled but were dry. Others, especially off the northern and southern flanks of the Khapchagay arch, are too deep to test with Soviet drills. Many probably have not been mapped because of the poor quality of the seismic data. Many of the structures that were tested were drilled on the basis of shallow mapping that could not be carried to deeper levels because of a total lack of reflections with the antiquated, single-fold techniques used.

The Khapchagay arch has been well tested along its crest (Fig.33), but the deeper zones below about 4,800 m are untested; in fact, most zones below 4,000 m are untested. The flanks of the Khapchagay arch are also undrilled, as are many parts of the basin flanks. Stratigraphic traps, a distinct possibility in the area, have not been sought because of the depth and expense of drilling the flanks of the Khapchagay arch.

The principal reserves are in the Upper Permian and Lower Triassic. Very small reserves are in the Lower to Middle Jurassic. Maksimov et al. (1979) have pointed out that the organic matter to 3,500-4,000 m is chemically immature, and that both the percentage (by weight) of

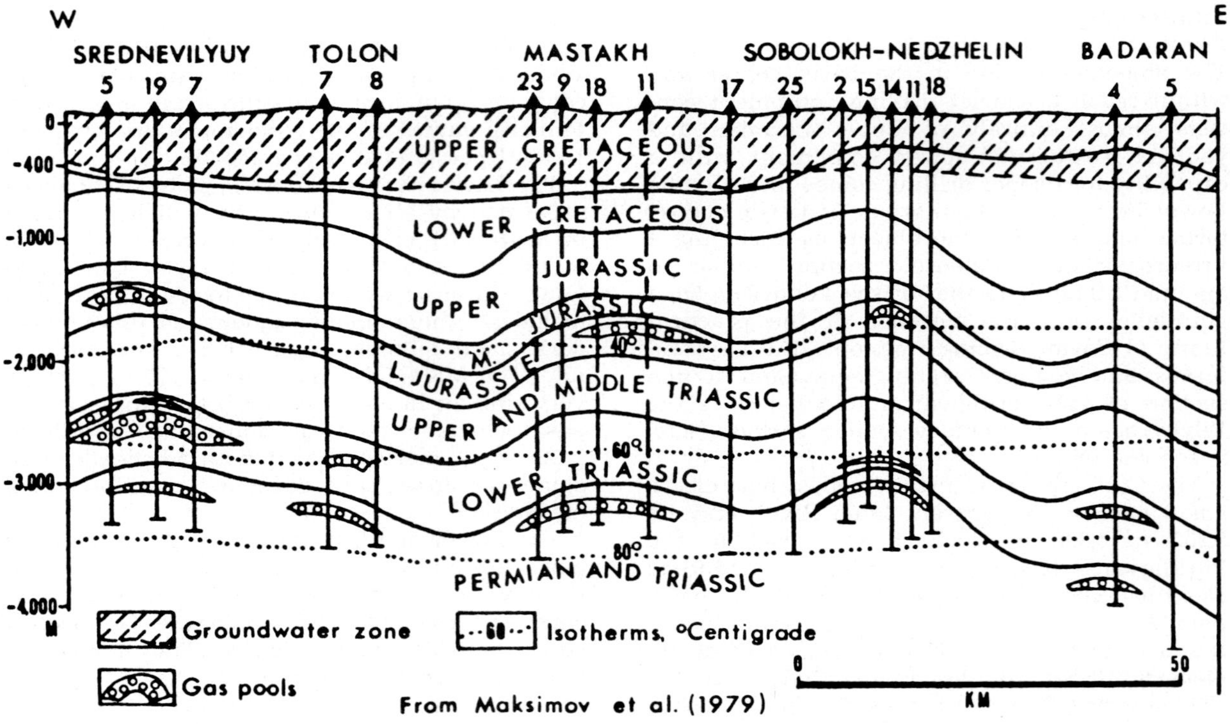


*Fig. 33* — W-E structural cross section along the crest of the Khapchagay arch, Vilyuy basin. Location is shown on Fig. 32. From Maksimov et al. (1979).

organic matter and of condensate increases downward. Therefore, Maksimov et al. wrote that the gas sources, which are of woody origin, are deeper than the drill has penetrated, because of the migration-differentiation effect upward (lightest fractions in the Jurassic; heaviest in the Permian).

The possibility of finding the Proterozoic-Devonian section to be prospective should not be overlooked. Devonian salt domes are present in the southwestern part of the basin. This older section will be extremely deep in the basin center, generally below 7,000 m. However, it could be prospective near the basin flanks where some shallow tests have found shows.

By the end of 1979, two fields had produced 60 Bcf for use in the city of Yakutsk and for the lumber and other industries south of Yakutsk. If large reserves eventually are established, they will be piped to the Amur-Ussuri (Vladivostok-Komsomol'-na-Amure) area of the Soviet Far East where 90 percent of the hydrocarbons now consumed in the Soviet Far East come from the West Siberian basin by rail. The remaining 10 percent comes from Sakhalin by pipeline (Meyerhoff, 1981). As of January 1, 1980, 17 Tcf of reserves had been proved; an estimated 19 Tcf of gas and 500 million bbl of liquids remain to be found (Table 1). The latter figure admittedly is conservative.

### Factors Bearing on Global Tectonic Problems

In hypotheses of plate tectonics, the problem of the location of the boundary between the North American and Eurasian «plates» has long been debated. Many persons who know little or nothing about the geology of Alaska and the Soviet Far East draw a «plate» boundary through the Bering Strait (e.g., Le Pichon, 1968). However, the continuity of the geology across the Bering Strait has long been known. Because of this, other hypotheses have been advanced. One of these is to place the boundary in the area of the present Lena Valley, between the Siberian platform and the Verkhoyansk Range. This is impossible because of the continuity of the Paleozoic-Mesozoic geology across the Lena Valley from the Lena-Anabar trough and the Vilyuy basin on the platform, to the Verkhoyansk fold belt east of the platform, a fact long recognized by Soviet geologists (see

Fradkin, 1969 for summary; see also Gusev, 1979). Therefore, Churkin (1972) proposed that the boundary between the two alleged «plates» is the zone between the Mesozoic Verkhoyansk Range and the Paleozoic Cherskiy Range. An unfortunate feature of Churkin's model is that it does not take into account the nearly identical stratigraphy of the Tournaisian and older beds beneath the Vilyuy basin, the Verkhoyansk Range, and the eastern side of the Cherskiy Range (Fig. 31). It stretches credulity somewhat to state that a «collision zone» must have existed between two areas which have nearly identical Tournaisian and older rock formations, climatic indicators (Devonian evaporites, Carboniferous coals), and faunas. The similarities which I list, however, are not absolute proof that some *narrow* (200-300 km) opening once did separate the areas of the present Cherskiy and Verkhoyansk Ranges. This is not a case like that proposed by Hamilton (1970), according to which the East European and Siberian platforms were once far apart and «collided» along the line of the present Urals. Hamilton's hypothesis had been disproved long before his paper was published, when it was realized that the Timan Ridge is an extension of the Baykalides and that the two orogenic belts are continuous *through* the Urals and beneath the West Siberian basin. Maksimov and Dikenshteyn's (1979) more recent study of the Nyurol'ka basin simply shows in detail how wrong Hamilton was. Gurev's (1979) work, as well as that by others before him, suggests that Churkin's (1972) hypotheses may be equally wrong, and for many of the same reasons.

## Lena-Anabar, Khatanga, and Yenisey Troughs

### Introduction

This group of troughs extends northward from the Vilyuy basin to the Arctic Ocean along the Lena River, between the Siberian platform and the Verkhoyansk Range; it turns abruptly westward just south of the Lena delta (Dikenshteyn, 1977) and extends along the Arctic coast to Nordvik, where it continues westward between the Siberian platform and the Taymyr fold belt (Fig. 34), which is a continuation of the Paleozoic fold belt of the Urals. The total area of the troughs is 430,000 sq km.

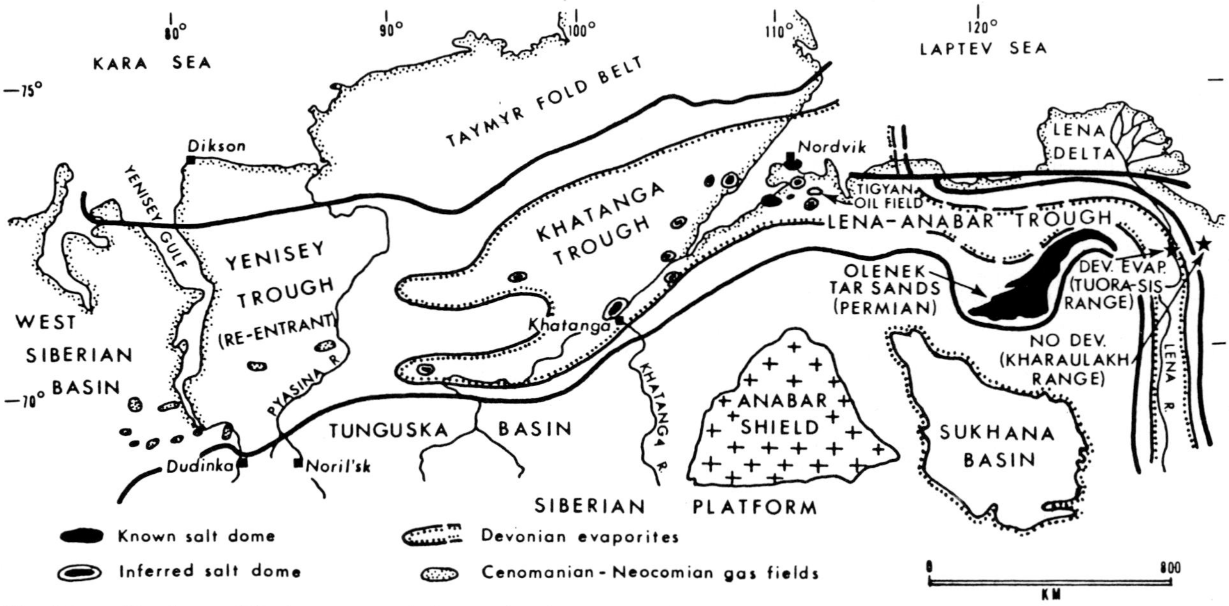


*Fig. 34* — Yenisey, Khatanga, and Lena-Anabar troughs, northern Siberia. Cret.-Jur. gas fields of western part of the Yenisey trough are shown, where the trough opens into the West Siberian basin. Devonian salt domes and evaporite basin are shown. Two of the known domes near Nordvik have produced oil since about 1912, and are shown in solid black. Location of the Permian Olenek tar-sand deposits is shown. Compiled from Vasil'yev (1968b), Vasil'yev and Zhabrev (1975), Dikenshteyn (1977), Gusev (1979), and Yaskevich et al. (1980).

The area was first explored for petroleum early in this century, and by about 1912, shallow cable-tool holes had been drilled on several salt diapirs in the region of Nordvik. Oil had been used locally for several decades from small surface pools fed by seeps. From 1912 until after the October Revolution, additional small quantities of oil were produced from shallow boreholes. Initial flow rates ranged from 3 to 10 bbl/d. In 1948, the small East and West Tigyan fields were found just southeast of Nordvik (Fig. 34). Initial flows ranged from 13 to 80 bbl/d. This oil still is produced for local use, although all of it's heavy oil — in the 11 to 17°API range (Vasil'yev, 1968a).

### Regional Geologic Framework

In the original Soviet literature — in fact until 1980 — the Yenisey and Khatanga troughs were lumped as a single unit, the Yenisey-Khatanga trough, with the dividing line between the two parts of the trough drawn just east of the town of Khatanga (Fig. 34). The trough east of the Nordvik salt-dome area was, and still is, called the Lena-Anabar trough. Recent work, however, has revealed that the Devonian salt and associated evaporites, which underlie all or most of the Lena-Anabar and Khatanga troughs, extend west of the Khatanga trough as originally defined — in fact, almost 900 km farther west to the region of the Pyasina River (Fig. 34; Yaskevich et al., 1980). Thus the Yenisey trough is little more than a re-entrant of the West Siberian basin into the area between the Taymyr Range and the Siberian platform (see Fig. 15). I would suggest that the term «Yenisey trough» be abandoned, and that the term «Khatanga trough» be applied to the belt from the Yenisey River to just east of Nordvik. If subsequent geologic studies demonstrate that the entire trough system from the Vilyuy basin to the Pyasina and/or Yenisey Rivers has essentially the same stratigraphy, a single name, such as «Lena-Khatanga trough,» should be applied, and the separate names abandoned.

### Stratigraphy

The stratigraphy is very similar to that of the Vilyuy basin. The oldest rocks penetrated in the center of the troughs are Devonian carbonates and evaporites. These are seen in the Vilyuy basin (Vasil'yev, 1968b), along the eastern bank of the Lena River in the Tuora-Sis Range (Gusev, 1979; see Fig. 34), and in the Nordvik area of the Khatanga trough (Vasil'yev, 1968a). Lower Carboniferous rocks are present mainly within the troughs, but are absent in some adjacent areas, such as the northern Verkhoyansk Range (Kharaulakh Range: see Fig. 34; Gusev, 1979). Tournaisian limestone, similar to that in the Dzhardzhan Valley farther south (Fig. 31), is present in the Nordvik area (Vasil'yev, 1968a; Gusev, 1979) and in Khatanga trough (Makarova et al., 1979). Upper Carboniferous terrigenous strata, largely marine, are present in some places, but commonly are absent (Fig. 31). The Carboniferous consists mainly of sandstone, siltstone, shale, and carbonaceous shale interspersed with coal seams.

The Permian through Jurassic section, like the Carboniferous rocks, is a thick (2,300 to 3,200 m in the Nordvik area) sequence of continental, deltaic, paralic, and coastal-marine sandstone, siltstone, and shale with much carbonaceous material and some very fossiliferous marine beds. Permian and Triassic sandstones produce heavy oil at the salt domes around Nordvik, most notably at the West and East Tigyan fields (Fig. 34). Porosity is 18 percent, but permeability is less than 20 md. The Olenek tar sands (Fig. 34) west of the mouth of the Lena River are mainly in very porous sandstone beds of Permian age, but the source of the oil is unknown. The Olenek deposits are little studied, but have been estimated to contain about 100 to 120 billion bbl of oil in place, with a recovery potential of 12 billion bbl (Meyerhoff, 1980a).

The Cretaceous forms the cover rocks in most of the area, with part of the Cretaceous removed (or not deposited) in some places. Total thickness of the section to the basement is about 7 km (this average is given by Dikenshteyn, 1977), but approaches 14 to 17 km where the thick West Siberian Jurassic and Cretaceous sections overlie the older section in the Yenisey reentrant at the western end of the Khatanga trough (Dikenshteyn, 1977; see Fig. 15).

### Petroleum Geology

The entire region except the Yenisey reentrant is poorly known and poorly explored. Verkhoyansk mountain-front folds are present in the Lena Valley. Platform and drape structures are found between the Lena River and Nordvik, but some of the folds strike east-west and appear to be compressional folds. Mountainfront folds are known in the pre-Cretaceous section of the Khatanga trough, and are reflected in the overlying Cretaceous beds east of the Yenisey re-entrant. Subthrust plays may exist beneath and at the southern edge of the Taymyr Range, which was thrust southward beginning in latest Triassic time (Dikenshteyn, 1977). A similar subthrust play can be expected along the full length of the Verkhoyansk Range, which was pushed westward against the Siberian platform. The Verkhoyansk Range, according to Dikenshteyn (1977), turns abruptly westward near the Lena delta, and parallels the Laptev Sea coast to Nordvik. Thus, a subthrust play could exist along the Arctic coast. Surface geology and geophysical data show that the entire Arctic coastal zone is faulted upward with respect to the Lena-Anabar trough. The boundary fault is shown on Figure 34.

In addition, deep plays in late Proterozoic through Devonian strata, most of which are carbonates, can be expected. They will be at depths (about 6,000 to 7,000 m in the centers of the troughs) that are too great to be economic, given the capacity of existing Soviet drilling equipment. However, the same Devonian and older strata presumably lie at shallower depths near the margins of the troughs, and stratigraphic pinchouts and reef plays can be expected.

Evaporites in this region are known in beds of Cambrian, Ordovician, and Devonian ages. Thus there are numerous potential planes of décollement, and thrusting of such ranges as the Taymyr and Verkhoyansk may have been extensive, and not limited to a relatively narrow zone along the fronts of these ranges.

The sheer size of the area (430,000 sq km excluding subthrust possibilities) and the volume of sedimentary rocks (2,600,000 cu km minimum) make it attractive for exploration. On Table 1, I have not included the Yenisey-Khatanga and Lena-Anabar troughs in my resource calculations. However, if one uses a recovery of 40,000 bbl/cu km for a thickness of 4 km of sedimentary cover (I exclude the upper 2 km of Jurassic and Cretaceous, because they are not important objectives), then the Yenisey-Khatanga and Lena-Anabar troughs have a resource potential of about 68 billion bbl of oil or gas equivalent. The great depths in much of this area suggest that the region will be gas prone. Drilling in the Vilyuy basin supports this suggestion. However, of the 68 billion-bbl resource, it is possible to deduct 10 to 12 billion in the Olenek tar sands. This leaves an estimated 56 billion bbl of oil or its gas equivalent to be found in strata that range in age from late Proterozoic through Late Triassic.

## Asiatic USSR — Offshore Basins of the Siberian Platform

### *North Kara Sea Basin*

The Kara Sea has an area of 777,000 sq km, and is divided into three geologic provinces (Fig.35). The southern half, about 350,000 sq km, belongs to the West Siberian basin. Deltaic to marine Cretaceous rocks crop out on small islands within and adjacent to the Kara Sea. Seismic data suggest that a 6- to 12-km thick section of Jurassic and younger sedimentary rocks occupies the southern part of the sea.

The West Siberian basin is separated from the North Kara Sea basin by the Novaya Zemlya-Taymyr (Urals) fold belt which was last deformed by compressive stresses at the end of the Triassic (Bolk et al., 1973; Fig. 42). The fact that the Taymyr fold belt is a continuation of the Urals fold belt now is well established.

The North Kara Sea basin has an area of about 350,000 sq km. The geology is poorly known. Bolk et al. (1973; Fig. 42) calculated a maximum depth to magnetic basement of 7 to 7.5 km. This 7-km thickness almost certainly includes some Paleozoic rocks. Gramberg et al. (1967) and Fedynskiy and Levin (1970) wrote that the basin is filled with Devonian carbonate and terrigenous clastic strata that have no petroleum potential. The basis for these statements seems to be Gramberg et al.'s opinion that much of Severnaya Zemlya, along the eastern margin of the basin, is underlain by Devonian rocks with a low organic content (Gramberg et al., 1967), an opinion not shared by Semenovich et al. (1973). Moreover, the numerous outcrops of Lower Cretaceous coal-bearing strata on the small islands in the area do not support the opinions of Gramberg et al. and Fedynskiy and Levin. Instead, it seems more likely that at least 6 km of the section consists mainly of Mesozoic and late Paleozoic strata similar to those in the Lena-Anabar trough.

Semenovich et al. (1973) wrote that the Devonian contains good reservoirs and a fair amount of organic matter; that the Devonian is 1,700 m thick; that Lower Carboniferous may be present; and that Triassic, Jurassic, and Lower Cretaceous more than 2 km thick are present. Thus the section would be Permian, Early Cretaceous. Mesozoic, Paleozoic, and Proterozoic rocks crop out in the Severnaya Zemlya and Taymyr areas, with rocks as young as Triassic being strongly folded in the Taymyr Range. Carboniferous

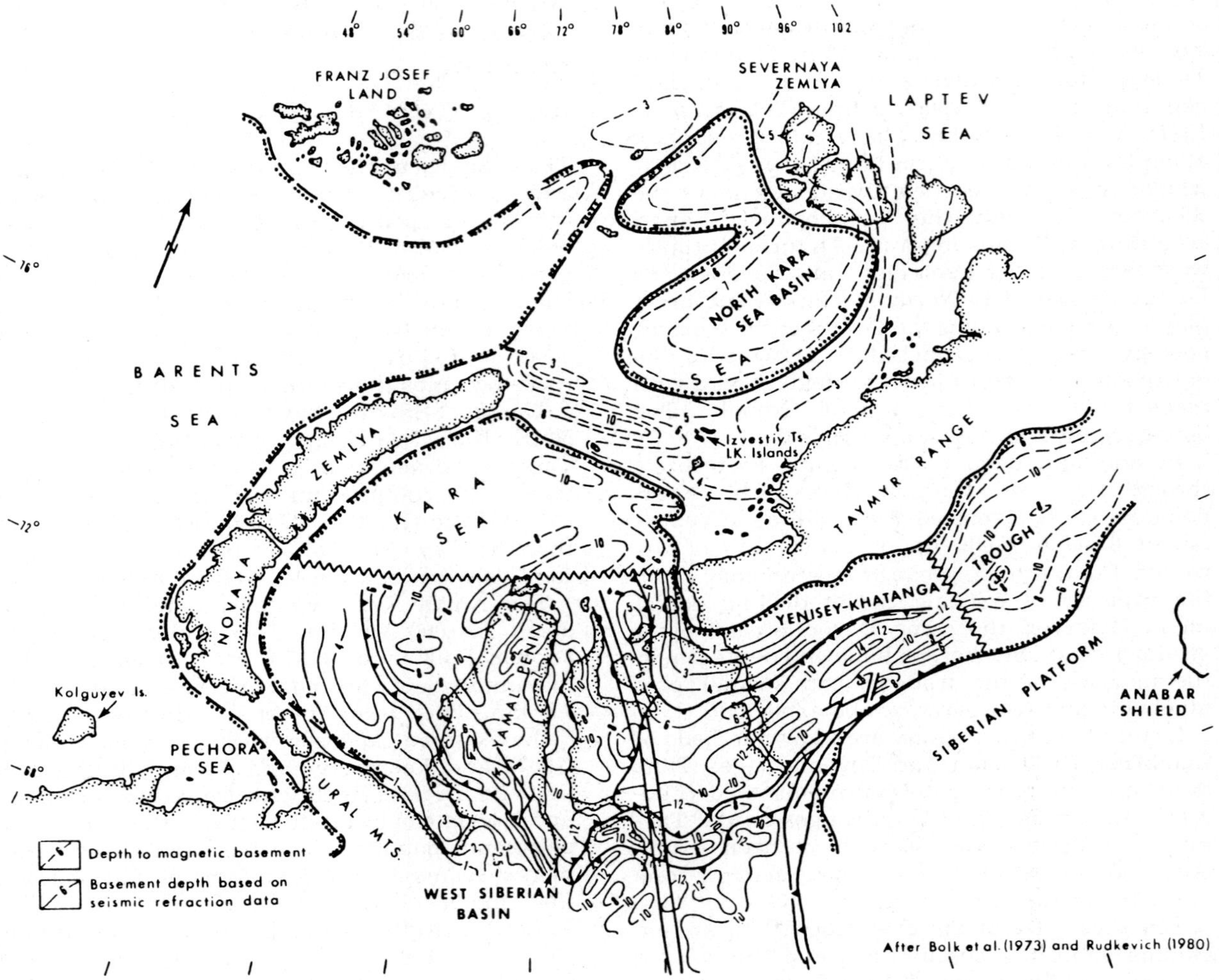


*Fig. 35* — Structural map of the Kara Sea. Southern half is the southern Kara Sea extension of the West Siberian basin, with depths to basement from seismic-refraction and subsurface data; northern half is depth to magnetic basement based on a 1972 air-magnetometer survey. Depths are in kilometers. Seismic data are from Rudkevich (1980); magnetic data are from Bolk et al. (1973).

through Early Cretaceous beds underlie Franz Josef Land on the west. Semenovich et al. (1973) postulated that the stratigraphy of the North Kara Sea basin is closely related to that of the Franz Josef Islands.

In Novaya Zemlya, folded Proterozoic through Paleozoic rocks of the Urals fold belt are present. The Ural trends continue to the Taymyr Range (Fig. 35). Proterozoic through Lower Cretaceous rocks crop out in the numerous islands north and west of the Taymyr Peninsula.

Gravity data indicate that the region of the North Kara Sea basin is underlain by denser rocks than those of the West Siberian basin. Therefore, many workers (e.g., Gramberg et al., 1967) have assumed that the North Kara Sea basin is underlain by Paleozoic rocks. The geological basis for this opinion is the widespread presence on Severnaya Zemlya of Ordovician, Silurian, and Devonian marine strata, mainly carbonates and up to 7,000 m thick (Ageyev, 1979). These same data have been interpreted by

Pol'kin et al. (1971) to indicate that 0.5 to 2.5 km of post-Jurassic strata underlie the Kara Sea, and that these are underlain by Jurassic, Triassic, and Paleozoic. The higher gravity values for the region may be the result of widespread Lower Cretaceous basaltic and doleritic flows and sills. Semenovich et al. (1973) stated that the entire Jurassic through Devonian section is prospective and rated the area very highly. I agree strongly with this assessment, even if Lower Cretaceous igneous rocks are fairly widespread. Much oil and gas are being found beneath igneous traprocks in many parts of the world (Kamen-Kaye and Meyerhoff, 1979).

It is very difficult to estimate hydrocarbon resources in the basin. If an average prospective thickness of 4 km is assumed, the sediment volume is about 1.4 million cu km. If a recovery of 40,000 bbl per cu km is assumed, then the resources below a depth of 2 km could be as high as 28 billion bbl or the gas equivalent. However, Semenovich et al. (1973) believe that the Lower Cretaceous contains numerous mafic sills and flows, which could have reduced the resource potential. Therefore, I have used a conservative estimate of ultimate recovery of 1 billion bbl of liquids and 50 Tcf of gas. The remaining 1 billion bbl of liquids and 450 Tcf of gas (Table 1) is believed to be in the West Siberian basin part of the Kara Sea.

### *Laptev Sea Basin*

This basin includes an area of 414,000 sq km, of which 250,000 sq km is in the south part, according to Grigorenko et al. (1975), is prospective (Fig. 36). Geophysical data indicate that the prospective area contains about 1 million cu km of strata with an average thickness of 4 km. The basin is bound by the Taymyr fold belt on the west, by the Lena-Anabar trough on the south, by the Verkhoyansk fold belt in the southeast, and by the New Siberia (Islands)-Chukotsk fold belt on the east.

The basin consists of two parts: a northern platform area underlain by about 4.5 to 6 km of block-faulted middle Proterozoic through middle Paleozoic section that consists principally of carbonate rocks (Semenovich et al., 1973; Grigorenko et al., 1975). The numerous faults make the area rather unattractive for exploration.

The southern part of the basin is the east-west-striking South Laptev trough or downwarp, in which geophysical measurements offshore and surface mapping onshore indicate that 2 to 3 km of Permian and Triassic and up to 3 km of Jurassic and Cretaceous may be present. Semenovich et al. (1973) stated that the stratigraphy (based on mapping of the southern edge of the basin where it crops out in the onshore area) is nearly the same as that in the Lena-Anabar trough farther south (Fig. 36). Thus the 250,000 sq km of the South Laptev trough and its flanks appear to be very prospective. The Upper Cretaceous-Tertiary section — largely continental, with some lava flows and marine beds — which forms a thin veneer of unconsolidated sediments above all of the older section is not believed to be prospective, partly because it is too shallow for hydrocarbons to have matured.

On Table 1, I have assigned 1 billion bbl of liquids and 50 Tcf of gas to this basin. These numbers are very arbitrary in view of the very small amount of knowledge available. I have considered mainly the South Laptev trough in my estimate. The numbers are conservative, if one considers that at least 970,000 cu km of sedimentary rocks is present in the South Laptev trough (Grigorenko et al., 1975).

## Asiatic USSR — Offshore Basins of the Pacific Mobile Belt

### *Stolbovoy Basin*

This basin occupies the southeastern part of the Laptev Sea, and its area is included in that of the Laptev Sea, given in the section on the «Laptev Sea basin.» The basin has been considered to be a part, or extension, of the South Laptev trough. Semenovich et al. (1973) wrote that the Stolbovoy basin and the Oloy basins (Fig. 36) are one and the same. The basin was called the Laptev-Yansk basin by Ivanov (1981). Meyerhoff (1980a) included the Stolbovoy basin in the Laptev Sea basin. Sevost'yanov (1971) included it in the South Laptev basin.

According to Ivanov (1981), the «basin» actually is a complex of semi-isolated grabens and depressions. Between 2 and 3 km of terrigenous Upper Jurassic and Lower Cretaceous marine to paralic sediments are exposed on Stolbovoy Island (Semenovich et al., 1973; Fig. 36). These are buried offshore by 0.5 to 1 km of younger rocks, presumably Upper Cretaceous sediments and volcanic flows and sills, and by Paleogene continental strata with some volcanic rocks. Both the Upper Cretaceous and Paleogene strata contain coal beds. Semenovich et al. (1973) wrote that oil and gas possibilities are best in the pre-Upper Cretaceous section (but above the Carboniferous and older basement); Ivanov (1981) stated that the best prospective section would be the Upper Cretaceous and Paleogene, in which gas generated from the associated coal measures could be anticipated. Surface outcrops show some zones of good porosity in the Upper Jurassic, Cretaceous, and Paleogene. However, no exploratory drilling has been conducted.

I have assigned no resource potential to the onshore section. My offshore potential is included in the figures assigned on Table 1 to the Laptev Sea. However, I do not rate the offshore Stolbovoy basin much higher than the onshore part. Oil seeps are present in the Paleozoic and Triassic sections of the nearby Kotel'nyy Island (Semenovich et al., 1973; Fig. 36).

## *North Laptev Basin*

The areal dimensions of this basin (Fig. 36) are included in the figure given for the Laptev Sea as a whole. This basin appears to be a part of Grigorenko et al.'s (1975) «North downwarp,» a part of their Novosibirsk basin which included the North Laptev basin, the Faddeyevskiy basin, the Oloy basin, and perhaps the western parts of the Northern East Siberian and Southern East Siberian basins (Fig.36). In any case, it is separated from the basins of the East Siberian Sea by the De Long arch, which strikes generally north-south through the New Siberia Islands. Grigorenko et al. (1975) estimated that the section includes a pre-Devonian basement, some upper Paleozoic objectives, an Upper Jurassic Lower Cretaceous prospective section, and a sequence of lava-bearing Upper Cretaceous and Cenozoic. Thus the section, as interpreted by Grigorenko et al., is similar to those of the Stolbovoy basin and the Laptev Sea area — especially the northern platform. Estimates of petroleum potential are very low, and are included in the overall figures for the Laptev Sea, as shown on Table 1. Oil seeps are present in the Paleozoic and Triassic section of the nearby Kotel'nyy Island (Semenovich et al., 1973).

## *Oloy Basin*

This basin is east of the De Long arch, and is about 40-45 percent onshore and 55-60 percent offshore in the East Siberian Sea. The East Siberian Sea, with an area of 968,000 sq km, contains other basins in addition to the Oloy. These are the Faddeyevskiy, Northern East Siberian, and Southern East Siberian basins. The resource potential of this area is considered conservatively to be 10 billion bbl of oil and condensate and 50 Tcf of gas (Table 1). I have not assigned specific estimates to each basin, although I consider the Northern East Siberian and Southern East Siberian basins to have the greatest potential.

The Oloy basin is part of the Novosibirsk basin of Grigorenko et al. (1975); it is the Novosibirsk basin of Ivanov (1981); and it is the western half of the Southern East Siberian basin of Sevost'yanov (1971). Meyerhoff (1980a), included it under the general term, «East Siberian Sea basin.»

According to Sevost'yanov (1971) and Semenovich et al. (1973), the basin is similar structurally and stratigraphically to the Stolbovoy basin, and is broken up by numerous horsts, grabens, depressions, and isolated faults. The primary objective, according to Semenovich et al. (1973), is the Upper Jurassic-Lower Cretaceous marine-terrigenous sequence. Ivanov (1981), however, stated that gas in the Upper Cretaceous-Paleogene coal measures associated with continental and volcanic rocks is the most important objective. Semenovich et al. (1973) observed that the section is thicker than that in the Stolbovoy basin, and that more surface manifestations of oil and gas are evident. They therefore regarded the basin as more prospective than the Stolbovoy.

## *Faddeyevskiy Basin*

This basin, named by Sevost'yanov (1971; Fig. 36), is just east of the New Siberia Islands (northern part). The existence of the basin was determined by surface geology and some geophysical data (Galponenko, 1971). It was included under the general name of «East Siberian Sea basin» by Meyerhoff (1980a). On the basis of surface geology, Sevost'yanov (1971) suggested that possible objectives include up to 2,000 m of Ordovician sandstone, shale, and limestone, about 1,000 m of Silurian and Devonian carbonates, and up to 1,200 m of terrigenous marine Triassic strata. The sequence is overlain by a thin section of Late Cretaceous, Paleogene, and Neogene continental and volcanic beds.

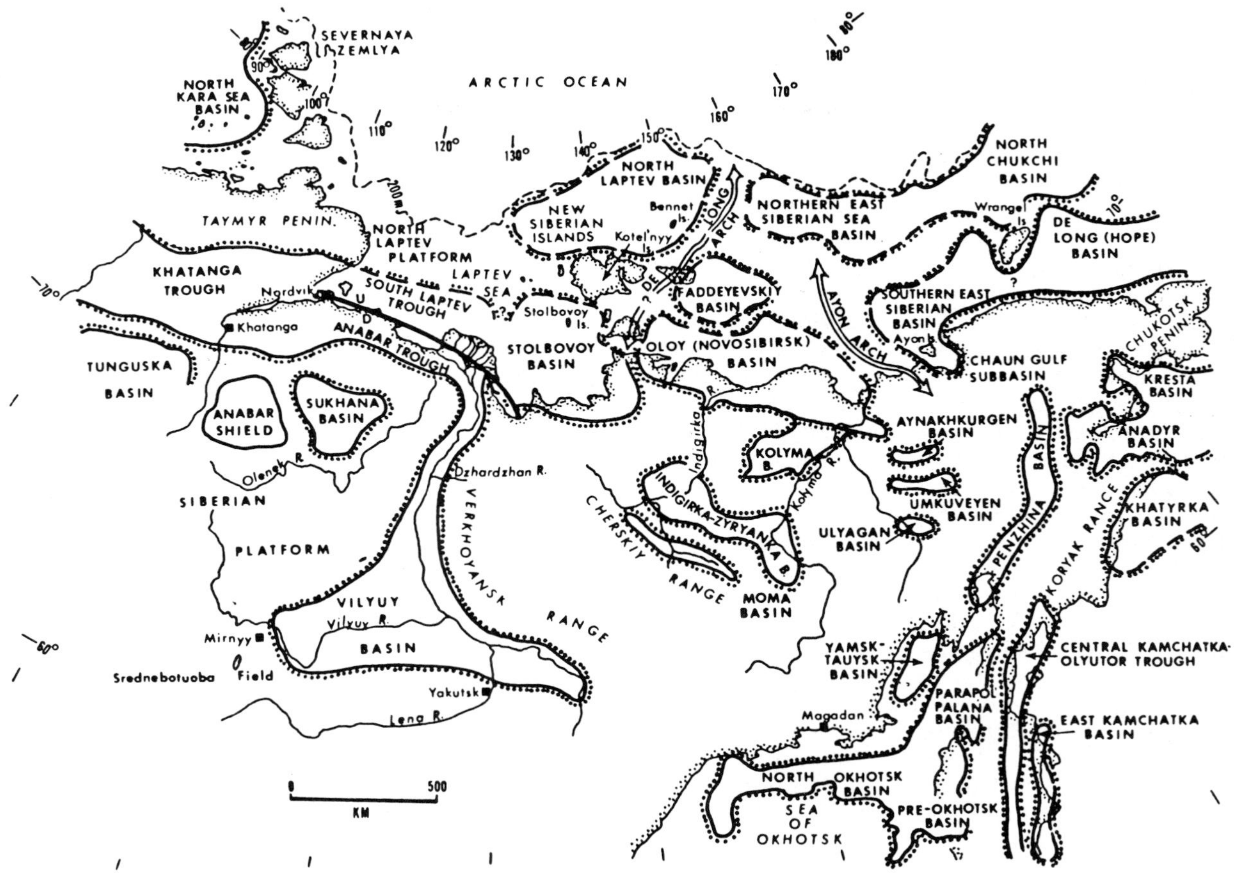


*Fig. 36* — Index map to the locations of 32 basins in the northeastern USSR. Outlines of the basins are based on geophysical information offshore, with some use of bathymetry. Onshore basins are based on geologic and geophysical studies. Only the Khatanga part of the Yenisey-Khatanga basin (Ust'-Yenisey basin) is shown. Soviet opinions of basin outlines and names differ significantly in the off- and onshore regions east of the Kara Sea. It is possible that the South Laptev trough and the Anabar trough originally were one basin which now is divided into two parts by a major fault zone along the northern coast. The South Laptev trough and the Stolbovoy basin may be one and the same. Semenovich et al. (1973) wrote that the Stolbovoy and Oloy (Novosibirsk) basins were a single basin. Grigorenko et al. (1975) included the North Laptev, Faddeyevskiy, Oloy (Novosibirsk), and apparently parts of the Northern East Siberian and Southern East Siberian basins under the term Novosibirsk basin. The Stolbovoy basin was termed the Laptev-Yansk basin by Ivanov (1981). The Indigirka-Zyryanka basin of Ivanov (1981) is the Moma-Zyryanka basin of Semenovich et al. (1973). Sevost'yanov (1971) included the North Laptev basin and the North Laptev shelf as a single basin, whereas Semenovich et al. (1973) and Grigorenko et al. (1975) treated the North Laptev shelf and South Laptev trough as one basin unit. The confusion, as can be seen, is nearly endless. Sources used to compile this map were Trofimuk (1971), Sevost'yanov (1971), Bolk et al. (1973), Semenovich et al. (1973), Grantz et al. (1975), Grigorenko et al. (1975), Vlasov and Tarasenko (1977), Mavrinskiy et al. (1980), Ivanov (1981), and Meyerhoff (1980a).

## *Northern East Siberian Basin*

### *(North Wrangel Basin of Eremenko et al., 1973)*

This basin also was named by Sevost'yanov (1971; Fig. 36). It is just east of Bennet Island, the northeasternmost of the New Siberia Islands group. The basin, although its existence is unproved by detailed geophysical data, is on strike with and may extend into the North Chukchi Sea basin of Grantz et al. (1975). A Cambrian-Ordovician and Middle Triassic through Cretaceous gray to dark gray carbonate and dark-gray to black terrigenous-clastic marine sequence is present on Bennet Island, and was assumed by Sevost'yanov (1971) to underlie the basin and to be a principal part of the objective section. However, Bennet Island is in the North Laptev basin, not the Northern East Siberian basin. Moreover, Grantz et al. (1975) presented strong evidence that the North Chukchi basin, on strike with the Northern East Siberian basin, is underlain by up to 5 km or more of Late Cretaceous through Neogene terrigenous-clastic strata intruded locally by shale (?) diapirs. If the two basins do join north of Wrangel Island (Fig. 36),one must consider the possibilities that (1) the stratigraphic interpretations (Sevost'yanov, 1971; Grantz et al., 1975) of the basins are wrong, or (2) the North Chukchi Sea basin is to the Northern East Siberian basin what the Yenisey trough (reentrant) is to the Khatanga trough (Fig. 34). In the latter case, the older (Devonian-Triassic) rocks of the Khatanga trough plunge westward beneath the thick Jurassic and Cretaceous rocks of the Yenisey trough (Fig. 35). A third explanation is that the two basins are separate and unrelated, but this does not seem to be too logical an explanation if one considers the location of the two troughs with respect to the Arctic Ocean shelf edge and one another. On the whole, it is more likely that the exposures on Bennet Island are unrelated to the objective section of the Northern East Siberian basin (except, possibly, at great depth), but instead are related to the objectives below the North Laptev basin. Thus, Grantz et al.'s (1975) opinion that the North Chukchi basin is underlain by 5 km or more of Late Cretaceous through Tertiary marine and associated strata most probably is true also for the Northern East Siberian basin.

## *Southern East Siberian Basin*

### *(South Wrangel Basin of Eremenko et al., 1973)*

What I call the Southern East Siberian basin is only the eastern part of Sevost'yanov's (1971) basin of the same name. It is included in Meyerhoff's (1980a) general term, «East Siberian Sea basin.» It includes the Chaun (Rauchuan) subbasin (Fig. 36). Trofimuk (1971) divided the basin into two parts — the Rauchuan and the De Long basin, with 52,000 sq km in the Rauchuan subbasin[1]

The Ayon arch borders the basin on the west, as I have shown on Figure 36. However, Ayon Island is in the northwestern part of the Chaun Gulf, and therefore either I have shown the position of the arch incorrectly, or it was poorly named. On the east, the basin continues into the De Long (Hope) basin (Galponenko et al., 1971). At least 3 to 4 km of Upper Jurassic through Tertiary section is present (Semenovich et al., 1973). The overall stratigraphy is believed to be similar to that of the Alaska North Slope (Grigorenko et al., 1975). There are organic-rich rocks with indications of petroleum in the Lower Cretaceous of Wrangel Island (Semenovich et al., 1973). The basin is considered to have good prospects. However, one well drilled onshore in the Chaun subbasin penetrated 700 m of Tertiary continental strata which overlies economic basement directly (Ivanov, 1981).

# **Basins of the Onshore Northeast USSR (Pacific Mobile Belt)**

## *Indigirka-Zyryanka and Moma Basins*

These basins originally were a single basin (Moma-Zyryanka basin of Semenovich et al., 1973). During Late Cretaceous and Tertiary deformation, the Il'in-Tasskiy arch was formed, and the original basin was divided into two parts as shown on Figure 36. The combined area of the two basins is about 115,000 sq km. Semenovich et al. (1973) regarded the prospects of both basins as «very promising.» Upper Jurassic

[1] The spelling of the Chaun subbasin or basin has been confused by many authors in the Soviet Union and elsewhere. The confusion is caused by the fact that the spelling of the gulf is «Chaun,» whereas that of the river just west of the gulf is «Rauchuan» (Trofimuk, 1971, p. 119).

marine terrigenous strata are particularly prospective; marine Lower Cretaceous strata are mainly shale which forms a seal for both of the basins.

Ivanov (1981) disagreed with the opinion of Semenovich et al. (1973). He regarded the sandstones interbedded with Aptian and Albian coal-bearing beds as the most prospective part of the section. In the southeastern parts of both basins, Ivanov stated that the organic-rich Carboniferous and Devonian marine clastic sediments are prospective. These Paleozoic strata are absent in the northwest where Callovian and younger rocks overlie economic basement directly. The unmetamorphosed section here is only 4.8 km thick (composite thickness), whereas the Devonian and younger section of the southeastern parts of the basins is 9 to 10 km thick (Ivanov, 1981). In the Moma basin, the Lower Cretaceous through Quaternary thickness is 3 to 4 km. All sediments are of miogeosynclinal or shallow-water type, with some continental strata.

### *Kolyma Basin*

The existence of this basin (Fig. 36) has been questioned by numerous workers, among them Ivanov (1981). Trofimuk (1971) wrote that the basin has an area of 260,000 sq km, but noted that it is badly broken up by faults. Dips generally are in the 10 to 20° range with local dips of 30° or more. Sevost'yanov (1971) described the rocks as a platform facies containing up to 2,000 m of Permian carbonate rocks; up to 250 m of Upper Triassic sandstone, shale, and limestone; perhaps 1,500 to 1,600 m of Lower and Middle Jurassic marine and continental terrigenous strata; 500 to 600 m of marine Upper Jurassic volcanic and associated rocks; and up to 6,000 m of Lower Cretaceous coal-bearing (including very thick Aptian-Albian) sections. The prospectivity of the basin is very low.

### *Aynakhkurgen, Umkuveyen, and Ulyagan Basins*

These basins (Fig. 36) have a total area of about 35,000 sq km. Up to 3 to 4 km of Triassic through Lower Cretaceous is present. Many of the rocks are intermediate-composition lava and associated tuff. Marine Triassic and Jurassic strata are present. The Lower Cretaceous consists of coal-bearing continental beds, whereas the Upper Cretaceous is 300-400 m of silicic lava and tuff. The Paleogene, where present, generally is basaltic (Trofimuk, 1971). Sevost'yanov (1971) wrote that the Albian sandstones could contain gas generated from the interbedded coals. No worker in the region appears to be wildly enthusiastic about the prospects of any one of these three basins.

### *Basins of the Amur and Ussuri Rivers Area*

These basins have been described by Trofimuk (1971) and Meyerhoff (1981; see Fig. 37). They include the Vladivostok shelf and Suyfun, Sikhote-Alin, Suchan-Daubikhe, Verkhne-Zeya, Ushumun, Zeya-Bureya, Verkhne-Bureya, Sredne-Amur, Bikin, and Khankay basins. The prospectivity of these basins is very low. Moreover, they are not truly arctic, in that all are generally south of 55°N lat and are removed from the permafrost zone. I therefore begin with the coastal and offshore basins of Sakhalin Island — at the same latitude — where tundra, taiga, permafrost, icebergs, and severe blizzards pose serious problems more akin to those of truly Arctic regions.

## Okhotsk and Japan Sea Basins (Pacific Mobile Belt)

### *East Sakhalin Basin*

#### Introduction

The East Sakhalin basin is the only commercial petroleum productive basin in the Soviet Far East. It contains about 84,000 sq km, of which 7,500 is onshore (Gnibidenko and Khvedchuk, 1983; Fig. 38). Although production is minuscule (42,000 bbl/d and 767,000 Mcf/d) in contrast to other USSR basins, this modest production is extremely important in plans for expanding the growth of the mainland industrial centers of the Amur and Ussuri Rivers. Intensive exploration with foreign help is underway to enlarge the petroleum-reserves base, particularly in the offshore.

The petroleum potential of the region was recognized early in the exploration of Tsarist Russia because of the large numbers of seeps, small asphalt lakes, and large surface areas

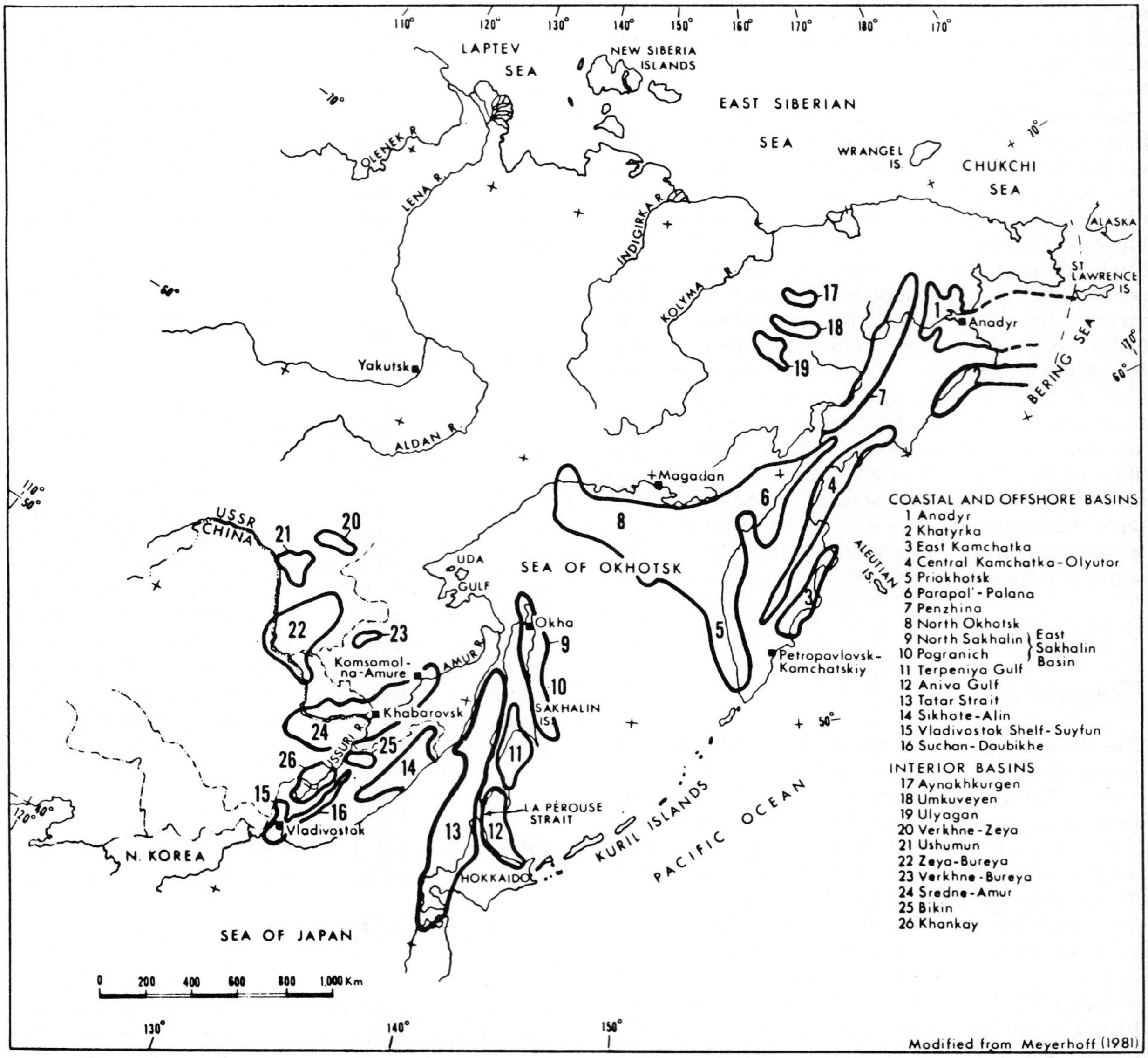


*Fig. 37* — Index map to petroleum-bearing and potential petroleum basins of the Soviet Far East. Modified from Meyerhoff (1981) and Gnibidenko and Khvedchuk (1983).

covered by tar mats. Studies directed toward exploiting the petroleum potential began as early as 1859. The first drilling was by a Russian private firm in 1889, and a small noncommercial find was made close to Okha (Fig.38). In 1897, a Chinese company drilled unsuccessfuly in the same area. Drilling continued intermittently on the many surface anticlines, but a commercial discovery was not made until the Japanese discovered Okha field in 1921. Okha turned out to be the largest onshore field with an estimated ultimate recovery of 150 to 200 million bbl. A concession was granted by the Soviet government to Sinclair in 1921, but the Japanese (who operated the Sakhalin fields) refused to let Sinclair geologists explore the area (Owen, 1975). Japan, in one of the strangest business partnerships in petroleum history, developed Sakhalin's

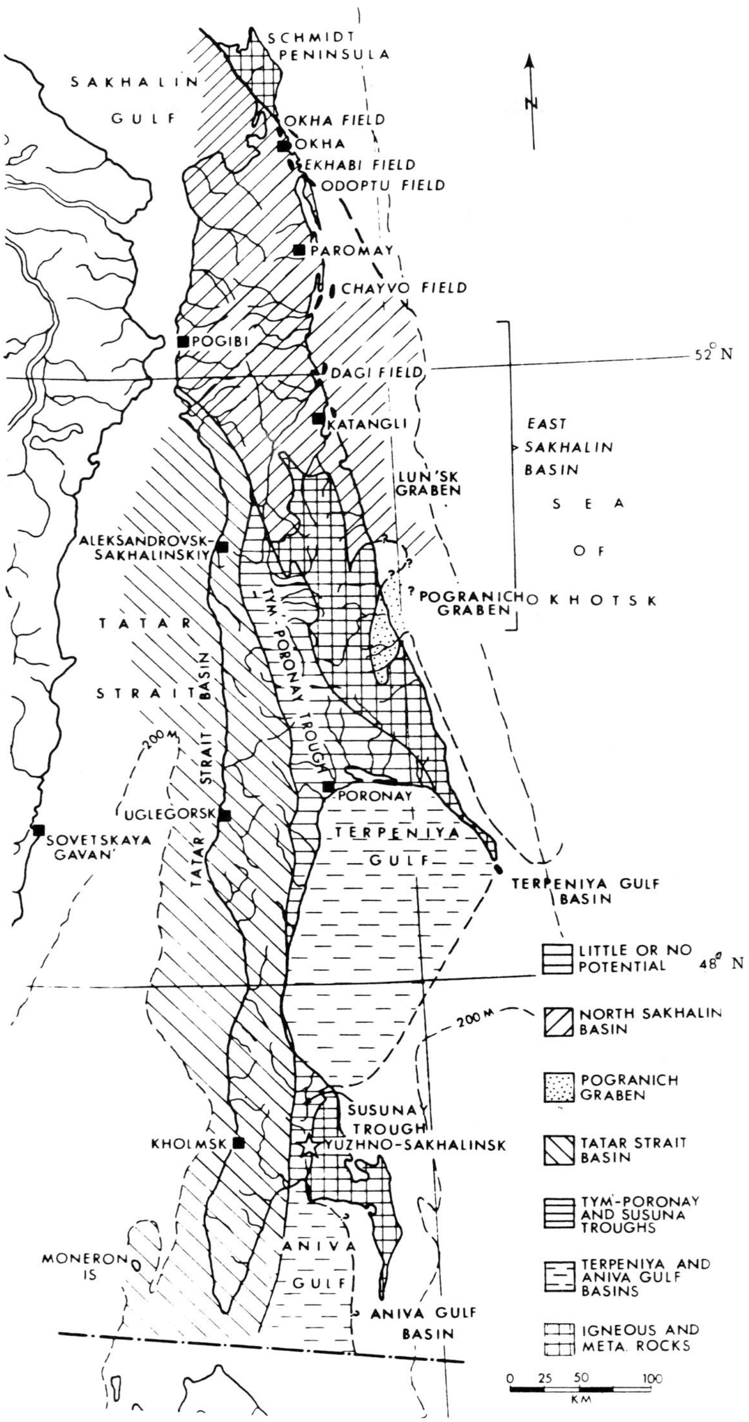


*Fig. 38* — Index map to Tertiary basins of Sakhalin **Island**. Oil and gas fields that are mentioned in the text are shown. From Meyerhoff (1981); offshore fields are from Nikolayev and Kleshchev (1981).

oil which it used to fight the Allies — and in part for use by the Soviets against the Germans — during World War II. The Red Army brought this partnership to a halt when it entered the Pacific war in August 1945. In 1971, the USSR invited Japanese and American companies to participate in offshore exploration, and in 1977, the first joint Japanese Soviet well was drilled offshore from Odoptu field (Fig. 38). Since 1977, the offshore part of Odoptu field has been partly developed, and two additional discoveries have been made, only one of which is commercial (Chayvo field; Fig. 38).

**Regional Geologic Framework**

Sakhalin Island is an extinct part of the Japan island-arc system, and is a northward extension of the geology of Hokkaido. The oldest dated fossiliferous rocks are Permian (Vereshchagin and Kovtunovich, 1970); these overlie older, more metamorphosed rocks. From an economic point of view, all pre-Campanian rocks are basement. Even pre-middle Eocene rocks, in most cases, are not prospective. In the East Sakhalin basin, Miocene generally overlies Cretaceous volcanic rocks, but in some areas (Schmidt Peninsula, Pogranich graben), Oligocene overlies the Cretaceous volcanics.

After major compression and tectogenesis during early and middle Eocene times, vertical movements predominated. These were followed by at least three more episodes of lateral compression — the late Oligocene to early Miocene Kuril tectogenesis, the middle to late Miocene Aleutian tectogenesis, and the late Pliocene and Pleistocene Sakhalin tectogenesis. During the late Pliocene, thrusting was from west to east. During Pleistocene time, thrusting ceased and vertical faulting commenced. Although the Pliocene folding and thrusting phase produced the major producing structures, the vertical faults which formed late during the Sakhalin tectogenesis allowed much of the oil and gas that accumulated to escape. The tectonic scheme of the island is shown on Figure 39, and the numerous folds of the East Sakhalin basin are indicated.

**Stratigraphy**

During Oligocene time, a gradual transgression of the basin began in both the northeast (Schmidt Peninsula) and southeast (Pogranich graben; Fig. 38). The sediment provenance area was the present mainland, and the sediments were carried and deposited by a paleo-Amur River system. In fact, the entire East Sakhalin basin may be regarded as the Oligocene through Pleistocene Amur delta, with deltaic, paralic, littoral, and nearshore neritic sequences of terrigenous sedimentary rocks. Offshore marine bars striking north-south characterize some eastern fields (Gololobov et al., 1981). The separation of Sakhalin from the mainland is very recent, and probably took place during the Sakhalin tectogenesis.

The Oligocene of the Schmidt Peninsula consists of a basal 400 m of coal bearing, gray, medium-grained sandstone, gray to dark-gray siltstone, dark carbonaceous shale, and conglomerate (Gladenkov, 1973). The upper 515 m consists of gray, well-bedded siltstone, mudstone, and sandstone. Tuff and claystone partings are numerous. The upper strata are more marine than the lower. In the Pogranich graben area, the Oligocene consists of several hundred meters of poorly sorted marine sandstone and siltstone with shale, tuff, and a basal conglomerate (Mitrofanova, 1978; Savitskiy et al., 1979).

The Miocene and Pliocene contain the principal source and reservoir rocks in the basin. The hydrocarbons are mainly in the upper Miocene Okobykay Suite (90 percent), but some are present in the middle to upper Miocene Dagi Suite and the lower Pliocene Nutovo Suite (Fig.41).

In general, the section ranges in thickness from 5,000 to 7,000 m (Alekseychik et al., 1963; Rusakov et al., 1967). The main upper Miocene (Okobykay) productive zone is 800 to 2,500 m thick. The unit is largely sandstone in the west, with a large volume of shale, some quite organic-rich, in the east. This fact is reflected in the productivity of the folds: the western area is largely barren of hydrocarbons (no seals are present), and the folds are tight, more compressed and severely faulted. Eastward the folds become more gentle and, where shale forms a large percentage of the section (and good source beds), the folds are productive. Productivity and the number of reservoirs increase eastward.

The sandstone is coarsest in the west and finest in the east. The color ranges from very light gray to medium gray and yellow gray. Shales and siltstone are various shades of gray. Coal appears in the basal part of the middle Miocene. The fauna is almost entirely benthonic in the center and east. The section is largely continental in the west. A Holocene cover, locally up to 1,000 m thick, is present in many areas.

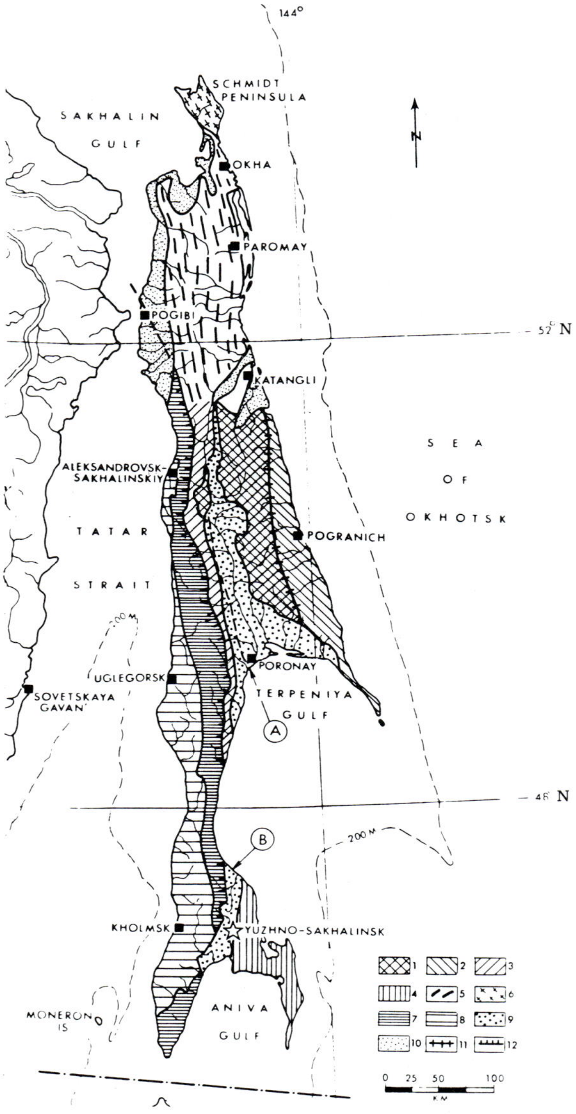


*Fig. 39* — Tectonic outline of Sakhalin Island.
1 = Sakhalin horst massif;
2 = East Sakhalin block-fault zone;
3 = Central Sakhalin graben or trough;
4 = South Sakhalin block-fault zone;
5 = East Sakhalin basin and fold zone;
6 = Schmidt Peninsula block-fault and fold zone;
7 = eastern margin of Tatar Strait basin (West Sakhalin Mountains), bounded on the east by the West Sakhalin thrust zone;
8 = axial part of Tatar Strait basin;
9 = superimposed post-early Eocene troughs;
10 = Quaternary depressions;
11 = East Sakhalin fault zone;
12 = West Sakhalin thrust zone.
From Pushcharovskiy (1964). Republished with permission of Scientific Press Ltd.

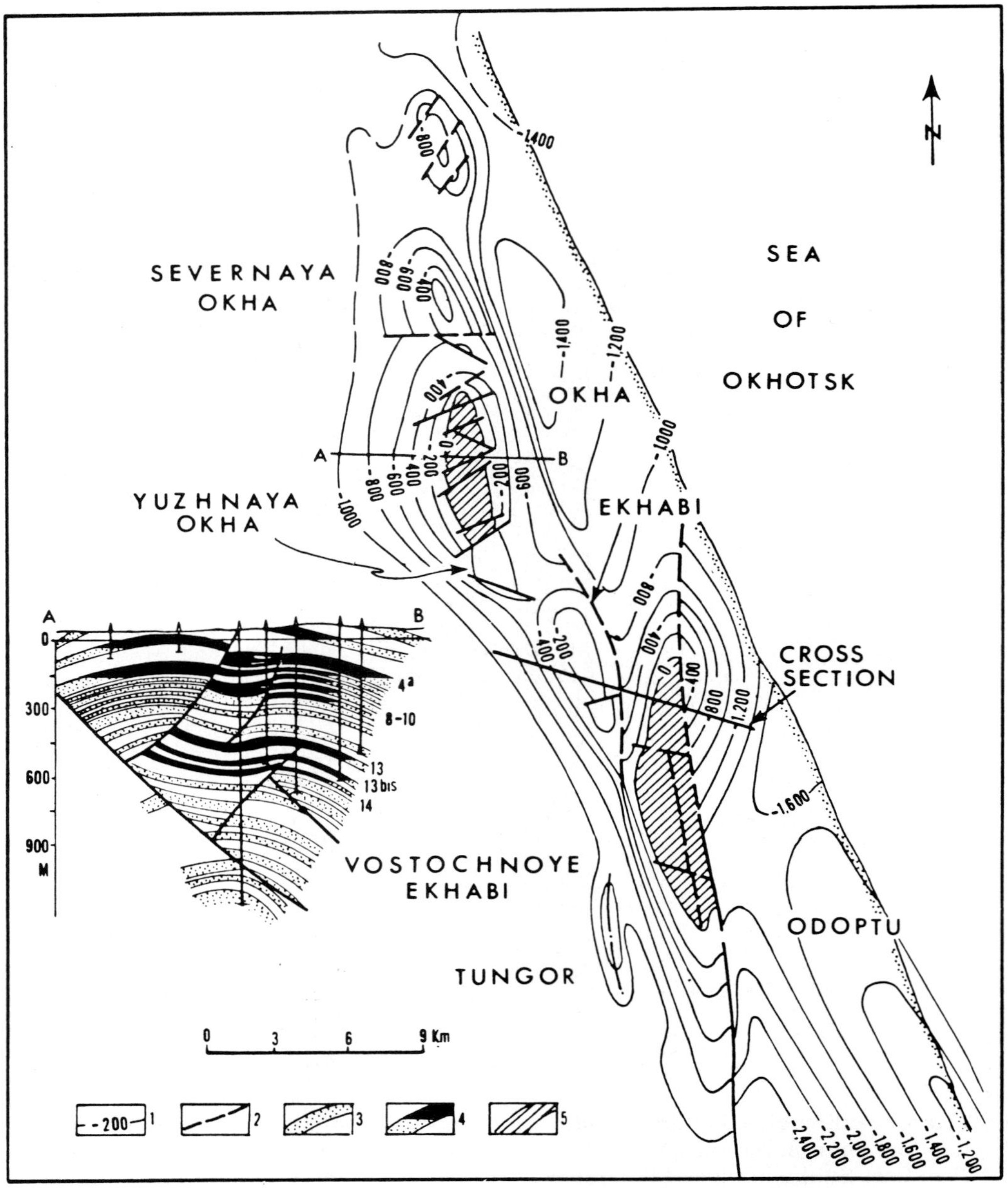


*Fig. 40* — Ekhabi anticlinal zone, with locations of Okha, Severnaya-Okha, Yuzhnaya-Okha, Ekhabi, Vostochno-Ekhabi, Tungor, and Odoptu fields. Structural datum is the top of the upper Miocene. Depths are in meters. Cross section is of the Ekhabi field.
1 = structural contours;
2 = faults;
3 = sandstone;
4 = oil-bearing reservoir;
5 = upper Miocene outcrops.
Horizontal and vertical scales are the same. Shows location of Fig. 41. From Vereshchagin and Kovtunovich (1974). Republished with permission of Scientific Press Ltd.

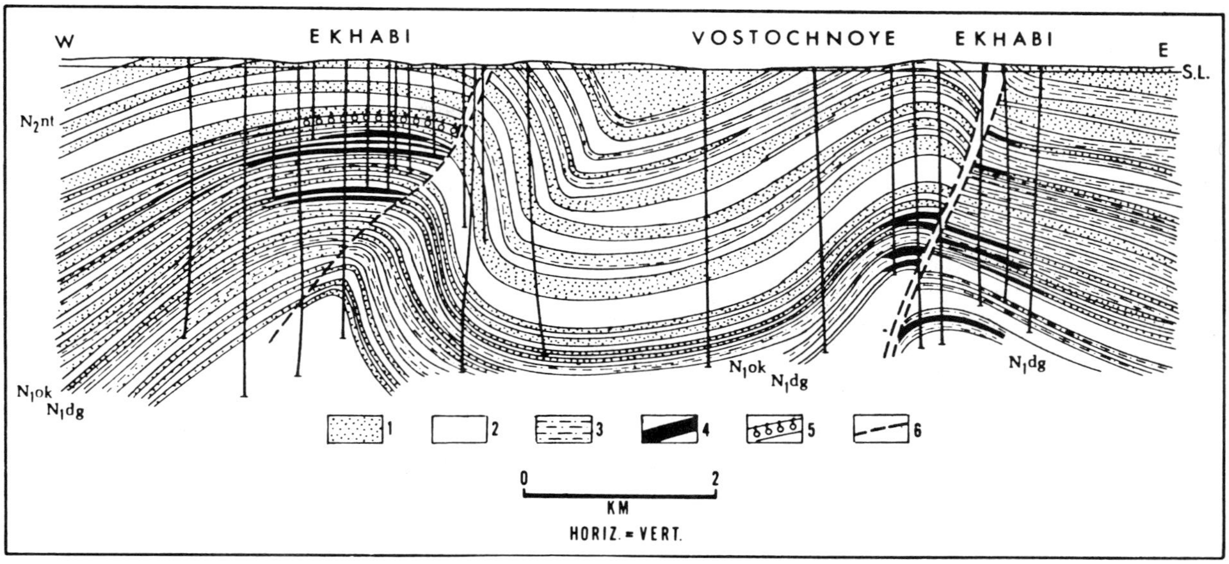


*Fig. 41* — West-east cross section through Ekhabi and Vostochno-Ekhabi fields. Location is on Fig. 40.
1 = sandstone;
2 = shale;
3 = siltstone;
4 = oil;
5 = gas;
6 = faults. $N_2nt$ is lower Pliocene Nutovo Suite;
$N_1ok$ is upper Miocene Okobykay Suite;
$N_1dg$ is middle and upper Miocene Dagi Suite. From Vereshchagin and Kovtunovich (1974). Republished with permission of Scientific Press Ltd.

**Petroleum Geology**

By the end of 1981, slightly more than 50 fields had been discovered with a total of approximately 34 productive zones. Two fields — Chayvo and Dagi — are offshore, and were discovered jointly by Japan and the USSR. Odoptu field, discovered in 1953, strikes offshore (Figs. 38, 40). Part of the offshore portion of the structure was developed by slant drilling that commenced in 1970 (Budnikov, 1975). Offshore drilling (with Japanese participation) from a mobile platform commenced in 1977. Several excellent wells with IPs up to 3,733 bbl/d of oil and 21,000 Mcf/d of gas have been found. (Dagi is now listed as noncommercial.)

All production in the basin is from fine- to medium-grained sandstone, with the best production in the east and the poorest in the west. The reason for this is that shale seals pinch out in a westward direction. Production ranges in depth from less than 20 m to 2,845 m. Source rocks are the marine shales that intertongue with the productive section. Nearly 88 percent of all reserves and 90 percent of the production is from the upper Miocene Okobykay Suite. The remaining reserves, 12 percent, are divided almost equally between the middle to upper Miocene Dagi Suite and the lower Pliocene Nutovo Suite. Although most traps are structural, many are wholly stratigraphic or a combination (Nikolayev and Kleshchev, 1981).

The largest fields (e.g., Okha) found onshore will produce no more than 150 to 200 million bbl. Offshore fields could turn out to be larger, because of greater volumes of source and sealing materials (Budnikov, 1975). Moreover, Pleistocene faulting caused much of the onshore oil and gas to escape, whereas orogenic movements were less intense onshore. Many of the

Miocene faults are growth faults (Alekseychik et al., 1963), but growth generally ceased before the end of Miocene time (Koval'chik et al., 1981). The onshore reservoirs are greatly undersaturated (less than 10 to 25 percent full, in many cases). The offshore fields probably will contain larger volumes of oil and gas in the reservoirs. A major problem is that the eastward extent of the sandstones is unknown. There is some evidence to suggest that they pinch out rather abruptly in some areas. However, reservoir quality is good, both onshore and off, with porosity values of 14 to 34 percent (values of 26-32 percent are common), and permeability values range from 90 to 1,500 md. Oil quality is best in the east, with density of 0.81, low paraffin, and low sulfur. Figure 40 is a structural-contour map of the Okha-Odoptu area (Fig. 39); Figure 41 is a structural cross section, and shows the typical structure style.

Proved reserves in mid-1974 were 87.6 million bbl of oil and 3.6 Tcf of gas. The oil was being depleted at a rate of 16 to 17 million bbl per year, with only minor gas production. Because of the huge cost of transporting oil to the Amur and Ussuri industrial centers by rail from West Siberia, Sakhalin has been explored more intensively. Gas, once a nuisance, is now being utilized more extensively. Reserves are now on the order of 24 million bbl of oil and condensate and 2.0 Tcf of gas. Resource potential is 1 billion bbl of liquids and 5 Tcf of gas. It will be several years before offshore production becomes significant. Although oil has been found in the structures drilled offshore, the largest offshore volumes found to date are gas and condensate at Chayvo field.

**Terpeniya Gulf Basin**

This basin has an area of 13,500 sq km (Fig 38). It and its onshore extensions, the Susunay trough on the south and the Tym'-Poronay trough on the north, are bounded on the west by the Central Sakhalin thrust zone (Fig. 39). The major thrusting, from west toward the east, took place during Miocene and later times, placing the eastern margin of the Tatar Strait basin on top of the western margin of the Terpeniya Gulf basin.

The Susunay and Tym'-Poronay troughs are narrow and severely deformed. Despite this fact, two small Miocene gas fields have been discovered (Naydenov, 1978). The Terpeniya Gulf may be a zone where the basin broadens, and parts may be more prospective, with longer, more gently dipping, and less deformed structures. The Miocene and younger section may not offer the only objectives in this area (Siryk, 1968). Gas and oil seeps are associated also with the Late Cretaceous section. Five mud volcanoes, whose gas originates in the upper Cenomanian-Senonian strata, are active in the area (Gorkun and Siryk, 1967). The main problem with the Upper Cretaceous is a lack of reservoirs, although good source beds are present.

**Aniva Gulf Basin**

This basin, with an area within the USSR of about 4,200 sq km, also is bounded on the west by the Central Sakhalin thrust zone. Several anticlines have been mapped onshore, and most of them have been drilled. Noncommercial quantities of oil and gas were found in several (drilled between 1968 and 1971). In May 1971, the Vostochno-Lugovskaya field was found in several upper Miocene sandstones between 927 and 1,182 m. A second field — Yuzhno- Lugovskaya — was discovered subsequently (Yurochko et al., 1981). The fields have tilted gas/water contacts.

The discovered reserves are small, but are large enough to justify their local use. The major remaining potential of the basin is offshore. The basin extends into eastern Hokkaido across the La Pérouse Strait. Descriptions of the geology were published by Siryk (1968), Bakayev (1973), Bakayev et al. (1973), Ovcharenko et al. (1973), and Taboyakov (1973).

**Tatar Strait Basin**

The prospective area of this basin has been given as 20,000 sq km. It occupies the western part of Sakhalin (Fig. 39) and the Tatar Strait, but does not extend onto the Soviet mainland. The basin is mainly a Late Cretaceous-Cenozoic successor type basin on older eugeosynclinal rocks.

Rodnikova et al. (1968) noted that the basin extends southward to include much of western Hokkaido, where 15 fields produce from upper Eocene, lower and middle Oligocene, middle and upper Miocene, and lower Pliocene. The fields are both oil and gas fields. Reservoir quality is poor, with very low permeability and porosity values, and a high volcanic ash and clay content. Only 7 million bbl has been produced in more than 25 years; average production per well ranges from a few liters to 10-12 bbl/d.

In western Sakhalin, the Upper Cretaceous through Paleogene section contains many source rocks but almost no reservoirs. The Oligocene through Pliocene contains some reservoir-quality strata, but has no source or seals. Moreover, much of the Miocene, considered by Rodnikova et al. (1968) to be a major objective in the basin, commonly is a sequence of andesite and basalt flows, tuffs, and diabase sills, as on Moneron Island and other structures drilled jointly since 1977 with the Japanese (Kudel'kin and Sheremet'yeva, 1981). Numerous seeps have been found, and some small oil and gas flows have been obtained from wells.

Overall the Tatar Strait basin is a highly unfavorable area for the exploration for petroleum. Offshore drilling since 1977, conducted jointly with the Japanese, simply reinforces this opinion. Some Soviet geologists disagree with this assessment, but the great amount of drilling and the wholly negative results speak for themselves (Vereshchagin and Kovtunovich, 1970, 1974; Budnikov et al., 1973, Melankholina, 1973, Zakharova, 1973, Ovcharenko et al., 1974; Budnikov, 1975; Kvedchuk et al., 1981; Kudel'kin and Sheremet'yeva, 1981). I have assigned no potential resources to the USSR part of the basin. (Meyerhoff, 1981, reviewed the stratigraphy and structure of the basin in detail, with drilling results.)

## *Priokhots and Parapol'-Palana Basins*

These are the basins of the West Kamchatka coastal plain and offshore shelf (Fig. 37). The configurations of the basins are not known with certainty, and are shown with many different orientations. Also, they have been given many different names by various authors (e.g., Trofimuk, 1971; Ovcharenko and Sevost'yanov, 1974; Vlasov and Tarasenko, 1977; Meyerhoff, 1981; Gnibidenko and Khvedchuk, 1983). For example, Trofimuk (1971) stated that the West Kamchatka and Priokhotsk basins are synonymous. Vlasov and Tarasenko (1977) divided the West Kamchatka basin into three parts: the Icha basin in the north; the Kolpakov basin in the center; and the Priokhotsk basin in the south (which Vlasov and Tarasenko interpreted to be an extension of the depocenters of the Kuril Islands). Trofimuk (1971) wrote that the Palana and Tigil' basins are one and the same. Ivanov (1981) used the name Pustoretskiy basin, instead of Palana or Tigil'. Ivanov also included the southern part of the Parapol' basin within the Palana or Pustoretskiy basin. Moreover, what Ivanov (1981) called the West Kamchatka basin actually is part of the Parapol' basin (northern part) and part of the North Okhotsk basin. Gnibidenko and Khvedchuk (1983) applied the term «West Kamchatka basin» to all of western Kamchatka, particularly the offshore.

This confusion is a direct result of the facts that, (1) even after drilling dozens of onshore wells, the dividing lines (if any exist) between basins are not apparent; and (2) few geophysical data, especially seismic, were available until Gnibidenko and his colleagues completed detailed studies offshore (Gnibidenko and Khvedchuk, 1983).

Four «discoveries» of gas have been recorded in the Priokhotsk and Parapol'-Palana basins. Only two possibly significant discoveries have been made (Kudryavtseva et al., 1979). These are at the mouth of the Liman River in the Priokhotsk basin (Liman anticline) and in the Tigil' area in the southern part of the Parapol'-Palana basin (Rossochino anticline). The Liman-1, discovered in 1973, flowed 385,000 cu ft/d of dry gas from a depth of 2,070-2,109 m in sandstones of Late Cretaceous age (according to Ovcharenko and Sevost'yanov, 1974) or early Paleogene age (according to Kudryavtseva et al., 1979). The Rossoshino-1 flowed 2,450 Mcf/d of dry gas from Paleogene sandstone (Kudryavtseva et al., 1979). Both structures are small; reserves are small; and the absence of a local market makes both discoveries noncommercial.

Hundreds of wells have been drilled onshore along the West Kamchatka coastal plain since the first structural and stratigraphic tests were drilled in 1935. Although the section onshore ranges in thickness from 3,500 to 4,000 m (above Senonian basement) and may be 6,000 m thick offshore, results to date have been exceedingly discouraging. Closures on folds are small; large and small faults are numerous; many folds (mainly compressional, but including some drape anticlines) are breached by faults and by erosion; reservoir quality is poor; and lateral reservoir continuity is doubtful. If there is any future for the basins of western Kamchatka, it is in the offshore (Vlasov and Belova, 1964; Trofimuk, 1971; Ovcharenko and Sevost'yanov, 1974; Vlasov and Tarasenko, 1977; Meyerhoff, 1981; Gnibidenko and Khvedchuk, 1983).

## *Severny Basin*

This basin strikes west-northwest for 600 km across the northern part of the Sea of Okhotsk (Fig. 37). More sophisticated and detailed seismic studies were published by Mavrinskiy et al. (1980) and Gnibidenko and Khvedchuk (1983).

The basin includes several troughs and deeps, separated by saddles and arches. The best study by far was made by Gnibidenko and Khvedchuk (1983; see also Burk and Gnibidenko, 1977).

Ivanov (1981) included his West Kamchatka basin in the eastern part of the Severny basin. Sevost'yanov (1971) included part of the western portion of the Severny basin in his northern extension of the East Sakhalin basin. He also included a part of the Yamsko-Tauysk basin of Ivanov (1981) in the East Sakhalin basin. Ivanov (1981) included the western part of the Severny basin in the Yamsko-Tauysk basin. (I have not included the Yamsko-Tauysk basin in this review, because the reality of its existence is doubtful.) Gnibidenko and Khvedchuk (1983) divided the Severny basin into three subbasins from west to east: Lisyanskiy, Severny, and Tinro.

Of all of the Sea of Okhotsk basins, including the East Sakhalin basin, this basin, with a minimum area of 72,000 sq km, shows the most promise. It is bounded on the northeast and southwest by faults. Objectives range from middle Eocene through lower Pliocene. Basement is Late Cretaceous (Burk and Gnibidenko, 1977). The minimum section thickness in the basin is 3,000 m; the average is 4,000; the deeps have slightly more than 6,000 m of section. No resource potential was assigned specifically on Table 1 to this basin, but it very likely could hold the equivalent of 6 (range of 5 to 12) billion bbl and 12 Tcf, provided that structural and other conditions favorable for generation, accumulation, and preservation are present.

## *Penzhina Basin*

This basin, 720 km long and 100 to 160 km wide, separates the Penzhina-Anadyr fold belt on the west from the Kamchatka-Koryak fold belt on the east. Its northern end is just west of the Tertiary Anadyr basin (Fig. 37). At least 8,000-9,000 m of post-Neocomian strata (mostly Cretaceous) are present (Ivanov, 1981).

More than 58 surface anticlines have been mapped. Some are large, 50 km long and 5 to 10 km wide. Many have been destroyed because the objective sections are exposed, or are so close to the mountains on either side that artesian waters have flushed possible hydrocarbon accumulations. However, many large and attractive anticlines, with 5 to 30° flank dips, remain. Oil and gas seeps are numerous, Kudryavtseva (1978) wrote that the best objective sections are sandstones in the Aptian-Albian, Senonian-Danian, and Paleogene. Porosity values are as great as 25 to 35 percent; permeability values are as great as 300-1,200 md. Trofimuk (1971) wrote that, of the 50,000 sq km that are prospective, about 15,000 sq km have really good potential. Sevost'yanov (1971), in contrast, did not believe the basin was worth listing in his discussion of prospective basins of eastern Siberia and the Chukchi Sea.

## *Central Kamchatka-Olyutor Trough*

This basin forms a graben-like depression that extends nearly 75 percent of the length of the Kamchatka Peninsula (Fig.37), or 950 km. The prospective area is 60,000 to 75,000 sq km, some of which is offshore (Trofimuk, 1971; Vlasov and Tarasenko, 1977). Some workers, such as Ivanov (1981), divide the depression into two basins — the Central Kamchatka and the Olyutor.

No intensive petroleum exploration has been conducted, although about 20 anticlinal structures have been identified, most of them 10-15 km long, 3-6 km wide, and with 5 to 10° flank dips. One gas seep has been known since 1913.

Several thousand meters of Paleogene-Neogene volcanogenic and marine strata are present. The stratigraphic history is related closely to that of the Kuril-Kamchatka island arc. The principal objectives and source rocks, on the basis of surface geology, appear to be in Miocene and Oligocene tuffaceous sandstones. The known surface stratigraphy suggests that the basin has little potential.

## *East Kamchatka Basin*

This basin is east of and parallel with part of the Central Kamchatka-Olyutor trough (Fig. 37). It has a prospective area estimated at 27,000 sq km,of which half is offshore (Trofimuk, 1971; Dikenshteyn, 1977; Vlasov and Tarasenko, 1977; Gnibidenko et al., 1980; Meyerhoff, 1981). It is close to the Kuril-Kamchatka deep-sea trench and is a part of the Kuril-Kamchatka island-arc system.

The onshore part of the basin is an area of active volcanism and mud volcanoes, with numerous oil and gas seeps that have been reported since 1755. Uzon volcano, in the southwestern part of the basin, has several groups of small but active mud volcano fields in its caldera; these small mud volcanoes or mudpots contain large pools and blobs of light-green oil (Beskrovnyy and Lebedev, 1971). The oil is believed to come from a Miocene reservoir (or reservoirs) at depth, reservoirs that are known to bear oil nearby.

The Bogachevskiy River seep was found in 1921, and flowed initially at a rate of 160 liters/d of pure oil. Active petroleum exploration began in 1941. Between 1941 and 1957, 88 shallow (450-500 m) and 5 deeper (1,400-2,790) tests were drilled. By 1970, 107 shallow holes, 28 deeper (mainly stratigraphic) tests, and 84 geothermal wells had been drilled in the basin. Small amounts of oil, exploited commercially since the 1920s, were hauled by cart or boat to local villages and the capital city of Petropavlovsk-Kamchatskiy.

Porosity values generally are low because of the clay and volcanic ash matrix (3-12 percent), and permeability values also are low. Small amounts of noncommercial production can be maintained for local use. The principal reservoirs are in the Oligocene and Miocene. Some wells have recovered up to 0.5 to 3.2 bbl/d of oil and 2,100 to 350,000 cu ft/d of gas. These initial flows generally drop very rapidly. As a whole, the onshore part of the basin has little promise except for geothermal energy. Development of geothermal energy, however, is several decades away. In contrast, the offshore has two depocenters where the Tertiary section is thicker — Kamchatskiy and Kronotskiy Gulfs. It is in these two offshore areas where the best petroleum prospects lie (Gnibidenko et al., 1980). The potential for large fields, however, is slight.

## Bering Sea Basins (Pacific Mobile Belt)

### *Khatyrka Basin*

#### Introduction

The Khatyrka basin has an area of at least 40,000 sq km, of which 3,500 to 4,000 sq km is onshore. The Khatyrka basin does not connect eastward with the Navarin basin (Figs.42, 43; Marlow, 1979; Marlow et al., 1983). This possibility has been suggested by various authors (e.g., Link and Meyerhoff, 1976).

Aeromagnetic, gravity, and field-geologic studies began in 1955. The oil potential was first studied in detail in 1959, although seeps had been known for more than 200 years. Reflection-seismic studies began in the early 1970s, and the first stratigraphic test was drilled in 1974. Flows of gas were obtained from the Oligocene and lower Miocene. In sandstones of the lower Miocene, flow rates of from 945,000 cu ft/d to 1,050 Mcf/d were recorded from the Onol'skiy R-30 well (Dikenshteyn, 1977). An oil discovery was reported during 1978, but nothing more has been written about this «discovery.»

#### Regional Geologic Framework

The Khatyrka basin is a successor basin on the eastern flank of the Koryak fold belt (Fig.43). Basement is the pre-Campanian Cretaceous, although the Campanian-Maestrichtian section includes large volumes of volcanic rocks, including lava flows. The principal objectives are in strata that range in age from latest Eocene through the Miocene. The Paleocene and most of the Eocene appear to be absent in large parts of the basin, with the result that the Oligocene covers the Cretaceous directly. There is major buried topographic relief on the pre-late Eocene surface.

#### Stratigraphy

The Tertiary section ranges in thickness from a feather edge updip to more than 5 km near the coast line. Gnibidenko and Svarichevskiy (1974) suggested that the average thickness is about 4 km. Burlin et al. (1974) agreed with this opinion. The Oligocene section contains thick units of dark-colored organic-rich shale with lenses of sandstone, siltstone, and beds with marly concretions. The Miocene was deposited in a shallower water to nonmarine environment near the basin margin, where coal beds (lignite) are present. The section becomes more marine seaward. Conglomerate, sandstone, siltstone, and shale comprise almost all of the section. The Miocene appears to be mainly middle and late Miocene; early Miocene seems to be absent (Trofimuk, 1971).

#### Petroleum Geology

At least 40 structures have been mapped in the basin, some of them offshore. Clay diapirs intrude some of these structures (Fig. 43). Most of the onshore folds are about 7-8 km long and 2-3 km wide. Offshore anticlines appear to be somewhat larger.

The onshore part of the basin is divided into two parts by a basement ridge. Numerous gas and oil seeps are present,as well as more than

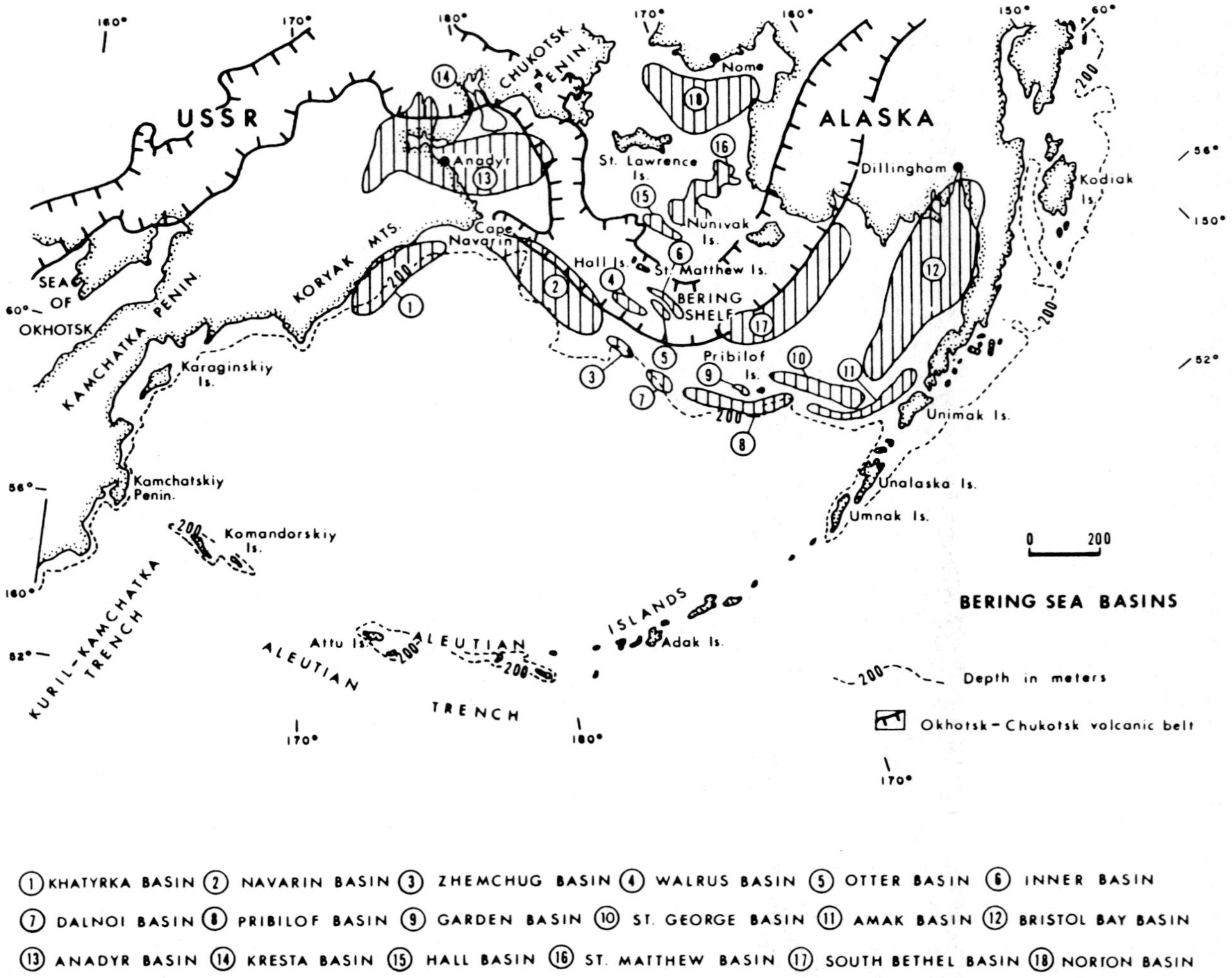


*Fig. 42* — Index map to prospective Tertiary basins of the Bering Sea Shelf. The Mongol-Okhotsk (Okhotsk-Chukotsk) volcanic belt is shown. Modified from Gnibidenko and Svarichevskiy (1974), Link and Meyerhoff (1976), and others.

200 $H^{2}S$ seeps. The Onol'skiy R-30 well recovered 945,000 cu ft/d of gas during one test from a lower Miocene zone, and 1,050 Mcf/d from a later test of the same zone (Rovnin et al., 1976; Dikenshteyn, 1977).

The onshore area of the basin is structurally and stratigraphically complex. It is reasonable to assume that, in the offshore region farther from the Koryak fold system, the basin may be more simple, both structurally and stratigraphically.

## *Anadyr Basin*

### Introduction

This basin occupies an onshore area of 30,000 to 35,000 sq km at the northern end of the Koryak Range (Fig. 44), and extends offshore toward Alaska (Meyerhoff, 1972, 1981; Link and Meyerhoff, 1976; McLean, 1979). The offshore extension (East Anadyr trough), according to McLean, occupies an addition aral 60,000 sq km.

The first petroleum-oriented field work commenced in 1959. Refraction- and reflection-seismic lines were shot during 1963-1966 (Dolzhanskiy et al., 1966). Structural corehole drilling began during 1963 and, before the first deep wildcat well was drilled during 1967 and 1968, 32 stratigraphic and structural tests had been drilled (Dikenshteyn, 1977). Gravity, magnetic, and single-fold seismic lines were run over both the on- and offshore parts of the basin (Verba et al., 1971).

The first exploratory well, completed in 1968, was dry, but the second, completed in 1970, tested between 7,000 and 10,000 Mcf of gas. A second discovery was made nearby during 1972, and produced 7,000 Mcf/d. A third discovery, in 1975, tested 3,500 Tcf/d (Rovnin et al., 1976). Drill-stem tests of the first two discoveries, conducted during the spring and summer of 1972, produced gas for a few hours before both wells went to water (oral commun., Ministry of Geology). Exploration efforts in the basin declined notably, until a resurgence took place beginning in 1980. A recent (1981) oil discovery has been described in the Soviet press.

## Regional Geologic Framework

The basin overlies folded, faulted, and metamorphosed Mesozoic rocks of the Koryak-Kam-

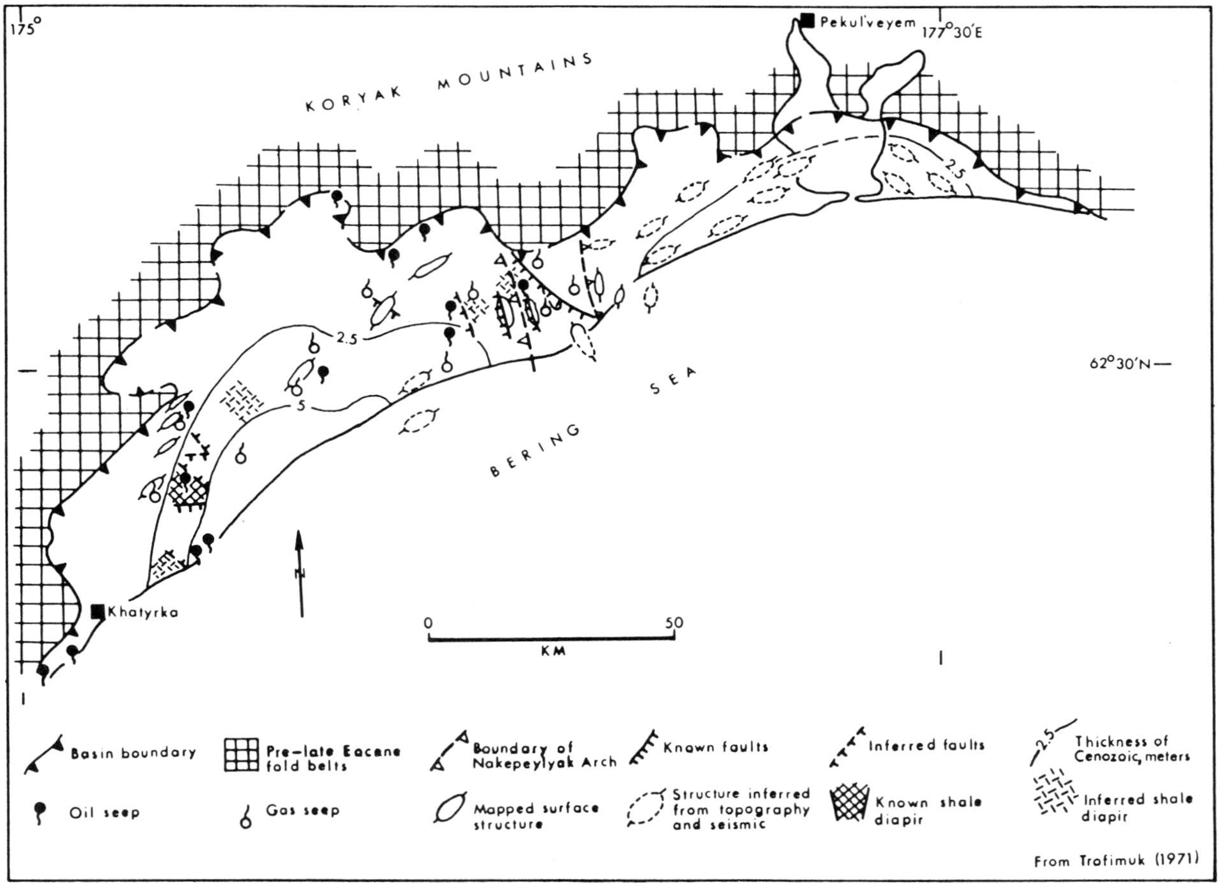


*Fig. 43* — Sketch map of the Khatyrka basin, western Bering Sea. Known and inferred structures are shown, together with known and assumed clay diapirs, faults, and oil and gas seeps. Some seismic thickness values for the Tertiary section are given. From Trofimuk (1971), as modified by Burlin et al. (1974). Republished with permission of Scientific Press Ltd.

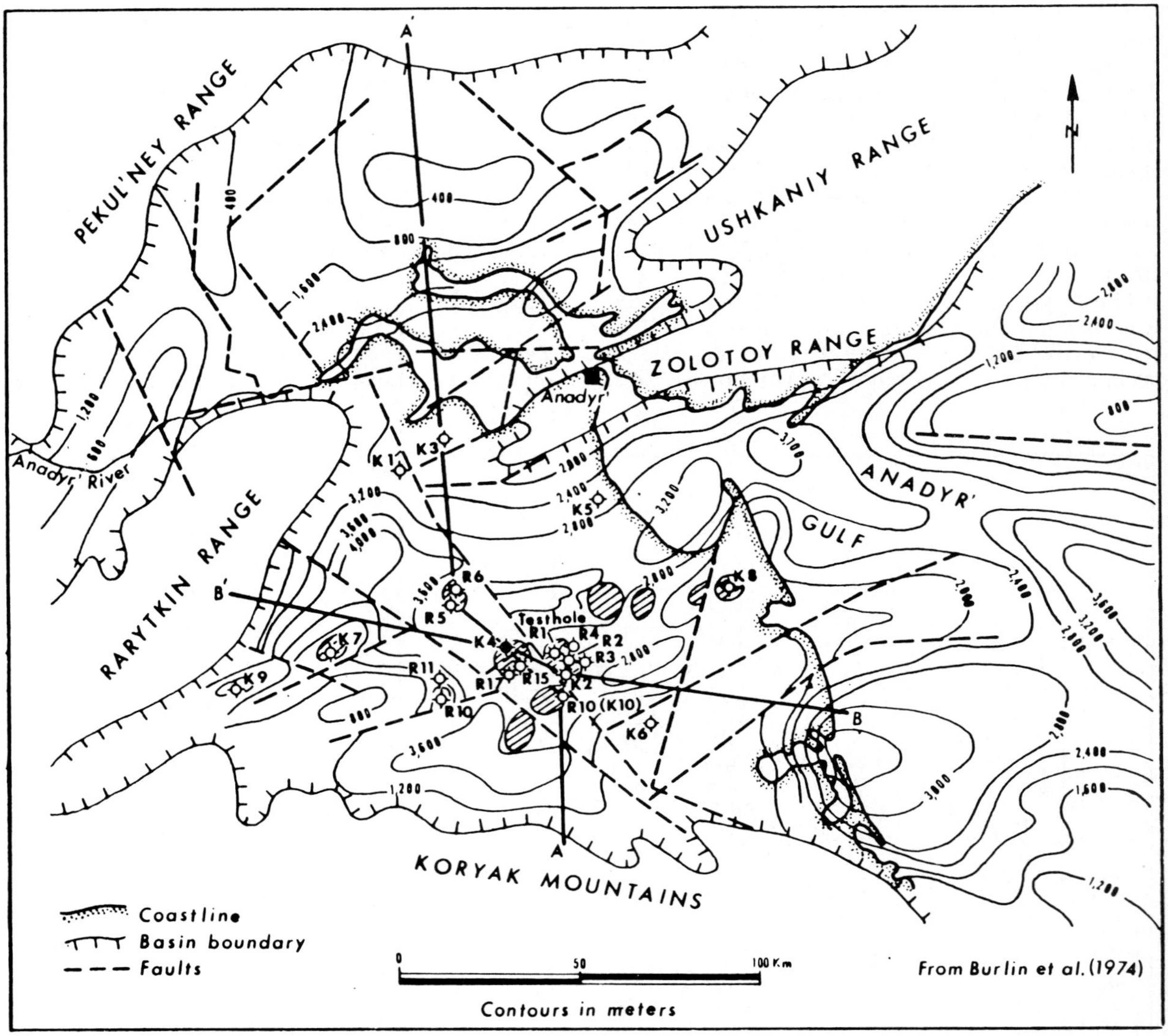


*Fig. 44* — Structural-contour map, Anadyr basin, Soviet Far East. Depths in meters. Structural datum is top of the basement. From Burlin et al. (1974), as modified by Link and Meyerhoff (1976). Location of Figure 45 is shown.

chatka fold belt. The Koryak Range bounds the southern part of the basin; the Pekul'ney Range separates it from the Penzhina basin; and the Mongol-Okhotsk (Okhotsk-Chukotsk) volcanic belt borders the basin on the north (Fig. 44). The Anadyr is a successor basin of mainly Campanian (Senonian) through Cenozoic age.

**Stratigraphy**

The Aptian and Albian succession, as in so many parts of the Soviet Far East, overlies and is overlain by a typically geosynclinal sequence. However, some bituminous carbonate rocks are present in that sequence. The Aptian and Albian, in contrast, are a shallow-water facies which, in a few wells, has gas and/or oil shows in moderately clean sandstones associated with organic-rich shales. Although the Aptian-Albian section is considered to be prospective, the intense tectogenesis which has affected this entire region probably has eliminated this part of the section from serious consideration as a primary drilling objective. The Neocomian rocks are largely chloritized tuff, sandstone, and andesite or basalt lavas.

A major unconformity separates the Upper Cretaceous from the Lower, and marks (approximately) the Turonian-Senonian boundary. The Senonian, which has some gas shows, is a terrigenous-clastic, molasse-like, shallow-marine sequence in the east. Some coal is present. Increasing amounts of coal appear westward, and the section becomes mainly nonmarine. Some diabase sills are present.

The Senonian is absent on many structures as a result of post-early Eocene erosion. The section is broken by numerous faults. The modern structures, which are partly compressional anticlines, but largely block-faulted drape anticlines, began to form during late Senonian time and continued to form through Miocene time. Dips range up to 40° in the Senonian strata and decrease upward.

The Senonian is believed to be conformable with the overlying Paleocene — early Eocene, which is present only along the northern margin of the basin. Here the Paleocene and early Eocene are mainly intermediate to mafic effusive rocks and their tuffs. The section wedges out southward. Oligocene shallow-water marine sandstone, siltstone, and shale are present in all of the basins, but are preserved mainly in grabens and depressions.

The Miocene is unconformable on all older rocks. Locally it is as thick as 3,000 m but it generally thins over the basement horsts and related block structures. It is overlain by 400 to 500 m of Pliocene and Quaternary. The early Miocene, locally absent, consists of conglomerate, sandstone, and shale, locally with coal beds. It grades upward into middle and upper Miocene littoral to marine sandstone, siltstone, and shale. A shale seal covers the basin and is present in the upper part of the Miocene section. The major gas shows encountered in the basin are in sandstones of middle to late Miocene age below the major shale sequence which seals most of the basin. The total thickness of the Tertiary section offshore is about 6 km (Verba et al., 1971).

**Petroleum Geology**

The Aptian, Albian, and lower Cenomanian sandstones are fairly clean, and similar to those found in the Penzhina basin. Porosity values average about 18 percent (9-25 percent range) in the cleanest units, and permeability values range from nil to 250 md. In the Miocene, Oligocene, and lower Miocene sandstones, porosity values are 10-13 percent, but permeability is very low. In the basal part of the middle to upper Miocene, porosity values are 18 to 26 percent; permeabilities normally are 92-98 md, but locally range up to 1,000 md.

More than 60 anticlinal structures have been mapped in the onshore part of the basin. Some may have originated during Late Cretaceous time as compressional structures, but movements during early and middle Tertiary times were largely vertical. As a result, most of the existing structures are drape anticlines. Fault seals probably are abundant. Most of the structures are 9-10 km long, 4 km wide, and have 2 to 5° flank dips. Almost all are syndepositional arches, and the bounding faults are growth faults. This fact is seen clearly on Figure 44, taken from Burlin et al. (1974). A negative feature of the repeated movements on the individual faults is that many structures are «bald-headed,» with the objective section either removed or never deposited. On such structures, the Pliocene overlies the Lower Cretaceous directly.

At least 42 wells have been drilled in the basin, 32 of them for stratigraphic and structural control. The deepest test was drilled in 1978 to 3,301 m. Almost all results have been discouraging. The sandstones are tight and «dirty.» However, they become cleaner upward in the section, and there is some evidence that they become cleaner toward the present Bering Sea. At least three gas discoveries have been made in Oligocene and middle to upper Miocene sandstones (3,500 to 10,000 Mcf/d) but the first two went to water during extensive testing. Up to 2 or 3 bbl/d of oil has been tested in the low-porosity, low-permeability sandstones of the Oligocene which, in the onshore area at least, appears to be in the oil window (the Miocene gases are largely biogenic).

The Soviet press reported an oil discovery (1,340 bbl/d) during 1981. Location of the well was not given, but its T.D. is 1,650 m, and severe pressure problems were reported. An offset well is being drilled.

Gas and oil shows have been found in post-

Turonian sandstones of the Upper Cretaceous. Detailed descriptions of test results and generalized summaries of the stratigraphy found in the wildcat wells, and in some of the stratigraphic tests, were given by Link and Meyerhoff (1976).

Results to date, although not encouraging, indicate that the offshore region may have been more stable; that fault movements were not so great; and that the sandstones are cleaner. I believe that commercial fields eventually will be found, mostly offshore, with mainly gas in the middle and upper Miocene, and oil and gas in the lower Miocene and Oligocene. My estimates of the reserve potential are shown on Table 1, where I have combined my estimates for the Khatyrka and Anadyr basins — 10 Tcf of gas and 100 million bbl of liquids.

## *Kresta Basin*

This basin occupies an area of about 5,000 sq km (Fig. 42). It is in Soviet waters, north of the Anadyr basin, but is separated from it by the Anaut uplift (Marlow et al., 1983). The Kresta basin contains up to 5 km of post-early to middle Paleocene sediments. The section presumably is similar to that in the Anadyr basin. Nothing more is known at this time.

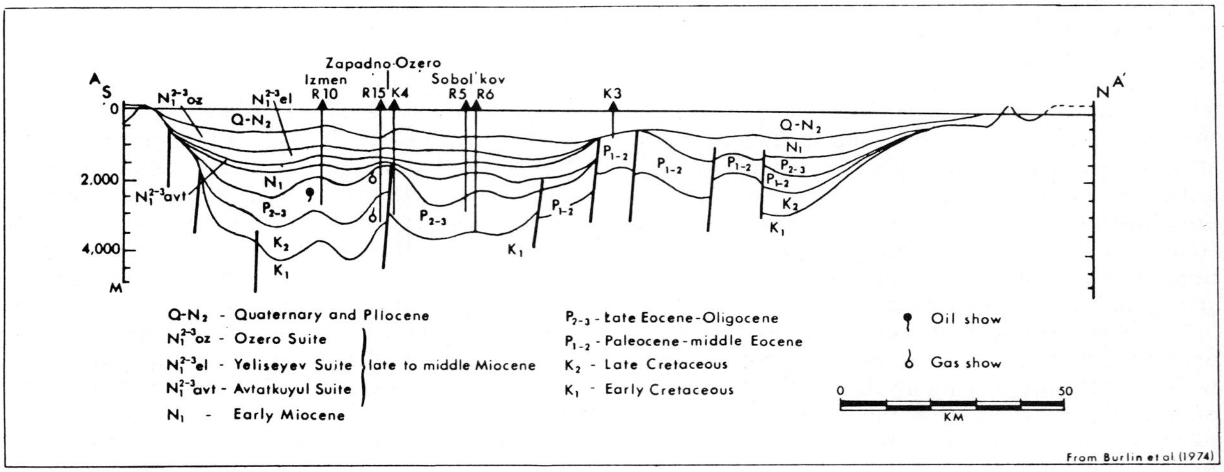


*Fig. 45* — South-north structural cross section of the Anadyr basin. Location is shown on Figure 44. Note that most of the faults are growth faults; note also the thinning of the Tertiary formations over the tops of the structures, a fact which indicates the synsedimentary nature of the structures. From Burlin et al. (1974).

# COAL

## Introduction

The major areas which produce coking and other high-grade coals are the Timan-Pechora basin (specifically, the northeastern part around Vorkuta and Inta), the Kuznetsk area, and the South Yakutia area (Fig. 1). In 1978, the Timan-Pechora basin supplied 4 percent of the nation's total coal production: the Kuznetsk area produced 20 percent; and the South Yakutia area, less than 0.5 percent (Dienes and Shabad, 1979). However, because of the high quality of South Yakutian coals, the mining facilities at Neryungri and Chul'man were being expanded at a rapid rate. Only the Kansk-Achinsk and Ekibastuz areas (4.4 and 5 percent respectively) also were important producers in the region. All but one of these major coal-mining areas are concentrated close to the Trans-Siberian railroad. The Timan-Pechora mining area has its own rail-transport line to Leningrad.

## Timan-Pechora Basin

### Introduction

Coal has been known in the northern Urals since the 1840s, and perhaps earlier (Wright, 1978). An application for a mining permit was made for the first time in 1864, but was refused by the Tsarist government. Detailed surveys were begun in the 1920s. Core drilling in the kuta River valley proved the presence of large reserves. Uneconomic mining began in 1929, but it was not until 1932 that the first economic mine was opened, 100,000 tons during the first year. Forced labor played a major role in developing the mines, and still does in some mines. At present, the area contains 28 mines (not all are producing) which yield 28.5 million tons per year. The Vorgashor mine, opened near Vorkuta in late 1975, is the largest shaft mine in the European USSR, with a designed annual capacity of 4.5 million tons. The coal is shipped by rail (the railroad was completed in 1941) to the Cherepovets iron and steel plant, 400 km east of Leningrad (Dienes and Shabad, 1979). Important volumes of steam coal are shipped to Leningrad, Arkhangel'sk, and Murmansk (Shabad, 1969).

### Geology

The oldest exposed strata are Early Carboniferous carbonates, mainly limestone (Nalivkin, 1973). Disconformable on the Lower Carboniferous are the *Schwagerina*-bearing limestones of the Asselian (earliest Permian) stage. Upper Carboniferous rocks are absent, as in most of the basin (e.g., see Fig. 10), and in most of the coal-mining area, both the Asselian and Sakmarian are absent (Nalivkin, 1973). The Yun'yaga Suite (late Asselian-Artinskian) is a marine sequence, 800 to 1,900 m thick, of shale, siltstone, sandstone, and some limestone. It becomes more calcareous westward. Above it is the main coal-bearing unit of the basin, the Kungurian and Ufimian Vorkuta Suite. This suite is 500 to 2,400 m of mainly continental deltaic, swamp, and lagoonal strata, although marine beds are present at the base. According to Kabalov (1979), it is a molasse. Above it is the Upper Permian Pechora Suite. It also is continental, and contains much coal, but of poorer quality than that in the Vorkuta Suite. The Pechora grades upward into the Triassic Kheyyakha Suite.

In the Vorkuta and Pechora Suites, there are more than 100 seams of coal, 30 of which are commercial. The commercial beds have an aggregate thickness of 30 m (Meyerhoff, 1982a). The coal sequences are cyclic, and in many respects resemble cyclothems.

### Resources

The Timan-Pechora coal deposits grade from west to east through nine ranks of coal, with the highest ranks on the eastern slope of the Urals, adjacent to the Kara Sea (Wright, 1978). Hard brown coal (=Russian brown coal) is in the west along the Chernyshev Ridge (Fig. 46). The brown coal grades eastward into low-rank bituminous coal (flame coal = lowest rank; gas coal; fat coal), high-rank bituminous hard coal (coking coal = lowest rank; sintering or coking coal; lean coal), semi-anthracite, and anthracite.

Dienes and Shabad (1979) wrote that resources are small. Wright (1978), after a thorough study of the basin, found this not to be true. The principal problems are the high degree of folding and faulting of the beds in the Urals and Pay Khoy; and the large amount of poorer quality coals in the Upper Permian. Resources are not less than 303,000 million tons (344,000, according to Semenov and Shcheglov, 1968), of which 20,000 are measured and proved. This volume of coal, by any standard, is very large. Of the 20,000, 7,900 million is in the proved category, and 3,000 million of this is coking coal.

## Eastern Urals

Several coal districts are present in the southwestern part of the West Siberian basin, along the eastern flank of the Urals. These districts lie in a belt about 600 km long, from Kopeysk south of Chelyabinsk, to Volchansk north of Chelyabinsk.

Lower Carboniferous anthracite is present in small amounts near Artemovskiy (Fig. 1). In

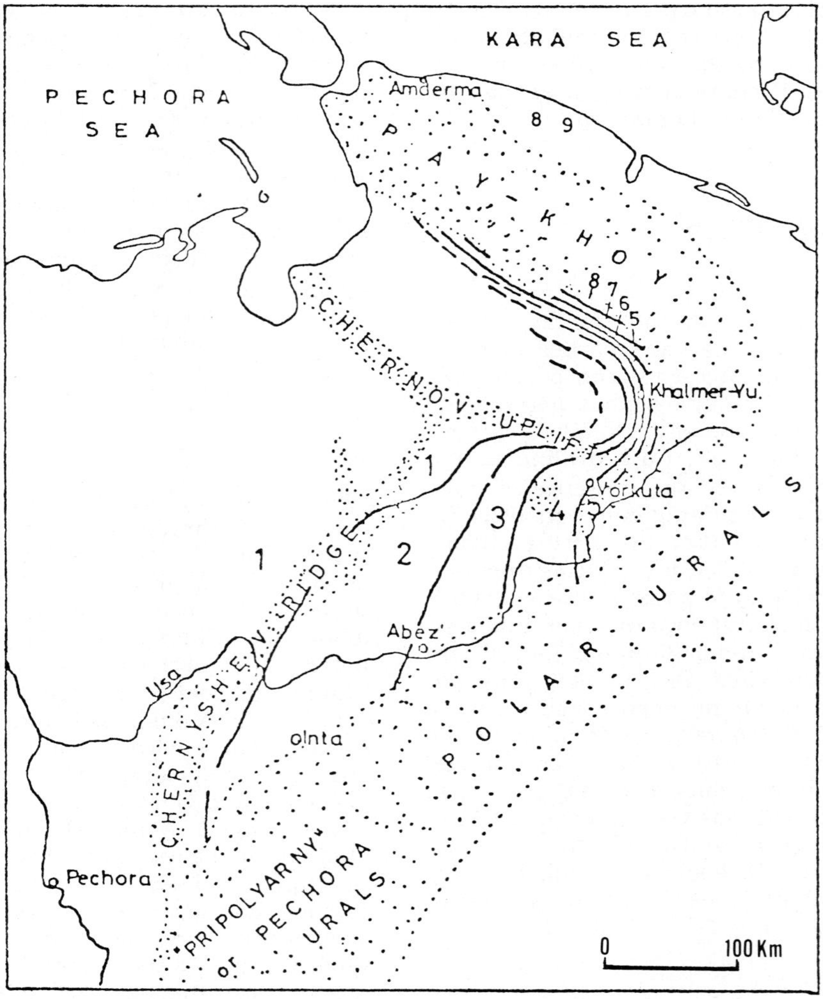


| *Russian terminology* | *Rank stages* | |
|---|---|---|
| 1 Brown coal | Hard brown coal | |
| 2 Flame coal | | |
| 3 Gas coal | Low rank | |
| 4 Fat coal | | Bituminous |
| 5 Coking coal | | hard |
| 6 Sintering or caking coal | | coal |
| 7 Lean coal | High rank | |
| 8 Semi-anthracite | | |
| 9 Anthracite | Anthracite | |

*Fig. 46* — Map of Vorkuta and Inta coal districts, Timan-Pechora basin, showing changes in coal ranks from west to east. From Wright (1978).

places, the coal has been converted into graphite near igneous intrusions. The original proved reserve was only 2,000 million tons.

The coals found from Volchansk to Kopeysk are lignites which crop out in Liassic (Early Jurassic) and Rhaetic (latest Triassic) continental, lacustrine, and lagoonal deposits of brown and gray sandstone and claystone. Several seams are present, one up to 100 m thick. They have been produced commercially since 1925. The total reserves are small, and are nearing depletion. Original reserves were 2,872 million tons (Cressey, 1945).

## Taymyr Peninsula

Bituminous coal has been mined in the Taymyr Peninsula (see Figs. 1 and 34 for location) for many years as a local fuel. Detailed surveys of the region have not been carried out. Nalivkin (1973) wrote that most of the reserves and resources are in the Upper Permian of the western part of the peninsula. Actually, thin seams are present in the Artinskian and Kungurian, with the most numerous and thickest seams in the Ufimian, Kazanian, and Tatarian (Wright, 1978). Estimated resources are not less than 75,500 million tons, whereas proved reserves are only 700 million tons (Nalivkin, 1973). Forty workable seams are known, some of them 6 to 12 m thick. (Other resource estimates were summarized by Miles and Wright, 1978.)

In eastern and northeastern Taymyr, some Permian seams also are known. In addition, brown coals (lignites) are known in Lower Cretaceous continental deposits along the northern coast. Some of these seams are 4 to 5 m thick. These coals probably never will be mined because of their remoteness. This opinion holds true for both the Permian and the Lower Cretaceous coals.

## Ekibastuz District

Ekibastuz (Fig. 1) had its beginnings as a primitive mining operation early in the present century. However, the modern mines originated in the early 1950s in a forced labor camp described by Aleksandr Solzhenitsyn in the third volume of the *Gulag Archipelago*. The first modern mine opened in 1954. The coal is sub-bituminous with a high ash content (up to 40 percent). The age is late Viséan, and the deposit occurs in a syncline 13 km long and 6 km wide (Nalivkin, 1973). A unique feature of the deposit is the presence of one seam 150 m thick (this thickness does include some very thin shale intercalations). Resources are 10,000 million tons, with 3,700 million proved.

## Kuznetsk (Kuzbas) Deposits

Mining began here at the turn of the century to supply coals for the newly opened Trans-Siberian railroad. This is the second largest coal-producing area in the USSR (Fig. 1), with 162 million tons in 1978. Ninety percent of the area's resources are above 600 m (252,000 million tons, according to Nalivkin, 1973, but believed to be at least 324,000 million, according to Semenov and Shcheglov, 1968). Ninety-four percent of the resources are above 100 m. Proved reserves are 33,000 to 35,000 million tons. Most of the coal is low-ash, high-rank bituminous coal, but at least 54,000 million tons of anthracite is present.

There are three coal-bearing units: (1) a Late Carboniferous-Early Permian formation; (2) one of Late Permian age; and (3) another of Early Jurassic age. Thickness of (1) is 800 to 2,700 m; it has 28 seams totaling 78 m in thickness. Thickness of (2) is 2,600 m with 45 workable seams having a thickness of 75 m. The Lower Jurassic has 10 seams totaling 18 m. The Jurassic is not being mined. Unlike many mines, about half of the Kuznetsk mines can be worked from the surface by strip mining; the other half are mined underground. Coking coal accounts for 40 percent of production, and bituminous steam coals, 60 percent. Large areas still are undeveloped (Shabad, 1969; Dienes and Shabad, 1979).

## Kansk-Achinsk District

The first mine in the Kansk-Achinsk basin (Fig. 1) was opened in 1940, after preliminary mining in the late 1930s. The coal-bearing zone is a Lower and Middle Jurassic lignite complex, with 44 seams, some of them 40 to 80 m thick. The lignite-bearing unit thickens from east to west — 40 to 200 m near Kansk and 1,200 to 1,300 m in the Chulym-Yenisey trough near Achinsk. Resources are 1,208,000—1,220,000 million tons; proved reserves are 73,000 million. Some layers are subbituminous or lignite. Coals of the same age, plus Neogene coals, are present in the Irkutsk area (Fig. 1).

## Tunguska Basin-Lena Trough

Figures 1, 23, and 34 show the location of the Tunguska basin, the Vilyuy basin, and the Lena trough. The coals of the Tunguska province are Carboniferous, Late Permian, and Triassic. In the east and southeast, Jurassic, and even younger (Early Cretaceous) coals are present. Many are associated with the Siberian traprocks which are Permo-Triassic and cover 1,000,000 sq km of the Siberian platform. Knowledge of the area is insufficient to differentiate among the different districts, even to determine where one district begins and another ends. For example, in some places the Kansk-Achinsk coals overlie Carboniferous coals (too thin to mine) of the Tunguska province. Semenov and Shcheglov (1968) estimated that the two provinces — Tunguska and Lena, covering the whole Siberian platform, contained resources of not less than 4,400,000 million tons, the largest near-surface coal field on earth. (Jurassic coals underlie huge parts of the West Siberian basin, thus forming an even larger coal basin, but they are too deep to mine.)

The Tunguska Permian coals have been mined for fuel at Noril'sk (consequently this is the only part of the Tunguska deposit shown on Fig. 1). This fuel was supplanted by hydroelectric power generated at the Khantayka River dam east of Igarka (Fig. 1). When the gas line from the West Siberian fields to Noril'sk was completed during 1969, the Khantayka power was diverted elsewhere and the mines at Noril'sk were kept as a reserve for gas pipeline failures. The coal still is used as a backup fuel during bad weather for the nearby Snezhnogorsk hydroplant.

Nalivkin (1973) estimated that the Tunguska basin proper is covered by formations containing about 1,516,000 million tons of coal, of which 429,000 million tons lie above 300 m. He estimated a resource of 2,400,000 million tons in Upper Jurassic and Lower Cretaceous rocks of the Lena trough. Thus his conclusion was that nearly 4,000,000 million tons underlie the Tunguska-Lena area, a result similar to that reported by Semenov and Shchlegov (1968). Regardless, the size of the deposits in this huge area of more than 2.5 million sq km is unknown, but undoubtedly is huge.

The Upper Jurassic and Lower Cretaceous coals of the Vilyuy basin also belong to this area, as do the mines at Kangalassy, 32 km north of Yakutsk, and at Sangar, 200 km farther northwest (Fig. 1). Both mines have high-grade bituminous Jurassic and Lower Cretaceous coals that are used to fuel power stations for the mining industries of the region. The mining organizations, however, are converting to natural gas, and the coals are now the subject of a possible trade/development deal with Japan.

## South Yakutia Coal Basin

The location of this basin is shown on Figure 1. It is part of the Soviet Japanese trade deal to develop the coals of Yakutia (Japan had been importing large amounts of coal from the more distant Kuznetsk basin).

The coals are high-grade bituminous coals of Jurassic age. The size of the deposit is unknown, but 40,000 million tons are inferred and 2,600 million is proved. The coal makes excellent coking coal. Steam coal also is produced. Seams up to 150 m thick lie close to the surface, so that strip mining is the cheapest way to develop the basin (Shabad and Mote, 1977).

## Miscellaneous

Within the Pacific mobile belt, there are numerous coal deposits. These occur generally in four age groups (Nalivkin, 1973): Upper Carboniferous; Lower Triassic; Upper Triassic-Lower Cretaceous (strata of this age group contain the largest industrial coal deposits in the Pacific mobile belt); Upper Cretaceous-Paleogene; and Neogene-recent.

The Indigirka-Zyryanka and Moma basins contain a resource in Aptian-Albian strata of 102 billion tons; the Suyfun basin (Fig. 36) produces coal from the Aptian-Albian; the Suchan-Daubikhe basin has proved reserves of 200 million tons in 13 seams within the Aptian, Albian, and, possibly, Cenomanian. The Verkhne-Bureya basin has 25,000 million tons of resources, proved reserves of 5.4 billion tons, and produces from the Jurassic, and possibly some beds of Late Triassic age.

Even smaller deposits are mined. Many of these small deposits are in fault blocks and folds within the various mining districts of the Soviet Far East. The Paleogene coals of the Tatar Strait basin (Fig. 36) of western Sakhalin Island are mined. Sakhalin proved reserves are 3,600 million tons out of a resource of 19,400 million tons. Without this resource, the industries on Sakhalin would find it difficult to survive.

It is clear then that coal is extremely important in the Soviet Far East and Maritime provinces. The reasons are (1) lack of roads for transporting fuel, (2) lack of forests for wood burning, and (3) lack of natural gas and oil. For example, in Magadan oblast in the northeast, one may observe the following:

**a.** The Susuman and Yagodnoye settlements and mines on the upper Kolyma River get their power from a 50,000-kw thermal plant (Arkagala plant) using local Kolyma valley coal.

**b.** The same plant supplies the Ust'-Nera district, 180 km farther northwest.

**c.** A plant in the Zyryanka area, farther down the Kolyma valley, uses the local Aptian-Albian coal to produce electricity for the gold and tin camps of northern Magadan.

**d.** In the Anadyr basin (Fig. 44), Aptian-Albian coals are mined at the Beringovskiy deposit in the southeasternmost part of the basin. The energy is used for providing electricity to tungsten and tin operations along the southern shore of the Chukotsk Peninsula, and also is used to fuel ships.

**e.** At Galimyr, 350 km east of Yagodnoye and 100 km from the Sea of Okhotsk coast, a coal-fueled plant keeps the tin operations running at Omsukchan.

These small examples are a few of many which enable the Soviet mining industry to develop in the treeless tundra regions of the Pacific mobile belt area. For this reason, geological field parties explore in large numbers every year to find new deposits to fuel local industry and to heat the plants and homes of the workers.

## OTHER ENERGY SOURCES

Oil shales are found in many places, but are not economic at present. The Olenek tar sands (Fig. 1) near the Arctic Ocean coast are large, but development costs are too great to justify their development at the present time, despite the fact that 10 to 12 billion barrels of oil may be recovered. Nuclear plants are being built in several areas, and some have been operating for years. Two are at Polyarnyy and Kola, near Murmansk, close to the border with Finland. In 1977 the Soviets built the small reactor at Lovisa, just inside of Finland. Three small 12 MW reactors are in the Kolyma valley near the gold-mining center of Bilibino. Another is being installed along the northern coast of Magadan oblast. Nuclear reactors in these remote areas are built only where other fuel supplies are lacking.

Hydroelectric power has not been developed widely, but plants are present, or are being built, in several areas. Fifty-six percent of the USSR's hydroelectric potential is in the northeastern part of the country alone. However, the economic worth of a multitude of hydroelectric facilities is debatable if other, cheaper forms of energy are at hand.

## CONCLUSIONS

The Soviet arctic and subarctic regions contain less than 25 million persons out of a total population of 262 million. Yet increasing amounts of energy for all of the USSR must come from Siberia, and the Soviet Far East and Maritime provinces. With manpower and capital in short supply, without roads and rails, and without a well-developed infrastructure to support future development, the Soviets are going to have to redouble their efforts to develop this vast region and to extract its natural wealth, not just to better its own economy, but also to better the lot of its people. The resources are there. The question is whether they can be developed in sufficient time to keep the country's unhealthy economy alive. An obvious solution to the problem is to introduce increasing amounts of foreign help and know-how. The Soviets have done this with the Finns on the Kola Peninsula, at Noril'sk, and other arctic areas; they have done this with the Japanese in offshore Sakhalin and to a lesser extent in southern Yakutia, farther west. The future of the region, therefore, depends not only on the USSR's ability to utilize its existing manpower wisely, but also to welcome more foreign help — a step which this sleeping giant always has appeared very hesitant to take.

## References


Abdullin, R. A., 1980, Priroda fil'tratsionnykh svoystv porod Bazhenovskoy svity Salymskogo mestorozhdeniya Zapadnoy Sibiri: Geol. Nefti Gaza, no. 11, p. 19-25.

Afanas'yev, Yu. T., Gaydeburova, Ye. A., and Mikulenko, K. I., 1966, Perskpektivy neftegazonosnosti triasovykh i yurskikh otlozheniy Vilyuyskoy gemisineklizy i Predverkhoyanskogo progiba: Geol. Nefti Gaza, no. 2, p. 21-25.

Ageyev, K. S., 1979, Blokovaya tektonika Severnoy Zemli, *in* Pogrebitskiy, Yu. Ye., ed., 1979, Tektonika Arktiki: razlomy materikovoy otmeli i okeana: Leningrad, Minist. Geol. SSSR, Nauchno-Issled. Inst. Geol. Arktiki, p. 33-54.

Alekhin, S. V., and Verba, M. L., 1979, Dinamika tektonicheskikh protsessov i razlomy Yuzhno-Barentsevskoy vpadiny, *in* Pogrebitskiy, Yu. Ye., ed., 1979, Tektonika Arktiki: razlomy materikovoy otmeli i okeana: Leningrad, Minist. Geol. SSSR, Nauchno-Issledovatel'-skiy Inst. Geol. Arktiki, p. 18-26.

Alekseychik, S. M., Gal'tsev-Bezyuk, S. D., Koval'chuk, V.S., and Sychev, P. N., 1963, Tektonika, istoriya geologicheskogo razvitiya i perspektivy nefte gazonosnosti: Leningrad, Vsesoyuznogo Neftyanogo Nauchno-Issledovatel'skogo Geologorazvedochnogo Inst. (VNIGRI) Trudy, v. 217, 275 p.

Anatol'yeva, A. I., Zharkov, M. A., and Sovetov, Yu. K., 1966, O korrelyatsii krasnotsvetnykh tolshch venda i nisov nizhnego kembriya yugo-zapadnoy okrainy Sibirskoy platformy: Akad. Nauk SSSR Doklady, v. 166, no. 2, p. 413-416.

Anonymous, 1981, Oil strikes in Poland, West Germany fuel new interest in Baltic: Oil and Gas Jour., v. 79, no. 23, p. 102, 105, 108, 111-112.

Antsiferov, A. S., 1980, Gazovaya i khimicheskaya zonal'nost' plastovykh vod neftegazonosnykh otlozheniy Yuga Sibirskoy platformy: Geol. Nefti Gaza, no. 1, p. 5-9.

Bakayev, G. A., 1973, Rol' uglenosnykh otlozheny v formirovanii gazovykh mestorozhdeniy Sakhalina: Neftegazovaya Geol. Geofiz., no. 5, p. 24-26.

Bakayev, G. A., Tyutrin, I. I., and Taboyakov, A. Ya.. 1973, Perspektivy gazonosnosti kaynozoyskikh otlozheniy Lyutogskoy sinklinal'noy zoni, Yuzhnogo Sakhalina: Geol. Nefti Gaza, no. 1, p. 67-73.

Bakirov, A. A., ed., 1979, Neftegazonosnyye provintsii i oblasti SSSR: Moscow, Nedra, 456p.

Balabanova, T. F., 1981, Terrigennyye porody paleozoya yugo-vostochnoy chasti Zapadno-Sibirskoy pliti: Akad. Nauk SSSR Izv., ser. geol., no. 4, p. 131-136.

Barentsov, M. E., Volkov, E. V., and Yushinskaya, Z. N., 1968, Tektonika i neftegazonosnost' Khantymansiyskoy vpadiny, *in* Sarkisyan, S. G., ed., 1968, Geologiya i neftegazonosnost' tsentralnoy chasti Zapadno Sibirskoy nizmennosti: Moscow, Nauka, p. 5-23.

Bazanov, E. A., 1973, Geologicheskoye stroyeniye Yaraktinskogo neftyanogo mestorozhdeniya Irkutskoy oblasti: Geol. Nefti Gaza, no. 7, p. 15-18.

Bazhenova, T. K., Platov, Yu. I., Makarov, K. K., and Shumenkova, Yu. M., 1980, Analiz paleotektonicheskikh dvizheniy Sibirskoy platformy v svyazi c otsenkoy generatsii i akkumulyatsii uglevodorodov i konservatsiya ikh zalezhey v domezozoyskikh otlozheniyakh, *in* Luchitskiy, I. V., 1980, Tektonika i nefte gazonosnost' Sibiri i Dal'nego Vostoka, Tektonika Sibiri, v. IX: Novosibirsk, Nauka (Sibir. Otdel.), p. 55-63.

Belen'kiy, V. Ya., and Kunin, N. Ya., 1978, Puti povysheniya effektivnosti seysmorazvedki pri podgotorke struktur v Zapadnoy Yakutii: Geol. Nefti Gaza, no. 5, p. 22-30.

Beskrovnyy, N. S., and Lebedev, B. A., 1971, Oil shows in the caldera of Uzon Volcano, Kamchatka: Acad. Sci. USSR Doklady, v. 201, p. 266-268 (trans.by Am. Geol. Inst.).

Bodunov, Ye. I., Izosimova, A. N., Zuyeva, I. N., Chalaya. O. N., Utkina, N. A., Trushcheleva, G. S., Tel'ye, A. V., Lebedev, V. S., Lebedev, Val. S., Bakin, V. Be., Tokin, V. V., and Tomskaya, Ye. S., 1981, Geokhimicheskaya kharakteristika gazov i neftey Botuobinskogo rayona Yakutii: Geol. Nefti Gaza, no. 8, p. 20-28.

Bolk, V. E., Ivanov, S. S., and Shimaravev, V. N., 1973, Metodika interpretatsii aeromagnitnykh dannykh pri izuchenii zemnoy kory na primere issledovaniya shel'fov severnykh morey SSSR: Leningrad, Minist. Geol. SSSR, Nauchno- Issledovatel'skiy Inst. Geol. Arktiki, 87 p.

Borisov, A. V., Leshchenko, V. Ye., and Nosovskiy, V. I.. 1976. Novoye napravleniye poiskovo razvedochnykh rabot na neft' i gaz v Verkhne-Pechorskoy vpadine, Predural'skogo progiba: Geol. Nefti Gaza, no. 8, p. 1-5.

Budnikov, N, P., ed., 1975, Shel'f Sakhalina — geologicheskoye stroyeniye perspektivy neft-

egazonosnosti i puti osvoyeniya resursov nefti i gaza: Moscow, Nedra, 192 p.
Budnikov, N. P., Ovcharenko, A. V., Taboyakov, A. Ya., and Tyutrin, I. I., 1973, Perspektivy neftegazonosnosti shel'fa Sakhalina: Geol. Nefti Gaza, no. 11, p. 9-12.
Burk, C. A., and Gnihidenko, H. S., 1977, The structure and age of acoustic basement in the Okhotsk Sea: Am. Geophys. Un. Maurice Ewing Ser., v. 1, p. 450-461.
Burlin, Yu. K., Dontsov, V. V., and Pastukhova, T. N., 1974, O dal'neyshem napravlenii neftepoiskovykh rabot na Severo-Vostoke SSSR: Geol. Nefti Gaza, no. 5, p. 17-22.
Churkin, M., Jr., 1972, Western boundary of the North America continental plate in Asia: Geol. Soc. America Bull., v. 83, no. 4, p. 1027-1036.
Cressey, G. B., 1945, The basis of Soviet strength: New York, McGraw-Hill Book Co., Inc., 287 p.
Dahl, T., 1982, Swedish group's small-hole shallow-drilling technique cuts costs: Oil and Gas Jour., v. 80, no. 16, p. 98-100.
Danilkin, S. M., 1980, O barovom tipe Markovskoy i Yaraktinskoy zon neftegazonakopleniya: Geol. Nefti Gaza, no. 6, p. 14-20.
Dienes, L., and Shabad, T., 1979, The Soviet energy system: resource use and policies: Washington, V. H. Winston and Sons, 298 p.
Dikenshteyn, G. Kh., ed., 1977, Neftegazonosnyye provintsii SSSR: Moscow, Nedra, 328 p.
Dolzhanskiy, B. G., Ivanov, V. V., Dontsov, V. V., Burlin, Yu. K., and Roshchin, G. I., 1966, Novyye dannyye o glubianom stroyenii tsentral'noy chasti Anadyrskoy vpadiny: Geol. Nefti Gaza, no. 10, p. 15-21.
Drobot, D. I., and Isayev, V. P., 1966, Novyye dannyye o sostave i svoystvakh nizhnekembriyskikh neftey Prilenskogo rayona Irkutskogo neftegazonosnogo basseyna: Novosibirsk, Geol. Geofiz., no. 10, p. 32-41.
Dubatolov, V. N., ed., 1981, Paleozoy Zapadno-Sibirskoy nizmennosti i yeye gornogo obramleniya: Novosibirsk, Nauka, Sibir. Otdel., 144 p.
Dubronin, M. A., 1976, Usloviya obrazovaniya solyanykh struktur Angaro-Lenskogo progibov: Novosibirsk, Geol. Geofiz., no. 6, p. 60-67.
Emelyanov, E. M., Litvin, V. M., Levchenko, V. A., and Martynova, G. B., 1971, The geology of the Barents Sea, *in* Delany, F. M., ed., 1971, The geology of the East Atlantic continental margin: London, Inst. Geol. Sci. Rept. no. 70/14, p. 1-15.
Eremenko, N. A., Malovitskiy, Ya. P., Gramberg, I.S., and Lebedev, L. I., 1973, Geologic structure and oil and gas potential of USSR continental shelf: Am. Assoc. Petrol. Geol. Bull., v. 57, no. 2, p. 235-243.
Fedynskiy, V. V., and Levin, L. E., 1970, The tectonics and oil and gas potential of the marginal and inland seas of the USSR: Lamont-Doherty Geol. Observatory, Columbia Univ., 301 p.
Filanovskiy, V., 1980, Zapadno-Sibirskiy neftegazovyy kompleks: resul'taty i perspektivy: Planovoye Khozyaystvo, no. 3, p. 19-26.
Fradkin, G. S., ed., 1969, Leno-Vilyuyskaya neftegazonosnaya provintsiya: Moscow, Nauka, 278 p.
Galponenko, G. I., 1971, Ispol'zovaniye rezul'tatov regional'nykh geologo-geomorfologicheskikh i geofizicheskikh issledovanii pri neftegazopoiskovykh rabotakh na Arkticheskom shel'fe Vostochnoy Sibiri i Severo-Vostoka SSSR, *in* Demenitskaya, R. M., Bolk, V. E., Gusev, B. V., Karasik, A. M. Litvinov, E. N., and Ivanov, S. S., eds., 1971, Geofizicheskiye metody razvedki v Arktike, v. 6: Leningrad, Minist. Geol. SSSR, Nauchno-Issledovatel'skiy Inst. Geol. Arktiki, p. 58-63.
Galponenko, G. I., Bolk, V. E., Demenitskaya, R. M., Zatsepin, Ye. N., Ivanov, S. S., Puminov, A. P., Shimarayev, V. N., and Gorodintskiy, A. M., 1971, Skhema elementov interpretatsii geofizicheskikh dannykh po Arkticheskomu poberezh'yu i shel'fu SSSR kak osnova prognozirovaniya neftegazovykh i rossypnykh mestorozhdeniy, *in* Deminitskaya, R. M., Bolk, V. E., Gusev, B. V., Karasik, A. M., Litvinov, E. M., and Ivanov, S. S., eds., 1971, Geofizicheskoye metody razvedki v Arktiki, v. 6: Leningrad, Minist. Geol. SSSR, Nauchno-Issledovatel'skiy Inst. Geol. Arktiki, p. 6-8.
Gavrilov, Ye. Ya., Kulibakina, I. B., and Teplinskiy, G. I., 1971, O formirovanii zalezhey uglevodorodov Markovskogo mestorozhdeniya: Geol. Nefti Gaza, no. 2, p. 30-31.
Geodekyan, A. A., ed., 1976, Geologicheskoye stroyeniye i perspektivy neftegazonosnosti tsentral'noy Baltiki: Moscow, Nauka, 110 p.
Geodekyan, A. A., Dubovskoy, I. T., Kleshchev, K. A., Mazur, V. B., Ostrovskiy, M. I., Sakalanskas, K. A., Ul'mishek, G. F., and Khubl'dikov, A. I. 1978 Neftegazonosnyye komplek-Nauk SSSR Izvestiya, ser. geol., no. 8, p. 128-137.
Gladenkov, Yu. B., 1973, Neogene Yoldias of northern Sakhalin: Acad. Sci. USSR Doklady, v. 208, p. 109-112 (trans. by Am. Geol. Inst.)

Gnibidenko, H. S., and Khvedchuk, I. I., 1983, The Cenozoic sedimentary basins of the Okhotsk Sea: Am. Assoc. Petrol. Geol., in press (Third Circum-Pacific Energy and Mineral Resources Conf.)

Gnibidenko, G. S., and Svarichevskiy, A. S., 1974, Struktura i perspektivy neftegazonosnosti akvatorii Beringova morya: Sovet. Geol., no. 1, p. 89-96.

Gnibidenko, H. S., Bykova, T. G., Veselov, O. V., Vorob'yev,V. M., Kim Chun Un, and Tarakanov, R. Z., 1980, Tektonika Kurilo-Kamchatstogo glubokovodnogo zheloba: Moscow, Nauka, 179 p.

Gol'dberg, I. S., Lebedev, B. A., and Frolov, B. M., 1981, Razdel'nyy prognoz razmeshcheniya gaza, nefti i bitumov na Sibirskoy platforme: Geol. Nefti Gaza, no. 2, p. 22-26.

Gololobov, Yu. N., Klimov, V. G., Nikolayev, I. Ye., Novikov, G. N., and Ostistyy, B. K., 1981, O lovushkakh barovogo tipa v Daginskoy svite Severnogo Sakhalina: Geol. Nefti Gaza, no. 2, p. 13-18.

Gorkun, V. N., and Siryk, I. M., 1967, Calculation of depth of deposition and volume of gas expelled during eruptions of mud volcanoes in southern Sakhalin: Internatl. Geol. Review, v. 10, no. 1, p. 4-12.

Gramberg, I. S., Kartseva, G. N., Sokolov, V. N., and Sorokov, P. S., 1967, Perspektivy neftegazonosnosti tsentral'nogo sektora Sovetskoy Arktiki: Leningrad, Minist. Geol. SSSR, Nauchno-Issledovatel'skiy Inst. Geol. Arktiki, reg. geol. ser., v. 2, p. 27-40.

Grantz, A., Holmes, M. L., and Kososki, B. A., 1975, Geologic framework of the Alaskan continental terrane in the Chukchi and Beaufort Seas: Canadian Soc. Petrol. Geol. Mem. 4, p. 669-700.

Grigorenko, Yu. N., Makarov, K. K., Gramberg, I. S., and Kolakov, Yu. N., 1975, The oil and gas producing (sic = prospective) basins of the shelves of northeastern Asia: Canadian Soc. Petrol. Geol. Mem. 4, p. 663-668.

Gusev, G. S., 1979, Skladchatyye struktury i razlomy Verkhoyano-Kolymskoy sistemy mezozoid: Moscow, Nauka, 208 p.

Hamilton, W., 1970, The Uralides and the motion of Russian and Siberian platforms: Geol. Soc. America Bull., v. 81, no. 9, p. 2553-2576.

Ivanov, V. V., 1981, Neftegazonosnost' osadochnykh basseynov Severo-Vostochnoy Azii, *in* Surkov, V. S., ed., 1981, Neftegazonosnost' Sibiri i Dal'negi Vostoki: Novosibirsk, Nauka (Sibir. Otdel.), Inst. Geol. Geofiz., Trudy, v. 513, p. 131-136.

Kabalov, V. K., 1979, Usloviya formirovaniya uglenosnykh otlozheniy Urala: Sovet. Geol., no. 10, p. 39-51.

Kalinkina, L. I., 1981, Neftegazonosnost' rifogennykh otlozheniy nizhne-sredne-Motskoy podsvity Nepsko-Botuobinskoy anteklizy: Geol. Nefti Gaza, no. 4, p. 12-20.

Kamen-Kaye, M., and Meyerhoff, A. A., 1979, Flood basalts: implications for global-tectonic hypotheses and petroleum exploration: Jour. Petrol. Geol., v. 1, no. 4, p. 29-37.

Kanareykin, B. A., and Yashkov, G. N., 1980, Povysheniye effektivnosti rabot MOGT pri izuchenii paleozoya na yugo-vostoke Zapadno-Sibirskoy plity putem predbaritel'nogo vychitaniya kratnykh voln: Novosibirsk, Geol. Geofiz., no. 6, p. 120-128.

Khabarov, V. V., Nelepchenko, O. M., Volkov, Ye. N.. and Bartashevich, O. V., 1980, Uran, kaliy i toriy v bituminoznykh porodakh Bazhenovskoy svity Zapadnoy Sibiri: Sovet. Geol., no. 10, p. 94-105.

Khvedchuk, I. I., Yefremenkov. A. F., and Balabko, G. F., 1981, Sovremennyy strukturnyy plan i geologicheskoye stroyeniye dna Tatarskogo proliva, *in* Kornev, O. S., Svarichevskiy, A. S., and Sheremet'yeva, G. N., eds., 1981, Rel'yef i struktura osadochnoga chekhla akvatorial'noy chasti Dal'nego Vostoka: Vladivostok, Akad. Nauk SSSR, Dal'nevostochniy Nauchnyy Tsentr, Sakhalinskiy Kompleksnyy Nauchno-Issledovatel'skiy Inst., p. 70-80.

Kontorovich, A. E., Nesterov, I. I., Salmanov, F. K., Surkov, V. S., Trofimuk, A. A., and Erv'ye, Yu. G., 1975, Geologiya nefti i gaza Zapadnoy Sibiri: Moscow, Nedra, 679 p.

Korzh, M. V., 1978, Paleogeograficheskiye kriterii neftegazonosnosti yury Zapadnoy Siberi: Moscow, Nauka, 135 p.

Koval'chik, V. S., Kuklich, L. A., Mishakov, G. S., and Tronov, Yu. A., 1981, Osobennosti stroyeniya i formirovaniya lokal'nykh struktur Severnogo Sakhalina: Geol. Nefti Gaza, no, 2, p. 6-12.

Krems, A. Ya., Vasserman, B. Ya., and Matviyevskaya, N. D., 1974, Usloviya formirovaniya i zakonomernosti razmeshcheniya zalezhey nefti i gaza (o tektonicheskom stroyenii i osnovnykh etapakh geologicheskogo razvitiya Timano-Pechorskoy neftegazonosnoy provintsii): Moscow, Nedra, 338 p.

Kudel'kin, V. V., and Sheremet'yeva, G. N., 1981, O stratigrafii kaynozoya i istorii razvitiya rayona ostrova Moneron (Tatarskiy proliv), *in* Kornev, O. S., Svarichevskiy, A. S., and Sheremet'yeva, G. N., eds., 1981, Rel'yef i struktura osadochnoga chekhla akvatorial'noy chasti Dal'nego Vostoka: Vladivostok, Akad. Nauk SSSR, Dal'nevostochniy Nauchnyy Tsentr, Sakhalinskiy Kompleksnyy Nauchno-Issledovatel'skiy Inst., p. 81-88.

Kudryavtseva, Ye. I., 1978, Gazoproyavleniya v Podkagernom rayone Zapadnoy Kamchatki: Geol. Nefti Gaza, no. 10, p. 41-44.

Kudryavtseva, Ye. I., Andreyeva, Z. A., Bogomolov, A. I., Voronkov, Yu. S., Kostromin, G. A., Sayfutdinov, Z. S., and Shchelovanov, Yu. S., 1979, O novykh neftegazoproyavleniyakh iz mesokaynozoyskikh otlozheniy Zapadnoy Kamchatki: Novosibirsk, Geol. Geofiz., no. 10, p. 145-149.

Kushnareva, T. I., and Rasskazova, N. B., 1977, Ordovik Pechorskoy sineklizy: Akad. Nauk SSSR Izv., ser. geol., no. 6, p. 81-92.

Lalayants, A., 1980, Neftegazovyy kompleks: segodnya i zavtra: Sotsialisticheskaya Industriya, 21 May, p. 2.

Le Pichon, X., 1968, Sea-floor spreading and continental drift: Jour. Geophys. Research, v. 73, no. 12, p. 3661-3697.

Leshchenko, V. Ye., 1973, Rol' gidrodinamicheskikh faktorov pri formirovanii i razrushenii zalezhey nefti i gaza v Timano-Pechorskoy provintsii: Geol. Nefti Gaza, no. 12, p. 31-35.

Link, P. K., and Meyerhoff, A. A., 1976, Bering Sea: geology and hydrocarbon potential: Geneva, Petroconsultants, S.A., 2 vol.: v. 1, 205 p., 15 tables, 53 figs.; v. 2, 6 pl., 17 seismic lines.

Makarova, Sh. D., Menner, V. V., Spasskiy, N. Ya., Stulov, A. T., and Chernyak, G. Ye, 1979, Nizhnekamennougol'nyye otlozheniya na severo-vostochnom borte Tungusskoy sineklizy: Novosibirsk, Geol. Geofiz. no. 10. p. 136-138.

Maksimov, S. P., and Dikenshteyn, G. Kh., 1979, O vozraste skladchatogo osnovaniya Zapadno-Sibirskoy plity i vozmozhnom rasprostranenii platformennykh neftegazonosnykh kompleksov paleozoya v yeye predelakh: Geol. Nefti Gaza, no. 7, p. 6-11.

Maksimov, S. P., Larkin, V. N., Brindzinskiy, A. M., and Bakin, V. E., 1979, Usloviya i vremya formirovaniya gazokondensatnykh zalezhey v. permo triasovykh otlozheniyakh Khapchagayskogo vala: Geol. Nefti Gaza, no. 4, p. 22-27.

Marlow, M. S., 1979, Hydrocarbon prospects in Navarin basin province, northwest Bering Sea shelf: Oil and Gas Jour., v. 77, no. 44, p. 190, 192, 194, 196.

Marlow, M. S., Cooper, A. K., and Childs, J. R., 1983, Tectonic evolution of the Gulf of Anadyr and formation of Anadyr and Navarin basins: Am. Assoc. Petrol. Geol. Bull., v. 67, in press.

Mavrinskiy, Yu. S., Baboshina, V. A., Kulikov, N. V., and Kharakhinov, V. V., 1980, O tektonike Okhotskogo morya, *in* Luchitskiy, I. V., ed., 1980, Tektonika i neftegazonosnost' Sibiri i Dal'nego Vostoka, Tektoniki Sibiri, v. IX: Novosibirsk, Nauka (Sibir. Otdel.), p. 119-127.

McLean, H., 1979, Review of petroleum geology of Anadyr and Khatyrka basins: Am. Assoc. Petrol. Geol. Bull., v. 63, no. 9, p. 1467-1477.

Melankholina, Ye. N., 1973, Zapadno-Sakhalinskiy geosinklinal'nyy progib i yego gomologi v Tikhookeanskom poyase: Moscow, Akad. Nauk SSSR, Geol. Inst. Trudy, v. 244, 174 p.

Meyerhoff, A. A., 1972, Russians look hard at the Anadyr basin: Oil and Gas Jour., v. 70, no. 43, p. 124, 129, no. 44, p. 84, 89.

Meyerhoff, A. A., 1980a, Petroleum basins of the Soviet Arctic: Geol. Mag., v. 17, no. 2, p. 101-186.

Meyerhoff, A. A., 1980b, Geology and petroleum fields in Proterozoic and Lower Cambrian strata, Lena-Tunguska petroleum province, Eastern Siberia, USSR: Am. Assoc. Petrol. Geol. Mem. 30, p. 225-252.

Meyerhoff, A. A., 1981, The oil and gas potential of the Soviet Far East: Beaconsfield (England), Scientific Press Ltd., 176 p.

Meyerhoff, A. A., 1982a, The USSR northern and far eastern coasts: petroleum geology and technology, mining activities, and environmental factors: Ottawa, Dept. Indian Northern Affairs, 223 p.

Meyerhoff, A. A., 1982b, Carboniferous oil and gas production in the Eastern Hemisphere: Am. Assoc. Petrol. Geol. Studies in Geology, in press.

Mikulenko, K. I., and Gaydeburova, Ye. A., 1966, K voprosu o sopostavlenii gazoneftenosnykh gorizontov nizhneyurskogo vozrasta Vilyuyskoy gemisineklizy i Predverkhoyanskogo progiba: Geol. Nefti Gaza, no. 2, p. 43-47.

Miles, P., and Wright, N. J. R., 1978, An outline of mineral extraction in the arctic: Polar Record, v. 19, no. 119, p. 11-38.

Mitrofanova, L. I., 1978, Bentosnyye foraminifery neogena Yuzhnogo Sakhalin: Vestnik Mosk. Univ., no. 5, p. 97-100.

Nalivkin, D. V., 1973, Geology of the U.S.S.R.: Edinburgh, Oliver and Boyd, 855 p.

Naydenov, V. N., 1978, Perspektivy neftegazonosnosti miotsenovykh otlozheniy Yuzhnogo Sakhalina: Geol. Nefti Gaza, no. 2, p. 21-25.

Nesterov, I. I., Salmanov, F. K., and Shpil'man, K. A., 1971, Neftyanyye i gazovyye mestorozhdeniya Zapadnoy Sibiri: Moscow, Nedra, 463 p.

Nikolayev, I. Ye., and Kleshchev, A. A., 1981, Tipy neantiklinal'nykh lovushek nefti i gaza i ikh rasprostraneniye v predelakh ostrovnoy chasti Severnogo Sakhalina: Neftegazovaya Geol. Geofiz., no. 1. p. 17-19.

Ovcharenko, A. V., and Sevost'yanov, K. M., 1974, Perspektivy neftegazonosnosti i napravleniye dal'neyskikh neftegazopoiskovykh rabot na Kamchatke: Geol. Nefti Gaza, no. 11, p. 6-11.

Ovcharenko, A. V., and Bakin, V. Ye., 1979, Perspektivy poiskovo — razvedochnykh rabot na gaz v Yugo-Zapadnoy Yakutii: Geol. Nefti Gaza, no. 1, p. 1-7.

Ovcharenko, A. V., Taboyakov, A. Ya., and Tyutrin, I. I., 1973, Gazonosnost' Anivskogo progiba Yuzhnogo Sakhalina: Geol. Nefti Gaza, no. 7, p. 8-13.

Ovcharenko, A. V., Tyutrin, I. I., Taboyakov, A. Ya., Bakayev, G. A., and Naydenov, V. N., 1974, Geologicheskoye stroyeniye i perspektivy nefte gazonosnosti Yugo-Zapadnogo Sakhalina: Geol. Nefti Gaza, no. 5, p. 12-17.

Owen, E. W., 1975, Trek of the oil finders: a history of exploration of petroleum: Am. Assoc. Petrol. Geol. Mem. 6, 1647 p.

Pettijohn, F. J., Potter, P. E., and Siever, R., 1972, Sand and sandstone: Berlin, Springer-Verlag, 618 p.

Pol'kin, Ya. I., Zatsepin, Ye. N., and Rybkin, G. S., 1971, Tektonika mesokaynozoyskogo chekla shel'fa Karskogo morya i perspektivy ego neftegazonosnosti po dannym kompleksnykh geofizicheskykh issledovaniy, *in* Demenitskaya, R. M., Bolk, V. E., Gusev, B. V., Karasik, A. M., Litvinov, E. M., and Ivanov, S. S., eds., 1971, Geofizicheskiye metody razvedki v Arktike, v. 6: Leningrad, Minist. Geol. SSSR, Nauchno-Issledovatel'skiy Inst. Geol. Arktiki, p. 44-47.

Pushcharovskiy, Yu. M., 1964, The tectonics of Sakhalin: Internatl. Geol. Review, v. 7, no. 12, p. 2184-2196.

Rice, D. D., and Claypool, G. E., 1981, Generation, accumulation, and resource potential of biogenic gas: Am. Assoc. Petrol. Geol. Bull., v. 65, no. 1, p. 5-25.

Rodnikova, R. D., Sevost'yanov, K. M., and Taboyakov, A. Ya., 1968, Structural formation connections between southern Sakhalin and Hokkaido Island in relation to hydrocarbon potential: Internatl. Geol. Review, v. 10, no. 7, p. 749-760.

Rönnevik, H. C., 1981, Geology of the Barents Sea, Chap. in Il 36, *in* Illing, L. V., and Hobson, G.D., eds.,1981, Petroleum geology of the continental shelf of north-west Europe: London, Heyden and Son, Ltd., p. 395-406.

Rovnin, L. I., Dvurechenskiy, V. A., and Ovcharenko, A. V., 1976, Rezul'taty geologorazvedochnykh rabot na neft' i gaz po Ministerstvu geologii RSFSR za 1971-1975 gg. i osnovnyye napravleniya na desyatuyu pyatiletku: Geol. Nefti Gaza, no. 7, p. 1-7.

Rudkevich, M. Ya., 1980, Tektonika i genezis Zapadno-Sibirskoy plity v svete novykh geologo-geofizicheskikh dannykh, *in* Luchitskiy, I. V., 1980, Tektonika i neftegazonosnost' Sibiri i Dal'nogo Vostoka, v. IX, Tektonika Sibiri: Novosibirsk, Nauka (Sibir. Otdel.), p. 23-30.

Salmanov, F. K., 1974, Zakonomernosti raspredeleniya i useloviya formirovaniya zalezhey nefti i gaza: Moscow, Nedra, 280 p.

Samsonov, V. V., 1975, Irkutskiy neftegazonosnyy basseyn: Irkutsk, Minist. Geol. SSSR, Vostochno-Sibirskoy Knizhnoye, 196 p.

Samsonov, V. V., and Tyshchenko, L. F., 1970, O geneticheskoy svyazi gazov zakrytykh por i gazov produktivnykh plastov: Geol. Nefti Gaza, no. 8, p. 33-36.

Savitskiy, V. Ye., Boldyreva, V. P., Danchenko, R. V., and Mitrofanova, L. I., 1979, Oligotsen-miotsenovyye otlozheniya Yuzhnogo Sakhalina (Pogranichnyy progib): Vestnik Mosk. Univ., no. 1, p. 81-86.

Semenov, A. I., and Shcheglov, A. D., 1968, Geologicheskoye stroyeniye SSSR, tom IV, Osnovnyye zakonomernosti razmeshcheniya mestorozhdeniy poleznykh iskopayemykh na territorii SSSR: Moscow, Nedra, 504 p.

Semenovich, V. V., Gramberg, I. S., and Nesterov, I. I., 1973, Oil and gas possibilities in the Soviet Arctic: Am. Assoc. Petrol. Geol. Mem 19, p. 194-203.

Sevost'yanov, K. M., 1971, Tektonicheskoye rayonirovaniye i perspektivy neftegazonosnosti akvatoriy Vostochno-Sibirskogo i Chukhotskogo morey: Geol. Nefti Gaza, no. 6, p. 16-20.

Shabad, T., 1969, Basic industrial resources of the U.S.S.R.: New York, Columbia Univ. Press, 393 p.

Shabad, T., and Mote, V. L., 1977, Gateway to Siberian resources (the BAM): Washington, Scripta Publ. Co., 189 p.

Shimarayev, V. N., Galponenko, G. I., Martynov, V. S., and Plyakhina, I. L., 1971, Vyyavleniye osobennostey glubinnogo stroyeniya yugovostochnoy chasti Barentseva mory a po aeromagnityn dannym: Leningrad, Minist. Geol. SSSR, Nauchno-Issledovatel'skiy Inst. Geol. Arktiki, Geofizicheskiye Metody Razvedki v Arktike, v. 6, p. 51-54.

Shnip, O. A., Gibshman, N. B., and Vasil'yeva, M. Yu., 1981, Karbonatnyy razrez paleozoya yugo-vostoka Zapadnoy Sibiri: Neftegazovaya Geol. Geofiz., no. 7, p. 17-18.

Siryk, I. M., 1968, Neftegazonosnost' vostochnykh sklonov Zapadno Sakhalinskikh gor (Yuzhnyy Sakhalin): Moscow, Nedra, 138 p.

Skorobogatov, V. A., 1980, Usloviya formirovaniya uglevodorodnykh skopleniy v verkhneyurskikh otlozheniyakh tsentralnykh i severnykh rayonov Zapadnoy Sibiri: Geol. Nefti Gaza, no. 11, p. 25-32.

Taboyakov, A. Ye., 1973, Ob otkrytii promyshlennogo mestorozhdeniya gaza na Yuzhnom Sakhalina, *in* Novikov, G. N., et al., eds., 1973, Voprosy geologii i neftegazonosnosti Sakhalina: Leningrad, Akad. Nauk SSSR, Geograficheskoye Obshchestvo SSSR, Sakhalina Otdela Izvest., v. 4, p. 182-184.

Tikhomirov, Yu. P., Konyukhov, V. I., and Storbin, Yu. A., 1978, Opyt razvedki slozhnykh po fazovomu sostoyaniyu uglevodorodov zalezhey na Var'yeganskom mestorozhdenii, *in* Tikhomirov, Yu. P., ed., 1978, Metodika razvedki neftegazokondensatnykh mestorozhdeniy v Zapadnoy Sibiri: Tyumen, Minist. Geol. RSFSR, Glavtyumen'geologiya Trudy, v. 134, p. 46-68.

Toporskaya, L. P., 1970, Kolvinskiy val — novaya zona neftegazonapleniya na Severe Timano-Pechorskoy provintsii: Geol. Nefti Gaza, no. 12, p. 5-7.

Trebin, Yu. A., and Rubtsova, N. M., 1978, Zakonomernosti rasprostraneniya zalezhey i mestorozhdeniy gazokondensatnoneftyanogo (GKN) i gazoneftyanogo (GN) tipa, *in* Tikhomirov, Yu. P, ed., 1978, Metodika razvedki neftegazokondensatnykh mestorozhdeniy v Zapadnoy Sibiri: Tyumen, Minist. Geol. SSSR, Glavtyumen'geologiya Trudy, v. 134, p. 3-10.

Trofimuk, A. A., ed., 1971, Neftegazonosnyye basseyny Dal'nego Vostoka SSSR: Moscow, Nedra, 184 p.

Trofimuk, A. A., and Vyshemirskiy, V. S., 1976, Problema neftegazonosnosti paleozoya na yugo-vostoke Zapadno-Sibirskoy nizmennosti: Novosibirsk, Nauka, Sibir. Otdel., 237 p.

Trofimuk, A. A., Vasil'yev, V. G., Karasev, I. P., Kosorotov, S. P., Mandel'baum, M. M., Mustafinov, A. M., and Samsonov, V. V., 1964, Main problems of prospecting the Markovo oil field in eastern Siberia: Geol. Nefti Gaza (Eng. trans., in *Petroleum Geology*, 1969, McLean, Va., v. 8, no. 2, p. 13-18).

Vasil'yev, V. G., ed., 1968a, Geologiya nefti, spravochnik, tom 2, kniga 1, Neftyanyye mestorozhdeniya SSSR: Moscow, Nedra, 764 p.

Vasil'yev, V. G., ed., 1968b, Gasovyye mestorozhdeniya SSSR, spravochnik: Moscow, Nedra, 687 p.

Vasil'yev, V. G., and Zhabrev, I. P., eds., 1975, Gazovyye i gazokondensatnyye mestorozhdeniya, spravochnik: Moscow, Nedra, 527 p.

Vedernikov, G. V., Yashkov, G. N., and Firsova, T. K., 1980, K voprosu ob izuchenii paleozoyskikh otlozheniy yuga Zapadno-Sibirskoy plity metodami seysmorazvedki: Novosibirsk, Geol. Geofiz., no. 5, p. 85-96.

Verba, M. L., Galponenko, G. I., Ivanov, S. S., Orlov, A. N., Timofeyev, V. I. and Chernenkov, Yu. F., 1971, Glubinnoye stroyeniye i perspektivy neftegazonosnosti severo-zapadnoy chasti Beringora morya: Leningrad, Minist. Geol. SSSR, Nauchno-Issledovatel'skiy Inst. Geol. Arktiki, Geofizicheskiye Metody Razvedki v Arktike, v. 6, p. 70-74.

Vereshchagin, V. N., and Kovtunovich. Yu. M., 1970, Geologiya SSSR, Tom xxxiii, Ostrov Sakhalin, Geologicheskoye opisaniye: Moscow, Nedra, 432 p.

Vereshchagin, V. N., and Kovtunovich, Yu. M., 1974, Geologiya SSSR, Tom xxxiii Ostrov Sakhnlin, Poleznyye iskopayemyye: Moscow, Nedra, 208 p.

Vlasov, G. M., and Belova, M. B., eds., 1964, Geologiya SSSR, Tom xxxi, Kamchatka, Kuril'skiye i Komandorskiye ostrova, chast' I, Geologicheskoye opisaniye: Moscow, Gosudarstvennyy Geol. Komitet SSSR, Nedra, 733 p.

Vlasov, G. M., and Tarasenko, T. V., eds., 1977, Geologiya SSSR, Tom xxxi, Kamchatka, Kuril'skiye i Komandorskiye ostrova, Poleznyye iskopayemyye: Moscow, Nedra, 351 p.

Vlasov, G. M., and Tarasenko, T. V., eds., 1977, Geologiya SSSR, Tom xxxi, Kamchatka, Kuril'skiye i Komandorskoye ostrova, chast' I, Geologicheskoye opisaniye: Moscow, Gosudarstvennyy Geol. Komitet SSSR, Nedra, 735 p.

Vorob'yev, V. N., 1980, Tipy zalezhey nefti i gaza v dokembriyskikh i nezhnepaleozoyskikh otlozheniyakh Sibirskoy platformy: Geol. Nefti Gaza, no. 1, p. 1-5.

Wright, N. J. R., 1978, Coal deposits of the Pechora basin, northwestern USSR: Cambridge (University) Arctic Shelf Programme (CASP) Rept., 112 p.

Yanshin, A. L., 1962, Perspektivy otkrytiya mestorozhdeniy kaliynykh soley na territorii Sibiri: Novosibirsk, Geol. Geofiz., no. 10, p. 3-22.

Yaskevich, V. I., Yakovlev, Yu. K., Chetvergov, A. P., Klyuchko, V. P., and Stepanov, V. I., 1980, Resultaty: problemy izucheniya tektoniki zapadnoy chasti Sibirskoy platformy i Yenisey-Khatangskogo progiba po geologo-geofizicheskim dannym, *in* Luchitskiy, I.V., 1980, Tektonika i neftegazonosnost' Sibiri i Dal'nego Vostoka, Tektonika Sibiri, v. IX: Novosibirsk, Nauka (Sibir. Otdel.), p. 79-84.

Yurochko, A. I., Svitenko, V. S., and Borisov, B. D., 1981, Vliyaniye gidrodinamicheskikh usloviy na razmeshcheniye skopleniy gaza v Anivskom progibe Yuzhnogo Sakhalina: Geol. Nefti Gaza, no. 12, p. 37-40.

Zakharova, M. A., 1973, Litologiya paleogenovykh otlozheniy Sakhalina i usloviya ikh obrazovaniya: Novosibirsk, Akad. Nauk SSSR, Dal'nevostochnyy Nauchnyy Tsentr, Sakhalinskiy Kompleksnyy Nauchno-Issledovatel'skiy Inst., 108 p.

Zapivalov, N. P., 1979, Geologicheskiye predposylki i metodika poiskov zalezhey nefti v paleozoye na yuge Zapadnoy Sibiri: Sovet. Geol., no. 3, p. 22-37.

Zapivalov, N. P., Zubarev, Ye. M., Min'ko, V. A., Pluman, I. I., Rozhok, N. G. Serdyuk, Z. Ya., and Sherikhova, V. Ya., 1975, Paleozoyskaya neft' v Novosibirskoy oblasti: Geol. Nefti Gaza, no. 2, p. 14-19.

Zapivalov, N. P., Serdyuk, Z. Ya., Zalazayeva, L. V., and Yashina, S. M., 1978, Neftegazonosnost' karbonatnykh porod silura-nizhego karbona Zapadnoy Sibiri: Geol. Nefti Gaza, no. 1, p. 25-32.

Zapivalov, N. P., Botneva, T. A., Pankina, R. G., Shulova, N. S., and Guriyeva, S. M., 1980, Geokhimicheskaya kharakteristika paleozoyskikh i mesozoyskikh neftey yuga Zapadnoy Sibiri: Geol. Nefti Gaza, no. 9, p. 8-14.

Zharkov, M. A., 1969, Ob ob'yemakh solenakopleniya v kembriyskuyu epokhu: Akad. Nauk SSSR Doklady, v. 184, no. 4, p. 913-914.

Zheleznova, A. P., and Kul'kova, A. A., 1981, Sravnitel'naya kharakteristika litologo-petrofizicheskikh parametrov peschanykh plastov Markovskogo i Srednebotuobinskogo mestorozhdeniy: Geol. Nefti Gaza, no. 3, p. 40-44.

Zolotov, A. N., Tyshchenko, L. F., Obchenkov, L. K., and Antsiferov, P. S., 1968, Stroyeniye gazokondensatnoy zalezhi Parfenovskogo gorizonta Markovskogo mestorozhdeniya: Geol. Nefti Gaza, no. 6, p. 26-30.

*Cold Regions Science and Technology*, 7 (1983) 167—177
Elsevier Science Publishers B.V., Amsterdam — Printed in The Netherlands

# Coal Resources in Arctic Areas Outside USSR

*Alv Orheim*

Store Norske Spitsbergen Kulkompani A/S
Longyearbyen

## Abstract

*The Arctic region contains several important coal basins, and coal is present in all parts of the Arctic.*

*They are inferred to represent more than 50 % of all coal resources of North America. In most areas, however, the exploitation of these resources have been hampered for a long period of time because of their remoteness. Today the only mines in operation are in Svalbard and at one locality south of Fairbanks, Alaska.*

*Polar conditions compose several restrictions to exploration and mining. This requires a re-ordering of economic determinants, taking into account elements like permafrost, lack of water for dust suppression, semi-annual shipping and the sensitive environment. The presence of mining is a proof that the coal resources are economic viable. The development of other Arctic geo-resources will be favourable to the coal industry as well. During the last 3—5 years there has been a dramatic and positive change in the attention given to Arctic coal resources.*

## Introduction

We do not know when coal was first used by the indigenous population of the Arctic. Presumeably it was a long time prior to the first written account of Alaskan coal by Nathaniel Portlock in 1786. In Svalbard coal, and its potential use to whalers was reported by Jonas Poole in 1610, one year before the first whale was killed.

The method of extraction during these early years was to scavenge the outcrops close to the seashore. The first serious attempts for commercial utilization began in 1855 when Siberian fur traders opened a coal mine at Port Graham, Alaska. The coal was intended for an export market which never materialized and the miners settled for local and maritime demands. Later several other mines started on the same assumption: to supply a growing export market. However, only the mines in Svalbard have achieved this goal. In fact the coal produceres of Alaska, the Yokon Territory and Greenland hardly survived the global transition from coal to oil after World War II. Presently there are only six mines in operation outside the USSR. Four of these are in Svalbard, one in Alaska and one in Yokon (Figure 1).

Several localities are now being evaluated for coal production. This paper discusses firstly some of the considerations pertinent to coal mining within subzero environment. Thereafter the most significant fields of coal production and resources are described.

## Considerations

### Mining

The many constraints facing any industrial activity in the Arctic are well documented. Severe climatic conditions, frost heave, lack of water, remote locations and difficult communications are some of the factors that make any activity expensive.

On the other hand, the chances of finding flat-lying coalbeds, near surface and of good quality and thickness, might be assumed to be greater here than in more well developed mining districts.

0165-232X/83/$03.00  

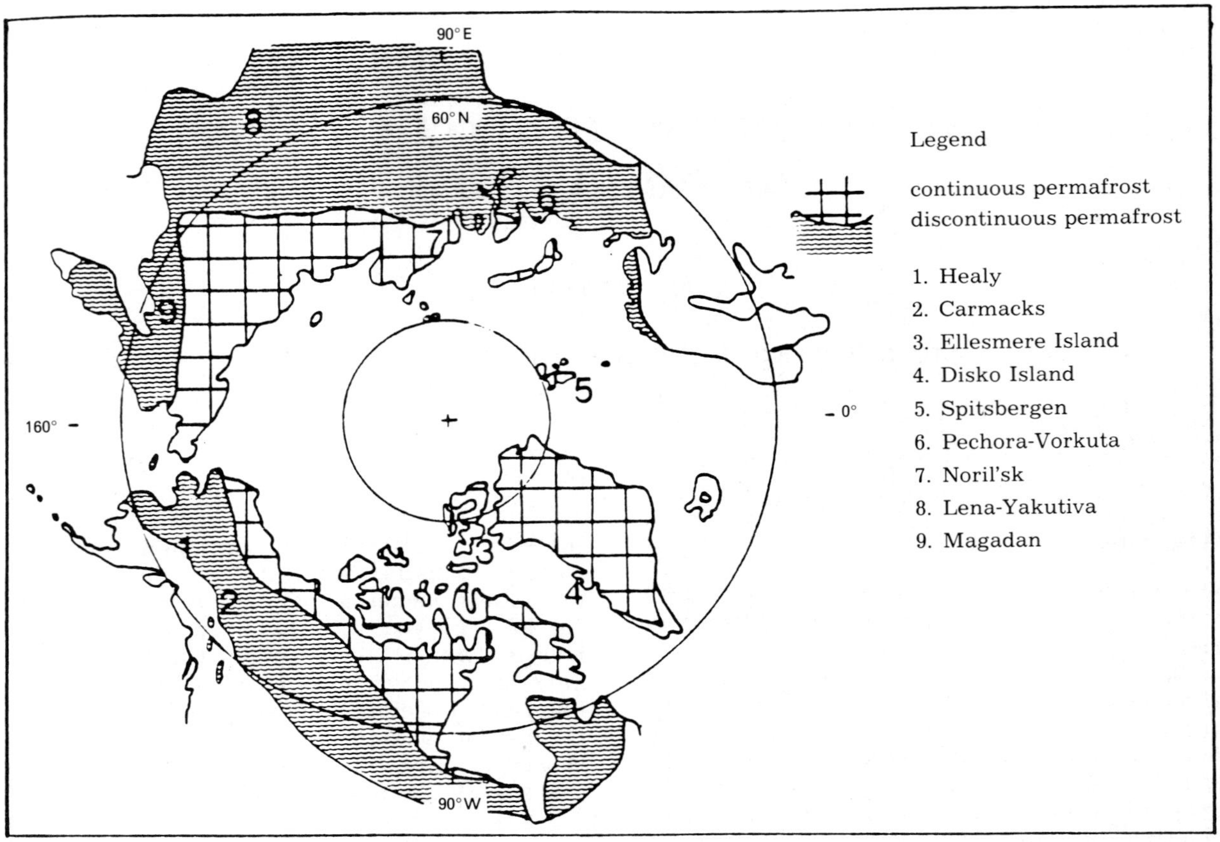


FIGURE 1. *Distribution of permafrost on the Northern Hemisphere and circumpolar coal resources.*

During the last 15 years there has been major shifts in mining techniques, towards fully mechanized mines. This has greatly improved productivity and reduced the hazards. However, every improvement on the technical side will increase the capital costs. Thus, the marginal tonnage of coal to support investments in new mines is rising. This general development will also have impact on the evaluation of Arctic mine prospects. Another similar trend is the increasing emphasis on high quality coal. It is likely that this will bring more attention to coal in remote areas.

*Mining within permafrost* faces numerous unparalleled conditions. Briefly they are summarized as: cool, dry environment underground, stable rock mechanic conditions, difficult dust suppression because of sub-zero temperatures, and particular precautions must be taken during the summer when humid, warm ventilation air disrupts the rock-mechanical state.

The overall net effect of permafrost on the underground mining activity itself is 10 — 30 percent increase in production costs.

*Methane gas and water* may accumulate in the area just below the permafrost level. When all fissures and pores are filled with ice, permafrost acts as cap rock for migrating gas and water under hydrostatic pressure.

As shown in Figure 2 it is of particular importance to consider these factors when drilling or mining at the lower part of steep slopes. If the mountains nearby is much higher than the maximum thickness of permafrost the interaction of permafrost and ground water will create an effect similar to Artesian wells.

*Weathered coal* is a significant problem in the Arctic. This is due to extensive frost action over the last few millions of years.

At Spitsbergen the result of this action can be seen as far as 150 meters below surface. Open cracks, now filled with ice and parallel to the mountainside are typical. Free circulating water have oxidized pyrite and other ash components. At the same time repeated freezing and thawing of inherent water in the coal has disrupted the coal structure itself. To analyse such coals for sulphur, ash, volatile matter or coking properties is therefore meaningless unless the results are put in proper context.

**Exploration.**

It must be remembered that coal exploration in the Arctic still is in its early juvenile stage. Only very few areas have been covered and small amounts of money and time have been spent. Most work has used relatively small, traditional types of equipment and methods. This situation reflects both the economic and environmental constraints that have faced the explorationists.

Many of the problems encountered in coal exploration, i.e. core recovery, may be overcome by using heavier, more oil-field type equipment — developments similar to what is seen in more temperate areas. However the economic and logistic consequences of applying these techniques in Arctic coal exploration are substantial. In most cases it is therefore assumed that smaller scale exploration must be completed prior to any commitment to large scale development work.

*Core drilling* is a widely used exploration tool, and also the most difficult. One basic question to ask when planning a drilling program is whether to use heated drill fluid, i.e. to suppress the permafrost in the borehole wall, or to use a fluid with freezing point below expected wall temperature.

Heated fluid is mainly chosen in cases where surface water is readily available, offering substantial savings in transportation costs.

For relatively shallow boreholes, maximum 150 meters, or when depth of permafrost is moderate, warm water gives satisfactory results. However, at greater depths the method involves high risk for freezing of drill rods, and total failure.

To retain permafrost in the borehole wall, and use a drill fluid with freezing point below the minimum ground temperature, is therefore the only solution to drilling deeper holes. The stability of the wall is maintained, additional thawing in the active layer is reduced, and the

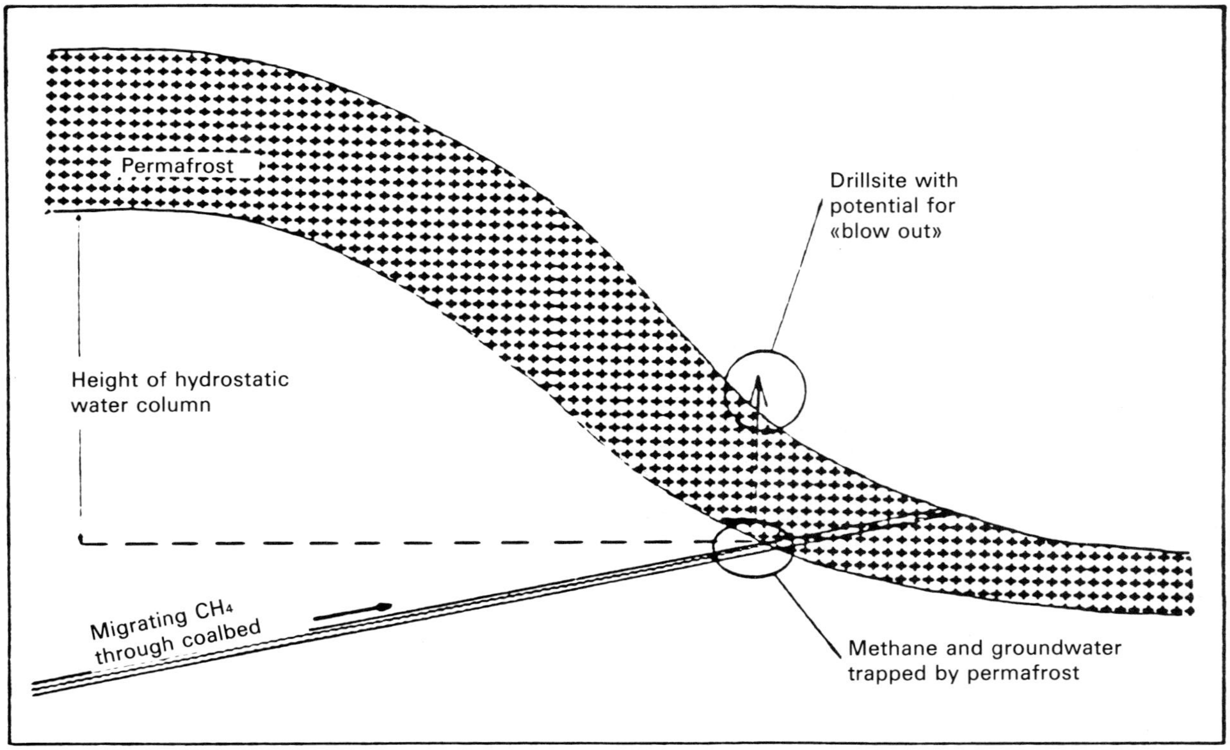


FIGURE 2. *Potential drilling problems when permafrost acts as caprock for methane gas and ground water.*

fluid characteristics offer some leeway for circulation breaks without immediate risk of freezing.

*Trenching* i.e. to expose the coal outcrop by removing the overburden, is an important exploration technique. One effect of permafrost to be aware of here is solifluxion. Through this, outcropping coalseams may be moved several meters down the hillside, Figure 3. Because the overburden is frozen below the active layer it may be difficult to determine exactly where the proper outcrop is. Common errors are therefore to start the trenches too low, to report too high seam thickness, and to «invent« false faults.

**Economic**

The object of exploration is to prove workable resources, and identify the most profitable exploitation. The relative abundance of coal deposits within the Arctic Circle in spite of the paucity of observations so far, implies considerable opportunities in the region.

On the other hand, activities here are faced with much higher risks compared to further south.

The most important factors in this picture are the vast distances, and the remoteness of potential mining areas. Consequences of the situation are:

— long lead time from initial discovery to production, i.e. high capital costs,
— higher investments because of infrastructure, housing, roads, etc.,
— particularly high costs during exploration,
— relatively high operational costs, e.g. higher wages, more expensive shipping.

Other constraining and economic factors relate to the protection of the vulnerable arctic environment. This need for prudence imposes restrictions especially during exploration.

The economic impact of the preceding factors will vary from one region to another. Within the same economic framework, and under otherwise equal conditions, cost of living is probably 10 percent higher in Arctic regions compared to further south. Similarly, production costs will most likely be 20 — 30 percent higher while the costs of constructions are 1 1/2 to 4 times higher. The most noticable difference is found in exploration costs, where work in the Arctic regions has proved to be 3 to 10 times more expensive than further south.

## Description of Resources

**Terminology**

In the most general description a resource or resource base is said to contain all deposits of a

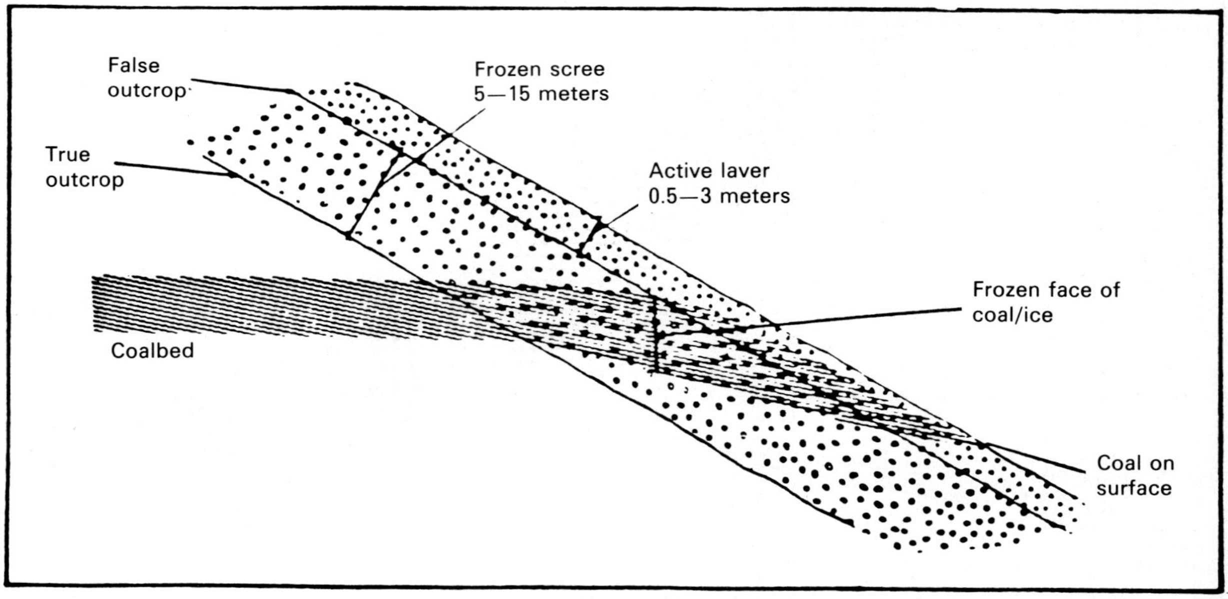


FIGURE 3. *Outcrop of coalseam slid downward through solifluction.*

potential extractable mineral or rock. The reserve of a deposit is said to be the mapped and economic mineable portion of the resources. A special case is made of the subeconomic discovered deposits. New technology, or market development might shift them into reserves, and the significance of already knowing the geology of the deposit is such that it can not be grouped with unknown resources. Internationally it is suggested to use the following areal limitations for identified resources (Figure 4):

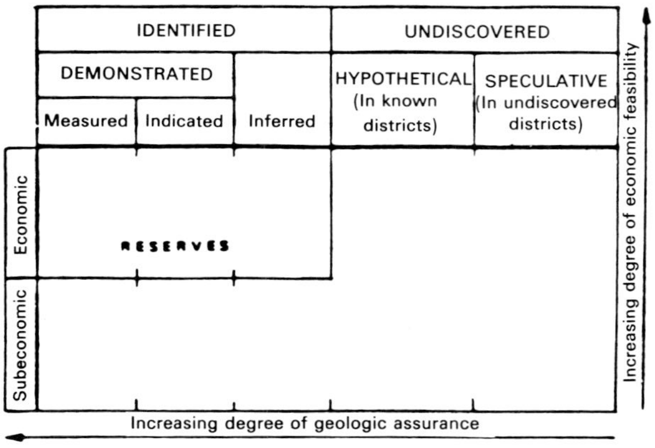


FIGURE 4. *Resource terminology adopted in this paper.*

- measured reserves: less than 0.5 kilometre from known observation (borehole, adit or outcrop)
- identified reserves: 0.5 — 1.0 kilometres from known observation.
- inferred reserves: from 1 to 5 kilometres distance of known observation.

Accordingly, hypothetical resources are at a distance greater than five kilometres from observed coal. These recommendations correspond in principle to the practice in Svalbard, but slightly stricter rules with regard to areal extent are practised by Store Norske there. This reflects the necessity of caution in reserve estimates under Arctic conditions, and it is recommended to adopt these narrower limits for all Polar regions. The present study is a compilation from several published and unpublished sources. As far as possible the terminology above was adopted in the summarized table of resource estimates (Table 1).

**Table I:**

*Arctic coal resources outside of USSR. Estimates summarized. (Billion tons).*

| | Identified | Undiscovered |
|---|---|---|
| Alaska | 14.000 | 3.000.000 |
| Svalbard | 0.144 | 26.780 |
| Canada | 1.000 | 200.000 |
| Greenland | 0.180 | 0.700 |
| | 15.324 | 3.227.480 |

**General overview**

The total land area of Alaska, Arctic Canada, Greenland and Svalbard is 6,000,000 km$^2$, of which 1/3 is covered by glaciers. Significant parts of Canada and Greenland have igneous or methamorphic rocks with no coal potential. In other areas, notably Alaska and the Canadian Arctic Archipelago large sedimentary basins are identified and coalresources are inferred. The geological mapping has achieved its most advanced state in Svalbard, partly due to fairly restricted area and partly due to the fact that there is hardly any vegetation cover. A general overview was therefore easily established. In Alaska mapping is also at an advanced state, leaving little doubt as to the overall coal resources.

Petroleum and coal are closely related in origin, and it is no surpise to encounter coal formations in the course of petroleum exploration. This has happened in several fields of the Norwegian North Sea, where so far 44 wells have reported coal. As of today there is no official documentation published of these observations. They are said to consists of several beds, with maximum thickness of 12 m. Similar situations are easily envisioned in the petroleum basins of northern North America. The potential resources at each of these fields must be in the order of hundreds of million tons.

Presently the technological challenges for extracting subsea coal, either by mining or by in situ gasification, are numerous and seemingly insuperable. Although this is not a rigid situation, the economical incentive to develop the technology is clearly not present today. Thus it will here suffice to acknowledge the presence of large coal resourses offshore, available when or if exploitation is economically feasible.

**Alaska**

The coal deposits are subdivided into 3 major fields and several smaller ones, Figure 5. Total

FIGURE 5. *Coal areas of Alaska.*

resources are in the order of two to five million tons of coal. These figures suggest that Alaska have 40 % of all USA coal resources. Every coal quality, from anthrasit to lignite is present within the state. The North Slope-area may constitute the worlds largest coal province, it is also one of the least known and one of the least probable for large-scale utilization. The coal are of late Cretaceous, early Tertiary age. Beds with thickness six to ten metres are not uncommon, and a total of 150 beds are identified.

The Nenana field, and numerous satelite fields, consits of Tertiary coals in an area 150 miles long, 30 miles wide. The only Alaskan mine in operation is found at Healy, in the central part of the field. Production at Healy is 0.7 million tons (1980), all of which is used in utilities in the Fairbanks district. Plans for expanded production and export to Korea will come into effect soon.

Cook Inlet — Susutna Basin is also of Tertiary age. It occupies both sides of Cook Inlet, and there are assumed large subsea resources. Due to the proximity to good habour facilities several of the individual fields have been evaluated as economical, with a proposed market in Asia.

Of the total inferred coal reserves of Alaska less than 1 % represents measured reserves, In Alaska the chances of exploitation will develop along two different paths. The incentive to use coal on a small scale will increase as oil prices go up. Programmes to substitute oil by coal is proposed for several of the small villages and also for Fairbanks and Anchorage. A full scale development along these lines would quadruple the present level of production.

The other path for coal development in Alaska will be to export the coal. Alaskan coal is generally very low sulphur, sub-bituminous to bituminous. Normally such coals will have good prospects within the energy-market, although the moisture content is somewhat high. The resources also include possible metallurgical coals and lignites. The decisive economic element in evaluating the export market is however neither quantity nor quality, but the costs of infrastructure, transportation etc. So far these elements have hampered all plans for coal export. Negotiations are in process for several different mines, and the situation to day have several positive facets.

**Svalbard**

In Svalbard two Russian and two Norwegian coal mines are in operation. These mines have a total annual production of 0.4 million tons from their two settlements, Barentsburg and Pyramiden.

Longyearbyen is the administrative center for the islands, and is also the center for the Norwegian mining company Store Norske Spitsbergen Kulkompani A/S (SNSK). The total population of Longyearbyen is 1200. The annual coal production is 0.4 millions tons.

50 km SE of Longyearbyen a new mine, Svea, is now being developed. Small scale production and larger exploratory works are carried out, while plans for a 0.25 million tpy production are being considered.

There are several coal formations on Svalrepresent the northern most industrial activity of the world. (Figure 6, Table 2). As there is no native population to Svalbard, the coal mining has always been geared at export. The mining was started by the American J.M. Longyear in 1906. Later several other nationalities began mining, but during the difficult economic times of the 1920's most settlements were abandoned.

Today the Russian company Trust Arkitikugol («Arctic Coal») employs 2200 people, and bard. (Figure 6, Table 2). However, only the coals from the Carboniferous period (north of Isfjorden) and the lower Tertiary in the central basin, are of any importance. The Tertiary coals represent the most promising reserves for the future.

So far explorations and mining have been restricted to an area 10 — 20 km wide along the outcrops. The whole basin, and in particular the eastern part from Longyearbyen to Svea, has favourable conditions for coal.

There are five coal seams in Longyearbyen, of which only the Longyear seam, has satisfactory thickness and quality for extraction. It is subbituminous with low ash and sulfur content,

FIGURE 6. *Location of coalmines in Svalbard.*

**Table 2.**

Coalbearing Strata

| Age (mill. years) | Formation thickness (m) | Period | Mining sites |
|---|---|---|---|
| 40 | | Upper | |
| | 800 | Tertiary | |
| 63 | | Lower | Barentsb. Ny-Ålesund Longyearb. Svea Grumant |
| | 500 | | |
| 120 | | Cretaceous | Advent City |
| | 1200 | | |
| 200 | | Triassic | |
| | 1400 | | |
| 330 | | Carboni-ferous | Pyramiden Bjørnøya |
| | 1100 | | |
| 370 | | Devonian | Bjørnøya |

well-suited both as metallurgical coal and steam coal. The thickness varies between 0.8 m and 1.8 m. The main seam at Svea has a maximum height of 5.5 m with quality similar to the Longyear seam.

Allthough both Svea and Longyear seams have satisfactory quality they are distinctly different in technological behaviour. This is mainly due to different chemical and physical environment during peat accumulation. This accumulation is the crucial factor both with regard to bed thickness and quality. The two seams are thus not interchangeable without some modifications of technology.

In the resource estimates this fact has led to different evaluation of coal observations, depending on whether a particular bed in question has been regarded as Longyear-type or Svea-type. One characteristic property of the Svea coalbed is a lateral discontinuity, with great variation in thickness. In the course of exploration it therefore becomes more difficult to have success. On the other hand, locally the bed is developed with a favourable thickness, well suited for modern equipment.

Systematic exploration on behalf of Norwegian interests, startet 15 years ago, both on a detailed and regional scale. The results of this

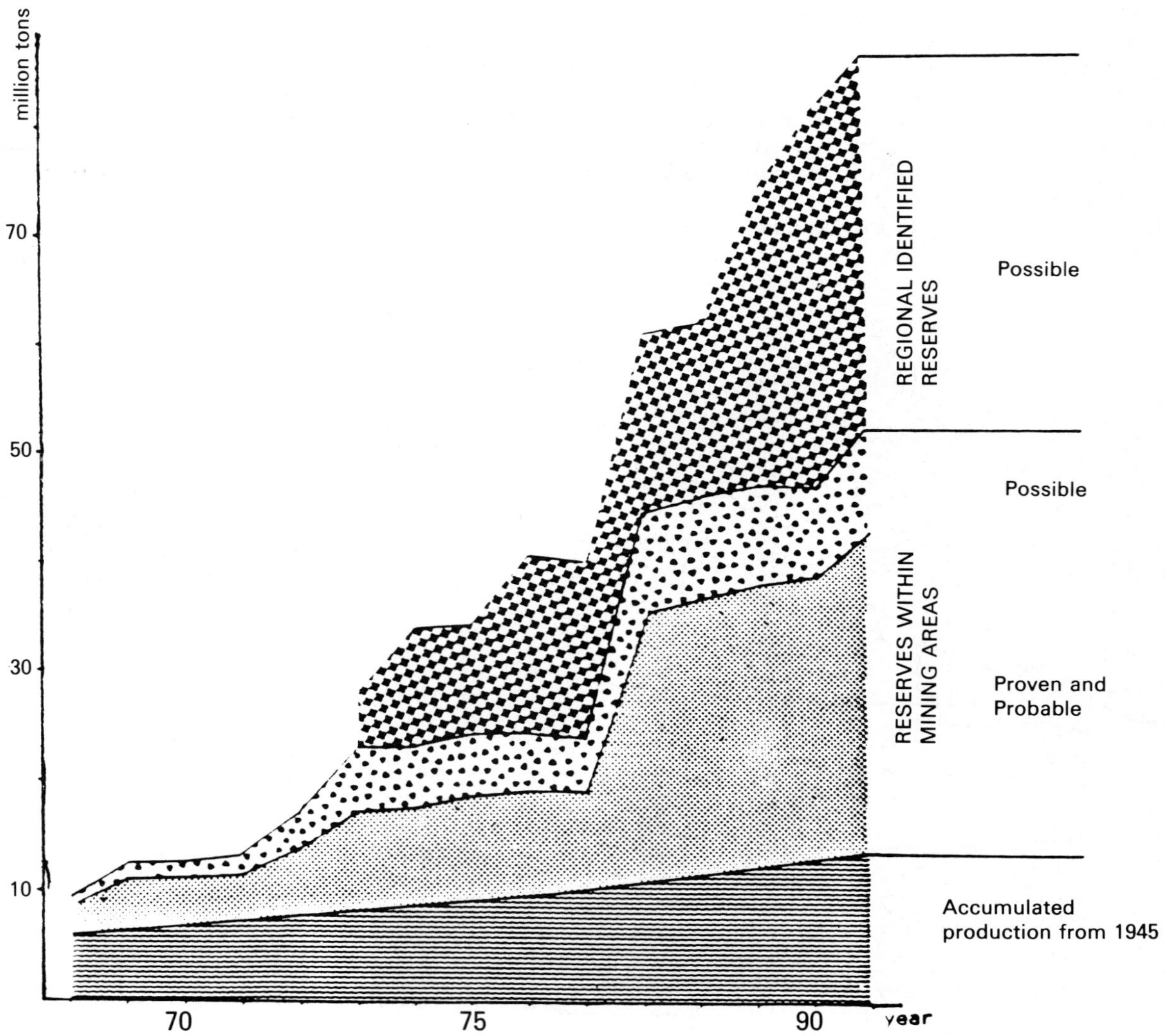


FIGURE 7. *SNSK resource estimates for the exploration period 1967-1981.*

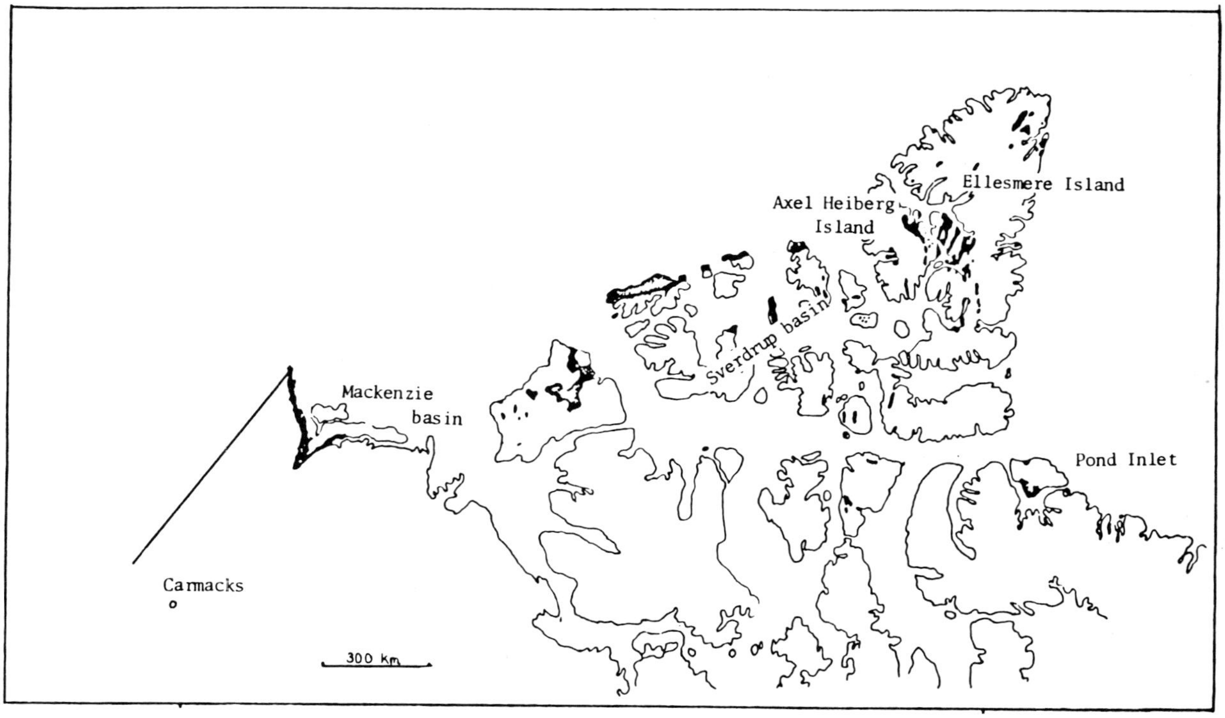


FIGURE 8. *Cretaceous and Tertiary outcrops with coal, Arctic Canada.*

activity, superimposed on the accumulated production of SNSK is presented in Figure 7. The substantial increase in identified, economical reserves is proof of the favourable exploration potential of Svalbard. Obviously this is a result of a historically low rate of production versus the resource base.

The total inferred coal resources of Svalbard are 144 million tons, with an undiscovered resource of possibly 26 billion tons.

**Canada**

The Canadian Arctic have numerous coal depesits (Figure 8). However mining has to this day been on a limited scale. At Carmacks in Yokon Territory and at Pond Inlet, nothern Baffin Island two small mines have operated. The coal has been used locally.

At both localities the coal are of Cretaceous age, corresponding to some of the very large deposits lately discovered in the Mackenzie and Sverdrup basins. Within the Arctic Archipelagos coal has been discovered in several geological formations. Most promising are the Isachsen Formation (Cretaceous) with coal seams 1.8 m, and Eureka Sound Formation (Tertiary) with several coalseams of 4.5 m or more.

The latter type is particular well developed on Ellesmere Island and Axel Heiberg Island. Several of the seams show a lateral discontinuity similar to the coals of Svalbard. It is belived that the sedimentary basins of these areas had related evolution some 50 to 100 million years ago.

The development of the large coal resources of Northern Canada is now just commencing. As a result of exploration efforts for petroleum significant coal resources have been recognized. Three different groups are now preparing programmes to evaluate the coal. Previous feasibility studies have not taken into account the beneficiary effect of other development in the area. The opening of lead-zinc mines at Little Cornwallis Island and at northern Baffin Island and petroleum production from the Mackenzie and Sverdrup basins, will effectively change the logistic situation of the whole area.

The total coal resources of the Arctic Canada have only been surmised on a very rough basis. Total resources may be in the order of 200 billion tons, of which only 0.5 % is inferred or measured. As exploration programmes are expanded these figures are however rapidly changing. This is illustrated by the results at

Bonnet Plume, Yokon, where two years work resulted in 380 million tons of measured, indicated and inferred reserves.

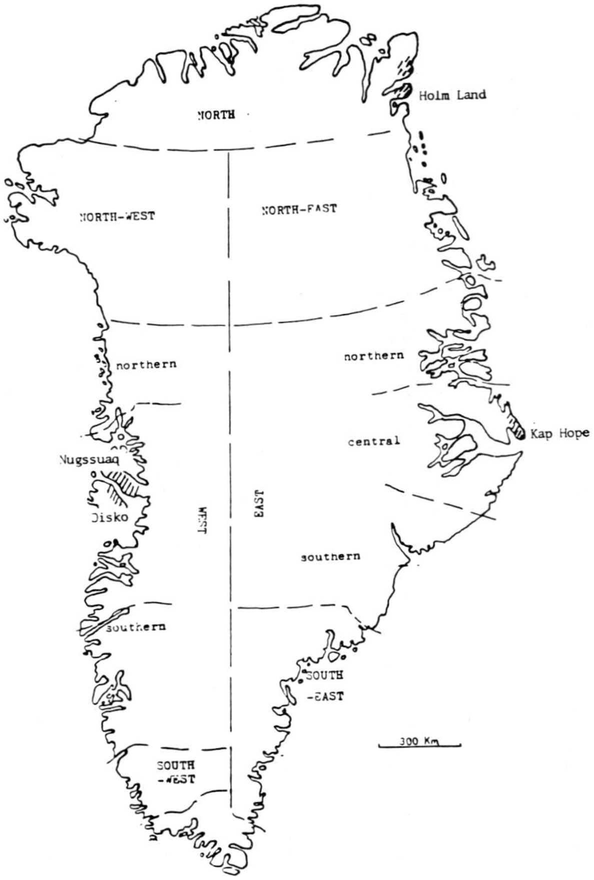


FIGURE 9. *Location of coal deposits in Greenland.*

### Greenland

This region seems to be the one with least coal potential. The most significant deposits are at the Disko and Nugssuaq areas on the west coast. (Figure 9). The coals there are of Cretaceous age, with maximum seam thickness 2 — 3 m. The quality characteristics are subbituminous, low sulphur, low calorific value. Similar coals are also found on the east coast of Greenland, at Kap Hope and Kap Stewart, but only in very small quantities.

Younger, Tertiary, coals are quite often found above the Cretaceous coals, but they have limited lateral extend. Early Carboniferous coals are found on the Eastern and Northern coasts. With maximum thickness of 0.5 m these may safely be disregarded as potential coal resources even though the quality is excellent.

Coal mining at Nugssuaq commenced in 1924, and terminated in 1972. The coal was only shipped to local users in Greenland. Following the oil embargo in 1973 a renewed interest for the deposits lead to futher explorations. The final results from this work have not been published. The resource base for all of Greenland is ca. 700 million tons within non-glaciated areas. Approximately 25 %, 180 million tons, are identified.

## Concluding Remarks

Coal mining and exploration have been part of the development of arctic regions through the last 8 decades. With some modifications the techniques and approaches are the same here as south of the 60th parallel. It is highly probable that mining coal gradually will become more intensive in Arctic regions. Partly this is a result of general expanding exploitation of resources in these areas. An increased level of coal production is also due to the recognition that this part of the world has large discovered and undiscovered coal resources. The potential result of futher exploration is demonstrated by the fact that the Arctic region (excluding Greenland due to its high precentage of icecoverage) have a resource density factor of 1100 tons per square kilometre, compared to global average of 81000 tons per square kilometre. The main reason for the discrepancy of these values must be the lack of knowledge of the coal deposits within the Northern areas.

## References


### Alaska

Rao, P.D. & Wolff, E.N. (ed) 1981: «Focus on Alaska's Coal '80.
Proceeding of the Conference, University of Alaska, Fairbanks».
MIRL report number 50. 518 pp.

### Canada

Fortier, Y.O. et al 1963: «Geology of the North Central part of the Arctic Archipelago. Northwest Territories».
Geol. Surv. Can. Mem. 320.

Tozer, E.T. & Torsteinson, R. 1964: «Western Queen Elisabeth Islands, Arctic Archipelago».
Geol. Surv. Can. Mem. 332.

Young, F.G. 1975: «Upper Cretaceous stratigraphy, Yokon,
Coastal Plain and north western Mackenzie Delta».
Geol. Surv. Can. Bull 249, XII, 83 pp.

Jackson, G.D. et al 1975: «Geology of the Pond-Inlet Map area,
Baffin Island, District of Franklin».
Geol. Surv. Can. Pap. 74 — 25.

**Svalbard**

Harland, W.B. et al 1976: «Some coal bearing strata in Svalbard».
Norsk Polarinstitutt Skrifter No. 164.

Orheim, A. 1979: «Coal mining in Svalbard».
Report, 5. Int. Conf. on Port and Ocean Engineering.
Norw. Inst. of Technology.

Orheim, A. 1982: «Undiscovered Tertiary coal resources of Svalbard. An assessment using Monte Carlo simulation».
Proceedings 3rd Int. Symp. on Arctic Geology.
Geol. Surv. Can. (in press).

**Greeland**

Schiener, E.J. 1976: «West Greenland coal deposits: distribution and petrography»
Geol. Surv. Green. Rpt. 77.

Schiener, E.J. 1976: «Coal Geology»
in: Geology of Greenland (eds A. Escher & W.S. Watt).

Surlyk, F. et al 1973: «Stratigraphy of the Jurassic-lower Cretaceous sediments of Jameson Land, East Greenland».

# Section III

*Technological/Economic Aspects of Exploration/Exploitation of Arctic Energy Resources*

*Cold Regions Science and Technology*, 7 (1983) 181—193
Elsevier Science Publishers B.V., Amsterdam — Printed in The Netherlands

# Exploration Methods in the Canadian Arctic

*J. Hnatiuk*

Gulf Canada Resources Inc.
Calgary

## Abstract

*Exploration for hydrocarbons in the Canadian Arctic is very challenging and expensive due to the unique and very harsh environments. Sea ice and icebergs present the most formidable problems, although sea bottom conditons such as ice gouging, permafrost and gas hydrates also make drilling more difficult. In the Beaufort Sea, artificial islands have been used out to water depths of 20 m, a concrete caisson is in place on a subsea berm in 22 m of water and four ice-reinforced drillships have been used during the short 110-day open water season in water depths of 18 m to 68 m. The latest generation of floating vessel is a round conical-shaped drilling barge which will commence drilling in 1983 and will extend the drilling season. Two other innovative systems to be used soon are a fully-integrated floating steel caisson and a used crude oil tanker, both of which will be held on location by the weight of the sand or water which will be pumped into their interiors.*

*In the Arctic Islands offshore areas, drilling is conducted off thickened sea ice in water depths of up to 370 m. Added flotation is provided by plastic foam which is incorporated into the ice platforms. Drilling is only conducted in Winter and only in areas where the ice does not move.*

*On the east coast where ice is a problem, drilling is seasonal using primarily dynamically positioned drillships which can quickly move if an approaching iceberg cannot be towed.*

*On land, the primary problems are presented by the sensitive tundra which cannot be disturbed since the ice rich soils would thaw and subside. Most drilling is done in Winter unless supplies are stockpiled and helicopters used for personnel and minor supplies.*

*Geophysical techniques differ in the Arctic Islands because often seismic surveys must be conducted from the ice surface. In the Beaufort Sea, seismic is by marine vessel. On land, seismic surveys are usually in Winter because of the sensitive tundra. Interpretation of geophysical data is often complicated by the presence of permafrost.*

*In spite of the challenges and difficulties presented, some 250 wells have been drilled in Canada's Arctic regions in the last twelve years, resulting in many major oil and gas discoveries.*

*The experience gained during exploration both offshore and onshore will be valuable in designing and operating future production systems.*

## Introduction

Much of Canada's Arctic area includes attractive sedimentary basins which may contain commercial oil and gas reserves. Drilling to date has indicated good potential in the Mackenzie Delta, Beaufort Sea, Arctic Islands, Labrador Sea and Grand Banks as shown on Figure 1.

The Labrador Sea and Grand Banks are not strictly Arctic but both of these areas are faced with iceberg and sea ice problems and for these reasons are included. The Norman Wells field at the Arctic Circle was discovered in 1920, is still on production and will soon be expanded nearly tenfold.

0165-232X/83/$03.00  

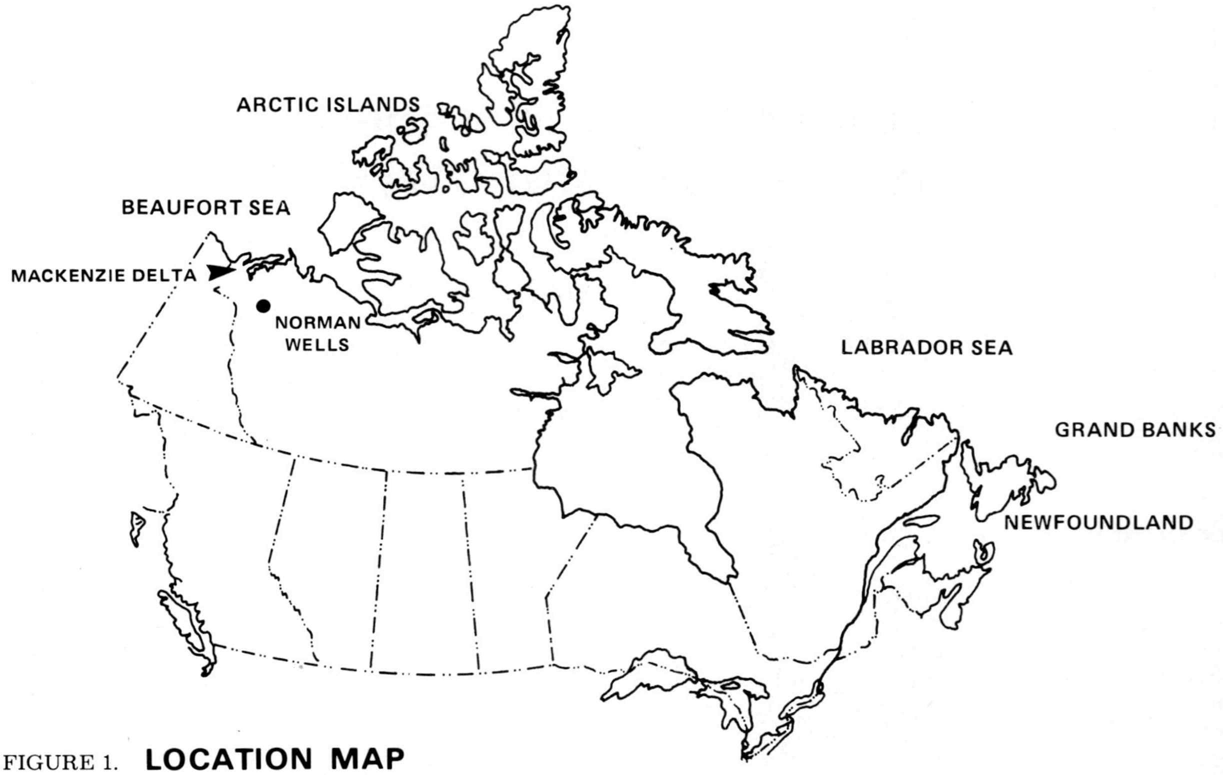


FIGURE 1. **LOCATION MAP**

The paper will emphasize the technology involved in new innovative drilling platforms for the Arctic and will thus concentrate on the Beaufort Sea where most of the new concepts are currently being developed.

The Beaufort Sea, Arctic Islands, Labrador Sea and Grand Banks all present hostile and challenging environments. Drilling technology has been adapted and advanced in order to drill safely and as efficiently as possible in all these areas. In the process of exploratory drilling, technology has been developed which can be applied to the production phase in the Arctic areas.

## Environmental Conditions

### Mackenzie Delta — Norman Wells

The Mackenzie River Delta consists of low lying active delta, much of which is flooded from five major river channels during Spring run-off. This area is bounded on the west by mountains and the north by the Beaufort Sea. The old delta to the east has elevations to 200 m and is dotted by many small lakes. The soil in both the old and new delta is ice rich permafrost and may contain massive ice lenses up to 10 m in thickness. Much of the land is patterned with polygonal ice wedges. Most of the delta is covered with an insulating layer of vegetation and humus which prevents the permafrost from melting. Permafrost thickness reaches 600 m and is thinner near the river channels. Permafrost may be absent beneath the rivers and the deeper lakes.

The environmental conditions in the Norman Wells area are similar to the Mackenzie Delta but permafrost near the river and under the islands is of lesser thickness. Although mostly wooded, the terrain in the Norman Wells area is also protected by an insulating layer of growing or dead vegetation.

Both areas suffer from extreme cold, lack of daylight in winter, blowing snow and fog.

### Beaufort Sea

The primary environmental constraint in the Beaufort Sea is sea ice, since the area is ice-covered for some nine months of the year. Freezeup commences in October and progresses seaward with the growth of landfast ice. A smooth area of landfast ice extends 25 km to the 10 m water depth and beyond that a rough pressure ridged area develops out to about the 20 m water depth contour as shown in Figure 2. Ice islands of glacial origin may ground about the edge of the landfast ice. A recurring open water lead is found along the water edge of the landfast ice which opens and closes depending on wind duration. Beyond the landfast ice is the shear zone of broken ice which is usually in motion and deforming when pressured. The polar pack ice, which is primarily multi-year ice, lies beyond the shear zone. It too is usually in a clockwise motion and usually retreats northward by freezeup.

FIGURE 2. *Winter ice conditions in the Beaufort Sea.*

The sea ice grows to a thickness of 2 m by May. The pressure ridges however can exceed 30 m in thickness. The multiyear ice will reach equilibrium thickness of some 3.5 m but ridges can approach 30 m in thickness.

Breakup occurs in late June or early July. Summer ice conditions are variable depending on the direction of the prevailing winds. Figure 3 shows the number of days of open water by years at a typical drill site in the Beaufort Sea in 30 m of water. Some years no open water existed, while other years there have been 120 days. This graph does not include those periods of thin ice during which drilling vessels can also operate with icebreaker support.

Waves are generally minor because of reduced fetches and presence of sea ice. Significant waves however can reach 5 m. Currents are less than 1 kt.

Subsea permafrost and gouging of the sea bottom by ice features are unique problems in the Beaufort Sea.

Total darkness in Winter, extended periods of extreme cold, drifting snow and fog hamper operations in the Beaufort Sea.

### Arctic Islands

The northwestern portion of the Arctic Islands, which is known as the Sverdrup Basin, is the most promising for hydrocarbons. The most severe environmental problem here is the sea ice. In this area, multi-year ice is present nearly the entire year. Some years, breakup occurs and some drifting of the pack ice takes place for two months in Summer. Open water periods are longer and more predictable in the eastern and southern portions of the Arctic Islands, but drilling is not taking place in those areas.

The period of darkness or limited daylight is even a greater problem in the Arctic Islands. Although precipitation is less in this area, blowing snow is still a severe operational problem.

### Labrador Sea

The ice cover in the Labrador Sea is characterized by very rough first-year ice and an abundant supply of icebergs and multi-year ice. Freezeup begins in November or December and ice extends 200 km seaward. During winter, the bergs and sea ice are transported southward by winds and strong currents. Ice drift is in the order of 17 km/d. Breakup and melting of the ice occurs from April to June depending on the latitude. Ice thickness grows to 2 or 3 m and multi-year ice somewhat thicker.

The icebergs originate primarily in Greenland. The bergs may reach tens of million of tons and are various shapes from tabular to

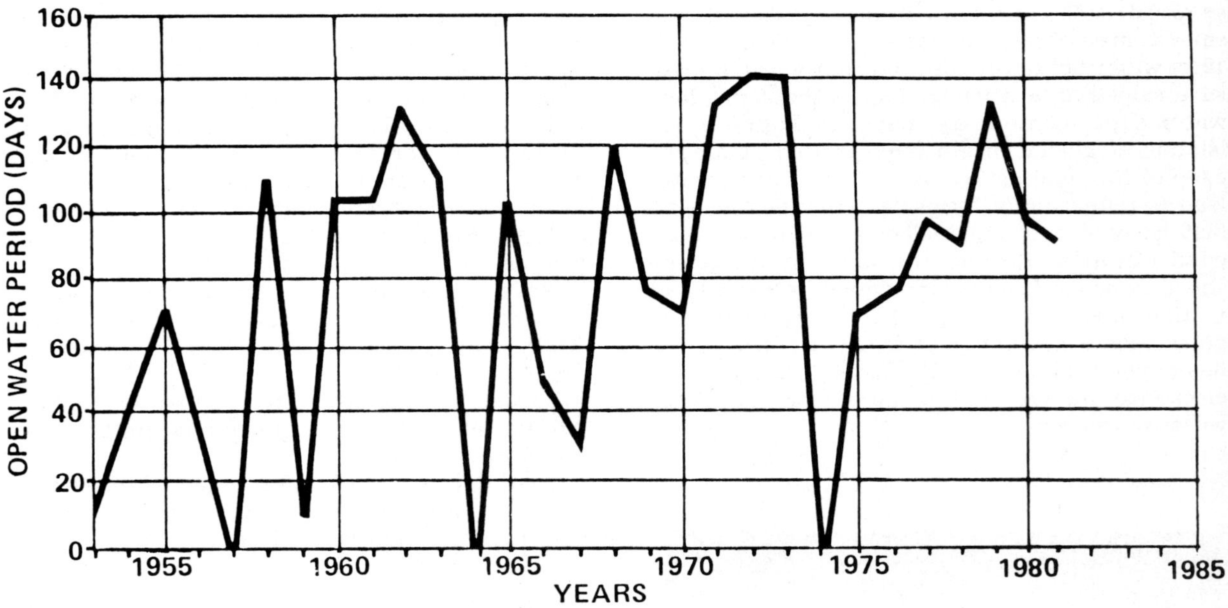


FIGURE 3. *Beaufort Sea; open water days per year lacation 70° 10'N, 133° W*

pinacled. In open water, they drift southward at about 0.5 kts under the influence of the strong current. The bergs often ground during their journey. In winter, sea ice impedes their progress. Bergs are most frequent in early Spring and fewest in the Fall.

Strong winds and high waves are also severe operating problems as is frequent fog and icing.

**Grand Banks**

The environmental conditions in Summer on the Grand Banks are similar to the Labrador Sea or the North Sea. In Spring however, numerous icebergs appear ranging up to ten million tons. Occasionally, sea ice will drift into the area from the north during the first five months of the year. Fog, icing and high waves are also environmental problems in this area from Fall to Spring.

## Pre-Drilling Exploration

Initial exploration in the Arctic, as in other areas, includes surface geology, magnetometer surveys and gravity meter surveys. Regional mapping of outcrops narrow down the prospective areas considerably onshore. Aerial photography has now been supplemented with satellite imagery to assist in interpretation and analysis. Geochemical and geothermal studies now also play a part in geological surveys.

It is the seismic surveys however that must be conducted in considerable detail to identify drillable anomalies. This can be accomplished in the Arctic using tracked vehicles and seismic trains on land. A major difference however is that most of the seismic work must be conducted in Winter to avoid disturbing a sensitive layer of vegetation which protects the underlying ice rich frozen soils.

Offshore seismic surveys are complicated by the presence of sea ice during all or a larger portion of the year. In those areas where ice breaks up and clears briefly, conventional seismic surveys can be run as ice conditions permit.

In those areas in the Arctic Islands such as the Sverdrup Basin where the ice only breaks up in occasional years, seismic surveys must be conducted on the ice in Winter. The ice surface is often rough or uneven and necessitates careful routing. The multi-year ice through which shot holes must be drilled, may be many meters thick. Extreme cold, prolonged darkness and drifting snow add to the problems of obtaining seismic records in Winter in the Arctic.

Permafrost, which may reach a depth of 600 m on both land and offshore in the Arctic, complicates interpretation of the records because of the increased velocities of propogated shock waves in ice bonded material. Both the depth of permafrost and the velocities may be unknown. Multi-year ice on which geophones are laid in the Sverdrup basin reduce the quality of the data.

New sophisticated interpretation techniques are applied to seismic records from the Arctic as for other areas. These include threedimensional analysis, searching for «bright spots» and high fold stacking. However, even having accumulated and interpreted great volumes of expensive records, the only means for determining the presence of oil and gas underground is to penetrate the prospect with a well bore and flow test fluids from the porous formations.

In the Beaufort Sea, the most modern seismic data acquisition and processing techniques are employed. High fold shooting and near and far trace offsets are used in shooting seismic. In processing, such techniques as velocity/dip filtering, seismic designature and wave equation migration are used to improve the geologic picture. In addition, three-dimensional data acquisition and processing techniques enable the delineation of subtle stratigraphic and structural features at oil or gas discoveries. This results in a more judicious placement of the expensive delineation wells.

## Arctic Drilling Technology

### Mackenzie Delta

The first well in the Mackenzie Delta was drilled in 1965, since which approximately 100 wells have been drilled. Two major gas pools and some smaller oil and gas pools have been discovered. Additional drilling in the Mackenzie Delta awaits pipeline transportation systems.

Drilling is generally restricted to the winter months to avoid disturbance of the insulating tundra which protects the ice rich underlying permafrost soils. If drilling is to be carried into the summer, the site must be heavily gravelled to insulate the permafrost and the drilling equipment must be set on wooden piles. All access is by helicopter in summer. Unless moved by helicopter, the drilling equipment must remain until freezeup before being moved.

Care must be exercised when drilling through the permafrost until casing is set. The upper portion of the hole is often refrigerated and surface casing is set quickly before there is much thermal degradation of the permafrost. Special cements are used to reduce generation of heat. The wells are monitored to ensure that possible gas hydrates are not permitted to decompose and release gas. Overpressured formations at depth cause some problems but for the most part, exploratory drilling in the Mackenzie Delta has become routine.

Production wells will be drilled directionally from drilling pads to reduce the number of roads, gathering lines and environmental disturbance. The production tubing may have to be insulated to prevent thawing and subsidence of the permafrost in some areas.

### Norman Wells

Current production wells, which total some 60, have been drilled along the Mackenzie River and on natural islands. Permafrost is present but due to the proximity of the river has been less thick and not a major drilling and production problem. Flooding during ice breakup on the river and Spring flooding due to run-off necessitates temporary removal of some wellheads.

Recently, in preparation for an increase from the current 3,000 barrels per day of production to 26,000 BPD into a pipeline to the south, Esso have constructed and drilled an exploratory well from an artificial island. Four of six artificial islands, 45 m x 80 m, will be constructed by the Fall of 1983 for directional drilling of production wells. Development drilling is in progress on the mainland. Norman Wells is the only producing oilfield in the Canadian Arctic and has been on production for over sixty years.

### Beaufort Sea

*Artificial Islands*

The first offshore drilling in the Beaufort Sea was undertaken in 1973 from a dredged gravel artificial island constructed in 3 m of water over a period of two Summers. To date, 24 wells have been drilled from 21 artificial islands, three of the wells having been drilled directionally. Three other islands are being completed to receive steel caissons in the near future. The artificial islands without caissons have been constructed by various techniques with water depths increasing with time and construction techniques becoming more efficient. The island building techniques have included:

1) Dredging through floating pipe and pipe mounted on wooden piles.
2) A sandbag dyke, later filled with silt and gravel.
3) Gravel trucked from land over the ice.
4) A barge-mounted drilling rig sunk within a dyke.
5) Sand hauled by dump scows from a dredge.
6) Stationary suction dredge and floating line.
7) Stationary cutter suction dredge.
8) Mobil suction hopper dredge.

Capability has increased from 3 m water depth at Immerk, the first island, to 20 m at the Issungnak oil discovery. A concrete caisson island, Tarsiut, was completed in 22 m of water and is discussed later. An underwater berm is being prepared in nearly 50 m of water in 1982 for drilling next year.

Prior to the Tarsiut caisson island well, all islands were built in landfast ice. However, even landfast ice can experience significant motion due to temperature change or wind. In order to protect these islands from ice overriding, a shallow sacrificial beach is built on which the ice can ground and form a protective annular ring of grounded ice. Most of the islands used sandbag protection as shown in Figure 4 to prevent Summer wave erosion. Some earlier ones were built only to provide a platform for winter exploratory drilling and were then allowed to self-destruct.

FIGURE 4. *Beaufort Sea Artificial Island*

The side slopes of dredged artificial islands have been in the order of 1:15. At islands for which sand had to be hauled long distances, there was a great economic incentive to reduce the island volumes through steeper slopes. Steeper slopes have now been achieved through controlled placement of the sand through stingers or dredge arms.

Another reason for the high cost of artificial islands has been difficulty in completing the above-water portion without the crown being washed away by waves and swell. Sometimes this crown or cap could not be achieved until commencement of freezeup, leaving little time to move drilling equipment onto the site. The use of sand filled caissons on subsea berms both reduces the volumes of sand required and solves the problem of the crown washing away. Tarsiut caissons placed in 1981, were the first used in the Beaufort Sea. In 1982, two steel caissons are being brought into the Beaufort Sea by Esso and Dome.

*Ice Reinforced Drillships*

In 1976, two existing vessels were ice reinforced and converted to drillships and a third Norwegian drilling vessel was ice reinforced. Two were brought into the Beaufort Sea around Alaska and a third through the Arctic Islands. Later, a fourth converted vessel was ice reinforced and brought into the Beaufort Sea. With the support of the 16,400 H.P. icebreaker Kigoriak and several large icebreaking supply boats, these drilling vessels are able to drill during the open water periods indicated in Fig. 3 and into the freezeup period while the ice is still thin. The average drilling periods have been 110 days. Drilling some years, has commenced in late June and has continued as late as November 29.

Should the ice concentration become excessive or the polar pack blow in from the north in summer, the drillships secure the wells and move off location until ice conditions improve. When drilling in water depths of less than 47 m, it is necessary that an excavation be dug for the blowout preventors. When the drillship is forced to leave the location, four blowout preventors remain below sea bottom where they will not be damaged by grounding sea ice.

To date, these drillships have drilled some twenty wells in water depths ranging from 18 m to 68 m. There have been four significant oil and two gas discoveries from wells drilled by drillships. Often, it takes more than one open water season to drill a well to total depth and most of an open water season to fully test a well. For this reason, a well drilled from a drillship which discovers oil or gas costs over $100 million. The main reason for the high cost is that the very expensive drillship and its support system are idle for eight or nine months of the year due to ice conditions. The next step in floating drilling vessel technology for the Beaufort Sea is to build a vessel which can, with the aid of support vessels, extend the drilling season to some six months or more.

*Tarsiut Caisson Retained Island*

The first caisson island in the Beaufort Sea was built for Gulf Canada and its partners as a drilling platform for the Gulf et al Tarsiut N-44 delineation well. The site was excavated to a depth of 2 to 3 m and construction commenced in the Fall of 1980. Carefully controlled placement of the sand enabled berm slopes of 1:5 to be achieved compared to the 1:15 common at previous artificial islands. The reasons for choosing the caisson over the artificial island were to reduce the amount of fill which had to be hauled some 100 km and to eliminate the problems encountered in placing the final cap on artificial islands. The underwater berm was completed in the summer of 1981, gravelled and carefully levelled. It was built in 22 m of water and reached to within 6 m of the water surface.

The concrete caissons were built on the west coast of Canada and moved to the Beaufort Sea on a specially built barge. This submersible barge was sunk, the concrete caissons floated onto the barge, and then the barge raised and towed to the Beaufort Sea. Each of the four concrete caissons shown in Figure 5, was 11 m high, 80 m long and 15 m wide and each weighed 5,300 tons. The floatable caisons were compartmentalized so that cells could be filled with water for ballasting and later filled with sand on site. The caissons were not fastened together when placed, but the gaps at the corners were closed with steel doors and filter cloth to prevent the sand from leaking out of the core. It is the weight of the sand filled caissons and the sand core that keeps the structure from sliding when subjected to severe ice loads. The core was built up 2.5 m above the caissons, levelled to receive a specially built drilling rig, and the slopes were protected with sandbags. This provided a freeboard of 7.5 m. It became evident during the storms that occurred near the end of the open water season that wave overtopping and spray would be a problem the following Summer.

FIGURE 5. *Concrete Caisson Island under construction in the Beaufort Sea.*

In Winter, as planned, the sea ice grounded on the underwater berm and formed an annular rubble pile around the caisson protecting it from direct impact and extreme ice forces. In the case of Tarsiut, the regulatory agencies required relief well capability before penetrating hydrocarbon bearing zones. This required flooding of the ice rubble pile to build a grounded ice island 150 m from the well and 100 m in diameter, large enough to accept a helicopter transportable drilling rig and supplies. This pad was built with an 8 m freeboard to prevent sliding under lateral ice loading.

The concrete caissons, the sand fill and the surrounding ice were extensively instrumented to monitor pressures and movements. This monitoring served to assure the safety of the operation as well as to collect valuable information for application to future drilling and production platforms.

The well drilled from late December 1981 until mid-April 1982, after which extensive flow testing was carried out. In preparation for the open water season, upgrading of the surface of the island was undertaken, commencing in April 1982. One thousand two-ton gabions or rockfilled wire baskets were airlifted to the island by helicopters and set on the caissons in place of the sandbags. At the corners, sheet piling and pipe piling were driven to prevent waves from washing sand through the doors between caissons. This work was completed by May 31, on schedule. As soon as the ice had broken up and cleared, another thousand gabions were barged to the site and placed to increase the freeboard by another three meters for a total freeboard of ten metres. Also, additional pipe piles were driven and rock up to 1.5 m in diameter was placed in the water around the toe of the caisson. The caisson has been hit by severe storms and has withstood the 3.5 m significant waves, although the island was, on one occasion to date, evacuated to ensure the safety of personnel.

*Steel Drilling Caisson*

Dome Petroleum is bringing into the Beaufort Sea from Japan during the Summer, a caisson called the Steel Drilling Caisson (SDC) which will be placed on an underwater berm in a 30 m water depth at the Uviluk site. The steel caisson in this case is a half of a supertanker as shown in Figure 6. It has been reinforced to resist ice pressure. It has a double hull with concrete in between. It will be ballasted with water and not filled with sand as are the other caissons. It too is re-usable at other sites, in fact a second berm is being built for it this Summer at Nerlerk. The Uviluk berm on which it will first be used required only about 500,000 $m^3$ of sand. The caisson is about 160 m long, 50 m wide and is about 25 m high. It will be set on a berm 9 m below water surface. The high freeboard protects against wave overtopping, spray and ice ride-up. Should a relief well be required, a helicopter-transportable drilling rig could be placed at the opposite end of the SDC. This caisson will be used in that portion of the Beaufort Sea beyond landfast ice as were the concrete Tarsiut caissons. In this region, pack ice is in motion most of the winter.

*Esso Caisson Retained Island*

Esso will bring its first steel caisson from Japan into the Beaufort Sea during the Summer of 1982. It consists of eight steel segments each weighing 1,000 tonnes and measuring 43 m long. The eight segments will be held together by 16 steel cables, 76 mm thick, to form a large «doughnut» 100 m in diameter. The steel caisson has a height of 17 m and walls 13 m thick at the base and 8 m thick at the top. The $55 million caisson will be assembled in calm waters and then towed to the Kadluk site in 14 m of water and ballasted down onto a subsea berm as shown in Fig. 7. Some 80,000 $m^3$ of dredged sand will be placed into the interior of the caisson to provide the sliding resistance against the ice forces which may develop. This caisson can be refloated and used at other locations. Other benefits of the system are the reduced volume required for the island and the ability to pierce the water with a platform which will not wash away during construction. Esso, as a pioneer in island building, has advanced the technology considerably over the past ten years.

*Gulf Conical Drilling Unit*

Gulf Canada Resources has under construction a floating drilling vessel called a Conical

FIGURE 6. *Canmar Singe Steel drilling caisson*

Drilling Unit (CDU) which will enter the Beaufort Sea in the Summer of 1983. It is shaped in such a way so that ice will be broken in bending by the downward pressure of the vessel, as shown in Figure 8.

The unit, which is not self-propelled, is moored with twelve 89 mm anchor lines. Most of the broken ice will move around the hull but pieces which are forced downward are deflected away from the mooring lines by the unique

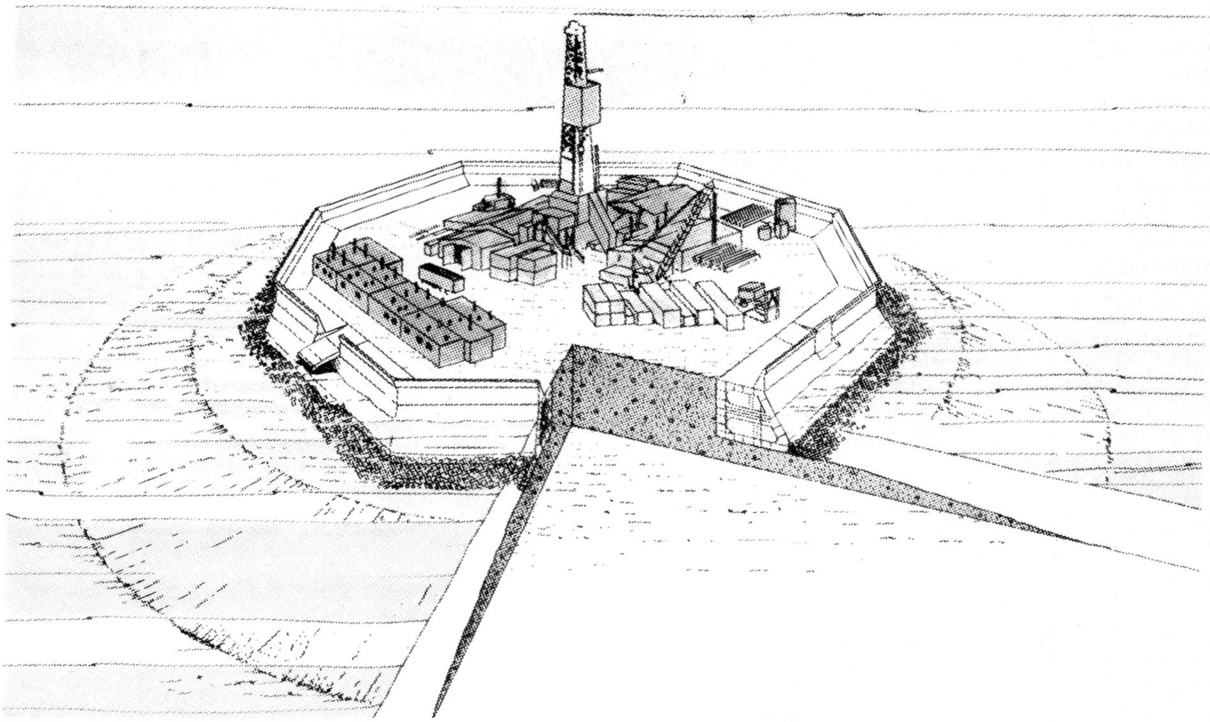

FIGURE 7. *Esso caisson retained island*

FIGURE 8. *Gulf conical drilling unit*

lower shape of the hull as shown in Figure 8. The CDU is 81 m in diameter and is 24 sided or nearly circular. It will likely be operating in water depths of 40 m to 60 m, beyond the land-fast ice. It is designed to withstand a moving ice sheet 1.2 m in thickness. Ice management will be by two 18,600 H.P. Class IV icebreakers and two 14,000 H.P. Class IV icebreaking supply boats. These vessels will break first-year pressure ridges and thick ice in order to extend the season well beyond the current 110 days. These vessels will also break ice to bring the CDU to its Winter harbour at the end of the season. The Class IV vessels are permitted to operate from June 1 to January 31 in Canadian waters, but it is unlikely that drilling from the CDU will ever by possible during that entire period. In the event of multi-year ice intrusions, the CDU will move off location. Extension of the drilling season however permits a much better utilization of very expensive drilling equipment. The CDU is one component of a 674 million drilling system which Gulf has under construction for use on lands it operates in the Beaufort Sea.

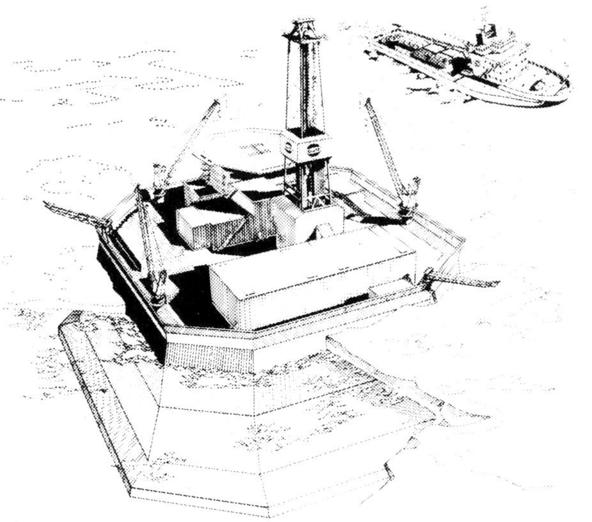

FIGURE 9. *Mobile arctic caisson*

*Gulf Mobile Arctic Caisson*

As the second major component of Gulf's new Beaufort Sea Drilling System, a large steel floatable caisson shown in Figure 9, is being built for use in the Beaufort Sea.

It will arrive complete with drilling equipment ready to drill in 1984. It will likely be used in water depths of 20 m to 40 m, beyond the landfast ice. A berm will first be built to within 21 m of sea level using dredged sand. After careful levelling of the berm, the caisson will be ballasted down onto the berm and the central core of the caisson filled with 115,000 $m^3$ of dredged sand. It is primarily the weight of sand within the caisson which will keep it from sliding under the extreme forces exerted by large pressure ridges and multi-year ice. The caisson itself weighs 33,000 tonnes. The hull will be insulated to prevent the sand core from freezing during the long cold winters. The shape of the caisson shown in Figure 9 was chosen based on model tests to prevent ice or waves from reaching the deck and to minimize spray from waves. The diameters of the caisson are 111 m at the bottom and 87 m at the top. The height is 29 m and the freeboard 8 m.

The Mobile Arctic Caisson (MAC) would only be moved during Summer. It could drill and fully test one well per year and if the well were a discovery, a second directional delineation well could be drilled without re-supplying the caisson with drilling supplies. The MAC can be repositioned to another site quickly because the drilling equipment need not be moved off and berm volumes will not be large.

## Arctic Islands

Drilling in the Arctic Islands from 1961 to 1973 had all been on land except for several near shore stratigraphic tests drilled off the ice in 1973. Each year since 1974, Panarctic, acting on behalf of itself and other operators has successfully drilled off thickened ice in water depths up to 370 m. In each case, flooding of either the first-year or multi-year ice was undertaken as early as possible after freezeup. The ice pad was built up to a thickness of some six meters as shown on Figure 10. For the first offshore well, an emergency kill line penetrated the ice surface at a point remote from the well. Initially, all but one blowout preventor was at the ice surface and connected to the sea bottom by high

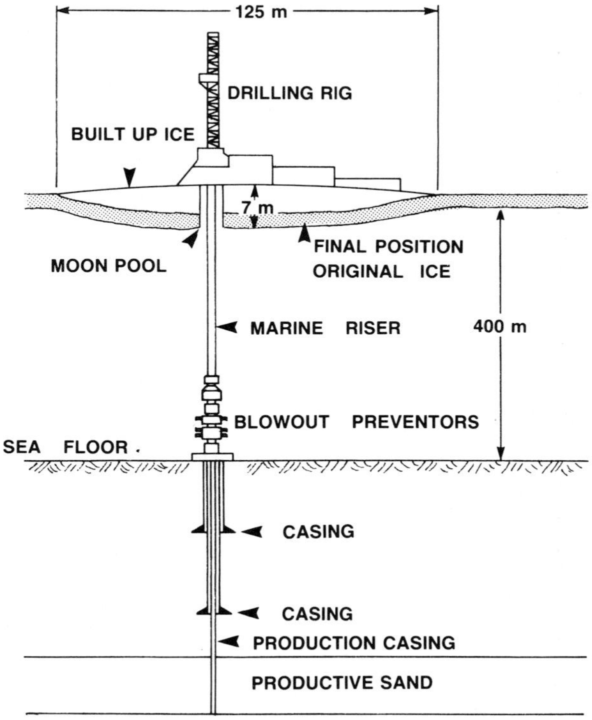


FIGURE 10. *Ice platform for drilling in the artic islands*

pressure casing. Now, as shown on Figure 10, all blowout prevention equipment is on the sea bottom, as is the case in conventional offshore wells. A second ice pad for relief well purposes is built nearby after the first pad is complete.

The latest innovation is the use of large urethane foam blocks within the ice island. With a density of only 30 km/m$^3$ compared to 920 kg/m$^3$ of ice, the urethane increases the load carrying capacity and reduces the construction time. In the use of urethane, some 275 m$^3$ of foam cells, 3 m square and 1.5 m deep are built in formers. The foam does not allow water penetration and has the advantage of being disposable by burning the following Summer. Construction of ice island pads commences in November when the first-year ice permits light vehicles and pumps to reach the site. Drilling must commence as early as possible in order that the well be completed by April. Termination date is set so that ample time for a relief well remains prior to breakup.

An important step prior to drilling off the ice is the measurement of ice movement over at least two Winters. A lateral movement of only 5 % of water depth can be tolerated by the drilling equipment. Ice movement can be measured by various methods but the most common are automated computerized stations using overhead satellites for positioning. Only one well to date has been re-spudded due to excessive ice movement.

One test well was actually drilled off the ice, completed with a wellhead and gathering line to shore and flow tested, all in one winter. The entire operation could have been completed off the ice without the use of divers. As a test well, the Drake F-76 subsea completion was in only 60 m of water and 1300 m from shore but proved new technology for the Arctic Islands offshore areas.

Other concepts have been proposed for offshore drilling in the Arctic Islands. A drilling rig supported on a towed air cushion vehicle has been proposed for some ten years and a prototype was built and tested. This drilling platform would be used at sites where Summer drilling was impossible and there was too much ice movement for drilling from a thickened ice pad. Two possibilities exist to compensate for ice movement. In one case, the air cushion barge would use waste heat to keep a pool melted around the vessel. Another possibility would be to cut and remove blocks of ice from around the drill hole as the ice moves. The vessel would only hover on air for purposes of moving and would sit on the ice surface or float in a pond while drilling.

Another innovative proposal was the use of a dynamicallypositioned ice cutting semi-submersible drilling rig. It was to cut ice while making way to location and would cut ice to maintain location while drilling, if the ice moved. Several million dollars worth of laboratory and Arctic field work were undertaken to investigate this technology.

Since the water depths at the offshore oil and gas discoveries are in the order of 300 m, it is unlikely that a bottom founded or gravity platform will be applicable in much of the Arctic Island offshore areas.

**East Coast**

All drilling in the Labrador Sea and on the Grand Banks has been exploratory or delineation drilling. In the Labrador Sea, drilling is seasonal due to sea ice and icebergs. Drilling off Labrador commences about July 1 using dynamically-positioned drillships which are not anchored and which can quickly move out of the way of an approaching iceberg if it cannot be towed aside. Iceberg towing has been routine and tows of bergs exceeding 10 million tons have been undertaken. One summer, a semi-submersible with quick release anchor lines was successfully used off Labrador. The problem with ship shape vessels is the inefficiencies due to rough seas in the Fall. Costs of drilling off Labrador are increased because of the mobilization and demobilization over long distances each season.

On the Grand Banks, nearly all drilling has been done with semi-submersibles which drill year round. The Hibernia discovery however, was made using a drillship. The high waves the majority of the year preclude drillships as a year round vessel. On the Grand Banks, drilling vessels must be prepared to move out of the way of encroaching sea ice, which occurs only every few years. Any time of the year, but particularly in Spring, icebergs are a threat and must be towed aside or the drilling vessel must move.

A floating production system for Hibernia is being considered to permit movement away from sea ice or icebergs if necessary.

## Summary and Conclusions

1. Major advances in technology for drilling in the Canadian Arctic offshore areas have been made over the past decade.

2. Twenty artificial islands have been used succesfully as drilling platforms in the Beaufort Sea out to water depths of 20 m. Caissons on underwater berms are now being employed for deeper water.
3. A concrete caisson island system in the Beaufort Sea has successfully withstood ice forces and wave action over the past year.
4. Three steel caissons for placement on subsea berms in the Beaufort Sea are either under construction or have arrived in the Beaufort Sea.
5. Four ice reinforced drillships have drilled successfully in the Beaufort Sea since 1976, but a new generation of floating vessel which is conical in shape is under construction and will be drilling in 1983.
6. Offshore drilling in winter in water depths up to 370 m. has been successfully undertaken since 1974 in the Arctic Islands using artificially thickened sea ice.
7. Dynamically-positioned drillships are used in the Labrador Sea because of the iceberg threat and drilling is only during the Summer months due to sea ice.
8. Most drilling in the Canadian Arctic is seasonal; in Winter on land and in the offshore Arctic Islands; and in Summer using floating vessels in the Beaufort Sea and Labrador Sea. Drilling from artificial islands and caisson islands could be conducted year round if relief well capability were not required.
9. All drilling in Canada's Arctic has been exploratory or delineation but valuable technology has been developed which can be applied to development drilling and production platforms.
10. Having been faced with hostile physical environmental conditions in the Canadian Arctic, industry has conducted research and engineering studies, built equipment and gained operational experience in a stepwise manner to meet the challenges of exploration drilling. The technology is also available to meet the challenge of production of hydrocarbons in most areas of the Canadian Arctic.

## Acknowledgement

The author wishes to express his thanks to Gulf Canada Resources Inc. for permission to publish this paper.

*Cold Regions Science and Technology*, 7 (1983) 195—199
Elsevier Science Publishers B.V., Amsterdam — Printed in The Netherlands

# Development of Petroleum Reserves in Arctic Alaska

## A Brief Discussion

*Robert E. Smith*

ARCO Oil and Gas Company
Dallas

## Abstract

*The development of petroleum reserves in arctic Alaska is reviewed. This review includes development activities from the year 1969, immediately following the discovery of Prudhoe Bay Oil Field, through to the present activities of the Kuparuk Oil Field. Primary and secondary recovery efforts are summarized. Probable methods for developing reserves found beneath the coastal waters of North slope Alaska are described.*

*Fiscal and technical aspects are presented. Predicted and actual economics are compared. Several technical problems which are related uniquely to the arctic environment are discribed. The research, testing and design efforts expended to determine design solutions are described briefly. These problems include thaw-movement of permafrost adjacent to production wells, settlement of piling in permafrost due to creep under sustained loading, marine transport of supplies and facilities from the contiguous 48 states and orient through the often ice covered waters offshore western and northern Alaska, low radiation large capacity flares over the thermal sensitive permafrost, and social and emotional health of a large work force under harsh confining arctic conditions.*

*The basic objective of this paper is to communicate the complexity and uniqueness of petroleum development in the Arctic. An exhaustive technical and fiscal treatment is not intended.*

## Introduction

Design studies for the development of petroleum reserves in arctic Alaska commenced in earnest following the discovery of the huge Prudhoe Bay field in February 1968. This discovery was preceded by continuous exploratory drilling in northern Alaska from 1963 to 1968. Nine dry holes were drilled during this period. The Kuparuk field, approximately 50 kilometers west of Prudhoe, was discovered in May 1969.

The subsequent development and exploration activities in arctic Alaska and Canada have been well documented in open literature. The petroleum industry as well as financial and governmental groups and the general public have followed closely these activities. Several papers at this conference will comment on some of these activities and the potential of oil and gas in the Arctic.

The objective of this paper is to communicate by illustrative examples the complexity and uniqueness of petroleum development in arctic Alaska and how safe, profitable production has been achieved. This will be accomplished by briefly reviewing three technical problems selected from the author's experience during the development of Prudhoe Bay and Kuparuk fields. An exhaustive technical treatment obviously is not intended. We welcome discussion and questions as this paper is being given or at any other point during this timely conference.

To facilitate our discussion, may we first describe very briefly the Prudhoe Bay and Kuparuk fields. Then we will take a look at the three selected problems.

0165-232X/83/$03.00  

## Prudhoe Bay and Kuparuk Fields

The Prudhoe Bay and Kuparuk fields are on the northern coast of Alaska. The centers of these fields are approximately 50 kilometers apart.

The Prudhoe Bay field delivered oil to the Trans-Alaska Pipeline System (TAPS) in June 1977. Kuparuk delivered oil to TAPS December 1981.

Some basic statistics are given in *Table 1.*

Due to the arctic environment of North Slope Alaska and the remoteness of this region, the arrangement and design of the production facilities, wells, roadways and personnel quarters are different from those of oil fields in more temperate, accessible regions.

Buildings and pipelines are elevated on piling to protect the 600 meters thick permafrost. Load capacities of these piles depend on adfreeze, not friction or cohesion.

Buildings (modules) are fabricated and equipment assembled under more amenable conditions in the continental United States and then transported to the North Slope by marine barges.

Gravel roadways connect production facilities and gravel drill pads. Gravel is a nominal 2 meters thick to protect the permafrost and tundra.

Large, comfortable quarters are provided. A majority of the work force is rotated on a regular basis.

Massive resupply of bulk material and delivery of new facilities occur only on an annual basis due to the ice conditions offshore northwest and north Alaska.

**Table 1**

| Item | Prudhoe Bay | Kuparuk |
|---|---|---|
| 1. Estimated Recoverable Reserves (Billion Cubic Meters) | | |
| Oil — Oil and Condensate | 1.5 | .24 |
| Gas — In solution | 230 | — |
| In gas cap | 5.0 | — |
| 2. Estimated Dimensions of Reservoir | | |
| Depth (Meters) | 2,600 | 1,900 |
| Thickness (Meters) | 180 | 15 |
| Area (Square Kilometers) | 500 | 500 |
| 3. Estimated Oil Production Rates (Cubic Meters/Day) | | |
| Typical Well in 1982 | 1,600 | 240 |
| Field Total in 1982 | 240,000 | 14,000 |
| Field Total in 1987 | 240,000 | 42,000 |
| Field Total in 1992 | 113,000 | 37,000 |
| 4. Estimated Number of Wells (Oil and Gas Producers, Water and Gas Injectors) | 650 | 800 |
| 5. Estimated Total Project Cost (U.S. 1981 Billion Dollars) (Project cost does not include costs of Trans-Alaska Pipeline System and Marine Transportation System) | 13.0 | 8.0 |

Facilities are enclosed to protect personnel and equipment from the arctic weather.

Now that we have a feel for these arctic oil fields, may we address three technical problems due primarily to the arctic conditions.

## Sea Ice Forecasting

The decision to transport material and facilities by annual «sea lifts» during the arctic summer was made early in the development of Prudhoe Bay. This decision was made with full appreciation of the possibility of loss due to storms and ice conditions and the possibility of total blockage if the ice did not clear from the coast of Alaska. Discussion with U.S. and Canadian arctic scientists, natives, arctic marine transporters, and defense personnel indicated that the probability of total blockage due to ice conditions in any given year was less then 5 %. The probability of a major loss of material or damage in any given year was judged to be in the same range.

During the 1975 Sea Lift some damage to several barges and their cargo did occur during a sudden storm off the northwest coast of Alaska. This same year the ice cleared Point Barrow late and caused the sea lift to arrive late at Prudhoe Bay. Insufficient time remained to offload several major modules prior to freezeup or to return some of the barges to southern waters.

With rapid, creative planning on the part of field personnel and very prompt permitting action by regulatory agencies, we were able to offload these barges during the fall and early winter of 1975. Considerable extra expense and some delays in construction were experienced.

This 1975 experience caused us to rethink our basic decision, but after reevaluation our conclusion was the same: «marine barge transport was the way to go».

During this reevaluation, the need for improvement in our weather and ice forecasting was recognized. Technical personnel from Prudhoe Bay Unit companies interviewed several oceanographic firms, arctic ice researchers, and retired U.S. Navy personnel. The advice of ice forecasters in Canada was sought and freely given.

During late 1975 and early 1976 an improved ice forecasting team was established. This team was a blend of retired U.S. Navy personnel, with arctic and subarctic experience, and arctic ice researchers. The forecasting office is located in the Washington, D.C. area. Currently, the U.S. National Weather Service, NOAA, U.S. Navy, and Canada Ice Central-Ottawa are cooperating on ice forecasting for the annual sea lifts.

Following is a brief summary of this forecasting effort.

Basic Data Sources:

1. Weather permitting, a trained ice observer flies the coast line daily. His reports are shared with U.S. and Canadian groups.
2. Canada Ice Central flies the coast line weekly and their reports are also shared.
3. Monthly U.S. Navy flight reports are shared.
4. VHRR visual mode satellite imagery is obtained on a daily basis.
5. Weather charts from the National Weather Services are obtained four times daily.

Service from Ice Forecaster:

1. An ice forecaster is available 24 hours a day during transit of sea lift.
2. Daily advisories are given from the forecaster to sea lift during transit of sea lift.
3. Five-day ice forecasts are made twice weekly during transit of sea lift.
4. Thirty-day ice forecasts are made twice monthly during transit of sea lift.
5. An advisory of expected ice conditions for the forthcoming sea lift is issued each year in December.
6. A season forecast is issued May 1 each year with a predicted first entrance date into Beaufort Sea; that is, a prediction of the date the sea lift can pass Point Barrow and move onto Prudhoe Bay.
7. An annual critique is made of the past year ice forecasting effort.

Information is transmitted by telecopy to the sea lift and other groups on a real time basis. Telecopy service has been satisfactory.

The owner companies have been pleased with this forecasting service. The cooperation of U.S. and Canadian government groups contributes greatly. It is felt that this forecasting effort reduces our risks appreciably by enabling us to make decisions based on better prediction of the ice and weather conditions. The reaction time to the «sudden storms» of the arctic are lengthened with this directed, single purpose, ice forecasting service.

## Design of Wells Through Permafrost

In the oil and gas industry the term «Frontier Area» has been used often during the past decade as a label for areas where extreme environmental conditions require new technology or an unusual mix of existing technology to find and produce oil and gas. In this sense Prudhoe Bay and Kuparuk fields are in a Frontier Area.

The existence of approximately 600 meters of permafrost (ground which remains frozen for many years) at Prudhoe and Kuparuk is an extreme environmental condition. Deep permafrost is a new design factor for the oil industry.

Immediately following the discovery of the Prudhoe Bay field, a concentrated study was made to identify and then solve problems related to thick permafrost. Our initial step was to study existing literature and to interview knowledgeable arctic scientists and engineers wherever they could be found. For the most part, these people were not associated with the petroleum industry. We found a small, but knowledgeable, group of people in several U.S. and Canadian universities and at the U.S. Cold Regions Research and Engineering Laboratory, the National Research Council of Canada-Building Research Board, and the U.S. Geological Survey.

These interviews and the literature study gave us a much better understanding of the arctic environment and, specifically, of permafrost. We used this understanding to identify and then solve problems of drilling and operating wells through the thick permafrost of North Slope Alaska. Following are descriptions of two such problems and their solutions.

Wells in the Prudhoe Bay or Kuparuk fields produce oil and gas which are much warmer than the surrounding permafrost. A conventionally completed well will thaw the permafrost and cause movement of the surrounding soil. This movement will stress the well casing and possibly disturb surface equipment. The magnitude of movement, casing stress, and surface disturbance depend on several factors such as permafrost properties, well temperature and properties of material in casing annuli, and the production history of the well.

Our initial predictions of possible movement and casing stress for conventional wells were very broad; that is, they could be very small and insignificant or quite large and major depending mainly on the thermal and physical properties of the permafrost. Prior to the 1970's, very little was known about the thick permafrost on the North Slope. Only a few shallow (several meters) cores had been taken.

Freezeback was another identified problem of wells through the thick permafrost. As stated above, a conventionally completed well will thaw permafrost. A thaw zone can reach a diameter of one meter or more in a short time. If flow is stopped, the freezable fluids inside the casing annuli and the water outside the external casing will freeze. Pressures will be created during this freezeback. The magnitude of this pressure depends on several factors such as permafrost properties, fluid properties in casing annuli, temperature of the permafrost, and length of the shut down. We found only one short paragraph in our literature search which mentioned a possible freezeback pressure problem which caused a casing collapse. However, our initial studies did indicate that freezeback could be a serious problem and needed to be studied.

Shortly after identifying these problems and developing several candidate solutions, we briefed the state regulatory body at a public hearing. Our plans to continue design studies were presented at this public hearing.

Approximately 10 million U.S. dollars were spent in five years to develop and test our designs. At the outset there was considerable difference, monetarily and technically, in each company's assessment of these problems and their design solution. After much study, many meetings, deep coring of permafrost, and large-scale field testing, design solutions were developed and accepted by all companies and the state regulatory body.

During this design study, we attempted, by written and verbal inquiries to the Soviets, to learn of their experience and approach to these two problems. Specifically, the cause of a major blowout in one of their arctic gas fields was of prime interest. We suspected this blowout was due to freezeback. No formal Soviet reply was received. We were given only a verbal suggestion during informal conversations at a technical conference in the Soviet Union that we were working on a «real problem». We shared our research findings with the Soviets by publishing papers at two technical conferences in the Soviet Union and a dozen or more papers in the West.

Some damage did occur to several wells completed early in development drilling of Prudhoe Bay, but we were aware of this potential problem and our safety checks revealed this damage. The damage was repaired before producing these wells.

For Kuparuk and Prudhoe Bay wells the design approach for freezeback is two fold. We do not leave freezable fluids in casing annuli. Considerable effort is made to displace freezable fluids. We design the external casing to withstand, with a margin of safety, the maximum predicted external freezeback pressures.

Our design approach for thaw movement for Prudhoe Bay and Kuparuk wells is straight forward. Gelled fluids are placed in casing annuli for insulation, thereby reducing thaw. External well casing has couplings which are strong enough to yield the casing. This arrangement permits the casing to yield and move with the movement of the thawed permafrost without loss of pressure integrity.

## Low-Radiation, Large-Capacity Emergency Flares

During emergency shut downs large volumes of gas have to be flared. A low-radiation, multiple-tip flare system was designed to minimize the thermal impact on the permafrost. This design called for these systems to be placed over several small lakes on the North Slope.

Locating these flares over natural lakes increased the probability of pollution of these lakes but reduced the thermal loading on the surrounding permafrost to an insignificant amount. Surface drainage was not allowed through these lakes. This system was proposed to the state regulatory group. Its advantages and disadvantages were openly discussed. Permission was received to install this unique flare system on the North Slope.

## Concluding Remarks

The remoteness of the arctic oil fields, coupled with the arctic environment, presented the owner companies a myriad of unique design problems. Solutions to these design problems were required prior to production. These solutions had to be both safe and profitable. A wise balance had to be found between profitable production and the desire to minimize physical and ecological impact.

It is this author's opinion that our design studies and field testing, coupled with open and objective discussions between owner companies, regulatory groups, academia, federal and state arctic groups and a concerned citizenry, enabled this wise balance to be struck. Experience to date supports this opinion. We trust the future will confirm.

The pool of arctic knowledge found in federal and state arctic groups and academia was an excellent resource to the petroleum industry.

Tax dollars spent by federal and state governments in frontier areas should be to gain knowledge of the environmental conditions, both extremes and norms. Basic environmental knowledge is helpful for any region, but it is essential for frontier areas. Without it, costly and needless mistakes will be made.

*Cold Regions Science and Technology*, 7 (1983) 201—216
Elsevier Science Publishers B.V., Amsterdam — Printed in The Netherlands

# Structures in Ice Covered Waters

*Svein Fjeld*

Det norske Veritas
OSLO

## Present oil activities in ice infested waters

The main ice infested areas in which structures for oil and gas activities are presently realized or are expected to be realised in this century, can be categorized as follows:

Sheet ice infested areas:

| | |
|---|---|
| Cook Inlet | 20-40 m depths |
| Sakhalin Area | 30 m depths |
| Po Hai Bay | 20 m depths |

Iceberg infested areas:

| | |
|---|---|
| Labrador Coast | Large depths |
| Newfoundland | Planned production |
| | < 100 m depth. |

High arctic areas:

| | |
|---|---|
| Beaufort Sea | 0-60 m depth (so far explored) |
| Barentz Sea | |

Depending on time and location the high arctic area is exposed to ice conditions of varying severity: Sheet ice, first-year and multi-year ridges and floes, pack ice and ice islands. The significance of the different types of ice to structural design is briefly summarized below.

## Ice properties in general

Ice is a material with rather unique strength properties. Important parameters are:

- crystal structure
- salinity and impurities
- temperature
- strain rate

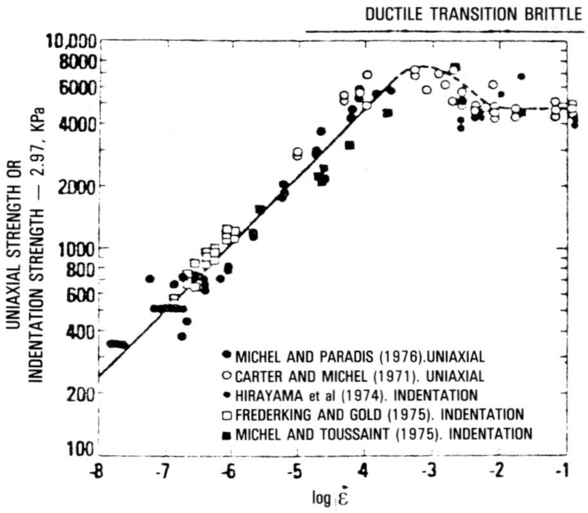


FIGURE 1. *Compressive strength of sea ice as a function of strain velocity* $\dot{\varepsilon}$.

The compressive strength of sea ice is shown in Fig. 1. Sea ice subject to confinement has much greater strength. Therefore very high localized contact stresses can be observed between structures and icebergs, multiyear ridges etc.

The flexural strength of ice is considerably lower than the compressive strength, Fig. 2. This fact is important for the configuration of structures with sheet ice. Fig. 2 also shows the strength dependence on the brine content. Newly frozen sea water contains approx. 3 % salt. The ice gradually rejects the brine. Three years old ice can be considered free of brine. Therefore, multi year ice is considerably stronger than first year ice.

0165-232X/83/$03.00  

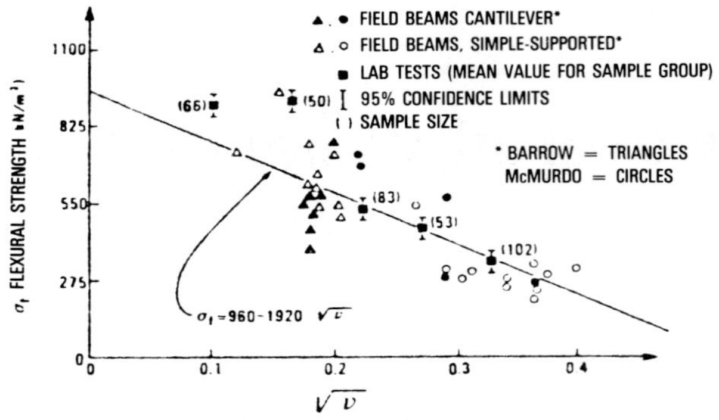


FIGURE 2. *Flexural strength of ice as a function of brine content.*

Due to the large number of parameters affecting the mechanical properties of the ice, laboratory and field measurements may show significant scatter from one test series to another and even from one sample to another.

Typical «higher» values of the E-modulus and the failure strengths of sea ice are shown below:

| | |
|---|---|
| Tension | = 0.8-1.0 MPa |
| Bending | = 1.0-1.2 MPa |
| Shear | = 0.3-1.0 MPa |
| Uniaxial compression | = 5.0-10.0 MPa |
| E-modulus | = 2.0-5.0 x $10^3$ MPa |

Strictly, the given values apply only to ice exposed to short load durations (in the order of some seconds). Under prolonged load durations or very low strain rates, creep effects may drastically reduce the apparent properties. This situation is, however, of little relevance to most structures in the Arctic.

## Sheet ice

The thickness of the sheet ice recommended for design vary from less than 1 m in Po Hai Bay to more than 2 m in the southern Beaufort Sea. Anchored to the shoreline and may be also by grounded ridges, sheet ice remain fairly stable, although motions up to 100 meters within a few hours have been recorded.

## Ice crushing loads

The horizontal force exerted by ice crushing against a structure may be calculated as:

$$F = I\, f_c\, C_x\, D\, t$$

where F = horizontal ice force.
I = indentation factor.
$f_c$ = contact factor.
$C_x$ = unconfined compressive strength of the ice.
D = diameter or width of the structure at the region of ice contact.
t = ice thickness.

The strain rate $\dot{\varepsilon}$ for the ice is a function of the ice approach velocity, v and the structure dimension. The following relationship has been used

$$\dot{\varepsilon} = \frac{v}{2D}$$

Fig. 3 gives examples of indentation factors for laboratory-grown columnar fresh water ice. These curves have also been used for columnar saline ice. For granular sea ice an indentation factor of 1.2 has been used for a large structure/sheet ice thickness ratio.

The contact factor, $f_C$, depends on:
- ice movement rate
- local geometric effects
- active defensive mechanisms

Analysis of ice forces on Cook Inlet structures suggest a maximum value of 0.55 for the product of I x $f_C$. Assuming an identation factor of 1.2, the calculated contact factor of 0.45 appears to be consistent with the relatively continuous ice movements and non-frozen-in conditions of Cook Inlet.

The ice pressure on local load areas may be several times higher than the average pressure on a total load area. Simultaneous ice crushing can be considered taking place in a limited number of contact points only. At these prints a multiaxial compression state with high crushing strength results. The following area dependence has been suggested

$$P_l = 6.0 \left( \frac{A}{0.35\ m^2} \right)^{-1/6} \sigma_c$$

where

$P_l$ = local pressure
A = local load area $m^2$

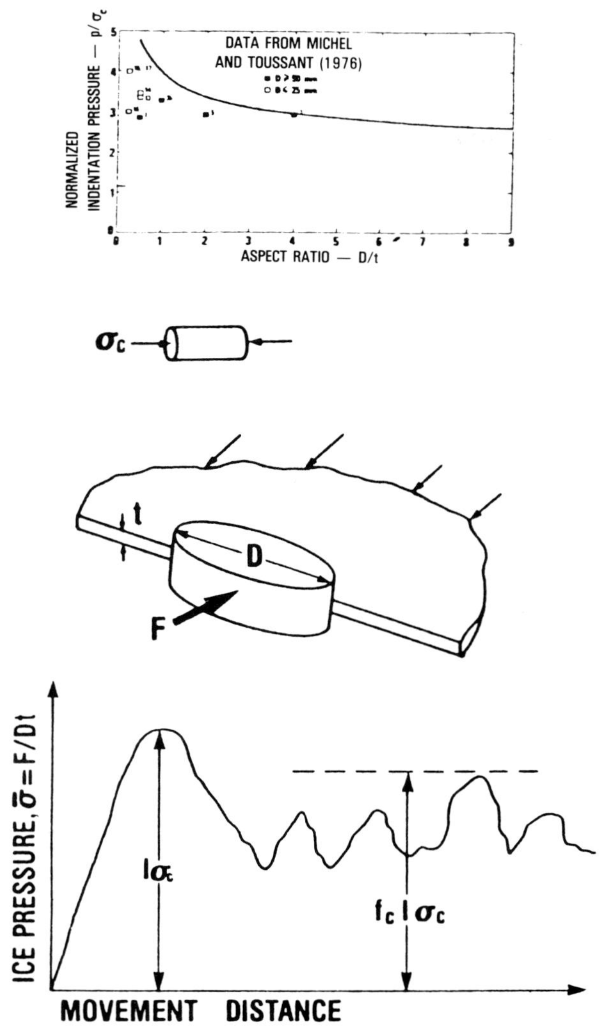


FIGURE 3. *Ice crushing strength with indentation factor.*

## Failure in bending

For sloping-sided structures, sheet ice moving against a structure may fail in bending at much lower horizontal force than if crushed. Forces may be calculated as (See Fig. 4):

$$R_H = [A_1 \sigma_f t^2 + A_2 \rho_w g\, t\, D^2 + A_3 \rho_w g\, t_R\, (D^2 - D_T{}^2)]\, A_4$$

$$R_V = B_1 R_H + B_2 \rho_w g\, t_R (D^2 - D_T{}^2)$$

where $R_H$ = horizontal force on the cone,

$R_V$ = vertical force on the cone,

$\rho_w g$ = unit weight of water,

$\mu$ = ice-structure friction coefficient,

$\sigma_f$ = sheet ice bending strength,

$t$ = ice sheet thickness,

$t_R$ = ice ride-up thickness,

$D$ = waterline diameter of the cone,

$D_T$ = top diameter of cone, and

$a$ = cone angle, as measured from the horizontal.

## Ice buckling

When the ice thickness is small, the ice sheet may buckle rather than crush or fail in bending.

The failure load due to buckling of a drifting ice sheet can be estimated from the formula

$$P_b = K_b \cdot \gamma_w \cdot L_i{}^2 \cdot b$$

$P_b$ = total buckling load of moving ice cover

$K_b$ = coefficient depending on the boundary conditions along the interaction line of the cylinder and ice cover, and the aspect ratio b/Li

$\gamma_w$ = specific weight of water

$L_i$ = characteristic length of the ice cover

$b$ = breadth of structure at waterline

$$L_i = \sqrt[4]{\frac{E \cdot h^3}{12\,(1 - v^2) \cdot \gamma_w}}$$

$E$ = E-modulus of ice

$h$ = ice thickness

$v$ = Poisson's ratio

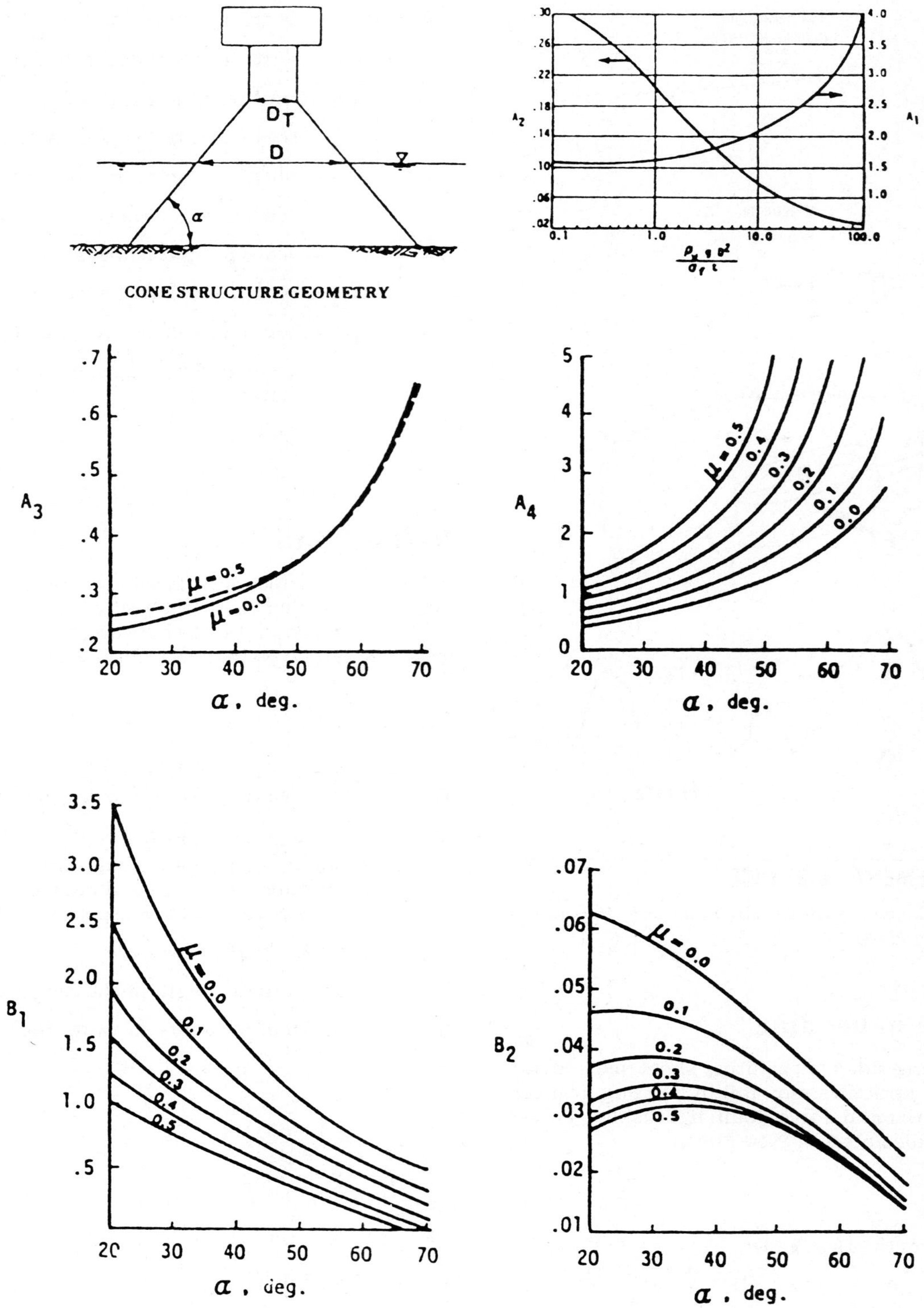


FIGURE 4. *Ice force coefficients for bending.*

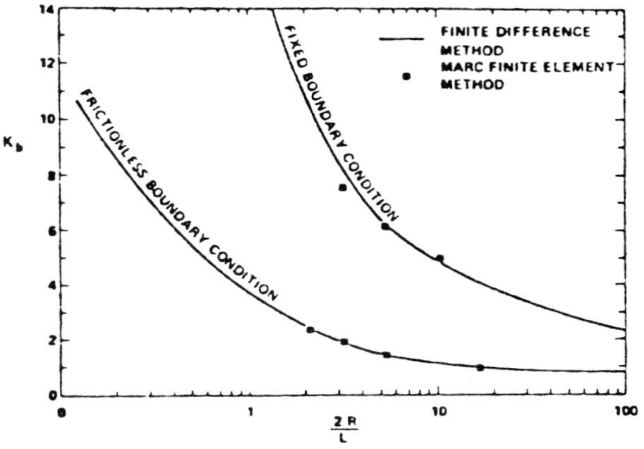


FIGURE 5. *Buckling failure coefficient* $K_b$

$$P_b = K_b \cdot \gamma_w \cdot L_i^2 \cdot b$$

Values of $K_b$ can be determined from Fig. 5, where $2R = b$ and $L = L_i$.

The fixed boundary condition represents an upper bound for the buckling load, whereas the friction-less condition is a lower bound.

For aspect ratios, $b/L_i$ less than approximately 5, the buckling load $P_b$, usually becomes larger than the corresponding crushing load, $P_c$, and is therefore of minor interest.

Multileg platforms may be found to experience much higher loads than anticipated as the sum of loads on each individual leg. The legs adjacent to the leg under consideration may significantly alter the ice failure mode. On the other hand the load peaks on the individual columns need not occur at the same time (see Fig. 6). Therefore, whether ice is anticipated to fail in bending or crushing, leg spacing should be reviewed to ascertain whether the failure mechanism is influenced by the presence of adjacent legs. In addition to analytical procedures to assess the effects of leg spacing on the forces exerted on the structure as a whole and the individual legs, model tests may be used.

The possibility that ice may be entrapped inside the legs should either be prevented or accounted for in the load calculations.

In conditions associated with continuous ice movements at relatively high velocity, ice fracture may occur at uniform time intervals. The fracture frequency or periodicity may be at or close to a fundamental frequency of the structure, resulting in the phenomenon of «ratcheting». A similar phenomenon is also theoretically possible in the case of cone-shaped structures.

## First year ridges and rubble field.

First-year ridges are usually composed of loosely consolidated blocks. Therefore crushing strength of ice rubble is assumed to be considerably lower than the crushing strength of individual ice pieces making up the rubble. The low winter ice velocity further tends to reduce the resulting ice force.

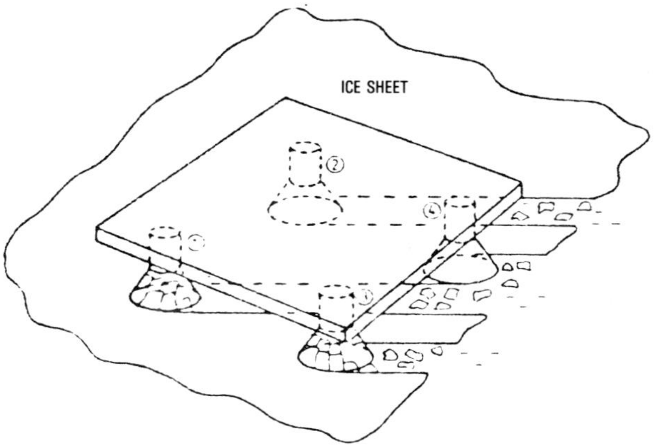


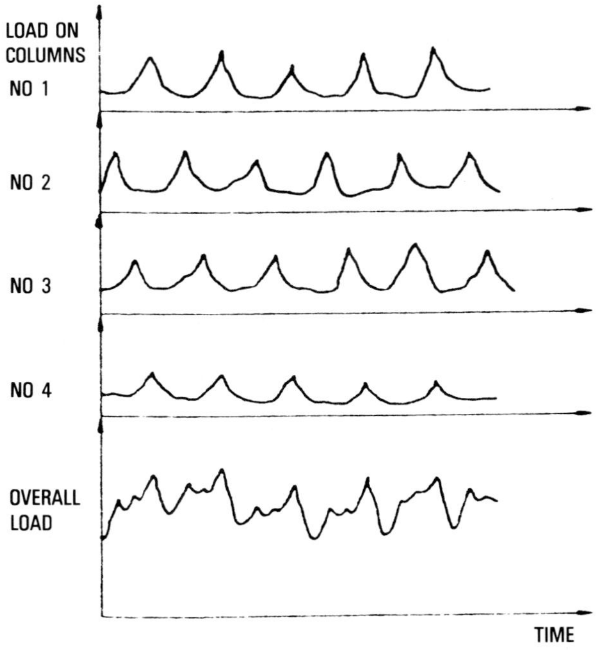


FIGURE 6. *Ice load histories on single columns and on complete structure.*

The shear strength of ice rubble appears to be a function of the cohesion between ice rubble pieces and the degree of confinement of the shear plane. Ice rubble behaves like a granular material. Measured values for the angle of internal friction of approximately 50 degrees were obtained and a full-scale cohesive strength was inferred to be from 5 to 10 psi, depending on the arrangement and size of the ice pieces in a ridge.

## Multiyear floes and summer ice.

Maybe somewhat paradoxically design ice forces in summer conditions are normally larger than for winter conditions. The summer ice is characterised by deteriorated multiyear ice floes, ridges or rubble.

These floes, compacted and refrozen to achieve high strength have weathered, rounded, consolidated sails and relatively solid keels approximately 3-4 times their sail height.

Multiyear ridges may either fail by crushing, bending or shearing against the structure or may come to rest at the structure without failing due to an insufficient driving force. The preferred mode of multiyear ridge failure is normally bending, since this may result in the lowest load from such a ridge. This means that short stocky ridges result in the highest forces, a force which is limited by the crushing bending strength of the ice.

In determining ridge forces, the geometry, ice properties, velocity and orientation of the ridge should be considered, along with the environmental force available to drive the ridge against the structure.

The pressure distribution of ice features in contact with a structure should be determined for local structural design considerations. Ice contact failure pressures are anticipated to be considerably larger than the uniaxial ice strength because of the confinement of the ice in local contact with the structure.

Inertial ice forces can occur when ice features of large mass at low velocities, e.g., ice islands and icebergs, or smaller features at high velocities e.g., summer storm driven floes, encounter a structure. The forces necessary to arrest or deflect the ice features should be evaluated and compared to the forces necessary to fail the ice features considering the higher ice strength corresponding to the relatively high velocity and thus strain rates. Allowance should be made for added mass effects from the water. For these types of collisions capacity to absorb impact energy can be a more vital safety criteria than brute strength.

## Arctic pack ice

The only offshore activities so far conducted in the pack are experiments with ice breakers, and this is likely to remain so for a long time.

The Polar pack is a vast assembly of ridged and hummocky floes averaging about 4 m thick, largely of multi-year blue ice.

Summer melting means that the oldest ice in the pack seldom exceeds 12 years.

Although day to day motion can be highly irregular, in the long term the pack ice of the Beaufort Sea grinds relentlessly round in a wind-generated clockwise circulation known as the Beaufort gyre, centred on a point roughly at 80° N, 150° W, 1,000 km in from the pack ice edge.

Ice drift speed on the southern rim of the gyre averages 2.5 km/day from east to west, although in spring it can approach 25 km/day.

Even in winter, as much as 5 % of the pack may be open water.

The most extreme ice conditions in which exploration drilling is taking place, is the transition zone.

At the winter boundary between the relatively static landfast ice and the circulating polar pack is a transition zone of rapidly deforming and heavily ridged ice. Ridges 20 m deep are not uncommon. The zone extends from wherever the landfast ice has reached that year, on roughly to the edge of the continental shelf — typically 100 km wide in the Canadian Beaufort.

Scattered in its predominantly first year ice are multi-year floes and fragments of ice islands. Whole ice islands are also possible — rare, but not to be ignored — with a frequency of appearance at a given location in the Canadian Beaufort calculated at 1 in 100 years.

They are the nearest that the Beaufort Sea comes to experiencing icebergs, originating generally from glaciers of the high Arctic islands where large tabular slabs of ice up to 60 m thick and 15 km across can calve off and circulate.

## Icebergs

Icebergs jeopardizing offshore energy activities are mainly encountered on the east coast of Canada, from the Davies strait to the Grand

Banks. Drifting ice objects can have a large variety of magnitudes from growlers with mass less than 100 tons, bergy bits 100 — 5,000 tons and real icebergs of size up to 10 — 20 mill. tons as far south as on the Grand Banks. The drift velocity of large icebergs is estimated to approx. 0.5 m/s. However, the largest ice bergs are not considered the greatest risk. The impact energy of an iceberg of approx. 10,000 tons in large sea states is still very large and these ice pieces are by far more difficult to observe by the monitoring systems.

## Further general design considerations for arctic structures

Several concepts for arctic structures are fitted with cones in the water line (Fig. 7). The cones are intended to fail the ice in bending and thereby reduce the horizontal forces. However, this reduction is not obvious.

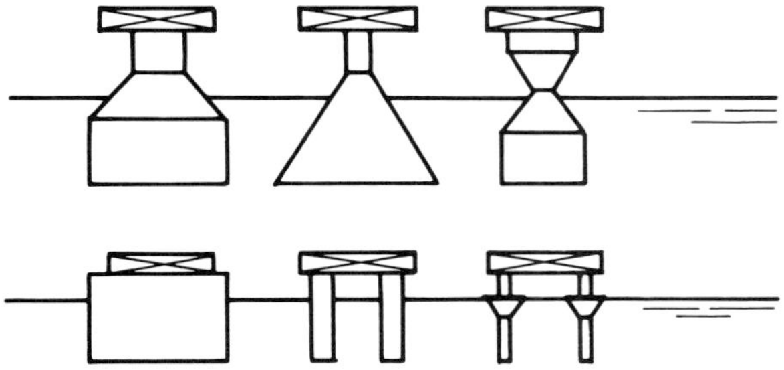

FIGURE 7. *Proposed principle configurations for ice resisting structures.*

- The ice will be broken upwards and compile a heap of rubble of considerable height. The forces necessary to clear off this rubble may be significant.
- The cone inevitably has an upper end. When the broken ice creeping up the cone reaches this top there still may be a failure of the ice in crushing.
- The cone will result in a significantly increased wave run-up.

Present design philosophy doubts the benefit of the cone. If a cone should still be included it should break the ice downwards rather than upwards. This shape will relieve all the above mentioned disadvantages.

«The waves are our enemy the ice is our friend». This, somewhat surprising saying among designers for arctic structures should remind us that construction and operation of platforms in ice infested areas are also hampered by serious problems beyond the ice conditions. These problems will just briefly be mentioned in the following remarks.

It is a fact that structures and civil works in the arctics have so far been impaired to larger extent by waves than by ice. We have already mentioned cones resulting in increased wave run-up. This is an example that many measures taken to reduce ice problems also reduce the wave resistance.

The environment of the arctics is furthermore characterized by

- temperatures down to —50°C resulting in steel material problems
- dense and frequent fog
- icing
- large tides (> 4 m) due to shallow waters
- seismic exposure
- breaking waves
- relic permafrost hampering pile driving and also jeopardizing gravity structures with wells heating the soil
- soil conditions, to a large extent consisting of fine silt unfeasible as material for gravel islands and foundation for gravity structures.

The latter item is an example of the complete lack of resources for civil engineering work, which, in combination with its remoteness, result in enormous construction costs. Even a material as stone for rip-rap has been reported to have a cost of 400 $/ton.

For this reason any construction work should be accomplished in the south if possible. The platform with complete production facilities should be precommissioned prior to the transportation to the Beaufort Sea. The platform would thus be built up in a practical and efficient manner, in the south, utilizing existing, well established construction techniques.

The facilities in terms of tugs etc. necessary to tow the gigantic platform concepts proposed to their site through the ice infested Beaufort Sea should not be underestimated. The ice at Point Barrow will seldom withdraw to a water depth more than 10 m, a fact which imposes severe weight restrictions. Even in the open sea periode ice cover in the Beaufort sea can be 2/10 to 4/10 and the towing season, normally from July to October, can in certain years be restricted from mid-August to early September. Finally fog and heavy snowfall can reduce the visibility considerably, a fact which can further complicate the marine operations.

## Ice platforms

A basic principle for structural design in hostile environment is — to the extent possible — work with nature instead of struggling against it. An elegant application of this principle is the establishment of fixed or floating ice platforms. Different methods have been applied to thicken the landfast ice artificially until grounding, 5-10 cm increasing thickness a day. Diameters of 300-400 m are not unusual for such platforms. Depending on the water depth the island will be completed approximately in January and will not break up till late summer.

The flooding and freezing of a floating ice platform take some two months. In one case a floating ice island survived the summer.

Floating ice islands techniques have changed little since the pilot installation in 1974. The ice is artificially thickened to about 7 m in the centre tapering to 1.5 m thick natural ice at the edge. Access with light vehicles and pumps is normally possible across 300-600 mm first year ice by early November so that flooding starts soon after.

Marine risers are designed for ice movement equivalent to 5 % of water depth.

A recent innovation is inclusion of large urethane foam blocks in the ice islands. With a density of only 30 kg/m$^3$, compared to the 920 kg/m$^3$ of ice, the urethane adds to the load-carrying capability of these floating platforms with out lengthening the construction period. The polyurethane inclusion is introduced because the time taken to thicken the natural ice by flooding and freezing to support a drilling rig has been becomes increasingly critical as rig loads have crept up to 1,630 tons and more.

A special ice island concept is the so called «Norwegian Production Gravity Ice Island». However, this platform concept should rather be grouped in the category «Large volume caissons» inasmuch as the ice is produced and contained inside an external vessel of steel or concrete.

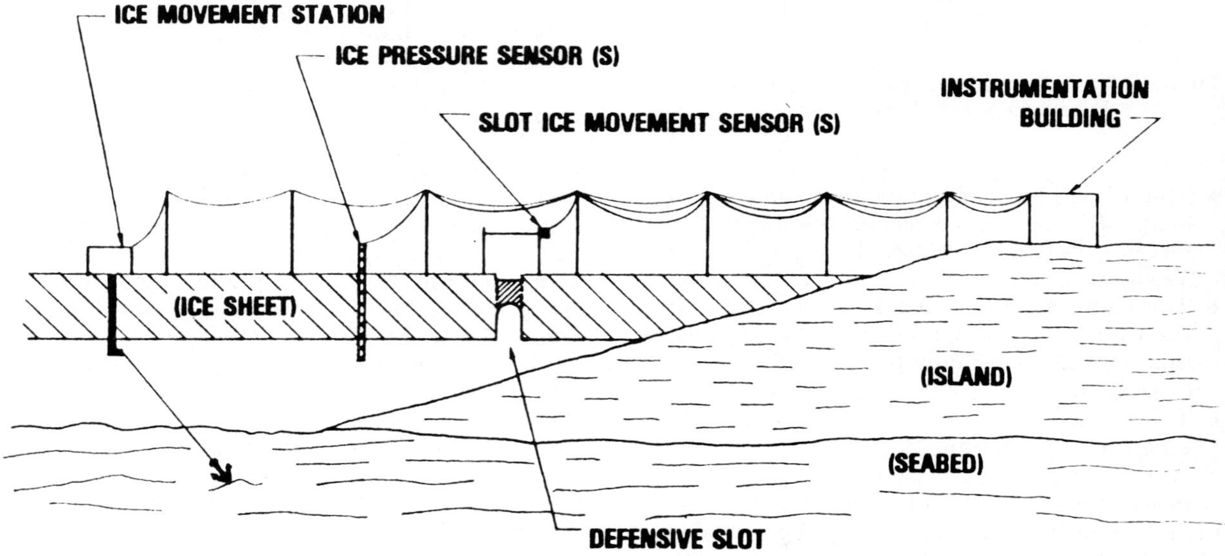


FIGURE 8. *Ice monitoring and defense system.*

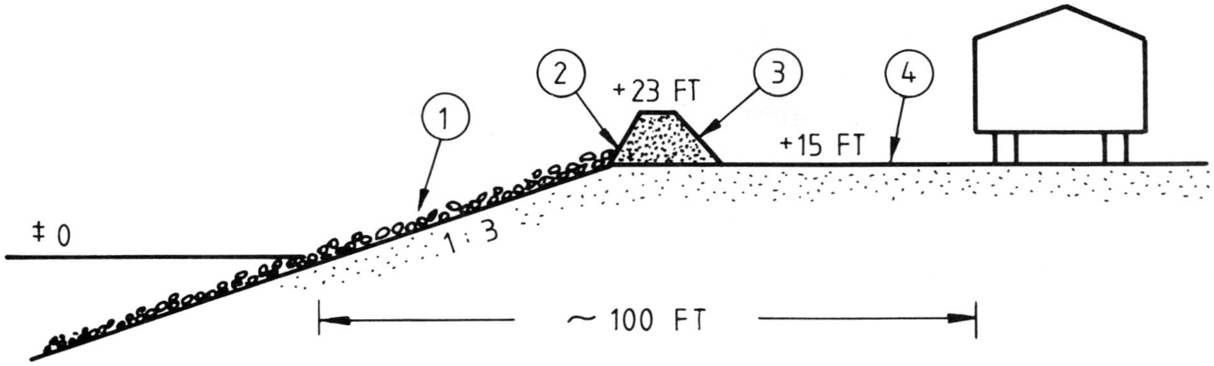


FIGURE 9. *Ice overvide protection features.*

1. *Rough slope to resist sliding*
2. *Steep bluff to cause jamming*
3. *Berm to trigger pile-up*
4. *Buffer zone to accommodate override*

## Gravel islands

This is by far the most common type of platforms in the Beaufort Sea.

Granular fill to be used for island construction is distinguished as dredged fill and trucked fill.

Dredging operations are likely to be conducted only in summer. During the summer open-water season, fill material is removed from the ocean floor by dredging transported by barge or pipeline, and deposited at the island construction site. This fill is likely to contain a significant portion of sand in addition to gravel. Considering the wave action during the summer construction and small grain size of the fill, the submarine slope of the dredged fill islands built to date used to be 1:15.

Trucked fill is used primarily for winter construction. Land fill material is trucked over the ice cover during the winter and deposited through a hole cut in the ice at the island construction site. This fill is generally coarse, composed mostly of gravel from a carefully selected borrow site. Considering the absence of wave action during the winter construction and generally larger grain size of the trucked fill, the submarine slope of the trucked fill island is taken as 1:3.

Lateral ice load resistance during winter is assumed as certained by freezing the fill. To escape ice conditions exceeding design assumptions, monitoring facilities are normally installed to provide indications of unusual ice activity. (Fig. 8).

Otherwise the design of the gravel island addresses the following protective measures to avoid ice piling or override (Fig. 9)

— slope angle
— freeboard
— berm

Wave erosion is avoided by appropriate

— slope protection (sand bags or stone)
— possibly sacrificial beach to absorb the wave energy at some distance.

The gravel island techniques open the possibility for gradual extension from a temporary wildcat exploration island to a permanent production island. (Fig. 10)

Production islands have also been conceived in the form of large loading atolls (APLA) which include a quay with sheltered waters for the loading tanker. (Fig. 11)

## Caisson retained gravel islands

In spite of the slope protection the ice and wave zone will be subject to some maintenance

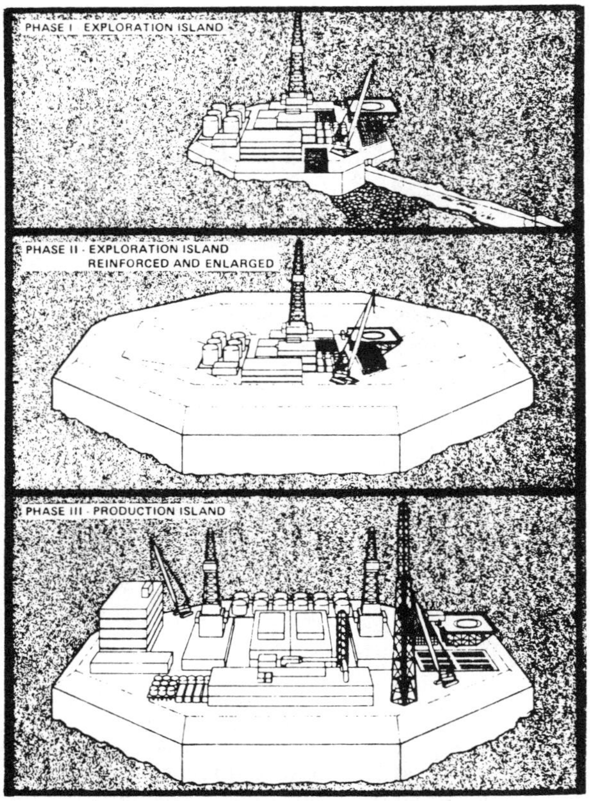


FIGURE 10. *Conversion from exploration to production island.*

work. Furthermore the amount og fill necessary to establish a gravel island of given slope inclination will increase with the height in approximately third power. Therefore, in deeper waters more than say 20 m, there is a recent trend to top the island with ring of steel or concrete caissons, containing the gravel fill in side. (Fig. 12)

In one project four light weight concrete caissons, each 80 m long and 11 m high, were built by forming the bases in a dry dock and slipforming to full height in floating condition. The caissons were towed from Vancouver round point Barrow. In another project steel caissons have been fabricated in yards in Japan and transported to the Beaufort Sea.

These platforms have been fitted with a lot of instrumentation to improve the knowledge about local and global ice forces.

## Grounded tanker

A main disadvantage of caisson retained gravel islands is their lack of mobility compared to conventional drilling platforms, i.e. drill ships, jack-ups and semisubmersibles. This summer an unique platform intended to overcome this inconvenience will be installed in the Beaufort Sea. A laid up tanker has been cut and the forward half is fitted to drill and test wells. The tanker is towed from the yard in Japan to the

FIGURE 11. *Arctic production and loading atoll.*

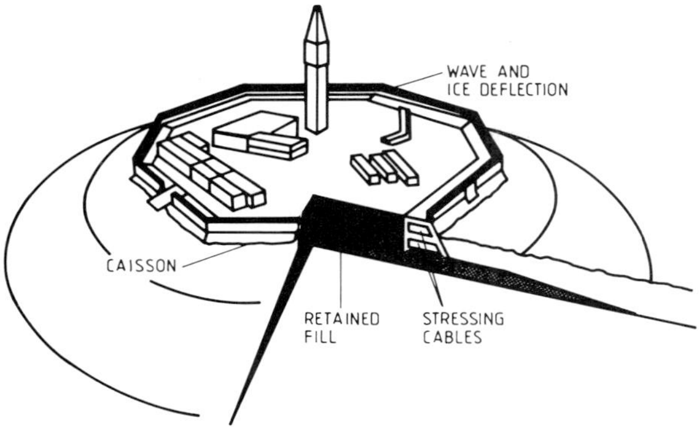


FIGURE 12. *Caisson retained gravel island*

Beaufort Sea site and settled on to a predredged 16 m high berm. Next summer the tanker will be raised by deballasting and moved to another field.

The 25 mm thick hull side plate has been backed by 1 m of concrete and the transverse hull frame significantly strengthened to resist the ice forces. As the summer ice is assumed more severe than the winter conditions, the rubble field of failed ice built up during the winter will be artificially thickened and consolidated along each side of the tanker by spraying water. This procedure significantly improve the sliding safety factor.

## Large volume caissons

(Fig. 13). Moving into deeper waters solutions based on dredged berms tend to be prohibitive from an economic and feasibility point of view. Large volume caissons are considered a more feasible alternative. Several large volume platform types, piled and gravity type, steel and concrete, have been conceived for the Beaufort Sea. However, so far none has been realized. These structures have smaller water line area and thus ice loads than the gravity islands. However, still the forces amount to several hundred thousand tons. The most feasible ice force resisting mechanism is large weight and bottom friction which again requires a large weight.

In case of soft clay this type of structures might turn out unfeasible due to inadequate global ice load resistance and excessive settlements irrespective its effective weight.

As concrete structures have no problems from thermal brittleness point of view, and as large dead weight should be considered an advantage, concrete has been considered a very feasible material for in resisting gravity structures.

These platforms can be made with vertical or sloped walls intended to reduce the ice loads. One of the advantages of large volume platforms is that substantial oil storage is possible within the structure. Another advantage is that, dependent upon structural refinements, a minimum of ice rubble fields will grow around it, thereby allowing tanker access to the caisson on the «down-stream» side of ice movement. The exact shape of a large volume caisson could vary; it might be simply a pillar with straight sides, or it could have a flanged top and bottom to provide a larger working surface and base. The base of such a large production caisson might measure 250 metres across, the neck of the caison being 120 metres and the working platform 160 metres

Also anchored floating structures have been conceived for the arctics. The large ice forces are a major challenge to the anchor system design. However, these concepts are assumed fitted with cones at the water line. The ice can then be broken actively by raising and lowering by the ballasting system. In this way the ice forces can be significantly reduced (Fig. 14).

## Ice breaking multileg and monopod structures

Moving out of the Beaufort Sea at the same time means moving into areas of more modest iceproblems. The Cook Inlet has lead the way in developing ice resistant platform types feasible for sheet ice areas. In addition to ice Cook Inlet is notorious for unusually large tides (up to 10 m) and associated fast currents (3 m/sec).

11 of altogether 14 platforms in Cook Inlet have four 4-5 m diameter legs housing conductors and flowlines with cross bracings kept well clear of ice levels. Two platforms are similar tripods Fig. 15 and one platform is a monopod with 9 m diameter column. The ice design criteria included a 2.1-MPa force on the structure, however, instrumented monitoring indicate the effective pressures actually experienced to be a small fraction of this only.

The large diameter legs protect the conductors etc. from ice impugnments which might

*Conceptual gavity caisson for future production in deeper water. The working platform may measure over 500 feet (160 metres) across. The base of the caisson would be used for oil storage, and would be buried deep within a sand berm, which serves to anchor and protect the platform from ice islands.*

*Exxon monocone for Alaskan Beaufort*

*Concrete monotower with icebreaking cone for depths of 200 m in the Navarin basin.*

FIGURE 13. *Conceptual caisson designs.*

otherwise destroy them. It also avoids additional forces on the platform due to ice build up if the conductors had been installed outside the platform legs.

The Cook Inlet ice contains large amounts of silt. The grinding action of the silt-laden ice pans results in extensive abrasion. Upon bad experiences with the pioneering platforms the ice abrasion is combated by thickening steel plate in columns and double skin with a grout filled annulus to resist impact.

In the Po Hai bay Chinese companies have installed several platforms of relatively conventional jacket type. The piles are designed to resist on ice pressure of 2-MPa from 70 cm thick ice.

## Platforms with gravity fenders

Where icebergs are encountered these are a dominating hazard to be considered. The maximum icebergs, e.g. offshore Newfoundland, must be considered to have a displacement of tens of millions of tons and a scouring depth approaching 10 m.

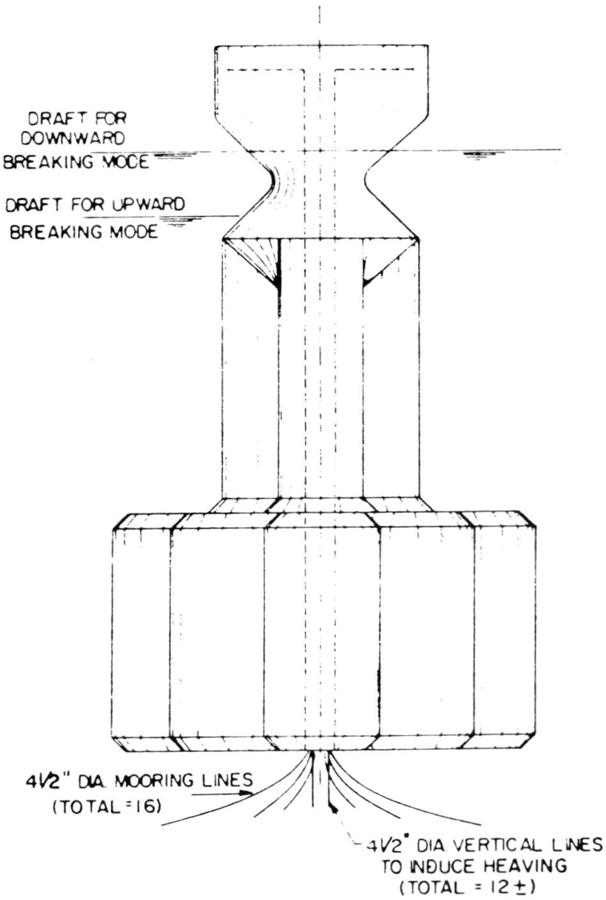


FIGURE 14. *Ice-breaking floating arctic caisson.*

The major problem as far as platforms are concerned is related to the impact energy. After the energy is absorbed, the drift and drag forces on a barge resting on a platform will be in the order of magnitude of 10,000 tons. If the iceberg is surrounded by a thick ice sheet, this force might be significantly larger but for conditions to be expected at Newfoundland still not excessive. The force can easily be taken e.g. by a fixed gravity structure.

Different impact protection philosophies have been proposed (Fig. 16). The naked platform can hardly be designed to resist the iceberg impact. A surrounding protective gravel berm suffers from several shortcomings.

— erosion problems
— inadequate height to stop dangerous bergs
— the iceberg cannot be brought to complete rest, but will rather turn in most unpredictible ways

The impact energy, which amounts to the order of magnitude one million *tm*, requires an unique fender system. If a design basis impact can be identified it seems in no way impossible to construct such a fendering system. A heavy ring around the platform with an effective weight of several hundred thousand tons can absorb the energy in friction at its base. After an impact which moves the fender ring it can be relocated by deballasting. This solution requires a relatively large structure and its economic feasibility depends on the productivity of the oil field.

An advantage of the fender platforms is their ability to withstand any ice loading, also bergy bits etc. which might escape the attention of the monitoring system.

## Structures designed to escape in case of approaching ice bergs

Towing of icebergs in the order of millions of tons does not appear to be a reliable solution considering the relatively strong current, wind and high waves in the iceberg season. The randomness in the drift direction might also complicate the planning and execution of such operations.

In iceberg areas the alternative to fendered platforms are floating platforms intended to escape in case of approaching platforms. Current semisubmersibles can be used as production platforms and possible storage requirements can be met by an anchored tanker or purpose-built vessel.

For all year production the platforms will have to be anchored to resist all forces from landfast ice and drifting ice, which can be of large magnitude. Further the safety is completely dependent on the reliability of the iceberg monitoring system. In this context it should be underlined that small bergy bits can strike the platform with high velocities and a great potential to damage a conventional floating platform.

It is essential that the mooring system and the sea floor well connections can survive passing iceberg and drifting ice without significant damage. Otherwise it would be impossible to reconnect the platform and bring it on stream

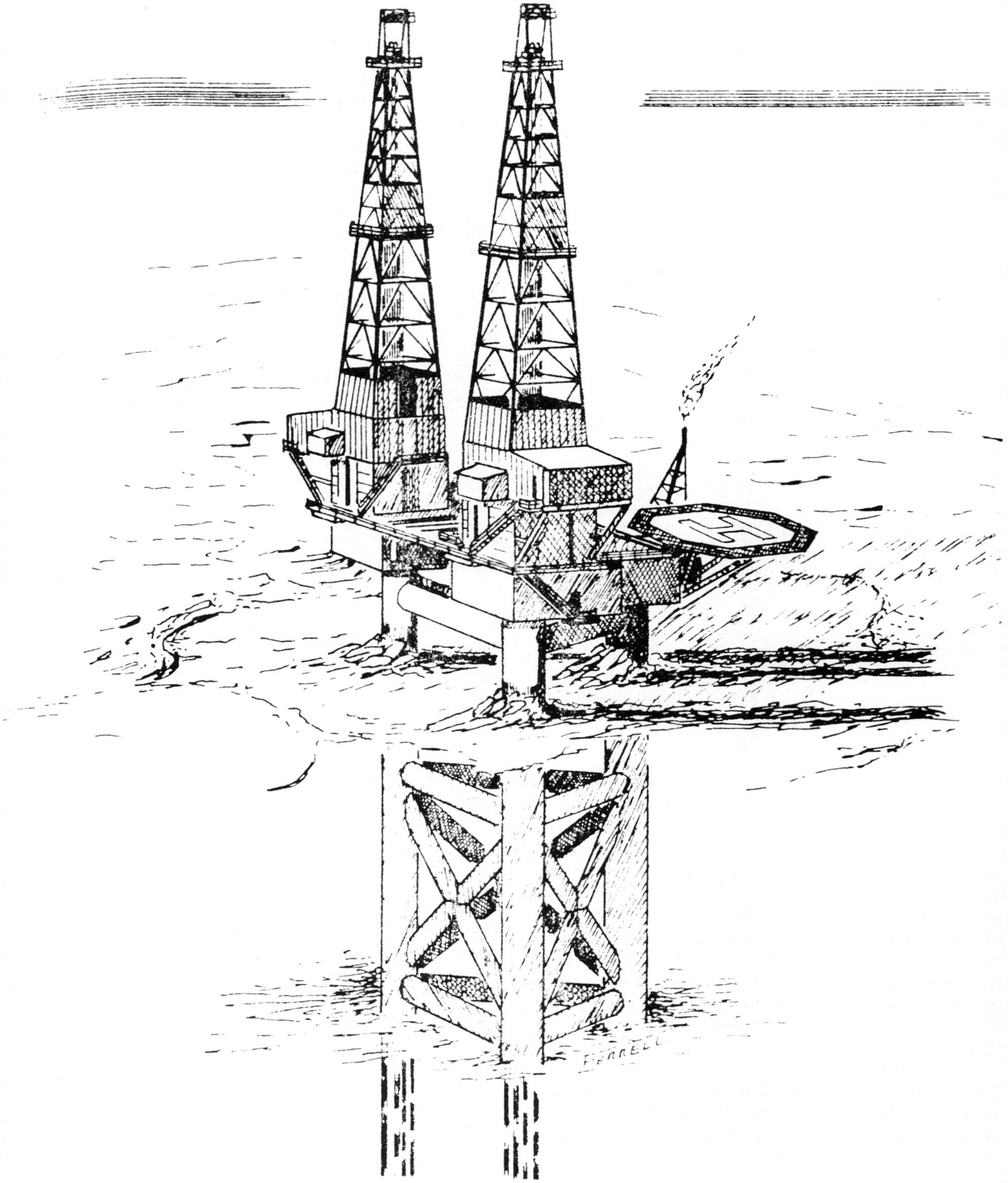

FIGURE 15. *Typical Cook inlet platform*

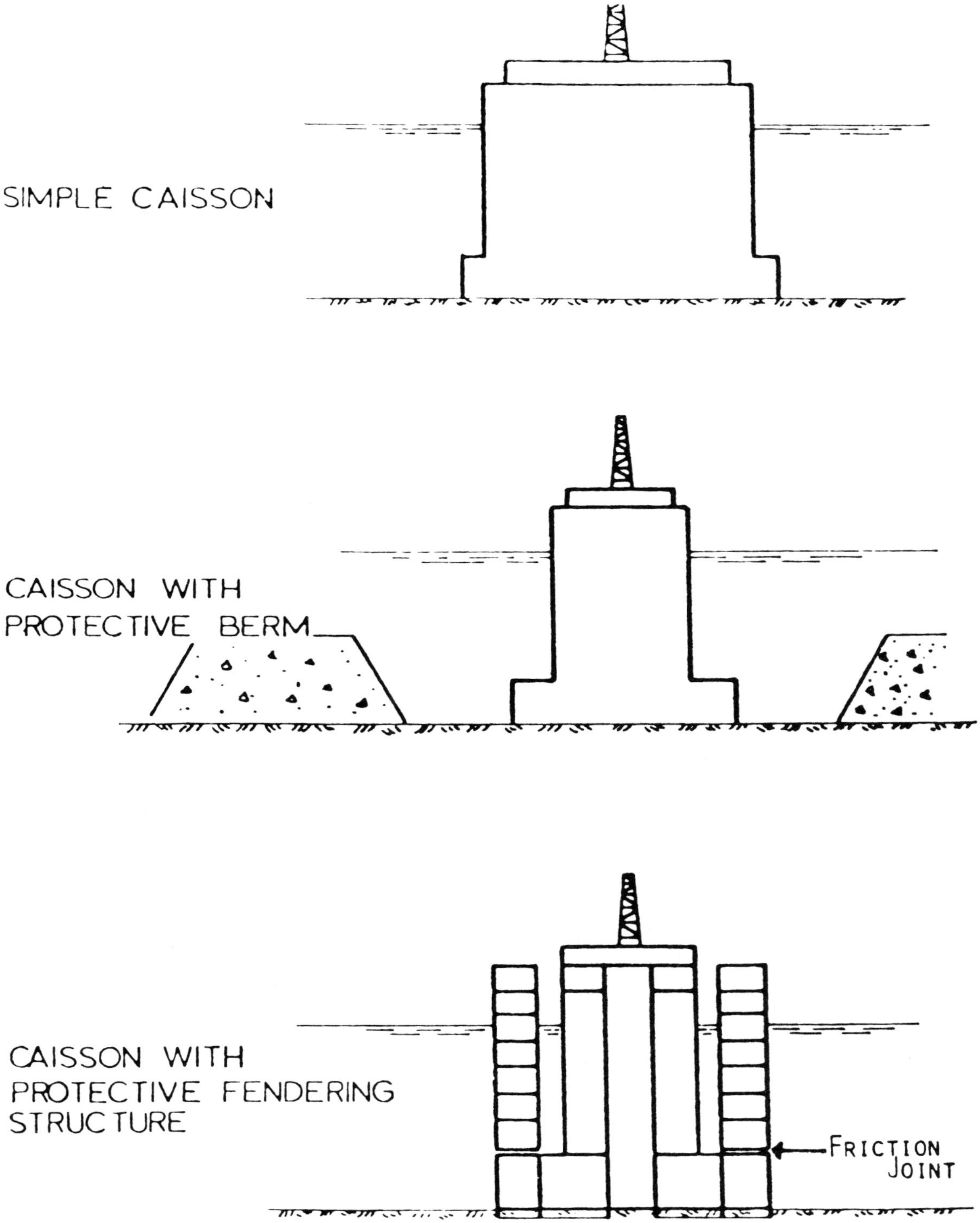


FIGURE 16. *Iceberg protection principles.*

again without significant delays. The scouring potential of the icebergs is an essential issue in this context. Production on floating semisubmersible platforms was an alternative for the development of the Hibernia field, offshore Newfoundland possibly favoured by the operator. Whether this field development plan will be maintained after the Ocean Ranger disaster remains to be seen.

A few weeks ago the Energy Minister of Newfoundland declared that all offshore platforms under his jurisdiction should be concrete gravity structures.

## Ice management systems

As mentioned an effective iceberg warning system is a prerequisite for a floating platform scheme. Ice surveillance will have to be carried out by work-boats, by air-craft reconnaissance and by radar. Each platform must have ice observers on board who are responsible for plotting iceberg and ice floe movements on a daily basis, or hourly basis, once ice moves into the immediate vicinity.

Radar can usually be relied upon to pick up a small to medium iceberg at 4 to 15 nautical miles except in the worst sea conditions. However, because seawater reflects radar better than freshwater iceberg ice, sea clutter seriously degrades detection range in very heavy seas.

Ice berg areas such as the Grand Banks are very foggy. This fog is worse in the spring months, coincident with the iceberg season; which is also a time of severe storms and high seas.

Consideration has also been given to synthetic-aperture radar (SAR) mounted in satellites or aircraft.

This appears to be a useful, though expensive, method for tracking the large-scale motion of ice, as well as individual large bergs. Other sensors considered for ice detection are sonar, low level light television (LLTV) and infrared detectors (IR). Both LLTV and IR outperform the eye in darkness but are little or no better in fog, rain and haze. On special importance is problems in detection of small bergy bits.

In severe sheet and ridge ice environments, structures may rely on active ice defence measures for the protection of the structure. The following considerations may serve as examples for reducing ice forces on structures in areas where low ice movement rates are expected:

- A moat can be cut into the ice around the structure. Although ice will refreeze across the moat, the moat creates a weakened zone, limiting the forces on the structure to those necessary to fail the ice. The moat should be wide enough to accommodate the maximum expected continuous ice movement.
- A narrow slot can reduce the contact factor thereby reducing the magnitude and the ice crushing force.

## Concluding remarks

The aim of the present paper is to give a brief over-view of present published state of the art of arctic structures. It should, however, not be overlooked that important aspects of this art cannot at present be published.

A variety of interesting new platform concepts, mainly in concrete, has been presented for these conditions to Veritas for precertification review. However, the designers have required a complete confidentiality and further details cannot be presented here.

In our cooperation with oil companies interesting new data about ice forces have been presented. This too will have to be kept confidential.

For these reasons it seems fair to state that present state of the art is not identical to published state of the art. This, in some ways discouraging, feeling can also lead us to the positive conclusion that safe and economic structures for the exploitation of the Arctic energy resources are probably nearer to their realization than we might imagine.

*Cold Regions Science and Technology*, 7 (1983) 217—226
Elsevier Science Publishers B.V., Amsterdam — Printed in The Netherlands

# A Polar Gas Pipeline for the Canadian Arctic

*Olavi M. Kaustinen*

Polar Gas Project
Toronto

## Abstract

*The Polar Gas Project was established in 1972 to determine the best means of moving frontier natural gas reserves from Canada's High Arctic to southern markets. The current participants in Polar Gas include TransCanada PipeLines Limited, acting as Project Manager, Panarctic Oils Limited, Tenneco Oil of Canada Ltd., the Ontario Energy Corporation and Petro-Canada. This paper summarizes the pipeline routings investigated, present and future reserves in the Canadian Arctic and outlines methods for installing the pipelines in the Arctic channels.*

*All the various pipeline routings investigated would be partly in continuous permafrost regions, partly in discontinuous permafrost and most cases only a portion would be in non-permafrost soils. The preferred route would be approximately 3,100 miles in length and would be buried throughout its length. The cost of constructing the Polar Gas pipeline is estimated to be $10,6 billion expressed in 1980 Canadian dollars excluding any interest charges.*

*Even though energy exploration in these areas is relatively only a few years old, substantial natural gas reserves have been identified — approximately 15.7 Tcf in the Arctic islands excluding 1981 discoveries and 5.8 Tcf in the Mackenzie Delta excluding any gas reserves in the Beaufort Sea. The total marketable reserves of 21.5 Tcf which have already been identified in these areas are more than sufficient to make the Polar Gas Project a feasible and economic venture.*

*In order to connect the natural gas reserves in the Canadian Arctic Islands, Polar Gas will have to install pipelines in two major marine crossings in the Arctic waters in the initial construction phase of the project. The two marine crossings are located at either end of Victoria Island — at Dolphin and Union Strait and at M'Clure Strait. Dolphin and Union Strait between Victoria Island and the Canadian mainland is about 19 miles wide and has a maximum depth of 400 feet. M'Clure Strait is between Victoria Island and Melville Island and is 76 miles wide with a maximum depth of 1650 feet. Different methods of construction are proposed for the two channels and these methods are described in the paper as well as outlining the technique for protecting the pipelines from potential ice scour.*

*Polar Gas and its consultants have designed a full scale demonstration of the one-atmosphere welding technique closely parallelling the underwater environment faced at M'Clure Strait to demonstrate that the tie-ins which would be necessary for the actual construction of the marine crossings could be satisfactorily undertaken. The equipment required for the demonstration is outlined in the paper and the general procedures are described.*

*The planned pipeline demonstration would significantly enhance the current state-of-the-art for deepwater pipelining worldwide as well as providing the project with invaluable information which can be used for the final design and construction of the offshore marine facilities. Therefore the demonstration should be of interest not only to participants in the Polar Gas Project but more generally to all energy companies concerned with deepwater tie-ins and marine pipeline repair.*

0165-232X/83/$03.00  

## Introduction

The significant increase in world prices for hydrocarbon supplies over the past decade has had the effect of reducing consumer demand through conservation and at the same time has resulted in increased activity in exploration for new sources of crude oil and natural gas. Although these trends have resulted in a current temporary surplus of hydrocarbon supply, the long term outlook indicates that additional sources will be required to meet the projected energy requirements in the future. In addition, there are a number of advantages for the North American continent to work towards self-sufficiency in energy to protect itself against potential uncertainties of outside supplies. In this regard the large potential of natural gas and crude oil in the Canadian Arctic frontier areas could be made available to achieve this objective.

Natural gas was first discovered in the Canadian Arctic Islands in the early 1960's and a major discovery was announced in 1969 at Drake Point on Melville Island. In succeeding years a number of fields were discovered in the Melville Island and King Christian Island areas. As the volume of natural gas reserves increased, it became increasingly evident that the development of a transportation mode was necessary in order to bring these reserves to markets in the south.

As a result the Polar Gas Project was formed in 1972 to determine the technical and economic feasibility of connecting these natural gas reserves from the frontier areas of northern Canada to markets in southern Canada and the U.S.A. Polar Gas is a consortium of five companies including: TransCanada PipeLines Limited, the largest Canadian natural gas transmission company, acting as Project Manager; Panarctic Oils Ltd., the major exploration company in the Arctic Islands; Tenneco Oil of Canada Ltd., a subsidiary of Tenneco Inc.; the Ontario Energy Corporation, a company wholly-owned by the Ontario Provincial Government and Petro-Canada, an integrated oil and natural gas company wholly-owned by the Federal Government of Canada. Over the past ten years Polar Gas has spent more than eighty million dollars studying, researching and engineering its transportation system.

In its early studies Polar Gas examined a wide range of transportation alternatives for moving the gas out of the Arctic. Figure 1 compares the energy efficiency of the various

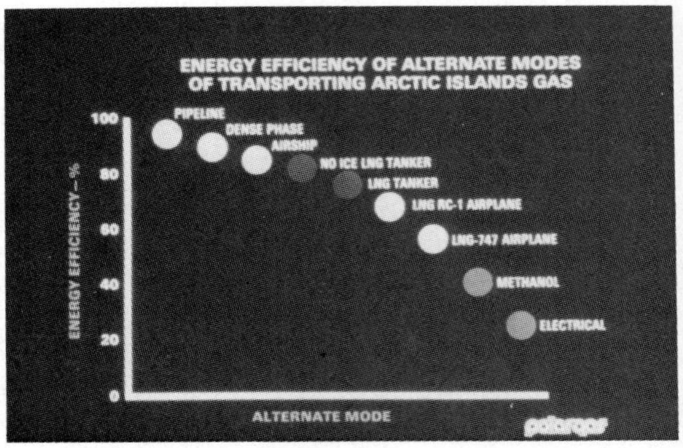


FIGURE 1.

systems considered and, as will be seen, the pipeline, which is close to 95 efficient, is the most preferred method of transporting large volumes of natural gas over long distances. As a fully buried pipeline it also has the advantage of being virtually impervious to the extremes of the Arctic weather.

A number of alternative pipeline routings have been studied all of which are «Y» shaped to draw gas from the two major supply areas, the Mackenzie Delta/Beaufort Sea Region and the Sverdrup Basin of the Arctic Islands. The preferred route shown in Figure 2 is called the «Direct» route and has the lowest capital cost and cost of transportation of all the alternative routes studied.

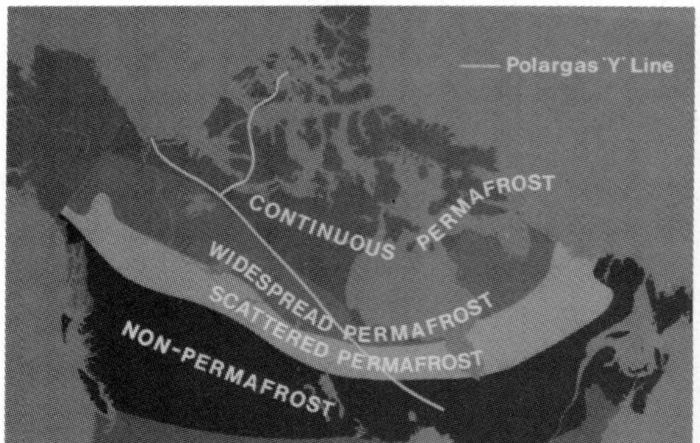


FIGURE 2.

The pipeline is approximately 3,100 mi (5,000 km) in length and passes through regions of continuous permafrost, widespread and scattered permafrost and non-permafrost terrain. Approximately two-thirds of the pipeline will be buried in continuous permafrost where it will be refrigerated below 32° F (0° C) to ensure a stable pipeline system at all times. A further twenty-five percent of the pipeline will be buried in discontinuous permafrost and only ten percent in the so-called conventional terrain and in these latter two zones the gas will be at a temperature above 32° F (0° C).

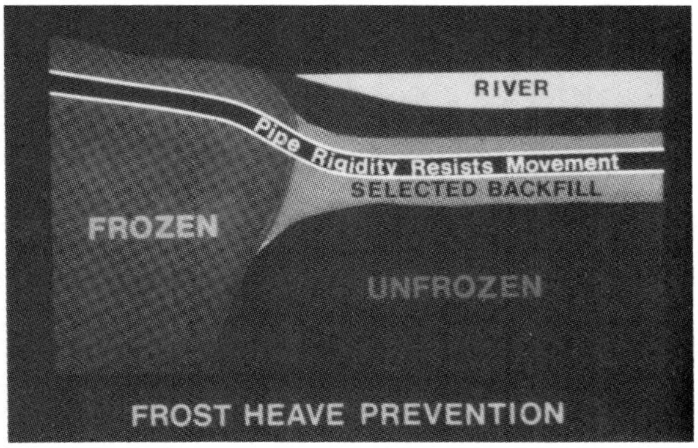


FIGURE 3.

Polar Gas has undertaken a number of field programs to determine the type and ice content of soils along the pipeline routing. Although it is not anticipated that frost heave will occur along the land portion in the continuous permafrost zone, there is a potential for frost heave under a number of the larger rivers which flow year round and therefore have unfrozen soils beneath the river beds. In these cases, where there is a potential for frost heave, it is planned to bury the pipeline in a trench surrounded by selected granular fill materials, insulate the pipeline and provide heat tracing in order to prevent any detrimental frost heave (Figure 3.)

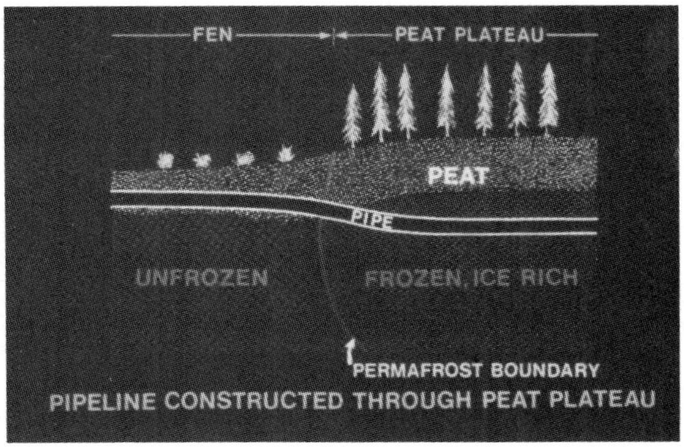


FIGURE 4.

Depending upon the ice content of the soils in the discontinuous permafrost regions, special construction measures (Figure 4) may be necessary to avoid thaw settlement effects and to maintain pipeline stresses within acceptable limits. Such measures would include one or more of: deep trenching, selected backfilling and below ground pipe supports.

## Marine Pipeline Crossing

The scope of the project in terms of both scale and the construction environment makes it a most interesting and unique undertaking and the major challenge, from an engineering perspective, is the design and construction of the major marine crossings which are required for the Arctic Islands lateral.

FIGURE 5.

The two marine crossings located at either end of Victoria Island are the Dolphin and Union Strait and at M'Clure Strait and are shown in Figure 5. The conditions are considerably different at each of these locations and therefore the construction techniques required are quite different. Polar Gas has carried out comprehensive field research programs and engineering studies and has concluded that it is feasible to construct these crossings by modifying existing marine construction methods to cope with the Arctic environment.

Dolphin and Union Strait between Victoria Island and the Canadian mainland is about 19 mi (31 km) in width and has a depth of 400 ft (122 m). Winter ice usually averages 6 ft (1.8 m) in thickness whereas the summer months normally ensure two or three months of ice-free open water. Under these climatic conditions the pipeline could be placed on the channel bottom using modified laybarge techniques in a similar manner to many marine pipeline installations that have been constructed in many parts of the world but using ice strengthened vessels to cope with potential intermittent ice floes.

To prevent potential damage from ice ridges that normally occur during the winter months in Dolphin and Union Strait, each end of the pipeline will be constructed inside foreshore tunnels to a water depth location of 80 ft (24 m).

The more northerly crossing at M'Clure Strait is somewhat more difficult since geographic and climatic conditions are more severe. M'Clure Strait is approximately 76 mi (122 km) wide and 1650 ft (503 m) deep. This body of water is normally never completely free of ice at any time of the year and during the winter months ice thicknesses range up to 39 ft (12 m). To protect the pipeline in the foreshore areas where there is potential ice ridging and scouring from ice islands that occasionally migrate through this Strait, the pipeline would be protected in a tunnel similar to that in Dolphin and Union Strait but in this case the tunnel would be installed to a water depth of 180 ft (55 m) as shown in Figure 6.

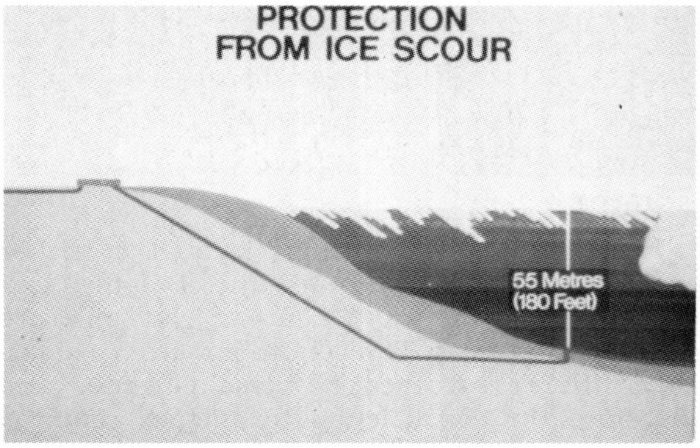


FIGURE 6.

The absence of a predictable weather «window» in the summer and the wide variability of ice thicknesses during winter in M'Clure Strait has ruled out the construction of this crossing by a laybarge or by continuous ice trenching methods. As a result a unique method, shown in Figure 7, has been developed for installing pipelines across M'Clure Strait and has been called the «ice hole bottom pull technique».

Actual construction would begin with pipe preparation at an onshore make up yard where

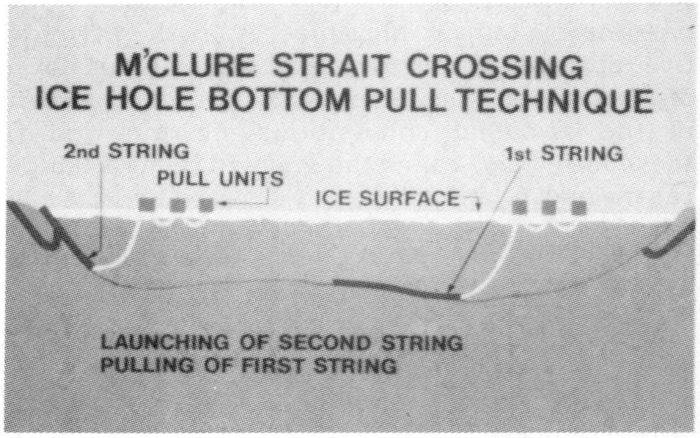


FIGURE 7.

individual 36" (914 mm) diameter pipe joints would be welded together into pipe «sections» and launched sequentially. When a sufficient number of sections have been welded together, the entire pipe «string» which would be several miles long, would be sealed and pulled across the bottom of M'Clure Strait.

From shore to shore each of the two 36" (914 mm) diameter pipelines would require 6 pipe strings.

Pulling of each pipe string from the onshore make up yard would be undertaken by large cable pulling units set up over the ice holes as shown in Figure 8. These units would provide a

FIGURE 8.

linear pulling capability thereby enabling the pull cable to be fed ahead under the ice to the next station as it follows the pipe string from behind.

Once all of these separate pipe strings have been laid across the bottom of M'Clure, they must be tied together to form a single pipeline.

This would be carried out by the use of a special deep sea welding technique which allows welders to work at normal atmospheric pressure during the entire pipe joining operation.

To bring the pipe ends together for the welding operation, alignment and receiver sleds mounted on adjacent pipe strings would be winched together inside the welding chamber and the pipe ends accurately lined up for welding. The welding crew would then be lowered from the surface in a personnel transfer vehicle which mates with the welding habitat and permits the welders to transfer into the habitat and carry out the welding under normal atmospheric conditions as illustrated in Figure 9.

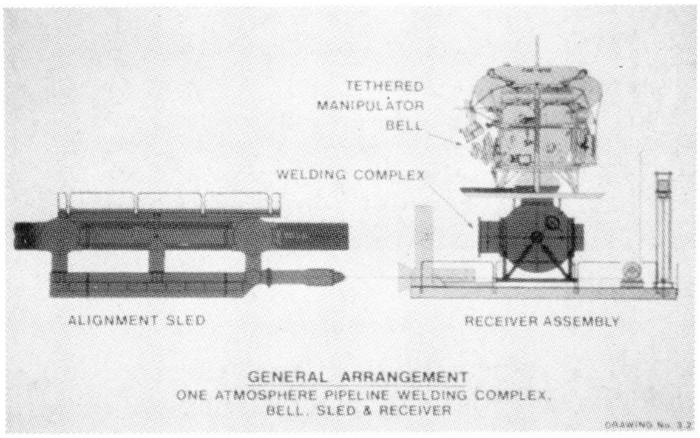


FIGURE 9.

## Logistics

To build the Polar Gas pipeline would require a very significant logistics effort in moving almost 5 million tons (4.2 million tonnes) of equipment, material and supplies to some of the most remote areas of Canada. The logistics requirements shown in Figure 10 illustrates the breakdown of the overall tonnage to be handled for the 3 separate zones. A total of 15 staging and transfer sites would be required to handle the total tonnage.

South of the 60th parallel existing road and rail systems would be used to move 1.7 million tons (1.5 million tonnes) of materials with almost half of this tonnage being moved in the 2nd year of construction.

Further north, river and ocean barges would be used to transport more than 2.3 million tons (2.1 million tonnes), again peaking in the 2nd year of construction.

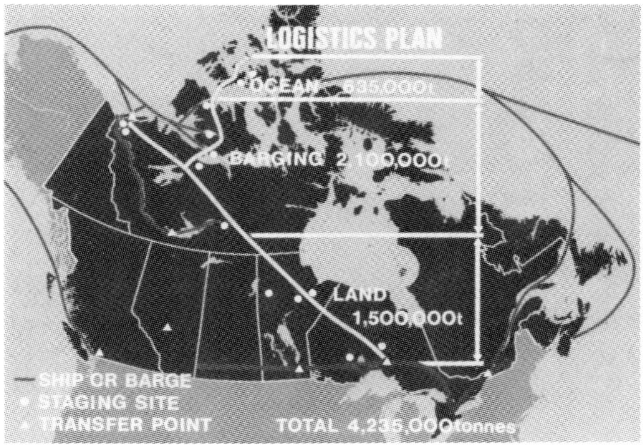


FIGURE 10.

In the High Arctic however, only ice class ships can reliably be used to transport the material through the Arctic seas. Figure 11 shows the tonnages to be shipped and number of ships required over the construction period. These ships would transport close to 700,000 tons (630,000 tonnes) of equipment and material from ports in eastern Canada and elsewhere and deliver their cargoes to Melville and Bathurst Islands. Since the shipping season to Bridport, on Melville Island, is usually only about one month duration each year all of the ships would require ice breaker escorts particularly west of Resolute. It would therefore be necessary to assemble sufficient ice class vessels with ice breaker escorts to ensure safe, thorough and cost effective material movement to meet the construction schedule.

**HIGH ARCTIC SHIPPING**

| | CONSTRUCTION YEAR | | | | | | | |
|---|---|---|---|---|---|---|---|---|
| | 1 | 2 | 3 | 4 | 5 | 6 | 7 | 8 |
| SHIPS | 10 | 21 | 28 | 15 | 3 | - | 1 | 3 |
| TONNES (000) | 67 | 158 | 242 | 126 | 18 | - | 4 | 15 |

TOTAL TONNAGE - 630,000 TONNES

FIGURE 11.

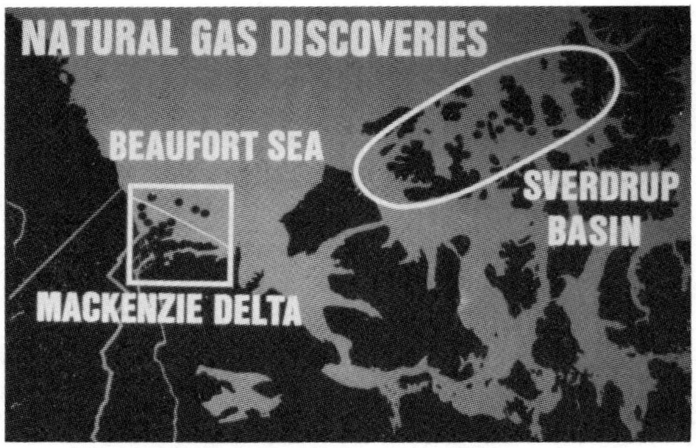


FIGURE 12.

## Natural Gas Supply and Demand

The major supply areas in the Canadian Arctic are the Mackenzie Delta/Beaufort Sea region and the Sverdrup Basin of the Arctic Islands shown in Figure 12. Although energy exploration in these areas is relatively recent, substantial proven natural gas reserves have been identified and total approximately 21.5 trillion cubic feet (609 x $10^9 m^3$) excluding any gas reserves in the Beaufort Sea area (Figure 13). These marketable reserves already identified represent about 10 years' supply for Canada at current consumption rates and are sufficient to make the Polar Gas pipeline a feasible and economic venture.

**ESTIMATE OF CURRENT MARKETABLE NATURAL GAS RESERVES**

($10^9 m^3$ @ 1atm & 15.5°C)

| | |
|---|---|
| Arctic Islands | |
| Melville Area | 249.3 |
| Lougheed Area | 85.0 |
| Ellef Ringnes Area | 110.5 |
| Sub-total | 444.8 |
| Mackenzie Delta | 164.3 |
| TOTAL | 609.1 |

(Excludes 1981 gas discoveries)

FIGURE 13.

With each new drilling season, proven reserves of natural gas grow substantially and the ultimate potential in the Arctic of 200 to 285 tcf (5.7 to 8 x $10^{12} m^3$) is equivalent to 3 to 4 times the current remaining natural gas reserves in western Canada.

The overall usage of natural gas in Canada is monitored by the National Energy Board (NEB), a regulatory arm of the Federal Government, and reserves surplus to Canadian requirements may be authorized for export by the NEB. In the Board's latest report issued in September 1981

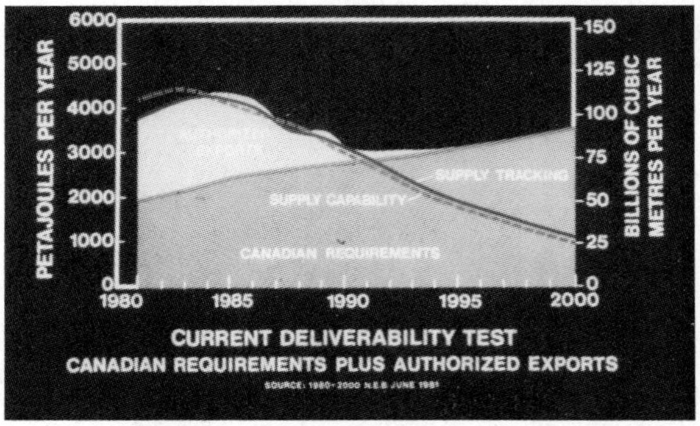


FIGURE 14.

the NEB indicated that the current supply would be adequate to meet Canadian requirements and existing authorized exports to the United States until the year 1992. In the forecast it is assumed that no new exports are authorized and therefore the exports to the U.S.A. taper off after 1985 and there would be virtually no exports after 1990 (Figure 14). The report also forecasted future reserve additions in conventional areas which would improve the supply capability and permit some additional exports

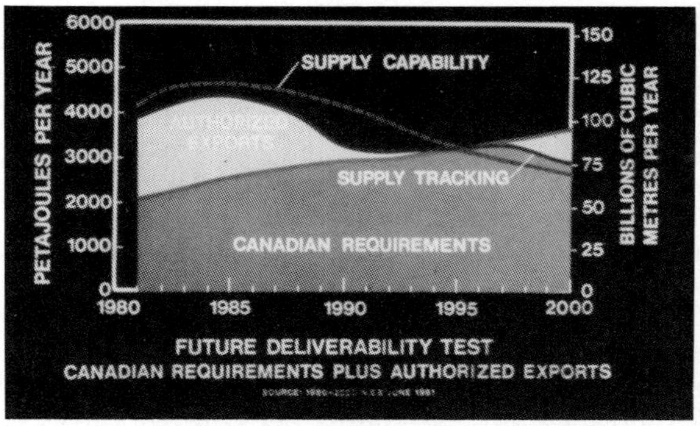


FIGURE 15.

up to the year 1995. In this scenario if no new exports are authorized and with demand tracking supply, Canadian requirements could be met until the year 1998 as shown in Figure 15. These projections do not, however, include any reserve additions from the frontier regions such as the east coast or the Canadian Arctic. In fact, the NEB indicated that these reserves will not be considered in the total supply scenario until a transportation system has been put in place.

The NEB also considered different supply scenarios and demand rates for natural gas. Looking at varying the demand but maintaining the base supply, the projections indicate that Canadian demand can be met until 1992 with the projected high demand, 1997 with the middle demand rate and 1999 at the low demand rate (Figure 16).

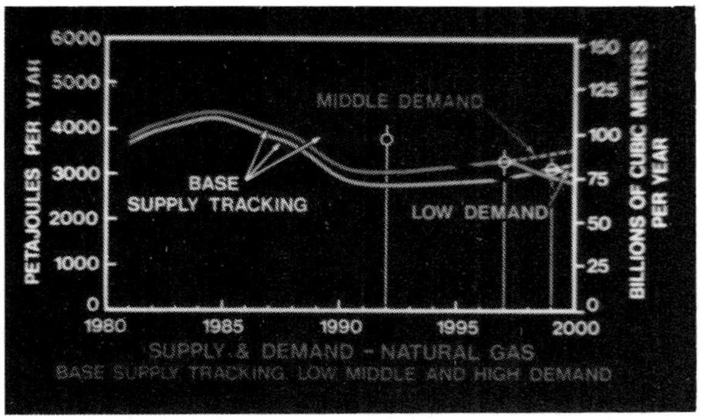


FIGURE 16.

Although the NEB recently revised its position in surplus of Canadian requirements by deleting the current deliverability test, based on the latest demand rates projected, a new source of gas is still required to meet Canadian requirements in the 1990's unless large new supplies are found in conventional areas in western Canada.

In the previous figures, it was shown that Canadian exports would be terminated when the licences become due. However, the graph on Figure 17 shows Tenneco's latest (August, 1981) projected U.S. gas supply by source to the year 2000. It should be noted that the U.S. are expecting to maintain conventional pipeline imports during this entire period. In fact the U.S. are planning to increase the yearly volumes starting in the 1980's and if Canada does not authorize a continuation of these volumes the supply must come from other sources or alternative energy projects. In any event, although there is a current deliverability surplus in the U.S. very similar to that in Canada, the figure indicates that there is a large market for natural gas south of the border.

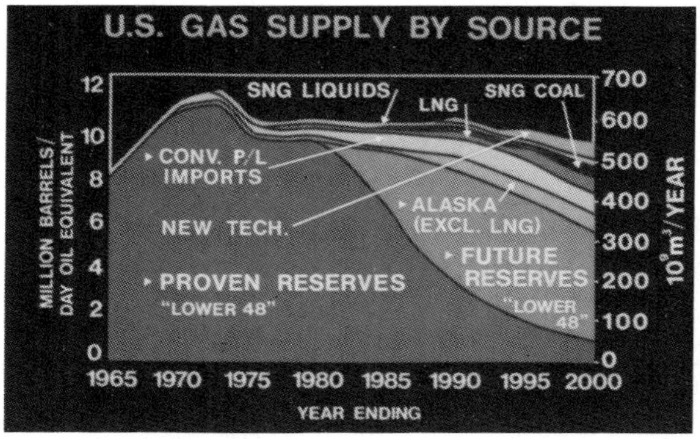


FIGURE 17.

## Economic Overview

The following brief overview clearly shows the economic viability of the project.

The cost of constructing the Polar Gas pipeline is high as would be expected for any frontier transportation system and is $10.6 billion expressed in 1980 Canadian dollars excluding any contingencies or interest charges (Figure 18). TransCanada would have to add a further $1 billion to expand its system from

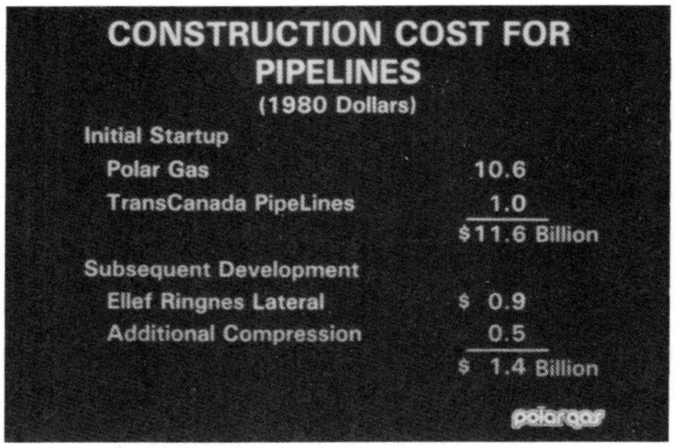


FIGURE 18.

northern Ontario to the markets in southern Ontario and for connection to potential export systems. Thus the total construction cost of the pipelines for the initial throughput of 2.15 bcf/d (60.9 million $m^3$/d) would be some $11.6 billion. Expanding the pipeline system to tap the reserves in the Ellef Ringnes area would add another $1.4 billion.

Individual producing fields utilizing currently identified marketable reserves, will be sequentially integrated into the Polar Gas transportation system as indicated in Figure 19. The

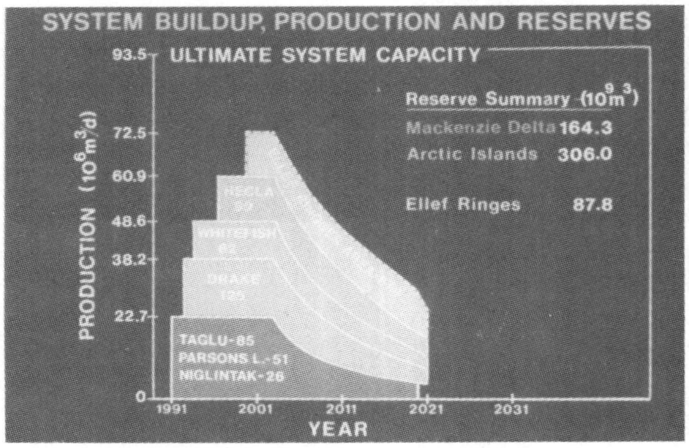


FIGURE 19.

Mackenzie Delta gas will be the first brought on line and different fields in the Arctic Islands will follow in subsequent years. Although project economics have been based on connecting only the Drake, Whitefish and Hecla fields, lateral lines to other areas could be connected in a similar manner when the reserves have been identified and as they are needed.

ENERGY PRICE ASSUMPTIONS

| Year | Persian Gulf Oil FOB ($/bbl. U.S.) | Persian Gulf Oil CIF (C$/bbl.) | Domestic Price Frontier Gas @ Toronto (C$/GJ) | Arctic Frontier Gas Average Price (C$/GJ) |
|---|---|---|---|---|
| 1981 | 33.70 | 41.70 | | |
| 1991 | 87.70 | 99.50 | 12.40 | 16.00 |
| 2000 | 176.70 | 191.20 | 24.60 | 28.70 |
| 2001 | | | 26.60 | 26.60 |
| 2010 | | | 53.00 | 53.00 |

polargas

FIGURE 20.

It is expected that in each of these areas, the reserves will have increased by the time the pipeline is in place. This would substantially improve the economics of the project since the ultimate capacity of the Polar Gas pipeline, as indicated at the top of Figure 19, is 3.3 bcf/d (93.5 million $m^3$/d).

The attractiveness of connecting these supply sources from a producer point of view has been determined based on energy price projections as shown in Figure 20.

For the calculations, the export gas price has been assumed to be equivalent to the cost of Persian Gulf oil laid down in Canada and domestic gas price for frontier gas has been assumed to be equal to the delivered cost of oil sands crude in Toronto. (The latter is currently perceived to be approximately 15 % less than the equivalent world price.) For future prices, as indicated in Figure 21 it has been assumed that the world price of crude oil and export gas price would increase at the rate of 9.1 percent each year for the period 1981 to 1991 and 7.5 percent

MARKET PRICE CHANGES FOR FRONTIER GAS (% per YEAR)

| Year | Projected Persian Gulf Oil FOB | Projected Persian Gulf Oil CIF | Domestic Price Frontier Gas @ Toronto | Arctic Frontier Gas (Average) |
|---|---|---|---|---|
| 1981-1991 | 10.0(1) | 9.1 | 8.0(2) | — |
| 1991-2000 | 8.1(1) | 7.5 | 6.8(2) | 6.8 |
| 2001-2010 | | | | 8.0 |

(1) U.S. GNP deflator + 2%
(2) Canadian CPI deflator

polargas

FIGURE 21.

per year for the period 1991 to 2000. This is based on the U.S. Gross National Product deflator (escalation) plus 2 percent. For the oil sands and domestic gas price in Toronto, the escalation rates have been taken as 8 percent per year for the period 1981 to 1991 and 6.8 percent for the period 1991 to 2000.

Based in the above, the average price for frontier gas would be $17.16/mcf ($16/gigajoule) in 1991 and rising in succeeding years as outlined in the preceding paragraph. The prices for each year are shown in Figure 22 on a logarithmic scale. The solid line shows the world price, the broken solid line shows projected frontier gas domestic price while the dotted line indicates the average price. Therefore, the average price shown as a dotted line is at the world level initially and decreases to the domestic number when exports are terminated. The remaining two lines show transportation costs by year from the two areas. The cost of transporting Arctic Islands' gas is somewhat higher in that it is assumed that each of the areas would pay for its own lateral line to the junction point near Coppermine in the Northwest Territories, and therefore, the Arctic Islands' lateral includes a longer line and the higher cost of the marine crossings. On this basis, although there would be a slightly lower netback for the Islands' producers, the return would be sufficiently high to invest in the necessary production facilities for the pipeline connection.

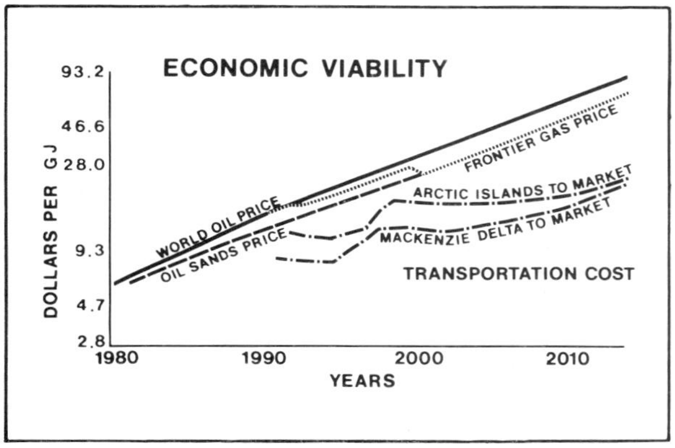


FIGURE 22.

## Benefits

The benefits to all sections of the Canadian economy are substantial on a national basis, and Figure 23 summarizes some of the advantages. When comparing supply from the various Canadian energy projects, it can be readily seen that Polar Gas would supply much more than any of the other projects recently completed, under construction or planned. For example, Polar Gas operating at its capacity of 3.3 bcf/d (93.5 million $m^3$/d) would provide four and one-half times as much energy as the massive James Bay hydro-electric complex in northern Quebec which cost close to $16 billion to build. All other projects shown in this figure, with the exception of the Dempster Lateral, supply even less energy than the James Bay Project.

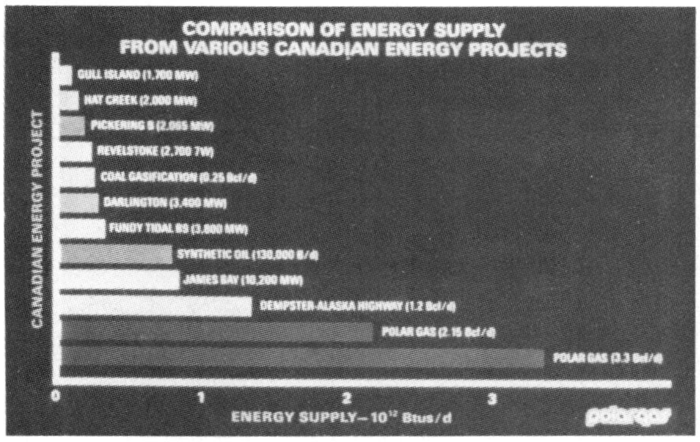


FIGURE 23.

A COMPARISON OF
MAJOR ENERGY OPTIONS
AVAILABLE TO CANADA
($ per GJ)

| | DELIVERED TO EASTERN MARKETS |
|---|---|
| GAS PROJECTS | |
| POLAR GAS | 3.70 |
| Synthetic Gas | 6.10 |
| OIL PROJECTS | |
| Synthetic Oil | 4.70 |
| Imported Oil | 11.70 |
| ELECTRIC PROJECTS | |
| Thermal—Coal | 14.90 |
| Nuclear | 9.30 |
| Hydro | 11.20 |
| Tidal | 23.80 |

FIGURE 24.

A comparison of the major energy options available to Canada to meet projected future thermal demand is shown in Figure 24. On a resource cost basis (expressed in 1985 Canadian dollars), that is, excluding taxes and royalties, frontier gas delivered by pipeline to eastern markets is less than the other options. For instance, imported oil would cost $12.50/mmbtu ($11.70/gigajoule) or over three times the delivered cost by Polar Gas. Electric projects for thermal demand could be even higher on an average ranging from $10.00 to $25.50/mmbtu ($9.30 to $23.8/gigajoule) for tidal power. These higher costs result from the fact that electrical utility companies in Canada do not pay income taxes or royalties.

Figure 25 shows that although the cost of the pipeline, expressed in 1980 dollars, is substantial, the cost of the system and related production facilities is relatively small when you consider the cash flow from the project.

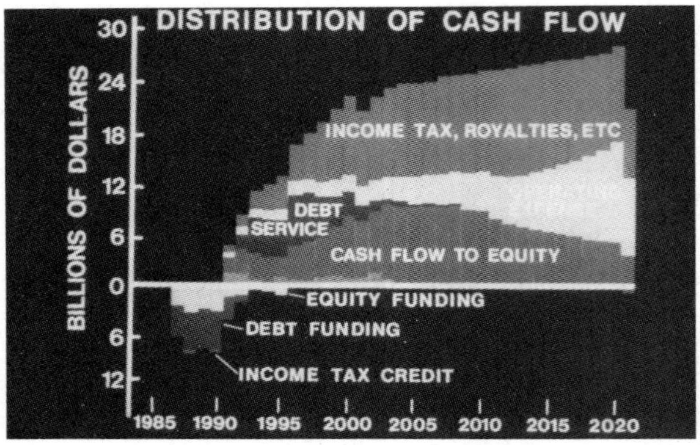


FIGURE 25.

In this figure, the equity and debt funding is shown below the line and the positive cash flow to equity, debt servicing, operating expenses and government revenue are shown above the line. Total cash flow during the operating life is some $650 billion undiscounted or $22.5 billion discounted at a rate of 14 percent. Figure 26 shows the distribution of net benefits. The transmission company only gets some 8 percent of this total. In escalated terms, the consumers benefit some $130 billion or 20 percent resulting from the difference in energy cost between frontier gas and imported oil. The producers receive some $143 billion or 22 percent of the total after deducting all costs of production. However, by far the largest beneficiaries are the governments who receive 50 percent of the total or some $325 billion. On a discounted basis this would amount to $11.3 billion.

## Project Benefits to the United States

There are also substantial benefits to the United States as well as the enormous benefits to Canadians as shown in Figure 27. First, it will bring in a new supply source of which a substantial portion must be directed to American markets in the early years. Second, a new secure large additional supply of energy will become available for a minimum of ten years which could be extended for several years beyond the initial term provided that exploration continues to prove up the huge potential in the Canadian Arctic frontier. Third, the import of Canadian frontier natural gas would serve to replace more costly imported LNG. Fourthly, the project would transport the natural gas directly to the northeastern states where the large demand exists. This would result in the lowest overall transportation costs as well as providing substantial savings in construction of additioial pipeline facilities in the United States. And, lastly, the imported natural gas could mean a substantial saving to U.S. consumers when compared to the laid-down cost of alternative fuels.

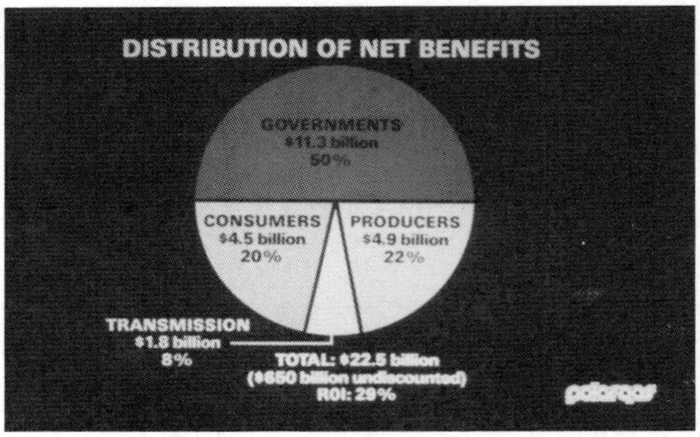


FIGURE 26.

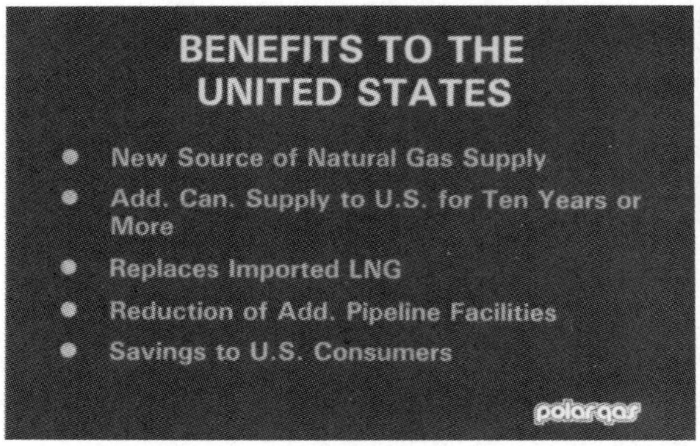


FIGURE 27.

## Summary

Sufficient reserves of natural gas have already been discovered in the Canadian Arctic to make this pipeline project economically attractive, and there is no doubt that Canada and the U.S. will need the gas by the early nineties of not earlier. The pipeline project will take ten years to build and the sooner it is built, the lower the real cost of construction will be in addition to providing significant boosts to the economy of both countries and the ability to look after our respective energy needs.

*Cold Regions Science and Technology*, 7 (1983) 227—230
Elsevier Science Publishers B.V., Amsterdam — Printed in The Netherlands

# Arctic Transportation Problems and Solutions

*Lindsay J. Franklin*

Panarctic Oils Ltd.
Calgary

## Abstract

*The explorer seeking oil and/or gas in the Arctic Islands is faced with problems relating to transportation and logistics on two levels. Firstly, he must deal with the immediate problems of moving men, equipment, materials and supplies to support the exploration venture. The problems are defined and the methods used to solve them are described, with appropriate comment on the success/failure of various innovative procedures utilized.*

*Secondly, the Arctic Islands explorer is faced with economic evaluation of discovered hydrocarbon accumulations. such evaluation is dependent upon the conception and predicted development of unique and unproven transportation systems. The relativemerits of several systems including ice breaking tankers, submarine tankers and pipelines are reviewed.*

## Artic Transportation Problems and Solutions

Have you ever wondered — why would any person decide to explore in the far northern portions of Canada where it is so cold — dark — isolated that conditions are probably more difficult than any other area of the world?

It has been suggested that it has something to do with a variation of Murphy's law. If the conditions are such that exploration is facilitated and the surroundings are pleasant, the probability of finding significant oil and gas is very low. Conversely, the Arctic Islands area of northern Canada must surely hide great and wonderful rewards.

The preliminary results from the small amount of exploration conducted in this forbidding and inhospitable area appear to confirm our feelings about the adversity of nature. In a prospective area measuring some 1200 kilometres by 600 kilometres there have been drilled some 160 exploration tests over the last 21 years.

This miniscule amount of exploration activity, relative to the area under examination, has resulted in the definition of gas reserves (circa 16 tcf) equal to some 20 % of the volume discovered in western Canada following exploration activity of a magnitude many hundreds of times greater than that in the arctic archipelago. Additionally, recent findings have revealed that the large structures, originally deemed prospective for gas accumulation, may contain very significant volumes of oil. The cisco structure for example, has been found to contain oil estimated to provide recoverable reserves of the order of 450 millions of barrels.

The indications of substantial reward are of sufficient magnitude that aggressive exploration will continue and will accelerate in the Canadian arctic archipelago, in spite of the formidable problems one must face, particularly with respect to transportation.

The problems of transportation and logistics must be dealt with on two levels. Firstly, one must develop and implement systems to move efficiently and safely men, equipment, materials and supplies to support the exploration activity.

Secondly, the arctic explorer is faced with evaluation of discovered hydrocarbon accumulations, and such evaluation is dependent upon the conception and predicted development of

0165-232X/83/$03.00  

unique and unproven transportation systems.

In dealing with transportation and logistics problems on the first level Panarctic has attempted to operate in harmony with the environmental conditions which prevail rather than confront them.

One selected as a major and permanent operating base Rea Point on the east end of Melville Island. This location was chosen as the spot reasonably central to expected activity, accessible to sealift each summer with existing and available marine equipment, and exhibiting a shoreline permitting direct offloading to shore without requirement for wharf or dock facilities.

Rea Point has proven to be a furtuitous choice. A short «two week window» presents itself each year from late August to early September. Each year for the past 13 years there has been successfully deposited at Rea Point the equipment and supplies and fuel to support the exploration activity of the succeeding year. Sealift volumes have been as high as 40,000 tons of dry cargo and bulk fuel with a requirement for involvement of five ships plus federal ice breakers which assist when required.

Equipment and supplies not provided by sealift (sometimes due to late delivery or to change in proposed operation) must be delivered by heavy-lift aircraft at relatively high transportation costs.

In recent times the high cost of capital has occasionally made it economically attractive to airlift certain high cost items having application late in winter drilling programs. The cost of capital tied up for some nine months may offset the higher cost of airlift.

It was initially anticipated that maximum work would be conducted during that period of the year when one might take advantage of the extended periods of daylight and attractive temperatures that facilitate the operation. It was soon evident that the warm temperatures and continuous daylight of June through September preclude operational activities more stringently than the darkness and cold of mid-winter.

Summer drilling operations can be conducted successfully but they require substantial planning and preparation. All material and equipment for use over the summer must be stock-piled at the location in spring while the frozen surface provides either overland transportation corridors or (more commonly) 1800 metre airstrips for the Hercules Aircraft. Stockpiling must make reasonable allowance for contingency situations such as loss of circulation that might consume major volumes of mud and chemicals, or the requirement to run intermediate casing not included in original well program.

After the surface has thawed a short airstrip (circa 600m) may be available for stol aircraft, but transport between such strip and the drillsite commonly will be by helicopter.

At a summer drilling location remote from Rea Point, the requirement to replace rig components too large for a Twin Otter will involve the substantial expense of a heavy-lift helicopter to sling such components from Rea Point to the drilling location with all the attendant problems of fuel caches at intermediate points, frequent foggy conditions generated by the open water, etc. In winter one would simply deliver the required part by heavy-lift aircraft to the drillsite where routine pickup by truck is effected for transport to the rig.

When drilling at a location is completed in the June-September interval there exist no alternative other than to await for frozen surface conditions in October for movement of a rig and equipment to the succeeding location.

Do not be fooled by pretty pictures of midsummer arctic drilling; there are many attendant problems principally associated with the knee-deep mud that is seldom shown in the pictures.

Now let us look at the cold portion of the arctic season where the problems of soft mud no longer inhibit activity. One must admit that there are substantial problems associated with working in the cold and the dark. One of the problems is hard water. If you think you have hard water problems consider what this fellow is facing.

In the area of Panarctic operations about 75 of the surface area is water covered and, accordingly, many of the attractive structures are in the offshore area. Following a policy of attempting to work in harmony with the environmental conditions of the Arctic, Panarctic elected to do exploratory offshore drilling utilizing the ice rather than fighting it in the more conventional manner. This ice is used as a stable platform which can be artificially thickened, strengthened, and smoothed. Work on the ice platform is initiated with equipment delivered to the site by helicopter as soon as the ice is of adequate thickness (circa 0.5 m) to support men and machinery. Helicopters slinging camp units, tractors, pumps etc. are utilized to gain precious time. Reliable surface access to a drilling site over the ice from land would not normally be available until several weeks later. Following occupation of the selected site flooding operations are quickly commenced and the buildup of the platform and a 1800 m airstrip progresses relentlessly but at a disappointingly slow average buildup rate of 10 cm per day. Over-zealous flooding rates produce unfrozen pockets of salt water that become more saline as the surrounding ice forms: such saline pockets, when insulated from the deep cold of the atmosphere, will never freeze, and they weaken the platform or the airstrip unacceptably. Patience is a virtue in the flooding operation.

When the airstrip on the ice is ready to accept the Hercules aircraft, movement of rig, camp, and supplies is energetically progressed. Loads have been assembled to utilize the Hercules effectively with respect to volume and weight lifting capacity. On a twenty-four hour basis the multiple Hercules are loaded and unloaded in a smooth and efficient manner to minimize aircraft time on the ground. from approval of the airstrip for Hercs to the point of spudding the well, a period of some 25 to 30 days, there will be of the order of 150 Herc loads delivered.

When ice platform drilling is in progress air support normally consists of a visit of a boeing 727 cargo/passenger combination once weekly for crew change and delivery of fresh food and produce, plus a Hercules about every second day with fuel and/or other consumable materials in support of the operation. The heavy aircraft services are supplemented by Twin Otters commonly making one or more visits daily to each location.

Now having looked at the mundane side of arctic transport problems — the everyday kind of support needed for an ongoing exploration activity — one is left with the future transportation problems associated with production activities. Firstly, one might expect to look at transport of the product to markets: However, this particular conference includes papers dealing with:

- *The Arctic Pilot Project,*
- *A polar gas pipeline for the arctic islands and*
- *Submarine transportation of hydrocarbons from the Arctic.*

These papers provide substantial information on the principal methods being considered for transport of hydrocarbons from the Arctic Islands and there is, accordingly, no need to comment further thereon. The three papers referred to will deal effectively, precisely, and exhaustively on their particular transportation mode.

One other aspect of future transportation problems of concern, and on which I will comment briefly, relates to the installation of the production facilities that will be required to process and deliver oil or gas to the pipeline, tanker or submarine that will transport it to market.

The Arctic Pilot Project, proposed for operations starting in the late 1980's will take some nine million cubic metres of gas daily from the Drake Point field on the Sabine Peninsula of Melville Island. The production facilities and processing plant suitable for delivery of such volume of gas having pressure, temperature and water content suitable for pipeline transmission will be installed at Drake Point some 180 km beyond Rea Point, the closest spot for which there is some reasonable certainty of reaching with sealift.

For the transport of materials and equipment from Rea Point to the field installation site the most promising systems or combinations were evaluated with reference to the highest probability of success and the lowest overall operating cost. The high cost of site assembly of numerous small units (i. e. Air transportable) versus the economies resultant from an ability to move large modules from fabrication site to erection site without dismantling, constituted an important factor in the evaluation of alternatives.

Our studies revealed that the requirement to move some 15,700 tonnes in two consecutive winter seasons could be accomplished most effectively via an over-ice route utilizing an air cushion vehicle (acv) having a payload capacity of 136 tonnes, towed/restrained by two rolligon vehicles and supported by one light rolligon support vehicle. This over-ice transport system would be supplemented by Hercules aircraft which would transport all cargo of size and weight appropriate for air transport. The acv/rolligon combination has remarkable tolerance to accommodate irregularities in the surface of the ice and has excellent load distribution characteristics. Safe ice thickness for the acv with 136 tonne payload is found to be 1.04 metres and such ice thickness over the selected route has probability of 50 % by February 1 and 80 % by May 1.

The acv rolligon transport facility is not deemed high technology, but there are sufficient unknowns concerning operation under arctic conditions that one year of operating experience is planned prior to direct application of the system for module transport.

In the summer of 1982 a five-ton boeing vertol 107 heavylift helicopter equipped with a 600-foot line towed a 4,000 ton barge across the water of Puget Sound. A towing speed of 10 knots was achieved within minutes, and a line tension of 10,500 pounds was measured with load cells installed in the line. This demonstration, conducted for an operation different and far removed from Panarctic operations, suggests a combination of towing vehicle and acv that might safely be utilized to move heavy loads across areas combining open water and ice-covered water where the ice cover cannot be depended upon for support of wheeled or tracked towing vehicles. The helicopter towing procedure might be expected to be a high cost option, but it makes application of an acv practical under conditions that could preclude the use of wheeled or tracked prime movers.

## Subsea Production Systems

To this point we have been considering some transportation problems associated with land or ice platform drilling and with transportation for land based production facilities. Now as we proceed to consider subsea production facilities under ice covered waters, we are confronted with transportation problems of an increased order of magnitude.

Let us look at the area east of Lougheed Island 77°N and 108°W where we find the Whitefish gas field and the disco oil field lying under some 300 m of ice covered water and 20-30 km from shore.

The favoured production system envisions two groups of five wells producing to subsea manifolds which in turn would produce to a single manifold for comingling and transport to shore by one flowline. It may sound reasonably straightforward, but when one looks in to the complexity and size of the subsea manifolds involved. The dimensions of the problem become apparent. Two «small» manifolds at 200 t each and a larger one, at some 400 t, must be placed on bottom. In August/September, when the area is sometimes clear of ice placement from floating vessels could be relatively routine; however, the area may not clear of ice in the next ten years and some alternative placement method must be developed.

Three alternatives are suggested for consideration:

1. Develop subsea modules that can be delivered to the installation site in air transportable component parts, and assemble components on site.
2. Develop an acv transport unit of dimension/capacity adequate for over ice transport of the large loads.
3. Develop subsea modules that could be towed under the ice at a controlled submarine elevation while imposing minimal load on ice cover.

Alternative no. 1 would involve the assembly and testing of complex and sophisticated units under the adverse conditions of arctic winter on the ice, and such prospect does little to commend this procedure. At this juncture it is suggested that it will be necessary to have an ability to deliver and place subsea units without disassembly after fabrication and testing at point of fabrication.

The relative merits of alternative 2 versus alternative 3 (development of large acv unit versus development of a system to tow the manifolds under the ice) will require substantial study and investigation. Cursory evaluation of the latter system indicates it to offer maximum ability to perform under the widely variable environmental conditions which may confront the operator at time of installation.

In a year of open water the manifolds could be transported directly to installation site on floating transport, and placed on bottom. If ice cover precluded marine access to location, unloading and sub-ice transport could be effected from the open water closest to the installation site. In the event that sub-ice transport were possible, only after extensive and prolonged preparation of the ice surface for towing vehicle, the subsea manifolds could be stored on or close to ocean bottom for recovery and towing when surface conditions were appropriate.

It is obvious that much work needs to be done to develop optimum solutions to the problems related to transport for sub-ice production systems. There have been presenteo for your consideration many more problems than solutions. But if this presentation stimulates any portion of an audience of knowledgeable people to think about these problems and potential solutions, then our time together is well spent.

In the meantime we will continue our exploration activities. As we define more and more energy reserves, as I am confident we will do, the increasing incentive to develop solutions to the transport problems here reviewed will undoubtedly lead to the optimum solutions.

*Cold Regions Science and Technology*, 7 (1983) 231—249
Elsevier Science Publishers B.V., Amsterdam — Printed in The Netherlands

# Maritime Transport in Support of Arctic Resource Development

*Donn K. Haglund*

The University of Wisconsin
Milwaukee

## Abstract

*Resources are available in the Arctic that will utilize maritime services as economic forces dictate their development. The clear immediate opportunity is for alternative transportation of oil and gas products although future exploitation of additional Arctic coal deposits (as well as other mineral resources) may generate additional maritime transport service demands. Living resources of the Arctic, although significant, are not expected to create development of maritime tranportation systems beyond a possible increase in fishing vessels.*

*Transportation technology is either already in existence for Arctic resource development or can be developed when economic and institutional factors are favourable, provided that appropriate research programs are instituted to provide the necessary fundamental knowledge.*

## Introduction

Recent years have seen the developed nations looking farther and farther afield for sources of essential commodities. The polar regions of the earth are among the planet's major remaining resource frontiers. The Antarctic and especially the Arctic realms have attracted scientific and resource investigators at an accelerating rate. Since World War II there has also been a military interest in the Arctic. The need for transport services in these remote regions has expanded with increasing interest in the far north and far south.

Among the three polar interest communities, scientific, military, and resource development, the present study addresses itself solely to the latter. It is assumed that the military services will continue to assess their own requirements and that they are not likely to generate a demand for substantial increases in transportation services except in an emergency. Similarly, transportation requirements of basic scientific investigation are not expected to extend beyond research vessels nor commercial-military aircraft use.

That there will be a significant increase in demand for transportation services in polar resource development is demonstrated by present events and those of the recent past. Although seekers of furs, fish, sea mammals, and precious metals have extracted wealth from the polar lands and seas for centuries, it was not until the 1960's that concerted efforts were made to develop such industries as petroleum and natural gas extraction on the polar frontier. The developed world, especially the United States, was formerly able to rely on domestic or more readily accessible foreign sources for such basic commodities. Production of the first Arctic coastal Alaskan oil and construction of an elaborate transport system, especially the Trans-Alaska Pipe Line System (TAPS), to deliver its product to market is a prime example of things to come.

While there are both actual and potential roles for various transport modes in polar resource development, the contribution that maritime services can make forms the basis of the present study.

Recognizing the need for a broad study of the role which maritime transportation can be expected to play in polar regions resource development, the United States National Academy of Sciences, through its Maritime Transportation Research Board, established, in late 1979, the Committee on Maritime Services to Support Polar Resource Development. The present communicator was asked to serve as its chairman. Twelve additional members were appointed who represented a variety of interests; academic, shipbuilding and ship operating, maritime labor, polar technology and port administration. Both Canadian and American (including Alaskan) specialists accepted appointments. Liaison officers from the U.S. Coast Guard, the U.S. Maritime Administration and the U.S. National Oceanic and Atmospheric Administration were assigned to the study group. Funding was provided by the U.S. Departments of Commerce, Defense and Transportation under provisions of a contract between the National Academy of Sciences and the U.S. Office of Naval Research entitled «Maritime Services to Support Polar Resource Development».

The final report was published in the latter half of 1981 by the National Academy Press, Washington, D.C. This presentation is intended to highlight, summarize, amplify and further develop the contents of that research.

While the twenty-first century may see major resource development activity in both the North and South Polar Regions, it is the judgment of the Committee that such efforts will be limited to the Arctic during the final two decades of this century, the time period to which this study is limited. Continuing and potential marine transportation needs for the Antarctic region are to support scientific research with only limited potential need, in the long run, for tourism and commercial development of living and mineral resources.

The basic emphasis of the present study therefore is on the Arctic. It is concerned with the special role of maritime transportation, both inbound and outbound, and it includes its ancillary facilities requirements in the movement of extractive resources of the far north. In order to assess this role it is necessary to describe the physical environment in which this transport will take place and then proceed to the technical considerations that the severities of the Circumpolar North impose upon maritime transportation systems.

## Definition

The Arctic is defined in a variety of ways, depending largely on the discipline, audience, or user addressed. To botanists or geologists it may be the area north of the tree line or where permafrost occurs near mean sea level; to cartographers it may be the area north of the Arctic Circle; to climatologists it may be where the average for the warmest month is 10°C or 50 °F. But to the marine community, and for our purposes here, it is that region where bordering seas usually have a sea ice cover for some part of the year or season. This may be mostly winter ice cover, i.e., the ice cover may retreat or melt completely for a significant part of the year (navigation season) as in the seas of the northern sea route of the Eurasian Arctic, the Bering Sea, the Chukchi Sea, the near shore areas of the Beaufort Sea, Baffin Bay, the Labrador Sea, the southern East Greenland Sea and the Barents Sea. On the other hand, in some areas such as seaward of the North Alaskan Slope (Beaufort Sea), in regions within the Canadian Archipelago north of the Parry Channel (the classical Northwest Passage), off the northern coasts of Greenland, and the broad, deep expanses of the central Arctic Ocean itself the sea ice cover is perennial. In these areas, ice cover is usually from three-quarters to complete cover, even in the warmest month of the year. It is this deep ocean area, with its thick sea ice cover, surrounding the North Pole that mainly distinguishes the Arctic from the Antarctic environment. Because of this marine character, extreme air temperature minimums rarely fall below −40° to −50°C over the vast expanse of perennial sea ice. In summer, this marine effect predominates even more. Variability at this time about a mean of 0°C is very slight.

# ARCTIC RESOURCES

The Arctic consists of portions of Alaska, Canada, Scandinavia and the U.S.S.R. plus Greenland and the adjacent water bodies. There is a wide range of resources, fossil fuels and metallic areas in both northern U.S.S.R. and Scandinavia and in undersea settings offshore of these nations. It must be assumed, however, that their exploitation will be by these countries. Similarly what Greenlandic resource development takes place, notwithstanding a single mining operation under Canadian lease at present, will likely be under Danish or local Green-

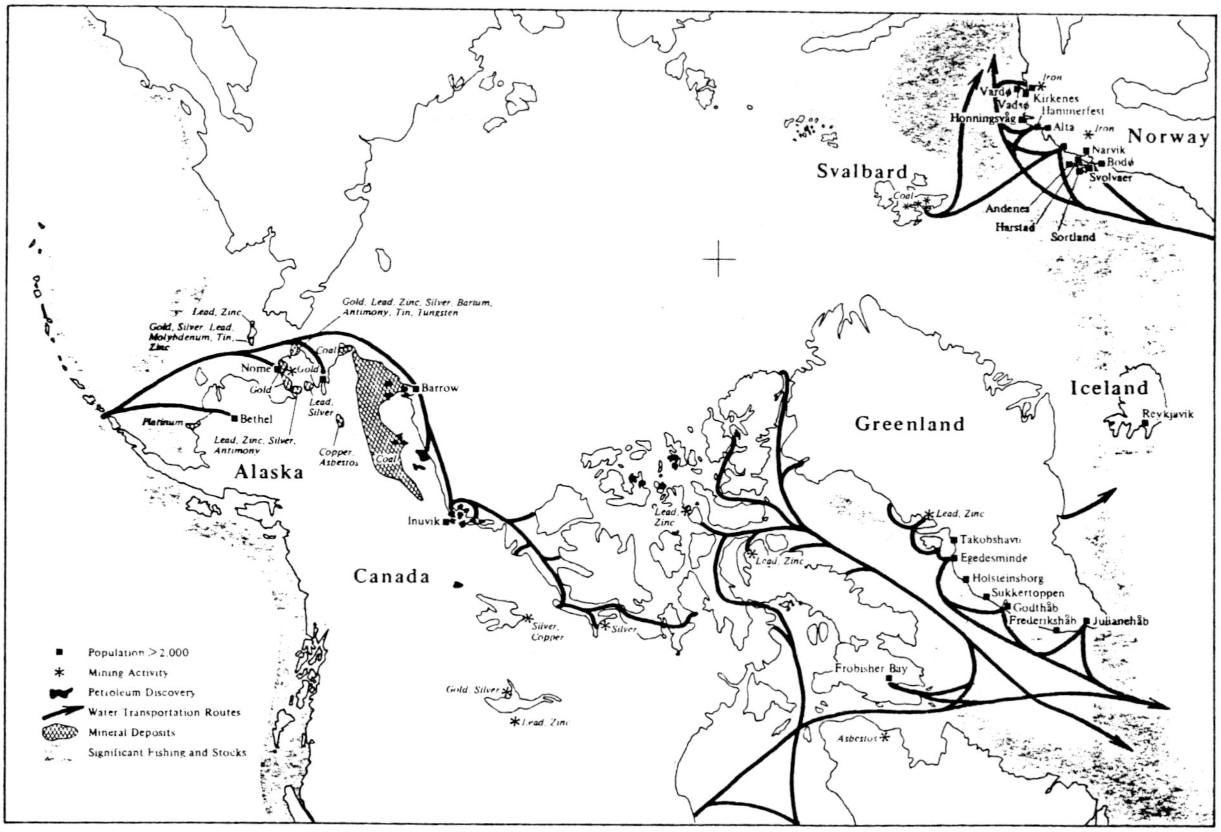


FIGURE 1. *Western Arctic Resources.*

landic control or both. Thus, Arctic maritime transportation opportunities that evolve for North America or organizations, whether in the public or private sector, in the forseeable future are expected to be associated with Alaskan and northern Canadian resources when economic forces dictate and not before. Consequently the study limited its circumpolar resource considerations to Alaska and northern Canada. Figure 1 is a map of western Arctic resource locations.

## Alaskan Mineral Resources

While Alaska has been frequently described as a vast treasure house of resources, from the standpoint of Arctic maritime transportation development, the range and distribution of these resources would appear to be limited to a few commodities in the northern and northwestern portions of the state. The present and potential demand for maritime services seems to be linked to selected mineral deposits situated in the regions from Bristol Bay northward and on the Arctic coastal plain, commonly referred to as the North Slope. Petroleum, natural gas, coal and a few metallics, notably copper, lead and zinc offer greatest promise for marine transport demand.

While the extent of both oil and gas reserves in Arctic Alaska, on-shore and off-shore, are still unknown, they are potentially very large. The Prudhoe Bay reserves alone are conservatively estimated at 9.3 billion barrels of oil and 764 billion cubic meters (27 trillion cubic feet) of gas. (*Oil and Gas Journal* Vol. 75, No. 14, April 14, 1977, p. 56). Alaska is also known to have extensive deposits of generally low sulfur content coal, aggregating some 130 billion tons, the energy equivalent of 350 billion barrels of crude oil. (E.F. Barnes, *Coal Resources of Alaska* U.S. Geol. Survey Bulletin 1242-B, 1967). Prudhoe Bay reserves of petroleum approximately equal two years of current U.S. consumption of oil from all sources, including both domestic and imports. Its gas reserves, if somehow called upon to meet total American demands, would provide a one and half year supply at current levels. Alaska's coal reserves, on the other hand, exceed five hundred times present national annual useage. Conversion of all recoverable Alaska coal to petroleum, were it feasible, would provide some one hundred oil years for the United States at present consumption rates. Over ninety percent of these reserves occur north of the Brooks Range and thus provide a major potential source of Arctic maritime transportation demand. Markets in Japan and other Far East areas, as well as those in the lower 48 states may be assumed to be the eventual utilizers of this marine transport opportunity.

High grade copper deposits near Kobuk and lead and zinc occurrences north of Noatak, both in northwest Alaska, indicate further potential Arctic maritime shipping. Much more research and exploration is needed before an estimate of the total metallic resource potential of north Alaska can be made. There is reason to believe, however, that extensive lead, zinc, copper and possibly other metals, in either concentrate or direct shipment form, should be anticipated as an eventual opportunity for maritime transportation.

### Northern Canadian Mineral Resources

Northern Canada, herein operationally defined as the Yukon and Northwest Territories including their off-shore area, stands as one of North America's resource frontiers. While the ancestors of contemporary northern native peoples have utilized renewable resources; animals, aquatic life, trees, plants, and the like, for many years, extractive energy sources and minerals are attracting today's commercial attention. Economic and geographic limitations preclude large scale commercial (and transport generating) development of forestry, fisheries, agriculture and grazing industries in Canada's north lands.

Mining, including oil and gas, is another matter. Gold, radium, and relatively small scale oil production have been part of northern Canada's economy for decades. Post-World War II developments in lead, zinc, copper, silver, nickel, tungsten, cadmium, and asbestos have followed. A combination of rail, truck, river barge, air and pipeline transport facilities have met all needs of the minerals industry of northern Canada until very recently.

In the late 1960's, as in Alaska, major exploration efforts for oil and gas came to the Canadian north. There is guarded optimism that the area will become, like Alaska's North slope, a significant production region within the foreseeable future.

Oil reserves in Canada's north at the 50 percent probability level, were estimated in 1975 at 18.8 billion barrels. This figure is still in use in 1982 by the Geological Survey of Canada (GSC). While this is a substantial total, it does fall significantly short of earlier optimistic industry estimates which ranged as high as 200 billion barrels. The 1975 GSC estimate for natural gas resources, also at the 50 percent probability level, stands at three and a half trillion cubic meters (123 trillion cubic feet), again much less than original industry estimates which were some two and a half to five times as great. Canada's total consumption of petroleum currently totals some 681 million barrels per year and annual natural gas consumption is 43.3 billion cubic meters. The far north could therefore, theoretically, serve as the nation's sole supplier of petroleum for almost 27 years and of natural gas some eighty years, based on GSC estimates of reserves.

Extensive reserves (in excess of 100 million tons) of direct shipment iron ore have been located in northern Baffin Island, as well as 7 million tons in lead and zinc deposits, from which production began during 1977 when the first concentrates were shipped to European markets.

At Coppermine a three-million ton copper deposit has been identified as well as substantial reserves at existing mine sites throughout the territories.

While various other mineral resources have been identified throughout the Canadian north, the cited examples are those for which ocean-going vessels are a logical and foreseeable transport mode.

## FACTORS IN ARCTIC RESOURCE DEVELOPMENT

### Existing Arctic Transportation

Extensive petroleum and natural gas reserves situated on the North Slope of Alaska plus potential undersea deposits in the Bering,

Chukchi and Beaufort Seas areas are of great interest to maritime shipping. Although the tankship *Manhattan's* transits of the Northwest Passage in the late 1960's did not lead to inauguration of maritime delivery of Arctic Alaskan oil, it did have positive results and was a most valuable research exercise. Present petroleum movement southward is entirely via the Trans-Alaska Pipeline System (TAPS) to trans-shipment. Future transport needs from expanded petroleum production on the North Slope cannot be satisfied by TAPS with its present pump station capabilities. There is no system, pipeline, liquid natural gas tanker, or other, for transporting the extensive natural gas resources of northern and northwestern Alaska to temperate zone markets, although a natural gas pipeline is presently being designed with the major engineering effort underway.

Oil and gas production in northern Canada began at Norman Wells on the Mackenzie River in the 1930's. Aside from the war years the limited production of this small field has been consumed regionally and has moved to local markets by barge. Most Arctic oil and gas reserves in northern Canada are, by their geographical locations, capable of being moved to market by maritime transportation.

## Social and Institutional Constraints

Although it is possible to analyze the impact of technological developments in the near future with some degree of confidence, it is with far less certainty that institutional, social, and economic trends may be approached.

Perhaps the most important issues that will affect polar development and maritime transportation are political and jurisdictional — both international and domestic. Foremost among these is the lack of agreement over the jurisdiction and control of waters and submarine resources. At present only Svalbard, the Greenland-Canada boundary, and the U.S. Russia convention line of 1867 exist under international treaty. Beyond these limited areas there is likely to be general disagreement between a sector principle, an equidistant line from coastal points, or control over the continental shelf. Until an agreement is reached, possibly emanating from the UN Conference on Law of the Sea (UNCLOS III), conflicting claims are likely to restrict exploitation of resources in disputed waters while each nation presses for its own advantage.

A similar problem exists in the designation of international waterways. Important to the United States is Canada's claim that the Northwest Passage is an internal water route and does not meet the standard international legal definitions of an international strait. Though open to the vessels of all nations, they must conform to Canada's various shipping and environmental acts. The United States disputes this jurisdictional claim.

Environmental considerations will certainly have important bearing on planning for polar development, too. At present Canada and the United States, cognizant of the fragility of Arctic ecosystems, are developing more stringent controls for shipping and off-shore operations. For Alaska these controls and procedures now require lead times of more than two years to meet environmental requirements before a project can begin and thus will increase uncertainties in planning and financial backing. Similarly Canada's Arctic is protected under several acts including the Arctic Waters Pollution Prevention Act, Navigational Waters Protection Act, Canada Shipping Act, and Wildlife Act among others.

A further ramification of federal regulations is likely to be the Jones Act in the United States and a similar clause of the Canada Shipping Act, both of which will restrict the use of foreign-built vessels to international voyages. Likewise, export controls in both countries may well inhibit the distribution of raw materials to foreign producers and consumers. Federal regulations may well conflict with state or territorial regulations over environmental safeguards or export controls.

Native peoples will be an important factor in planning for polar development. Through their corporations, they may prove to be a significant source of investment. They can also contribute a limited amount of manpower to construction projects. There is no doubt that increasing power and sense of self-control of native groups will require that northern developers carefully consider Natives' demands for compensation both for losses of hunting areas and for qualitative changes in their lives (biological, social, and economic) resulting from polar development. It is likely, further, that they will demand a direct stake in development and may attempt to impose taxes on real property, vehicles, vessels, and incomes within their legal boundaries.

# ARCTIC ENVIRONMENT

## Sea Ice

Sea ice means different things to the various users that must work with it. This is especially true of the expanding community exploring for and developing polar resources. Those conducting exploratory drilling for oil and gas are interested in the nearshore ice conditions. Specifically they are interested in the so-called seasonal sea ice zone (SSIZ), the area within the average maximum and minimum annual limits of the pack ice extent. In winter they may be primarily concerned with the extent or boundary and thickness of the fast ice. Fast ice is the ice that forms first in shoal areas near the coasts and grows seaward. If drilling is to be done on fast ice, its thickness and strengths must be known. Although termed fast ice, small scale movements induced by such phenomena as wind, tides, storm surges, spring runoff and thermal expansion and contraction may occur. During summer — or in all seasons beyond the extent of fast ice — other ice features and processes are more important. Movement and dynamics of the ice become dominating influences. Such factors as the stage of development of the ice must be considered. An impingement of old ice that has survived one or more summers'melt could obstruct operations. Deformation features, such as ridging are also major problems. Ridging usually occurs through alternate processes: convergent stresses cause fracturing, new ice forms in the fracture, subsequently crushing or shearing. As a result ridges attain thicknesses more than ten times that of undisturbed sea ice, whose maximum thickness is generally limited by thermodynamic ocean and climatic processes to about ten to twelve feet. These ridges can move rapidly, interacting with drilling structures and feeder lines from drilling structures to shore. Specific aspects of sea ice ridges that are important to drilling and other activities are

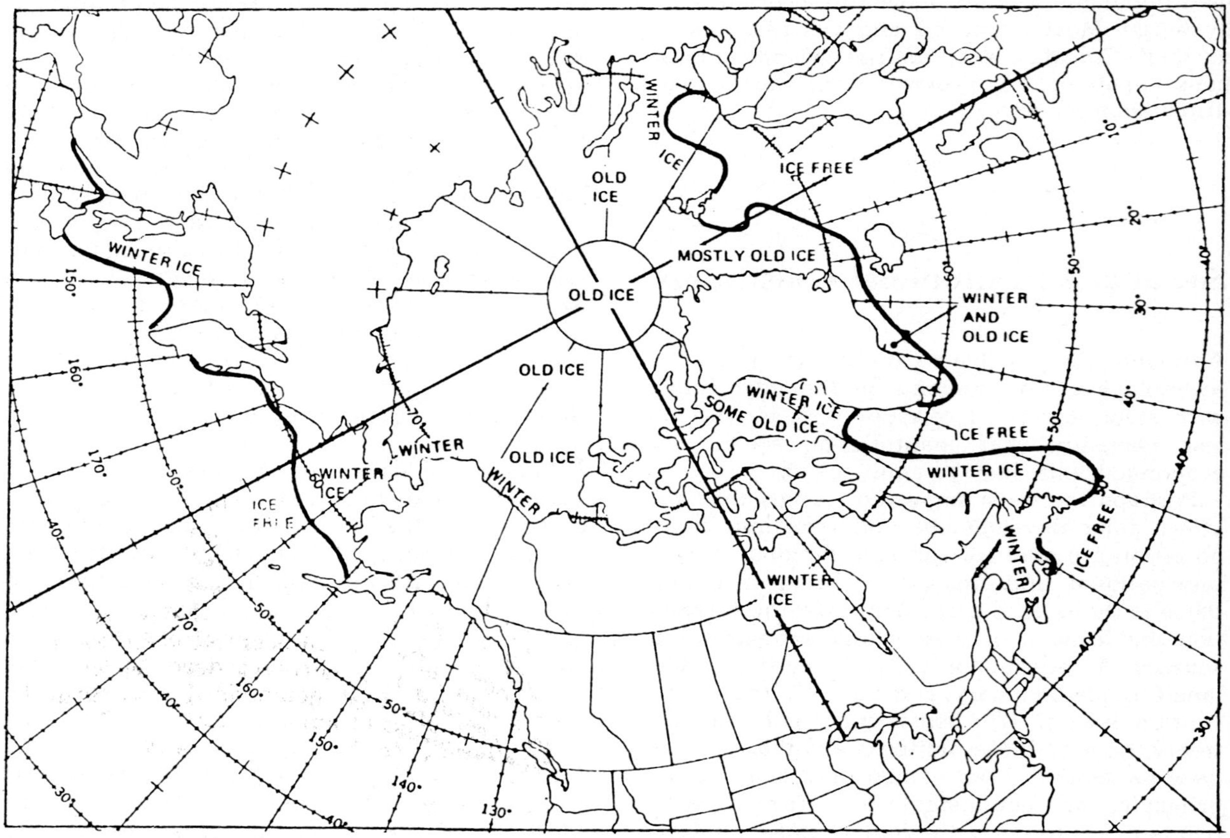


FIGURE 2. *Average Maximum Extent of Sea Ice.*

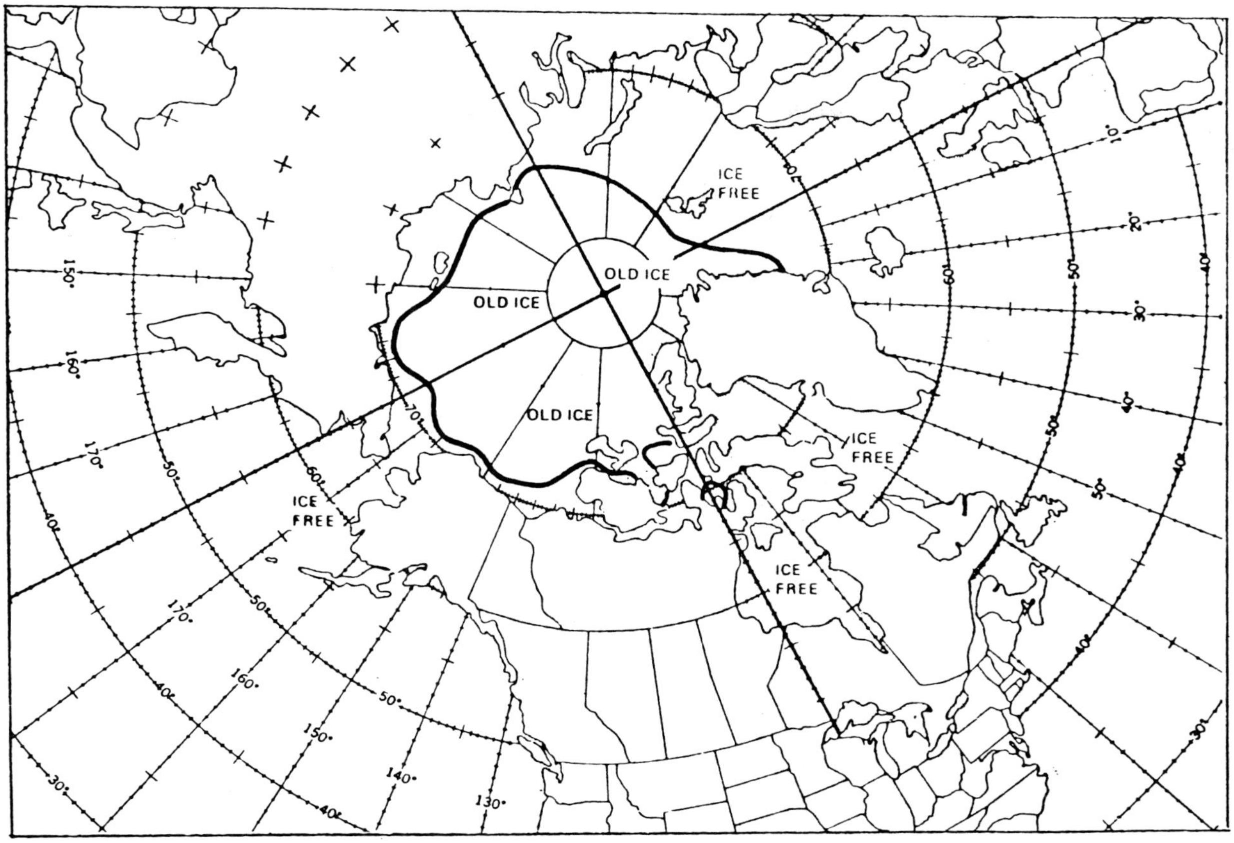


FIGURE 3. *Average Minimum Extent of Sea Ice.*

their frequency, shape, height-to-draft ratio and whether they are grounded or free floating.

Transportation of resources depends on expeditious and safe transit to and from a commodity source. Concentrations of ice (amount of ice versus ice-free water in an observational area) then become of overriding importance along with the existence of and conditions within the pack, shore, and flaw leads. A shore lead is an area of largely open water between the pack ice and the shore. A flaw lead is an area of largely open water between pack and fast ice. These leads may or may not be refreezing or refrozen. Of course, ice thickness, strength, roughness, or deformation features are also of concern as are other features, such as floe size, stage of melt, snow cover and presence of glacial or ice island fragments. The seriousness of problems caused by the features and processes mentioned is related to the size, design, strength, and power characteristics of the carrier itself. Also of great importance is the position of the highly variable ice edge.

The maximum extent of sea ice varies widely from year to year and from locale to locale. The average maximum is shown in *Figure 2.* It is noteworthy that when the eastern Bering Sea is enjoying light ice conditions, the western Bering Sea and the Sea of Okhotsk may have heavy ice. It is also noteworthy that light ice conditions in the Bering Sea do not necessarily signify light ice conditions in the Beaufort Sea. Indeed, the reverse may be true in many years.

The chart of average minimum extent of sea ice, *Figure 3,* depicts the position of the sea ice pack in its usual position along the north coast of Alaska and in the Parry Channel. The Beaufort Sea itself can be broken into two distinct areas of relative sea ice severity; west of Prudhoe Bay and east of Prudhoe Bay.

Insofar as polar resource development is concerned, its effectiveness is completely coupled with understanding of sea ice behavior and distribution. Sea ice affects operations of the wide variety of current vessels and carriers and also affects development of marine structures and

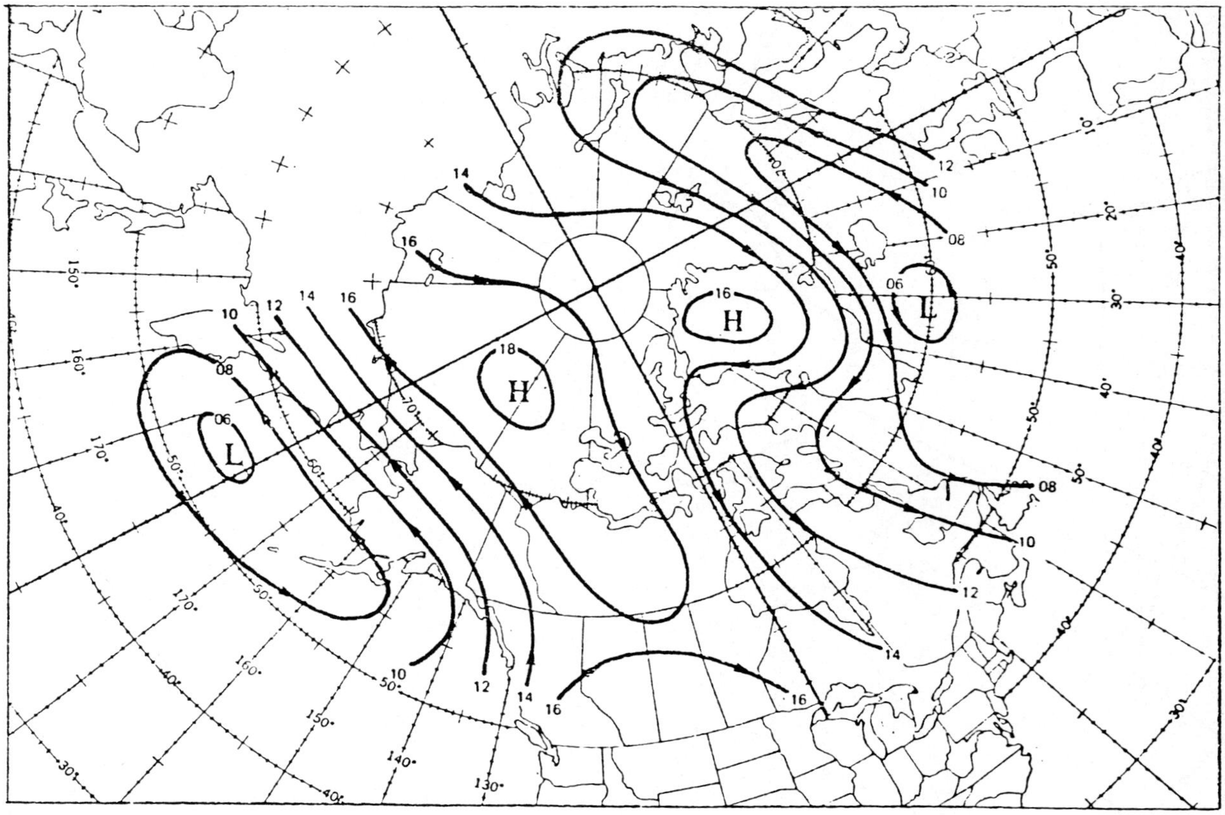


FIGURE 4. *Mean Annual Sea Level Pressure (In Millibars over 1,000)*

architecture. Now vehicles for transiting sea ice are being developed by all countries bordering upon polar seas. Terminals, drilling mechanisms, and structures are also being developed.

## Atmosphere and Currents

The mean annual sea level pressure chart, *Figure 4,* illustrates the primary centers of meteorological action in the lower atmosphere; the Aleutian and Icelandic lows. It also illustrates the secondary centers of action; the Pacific or Beaufort Gyre and the Greenlandic High. The position of these centers appears to be important to forecasting relative severity of ice in different areas during the coming ice season.

*Figure 5* illustrates the major oceanic circulation in the polar areas. Note, that the primary and secondary centers of action in the lower atmosphere coincide with similar circulations in the ocean. There are other currents in the Arctic but they are not as pronounced and can quickly become wind driven currents when the wind speed is sufficient. In general these lighter currents circulate in a clockwise fashion around land masses or the bigger islands. Although the sea ice extent will vary from year to year, the same general contour is usually followed. This contour appears to follow the oceanic and lower atmosphere circulations.

## Permafrost

Permafrost occurs commonly in high-latitude and high-altitude land environments and is found also in the sea floor along Arctic coasts. Because the characteristics of permafrost-soil structures are readily altered by human activity, permafrost poses many problems for construction engineering in regions that are prospects for resource development.

Much less is known about offshore permafrost than is known about land permafrost. Ability to predict the probable distribution of offshore permafrost is limited. The interdependence of thermal, hydrologic, and sedimentary factors affecting offshore permafrost is not well understood. Even if the role of these factors were better known, our knowledge of the general distribution of offshore permafrost would be limited by our inadequate knowledge of the distribution of bottom-water temperatures and bottom currents in the Arctic Ocean.

Scientific assessment of offshore permafrost has shifted from highly speculative to more specific in the past 20 years. Although the literature on offshore permafrost is sparse, it shows evidence of substantial permafrost offshore.

Exploration, extraction, and transportation of hydrocarbons will raise problems of engineering and construction related to the existence of permafrost in the off-shore areas and sea beds. Wells, offshore structures, and port facilities can affect the characteristics of permafrost-soil structures and may require special designs. Offshore operations can proceed with greater confidence when more comprehensive knowledge of offshore permafrost is obtained.

## Normal Navigation Seasons

The period of normal navigation by unescorted vessels varies widely and is dependent on several factors: (1) the strength of the vessel, (2) the draft of the vessel, (3) the experience of the master, (4) the support services (aerial reconnaissance and sea ice forecasting) provided, and (5) the ice conditions themselves. Some vessels may be safe operating in four tenths coverage of ice while other vessels may not enter ice with a concentration of over one tenth. *Table 1*, therefore provides only generalities.

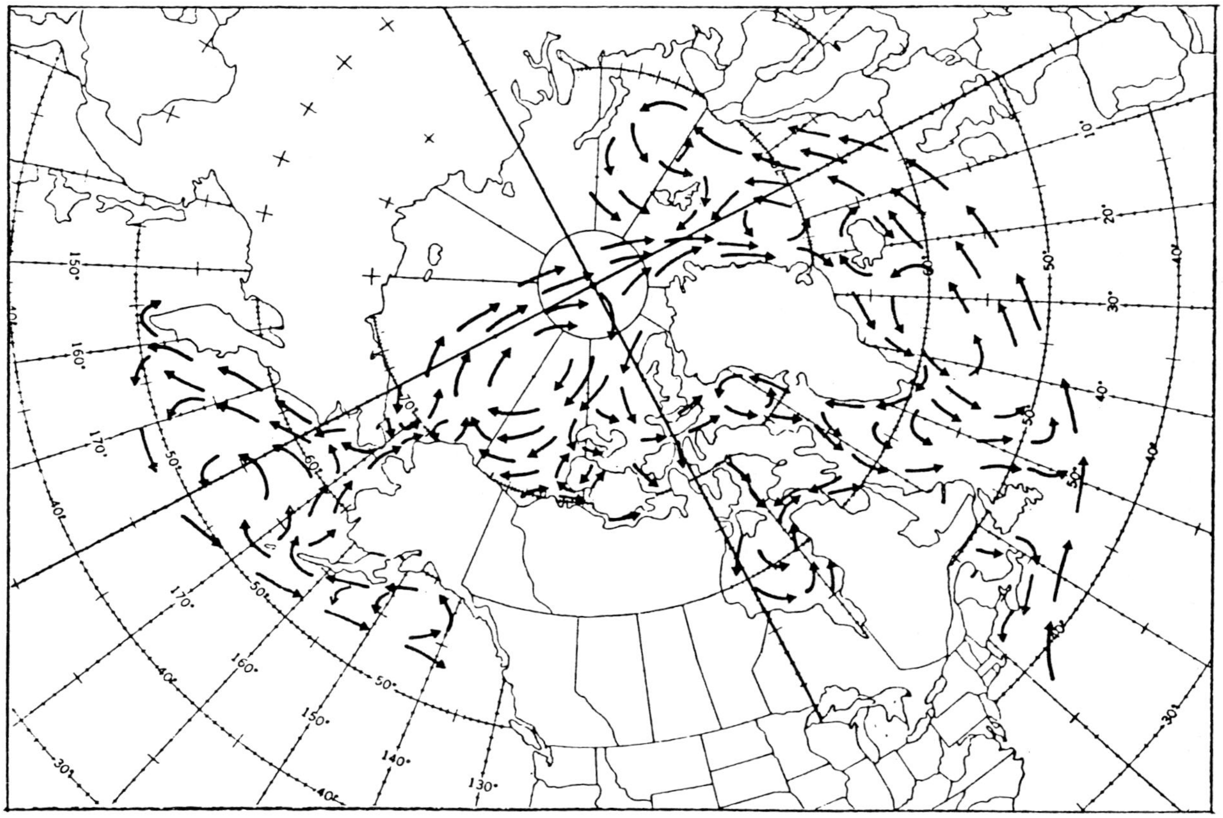

FIGURE 5. *Major Ocean Currents.*

*Table 1*
**Normal Navigation by Unescorted Vessels**
(*Varies vith severity of ice season*)

| | |
|---|---|
| Bering Sea | June through mid-November |
| Chukchi Sea | mid-July through mid-October |
| Beaufort Sea | August and September |
| High Canadian Arctic | Portions August and September |
| Lower Canadian Arctic | Portions mid-July through September |
| Hudson Bay, Labrador Sea | July through October |
| Baffin Bay | mid-July through mid-October |
| East Greenland | August and September south 70N |
| Barents Sea | August and September to near 80N |
| Arctic Ocean | closed to normal navigation |

Icebreaker support can extend the navigation season. Canadian vessels operate through November with icebreakers and support vessels assisting rigs in the southeastern Beaufort Sea and outer Mackenzie Bay. U.S. tug and barge operations can be extended into October, and possibly November, along the northern coast of Alaska by using icebreaking barges. U.S. icebreakers have been conducting Bering Sea winter patrols for several years. Soviet nuclear icebreakers have demonstrated their ability to penetrate the deep Arctic ice pack by reaching the North Pole on at least one occasion. Thus it is clear that Arctic shipping seasons could be extended both in spring and fall in all areas if powerful icebreakers were available.

## CURRENT ARCTIC MARINE SYSTEMS

### Ports

A port is defined as an interface between marine mode and any other transportation mode including marine, road, rail, pipeline, and air. If protection and anchorage for ships are provided as well, the term harbor is used.

As the port is an integral part of the marine transportation system, it will influence the economic, environmental, and demand factors of the total system. Present Arctic harbors, such as that at Churchill, Manitoba, use the same technology as warm water ports with minor modification, limiting their operations to the summer season when Arctic conditions are least severe. During this brief season, incremental costs at Churchill are approximately 35 percent above those of open water ports.

So far, northern ports have been located in natural harbors, such as bays or tidal estuaries, so that the land offers protection from winds, wave action, and currents. Artificial harbors protected by breakwaters can be constructed using existing technology. Costs of constructing artificial harbors would depend largely on local availability of construction materials. A protected harbor, either natural or manmade, is a requirement for either general cargo or bulk cargo loaded by a high speed loader. Either usually requires the ship to be alongside a conventional wharf.

Offshore single point moorings, designed to resist ice pressures, would be suitable for loading bulk liquids or slurries. Such moorings could be used for surface ships, submarines, or semi-submersibles where the water is sufficiently deep. Because less shelter is afforded, operations would be vulnerable to the severe wind and ice conditions that are normal in the Arctic. Detailed studies would be needed of each structure and site to satisfy all structural and environmental requirements.

Prior to choosing a port site, a study of navigation problems should be made. The study should cover the characteristic features of the ice regime including the river and sea approaches and routes. An estimate should be made of the increase in static ice thickness resulting from ship traffic in the harbor. The study of the ice and thermal regimes should be directed toward forecasting dates of ice break-up and freezing.

Research and development for Arctic terminals must consider the following in determining type of harbor facilities to be constructed.

1. Ice conditions and forces exerted by ice pressure.
2. Soil conditions, permafrost, allowable bearing pressure, settlement, etc. on shore and underwater.
3. Size and type of vessel to be berthed. Depth of water required in the dock and approach channel, together with appropriate ice-freezing arrangements.
4. Berthing forces and fendering; tug requirements modifications to what would be needed for open water conditions to allow for ice conditions, extreme temperatures, and isolation should be included in design especially for tugs.
5. Mooring forces — including factor for ice jamming between wharf and ship. Automatic system for line tending will have to be considered.

6. Wind and waves — wind velocity and direction should be measured at site.
7. Current and tides — current suitable for ice flushing should be sought and studied.
8. Earthquake forces where applicable.
9. Thermal forces.
10. Dead load on structure — important where low-loadbearing soils are encountered. In addition, change in soil conditions due to permafrost or possible modification of permafrost must be studied.
11. Live load due to machinery, vehicles, uniform deck load, or from various structures built to serve specific requirements.
12. Topographic and hydrographic profile of the site considered.
13. Acessibility and distance to the resources to be developed.
14. Ecological, environmental, social, legal, and economic aspects.

Due to the above factors, extensive exploration and studies are required to arrive at the most favorable type and location of harbor and dock. Hydraulic model studies of wind, waves, and currents; aerial and on-land observations; and photography of the movements of the ice for one or two years would be necessary to establish the most favorable location and orientation for the harbor and dock.

## Tugs and Barges

At the present time, tug and barge combinations, both ice reinforced and standard type, constitute the main Arctic marine transportation system that regularly, but mainly on a seasonal basis, serves the American and Canadian Arctic. This service started in the mid-fifties and now provides regular service to various government agencies and native communities scattered throughout the whole Arctic.

The barge and tug service has replaced almost completely the previous method of having either standard cargo ships escorted by ice-breakers or very small (50 to 200 ton) ships supply the area. Annual transport includes various general cargo, from needles to heavy construction equipment, and heating oil.

With the discovery of the North Slope Oil Field in 1967 and the on-going drilling programs in the Arctic, tugs and barges have come into their own to deliver living quarters, power stations, and service buildings of modular construction. This method is quite suitable as most Arctic sites are without wharves and other conveniences, and also have a limited water depth. Some modules are nine stories high and can weigh up to 1200 tons each. Using specialized equipment, they are «walked» on and off the barges. Recently, a barge with icebreaking capability has been introduced into this service.

## Arctic Icebreaker Operations

Until now it has been possible to meet Arctic cargo commitments during the summer season when sufficient open water appears in some locations to permit penetrations by shipping. In a large measure, then, the barrier posed by ice has been avoidable. But now, with accelerating commercial development, the situation is changing. Existing commitments will not only continue, they will expand. Important new ones are also appearing.

To meet this situation, the need to expand marine operations and extend the navigation season is imminent. A requirement for year-round operations into selected locations will not be far behind. Supporting all this activity is the icebreaker. It will continue to play a key role in the ongoing development of the far North. There are now too few icebreakers to meet present commitments and the fleet is aging. Some units suffer from inadequacies (e.g. low endurance) that lessen their effectiveness and none is able to extend the navigation season beyond its present limits to a significant degree.

Icebreakers are specialized vessels that fill an essential role. Without their services the present commercial shipping operations through ice during the summer or any season would be out of the question. Icebreakers are costly vessels because they require great strength, very high powers and long endurance. The trend in icebreaker design is toward even more powerful and larger ships than their predecessors.

The primary role of the existing icebreaker has been, and remains, escort of commercial shipping through ice. Such shipping consists of unstrengthened vessels and vessels with a degree of ice-strengthening from Lloyds Ice Class 1 to Ice Class 3 or equivalent rules of the American Bureau of Shipping and other classification societies. Their size can vary from 1,000 tons or so to as much as 42,000 tons or more. The upper limit is usually governed by the beam of the escorted ship for if it exceeds that of the icebreaker, and the escort operation must be conducted, for example, through fast ice, quite obviously the larger ship will experience difficulties and may even suffer damage to framing and shell plating. Such an escort role is well established and will continue to be used during the summer navigation season, whether extended or not, to conduct convoys to and from selected unloading sites in the north.

A secondary and largely Canadian role has been to deliver cargo to remote northern stations. This commitment has been downgraded in recent years because the quantity of cargo that could be embarked was not especially significant, better means became available and, most importantly, such commitments interfered with the primary role of the icebreaker. In addition, structural strength in an icebreaker is jeopardized somewhat by having a large void space for cargo in a location in the hull where strength is essential.

Not only do icebreakers escort shipping through ice in the summer on a planned basis, but they also are on call to assist unescorted ships that, for a number of reasons, get into difficulties and call for help. They also carry out scientific tasks (including all-important hydrography and oceanography) on a 'not to interfere' basis.

Success is ice breaking calls for both thrust and inertia. Thrust is the horsepower developed by the propellers and mass is the quantitative measure of inertia. What is required, then, is the best mix of horesepower and displacement for the area of operations and the season of the year.

Thrust is needed to drive a ship through a level sheet of fast ice on a continuous basis, while it is inertia that keeps the same vessel advancing through harder and thicker multiyear ice ridges and hummocks that are common features in ice fields.

*Manhattan*, at 155,000 tons and 43,000 shaft horsepower (shp), had the needed mass but too little thrust for other than an experimental voyage. The new U.S. Coast Guard icebreaker, *Polar Sea*, at 12,000 tons and 60,000/75,000 shp, would seem to be the opposite. Icebreaker tankers considered at one time by Humble Oil for shipping Prudhoe Bay oil through the Northwest Passage were to displace about 250,000 deadweight tons (dwt) and develop about 200,000 shp. Those ships would have been excellent icebreakers for, all other things being equal, they would have had what presently appears to be the ideal combination of power and mass.

Canada's «new and powerful» icebreaker, the *Louis S. ST. Laurent*, displaces 14,000 tons and develops 24,000 shp. During special trials on the second *Manhattan* voyage in 1970, this combination of power and mass failed to break a level sheet of fast ice four feet thick. The ship charged this obstacle and advanced about a ship's length before being brought to a standstill. The only way she could advance was by backing and ramming. In the same conditions, the *Manhattan* herself was equally unsuccessful.

Arctic first-year ice attains a thickness of approximately six feet. If there is to be an icebreaking capability over a longer period of the year, if not year-round, and over a wider area of operations than has been attempted up to this time, then icebreakers possessing materially better capabilities than those possessed by the *St. Laurent* are called for. In addition to first-year ice, there are also quantities of multiyear ice and ridging which present an even more redoubtable obstacle. The type, thickness and distribution of sea-ice in all its forms is dealt with in detail elsewhere in this report.

The United States Coast Guard embarked on an ice breaker construction program in 1966 because of the demonstrable inadequacies of their old Wind-Class ships. The need for better icebreakers became even more obvious during the Northwest Passage voyage of the *Manhattan*, in 1969, when the chief icebreaker role had to be assumed by the Canadian Coast Guard Ship *John A. Macdonald*.

Experience with the icebreaking tanker *Manhattan* demonstrated that large ships, designed as such, would make the most effective icebreakers. Such ships would combine the qualities essential for success in ice, namely, mass, and power, making possible year-round navigation on a routine basis into selected Arctic areas and extended seasons into a number of others. These ships would be icebreakers in nearly every sense of the term and their performance in ice would exceed by a wide margin that of any present government icebreaker but may not be as maneuverable. They should not be confused with ice-strengthened bulk carriers which would not possess the desired qualities.

Year-round Arctic navigation would seem to be justifiable on three counts; firstly, a requirement to remove very high tonnages of minerals such as iron ore; secondly, the removal of natural gas in liquid form if the economics of such a marine operation can be justified; and, lastly as an alternative mode of oil transportation.

One mining project that could require year-round shipping (10-12 months) is the Baffinland iron deposit where the amount of direct shipping ore to be moved would probably exceed five million tons. However, factors of economics and marketing are not now favorable for development of this deposit, although world demand for steel and the political stability of the emerging source countries could change this state of affairs very quickly.

There is, therefore, no requirement at this time for ice breaking bulk carriers. However, demand for energy will continue to spur the search for more gas and oil. The situation could change very suddenly, placing strong emphasis on Arctic resources and a need for icebreaking tankers.

## AUXILIARY SUPPORT SERVICES

While ports and vessels are the major components of any maritime transportation system, successful establishment and functioning of a transportation system depend upon a number of support services. The vessels and ports must communicate with each other as well as third parties. Vessels must be capable of proceeding in a safe and timely manner to their destinations. Resources in the form of personnel, equipment, and other supplies must be transported to establish and maintain port and production facilities and operations. Although some degree of self-sufficiency, in terms of these functions, is inherent in ports and vessels, auxiliary support services are generally relied upon to fulfill the bulk of these needs. This section addresses some of the more important support services for an arctic marine transportation system.

### Communications

Any Arctic marine transportation system will necessarily depend heavily upon reliable short-range (line-of-sight) and long-range communications. Short-range communications requirements can likely be satisfied by an expanded VHF-FM communications network and, perhaps, a shore network (e.g., marine operator) for ship-to-shore communications.

While short-range very high frequency (VHF) and ultra high frequency (UHF) communications are not impaired by polar-unique conditions, longer range high frequency radio communications are unreliable due to polar atmospheric conditions. For limited traffic levels, shoreside landline relay at remote sites may be sufficient; however, as traffic levels expand and/or as communications requirements dictate, it may be necessary to implement a communications system capable of rapid and reliable handling of not only operational traffic but also data transmissions between Arctic and temperate stations.

Communications satellite technology appears to be the most promising means for satisfying the long-range communications requirements. Existing maritime satellite services do not, however, provide coverage to the Beaufort Sea and northern Canadian Archipelago since current communication satellites are positioned in equatorial geostationary orbits to provide coverage of the major ocean areas. A network of satellites in polar or near-polar orbits would be necessary for adequate communications services for an expanded marine transportation system.

### Navigation

The practice of marine navigation, i.e., the determination of one's geographic position on the water, in the Arctic regions is impaired by several factors. Because of the remoteness and limited marine activity in these regions the waters are relatively poorly charted and coverage by either traditional or modern electronic aids to navigation is limited. The low relief and featureless terrain characteristic of much of the Arctic coastline hinders navigation by traditional radar and visual means. Moreover, application of visual piloting methods as well as celestial navigation is limited by the inhospitable Arctic weather. High latitudes also place restrictions on use of magnetic and gyrocompasses and on celestial navigation. Navigational light and buoy systems, the heart of navigation systems in restricted waters in temperate regions, are precluded, or at least limited, by the ice and environment of the Arctic. In short, the utility of traditional navigational methods in both restricted and open waters is significantly diminished in the Arctic.

To overcome these limitations on conventional methods, current Arctic marine traffic relies upon satellite navigation, OMEGA, and installed shore aid systems, such as the system of RACONS along the Alaskan North Slope, to supplement traditional navigational methods. Only satellite navigation provides coverage throughout the Arctic. As Arctic marine commerce expands, accurate and reliable marine navigation over larger portions of the Arctic will become critical. Consequently, the limited navigational capability in the Arctic will require upgrading to assure safe navigation. Existing systems (e.g., the Alaskan RACON System) may require expansion. Increased use of satellite navigation and possible expansion of LORAN C coverage will likely provide adequate navigation coverage to open water areas. The continued movement of tug and barge traffic along near-shore open water passages and the future emergence of deep draft surface or submarine Arctic carriers will place a premium on accurate and comprehensive bottom sounding information. Most of the existing hydrography in Alaskan waters north of the Aleutian Islands does not meet National Ocean Survey spacing criteria. Similarly, the quality of bottom profiling in the Canadian Arctic is generally far below that in temperate regions. The known presence of significant and relatively abrupt rises in the bottom of normally deep Arctic channels further emphasize the need for good sounding information.

Provision of adequate navigation in restricted Arctic waters and approaches to harbors or offshore terminals will require development of substitutes for the buoy and shore aid systems utilized in temperate waters. Candidate systems for such substitutional include:

1. high frequency radio systems (e.g., several commercially available systems are currently used for survey and experimental work);
2. low frequency radio systems (e.g., mini-LORAN C and DECCA systems);
3. acoustic systems (i.e.. bottom-anchored acoustic transponders interrogated by a vessel-generated acoustic signal); and
4. leader wire systems (i.e., electronic tracking along a buried or sub-surface wire).

All of these systems lie within state-of-the-art technology. However, adaptation of particular systems to the Arctic marine environment may generate research needs.

## Environmental Information Services

While icebreaker assistance is required to permit commercial vessels to deal with the ice conditions encountered, weather and ice information services strive to optimize vessel movements by helping the commercial carrier avoid adverse environmental conditions. An environmental information service performs three fundamental tasks: data collection, analysis, and dissemination. The product information is of two forms; (1) operational data that describe local conditions and is immediately applicable to ongoing operations, and (2) forecasts that are useful for vessel route planning and scheduling. Though different in the degree of detail required, the information requirements for ice navigation are generally the same for both operational data and forecasts. They may include the usual marine weather information as well as ice age and thickness distribution, ice ridging distribution, iceberg and floeberg information, grounded fast ice limits, and the location of leads, polynyas, and fracture zones.

## Logistics

Other than the resupply of military installations, polar marine logistic needs have been associated almost exclusively with scientific research efforts. The transport of relatively small quantities of scientific equipment, supplies, and personnel among or between ice camps, vessel platforms, and shore stations has been a major logistics activity. The principal vehicles for fulfilling these logistics needs have been icebreakers, helicopters, small aircraft, and conventional land carriers, such as tractors, trucks and snowmobiles. Each of these vehicles has obvious limitations. The use of large, expensive icebreakers is clearly cost-effective only for large-scale (in terms of cargo quantity and distance) logistic needs. Airborne carriers have limited payload capabilities and are highly weather-sensitive. Aviation is expected however to continue to provide the predominant personnel transport system in the Arctic. Land-carriers are generally limited to land-fast ice where open water areas and topography (ridging and hummocking) characteristic of the dynamic Arctic pack are not prevalent.

As Arctic marine scientific and commercial activity expands, logistic support requirements will also increase. Offshore platforms may require periodic deliveries of supplies and personnel from shoreside staging points. Regular and emergency servicing of offshore installations, including platforms and pipeline systems, may generate a need for utility vehicles capable of reliable service under the most adverse Arctic conditions. The limitations of traditional Arctic logistics vehicles will necessitate development of vehicles more capable of fulfilling these emerging needs.

Candidates for Arctic logistics vehicles include hovercraft air-cushioned vehicles (ACV) and surface effect vehicles (SEV); vertical/ short take-off and landing (V/STOL) aircraft; Archimedean screw (worm screw) vehicles, and mini-submersibles. Each, of course, has its strengths and weaknesses. Air-cushioned vehicles have been tested in the Arctic with some success. These vehicles can transit the ice-water continuum at considerable speeds with significant payloads. However, hovercraft are difficult to maneuver in high winds, are limited in their ability to negotiate pressure ridges, and ACV's are particularly vulnerable to skirt damage due to ice abrasion.

V/STOL aircraft, as well as hovercraft, have been evaluated for Arctic application. They offer the advantage over conventional fixedwing aircraft of a limited take-off and landing area requirement while providing a generally larger payload capacity than conventional helicopters. However, as with other airborne carriers, V/STOL aircraft operations must be accommodated to the particular environmental conditions of the Arctic.

Archimedean screw vehicles are a relatively new development for polar regions, still in the developmental stage. Initial research indicates that these vehicles can operate effectively in the ice-water continuum; however, their speed does not approach that of other candidate vehicles. They may also be restricted in their ability to negotiate ice ridges. Though very limited in payload capacity, they are apparently very maneuverable and adaptable for towing and other utility functions.

Since they avoid ice and weather, mini-submersibles may be very attractive as all-weather logistic vehicles. Submersibles provide a capability for servicing sub-surface pipelines and other structures not present in other logistics vehicles. However, their safe operation requires extensive knowledge of the acoustic properties of the operating waters and good sounding data — two criteria not generally satisfied in polar waters. Moreover, operation of submersibles in ice-covered waters will present special problems in vehicle docking and emergency recovery.

Future Arctic marine logistic services will no doubt be provided by a range of vehicles. Research in this area should concentrate on development of new vehicles to overcome the limitations of conventional logistic vehicles. In particular research directed at eliminating or minimizing the shortcomings of the candidate vehicles discussed above could effectively increase the options available to future designers of logistics support systems for Arctic commercial and scientific ventures.

# NEED FOR MARITIME SERVICES

## Marine Services for Transportation

Many studies have been conducted in the last few decades concerning resources of the polar regions. These studies identified significant quantities of valuable materials in the north polar region in the form of oil, gas, coal and ores that would be useful to the world at large if they could be extracted and transported outside the Polar area.

The worldwide shortage of oil and gas and, to a lesser degree, many other minerals has sparked a recent serious interest in the natural resources of the north polar regions. In the past, economics and severe environment have hampered commercial development of these resources, but today highly inflated costs from traditional sources make such a development economically attractive.

Extraction of Alaskan oil became economic in the last decade. As a result, many problems peculiar to Arctic operations were solved, a pipeline system was constructed, and is now delivering over one million barrels of oil per day to the lower forty-eight states. Natural gas in the Prudhoe Bay area has also reached an economic trade-off point and, it appears at this time, another pipeline will be added in the near term to handle this product. So, in the next few years, two products will he moving from two locations in the Arctic to locations outside.

An additional approach is needed if other Arctic resources are to be developed in the near to midterm future and added to the world supply of useful products. This is not to say that the pipeline approach is wrong or should not be further considered. Indeed, for those products that can be pumped (oil, gas, and coal slurries) the pipelines and marine transportation systems may complement one another or, in some cases, be in competition. For those products that cannot be pumped, the marine transportation system offers the only economically viable means of transportation.

Waterborne transport of bulk cargo in temperate regions is known to be the cheapest form of transportation today; for this reason it is natural that marine transportation systems be given serious consideration for use in the Arctic. A marine system, with due consideration for the extra costs resulting from the severe and unusual environment, may still enjoy an appreciable advantage over other systems.

## Transportation Systems

The fundamental criterion on which any transportation system is judged is the unit cost of transporting cargo over the expected life of the transportation system. Calculation of these costs involves numerous assumptions, each of which might be assigned a probability of accuracy or assigned a range of application. As an example, a proposed Arctic marine transportation system might consist of a fleet of large tankers that will have to transit many miles of icecovered waters. To compute the annual throughput of this fleet of tankers, one must be able to compute voyage times throughout the year. This will involve predicting speeds along each leg of the route, which in turn will require information on ice conditions and performance of the ship in ice. The latter information will involve assumptions that can greatly influence the unit cost of transporting the cargo.

The accuracy of the assumptions made in computing unit transportation cost depends heavily upon practical operating experience with the transportation system. The TAPS pipeline for example will provide valuable experience in computing the unit transportation cost of transporting oil and gas in Arctic regions by pipeline. Arctic marine transportation systems on the other hand do not currently enjoy similar operating experience. Costs associated with constructing and operating such a system could be perceived to be more risky than similar costs associated with a pipeline system. As a consequence, the unit transportation cost of an Arctic marine system would have to be substantially lower than that of a pipeline system to offset its associated risks.

Another important criterion for judging a transportation system is its potential effect on the physical environment. Factor & Grove (1979) describe the legislative requirements for environmental protection that had to be met before the TAPS pipeline could be constructed. They estimate that \$3.2 billion of the \$8.0 billion construction costs were a direct cost of delays caused by environmental litigation. In addition, they estimate that construction was delayed about four years by these claims and the Alaskan Native land claims. Only through a special act of Congress were permits finally granted for construction of the pipeline. The nature and extent of litigation regarding environmental protection requirements that would have ensued had an Arctic marine transportation system alternative been selected by the oil companies is unknown. However, from the environmental litigation surrounding the tanker operations into Valdez and Puget Sound, one might expect it would have been considerable. The greatest difficulty in applying environmental criteria to selection of a transportation system lies in quantifying environmental effects.

## Marine Transportation System Advantages

Some of the marine system advantages, other than cost, that enhance its attractiveness are:

1. almost any product can be transported;
2. a ship has flexibility in routing, since it can load or unload at any location having suitable facilities;
3. a ship can move to a more temperate location for routine and periodic inspections, maintenance and overhauls, thereby decreasing the cost and inconvenience of these operations;
4. a marine transportation system would disturb the sensitive tundra, vegetation, and wildlife, less than some other systems;
5. the capacity of the system can be increased (or decreased) as needed by varying the number of ships used or the number of annual trips per vessel;
6. a marine system is the most practical way to provide life support to an enlarged population in the barren areas.

A marine transportation system for the Arctic comprises more than ships. At present, there are no deep-water terminals and few aids-to-navigation. Communication, search-and-rescue, and ship repair facilities are not adequate for full scale operations. The transportation system must, therefore, include a fleet of ships, an Arctic loading terminal, a southern of floading or trans-shipment terminal (unless existing facilities can be used), adequate navigation and communication facilities, and support facilities such as line-handling launches, tugboats, harbor or route icebreakers, etc. Major units in the system (ships and terminals) must be configured for the cargo to be handled, but the other units would be adaptable to any cargo.

## Marine Services for Resource Development

Up to this point, emphasis has been on transportation of Arctic resources to other areas. Another important consideration is offshore marine structures, such as drill platforms for exploration and production, and various kinds of processing plants for preparation or reduction of the products for transport. As the frontier expands there will be a need for more and more of these structures and for life support systems, which will surely be largely marine oriented.

There is much common knowledge of the various marine structures associated with oil and gas exploration and production. The rigorous environment will require special attention in their design and material selection, and care must be taken in their movement, placement, and operation to prevent damage to the environment.

The growing need for various processing plants will expand as resources are developed. These plants include petroleum oil degassing and pre-treatment facilities to condition the oil for shipment, liquefaction facilities to convert petroleum gas to liquified natural gas or methanol, and processing facilities to refine ore into the desired minerals. Due to the severe environment and the lack of local facilities it is likely that such plants will be constructed and barge mounted away from the Arctic and then moved into place by marine systems. Their successful placement and operational support would be greatly enhanced by a yearround marine transportation system.

# CONCLUSIONS

## The Committee has concluded that:

1. Many resources are available in the Arctic that will utilize maritime services in the future as economic forces dictate their development. The clear immediate opportunity is for alternative transportation of oil and gas products. Living resources, although important, will not drive development of marine transportation systems. Requirements for fishing vessels, however, may increase.

2. Transportation technology is available for resource development, or can be developed when economic and institutional factors are favorable, if appropriate research and development programs are instituted to provide the necessary fundamental knowledge.

3. Economic and institutional constraints so far have made movement of Arctic resources by marine transport impractical. Although all of the problems of pipeline operation have not been solved, the Trans-Alaska Pipeline now in use provides a baseline against which costs and risks of other transportation system can be compared. No similar cost data are available for marine transportation systems. A pilot-project marine transportation system could develop the data and operational experience for a full scale system when needed and provide, a partial back-up to the Trans-Alaska Pipeline System in the event of an interruption of its operation.

4. Arctic marine transportation development and associated industrial development will generate opportunities for the maritime industry. Opportunities may open for construction not only of ships, but also of offshore drilling platforms and resource conversion plants to be transported to the Arctic.

5. While Antarctic resource development will be many years behind Arctic resource development, techniques perfected for the Arctic may be used in the Antarctic.

# BIBLIOGRAPHY

M.A. Adelman, Paul G. Bradley, and Charles A. Norman, *Alaskan Oil: Costs and Supply* (Praeger Publishers, New York, 1971).

Arctic Ice Dynamics Joint Experiment, Division of Marine Resources, University of Washington, *AIDJEX Bulletins Nos. 1-40* Seattle, WA.

Alaska Field Operations Center, U.S. Bureau of Mines and C.C. Hawley, *Alaska's Mineral Potential, 1978* (Alaska Field Operations Center, U.S. Bureau of Mines, Juneau Alaska).

Arctic Research Subcommittee, Task Group or Ice, Wind and Waves, *Wind and Wave Forces* Final Report (Alaska Oil and Gas Association, 1976.

Per F. Anderson, *Ice Free Harbors* (Engineering Journal, July-August 1972, Engineering Institute of Canada).

Arctic Environmental Information and Data Center, *Current Research Profile for Alaska, Fourth Edition* (University of Alaska, Anchorage, 1977).

Bechtel, Inc. *Arctic Terminal for Icebreaking Tankers, Conceptual Design* (For the Maritime Administration, U.S. Departmental of Commerce, 1979).

Mary Clay Berry, *The Alaska Pipeline: The Politics of Oil and Native Land Claims* (Indiana University Press, Bloomington, 1975).

Ye P. Borisenkov and I.G. Pchelko, Eds. *Indicators for Forecasting Ship Icing* (Draft translation by Cold Regions Research and Engineering Laboratory. Corps of Engineers, U.S. Army, Hanover, NH, 1975).

Canadian Arctic Resources Committee, *Northern Transitions* 2 vols, Proceedings of Second National Workshop on People, Resources, and the Environment North of 60° (Canadian Arctic Resources Committee, Ottowa, Ontario, 1978).

Central Intelligence Agency - *Polar Regions Atlas* May 1978.

S. T. Culshaw and B. R. LeDrew, *Labrador Sea Ship-in-the-Ice: A Pilot Study* (Proceedings of the Fourth International Conference on Port and Ocean Engineering Under Arctic Conditions, St. John's Newfoundland, 1977: pp 1122-1131).

J. W. Devanney III and J. B. Lassiter III, *Transportation of Arctic Petroleum Resources* (Maritime Administration, U. S. Department of Commerce, Washington, D.C., 1973).

Edgar J. Desner, *Arctic Seas: Marine Transportation and High Arctic Development* (Canadian Arctic Resources Committee, Ottawa, Ontario, 1978).

John M. Drury, Stewart French, and John C. Reed, *An Arctic Submarine Port Facility: The Potential Participation of Alaskan Natives in the Establishment and Operation of a Submarine Oil Port Facility in Arctic Alaska,* (Arctic Institute of North America for the Maritime Administration, U.S. Department of Commerce, Washington, D.C., 1973).

Energy, Mines, and Resources Canada, *Oil and Natural Gas Resources of Canada 1976,* Report EP-77-1, (Supply and Services Canada, Ottawa, Canada, 1976).

Stanley Factor and Sandra Grove, *Alaskan Transportation: An Overview of Some Aspects of Transporting Alaskan Crude Oil* Marine Technology, Volume 16, No. 3., July 1979, pgs. 211-224.

Robert Faylor, et. al., *Arctic Marine Commerce Study, 3 Volumes* (Arctic Institute of North America, for the Maritime Administration, U.S. Department of Commerce, Washington, D.C., 1973).

Global Marine Engineering Company, *A Preliminary Feasibility Study of Tanker Transportation Systems Serving the Northwest Coast of Alaska, (Maritime Administration, U.S. Department of Commerce, 1978).*

Lulijian and Associates, Inc., *Environmental Considerations for Submarine Crude Oil Tankers* (for the National Oceanic and Atmospheric Administration, U.S. Department of Commerce, 1973).

Joseph Macknis, *United States Policy in Antarctia, Marine Policy Reports, Vol. 2 No. 2* (Sea Grant College Program, University of Delaware, 1979). Marine Board, National Research Council, Engineering at the ends of the Earth: Polar Ocean Technology for the 1980,s, (National Academy of Sciences, Washington, D.C., 1979).

Maritime Research Information Service, Transportation Research Board, *Marin Systems in Frigid Environments, Special Bibliography* (National Academy of Sciences, Washington, D.C., 1978).

Howard A. Meyerhoff and A.A. Meyerhoff, *Arctic Geopolitics* (Memoirs of the American Association of Petroleum Geologists, 1973).

Patrick Moloney, Submarine Tanker Concepts and Problems, (A Kings Point Scholar Project, U.S. Merchant Marine Academy. Kings Point, NY, 1974).

D. B. Muggeridge, Editor *POAC 77 Proceedings of the Fourth International Conference on Port and Ocean Engineering Under Arctic Conditions,* (Memorial University of Newfoundland, St. John's Newfoundland, Canada, Sept., 1977).

National Science Foundation, *MOSAIC Vol. 9, No. 5, Research in the Polar Regions* (Government Printing Office, Washington, D.C., Sept/Oct, 1978).

Newport News Shipbuilding and Drydock Co., *Arctic Submarine Transportation System* 7

Vol's. (for the Maritime Administration, U. S. Depart ment of Commerce, Washington, D.C., 1975).

Everett Peterson and Janet B. Wright, editors, *Northern Transitions, Volume I, Northern Resource and Land Use Policy Study* (Canadian Arctic Resources Committee, Ottawa, Ontario, 1978).

Max G. Pitcher, Ed. *Arctic Geology: Proceedings of the Second International Symposium on Arctic Geology* (American Association of Petroleum Geologists, Tulsa, OK., 1973).

Polar Research Board, National Research Council, *Problems and Priorities in Offshore Permafrost Research* (National Academy of Sciences, Washington, D.C. 1976).

A. G. Ronhovde, *Arctic Operations Technical Needs, Vol. II to Technical Report SRC-71-TR-N3501* (Systems Research Center, Washington, D.C., 1971).

John E. Sater, *The Arctic Basin, Revised edition* (The Arctic Institute of North America, Arlington, VA., 1969).

John E. Sater, A.G. Ronhovde, and L. C. Van Allen, *Arctic Environment* and Resources, (The Arctic Institute of North America' Arlington, Va., 1971).

Myron A. Sayles, Knut Aagard, and L. K. Coachman, *Oceanographic Atlas of the Bering Sea Basin* (U.S. General Accounting Office, Washington, D.C., 1980).

Scientific Committee on Antarctic Research - *Preliminary Assessment of the Environmental Impact of Mineral Exploration/Exploitation in Antarctia* - (August, 1977).

Harold V. Sverdrup, Dynamics of tides on the North Siberian Shelf, *Geofysiske Publikasjoner* Vol. 4, No. 5, Oslo, 1927.

A. F. Treshnikov, editor, *Problems of the Arctic and the Antarctic* (National Science Foundation, Washington, D.C., I973).

James H. Zumberge, *Mineral Resources and Geopolitics in Antarctica* American Scientist, Volume 67 - January-February, 1979.

*Cold Regions Science and Technology*, 7 (1983) 251—257
Elsevier Science Publishers B.V., Amsterdam — Printed in The Netherlands

# The Northern Sea Route Today

*Terence Armstrong*

Scott Polar Research Institute,
Cambridge

## Abstract

1. *Outline of typical recent season's operations; freighting pattern, equipment employed, length of season.*
2. *Current uses for the route: routine resupply, including to some inland areas; bringing out raw materials (minerals, timber); bringing in fuel and manufactured goods, including river craft; absence of significant through traffic.*
3. *Future prospects: greater use for resupply, including to inland areas where routes from the south cannot be expanded; enlarging through traffic; possibility of year-round navigation; prospects in the central polar basin opened by* **Arktika** *voyage to North Pole in 1977; importance for offshore oil recovery in Kara Sea; possible effect of change in ice conditions, arising either naturally or through change in river run-of.*

## Introduction

The Northern Sea Route may be defined as the system of sea lanes north of Asia, between the straits joining the Barents and Kara Seas in the west and Bering Strait in the east (Fig.1). An accurate picture of a recent season's operations in these waters is not possible to obtain, because no reports that could be called in any way complete are made public. All that can be pieced together is an impresssion of each season's activities derived from news items reported in the popular and specialist press and on the radio. For the present paper, these sources are *Pravda*, *Vodnyy Transport*, *Rechnoy Transport* and the BBC's *Summary of World Broadcasts. Part 1. USSR.* On that admittedly poor basis, let us construct what account we can of the 1981 season (the 1982 season is still in progress, and it would be premature to comment)[1]

## The Freighting Pattern

The freighting pattern seems to have remained substantially the same as it has been for some time. The busiest sector is that between the western terminus at Murmansk and the Yenisey river, where both the mineral industry and the timber industry are strongly represented, at Noril'sk and Igarka respectively. This is the region where there has been a most notable lengthening of the operating season (to be considered below), permitting significant increase in the mineral export. The volume of freight passing through Dudinka, the port for Noril'sk, was expected to be 5.5 million tonnes in 1981, but over half of this was to be river-borne. The target was probably reached, and may have been exceeded. Supposing two million tonnes travelled by sea,

[1] A somewhat parallel account, for 1978,was published by the author J.B.Wright, ed. in *Marine transportation and High Arctic development: Policy Framework and Priorities* (Ottawa, Canadian Arctic Resources Committee, 1979), p.210-22: and another, with more historical background, in *Ymer* 1980, p.86-129. Short summaries of each season's activities are published annually in *Polar Record*.

0165-232X/83/$03.00  

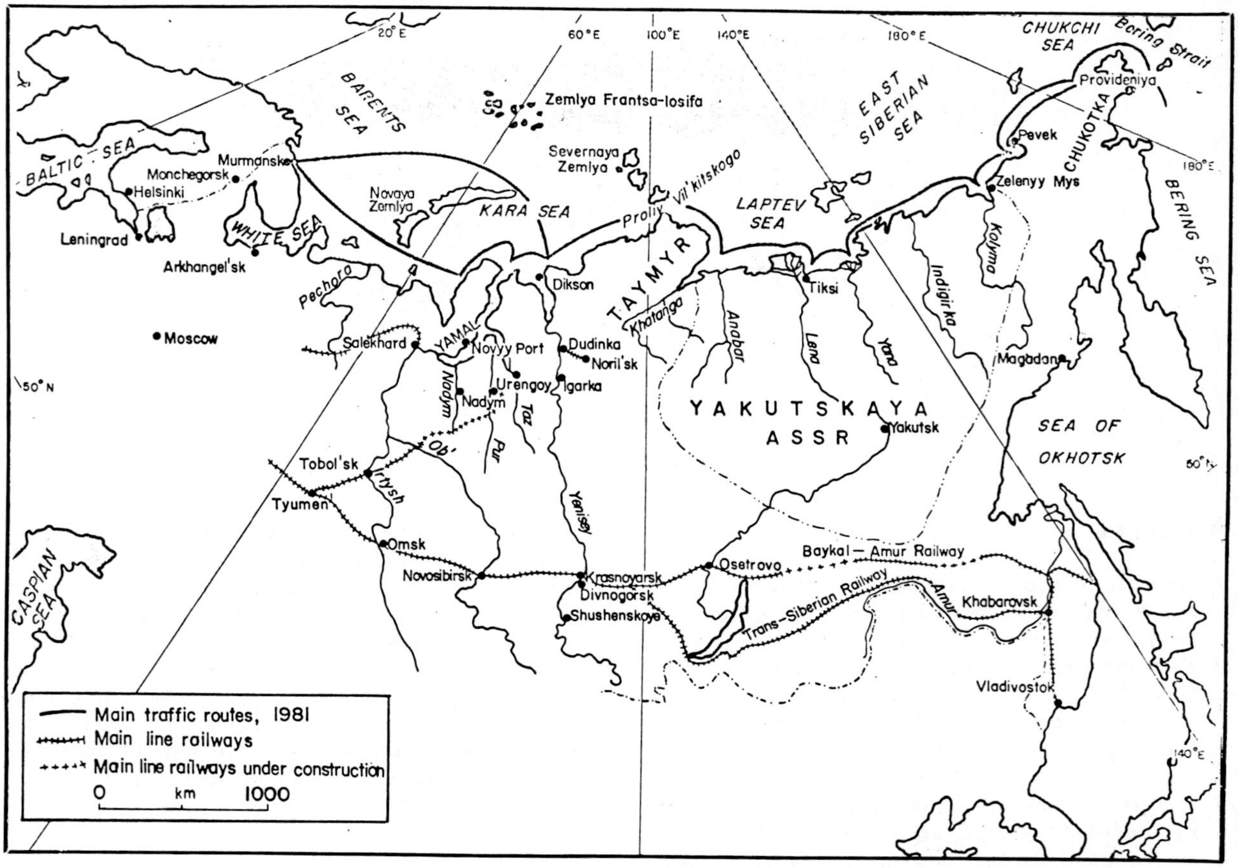


FIGURE 1. *Main traffic routes through the Northern Sea Route 1981.*

this would imply 100 cargoes of 20,000 tonnes, which is the capacity of the biggest ore ships operating on this route. There are probably only 20 or so ships in this class, and they may not be able to operate up the river at full load, so they could carry two million tonnes only if each ship made over five voyages. This is possible but perhaps not probable, so the number of cargoes carried would probably have been larger than 100. The timber port of Igarka had loaded 150 sea-going ships by mid-October, carrying a million cubic metres of timber (about 550,000 tonnes). These figures may have been exceeded before the season ended some days later. The ships trading out of Igarka are necessarily smaller than those stopping at Dudinka. But the two operations between them are likely to call for at least 300 cargoes and account for 2.5—3 million tonnes of freight.

By contrast, the adjacent estuary of the Ob' river carried much less traffic, although the volume increased from the level of preceding years. It was linked with the oil and gas development round the estuary itself. The plan called for bringing in 150,000 tonnes of gas pipe to Novyy Port, which is twice the total for 1980, but it is not clear if it was fulfilled. The Ob'-Irtysh river fleet was very active, however, and was receiving reinforcement with ships able to operate in the estuary. The oilfield expansion calls for use of the smaller rivers, such as the Nadym, Taz and Pur, which drain into the estuary and thus must be reached through it.

The special late-winter operation to the west coast of Yamal, where the fast ice is used as a quay on which to land freight for the oil industry, took place again in 1981, but without any indication of the scale. The fast ice was late forming, so the first ships arrived only on 20 April. The operation lasted until 5 June, and was supported by four icebreakers, including two nuclears. In 1982 the season seems to have been even later, with the first ship arriving on 7 May.

A third distinct freighting operation links

ports in the eastern sector, notably Zelenyy Mys at the mouth of the Kolyma, and Pevek, with Bering Strait and the Pacific. There was little information on the functioning of this route in 1981, partly because the ice was bad. Traffic is certainly much smaller than into the Yenisey. The last figure for freight lifted in this sector was 40,000 tonnes in 1978, and it seems unlikely that this will have been exceeded in 1981.

The remaining sector of the route - that centred on the Laptev Sea - may be served from either end, but more commonly is reached from the west, which was the case in 1981. Coastwise traffic radiates out of Tiksi, at the mouth of the Lena, to the smaller rivers of this basin, and this year increased attention was paid by the press to these rivers (Khatanga, Anabar, Yana, Indigirka). The volume of freight arriving by sea at Tiksi was to grow by a factor of 2.4 over the 1980 level; but it is not known what the absolute value was, nor whether the increase was achieved.

There was, as usual, virtually no reported use of the through route by ships making the transit from Atlantic to Pacific or vice versa.

## Ships

Of the eighteen major icebreakers (i.e. over 10,000 shp) known to be in recent use on the Northern Sea Route, fifteen were mentioned in the press as operating in 1981. This total included the three nuclear-powered ships (see Fig.2). The three absentees were probably undergoing refits, but one of them, a Finnish-built ship dating from 1954, may have been withdrawn from service. A new nuclear icebreaker, *Rossiya*, apparently of the *Arktika* class, is being built in Leningrad and three smaller conventional icebreakers designed primarily for the Baltic and other areas of light ice, are on order from Finland. The Minister of the Merchant Fleet had in 1977 expressed the intention to build more, so there may be others, especially non-nuclear, at the design or even construction stage.

The fleet or ice-strengthened freighters is also growing. The intention stated a few years ago is to increase it from 600,000 dwt in 1977 to 1.5 million dwt by 1990. The number of ships in the two strongest ice classes (ULA and UL) of the six

FIGURE 2: *Icebreakers of 10,000 shp or more known to operate on the Northern Sea Route.*

| Name | Where built | Power shp | Displacement Tonnes | Entered service | Remarks |
|---|---|---|---|---|---|
| *Kapitan Belousov* | Finland | 10,500 | 5,360 | 1954 | |
| *Kapitan Voronin* | Finland | 10,500 | 5,360 | 1955 | |
| *Kapitan Melekhov* | Finland | 10,500 | 5,360 | 1956 | |
| *Lenin* | USSR | 44,000 | 16,000 | 1959 | nuclear |
| *Moskva* | Finland | 22,000 | 13,290 | 1960 | |
| *Leningrad* | Finland | 22,000 | 13,290 | 1961 | |
| *Kiyev* | Finland | 22,000 | 13,290 | 1965 | |
| *Murmansk* | Finland | 22,000 | 13,290 | 1968 | |
| *Vladivostok* | Finland | 22,000 | 13,290 | 1969 | |
| *Yermak* | Finland | 36,000 | 20,240 | 1974 | |
| *Admiral Makarov* | Finland | 36,000 | 20,240 | 1975 | |
| *Arktika* | USSR | 75,000 | 23,400 | 1975 | nuclear |
| *Krasin* | Finland | 36,000 | 20,240 | 1976 | |
| *Sibir'* | USSR | 75,000 | 23,400 | 1977 | nuclear |
| *Kapitan Sorokin* | Finland | 22,000 | 14,900 | 1977 | shallow-draught |
| *Kapitan Nikolayev* | Finland | 22,000 | 14,900 | 1978 | shallow-draught |
| *Kapitan Dranitsyn* | Finland | 22,000 | 14,900 | 1980 | shallow-draught |
| *Kapitan Khlebnikov* | Finland | 22,000 | 14,900 | 1981 | shallow-draught |
| **Under construction** | | | | | |
| *Dikson* | Finland | 10,000 | | 1982-83 | not primarily for the Northern Sea Route |
| *Magadan* | Finland | 10,000 | | 1982-83 | » » |
| *Mud'yuga* | Finland | 10,000 | | 1982-83 | » » |
| *Rossiya* | USSR | 75,000 | | 1985-86? | nuclear |

distinguished in the merchant shipping register was 287 in 1977 - the average cargo capacity was thus of a little over 2000 tonnes. There was a notable shortage of larger ships in the strongest class, but this gap is being filled, for example, by orders for 14 ships of 20,000 dwt from Finnish yards for delivery in 1982-84. Another class of 20,000 dwt ships are the ore-carriers built in the German Democractic Republic. One was delivered in 1981, with two more expected soon afterwords, and this would bring the total of this class to at least 14. Thus these two classes will provide by 1984 well over 500,000 dwt.

A particularly interesting development is the announced proposal to build a nuclear-powered lighter-carrier (LASH) for Arctic use. She will carry 74 lighters with a load capacity of 360 tonnes each, or 1336 containers. She will be much the biggest freighter to operate in Soviet Arctic waters, and will presumably be able to do so because her special role removes the need to enter shallow ports and estuaries. Another Soviet-built LASH ship, not nuclear-powered, is the *Aleksey Kosygin*. She is due for delivery by 1985, and may be used to supply settlements in Chukotka.

## Length of Season

For many years the navigating season was accepted as from two and a half to four and a half months, between July and November. But strenuous efforts in the 1970's led to notable extension. In particular, the Yamal operations already mentioned were conducted in early spring (generally starting a month or more earlier than in 1981): and then the Dudinka operations, aided by the new shallow-draught icebreakers specially designed for use in the Yenisey estuary, were extended at both ends of the season until in 1979 they were claimed to be year round. In fact there was, and remains, a gap of at least six weeks in May-June, when the river ice is going out. But the achievement is very considerable. This extension relates only to the southwest Kara Sea, however, and the rest of the route, chiefly east of Mys Chelyuskina, remains in use for around three months. (A report of spring 1982 mentions a season of 75-90 days at the mouth of the Kolyma.) It should be added that a voyage was apparently made to Severnaya Zemlya, in the northeast Kara Sea, in January 1982 - the first winter voyage to that area. The purpose was unspecified.

## Current uses

There are two major components in the freighting between northern Siberia and the route's eastern or western terminals: on the one hand bringing in fuel, food, and manufactured goods, either as an exceptional operation in support of some major construction project, or as part of the pattern of routine resupply of a number of settlements on the islands, the coast, or up the rivers; and on the other hand bringing out raw materials, of which minerals and timber are much the most important.

The import component appears to be the smaller in volume. This is inference rather than stated fact, and if true it follows that many ships inward bound to Siberia sail in ballast. The routine resupply of small settlements is a time- and effort-consuming exercise, in that many of the places to be supplied do not have adequate harbours or facilities. This is where the LASH ships are expected to prove specially useful. There is an annual import of river craft, built in European USSR or eastern Europe and due for delivery to Siberian rivers. These craft, which may include barges, dredgers, floating cranes and other specialised floating stock, pose a navigational problem in ice-filled seas, but their delivery to the rivers designated has been apparently safely achieved. In 1981 convoys reached the Ob', the Yenisey, and the Lean. In the past, some convoys have made the full west to east transit, being bound for Pacific rivers such as the Amur. But perhaps significantly, river ships bound for Pacific destinations did not take the Northern Sea Route in 1981, but followed a south-about course through the Indian Ocean during the winter of 1981-82.

Among the more notable imported freights were two more turbines for the Sayano-Shushenskoye hydroelectric station on the upper Yenisey; they were trans-shipped at Dudinka, but in previous years the sea-going ship which brought them from Murmansk ascended the river much further, to Divnogorsk near Krasnoyarsk. On the other hand, the Ob' was successfully ascended by ships bringing fractionating columns, 90 m long and weighing 600 tonnes each, from Leningrad to Tobol'sk.

Freight outward bound from northern Siberia is the most important component, in the sense that it is the basic reason for operating the route. The Yenisey is the major single outlet, with the two industries: the copper nickel mines of the Noril'sk area, linked to the Yenisey by rail to the river port of Dudinka; and the timber milling centre at Igarka, 230 km upstream. The two freighting programmes seem to be kept quite separate, presumably for obvious practical reasons of difference in freight and in port location. The link provided by the sea route for transport of nickel are is very important for the industry, since the smelter at Noril'sk is unable to handle all the production, while there is surplus smelter capacity at Monchegorsk, near Murmansk. The mining concern is evidently prepared to pay the considerable costs involved in keeping this route open in winter (though one may doubt if the accounting system obliges the charterer to meet anything like the full costs of icebreaker assistance). The Igarka timber traffic is smaller in volume, but it is the only raw material from northern Siberia to be exported abroad direct from the point of origin.

The Yenisey is not, of course, the only source of raw materials. The next most important area is the shore of the East Siberian Sea, including the mouth of the Kolyma and the port of Pevek. These places serve a hinterland with important gold and other mines. Little is known of the volume of freight taken out, but it is probably rather small. Gold ore is much too bulky to be refined elsewhere than on the spot, and the refined product is not large. The Lena, as a major river giving access far into the interior, is also not a significant route for taking out raw materials. At times in the past timber has been exported from the Lena to Japan, but this never flourished and has long been discontinued.

The service area of the Northern Sea Route, whether for imports or exports, extends far inland where navigable rivers exist. The route's main competitor, for heavy freight, is a combination of the Trans-Siberian railway and river transport, and this southerly route is particularly advantageous, of course, where the railway intersects major north-flowing rivers like the Irtysh, Ob', Yenisey, and Lena. Thus the southern boundary of the service area is very convoluted. It also varies with the commodities carried (perishable goods cannot be risked in the same way as unperishable).

The virtual absence of through traffic (apart from possible naval activity, which is never publicised) indicates that the route is not a successful competitor with the railway as a link between Atlantic and Pacific. In the early days of realisation that a northeast passage was navigable, this had seemed a main objective. But the strong seasonality of the route, and the risks of delay and damage, have up to now prevented its achievement.

## Future prospects

The Soviet government clearly intends to expand shipping operations on the Northern Sea Route. The level of investment in ice-going ships, if no other factor, provides ample evidence of this. What are the possible directions that this expansion could take?

1. There may be growth of traffic on the present freighting pattern. To the extent that the sea route is the cheapest option, this will tend to take place anyway as mining develops. But there are also some areas where another factor plays a part. For instance, as soon as a spur of the Trans-Siberian railway intersected the Lena, in 1951, Yakutskaya ASSR received less goods by sea, and much the greater part by way of the railway and river. But in recent years traffic using the port of Osetrovo at the rail-river intersection has been unable to expand further because of the limitations imposed by water depth. This resulted in planning for the sea route to take a larger proportion of the incoming freight, and the 1981 plan called for a continuation of this. Goods bound for the Khatanga river from Krasnoyarsk were routed down the Yenisey to Dudinka and thence by sea, rather than by rail to the Lena, down it, and along a shorter sea passage. It seems, however, that this planned expansion has not been wholly satisfactory (the level of fulfilment of the plan has not been published). An article in *Ekonomicheskaya Gazeta* 1982, No.11, p.6, by the Party Secretary for Yakutskaya ASSR and the Chairman of the Academy of Sciences branch there came out strongly in favour of extending the Baykal-Amur railway (BAM) to Yakutsk (a project mentioned in the 11 th five year plan, 1981/85), and mentioned that the Northern Sea Route was difficult and expensive. This is the more interesting in that the same Party Secretary had earlier been one of those calling for greater use of the sea route. The situation seems to be that the sea route is acceptable only if no other route can be used — but there are and will remain many areas for which this is true, and an expanding economy in these areas will assure the sea route of more use.

2. Through traffic may be expanded. The shortest route between Murmansk and Bering Strait lies not along the coast but across the middle of the Arctic Ocean, passing close to the Pole. This route also has greater water depth, minimising the constraints imposed by the shallow waters of the continental shelf. The region is heavily icebound all the year round, but its

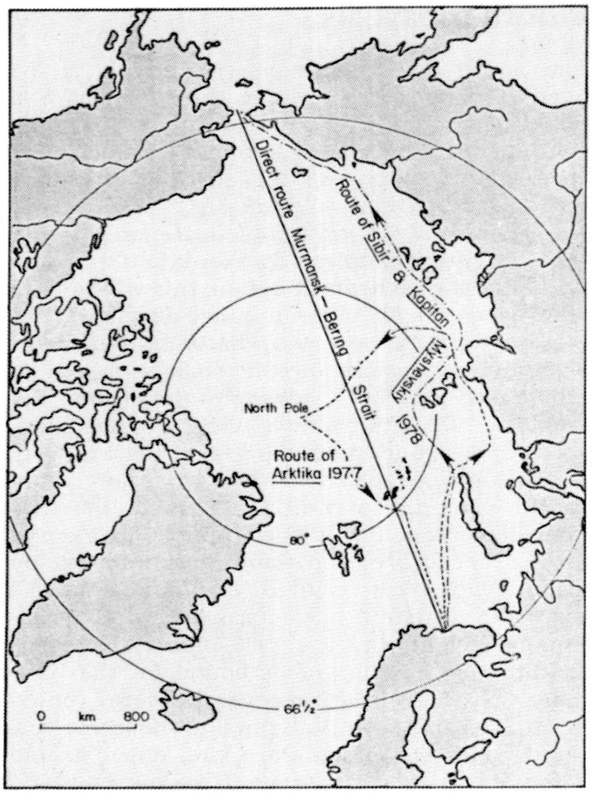


FIGURE 3: *Arctic Ocean and adjacent waters.*

penetration in August 1977 by the Soviet nuclear icebreaker *Arktika* opened up new possibilities. *Arktika* made a speedy voyage to the Pole and back, becoming the first surface vessel ever to reach it. The following year her sister ship *Sibir '* escorted a freighter, *Kapitan Myshevskiy,* over a route north of the island groups in the more difficult months of May and June. The stated object of these voyages (Fig.3) was to explore the possibility of trans-Arctic Ocean navigation — which might be Soviet or non-Soviet. Since 1978, however, there has been no further activity of this kind, and little more mention of a trans-Arctic route.

It should be noted that neither in the USSR nor elsewhere in the world is there a perceived need to carry freight over such a route. K. N. Chubakov, head of the Administration of the Northern Sea Route, suggested in 1979 that the route might provide relief for the Trans-Siberian railway — but he was speaking of a general intention in the indefinite future. No major development is being held up by the lack of a through route (which is not to say that if one existed, freight would not be found). Nor does any nation possess enough of the kind of ships which would be required: the relatively small ice-strenghened freighters at present available cannot take advantage of the deeper water. So the question arises, why is the USSR interested? One possible explanation is that the USSR sees itself, rightly, as the world leader in ice navigation. She will have noted the American intention in the early 1970s to build two new icebreakers, *Polar Sea* and *Polar Star,* of novel design and comparable power to *Arktika* and *Sibir',* and may therefore have concluded that if she did not show her paces in the central Arctic Ocean she might be beaten to it. She was not to know at that time that the new American ships would prove dismal failures. Having ascertained this later, she lowered the priority of developing a trans-Arctic route. But the possibility of pioneering one at a later stage remains open, and may well become important in the future.

3. The navigating season may be lengthened. Very impressive results have been obtained in the southwest Kara Sea in keeping access to the lower Yenisey for perhaps ten months of the year. The stimulus for this came from the nickel mine. The achievement of nearly year-round operation in one sector invites the thought that it might also be possible elsewhere, if a sufficient motive were there. The influential economist Academician Aganbegyan said in 1980 that year-round operation on the Northern Sea Route was a major programme for the future, and the First Deputy Minister of the Merchant Fleet, V. Tikhonov, was reported on 14 December 1981 as saying that it was under study. Ice conditions are on the whole more severe farther east, so the task of lengthening the season would be harder. But it is very probable that a significant lengthening could be achieved.

4. This brings us to consideration of the physical environment, in particular the ice. It is now clear that the climate, as reflected in the ice situation, has been worsening since the 1940s. Up to now advances in technology and science — ship design and power, communications, ice forecasting techniques — have offset, and more than offset, deteriorating ice conditions. The deterioration is expected to continue, but this is only a forecast. What is certain is that the summer seasons of 1978, 1979 and 1981 were all exceptionally bad. There is also the question of whether man-made changes in the flow of northward-running rivers could further exacerbate the situation. This is a topic much discussed recently in the context of proposed

irrigation schemes by which water from these rivers will be directed to the Caspian Sea and central Asia. How much longer, then, can technology ensure the movement of ships in the waters of the Northern Sea Route? This is a hazardous question to try to answer. My own forecast would be that since technology may be expected to continue improving, with the impetus of resource exploitation behind it, nature may be kept at bay for some time yet, unless a catastrophic deterioration occurs; and the river change plans at present envisaged are likely to have too small an effect to make any significant difference.

5. Finally, to revert more directly to the subject of this conference, Arctic energy resources, let us consider possible expansion in that context. Oil and gas prospects in the Kara Sea are believed to be good: the northwest Siberian petroliferous zone (which at present produces more than ten per cent of the *World's* oil and gas) very likely extends offshore. Prospects are also good in the Barents Sea, where a drilling ship was reported to be active in May 1982. No finds have been announced, and it is possible that in the Kara Sea no exploration has been done. But ice strengthened drilling ships are being obtained from Finland, so work can start at any time (it may be one of these ships which is now active in the Barents Sea, although the ice hazard is likely to be much less there). At all events, when offshore oil production gets going, especially in the Kara Sea, ability to navigate in these waters will be crucially important.

## Conclusion

The Northern Sea Route is far ahead of any other sea route through the Arctic Ocean and its adjacent seas in respect of traffic carried. Its Soviet operators have more of the best equipment and a much larger number of ice-trained crews than any other country involved in this work. They are in a good position to expand their operations, and they seem certain to do so. Existing trade links can be exploited more intensively, new routes can be opened up, the season can be lengthened. These things will be done primarily to benefit the domestic economy of the USSR. The probable oil and gas development is one of these. But they may also have a wider impact, since the primacy of Soviet expertise could allow the USSR to sell its services, if it so wished, to western developers faced with similar problems. This aspect has not yet properly emerged (there was a brief and half-hearted attempt to sell escorted passage through the Northern Sea Route in 1967), but it is one which may play an important part in the future.

*Cold Regions Science and Technology*, 7 (1983) 259—272
Elsevier Science Publishers B.V., Amsterdam — Printed in The Netherlands

# The Arctic Pilot Project

*Ricard A. Bailey*

Petro-Canada
Calgary

## Introduction

My presentation has been divided into three sections namely: an overview of the Arctic Pilot Project (APP); some of the specific Project requirements that we feel may be of interest to suppliers; and some of the many benefits the project offers for the Canadian economy.

Before providing an overview of the Project I would like to briefly comment on some of the changes that have occurred in the energy sector over the past decade. With declining reserves and production of conventional energy sources, Canada has been forced to consider new and previously uneconomical energy reserves.

A variety of new techniques are now being employed to squeeze additional production from wells exhausted by primary recovery. In addition, a variety of projects at various stages of completion are being undertaken to recover energy from heavy oil reserves, the tar sands, the East coast offshore and the Arctic. Naturally, the orientation of the Arctic Pilot Project is towards the extraction of energy supplies from Canada's far North. In our opinion, the Arctic Pilot Project is an essential prerequisite to the orderly energy development of the Arctic.

## SECTION I — PROJECT OVERVIEW

### The Project

The Arctic Pilot Project is a transportation system designed to deliver 270 million cubic feet/day of natural gas from Melville Island in the Arctic Islands to a terminal location in Eastern Canada on a year-round basis using icebreaking liquefied natural gas (LNG) carriers. LNG was chosen as the cargo because of the abundance of natural gas in the Arctic and the fact that LNG is a non-pollutant. The project will provide experience in the movement of resources from the Arctic on the smallest economic scale. Even though the project is small in scale compared to alternative energy projects, it will yearly provide enough gas to heat approximately 725,000 homes.

It is the sponsors' belief that the project is complementary to other energy transportation systems and is highly flexible in that this type of system could potentially be used to extract other forms of energy and minerals from any part of the Arctic. It does not commit the federal government to only marine transportation of resources.

### What is LNG?

Most of the world's natural gas is transported in pipelines. Although pipelines have provided an effective means of gas transportation, their use has been restricted by oceans and other geographical barriers. Furthermore, pipelines are expensive and permanent facilities that cannot be moved if gas supply or market conditions change. To a certain degree, these factors have limited pipeline construction and as a result, some natural gas reserves have been isolated from the energy-consuming public. To overcome this problem, other transportation systems have been investigated. To date, the most successful

0165-232X/83/$03.00  

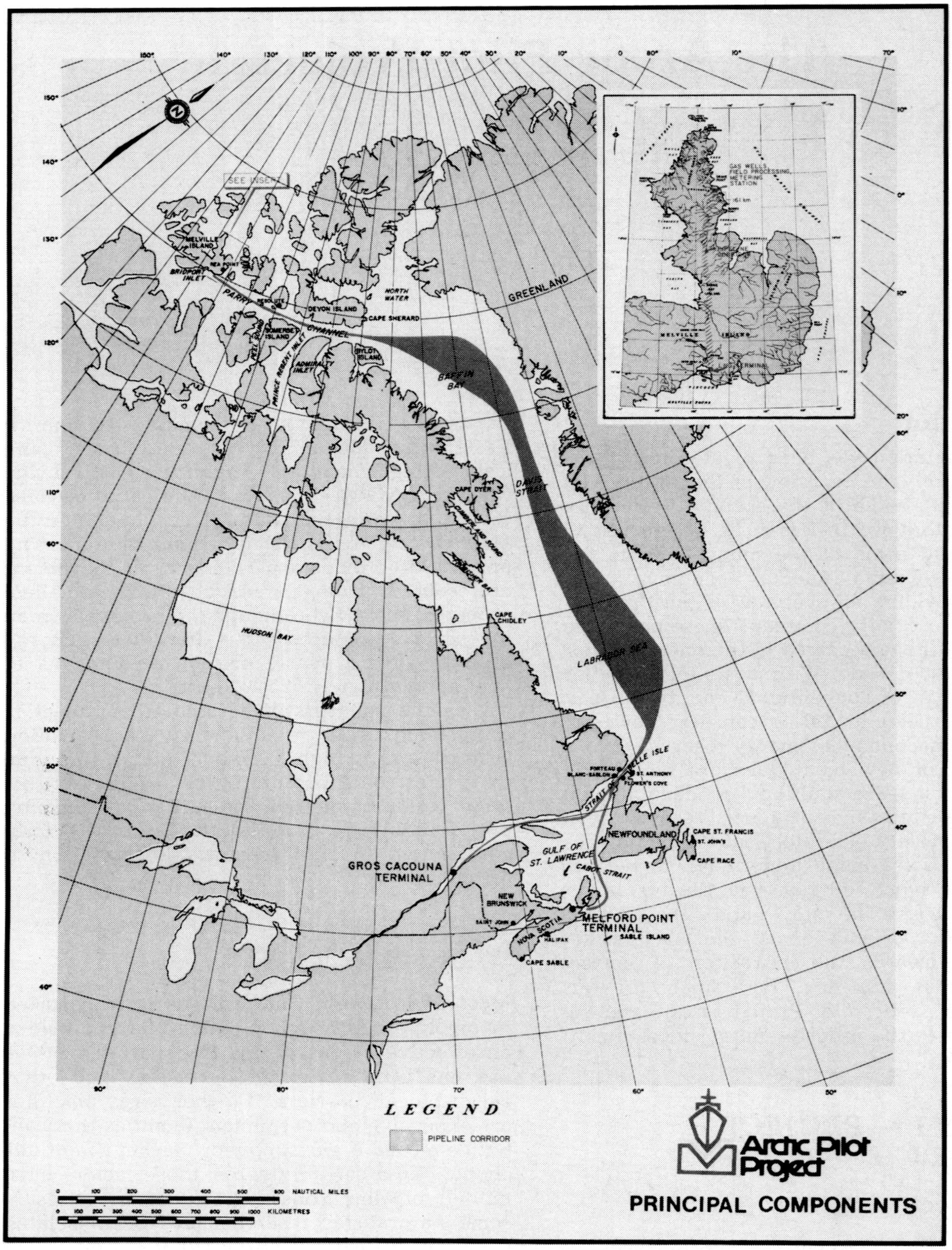
SEE INSET
MELVILLE ISLAND
BRIDPORT INLET
PARRY CHANNEL
DEVON ISLAND
SOMERSET ISLAND
CAPE SHERARD
NORTH WATER
GREENLAND
BYLOT ISLAND
ADMIRALTY INLET
PRINCE REGENT INLET
BAFFIN BAY
DAVIS STRAIT
CAPE DYER
CUMBERLAND SOUND
FROBISHER BAY
CAPE CHIDLEY
HUDSON BAY
LABRADOR SEA
STRAIT OF BELLE ISLE
PORTEAU
BLANC-SABLON
ST. ANTHONY
FLOWER'S COVE
NEWFOUNDLAND
CAPE ST. FRANCIS
ST. JOHN'S
CAPE RACE
GULF OF ST. LAWRENCE
CABOT STRAIT
GROS CACOUNA TERMINAL
NEW BRUNSWICK
SAINT JOHN
NOVA SCOTIA
HALIFAX
CAPE SABLE
MELFORD POINT TERMINAL
SABLE ISLAND
GAS WELLS FIELD PROCESSING METERING STATION
LEGEND
PIPELINE CORRIDOR
NAUTICAL MILES
KILOMETRES
Arctic Pilot Project
PRINCIPAL COMPONENTS

FIGURE 1.

alternative has been to liquefy the gas, reducing its volume about six hundred-fold. The liquefied natural gas is then transported by ship to the consuming areas, where it is regasified and delivered to local pipeline systems. Marine trials carried out in 1959 successfully demonstrated the technical and economic feasibility of transporting natural gas in this way. Since then, LNG technology has advanced significantly and today liquefied natural gas is shipped on over a dozen international trade routes.

The process of producing liquefied natural gas begins by cooling the gas until, like most gases, it eventually condenses into a liquid. During the cooling process natural gas achieves the equivalent density of air at $-104°C$. Condensation into a liquid occurs at the ultra-cold (cryogenic) temperature of $-162°C$.

Once the gas has been liquefied on Melville Island, it will be stored as cold but unpressurized LNG in special tanks to await the arrival of the ships. The two ships will transport the LNG to a receiving terminal where it will be stored and regasified according to market demand. The conversion of LNG back to quality gas at the receiving terminal requires only the application of heat from the surrounding environment.

## Integrated Route

To date, considerable biological and physical data have been collected and analyses done to define the shipping corridor from Melville Island to eastern Canada for the APP transportation system. The shipping component, of course, is the most pioneering aspect of the APP, both in the area of ship design and operation, as well as in route selection.

The Arctic Pilot Project has prepared an Integrated Route Analysis which has been filed with various federal and provincial government departments in Canada. The Integrated Route Analysis provides a baseline that will prove to be the foundation for work to be conducted in the future by the APP. It defines the early stages in the evolution of the most desirable ship route for the project, considering ship safety, environmental considerations, and other factors.

The route chosen by the APP reflects a very careful consideration for the potential effects of the ships on the marine environment and on the peoples of northeastern Canada and Greenland.

Two examples of this consideration are the studies done on the potential impact of ship generated noise on the marine mammals of the regions traversed, and the effect of a ship track in ice on the movement of hunters.

The Arctic Pilot Project intends to develop and implement a long-term research and monitoring program for underwater acoustics. A major step towards developing such a program was the Arctic Pilot Project workshop on underwater noise held in Toronto, in February of 1981.

To date, investigations by the Project have shown that ambient noise levels in the Baffin Bay, Davis Strait and Labrador Sea areas are naturally high as a result of iceberg rollover and sea ice ridge creation. Even so, there will be an area to either side of the ship wherein received noise levels will be at or above ambient noise levels in the ocean.

Technical investigations indicate that the zone of potential noise influence will be such that the worst case location will receive ship noise greater than the ambient level in a narrow band, parallel to the ship track for only two to four hours for a given vessel passage. There are very few sea mammals in the high Arctic during periods where ice cover is most extensive and noise from icebreaking most intense. During open water periods when sea mammals are most prevalent, the vessel's noise will be less than many of the vessels which travel the St. Lawrence River. Perhaps through operational monitoring and experimental programs such as the APP Noise Research Plan, it may be possible to answer such questions as why the whale populations of the St. Lawrence River are able to co-exist with the very heavy shipping traffic activity of one of the world's most utilized sea-lanes.

In terms of potential effects of the Project on the peoples of the regions traversed, the primary concerns have been with the ability of hunters, particularly from Resolute Bay, to cross the track of the ship, and with the potential impact on the hunters of West Greenland.

To investigate ship track crossing capabilities, two trials were held at McKinley Bay in the Beaufort Sea. During the last trials, held in March, 1982, Inuit were able to walk across the ship track 1/2 hour after the passage of the icebreaker *M. V. Kigoriak*. One hour later a skidoo was driven over the track and, this was followed by a skidoo hauling a loaded komatik.

A ship track was also crossed in the harbour of Thunder Bay using the experimental vehicle the Archimedian Screw Tractor (A.S.T.). There are also plans to test a modified boat which will ferry a snowmobile across open water areas during the break-up season. The Arctic Pilot Project is confident that based upon the tests to date, it will provide a method that will ensure safe crossings.

In terms of external affairs, the APP has for over two years conducted negotiations with Greenland via Denmark and has established an APP Working Group to resolve contentious issues. As a result of these discussions the ship route has been moved further to the west than originally planned.

It now comes no closer than 100 km to the Greenland coast, thereby minimizing the impact on marine animal life and resource harvesting activities. In addition the Project, through the Working Group has committed to conduct research work into the question of the noise generated by ships operating in ice and the effects of this noise on marine mammals.

## APP Participants

The Arctic Pilot Project is a Joint Venture of four Canadian companies: Petro-Canada Exploration; NOVA, An Alberta Corporation; Dome Petroleum Limited; and Melville Shipping Ltd. Melville Shipping is a consortium of three shipping companies: Federal Commerce and Navigation; Upper Lakes Shipping and the CSL Group Inc. All of these large, Canadian companies have relevant experience to assist the Arctic Pilot Project in this northern development activity. Petro-Canada is the overall project operator and manager. With such a strong team we feel very confident of our ability to successfully carry out this most challenging northern development project.

The five major project segments are as follows:

(a) the DRAKE POINT FIELD FACILITIES to be owned and operated by Panarctic Oils;
(b) the MELVILLE ISLAND PIPELINE to be owned by the APP and designed and operated by NOVA;
(c) the BRIDPORT LNG FACILITIES also owned by the APP, and designed and operated by Petro-Canada;
(d) the TWO LNG CARRIERS owned by the APP, and designed and operated jointly by Dome, Petro-Canada and Melville; and
(e) the SOUTHERN TERMINAL to be owned and operated by TransCanada PipeLines (TCPL).

## Drake Point Field Facilities

The Drake Point Field facilities will consist of six or eight wells drilled from two pads into the on-land portion of the Drake Point field. Flowlines will connect the wells to a central gathering facility where the gas will be treated and metered prior to sale to the Project. Since the gas in the field is sweet and averages over 97 % methane, treatment will be relatively minor compared to that required for most other gas fields. Associated with the facilities will be a permanent camp for operating personnel and an airstrip for crew rotation and replenishment of consumables.

Risked gas-in-place at Drake, defined at 90 % of proven, plus 50 % of probable, plus 10 % of possible gas-in-place has been estimated at 113.3 billion cubic meters (4.0 trillion cubic feet). Marketable gas is estimated at 90 % of risked gas-in-place based on excellent permeability and good lateral flow characteristics. To date, eight wells have been drilled into the producing formations, with production flow rates averaging 195,000 cubic meters/day (6.8 million cubic feet/day) and calculated absolute open flow per well of 8.9 million cubic meters/day (317.5 million cubic feet/day) Based on this information producing flow rates of 1.1 to 1.4 million cubic meters per day per well (40-50 million cubic feet) are projected for the Project.

## Melville Island Pipeline

The Melville Island pipeline will originate at the Drake Point Field on the north end of Melville Island and will terminate at Bridport Inlet in the south. The 559 mm (22 in.) diameter pipeline will be 160 km (100 mi.) in length and will be completely buried to minimize disruption to the environment. The gas flowing into the pipeline will be chilled to below freezing temperatures to prevent melting of the permafrost.

The pipeline will be constructed during 6 to 8 week periods in each of the spring and autumn seasons over a two to three year period. This unusual construction schedule is dictated by the harsh Arctic winter conditions and by the 24 hour summer sunshine which can cause the upper 40 cm (16 in) of the ground to thaw to an impenetrable, yet fragile state. Construction will therefore take place when the ground is frozen yet the temperatures and winds are not so severe as to make unprotected work hazardous.

The pipeline will be constructed using an

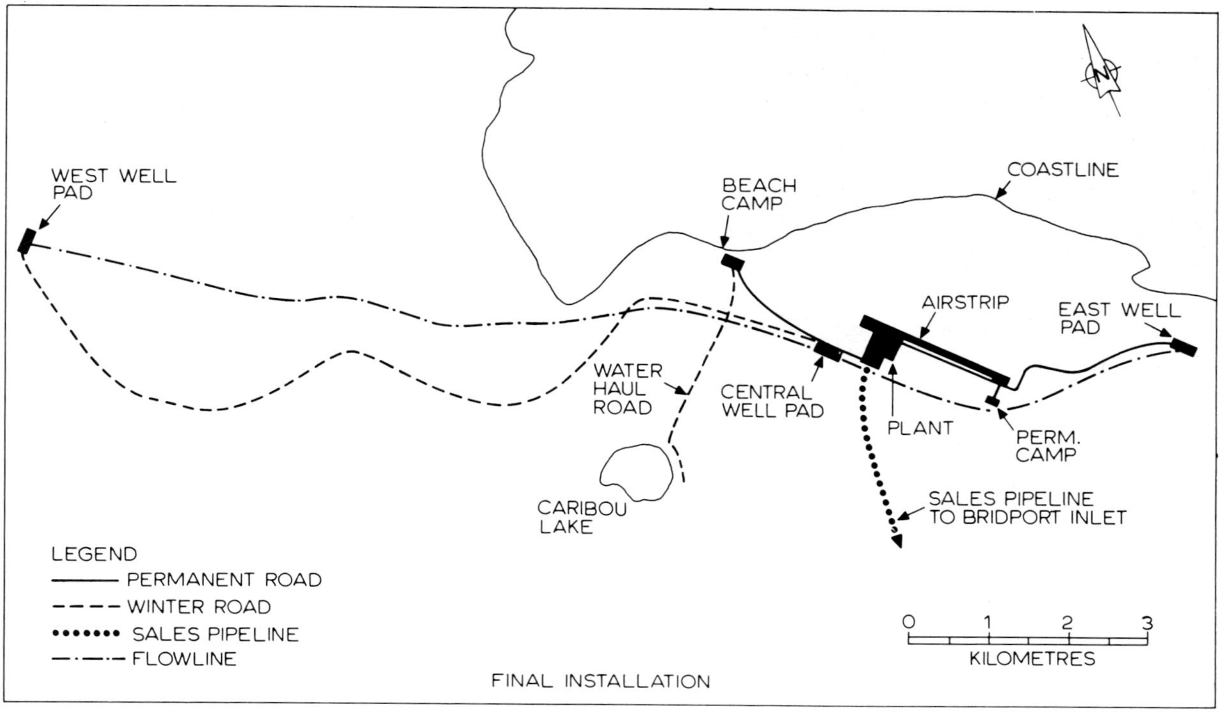


FIGURE 2.

Arctic-class ditcher. Preliminary trials have been held in 1981-82 during which such a specially designed machine was taken to Melville Island and digging trials were held. Research areas included ditchability rates with different tooth designs, metallurgies and orientations, with achieved results generally as-good-as or better than predicted.

## Bridport Facilities

At Bridport Inlet the natural gas will be liquefied and temporarily stored before being loaded onto the ships. The LNG plant and storage tanks will be barge-mounted. The three barges will be built in southern Canada where control over costs and construction schedule can be effectively maintained, and at the same time environmental impact at Bridport Inlet can be minimized.

The Bridport liquefaction plant will use the Technip-Snamprogetti precooled, two-fluid, mixed refrigerant liquefaction process. The incoming gas will first be treated to remove carbon-dioxide, mercury and water, and then condensed and subcooled in a large vertical spiral wound heat exchanger. The total LNG storage of 200,000 $m^3$ LNG capacity (4.4 billion cubic feet of gas equivalent) will be contained in two barges, each having two 50,000 $m^3$ (1.1 billion cubic feet of gas equivalent) tanks. The asic design of each storage barge will beessentially that of the cargo portion of an LNG carrier. The membrane containment system is currently the preferred type for the storage tanks due to lower costs, greater Canadian content and industrial benefits, and other reasons.

The Bridport facilities will also include accommodation for operating crews and an all weather air strip for crew rotations. An ice management system for reducing the growth of ice adjacent to the dock will be incorporated in the design. This system will utilize a portion of the heat of rejection from the liquefaction process.

## Kigoriak

This project will be the largest-scale operation to be undertaken north of the 75° latitude, an area which is inaccessible by existing ships for all but 8 weeks of the year. During this brief

FIGURE 3. *Bridport liquefaction factories (artist's rendering)*

period, from the latter part of August and until about the end of September, only ships with ice-strengthened hulls can safely venture into these waters. Vessels are at constant risk of being beset by wind-driven ice. In such a case they may only be extricated by a favourable change in wind direction or by the assistance of the powerful icebreaking ships operated by the Canadian Coast Guard.

The design of the APP ships is based on performance of Canadian Coast Guard Icebreakers which operate in the Arctic during the summer months, on commercial vessels such as Melville Shipping's *M.V. Artic*, and on Dome Petroleum's *M.V. Kigoriak*. This vessel is an ice-strengthened supply boat that has demonstrated year-round operating capability in the Arctic over the last three years. Many of the design features proposed for the APP ship's are based on refinements of the design of this ship.

## Ship Design

As a consequence of the ice conditions, the Arctic Pilot Project has undertaken to design the largest and most powerful ice-breaking vessels ever envisioned. These vessels would have the enormous strength and power necessary to voyage anywhere in the Arctic during any season. As such, they will incorporate design features that have been trial-proven in smaller-scale vessels such as the *M.V. Kiqoriak* Of particular interest is the spoon-shaped bow with reaming flanges that extend the bow-width beyond the beam of the mid-section of the hull to permit short-radius turns. Also, the use of water spray nozzles, to direct water upon the ice before the bows, provides lubrication and reduces friction when breaking ice.

Motive Power for the ships is generated by gas-turbine engines, which burn the LNG vapours which boil-off from the vessel's LNG cargo. Additionally, the vessels may use diesel fuel in substitute for the gas. The propulsion system employs shrouded, variable-pitch propellers which reverse through pitch adjustment. With the two shafts always turning in the same direction, power-train simplicity is preserved and the ship has the required power astern to back away from the ice-face, when necessary to ram concentrated ice ridges in heavy pressure formations.

The vessels will be equipped with the most advanced ice-detection systems in existance, including infra-red detection and low light level television. Also, they will be equipped with satellite navigation and communication apparatus and the conventional HF/VHF radio. The crew for each ship will be approximately 42 persons, each crew being changed about every two weeks.

These vessels will have several unique capabilities which offer great manoeuvring safety. They will be able to turn in a very short (less than 2 km) radius and can come to a complete stop in 1.5 times her length from crusing speed. The bridges will be located forward in the hull and the crews will have direct visibility of the waterline near the bows. While breaking ice, the underwater shape of the hull will force ice to submerge and then be directed outboard and under the ice alongside.

With these features, the ship can safely accomplish return voyages from the area of the St. Lawrence river to Melville Island in 33 days during winter and in 16 days during summer, a total distance of some 10,000 km. Since there will be two vessels in operation, a return trip is possible every 8 to 16 days.

These ships will be amongst the largest, and most sophisticated vessels ever built anywhere in the world for commercial trade. At 112 megawatts (150,000 horsepower) per vessel, the LNG carriers will be the most powerful non-military vessels in the world.

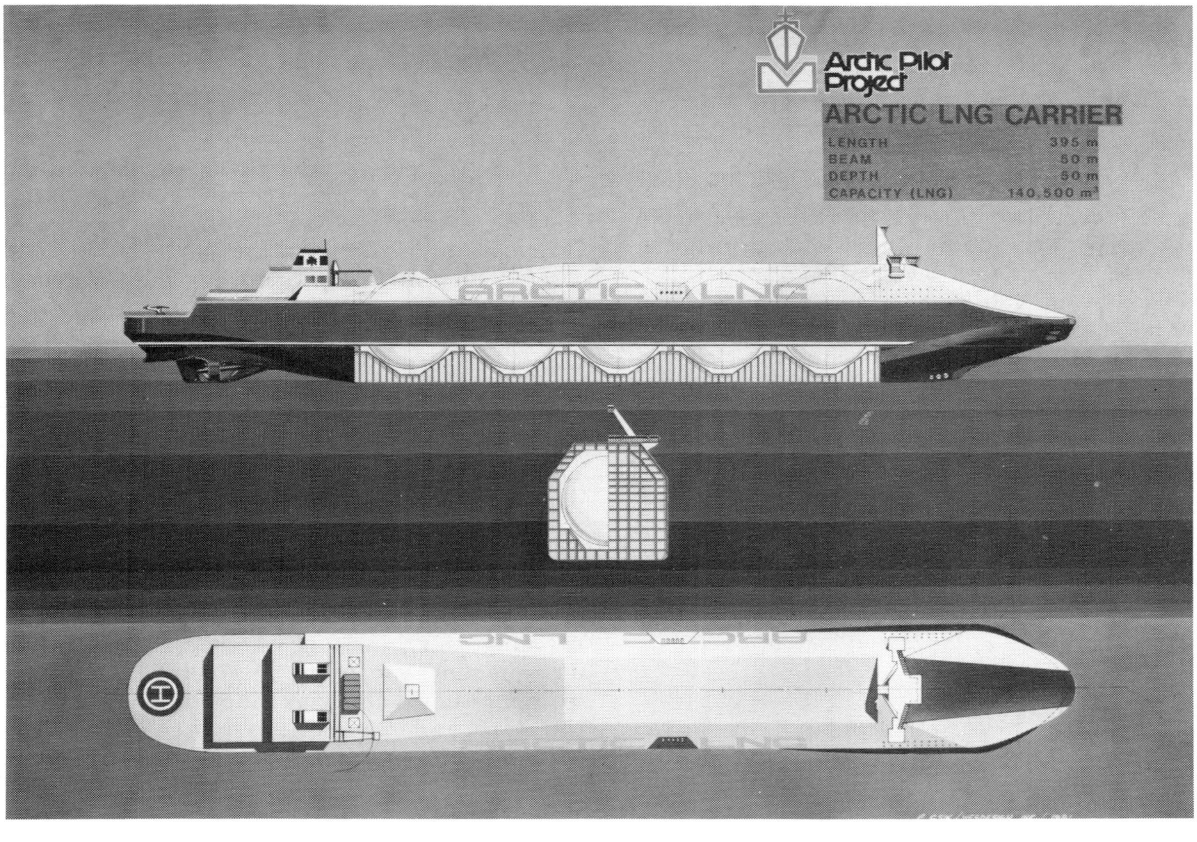


FIGURE 4.

FIGURE 5.

## Bow Design

The bow design features a broad spoon shape that incorporates lateral reamers which in total are some five meters wider than the parallel mid-body of the hull. The reamers allow the bow to break a track in the ice wider than absolutely necessary in order to reduce both the turning radius and the friction between ice and the parallel mid-body of the hull. The spoon shaped bow will break ice in the manner of least resistance; that is, in a downward direction.

The forward skeg located below the bow serves to contain the lateral thrusters used for docking manoeuvers. The wedge shape of the skeg acts to deflect pieces of ice broken by the bow away from the bottom of the hull and to the sides.

## Stern Design

The stern design features two controllable pitch propellers mounted in Kort nozzles, with twin, independently controlled rudders, all mounted on an ice island. Controllable pitch propellers allow for quicker response from full ahead to full astern power which is very useful in ice ramming manoeuvers. The Kort nozzles serve to channel the flow of water giving greater thrust at any level of power, and also prevent large pieces of ice from entering the propellers. The ice island feature will keep ice away from the propellers and rudders by virtue of its pointed, tear-drop shape and by the reverse slope on the sides which constrain ice by its natural bouyancy to the bottom of the hull. These features have been tested in ice model tanks and proven successful beyond our expectations.

## Sea Ice

A very important consideration in the overall shipping component is of course the marine environment. The Arctic Pilot Project will seek to minimize, to the maximum practical extent, the impact of the Project on the people, the environment and the animals of the regions traversed.

An equally important consideration to the project is the physical environment, particularly as it relates to the passage of the ships. For several years the Project has gathered relevant physical data in an effort to predict with confidence the technical performance and economic feasibility of the proposal.

To date the APP has been in existence for five years and the sponsors have spent a total of approximately $50 million on it, of which about $6.5 million has been spent on environmental and socio-economic studies and regulatory related matters.

## Southern Terminal

The last component of the Project is the Southern Terminal. Two locations are being considered; the first at Gros Cacouna Island which is adjacent to the south shore of the St. Lawrence River, near Riviere-du-Loup, Quebec; and the second is Melford Point on the mainland side of the Strait of Canso, opposite Port Hawkesbury, Nova Scotia.

At one of these locations, TransCanada PipeLines will build and operate the receiving and regasification terminal. At the chosen site, the LNG will be received and stored in liquid form in special tanks, regasified by applying heat and the gas distributed to the market through conventional pipelines. As a by-product, significant amounts of refrigeration could be produced, which would be useful in other commercial enterprises.

The gas will be marketed in Eastern Canada at the prevailing domestic price. An equivalent volume of Western Canadian gas that would otherwise have moved into this market will be exported into the United States market at the prevailing border price. Access to this higher price makes the Arctic Pilot Project economically attractive.

The Arctic Pilot Project is currently pursuing an alternate market in western mainland Europe. Discussions have been held with several companies in different countries that are interested in alternate supplies. From the Canadian perspective, gas supplies exceed demand by a considerable margin and at a national level there is some expressed desire to move away from the monopsony situation of only marketing exportable gas in the United States. Such factors as currency stability, the broadening of trade and geopolitics argue in favour of pursuing the European option. The facilities designed for the Project are capable of liquefying and transporting only marginally smaller quantities of LNG to Europe at very similar prices and economics.

FIGURE 6.

## Options

During this conference other proponents will describe alternate means of moving Arctic resources to market, such as large diameter pipelines for transporting natural gas, and concepts for submarine transport of liquids. The Arctic Pilot Project has carefully considered each of these options and concluded that surface ship transport of LNG offers the best combination of proven economics, technology, and consistency with energy demands and financial market capabilities.

With reference to large diameter pipelines for marketing natural gas it is our assessment that the following factors mitigate against the choice of a pipeline at this time:

- the as yet inadequate reserve base to support the economics of a large diameter pipeline,
- the relatively small size of the domestic gas markets compared to economic pipeline throughputs,
- the reluctance of government agencies to devote large quantities to export markets,
- the recently poor track record of the take of Canadian gas by the existing markets in the United States,
- the relatively large, up front capital cost of frontier pipelines, and
- the great difficulty of financing the pipeline investments.

Compared to this option the APP is relatively small, is in tune with both the reserve base and the market demand, will be largely constructed in the south for maximum technical and financial control, and by its very nature is both flexible and incrementable thereby reducing financial difficulties. Furthermore, by generating a cash flow at an early date to Arctic explorers, the APP will contribute to reserve delineation and will thereby enhance the future viability of the pipeline option.

With reference to submarines for the marketing of LNG it is our position that such proposals may be feasible in the future, provided that considerable time and effort are invested in the design concepts. This position is based on the assessment that acceptable answers must first be found to such fundamental questions as:

- navigation by large vessels in shallow waters which frequently contain deep ice ridge keels,
- precise control over depth and pitch angles to reduce likelihood of grounding or under ice contact,
- the inability of sonar under Arctic sea conditions to reliably provide adequate manoeuvring and hazard avoidance capability,
- the low Canadian content and industrial benefits, and insufficient maritime experience associated with submarines,
- the safety hazards posed by either liquid oxygen supported, or nuclear propulsion systems,
- the untested underwater LNG loading systems, particularly under Arctic conditions, and
- the inconsistency between the need for frequent, extensive and time consuming maintenance and the commercial desire to adhere to strict and relatively inflexible schedules.

Compared to this option, the APP will utilize conventional, demonstrated technology in a unique configuration, will use available and reliable satellite navigation and communications capabilities, and will have sufficient flexibility inherent in the delivery schedules to permit the small amount of unscheduled maintenance to be readily conducted.

## Project Master Schedule

It will require approximately four years after regulatory approval until the Project becomes operational in 1987. The schedule will allow sufficient time to carry out R & D, engineering and prototype development prior to final material procurement. While this schedule is tight, it is predicated on late 1982 or early 1983 approval, followed by an immediate start to detailed design, followed about one year later by start of construction. Major sealifts are planned for each of the years 1984 through 1986 with final construction tie-ins in 1987 just prior to start-up.

We quite often get asked the question, why not delay the project for several years as we do not need the gas in Canada at this time? In response, we feel we have an opportunity and a window in time to develop a transportation system that will connect Canada's northern resources for the future benefit of all Canadians. We are planning to develop a transportation system that in our opinion is vitally necessary to provide access to natural resources in the north such as gas, oil and minerals. Without a proven transportation link we will be unable to have the assurance that resources in the Arctic and other frontier areas can be moved to market when needed. If we are contemplating moving other resources from the Arctic it will not be until the late 1980's or early 1990's that the technology proven by the Arctic Pilot Project will be available.

We believe that we already are behind with this Project and to delay it further would in fact destroy the opportunity for the orderly development of a year-round Arctic transportation system. Society would then probably be faced with a need for moving larger volumes of energy from the Arctic in the last decade of this century without the vital technological development, experience, and knowledge we would have gained from this first research/development type project proposed the Arctic.

## Research and Development Program

An important aspect of the Arctic Pilot project is its commitment to an ongoing Research and Development program throughout the project duration. This program under the direct administration of the Arctic Pilot Project will cost an estimated $220 million in today's dollars, over the nominal twenty-year project life. It represents a significant commitment by the Project to conduct research into environmental, socio-economic and technical questions, as well as to monitor any effects that year-round transportation may have on the Arctic.

## Regulatory Processes

The Arctic Pilot Project has been extensively reviewed by government-appointed regulatory bodies. To begin with, early in 1980 the Project participated in a public environmental assessment and review process hearing for the facilities north of 60°. Approval with terms and conditions was granted late in 1980.

Secondly, in early 1981 the Project participated in two joint federal-provincial public environmental hearings; one each in Quebec and Nova Scotia. Approvals with terms and conditions were given late in 1981.

Thirdly, in 1981 the Project voluntarily undertook to have the Canadian Coast Guard examine in detail the ship design, operations and routing, and harbour design under the TERMPOL process. Later in the year the three separate TERMFOL reports all concluded that the Project was acceptable subject again to terms and conditions.

Lastly, the Project is currently being assessed by the National Energy Board through public hearings which began early in February. We anticipate completion of these hearings by Autumn, and a favourable recommendation by the Board sometime in the first quarter of 1983. Following this, our customers in the United States will seek approval for imports from the FERC and ERA regulatory processes. Approvals from these bodies should be received by the third quarter of 1983.

# SECTION II - PROJECT REQUIREMENTS

## Project Requirements

I would now like to address some of the Project's requirements which may be of interest to the attendees at this conference. As you will see the Project will require the application of many different types of technology in order to meet our objectives. Some of the requirements can be met with existing products and services. However, in other cases companies may need to apply existing knowledge and technology to new applications.

## Instrumentation/Control Systems

Every operational component will require control systems including the gas fields, the pipeline, the liquefaction plant, the carriers and the southern terminal. Some of these will be very complex and will require the extensive use of computers.

A variety of sensors will be needed as front ends for control systems, safety systems and data acquisition systems. Some of the sensors envisaged include pipeline line break detectors, carrier hull instrumentation, flow meters and environmental measurement devices.

## Data Acquisition/Safety Systems

Because of the pilot nature of the Project, a great deal of emphasis will be placed upon safety systems and the acquisition of operational data. There will be a need for a line break detection system which is capable of discovering a rupture of the pipeline and then quickly isolating that section by the closure of remotely located valves.

Gas detection and fire control systems will be

necessary for some of the pipeline facilities, the liquefaction plant, the carriers and the southern terminal.

Great care will be taken to instrument the carriers and collect data on all aspects of its operation. Data acquisition of land, ocean surface and subsea data will be required to support operations and determine the effects, if any, of the Project on the environment.

## Satellite Communications and Navigation

In the remote and harsh environment of the North, satellites play an important role in communications and navigation. The APP will be an important user of many of the satellites proposed in the Canadian space program.

First, and of primary importance, is the requirement for both voice and data communications amongst Bridport Inlet on Melville Island, the carriers and the southern based Project head office. A number of satellites are important potential candidates.

M-SAT would be very useful for automated remote data collection of environmental data. In the case of an emergency, stranded research parties could be located by a search and rescue satellite.

Northern navigation will be greatly aided by the use of positioning satellites.

## Earth-based Communications and Navigation

In addition to satellites, earth-based equipment will be required for communications and navigation. An LNG carrier-based system will be required for ice detection and collision avoidance. Surface-based and subsea navigational aids will be required to assist the carrier in docking at both the northern and southern terminals. The carriers will require an integrated navigational system capable of accepting inputs from a variety of navigational subsystems.

Low-power «Arcticized» radio transmitter/receivers will be necessary in both portable and unattended configurations. These will be used for voice and data transmission during environmental and sea mammal data collection. They will also be required for controlling line block valves in case of a pipeline rupture. Emergency locator transmitters will be required to ensure safety of research crews.

An all weather airport will be located at Bridport Inlet. This airport will require a variety of instrumentation and navigational aids including Radar, ILS, VHS and communications.

## Engineering, Modelling and Design

As I mentioned in the first section of this paper, the LNG carrier will employ improved designs for the hull and a propulsion system capable of operating at full power in extreme ice conditions.

Considerable work remains to be completed as regards heat transfer modelling. We also hope to increase our understanding of ice mechanics and improve our ability to forecast ice movements.

A variety of engineering challenges remain such as subsea power systems as well as searching for technically and economically feasible alternate energy scenarios for northern communities.

There will also be a need for a simulator suitable for training carrier crews. There is an apparent need for improved cartography methodology.

## Ice/Cold Ocean Engineering

Because the APP will operate year-round in Arctic waters, a great deal of ice and cold ocean engineering activity will be required. We will need to conduct additional research on ice characteristics and properties.

We have carried out a considerable number of ice model tests on the LNG carrier but need to do more testing as we refine the design. We will also want to use model testing for operational scenarios such as docking. The knowledge gained during this study may be useful in the development of a training simulator for ship crews. Speaking of modelling, we will be using simulation in many areas such as the vessel engine control system.

We will need to conduct coastal engineering for Bridport Inlet to ensure that this excellent harbour is optimally utilized and maintained. Naturally, we will require engineering for docking and harbour facilities at both Bridport Inlet and the southern terminal.

In order to choose the best route for the LNG carriers, it is desirable to be able to predict the creation and movement of difficult ice formations. Accordingly, ice modelling and ice movement forecasting should be useful for strategic carrier routing.

Besides being able to forecast difficult ice situations, we must also be able to employ ice management techniques to minimize the inconvenience of ice effects.

## Technology Development and Transfer

New technology requirements for the project have and will be further identified as engineering progresses. These requirements will be communicated to potential Canadian suppliers through briefings such as these and through regional business opportunity seminars. After Canadian capability has been determined, a decision will be made as to whether the technology should be developed in Canada or whether foreign technology must be utilized. If foreign technology is required, Canadian firms will be sought as suitable repositories for the technology. In this case the contract will be placed with the Canadian firm or the joint venture in which that firm participates.

## Technology Development

The Project is committed to spend $220 million dollars (1981 funds) on research and development over a twenty-year period. Although we cannot yet be absolutely certain of the breakdown of these expenditures, we expect that approximately $20 million dollars will be used for R & D capital expenditures and approximately $80 million dollars for technology and prototype development.

The allocation of substantial funds such as these is indicative of the Project's important role in supporting the development of technology to meet the specific needs mentioned earlier.

# SECTION III — BENEFITS

## Benefits to Canada

I would now like to conclude by providing a broad overview of the benefits to be derived from the Project.

The Project's estimated capital costs of $2.5 billion dollars, as spent, will ultimately result in a cumulative effect on the economy of more than $15 billion dollars. In addition to the economic benefits these expenditures will create, the Project will generate significant benefits in several other areas:

**1. Technology**

(a) The Project will pioneer and develop technological expertise in Arctic LNG production and transportation systems, especially where the inaccessability and small size of gas reserves makes a portable system appropriate.

(b) The Project will play an important role in utilizing many satellite systems for remote sensing and communications. In addition, the APP will be users of a variety of earth-based communications, navigation, data acquisition safety and control systems.

(c) The Project represents a chance to attain supremacy in the design and operation of icebreaking vessels, vital factors in a northern country such as Canada.

(d) The technology acquired and developed by Canadian firms will allow them to exploit export opportunities. Experience on this and other demanding projects will ehance their ability to complete effectively in world markets.

(e) The Project will provide an economically viable opportunity to adapt and refine conventional engineering, construction and operational procedures to Arctic conditions.

**2. Access to Arctic Resources**

(a) The Project will add several trillion cubic feet of natural gas to Canada's supply by demonstrating that the discovered Arctic gas can be brought to market economically.

(b) The APP will provide impetus for further development of the substantial resources in the extreme north such as oil and minerals by proving the viability of a year-round transportation system in the Arctic.

(c) The APP will provide a much needed cash flow for the search for hydrocarbons in the Arctic, and will be positive proof of deliverability, thereby encouraging greater exploration.

**3. Employment Opportunities**

The Project will create significant national employment opportunities and result in extensive investments which will in turn generate more effects. These employment opportunities will exist across the country with significant concentrations in the Arctic and East Coast areas, both of which are regions of relatively high unemployment.

**4. Knowledge**

The Project will add considerable knowledge in the fields of climatology, bathymetry, meteorology, ice management, communications, wildlife management and Canada's northern environment generally.

**5. Sovereignty**

We feel that the Arctic Pilot Project will provide substance to the Canadian claims of sovereignty in the high Arctic by sheer physical presence. In addition, we would establish technological sovereignty in the Arctic by moving to a position of number one in the world with respect to icebreaking knowledge and capability.

## Employment Benefits

The employment benefits to be derived from the APP are extensive and geographically wide spread. During the construction phase approximately 8700 equivalent person-years of employment will occur in Canada. The operations phase will require about 325 direct jobs, resulting in an induced employment effect on the Canadian economy of about 2,000 jobs for each of the nominal 20 years of operation. These jobs would likely not exist without this Project.

## Net Monetary Benefit

The net social benefit of the Project will vary between $1.4 billion and $6.5 billion (all in 1981 dollars, discounted at inflation plus 10 % per annum). The $1.5 billion value would apply if the economy were fully employed, whereas the $6.5 billion value applies more exactly to todays relatively high unemployment situation.

## Expenditure Effects

In 1981 dollars the direct plus induced expenditure effects will exceed $15 billion. More important than the magnitude of the estimate, however, is the broad and fairly even spread amongst various geographic regions.

## Total Cash Flow

Over the twenty-year operating life, the APP will generate approximately $72 billion (as spent dollars) worth of gross revenues. Because of the equivalent volumetric sale of gas to customers in the United States, there will be an equivalent, positive effect on Canadian balance-of-payments.

Of the $72 billion of gross revenue, approximately $38 billion will flow to government through taxation and royalty schemes. This value excludes personal income taxes on salaries and wages.

## Conclusion

In conclusion, I believe that the Arctic Pilot Project is an exciting transportation development program of immense significance. It is an important formative element in the development of a technological infrastructure capable of supporting this and other resource projects. It will serve to position industry for the future by providing the technology and experience needed to exploit new resource project opportunities. The Project will also serve as an important vehicle for forging links amongst the resource, manufacturing and service sectors of the economy.

We look forward to working closely with industry and government in jointly addressing the challenges ahead to the benefit of all parties.

*Cold Regions Science and Technology*, 7 (1983) 273—283
Elsevier Science Publishers B.V., Amsterdam — Printed in The Netherlands

# Submarine Transportation of Hydrocarbons from the Arctic

*Lawrence R. Jacobsen and James J. Murphy*

General Dynamics Corporation
Quincy

## Abstract

*A submarine tanker system is presented as a technically feasible, economically viable transportation system to move oil and gas from the Arctic to markets in Western Europe and North America. Proven by over two decades of nuclear submarine operations in the Arctic, submarines can safely and reliably transport cargo under ice over virtually any proposed shipping route in the Arctic, with assurance of year-round delivery at uniform, predictable schedule intervals.*

*To establish the technical feasibility of commercial cargo submarines, baseline designs for a 181,400 dwt submarine oil tanker and a 140,000 m³ submarine LNG tanker have been developed, using either nuclear or conventionally fueled propulsion plants.*

*The submarine oil tanker design is based on a contract design developed in the 1960's, which has been refined to comply with recent oil pollution regulations. The submarine LNG tanker is an adaptation of that basic design to accommodate cryogenic cargo containment and transfer systems. Approximate characteristics of both designs have been identified, including ship dimensions, cargo system arrangement, propulsion machinery, and operational capabilities. Both submarines are designed to load cargo at a submerged terminal, and to offload cargo in the surfaced mode.*

*Approximate transportation costs are provided for both the oil and LNG submarine tanker systems.*

## SUBMARINE TRANSPORTATION IN THE ARCTIC

### The Arctic Ice Cover

The presence of the Arctic ice cover is a compelling reason for consideration of a submarine transportation system for development of Arctic hydrocarbon resources. Figure 1 shows the extent of the average ice cover and ice concentrations occurring in April and September, representing the least and heaviest coverage, respectively. Even under conditions of least ice coverage, the waters of the Beaufort Sea and around the Arctic Islands typically have 0.8-1.0 coverage.[1] The maximum seasonal average thickness of the ice in these areas is 3.5 to 4 meters.[2] Movement of the ice creates pressure ridges with drafts of between 18 to 24 meters in summer, and 21 to 30 meters in winter, which typically cover 13 to 18 percent of the ice area, according to ice profile data obtained by nuclear submarines. From U.S. aerial reconnaissance observations, pressure ridges occur at a frequency of between 300 and 1,200 every 48 kilometers.[3] Recent Canadian stereo photographic data of the Beaufort Sea ice cover showed the average frequency of pressure ridges to vary between 7.8 per km in February to 17.5 per km in May.[4] The location of these ridges constantly varies, according to the movement of the ice. It is an inescapable fact that, during any season of the year, ice at least 2 meters thick will be encountered, and pressure ridges will be present in any areas of heavy ice concentration.

0165-232X/83/$03.00  

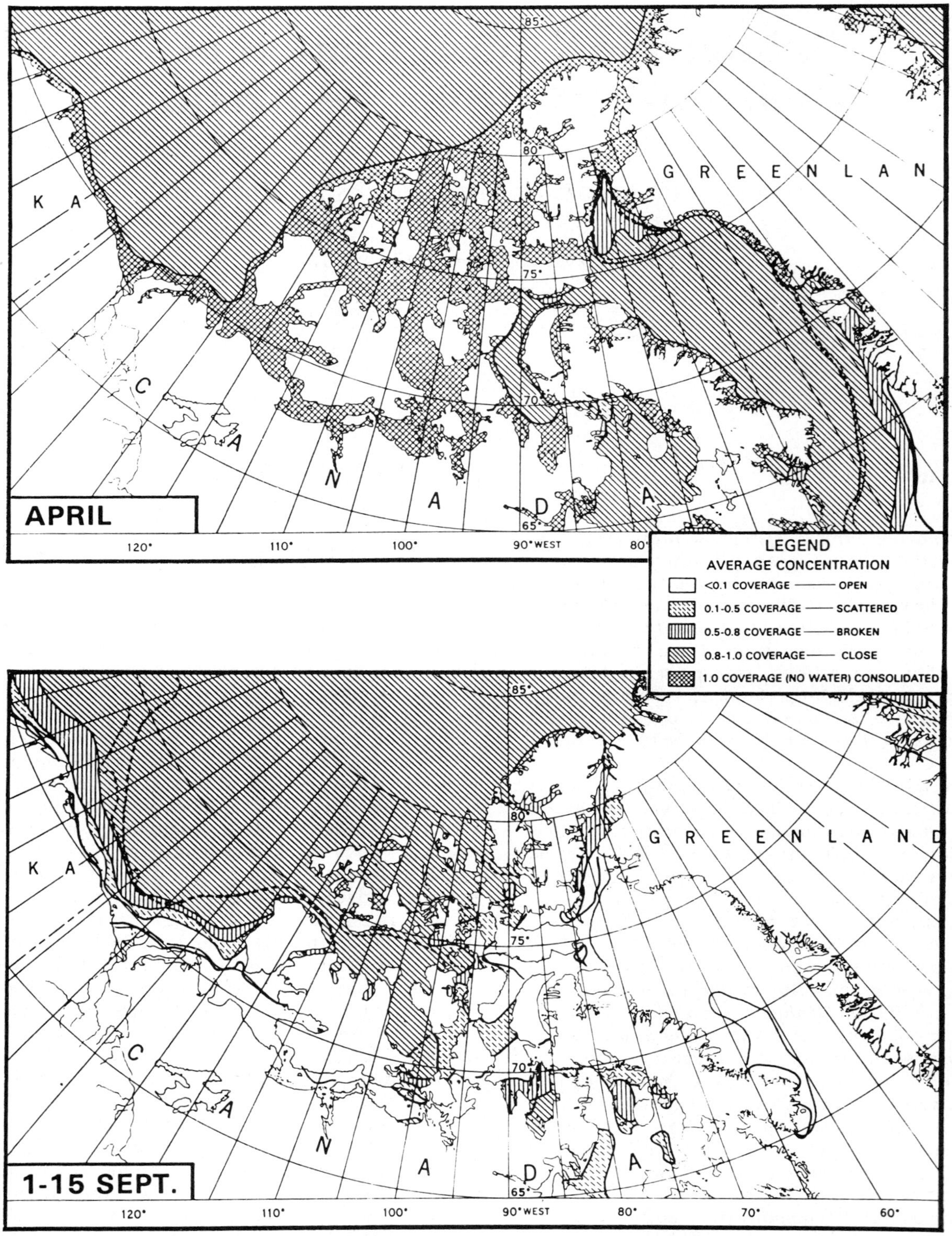


FIGURE 1. *Average Ice Cover and Concentrations, April and September*[1]

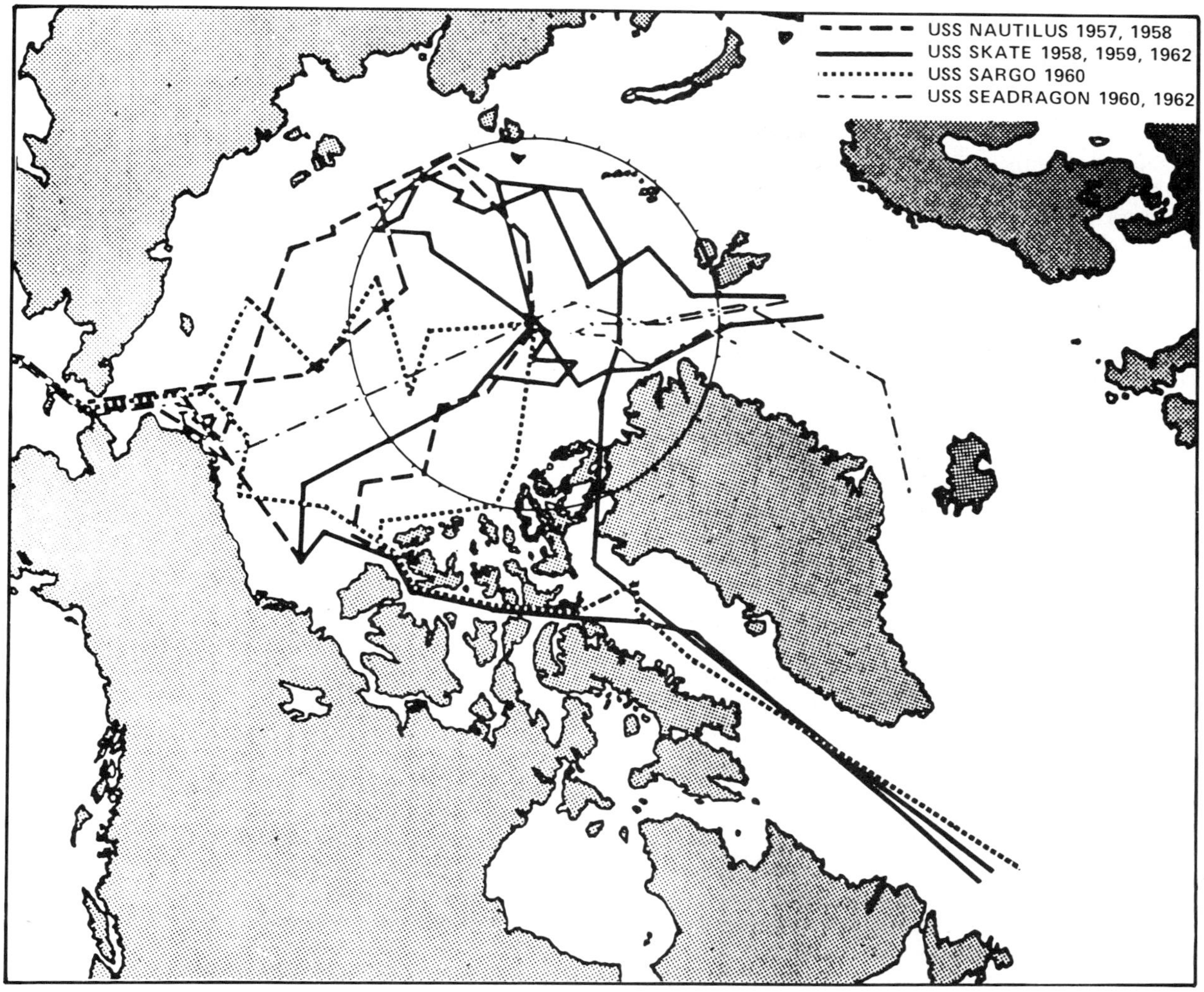


FIGURE 2. *Nuclear Submarine Operations in the Arctic — 1957—1962.*

## Advantages of a Submarine System

In the face of the Arctic ice barrier, a submarine transportation system offers significant advantages. First and foremost, nuclear submarine operational experience in the Arctic has established that submarines can safely conduct extended submerged voyages under the ice in virtually any area of the Arctic regardless of seasonal ice conditions. As shown in figure 2, the initial, highly publicized exploratory voyages of the *Nautilus*, *Skate*, *Sargo* and *Seadragon* opened up undersea lanes over most of the shipping routes proposed for transportation of hydrocarbons to North America and Europe. Second, the submarine has an inherent safety advantage in that the ship and its cargo are not exposed to the hazards of surface ice and weather conditions. Third, an undersea transportation system capable of avoiding the ice barrier rather than breaking through it poses the least threat to the fragile Arctic ecology. Fourth, and most important in terms of economic viability, a submarine system offers reliable all-season delivery of constant cargo baseloads at uniform, predictable schedule intervals.

With these advantages, a submarine transportation system is a string candidate for safe, reliable transportation of hydrocarbons from the Arctic islands and offshore areas.

## Cargo Submarine Technology Development

Although cargo submarines have not previously been built or operated for commercial service, the use of submarines as cargo carriers is not a novel concept entirely without precedent. The first cargo submarine was built in Germany during World War I with private capital. This submarine, the *Deutschland*, entered service in 1916, and made several voyages to the United States, returning with cargoes of rubber, nickel, and tin. During World War II, military submarines were used by the German, British, United States, and Japanese navies to carry oil and other supplies in areas where surface transport was too vulnerable.

This experience established the usefulness of submarines as covert cargo vessels, although not on a commercially economical basis. Not until the advent of nuclear power were submarines capable of the submerged endurance necessary for sustained voyages under ice. Following the success of nuclear submarine operations in the Arctic, discoveries of major oil and gas reserves in the Alaskan and Canadian Arctic spurred a series of studies of the potential for a submarine Arctic transportation system, which pointed to the submarine as a technically and economically viable transportation alternative to icebreaking tankers and pipeline systems.[5,6] General Dynamics has had a long-standing interest in the development of a commercial Arctic submarine transportation system, and in the late 1960's proposed contract designs for a family of nuclear-powered submarine oil tankers ranging from 170,000 dwt to 300,000 dwt as the basis for a technically and economically feasible transportation system for Arctic oil.[7] Recently, General Dynamics has also taken the lead in design development of a submarine tanker capable of transporting LNG cargo, in response to the current active interest in development of the vast known reserves of Arctic natural gas. In this effort, a non-nuclear propulsion system using existing technology has been proposed as a technically feasible alternative to nuclear power for sustained submerged voyages of up to 6,400 nautical miles (11,850 km).[8]

In short, the technology is currently available for development of an Arctic commercial submarine transportation system economically competitive with other transportation alternatives.

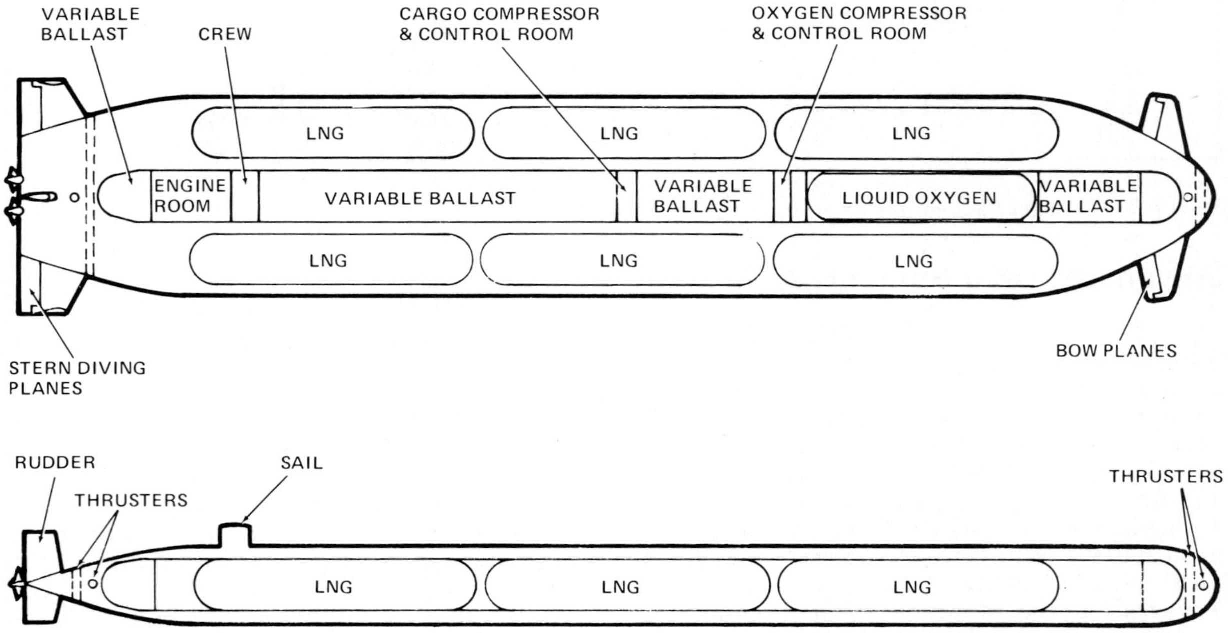


FIGURE 3. *Baseline 140,000 m³ Submarine LNG Tanker Design (Non-Nuclear Propulsion).*

# BASELINE SUBMARINE TANKER DESIGN

The objectives of General Dynamics' submarine tanker design development effort have been to: (1) provide a submersible vehicle capable of safe and reliable transport of cargo and crew in the Arctic environment; (2) minimize developmental risk by the use of existing proven technology; and (3) establish with regulatory agencies the technical basis for acceptance of a cargo submarine for commercial service.

This paper presents the basic features of General Dynamics' submarine and oil tanker designs at the current stage of development. The initial submarine LNG tanker concept described in a paper presented at the Gastech '81 LNG/LPG Conference and Exhibition has been refined to ensure safe containment of LNG cargo and, for the non-nuclear propulsion plant, safe containment of liquid oxygen used to support combustion. The submarine oil tanker design initially developed in the late 1960's has also been revised to meet recent Arctic pollution requirements, and has been reexamined in view of current regulations for oil cargo ships.

Figures 3 and 4 show the basic configuration and arrangement features; Table 1 gives the principal ship characteristics.

Ship size and cargo capacities have been predicated on cargo lift and voyage profile considerations for a large-scale development

**Table 1. Baseline Ship Characteristics**

*140,000 m³ Submarine LNG Tanker*

| | Non-Nuclear | Nuclear |
|---|---|---|
| Length Overall | 1470 ft (448 m) | 1290 ft (393 m) |
| Beam | 243.7 ft (74.3 m) | 243.7 ft (74.3 m) |
| Depth | 94 ft (28.6 m) | 94 ft (28.6 m) |
| Shaft Horsepower | 50,000 | 75,000 |
| Speed | 12.3 knots | 14.8 knots |
| Operating Depth | 600 ft (183 m) | 600 ft (183 m) |
| Submerged Displacement | 727,400 L.T. (739,038 M.T.) | 660,800 L.T. (671,373 M.T.) |

*181,400 dwt Submarine Oil Tanker**

| | Non-Nuclear | Nuclear |
|---|---|---|
| Length Overall | 1020 ft (311 m) | 1020 ft (311 m) |
| Beam | 170 ft (51.8 m) | 170 ft (51.8 m) |
| Depth | 90 ft (27.4 m) | 90 ft (27.4 m) |
| Shaft Horsepower | 50,000 | 75,000 |
| Speed | 14 knots | 16 knots |
| Operating Depth | 600 ft (183 m) | 600 ft (183 m) |
| Submerged Displacement | 370,960 L.T. (376,895 M.T.) | 370,960 L.T. (376,895 M.T.) |

* Cargo capacity for non-nuclear submarine oil tanker will be less than 181,400 dwt, depending on the size of the fuel oil and liquid oxygen tanks.

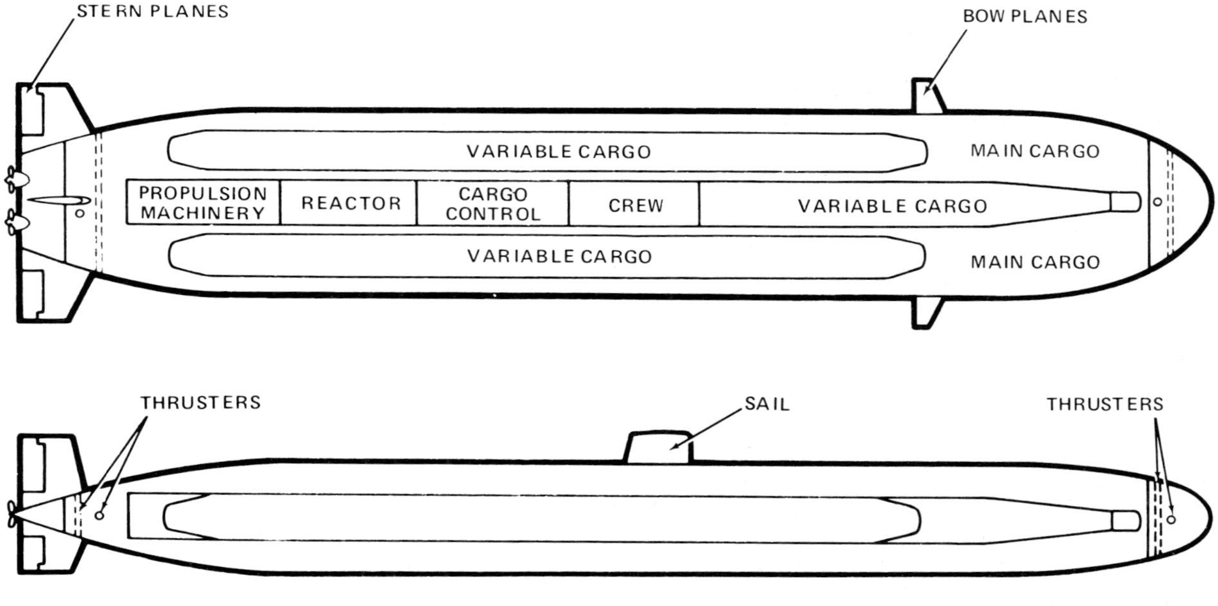


FIGURE 4. *Baseline 181,400 DWT Submarine Oil Tanker Design (Nuclear Propulsion).*

project involving a 6,400 nautical mile (11,850 km) round-trip submerged voyage over one of the longest projected shipping routes. However, the baseline design features would apply equally to a range of cargo submarine sizes optimized for various cargo lift and route distance requirements.

For efficient carriage of cargo with minimal draft, a rectangular outer hull has replaced the body of revolution hull shape typical of modern

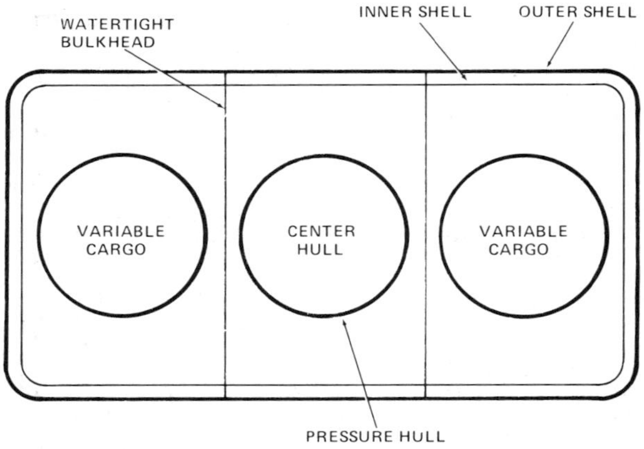


FIGURE 5. *Typical Cross Section, Submarine Oil Tanker.*

nuclear submarines. The outer hull for the submarine oil tanker is double-walled, in order to satisfy regulatory requirements for pollution control. (See figure 5.) The non-pressure outer hull structure of both the submarine oil and LNG tankers is designed to resist hydrostatic, hydrodynamic, and ice impact loading.

The outer hull encloses cylindrical pressure hulls containing cargo, crew, and machinery. Cargo containment hulls are arranged on either side of a center hull that contains crew spaces, the cargo control room(s), and propulsion machinery. These pressure hulls are structurally designed to withstand full collapse depth sea pressure. The location of crew spaces and vital ship systems in the center hull offers maximum protection in the event of a collision. Within the center hull, crew and control spaces are further protected by watertight bulkheads designed to withstand collapse depth sea pressure.

# SHIP SYSTEMS

## Main Ballast System

The ship's main ballast tanks, which are flooded with sea water to submerge the ship and are emptied to surface the ship, are located in the

FIGURE 6.

space between the inner cylindrical pressure hulls and the rectangular outer hull. The ballast tanks are subdivided by watertight bulkheads in order to maintain sufficient reserve buoyancy to surface the ship if a number of tanks should be ruptured while the ship is submerged. If the ship must be brought rapidly to the surface in an emergency, the ballast tanks are blown with a dedicated supply of high-pressure air. The ship has sufficient buoyancy and topside structural reinforcement to permit emergency surfacing through ice up to 12 feet thick. The emergency ballast tank blow system ensures that the ship can always come to the surface when required.

### Ship Control

The ship has twin screws for reliable under-ice operation. The control surfaces (rudder and planes) are sized for positive ship control through the full range of operational speeds. Thrusters at the bow and stern provide maneuverability during docking. A sail on the main deck houses the periscopes and antennas, and has a bridge for conning the ship on the surface. Hydrodynamic tests conducted on a model of General Dynamics submarine oil tanker design (Figure 6) confirmed the ship's control characteristics.[9]

Redundant hydraulic systems, emergency operation of hull closures and control surfaces by an independent stored energy supply, and emergency ship control systems have been incorporated in accordance with standard submarine design practice to ensure the safety of the ship and crew under emergency conditions.

### Navigation

Based on operational experience, the submarine's navigation systems have been specifically selected for safe transit under ice, for precise course-keeping, and for docking at a submerged terminal. Navigation equipment includes inertial and dead reckoning navigation systems, omega radio navigation signal reception capability via retractable floating wire antenna, doppler sonar, search and profile sonar, long-range collision avoidance sonar, depth sensors, and acoustic arrays for reception of accoustic signals from bottom-mounted transponders.

### Atmosphere Control Systems

Two independent and isolated ventilation and air revitalization systems provide a safe, contaminant-free atmosphere for the crew. One system serves the crew compartment, which contains ship control, navigation, and habitability spaces; the other serves the engine room and cargo control spaces. The crew compartment is thereby effectively isolated from potentially hazardous contaminants originating in machinery or cargo spaces.

The main supply of oxygen for air revitalization is from onboard flasks; backup oxygen is obtained by burning sodium chlorate candles in chlorate furnaces located in the engine room and crew compartment. Both sources are adequate to support 27 days of submerged operation, per ABS requirements.

Chemical absorbent beds and blowers in the crew compartment and engine room maintain carbon dioxide levels at or below ABS requirements. For backup carbon dioxide removal, the chemical absorbants, lithium hydroxide and barralyme, are spread out on flat surfaces.

Emergency air breathing equipment with a dedicated air supply is located in all areas of the ship for use in emergency situations.

## PROPULSION SYSTEMS

Either nuclear or non-nuclear steam propulsion can be used for the baseline ships; both plants use existing technology or components, and can be adapted to hull diameters between 50 feet (15.2 m) and 64 feet (19.5 m). Both plants are designed with sufficient backup propulsion capability and component redundancy to ensure safe, reliable under-ice operation.

### Nuclear Propulsion

A steam generator supplies steam via two independent piping loops to two cross-compound propulsion turbines and four turbine-generators sized to meet all ship's electrical load requirements with reserve capacity. Twin shafts and screws provide the added element of reliability appropriate to a commercial submarine operating under ice and conducting frequent under-ice docking maneuvers.

The reactor plant has double containment; that is, the reactor core, steam generator, coolant loop, and main coolant pumps are contained in a separate pressure vessel within the pressure hull. Double containment, together with the application of proven radiation shielding techniques, provides the margin of safety necessary for commercial operation.

Because both propulsion shafts are driven by a single power source, secondary electric propulsion motors are connected to the reduction gearing for each shaft to provide emergency propulsion. These motors are driven either by a diesel generator during surface operation, or by large lead acid batteries when the ship is submerged. Both the diesel generator and battery systems are sized to give the submarine a specified propulsion endurance, to support nuclear plant start-up operations, and to meet critical habitability, communication, and ship operational requirements. An on-line DC-AC motor generator system gives the submarine an uninterrupted power supply for critical electrical loads in the event of loss of power from the turbine-generators.

## Non-Nuclear Propulsion System

The non-nuclear steam propulsion plant has three independent supercharged boilers that are fired with methane or distillate fuel oil. In the submarine LNG tanker, methane fuel is drawn from the cargo tanks; in the submarine oil tanker, a separate distillate fuel oil tank is provided.

While the ship is submerged, the boilers operate with recirculating exhaust gases injected with oxygen gas obtained from an onboard liquid oxygen tank. When the ship is surfaced, turbine-driven superchargers allow the boilers to use air for combustion. Since the three boilers provide independent sources of steam for the propulsion turbines and turbine generators, there is no need for secondary propulsion motors. However, diesel generator and battery systems are provided for in-port or emergency electric power generation.

The liquid oxygen used as oxidizer for the non-nuclear plant is stored in an insulated tank within a separate compartment in the center pressure hull, where it is effectively isolated from crew and machinery spaces by structural bulkheads and ballast water. Within the tank compartment, the liquid oxygen tank is surrounded by an inert atmosphere. The tank itself is internally pressurized to permit operation of the liquid oxygen pumps, which, for safety reasons, are located outside the tank in the compressor room. Oxygen is supplied to the engine room boilers in gaseous form, pressurized to 6 atmospheres absolute. Oxygen gas is run outside the pressure hull in pressureproof piping in order to segregate the oxygen from potential contact with combustible materials or gases. Double-walled inerted piping is used in the engine room and in any other areas of the ship where effective isolation of oxygen piping is required to ensure safety.

For the non-nuclear submarine oil tanker, the addition of fuel oil and liquid oxygen tanks will reduce cargo capacity in proportion to the size of the tanks required for the selected voyage profile.

# CARGO SYSTEMS

The foregoing discussion has described the basic design features of a submersible vessel for safe and reliable transportation of hydrocarbon cargo in the Arctic environment. The basic configuration of the ship is essentially similar for either LNG or oil cargo; there are, however, major differences in the cargo systems.

## LNG Cargo System

In the submarine LNG tanker, LNG is carried in six identical cylindrical cargo tanks, each enclosed by a pressure hull. These cargo containment hulls are arranged in groups of three on either side of the center pressure hull. When the ship is submerged, the cargo hulls are surrounded by sea water ballast. Within the cargo hulls, the space between hull and tank is filled with inert gas.

Because the cargo tanks are not ballasted with sea water, there are variable ballast tanks in the center hull, which contain sufficient additional sea water to submerge the ship when the cargo tanks are empty.

The cargo system is essentially similar to the proven system used on General Dynamics' surface LNG tankers. The cargo tanks are insulated for a design boiloff rate of 0.2 percent per day. Internal cargo tank piping is contained within a pipe tower with an access dome at the top and two submerged cargo pumps at the bottom. From each dome, cargo piping is run within pressureproof tunnels to a central loading/unloading station. To enable both surfaced and submerged cargo transfer, the cargo piping connects to a common manifold leading to access trunks at the top and bottom of the center pressure hull. The lower trunk is designed to interface with a submerged terminal; the upper trunk has piping connections for cargo transfer at a surface terminal.

For the nuclear tanker, LNG boiloff from the cargo tanks is reliquefied and returned to the tanks; on the non-nuclear tanker, normal boiloff is burned as fuel. In both cases, a backup system is provided to compress excess boiloff for overboard discharge at submergence depth.

For carriage of LNG cargo, the submarine has an inherent safety advantage, in that it operates in a seawater environment that does not support combustion of methane gas. The lack of access to the atmosphere, however, poses design requirements unique to a submarine tanker in order to ensure safe containment of LNG cargo under all operational conditions. To meet these requirements, the cargo system incorporates the following design features:

- The cargo tanks are effectively isolated within independent pressure hulls and are surrounded by an inert atmosphere.
- Cargo system piping is run within pressure-proof piping tunnels outside the center pressure hull compartments containing crew and machinery.
- All cargo system piping is run within an inert atmosphere except in the cargo compressor room.
- All spaces subject to potential methane gas leakage are enclosed by gastight boundaries.
- Methane gas leakage can be evacuated and discharged overboard at ambient sea pressure.
- The cargo tanks are designed to permit under-ice venting on loss of propulsion power.

## Oil Cargo System

In the submarine oil tanker, cargo is carried in two types of tanks: (1) main cargo tanks between the inner cylindrical pressure hulls and the double-walled outer hull, and (2) pressure-resistant variable cargo tanks at the forward end and on either side of the center pressure hull. When the cargo is offloaded, the main cargo tanks are flooded with sea water to compensate for the weight of the cargo. The variable cargo tanks carry only oil, and are filled as required to compensate for variations in cargo density. Three main ballast tank groups, located forward, aft, and amidships, carry only clean sea water ballast; these tanks are flooded to submerge the ship, and are emptied as required to surface the ship.

The shipboard cargo handling system for the submarine oil tanker is self-compensating; that is, the ship's cargo tanks are always full of oil or full of water or some combination of the two. This system enables the tanker to be loaded or offloaded at a submerged terminal facility. During the loading operation, cargo oil is pumped from the terminal into the main cargo tanks, forcing ballast water from the bottom of the tanks. When the oil/water interface approaches the discharge strum, loading is slowed and the ballast water is passed into an expansion tank to allow separation to take place. The ballast water emptied from the cargo tanks is pumped to a separation facility at the loading terminal.

Unlike a surface tanker, the submarine tanker's main cargo tanks are not exposed to air, which eliminates the presence of combustible volatile gases in the tanks and greatly reduces the potential for the formation of oil clingage and sludge. Any sludge which may collect on the tank surfaces will be removed by a washing system provided in accordance with IMCO regulations.

An expansion tank accommodates variations in cargo volume caused by temperature differences occurring in route; sea water is admitted to or discharged from the tank to compensate for changes in cargo volume. The tank is sized to accommodate the maximum variation in cargo volume, with an adequate margin to prevent discharge of oil cargo to sea.

An inert gas system in accordance with IMCO regulations is provided for all cargo tanks.

Like the submarine LNG tanker, the oil tanker cargo system is completely isolated from crew and machinery spaces and from other ship systems. The main ballast tanks which are emptied directly to sea contain only clean sea water. Compensating sea water ballast used in the cargo tanks is discharged only to a shore facility, never directly to sea.

## Suitability for Other Hydrocarbon Cargo

The successful adaptation of the baseline submarine tanker for cargoes as diverse as oil and LNG demonstrates that the basic vehicle is equally capable of transporting other types of hydrocarbon cargo, including methanol, MBTE, distillate, and LPG.

## REGULATORY REQUIREMENTS

To ensure that the design features for ship, crew, and cargo safety satisfy basic regulatory requirements, the submarine LNG tanker design has been exposed to regulatory agencies in technical briefings with the U.S. Coast Guard and the American Bureau of Shipping. As a result, the technical groundwork has been established for development of regulatory requirements applicable to LNG cargo submarines in those areas where existing regulations for surface ships do not apply or must be modified. Similar briefings were held during the initial design development of General Dynamics' submarine oil tanker; since then, however, significant changes in the regulations for oil cargo carriers have been made. The submarine oil tanker design has therefore been reviewed to identify areas where reexamination by regulatory agencies is required. Recent Arctic pollution regulations stipulate that an oil cargo carrier have an inner and outer skin. To meet this requirement, a double-walled shell has been provided for the entire outer hull of the submarine oil tanker. Another area of regulatory concern is the potential for overboard discharge of oil. Since the ship's main ballast system used to submerge or surface the ship is completely independent of the cargo ballast system, and since all cargo tank ballast water is pumped to a shore facility during loading, the submarine tanker cargo system meets the intent of the regulations that impose strict controls and limits on the quantity of oil discharged overboard at sea. As in the case of the submarine LNG tanker, there are areas where specific requirements applicable to a submarine rather than a surface tanker must be developed and accepted by regulatory agencies.

In the event that licensing problems constrain the application of nuclear power to a commercial cargo submarine, a technically feasible non-nuclear plant has been proposed. In the recent technical briefings with the U.S. Coast Guard and the American Bureau of Shipping, safe containment of liquid oxygen and the isolation of oxygen piping from the fuel system and from other areas of the ship were addressed as the major safety issues. The solutions to these problems incorporated in the baseline LNG submarine design were considered a technically acceptable approach.

Since regulatory requirements are essentially concerned with the safety of the ship and crew and safe containment of cargo, it should be emphasized that the submarine has a significant safety advantage for transportation of hydrocarbons in the Arctic. The ship and its cargo are not continuously exposed to the hazards of ice impact damage in transit through the ice cover; rather, the submarine travels in a stable undersea environment ideally suited for safe navigation and ship control.

## CONCLUSIONS

Three main conclusions can be drawn from this examination of the potential for a submarine transportation system for development of Arctic hydrocarbon resources:

- The presence of the Arctic ice cover favors an undersea transportation system.
- The technology is available for commercial development of cargo submarines for safe and reliable transportation of hydrocarbon cargo.
- Submarine LNG and oil tanker design has been brought to the point where further development can proceed at minimal technical risk, with confidence that regulatory requirements can be met.

## References


1. *Oceanographic Atlas of the Polar Seas, Part II - Arctic* U.S. Navy Hydrographic Office, Washington, D.C.: 1958 (Reprinted 1968).
2. Markham, W. E., *Ice Atlas - Canadian Arctic Waterways.* Minister of Supply and Services Canada: 1981.
3. Wittmann, W. I. and J. J. Schule, Jr., «Comments on the Mass Budget of Arctic Pack Ice», *Proceedinqs of the Symposium on the Arctic Heat Budget and Atmospheric Circulation,* Memorandum RM-5233-NSF., The Rand Corporation, Santa Monica, California: December 1966.
4. Wright, B. D., D. L. Schnab, and D. McGongal,

«Ice Conditions affecting Navigation in the Beaufort Sea.» Presented at SNAME Spring Meeting, Ottawa, Ontario, June 17-19, 1981.

5. *Maritime Submarine Transportation System Conceptual Design, Study* Office of Commerical Development, U.S. Maritime Administration: January 1975.
6. *National Petroleum Reserve - Alaska: Marine Transportation System Analysis.* Office of Maritime Technology, U.S. Maritime Administration: October 1980.
7. Jacobsen, L. R., «Subsea Transport of Arctic Oil - A Technical and Economic Evaluation.» Presented at the Offshore Techonology Conference, Dallas, Texas, 1971.
8. Veliotis, P. T. and S. Reitz, *A Submarine LNG Tanker Concept for the Arctic.* Revised 1 October 1981. Presented at the Gastech '81 LNG/LPG Conference and Exhibition, Hamburg, Germany, 20-23 October 1981.
9. Kohl, R. and A. Goodman, *Experimental Determination of Stability and Control Derivatives of a Proposed Submarine Oil Tanker from Captive Model Tests.* Hydronautics, Inc. Technical Report 7034-1: May 1970.

*Cold Regions Science and Technology*, 7 (1983) 285—291
Elsevier Science Publishers B.V., Amsterdam — Printed in The Netherlands

# Thyssen-Waas Icebreaker Concept
## Model Tests and Full Scale Trials

*H. Wilckens, A. Freitas*

Thyssen Nordseewerke GmbH
Emden

## Abstract

*An icebreaker hull form based on a fundamentally new concept has been developed by Thyssen Nordseewerke in cooperation with the Hamburg Ship Model Basen (HSVA). The concept, originated by Dr. H. Waas, employs a novel icebreaking mode which leads to a considerable reduction in ice resistance and ice loads on hull and propeller. The ice is broken in a defined manner, initially by shearing followed by bending fracture. The broken floes are displaced in such a manner that the propeller area remains free of ice.*

*The new icebreaker forebody has been developed within an extensive research program embracing systematic model tests in ice and open waters and full scale trials with an experimental icebreaker. The objective of the research work is the application of the concept to various marine operations.*

*The research program was carried out in three stages: the definition phase, the optimization phase and the full scale trials. The first phase concerned the determination of the required form characteristics and parameters which were then optimized in the second phase.*

*The experimental vessel used for the full scale trials is the converted icebreaker MAX WALDECK (55 m length, 3100 HP). A foreship section according to the Thyssen-Waas concept was fitted onto the existing hull.*

*The trials were made in the Gulf of Bothnia during March/ April 1981 and 1982. The program encompassed test runs in level ice, in brash and mush ice and through ridges. The trials confirmed the model basin tests in all respects.*

*The modified MAX WALDECK demonstrated impressive performance under all ice conditions. The vessel broke ice upto a thickness of 90 cm in the continuous mode; the power take-off was about one third that required by conventional icebreakers of comparable size. Ridges with a total depth of about 8 m were negotiated in only two runs.*

*The low power requirements, the effectiveness in irregular ice formations and the reduced loads on hull and propeller makes the Thyssen-Waas concept a technical and economical viable system for marine operations in arctic regions.*

## Introduction

The hull shape of existing icebreaking vessels have been evolved from the classical ship form. Though various improvements have been made to reduce ice resistance, to withstand ice pressure and to keep the ice away from the propellers, the basic shape of the sea-going vessel has been maintained. The conventional bow acts as a wedge resulting in the crushing of the ice and high friction losses.

The Waas concept takes a fundamentally new approach to the problem of breaking ice efficiently. The concept, which is principally a new hull configuration, was developed by studying the mechanics of ice fracture and the kinetics of the broken ice.

Thyssen Nordseewerke has been working for several years on a research and development program to use the Waas concept for marine operations in ice covered waters. The work has been done in collaboration with the Hamburg

  

Ship Model Basin (HSVA) and has been sponsored by the German Ministry for Research and Technology.

The remarkable performance of the new concept as established in the model tests was incentive enough to try out the system at full scale under real conditions. The icebreaker MAX WALDECK was converted accordingly and tested in the ice fields of the Gulf of Bothnia.

Both model and full scale tests have proved the superior capabilities of this new icebreaker concept.

As the Waas concept has been published often, the description here is limited to its main features.

The forebody has a pontoon-shaped bow with a flat bottom and sharp edges at the sides which may be accentuated by fitting runners (Fig. 1). The ice is thereby loaded uniformly and broken mainly through shearing. The widest part of the ship is right forward at the shearing edges; thus a channel is created which is wider than the ship's mainbody.

Aft of the pontoon-shaped bow, the flat bottom gradually goes over to a V-shape. Here the broken floes are displaced to the sides so that no ice is ingested in the propeller area. If desired it is also possible to create an ice-free channel behind the ship.

The icebreaking mode is characterized by the following features:

- The ice is broken by shearing and cantilever bending. The ultimate stress of these failure modes is lower than that by other forms of bending fracture.
- The ice is broken in large floes; the work done in breaking the ice is thus reduced.
- The movement of the ice floes under the ship is defined; tumbling and jamming of the ice pieces with corresponding energy losses are avoided.
- As the ship is widest at the shearing edges, there is no crushing of the ice which occurs at the shoulders of the conventional ship form.
- The channel between the ship's sides and the ice cover is free of ice rubble so that there are no friction losses here.
- The greater part of the broken ice is displaced to the sides under the ice cover. Thus friction between ice and the ship's hull and icemilling in the propellers are practically eliminated.

All these factors lead to a reduction in ice resistance by about 30 % and in power requirements by 50 % to 70 % depending upon the ice thickness.

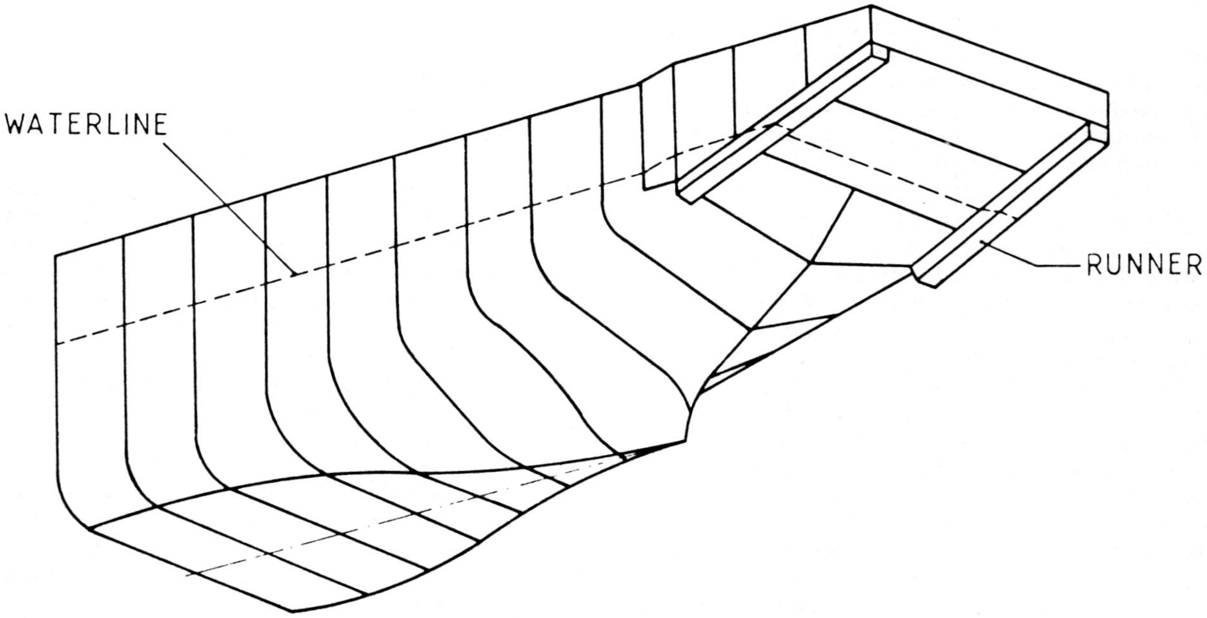


FIGURE 1. *Waas Icebreaker Forebody*

## Model Tests

In the definition phase of the research program the form characteristics and parameters influencing performance were determined in tank tests with models of following ship types:

— large polar tanker
— polar research vessel
— icebreaking pusher-tug.

The program encompassed tests in saline ice, smooth waters and in waves.

Resistance and propulsion tests were made in level ice of various thicknesses and strengths. Besides these there were also propulsion tests in brash ice and with pressure ridges. Fig. 2 shows the results of the propulsion tests in level ice of one of the tanker models (No. 2) in comparison to those of a corresponding model with conventional lines.

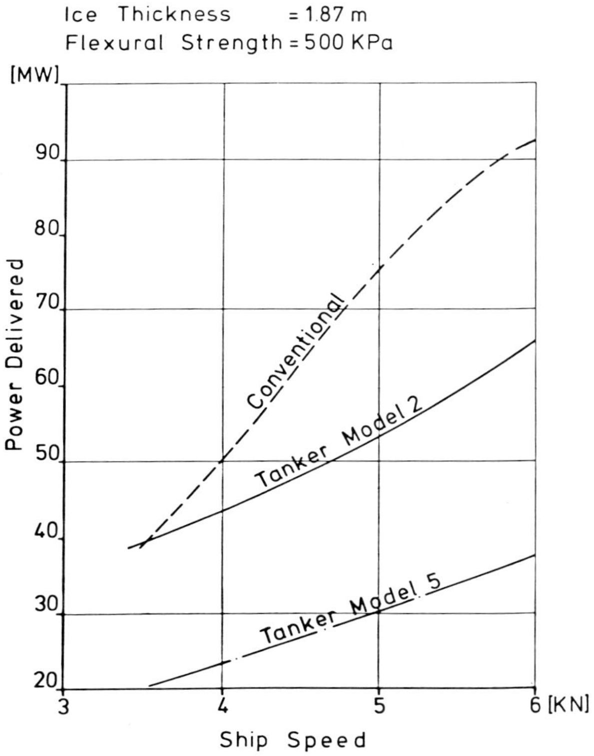


FIGURE 2. *Propulsion-Tests in Ice (Full Scale Values).*

The tests in open waters demonstrated that the Thyssen-Waas forebody form had no negative effect on the hydrodynamic performance. The speed/power relationship is comparable to that of normal sea-going vessels.

The tests in waves, however, showed that the new forebody has very good seakeeping characteristics. The roll and pitch amplitudes as well as the vertical accelerations at the foreship were lower than those of conventional ships of similar dimensions. This is due to the dampening effect of the increased water-plane area forward and the sharp edges of the bow.

In the second phase of the development program the Thyssen-Waas forebody was optimized with respect to performance both in ice and in open waters. The work during this phase was concentrated mainly on the polar tanker project.

Three further variants of the tanker forebody were tested. With the best of these, Model No. 5, a further improvement in performance of about 40 % over Model No. 2 of Phase 1 was achieved (see Fig. 2).

## Full Scale Trials

The full scale tests were performed with a converted conventional icebreaker. A bow section according to the Thyssen-Waas concept was fitted onto the forebody of the icebreaker MAX WALDECK owned by the German Ministry of Transport.

The MAX WALDECK had prior to modification a conventional bow with circular shaped frames. She has a single screw and a 3100 HP diesel-electric propulsion plant. The following are the dimensions of the vessel after conversion:

| | |
|---|---|
| Length (o.a.) | 55.34 m |
| Beam (midship) | 11.90 m |
| Beam (bow) | 13.10 m |
| Draft | 4.95 m |
| Displacement | 1295 ton |

The shape of the added foreship section was determined in tanks tests with a model of the original vessel and two model variants with the Thyssen-Waas forebody. The results of the tests showed that at full power an increase in speed by almost 100 % was achieved with the better of the two forebody variants over the model of the original ship.

The modified MAX WALDECK was w.r.t. the hull form far from ideal. The transition from the almost rectangular frames of the new forebody to the circular frames of the existing vessel had to be made over a short length resulting in a step between the new and old hull. Further due to trim problems, the forebody beyond the forward perpendicular was cut short,

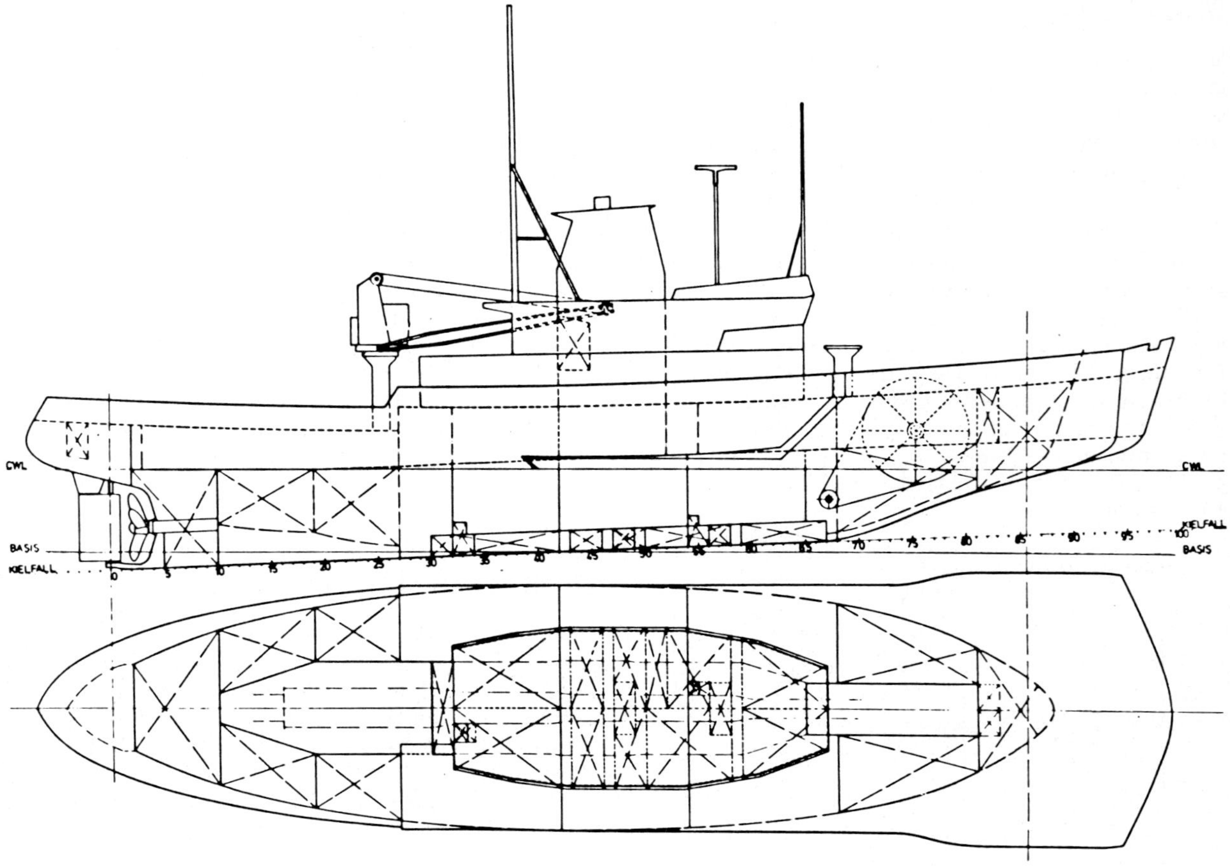


FIGURE 3. *General Arrangement Sketch of the modified MAX WALDECK*

resulting in a very blunt bow (see Fig. 3). The MAX WALDECK is therefore not a prototype of the Thyssen-Waas icebreaker form. Inspite of the compromises made, the performance of the icebreaker was quite remarkable.

The first expedition took place in spring 1981 in the archipelago off Vaasa, Finland. The thickness of the level ice varied from 20 cm to 50 cm with snow cover between 5 cm and 30 cm. The test runs were made at different speeds. The highest speed measured was 6.2 kn in 40 cm thick ice at 1450 kW power level. The MAX WALDECK also broke fast frozen rafted ice 90 cm thick in the continuous mode. A fire on board put a premature end to the expedition.

After repairs a second expedition was started in March 1982 in the same area. The level ice was quite uniform in this year, about 65 cm on the average with very little snow. The flexural strength was however, lower: 350 kPa as against 500 kPa in 1981. The maximum speed measured was 6.7 kn at 2050 kW.

The new forebody was also quite effective in brash and mush ice. Test runs were made in the channel leading to Vaasa, the depth of ice mass being 1.0 m to 5.0 m. As in the case of fast ice, the brash ice was displaced under the onmoving bow so that even large ice floes were negotiated without any impact.

Breaking pressure ridges presented no problems to the MAX WALDECK. Pressure ridges with a keel depth of 7.0 m were broken in only two runs. The first run was started 100 m before the ridge crest; the second run in the broken channel, was made over a length of 50 m. Fig. 4 shows the profile of one of the larger ridges broken.

Fig. 5 shows the MAX WALDECK in action.

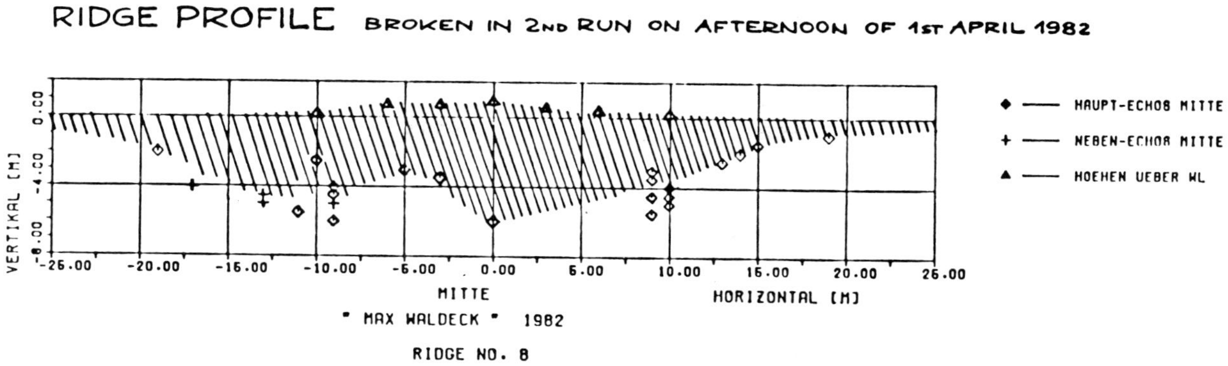


FIGURE 4. *Profile of a Pressure Ridge*

FIGURE 5. *Experimental Icebreaker MAX WALDECK*

FIGURE 7. *Manoeuvre Test*

FIGURE 6. *Channel broken by the MAX WALDECK*

FIGURE 8. *Widening the channel*

The typical channel with the clean cut edges can be seen here as also in Fig. 6. At lower speeds it is possible to create a completely ice-free channel.

The manoeuvrability of the icebreaker is demonstrated in Figure 7. Turning circles with diameter of 8 ship lengths were measured in 70 cm thick level ice. Figures 8 and 9 show the MAX WALDECK while widening an existing channel. This is a manoeuvre that icebreaker with a conventional bow cannot easily perform.

Figure 10 shows the icebreaker passing through the crest of a ridge.

FIGURE 9. *Widening an existing channel*

FIGURE 10. *Breaking through a pressure ridge*

Fig. 11 gives a comparison between the performance of the MAX WALDECK and the WERDERTOR. The WERDERTOR is a 10500 HP offshore supply vessel with a icebreaker forebody. The expedition with WERDERTOR took place in 1977 near Spitzbergen. The dimensions of the two vessels as well as the ice conditions in which they were tested are similar

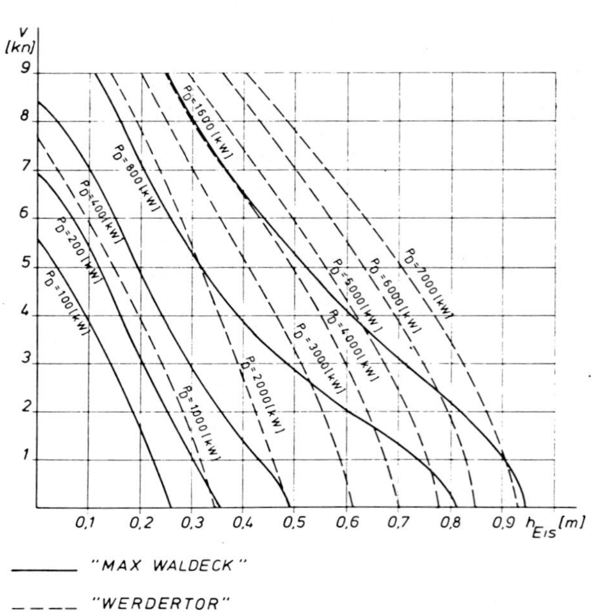


FIGURE 11. *MAX WALDECK/WERDERTOR Performance Comparison*

enough to justify direct comparison of the results of the two expeditions. It can be seen from the diagram that the power requirement by the MAX WALDECK is around 65 % — 75 % lower than that needed by the WERDERTOR.

The MAX WALDECK returned from both expeditions without any damage to the hull or the propeller. Subsequent dockings showed that the paint (standard anti-fouling), with the exception of the shearing edges, was still on. This is an indication this new icebreaking mode reduces loads on the hull to an appreciable extent.

## Conclusion

The development of arctic navigation has been hindered by two significant factors:

- Due to the high power requirement and fuel consumption, the economical feasibility of arctic marine transport systems could so far not be guaranteed.
- The heavy loads on hull and propeller and the possibility of getting stuck in heavy ice has made arctic navigation a hazardous undertaking.

The Thyssen-Waas icebreaker concept could offer, as demonstrated in both model and full scale tests, a solution to these problems. The low power needed, the reduced loads on the ship, the ability to negotiate all kinds of ice formations and the good manoeuvrability make this new concept a technically and economically viable system for ice navigation.

---

## References


(1) Waas, H.: «Arctic Merchant Vessels of the Second Generation». Yearbook of the Schiffbautechnische Gesellschaft, Vol. 72, 1978.

(2) Schwarz, J.; Freitas, A.; Weiss, F.: «Ice breaking Tests in Model and Full Scale». Status report «Developments in Marine Technology», 1981 of the German Ministry of Research and Technology.

(3) Freitas, A; Wilckens, H.: «A Novel Icebreaker Concept». INTERMARITEC 80, Hamburg.

(4) Freitas, A.: «Experimental Investigation of a Novel Icebreaker Concept». ICETECH 81, Ottawa.

(5) Schwarz, J.; Hoffmann, L.: «Icebreaking in Model and Full Scale». Yearbook of the Schiffbautechnische Gesellschaft, Vol. 72, 1978.

*Cold Regions Science and Technology*, 7 (1983) 293—295
Elsevier Science Publishers B.V., Amsterdam — Printed in The Netherlands

*A short note to section III:*
*Technological/economic aspect of exploration/exploitation*

# Special Hazards in Open Waters at High Latitudes

*Torkild Carstens*

The Norwegian Hydrodynamic Laboratories
Trondheim

Long before reaching the high arctic with its icepack and associated design and operational problems, marine transportation and other off-

FIGURE 1.

FIGURE 2.

FIGURE 3.

shore activities encounter the special hazards of open waters at high latitudes. The worst threat to fishermen, whalers and sealers for centuries, superstructure icing, must now be faced by the oil industry as it moves into northern waters. Figs 1 and 2 show ship icing, while Fig 3 is a photo of a semisubmersible in the Gulf of Alaska after an icing event. The load added by the ice is less likely to impair the stability of a rig and cause it to capsize as it does to ships. What the ice most definitely does, however, is to interfere with operations. Fig 4 is a close-up where we can see that clogged air intakes have already been de-iced, while other deck equipment is allowed to remain inoperational.

FIGURE 4.

The pictures bring home the point that virtually all exposed surfaces are covered and all exposed equipment is knocked out for a shorter or longer time. Today's primitive de-icing technology makes sure that this time will be long rather than short. The things to the left in Fig 4 may not be essential to either economy or safety, but the life raft to the right does not meet the most elementary requirements to emergency equipment.

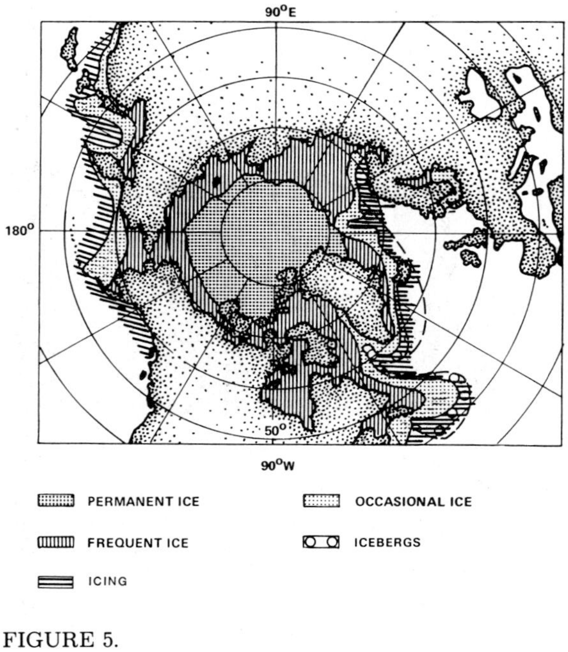

FIGURE 5.

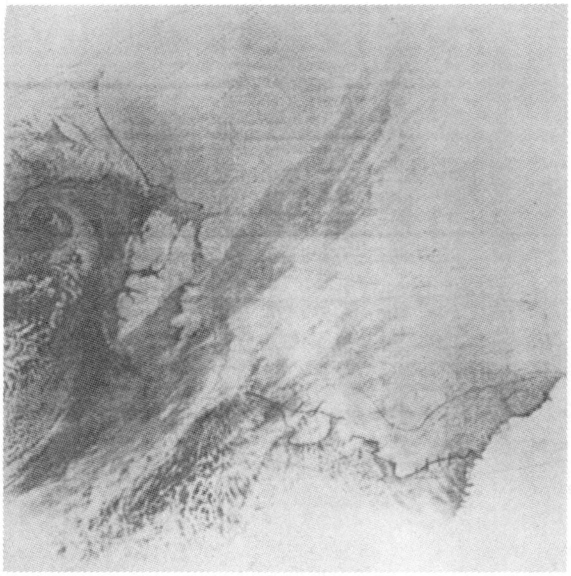

FIGURE 6.

Fig 5 shows the areas in which icing occurs: the open waters south of the ice edge, or outside cold continents. The icing events are outbreaks of cold air from N, NW or NE, and the icing-prone belt extends southwards for 1000 or 2000 km until the air is heated by the warm ocean to temperatures above freezing.

Embedded in such massive outbreaks of cold air one sometimes finds the other major hazard of high latitude open waters: the polar lows. These arctic storms are small but intense cyclones that are generated in the destabilized cold air heated from below. They may appear close to the ice edge as the storm west of Spitzbergen in Fig 6, in mid-ocean as the low near Bear Island in Fig 7, or on the exposed coasts as the low off the Lofoten Islands in Fig 8. Even the North Sea may be raged by an occasional polar low.

In many respects these arctic cyclones are smaller relatives of the tropical cyclones. The wind speed may reach 60, 80 or 100 knots and the air temperature may be −10°, −20° or −30°C. In a quadrant to the right of the edge, when facing the direction of storm movement, there are trapped waves 5, 10 og 15 m high. Visibility may be 100 m, 50 m or even lower.

Because of their origin in remote areas with few meteorological stations these storms usually remain undetected, arrive suddenly and without warning. Nice weather may turn into full storm, complete with high winds and waves, in 10 minutes.

FIGURE 7.

We have recently, in the «Ocean Ranger» disaster, been reminded of what a sudden storm can do. That storm was probably not a polar low, but it was not predicted.

My plea is for an upgrading of the research needed to meet the problems of open waters at

FIGURE 8.

high latitudes. Rather than sit and wait for more disasters to trigger such efforts, we should give high priority now to research and development for

a) anti-icing and de-icing technology,

b) monitoring and forecasting of polar lows.

# Section IV

## *Environmental and Social Impact*

*Cold Regions Science and Technology*, 7 (1983) 299—302
Elsevier Science Publishers B.V., Amsterdam — Printed in The Netherlands

# The Impact of Industrial Development on the Arctic Environment

*Geoffrey Larminie*

British Petroleum Company
London

— This paper does not represent the petroleum industry's viewpoint; it is based on the author's personal experience and involvement with the one existing major petroleum development in the non-Communist sector of the Arctic, namely the discovery and development of the Prudhoe Bay oil field and the Trans-Alaska pipeline. It is salutary to reflect that, to date, virtually all the available technical know-how and operational experience of large scale oil and/or gas production operations in the Arctic comes from Alaska and the USSR. As we do not as yet have the environmental equivalent of Art Meyerhoff*, we have virtually no information on impacts from the Soviet Union and almost all our scientific and technical data on the environmental aspects of development in high latitudes comes from Canada, Scandinavia and Alaska. One final point: most of this data relates to onshore production operations; to date there are no large offshore fields in production in the Arctic Ocean, and the oil industry's practical experience of operations in these waters is derived from its exploration work. So far there have been no serious oil spillages associated with the exploration drilling so that contingency plans for dealing with such incidents under or adjacent to ice are largely hypothetical. However, that is not to say that these plans are without foundation for we have some experience from incidents in cold waters in high latitudes (e.g. northern Norway), and a number of major co-operative research projects between industry and government are in progress of which the Baffin Island Oil Spill Project is the best example.

* See Meyerhoff's paper; «Energy Resources of the Soviet Artic» in section II

When the Prudhoe oil field was discovered and the decision to transport the oil by pipeline from Prudhoe to Valdez was announced the conservationists predicted the direct consequences. The oil field would destroy the last great wilderness and the pipeline would result in 800 miles of devastation extending from the caribou breeding grounds of the Arctic Coastal Plain in the North to the rich fishing grounds of the Valdez Arm and Prince William Sound in the South. The construction crews would destroy hundreds of spawning streams in the Interior; the hot oil would melt the permafrost, degrade the tundra and destabilize the hillslopes; tankers and terminal operations would spill oil into the Valdez Arm and its approaches — in short, the Alaskan wilderness was doomed.

In the event, the development of the Prudhoe Field was subject to serious delays consequent upon the need to satisfy stringent conditions set down by the U.S.Government in response to a powerful, albeit emotional, environmental lobby, which itself was a unique (both in time and place) reflection of a sudden world-wide upsurge of concern for conservation of the environment. The matter was further complicated by a Federal injunction against any land leasing pending settlement of the Alaska Native Land Claims issue. Although the participants in the Prudhoe oil field and the Alyeska Pipeline Company were poised to start development in 1969 — and all 800 miles of 48 inch diameter pipe for the pipeline had been delivered to Alaska by October 1971 — it was not until April 1974

0165-232X/83/$03.00  

that construction started. All through the period of delay work continued on further refining the design, compiling base-line data on the environment, additional testing of the pipe itself, detailed soil studies along the route, and much else. It is estimated that these pre-construction costs of the Trans Alaska Pipeline alone amounted to 100 million dollars, and Alyeska have estimated the total delay costs at over two billion dollars.

It is fair to say that by and large the prophets of doom have been confounded. The prediction by some naturalists that the oil field developments at Prudhoe would force the caribou to abandon the area was wrong and studies by the U.S. Fish and Wildlife Services confirm that this has not happened. Valdez Arm and its approaches have not been polluted by oil and the terminal has a very good operational record best exemplified by operations during 1981 when 735 tankers were loaded with almost 23 billion gallons of oil and only 8 gallons were spilled. However, it must be admitted that there have been several incidents on land of which those in the Atigun Pass, and North of Fairbanks were the most serious. At Atigun thaw resulted in unstable ground conditions which disturbed a buried section of the line set in a concrete «coffin» (a construction technique imposed on the owners) and as a result the pipe ruptured spilling an estimated 1500 gallons before the leak was detected. The spill was cleaned up, the line was repaired, and the site successfully restored. The spill near Fairbanks was the result of sabotage in 1978 when a hole was blown in the line and ca. 550,000 gallons of oil was spilled. The resulting massive clean-up operation proved highly effective, grass was grown, the site recolonized and the alders have now grown up to the mid-line of the pipe and there are birds everywhere. Thus it has been demonstrated that the consequences of the inevitable accidents are at worst short-term and, provided there are first class contingency plans in place backed up by trained manpower and all the necessary resources of equipment and expertise in site rehabilitation, there is no inherent reason to predicate that the exploitation of oil and gas in the Arctic will result in irreversible damage to the environment. Hand in hand with the developments in high latitudes, coupled with the consequent stimulus to extensive scientific study of the ecosystem, has come the realisation that the tundra is more robust than many erstwhile experts predicted. In this context it is interesting to note that in the past eighteen months the author has attended a number of major meetings and workshops (none of which were sponsored by the oil industry) without once hearing the epithet «fragile» applied to the tundra — yet this word seemed to be on everybody's lips in the ten years following the discovery of oil at Prudhoe.

New developments on land will essentially follow the pattern pioneered in Arctic Alaska for the Prudhoe and Kuparuk oil fields and the Trans-Alaska pipeline and the major potential impacts associated with exploration and production are now pretty well appreciated. Geoglogical surveys, seismic surveys, gravel extraction for pad construction (and later for road and site construction if hydrocarbons are discovered in commercial quantities), exploration drilling, production drilling, construction of production facilities, are all conducted in the context of a proper appreciation of the consequent stresses on the ecosystem. The techniques of remote sensing coupled with geobotanical mapping and appropriately designed ground-surveys which were developed from the Tundra Biome sector of the International Biological Programme (part of which was based on Prudhoe) are particularly applicable to the layout and siting of the surface facilities associated with the discovery of oil and gas onshore in the Arctic, and of course this also applies to the onshore installations which may be required in connection with an offshore discovery.

The industry is now looking to the search for and production of hydrocarbons offshore in the Arctic and others in this proceedings have described the techniques of exploration drilling through artifically thickened ice and from gravel islands in considerable detail. The problems of producing oil from offshore fields in the Arctic (particularly in the deeper waters seawards of the land-fast ice) are considerable and many of the problems have been identified and data-collection and/or research directed towards solving them is now in hand.

What sort of information is required by an operator in the Arctic Ocean? The answer would include, but not necessarily be limited to, the following main headings:

— ice and ice conditions
— meteorology
— oceanography
— coastal geomorphology
— biological data
— archaeology
— anthropological and sociological data on the indigenous human population

There seems to be much unnecessary and unproductive argument current as to who should be responsible for providing this sort of base-line data which is an accepted pre-requisite for environmental impact assessments. Governments are accused of paying inadequate attention to their northern marches, of devoting insufficient resources for the scientific study of the Arctic, tantamount to neglect of their stewardship; oil companies are under pressure to provide and pay for all the missing information and all too quickly polarisation sets in and the end result is an unproductive confrontation. In the opinion of the author both parties have responsibilities and the most efficient way to discharge these is by co-operation. The BIOS (Baffin Island Oil Spill Study) previously referred to is a good example of what can be done in a specific field, and in the broader field of environmental base-line studies the EAMES Programme (Eastern Arctic Marine Environmental Study) is an outstanding example.

EAMES began in 1977 in response to oil industry interest in exploring for hydrocarbons in the eastern Canadian Arctic for which the Minister for Indian Affairs and Northern Development required a comprehensive environmental assessment before drilling would be permitted. In 1981 Indian and Northern Affairs and the oil industry participants published a booklet entitled «The Achievements of an Environmental Programme» describing the project, and the scientific results were published in a series of reports, and the whole of the March 1982 issue of Arctic (vol 35 No 1) was devoted to the results from the northern sector (Baffin Bay — Lancaster Sound).

The organisation and objectives of the programme are set out in the 1981 booklet which states that «scientists mainly from the Federal Departments of Fisheries and Oceans, the Environment, and Indian Affairs and Northern Development co-operated with industry and local communities to develop a program of environmental field studies. The purpose of the program was to investigate the offshore and nearshore waters and the marine life of this biologically-important area, to obtain sufficient data to prepare Environmental Impact Statements (EIS)». These documents would be submitted for review by panels established by the Federal Environmental Assessment Review Office (FEARO). Once under way the programme was managed by a joint federal/territorial government and industry committee under the chairmanship of the Department of Indian Affairs and Northern Development. In addition an EAMES Advisory Board was set up comprising representatives of the eastern Arctic Communities and non- government specialists to liaise with northern peoples and to provide an independent assessment of the scientific work. All concerned with the future orderly development of Arctic resources would be well advised to familiarise themselves with this project which must rank as a very important precedent for environmental impact assessment studies concerned with development anywhere in the Arctic whether by government or industry.

To work in the Arctic involves a continuous learning process, and the lessons derived from field and operational experience must be applied and widely disseminated (this is why the rapid publication of the EAMES results is such a landmark).

As base-line data, impact assessments and associated studies accumulate some common denominators will be established to which the site-specific components can be added for individual developments. Above all it is important to accept that an environmental impact assessment is not a penalty (regretably there is still a corpus of opinion which regards it as an unwarranted imposition), but an important management tool. By itself it is not enough and must be accompanied by a well designed monitoring programme during construction and the early stages of operation, whereby one can make a match between performance and prediction. Experience in monitoring North Sea production operations (e.g. the Forties Field and the Sullom Voe Terminal in Shetland) have shown that effective low-cost monitoring can be achieved both offshore and in coastal waters. As time goes on, the objective of responsible operators in all industries, not just in the oil industry, must be to make the environmental element as endemic in management philosophy as safety, so that such tools as impact assessment, monitoring, performance review and audit, and so on, are the rule rather than the exception. All of which involves an appreciation of the fact that man is a part of not apart from the ecosystem and as a potential modifer must take cognizance of his role in the flow of energy and nutrients which sustain the total system. That being so I believe it is quite within our capacity to undertake major development projects in an environmentally safe and acceptable manner in the Arctic and be justly proud of the results.

In conclusion, it is worth reiterating and

expanding on a point touched on above. All who work in the Arctic, or indeed anywhere in the world, have a responsibility to share the environmental component of the knowledge of how things should or should not be done, and it is in the interests of us all that this environmental expertise should never knowingly be withheld on spurious grounds of confidentiality. It is non-competitive information and it is a measure of the importance of this concept that so much information has been given so freely at this conference. Long may it continue.

*Cold Regions Science and Technology*, 7 (1983) 303—312
Elsevier Science Publishers B.V., Amsterdam — Printed in The Netherlands

# Beaufort Sea Energy Production and Environmental Protection

*Richard A. W. Hoos*

Dome Petroleum Limited
Calgary

## Abstract

*Exploration permits in the Beaufort Sea-Mackenzie Delta Region were issued to the petroleum industry by the Federal Government beginning in 1965. The purpose of these permits was to encourage exploration for hydrocarbons in this region. To retain the permits, industry was obliged to carry out a certain minimum level of exploration-related activity. Since that time, almost $ 2 billion has been spent on exploration. Drilling results to date, and particularly the recent oil discoveries at Kopanoar, Tarsiut, Issungnak and Koakoak has led the major operators to commence the design of systems to extract the hydrocarbons and deliver them to market.*

*Permanent islands built of sand, concrete and steel in waters up to 200 feet deep will provide the offshore platforms for producing and processing the hydrocarbons. Initially oil will be of primary interest, with the plan being to transport it via Arctic icebreaker tankers and/or pipeline(s) to markets. As the demand for gas develops, it will also be transported to market, most likely by pipeline. A significant shorebase infrastructure complete with quarries, roads, harbours, airports, maintenance facilities and other services will be required to support the impending development.*

*All of the foregoing activities must be carried out in an environmentally and socially responsible manner. The proposed developments are presently being examined under the auspices of Canada's Federal Environmental Assessment and Review Process. This process is assisting by involving the public in the early identification of the major environmental and socio-economic concerns respecting proposed developments. The outcome of this effort, along with more specific reviews to be conducted by the appropriate regulatory agencies, should ensure that the final developments which take place will conform to the high environmental protection and socio-economic standards expected by all Canadians and the international community.*

## Background

Petroleum activity in the Beaufort Sea and Mackenzie Delta area began in the early 1960's with seismic exploration and the drilling of exploration wells by a number of operators. The prelude to offshore exploratory drilling occurred in 1971 when Imperial Oil Ltd. applied for permission to drill in shallow water from man-made islands. Approval was granted for the first such island, Immerk, in 1972, from which a well was drilled soon afterwards. Since then 18 man-made, or artificial islands have been built in waters off the Mackenzie Delta (Figure 1). Industry has gained valuable operating experience both on land in the Delta and offshore in the Beaufort Sea and more than 130 exploration wells have been drilled.

Dome, through its subsidiary Canadian Marine Drilling Ltd. («Canmar») began its exploratory drilling program in the Beaufort Sea during 1976 with three ice-reinforced drillships (Figure 2). A fourth drill ship was added in 1979. In addition to the drillships, the present drilling fleet includes eight supply vessels four of which

0165-232X/83/$03.00  

FIGURE 1. *Issungnak, site of one of the more recent offshore oil and gas discoveries, was built by Esso Resources Canada Limited. It is located about 30 kilometres north of the nearest Arctic coastline in 60 feet of water. Issungnak required the dredging of five million cubic metres of sand from the sea floor.*

are ice class, one Class 4 icebreaker, four ice class support tugs, an ice reinforced fuel tanker, an offshore crane barge, a floating drydock and six specialty barges.

FIGURE 2. *Three of the ice-reinforced drillships, several supply ships and at the top of the photo, the icebreaker Kigoriak, while moored in McKinley Bay during the winter of 1979-80.*

To increase the pace of offshore drilling activity, in 1980 Dome commenced the construction of exploration drilling islands in the offshore Beaufort at the Kaglulik and Tarsiut sites. The Tarsiut island, operated by Gulf, was completed in 1981 and drilling at this site is presently underway (Figure 3). Tarsiut is the world's first caisson retained island built in Arctic waters.

FIGURE 3. *Tarsiut, the first caisson-retained island in the Beaufort Sea, was completed during 1981. A successful delineation well was drilled from the island during the winter of 1981-82, and further wells are planned including one currently underway.*

This type of island construction represents a significant step forward in the design and construction of future Arctic offshore platforms. Unlike the more conventional sacrificial beach islands built in the shallower waters of the Beaufort, Tarsiut required much less gravel and sand to complete the structure. It was built with a more steeply sloped berm foundation, with a 1 in 5 slope, compared to the more conventional 1 in 15 for previous islands. In addition, because of the use of caissons to break through the water-air interface, the subsea berms end approximately 6 metres (20 ft.) below the waterline.

Tarsiut also serves as a research laboratory, having been equipped with more than a million dollars worth of instrumentation to measure ice and other forces throughout the life of the island. This preliminary experience, together with the research data generated, are being directed into the design of other exploration islands projected for the Uviluk and Koakoak areas, and will make a major contribution to the design of future production facilities.

To support the exploratory drilling effort, sizeable shorebases are operated by Dome, Esso and Gulf (being built) at Tuktoyaktuk. The base facilities include docks, staging areas, accommodation facilities, administration complexes, warehouses, light steel fabrication and machine shops.

The bases are serviced by air from the Tuktoyaktuk airport, and a small strip at the Esso base. These facilities accommodate the variety of aircraft including Boeing 737's, Hercules transporters, helicopters and STOL planes needed to transport personnel and equipment.

During the first three years of operation in the Beaufort, the drillships were anchored through the winter at natural harbour sites located at Herschel Island and at Cape Parry. This posed various operational difficulties and in 1979 permission was received to create a new winter anchorage at McKinley Bay on the Tuktoyaktuk Peninsula. A large cutter suction dredge known as the Aquarius was brought in, and by the end of the 1979 open water season, had dredged a navigation channel and basin in McKinley Bay (Figure 4).

FIGURE 4. *The cutter suction dredge Aquarius was brought to the Beaufort in 1979 to dredge an overwintering harbour at McKinley Bay. Since then it has been used to perform a variety of jobs including enlarging the McKinley Bay harbour, providing sand for the Tuktoyaktuk water reservoir, and to help with the building of Tarsiut.*

During 1980 and 1981 the McKinley Bay harbour was enlarged to accommodate the expanding drilling fleet. In the process, an island was built on the north side of the mooring basin. This island, which serves to protect the ships anchored in the basin, has grown to 63 hectares in size. Geotechnical studies conducted on the island have shown it to be very adequate as a foundation for future support base facilities and activities.

Plans are presently being developed to use the island as a major support base to service year-round exploration drilling and the initial future production development activities.

## Drilling Results to Date

The exploration effort has been very successful in corroborating the expectations of geologists who have long predicted that the Mackenzie Delta-Beaufort Sea area would be a major oil and gas producing province (Figure 5). On-land in the Delta, over 100 wells have been drilled and three major gas fields have been confirmed at Taglu, Parsons Lake and Niglintgak. Proven gas reserves are estimated at 6 trillion cubic feet. Offshore, the drillship operations have completed 15 wells and are credited with four of the offshore oil discoveries and two gas discoveries. Likewise oil has been discovered at several offshore island sites with the most promising locations being Issungnak in 19 metres of water and Adgo, in very shallow water.

Delineation drilling is required to establish the commerciality of the offshore discoveries. One successful delineation well was drilled in 1981 offsetting the Kopanoar discovery, and a successful delineation well was completed from the Tarsiut island during the winter of 1981-82. Two further delineation wells (1 from island, 1 by drillship) are presently being drilled at Tarsiut; delineation drilling of the Koakoak discovery is planned for 1983; and further drilling at Kopanoar for 1983 and 1984. The Tarsiut discovery is located in the shallowest water and therefore has the greatest potential for early production from the Beaufort.

Estimates for Beaufort hydrocarbon potential vary over a range from 6.9 to 34 billion barrels of oil and up to 60 trillion cubic feet of gas. However, the precise potential of the Beaufort should not be an issue at this stage of evaluation since only a few fields would be fully developed by the end of this century. The industry expects to confirm 4.5 to 7.5 billion barrels of oil in the Beaufort by 1990.

Consultants estimates for recoverable oil reserves at Kopanoar range from 270 million barrels to 1.8 billion barrels and at Koakoak from 300 million barrels to 2.0 billion barrels. Estimates have not been released for other discoveries.

## The Development Objective

The Beaufort Sea-Mackenzie Delta region has the potential to help Canada become self-sufficient in oil production within this decade and in the process could stimulate considerable economic and employment activity in Canada.

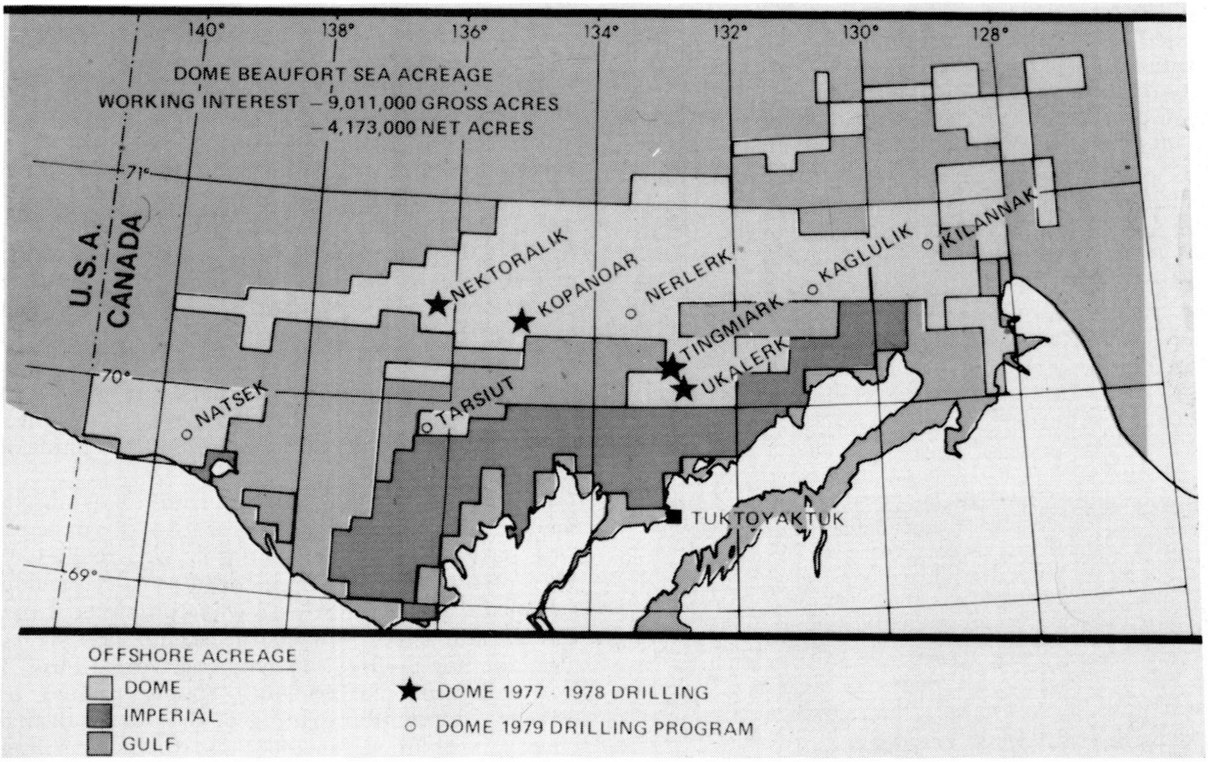


FIGURE 5. *Acreage holdings of Dome, Esso and Gulf in the offshore Beaufort Sea region.*

The objective of commencing oil production from the Beaufort Sea is consistent with the national interests and goals of all Canadians; oil self-sufficiency by 1990, regional benefits where possible and increased technological infrastructure to ensure employment opportunities. However, the total potential of the Beaufort Sea-Mackenzie Delta region for oil and Canadian benefits cannot be realized unless production is allowed to begin. Basic building blocks must be put in place now to achieve results in the future.

Recognizing the difficulty of forecasting demand, the long lead times required to develop the supply alternatives, and the potential of sudden interruptions in foreign supplies, the only prudent direction that the country can follow at this juncture is the development of all alternative energy sources. Once production has commenced from frontier areas such as the East Coast and the Arctic the government will be in a much better position to match supply and demand by controlling the pace of further development in these areas.

## Development Schedule

The development schedule leading to future production from the Beaufort Sea is dependent on numerous factors. They include: the time taken to delineate a discovery; water depth and location of the field; the transportation mode employed, particularly in the early stages of development; the complexity of the production scheme selected; and government approvals.

Shallow water developments (less than 25 m) will be simpler to bring on production than deep water developments (30-60 m). This is because much less dredged material will be required to build the necessary islands, and the islands themselves can be designed for lower levels of ice forces, since most of the larger ice features will be prevented from moving into this area due to draft considerations. Likewise, finding the oil in shallow zones (less than 2,000 m) will make such fields easier to develop than those with deeper reservoirs because the well pressures will be lower and drilling time will be reduced.

For this presentation, I will examine in fair detail the current plan for developing the Tarsiut field although similar plans have been developed for other fields such as Kopanoar, Koakoak and Issungnak. The Tarsiut discovery holds great promise for early production from the Beaufort because the water is relatively shallow, ranging from 17 — 22 metres deep, and the oil discovery zone is at roughly 1,500 metres. Furthermore, the Tarsiut structure (like most structures in the Beaufort) is relatively simple so only a few delineation wells will be required around each prospective island location to justify proceeding with development. As an example, Prudhoe Bay, 300 miles to the west, was evaluated with only 2 wells. The Tarsiut structure, illustrated in Figure 6, is approximately 20 miles long, by two miles wide. It indicates the potential ultimate development scheme with five development islands including one Artificial Production and Loading Atoll (APLA). The present island could be modified to accommodate an early production scheme.

Early production systems were first used in the North Sea where there was a long construction time for permanent production platforms. Wells drilled from semi-submersible rigs (frequently delineation wells) were completed with subsea wellheads and connected to a semi-submersible floating rig with flexible riser pipes. A storage tanker equipped with a production process system was moved to a single point mooring system and wells were placed on production. Tankers loaded directly at the storage tanker. Similar systems are being employed in other parts of the world including the Philippines (Figure 7).

At Tarsiut, by the end of 1982 there could conceivably be four or five wells capable of production and more wells could easily be drilled from the existing island. Thus, if a floating production and storage system was available along with suitable transportation, this island could theoretically be capable of producing in the summer of 1984. Recognizing this possibility, conceptual plans have been developed for an

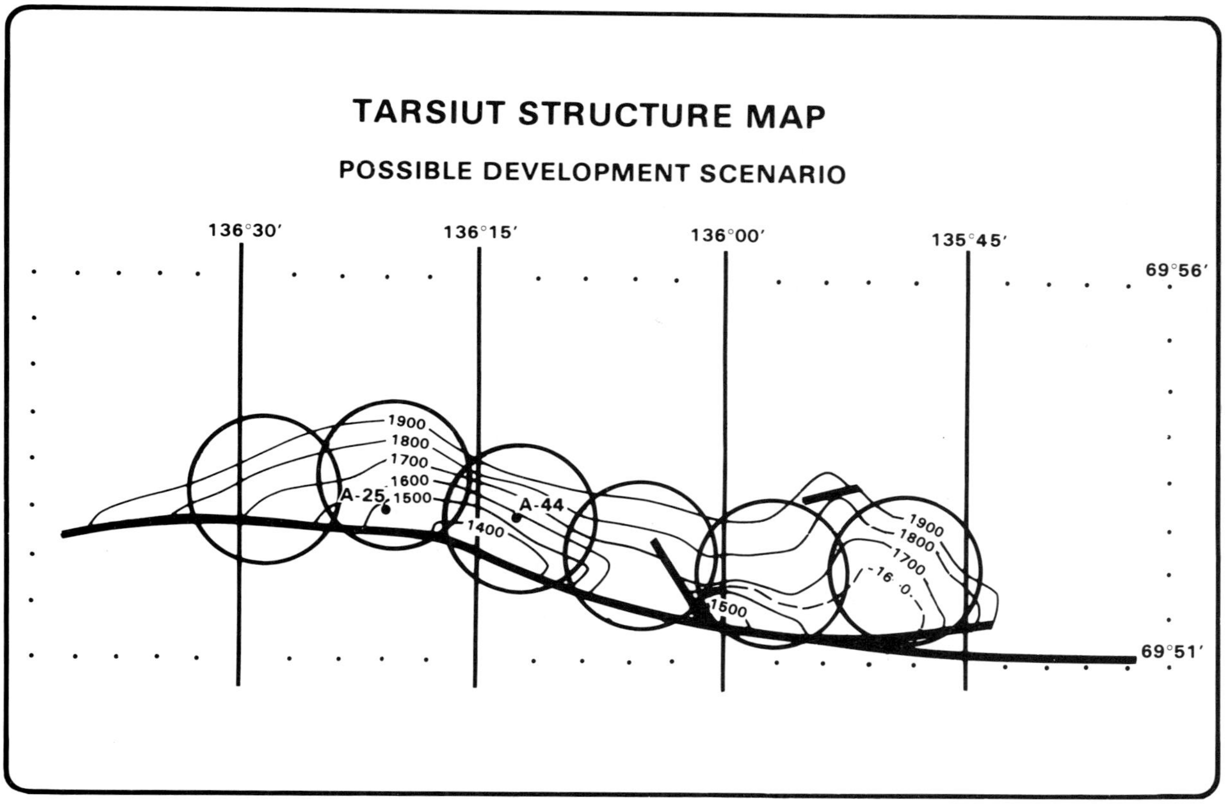


FIGURE 6. *The structure map for the Tarsiut field. The structure is 20 miles long by 2 miles wide and would require up to 5 development islands to efficiently produce the recoverable oil.*

early production system for Tarsiut. This system would be portable so after permanent systems had taken over at Tarsiut, it could be moved to another site such as Koakoak. It also provides a technique for bringing the Beaufort on production slowly.

FIGURE 7. *This photo shows one of the newest early production systems (EPS) presently in use in the Philippines. An EPS is a crude oil processing and handling system which allows the earliest possible delivery of oil, before the permanent production system comes on line. Subject to approval, a similar system using an ice-reinforced tanker, could be used in the Beaufort as early as 1985.*

Figure 8 shows how Tarsiut could be produced during open water and moderate ice conditions. The island would be connected to a single point mooring system located a few kilometres away. The processing and storage tanker would be similar to that used in the Philippines example except the vessel would be ice-strengthened and all of the facilities will be enclosed.

To convert the first island from an early production system to a full scale year-round operation, the island would be expanded over a period of approximately two years, and during the process, transformed into an Artificial Production and Loading Atoll (APLA) (Figure 9). The APLA provides a berth for the floating processing and

FIGURE 8. *Earliest production from the Beaufort could be carried out as illustrated in this artist's rendering. Oil from a drilling island (eg. Tarsiut) would be pipelined into deeper water where a marine riser would bring it into an EPS, much as the Phillippines example, for processing, prior to loading it into a tanker.*

A CONCEPTUAL DESIGN FOR TARSIUT ISLAND No.1

FIGURE 9. *The most recent artist's rendering of the Arctic Production and Loading Atoll (APLA) being designed for the Tarsiut field is illustrated here. It will contain permanent production, crude oil storage and processing, and loading facilities to transfer oil to Arctic tankers.*

storage facilities as well as providing protection for tankers during loading operations.

The additional production islands in this field can be constructed using conventional dredging equipment at a rate of one per year. Because of the shape of the Tarsiut field all production locations will be connected to the APLA by a subsea pipeline. This will be carried out in phases with islands being connected when they become completed.

Dome proposes to transport at least the first Beaufort oil to market with double-hulled class 10 icebreaking oil tankers (Figure 10). The first ship of this new generation is likely to be fairly small, in the order of 80,000 DWT (Figure 10). It will serve to demonstrate the technical feasibility of the plan as well as continue the process of evaluating the environmental and other implications associated with year-round shipping. Subsequent vessels will range to 200,000 DWT, with an oil carrying capacity of 1.5 million barrels. These ships will combine the best features of both icebreakers and tankers with the final product being the strongest, safest tankers that may ever be built. Eventually, when much greater oil reserves have been delineated in the Beaufort and daily oil throughput approaches 1.0 million barrels, we believe the use of a pipeline becomes a much more viable option and could become realized.

FIGURE 10. *The tankers being proposed to transport oil through the Arctic seas will be ice Class 10 double-hulled ships. The first ship of this new generation will be fairly small, at approximately 80.000 DWT. Subsequent ships will range to 200.000 DWT with an oil carrying capacity of 1.5 million barrels.*

## Beaufort Production and the Environment

The preceding discussion examined some of the plans and engineering hardware that will form part of the Beaufort development picture. However we believe it is the view of all Canadians, that whatever activities take place in the region, they should do so with preferably no, or at most minimal impacts to the Arctic environment. This as you will recognize, is also quite a challenge, but we believe firmly, can be met.

Many of the answers to technical problems have gone hand in hand with environmental and socio-economic solutions. Most of these solutions have been developed through the course of lengthy and diligent research, stepwise and cautious development, and the sensible application of real world experience. The oil industry's expertise, built up over more than 15 years, has given it invaluable insight into both the technical and the environmental concerns. A sensible approach to these concerns is reflected in the excellent safety record achieved in the Beaufort Sea and Mackenzie Delta. There have been few serious mistakes, no well blowouts, and little ecological damage caused by exploration.

Scientists in general have become acutely aware of the sometimes delicate balance between climate, geography, and nature. Since some of the interactions between living things in the Delta and the Beaufort Sea may be disrupted by exploration and development, the industry has been especially cautious in applying new technology to the region.

Frequently small scale experimental models are constructed and tested by the industry before any full scale activity is undertaken. For example, model islands were built first to test the impact of ice on these structures, and thereby establish safety criteria which have been demonstrated in the building of islands in progressively deeper water. The Kigoriak Class Four icebreaker is a fore-runner of a wholly new class of icebreakers including the 200,000 dead weight ton tankers.

The possible influence of year-round icebreaking through the Northwest Passage, and the building of artificial islands in coastal waters, have been examined through ongoing research and development programs. The experimental program with the John A. Macdonald icebreaker, the ongoing Kigoriak research program, the comprehensive ice tank testing, plus the cumulative experience acquired by operating in the Arctic environment year after year, in all seasons, are providing a backlog of data which are showing that environmentally acceptable solutions are available.

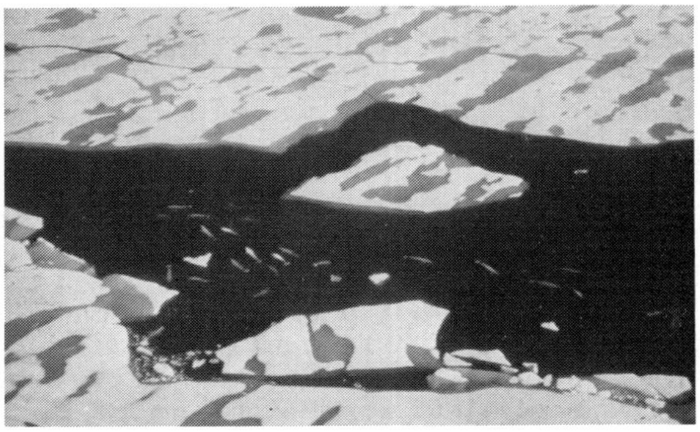

FIGURE 11. *Up to 7.000 beluga whales have been counted in the shallow waters of the Mackenzie River estuary during the summer season. Industry conducts aerial monitoring programs every year to determine where the whales are, and to redirect ship traffic if necessary to ensure that possible interference to the whales is minimized.*

Industry and government agencies have conducted research on the impact of ship traffic upon ice and the sea life that make their home on, or under that ice (Figures 11 and 12). For example, there have been noise studies conducted with Kigoriak to determine the impact, if any, of tanker passage on animal life, their habitats, and migration patterns. With this information, shipping routes can be selected to avoid ecologically vulnerable areas. Shipping corridors will be surveyed to obtain adequate bathymetric and sea ice data, and a preliminary study has been completed on the ability of hunters to cross the broken ice left in the track of an icebreaker (Figure 13).

Research and development has been conducted on a variety of other technical issues relevant to environmental protection matters and solutions are evolving to handle them. Examples include the detailed work on assessing methods to prevent oil spills, or clean them up should they occur; the effects of artificial island construction dredging and drilling operations; and the installation and operation of safe sub-sea well control systems, pipelines and other submarine hardware.

Valuable information has been obtained on the behaviour of oil under and in sea ice, and ways and means for cleaning up the oil. A study conducted in 1980 on the spread of oil under ice showed that up to 80 percent of the oil could be removed while the remaining 20 percent dispersed naturally with negligible impact on the environment. Surprisingly, ice has proven to be more of an ally than a foe when it comes to restraining and cleaning up oil spills. What we have found is that ice edges form natural booms to collect oil against, and in the case of landfast ice, which is generally the last to leave the Beaufort region, it protects the important foreshore coastline. During the lengthy winter period ice continues to grow and encapsulates oil, keeping it away from contact with marine mammals and other biota. During this process of «cold storage» ample time is available for trained people to plan an effective clean up strategy, to transport and set up equipment, and to handle the oil once it begins to be liberated from the ice in the spring. The freezing of the oil during the winter months preserves its most volatile components, which helps tremendously with burning during the spring. Although industry and government work in this important area continues, the statement «ice is nice» is becoming quite appropriate in the field of Arctic oil spill cleanup research.

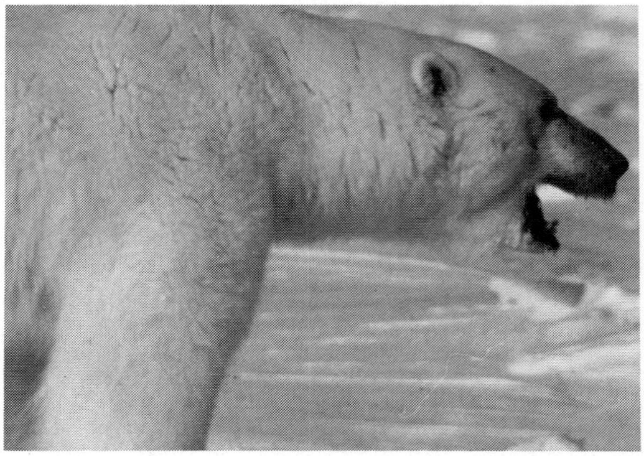

*FIGURE 12. Polar bears, a valuable species to the Inuit, frequent the offshore drilling area during winter and have been known to come onto offshore drilling islands.*

Artificial islands constructed in the offshore Beaufort must be safe and able to withstand ice forces within the realm of probabilities. Much experience has been gained in the construction of shallow water islands, and new innovative technologies were applied in the construction of the latest island, Tarsiut. Methods have been developed and tested to increase the slope of the subsea mound used as the base for mounting the large retaining walls. Increasing the slope means less material is required, which, of course, reduces the extremely expensive construction costs. However, it also results in reduced habitat alterations, both at the dredge sites and in the vicinity of the island. The retaining walls, or caissons, further reduce material required, and are also designed to be portable and reusable. The multi-million dollar research program which is investigating factors such as ice, erosion, and earthquake forces and their possible influence on the integrity of the island, as well as numerous other programs, will play a very important role in the design of safe future production platforms in the Beaufort region.

*FIGURE 13. The icebreaker Kigoriak sails out of the Arctic twilight during the experimental icebreaker track research program conducted at various times during the winter of 1981-82 in the Beaufort Sea.*

The sub-surface well heads, manifolds, and control systems employed on the ocean floor have been designed with safety and stability as paramount ingredients. They have to function flawlessly in cold waters, be maintainable under the oceans, and be protected from moving ice. Drillships are equipped with rapid disconnecting systems which permit a shut down of exploratory drilling operations, thereby minimizing the risk of a blowout. This rapid disconnect is necessary should severe ice conditions move into the drilling area.

Further research is being conducted on the laying of sub-sea pipelines on the seafloor in areas of ice scouring. We have concluded that should pipelines be laid to shore, they will be buried beneath the depth of maximum probable ice scouring in areas where this occurs.

## The Environmental and Socio-Economic Review Process

The foregoing has described our general plan for developing the hydrocarbon resources of the Beaufort Sea region as well as some of the research and monitoring programs being undertaken to evaluate and minimize impacts of present and future activities in the region. To assist in ensuring that future development takes place in a socially and environmentally acceptable manner, in July, 1980, the Minister of Indian and Northern Development referred all development proposals in the Beaufort region to the Minister of the Environment for a formal review under the Environmental Assessment and Review Process (EARP).

The Minister of Environment, in turn, established an independent panel comprised of 7 members to carry out the necessary public review. The Panel's mandate is to identify major developmental effects, both positive and negative, upon the physical, biological and human environments and recommend ways and means of dealing with them. As part of this mandate, the Panel will review and assess all relevant information associated with the physical, biological and socio-economic considerations of the proposal, amd should be prepared to address possible community changes such as effects on existing life-style and conditions. The Panel review is to include all related activities north of 60° of the proponent's proposal associated with or resulting from the commercial production and shipment of hydrocarbon resources from the Beaufort Sea area. This includes possible onshore and offshore oil and gas production facilities in the Canadian Beaufort Sea area and subsequent transportation of oil and gas to southern markets by icebreaking tankers or pipeline(s) or by both means.

The Panel's first task was to develop a set of guidelines for the Industry which would be used to prepare the necessary documentation

(Environmental Impact Statement). Following a series of public meetings held throughout the north in November-December, 1981, to solicit input, guidelines were issued to the Industry in February, 1982. Upon receipt of the guidelines, the Industry consortium, spearheaded by Dome, Esso and Gulf, began to finalize the EIS.

The Environmental Impact Statement consists of 7 primary volumes. Each volume covers specific areas of information needed to address developmental, environmental and/or socio-economic issues. The basic information needs are summarized in Figure 14. Each of the 7 volumes responds to specific information needs. The master index illustrated in Figure 15 provides the titles of these volumes.

| KEY QUESTIONS THE EIS MUST ADDRESS |
|---|
| • WHAT IS THE DEVELOPMENT PLAN? |
| • WHAT ARE THE REGIONS LIKE NOW? |
| • WHAT WILL THE REGIONS BE LIKE IN THE FUTURE? |
| • WHAT IS BEING DONE ABOUT UNKNOWNS? |
| • WHAT IS INDUSTRIES' RESPONSE TO SPECIFIC ISSUES? |

FIGURE 14. *The key types of questions that the Environmental Impact Statement must address.*

| | |
|---|---|
| VOLUME 1 | SUMMARY |
| VOLUME 2 | DEVELOPMENT SYSTEMS |
| VOLUME 3A | BEAUFORT SEA-DELTA SETTING |
| VOLUME 3B | NORTHWEST PASSAGE SETTING |
| VOLUME 3C | MACKENZIE VALLEY SETTING |
| VOLUME 4 | BIOLOGICAL & PHYSICAL EFFECTS |
| VOLUME 5 | SOCIO-ECONOMIC EFFECTS |
| VOLUME 6 | ACCIDENTAL SPILLS |
| VOLUME 7 | RESEARCH AND MONITORING |

*FIGURE 15. The Environmental Impact Statement has been structured to provide all the necessary background information and to address the key questions in a series of 7 volumes. Information is presented in summary form, by discipline, and by geographic regions.*

The EIS has now been completed and distributed to the panel, interested public and government agencies for review. Following the review period of approximately 90 days, public hearings will again be held throughout the north in communities which are mostly likely to be affected by future development of the Beaufort region's hydrocarbon resources and the transportation thereof to southern markets. More thorough technical hearings will be convened in larger southern centres. Upon completion of the hearing process, the panel will prepare a report on their findings and recommendations for consideration by the ministers of Environment and Indian and Northern Affairs.

Through this process, the views of the public and factors influencing the environmental and social well-being of the north will have been presented in a clear and thorough manner. As a result, it will make a major contribution to future decisions affecting the sound and beneficial development of the Beaufort Sea-Mackenzie Delta hydrocarbon resources.

*Cold Regions Science and Technology*, 7 (1983) 313—341
Elsevier Science Publishers B.V., Amsterdam — Printed in The Netherlands

# Arctic Marine Oil Spill Research

*H.R. Hume, I. Buist, D. Betts and R. Goodman*
Petro Canada
*Calgary*

## Abstract

*Canada, with its immense arctic and sub-arctic regions, wants to develop the promising hydro-carbon potential of the Beaufort Sea, the high arctic islands, the eastern arctic and Labrador. Of necessity, Canada has gradually become a world leader in arctic marine oil spill research.*

*Some of this research has been done by the Canadian federal government through AMOP, the Arctic Marine Oilspill Project. There has been joint oil industry/government research, such as the Beaufort Sea Project and the Baffin Island Oil Spill Project. In oil company initiatives, Dome Petroleum has been a leader in the Beaufort Sea region. Finally, a cooperative oil industry organization, the Arctic Petroleum Operators' Association and its research group, the Canadian Offshore Oil Spill Research Association, have made many important studies.*

*Initial efforts aimed to strengthen existing containment and recovery equipment to operate in light ice conditions. When the limits of this technology were rapidly reached, research turned to developing burning techniques, including fireproof booms,* in situ *burning against ice edges and in spring melt pools, air deployable igniters and portable burners and incinerators. New dispersant technology was developed for aerial application, for cold water dispersion and for solidification. The effects of dispersed oil in the arctic nearshore and onshore environment were assessed. More recently, subsea containment has become an important area of study.*

*Future research will include methods of mechanical recovery of oil in ice, oil/water emulsions in and under ice, and the mechanism of oil migration upward through pack ice.*

## Introduction

When northern nations began to develop the hydrocarbon potential of their arctic and sub-arctic regions, it became apparent that oil spill countermeasures that were appropriate for temperate and tropical waters were unsuitable for colder northern waters. Ice and icebergs, cold sea and air temperatures, and fluctuating salinities all severely limited existing southern oil spill technology.

This paper reviews the work done in Canada by government, industry and academics to understand and minimize the impact of hydrocarbon spills in the northern environment. The first section summarizes the historic and organizational contributions to oil spill research. The major Portion of the paper covers the various research themes that have been pursued by Canadian organizations and the final section summarizes areas of future research.

Similar research is being undertaken by the U.S.A., Norway, the U.S.S.R., and the U.K. There are strong affiliations between Canada and the U.S.A. in oil spill research in offshore Alaska and the Yukon to prevent duplication of effort and to share information. The U.S.A., Norway and Britain are currently participating in a major research project in Canada's eastern arctic.

0165-232X/83/$03.00  

Arctic oil spill research in Canada has been undertaken by the federal government through the Arctic Marine Oil Spill Project. Joint oil industry/government research includes the Beaufort Sea Project and the Baffin Island Oil Spill Project. Oil industry research has been done either by individual companies such as Dome Petroleum in the Beaufort Sea, or cooperatively through the Arctic Petroleum Operators' Association and its research group, the Canadian Offshore Oil Spill Research Association.

## Agencies Conducting Oil Spill Research in Arctic Canada

Various government and oil industry agencies have conducted arctic marine oil spill research in Canada, both individually and jointly since 1970. Oil industry research, unless it involves commercial advantage, is normally done cooperatively both to share the cost and to overcome the problems of confidentiality when done in isolation. The major exception was work conducted by Dome Petroleum in the mid-1970's where because their offshore drilling activities were ahead of the rest of industry in the western arctic, they were obliged to conduct research in isolation. Joint research projects were done initially through APOA for arctic research, where interested companies shared the cost. More lately joint projects are done through COOSRA, where the cost is spread across the entire offshore oil industry. The arctic oil spill research arm of government has been AMOP since 1977.

The larger research projects, because of the high costs involved have been done jointly by government and industry (e.g. the Beaufort Sea Project and BIOS).

## Arctic Marine Oilspill Program (AMOP)

In 1977, following the government decision to allow drilling for oil in the Canadian Beaufort Sea, the Department of the Environment initiated the $7.5 million Arctic Marine Oilspill Program (AMOP) to develop oil spill countermeasures for ice-infested waters. The eight-year program is managed by the Environmental Protection Service (EPS), Environment Canada, with the assistance of an interdepartmental committee of government and oil industry specialists.

AMOP consists of engineering projects in such areas as oil spill recovery and containment devices, combustion and incineration systems, dispersant equipment and strategies, remote sensing equipment and shoreline cleanup methods. AMOP also studies the fate and Physical effects of oil and the movement and behaviour of spills in ice-infested waters. Finally, AMOP includes biological assessment, particularly that related to spill-treating agents like chemical dispersants.

Presently, approximately 100 projects have been conducted under contract by consultants, university research. ers and government scientists, and most of this work is performed by private companies. Much of the work is jointly funded by other government agencies, by the oil industry and sometimes by other Arctic nations.

AMOP reports are published as government technical reports, which are often summarized in Spill Technology Newsletter. Also, project updates are provided at annual AMOP seminars, which are an international review forum. Appendix B lists published AMOP research reports.

## Beaufort Sea Project

The Beaufort Sea Project was a set of studies jointly funded and managed by government and the oil industry that investigated wildlife, marine life, oceanography, meteorology, sea ice and oil spill countermeasures. When the federal cabinet granted Approval-in-Principle for exploratory drilling in the Beaufort Sea in 1973, it stipulated that the Drilling Authority would be issued subject to operating conditions developed by this project.

The oil spill studies were based on a need to react to a large blowout in which thousands of tonnes of oil could escape into the ice-infested sea and that the blowout could continue for a year or longer. Innovative work was conducted on the underwater dynmics of a blowout plume (Topham 1975). Results of these studies have been published in the Beaufort Sea Project Technical Report Series and in the APOA report series 72 (see Appendix A). Ross *et al* (1977) have also summarized this work.

## The Baffin Island Oil Spill (BIOS) Project

In the late 1970's, AMOP undertook responsibility for identifying and coordinating research into experimental oil spills in cold waters. Some of this research was integrated into a single four-year project called the Baffin Island Oil Spill (BIOS) project.

This $5 million project is administered by the Environmental Protection Service, and directed by an international management committee of representatives of the funding agencies. These include the Canadian Offshore Oil Spill Research Association, the Department of the Environment, the Department of Indian and Northern Affairs, the Department of Fisheries and Oceans, the Canadian Coast Guard, the U.S. government, the Norwegian government, and British Petroleum.

The primary project objectives are to determine whether the use of chemical dispersants in the Arctic nearshore marine environment will reduce or increase the environmental effects of spilled oil, to assess the fate of spilled oil, and to compare the relative effectiveness of other shoreline protection and cleanup techniques.

The site of the BIOS project is Cape Hatt, Baffin Island. Baseline environmental information was collected throughout the study area in 1980. In the summer of 1981, one small bay was treated with an experimental oil slick, the subsurface of two bays were treated with different doses of oil and dispersant mixture, and another bay remained an uncontaminated control. During the springs of 1981 and 1982, the effects of oil and chemical treating agents on the under-ice biology was examined. Monitoring will continue until 1984 on the physical and chemical fate of the oil and on its biological effects. Besides dispersants, other experimental shoreline protection and cleanup techniques are being evaluated on other nearby beaches.

Technical reports are published annually and the research is discussed at annual BIOS Project workshops. Final reports will be available in 1985.

## Dome Petroleum

Before starting its drilling program in the Beaufort Sea in 1976, Dome Petroleum Ltd. undertook an oil spill research program to develop appropriate oil spill countermeasures. Besides developing containment, recovery and disposal techniques and systems for the Beaufort Sea, Dome initiated studies in the areas of contingency planning, *in situ* burning, tracking and monitoring, shoreline protection and cleanup, and bird and wildlife protection techniques. Research is conducted every year to develop new technologies and to improve the existing oil spill cleanup capabilities in the Beaufort Sea. Recently, much of this work has been done through the Arctic Petroleum Operators Association (APOA) and the Canadian Offshore Oil Spill Research Association (COOSRA).

## Arctic Petroleum Operators' Association

The Arctic Petroleum Operators' Association (APOA) was formed in 1970 to coordinate arctic research and avoid duplication of effort (Hnatiuk *et al.* 1980). APOA's area of concern reaches from the Alaska-Yukon border across the arctic archipelago to Davis Strait.

Much of APOA's research involves environmental and engineering studies. Together with the Eastcoast Petroleum Operators' Association (EPOA), APOA has conducted a series of oilspill studies from 1970 to 1979. These are listed by Pallister (1982) and are available in Photocopy or microfiche. Appendix A lists APOA research projects.

Since 1980, the APOA/EPOA Oil spill Committee has limited its activities to operational and contingency planning, while oil spill research has been undertaken by a specialized research agency, the Canadian Offshore Oil Spill Research Association.

## Canadian Offshore Oil Spill Research Association (COOSRA)

The Canadian Offshore Oil Spill Research Association was formed in 1980 as the oil spill research arm of the Arctic Petroleum Operators' Association and the Eastcoast Petroleum Operators' Association. Initially it was to provide the oil industry's contribution to the Baffin Island Oil Spill Program.

COOSRA originally had 16 full members and two associate members; one associate has since become a full member. Each member's costs are shared on the basis of offshore acreage and drilling days for each year. Membership in COOSRA overcomes the problems of confidentiality and duplicated research effort that could hamper spill research.

COOSRA has a technical advisory committee of approximately 12 members who review all research proposals, assess priorities to projects, and assign a regional applicability factor to each project for assessing an individual company's cost share. The recommendations of the technical advisory committee are then passed to a management committee, and finally to a members' committee for acceptance or modification.

All COOSRA Projects to date have been either wholly or partially applicable to the arctic. Appendix C lists COOSRA projects from 1980 to 1982.

The above organizations have individually and in combination funded arctic marine oil spill research from the early 1970's up to the present. Since the mid 1970's, there has been a conscious effort to coordinate government and industry research to prevent duplication and to endeavour to approach oil spill research in a systematic manner. One of the major advantages Canada has in this field of research is the close working relationship between government (federal, provincial and territorial), oil industry and academic operational and research personnel. Although a complete systems approach to oil spill research in arctic waters has not been achieved, the following section describes the main categories of research themes that have been pursued to date.

## Oil Spill Research Themes

Arctic marine oil spill research since the early 1970's has followed several well defined themes. These include fate and behaviour studies, which help to improve the effectiveness of countermeasures such as monitoring and surveillance, mechanical containment and recovery, combustion, subsea containment, shoreline cleanup, and chemical dispersants.

Initially, mechanical containment and recovery was the only proven technology for open water. However, such definite countermeasures are ineffectual in high seas and in ice-infested waters. Since the mid-1970's, considerable progress has been made in improving combustion and shoreline cleanup techniques and in reducing dispersant toxicity. Major efforts are currently underway in the recovery of oil from ice, in more effective cold water dispersants, in chemical solidifying agents, and in designing and testing feasible subsea containment systems.

## Fate and Behaviour

A basic understanding of oil, its chemical and physical changes, and its interaction with open water and ice is of primary importance for spill response. Both mechanical recovery and chemical dispersion depend upon the nature of the oil. Since the behaviour of an oil slick is highly dependent on the type of oil and ambient climatic conditions, studies have been carried out on both under ice and open water conditions.

### Ice Conditions

Much effort has been devoted to the fate and behaviour of oil under sea ice, commencing with the Beaufort Sea Project. One study in a small bay southwest of Cape Parry N.W.T. (NORCOR 1975) involved nine controlled discharges of approximately 45 $m^3$ of two types of crude oil under fast ice at various stages of growth. Fundamental data on the interaction of oil with Arctic ice was collected to study the mechanics of transport, the area of contamination and albedo modification. Rosenegger (1975) also studied the mechanics of movement of oil in ice-covered water in the laboratory. Results of these two studies are as follows:

i) throughout the winter, oil penetrates only between 5 to 10 cm into the loose skeletal layer on the bottom of an ice sheet;
ii) penetration of the oil into the ice sheet is not normally expected until brine drainage channels are open;
iii) the oil's effect on ice growth tends to be relatively insignificant in comparison to natural variations;
iv) the equilibrium thickness of two crudes under ice was 0.80 and 0.88 cm;
v) oiled areas are likely to be free of ice between one and three weeks before the gross failure of the sheet;
vi) data enabling calculation on minimum currents required to move an oil bubble are presented.

These studies recommended that further work be carried out on the weathering and degradation of oil, on entrainment and migration of oil in multi-year ice, effect of oil on sea ice formation, on the pumping action of leads that open then close, and on ice movement.

Walker (1975) discussed the effects of a blowout on the Arctic ice. He concluded that the formation of ice would not be much delayed, because wind would push oil into leads and these would be quickly covered over by snow. In late 1982, COOSRA will study oil in pack ice to understand the mechanisms of oil moving in leads to simulate freeze-ups and break-up conditions.

Milne *et al* (1977) found that the porosity of multi-year ice was 0.2 % to 0.03 % and anticipates that most, if not all, of any light to medium crude oil trapped in and under multi-year ice would migrate upward through the ice. In another study of the fate of oil in multi-year ice, the mean maximum thickness of oil was 0.6 cm if gas venting occurred (NORCOR 1976b).

The spreading of oil beneath sea ice depends on the undersurface characteristics and oil and gas effects. Purves (1978) explained that the configuration of the ice/oil/gas/ water system that will be energetically favoured would be determined by various interfacial tensions. Purves concluded that oil layers as thin as 0.2 cm would probably occur when well blowout products accumulate under ice. The presence of gas would greatly increase the area contaminated, but would have little or no effect on the timing of release of oil trapped under first-year ice. The time of release of oil and gas might be predicted by ice thickness, salinity profile and meteorological data.

In the winter of 1979-80, Dome Petroleum Ltd. (1981a) experimentally simulated an oil and gas blowout under shorefast ice near McKinley Bay, N.W.T. The major study objectives were:

1) to understand further how oil and gas behave when discharged under ice;
2) to investigate water-in-oil emulsion formation;
3) to field test the effectiveness of *in situ* burning of oil on sea ice;
4) to assess the capability to clean up a blowout under ice.

Oil formed buoyant droplets that adhered to the undersurface of the ice. Oil discharged under new flat December ice occurred mainly as droplets, but oil released in April-May collected in pools in undulations beneath the ice. The amount of gas released greatly affected the area of the ice undersurface contaminated by oil. Oil was encapsulated by the growing ice sheet within 24 to 48 hours. A sample of oil from the undersurface of the ice showed loss of lighter fractions up to $C_{12}$. This loss was due to the interaction between air and oil, but would not take place during a real blowout. *In situ* burning efficiencies averaged 80 %.

Oil released in moving sea ice has a very different fate from oil under stable fast ice. A 7000 ton Bunker spill off Nova Scotia in March 1979 left an average oil-in-ice concentration of 100 ppm (Reimer 1980). Due to large scale mixing, the oil was dispersed into progressively smaller particles, so that burning and mechanical removal were unfeasible. Crude oils from a blowout would rise through the frazil and brash ice, forming an emulsion that would be distributed vertically as was the bunker oil.

Some years ago, there were concerns that oil pollution of arctic waters could change the earth's climate by altering the surface albedo and consequently reducing ice cover (Milne 1978). NORCOR (1975) studied the effects of entrained and free oil on the surface heat exchange. Although there were measured changes in albedo, no appreciable effect on the growth of an ice sheet was demonstrated.

## Open Water

Early mathematical models of the behavior of surface oil slicks have been developed by Mackay and Leinonen (1977). These have subsequently been refined to include calculations for dispersion and solution of oil in the water column, emulsification, oil properties, and blowout versus instantaneous release conditions (Mackay and Paterson 1980a, 1980b). Figure 1 is a mass balance of oil on open water, illustrating the modelling attempts of Mackay and his coworkers. Spreading equations have been based on Fay's surface tension-viscous equation. These are solved numerically to accommodate the complex behaviour of oil spills. The models have a thick slick (1 mm to 1 cm) surrounded by a thin slick (1 μm to 10 μm) eight times the size (Mackay, Buist, Mascarenhas and Paterson 1980). Allowance has been made to simulate the spreading of the thick portion of the slick to form the thinner portion of the slick. Figure 2 illustrates the spreading of oil on water as a function of time and volume.

Evaporation changes the properties of oil slicks, such as volatility, solubility, density, viscosity and surface tension. Mackay and his co-workers have quantified these changes in relation to temperature, area, time and wind speed. This work is a behavioural model that not only predicts location with time, but also incorporates weathering processes. Such modelling provides exposure data that can be linked to an ecosystem model. Tebeau (1982) has shown that these behavioural models can be used in slick trajectory models to predict the weathering of crude oil under Arctic conditions.

Oil rising in a blowout plume consists of droplets of 0.5 mm - 1.0 mm diameter (Topham 1975). The hydrodynamics of the plume are controlled by the volume of natural gas. LeDrew

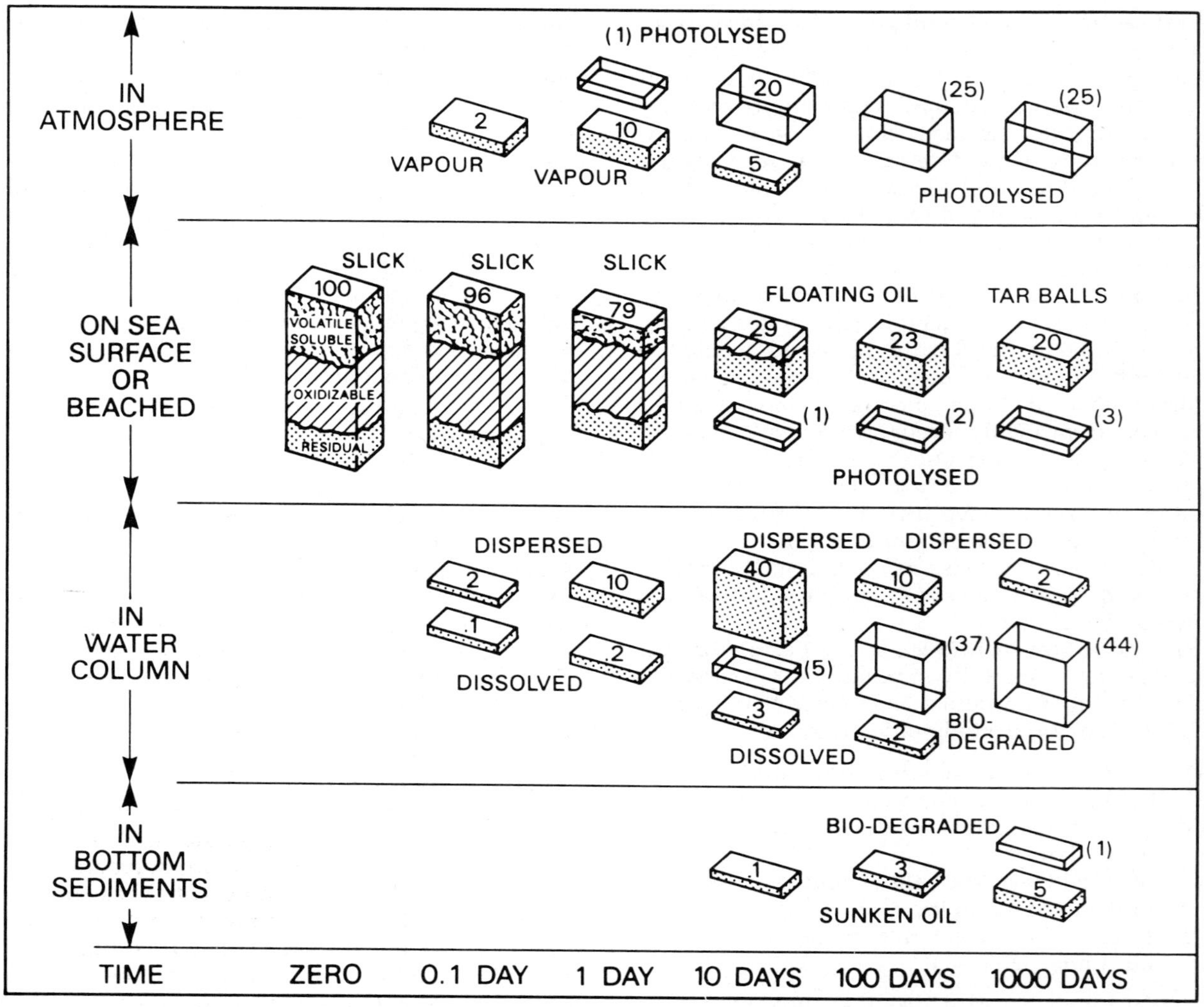


Note: Empty unshaded boxes represent oil converted to another chemical form.

FIGURE 1. *Speculative mass balance for batch spill. Source: Mackay et al 1980.*

and Gustajtis (1979) speculated on the interactions between pack ice and a hypothetical blowout plume in the Labrador Sea. They expected that, under calm conditions, the 0.5 m/s induced currents would largely prevent ice floes from crossing the wave ring at the tope of the plume. During turbulent conditions, however, the wave ring would be covered by ice, and the oil would become entrained in a variable mixture of water and ice.

The importance of understanding why certain oils form emulsions and their rate of water uptake was illustrated by the IXTOC 1 well blowout, where countermeasures for recovering the oil were severely limited by the changing properties of oil/water emulsion.

The tendency of oil to form water-in-oil emulsions has been studied by Mackay and Zagorski (1981, 1982). It has been hypothesized that the turbulent mixing within a blowout causes emulsions to form. Mackay suggests that the presence of certain chemical compounds influence the stability of an emulsion. Stability is hypothesized to be controlled by a combination of wax and asphalt content (Figure 3).

## Shorelines

During the ongoing BIOS experiment, 15 m$^3$ of oil were spilled into one test bay and 15 m$^3$ of dispersed oil in others. The bulk of the untrea-

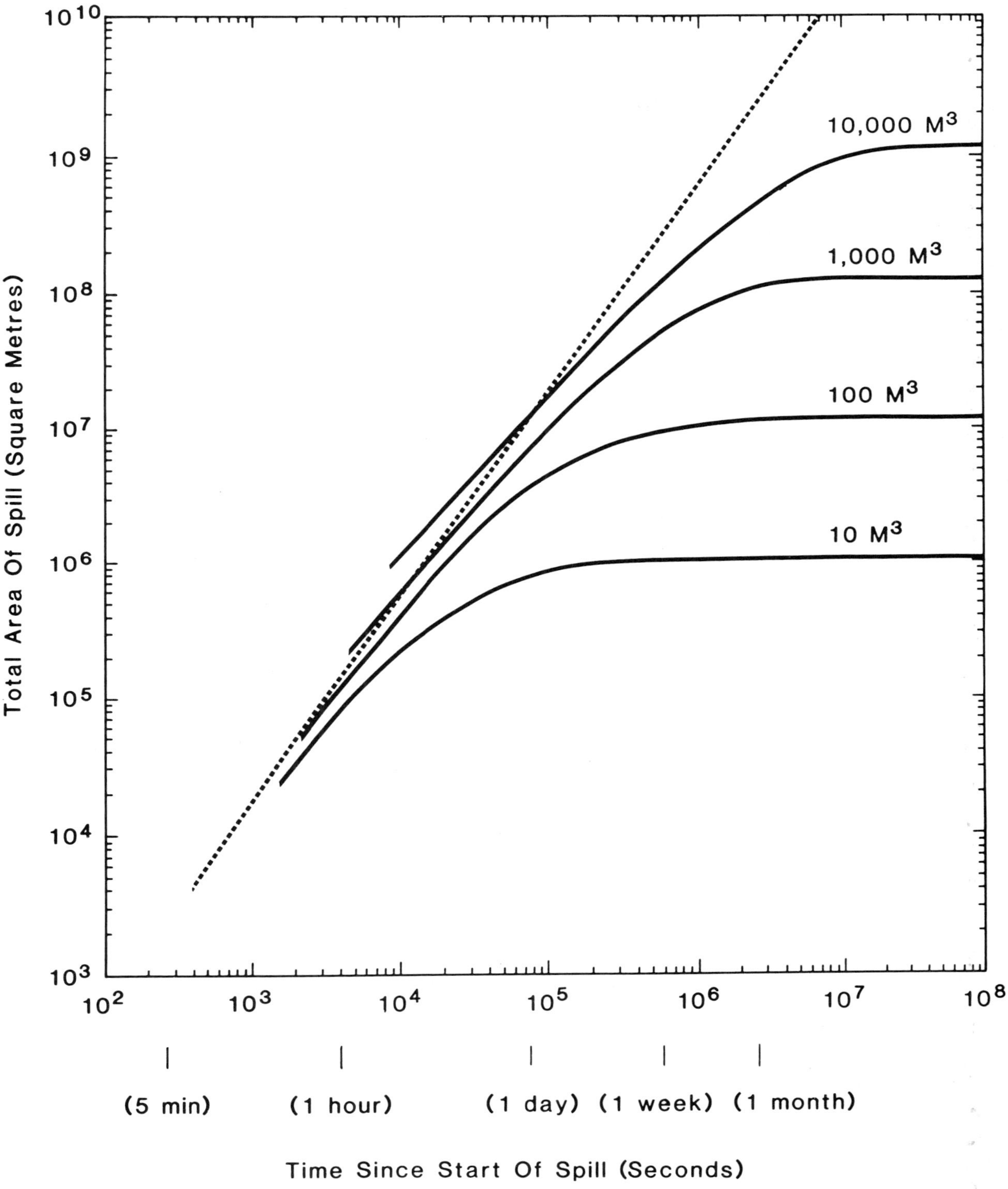


FIGURE 2. *Area vs Time Per Spill Volume. Source: Oil Spill Processes* ***and Models*** *(Mackay et al 1980).*

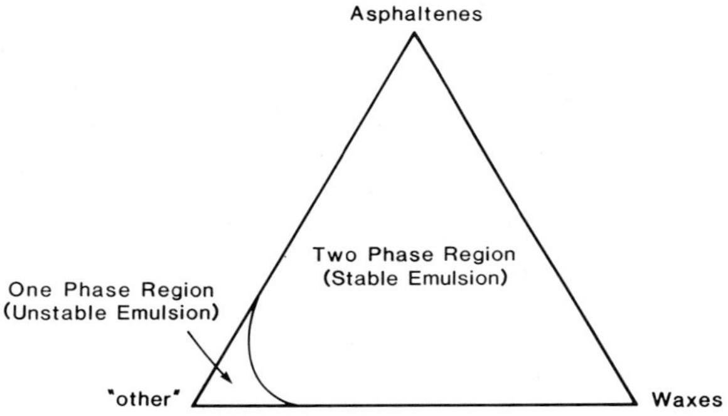


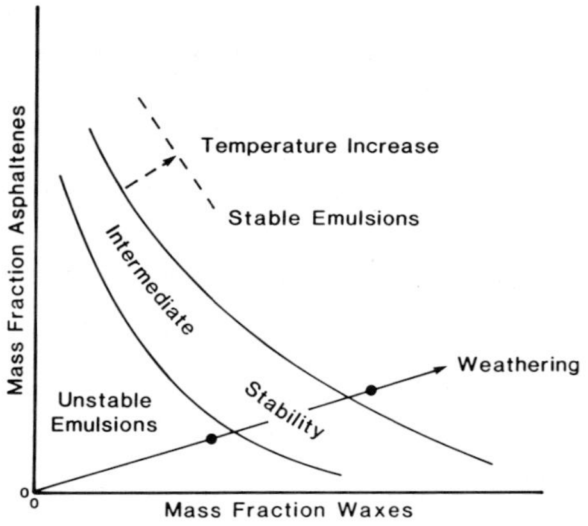


FIGURE 3. *Triangular and rectangular phase diagrams showing stability regions. Source: Mackay et al 1982.*

ted oil washed ashore on a rising tide and the remaining oil was mechanically collected from the water surface. Preliminary unpublished results (D. Thornton, pers. comm.) indicate that oil concentrations of a few ppm occur in the sublittoral sediments, perhaps as a result of leaching from the beach. Stranded oil weathered and bonded to the beach within 24 hours. Photos taken after the ice receded in the summer of 1982 show a large amount of oil still present. The dispersed oil had a small but rapid sediment penetration rate and contaminated the adjoining strait to 10 m depth within a few days.

## Monitoring and Surveillance

Once an oil spill has occurred, one of the main concerns is a determination of the motion, size and thickness of the spill. This information is essential to optimize spill response and to determine potential impacts on various ecological communities. A major component of any integrated oil spill mitigation package is remote sensing of the oil spill, tracking it, and predicting its motion. This information is useful, not only for the On-Scene Commander and his support staff in the technical planning of the response, but it is essential to give adequate public information concerning the spill. While remote sensing, tracking and trajectory prediction are independent, this information is a complementary set of data for a fuller understanding of the oil spill. The problems of remote sensing, tracking, and prediction of oil spill trajectories are further complicated by the presence of ice during most of the year in northern waters.

### Remote Sensing

Oil on the water surface is most readily detected by aerial reconnaissance. Should weather conditions permit, the cheapest and best method of detecting oil on the sea surface is to see it from an aircraft. Aerial reconnaissance is routinely used in most oil spill responses to give the On-Scene Commander a first quick look at the size of the oil spill and of his task. A trained observer is required because there are many other sea features that can resemble floating oil, including cloud shadows, beds of seaweed, or suspended particles in nearshore waters. The experienced observer can distinguish these features and positively identify an oil slick. However, in fog, cloud, rain or at night, visual observation cannot be used. Such conditions occur frequently in Canadian coastal waters and darkness is prolonged during winter in Arctic Canada. Canadian researchers have tried to develop an all-weather, day-and-night operational remote sensing system to detect oil spills.

The Arctic Marine Oil Spill Program (AMOP) supported a project of the Canada Centre for Remote Sensing (CCRS) to determine the most suitable integrated sensor package for Canadian requirements. Sensors include: Side Looking Airborne Radar (SLAR), UV/IR dual channel line scanner (DCLS), a laser fluorosensor, a low-light-level television (3LTV) and annotated photographic cameras. Each of these sensors was independently evaluated. Subsequently, an integrated system consisting of suitable sensors and a real time display with operator interaction was developed. The system was tested in four experiments: around Montreal Island, the site of major refining activities; Scott Inlet, a natural oil seep in the Canadian Arctic; the Wallops Island experimental oil spill sponsored by NOAA; and the Kurdistan spill, which occurred in ice in the spring of 1979 (O'Neil 1982). Particular requirements for all-weather operations, detailed identification of the spill source, and operation in ice were unique features investigated under the AMOP-CCRS Program.

The recommended combination of sensors is a laser fluorosensor, dual channel line scanner, side-looking airborne radar, low-light-level television, and conventional photographic cameras with suitable filters. The output of the first three of these sensors is fed into an on-board computer system that allows analysis, displays the information, and integrates it with positional information derived from the navigational equipment aboard the aircraft. Total weight of the system, including operational personnel, is approximately 1,850 kg.

The laser fluorosensor gives a positive identification of the presence of oil, using a cross-correlation technique with previously recorded spectra. This sensor helps the operator identify the type of oil and its thickness. The dual channel line scanner operates in the UV (300 - 370 nm) and in the infrared (8.5 - 12.5 nm) band, and can discriminate between the thick and thin components of the oil. The UV channel is particularly sensitive to the sheen, while thermal differences associated with differential adsorption of solar radiation by thick oil are detected in the IR channel. Unlike the laser fluorosensor, the dual channel line scanner only detects oil shape with some indication of thickness, and presents no information on the nature of the oil. The SLAR can be used for quick coverage of large areas. The system has been used by the U.S. and Swedish Coast Guards (U.S.C.G. 1979) and is incorporated in the Canadian system. SLAR requires a dedicated aircraft, which limits the selection of aircraft.

A low-light-level TV (3LTV), a small portable video recorder and annotated photographic cameras with suitable filters aid visual observations in difficult lighting conditions and improve the ability to discriminate between oil and other features. These two systems have been extensively developed and tested as part of the Canadian Remote Sensing Program.

The ability to detect and operate under low illumination during an oil spill is another project of AMOP. A night light, consisting of a commercially available UV light with suitable filters, aids visual detection (Fingas 1982). The system is hand portable and readily operable from a small boat. While such a system is less complex than aerial remote sensing equipment, it helps to extend the operational time of an oil spill response.

## Tracking

If sophisticated remote sensing equipment is not available or when bad weather interferes with aircraft operation, an alternate method of tracking an oil spill is required. Initial experiments with a wide variety of drifting buoys, drift cards, plywood sheets and other surface tracking devices have been carried out in Canada to determine the system that follows the track of the oil accurately (Fingas and Lea 1981). This is not a simple problem, since oil may be at several depths from the surface to several metres below it, and all oil must be tracked. No one single piece of equipment will follow thin oil, thick oil, and oil in the water column. A series of buoys has been developed to track the various components of an oil spill.

A 1976 AMOP study evaluated the need for various types of buoys. Two types of buoys, one capable of close support for strategic decisions and another to determine the long range trajectories, were required (Goodman 1978). These two systems, a micro and a macro, have been independently developed. The micro system has been developed by Orion Electronics in Nova Scotia with funding by the AMOP. This system is a top hat buoy containing a transmitter. The buoy shape was choosen to remain in the surface sheen. This system is used in a bearing mode, and a fixed wing aircraft or a boat is used to find the target buoy. The location is determined by the navigation system of the search vessel. Such buoys are widely used in Canada for tracking spills.

A more sophisticated system has been developed by Esso Resources Canada Limited and Orion. It uses a transmitter-receiver in the buoy and a range/bearing system, so that the location of the buoy may be determined from a single land location. The system has been tested and found effective (Roddis 1982). Both the top hat-tracking-buoy system and the range/bearing tracking systems are limited to line of sight characteristic of the operating frequencies of this equipment.

A joint industry-government experiment to determine the tracking characteristics of several buoy types has been conducted by the Bedford Institute of Oceanography in Dartmouth. Various sizes of masts and a variety of buoy shapes have been studied, to determine the drift characteristics of the buoys and their reaction to both wind and surface currents (MacLaren Plansearch 1982). These experiments indicate the nature of ocean surface currents and the trajectories of oil spill buoys. These measurements indicate the nature of the horizontal dispersion process and thus, the accuracy of oil spill trajectories.

### Oil Spill Trajectories

Canada has had a major role in the use of computer simulation to determine oil spill trajectories. Eight of the 35 oil spill simulation models reviewed in a recent API publication (Huang and Monastero 1982) were Canadian. Most trajectory models have three components: advection, spreading, and state of the oil. The advective component is generally physically based and uses a wind drift of 2-4 % of the wind speed with a direction of between 0-30° to the right of wind direction (northern hemisphere). Typical of these models is one that has been developed by the Atmospheric Environment Service of the Government of Canada for use in Canadian oil spill trajectory prediction. This model uses a wind factor of 3.5 and a Markov chain for predicting future wind speeds. The model accounts for spreading, dispersion and advection. It has recently been updated by a Canadian firm (MEP 1982) to include other processes and to use forecast rather than Markov-predicted winds. A ocean current component of the advection is presently being added.

A major research effort on the fate of oil has been led by Dr. Don Mackay (Mackay *et al* 1980) of the University of Toronto. He has developed elaborate models for spreading theory, emulsification, natural dispersion, and other fate parameters of an oil spill. These trajectory models have recently been incorporated into a biological model to determine the effect of oil and oil dispersants in the marine environment (Trudel 1982). This work has been supported by both the Government of Canada and the Canadian oil industry.

The use of trajectory models for both impact assessments of oil spills and for the real time support of an oil spill response has been reviewed by a working group of federal government and oil industry scientists, consultants and academics. This group evaluated many models for both requirements, and their report has been issued (Bedford Institute of Oceanography 1980). Suitable models have been chosen for both scenario work and real time support.

The evaluation of new models, in terms of advection, will be undertaken using the calibrated drifter buoys that were developed as part of the buoy tracking experiment.

Canada, through both the petroleum industry and government involvement, has undertaken numerous experiments in remote sensing of oil spills, tracking of oil spills using buoys, and the prediction of oil trajectories in computerbased models. This information has been integrated with work in other nations to enable Canada to respond to any oil spill that might occur there. Many areas still require development including the detection of oil under and on ice, improvement of modelling capabilities in ice-infested waters and improvement of the ability to track oil in pack ice. These are areas under active investigation by the Canadian research community, with the support of both government and the Canadian oil industry.

## Mechanical Containment and Recovery

Oil can be mechanically contained on the sea surface by floating barriers or booms, which prevent the spreading of an oil slick and facilitate the recovery of contained oil with devices that skim oil from the water surface. This is a tried and tested technology in temperate waters, in wave heights up to approximately 2 metres.

Many boom and skimmer systems have been developed for ice-free coastal waters (Solsberg 1982). An Arctic oil spill in summer open water conditions would be similar to a spill in more temperate regions, except that low water temperature would increase oil viscosity. During winter, recovery would be severely limited by landfast ice, first-year pack ice or multi-year pack ice. AMOP, APOA and COOSRA have supported many projects to develop containment and recovery devices for cold, ice-infested waters. A special workshop on the recovery of oil in ice will be sponsored by COOSRA in 1982.

## Ice Conditions

Problems with conventional containment booms in ice-covered waters were identified by Logan *et al.* (1975). These problems were:

i) most equipment had not been tested in ice-infested waters, and so it was impossible to predict its effectiveness;
ii) booms with internal tension members fail by rotating onto the ice;
iii) booms with external tension members eventually undergo structural or mooring failure;
iv) the connectors and end plates of all booms should be improved.

Of the available skimmers, disc-drum or oleophilic belt recovery devices are the most Promising for development.

The effectiveness of open water containment and recovery devices is limited when ice concentrations exceed 10 to 20 %. Handling, blockage and damage to oil booms are serious problems in brash ice. Large floes make booms ineffective. AMOP developed an ice-oil boom which yields, reduces its angle to the ice, and then swings back under its restoring moment (Tsang and Vanderkooy 1979). This concept is based on the glance boom principle used by the pulp and paper industry.

Canadian Marine Drilling Ltd. (CANMAR) developed a solid flotation boom (Nadeau *et al* 1976). This boom can contain oil in waves up to 1.5 m, and is deployed with a modified Lockheed R-2003 Skimmer (Schultz 1976). Some of the modifications are: ice-processing pumps, (Western Canada Hydraulic Ltd. 1979), greater oil recovery ability, ice fenders, and a space heater.

Little recovery can occur in fast ice. Although trenching is successful on ice-covered rivers (Canadian Petroleum Association 1978), specialized equipment is necessary to cut through thick ice. Blasting of trenches is successful in the High Arctic (Van Ieperen 1980). Oil trenches with a 1 to 2 km circumference can be blasted around offshore rigs in strong multi-year ice in

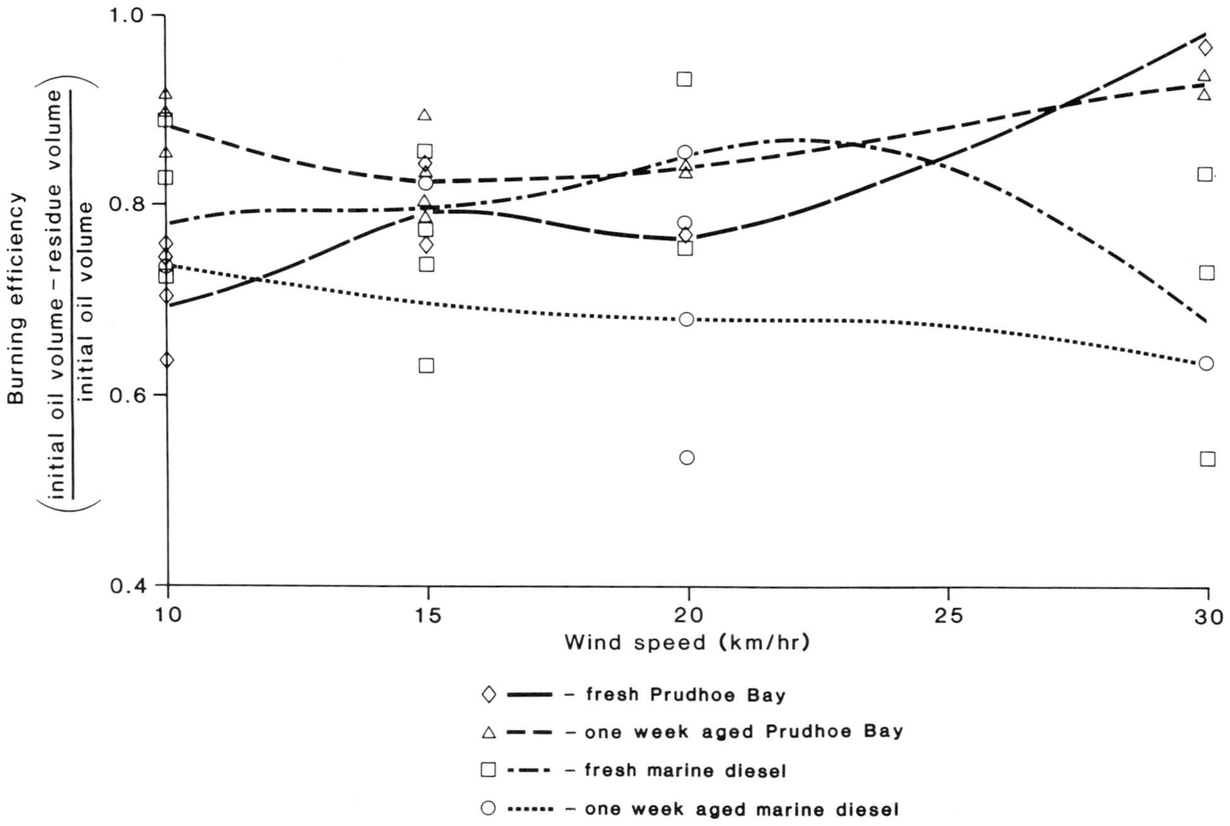


FIGURE 4. *Burning Efficiency vs Wind Speed.*

two to three days. Best results are with holes drilled to within 0.7 m of the under surface of the ice, spaced 1 m apart, and with 4.2 kg of explosive per metre of trench length.

## Open Water

AMOP developed an Arctic Boom for oil recovery in remote Arctic localities (Meikle 1981c).

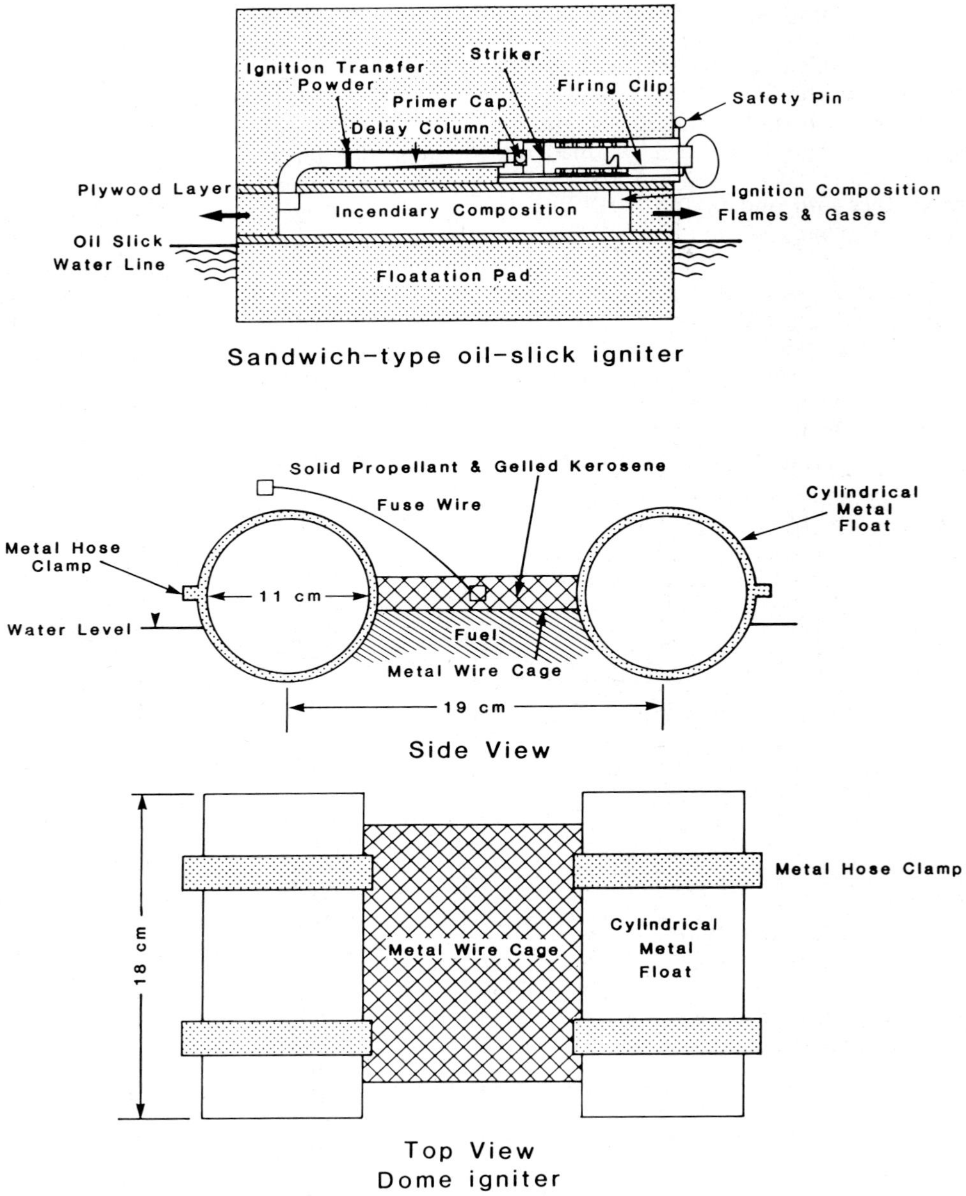


Both the igniters shown above are designed to be dropped from helicopters onto oiled melt pools on ice. They have been successfully field tested (Dome, 1981).

FIGURE 5. *Igniter designs.*

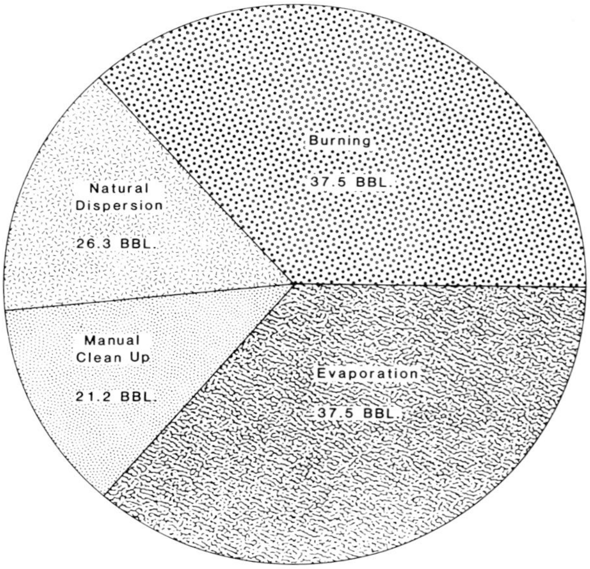


FIGURE 6. *Mass balance for oil (122.5 BBL.) discharged under ice.*

Due to requirements for light weight and for minimum mechanical support systems, this boom is self-inflating (McAllister 1980). Tests of Arctic Boom and a compacting Zoom Boom compared favourably to the VIKOMA Ocean Pack system (Purves *et* al 1980).

A skimmer with a skimming head that moves at the same speed as the surface current was developed and tested by AMOP for cold ocean waters (Abdelnour *et al* 1980). It will be used with the Beaufort Sea Co-op oil recovery barge, which has a separator, heat treater and burner capable of Processing 800 $m^3$/d (Pistruzak 1981).

## Burning

This section discusses Canadian research and development into burning Arctic marine oil spills in ice-infested waters and in cold open water. This section also discusses research to develop disposal burners for recovered oil.

### Burning of Oil Entrapped in Ice

In 1975, APOA and Environment Canada studied a controlled spill of crude oil under first-year Arctic sea ice (NORCOR 1975). This study found that most of the oil could be disposed of by *in situ* burning the following spring. This led to research and development studies of the *in situ* combustibility of various types of oils under Arctic conditions and to the development of oil slick ignition devices.

Research on the *in situ* combustibility of spilled oil in Canada initially concentrated on only crude oils (Arctec 1977, Energetex 1977, NORCOR 1976a). Further work investigated weathered and emulsified crude oils and other types of oil (including diesels and bunker fuels) under simulated Arctic environmental conditions (Energetex 1980a, 1981a). The major conclusions of this work were as follows. It is possible to burn oils *in situ* in ice conditions. The minimum ignitable thicknesses of various oils are approximately: 1 mm for fresh crude oil and diesel, 3 to 5 mm for weathered crude oil, up to 10 mm for bunker oils and heavily emulsified crude oils. It is possible to burn oil *in situ* at wind speeds up to 45 km/h, at ambient temperatures as low as —40°C and with entrained snow and slush contents as high as 60 % by weight. Burning efficiencies as high as 98 % have been achieved, with an average for weathered crude oil under wind herding conditions being in the 80 % range. Burn efficiency (on a volume basis) is generally lower than that for fresh crude oil, for weathered oils, emulsified oils, and heavy oils. Thicker slicks burn more completely. In concentrations below 30 % by weight, entrained snow increases combustion efficiency, although higher snow concentrations tend to reduce burn efficiency.

Figure 4 shows results of some research into *in situ* burning in wind conditions.

There has also been much research into techniques to ignite oil *in situ*. These efforts have concentrated on developing devices suitable for use from aircraft. Early development focused on the selection of suitable materials for an oil slick igniter (Energetex 1977, 1978). Solid propellant was the most appropriate and reliable ignition agent. Subsequent work concentrated on the design and testing of air deployable igniter packages (Meikle 1981a, Energetex 1980b, Dickins 1979). Figure 5 shows the two igniter designs selected.

Dome Petroleum Ltd. (1981a) recently simulated an oil blowout under ice in the winter of 1979-80. The oil released in this experiment appeared on the ice surface in pools the following spring. Approximately 80 % of the oil discharged was burned before breakup (Figure 6). One-third was burned *in situ* using igniters dropped from a low-flying helicopter (plate 1). Individual Pool burning efficiencies averaged 80 % (Dome Petroleum Ltd. 1981a).

Investigations into the logistics of helicopter

PLATE 1: *Helicopter deploying igniters onto oiled melt pools.*

deployment and into ice drift trajectories indicate that most of the oil from a major winter subsea blowout in the Beaufort Sea could be disposed of by igniters dropped in spring (Ross Environmental Research Ltd. 1981). Thus, a novel countermeasure has been developed for ice-infested waters. Tracking oiled ice over winter, burning oil that appears in leads and ice cracks and deploying igniters from helicopters in the spring would enable cleanup of much of a major oil spill before breakup.

## Open Water

The IXTOC 1 incident and the sinking of the *Atlantic Empress* provide evidence that it is possible to burn oil on open water with some success (Ross *et al* 1979, Horn and Neal 1981). Canadian industry and government research has shown that ignition of oil on water requires a certain minimum oil thickness.

Prototypes of fireproof booms have been developed to thicken floating oil and contain it during combustion. Work has been conducted on several design concepts (e.g. Roberts and Chu 1978, Purves and Daoust 1978, McAllister 1979, McAllister and Buist 1980). Stainless steel booms (Plate 2) can contain burning crude oil for long Periods, in waves equivalent to seastate 2-3 (the highest tested to date) and in currents up to 0.5 m/s (McAllister and Buist 1980, Dome Petroleum Ltd. 1981b). Final offshore durability trials of this boom are underway. It is possible to collect and dispose of large volumes of oil from a blowout with only moderate lengths of the boom and minimal logistical support (Figure 7). A less robust fireproof boom design is also being tested (Meikle 1982).

The completion of work on fireproof booms and their eventual stockpiling in Canada will be a valuable complement to conventional oil spill response technology.

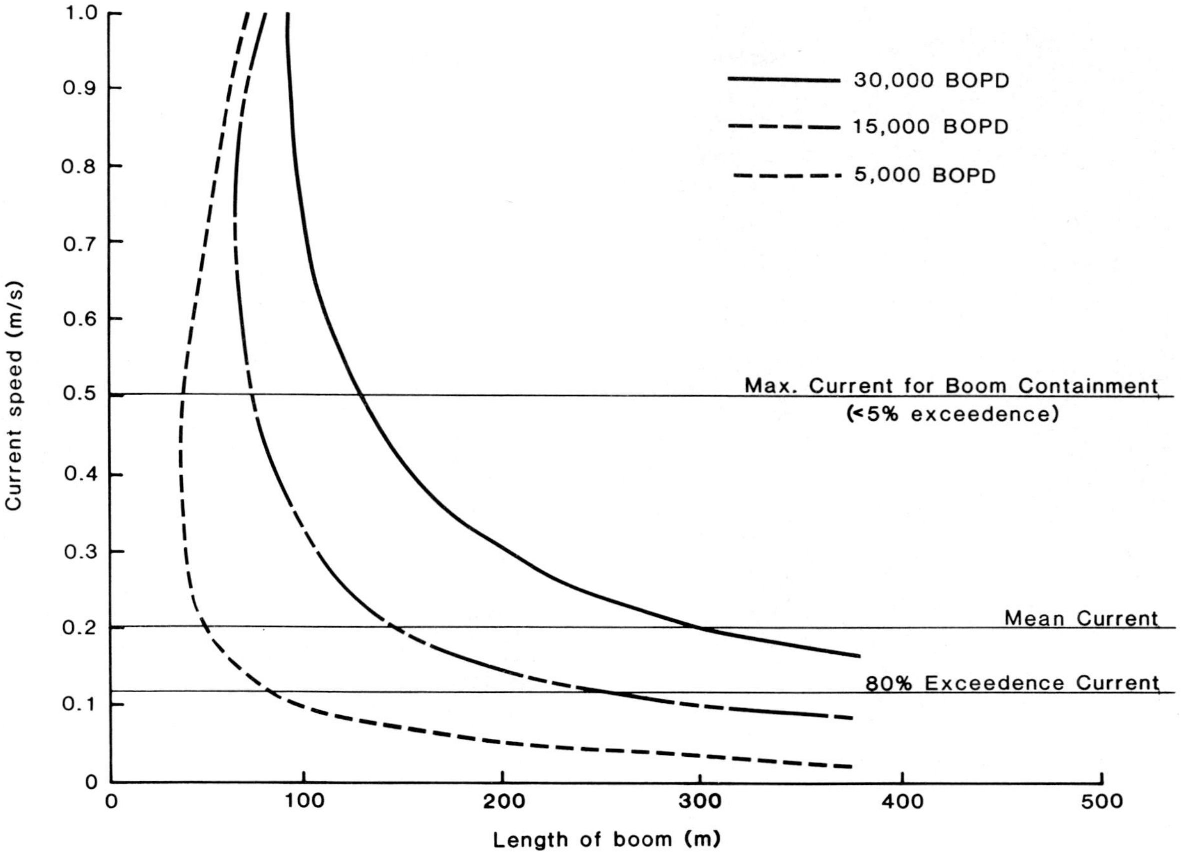


The graph illustrates the lengths of fireproof boom required to burn in-situ certain volumes of oil per day. The higher the current the boom is required. Also shown are the range of currents in Canada's Beaufort Sea.

FIGURE 7. *Length of boom required.*

## Disposal of Recovered Oil

A problem in any oil spill cleanup operation is the disposal of recovered oil and oiled material, particularly in remote areas and on rocky or tundra coasts.

Canadian research has focused on controlled burning techniques to dispose of oily material collected in remote Arctic and sub-Arctic sites. The basic approach has been to divide recovered material into three categories: fluid oils and emulsions with viscosities less than 500 cS; oiled debris (driftwood, seaweed, sorbents) and viscous emulsions; and oiled beach material and tar balls. The Primary design criterion for burners is portability by medium lift helicopter.

Research on fluid oil burners has concentrated on atomization techniques and system design. Early work showed that simple pneumatic atomization burners were suitable for light oils with low water content (Nikolaychuk and Thornton 1976, Pistruzak 1981). However, auxilliary equipment, particularly the compressors required for these systems, limited transportation by helicopter. One such burner does form an integral part of the Beaufort Sea Co-op response barge (Plate 3). The barge carries oil separators, heaters and the necessary auxilliary equipment to burn as much as 800

PLATE 2: *The stainless steel fireproof boom, in a test tank.*

m³/d of recovered oil with as much as 40 % water content by volume (Pistruzak 1981).

Another technique is mechanical atomization. A burner, portable by helicopter, has recently been developed for shoreline land nearshore cleanups (Dome Petroleum Ltd. 1982). The burner has a rotary cup to atomize the oil. The system, shown in Plate 4, can burn up to 80 $m^3/d$ of a wide variety of oils, including emulsions with as much as 60 % water content by volume. This system is also routinely used for slop and waste oil disposal.

PLATE 4: *The heli-portable burner is capable of disposing of up to 80 $m^3$/day of recovered oil.*

Two types of incinerator have been developed to dispose of oiled debris and heavy emulsions during shoreline cleanup. The Air Curtain Destructor is a pit dug in the ground, with a blower and ducting to force air into the pit from one side (PROSCARAC 1980). The resulting vortical air flow burns large amounts of oiled debris very efficiently (Plate 5). However, this type of incinerator is not suitable for ice-rich soils.

PLATE 3: *The Beaufort Sea co-op response barge.*

PLATE 5: *The air curtain destruction.*

A second incinerator is the box type, consisting of components light enough for helicopter transportation (Plate 6). The device has been tested with a wide range of oiled debris and has a capacity of one tonne per hour (Twardus 1979). Work is continuing to improve the design of box incinerators, to reduce weight for helicopter transport and to minimize field assembly.

PLATE 6: *Heli-Portable box incinerator*

A reciprocating kiln has been developed to clean and replace oiled beach sediments (Gill and Stevens 1980). The device, shown in Plate 7, is made of empty oil drums. Oiled sand is shovelled in one end and tumbled by the reciprocating action. Burning wood or waste oil in a chamber beneath the kiln preheats the oiled sand. Oil in the sand is burned off and clean sand exits. The entire unit can be carried

PLATE 7: *Heli-Portable reciprocating kiln*

by helicopter. Work is continuing to extend the operating life of the kiln by substituting rolled stainless steel for oil drums. The entire unit can be carried by helicopter.

## Dispersants

Chemical dispersants may be the only practical oil spill countermeasure in many arctic and subarctic areas during break-up, the open water period and freeze-up. Remoteness and the danger to field personnel may prevent physical containment and recovery of oil, while the thinness of oil slicks may prevent *in situ* burning.

Using chemical dispersion of oil to accelerate natural dispersion entails five problem areas:

i) *logistics:* how to get the right dispersant to the right place at the right time in the right quantities;
ii) *application techniques:* how to apply the chosen dispersant most effectively to the oil slick at the best application rate and correct droplet size;
iii) *effectiveness:* of the dispersant on fresh and aged oil at the ambient temperatures and salinities;
iv) *toxicity:* the lethal and sublethal effects of dispersant that can be expected and tolerated;
v) *authority to make decisions:* how and when decisions to use dispersant are made by the operator and the regulatory authorities.

### Logistics

Betts and Golob (1982) assessed the logistics needed to develop a dispersant response plan and to systematically prepare for dispersant application. He outlined three planning and three implementation steps for a response plan capability. He looked at two recent case histories, and concluded that systematic planning could help avoid a repetition of past problems.

### Application Techniques

Hildebrand *et al* (1977) studied the feasibility of various methods of dispersant application in the Beaufort Sea. These included application by fixed-and rotary wing aircraft, and by air cushion and surface vessels. Using a hypothetical oil spill scenario, the study concluded that any method would be very expensive, but that fixed and rotary wing aircraft could be suitable for arctic dispersant application.

A series of trials of aerial application of dispersant have been carried out by an oil industry/government group, the Canadian Offshore Aerial Applications Task Force (COAATF). The trials used a Conair DC-6B aircraft. There were land-based trials at Riviere du Loup, Quebec in 1978; Abbotsford, B.C. in 1979; Suffield, Alberta in 1980; and a marine trial off St. John's, Newfoundland in 1981. A landbased helicopter trial is scheduled for Dartmouth, Nova Scotia in October 1982. The last two trials are COOSRA projects. These trials measured the effectiveness of aerial application and measured dispersant drift and swath width, droplet size, application rates and overall coverage (Gill and Ross 1980, 1982). The Abbotsford and Suffield trials achieved a dosage rate of approximately 190 l/ha using nozzles or pipes, depending upon atmospheric conditions. These trials also showed that a decrease in dispersant viscosity may be associated with a reduction in droplet size (Gill 1981).

## Effectiveness

Because dispersant use is an expensive oil spill countermeasure, one must know whether chemical dispersants will in fact disperse different types of crude oil, at different stages of weathering and emulsification, and at various temperatures and salinities.

Much of the early work in the effectiveness of dispersants in Canada was conducted at the microscale level in laboratories by Mackay and co-workers. (Mackay *et al* 1978, Wells and Harris 1979, Harris and Wells 1980, Mackay *et al* 1980b; Mackay and Wells 1982). They tried to devise a satisfactory method of rating dispersant effectiveness both in sea water and in simulated ice conditions. To coordinate this work with other international research, a symposium was held in Toronto in 1980, where several papers were given on laboratory tests by U.S., British, French and Dutch researchers (Mackay *et al* 1980).

To test dispersant effectiveness in a cold open water environment, COAATF performed the St. John's experimental oil spill trial using Corexit 9527 and fresh Lago Medio crude. Preliminary reports indicate problems in sampling the water column for dispersed oil. There were good results from remote sensing, which showed a brown cloud of oil-in-water emulsion that lasted for 3 hours (Gill and Ross 1982). To obtain better water samples, the 1982 COAATF trial will be conducted at an air base in Dartmouth, Nova Scotia with water tanks and application by helicopter.

A standard dispersant effectiveness test is needed for two reasons. New and existing commercially available chemical dispersants must be tested by regulatory agencies. Secondly, operators must have a quick and easily conducted test to use during oil spills, to establish whether the available dispersant is effective at that particular time and place on the type of crude oil involved. Such a test would determine whether the operator should regard dispersants as a feasible oil spill countermeasure.

Ongoing research includes laboratory studies and field trials in a Project of British Petroleum and Petro-Canada, and studies by Exxon to develop an effective cold water dispersant for arctic and subarctic regions.

## Toxicity

Following the *Torrey Canyon* oil spill in 1967, there has been much study of the acute and sublethal toxicity of dispersants and dispersed oil to phytoplankton, zooplankton, the benthos, fish, birds and marine mammals. Little of this work deals with arctic conditions, because it is difficult to simulate an arctic environment in the laboratory. Good summaries of toxicity research are in Wells and Harris (1980), Mackay and Wells (1981), and Sprague *et al* (1982). Blackall and Sergy (1981) describe the preliminary design of the BIOS project, which will investigate the toxicity of dispersed oil to nearshore organisms.

## Dispersant Use Decisions

In Canada, unless there is a danger to human life or safety at a spill site, approval for the use of chemical dispersants must be sought from the Environmental Protection Service. The present «Guidelines on the Use and Acceptability of Oil Spill Dispersants» (Environment Canada 1973) are based on freshwater toxicity tests with fingerling rainbow trout. These guidelines will be replaced in 1982 by new effectiveness, toxicity and use standards (Mackay and Wells 1982).

Because of the need to plan the logistics of a large scale chemical dispersant application, the

Canadian oil industry must know in advance the conditions under which approval to use dispersant would be given. Such approvals would involve an evaluation of the relative risk of driving dispersed oil into the water column, where it may affect pelagic and benthic organisms, versus leaving untreated oil on the water surface, where it could affect shorelines, seabirds and marine mammals (Trudel 1980). Scenario modelling can aid such evaluations. Trudel (1980) used the oil fate model of Mackay and Paterson (1980a) to assess the impact of oil to a common murre colony and to Atlantic cod stocks off southeastern Newfoundland. This model has since been modified to include several additional species and applied to the mobil Grand Banks environmental data set as a COOSRA project. It is anticipated that once this model is operational, a series of government/ industry workshops will be conducted to establish criteria for approvals to use dispersants.

## Subsea Containment

The concept of a subsea device that would contain the oil and gas escaping from a blowout has been considered for years. Chen (1979) and Abdelnour *et al* (1977) have studied the engineering feasibility of subsea containment. As well, a joint industry/government workshop in 1981 reviewed various aspects of subsea containment and identified areas for further research (Meikle 1981b).

Since the 1981 workshop, Canadian government and industry investigations into subsea containment systems for Arctic marine waters have focused on six areas:

i) the flow of gases and liquids in risers;
ii) collector and riser model testing;
iii) environmental loads,
iv) *in situ* incineration of the oil and gas at the water surface;
v) preliminary design and feasibility testing of a floating cylinder incinerator system; and
iv) preliminary design and feasibility testing of a converted oil tanker system for separation and recovery of the oil.

Each of the areas of research concentrated on one component of the conceptual subsea containment system shown in Plate 8.

Topham *et al* (1982) investigated two phase flow (of liquids and gases) in risers, using computer simulation techniques. They confirmed that the water: oil ratio in the fluid collected strongly depends upon riser resistance, collector/source separation and pipe diameter. They calculated that, for a blowout of 4800 $m^3$ oil/day with a gas to oil ratio of 160:1, a 75 cm diameter riser could handle the flow up to a water : oil ratio of 50:1 to 70:1.

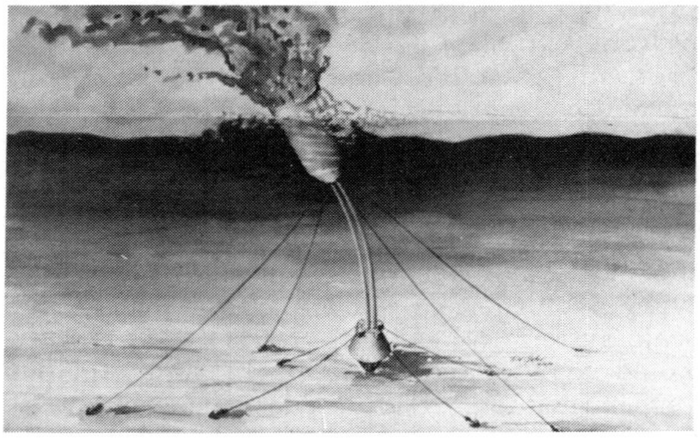

PLATE 8: *The conceptual arctic sub sea containment system*

Small scale model tests by C-CORE (1981) of riser and collector types showed that, if a riser could be brought close enough to a blowout source, no collector would be required. The air-lift action of the gas in the riser would draw in all the escaping fluid at the bottom. Flexible and semi-permeable risers were also evaluated. These were very susceptible to fluttering, which in actual conditions may produce unacceptable wear.

CanOcean (1981) analyzed the environmental loads on a system such as the one shown in Plate 1. They made separate calculations of the forces exerted on the collector, riser(s) and incinerator by each of waves, wind, currents, level sea ice, small ridges and large ridges. Since no mooring system could withstand large ice masses, a limiting ice criterion was set, whereby the incinerator would be submerged by a large ice mass and break through thinner ice once the large mass had passed. The results of the environmental load study were reduced to simple equations and nomographs, so that the effect of system design changes could be easily and quickly examined.

A concept for disposal of the collected products of an Arctic subsea blowout is to burn the oil and gas *in situ*. To evaluate the feasibility of this concept, Energetex Engineering (1981b) tested both an atomizing tip and a pool type burner. Oil, gas and water were Pumped into the burners to simulate gas-to-oil ratios of from 200:1 to 400:1 and oil to-water proportions of from 0 to 100 %.

These tests indicated that the atomizing burner was more effective for water contents up to 50 % and it produced a clean burn. If water content was more than 50 %, the pool burner was more effective. Since the water content of the collected product could vary considerably, the incinerator design should incorporate both types of burner, as shown in Figure 8. A full scale system to deal with a 2,400 $m^3/d$ blowout would require an atomizing tip approximately 1 m in diameter and a pool burner approximately 30 m in diameter.

CanOcean (1982) made a preliminary system design that combined all these features to determine the feasibility of an Arctic subsea containment system. CanOcean examined the structural design; materials of construction; and the availability and deployment of the collector, riser and incinerator (Figure 9). An important criterion was the capability for assembling most of the system within three weeks, thus eliminating expensive pre-construction and stockpiling. Rapid assembly would also allow the choice of appropriately sized components for the flow rates observed at the blowout site. CanOcean concluded that, for severe ice conditions, such a system would be extremely expensive and require a long time to construct. In particular, the mooring system required to resist the design ice load would be difficult to obtain. The major limitation on the mooring strength is the anchor holding power required to resist ice loads. The deployment of such a massive system would be immensely difficult, if not impossible,

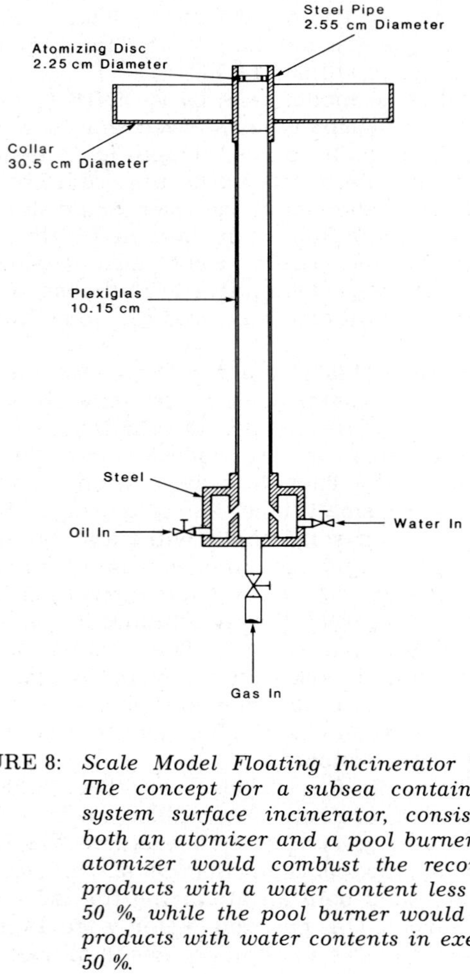


FIGURE 8: *Scale Model Floating Incinerator Tests. The concept for a subsea containment system surface incinerator, consists of both an atomizer and a pool burner. The atomizer would combust the recovered products with a water content less than 50 %, while the pool burner would burn products with water contents in exess of 50 %.*

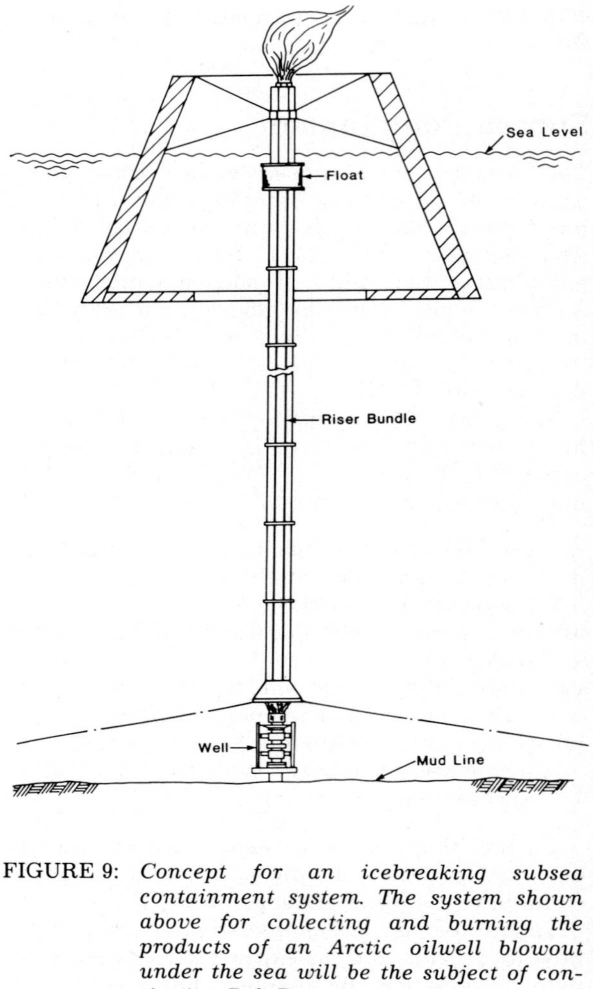


FIGURE 9: *Concept for an icebreaking subsea containment system. The system shown above for collecting and burning the products of an Arctic oilwell blowout under the sea will be the subject of continuing R & D.*

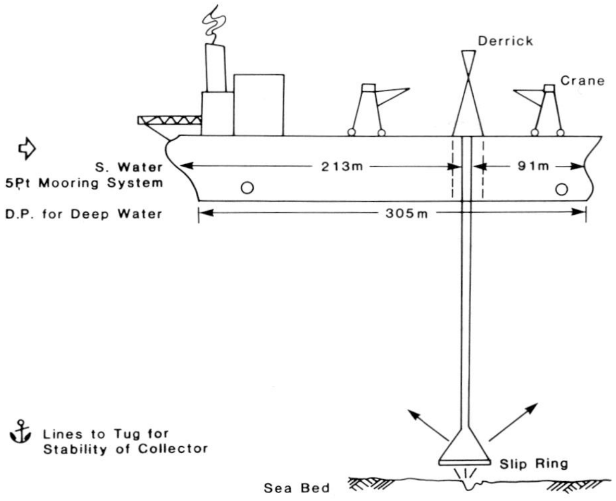


FIGURE 10: *Tanker subsea containment system. The tanker system for subsea blowout containment shown above was suggested by Capt. T. Robinson of Dome at the recent workshop on subsea containment. Further analysis has indicated that the use of a converted VLCC could be possible.*

and it would be beyond existing marine technology to remotely pretension the anchor chains.

An alternative to a collector and incinerator system is to collect the blowout products and pass them to a platform for separation and recovery of the oil. German and Milne Inc. (1982) studied the feasibility of converting a conventional VLCC to an appropriate subsea containment platform.

This concept entails cutting a moon pool into the bow section of a tanker through which a collector and riser could be lowered (Meikle 1981b). The tanker would move over the blowout, position the collector over the source, recover the collected fluids, separate oil, water and gas in the cargo tanks, store the oil, flare the gas and discharge the water (Figure 10). Initially, the concept seems feasible and there are no classification society rules or regulations that could not be met. Inert gas systems, water sprays and other systems can be installed with present technology. The positioning of the tanker accurately over the blowout would require some further investigation. The system would operate only in open water conditions.

## Shoreline Protection and Cleanup

The Canadian approach to the protection and cleanup of its vast arctic coastlines has three components. Coastal areas are being catalogued with standardized shoreline classification systems. Protection and cleanup strategies are being designed to be consistent with the various shoreline types and logistical limitations. Techniques and hardware are being developed to clean up polluted coastal areas and to dispose of oil and oiled debris in remote areas of the arctic.

### Shoreline Classification

One of the initial projects in this field was the Beaufort Sea Shoreline Study (Worbets 1979). This study was a sensitivity mapping of all coastal regions between the Alaska-Yukon border (141°W) and the Baillie Islands (128°W), and was based on two seasons of extensive field work. Combined with the Beaufort Sea Atlas, the shoreline study was used to design countermeasure options for the southern Beaufort Sea coastline.

In the late 1970's, further classifications were conducted by AMOP and the Geological Survey of Canada, designating coastal types of Labrador and Baffin Island. Since this time, there have been major advances in the methods of collecting and presenting coastal geomorphological information. The Victoria B.C. office of Woodward-Clyde Consultants has used a video-taping technique for much of the arctic and subarctic shoreline, for both Dome Petroleum and Petro-Canada. The raw footage is annotated and used to produce a physical shore zone analysis manual.

### Oilspill Response Strategies

The results of these shoreline classification studies (Worbets 1979, Robilliard *et al* 1980, Robilliard and Owens 1981) are then incorporated with the shore zone analysis into oil spill cleanup manuals. The manuals rank geological, biological and human sensitivities and areas of potential disturbance. The manuals also outline response actions and potential cleanup methods for specific geographic areas and sometimes indicate local and regional logistic facilities available.

The pioneering of these methods is legitimate oil spill research, although further development will emphasize operational techniques.

## Shoreline Cleanup and Waste Disposal

Hardy Associates (1979) investigated oiled debris disposal and storage sites along the Beaufort Sea coast and the problems involved in site selection and construction in permafrost areas. The report, however, was not applicable to the central and western arctic, where the shorelines were sufficiently different to make this approach impractical.

Twardus (1980) evaluated *in situ* combustion as a shoreline cleanup technique. This work indicated that low viscosity oils rapidly penetrated beach sands, making *in situ* combustion impossible. However, the same oil, when emulsified by rain, was combustible because of its reduced viscosity. The addition of peat moss also improved *in situ* combustion.

Dickins (1981) performed a follow-up study on the work by NORCOR (1975) in Balaena Bay. He found that, although the initial spill was considered well cleaned up, approximately 5 % of the applied crude oil remained at the experimental salt marsh site after six years. Oil residues left on the test core shoreline were reduced in volume from about 1 $m^3$ to less than 0.2 $m^3$ since August 1975.

The most recent work on shoreline countermeasures in the Canadian Arctic has been the shoreline component of the BIOS project. In 1980, eight test plots were oiled using aged Lago Medio crude oil and an emulsion of water in aged Lago Medio crude oil. The oils were spilled on both backshore and intertidal test plots in both high and low energy areas of the beach. At the high energy beach plots, 50-90 % of the oil was removed in 24 to 48 hours; at the low energy or sheltered plots, 30-90 % of the oil was removed by tidal action (Woodward-Clyde Consultants 1981).

This work provided the baseline data for the 1982 field season (Owens *et al* 1982), when the following counter-measures were investigated in a medium energy shoreline:

a) *in situ* combustion using the DREV igniter;
b) chemical surfactants designed to disperse oil;
c) mechanical mixing of oil-contaminated sediments;
d) application of a surface solidifying agent.

Results indicated that the incendiary device was not effective in igniting either the aged crude or the emulsified oil. The water-based dispersant effectively reduced the total oil loading, partly by driving the oil deeper into the sediments. This dispersant appeared to be less effective on the emulsified oil plot. The oil-based dispersant also drove the oil more deeply into the sediments, but by day 40, all the total hydrocarbon values were reduced to trace concentrations. This may be significant where it is necessary to remove stranded oil from the beach surface in shorter periods than could reasonably be expected from natural cleaning processes.

The mixing operation, not suprisingly, reduced the surface concentrations of stranded oil, but increased the subsurface concentrations.

The solidifying agent caused much of the oil to form a bonded compound that retards sediment reworking, but facilitates physical recovery.

In 1982, it is planned to determine if oil can be more readily removed from low energy environments, mainly by the application of dispersants. These low energy environments make up almost 40 % of the total shoreline in the Arctic, and they tend to be biologically productive. It is therefore important to have some basic understanding of the persistence of oil in these areas.

In general, little research has been conducted on arctic shorelines, because of the logistical problems of conventional shoreline cleanup. BIOS remains the best practical field study in shoreline cleanup countermeasures.

# Future Research Directions

At the present time arctic marine oilspill research is being conducted by AMOP and COOSRA. These two organizations are the major funders of BIOS, which is scheduled to be completed in 1983. AMOP was planned as an eight-year study from 1977-1984. COOSRA was conceived an industry association with an initial four-year life (1980-1983) with the option of continuation or termination in 1983. Recent legislation in Canada has introduced the concept of two Environmental Studies Revolving Funds (one north of 60° N Lat. and one south of 60° N), both funded by the oil industry and managed by the federal government with input from industry representatives. The revolving funds may eventually replace or incorporate COOSRA.

Major areas of future arctic oil spill research are seen as:

### A. *Monitoring and Surveillance*

Improved oil spill following buoys both for surface and subsurface tracking of oil. Following the Kurdistan oil spill, it was found that thick weathered bunker-C oils were found to travel at subsurface depths below the ice and finally contaminated beaches tens of kilometres away.

A promising technique for detecting oil under ice is sonar penetration from the upper ice surface.

Modelling of oil-under-ice trajectories is an area of research that requires further investigation.

### B. *Mechanical Containment and Recovery*

A workshop on the recovery of oil in ice is scheduled for Calgary in October 1982. It is expected to look at past technology and make recommendations for future research in an area where there is currently little available effective equipment.

### C. *Dispersants*

From the BIOS results, it is anticipated that guidelines for dispersant use, shoreline protection and cleanup will be produced for arctic waters.

Further work on the B.P./Petro-Canada and Exxon cold water dispersants can be expected. Effectiveness and toxicity tests and field trials can be expected.

The B.P. polymer solidifying agent was tested at Cape Hatt in the spring of 1982. Because this seems to be a promising countermeasure for both oil under ice and arctic shoreline cleanup, further developmental research is expected. AMOP is considering a series of applied field trials in arctic conditions.

### D. *Subsea Containment*

Further work on subsea containment systems in both open water and light ice conditions can be expected to reduce the mooring system stresses and loads (Buist and Potter 1982).

### E. *Miscellaneous*

Because of the high cost of conducting arctic marine oil spill research in the field, both AMOP and some oil companies have set money aside so that oil spill scientists and countermeasure specialists may visit accidental spills in the arctic to better understand the fate, behaviour and effect of oil in the arctic environment.


## Acknowledgements

The authors would like to express their appreciation of valuable contributions to this paper by D. Thornton of Environmental Emergency Branch, E.P.S., W. Pistruzak of Dome Petroleum, and W. Robson of Petro-Canada. Encouragement for this paper was given by J. Hnatiuk, Chairman of the Arctic Petroleum Operators Association and J. Gainer both of Gulf Canada Ltd.

Finally thanks is due to D. Karasiuk for the difficult task of editing the various diverse contributions and to Ms. S. Herman and G. Leduc for the typing of the manuscripts.


## Appendix A

A.P.O.A. Oil Spill Projects
(including Beaufort Sea Projects)

72-27 *The Interaction of Crude Oil with Arctic Sea Ice* NORCOR Engineering and Research Ltd. Department of the Environment, 1975.

72-28 *Movement of Oil Under Sea Ice* L.W. Rossenegger. Department of the Environment, 1975.

72-29 *Light Intensity and Primary Productivity Under Sea Ice Containing Oil* W.A. Adams. Department of the Environment, 1975.

72-31b *Oil Spill Countermeasures for the Beaufort Sea: Appendix* W.J. Logan, D.E. Thornton, and S.L. Ross. Department of the Environment, 1975.

72-33 *Hydrodynamics of an Oil Well Blowout* D.R. Topham. Department of the Environment, 1975.

97-1 *Tests of the Arctic Boat Configuration of the Lockheed Clean Sweep Oil Recovery System in a Broken Ice Field* L.A. Schultz, ARTEC Canada Ltd., 1976.

100-1 *Proposal to Evaluate an Oil Containment Boom for Use in Ice-Infested Waters* ARCTEC Canada Ltd. For CANMAR Drilling Ltd., 1975.

100-2 *Evaluation of the Bennett-Canmar Oil Contaiment Boom* J.R. Nadreau, W.G. Wallace, R.Y. Edwards Jr., ARTEC Canada Ltd., 1976.

107-1 *Some Aspects of Weathering and Burning of Crude Oil in a Water-*

*and-Ice Environment* J. Belicek & J. Overall, NORCOR Engineering, 1976.

108-1 *In-Situ Burning of the Products of an Sub-Sea Blowout* ARCTEC Canada Ltd., T. Brzustowski (University of Waterloo), K. Aziz (University of Calgary), 1977.

114-1 *Preliminary Tests of Bird-Scare Devices on the Beaufort Sea Coast* P.L. Sharp, LGL Limited, 1978.

119-1 *Detection of Oil Trapped Under Ice Using Impulse Radar, Phase II Report* J.R. Mann, Geophysical Survey Systems Inc., with CRREL, 1976.

149-1 *Sliktrack Simulations — East Coast* D. Bradfield, M. Metge, A. Telford, Imperial Oil Limited, 1977.

149-2 *Oil in Pack Ice Coldroom Tests* M. Metge, Imperial Oil Limited, 1978.

149-3 *Deep Sea Dispersion Analyses* NORDCO Ltd., with R.H. Loucks Oceanology Limited, 1977.

# Appendix B

AMOP PROJECT PUBLICATIONS
1977 — 1982

1977 — Probable Behaviour and Fate of a Winter Oil Spill in the Beaufort Sea — EPS-4-EC-77-5
— Novel Countermeasures for an Arctic Offshore Well Blowout — EPS-3-EC-77-14
— The Feasibility of Oil Spill Dispersant Application in the Southern Beaufort Sea — EPS-3-EC-77-16
— A Field Study of the Permeability of Multiyear Ice to Sea Water with Implications on its Permeability to Oil — EPS-4-EC-77-11
— Field Evaluation of Oil Mop and Preheat Unit — EPS-4-EC-77-12

1978 — Pumps for Oil Spill Cleanup — EPS-4-EC-78-3
— Field Evaluation of the Super Seahawk and Marco Class V Oil Skimmers — EPS-4-EC-78-2
— Field Evaluation of Eight Small Stationary Skimmers — EPS-4-EC-78-5
— The Interaction of Crude Oil and Natural Gas with Laboratory-Grown Saline Ice — EPS-4-EC-78-9
— Probabilities of Blowouts in Canadian Arctic Waters — EPS-3-EC-78-12
— Review of Ice Movement Buoys for Tracking Oil Spills — EPS-3-EC-78-7
— Arctic Oil Spill Countermeasures Logistics Study: Analysis Report — EPS-3-EC-78-9
— Arctic Oil Spill Countermeasures Logistics Study: Summary Report — EPS-3-EC-78-8
— Testing of Air-Deployable Incendiary Devices for Igniting Oil on Water — EPS-3-EC-78-11

1979 — Theory, Development and Testing of an Ice-Oil Boom — EPS-4-EC-79-2
— Evaluation of Pumps and Separators for Arctic Oil Spill Cleanup — EPS-4-EC-79-3
— Characteristics of Atmospheric Emissions from an in-situ Crude Oil Fire — EPS-4-EC-79-1
— Oil Spill Scenario for the Labrador Sea — EPS-3-EC-79-4
— Oiled Debris Disposal and Storage Sites: Beaufrot Sea Coast — EPS-3-EC-79-3
— A Review of Oil Slick Combustion Promoters — EPS-3-EC-79-8
— Feasibility of Surface Detection of Oil under Ice — EPS-3-EC-79-11

1980 — Microwave Systems for Detecting Oil Slicks in Ice-Infested Waters: Phases II and III — EPS-3-EC-80-3
— Effect of Surface Temperature on the Remote Sensing of Oil Using Laser Fluorsensors — EPS-4-EC-80-2
— Thin Oil Film Detection Limits of Laser Fluorsensors — EPS-4-EC-80-3
— A Selected Bibliography on the Fate and Effects of Oil Pollution relevant to the Canadian Marine Environment — EPS-3-EC-80-3

1981 — Microwave Systems for Detecting Oil Slicks in Ice-Infested Waters: Phase I — Literature Review and Feasibility Study — EPS-3-EC-81-3
— The Use of Satellite Imagery for Tracking the Kurdistan Oil Spill — EPS-4-EC-81-5
— An Evaluation of Oil Pumps and Skimmers — EPS-4-EC-81-4

1982 — Deep Water Blowout Trajectory Models for the Lancaster Sound Region — EPS-4-EC-82-2

# Appendix C

Canadian Offshore Oil Spill Research Association (COOSRA) Projects

1980
1. Baffin Island Oil Spill Project
2. Beaufort Sea Oil and Gas Under Ice Experiment
3. Portable Burner Development
4. Fireproof Boom Development
5. *In situ* Combustion Against Ice Edges

1981
1. Baffin Island Oil Spill Project
2. Offshore Dispersant Trial
3. NOAA Real time Oilspill Trajectory Model Conversion
4. Fireproof Boom Testing at the OHMSETT facility
5. Dispersant Use model
6. Ring Containment System Design
7. Return to Balaena Bay

1982
1. Baffin Island Oil Spill Project
2. Fireproof Boom Field Trials
3. Subsea Containment Field Trials
4. Emulsions in Ice Study
5. Oil in Pack Ice Study
6. Dispersant Effectiveness Trial
7. Mechanical Recovery in Ice Workshop
8. Portable Incinerator Development

---

# References


Abdelnour, R., A.M. Nawwar, P. Hildebrand and W.F. Purves. 1977. Novel Countermeasures for an Arctic Offshore Well Blowout. Environment Canada Report EPS-3-EC-77-14. Ottawa, Canada.

Abdelnour, R., B. Roberts, W.P. Purves and W. Wallace. 1980. Field Evaluation of Oil Skimmers. Proc. 4th AMOP Technical Seminar. Environment Canada. Edmonton.

Arctec Canada. 1977. *In Situ* Burning of the Products of a Subsea Blowout. APOA Project 108. Calgary, Canada.

Bedford Institute of Oceanography. 1980. Review of Real-Time Oil Spill Trajectory Models and Recommendations for Real-Time Oil Spill Model Services in Canada. First Draft by the Working Group on Oil Spill Trajectory Modelling.

Betts, D.A. and R.S. Golob. 1982. Planning for a Dispersant Response. 5th AMOP Technical Seminar. Edmonton, Alberta. pp. 285-294.

Blackall, P.J. and G.A. Sergy. 1981. The Baffin Island Oil Spill Project. Proc. 4th AMOP Technical Seminar. Edmonton, Alberta. pp. 705-741.

Buist, I.A. and S.G. Potter. 1982. Subsea Containment: COOSRA Research to Date. Proc. 5th AMOP Technical Seminar. Environment Canada. Edmonton, Canada.

Canadian Petroleum Association. 1978. Oil Recovery From Under River Ice. Calgary, Alberta.

CanOcean Resources Ltd. 1981. Subsea Containment Study: Phase 3, Environmental Loads. COOSRA report CS8V3. Calgary, Canada.

CanOcean Resources Ltd. 1982. Subsea Containment Study: Phase 4, Preliminary System Design. COOSRA report CS8V4. Calgary, Canada.

C-CORE. 1981. Subsea Containment Study: Phase 1, Preliminary Feasibility. COOSRA report CS8V1. Calgary, Canada.

Chen, K.W. 1979. The Feasibility of Underwater Containment of Subsea Oil Spills in Arctic Waters. Proc. 2nd AMOP seminar. Environment Canada. Edmonton, Canada.

Dickins, D.F. 1979. Air-Deployable Oil Spill Igniter Tests — Yellowknife. APOA Project 164. Calgary, Canada.

Dickins, D.F. 1981. Return to Balena Bay. COOSRA Report LS7. Calgary, Canada.

Dome Petroleum Ltd. 1981a. Oil and Gas Under Sea Ice Study. COOSRA Report CS4V1. Calgary, Canada.

Dome Petroleum Ltd. 1981b. Fireproof Boom Development — OHMSETT Trials. COOSRA report CS3. Calgary, Canada.

Dome Petroleum Ltd. 1982. The Development and Testing of a Helicopter Portable Burner. COOSRA report CS5. Calgary, Canada.

Energetex Engineering. 1977. Ignition and Burning of Crude Oil on Water Pools and Under Arctic Spring Time Conditions. APOA Project 141. Calgary, Canada.

Energetex Engineering. 1978. Testing of Air-Deployable Incendiary Devices for Igniting Oil on Water. Report to Environment Canada. Ottawa, Canada.

Energetex Engineering. 1980a. A Study to Evaluate the Combustibility and Other Physical and Chemical Properties of Aged Oils and Emulsions. Report to Environment Canada. Ottawa, Canada.

Energetex Engineering. 1980b. Improvement of Air Deployable Oil Slick Igniters. APOA Project 165. Calgary, Canada.

Energetex Engineering. 1981a. Burning of Crude Oil Under Wind Herding Conditions. COOSRA Report No. CS2. Calgary, Canada.

Energetex Engineering. 1981b. Subsea Containment Study: Phase 2, Turbulent Burning. COOSRA report CS8V2. Calgary, Canada.

Environment Canada. 1973. Guidelines on the Use and Acceptability of Oil Spill Dispersants. Environment Canada, Environmental Protection Service. Report EPS-1-EE-73. 54 p.

Fingas, M.F. 1982. A Simple Night-Time Oil Slick Detector Spill Technology Newsletter: 7 (1).

Fingas, M.F., and B. Lea. 1981. Evaluation of three Oil Spill Tracking Buoys. Spill Technology Newsletter: 6 (6).

German and Milne Inc. 1982. Subsea Blowout Containment System Concept Analysis. Report to Environment Canada. Ottawa, Canada.

Gill, S.D. 1981. The Suffield Trial of Aerially Applied Oil Spill Dispersants. PACE report No. 81-6.

Gill, S.D. and C.W. Ross. 1980. Aerial Application of Oil Spill Dispersants. Proc. 3rd AMOP Technical Seminar. Edmonton, Alberta. pp. 328-339.

Gill, S.D. and C.W. Ross. 1982. 1981 Dispersant Application Field Trial St. John's, Newfoundland. Proc. 4th AMOP Technical Seminar. Edmonton, Alberta. pp. 255-263.

Gill, S.D. and G. Stevens. 1980. Use of the TRECAN-PACE Oil Drum Incinerator During the Kurdistan Clean-up Operation. Spill Technology Newsletter: 5(1). Ottawa, Canada. pp. 9-14.

Goodman, R.H. 1978. Arctic Marine Oil Spill Program Ice Movement Buoys for Tracking Oil Spills. In: Arctic Marine Oil Slick Program Technical Seminar. Preprints. Calgary, Alberta.

Hardy Associates. 1979. Oiled Debris Disposal and Storage Sites: Beaufort Sea Coast. Report EPS-3-EC-79-3. Ottawa.

Harris, G.W. and P.G. Wells. 1980. Laboratory Effectiveness Testing of Oil Spill Dispersants. Proc. 3rd AMOP Technical Seminar. Edmonton, Alberta. pp. 305-316.

Hildebrand, A., A. Allen and C.W. Ross. 1977. The Feasibility of Oil Spill Dispersant Application in the Southern Beaufort Sea. E.P.S. 3-EC-77-16. D.0.E. Ottawa, Canada.

Hnatiuk, J., B.D. Wright and E.H. Kustan. 1980. The Arctic Petroleum Operators' Association and its Role in Arctic Research. Petroleum & Marine Environment Conference. Monaco 1980.

Horn, S.A. and P. Neal (Mobil Oil). 1981. The *Atlantic Empress* Sinking — A Large Spill Without Environmental Disaster. Proc. 1981 Oil Spill Conf. EPA/API/USCG. Atlanta, GA, U.S.A.

Huang, James and Francis C. Monestero. 1982. Review of the State-of-the-Art of Oil Spill Simulation Models. Final Report Submitted to American Petroleum Institute.

LeDrew, B.R. and K.A. Gustajtis. 1979. Oil Spill Scenario for the Labrador Sea. E.P.S. 3-EC-79-4. Department of Environment. Ottawa, Canada.

Logan, W.J., D.E. Thornton and S.L. Ross. 1975. Oil Spill Countermeasures for the Southern Beaufort Sea. Beaufort Sea Project No. 31. Department of Environment. Victoria, B.C. Canada.

Mackay, D., I. Buist, R. Mascarenhas and S. Paterson. 1980. Oil Spill Processes and Models. Fisheries and Environment Canada. E.P.S., EE-8 1980.

Mackay, D. and P.J. Leinonen. 1977. Mathematical Model of the Behaviour of Oil Spills on Water with Natural and Chemical Dispersion. Environment Canada, EPS-3-EC-77-19. Ottawa, Canada.

Mackay, D., R. Mascarenhas, K. Hassain and T. McGee. 1980. The Effectiveness of Chemical Dispersants at Low Temperatures in the Presence of Ice. Proc. 3rd AMOP Technical Seminar. Edmonton, Alberta. pp. 317-327.

Mackay, D., J.S. Nadeau and C. Ng. 1978. A Small Scale Laboratory Dispersant Effectiveness Test. *In*: McCarthy, L.T. Jr., G.P. Lindblom and H.F. Walter (eds.). Chemical Dispersants for the Control of Oil Spills. American Society for Testing and Materials. STP 659. pp. 35-49.

Mackay, D. and S. Paterson. 1980a. Development of an Oil Spill Model. Proc. 3rd AMOP Technical Seminar. Edmonton, Canada. pp. 114-119.

Mackay, D. and S. Paterson. 1980b. Program for Simulating Oil Property Changes During Weathering. Proc. 3rd AMOP Technical Seminar. Edmonton, Canada. pp. 128-138.

Mackay, D., S. Paterson and K. Trudel, 1980. A Mathematical Model of Oil Spill Behavior. Dept. of Chem. Engineering and Applied Chemistry, Univ. of Toronto. Report submitted to Environmental Protection Service.

Mackay, D., P.G. Wells and S. Paterson. 1980. Chemical Dispersion of Oil Spills, and International Research Symposium Report EE-17. Inst. Environ. Studies, University of Toronto.

Mackay, D. and P.G. Wells. 1981. Factors Influencing the Aquatic Toxicity of Chemically Dispersed Oils. Proc. 4th AMOP Technical Seminar. Edmonton, Alberta. pp. 445-468.

Mackay, D. and P.G. Wells. 1982. Recent Advances in Chemical Dispersion of Oil Spills. Proc. 5th AMOP Technical Seminar. Edmonton, Alberta. pp. 275-284.

Mackay, D. and W. Zagorski. 1981. Studies of the Formation of Water-in-Oil Emulsions. Proc. 4th AMOP Seminar. Edmonton, Canada.

Mackay, D. and W. Zagorski. 1982. Water in Oil Emulsions: A Stability Hypothesis. Proc. 5th AMOP Seminar. Edmonton, Canada.

MacLaren Plansearch, 1982. Research and Evaluation of Oil Spill Tracking Buoys. Phase 1 Summary and Phase 2 Proposal. Prepared for COOSRA.

McAllister, I.R. 1979. Development and Testing of a «Quickie» Fire Resistant Oil Containment Boom. Report to Dome Petroleum Ltd. Calgary, Canada.

McAllister, I. 1980. Development of an Offshore Self-Inflating Oil Containment Boom for Arctic Use. Proc. 3rd AMOP Seminar. Edmonton, Canada.

McAllister, I.R. and I.A. Buist. 1980. Fire Proof Boom Development, Phase III - Prototype Construction and Testing. COOSRA Report CSl. Calgary, Canada.

Meikle, K.M. 1981a. Incendiary Device for Oil Slick Ignition. Proc. 4th AMOP Seminar. Environment Canada. Edmonton, Canada.

Meikle, K.M. 1981b. Subsea Containment Workshop. Proc. 4th AMOP Seminar. Environment Canada. Edmonton, Canada.

Meikle, K.M. 1981c. Development and testing of the AMOP Boom. Proc. 4th AMOP Seminar. Edmonton, Canada.

Meikle, K.M. 1982. Pers. comm. re Gemeng concrete fireproof boom. Ottawa, Canada.

MEP. 1982. System for Oil Spill Motion Prediction. SOS Users Guide. Prepared for A.E.S. by Meteorological and Environmental Planning Limited. Downsview, Ontario.

Milne, A. 1978. Oil, Ice and Climate Change, The Beaufort Sea and the Search for Oil. Department of Environment, Institute of Ocean Sciences. Sidney, British Columbia.

Milne, A.R., R.H. Herlinveaux and G. Wilton. 1977. A field Study on the Permeability of Multiyear Ice to Sea Water with Implications on its Permeability to Oil. EPS-4-EC-77-11. D.F.E. Sidney, B.C.

Nadeau, J.P., W.G. Wallace and R.Y. Edwards. 1976. Evaluation of the Bennett-Canmar Oil Containment Boom. APOA Project 100. Arctec Canada Ltd. Montreal, Canada.

Nikolaychuk, J. and D.E. Thornton. 1976. Technique for Burning Oil Safely Where Recovery is not Feasible. Report to PROSCARAC. Calgary, Canada.

NORCOR. 1975. The Interaction of Crude Oil with Arctic Sea Ice. Beaufort Sea Project Report No. 27. Institute of Ocean Sciences. Sidney, Canada.

NORCOR. 1976a. Some Aspects of Weathering and Burning Crude Oil in a Water-and-Ice Environment. APOA Project 107. Calgary, Canada.

NORCOR. 1976b. Behaviour and Fate of an Oil Spill in the Beaufort Sea. Report to Department of Environment. Ottawa, Canada.

O'Neil, R.A. 1982. The Arctic Marine Oilspill Program (AMOP) Remote Sensing Study. Draft Report.

Owens, E.H., C.R. Foget and W. Robson. 1982. Arctic Shore line Countermeasures. Initial Results and Implications from the BIOS Experiment. Proc. 5th AMOP Technical Seminar. Edmonton, Alberta. 349-362 pp.

Pallister Resources Management Ltd. 1982. Catalogue of Research Project Reports. Calgary, Alberta.

Pistruzak, W.M. 1981. Dome Petroleum's Oil Spill Research and Development Program for the Arctic. Proc. 1981 Oil Spill Conference. EPA/EPI/USCG. Atlanta, GA. USA.

PROSCARAC. 1980. Air Curtain Incinerator. Report to Environment Canada. Ottawa, Canada.

Purves, F. 1978. The Interaction of Crude Oil and Natural Gas with Laboratory - Grown Saline Ice. EPS A-EC-78-9. D.F.E. Ottawa, Canada.

Purves, W.F. and A. Daoust. 1978. Booms for In-Situ Burning of Oil Spills. Report to Environment Canada. Ottawa, Canada.

Purves, W., R. Abdelnour and B. Roberts. 1980. Booms Offshore. Proc. 3rd AMOP Seminar. Edmonton, Canada.

Reimer, E. 1980. Oil in Pack Ice: The *Kurdistan* Spill. Proc. 3rd AMOP Seminar. Edmonton, Canada.

Roberts, D. and D.K.T. Chu. 1978. Development of Oil Spill Burning Equipment. Report by Bennett Pollution Controls. Vancouver, Canada.

Robilliard, G.A., E.H. Owens, R. Castle and C. Foget. 1980. Arctic Shoreline Protection and/or Cleanup - How to Decide What to Do When and Where. Proc. 3rd AMOP Technical Seminar. Edmonton, Alberta. 342-378 pp.

Robilliard, G.A. and E.H. Owens. 1981. An Integrated Biological and Physical Shoreline Classification Applicable to Oil Spill Countermeasure Planning. Proc. 4th AMOP Technical Seminar. Edmonton, Alberta. 517-535 pp.

Roddis, Hugh A.J. 1982. A VHF Radio System for Tracking Drafting Buoys Automatically. *In* Proc. 5th Arctic Marine Oil Spill Technical Seminar. Edmonton, Alberta.

Rosenegger, L.W. 1975. Movement of Oil Under Sea Ice. Beaufort Sea Technical Report No. 28. Department of Environment. Victoria, B.C.

Ross Environmental Research Ltd. 1981. The Use of Aerially Deployed Igniters for a Blowout in the Southern Beaufort Sea. Report to Dome Petroleum Ltd. Calgary, Canada.

Ross, S.L., W.J. Logan and W. Rowland. 1977. Oil Spill Countermeasures: the Beaufort Sea and the Search for Oil. Beaufort Sea Project, Victoria, British Columbia.

Ross, S.L., C.W. Ross, F. Lepine and E.K. Langtry. 1979. IXTOC 1 Blowout. Spill Technology Newletter: 4 (245).

Schultz, L.A. 1976. Tests of the Arctic Boat Configuration of the Lockheed Clean Sweep Oil Recovery System in a Broken Ice Field. APOA Project 97. Arctec Canada Ltd. Montreal, Canada.

Solsberg, L.B. 1982. A Catalogue of Oil Skimmers, Environment Canada. Environmental Emergencies Branch. Ottawa, Canada.

Sprague, J.B., J.H. Vandermeulen and P.G. Wells. 1982. Oil and Dispersants in Canadian Seas - A Research Appraisal. Report EE-1, Environment Canada. Ottawa.

Tebeau, P.A., D. Mackay, Wan Ying Shiu, W. Stiver, K. Hossain, D. McCurdy and S. Paterson. 1982. Oil Weathering Under Arctic Conditions. Proc. 5th AMOP Seminar. Environment Canada. Edmonton, Alberta.

Topham, D.R. 1975. Hydrodynamics of an Oilwell Blowout. Beaufort Sea Project No. 33. Environment Canada. Victoria, B.C.

Topham, D.R. 1981. The Modelling of Deep Oilwell Blowout Plumes, Including the Effects of Hydrate Formation. Proc. 4th AMOP Technical Seminar. Edmonton, Alberta. pp. 87-112.

Topham, D.R., G.A. Gregory and M. Fogarasi. 1982. Calculations of the Flow in the Marine Riser Section of a Sub-Sea Oilwell Blowout Collection Device. Proc. 5th AMOP Technical Seminar. Environment Canada. Edmonton, Canada.

Trudel, K. 1980. Modelling the Biological Impact of Spilled Oil. Proc. 3rd AMOP Technical Seminar. Edmonton, Alberta. pp. 140-143.

Trudel, K. 1982. Methods for Predicting the Ecological Effects of Untreated and Chemically Dispersed Oil Spills. *In*: Proceedings of the Fifth Arctic Marine Oil-Spill Technical Seminar. June 15-17, 1982. Edmonton, Alberta.

Tsang, G., and N. Vanderkooy. 1979. Theory, Development and Testing of an Ice-Oil Boom. E.P.S. 4-EC-79-2. Environment Canada. Ottawa, Canada.

Twardus, E.M. 1979. Evaluation of a Portable Open-Pit Incinerator for the Disposal of Oil-Soaked Debris. Report to Environment Canada. Ottawa, Canada.

Twardus, E. 1980. *In situ* Combustion: An Oilspill Counter measure for Arctic Shorelines. Proc. 3rd AMOP Technical Seminar. Edmonton, Alberta. 385-401 pp.

U.S.C.G. 1979. Airborne Remote Sensing of Oil Spills in Coastal Waters. Proceedings of the First Workshop Sponsored by Working Group I of the French Pilot Study for the NATO Committee on the Challenges of Modern Society, held at Washington, D.C., April 18-20, 1979.

Van Ieperen, M.P. and G.L. Hood. 1980. Blasting of Oil Containment Trenches in Multi-Year Ice. Proc. 3rd AMOP Seminar. Edmonton, Canada.

Walker, E.R. 1975. Oil, Ice & Climate in the Beaufort Sea, Beaufort Sea Technical Report No. 35. Department of Environment. Victoria, B.C.

Wells, P.G. and G.W. Harris. 1979. Dispersion Effectiveness of Some Oil Spill Dispersants: Tests with the «Mackay Apparatus» and Venuzuelan Lago Media Crude Oil Spill Technology Newsletter: 4(4).

Wells, P.G. and G.W. Harris. 1980. Biological Effects of Dispersants. Proc. 3rd AMOP Technical Seminar. Edmonton, Alberta. pp. 144-157.

Western Canada Hydraulic Ltd. Evaluation of Pumps and Separators for Arctic Oil Spill Cleanup, E.P.S. 4-EC-79-3. Environment Canada. Ottawa, Canada.

Woodward-Clyde Consultants. 1981. Shoreline Countermeasures - 1980 Study Results. (BIOS) Baffin Island Oil Spill Working Report 80-4: 83 p.

Worbets, B.W. 1979. Shoreline Oil Spill Protection and Cleanup Strategies: Southern Beaufort Sea. Report on APOA Project No. 136.

*Cold Regions Science and Technology*, 7 (1983) 343—348
Elsevier Science Publishers B.V., Amsterdam — Printed in The Netherlands

# Maarmorilik
# Environmental Impact of an Arctic Mine

*Erik Sprunk-Jansen**

Greenex A/S
Copenhagen

## Abstract

*Greenex A/S is a Danish Limited company, which since 1973 has operated a lead/zinc mine at Maarmorilik on the west coast of Greenland, 500 km north of the Polar Circle.*

*In 1981, 650,000 tons of lead and zinc ore were mined and concentrated to 38,000 tons of lead, 144,000 tons of zinc concentrate and 468,000 tons of waste — called tailing-containing minerals of no value and traces of lead, zinc and cadmium.*

*After chemical treatment, the tailings are discharged in a sea water suspension into a threshold fiord in 40 m depth via an especially constructed outled system.*

*When the tailings are mixed with sea water, cadmium, lead and zinc are partly dissolved and spread in the marine environment by natural transport processes.*

*Determined efforts have been made over the years to protect the natural environment, aimed especially at reducing the metal content in the tailings, and in particular the content of soluble metals.*

*Good results have been achieved from these comprehensive endeavours which have been undertaken at a cost of 40 mill. Dkr.*

*Since 1977, the biology and marine chemistry as well as the oceanography, including mixing and transport mechanisms in the sea water around Maarmorilik, has been studied intensively by Greenex in cooperation with a great number of both international and Danish experts.*

*Today, the marine environment around Maarmorilik is one of the best studied areas in the world.*

* This paper was presented by Prof. P. Søltoft.

## Introduction

Greenex A/S is a Danish Limited company, which since 1973 has operated a lead/zinc mine at Maarmorilik on the west coast of Greenland, 500 km north of the Polar Circle.

Long narrow fjords with steep mountaineous borders characterize the area. At the corner where the fjord Agfardlikavsa joins the side of the greater fjord Qaumarujuk stands the mountain «The Black Angel» which in a height of about 700 meters contains the rather rich ore.

In 1981, 650,000 tons of lead and zinc ore were mined and concentrated to 38,000 tons of lead, 144,000 tons of zinc concentrate and 468,000 tons of waste - called tailings - containing minerals of no value and traces of lead, zinc and cadmium.

By the chemical treatment in the concentrator plant the ore is depleted of zinc and lead compounds as completely as practice allows. Small unavoidable remains in the tailings, however, represent a potential source of pollution when the tailings are discharged in a sea water suspension into the Agfardlikavsa in 40 metre depth.

## Greenex and the Environment

Environmental protection is a natural matter of concern to the management of a modern industrial enterprise. Through the years, Greenex has devoted ever growing efforts in this field, and spent a considerable amount of money. In this way Greenex has succeeded - in cooperation with consultant engineers - in elucidating some of the environmental difficulties and in pointing out ways of control which have proved to be efficacious, and this leads to a quite relieving view on the future.

To a great extent the environmental work has been pioneer work. Although several other mines extract lead and zinc, their experiences have not been directly usable at Maarmorilik, where salt water, very unconventionally, had to be used in the concentrator.

The environmental risk emerges when heavy metals (lead, zinc and cadmium) are dissipated in the surroundings, especially in the fjord water around Maarmorilik. These are the metals which are recovered from the ore by flotation in the concentrator affiliated to the mine, but unfortunately the recovery is not complete. The residue from the flotation — the so-called tailings — consisting of less valuable minerals like limestone and pyrite still contain small quantities of lead and zinc ore.

## The Troublesome Tailings

It has been chosen to deposit the tailings, approximately 468,000 tons per year, at the bottom of the Agfardlikavsa fjord. It is practically impossible to place the tailings on land because of the steep mountains. From a chemical point of view depositing in water should be harmless, as the ore's lead and zinc content consists of sulphides (galena and zinc sulphide), which are insoluble.

In practice things went different. For example it may be assumed that in 1978 about 25 kg lead and about 70 kg zinc have been dissolved each day. The explanation is that not all the zinc and lead present are bound as insoluble compounds. A certain fraction — ordinarily only one percent — is made up of more soluble substances. Their nature is difficult to identify because they constitute only 0.1 — 0.2 g/kg of tailings.

Two important goals in the current environmental protection program are to keep down the tailings' total content of the metals examined and, perhaps even more important, to curb the content and the effect of the soluble substances.

## Lower Metal Content in the Tailings

A two-fold purpose is served if the recovery of the lead and zinc concentrates in the concentrator can be increased. The profitability goes up, and the metal content in the tailings goes down. For this reason great effort has been made to keep the concentrator operation under control and to improve the yield of concentrates. At the same time it has been possible from year to year to reduce the zinc and lead content in the tailings.

*Figure 1* shows the average content of lead and zinc in the tailings in the years 1978 and 1981. As may be seen the reduction of the concentration is about 36 % for lead and 47 % for zinc.

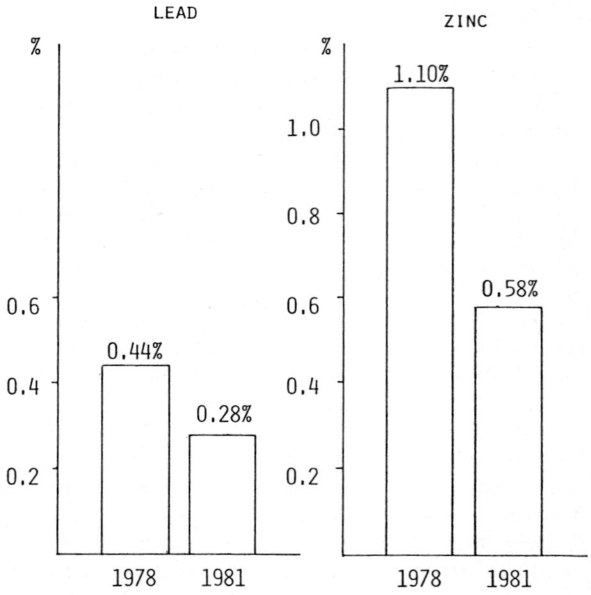


FIGURE 1. *Lead and zinc in tailings solid 1978 and 1981.*

## The Soluble Substances

Many of the measures put into use for reduction of the total metal content of the tailings have simultaneously curbed the content of the accompanying soluble substances, although they have not been completely eliminated.

In *Figure 2* the average content of the soluble lead and zinc compounds in the tailings in the years 1978 and 1981 is illustrated. It should be observed that the scale is gramme per ton tailings which is 10,000 times less than the percent scale used in *Figure 1.*

It is seen that the reduction of soluble lead compounds is about 65 % while soluble zinc compounds are reduced about 70 %.

Only a minor part of the soluble substances originates from the ore in the mine. Additional small quantities are created after the extraction of the ore by the influence of the oxygen in the air. Further increases of the content, especially of the soluble zinc compounds, occur in the concentrator when zinc sulphate and copper sulphate (flotation reagents) are added during the flotation. It has not yet been possible to dispense completely with these reagents — only to limit their use. It might therefore appear that we have to live with a certain content of soluble metal compounds in the tailings.

But to a great extent, these can be neutralized. For this purpose a method has been developed by which the tailings undergo an aftertreatment with aluminum sulphate and calcium hydroxide. Also a flocculating agent is added. The result has been that the metals compound into a form with a strongly reduced solubility. After some very convincing effects had been obtained in the laboratory the method was introduced in full scale.

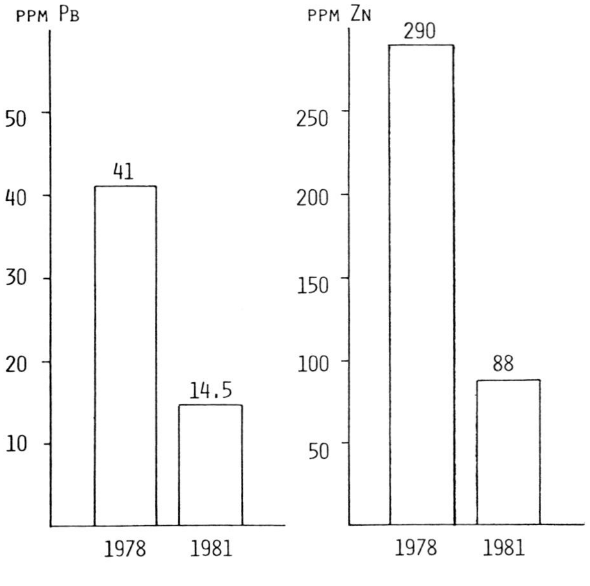


FIGURE 2. *Available soluble lead and zinc from tailings solid in sea water at pH = 7 in 1978 and 1981.*

## Experience from Operation

Lime addition started already in the middle of 1978. From the beginning of 1979, the tailings are treated with both aluminum sulphate and lime before deposition in the fjord, and shortly later also a flocculant was added in order to increase the settling velocity of the tailings.

Little by little these improvements demonstrate their success in Agfardlikavsa's 80 million cubic meters of water, the metal content of which is steadily decreasing. A similar amelioration occurs in Qaumarujuk's water.

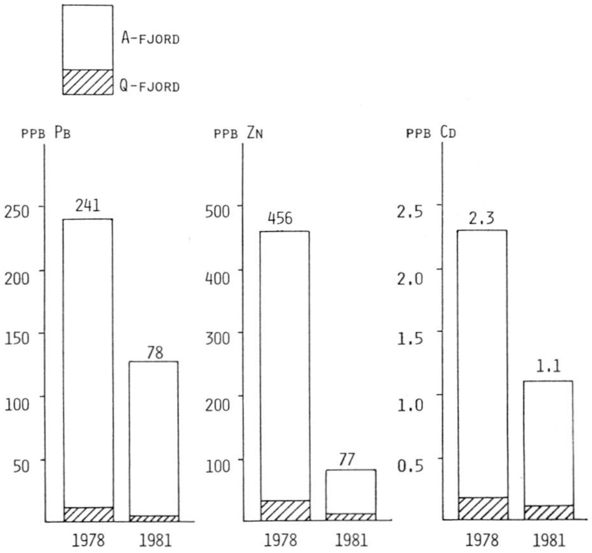


FIGURE 3. *Concentrations of dissolved lead, zinc and cadmium in A- and Q-fjord in 1978 and 1981.*

*Figure 3* compares the concentration of dissolved metal in the fjords in 1978 with the corresponding values in 1981. Again a very encouraging improvement can be noted. The concentration of lead is lowered considerably. The concentration of zinc has decreased even more. The concentration of cadmium, the extent of which is only around one hundredth of the other concentrations, is lowered although to a smaller degree. It may be inferred that some of the cadmium comes from sources other than the tailings.

The improvement of the water quality can be followed more detailed by analysis of the water in the deeper part of the Agfardlikavsa which receives the tailings discharged from the concentrator. *Figure 4* illustrates the variation of the concentration of lead, zinc and cadmium in this bottom water from 1977 to 1981. In spite of the violent seasonal fluctuations the decreasing tendency of all three concentrations is clearly discernible.

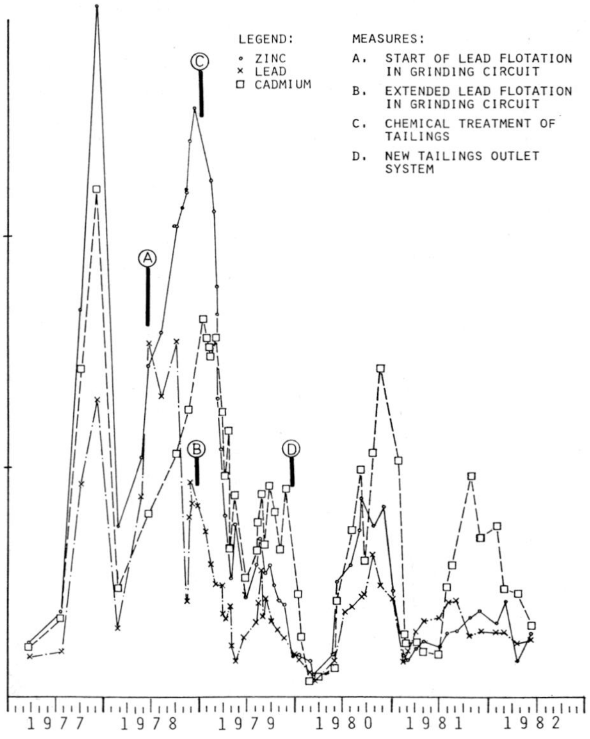


FIGURE 4. *Concentration of dissolved lead, zinc and cadmium in A-fjord bottom water and the introduction time of important measures.*

By arrows «A» and «B» is indicated the time when a couple of flotation cells were introduced into the grinding circuit, respectively when this arrangement was further developed. This effected a strong decrease of the lead concentration.

Arrow «C» refers to the introduction of the final treatment of the tailings with chemicals. Especially the zinc and cadmium concentrations are lowered by this mean.

Finally, arrow «D» indicates that a new outlet arrangement for tailings is put into use. So far, it had been a nuisance that air was sucked into the outlet tube giving rise to dissipation of tailings particles in the fjord. This is now avoided and the tailings sedimentation takes place unhindered.

## Nature Helps

It has been quite a task to obtain a comprehensive survey of the changes in the fjords. A vast number of measurements forms the basis for a report, which presents a record of the input of metal into the Agfardlikavsa fjord in 1978 and the further transportation to the Qaumarujuk fjord. It turned out that not all the metal is disposed of in this way — part of it disappears in another way, not putting any strain on the surrounding environment.

A small torrent falls into the inner part of the Agfardlikavsa fjord. During one year, it carries about 1600 tons of very finegrained glacier silt, which for some time is floating around in the fjord water. The enormous surface of the silt will absorb part of the metal content in the water and carry it until it settles as sediment.

This natural purification process inspires confidence that the disadvantageous environmental influence can be avoided in future, even though it may not be possible to make all the metal in the tailings insoluble.

## Supervision of the Environment

The environmental situation in the fjord areas is regularly followed via samples which are chemically analysed to control the metal content. Water samples are taken at chosen places in the Agfardlikavsa fjord and in the Qaumarujuk fjord. From each place samples from many depths are analysed. The salt content and the temperature are also measured, and samples are also taken from the sediment at the bottom of the fjords in order to find out how far the tailings have spread from the place of depositing.

Every spring and autumn samples of plants and animals in the area are collected in order to check if the same species still live there, and if their numbers have changed. For many species the metal content is determined by analyses of a greater number of each in order that natural variations shall not present a false picture of the situation. Among the closely observed species may be mentioned seaweed, mussels and the three fish: halibut, fjord cod and catfish.

Through these current environmental analyses a comprehensive material is collected to evaluate the environmental influence.

## Plants and Animals in the Surroundings

Seaweed and mussels, well known for their fast accumulation of metals from their surroundings, show a clearly increasing metal content as a consequence of the dissipation in the fjords. Meat from halibut, fjord cod and catfish shows that the metal content is still at the same level as before the start of the mine, but the liver shows a small tendency to increase, at least in the catfish. The ringed seal from the nearby settlement, Uvkusigssat, has not been affected by metals from Maarmorilik, but contains rather much cadmium, as do seals from other parts of Greenland. The cadmium is thus not a consequence of the mining operations, but is due to natural sources.

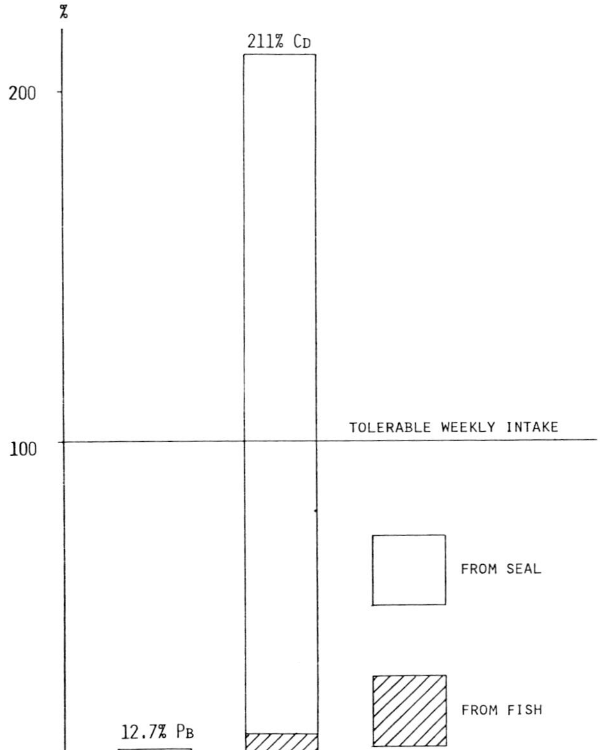


FIGURE 5. *Normal food in West Greenland and the metal load in relation to the tolerable weekly intake — cf. FAO/WHO (1972).*

The ordinary diet of the hunters in West Greenland comprises meat and liver from both fish and seal, while mussel is not a common ingredient. If this food (of average composition) is found in the Qaumarujuk fjord area, the influence of the most important metals, lead and cadmium, on the inhabitants can be illustrated as shown by *Figure 5*

The metal concentration is compared with the tolerable weekly intake, as stated by FAO/WHO (1972). The lead intake is only approximately 13 % hereof, whereas the cadmium intake is approximately twice as high, but as it derives almost exclusively from seal it cannot be traced back to the mining operations. However, the investigation will naturally be an incentive to watch the cadmium outlet with the greatest attention.

As it has recently been observed that a few inhabitants want to include mussels in their diet it has been deemed necessary, in view of their rather considerable lead content, to prohibit their collection for food.

## The Sediments at the Bottom of the Fjord

When the tailings have settled as sediment at the bottom of the fjord, it will release only small quantities of metal. The greater part of the minor content of soluble substances in the tailings has been released already while the particles were floating in the water. Additional small quantities may be released at the bottom of the fjord, although very slowly, when the insoluble sulphides in the tailings react with oxygen carried to the bottom dissolved in the water.

Therefore, if the mining operations, and thus the tailings outlet, should come to an end, the dissipation of metals would decrease to a fraction of what it has been up till now. In time, even further decreases may be expected, partly because the soluble substances are washed away and partly because the bottom of the fjord is covered by natural sediments.

## Covering the Tailings

If, after the cessation of mining operations, it is still desirable, enhanced safety can be obtained by covering the deposited tailings with waste rock that has been crushed (stones from the mine without ore content), and if necessary also with natural sediments from suitable parts of the bottom of the fjord. It will be appropriate to treat the waste with aluminum sulphate and lime. According to laboratory experiments a drastic reduction of the metal dissipation will be obtained. Conditions, similar to the situation before the start-up of the mining operations, can be created and there is no risk that the surrounding fauna and flora will suffer any damage.

*Cold Regions Science and Technology*, 7 (1983) 349—355
Elsevier Science Publishers B.V., Amsterdam — Printed in The Netherlands

# Arctic Third World: Indigenous People and Resource Development

*Franklyn Griffiths*

University of Toronto
Toronto

Resource scarcity, technological innovation and the growing reach of governments, corporations and peoples are combining to alter the face of the Arctic. The transformation is occurring along North-South and East-West lines. A region of the world that was long regarded as inaccessible is now being linked by an ever more complex network of ties to the industrial civilization that lies to the south. And within the Arctic itself, processes of economic, social and political change are beginning to spill across national frontiers that not so long ago marked separate national jurisdictions. The rapidity of change in the Arctic may be in doubt, depending as it does on the OPEC oil price and the state of the world economy. But the fact of change cannot itself be doubted.

With change come deprivations as well as advantages. Governments and corporations engaged in Arctic energy resource operations endeavour of course to plan their activities to best advantage. Though the region presents a novel and perhaps uniquely unforgiving environment, the physical and economic planning of Arctic energy extraction and transportation poses problems that are manageable to the extent that they are properly quantifiable. The situation is different where social and political planning is concerned. Here the essential variables are not quantifiable and not likely to become so. Science or an acceptable way of proceeding is not available. And yet there is a very real need to plan for social and political change. One has only to consider the controversies that have been prompted by overland pipeline routes, the emergence of Greenlandic opposition to Canada's proposed Arctic Pilot Project, or the furor that has arisen over the export of natural gas from the Soviet Arctic to western Europe, to recognize that a preoccupation with the physical and economic variables in Arctic energy undertakings risks unnecessarily severe conflict and diminished net benefits.

As in any political exercise, there are inherent limitations in achieving a fair distribution of advantages and deprivations within and between nations engaged in the development of Arctic energy resources. We should nevertheless be paying greater attention to the anticipation of social and political impacts of Arctic resource megaprojects. We should also be seeking to elaborate decision-making and implementation procedures that are acceptable to those most directly concerned. This applies in particular to the indigenous peoples of the Arctic, who are likely to be the most immediately affected by major Arctic resource development schemes.

In the brief probe that follows, I would like to suggest a framework for consideration of the current and possible future impact of Arctic energy resource development on the Arctic aboriginal peoples — the Inuit of Alaska, Canada and Denmark/Greenland, the Sami population of the Nordic countries, and the «small northern peoples» of the Soviet Union. Clearly there is considerable variation in the character of these communities, in the degree to which they are or may be affected by large-scale energy projects, in their relationships with central authorities to the south, and in the way they are viewed by the predominant southern population in each country. These differences notwithstanding, Arctic energy development

0165-232X/83/$03.00  

can benefit from a comprehensive view of the native populations.

From a social and political planning perspective, certain advantages are to be had from the adoption of an integrated or pan-Arctic assesment. For one thing, it should help to provide a common basis for the exchange of experience across separate jurisdictions, thereby improving immediate performance in project adaptation to the social and political as well as the physical environment of the Arctic region. Over the long haul, the use of a pan-Arctic mode of interpretation may offer an improved basis for expert discussion, thereby contributing to the development of region-wide norms and practices that stand to facilitate projects requiring international cooperation and the construction of international resource management regimes. As well, the adoption of an integrated perspective could be of assistance to the indigenous peoples in orienting themselves to industrialization and modernization in the Arctic. This much granted, the question remains as to whether a viable pan-Arctic perspective can be identified for the interpretation of the indigenous peoples' role in the exploitation of Arctic energy resources.

At this early stage in the development of systematic thought about the Arctic native peoples as a collectivity, we are in a position to offer hypothesis and not verities. The utility of a hypothesis lies not so much in whether it is right or wrong, as in its value in organizing the data and providing guidance for policy. With these caveats in mind, I suggest that we regard the Arctic indigenous peoples and the lands and adjacent waters they occupy as part of the Third world. (Ref. 1).

The initial reaction to the hypothesis may be one of disbelief, «Third world» suggesting tropical heat waves and newly independent or neo-colonial status. Upon reflection, however, it will be recognized that the term also connotes underdeveloped and developing areas whose evolution is substantially affected by decisions issuing from remote metropolitan centres, and whose populations experience both an erosion of traditional culture and a sense of powerlessness to determine their own future in the midst of externally generated change. Again, there will be variation in the applicability of any single hypothesis to the diverse situations of the Arctic native peoples. Nevertheless, there would appear to be some degree of fit between the hypothesis and the empirical data. Let's see what can be done with the idea.

Much of what has been reported to this conference serves to confirm the fact that the Arctic is an underdeveloped area of the world that is now undergoing accellerated development. The process is most advanced in the Soviet Arctic, where the establishment of large urban and industrial complexes has been accompanied by a massive population influx from elsewhere in the country. These changes, which have been going on for some decades, appear to be accompanied by an effective social policy that allows the indigenous peoples the option of remaining on the land or joining the industrial society. (Ref. 2). In the Arctic areas of the Nordic countries, economic development has been less intense and the Sami, while to some extent able to pursue a traditional way of life are frequently indistinguisable from the larger populations of which they have long formed part. On the other hand, as of the past decade, population retreat from the rural counties of northern Norway, Sweden and Finland was the order of the day as conditions of life deteriorated relative to the more prosperous areas to the south. A comparatively high Sami communist vote also suggests dissatisfaction with prevailing conditions. (Ref. 3).

In Greenland, the transition to an industrialized and modern way of life has been abrupt and far-reaching as a consequence of postwar Danish social policy. Copenhagen aimed to improve the living conditions of Greenlanders and to meet their desire to partake of the modernity to which they had been exposed during World War II. Culture shock has accordingly been part of the Greenlandic experience in recent times. Though home rule was granted in 1979, limited job and career opportunities have engendered some feelings of frustration and powerlessness among the educated youth who constitute a major portion of the population. (Ref. 4). As to Canada, a well-intended postwar social policy that aimed to integrate the Inuit into the fabric of national life succeeded in detaching them very largely from their traditional ways, while failing to provide much in the way of alternative means of subsistence aside from welfare. At the same time, the infrastructure left from wartime operations was enlarged for use in northern economic development, which saw first the expansion of mining activity and then a major high-technology effort to discover and extract hydrocarbons from the Arctic. (Ref. 5).

Finally, in Alaska, economic development, the achievement of statehood in 1959, and the land claims settlement of 1971, have contributed to a situation that is reflected in the following Inuit assessment:

> Too much is happening to the people. Too many outside pressures are forcing in on us. Changes are coming too fast, and we are being pushed in all different direction by forces that come from someplace outside. People thought that the land-claims settlement was the end of our problems, that it meant the future was secure; but it was only the beginning. Even before the lands were all selected the government wanted pipeline easements and road corridors right through our territory. These would take away strips miles wide, cutting right across our land. And instead of open access to our land, the Eskimos might be surrounded by huge pieces of country that are declared national resources for all of the people. Land that has always belonged to the natives is being parcelled up and divided among the takers. (Ref. 6).

Alter the references to reflect local conditions and modify the strength of feeling to correspond to local attitudes, and the foregoing statement may be taken to reflect an outlook that is shared in one form or another by many of the aboriginal inhabitants of the Arctic region.

That the advent of industrial society has been unsettling and disruptive, sometimes profoundly so, for the peoples of the Third World, is a commonplace that deserves repeating in the Arctic context. The Arctic indigenous populations have lost as well as gained in dealing with change. Here, as elsewhere in the less developed areas, change has for the most part been externally generated. It has occurred in thinly populated areas in which a renewable resources economy and a traditional way of life were the rule until only quite recently. Viewed in the ensemble, the effects of economic and social development on the native peoples of the Arctic have much in common with those experienced in developing areas elsewhere in the world. (Ref. 7).

Aside from the economic and social dimensions of change, the indigenous inhabitants of the Arctic share with the peoples of the Third World a shortage of capabilities in relation to political needs. Elsewhere, the principal political need has been self-determination. Stated most frequently in the demand for independence and now for a new international economic order to permit the elimination of more subtle forms of dependence, the urge for self-determination may also be expressed as a desire for more effective participation in the existing political structure, or for the acquisition of delegated powers of decision under home rule arrangements. While at present there are no indepenence movements in the Arctic, the indigenous-peoples of the region do exhibit a varying commitment to self-determination. This is to be seen in the self-governing activities of the North slope Borough in Alaska; in the demand for constitutional change, land claims settlements and a separate Inuit territory in Canada's eastern Arctic; in the cautious efforts of the Greenlanders to enlarge the competence granted under the terms of home rule, as for example by their ventures into the foreign affairs; or in the opposition of a minority of the Norwegian Sami to the adverse environmental effects of hydroelectric development on the Alta river. (Ref. 8). Where the Soviet Arctic is concerned, the official view is that self-determination is fully ensured under conditions of socialism; though evidence to the contrary is not-available, it would be most unusual if the smaller northern peoples had wholly failed to share-the desire for autonomy of action that is displayed by the larger national minorities of the USSR.

As is the case in the temperate areas of the Third World, political and technical capabilities are required to realize the diverse aspirations for self-determination in the Arctic. Among these are leadership, legitimate organization, media of communication, and the capacity to offer benefits and impose deprivations on central decision-makers whose actions may have far reaching local effects. Equally important but less well appreciated is the need to be able not only to speak to the issues, but to take an active part in defining them. For this, a technical competence is required that is frequently beyond the ability of the Arctic aboriginal peoples to provide. Whether the issue concerns the taking of bowhead whales, a proposal to alter the flow of an Arctic river to run southward, or a marine transportation scheme that may affect the behaviour of marine mammal populations, the problem is likely to be defined not so much on the basis of indigenous peoples' judgements, as on the technical and scientific findings of southern specialists. The Canadian Government, for example, has responded to this problem by providing funds to to the Inuit Tapirisat of Canada to secure legal and expert counsel in contesting government and corporate proposals. The ITC is nevertheless hard put to provide adequate staff support for the full range of issues with which it must deal. Nor is it making much headway in asserting the value of traditional knowledge. In varying measure, a shortage of political and also technical capabilities would seem to characterize the situation of the aboriginal peoples in the Arctic region, as it does in the Third World more generally.

Disparities between the desire for self-determination and the capacity to achieve it seem likely to yield a sense of injury and powerlessness in relation to southern centres of administration and decision. Again, we would expect to see variation in the strength of such feelings across the Arctic region, just as there is a range of attitudes among the developing countries toward the state of their relations with the First and Second worlds. There again being no evidence to the contrary, the Soviet Arctic peoples must provisionally be regarded as having achieved the most favourable accommodation with the central authorities or as having been most effectively integrated into the existing socio-political system. After them may come the Sami, whose comparatively high degree of integration is evident. Less fully integrated are the Greenlanders and Alaskan Inuit, who have found it necessary to seek varying degrees of self-government in asserting control over their own affairs. To rid themselves of what is seen to be an alien power of decision, the Greenlanders have as well voted to opt out of Danish participation in the European Economic Community. At the bottom of this provisional list are the Canadian Inuit, whose present condition may be regarded as one of colonial status. (Ref. 9).

Comparative field research might well alter the details of the foregoing ranked assessment of the situation, attitudes and self-images of the various Arctic native populations. But the research would be highly unlikely to yeild the finding that these populations and their politically active minorities are generally inclined to view with satisfaction their ability to see to it that the development process is properly adapted to local needs. Rather, the underlying attitude is likely to be a dualistic one in which a readiness to utilize the benefits of development is accompanied by a determination to preserve their cultural identity and to improve the conditions of an indigenous way of life. As occurs in the Third World more broadly, the aboriginal communities of the Arctic are poised between tradition and modernity. Caught up in processes over which they have little or no real power to control, theirs seems to be the choice between self-determination and assimilation, between survival or decay as separate peoples.

Let us assume, then, that it is indeed appropriate to regard the Arctic as part of the Third World where the indigenous peoples are concerned.* What consequences might the hypothesis lead us to expect from more intense development of Arctic energy resources? As the rate of industrialization and modernization picks up in the Arctic, the region will experience some of the same phenomena as have occurred in other developing areas. Among other things, we may anticipate a heightened commitment to self-determination, the appearence of indigenous equivalents of nationalism, increased transnational collaboration among native populations, new efforts to alter the terms under which development occurs, and a heightened readiness to block specific development projects.

To begin with, the choice between selfdetermination and decline will be sharpened. Some may opt for acquiescence in the face of change that alters the terms of existence without providing means of survival as a distinct people. Others will seek to assert themselves. The Greenlanders, for example, have stated in no uncertain terms that they consider the Arctic Pilot Project (APP) to be a threat to their survival. Whatever may be thought of their evaluation of the environmental and social con-

* Aside from the evident difficulties posed by variation in the situation and outlook of the Arctic indigenous peoples, the use of a Third world or more particularly a dependency framework may invite objection on other grounds. For one thing, it may be argued that the region contains a substantial European population whose presence effectively negates a Third World comparison. This point of view, which is likely to be shared by a good many Arctic residents as well as southern observers, may be accompanied by a preference to view the difficulties of the indigenous populations as part of a larger problematique that is specific to the Arctic and concerns all inhabitants of the region. To my way of thinking, the adoption of an Arctic-specific point of view, while ultimately preferable, is likely, at this point, to understate the specific problems of the native peoples which are better understood by reference to the larger Third World experience of development, dependence and selfdetermination.

sequences of repeated tanker transits through Baffin Bay and Davis Strait, the plain political fact is that an Arctic resource megaproject is viewed as facing them with life-and-death decisions. Less pointed but broadly similar assessments of Arctic resource development projects, including the APP, are shared by active minorities among the Canadian and Alaskan Inuit. Though equivalent attitudes and strength of feeling do not seem to exist among the indigenous populations of the Nordic countries and the Soviet Union, it would be surprising if hydro-electric development and an intensification of on-shore and off-shore hydrocarbon exploration, extraction and transportation operations accompanied by significant immigration and cultural change failed to evoke a sense of apprehension in these parts of the region as well. If the experience of the Third World is any guide, it suggests that arousal and mobilization will be the predominant initial reactions of native populations faced with a new wave of development stimulated by southern energy needs.

Mobilization to deal with externally induced change has formed part of the process of nation-building and nationalism in Third World countries, some of whose populations are comparable in size to those of the larger Arctic indigenous peoples. The population of the Seychelles, for example, is roughly equivalent to that of Greenland. Whether nationalism and demands for independence will arise in the Arctic is doubtful, if only because of the high cost of essential services in this part of the world. Short of independence, however, self-determination movements may confine themselves to dealing with specific issues and stop there. Or the concern with specifics may be interwoven in greater or lesser measure with an effort to secure and utilize a broader grant of local decison-making powers. Whatever the course chosen in seeking to shape the conditions of existence, the endeavour will probably be accompanied by some welding together and some reaffirmation of shared identify within communities that see themselves as standing apart from the bearers of development and modernity. In short, an acceleration of Arctic energy development activities seems likely to provide the aboriginal peoples with issues that may be acted upon to affirm their solidarity and autonomy. In practice this may mean higher levels of opposition and conflict between the indigenous peoples of the Arctic and the central authorities to the south.

The experience of the developing countries suggests as well that the Arctic native populations will seek to improve their prospects by means of cooperation. Though exact Arctic equivalents of non-alignment, the formation of the Group of 77, or the use of the United Nations General Assembly are improbable, the tendency to band together is already well advanced. Indeed, the common purpose on Arctic affairs that has been achieved by certain of the indigenous peoples may well exceed that of their own governments. Transnational cooperation appears to be most extensive between the Alaskan and Canadian Inuit and the Greenlanders in the setting of the Inuit Circumpolar Conference. Launched in 1977 in Barrow, Alaska, the ICC provides a framework for mutual assistance in preserving a common culture, enhancing Inuit control over the development process in their part of the Arctic region, and in lending support on specific issues that may face any of the constituent groups. In addition, an appeal that is so far unanswered has gone out to the Soviet Government to allow the participation of the Siberian Inuit in the ICC. At the same time, the Inuit of North America have begun to build ties with the Sami population of Scandinavia, lending moral support, for example to the Norwegian Sami in connection with the Alta river project. Furthermore, there appears to be a growing interest in cooperation with indigenous peoples in other areas of the globe, as was expressed by Inuit participation in the World Assembly of First Nations in Regina, Saskatchewan in 1982. Yet another indication of readiness to reach outward for support is to be had in the declared readiness of the Greenlanders to appeal to the United Nations in the event of a Canadian decision to move ahead with the Arctic Pilot Project, while the achieved level of cooperation should not be overstated, and while the future evolution of the transnational operations of Arctic indigenous peoples remains uncertain, the pattern of activity to date suggests a growing capacity and willingness both to cooperate across national frontiers and to internationalize Arctic energy issues. In all of this, a basic aim seems to be an improved bargaining position vis-a-vis government and corporations.

Finally, it should be evident from the foregoing that large-scale Arctic energy undertakings risk becoming targets for the expression of dissatisfactions, claims and ambitions that may have little to do with the character and immediate implications of the projects themselves. This is not to derogate indigenous attitudes toward specific projects, which may be well founded. Rather it is to recognize that, as occurs more broadly in relations between the developed and less developed worlds, energy development issues are likely to become increasingly politicized and therefore increasingly difficult to manage. Again, politicization and the creation of lightning-rod effects promise to be international as well as national phenomena, owing to the employment of modern means of communication to publicize issues, to create attitudes and networks where none existed before, and to invoke broader participation.

In sum, a Third World perspective on the Arctic suggests that some trouble lies ahead for energy resource development. To the extent that projects move forward rapidly and without effective consultation with the region's native peoples, they will stimulate demands for self-determination, create controversy, and serve to attract ambient political energy. A wider array of political actors, both within the Arctic states and internationally, would be drawn into the resolution of these issues. The end result would be increased delay, cost and uncertainty in Arctic energy development, and increased instability in the political life of a region that continues to have militarystrategic significance. And yet the possibility or hope remains: properly implemented, Arctic development should yield substantial benefits to all concerned.

If this version of what may lie ahead is well founded, it suggests a numher of things that might be done to improve the outlook for resource development undertakings. First, the existence and persistence of indigenous peoples' communities in the Arctic should be regarded as values to be supported by development, not as impediments that are somehow to be overcome in the course of project development and implementation. The value of indigenous communities lies ultimately in their capacity to lend dignity and autonomy to the individual-objectives that are central to the purposes of all circumpolar societies. (Ref. 10). To act on these purposes with greater effect requires a heightened awareness of social and political variables in Arctic resource planning. At present, social and political factors tend to be handled as «add-on» considerations, much like environmental impact assessments, in what is essentially an economic and hardware planning process. As far as can be determined, neither private-sector not state-owned energy corporations nor central government agencies themselves accord a high place to Arctic social and political planning units, insofar as they exist. This will have to change, together with prevailing views of the value of indigenous communities as such, if Arctic energy megaprojects are to be designed and managed in a way that maximizes the support of the native peoples.

Secondly, a greater emphasis on social and political planning should allow for a wide-angle view of the field, as well as more vigorous projectspecific efforts to anticipate and avert adverse impacts. For example, to attenuate the lightning-rod effect, resource development sponsors could well seek to act on behalf of indigenous peoples in matters of cultural policy, land claims settlements and other issues of concern. The reaction to this proposal is likely to be, «we have enough trouble as it is, without getting involved in a still larger agenda on which our right to speak is open to question.» The fact is, however, that if the trouble isn't taken to reduce the generalized sense of deprivation among native populations experiencing industrialization and modernization, trouble will be encountered in the implementation of specfic energy projects.

Most important, the indigenous peoples must be allowed a fuller measure of participation in project planning and implementation. This is easier said than done. It will require time as well as experimentation that is adapted to the varying institutional structures and customs of the ice states. Without effective participation, the native peoples will be inclined to regard themselves as being acted upon. In such circumstances their propensity to self-determination is more likely to take the path of opposition to southern-based energy schemes. This is not to suggest that they be accorded veto rights. Rather they should be enabled to play a fair and equal role in affecting the outcomes of corporate as well as governmental decisions. This means finding a way of assimilating traditional knowledge and modes of discourse in planning activities. It also means making specialists available to native peoples while educational incentives are provided to encourage the development of indigenous expertise in technical discourse. Ways should be sought to associate Arctic indigenous inhabitants with all phases of the project process from initial design of specifications through to implementation. Where implementation in particular is concerned, they should be accorded a direct role in the monitoring of environmental and social impacts.

Finally, we in the south should think once, think twice, and think three times before imposing an Arctic energy project upon an unwilling indigenous population. Rather than strengthening a Third World pattern of development in the Arctic, northern native peoples and southerners are called upon to cooperate in avoiding the reproduction of North-South cleavages in the north. Rejecting confrontation and assimilation alike, we are summoned to invent new means of accommodation and ultimately new ways of life that serve to reconcile indigenous culture and the culture of development.

## Reference


1. Maurice Strong, former Director of the United Nations Environment Programme, has advanced a similar proposition. See Maurice F. Strong, «The Arctic: Canada's Third world,» notes for a lecture delivered at the Ontario Science Centre, Toronto, October 22, 1978.
2. A recent discussion of Soviet policies is to be found in Frances Svensson, «Comparative Ethnic Policy on the American and Russian Frontiers» *Journal of International Affairs* Vol. 36, No. 1 (Spring/Summer 1982), pp. 83-103.
3. Terence Armstrong, George Rogers and Graham Rowley *The Cicumpolar North* (London: Methuen, 1978), pp. 217-219. See also the chapters by Hansen, Bykelund and Palomaki in William C. Wonders, ed., *The Artic Circle* (Toronto: Longmans, 1976), and Enio Siuruainen, «Natural Resources and Human Activity in Northern Finland in Post-war Times,» in Milton M. R. Freeman, ed., *Proceedings of the First International Symposium on Renewable Resources and the Economy of the North* (Ottawa: ACUNS, 1981), pp. 99-117.
4. Jørgen Taagholt, «Greenland — seen from a historic and security point of view,» unpublished paper, November 1981, pp. 4-8.
5. Armstrong *et. al., op. cit.*, pp. 88-99.
6. Quoted in John McPhee, *Coming into the Country* (New York: Farrar, Straus and Giroux, 1977), p. 36.
7. In particular the situation of other aboriginal communities is suggestive. See the issue of *Journal of International Affairs, op. cit .*, which is devoted to the human rights of indigenous peoples.
8. Fune Stormo and Eva Solem, «Sami Rights and the Alta River» *Northern Perspectives* Vol. 9. No. 3, 1981, pp. 2-6 (published by the Canadian Arctic Resources Committee, Ottawa).
9. Strong, *op. cit.*, p. 12.
10. Cf. Svensson, *op. cit.*, p. 84.

*Cold Regions Science and Technology*, 7 (1983) 357—361
Elsevier Science Publishers B.V., Amsterdam — Printed in The Netherlands

# The Developing Arctic

*Finn Sollie*

The Fridtjof Nansen Foundation
Lysaker

The Arctic is a vast region where natural conditions have remained major obstacles to penetration, settlement and development. Today, with the aid of modern technology and inspired by new demands for natural resources, man is finally able to invade the Arctic to include even that part of the world in his economic and industrial domain. In introducing his technical civilization into an environment that is so harsh and yet so vulnerable, man is opening new opportunities for development. In so doing, he is also taking upon himself an awesome responsibility for the effects of his actions.

In the global perspective, the Arctic may well be regarded as one of the world's largest underdeveloped regions and, in this respect, as one part of the underdeveloped Third World which exists inside some of the world's most advanced industrial nations. That perspective serves to remind us that vast and desolate regions in the far north which are devoid of advanced technical civilization, have long been a homeland for northern native peoples who have adjusted to the harsh climate and who have developed unique, social patterns and cultures in harmony with the arctic environment.

The Third World perspective may be helpful by forcing us to regard arctic development as part of a broader process where the demands of industrial civilization for change and development collide with established cultural values and social patterns. On the other hand, if we adopt the idea that in some respects the Arctic may be considered as a northern equivalent of the southern, underdeveloped Third World, it is important to remember that in other respects the Arctic is distinctly different from Third World countries and that there is no uniform and general solution to Third World problems.

First of all, all arctic areas are part of established, industrial states. In political and in legal terms, the underdeveloped Arctic is entirely different from underdeveloped countries which exist as separate political and constitutional entities. In the case of Third World countries, one may talk of an international conflict of interest between them on the one hand and, on the other hand, the industrialized countries. In the case of underdeveloped arctic regions, it is a question of conflicting interests inside and within single state systems. The difference between these international and internal conflicts is fundamental and calls for different solutions and different procedures.

In some respects, the problems that arise between an advancing industrial civilization and traditional societies in the far North may appear to be similar to problems that exist in the relationship between the underdeveloped South and the industrial states of the Northern Hemisphere. Nevertheless, the basic nature of the conflicing interests may be as different as are the formal differences between an international dispute and an internal conflict. Thus, the very development that an arctic people may resist as a threat against its traditional life-style, may be regarded as highly desireable in most Third World countries' and consequently, will be welcome there. In this respect, the North-South issues between the Third World and the industrial states appear to be primarily a conflict between competing economic interests, with the underdeveloped South demanding a larger share of the technological and industrial

0165-232X/83/$03.00  

development in order to improve the economic lot of its peoples. In the far North, on the other hand, the issue usually is presented as a conflict between different cultures within the national economic and political system.

Resistance to development to avoid change is quite different from a struggle for a just and equal share of the benefits of development. With both types of interest involved, the problems and conflicts that arise from development become complex and intricate and easy simplifications and generalizations serve no practical purpose. The clash of cultures that may be seen in arctic development would seem to have an equal share of both elements. On the one hand, there is no doubt a genuine desire to preserve and to strengthen the uniqueness of native societies and cultures. On the other hand, there are also desires to strengthen the economic base for those cultures and they require development. In this respect, there may be a conflict of interest within the minority groups and native peoples that is quite similar to conflicts between proponents and opponents to change and development in the national community at large.

We find here a problem of priority of values that is in one sense universal and that cuts across ethnic distinctions and cultural differences. In some cases a people may be so united in its preferences that a choice for or against development is simple. In other cases, the choice may be difficult and, furthermore, preferences will change over time. Finally, while it is true that in many parts of the Arctic native peoples may be the dominant population and that a large majority of the group lives in separate communities from other ethnic groups, in other instances communities are mixed and members of one ethnic group may live under different economic and social conditions in different communities. Consequently, members of one and the same ethnic group may have different economic and social preferences. In such instances the definition of values that should guide decisions in regard to development and change becomes extremely difficult and cannot be made on the basis of ethnic criteria.

As an example we may think of the various communities in the northernmost parts of Norway, where a minority of the laplanders remain as migrating reindeerhunters, while a majority are «sea-laps» and live on the coast as small farmers and fishermen under material conditions and with a life-style that is identical to those of the ethnic Norwegians in the same region. There is no doubt that the advent of large-scale offshore oil development in the Norwegian North will lead to changes that will affect the people in the coastal communities far more than the inland laplanders. Neither is there any doubt that the changes will be similar for all ethnic groups who share the economic and technological conditions along the coast.

Under such circumstances, the Third World parallel is not relevant, nor does the problem of change for native minorities appear significantly different from the general problem of change in society.

These questions about the effects that the development of arctic resources may have upon the native peoples of the North are not asked in order to minimize the problem. Rather, the purpose is to remind ourselves that there is not one, but several interlocking problems that vary from case to case and from area to area, and that there is no simple formula or reference to similar issues in other parts of the world that will make it less difficult to find the right balance between those interests that call for new technological and economic development and those interests that call for reticence in development to avoid social change and to preserve traditional ways of life.

In this context, a final thought might be offered as a reminder that adjustment to change may be as important as is change itself. The thought was inspired by an experience with the Alaska pipeline. To avoid interference with the migration of caribou (reindeer), the design of the pipeline allowed for appropriate passages at points where the deer herds were known to have their normal treks. However, wolves soon learned that these passages served as chokepoints where the reindeer queued up to pass and thus became easy prey to packs of wolves that assembled at the passage points. Eventually the caribou learned its lesson and stopped using the man-made passages by adjusting to pass under or over the pipeline in any place where conditions made it possible and without regard to the passages built for easy crossing. An example of well-intended but unsuccessful technical adjustment to natural patterns of behaviour, the experience is also one of successful natural adjustment to man-made change. The conclusion would seem to be that change and adjustment make a double process, where technology may be adjusted by design to satisfy observed patterns of behaviour, and where group — or social — patterns are adjusted to neutralize undesired effects. Surely, human progress in terms of culture as well as in material terms is a result of man's unique ability to adapt and adjust in combination with his capacity for invention and development.

The selection of the right and the necessary criteria and values to measure the benefits and the costs of development is difficult and important. Some but not all of the costs and the benefits can be easily measured in terms of the «monies» that are involved: the direct expenses and profits of exploring and exploiting new resources. In addition to such material values, there are a number of non-measurables that concern the natural and the social environment, where losses from polution and change cannot be easily measured against the advantages of growth and development. These costs and benefits are so widely discussed that they call for no special attention here. However, there also is a third dimension to change and development in the Arctic where comment may be called for, namely the many international «imponderables» that concern the interaction and relations between states, rather than between individuals and groups within states. These are the legal and political developments that are related to or caused by technical and economic change, as well as the shifts in the strategic role and importance of the Arctic that will result from the development of resources and increased activities in the region.

In this context, it should be remembered that the Arctic is a very large area which includes vast land areas as well as an entire ocean, the Arctic Ocean; and numerous rim seas, where we find some of the largest and shallowest continental shelf areas in the world. In many respects this huge and genrally undeveloped area can be considered as the last frontier for human expansion. Apart from the Antarctic Continent, the Arctic is the last major area in the world where man is imposing the imprint of his civilization.

The general lack of development in the Arctic is not a result of lack of interest in using the region. It is more than four hundred years since the first determined effort was made by a developing, progressive commercial nation to penetrate the Arctic. In 1553, a group of London merchants sent the first expedition, under the leadership of Sir Hugh Willoughby on a search for a northern sea route to the far East, to reach the fabled riches of Cathay, or China. «Monies» were the issue and the search for wealth in distant lands was conducted in fierce competition between established and developing maritime states.

At that time, in the sixteenth century, Spain and Portugal were dominant maritime powers who had discovered and monopolized the southern sea routes (around Africa and around South America) to the Far East. The London merchants were commercial leaders in a new and rapidly expanding maritime power and looked for new searoutes to secure a share of the eastern trade. The Dutch, who were the leading shipping nation in Northern Europe at that time and who were struggling to secure independence from their Spanish lords (the Netherlands were part of the Habsburg Empire and with the division of the Empire in 1555 became subject to Spain. The Dutch revolt in 1572 resulted in protracted war with Spain that did not end until 1648), were equally eager to find new sea routes and they too looked to the north for a new passage. This is why so many of the place names in the Atlantic parts of the Arctic are Dutch or English. They were given by sailors looking for a passage to the Far East and by whalers and sealers who followed the commercial explorers when they learned about the riches of the northern waters.

The important point is that early exploration of the Arctic began as an effort to find a new sea route and to circumvent the opposition of established maritime powers with a desire to maintain monopoly of power in the southern routes: early exploration of the Arctic was inspired by commercial interest and directed by strategic considerations and political interest.

As it happened, the failure to establish a northern sea route did not prevent the growth of English and Dutch maritime initiative and power. With the rapidly growing strength of the northern states and the sapping of the power of Spain and Portugal, the Iberian monarchies were not able to maintain control of the southern waters, which they considered to be their own by right. Instead, the new maritime powers were able to establish new principles of the freedom of the sea (*Mare Liberum*, Hugo Grotius, 1609). Consequently, and fortunately for the new shipping nations, strategic motives for penetration of the Arctic Ocean became irrelevant just at the time when it became obvious that ship technology would not allow the development of a northern route to circumvent the naval opposition of Spain and Portugal.

Today, modern technology can provide the necessary means to pass through — or underneath — the ice and maritime traffic is graduly increasing in arctic waters. In the sixteenth century, the purpose of exploration was to find a passage through the Arctic. In the twentieth century, the primary purpose of arctic shipping is to provide transport for arctic development and shipping of materials from the Arctic. Nevertheless, with the continuing development of arctic shipping technology, the dream of the London merchants in the sixteenth century for a safe, northern sea route between the Atlantic and the Pacific may soon become a reality. In that development, the same forces that moved the London merchants in their early initiative will still be at work. In the arctic development of resources and of transportation, the same elements of economic interest and of strategic and political considerations once again are active forces that direct development. Now, when earlier goals can in fact be reached, the strategic implications of trans-arctic shipping also are reinforced by the advantage of access to and and control of resources that will be scarce and that will have strategic as well as economic value.

It is not possible now to say how rapid this development will be. What we do know, is that the Arctic is rich in some of the most essential resources, and notably in energy resources, and that those resources will be developed. The speed of the development will depend on national policies and on market conditions that may speed up full-scale development to begin before the end of this century, or that may delay development at a large scale until the middle of the next century. Nevertheless, we do recognize that the process has started. Consequently, it is indeed necessary to establish a sufficient scientific base to predict and to master its potential effects.

We know that the Arctic is a region where the environment is particularly vulnerable and that large-scale development involves risks of pollution and disturbances that may have long-term and far-reaching effects. Consequently, it is a matter of necessity, and possibly one of urgency, to carry out essential research to avoid and to prevent major negative effects. This is a question of scientific investigation of the Arctic itself, as well as of research to develop necessary safeguards, technical precautions, essential monitoring procedures and information exchanges. There can be no doubt that there is a need for an increased scientific effort in preparation for the development that is in progress for the development of arctic resources and for the use of arctic waters for transportation and other activities.

This will be a massive scientific undertaking that no single nation may be expected to be able to carry out alone. The size of the area and the need for data from all parts of the Arctic to understand the interconnections of phenomena and natural processes make it essential to pool resources for more effective investigation, analysis and exchange of essential information. One might, in short, wish for a system of international scientific cooperation in the Arctic similar to that which has long existed in the Antarctic.

As development in the Arctic goes on and picks up momentum in the exploration and exploitation of resources, the need for scientific study and understanding increases proportionally and the need for harmonization and coordination of scientific programs becomes a matter of great concern and of urgency. In this connection two major obstacles must be negotiated.

One difficulty is the ingrained tendency of secrecy in industry with regard to research and scientific results. Scientific investigations in the Arctic can be expected to be increasingly performed under the aegis of or with economic support from industries with a direct commercial interest in the results. Freedom of information and exchange of scientific result is not an established practice in business, but in the Arctic sharing and pooling of data and information may be essential to the achievement of safe and balanced devlopment and, indeed, to effective preparation and execution of development projects.

A second difficulty is that of scientific coordination and sharing of information between nations. While international cooperation has been conducted with a fair degree of success in the Antarctic, cooperation in the Arctic leaves much to be desired. Yet we know that for optimum results, international cooperation and coordination in scientific research may be essential. This leads up to the important question about the role and attitude of the Russians in the development of the Arctic. This is a difficult question where political and strategic considerations interfere with the traditional attitudes of scientists for open and free communication and cooperation and where national policies restrict the pragmatism of technical and economic cooperation.

The importance of the Soviet Union in arctic affairs is undisputed. The Soviet Union includes nearly half of the land areas and of the continental shelf and ocean areas subject to national jurisdiction in the Arctic. Furthermore, nearly half of the Soviet land territory is arctic or subarctic land and, as far as we can tell today, very large proportions of Soviet energy resources are located in the arctic and subarctic region. As a result, the Soviet Union more than any other nation is dependent upon development of its northern regions in support of its industrial growth and economic development. Over the last few years Soviet production of oil and natural gas has been concentrated increasingly in its subarctic and arctic regions and in coming decades exploration and, eventually, exploitation of petroleum resources will move into the vast continental shelf areas in the north. Large-scale development of northern resources in the Soviet Union — and not only of hydrocarbons, but of other minerals and resources as well — is dependent upon an effective transportation system. With its long northern coastline and with its large northern rivers, the Soviet Union has long emphasized arctic maritime transport as a mainstay in its northern development. Furthermore, the Soviet Union with its highly advanced nuclear icebreaking technology appears to be the only nation which is now making a determined and systematic effort to develop the northern sea route for regular freight connection between Atlantic and Pacific ports.

For these reasons — size of area and scale of effort — it is Soviet development that will have the greater share of the total impact upon the Arctic. This is an impact that may be felt also in the other parts of the Arctic. Thus the Siberian rivers provide the major part of the sweet water flow into the Arctic and the conditions of the water in the Soviet Arctic will influence water conditions around the entire Arctic. For these reasons scientific cooperation and exchange of data with the Soviet Union will be important to the understanding and interpretation of natural processes and phenomena the Arctic at large. There can hardly be any doubt that better arrangements for international scientific cooperation, including Soviet scientists, would be a benefit for all parties and for the total effort to establish the foundations of scientific knowledge and understanding that is so essential for safe and effective development of the Arctic.

Unfortunately, political and strategic considerations make cooperation with the Soviet Union in arctic development a difficult issue. And yet, we know that a combined pool of arctic knowledge may make everybody better prepared to meet the challenge of arctic resource development.

Arctic development presents a many-faceted challenge. It is essential to understand the Arctic in all its aspects and not only in terms of the technological and economic questions involved. There is also a question about social and cultural values that will be affected by change caused by development and there are political difficulties that arise in a region which essentially is one continuous natural environment, but where separate national jurisdictions and strategic interests determine national policies. Above all, there is the question about a highly vulnerable environment that calls for very special care that accidental disturbances be avoided and that all parties that are involved or that may be affected if major disturbances do occur, be prepared to cooperate and to coordinate their efforts to minimize the negative effects.

# SUBJECT INDEX

# Cave and Karst Systems of the World

**Series Editor**

James W. LaMoreaux, P. E. LaMoreaux and Associates, Tuscaloosa, AL, USA

This book series furthers the understanding of cave and karst related processes and facilitates the translation of current discipline-specific research to an interdisciplinary readership by dealing with specific cave or karst systems. Books in this series focus on a specific cave or karst system, on the cave or karst systems of a specific region, on a specific type of cave or karst system, or on any other perspective related to cave and karst systems of the world. The book series addresses a multidisciplinary audience involved in anthropology, archaeology, biology, chemistry, geography, geology, geomorphology, hydrogeology, paleontology, sedimentology, and all other disciplines related to speleology and karst terrains.

More information about this series at https://link.springer.com/bookseries/11987

Philip J. Hobbs · Harrison Pienaar ·
Eddie van Wyk · Yongxin Xu
Editors

# Anatomy of a South African Karst Hydrosystem

## The Hydrology and Hydrogeology of the Cradle of Humankind World Heritage Site

Springer

*Editors*
Philip J. Hobbs (deceased)
Pretoria, South Africa

Eddie van Wyk
Bloemfontein, South Africa

Harrison Pienaar
Smart Places—Water Centre
Council for Scientific and Industrial Research
Pretoria, South Africa

Hebei University of Engineering
Handan, China

Yongxin Xu
Department of Earth Sciences
University of the Western Cape
Bellville, South Africa

ISSN 2364-4591 ISSN 2364-4605 (electronic)
Cave and Karst Systems of the World
ISBN 978-3-030-95828-2 ISBN 978-3-030-95829-9 (eBook)
https://doi.org/10.1007/978-3-030-95829-9



This Springer imprint is published by the registered company Springer Nature Switzerland AG
The registered company address is: Gewerbestrasse 11, 6330 Cham, Switzerland

# Declaration

I, **Philip J. Hobbs**, declare that this publication is my own work and has not previously been submitted by me for publishing at this or any other institution.

P J Hobbs.

Panoramic view, looking south, of the COH WHS 'core' area in the John Nash Nature Reserve showing the valley carved by the Grootvlei Spruit from left (south-east) to right (north-west) across the landscape; also visible is the early winter smog layer over Johannesburg on the horizon (*Photo* P. Hobbs, date 19/05/2010)

# Acknowledgements

It is with gratitude that I acknowledge Prof. Pat Eriksson, former Dean of the Faculty of Natural and Agricultural Sciences, and Prof. Louis van Rooy, Head of the Department of Geology, at the University of Pretoria. It is also fitting that the contributions of the Management Authority of the Cradle of Humankind World Heritage Site (COH WHS), and in particular Mr. Peter (Spike) Mills, Deputy Director Integrated Environment and Conservation Management for the COH WHS and Dinokeng Project, be acknowledged. The Management Authority is thanked for entrusting me with the task of improving the understanding of the water resources environment that contributes to the outstanding universal value of this globally treasured landscape. It is my hope that this publication will serve the Management Authority well in its task of managing and protecting also the water resources component of its UNESCO-entrusted mandate into the future. Peter Mills is thanked for his companionship and support on many excursions into the field.

The contribution of numerous other individuals to the work reflected in this dissertation is acknowledged separately at the end of the text. Many of these are landowners in the study area, and it is my hope that as stakeholders they will benefit from the material and knowledge presented in this dissertation. Others are professional colleagues in the employ of such organisations as the Department of Water and Sanitation (formerly the Department of Water Affairs), the Council for Geosciences and the Council for Scientific and Industrial Research. My thanks go to these individuals for the contribution of their time and effort.

In conclusion, I am thankful for the premeditated and fortuitous factors that a universal intelligence has considered fit to inform my professional career in a scientific discipline that offers so much unsolicited rich return. It is a privilege to contribute towards a better understanding of a complex and largely unseen hydrosystem such as underlies 'The Cradle'.

***"Entia non sunt multiplicanda praeter necessitatem."***
***(Entities must not be multiplied beyond necessity.)***

***John Punch***
***(Irish theologian 1603-1661)***

***aka***
***The law of parsimony (L. Lex parsimoniae)***
***as formulated by Occam's Razor***

View of surface flow and water quality monitoring station A2H049 at the lower end of the Bloubank Spruit at Zwartkop showing hut housing automated stage gauging instrumentation at left, and vertical stage gauge plates in middle and right foreground for visual observation; the blockage by vegetation and debris of the left flank of the weir (right of picture) is not ideal; the northern slope of the 1626 m amsl Zwartkop peak forms the backdrop to this view (*Photo* P. Hobbs, date 05/02/2010)

# Historical Timeline of Key Events Relevant to this Dissertation

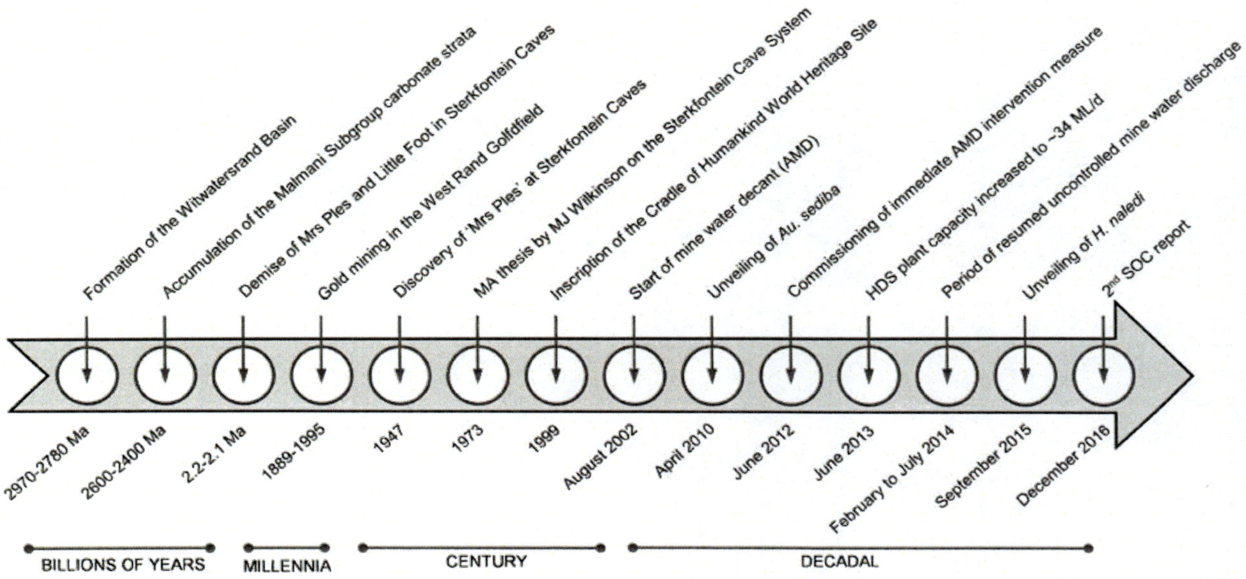


**Timeline 1.** Historical timeline of key events relevant to the study area. Acid mine drainage and its impact on receiving surface water and groundwater resources is a dynamic phenomenon that is continually evolving in response to both controlled (engineered) and uncontrolled (natural) circumstances. The immediate and short-term intervention measures implemented by the Department of Water and Sanitation (DWS) to control and manage acid mine drainage in the West Rand Goldfield (aka the Western Basin) were commissioned in June 2012. The impact of these measures on the receiving water resources is first manifested in August 2012. This marked the commencement of a new evolving dynamic in the study area, with the termination date for this study of September 2017 representing ~5 years of 'new dynamic' observation

Dolomite pinnacle protruding from a doline formed following flooding by stormwater of a soil borrow-pit alongside Dolomite Road (*Photo* P. Hobbs, date 04/11/2009)

# Recent Timeline of Key Events Relevant to this Dissertation

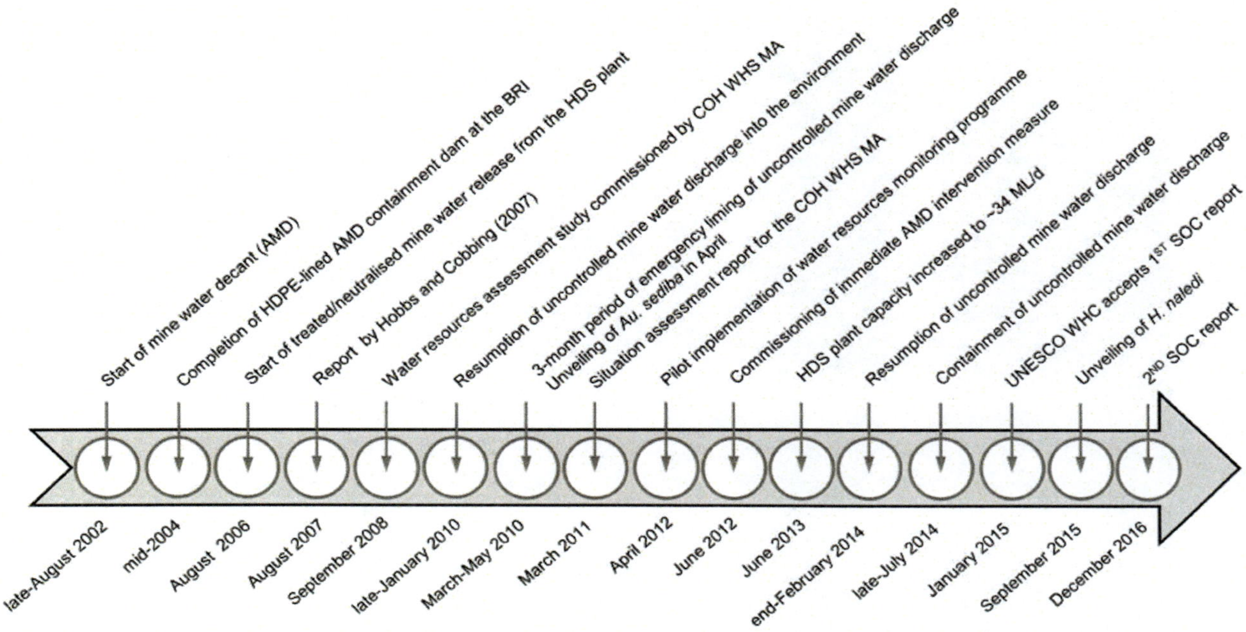


**Timeline 2.** Recent timeline of key events relevant to the study area. August 2017 marks 15 years since the phenomenon of acid mine drainage first appeared in the Western Basin. The impact of the intervention measures implemented in mid-2010 by the DWS to control and manage AMD in the immediate and short-term started manifesting a positive impact on the downstream surface water resources in August 2012. This represents a key event in the evolving dynamic response of the receiving water resources environment. Under circumstances where much of the hydrological analyses set out in this dissertation use a hydrological and not a calendar year as base temporal unit, this combination of factors signify September 2017 (the most recent complete hydrological year) as an appropriate nominal termination date for the material presented and discussed in this work

Surface expression of the epikarst at the Swartkrans fossil site (*Photo* P. Hobbs, date 18/05/2010)

# Extended Summary

## Introduction

The fossil hominin sites of Sterkfontein, Swartkrans, Kromdraai and environs (the so-called Cradle of Humankind) were inscribed by UNESCO in 1999 for protection of their cultural heritage in terms of the World Heritage Convention Act (Act No 49, 1999). The Management Authority (MA) of the Cradle of Humankind World Heritage Site (COH WHS) property exercised its mandate to protect also the aquatic environment of the property by commissioning a study aimed at establishing a monitoring system for surface water and groundwater resources in its area of jurisdiction.

The implementation of an appropriate integrated hydrologic and hydrogeologic monitoring programme is crucial to the successful management of the water resources in the COH. An effective routine monitoring programme serves to measure and demonstrate the success or failure of management efforts by the MA to protect the aquatic environment of the property. Such protection is not only required for the preservation of the karst environment and its palaeo-anthropological wealth, but also for the water users who reside in the area and depend on local water resources (primarily groundwater) for their livelihood.

## Surface Water Resources

### *Quantity*

The Skeerpoort River system is in a nearly pristine condition. Its perennial nature is sustained mainly by the combined discharge (>300 L/s ≈ 25.9 ML/d ≈ 9.6 $Mm^3/a$) of three high-yielding karst springs located in the John Nash Nature Reserve. The flow gauging record for the Skeerpoort River indicates a long-term median discharge of ~9.57 $Mm^3/a$ to Hartbeespoort Dam via the Magalies River. This represents ~5% of the net capacity (~190 $Mm^3$) of the dam.

The flow record for the heavily impacted Bloubank Spruit system indicates a long-term median discharge of ~22.7 $Mm^3/a$ to Hartbeespoort Dam via the Crocodile River. This represents ~12% of the net capacity of the dam. The Bloubank Spruit system experienced above average discharges in its upper reaches via the Tweelopie Spruit and the lower Riet Spruit following the resumption of uncontrolled raw mine water decant in late-January 2010. The combined discharge of raw and treated mine water realised quantified surface water losses of 20 to 32 ML/d to the karst aquifer from the lower Riet Spruit, equating to an infiltration rate of as much as ~90 L/s/km. Together with the discharge from the Percy Stewart Wastewater Treatment Works (WWTW) via the Blougat Spruit, these circumstances resulted in an unprecedented volume of surface water flow in the Bloubank Spruit system through the 2010, 2011 and 2014 winter seasons. The median discharge of ~4.7 $Mm^3/a$ (~13 ML/d) of treated sewage effluent to the Blougat Spruit from the Percy Stewart WWTW equates to ~21% of the long-term median annual discharge of the Bloubank Spruit system.

The Crocodile River flow gauging record at the confluence with its Bloubank Spruit tributary indicates a long-term median discharge of ~9.5 $Mm^3$/a (~5% of the Hartbeespoort Dam net capacity). As only 15% of this catchment falls within the COH, this discharge is excluded from the aggregate long-term contribution of ~18% (~34.7 $Mm^3$/a) delivered to Hartbeespoort Dam by the Skeerpoort River and Bloubank Spruit catchments. These catchments together represent ~71% of the study area.

## *Quality*

The Skeerpoort River system delivers a CaMg-$HCO_3$ water composition of excellent quality. Up until mid-2010, the impact of the poor quality associated with the abnormal combined discharge of treated and raw mine water in the upper reaches of the Bloubank Spruit system was mitigated by the contribution of treated wastewater effluent discharged by the Percy Stewart WWTW and the above average surface water runoff associated with the extremely wet 2010 summer. Since mid-2010, the increase in specific electrical conductivity (SEC) of Bloubank Spruit water from ~50 to >100 mS/m, together with a decrease in pH from 7.2 to 6.9 at the downstream end of the Zwartkrans Basin, reflects the increasing contribution of mine water to the middle reaches of the Bloubank Spruit.

The combination of long-term discharge and water chemistry records for the DWS gauging/sampling stations on the Skeerpoort River (A2H034), the Bloubank Spruit (A2H049) and the Crocodile River (A2H050) allow for an assessment of the total dissolved solids (TDS) loads associated with the respective drainages. This assessment indicates that the Skeerpoort River and (upper) Crocodile River deliver similar TDS loads of 2937 and 3249 t/a respectively, compared to the 10 173 t/a delivered by the Bloubank Spruit system. Again excluding the (upper) Crocodile River load, the values translate into contributions of 22% by the Skeerpoort River and 78% by the Bloubank Spruit system to the total TDS load of 13 110 t/a delivered to Hartbeespoort Dam by the COH drainages. In a regional context, this load constitutes only 13% of that entering the dam, being surpassed by the 51% (51 023 t/a) of the Jukskei River and the 28% (27 579 t/a) of the Hennops River. The balance of 8% (8085 t/a) is shared by the Crocodile River (3%) and the Magalies River (5%).

The quality of surface water resources in the Bloubank Spruit system is further compromised by bacterial contamination and associated elevated nitrate and phosphorus concentrations derived mainly from wastewater effluent. These circumstances also make it difficult to assess the agricultural impacts on the quality of surface water resources, as these are similarly associated with nutrient inputs. As a subset of the total salt load, the nutrient load entering Hartbeespoort Dam is of specific concern given the hypertrophic status of this impoundment. The sampling stations on the Bloubank Spruit system and the (upper) Crocodile River reflect median $NO_3$-N and $PO_4$-P loads of 129 and 2.9 t/a, respectively, to the dam in the period 1980 to 2013. A similar appraisal for the Jukskei and Hennops rivers indicates combined median $NO_3$-N and $PO_4$-P loads of 1245 and ~104 t/a, respectively, for the same period. In summary, the Crocodile River and Bloubank Spruit systems together contribute <10% to the median long-term $NO_3$-N load entering Hartbeespoort Dam, being overshadowed by the Jukskei River contribution of ~70% and the Hennops River contribution of ~20%. The $PO_4$-P load is dominated even more by the Jukskei and Hennops rivers, with the Crocodile River and Bloubank Spruit systems delivering <3% of this nutrient load to the dam annually.

A primary concern for the downstream environment is the impact of mine water, in particular the presence of trace/heavy metals, metalloids and radionuclides, on the quality of water in the Bloubank Spruit system. In the period of maximum likely impact, namely February 2010 to July 2012, median Fe and Mn levels of 0.013 mg Fe/L and 0.003 mg Mn/L in surface water at the lower end of the system on 33 sampling occasions, compare favourably with levels of 163 mg Fe/L and 65 mg Mn/L in composite mine water discharge in the upper reaches of the system on 129 sampling occasions. Mercury levels in surface water typically do not exceed 0.002 mg/L. Arsenic levels similarly seldom exceed the detection limit of 0.002 mg/L. Nickel presents as the most persistent trace metal in upper (headwater) reaches, the median concentration of 0.1 mg Ni/L from 41 sampling occasions exceeding the SANS (2015a) limit of 0.07 mg/L. As with Fe and Mn, Ni levels in the lower reaches of the system do not test the 0.07 mg/L limit.

Uranium levels in surface water nowhere and on no sampling occasion exceeded the analytical detection limit of 0.001 mg/L for this analyte. Radon ($^{222}Rn$) activity levels representative of the headwater reach in the mine area fall within the minimum detectable activity (MDA) of ~0.5 Bq/L. This compares favourably with the maximum contaminant level (MCL) of 11.1 Bq/L set by the USAs Safe Drinking Water Act (SDWA). Similarly, radium ($^{226}Ra$) activity levels do not exceed the SDWAs MCL of 0.185 Bq/L for this radionuclide in the extremely sparse set of available data.

The persistence of poor bacteriological quality as reflected in alarmingly high faecal coliform and *E. coli* values associated with surface water in the Bloubank Spruit, continues to represent a significant threat to the 'fitness for use' of this resource. This situation reflects the poor score achieved by the Percy Stewart WWTW in both the 2009 and 2011 DWS 'Green Drop' reports, and in particular the non-compliance in regard to the effluent wastewater quality metric. A thorough evaluation of this threat is thwarted by the non-disclosure of pertinent monitoring data by the local authority.

## Groundwater Resources

### *Quantity*

Springs are widely recognised as the most appropriate gauging, sampling and monitoring points in a karst environment. The study has enumerated eleven springs (excluding the seven located in the Krugersdorp Game Reserve) in the subregion. The total number of such features in the subregion is almost certainly greater. Some of the features represent groups of springs (and seeps) located in close proximity to one another. Nine of these drain dolomitic strata, the 'weakest' delivering ~2 L/s and the 'strongest' ~307 L/s. The total yield of these sources amounts to ~827 L/s (~71.5 ML/d ≈ 26.1 $Mm^3/a$). This equates to ~14% of the net capacity of Hartbeespoort Dam, and reflects the very important contribution of mainly good to excellent springwater to the water resources of the wider region. None of the enumerated springs are subject to regular and routine discharge measurements. Synoptic discharge measurements in this study have served to quantify the yield of many of these features for the first time.

Groundwater quantity is further represented by groundwater level data and information. The study has generated 117 groundwater level measurements from as many sources (18 springs and 99 boreholes). Each of these measurements has been translated into an absolute value representing a groundwater elevation above mean sea level. Together with the locations and elevations of the various springs, this information has led to an improved understanding of groundwater flow and movement especially in regard to the dolomitic strata. As a consequence, redefinition of the physical hydrogeologic environment recognises a degree of compartmentalisation that contributes significantly to a more informed understanding of the karst groundwater environment. A total of ten dolomitic compartments, two of which comprise subcompartments, are identified in the COH. Most of the compartments are drained by springs. Water budget calculations for the seven karst basins drained by springs with yields >20 L/s and factoring in their surface extent, indicates that 17 ± 5% of a mean annual precipitation of 710 mm provides a reasonable approximation of natural autogenic recharge from rainfall for the karst hydrosystem.

The behaviour of groundwater levels associated with the karst aquifer is reflected in the long-term water level records for 15 DWS monitoring boreholes dating back to 1985. In 11 of these instances, the record period extends to the present. An analysis of the data indicates a generally excellent agreement between the mean and median values. This reflects the large measure of constancy in this variable. Further, there is little correlation between the depth to groundwater rest level and the magnitude of water level variation; relatively small variations (<3 m) being associated with both 'deep' (>60 m bs) and comparatively 'shallow' (<30 m bs) water levels. The data set reveals a maximum water level variation value of ~12.2 m, with mean and median values of ~6.2 m and ~5.6 m, respectively. The slightly smaller differences associated with the 5%ile to 95%ile interval are characterised by a maximum value of 9.8 m, and mean and median values of 5.2 and 4.6 m respectively.

The very wet 2010, 2011 and 2014 summers precipitated an exceptional recharge of groundwater resources in the study area. A rise in groundwater rest levels by ~4.9 m on average testifies to these circumstances. Greater water level rises (by up to ~8 m) are attributed to artificial and allogenic recharge associated with the infiltration of surface water contributed from extraneous sources including mining and municipal wastewater effluent. This infiltration has amounted to as much as ~32 ML/d in the case of mine water, and ~7 ML/d in the case of municipal wastewater. The recent (since 2012) groundwater level (water table) elevations in the COH are the highest in the ~30-year record of monitoring.

### *Quality*

Groundwater quality in the COH is defined on the basis of chemical analyses carried out on water samples obtained from 51 sources (7 springs and 44 boreholes). The analytical suite include inorganic, organic and bacteriological variables, heavy/trace metals/metalloids and environmental isotopes as well as pesticide residue analyses employed selectively. In addition, numerous measurements of field variables (pH, EC, ORP/Eh and temperature) have been carried out on an ad hoc basis at a number of springs.

As might be expected, the hydrochemistry reflects a greater or lesser spatial variation depending on the position in the physical hydrogeologic environment. For instance, the subcompartments receiving water of compromised quality in terms of either trace/heavy metals/metalloids and elevated TDS loads associated with mine water, and/or elevated bacterial and nutrient loads associated with municipal wastewater (both representing allogenic recharge), reflect the poorest groundwater quality. Despite its location, however, the Lake water in Sterkfontein Cave continues to reflect an SEC of <70 mS/m as it did in June 2006. Karst basins receiving only autogenic recharge remain largely unaffected in terms of groundwater chemistry/quality.

## Conclusions

The understanding of the surface water and groundwater environments in the COH, also in regard to the inter-relationship between these resources, is considerably expanded by this study. This understanding extends as much to the water chemistry aspect as it does to the water quantity aspect. The platform built from historical data, and its integration with a wide range of rigorous and defensible newly-generated and interpreted hydrologic and hydrogeologic data and information, convincingly underpins the situation assessment of the surface water and groundwater environments. This, in turn, has provided the means to objectively gauge the impact of varied and numerous threats on the water resources in the study area, and to develop a coordinated, appropriate and cost-effective water resources monitoring programme. Outcomes of the study that are considered especially significant are summarised as follows.

- The quantification of surface water flow losses, especially those dominated by a mine water character in the lower Riet Spruit valley.
- The quantification of spring discharges.
- The definition of basins/subcompartments and corresponding groundwater resource units (GRUs) associated mainly with the karst formations in the study area.
- The development of semi-quantitative resource water quality objectives (RWQOs) to inform surface and groundwater resource directed measures for the karst portions of the study area.
- The derivation of a fossil site risk assessment that informs the vulnerability of each recognised fossil site and associated cave system in the context of its hydrogeologic setting.

A cause for grave concern is the unprecedented abnormally high flow conditions experienced in the Bloubank Spruit system in the more recent hydrological years, as this discharge is the result of abnormally high mine water decant driven by copious recharge associated with above average rainfall. This has already manifested itself as historical maximum SEC and sulphate values at the lowest end of the Bloubank Spruit system.

## Recommendations

The study has identified various concerns that give rise to the following general recommendations.

- The advisability of carrying out a gravimetric survey in the lower Riet Spruit valley extending from the confluence of the Tweelopie Spruit and the Riet Spruit down to the confluence of the Blougat Spruit and the Riet Spruit. The results of such a survey will indicate the measure of karst dissolution present in this important E–W corridor that hosts the N14 national road.
- The advisability of extending the hydrovulnerability assessment to other cave systems in the study area, together with a refinement of the applied assessment methodology.
- The establishment of a monitoring committee comprising a core of key stakeholder groupings, e.g. national, provincial and local government, environment and tourism, agriculture.
- The hosting (by the Management Authority) of a workshop or seminar to communicate the outcomes of the study to as wide an audience of stakeholders and interested and affected parties as are interested.
- The expansion of the mine water treatment capacity in the headwaters of the Tweelopie Spruit to accommodate a decant volume of ~60 ML/d, representing a 2-fold increase in the current treatment capacity.
- The establishment of additional mine water treatment facilities in the headwaters of the Tweelopie Spruit to further 'polish' the treated mine water that is generated by the expanded mine water treatment capacity and released into the environment.

# Contents

# Symbols, Acronyms and Abbreviations

| | |
|---|---|
| $\sim$ | Approximately |
| $\equiv$ | Equivalent |
| $>$ | Greater than |
| $>>$ | Much greater than |
| $\geq$ | Greater than or equal to |
| $\leq$ | Less than or equal to |
| $<$ | Less than |
| # | Number |
| $\pm$ | Plus-minus |
| $\delta$ | Delta (notation) |
| $\Delta$ | Change in |
| $\Sigma$ | The sum of |
| μg/g | Microgram(s) per gram |
| % | Per cent (parts per hundred) |
| ‰ | Per mil (parts per thousand) |
| %ile | Percentile |
| °C | Degree(s) Celsius (centigrade) |
| °C/m | Degree(s) Celsius per metre |
| °E | Degree(s) East (longitude) |
| °S | Degree(s) South (latitude) |
| $^{2}H$ | Deuterium |
| $^{3}H$ | Tritium |
| $^{18}O$ | Oxygen-18 |
| a | Annum |
| $a_h$ | Hydrological year |
| *A.* | Afrikaans |
| amsl | Above mean sea level |
| ABA | Acid base accounting |
| AET | Actual evapotranspiration |
| A.H. | Agricultural Holdings |
| aka | Also known as |
| Al | Aluminium |
| AMD | Acid mine drainage (or) decant (or) discharge |
| ARC | Agricultural Research Council |
| As | Arsenic |
| ASPT | Average score per taxon |
| ATSDR | Agency for Toxic Substances and Disease Registry |
| atm | Atmosphere(s) |
| *Au.* | Australopithecus |
| B | Boron |
| Ba | Barium |

| | |
|---|---|
| bc | Below collar |
| BE | Built Environment (a business unit of the CSIR) |
| bgl | Below ground level |
| BP | Before present |
| Bq/g | Becquerel(s) per gram |
| Bq/L | Becquerel(s) per litre |
| bs | Below surface |
| BSR | Bacterial sulphate reduction |
| bwl | Below water level |
| C | Carbon |
| $C_X$ | Concentration ($C_U$ = upstream; $C_D$ = downstream; $C_F$ = furrow; $C_S$ = spring; $C_G$ = groundwater) |
| $C_5$ | Concentration exceeded 5% of the time (equivalent to 95%ile) |
| $C_{50}$ | Concentration exceeded 50% of the time (equivalent to 50%ile) |
| $C_{95}$ | Concentration exceeded 95% of the time (equivalent to 5%ile) |
| ca. | Circa (about) |
| Ca | Calcium |
| Cd | Cadmium |
| CD | Compact disc |
| CDSM | Chief Directorate: Surveys and Mapping (in the Department of Land Affairs) |
| cfu | Colony forming unit(s) |
| CGS | Council for Geoscience |
| Cl | Chloride |
| CMB | Chloride mass balance |
| CN | Cyanide |
| Co | Cobalt |
| $CO_2$ | Carbon dioxide |
| COD | Chemical oxygen demand |
| COH WHS | Cradle of Humankind World Heritage Site |
| CoV | Coefficient of variation |
| CPOM | Coarse particulate organic matter |
| Cr | Chromium |
| CROSA | Cave Research Organisation of South Africa |
| CSIR | Council for Scientific and Industrial Research |
| CTR | Corrosion tendency ratio |
| Cu | Copper |
| d | Day(s) |
| dd.ddddd | Degrees latitude (or) longitude expressed to the 5th decimal |
| dd/mm/yyyy | Date format as day/month/year, e.g. 10/11/2012 $\equiv$ 10 November 2012 |
| DEA | Department of Environmental Affairs (formerly Department of Environmental Affairs and Tourism) |
| DEAT | Department of Environmental Affairs and Tourism |
| DED | Department of Economic Development (Gauteng Province) |
| DL | Detection limit |
| DLA | Department of Land Affairs |
| dm | Decimetre |
| DME | Department of Minerals and Energy |
| DMR | Department of Mineral Resources (formerly Department of Minerals and Energy) |
| DMS | Dissolved mineral salts |
| DO | Dissolved oxygen |
| DOC | Dissolved organic carbon |

| | |
|---|---|
| DPLG | Department of Development, Planning and Local Government |
| D:RQS | Directorate: Resource Quality Services (a directorate in the DWS) |
| DWA | Department of Water Affairs (formerly Department of Water Affairs and Forestry) |
| DWAF | Department of Water Affairs and Forestry |
| DWS | Department of Water and Sanitation (formerly the Department of Water Affairs) |
| EB | Electrical balance (aka ion balance or charge balance) |
| EC | Electrical conductivity |
| ECL | Environmental critical level |
| *E. coli* | *Escherichia coli* bacteria |
| EDC | Endocrine disrupting chemical |
| Eh | Free electron ($e^-$) activity defined as $-\log_{10}[e^-]$; also assigned the abbreviation $p\varepsilon$ |
| e.g. | *Exempli gratia* (for example) |
| EI&S | Ecological importance and sensitivity |
| EMF | Environmental management framework |
| EMS | Environmental management system |
| EoP | End-of-pipe |
| EPA | Environmental Protection Agency |
| Eq | Equation |
| eq | Equivalent |
| ERWAT | East Rand Water Care Company |
| ESE | East-south-east |
| ET | Evapotranspiration |
| EurepGAP | Euro-Retailer Produce Working Group Good Agricultural Practice |
| Fe | Iron |
| FFG | Functional feeding group |
| FIB | Faecal indicator bacteria |
| Fm. | Formation (geological term) |
| FPOM | Fine particulate organic matter |
| FSE | Federation for a Sustainable Environment |
| FSLTS | Feasibility study for a long-term solution (commissioned by the DWS) |
| G | Billion ($10^9$) |
| G1 | Gold1 (operating in conjunction with Rand Uranium) |
| g/t | Gram(s) per ton |
| Ga | Billion years |
| GC | Gas chromatograph |
| GDACE | Gauteng Department of Agriculture, Conservation and the Environment |
| GDARD | Gauteng Department of Agriculture and Rural Development (formerly GDACE) |
| Gg/a | Gigagram(s) per annum |
| GIS | Geographic information system |
| G.M. Co. | Gold Mining Company |
| GMU | Groundwater management unit |
| GNIP | Global network of isotopes in precipitation |
| Gp. | Group (geological term) |
| GPS | Global positioning system |
| GRA | Groundwater resource assessment |
| GRDM | Groundwater resource directed measures |
| GRU | Groundwater resource unit |
| h | Hour(s) |

| | |
|---|---|
| H | Weighting value |
| *H.* | Homo |
| $HCO_3$ | Bicarbonate |
| ha | Hectare(s) |
| HDS | High density sludge |
| HDPE | High density poly-ethylene |
| $H_i$ | Hazard index |
| HPC | Heterotrophic plate count (also referred to as total plate count) |
| *i* | Hydraulic gradient |
| I&AP | Interested and affected party |
| ICOMOS | International Council on Monuments and Sites |
| ICP | Inductively coupled plasma |
| IDP | Integrated development plan |
| i.e. | *Id est* (that is to say) |
| IGTT | Inter-Governmental Task Team (on AMD) |
| IGU | International Geophysical Union |
| IHAS | Integrated Habitat Assessment System |
| IHIA | Integrated Habitat Integrity Assessment |
| IMC | Inter-Ministerial Committee (on AMD) |
| iro | In respect of |
| iTLABS | iThemba LABS (a National Research Foundation facility) |
| IUCN | International Union for Conservation of Nature |
| JFA | Johan Fourie and Associates (Environmental Consultancy) |
| JNNR | John Nash Nature Reserve |
| K | Potassium |
| $K_X$ | Equilibrium constant for mineral X, where X = C = calcite, and X = D = dolomite |
| Ka | Thousand years |
| KFD | Koelenhof Farm Dam |
| kg/d | Kilogram(s) per day |
| $kg/km^2/a$ | Kilogram(s) per square kilometre per annum |
| $kg/m^3$ | Kilogram(s) per cubic metre |
| KGR | Krugersdorp Game Reserve |
| km | Kilometre(s) |
| $km^2$ | Square kilometre(s) |
| km/h | Kilometre(s) per second |
| kt | Kiloton(s) |
| *L.* | Latin |
| $LC_x$ | Concentration causing x% lethality |
| L/kg/d | Litre(s) per kilogram per day |
| LMDC | Leadership and Management Development Centre (Nedbank's Olwazini Estate) |
| LoD | Locus of decant |
| L/s | Litre(s) per second |
| LSC | Liquid scintillation counting |
| L/s/km | Litre(s) per second per kilometre |
| m | Metre(s) |
| M | Million ($10^6$) |
| m/Ma | Metre(s) per million years |
| $m^2/d$ | Square metre(s) per day |
| $m^3/km^2/a$ | Cubic metre(s) per square kilometre per annum ≡ mm/Ka |
| m/s | Metre(s) per second |

| | |
|---|---|
| $m^3/s$ | Cubic metre(s) per second (also referred to as 'cumec(s)') |
| MA | Management Authority |
| Ma | Million years |
| MAD | Mean annual discharge |
| MAP | Mean annual precipitation |
| $MA_hP$ | Mean hydrological year precipitation |
| MAPE | Mean annual potential evaporation |
| MAR | Mean annual runoff |
| MAT | Mean annual temperature |
| max. | Maximum |
| MCL | Maximum contaminant level |
| MCLM | Mogale City Local Municipality |
| MDA | Minimum detectable activity |
| MDB | Municipal Demarcation Board |
| MEC | Member of the Executive Council |
| med. | Median |
| MG | Mogale Gold (operating in conjunction with Mintails SA) |
| Mg | Magnesium |
| mg/kg | Milligram(s) per kilogram |
| MG/MSA | Mogale Gold/Mintails SA |
| mg/L | Milligram(s) per litre |
| mg/s | Milligram(s) per second (1 mg/s ≡ 0.000084 t/d) |
| min. | Minimum |
| mL | Millilitre(s) |
| ML | Megalitre(s) (1 ML ≡ 1 million litres) |
| ML/d | Megalitre(s) per day |
| mm/a | Millimetre(s) per annum |
| mm/Ka | Millimetre(s) per thousand years |
| mm/yyyy | Month/year (e.g. 01/2010) |
| $Mm^3/a$ | Million cubic metre(s) per annum |
| $Mm^3/m$ | Million cubic metre(s) per month |
| Mn | Manganese |
| MNR | Motsetse Nature Reserve |
| Mo | Molybdenum |
| MP | Management plan |
| MPN | Most probable number |
| MRD | Mine residue deposit |
| mS/m | MilliSiemens per metre |
| MSM | Monitoring system manual |
| MSRF | Morphologic suite of rising flow |
| mV | MilliVolt(s) |
| n | Count (of sample population) |
| n.a. | Not analysed |
| n/a | Not applicable |
| Na | Sodium |
| n.d. | Not determined |
| NAEHMP | National Aquatic Ecosystem Health Monitoring Programme |
| NE | North-east |
| NFEPA | National Freshwater Ecosystem Priority Areas |
| NGO | Non-governmental organisation |
| $NH_3$ | Ammonia nitrogen |
| $NH_4$ | Ammonium nitrogen |

| | |
|---|---|
| Ni | Nickel |
| n.m. | Not measured |
| nm | Nanometre(s) (1 nm = $1 \times 10^{-9}$ m) |
| No | Number |
| $NO_2$ | Nitrite nitrogen |
| $NO_3$ | Nitrate nitrogen |
| NOE | Nedbank Olwazini Estate |
| NRE | Natural Resources and the Environment (a business unit of the CSIR) |
| n/s | Not specified |
| NTU | Nephelometric turbidity unit(s) |
| O-$PO_4$ | Ortho-phoshate |
| ORP | Oxidation reduction potential; also simply referred to as 'redox' potential |
| OUV | Outstanding universal value |
| p | Page |
| P | Phosphorus |
| Pb | Lead |
| pCi/L | PicoCurie(s) per liter |
| $P_{CO2}$ | Partial pressure (activity) of $CO_2$ |
| p$\varepsilon$ | Free electron ($e^-$) activity defined as $-\log_{10}[e^-]$; also assigned the abbreviation Eh |
| pH | Free hydrogen ion activity ($\alpha_{H+}$) defined as $-\log_{10}\alpha_{H+}$ where $\alpha_{H+} = f \times [H^+]$ with $f$ the activity coefficient and $[H^+]$ the hydrogen ion concentration |
| $PO_4$ | Phosphate |
| pp | Pages |
| Pt | Petaton |
| Pt | Platinum |
| Ptn | PORTION |
| $Q_X$ | Flow or discharge ($Q_U$ = upstream; $Q_D$ = downstream; $Q_F$ = furrow; $Q_S$ = spring; $Q_G$ = groundwater) |
| $Q_{25}$ | Flow or discharge exceeded 75% of the time (equivalent to 25%ile) |
| $Q_{50}$ | Flow or discharge exceeded 50% of the time (equivalent to 50%ile) |
| $Q_{75}$ | Flow or discharge exceeded 25% of the time (equivalent to 75%ile) |
| $Q^n$ | Ranking factor |
| $Q_{min}$ | Minimum flow or discharge |
| $Q_{max}$ | Maximum flow or discharge |
| R. | River |
| $R^f$ | Reduction factor |
| RAIS | Risk Assessment Information System |
| REGM | Randfontein Estates Gold Mine |
| RET | Riparian evapotranspiration |
| RHP | River health programme |
| $RI_i$ | Risk intensity index |
| RMW | Raw mine water |
| RSA | Republic of South Africa |
| RWQOs | Resource water quality objective(s) |
| RU | Rand Uranium (successor to Harmony Gold) |
| RU/G1 | Rand Uranium/Gold1 (successor first to Harmony Gold and then Uranium1) |
| RWL | Rest water level |
| S. | Spruit |
| SABS | South African Bureau of Standards |
| SAC | Satellite Applications Centre |
| SAFF | Submerged aeration fixed film |

| | |
|---|---|
| SAIEEG | South African Institute of Engineering and Environmental Geology |
| SAKWG | South African Karst Working Group |
| SANAS | South African National Accreditation System |
| SANBI | South African National Biodiversity Institute |
| SANParks | South African National Parks |
| SANS | South African National Standard |
| SAR | Sodium adsorption ratio (a calculated chemical variable) |
| SASS | South African Scoring System |
| SAWS | South African Weather Service |
| Sbgp. | Subgroup (a geological term) |
| SC | Sterkfontein Cave |
| SD | Standard deviation |
| SDF | Spatial development framework |
| SDM | Synoptic discharge measurement |
| SDWA | Safe Drinking Water Act (administered and overseen by the US EPA) |
| Se | Selenium |
| SEC | Specific electrical conductivity (EC @ 25 °C) |
| SECL | Socio-economic critical level |
| SEM | Scanning electron microscopy |
| SHE | Standard hydrogen electrode |
| Si | Silicon |
| $SI_C$ | Saturation index of calcite expressed as $\log\{[Ca^{2+}][CO_3{}^{2-}]/K_C\}$ |
| $SI_D$ | Saturation index of dolomite expressed as $\log\{[Ca^{2+}][Mg^{2+}][CO_3{}^{2-}]^2/K_D\}$ |
| SMOW | Standard mean ocean water |
| SoC | State of conservation |
| SoE | State of the environment |
| $SO_4$ | Sulphate |
| Sp. | Spring |
| Spgp. | Supergroup (a geological term) |
| Sr | Strontium |
| SRB | Sulphate reducing bacteria |
| SRK | Steffen, Robertson and Kirsten (Consulting Engineers and Scientists) |
| SS | Suspended solids |
| S-S | Sibanye-Stillwater (successor to Uranium1) |
| StatsSA | Statistics South Africa |
| SW | South-west |
| t | Ton(s) |
| *T* | Transmissivity |
| t/a | Ton(s) per annum |
| t/d | Ton(s) per day |
| t/m | Ton(s) per month |
| t/ML | Ton(s) per megalitre |
| T.Alk. | Total alkalinity (as $CaCO_3$) |
| TC | Total carbon |
| TCLP | Toxicity characteristic leaching procedure |
| TCTA | Trans-Caledon Tunnel Authority |
| TDS | Total dissolved solids/salts |
| TIC | Total inorganic carbon |
| TMW | Treated/neutralised mine water |
| TOC | Total organic carbon |
| ToC | Table of Contents |
| TOL | Target operating level |

| | |
|---|---|
| TON | Threshold odour number |
| ToR | Terms of reference |
| $TR_i$ | Total risk index |
| TU | Tritium unit(s) (1 TU = 1 tritium in $10^{18}$ hydrogen atoms) |
| Turb. | Turbidity |
| TWQR | Target water quality range |
| U | Uranium |
| UIS | Union Internationale de Spéléologie (International Union of Speleology) |
| UNESCO | United Nations Educational, Scientific and Cultural Organisation |
| US(A) | United States (of America) |
| USACE | United States Army Corps of Engineers |
| USDA | United States Department of Agriculture |
| V | Vanadium |
| $V_i$ | Vulnerability index |
| VCR | Ventersdorp Contact Reef |
| vs. | Versus |
| *w* | Width |
| WARMS | Water authorisation and registration management system (a DWS database] |
| WBTWG | Western Basin Technical Working Group |
| WCPA | World Commission on Protected Areas |
| WGC | Water Geosciences Consulting |
| WGS84 | World Geodetic System 1984 (reference ellipsoid for Hartebeesthoek94 Datum) |
| WHC | World Heritage Centre (could also denote World Heritage Committee) |
| WHO | World Health Organisation |
| WMA | Water management area |
| WNW | West-north-west |
| WRC | Water Research Commission |
| WRDM | West Rand District Municipality |
| WWTW | Wastewater treatment works |
| XRD | X-ray diffraction |
| XRF | X-ray fluorescence |
| y | Year(s) |
| Zn | Zinc |

# List of Figures

## Introduction and Background



## Description of the Physical Environment

**Overview of Karst**



**Physical Hydrology**

### Chemical Hydrology

**Physical Hydrogeology**

**Chemical Hydrogeology**

# List of Tables

**Description of the Physical Environment**



**Overview of Karst**



**Physical Hydrology**

**Chemical Hydrology**

**Physical Hydrogeology**

**Chemical Hydrogeology**

# List of Plates

**Chemical Hydrology**

**Physical Hydrogeology**



**Chemical Hydrogeology**

# List of Text Boxes

# Précis

A poor understanding of the surface and groundwater resources of the Cradle of Humankind (COH) World Heritage Site (WHS) property has precipitated often alarmist reporting in the media regarding the negative impacts associated with various sources of poor quality water. The most notable of these is the acid mine drainage (AMD) threat to karst ecosystems and fossil sites across the property. These circumstances have generated wide and considerable concern for the preservation of the UNESCO-inscribed fossil sites and integrity of the water resources of the property. An assessment of the water resources environment has better informed this situation. It also formed the basis for the implementation of an integrated water resource monitoring programme in support of management efforts to protect the aquatic environment and assist in maintaining the outstanding universal value of the property. The combined results of these activities inform the multi-dimensional analysis set out in this work.

The surface water environment comprises the largely unspoilt Skeerpoort River catchment and the heavily impacted Bloubank Spruit system. These drain to the regionally important Hartbeespoort Dam via the Magalies River and the Crocodile River as respective main stems. The Skeerpoort River, with a long-term median discharge of ~9.6 $Mm^3/a$, is fed by karst springs delivering a similar quantity of excellent quality karst groundwater. This contribution represents ~5% of the ~190 $Mm^3$ full supply capacity (FSC) of Hartbeespoort Dam. The Bloubank Spruit system, with a long-term median discharge of ~22.7 $Mm^3/a$ (~12% of FSC), receives >4.4 $Mm^3/a$ of poor quality raw and treated mine water and >2.5 $Mm^3/a$ of treated municipal sewage effluent on average in its non-karst headwater reaches upstream of the karst environment. The balance is contributed by the karst aquifer delivering >16.6 $Mm^3/a$ of good to excellent quality dolomitic groundwater. The Crocodile River median long-term discharge of ~9.5 $Mm^3/a$ good quality surface water at the confluence with its Bloubank Spruit tributary, similarly represents ~5% of the dam's FSC. Expressed as a long-term (1980 to 2015) median annual salt load, these drainages contributed ~2.9 kt (Skeerpoort River), ~10.6 kt (Bloubank Spruit) and ~3.8 kt (Crocodile River) of total dissolved solids to the impoundment. The negative impact of a significant raw mine water discharge component for a large part of the 2010 to 2014 period on the Bloubank Spruit system, was partially mitigated by the commissioning in mid-2012 and further upgrades by mid-2013, of immediate and short-term mine water control and management interventions by the Department of Water and Sanitation. The persistence of poor to very poor bacteriological quality associated with surface water in the Bloubank Spruit represents a second significant and pernicious threat to the 'fitness for use' of the ambient water resources.

The karst hydrosystem is assessed as comprising eleven compartments (basins), two of which are subdivided into subcompartments. Most of the basins are drained by high-yielding (>20 L/s) springs for which quantitative discharges and hydrochemical composition are known. The aggregate discharge of eleven enumerated karst springs amounts to ~26.2 $Mm^3/a$ (~71.8 ML/d), or ~13% of the Hartbeespoort Dam FSC. Together with an additional ~4.1 $Mm^3/a$ (~130 L/s) associated with groundwater resurgence along effluent stream sections traversing dolomite, the total groundwater yield of ~30.3 $Mm^3/a$ (~83 ML/d) equals ~16% of the reservoir FSC. The combined long-term median discharge of the Skeerpoort River and

Bloubank Spruit amounts to ~34.7 $Mm^3$/a (~95.1 ML/d), a little over 18% of the dam's FSC. These circumstances reflect the important contribution of good to excellent quality karst groundwater rising mainly in the COH, to the water budget of the wider region.

Water budget calculations indicate that 17 ± 5% of a mean annual precipitation of 710 mm provides a reasonable approximation of natural (autogenic) recharge from rainfall for the karst hydrosystem. The very wet 2010, 2011 and 2014 summers resulted in exceptional recharge of groundwater resources in the study area. A rise in groundwater rest levels by ~3 m on average, also observed in the Sterkfontein Cave, reflects this. Water level rises of this magnitude and by as much as ~5 m are attributed jointly to allogenic recharge associated with the infiltration of surface water contributed in the form of mining and municipal wastewater effluent from upstream non-karst areas, and natural autogenic recharge. The allogenic recharge in the case of mine water has amounted to a minimum of ~5.1 $Mm^3$/a (~14 ML/d) rising to at least ~11.7 $Mm^3$/a (~32 ML/d), and from ~1.1 to ~2.6 $Mm^3$/a (~3 to 7 ML/d) in the case of municipal wastewater effluent. The mine water discharges have introduced a new set of drivers causing a resetting of the natural water resources environment. In the case of groundwater, this is immediately and most evident in potentiometric data. It is postulated that the impact on the physical hydrogeology will result in higher (by 10 to 15%) baseflows in the Bloubank Spruit system in the future.

A correlation between spring discharge, karst basin catchment area, springwater chemistry and rainfall recharge is demonstrated for most of the karst basins. This has provided an improved understanding of karst groundwater flow patterns in the study area that supports a plausible conceptual model of the hydrogeologic environment. Cognisant of the variation in climatological and physiographical factors that influence landscape development in the long-term, the mean solutional denudation rate of the COH karst landscape is placed at 9.5 mm/Ka within the range 5 to 14 mm/Ka. This represents a total rate of theoretical void development of ~2100 $m^3$/a. The average total carbon loading in the karst springwaters is found to be 44 mg C/L in the range 31 to 65 mg C/L. The average total carbon flux carried in these waters is ~442 kg/d (~161 t/a), with eight high-yielding springs delivering a combined total C-flux of ~3.1 t/d (~1.1 t/a). The average C-export rate for seven karst basins is calculated at ~4.6 t/$km^2$/a in the range ~2.8 to ~9.7 t/$km^2$/a. This is comparable to values reported in the literature for other karst terranes globally.

The analysis and interpretation of vertical hydrogeochemical profiles in the karst aquifer impacted by allogenic mine water recharge addresses an often neglected aspect and contributes materially to the multi-dimensional span of this work. Similar sentiments can be expressed in regard to the temporal assessment of net acidity of karst groundwaters, the environmental isotope characterisation of the water that characterises various sources, and the hydrovulnerability risk assessment of individual fossil sites.

The platform built from the integration of historical data with a wide range of newly-sourced and interpreted hydrologic and hydrogeologic data and information, convincingly supports a more robust understanding of the surface water and groundwater environments in the COH. This, in turn, has provided the means to objectively gauge the impact of varied and numerous threats to and on the water resources, and to develop an appropriate water resources monitoring programme. The unprecedented abnormally high flow conditions experienced in the 2010, 2011 and 2014 hydrological years in the Bloubank Spruit system precipitated adverse impacts attributable to abnormally high raw mine water discharges. Subsequent hydrologic and hydrogeologic observations have provided new insight into the response of the receiving karst groundwater resource. These circumstances have magnified the focus on the immediate and short-term intervention measures implemented in the Western Basin to manage and control mine water discharge. The impact of climate change on this situation in the long-term remains an unknown factor in quantitative terms.

# Synoptic Information on Sterkfontein Cave

Sterkfontein Cave is the flagship fossil site on the Cradle of Humankind World Heritage Site property. It is synonymous with the property to the extent that a common and mistaken perception exists that the cave and its immediate surrounds represents the whole of the World Heritage Site, when in fact it occupies only a very small portion in the south-western corner of the property. The property has a maximum N–S extent of roughly 27 km, and an E–W extent of some 30 km. The cave is the quintessential archaeological and palaeoanthropological excavation on view to tourists, based as it is on a long and distinguished history as a 'dig' dating back to the 1930s. Yet a Google® search for such basic information as cave air temperature and humidity, cave water temperature and depth, and depth below surface reveals a surprising paucity of information. Synoptic information that characterises the physical surface and subsurface environment is tabulated below. This has proven useful to Sterkfontein Cave tour guides who are faced with questions about these aspects from tourists. The information contributes to broadening the general knowledge of the cave environment beyond merely the well documented archaeological and palaeoanthropological aspects.

| Parameter/aspect | Value as a range and unit |
|---|---|
| Surface elevation | 1440 m amsl (river) to 1487 m amsl (hilltop) |
| Rainfall (wet) season | Summer (October to March) |
| Mean annual precipitation | 684 mm |
| Mean summer precipitation | 588 mm |
| Mean daily temperature (surface) | 20 to 22 °C (summer)<br>10 to 12 °C (winter) |
| Mean annual evaporation (surface) | 2200 to 2400 mm |
| Mean daily relative humidity (surface) | 66 to 68% (summer)<br>58 to 60% (winter) |
| Mean in-cave air temperature | 17 to 19 °C (summer)<br>15 to 17 °C (winter) |
| Mean daily in-cave relative humidity | ?? to ??% (summer)<br>?? to ??% (winter) |
| Cave depth (known) below surface entrance | 50 to 55 m below hilltop (1487 m amsl)<br>42 to 47 m below entrance (1479 m amsl) |
| Cave water temperature | 15.5 to 18.5 °C |
| pH of the cave water (in situ) | 7.2 to 7.9 |
| TDS (total dissolved solids) of the cave water | 475 to 525 mg/L |
| Cave water saturation indices (calcite and dolomite) | $SI_C$ –0.23; $SI_D$ –0.72 (slightly under-saturated) |
| Cave water depth (known) | 3 m (below 1434 m amsl)<br>8 m (below 1439 m amsl) |

(continued)

| Parameter/aspect | Value as a range and unit |
|---|---|
| Cave water level elevation | 1434 to 1439 m amsl (multi-year variation) |
| Cave water level fluctuation | 2 to 3 m (typical) over years<br>5 to 6 m (maximum) over decades and longer |
| Water table gradient and direction | 0.003 (30 cm per 100 m) to the ENE |
| Rainwater chemistry | pH = 5.1 to 6.5; TDS = 90 to 110 mg/L |
| Surface solutional denudation rate | 4.4 to 5.5 mm per 1000 years (local)<br>5.5 to 15.5 mm per 1000 years (regional) |
| Geological strata | Type: Dolomite of the Malmani Subgroup, Chuniespoort Group, Transvaal Supergroup<br>Age: ~2600 million years<br>Dip: ~20° to the NW (315°) |

Syenite boulders marking the position of the Plover's Lake Sill immediately downstream of the ~2 L/s Cradle Spring, discharging at back left of picture, in the Motsetse Nature Reserve (*Photo* P. Hobbs, date 21/12/2010)

***We are to admit no more causes of natural things than such as are both true and sufficient to explain their appearances. Therefore, to the same natural effects we must, so far as possible, assign the same causes.***

*Sir Isaac Newton*

***(English physicist, mathematician, astronomer, natural philosopher, alchemist and theologian 1642-1727)***

# Integrated Monitoring Approach

Harrison Pienaar and Philip J. Hobbs

The Cradle of Humankind World Heritage Site (COH WHS) enjoys international significance for its palaeoanthropological and archaeological fossil finds. A review of international efforts has not located another UNESCO-protected property in the world [#35 on the World Heritage List with karst of outstanding universal value (Williams, 2008)] that experiences the impact of acid mine drainage (AMD). The mine water threat was first manifested in late-August 2002 (van Biljon, 2006) with the start of mine water decant from the West Rand Goldfield, aka the Western Basin. This situation has generated wide and considerable concern for the preservation of the UNESCO-inscribed fossil sites on the property. The Environmental Management Framework/Plan (EMF/P) Status Quo Report (Beater and Kilian, 2009) identified information in regard to sources, extent and affect of AMD in the COH as being deficient. It is therefore unfortunate but not surprising that the paucity of hydroscientific information in regard to the property precipitated the publication of largely unsubstantiated (and often alarmist) media reports regarding negative impacts on the fossil heritage sites (e.g. Béga, 2008a; 2008b; 2010; 2013; Seccombe, 2008; Masondo, 2010; Groenewald, 2010a). Regrettably, this perception is afforded credibility by Wells et al. (2009) in the respected publication *Environmental Management in South Africa* (Strydom and King, 2009), and also in generic scientific articles such as by Durand et al. (2010), Durand (2012) and, more recently, by Abiye (2014).

The COH property could clearly benefit from a comprehensive assessment of the water resources system(s) to mediate an improved understanding of the integrated hydrologic (surface water) and hydrogeologic (groundwater) environments. In regard to the groundwater environment, this entails defining flow directions, boundary conditions and water chemistry (quality) aspects based on historical information, long-term monitoring data and new data and information. Similar considerations also apply to the surface water environment. The dynamics of interaction between surface water and groundwater resources in regard to both quantity (volumes) and quality (chemistry) is a crucial element of this understanding. This is especially important in the COH because of the risk to karst groundwater resources from threats such as over-abstraction associated with irrigated agriculture, mine dewatering and rewatering, and contamination/pollution from municipal wastewater, mine water discharge and on-site sanitation facilities on unserviced properties, many of which are smallholdings.

Reporting on the conflict between different communities over water resources of the karst that hosts the Choranche Caves and forms the Coulmes Plateau in the Vercors Mountains, France, Gauchon et al. (2006) emphasise that '…. water's place in the system must be defined'. The material presented in this publication provides such definition for the COH by building on and extending the collective knowledge and understanding of the water resources environment. This finds support in a conceptual hydrophysical model of the integrated surface water and groundwater environments that accounts for observed congruencies between rainfall, groundwater recharge, groundwater drainage patterns, spring discharges, and the surface extent of groundwater basins. Characterisation of the hydrochemistry associated with the various water resource components lends further support and rigour to the conceptual model of the hydrophysical domain.

The water resources system assessment underpins the implementation of an appropriate integrated water resources monitoring programme for the COH. This is crucial to the successful management of water resources and the impact on these resources in the study area. A focussed routine

P. J. Hobbs: Deceased

H. Pienaar (✉)
Council for Scientific and Industrial Research, Smart Places, Water Centre, Pretoria, South Africa
e-mail: hpienaar@csir.co.za

H. Pienaar
Hebei University of Engineering, Handan, China

P. J. Hobbs
Pretoria, South Africa



P. J. Hobbs et al. (eds.), *Anatomy of a South African Karst Hydrosystem*, Cave and Karst Systems of the World,
https://doi.org/10.1007/978-3-030-95829-9_1

hydromonitoring programme provides the means to measure and demonstrate the success or failure of management efforts to protect the aquatic environment in the COH. The water resources system assessment and derived conceptual model also facilitate the evaluation of ancillary issues that include the following:

- the status quo of water resources monitoring activities;
- the state of conservation (SoC) of the water resources environment;
- the establishment of resource water quality objectives;
- solutional denudation as an agent in the evolution of the karst landscape;
- carbon flux associated with karst spring discharges; and
- the vulnerability of the fossil sites to impacts on the (hydro)environment.

The material provides a foil, as demonstrated by Béga (2012a), albeit fleetingly,[1] van Schie (2013) and Vermeulen (2013), for the many and widespread misperceptions founded on ignorance, sentiment and/or emotion that have driven the public awareness of this World Heritage property. Most importantly, however, it informs and supports sound management actions in regard to responsible and effective governance of the water resources environment to also protect this aspect of the outstanding universal value (OUV) of the property in the interest and on behalf of all nations and humankind (WHC, 2013 and Text Box 1, p. 3).

View from inside Sterkfontein Cave upward through an aven extending ~45 m to surface with a sheltering tree visible through the surface opening; the near-circular geometry of the feature reveals the modification of a natural aven by early miners into a shaft for hoisting limestone to surface rather than via the steep and tortuous natural access route

[1] See the article published by the same journalist just two weeks later (Béga, 2012b).

**Text Box 1** Excerpt from the World Heritage Centre draft decision 37 COM 7B.44 (WHC, 2013) presented at the 37th Session of the World Heritage Committee in Phnom Penh, Cambodia, 16 to 27 June 2013

*In January 2011, the World Heritage Centre requested the State Party to undertake a study to address concerns that had been raised by various parties about a potential threat to the World Heritage property posed by effluent from abandoned and active mines in the vicinity. The study, commissioned by the Cradle of Humankind World Heritage Site Management Authority, Department of Economic Development, Gauteng Province, South Africa, was prepared in 2011 by the Council for Scientific and Industrial Research. On 14 March 2013, ICOMOS completed an assessment of this exhaustive study. This assessment is available online at the following Web address:* http://whc.unesco.org/en/list/915/documents.

*Over a number of years, concerns have been raised that acidic waters rich in iron and other minerals (Acid Mine Drainage, or AMD) from abandoned mines in the vicinity of the World Heritage property, plus the input of treated mine water (and, in recent years, the overflow of untreated and partly treated mine waters at times of heavy rainfall) from the workings of an active mine about 15 km distant into the hydrology system of the property, might have a harmful effect on the caves that underpin the Outstanding Universal Value (OUV) of the property. The 2011 study, "Situation assessment of the surface water and groundwater resource environments in the Cradle of Humankind World Heritage Site", attempts for the first time to give a comprehensive picture of the surface and sub-surface water regime across the property.*

*The World Heritage Centre and the Advisory Bodies suggest that the Committee congratulate the State Party for an excellent study. The work described appears to have been undertaken in accordance with best practice, and the results appear to reflect current scientific knowledge. The amount of field measurements to support the work is impressive. The study has successfully answered some of the uncertainties in knowledge about the hydrology and hydrogeology of the property.*

*The World Heritage Centre and the Advisory Bodies are of the view that the recommendations in the 2011 study should be acted upon. Cooperation with other agencies, especially the Department of Water Affairs, should be strongly encouraged, and there should be consultation and agreement on the proposed resource water quality objectives. Regular monitoring should be established, and a timeframe set to meet the agreed quality objectives.*

*The World Heritage Centre and the Advisory Bodies note that since the study was written, the Department of Water Affairs has initiated a "Feasibility study for a long-term solution to address the AMD associated with the East, Central and West Rand underground mining basins". A void assessment report was issued in draft form in March 2013, and a feasibility report proposes a system of active treatment of AMD and leachate from mine dumps for the West, Central and East Rand. If implemented in the West Rand, it should reduce the risks of AMD to the OUV of the property. They recommend that a joint World Heritage Centre/ICOMOS reactive monitoring mission be undertaken in order to assess the AMD impact to the property and propose recommendations to the State Party.*

View of the southern sidewall of Sterkfontein Quarry (part of the Bolt's Farm fossil site) showing the barricaded portal to Alladin's/Quarry Cave at centre right, the ~ 4 m thick epikarst zone (above the sub-horizontal dotted line) and the gentle boudinage-type folding of the carbonate strata exposed in the sidewall

# Introduction and Background

Harrison Pienaar and Philip J. Hobbs

## 1 Legal and Physical Framework

The UNESCO Convention concerning the Protection of the World Cultural and Natural Heritage (1971), ratified by South Africa in 1997 and incorporated into South African law in terms of the World Heritage Convention Act (Act No 49 of 1999), in 1999 inscribed the fossil hominid sites of Sterkfontein, Swartkrans, Kromdraai and environs (known as the Cradle of Humankind) for protection in terms of their 'collective' cultural heritage (Strydom, 2009). The Cradle of Humankind (COH) is located on the north-western margin of Gauteng Province in the central north-eastern part of South Africa (Fig. 1). It is among some 50 karst sites globally that have been inscribed by UNESCO (Hamilton-Smith, 2006), many of these for other values (e.g. cultural) than purely their karst landforms.

The management of a protected area is undertaken by the State Party, in this instance the Department of Environmental Affairs (DEA), that is answerable to UNESCO's World Heritage Committee (WHC). The DEA has delegated the management function to the Management Authority (MA) assigned administrative responsibility at a provincial level. The MA of the COH WHS is hosted by the Department of Economic Development, Gauteng Provincial Government, that is answerable to the DEA on issues such as the State of Conservation (SoC) of the property.

A little more than 1 billion people ($\sim$15% of the global population) live on karst (Kaufmann 2013), and karst aquifers are estimated to provide $\sim$ 1.5 billion people with potable water (Ford and Williams 2007; Bauer et al. 2008; Supper et al. 2008). Karst areas also count among the most vulnerable hydrogeologic environments to human impacts in the world. The distribution in South Africa of Neoarchaean ($\sim$2.65 to 2.50 Ga) major 'hard' sedimentary carbonate deposits is shown in Fig. 1. These are the Malmani Subgroup and the Campbell Rand Subgroup located east and west of longitude 25°E, respectively. Together representing at least 98% of carbonate strata (dolomite, limestone, calcrete, dolocrete, travertine, etc.) in South Africa, these formations cover $\sim$ 3% of the total surface area of the country (van Schalkwyk 1981). The CGS/SAIEEG (2003) guideline provides a concise description of the dolomite and limestone occurrences in South Africa. A more detailed description can be found in Martini and Wilson (1998) and Martini (2006). In regard to land and water resources issues, Vesper (2008) identifies the following characteristics unique to karst systems that require focussed study:

- close connections between surface and subsurface processes render karst systems highly vulnerable to impacts from surface activities;
- spatial heterogeneity hinders the ability to easily monitor and assess water quality and quantity;
- rates of physical processes are highly variable;
- subsidence may occur at an almost imperceptible rate or be nearly instantaneous;
- contaminants may be rapidly flushed through the system or trapped indefinitely; and
- water flow at springs may be consistent through time or change rapidly in response to storm events.

The International Union for Conservation of Nature and Natural Resources (IUCN) publication by Watson et al. (1997) lists some of the reasons for the protection of karst landscapes as their provision of the following 'goods and services':

P. J. Hobbs: Deceased

H. Pienaar (✉) · P. J. Hobbs
Smart Places, Council for Scientific and Industrial Research, Pretoria, South Africa
e-mail: hpienaar@csir.co.za

H. Pienaar
Hebei University of Engineering, Handan, China



P. J. Hobbs et al. (eds.), *Anatomy of a South African Karst Hydrosystem*, Cave and Karst Systems of the World,
https://doi.org/10.1007/978-3-030-95829-9_2

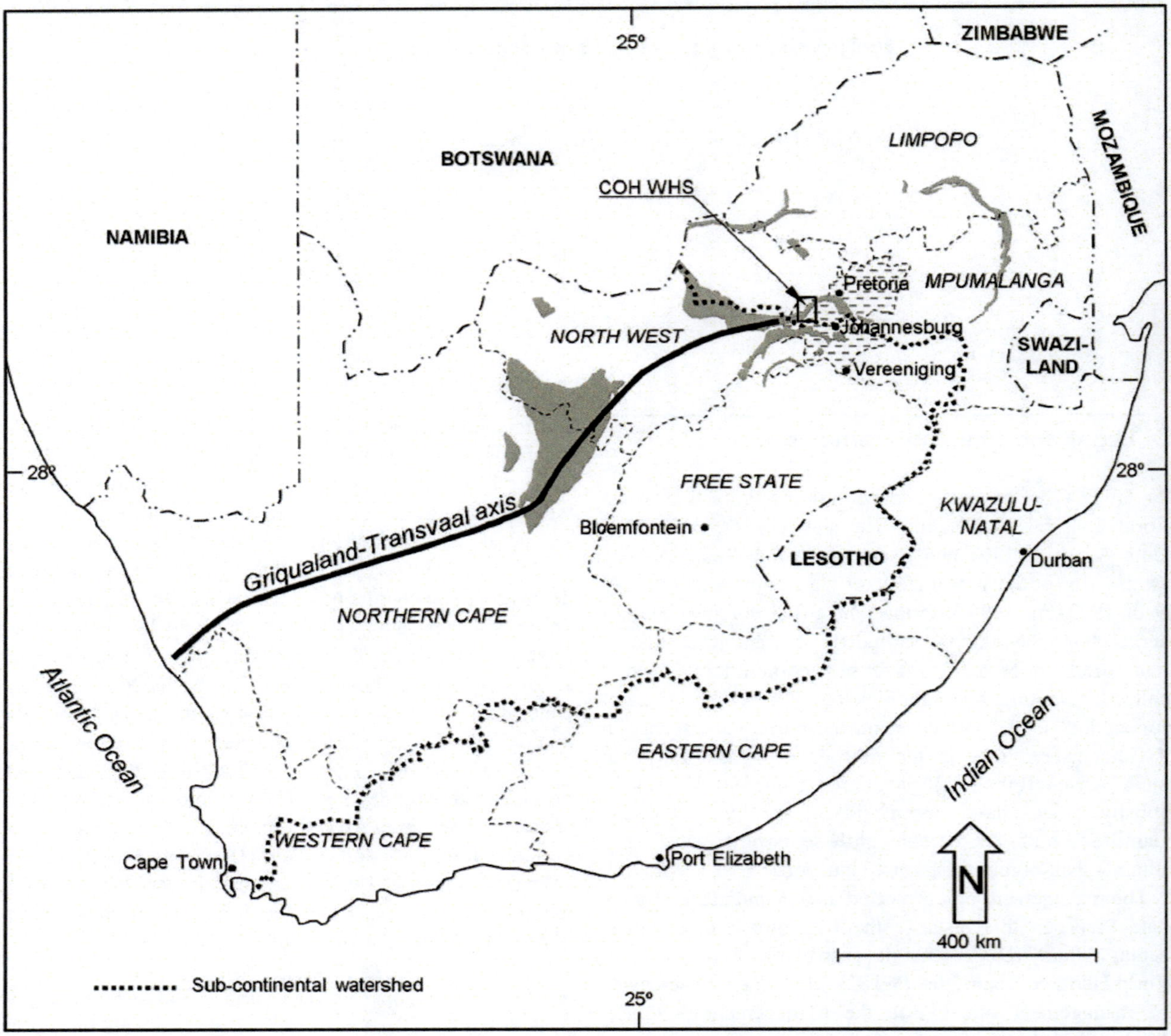


**Fig. 1** Location of the Cradle of Humankind World Hertigae Site (COH WHS) in Gauteng Province (light shaded area), South Africa, and in relation to the distribution of 'hard' sedimentary carbonate deposits (darker shaded areas) of the Chuniespoort Group in the South African interior (adapted from Martini and Wilson, 1998)

- habitat for endangered species of flora and fauna;
- sites containing rare minerals or unique landforms;
- important sites for the study of geology, geomorphology, palaeontology and other disciplines;
- culturally important sites, both historic and prehistoric;
- spiritual or religious features;
- specialised agriculture and industries;
- 'windows' into understanding regional hydrology and karst hydrogeology;
- sources of economically important materials, especially groundwater;
- tourism and its associated economic benefits;
- purely recreational areas, both scenic and challenging; and
- *role as carbon source and sink in the context of the global carbon cycle*[1].

Although the COH was inscribed for its cultural heritage represented by the reasons shown as **bold** text in the above list, it is evident that the property also meets many of the other criteria (shown as underlined text) in the list. The

[1] Italicised text added by author.

earlier activities of the IUCN-SA[2] Karst Working Group (Fourie 2005) are described in Text Box 1, and in the WRC/IUCN/SAKWG (2010) publication titled '*The karst system of the Cradle of Humankind World Heritage Site*'.

**Text Box 1** The IUCN-SA Karst Working Group

In May 2004, the South African country office of IUCN-The World Conservation Union, in collaboration with the Gauteng Department of Agriculture, Conservation and Environment (GDACE) and the Cave Research Organisation of South Africa (CROSA), established a Karst Working Group, made up of more than 30 individuals from diverse backgrounds and disciplines, including national, provincial and local government, industry, four national universities and a number of voluntary caving organisations. Their task is to develop a holistic monitoring and management system for karst systems in South Africa, and they have chosen the Cradle of Humankind World Heritage Site as a pilot study, in part to support the Cradle in meeting its reporting obligations to the World Heritage Authority. One of the group's main concerns is the threat of water pollution in the Cradle area, and the effect of such pollution on the karst ecosystems. The group is supported by the GDACE, the Water Research Commission and the IUCN World Commission on Protected Areas Cave and Karst Taskforce. It is also establishing partnerships with research institutions in Europe, as well as international NGOs Flora & Fauna International and Earthwatch.

From Fourie (2005).

The National Water Act (RSA 1998) posits the reminder that water resource protection is mandated in terms of Section 19 of the National Water Act (NWA), Act No 36 of 1998, with a particular focus on 'prevention and remedying effects of pollution'. Moreover, the NWA also provides mechanisms for promoting the sustainable use of inland waters (King and Pienaar 2011), as well as specific legal provisions to develop and implement groundwater protection measures, in particular the Resource Directed Measures (RDM) as described in Chap. 3 of the NWA (Pienaar et al. 2021). Section 19 of the NWA requires of a landowner, person in control of land or person occupying or using land to take all reasonable measures to prevent pollution of a water resource from occurring, continuing or recurring. It is perhaps even more pertinent to recognise the need to query the popular perception of a large-scale 'tragedy of the commons', as formulated by Hardin (1968), playing itself out regarding water resource pollution and consequential loss of fossil site integrity in the COH and, ultimately, loss of World Heritage status. This situation represents a 'wicked problem' as described by Davidson and Wei (2012).

[2] South African country office of the IUCN-The World Conservation Union.

## 2 Hydrophysical Framework

Ford and Williams (2007) report that karst resources are coming under increasing pressure worldwide and have great need of sustainable management. This has no greater relevance than for a World Heritage property, where the successful execution of administrative responsibilities is not sufficient to ensure that desired results in terms of environmental management outcomes are or will be achieved. Kasum et al. (2019) also argue that water resources are indissolubly linked to a wide range of human-made activities defined by terms such as *underwater exploitation* and *human health* (among other), which necessitates the development and advancement of models for forensic hydrography. Successful achievement of environmental objectives such as appropriate water quality and adequate biodiversity, require specific and directed interventions informed by a sound understanding of the water resources environment. Parrish et al. (2003) recognise monitoring programmes as crucial elements of any protected area management programme.

Knowledge of the current situation, trends of change, likely source(s) of undesirable constituents, threat status, etc. underpins the implementation of sound management actions directed at responsible and effective governance on the part of a Management Authority. The implementation of an appropriate integrated hydrologic and hydrogeologic monitoring programme is crucial to the successful management of water resources and protection of the aquatic environment in the COH. Such protection is not only required for the preservation of karst features and palaeoanthropological finds, but also for resident water users who depend on local water resources (primarily groundwater) for their livelihood. The Environmental Management Framework/Management Plan (EMF/MP) developed for the COH (Brown 2009), presents a first but understandably superficial description of the water resources of the property. An earlier study by Holland (2007) focussed on groundwater resources only.

## 3 Aims and Objectives

The primary objective of this study is to develop a sound and integrated conceptual model of the hydrologic and hydrogeologic environments that comprise the COH. Regarding

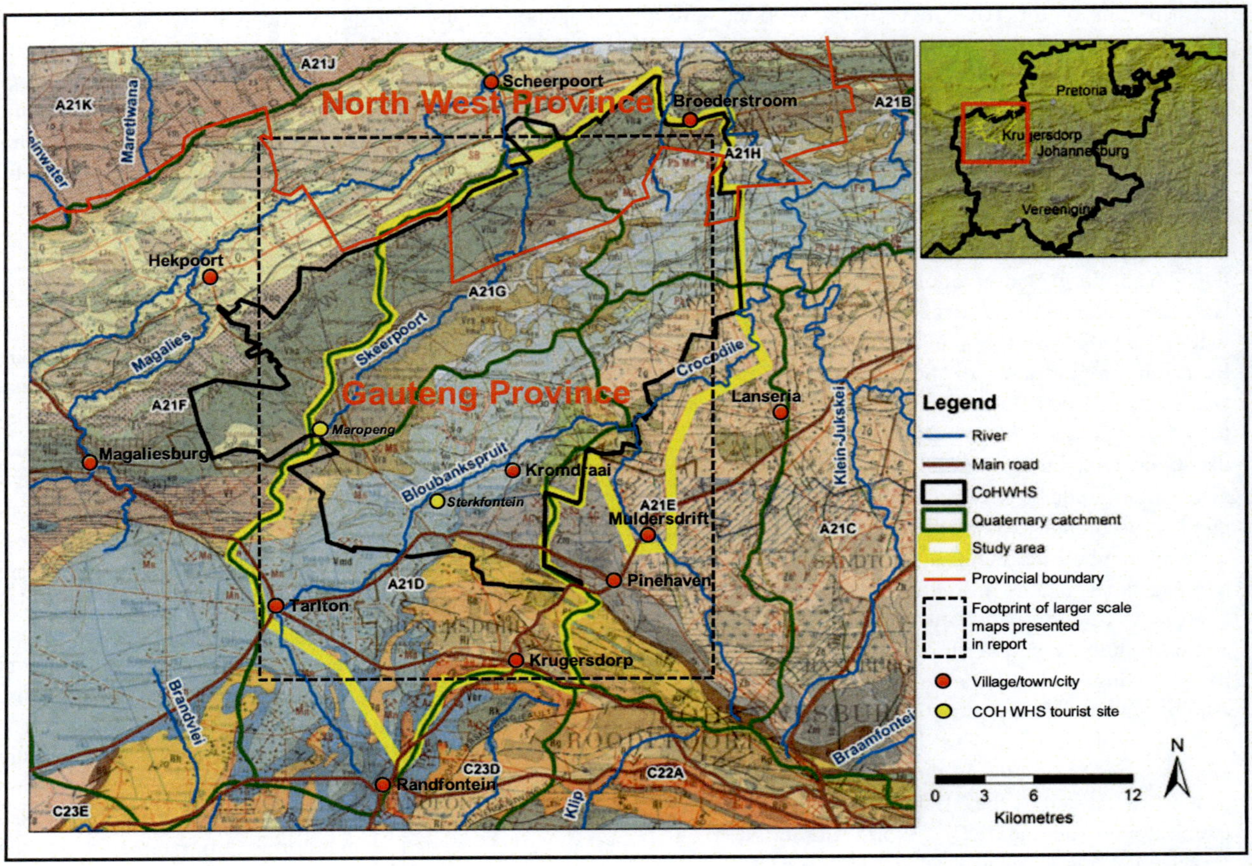


**Fig. 2** Definition of the study area in regard to geographic localities, geology and hydrology

the groundwater environment, this entails defining flow directions, boundary conditions and water chemistry (quality) aspects based on historical information, long-term monitoring data and new data and information. Although similar considerations also apply to the surface water environment, the dynamics of the surface and subsurface hydrologic interaction (aka stream – aquifer interaction) is considered a key element of this understanding. The successful outcome will provide a basis for implementing measures aimed at ensuring the sustainable functioning of the karst ecohydrology of the COH in the face of rapidly growing pressure from a variety of anthropogenic impacts.

An assessment of the water resources system (s) that provides an improved and rigorous understanding of the integrated hydrologic and hydrogeologic environments, also provides a more informed reference framework against which to gauge the veracity of concerns for the heritage value of the property. The integrated conceptual model of the surface water and groundwater environments developed in this study meets this requirement, and also identifies data and information gaps in the water resources environment. Further, the outcome elucidates ancillary aspects such as the following:

- the status quo of water resources monitoring activities;
- the state of conservation of the water resources environment (e.g. DEA 2014a, 2016);
- the establishment of resource water quality objectives for the water resources;
- solutional denudation as an agent in the evolution of the karst landscape;
- carbon flux associated with karst spring discharges; and
- the vulnerability of the fossil sites from negative impacts on the water resources.

## 4 Definition of the Study Area

The area defined in Fig. 2 represents the most appropriate demarcation based primarily on the harmonisation of the COH property boundary (as promulgated) with aspects such as drainage basin footprints, the distribution of geological strata with emphasis on the vulnerable karst hydrosystem, and the location of known threats associated with mining, industrial, urban and agricultural land uses. The measure of congruence between the study area boundary and the Quaternary surface water catchments is evident in Fig. 2.

The study area encompasses ~ 64 900 ha, compared to the ~ 52 000[3] ha of the COH property. The inclusion of the mining area in the extreme southern part of the study area between Krugersdorp and Randfontein (Fig. 2) recognises the importance of this activity and land use not only in terms of the mine water threat, but also the comparatively large amount of data and information related to this activity that already exists for this portion of the study area. The inclusion of a portion of the (upper) Crocodile River in Quaternary catchment A21E as far upstream as Muldersdrift in the south-eastern corner, incorporates the potential impact of the Driefontein Wastewater Treatment Works (WWTW) together with that of the Percy Stewart WWTW located in the upper reaches of Quaternary catchment A21D (Fig. 2), on the water resources environment.

Excluded from the study area is that portion of the property in Quaternary catchment A21F (the Magalies River catchment) along the north-western margin. Much of this area is located on younger geological strata (Sect. 4 in Chapter "Description of the Physical Environment"), and it does not share to any significant degree the sensitive groundwater environment that forms the core of the COH. The carbonate strata extending to the west-southwest beyond the study area boundary defines the eastern portion of the adjoining Steenkoppies Basin. This basin is drained by Maloney's 'Eye' (spring) (Sect. 1 in Chapter "Physical Hydrology"), the source of the Magalies River, and therefore also does not share the karst hydrosystem of the COH beyond purely geologic proximity/contiguity (Sect. 7 in Chapter "Physical Hydrogeology").

The Steenkoppies Basin has recently been studied in some detail (Holland et al. 2009) with a view to managing the high demand placed on the karst water resource for intensive irrigated agriculture. This demand has impacted negatively on the yield of Maloney's Eye, to the detriment of downstream water users in the fertile Magalies River Valley which supports intensive irrigated agriculture.

Irrigation of 'instant' (cultivated) lawn with treated municipal wastewater effluent (from the Percy Stewart WWTW) on dolomite in the south-western portion of the study area; compare the early spring brown natural vegetation typical of the Transvaal Highveld in the background with the lush green foreground

[3] The area of ~ 47 000 ha commonly reported, refers to the core area (i.e. excluding the buffer zone).

# Description of the Physical Environment

Harrison Pienaar and Philip J. Hobbs

## 1 Morphology and Drainage

The surface water drainage characteristics (physical hydrology) of the study area, including drainage pattern and mean annual runoff (MAR), are discussed in greater detail in Sect. 5.1 in Chapter "Physical Hydrology". The following discussion provides a synoptic general overview of this aspect.

The watershed that forms the continental divide between the Vaal River system to the south (draining westward to the Atlantic Ocean) and the Limpopo River system to the north (draining eastward to the Indian Ocean) (see Fig. 1 in Chapter "Introduction and Background"), also occupies the highest natural elevation ($\sim$1720 m amsl) in the study area. Extending to the north of this divide is a diverse landscape that includes undulating terrain with low to moderate relief along an SW–NE strike roughly concordant with the main drainages. This terrain is flanked to the south-east by prominent ridges incised at right angles by mainly ephemeral tributaries, and to the north-west by sub-parallel ridges and valleys. The flanking landscapes mark a transition in the geology across the landscape, with dolomite sandwiched between older underlying sedimentary strata (quartzite) in the south-east and younger overlying sedimentary rocks (quartzite and shale) in the north-west.

At an elevation of 1664 m amsl, the peak of Spioenkop on the farm Danielsrust 518JQ is the highest point in the study area. This is followed by Zwartkop peak at 1626 m amsl on the farm Zwartkop 525JQ near the south-eastern margin of the property. The lowest elevation of $\sim$1200 m amsl occurs in the vicinity of Broederstroom (Fig. 2 in Chapter "Introduction and Background"), a hamlet at the north-eastern end of the property immediately south of the regionally important (Sect. 5.1 in Chapter "Physical Hydrology") Hartbeespoort Dam storage reservoir (Fig. 1). This fits the description by Kruger (1983), who characterises the terrain morphology as undulating hills and lowlands with relief in the range 130 to 450 m, a drainage density of 0.5 to 2 km/km$^2$, a stream frequency of 0 to 6 per km$^2$, and 20–50% of the area having slopes of <5%. The property spans portions of five Quaternary catchments (Fig. 2 in Chapter "Introduction and Background" and Table 1) located in the Limpopo Water Management Area (WMA #1) (Fig. 1). Table 1 shows that basins A21D (the Bloubank[1] Spruit system) and A21G (Skeerpoort River) together cover $\sim$70% of the study area, followed in roughly equal proportions ($\sim$15%) by basins A21E (upper[2] Crocodile River) and A21H (lower Crocodile River).

## 2 Climate and Rainfall

The property falls within the warm temperate summer rainfall region that characterises the typical Highveld climate of the central north-eastern South African interior. The following synoptic information is sourced mainly from Schulze et al. (1997).

The mean annual temperature (MAT) falls in the range of 16 to 18 °C, with daily mean temperatures in the range of

P. J. Hobbs: Deceased

H. Pienaar (✉) · P. J. Hobbs
Smart Places, Council for Scientific and Industrial Research, Pretoria, South Africa
e-mail: hpienaar@csir.co.za

H. Pienaar
Hebei University of Engineering, Handan, China

[1] The spelling 'Bloubank' subscribes to that used on the most recent edition of the published 1:50 000 scale topocadastral map 2527DD Broederstroom (5th ed., 2001), and not to that of 'Blaauwbank' used on earlier maps and in other reports (e.g. Holland et al., 2009; SRK, 2009).

[2] The terms 'upper' and 'lower' are used in the context of the reaches above and below the confluence with the Bloubank Spruit upstream of Hartbeespoort Dam, and not in the context of the entire reach of the Crocodile River down to its confluence with the Marico River to form the Limpopo River (Fig. 3).



P. J. Hobbs et al. (eds.), *Anatomy of a South African Karst Hydrosystem*, Cave and Karst Systems of the World,
https://doi.org/10.1007/978-3-030-95829-9_3

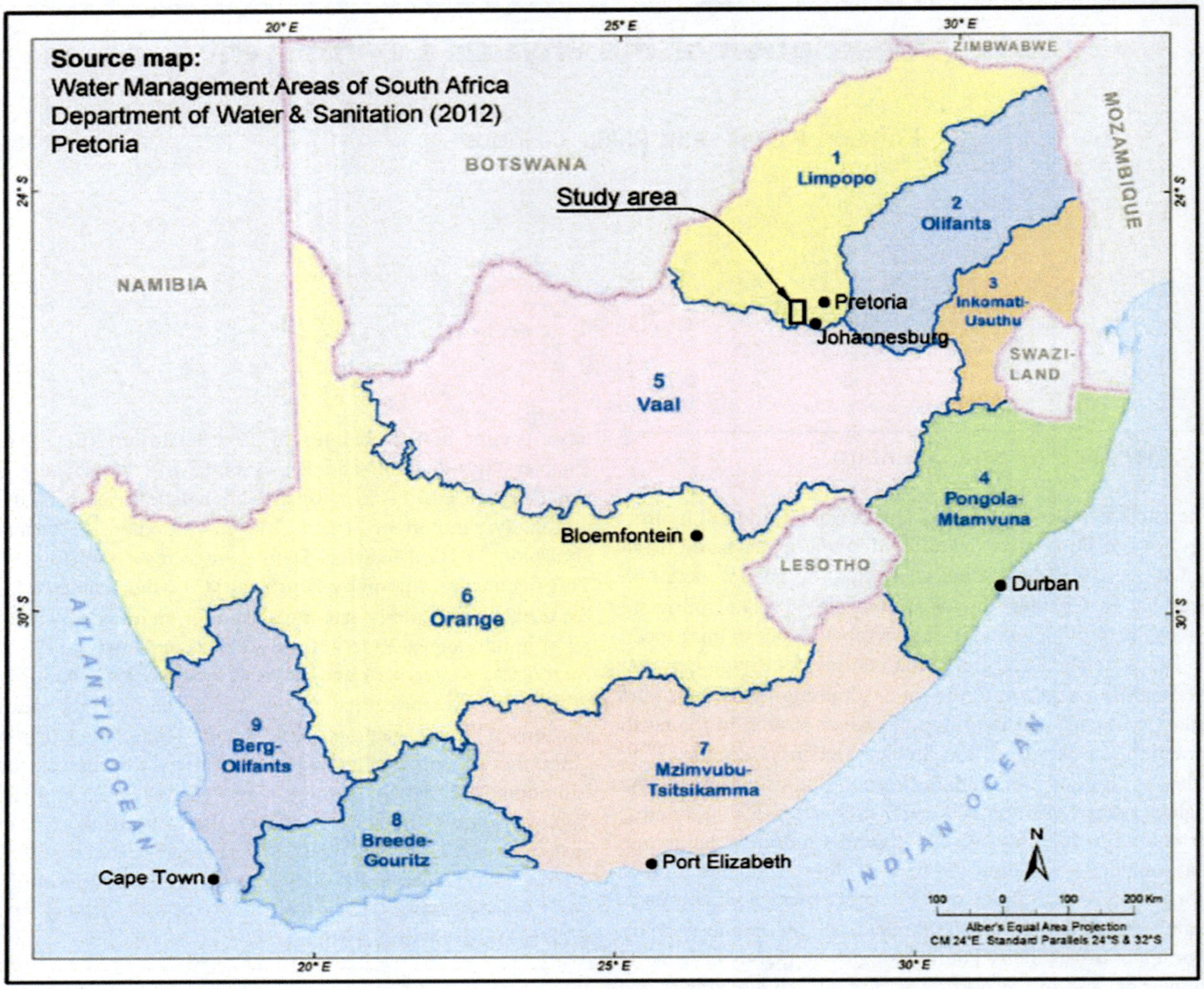


Fig. 1 Position of the study area within the national distribution of water management areas (WMAs; numbered and labelled) and the Limpopo WMA (#1) in particular

20–22 °C in summer (December/January) and 10 to 12 °C in winter (June/July). The daily mean relative humidity falls in the range 58–60% in winter and 66 to 68% in summer, with daily minima falling in the ranges 32–34% and 46–48% for these seasons, respectively. The A-pan equivalent mean annual potential evaporation (MAPE) falls in the range of 2200–2400 mm. The long-term A-pan evaporation recorded at OR Tambo International Airport located at 26.13°S—28.29°E, a distance of ~55 km east-southeast of the property, is 2160 mm/a. This study accepts the lower bound of 2200 mm/a as representative of the A-pan evaporation value for the study area, and a mean actual evapotranspiration of 1540 mm/a.

As the principal agent of groundwater recharge, rainfall is an integral component of the hydrological cycle, rendering knowledge of its magnitude and distribution an important aspect of a study such as this. The distribution of rainfall gauging stations in the study area and surroundings is shown in Fig. 2. The mean annual precipitation (MAP) associated with each of the Quaternary basins in the study area (DWA 2006) is shown in Table 1 and indicates a mean value of 696 mm/a (excluding catchment A21F representing 1% of the area).

The rainfall stations reflect a skewed geographic distribution that favours the south-western portion of the subregion. They are operated by various entities identified as follows:

- South African Weather Service (SAWS), generally located at police stations and municipal offices;
- Sibanye-Stillwater (S–S), located at various sites on the mine property;
- Mogale City Local Municipality (MCLM), located at the Percy Stewart WWTW; and

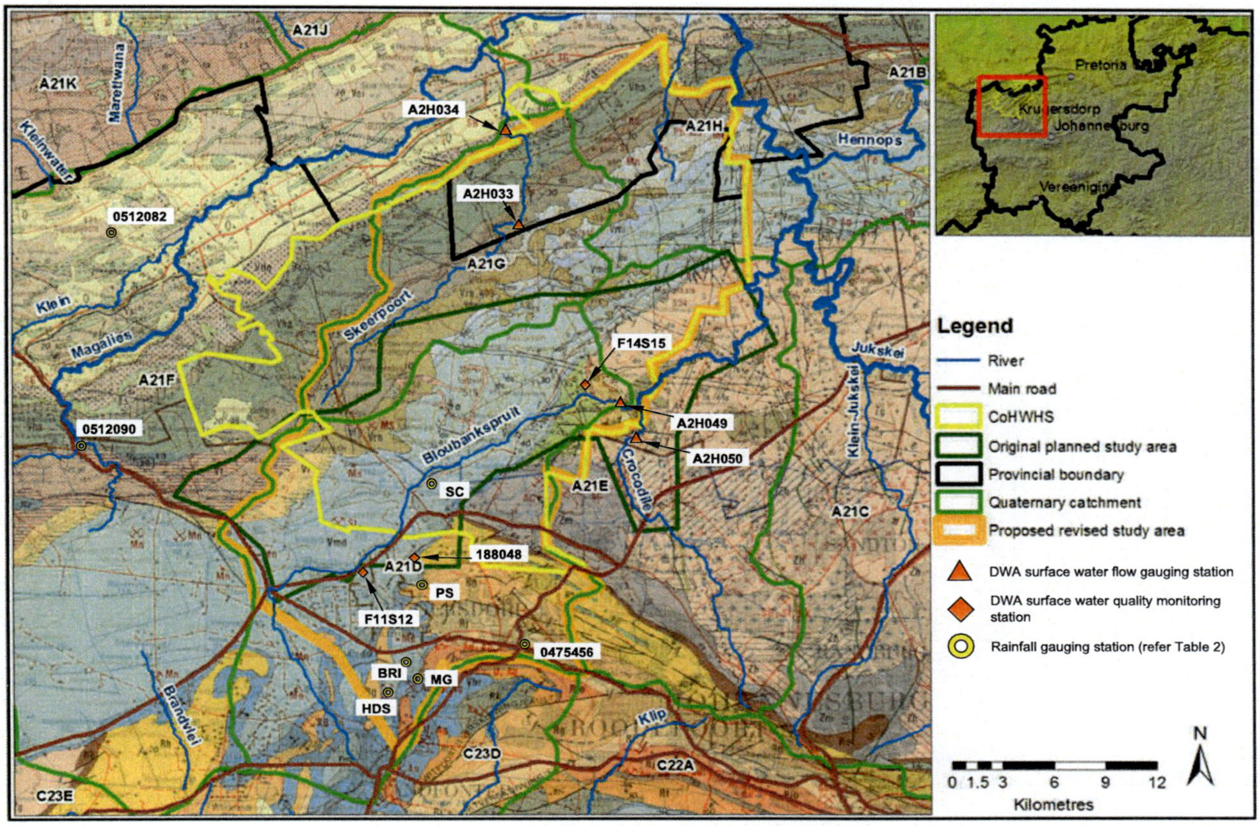


**Fig. 2** Definition of study area showing positions of rainfall, surface water flow and surface water quality gauging stations with hydrologic features superimposed on geology as backdrop

**Table 1** Geographic definition of quaternary catchments spanned by the study area

| Quaternary basin | Main drainageand relative latitude | MAP (mm) | Total area (km$^2$) | Proportion of basin in study area | |
|---|---|---|---|---|---|
| | | | | (km$^2$) | (%) |
| A21D | Bloubank spruit system south | 714 | 372 | 303 | 46.7 |
| A21E | Crocodile River (upper) south | 707 | 290 | 100 | 15.4 |
| A21F | Magalies river north | 677 | 1001 | 1 | 0.2 |
| A21G | Skeerpoort river central | 694 | 160 | 156 | 24.0 |
| A21H | Crocodile river (lower) north | 668 | 514 | 89 | 13.7 |
| Total | | | 2337 | 649 | 100 |

- Department of Water and Sanitation (DWS) located at Sterkfontein Cave and, since 2013, also on Ptn 36 of Vlakplaats 160IQ near Tarlton, Ptn 8/2 of Sterkfontein 173IQ north of the Krugersdorp Game Reserve, and at the HDS mine water treatment plant.

A daily precipitation record is maintained in most of the above-listed instances. The DWS stations comprise a cumulative (totalling) gauge developed by van Wyk (2010). These gauges have a rainfall equivalent capacity of ~450 mm and are therefore emptied every 2–4 months depending on season. The length of record and other information associated with each station is given in Table 2. It shows that rainfall monitoring by S–S (and its predecessors) in the mine area is comparatively recent. This reflects the importance of rainfall and associated recharge on the hydrostatic head in the subsurface mine workings and the more recent periodic recurrence of excess mine water discharge (Sect. 5.3.1 in

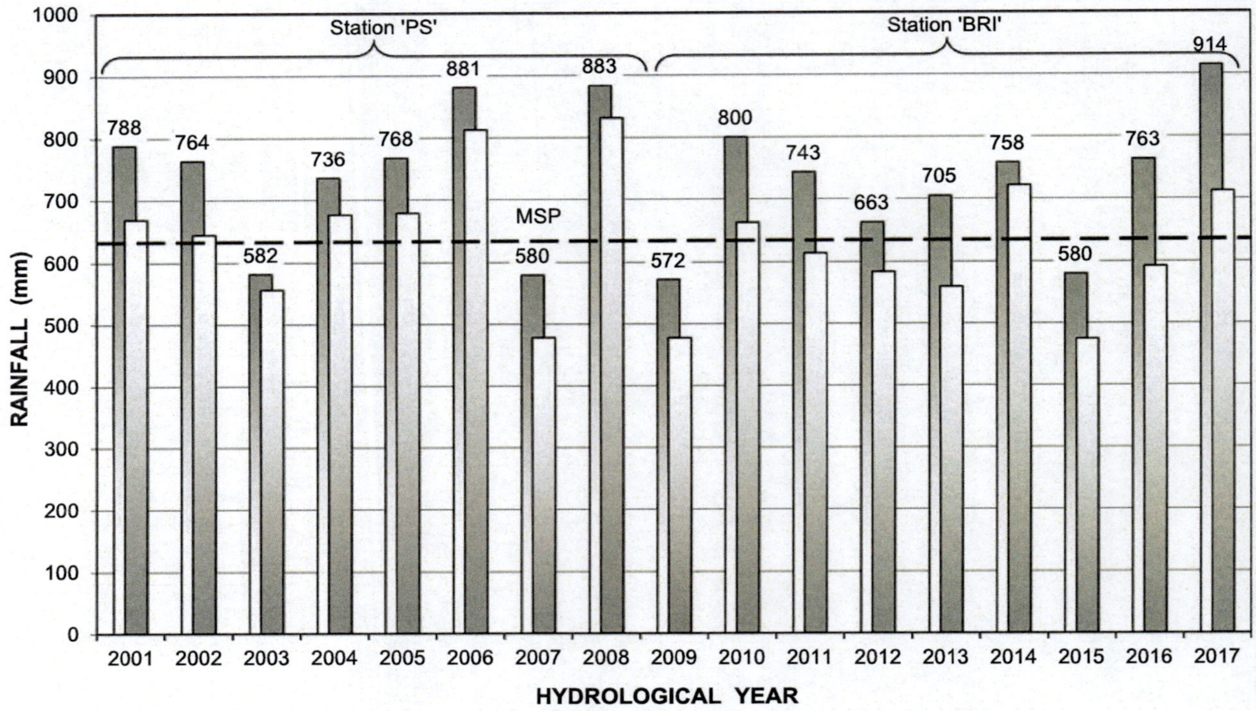


**Fig. 3** Composite annual and overlapping summer (wet season) rainfall record for stations PS and BRI; the mean summer precipitation of 632 mm is indicated by the dashed line labelled MSP

Chapter "Physical Hydrology"). Nevertheless, an analysis of the available precipitation record indicates that rainfall is distributed approximately 80–20% between the wet summer and the dry winter seasons.

The composite precipitation record of rainfall stations PS [for the 2001 to 2008 hydrological years[3] ($a_h$)] and BRI (for $a_h$ 2009 to 2017) shows the 'wetter' 2006 (813 mm), 2008 (831 mm), 2010 (663 mm), 2014 (721 mm) and 2017 (712 mm) summer rainfall seasons (Fig. 3). Both these stations are located north of the watershed (Fig. 2).

A comparison of the HDS and Sterkfontein Cave (SC) monthly rainfall data for the period of common record June 2010 to September 2017 returns the correlation shown in Fig. 4. The data set excludes months of no rainfall at both stations in order to remove the false correlation created by common null values. This shows that station SC experiences ~22% less rainfall on a monthly basis than does station HDS on the watershed ~13 km to the south. Wet months at both stations represent ~80% of the annual ($a_h$) record, equating to 9.6 such months per year. The regional reduction in rainfall northward that is evident in Table 1 is therefore echoed at the subregional scale. In summary, the long-term MAP for the northern third of the study area falls in the range 668–694 mm. The remaining ~67% comprising the central and south-western portion experiences ~710 mm (Table 1).

Considered of greater significance is the observation that ~52% of the 2014 summer rainfall at Sterkfontein Cave (396 out of 760 mm) and at station HDS (413 out of 799 mm) occurred in February and March 2014. These circumstances explain the resumption of uncontrolled mine water discharge from the mine area in late-February 2014 (Hobbs 2014a). This is similar to the situation that prevailed in the 2010 and 2011 wet seasons (Hobbs 2013a, b) that led to the severest mine water impact on the receiving water resources environment. The increased capacity of the mine water treatment plant to ~34 ML/d in June 2013 (**Timeline 2** p. ii and Text Box 1 in Chapter "Physical Hydrology", p. 61) remained insufficient to contain and treat the volume of mine water issuing from the flooded underground mine workings following recharge. Similarly, $a_h$ 2017 experienced an anomalous rainfall pattern marked by the following:

- A wetter than average summer (864 mm at station HDS and 712 mm at station BRI) yet in both instances typically representing ~80% of the $MA_hP$ of each station; and

[3] See **GLOSSARY.**

**Table 2** Rainfall monitoring stations in the study area and surrounds

| Station | Station name | Source | Coordinates[a] | | Elevation (m amsl) | Start date (mm/yyyy) |
|---|---|---|---|---|---|---|
| | | | Latitude | Longitude | | |
| 0475338 | Randfontein | SAWS | 26.13°S | 27.70°E | 1710 | 12/1954[b] |
| 0475456 | Krugersdorp kroningspark | | 26.10°S | 27.77°E | 1699 | 07/1904 |
| 0475605[c] | Wits botanical gardens | | 26.08°S | 27.85°E | — | 04/1988 |
| 0512082 | Hekpoort nooitgedacht | | 25.87°S | 27.55°E | 1463 | 01/1972[d] |
| 0512090 | Magaliesburg police station | | 26.00°S | 27.55°E | 1480 | 01/1969 |
| 0512783[c] | Pelindaba | | 25.80°S | 27.92°E | — | 01/1965 |
| PS[c] | Percy stewart WWTW | MCLM | 26.08°S | 27.73°E | 1590 | 01/2000 |
| BRI[c] | Black reef incline | S–S | 26.11566°S | 27.72319°E | 1662 | 10/2004 |
| HDS[c] | HDS plant | | 26.13384°S | 27.71579°E | 1714 | 10/2004 |
| SC[c, e] | Sterkfontein cave | DWS | 26.01566°S | 27.73413°E | 1480 | 07/2007[f] |

[a] Reported as decimal degrees in Hartebeesthoek94 Datum, WGS84 ellipsoid
[b] Closed March 2009
[c] Not used in WR 2005 (Middleton and Bailey 2008a, b)
[d] Closed November 2009
[e] Identified as STERK_RF by the DWS
[f] Data only available from June 2010

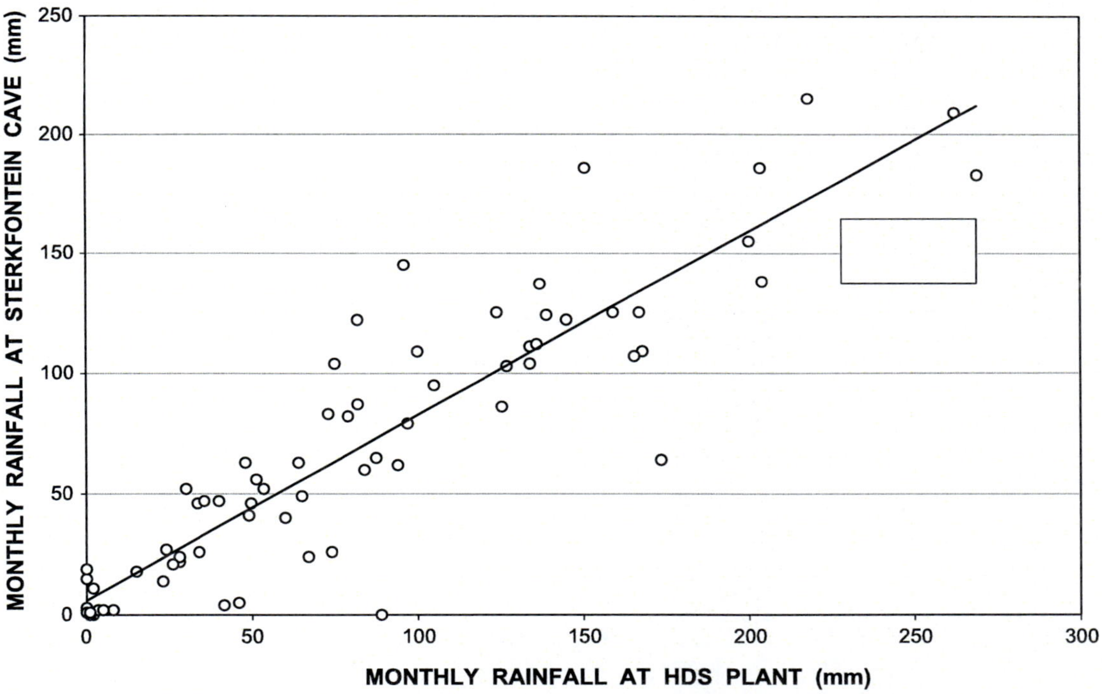


**Fig. 4** Correlation of monthly rainfall at Sterkfontein Cave with that at station HDS on the watershed for the period of common record June 2010 to September 2017; data set ($n$ = 71) excludes months of no rainfall at both stations ($n$ = 17)

- A very wet early winter (~200 mm at both stations HDS and BRI) representing ~20% of the $MA_hP$, i.e. equivalent to the typical proportional full dry season contribution.

An aspect not revealed in the preceding discussion is the correlation between individual high rainfall periods and flooding. For example, the rainfall event recorded for 16/12/2010, when 90 mm was measured at the HDS station and 106 mm at Sterkfontein Cave, precipitated the flooding in the region that contributed to a fish mortality event (Sect. 6.5.2 in Chapter "Chemical Hydrology") in January 2011 (Hobbs and Mills 2011).

A further aspect of precipitation that requires inspection for its potential relevance to groundwater recharge estimations, is the period considered for calculation of the annual mean. MAP values reported in the literature generally pertain to calendar years. In water resources assessments, however, it is the hydrological year that serves as base temporal unit, and it is, therefore, appropriate that a comparison be drawn between MAPs based on calendar and hydrological years. An inspection of the monthly data associated with the PS, HDS, BRI and SC stations aggregated to calendar and hydrological years, returns the results presented in Table 3. It is evident that the MAP values for calendar and hydrological years at each of the rainfall stations are similar, although the stations' PS and SC indicate hydrological years to be marginally wetter (by ~3%) than calendar years. The location of these two stations in more rugged terrain prone to orographic effects north of the watershed might explain this result. Nevertheless, it is concluded that natural groundwater recharge estimations based either on calendar or hydrological year averages will return similar results.

Middleton and Bailey (2008a) report an A-pan evaporation rate for the study area in the range of 2200 to 2600 mm/a. In broad terms, the mean annual potential evaporation in the study area exceeds the MAP by a factor of three. The drainages in the COH, however, are not characterised by large expanses of open water that promote evaporation losses. The dense arboreal vegetation along substantial reaches of the drainage channels provides a canopy that further inhibits evaporation, but more than likely promotes riparian evapotranspiration (RET).

Evapotranspiration (ET) has been identified in various studies (e.g. Hail and Prudic 1998; Harlow et al. 2005; Martini and Kavalieris 1976; Prudic et al. 1995; Schaefer et al. 2005) as representing the most important water 'loss' component of the water resources budget in karst environments. Rutledge and Mesko (1996) draw attention to the particular contribution of riparian evapotranspiration (RET) in this regard. The potential significance of RET as a water loss component in the hydro(geo)logic budget of the study area is evaluated in Sect. 7.4.1 in Chapter "Physical Hydrogeology".

Bredenkamp and van Staden (2009) report that the riparian tree zone is typically 10–20 m wide and the vegetation is dense and almost forest-like, dominated by indigenous trees and very dense forest-like bush on both banks of rivers, streams and creeks. Krige (2009) reports a calculated (R)ET value of 5 ML/d based on a total stream length of ~27.5 km on average 40 m wide, and subject to

**Table 3** Comparison of mean precipitation for full calendar and hydrological years at four rainfall stations

| Calendar year | Station rainfall (mm) | | | | Hydrological year | Station rainfall (mm) | | | |
|---|---|---|---|---|---|---|---|---|---|
| | PS | HDS | BRI | SC | | PS | HDS | BRI | SC |
| 2002 | 702 | | | | 2002 | 764 | | | |
| 2003 | 482 | | | | 2003 | 582 | | | |
| 2004 | 780 | | | | 2004 | 736 | | | |
| 2005 | 776 | | | | 2005 | 768 | | | |
| 2006 | 962 | | | | 2006 | 881 | | | |
| 2007 | 567 | | | | 2007 | 580 | | | |
| 2008 | 772 | | | | 2008 | 883 | | | |
| 2009 | | 737 | 612 | | 2009 | | 679 | 572 | |
| 2010 | | 903 | 823 | | 2010 | | 870 | 800 | |
| 2011 | | 885 | 711 | 750 | 2011 | | 967 | 743 | 824 |
| 2012 | | 735 | 714 | 691 | 2012 | | 660 | 663 | 606 |
| 2013 | | 714 | 662 | 596 | 2013 | | 752 | 705 | 702 |
| 2014 | | 805 | 739 | 808 | 2014 | | 832 | 758 | 812 |
| 2015 | | 451 | 490 | 462 | 2015 | | 596 | 580 | 566 |
| 2016 | | 1043 | 863 | 709 | 2016 | | 822 | 763 | 592 |
| Average | 722 | 784 | 702 | 669 | Average | 742 | 772 | 698 | 684 |

**Table 4** Classification of the vegetation of the study area (after Mucina and Rutherford 2006)

| Biome | Bioregion | Vegetation unit |
|---|---|---|
| Grassland (G) | Dry Highveld Grassland (Gh) | Carletonville Dolomite Grassland (Gh15) |
| | Mesic Highveld Grassland (Gm) | Soweto Highveld Grassland (Gm8) |
| | | Egoli Granite Grassland (Gm10) |
| Savanna (SV) | Central Bushveld (SVcb) | Moot Plains Bushveld (SVcb8) |
| | | Gold Reef Mountain Bushveld (SVcb9) |
| | | Gauteng Shale Mountain Bushveld (SVcb10) |
| | | Andesite Mountain Bushveld (SVcb11) |

an ET rate of 1674[4] mm/a. The significance of RET in the COH must, however, be tempered by the observed depth to groundwater level in and along stream reaches as discussed in Sect. 7.3 in Chapter "Physical Hydrogeology", except in those instances where the potentiometric surface approaches and intersects the stream channel (Sect. 7.4.1 in Chapter "Physical Hydrogeology"). Compared to RET, the mean annual actual evapotranspiration (AET) in the study area amounts to 1540 mm (Schulze et al. 1997).

## 3 Vegetation and Soils

The study area encompasses two biomes, namely the Grassland Biome and the Savannah Biome (Mucina and Rutherford 2006). As shown in Table 4, the Grassland Biome is represented by two bioregions, and the Savanna Biome by a single bioregion. Further subdivision of the bioregions recognises the vegetation units listed in Table 4 as occurring in the study area. A more comprehensive discussion of the vegetation types in the subregion is provided by Bredenkamp and van Staden (2009).

As the name implies, the Carletonville Dolomite Grassland (Gh15) vegetation unit covers the area underlain by dolomitic strata of the Malmani Subgroup (Table 5). It therefore also covers the largest expanse ($\sim$50%) of study area characterised mostly by the shallow Mispah and Glenrosa soil forms, with deeper red to yellow apedal soils of the Hutton and Clovelly forms occurring sporadically. The vegetation is characterised by species-rich grasslands forming a complex mosaic pattern with many species dominant (Mucina and Rutherford 2006). A notable characteristic of this vegetation unit is the occupation of sheltered valleys and sinkholes by trees such as *Celtis Africana* (white stinkwood, *A.* witstinkhout), *Kiggelaria Africana* (wild peach, *A.* wildeperske) and *Leucosidea sericea* (oldwood, *A.* ouhout) that represent traces of temperate or transitional forest (Acocks 1988). The Soweto Highveld Grassland (Gm8) vegetation unit occurs along the southern margin of the study area underlain by quartzitic strata of the Witwatersrand Supergroup. The deep, reddish soils on flat plains are absent in the presence of outcropping bedrock. The landscape supports short to medium–high, dense, tufted grassland dominated by *Themeda triandra* (red grass, *A.* rooigras).

The Egoli Granite Grassland (Gm10) vegetation unit occurs along the south-eastern margin of the study area where it is underlain by Halfway House Granite that builds the pluton known as the Johannesburg Dome. The landscape supports leached, shallow, coarsely grained, sandy and nutrient-poor soils of the Glenrosa form. Grassland vegetation is dominated by *Hyparrhenia hirta* (common thatchgrass, *A.* dektamboekiegras) with some woody species on outcropping bedrock.

The Savanna Biome is represented by the Gold Reef Mountain Bushveld (SVcb9) and Andesite Mountain Bushveld (SVcb11) vegetation units that occupy very small portions along the south-eastern margin of the study area, and the Moot Plains Bushveld (SVcb8) and Gauteng Shale Mountain Bushveld (SVcb10) vegetation units found along the north-western margin. The latter two units are readily defined by their association with the mainly sedimentary strata (quartzite and shale) of the Pretoria Group that overlie the Malmani Subgroup dolomite (Sect. 4). The soils vary from stony, colluvial, clay-loam, red-yellow freely drained forms to shallow Mispah soil forms. The low, broken ridges varying in steepness support shortly (3–6 m tall), semi-open thicket dominated by a variety of woody species including *Acacia caffra* (hook thorn, *A.* haakdoring) and *Cussonia spicata* (common cabbage tree, *A.* kiepersol).

## 4 Geology and Geophysics

The geology of the broader region is described in numerous texts, e.g. Clendenin (1989), Eriksson and Reczko (1995), Eriksson et al. (2006), Obbes (2001) and Robb and Robb (1998a). The geology of the study area is dominated by carbonate strata (primarily dolomite) associated with the Malmani Subgroup of the Chuniespoort Group within the

[4] Equal to 0.775 of the long-term A-pan equivalent of 2160 mm/a at OR Tambo International Airport.

**Table 5** Simplified lithostratigraphic subdivision of strata in the subregion

| Basic lithology[a] | Lithostratigraphic unit | | | Era (Age) | |
|---|---|---|---|---|---|
| Alluvium | Quaternary sediments | | | Late Cenozoic (<10 000 y) | |
| Dolerite/diabase/syenite [Jd] | Dyke/sill intrusive structures | | | Early Mesozoic (150 to 190 Ma) | |
| Andesite, basalt, subordinate shale [Vh] | Hekpoort formation | Pretoria group | Transvaal supergroup | (~2224 Ma) (~2555 Ma) | Vaalian |
| Ferruginous shale and quartzite, hornfels [Vt] | Timeball hill formation | | | | |
| Quartzite, shale, chert breccia [Vrs] | Rooihoogte formation | | | | |
| Dolomite [Vmd] | Malmani subgroup | Chuniespoort group | | | |
| Quartzite, shale [Vbr] | Black reef formation | | | (~2650 Ma) | |
| Graywacke, conglomerate, volcanics [R-Vk] | Kameeldoorns formation | Platberg group | Ventersdorp supergroup | (~2650 Ma) (~2780 Ma) | Randian |
| Quartzite, conglomerate [Rjo] | Johannesburg subgroup | Central rand group | Witwatersrand supergroup | (~2780 Ma) (~2970 Ma) | |
| Shale, quartzite [Rj] | Jeppestown subgroup | West rand Group | | | |
| Quartzite, greywacke [Rg] | Government subgroup | | | | |
| Ferruginous shale, quartzite [Rh] | Hospital hill subgroup | | | | |
| Mafic and ultramafic rocks [Zm] and granite [Zg] | Undifferentiated | | | Swazian (>3100 Ma) | |

[a] Lithology colours correlate broadly with those used in Fig. 2 in Chapter "Introduction and Background"
Ma = million years

Transvaal Supergroup succession of mainly sedimentary strata. The Malmani Subgroup is subdivided into four units identified as the Oaktree Formation at the base, overlain in turn by the Monte Christo, the Lyttelton and the Eccles formations. In a regional context, these strata dip to the north-west at an angle of between 15° and 30°, disappearing beneath younger Pretoria Group sedimentary rocks (quartzite and shale) roughly coincident with the SW–NE strike of the Skeerpoort River valley (Fig. 2 in Chapter "Introduction and Background" and Fig. 9).

The moderately rugged relief that occupies an elevation ranging between ~1400 and ~1600 m amsl between the Skeerpoort River draining the north-western portion of the property, and the Bloubank Spruit system draining the southern and south-eastern portions (Fig. 2 in Chapter "Introduction and Background" and Fig. 9), forms the escarpment-type karst morphology defined by Martini and Kavalieris (1976). These features reflect the degree of structural geologic influence on the landscape as recognised, amongst others, by Dirks and Berger (2013) as an agent in landscape dynamics in the Malapa and Gladysvale sections of the property (Table 6). The much flatter terrain occupying an elevation between ~1560 and ~1660 m amsl that characterises the landscape of the Steenkoppies Basin west of Tarlton (Fig. 2 in Chapter "Introduction and Background"), represents the plateau-type karst morphology of Martini and Kavalieris (1976).

The outcrop of individual karst formations is only defined north of latitude 26°S on published geological maps (RSA 1973a, b). South of 26°S, only the composite Malmani Subgroup is defined to the west as far as longitude 27°15′E (RSA 1986), beyond which the formations are again individually mapped.

Along the southern and south-eastern margin, the Malmani Subgroup dolomite rests on older sedimentary rocks of the Witwatersrand Supergroup and even older intrusive granitic and gneissic rocks forming the Archaean basement exemplified by the Halfway House Granite pluton. The stratigraphic relationship of the various lithologies described above is shown in Table 5.

The significance of karst formations such as underlie ~26 860 ha (~52%) of the COH, and ~27 850 ha (~43%) of the study area, is summarised by Hamilton-Smith (2006) who states that karst often has '.......

**Table 6** Salient hydrologic and hydrogeologic aspects associated with the most well-known fossil sites in the study area

| Fossil site | Coordinates[a] | Quaternary catchment | Main drainage | Groundwater basin | Synoptic cave morphology | SAAN[b] # with brief description and main significance |
|---|---|---|---|---|---|---|
| Bolt's farm | 26.03°S 27.72°E | A21D | Bloubank spruit | Zwartkrans | Near-surface and underground | SAAN—: comprises a complex of ~ 23 caves that include the oldest deposits in the Sterkfontein Valley on the basis of microfaunal dating; also the site of the recent *H. naledi* finds in the Rising Star Cave system |
| Swartkrans | 26.02°S 27.72°E | | | | Near-surface and underground | SAAN—0021: well-documented excavations and faunal analyses by CK Brain, hosts the largest collection of *Paranthropus robustus* fossils and evidence for the use of fire and bone and horn tools; has yielded >500 hominin fossils |
| Sterkfontein | 26.01566°S 27.73413°E | | | | Near-surface and underground | SAAN—0020: internationally recognised focus of the COH WHS and home of the *Au. africanus* specimens Mrs. Ples (Sts 5) and Little Foot (StW 573), has yielded >500 hominin fossils, and >25 000 identifiable fossils in total |
| Cooper's | 26.01°S 27.75°E | | | | Near-surface | SAAN—0023: specimens of *Paranthropus robustus*, including a face, uncovered; potential for more hominins and faunal dating |
| Kromdraai | 26.01°S 27.75°E | | | | Near-surface | SAAN—0022: one of the earliest sites studied, yielded the remains of at least three *Paranthropus robustus* individuals, and a few earlier Stone Age artefacts |
| Minnaar's | 26.01°S 27.74°E | | | | Near-surface and underground | SAAN—0004: known for the excavation of an extinct jackal skull, has the potential for the discovery of hominin fossils |
| Plover's lake | 25.98°S 27.78°E | | | Krombank | Near-surface and underground | SAAN—0025: hosting breccias <1 Ma old and revealing no hominin remains as yet, this site retains the potential for revealing the remains of early modern humans associated with Middle Stone Age artefacts |
| Wonder Cave | 25.97033°S 27.77143°E | | | | Underground | SAAN—0027: perhaps the best example of a tourist-centred (show) cave, as yet not known to host hominin fossils, but demonstrates the development of a recent talus cone containing semi-fossilised skeletal material |
| Drimolen | 25.97°S 27.76°E | | | Danielsrust | Near-surface | SAAN—0024: has yielded ~ 75 *Paranthropus robustus* specimens and five *Homo sapiens* species, as well as bone and horn tools |
| Gladysvale | 25.90°S 27.77°E | A21G | Skeerpoort River | Uitkomst | Near-surface and underground | SAAN—0001: has yielded several thousand Plio-Pleistocene mammalian remains including hominin (*Au. africanus*) teeth |
| Motsetse | 25.91°S 27.83°E | A21E | Crocodile river | Diepkloof | Near-surface | SAAN—0030: although not renowned for any significant discoveries, has research potential coupled with an attractive setting and visitor access |
| Haasgat | 25.86°S 27.83°E | A21G | Skeerpoort river | Diepkloof | Near-surface | SAAN—: has yielded 83 craniodental fossils of the Cercopithecidae Family (Old World monkeys) primate *Papio angusticeps*, extinct relative of *Papio ursinus* (Chacma Baboon) |

(continued)

**Table 6** (continued)

| Fossil site | Coordinates[a] | Quaternary catchment | Main drainage | Groundwater basin | Synoptic cave morphology | SAAN[b] # with brief description and main significance |
|---|---|---|---|---|---|---|
| Gondolin | 25.83°S 27.86°E | A21H | Crocodile river | Broederstroom | Near-surface | SAAN—0031: has yielded *Paranthropus robustus* specimens |
| Malapa | 25.90°S 27.80°E | A21G | Skeerpoort river | Diepkloof | Near-surface and underground | SAAN—: host to *Au. sediba*, the recent anthropological find in the COH |

[a] All coordinates except for Sterkfontein Cave and Wonder Cave truncated to the 2nd decimal in order to protect these sites from unwelcome visitation

[b] Site number from Berger and Brink (undated), together with a substantial amount of ancillary descriptive information including a mixed assemblage list of fauna

- Invaluable geological data (particularly in the cave floors);
- Important geomorphic structures and processes;
- Characteristic surface and often significant landscapes;
- Important surface ecosystems and even more important subterranean ecosystems; and
- Fossils and cultural heritage (pre-historic, historic and living)';

to which must be added.

- Significant quantities of fresh groundwater; and
- A capacity to function as a sink for carbon in the global carbon cycle.

The hydrogeologic significance of the chert-rich dolomitic strata of the Monte Christo and Eccles formations, versus that of the chert-poor Oaktree and Lyttelton formations, bears mention. The chert-rich formations are associated with significantly greater water-bearing properties than the chert-poor formations, due mainly to preferential dissolution of the dolomite that is interlayered with the chert bands (Martini and Kavalieris 1976). In outcrop, the insoluble chert bands protrude from the dolomite as is illustrated in Plate 1, Plate 1. Preferential dissolution is enhanced in the presence of fractures and joints that intersect the rock mass and facilitate the movement of infiltrating slightly acidic rainwater as primary speleogenetic agent, as observed by Wilkinson (1973) in Sterkfontein Cave.

The structural complexity of the geologic environment in the subregion presents a significant challenge. This challenge pertains, amongst others, to an understanding of the geologic control that structures such as faulting, folding and intrusions of dolerite, diabase or syenite in the form of sills (subhorizontal intrusions) and dykes (subvertical intrusions) exercise on the hydrogeologic regime. Staff and students of the Witwatersrand University (Wits) School of Geoscience and of the University of Johannesburg continue to carry out structural geologic mapping and other bio- and geo-environmental studies (including hydrogeology) in the area.

A number of geophysical surveys that help inform various aspects of the geological environment in the subregion and wider surroundings have been carried out in the past. The most significant of these is considered to be the gravimetric survey carried out by the then Department of Water Affairs and Forestry (DWAF) in the mid-1980s (Bredenkamp et al. 1986). This survey covered large tracts of the eastern portion of the Steenkoppies Basin and the western portion of the Zwartkrans Basin (refer to Sect. 7.2 in Chapter "Physical Hydrogeology" for compartment definition), spanning ~ 90 km$^2$ at a grid interval of 100 m between measurement stations.

The gravity survey was augmented with ground-based magnetic and electromagnetic surveys primarily to better define linear geological structures (mainly intrusive dolerite, diabase or syenite dykes) of possible hydrogeological significance. The magnetic surveys indicated that the dyke intrusions generally exhibit a negative magnetic signature, i.e. possess a magnetic field strength that is lower than that of the host rock intruded by these structures. The gravimetric data set has been converted to digital format (F Wiegmans, personal communication) which facilitates its further application and use in a GIS environment.

More recently (ca. 2005), the then DWA commissioned the Council for Geoscience (CGS) to carry out a regional aeromagnetic (airborne magnetic) survey covering the West Rand and Far West Rand goldfields. The limited ground-based application of the magnetic, electromagnetic and resistivity tomography survey methods to identify potential subsurface conduits of mine water drainage in the Krugersdorp Game Reserve (KGR) is reported by Coetzee et al. (2009). The CGS has also recently (in mid-2010) carried out an airborne electromagnetic survey of the area covered by the earlier aeromagnetic survey. Funded by the Department of Mineral Resources (DMR), the data generated by this survey has yet to be published and made

**Plate 1** Epikarst exposed in the form of a pinnacle ~2.5 m high in the Eccles Formation, showing the preferential dissolution of dolomite interlayered with less soluble and therefore more prominent chert bands, the slanting aspect of which also defines the local ~15° dip of the strata to the north-west (at bottom left of picture), with a partially wad-filled and vegetated vertical joint ('slot') separating the pinnacle from the overgrown rock mass at right of picture

available for wider application. It is imperative, however, that the additional information generated by this survey be applied to further inform the geologic and hydrogeologic understanding developed and presented in this manuscript.

A geophysical survey carried out as part of the water resources situation assessment for the MA (Hobbs et al. 2011a) has sought to better define features that inform the structural geologic setting of Sterkfontein Cave. This survey employed ground-based magnetic, electromagnetic and electrical resistivity surveys (Chirenje et al. 2010), and was successful in identifying several structures and their possible influence on the cave morphology.

## 5 Mining Geology

The threat posed to the COH by the legacy of mining in the West Rand Goldfield, in particular, that associated with mine water discharge, is an important aspect of the water resources environment in the study area. It is therefore pertinent that a discussion of this activity is included for the purpose of elucidating its position within this environment.

The mining geology of the West Rand Goldfield (aka the Western Basin) is described by Toens and Griffiths (1964) and Tucker and Viljoen (1986), amongst numerous others. The history of mining is discussed by Handley (2004), from which it would appear that gold mining in this field commenced in ca. 1889 with the opening of Randfontein Estates Gold Mining Company (REGM). Another mining activity in the area has included uranium (together with gold), limestone from caves (e.g. Sterkfontein) and excavations (e.g. Sterkfontein Quarry). Although primary extraction of these mineral commodities is now history, secondary extraction of gold is associated with the reworking of the older mine surface residue deposits in the form of sand dumps (Plate 2 and Fig. 5). The tailings reclaimed by Mogale Gold/Mintails SA (MG/MSA) have reported grades in the range of 0.26 to 0.3 g/t (Louw 2012), compared to average in-situ ore grades in the range of 4 to 19 g/t for the auriferous reefs in the West Rand Goldfield (Handley 2004). Total gold production in the 114 years to 2002 amounted to ~2.4 kt at an average grade of 5.09 g/t (Handley 2004). In the period 1952 to 1995, this goldfield also produced ~ 27.9 kt of uranium at an average grade of 370 g/t (Cole 1998), extracted primarily from the Main, White and Monarch reefs of the Johannesburg Subgroup. The REGM operations closed in 1999.

Surface features of mining-related significance in the Western Basin are identified in Figs. 5 and 6. The conceptual inter-relation of these features with the geologic environment in the subsurface is illustrated in Fig. 7. The illustrations show the association of the opencast mining activities (represented by the West Wits and Millsite pits) with the package of auriferous Kimberley Reefs, and the positions of the Black Reef Incline (BRI, Plate 3), #s 8, 9 and 9E Shafts, and #s 17 and 18 Winzes in the geologic environment. The position of the Main Reef in this environment, its exploitation by underground workings via shafts such as the ~450 m deep #8 Shaft, and its possible linkage to the Black Reef workings via the BRI and #17 and #18 Winzes, is also illustrated.

**Plate 2** Recovery of mine residue deposits on the northern side of Dump 20 (see Fig. 5 for location) destined for reworking in the Mogale Gold/Mintails SA plant; the excavated/eroded area in the foreground is the southern end of the Millsite Pit (photo W Bason, date 09/12/2009)

**Plate 3** Panoramic view of the HDPE-lined dam at the Black Reef Incline (under grid at front right-of-centre), showing the pump station at left-of-centre beyond the dam, from where the water is pumped to the HDS treatment plant located on the skyline (continental divide) at left of picture; the Tweelopie Spruit on the far side of the dam flows from left (south) to right (north) across this 'fish-eye' view (photo W Bason, date 09/12/2009)

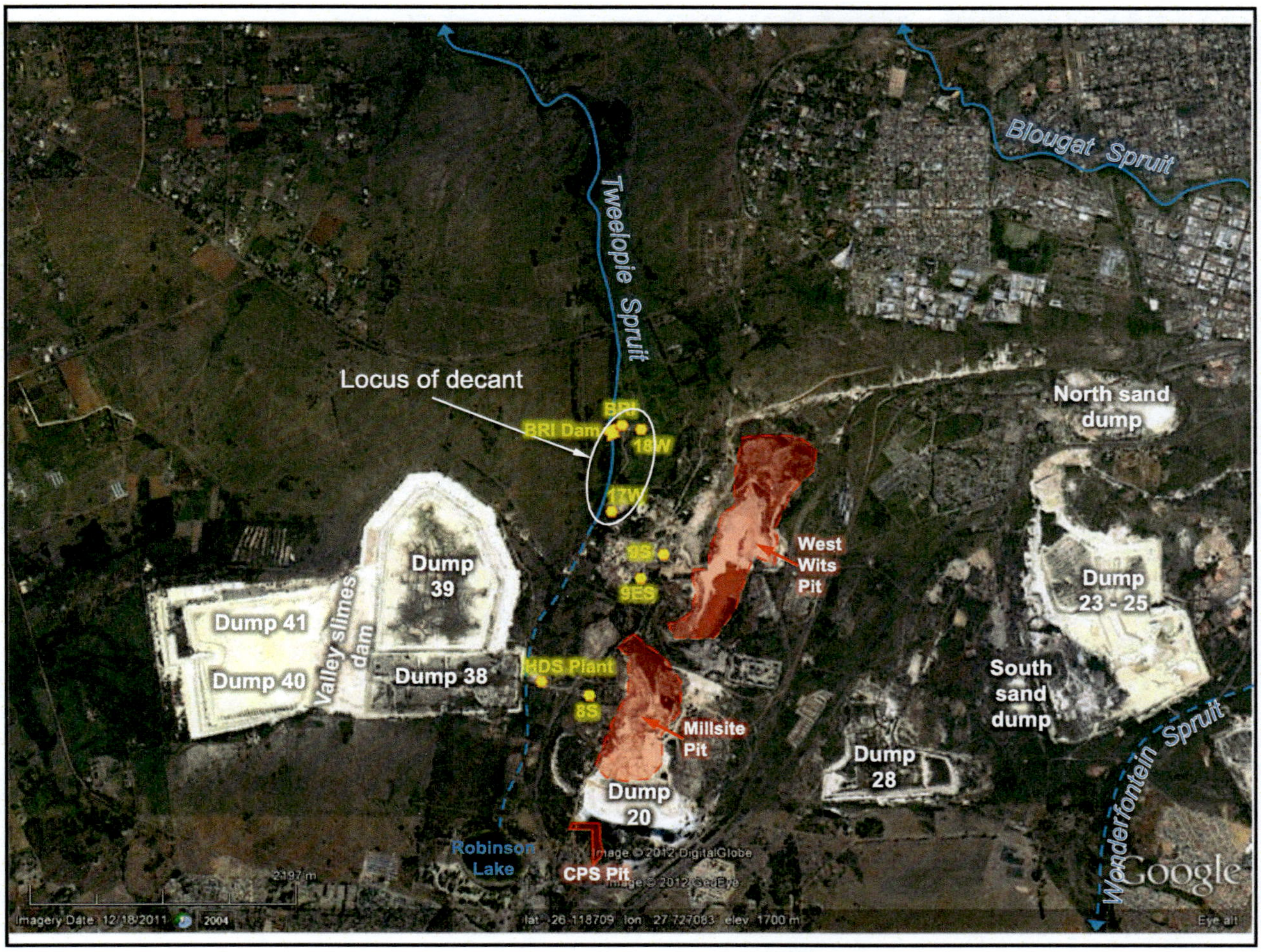


**Fig. 5** Google Earth® image showing surface features relevant to the mining environment in the Western Basin

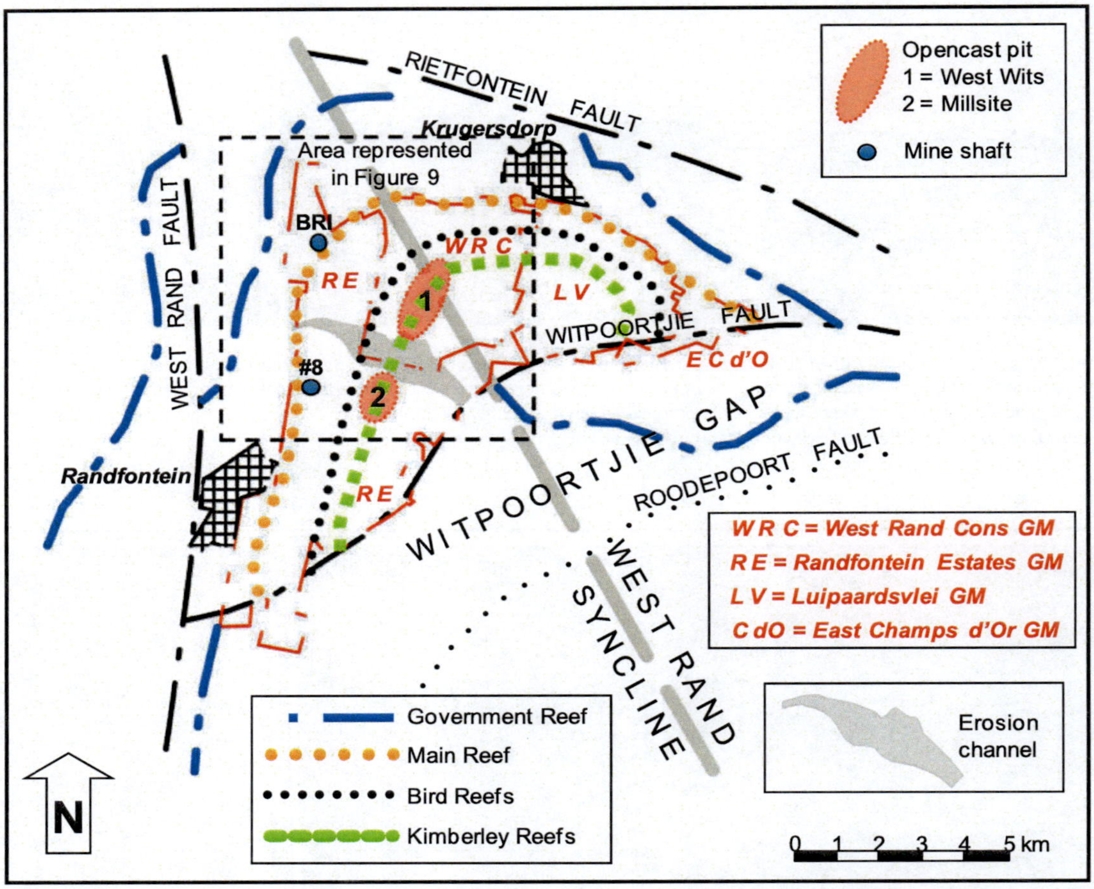


**Fig. 6** Diagram of mining-related features (e.g. geology, mine areas, shafts and pits) in the West Rand Goldfield. (Modified from Toens and Griffiths 1964; Tucker and Viljoen 1986)

## 6 Palaeontology, Archaeology and Ecology

The palaeontology and archaeology of the study area are intimately associated with the karst terrane of the subregion. Fossil sites often represent the erosional remnants of ancient cave systems and their contents (Esterhuysen 2009) and, for this reason, are dynamic features that continue to evolve in response to natural (mainly geomorphological) processes and anthropogenic impacts. Whereas the natural processes such as weathering and erosion occur over extended periods of time (typically millennia), the anthropogenic impacts are often manifested in centuries or even decades. Partridge (1973) derived estimates for when the Sterkfontein and Swartkrans caves became drained of 3.26 and 2.57 Ma, respectively. The difference of $\sim$0.7 Ma is surprising given the similar geomorphologic settings of these two cave systems located only $\sim$ 1100 m apart in the same dolomitic subcompartment. The estimated base elevation of Swartkrans Cave ($\sim$1465 m amsl) places it $\sim$25 m above the current ambient groundwater rest level elevation of $\sim$ 1439 m amsl (Sect. 7.3.2.2 in Chapter "Physical Hydrogeology"). Although this cave system is not accessible (at least as is known at present) to the same depth as that of the Sterkfontein system, it might be expected from the similar surface elevations occupied by both systems (1470–1487 m amsl) and from their shared hydrogeologic environment, that they would have drained contemporaneously.

The most well-known Australopithecine fossils associated with the Sterkfontein Cave site, namely the remains of 'Little Foot' (catalogue # StW 573) and of the *Australopithecus africanus* species 'Mrs Ples' (catalogue # Sts 5), have been assigned ages of $\sim$2.2 and $\sim$2.1 Ma, respectively (Pickering 2010; Pickering and Kramers 2010). The age of 'Little Foot' (*Au. prometheus*) was more recently set at $\geq$3 Ma (Bruxelles et al. 2014), and most recently at 3.67 Ma (Gardner 2015). The relevance of these ages in regard to landscape development in the region is examined in Sect. 8.12.1 in Chapter "Chemical Hydrogeology".

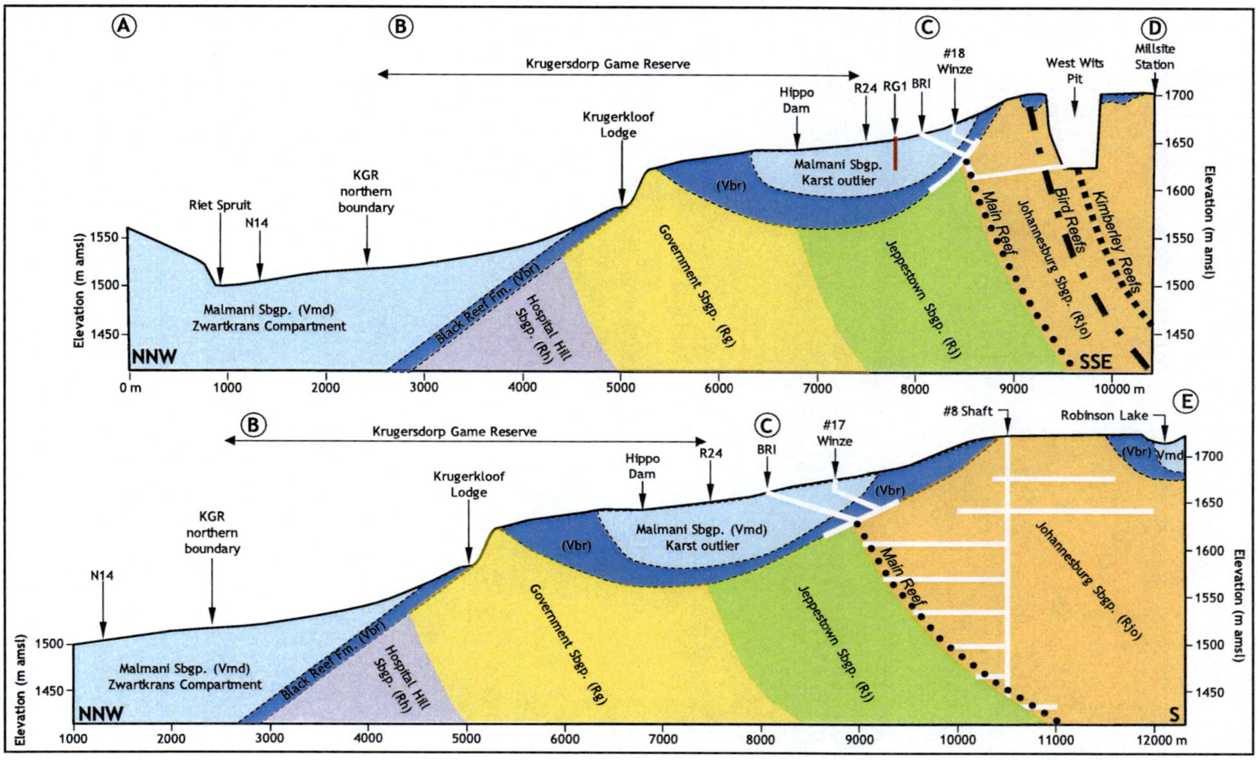


**Fig. 7** Schematic profiles of geologic and mining-related features in the northern portion of the West Rand Goldfield (see Fig. 8 for profile transects)

The Sterkfontein Cave site has yielded >500 hominin[5] fossils (Table 6) since the discovery of 'Mrs. Ples' in 1947 (Cooke 1969). Thirteen fossil sites in the COH (Fig. 9) have been declared National Heritage Sites in accordance with the National Heritage Resources Act (Act No 25 of 1999). Many fossil-bearing deposits and sites of palaeontological, archaeological, historical and cultural heritage importance (both categorised and uncategorised) are known to exist in the study area (Esterhuysen 2009). The COH has already yielded ~35% of the total record of human evolution in Africa (McCarthy and Rubidge 2005). The probable existence of yet undiscovered sites on the property that may yield material of palaeontological and archaeological value is substantial. This is exemplified by the unveiling on 08/04/2010 of the *Au. sediba* species fossils (Berger et al. 2010), dated at 1.977 ± 0.002 Ma (Pickering et al. 2011), found at the Malapa site in the north-eastern part of the COH (Dirks et al. 2010). An even more recent example is the unveiling on 09/09/2015 of the *Homo naledi* remains (a new species in the genus *Homo*) from the Dinaledi Chamber of the Rising Star Cave in the Bolt's Farm complex (Berger et al. 2015; Dirks et al. 2015) in the south-western portion of the property.

An important ecological aspect associated with the cave systems in the study area is their provision of a habitat for stygobitic fauna (Tasaki 2006). The aquatic cave-dwelling biota, typically blind amphipods, have been found in Sterkfontein Cave and two other caves in the study area (Koelenhof and Yom Tov) that intersect the water table. It is probable, therefore, that investigation of other more remote and less accessible caves in the area that intersect the water table might yield results of ecological significance also in this regard. This might alter the situation where no 'hotspot' of subterranean biodiversity is identified on the African continent (Culver and Sket 2000). Although it is beyond the scope of this study to comment on the suitability of the current subsurface habitat(s) that support this fauna, it does provide information (e.g. water chemistry data) that, in the hands of appropriate experts and applied with caution, might better inform this aspect. Some discussion on this topic is provided in Sect. 8.9.3 in Chapter "Chemical Hydrogeology". Further ecological significance is afforded by Durand and Peinke (2010) to the many cave features in the subregion that provide a habitat and roost for bat colonies.

[5] Use of the term hominin (as opposed to hominid) after Berger (2001).

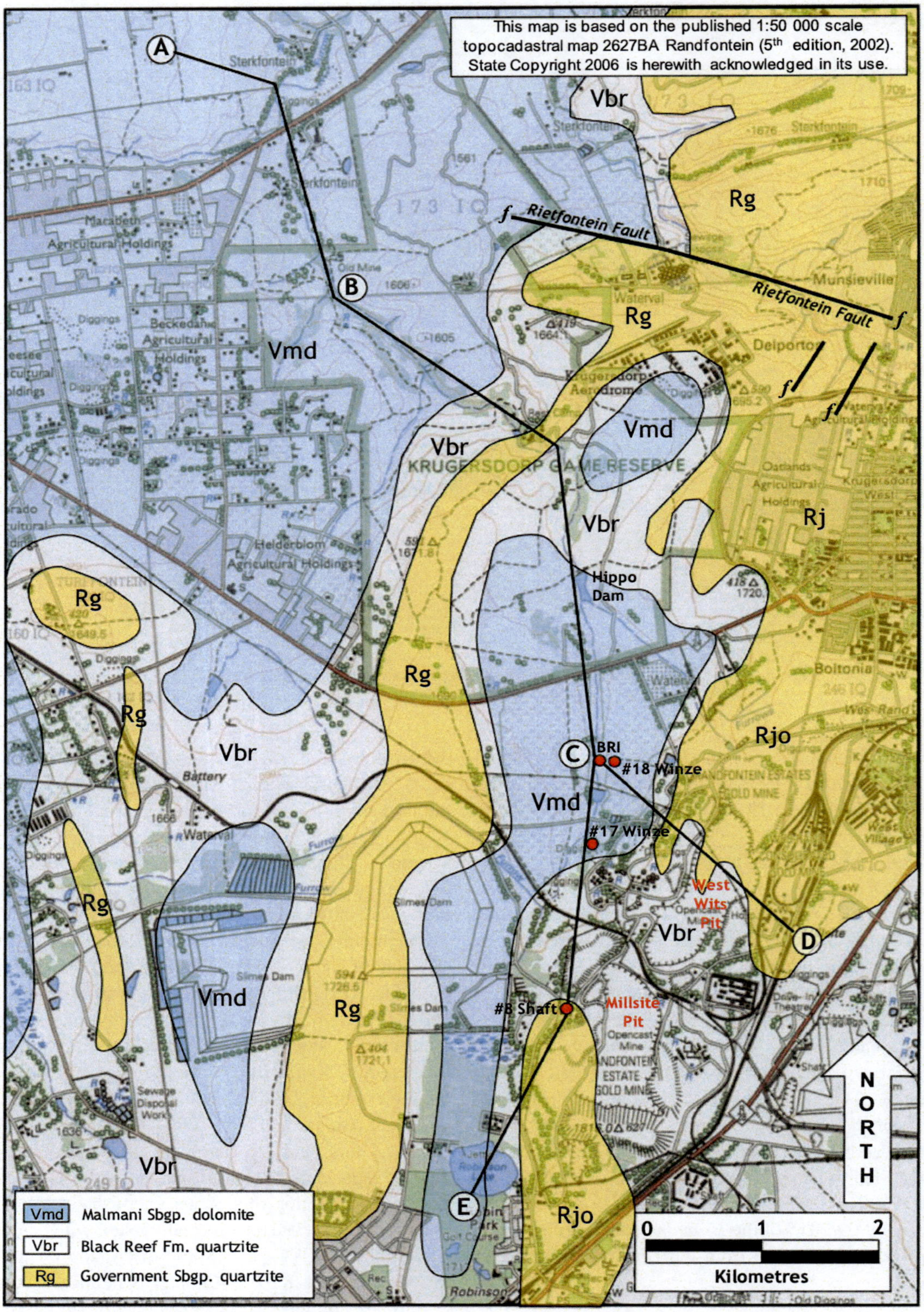


**Fig. 8** Geologic map of the north-western portion of the Western Basin showing surface features of mining-related significance and the position of the profile transects illustrated in Fig. 9 (original geology from Mellor 1917)

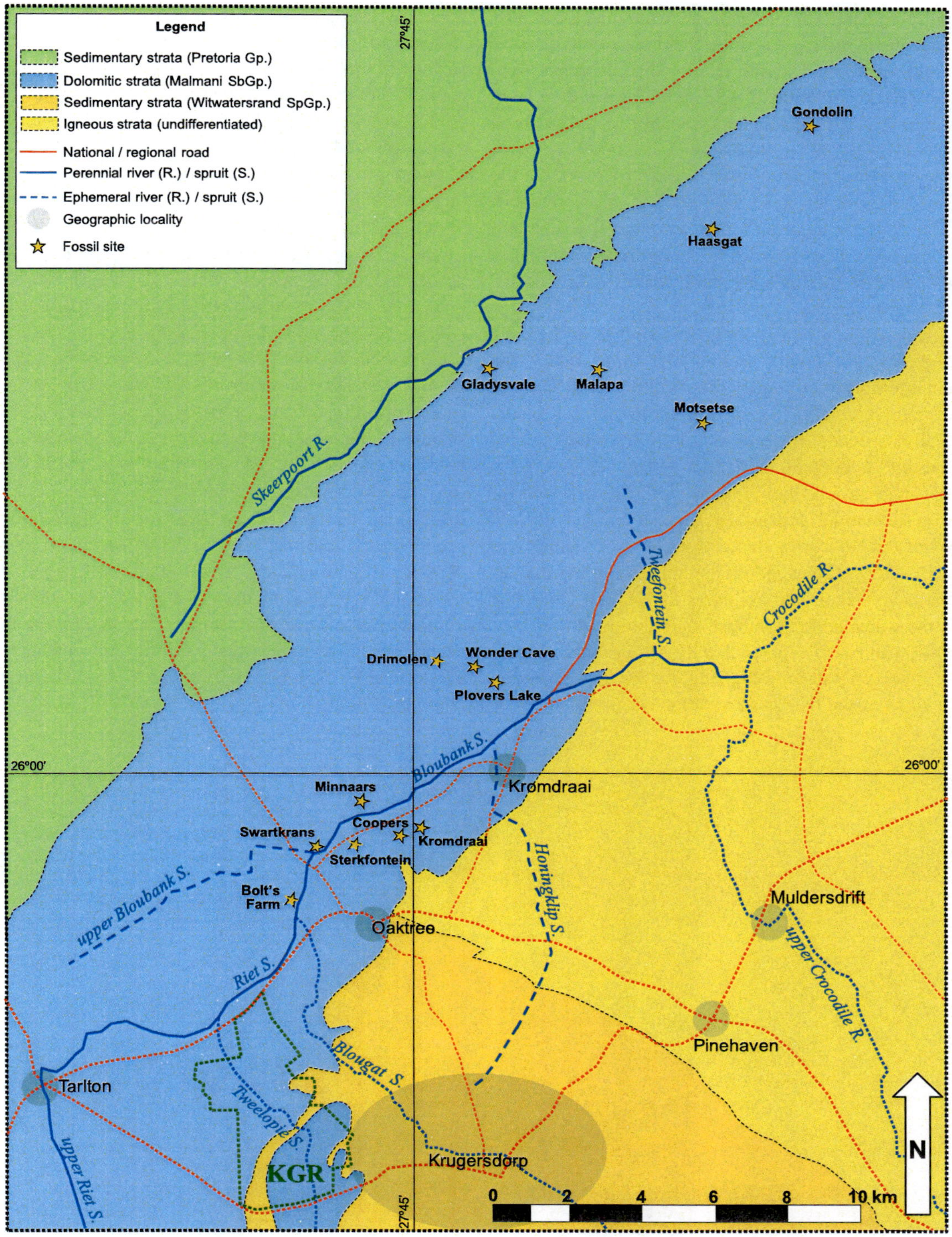


**Fig. 9** Locality map of the UNESCO-inscribed fossil sites in the study area superimposed on a simplified geological, surface water drainage and road map

The hydrological and hydrogeological settings associated with the inscribed[6] fossil sites in the COH are summarised in Table 6. Although some of the tabulated information (e.g. the groundwater compartment and groundwater rest level elevation information) is presented prematurely (i.e. without prior context in terms of this work), it is provided for completeness of this section of text. The observation that the geographic footprint of karst basins (e.g. the Diepkloof Basin, Table 6) spans adjacent (and hydrologically separate) surface drainage systems testifies to the hydraulic continuity that exists between surface water and groundwater systems in the COH karst terrane.

Included in Table 6 is a synoptic description of the cave morphology associated with each of the fossil sites. This description attempts to arbitrarily classify the morphology of a cave system in relation to its setting in the landscape. Simplistically, a near-surface morphology describes a cave system that does not extend more than ~5 m below surface, whereas an underground morphology describes a cave system with a substantial subsurface extension. The ultimate expression of the latter is intersection of the water table, e.g. Sterkfontein Cave. Caves of this nature are classified as water table caves (Ford and Williams 2007), and those that lie above the water table as vadose caves.

The CGS (undated) reports that in all instances in the COH, the important palaeontological finds are located within the upper 10 m of the host cave system. Whilst this may be generally true, at least in the case of Sterkfontein Cave the 'Little Foot' find is located a greater depth (~25 m) below surface (Partridge et al. 2003). Given a (maximum) surface elevation of 1487 m amsl (Sect. 7.1 in Chapter "Physical Hydrogeology"), this places the 'Little Foot' site at an elevation of ~1462 m amsl, at least ~22 m above the contemporary maximum attainable cave water level of ~1440 m amsl (Sect. 7.3.2 in Chapter "Physical Hydrogeology"). These descriptions have specific relevance to the vulnerability of each of the fossil sites (and their associated cave systems) where these are examined and discussed in the context of their hydrogeologic setting.

The Malapa excavation understandably suffers from a relative paucity of hydrogeologic information. This site is in the process of being listed with UNESCO as the fourteenth site on the property to be inscribed. Recently published material associated with the hominin remains being excavated at this site (Henry et al. 2012) affords the site ever-increasing importance. These circumstances inform the concern of the MA for the protection of the pre-historic 'treasures' of the COH as much as its concern for the current and future ecological sustainability of the property. This dictates the need for support of all components of the environment (including human habitation and settlement) that inform the outstanding universal value (OUV) of the area (P Mills, personal communication).

[6] Registered with UNESCO for protection in terms of their cultural heritage as part of the COH.

Accumulation of flood debris blocking the entrance to culverts of the N14 bridge over the Blougat Spruit near Oaktree; the hazard for potential flood damage to the national road is self-evident

# Overview of Karst

Philip J. Hobbs and Harrison Pienaar

## 1 General Overview of Global Karst

Dolomite [$CaMg(CO_3)_2$] bears the name of the French naturalist/geologist/mineralogist Déodat Gratet de Dolomieu (1750–1801) who first described this rock in 1791 from outcrops in the Dolomite Alps of northern Italy. The Serbian geographer/geomorphologist Jovan Cvijić (1865–1927) studied this rock in the Dinaric Kras region of Slovenia in central Europe, where he established its association with rock dissolution giving rise to the diagnostic landscape characterised by features such as dolines, sinkholes, caves and solutional conduit networks. Since this region represents 'classical' karst, in geologic terms the type area globally for such a landscape, the term karst (Germanic translation of *kras*) has become the internationally recognised name for any similar landscape[1] worldwide. Gams (1991), however, suggests that the word *karst* derives from the pre-Indoeuropean word *karra/gara* meaning stone.

The surface extent of carbonate strata across the globe (Fig. 1) is placed at ~13% of the total ice-free land (continental) surface area (Kaufmann 2013).

It is well known that karst aquifers form some of the most productive of all groundwater resources globally. This alone is testimony to their importance and value as a sustainable natural resource and, as this literature review shows, is equally true of a number of South African karst formations (Sect. 2). There is also a growing awareness of the potential role of karst in the global carbon budget as a carbon 'sink' (e.g. Gombert 2002; Larson and Mylroie 2013; Martin et al. 2013; Veni 2013; White 2013) for its carbon sequestration properties. Finally, the value of ecosystem services delivered by karst terranes is an often forgotten benefit that similarly is gaining awareness in the literature (e.g. Bonacci et al. 2009; Goldscheider 2012; Kiss et al. 2011; Zhang et al. 2011). The schematic diagram presented in Fig. 2 provides an indication of the diversity of elements that contribute to a karst hydro-ecosystem.

### 1.1 Age and Distribution

The majority of the considerable number of karst formations on earth (Fig. 1) formed in the Mesozoic Era 250–65 Ma before present (BP). The European carbonate strata that include the classic Dinaric Karst and the Italian Dolomites[2] belong to this era. The limestone formations of the English Midlands in the United Kingdom formed in the Cretaceous Period (144–65 Ma). The spectacular landscape of the Fengcong (cone) and Fenglin (tower) karsts of the Libo Karst cluster in the South China Karst[3] that covers large parts of the Yunnan, Guizhou and Guangxi provinces of southern China (Waltham 2008), date back to the Permian Period (298–250 Ma). The oldest of the numerous karst aquifers of North America are the Ozark Plateau Karst Aquifer in the east-central USA dating back to the Palaeozoic Era (542–251 Ma), and the Cambrian to Mississippian (542–318 Ma) karst of Alberta Province, Canada. It is evident, therefore, that these karst environments are not

P. J. Hobbs: Deceased

[1] The term karst is not limited to carbonate formations, since a similar landscape developed in other strata, e.g. the dissolution features formed in the Table Mountain Group sandstone, the quarzitic karst of South America, and the basalt lava fields that form the Pacific Northwest Pseudokarst Aquifer in the north-western United States, is typically referred to as pseudo-karst.

[2] Declared a World Heritage Site in 2009.

[3] Declared a World Heritage Site in 2007.

H. Pienaar (✉)
Council for Scientific and Industrial Research, Smart Places, Pretoria, South Africa
e-mail: hpienaar@csir.co.za

H. Pienaar
Hebei University of Engineering, Handan, China

P. J. Hobbs
Pretoria, South Africa





P. J. Hobbs et al. (eds.), *Anatomy of a South African Karst Hydrosystem*, Cave and Karst Systems of the World,
https://doi.org/10.1007/978-3-030-95829-9_4

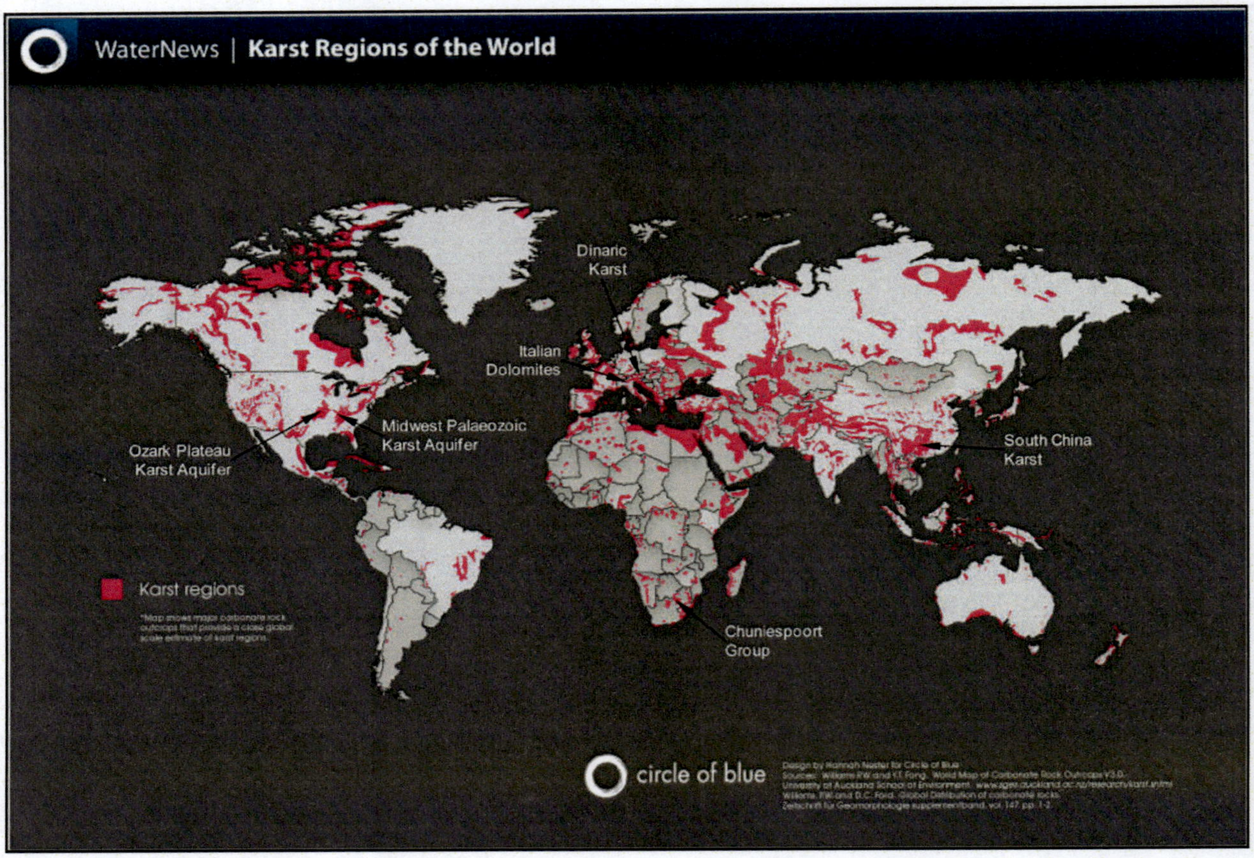


**Fig. 1** Distribution of karst regions worldwide, showing karst aquifers referenced in the text (sourced at http://www.circleofblue.org/waternews/ on 28/04/2012)

**Fig. 2** Schematic presentation of the components of an undisturbed karst hydro-ecosystem and its associated values and ecosystem services (modified after Goldscheider 2012)

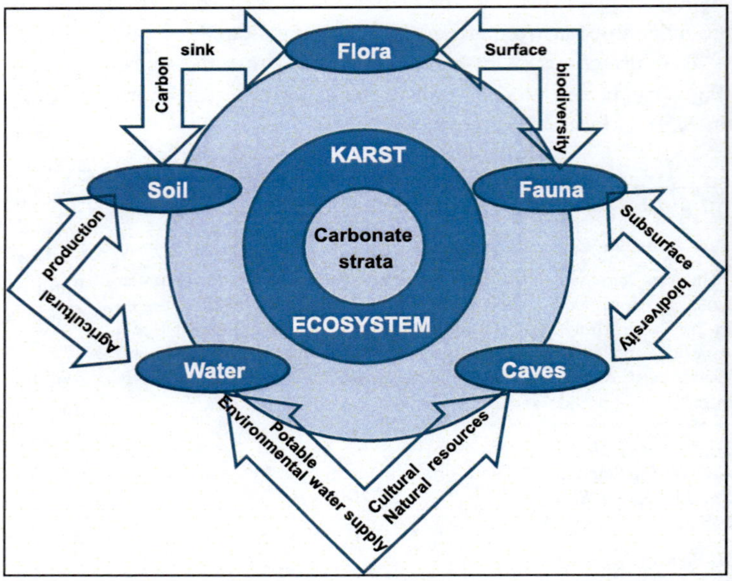

directly comparable to the local and regional 'hard' carbonate formations associated with the considerably older (by a factor of 5–10) Neoarchaean Vaalian strata (~2.55 Ga) of the Chuniespoort Group (Table 5 in Chapter "Description of the Physical Environment") that bracket the late Archaean and early Proterozoic. Apparent from Fig. 1 is the relative paucity of carbonate strata in South America, Australia and, indeed, the southern hemisphere. White and White (2009) do, however, describe the Bullita Cave system in the mid-Proterozoic (~1.6 Ga) dolomite that forms the Gregory Karst in the Northern Territory, Australia.

### 1.2 Karst and Acid Mine Drainage

There is comparatively little scientific material in the literature (at least in English publications) regarding karst resources that are impacted or threatened by acid mine drainage. The studies by Sasowsky and White (1993), Webb and Sasowsky (1994) and Sasowsky et al. (1995) describe the interaction between coal mine AMD and the Carboniferous Period karst formations (primarily limestone) of the Midwest Palaeozoic Carbonate Aquifer in the Obey River Basin in north-central Tennesee, USA. These studies reflect a situation where the acidity of mine water infiltrating a limestone aquifer survives a travel distance of ~8 km to where it exits the aquifer at a spring. The acid neutralisation results from dolomite dissolution and dissociation of carbonic acid ($H_2CO_3$) through $CO_2$ degassing (Sect. 8.12 in Chapter "Chemical Hydrogeology").

Observations made in the COH, where mine water has been infiltrating a portion of the karst aquifer since late-2002, have the potential to significantly advance the understanding of the interaction between acidic mine water and a receiving karst hydrosystem. The circumstances that describe this impact are further complicated by the simultaneous contribution of municipal wastewater discharge (albeit of lesser magnitude) into the same karst hydrosystem. This portion of the COH karst environment therefore represents a natural laboratory where the interaction between metal- and metalloid-compromised mine water discharge and bacteriologically and nutrient-compromised municipal wastewater effluent has occurred naturally and 'by default' in the subsurface over the last 1½ decades.

## 2 Overview of South African Karst Water Resources

### 2.1 Water Supply Aspect

Springs (also known locally as 'eyes') are as much a characteristic of karst environments as are dissolution features such as caves and sinkholes. In a country where the geology dictates that much of the hydrogeologic environment is characterised by generally low-yielding fractured and weathered-and-fractured secondary aquifers, it is no surprise that the high yielding cold springs drain karst aquifers.

Vegter (1995) lists 57 cold springs in South Africa yielding > 1 ML/d (>11.6 L/s). Most of these drain karst aquifers. Perhaps the best known example is the Kuruman Eye in the Northern Cape Province delivering ~18 ML/d (~208 L/s) (Martin 2006). The Tshwane Metropole in Gauteng Province obtains ~7% (22.7 $Mm^3$) of its total bulk input of 305 $Mm^3$/a (Lotter 2012) from dolomitic groundwater resources via springs and boreholes. Irrigated agriculture (excluding so-called 'improved grassland') in Gauteng Province alone covers 275 $km^2$, of which 95 $km^2$ (~35%) is located on dolomite. This testifies to the fact that portions of the karst landscape are admirably suited to large-scale irrigated agriculture, with the attendant threats of groundwater over-abstraction (Hobbs 2008a), pesticide/insecticide/herbicide contamination and nutrient-enriched agricultural return flow. An example in this regard is the situation that has developed in the Steenkoppies Basin immediately to the west of the COH, where unbridled expansion of irrigated agriculture based on the abstraction of karst groundwater has impacted adversely on the discharge of Maloney's Eye (Holland and Cobbing 2008; Holland et al. 2009), the source of the Magalies River.

It is not surprising, therefore, that karst aquifers have been the focus of detailed hydrogeologic studies aimed primarily at determining and utilising their water supply potential. This is exemplified by the groundwater investigations carried out in the 1980s by the then DWAF into the water supply potential of the carbonate strata in the Pretoria-Witwatersrand-Vereeniging (PWV or present-day Gauteng Province) area. These studies were precipitated by the prolonged drought experienced in the region from 1977 to 1986 (Roberts 1988), during which period the main source of water supply for the industrial heartland of the country, the Vaal Dam, emptied to only 7% of its capacity of ~2610 $Mm^3$. The karst aquifers were identified as an emergency source of potable water (Vegter 1988), and their potential exploitation for this purpose explored and evaluated on a large-scale commencing in 1983. These studies are documented in reports by, amongst others, Kok et al. (1985), Bredenkamp et al. (1986), Kafri et al. (1986), Leskiewicz (1986), Reynders (1988), Hobbs (1988a) and Kuhn (1989). At the time, Roberts (1988) put the cost of these studies to the State at R18 million.

In the period late-1983 to mid-1987, some 400 large diameter ($\geq$ 165 mm up to 381 mm) boreholes were sunk in the carbonate strata of the PWV area (Hobbs 1988b; Mulder 1988). Despite the successful establishment of numerous extremely high-yielding boreholes (many capable of yielding >25 L/s, and some even >100 L/s) in the course of these studies (Hobbs 1988b), and with the exception of the Tshwane Metro use mentioned earlier, exploitation was limited to land owners who later capitalised on these facilities for irrigated agricultural use and, in at least one instance near Olifantsfontein south of Pretoria, for municipal water supply purposes. Nevertheless, the legacy of these studies is two-fold, namely:

- The unprecedented deepening of the knowledge and understanding of the hydrophysical and hydrochemical characteristics of these groundwater resources; and
- The incorporation of many of the boreholes (297 in Gauteng Province alone, Table 1) in a network of groundwater monitoring stations that continues to generate invaluable hydrogeologic data some three decades later.

Prior to the above-mentioned studies, the dolomitic aquifers of the Far West Rand first attracted attention in a geotechnical and engineering geologic context (Sect. 2.2). These aquifers later drew the attention of Fleisher (1981) in a study of their hydrogeologic characteristics against the backdrop of their substantial dewatering by the deep gold mines in the Carletonville Goldfield (aka the Far Western Basin) (Sect. 2.2).

Further to the west, the karst aquifer of the Campbell Rand Subgroup on the Ghaap Plateau in north-central South Africa (Fig. 1 in Chapter "Introduction and Background") was mapped and described by Beukes (1978; 1987). Dewatering of this aquifer at Sishen Iron Ore Mine at ~36 ML/d on average (WS&E 2012) has resulted in water level drawdowns of >100 m in order to establish safe opencast mining conditions at the mid-2012 depth of 260 m below surface (bs), but raising concerns for the impact on small-scale groundwater users and the environment (van Dyk et al. 2007; WS&E 2012). In general, however, this groundwater resource has not received the same measure of attention as that of its eastern equivalent, the Malmani Subgroup.

The considerably younger (16–5 Ma) karst formations occurring mostly along the South African coastline (Fig. 3), comprise mainly limestone and calcarenites (calcareous sandstone). These formations make up the remaining 2% of carbonate strata. Many coastal towns and villages augment their surface water supplies with groundwater drawn from these strata, yet documentation of these instances remains fragmented and poorly quantified.

## 2.2 Geotechnical and Vulnerability Aspects

Studies of the geotechnical characteristics of the karst aquifers of the Far West Rand (Carletonville Goldfield) were precipitated by the severe and sometimes catastrophic sinkhole formation triggered by mine dewatering activities (Brink 1996). Most notable in this regard was the West Driefontein Mine sinkhole disaster of 12/12/1962. Drawdown in the Oberholzer Basin exceeded 300 m in the mid-1960s (Foose 1967) in the endeavour of West Driefontein Mine to 'make safe' its mine workings. Prior to this, the Venterspost Basin had been dewatered by the early-1960s when pumping rates peaked at ~ 57 ML/d (Brink 1996), and for which Swart et al. (2003) report a drawdown of 118 m in early-1974. Similar impacts were manifested in the Bank Basin in the early-1970s following

**Table 1** Geographic distribution of karst water level monitoring stations operated nationally by DWS (from Bertram 2008)

| Province | Geographic area | # of stations | | Karst hydrosystem |
|---|---|---|---|---|
| Gauteng | Centurion | 13 | 297 | Malmani Subgroup |
| | East Rand | 84 | | |
| | Kempton Park | 13 | | |
| | Natal Spruit | 67 | | |
| | Pretoria | 5 | | |
| | Rietvlei | 20 | | |
| | Tarlton | 37 | | |
| | West Rand | 58 | | |
| North West | Bo-Molopo | 122 | 191 | |
| | Dinokana | 16 | | |
| | Lichtenburg | 5 | | |
| | Polfontein | 11 | | |
| | Ventersdorp | 37 | | |
| Northern Cape | Danielskuil | 14 | 14 | Campbell Rand Subgroup |
| Total | | 502 | | |

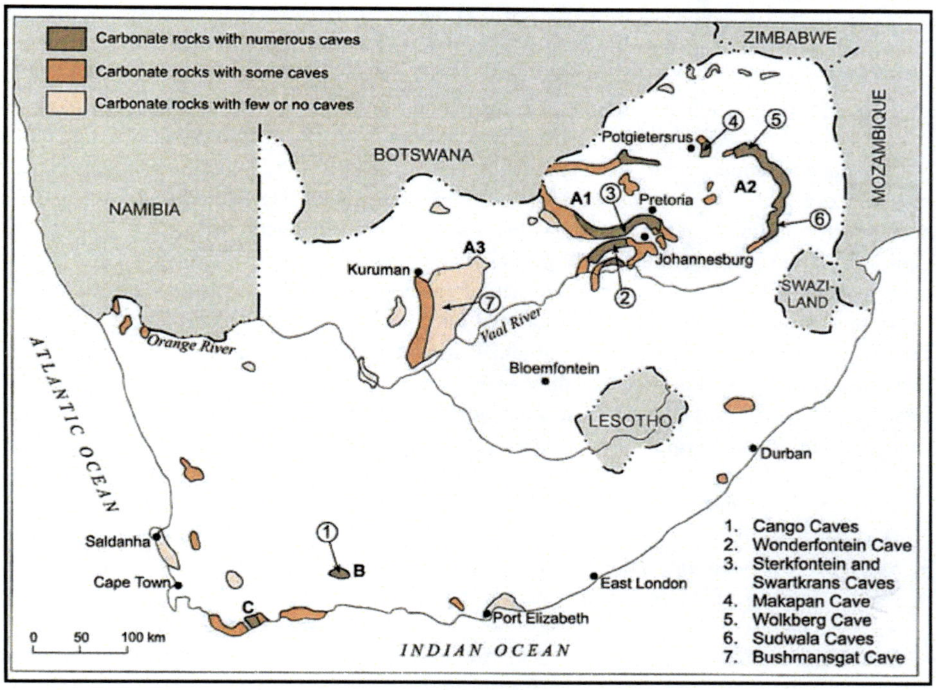


**Fig. 3** Distribution of carbonate strata in South Africa, showing the main areas of karst development as A (Transvaal Supergroup dolomite), B (Cango Caves Group limestone) and C (Bredasdorp Group limestone) (from Martini 2006)

dewatering at rates of up to 340 ML/d (Brink 1996). These studies helped build a sound understanding of the hydrogeotechnical and hydrogeologic characteristics of the karst formations that overlie the Ventersdorp Supergroup and gold-bearing Witwatersrand Supergroup strata. The voluminous data and writings generated in the course of these activities, and secured for posterity in the collection of Prof. EJ (Leslie) Stoch (deceased), has facilitated studies such as those by Schrader et al. (2014).

The vulnerability of karst aquifers to contamination is described extensively in the literature, and has resulted in the development of numerous mapping methods[4] to quantify their susceptibility to pollution, e.g. DRASTIC[5] (Aller et al. 1985), EPIK (Doerfliger et al. 1999), COP (Vías et al. 2003), PI (Goldscheider 2003), VULK (Cornaton et al. 2003), Localised European Approach (Dunne 2003), Time-Input Method (Kralik and Keimel 2003) and VUKA (Leyland et al. 2008). The VUKA method was developed for South African karst environments, and its application demonstrated in the COH (Leyland et al. 2008; Witthüser et al. 2010) and the Blyde River Canyon dolomite in Mpumalanga Province (Witthüser et al. 2010). An improvement on the VUKA method that incorporates a risk assessment as applied to the COH, is presented by Leyland (2010).

[4] Many as part of the COST Action 620 project (Zwahlen, 2003).
[5] Developed for groundwater resources in general, and not karst aquifers in particular.

## 3 Speleogenesis of Karst Aquifers

Following the earlier work by Choquette and Pray (1970), it has been recognised in the international karst literature that the development of karst systems progresses from syngenetic (eogenetic) forms through hypogenic (mesogenetic) forms to hypergenic (epigenetic) forms (Ford 2006; Ford and Williams 2007; Gunn 2004; Klimchouk et al. 2000; Klimchouk and Ford 2009; Palmer 2007; Dublyansky 2014; Klimchouk 2014). This 'evolution' is described by Klimchouk and Ford (2009) as follows.

- Eogenetic forms associated with coastal and oceanic sediments (primarily limestone) occurring in young rocks (Cenozoic Era) with high matrix porosity and permeability where caves form in the zone of mixing of marine and meteoric water.
- Hypogenic forms mainly associated with confined systems formed by burial of the eogenetic forms normally for very long periods of time, and where aggressive water enters the soluble formation from below driven by either solute or thermal density differences, creating cross-formational hydraulic continuity between laterally transmissive beds in heterogeneous soluble strata across cave-forming zones.
- Hypergenic forms associated with unconfined systems formed by geologic processes such as tectonic uplift and erosion (giving rise to so-called telogenetic karst) that

result in the gradual exhumation of hypogenic karst forms, and characterised by gravity-driven flow systems wherein downward meteoric water (rainwater) recharge drives a process such as carbonic acid dissolution further.

The listed sequence of karst development is readily recognised as correlating to the normal geologic, diagenetic and hydrogeologic sequence that describes the evolution of geologic strata (Klimchouk and Ford 2009). Equally recognisable is the potential for a later form of speleogenesis to inherit some measure of dissolutional structure and geometry from a preceding form. Ford and Williams (2007) recognise that the development of flow paths in karst aquifers depends on the availability and distribution of energy derived from the following:

- Throughput of water;
- Hydraulic gradient between recharge and discharge areas;
- Spatial distribution and nature of recharge; and
- Aggressiveness of the recharge water(s).

Bakalowicz (2006) draws attention to the different schools of thought represented by European and North American karst scientists (karstologists), respectively. Whereas the European view emphasised tectonic features, fractures and faults as the principal agents in karst development, the North American view assigned greater significance to bedding planes. These differing views are rooted in the different geological environments that characterise the carbonate strata of the respective continents. Compared to the relatively undisturbed nature of the North American strata, that of south-central and south-eastern Europe reflects severe deformation as a result of tectonic uplift. Perhaps the Neoarchaean karst strata of the South African interior represent a combination of the European and North American views on speleogenesis, varying from a bias toward the North American view in the plateau-type karst morphology, and toward the European view in the escarpment-type karst morphology (Sect. 3.4 in Chapter "Description of the Physical Environment").

## 3.1 Speleogenesis of South African Karst

### 3.1.1 Overview

An extensive review of international efforts has not revealed the assignment of South African karst systems to the above-mentioned forms of speleogenesis (karst development). Local opinion has followed the historical international trend of accepting the concepts and theories developed for unconfined karst systems (Klimchouk 2011), namely epigenic/hypergenic forms of speleogenesis, as characterising karst hydrologic and associated geomorphologic processes locally. The efficacy of these processes in the COH is discussed in Sect. 8.12.1 in Chapter "Chemical Hydrogeology". Klimchouk (2011) suggests, however, that Martini and Kavalieris (1976) and Martini et al. (2003) recognise a hypogenic form of speleogenesis in the so-called 'Transvaal Basin'. The following discussion explores the admittedly sparse evidence that provides hypothetical exploration of this question in a South African context. Although South African karst systems are addressed in broad terms, specific attention is afforded the COH.

Martini and Kavalieris (1976) recognise four periods of karstification in the ~2.5 Ga history of the Chuniespoort Group dolomite. These are the following:

- Pre-Pretoria Group period (>2.2 Ga), defined by the Rooihoogte Formation representing a palaeokarst residual breccia at the base of the Pretoria Group, and therefore developed on the weathered surface of the Chuniespoort Group carbonates;
- Pre-Waterberg Group period (>1.9 Ga), defined by dissolution cavities located at the unconformity with the overlying 1.9–1.5 Ga Waterberg Group;
- Pre-Karoo Supergroup period (>320 Ma), defined by palaeosinkholes filled with chert breccia, kaolinitic clay, carbonaceous sediments (including coal horizons) and, most tellingly, diamictite associated with the ~300 Ma Dwyka Group filling solution cavities in Transvaal Supergroup dolomite near Vereeniging in southern Gauteng Province (McCarthy and Rubidge 2005); and
- Tertiary to Recent Period (<65 Ma), defined by the present karst terrain formed after the removal of sufficient of the overlying Karoo Supergroup strata (following late Mesozoic uplift) allowed epigenic speleogenesis to sculpt the landscape, the geometry of preferential dissolution being informed (at least in part) by the overprint of regional WNW–ESE striking structural lineaments in the COH that cut through the NW dipping stratigraphic sequence of carbonate strata (Sect. 3.4 in Chapter "Description of the Physical Environment" and Dirks and Berger 2013).

### 3.1.2 Discussion

The distribution of carbonate (limestone and dolomite) strata in South Africa is shown in Fig. 3. This provides a basis for interrogating the speleogenesis of local karst systems in light of the forms described earlier in this section.

The 'young' Cenozoic shallow marine De Hoopvlei Formation (Bredasdorp Group) limestone deposits of middle- to late-Miocene age (16–5 Ma) along the southern Cape coast described by Malan (1990) readily represent a syngenetic/eogenetic system. Whereas caves that developed in the purer more indurated limestone are preserved, the softer more sandy limestone developed on marine platforms host only large depressions and shallow dolines (Martini 2006). Although currently of a terrestrial (land-based or continental) nature, their coastal setting accommodates a syngenetic form of speleogenesis associated with a coastline that in the last 120,000 years witnessed an ice-age related fall in sea level of ~130 m, recovering to its present position only in the last 10,000 years (McCarthy and Rubidge 2005; Roberts et al. 2006). Syngenetic/eogenetic speleogenesis in coastal settings, although created by unconfined circulation (i.e. epigenic karst speleogenesis), results in distinct porosity patterns because of the specific conditions associated with the mixing of different water chemistries at the halocline producing dissolution of porous, poorly indurated carbonates. It is for this reason that it is commonly distinguished as a separate category (Klimchouk 2011).

The much older 'hard' sedimentary carbonate deposits of Vaalian age (~2.5 Ga) in north-central and north-eastern South Africa undoubtedly host comparatively recent epigenic karst forms. Bearing in mind that the mapped distribution of carbonate strata represents their respective outcrop footprints, it is plausible that the much more extensive buried portions of these deposits might selectively host (even currently) the 'intermediate' hypogenic form of speleogenesis. There is therefore little reason to question the likelihood that the currently exposed and readily classified epigenic karst systems might earlier have represented a hypogenic form. The ramiform and spongework geometry that is a characteristic of hypogenic karst (Ford 2003; Klimchouk 2011) is evident in Sterkfontein Cave (Fig. 4), although Palmer (2011) has pointed out that maze caves may also result from hypergenic speleogenesis.

The hypogene mechanism of speleogenesis invokes the existence of a confined system within which the upward migration of 'aggressive' solutions results in selective dissolution of the carbonate strata and the formation of a significant secondary porosity. Klimchouk (2011) recognises thermal $H_2S$ solutions as a prime candidate for dissolution agent, which requires that sufficient sources of dissolved sulphate exist for reduction to $H_2S$ and, further, that the $H_2S$ can escape from its reducing environment. When rising $H_2S$-rich waters mix with infiltrated oxygenated $CO_2$-rich meteoric groundwaters at shallower depths, sulphuric acid dissolution manifests as a strong speleogenetic agent (Klimchouk 2011; Kirkland 2014).

The presence of palaeokarst features infilled with Karoo Supergroup sediments in the present karst landscape (and which defines the pre-Karoo period of karstification recognised by Martini and Kavalieris 1976) indicates the confining role played by the late-Palaeozoic Permian (298–250 Ma) and Mesozoic (250–65 Ma) strata prior to its removal by peneplanation following tectonic uplift, culminating by late-Mesozoic (Cretaceous) to early-Cenozoic (Tertiary) times (~65 Ma) in the African surface after the removal of >1200 m of material (Partridge and Maud 1987). Erosion in the post-African cycles subsequently exposed the pre-Karoo surface to the extent that it forms an important component of the present landscape in the southern and south-western portions of the Gauteng and North-West provinces. Dirks and Berger (2013) report a revised[6] erosion rate of 5–6 m/Ma for the plateau representing the African Erosion Surface south of the Malapa fossil site, and a ~ tenfold greater rate of ~55 m/Ma in the Skeerpoort River valley to the WNW of this site. The discussion presented in Sect. 8.12.1 in Chapter "Chemical Hydrogeology" suggests that solutional denudation in the COH averages 9.5 m/Ma in the range 5–14 m/Ma (see also Hobbs and Smith 2018). Klimchouk (2011) suggests that the hypogene speleogenesis of the Sterkfontein Cave system might have commenced in the early Miocene (18 Ma BP), but does not offer a potential source of 'aggressive' solutions that might serve as speleogenetic agent. The following discussion explores the possibility that hypogene speleogenesis of the Malmani Subgroup dolomite might date much further back to the early palaeo-Proterozoic (~2 Ga BP). It also offers potential sources of aggressive solutions from older Randian and Vaalian formations.

The carbonate strata of the Chuniespoort Group in the north-central and north-eastern parts of the country were formed in a marine environment developed on the stable Kaapvaal Craton (Eriksson and Reczko 1995) already carrying the 2.97–2.78 Ga Witwatersrand Supergroup and 2.78–2.65 Ga Ventersdorp Supergroup strata. By 2.7 Ga, these strata had been subjected to regional faulting and fracturing associated with the emplacement of the basaltic lavas of the Klipriviersberg Group (McCarthy and Rubidge 2005). Subsequent to the complete development of the Transvaal Supergroup by ~2.05 Ga, further structural disturbance of the accumulated Randian and Vaalian strata on the Kaapvaal Craton came in the form of:

- The Bushveld Igneous Complex (BIC) intrusion at ~2.05 Ga (Cawthorn et al. 2006) coeval with;
- Deformation resulting from a north-east verging compressive event which reactivated basement structures that were propagated through the Black Reef–Malmani–Rooihoogte succession (McCarthy et al. 1986; Obbes 2001); and
- The Vredefort meteorite impact event at ~2.02 Ga (Reimold 2006).

[6] Dirks et al. (2010) report an erosion rate of 3 to 5 m/Ma for the same landscape (see Sect. 8.12.1 in Chapter "Chemical Hydrogeology").

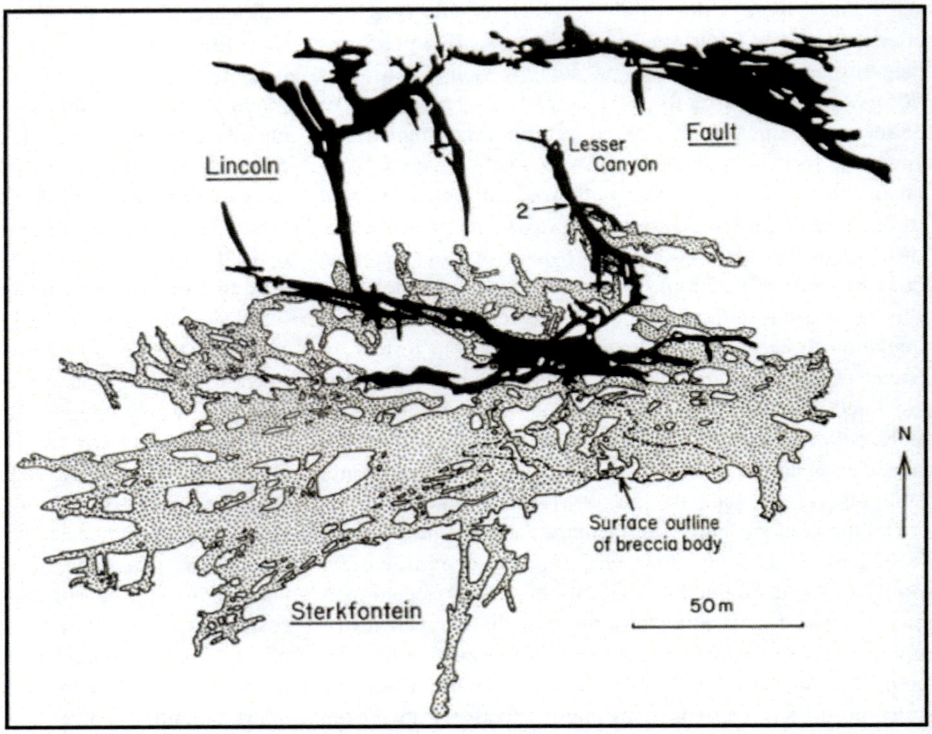


**Fig. 4** Map of the Sterkfontein Cave system showing the labyrinthic joint-controlled pattern of both the Sterkfontein and Lincoln cave sections (from Martini et al. 2003)

The geological setting of the Witwatersrand Basin as repository for 'the greatest of the world's gold fields' (John Hays Hammond as quoted in Pretorius 1986) has been intensively studied and hence well defined (see Robb and Robb 1998a and McCarthy 2006 and references therein). Duane et al. (1997) discuss the geochemistry of some deep gold mine waters (12 samples collected at a depth of 1050–3400 m bs in four mines[7]) from the Witwatersrand Basin, reporting the occasional encounter at depth of hot water (up to 40 °C). Zhao et al. (1994) note the enrichment of the Ventersdorp Contact Reef (VCR) with base metal sulphides such as pyrite ($FeS_2$), pyrrhotite [$Fe_{(1-x)}S$ (x = 0 to 0.2), e.g. FeS = troilite], galena (PbS), sphalerite (ZnS) and chalcopyrite ($CuFeS_2$). In a study of lead (Pb) isotopes associated with chalcopyrite and pyrrhotite in the VCR, Zartman and Frimmel (1999) note temperatures of up to 300 °C corresponding to the peak regional metamorphic temperature associated with burial of the VCR beneath 6500 m of Ventersdorp and Transvaal Supergroup strata.

The VCR separates the Witwatersrand clastic sediments from the overlying Ventersdorp volcanics (lavas). The carbonate formations of the Malmani Subgroup, sandwiched between the underlying Witwatersrand and Ventersdorp strata and the overlying Pretoria Group sediments (Table 5), could therefore not have escaped the intrusion of hot $H_2S$-rich connate groundwater rising from the still comparatively young buried bedrock strata. The invariable presence of carbon-rich material in the form of kerogen (bitumen) in association with the gold-bearing conglomerates (McCarthy and Rubidge 2005) is of further interest. Representing a form of hydrocarbon typically associated with $CO_2$-rich waters, this provides another source of thermal acidic fluid conducive to hypogene speleogenesis. Migration pathways for the hydrothermal brine might have taken the form of fractures, fissures and faults created by the afore-mentioned BIC intrusive and Vredefort impact events. Such circumstances clearly favour hypogenic speleogenesis of the carbonate formations by any of the mechanisms described in Text Box 1 (p. 45) long before their exhumation in the more recent geologic past. The mechanism of artesian transverse speleogenesis as identified by Klimchouk (2006) is therefore readily invoked in regard to a confined Malmani Subgroup karst hydrosystem both in the geologic past and, under specific conditions such as might be associated with the deeply buried dolomitic strata of the Chuniespoort Group basin, perhaps even at present. In this regard it is worth noting that numerous thermal springs in the southern Waterberg area of Limpopo Province, e.g. Warmbaths Spring at Bela-Bela (52 °C), and Die Oog (39 °C) and Rhemardo Spring (42 °C) at Mookophong (Olivier et al. 2008). These springs occur in the central part of the Bushveld Complex where sandstone of the Swaershoek

[7] Western Deep Levels, Vaal Reefs, Freddies and President Steyn.

Formation at the base of the Waterberg Group rests unconformably on felsites of the Rooiberg Group, and locally also on granites of the Lebowa Granite Suite (Bushveld Complex). Updomed fragments of Bushveld Complex floor rocks (dolomite) also occur in this area (Visser 1989) where structural disturbance is evidenced by regional faults and diabase dyke intrusions with which the springs are variously associated (Olivier et al. 2008).

Attention is also drawn to the recognition by Wilkinson (1973) of 'dome cavities' developed in breccia capped by a resistant travertine carapace (Fig. 5) in Sterkfontein Cave. Wilkinson (1973) attributes these features to current-induced erosion by water flowing through the cave system, the protrusion of fragile bones from the breccia surface suggesting gentle flow conditions. Palmer (2003) reports the enlargement of ceiling solution pockets by turbulent eddies (Slabe 1995 in Palmer 2003) or by elevated $CO_2$ partial pressure resulting from the compression of air by rising water (Lismonde 2000 in Palmer 2003). Such agents would occur in epiphreatic conditions during flooding of the cave system. Nevertheless, the Sterkfontein 'dome cavities' are reminiscent of the ceiling cupolas recognised by Klimchouk (2011) as a component in a suite of morphologic features created by transverse rising flow (aka 'morphologic suite of rising flow' or MSRF) associated with hypogene speleogenesis. The age (<3 Ma) of the breccia, however, is too young for the 'dome cavities' to represent ceiling cupolas formed by upward flow in a confined karst aquifer. Certainly closer inspection of the Sterkfontein Cave and other cave systems in the area for features such as described above is warranted.

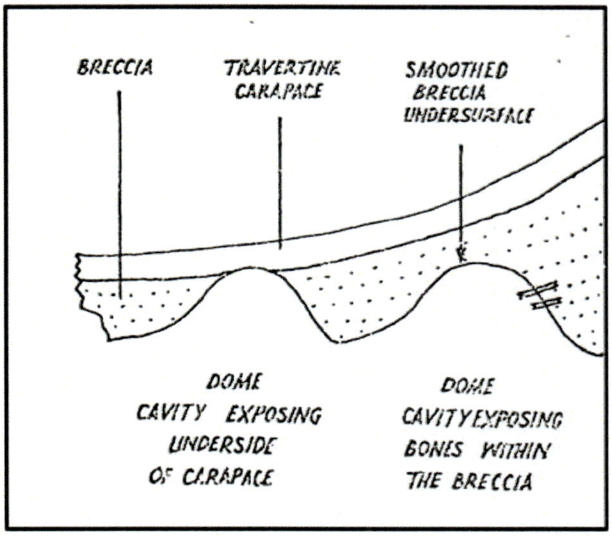


**Fig. 5** Sketch of 'dome cavities' in the Mound breccia of the Milner deposit beneath a protecting travertine layer in Sterkfontein Cave (from Wilkinson 1976)

## 3.2 The Epikarst

### 3.2.1 General Discussion

Epikarst is recognised as an integral component of the karst landscape by Williams (1983), who referred to this zone as the subcutaneous layer or 'skin' that overlies karst formations at the surface. Its significance derives from specific functions together serving as a regulatory subsystem at the top of the unsaturated zone within which autogenic recharge is stored, segregated into several components and spatially distributed to preferential ingress paths developed in the underlying vadose zone. In classical terms, stress release and weathering are the principal agents of formation of the incipient epikarst zone, and dissolution plays only a minor role. The processes of stress release and weathering (both mechanical and chemical) are common to virtually all rock types (including insoluble) exposed at surface. The regolith (weathered mantle) is formed as a consequence of the extension and opening of existing joints and formation of new joints, the accentuation and opening of bedding planes and micro-fissures, and the enhancement of fissure frequency and connectivity of fissure networks. Chemical weathering penetrates deeper into the bedrock, the resulting mineralogical alteration of the parent strata producing fines that may 'choke' the regolith and decrease its permeability. Controlling factors in this regard are the composition and structure of host rocks, climate, rate of unloading informed by the competition between uplift and denudation rates, and topography (Klimchouk 2004).

Klimchouk (2004) distinguishes between 'classical' weathering as described above, and karstification represented by the dissolutional removal of material accumulated in the regolith developed over carbonate bedrock. A key element in karstification is the development of greater porosity and an increasingly organised permeability focussed on vertical drains in the underlying vadose zone. The evolution of these properties progresses from an incipient epikarst on a freshly exposed carbonate surface, through a young epikarst to a mature soil-covered epikarst, and culminating in a barren old relict epikarst surface.

### 3.2.2 Epikarst of the COH

The epikarst of the COH represents primarily a soil-covered mature epikarst landscape characterised by dolines representing dispersed but evolving karstification favouring the concentration of autogenic recharge. This is illustrated in Plate 1, from which it is evident that denudation of the carbonate surface is an active process in the current landscape. The presence of soil enhances solutional enlargement of fissures in the epikarst by providing a ready source of $CO_2$. The landscape of the COH also hosts at least 412 sinkholes and dolines that have been classified as 'swallow'

**Plate 1** Examples of the epikarst in the COH showing **a** dolomite pillars intersected by cutters (dissolution 'slots') filled with soil and covered by a variable thickness of soil cover, **b** ~ 4-m thick epikarst horizon overlying the gently folded strata forming the vadose zone exposed in the southern sidewall of Sterkfontein Quarry, with detail of the exposed features illustrated in (Plate 2), **c** chert-rich dolomite outcrop in the form of a quasi-karren field, **d** chert-rich bedded and dipping dolomite outcrop, **e** dolomite outcrop with soil cover and infill at the Zwartkrans fossil site, and **f** exposed epikarst at the Gladysvale fossil site (all photos P Hobbs; date shown in upper right corner)

features after field verification of 734 suspected such features (R Leyland, personal communication). The question arises whether these features are also associated with epikarst evolution and therefore form part of the epikarst, or not. Klimchouk (2004) suggests that features such as point recharge dolines and collapse sinkholes are unrelated to epikarst morphogenesis, and simply represent holes in the epikarst. The author is of the opinion that these features are an intrinsic component of epikarst evolution.

As recognised by Klimchouk (2004), the epikarst and vadose zone fulfills crucial hydrologic functions and roles also in the regional karst system in the COH. Principal of these is the retardation of throughflow and mixing of recharge in its often lengthy and tortuous pathway to the aquifer. These circumstances account for the relatively constant flows—Martini (2006) refers to 'very regular discharges at resurgences'—that characterise springs draining the Neo-Proterozoic carbonate strata of the South African interior, compared to the 'flashy' nature of springs that drain the telogenetic and much younger European karst formations (Florea and Vacher 2006). It is posited that the local karst terrane, despite the absolute age of its carbonate strata, is in transition from a juvenile to a mature stage of evolution. The epikarst of the COH deserves considerably more study than it has received in the past. Unravelling the typology of this epikarst will contribute to the global understanding of karst morphogenesis and the transport of carbon through the vadose zone (Jones 2013).

**Text Box 1** Hypogenic speleogenesis

*Carbonic acid dissolution, which dominates overwhelmingly in epigenic carbonate speleogenesis, also operates as a hypogenic agent, though the origin of the acidity is different. It can be related to $CO_2$ generated from igneous processes, by thermometamorphism of carbonates, or by thermal degradation and oxidation of deep-seated organic compounds by mineral oxidants. The latter is common in hydrocarbon fields, where waters characteristically contain high $CO_2$ concentrations (Kaveev 1963). Hydrogen sulfide is another common hypogenic source of acidity where there are sufficient sources of dissolved sulfate for reduction and where the $H_2S$ generated can escape from the reducing zones. However, it is generally believed that creation of significant caves, where these acids provide the dissolution mechanism, depends mainly upon rejuvenation of aggressiveness by mixing or cooling. These conditions are commonly met in the ascending limbs of intermediate or regional flow systems, when they interact with shallower flow systems. Aggressiveness in hypogenic speleogenesis may also reflect acquisition of new sources of acid within the soluble formation itself, when groundwater flows transversely across it, or can be due to a number of* *mechanisms that rejuvenate the dissolutional capacity of fluids, such as the cooling of water, mixing of groundwaters of contrasting chemistry, and sulfate reduction and dedolomitization as mentioned above. In shallower conditions, where $H_2S$-bearing waters rise to interact with oxygenated meteoric groundwaters, sulfuric acid dissolution can be a very strong speleogenetic agent. It is recognised as the main speleogenetic process for certain large caves (e.g. caves such as the Carlsbad Caverns in the Guadalupe reef complex in New Mexico, south-western USA, and Frasassi Cave in Italy) and many smaller caves. Based on this, some researchers distinguish sulfuric acid karst/speleogenesis as a peculiar type (Hill 1996; 2000a; 2003a). Substantial sulfuric acid dissolution can also be caused by oxidation of iron sulfides such as pyrite and marcasite, where it is localised in ore bodies (Bottrell et al. 2000). Lowe (1992) suggested that oxidation of pyrite along certain horizons or bedding planes in carbonates ("inception horizons") may create preferential flowpaths that later will be inherited by epigenic speleogenesis. The increased solubility of calcite in cooling waters can cause dissolution along ascending flowpaths. The latter mechanism is commonly labeled as hydrothermal speleogenesis, occurring in high-gradient zones where ascending flow is localised along some highly permeable paths (Malinin 1979; Dublyansky 1980; Dubljansky Ju 1990, 2000a; Ford and Williams 1989; Palmer 1991; Andre and Rajaram 2005). Solutional aggressiveness can be renewed or enhanced by mixing of waters of contrasting chemistry and dissolved gas content (Laptev 1939; Bogli 1964; Palmer 1991), the effect widely cited in the karst literature as "mixing corrosion".*

*There are an increasing number of arguments and evidence suggesting that more than one process could be involved in many cases, operating either in combination or sequentially. Cross-formation flow is the main mechanism for hypogenic speleogenesis, which can integrate and trigger many cave-generating processes. Either carbonic acid or $H_2S$ dissolution can operate in hydrothermal systems, which are essentially ascending transverse phenomena common in many*

**Plate 2** Detail of epikarst and vadose zone features exposed in the southern sidewall of Sterkfontein Quarry; note that the lower image is contiguous with the right side of the upper image along the common chevron line

*basins. Mixing of contrasting waters is also commonly involved in hypogenic speleogenesis, at least at some stages. This, again, suggests that labeling types of karst and speleogenesis by a single dissolutional mechanism is misleading and should be avoided.*

From Klimchouk (2007).
See BIBLIOGRAPHY for references herein.

Tufa terraces across the channel of the Grootvlei Spruit ~600 m downstream of the Nouklip Spring

# Physical Hydrology

Harrison Pienaar and Philip J. Hobbs

## 1 Surface Water Drainage

The study area encompasses the Quaternary catchments and drainages identified in Table 1 in Chapter "Description of the Physical Environment", Sect. 3.1 in Chapter "Description of the Physical Environment". The water resources of the subregion are sparsely described in the literature, their extent militating against the regional scale at which studies are typically carried out, i.e. the effort expended in analysing the runoff characteristics of subcatchments such as straddle the study area is often not warranted. It is most often sufficient to simply describe the hydrology in terms of mean annual runoff (MAR) and catchment size. The MAR is reported either as a volume typically with units $Mm^3/a$, or as an equivalent depth of precipitation in millimetres (mm) over the catchment footprint area.

The study area spans the south-western quadrant of the Hartbeespoort Dam catchment. The net capacity of the dam, equivalent to its full supply capacity (FSC), is slightly less than its gross capacity of 195.1 $Mm^3$, which approximates the combined MAR of the contributing catchments. Middleton and Bailey (2008b) report an observed long-term MAR of 193.9 $Mm^3$ and a simulated MAR of 185.4 $Mm^3$ for Hartbeespoort Dam. Given this range of reservoir capacities, a nominal value of ~190 $Mm^3$ is used in this study to represent the FSC of the dam. It is a regionally significant reservoir for various reasons, including the following (in no specific order of importance):

- the attenuation of floods generated by rainfall on the highly urbanised Witwatersrand ridge that forms the continental divide which defines the southern boundary of the catchment;
- the provision of water for municipal water supply to towns and villages located downstream of the dam;
- the provision of water for irrigated agriculture, historically via a canal system, to riparian properties downstream of the dam;
- the support of a burgeoning property industry that has seen the expansion of residential areas and the proliferation of up-market security estates on the shores of the dam; and
- leisure and tourist activities.

Runoff in the study area is gauged at the four stations labelled A2H033, A2H034, A2H049 and A2H050 in Fig. 2 in Chapter "Description of the Physical Environment". A simplistic 'stick' diagram (Fig. 1) provides more readily identifiable components of the regional surface water drainage pattern and associated flow gauging stations.

Whereas the adjoining Steenkoppies Basin to the west-southwest is naturally drained by a single high-yielding spring,[1] the study area is drained by numerous and distributed such features. These feed either the Skeerpoort River or the Bloubank Spruit system. A more complete description of the discharge characteristics of these drainage systems is presented in the relevant sections that follow. Although the contribution of groundwater resources to the surface water hydrology of the study area is mentioned in the following sections (where relevant), the inseparable nature of the surface and subsurface environments that characterise the largely

P. J. Hobbs: Deceased

H. Pienaar (✉) · P. J. Hobbs
Smart Places, Council for Scientific and Industrial Research, Pretoria, South Africa
e-mail: hpienaar@csir.co.za

H. Pienaar
Hebei University of Engineering, Handan, China

[1] Maloney's Eye, with a long-term mean annual discharge of ~14.4 $Mm^3$ (~456 L/s), reduced to ~9.8 $Mm^3/a$ (~311 L/s) in the first decade of the 21st Century (Holland et al., 2009).



P. J. Hobbs et al. (eds.), *Anatomy of a South African Karst Hydrosystem*, Cave and Karst Systems of the World,
https://doi.org/10.1007/978-3-030-95829-9_5

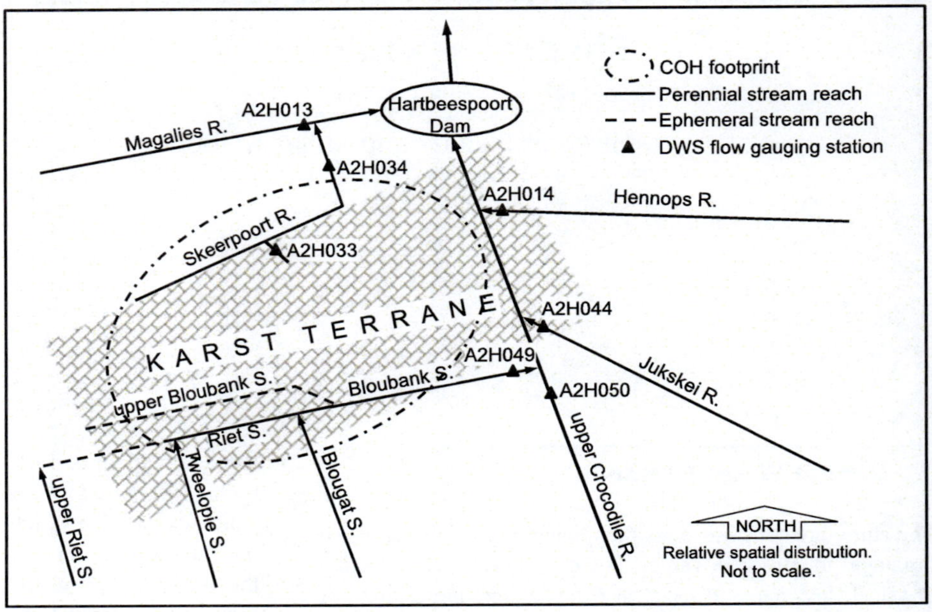


**Fig. 1** Schematic diagram of surface drainage and gauging network superimposed on the COH karst footprint; karst strata (not shown) extend to the west and east (see Fig. 1 in Chapter "Introduction and Background").

karst terrane of the study area is described in greater detail in Sect. 7.2 in Chapter "Physical Hydrogeology".

## 2 Skeerpoort River System

DWS flow gauging station A2H033 provides a ~50-year discharge record of the Grootvlei Spruit, a tributary of the Skeerpoort River, just before its confluence with the latter (Fig. 1). It yields the monthly discharge statistics presented in Table 1 and for which the annual discharge per hydrological year is shown in Fig. 2. The autumn month of March reflects the lowest coefficient of variation (CoV) of ~43%, and September the lowest monthly median discharge of 0.284 $Mm^3$ (~110 L/s).

Driven by the perennial Nouklip Spring discharge, the instantaneous flow pattern at station A2H033 up to September 2015 (the DWS record only extends up to November 2015) is shown in Fig. 3. This reveals a distinct step-like increase from ~0.05 to ~0.12 $m^3/s$ (~50 to ~120 L/s) in the 'base' values ca. October 1996. This step is also observed in the A2H034 record (Fig. 8 in Chapter "Physical Hydrology"). Although approached for an explanation, the DWS has not resolved this anomaly. Also evident in Fig. 3 are distinct recession curves following peak discharge events, for example ca. February 1978 and May 2000. Hydrograph segments of 260 days in 1999 (spanning the 1999 winter season) and of 220 days in 2005 (also spanning the winter season) were analysed graphically by Holland (2007), from which total potential groundwater discharges of 5.09 and 5.0 $Mm^3$, respectively, were estimated. These results prompted Holland (2007) to estimate, by extrapolation, a groundwater contribution to baseflow in 2005 of 8.3 $Mm^3$. This value significantly exceeds the 1997 to 2015 median discharge of ~5.4 $Mm^3/a_h$ [~170 L/s (text box, Fig. 2]. It also exceeds the maximum discharge value of ~7.5 $Mm^3/a_h$ [~238 L/s (text box Fig. 2] in the year 2000 of this period.

The Nouklip Spring synoptic discharge measurements (SDMs) of ~173 L/s in May 2010 and ~211 L/s in August 2014 following very wet summers (Fig. 3 in Chapter "Description of the Physical Environment"), equate to flows of ~5.4 and ~6.7 $Mm^3/a_h$ respectively. The former is in good agreement with the median discharge in the period 1997–2015 reported above, and exceeds by <10% the shorter segment values obtained by Holland (2007). This suggests that aggregating segmental recession-based calculations of groundwater contributions to an annual value is invalid in this instance.

An inspection of the data presented in Table 1 provides further insight into the Nouklip Spring discharge. Station A2H033 (Plate 1) is located ~525 m downstream of the spring, and therefore provides a measure of the springflow regime when very high flows dominated by storm runoff from the catchment above the spring are neglected. It is one of only eight DWS gauging stations in the country that record the discharge of a karst spring up to as recently as 2017. High yielding municipal water supply sources such as the Tshwane Metro's Pretoria Fountains, Grootfontein Spring and Sterkfontein Spring are gauged by the local authority. The long-term (whole record) median monthly flow ranges from 165 L/s in April (end of the wet summer

**Table 1** Statistical analysis of Grootvlei Spruit mean monthly discharge at station A2H033 in the period October 1964 to September 2015 (no data available since 13/12/2015)

| Variable | Month | | | | | | | | | | | |
|---|---|---|---|---|---|---|---|---|---|---|---|---|
| | Oct | Nov | Dec | Jan | Feb | Mar | Apr | May | Jun | Jul | Aug | Sep |
| n | 47 | 46 | 48 | 44 | 43 | 42 | 44 | 44 | 47 | 46 | 48 | 48 |
| Minimum | 0.160 | 0.159 | 0.165 | 0.166 | 0.142 | 0.162 | 0.155 | 0.160 | 0.150 | 0.151 | 0.150 | 0.145 |
| 5%ile | 0.170 | 0.166 | 0.178 | 0.185 | 0.162 | 0.179 | 0.194 | 0.176 | 0.161 | 0.165 | 0.166 | 0.160 |
| Mean | 0.382 | 0.400 | 0.409 | 0.431 | 0.410 | 0.411 | 0.411 | 0.404 | 0.366 | 0.372 | 0.358 | 0.321 |
| Median | 0.345 | 0.370 | 0.402 | 0.400 | 0.384 | 0.424 | 0.428 | 0.407 | 0.358 | 0.358 | 0.340 | 0.284 |
| 95%ile | 0.706 | 0.904 | 0.733 | 0.789 | 0.755 | 0.722 | 0.731 | 0.725 | 0.695 | 0.695 | 0.675 | 0.596 |
| Maximum | 1.642 | 1.756 | 1.162 | 1.02 | 1.118 | 0.822 | 0.816 | 0.881 | 0.829 | 0.891 | 0.821 | 0.692 |
| SD | 0.249 | 0.286 | 0.203 | 0.208 | 0.212 | 0.175 | 0.177 | 0.187 | 0.179 | 0.180 | 0.170 | 0.144 |
| CoV (%) | 65.1 | 71.6 | 49.5 | 48.2 | 51.7 | 42.6 | 43.0 | 46.2 | 49.0 | 48.4 | 47.6 | 44.8 |

All units are $Mm^3$ unless otherwise indicated. Analysis excludes months with missing and station rating exceedance data, but includes unaudited (recent) and estimated data

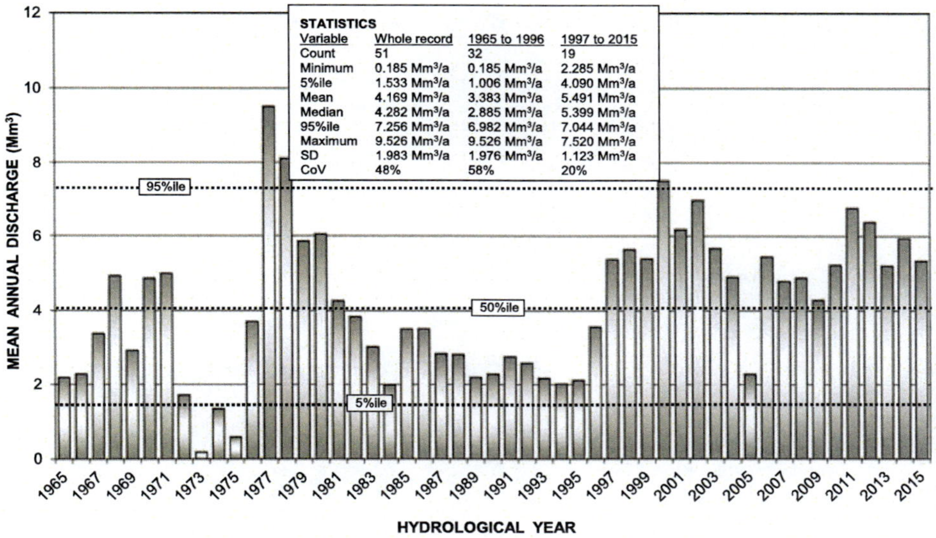


**Fig. 2** Pattern and trend of Grootvlei Spruit mean annual ($a_h$) discharge gauged at station A2H033 in the period October 1964 to September 2015

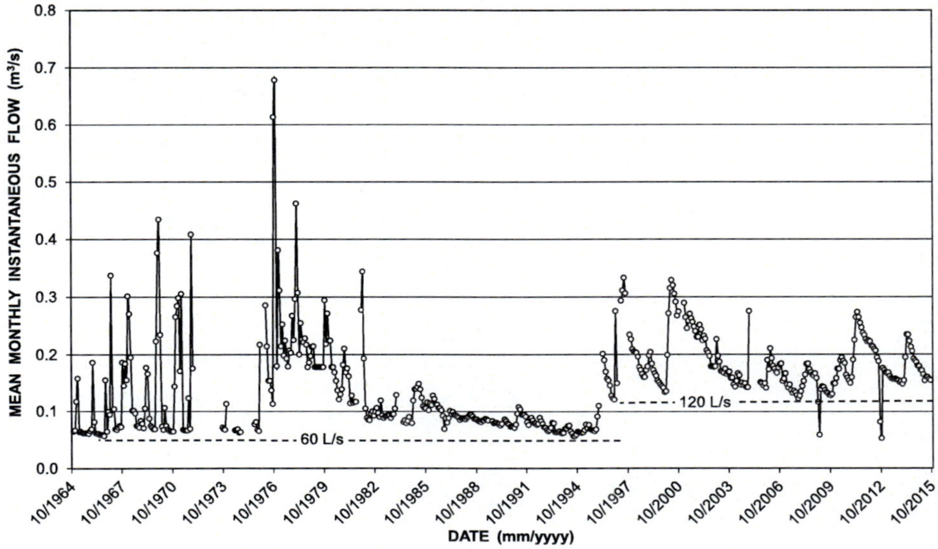


**Fig. 3** Long-term monthly hydrograph of the Grootvlei Spruit at station A2H033 for the period October 1964 to September 2015

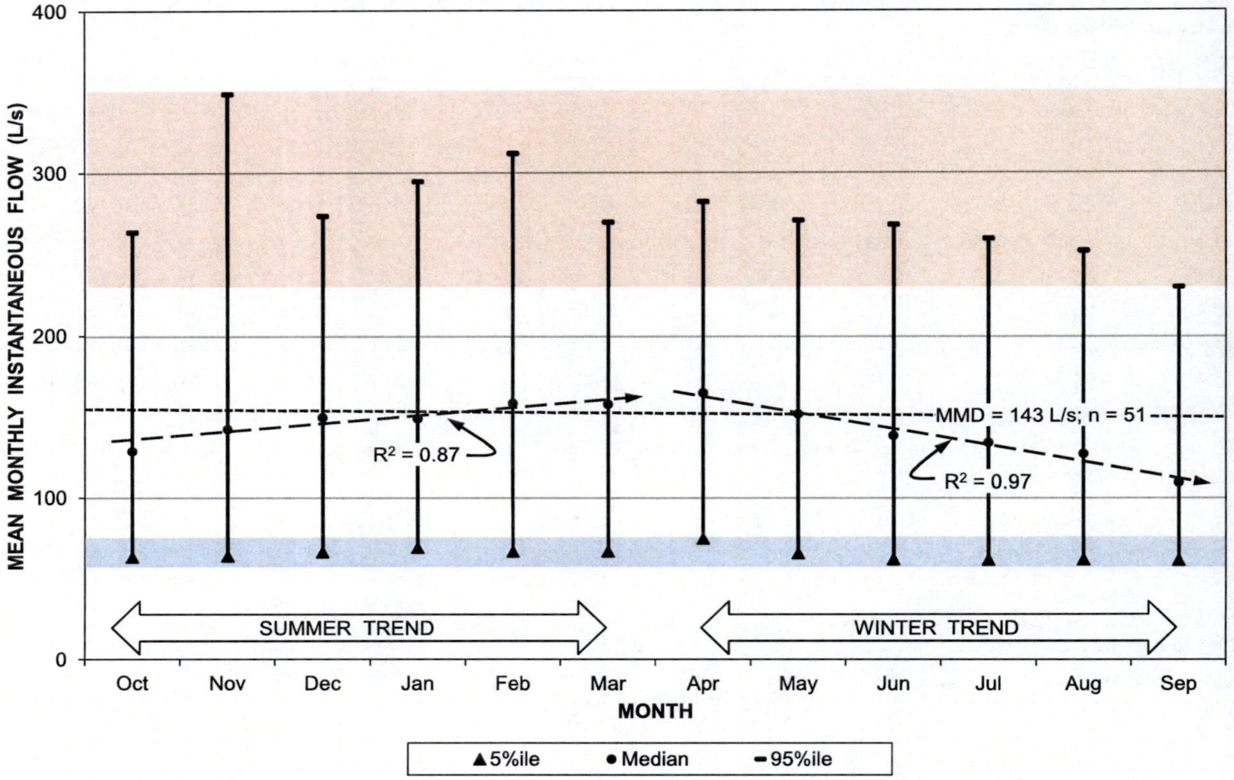


**Fig. 4** Long-term monthly discharge pattern of the Nouklip Spring, showing median values with 5%ile and 95%ile bounds and summer and winter trends as linear regression traces

**Plate 1** View from the right bank of the combination V-notch/rectangular weir flow gauging station A2H033 on the Grootvlei Spruit, showing the stage recorder housing (top centre) on the opposite bank

season) to 109 L/s at the end of winter in September (Fig. 4), bracketing mean and median annual values of 143 and 146 L/s, respectively. The 5%ile value occupies the comparatively narrow range 62–75 L/s, and the 95%ile value the broader range 230–349 L/s. The flow record indicates that a 2σ deviation at the 95%ile level exceeds the long-term mean monthly discharge (MMD) by ~40% to ~111%. These circumstances illustrate the Martini (2006) statement regarding the 'very regular discharges' that characterise karst springs of the Malmani Subgroup.

A similar analysis for the period 1997–2015 yields a median monthly flow ranging from 159 L/s in November to 187 L/s in March and May (Fig. 5), bracketing mean and median monthly values of 171 L/s. The 5%ile values occupy

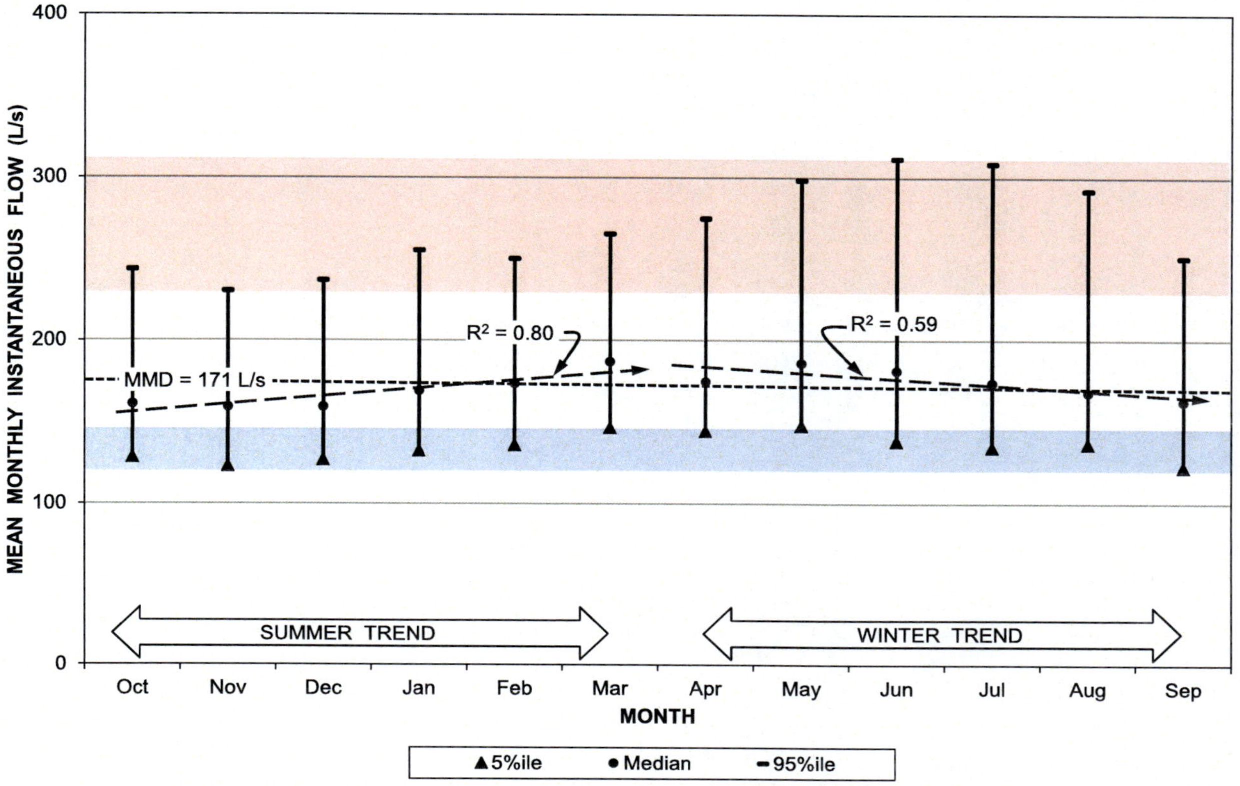


**Fig. 5** Monthly discharge pattern of the Nouklip Spring in the period 1997–2015, showing median values with 5%ile and 95%ile bounds and summer and winter trends as linear regression traces

the narrow range 123–147 /s, and the 95%ile values the range 230–309 L/s. The flow record indicates that a 2$\sigma$ deviation at the 95%ile level exceeds the mean monthly discharge (MMD) by ~35% to ~81%. The differences with the whole record analysis are apparent. The most interesting of these is the difference in the 95%ile pattern which, in Fig. 5, shows increasingly greater excursions from late summer through to late winter, peaking in June and July. As this pattern cannot be due to flood-driven runoff from high rainfall, it must reflect the hydraulic response time of the Diepkloof Basin (Sect. 7.2.7 in Chapter "Physical Hydrogeology"to groundwater recharge in the preceding wet season(s) (Table 2).

The 95%ile 'peaks' in June and July suggest a hydraulic response lag time of at least three months for the Diepkloof Basin karst aquifer. The mainly springflow-driven median values also reveal similar peaks of ~186 L/s in March and May. It is also evident that the decline in median discharge through the winter season continues into November and December with flows of 159 L/s. An aspect of the flow pattern that is not evident from the preceding discussion, is the fact that the spring discharge includes the autogenic recharge contribution of the hydrogeologically separated Tweefontein Spring (Sect. 7.4.6 in Chapter "Physical Hydrogeology"This was measured at ~30 L/s (~0.95 Mm$^3$/a) in May 2010, and ~26 L/s (~0.82 Mm$^3$/a) in August 2014. Clearly, this contribution would be included in the recession-based calculation of groundwater contribution to surface discharge as performed by Holland (2007), and would therefore not explicitly recognise the hydrogeologically separated Diepkloof and Tweefontein basins drained by the Nouklip and Tweefontein springs respectively. Against this background, the February 2018 SDM value of 105 L/s is inexplicably low, although Table 2 shows a minimum value of 0.142 Mm$^3$ (59 L/s) for February. A more detailed analysis of the springflow characteristics is presented in Sect. 7.4.7 in Chapter "Physical Hydrogeology".

It should also be noted that at its confluence with the Grootvlei Spruit, the Skeerpoort River already carries the discharge of the Nash Spring (Sect. 7.4.8 in Chapter "Physical Hydrogeology"). This spring is located a distance of ~4.9 km (measured along the sinuous river reach) upstream of this confluence. Its discharge was measured at ~130 L/s (~4.1 Mm$^3$/a) in May 2010, and ~208 L/s (~6.6 Mm$^3$/a) in August 2014. Together with the preceding discussion, these circumstances further inform the discharge

**Table 2** Statistical analysis of Grootvlei Spruit mean monthly discharge at station A2H033 in the period October 1997 to September 2015

| Variable | Month | | | | | | | | | | | |
|---|---|---|---|---|---|---|---|---|---|---|---|---|
| | Oct | Nov | Dec | Jan | Feb | Mar | Apr | May | Jun | Jul | Aug | Sep |
| n | 19 | 19 | 18 | 17 | 18 | 17 | 17 | 18 | 18 | 18 | 19 | 18 |
| Minimum | 0.342 | 0.318 | 0.324 | 0.313 | 0.142 | 0.374 | 0.368 | 0.384 | 0.357 | 0.356 | 0.355 | 0.212 |
| 5%ile | 0.344 | 0.319 | 0.339 | 0.355 | 0.328 | 0.392 | 0.374 | 0.395 | 0.358 | 0.359 | 0.365 | 0.317 |
| Mean | 0.459 | 0.434 | 0.463 | 0.485 | 0.438 | 0.518 | 0.513 | 0.547 | 0.519 | 0.525 | 0.504 | 0.441 |
| Median | 0.431 | 0.412 | 0.426 | 0.453 | 0.420 | 0.501 | 0.453 | 0.499 | 0.471 | 0.466 | 0.450 | 0.421 |
| 95%ile | 0.652 | 0.596 | 0.634 | 0.684 | 0.605 | 0.712 | 0.713 | 0.799 | 0.809 | 0.827 | 0.782 | 0.650 |
| Maximum | 0.733 | 0.607 | 0.736 | 0.736 | 0.698 | 0.726 | 0.816 | 0.881 | 0.829 | 0.891 | 0.821 | 0.692 |
| SD | 0.106 | 0.086 | 0.109 | 0.118 | 0.116 | 0.103 | 0.122 | 0.144 | 0.146 | 0.152 | 0.136 | 0.114 |
| CoV (%) | 23.1 | 19.9 | 23.6 | 24.3 | 26.4 | 19.8 | 23.7 | 26.4 | 28.1 | 29.0 | 27.0 | 25.9 |

All units are $Mm^3$ unless otherwise indicated. Analysis excludes months with missing and station rating exceedance data, but includes unaudited (recent) and estimated data

**Table 3** Statistical analysis of Skeerpoort River mean monthly discharge at station A2H034 in the period October 1965 to September 2017

| Variable | Month | | | | | | | | | | | |
|---|---|---|---|---|---|---|---|---|---|---|---|---|
| | Oct | Nov | Dec | Jan | Feb | Mar | Apr | May | Jun | Jul | Aug | Sep |
| n | 49 | 48 | 48 | 49 | 49 | 47 | 49 | 50 | 51 | 48 | 50 | 49 |
| Minimum | 0.266 | 0.349 | 0.353 | 0.397 | 0.333 | 0.334 | 0.321 | 0.362 | 0.405 | 0.452 | 0.401 | 0.334 |
| 5%ile | 0.399 | 0.393 | 0.491 | 0.448 | 0.381 | 0.388 | 0.374 | 0.383 | 0.465 | 0.462 | 0.447 | 0.364 |
| Mean | 0.782 | 0.867 | 1.011 | 1.206 | 1.239 | 1.348 | 1.091 | 1.050 | 0.939 | 0.865 | 0.847 | 0.770 |
| Median | 0.735 | 0.791 | 0.935 | 1.061 | 0.940 | 0.994 | 0.840 | 0.836 | 0.806 | 0.812 | 0.801 | 0.718 |
| 95%ile | 1.344 | 1.832 | 2.294 | 2.401 | 3.329 | 2.725 | 2.089 | 2.210 | 1.766 | 1.595 | 1.634 | 1.527 |
| Maximum | 2.078 | 2.25 | 3.498 | 3.16 | 6.625 | 8.55 | 5.581 | 6.231 | 4.396 | 1.972 | 1.902 | 1.843 |
| SD | 0.361 | 0.431 | 0.591 | 0.678 | 1.179 | 1.299 | 0.847 | 0.928 | 0.623 | 0.368 | 0.367 | 0.356 |
| CoV (%) | 46.1 | 49.8 | 58.4 | 56.2 | 95.1 | 96.3 | 77.6 | 88.4 | 66.4 | 42.6 | 43.3 | 46.2 |

All units are $Mm^3$ unless otherwise indicated. Analysis excludes months with missing and station rating exceedance data, but includes unaudited (recent) and estimated data

characteristics of the Skeerpoort River as gauged at station A2H034 located ~7 km downstream of station A2H033 (Fig. 1). At this location, the river has left the pristine environment of the John Nash Nature Reserve, traverses Pretoria Group sedimentary strata, and also supports agricultural activities in the vicinity of the confluence with the Witwatersrand Spruit.

The A2H034 flow record spans a period of 52 years. It provides the monthly discharge statistics presented in Table 3, and for which the annual discharge per hydrological year is shown in Fig. 6. The winter month of July reflects the lowest CoV value of ~43%, and September the lowest median discharge value of 0.718 $Mm^3$ (277 L/s). The data again reflect the constancy and perennial nature of the spring-driven autogenic discharge at this location. The combined discharge (>300 L/s) of the karst springs that sustain the Skeerpoort River (Sects. 7.4.6, 7.4.7 and 7.4.8 in Chapter "Physical Hydrogeology") equates to a discharge of at least 9.5 $Mm^3/a_h$, similar to the long-term median value of ~10.2 $Mm^3/a_h$ (text box, Fig. 6). These values represent ~5% of the ~190 $Mm^3$ FSC of Hartbeespoort Dam (Sect. 1), testifying the important groundwater contribution delivered by this portion of the COH to the regional water resources budget.

The truncated period statistics presented in Fig. 6 for station A2H034 mimic those of Fig. 2 for station A2H033 in order to provide a direct comparison under circumstances where the A2H034 record extends to September 2017, whereas that of A2H033 only extends to November 2015. This comparison is taken further in Fig. 6, which shows the cumulative discharge at stations A2H033 and A2H034 over the truncated period of common record.

The discharge pattern at station A2H034 for the complete record is shown in Fig. 8. Although less evident than in Fig. 3, distinct recession curves following peak discharge events, for example ca. October 1977 and October 2000, are also discernible in the hydrograph. Evident at the end of this record are the exceptional discharges (some of the highest on

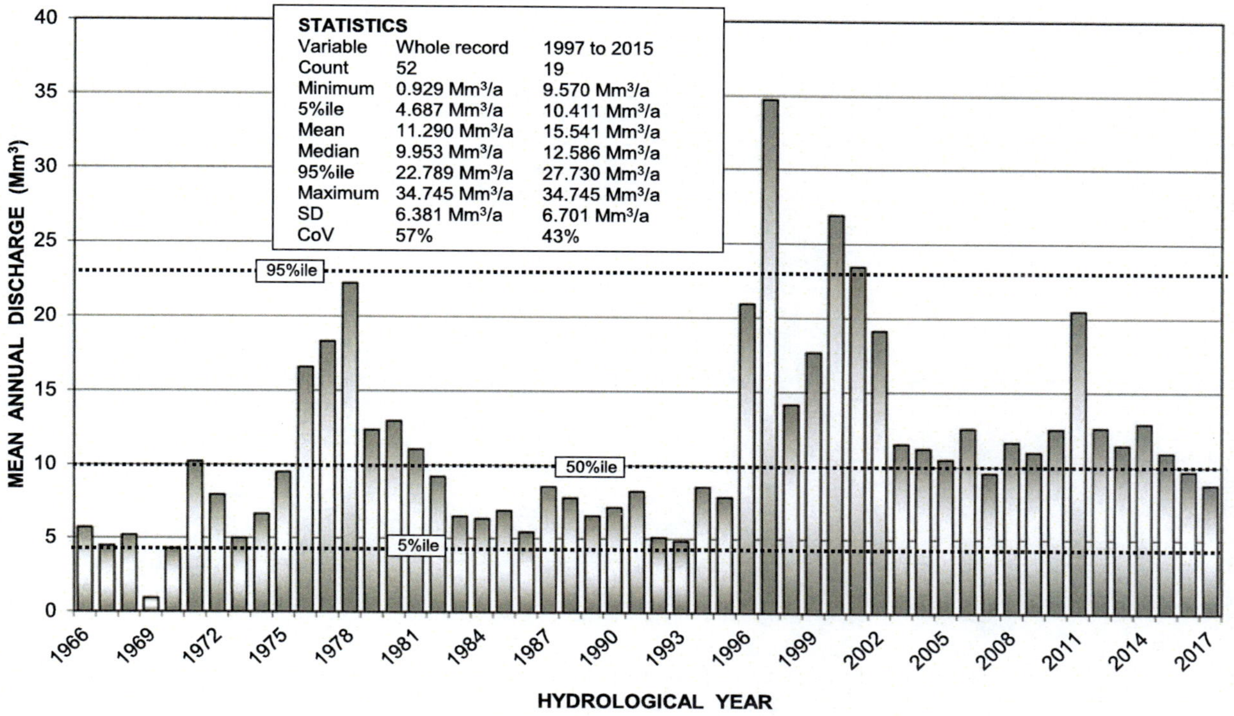


**Fig. 6** Pattern and trend of Skeerpoort River mean annual ($a_h$) discharge gauged at station A2H034 in the period October 1965 to September 2017

record excluding the uncensored anomalous data) recorded in January and March 2011 and March 2014. The aftermath of such events is shown in Plate 2 taken on the farm Dwarsvlei 503JQ in the upper reaches of the Skeerpoort River valley following a flash flood event in October 2009. Notable, however, is the absence in Fig. 9 of a flow peak associated with this `response in Fig. 9, illustrating the degree to which singular events are absorbed in monthly and annual values. As indicated previously, station A2H034 is located a substantial distance downstream of its principal perennial sources at a location where agricultural impacts in the form of abstraction for irrigation, as well as the contribution of the ephemeral Witwatersrand Spruit tributary entering from the east, are also likely manifested. These circumstances negate attempting a correlation between spring discharge and rainfall as has been attempted by Holland et al. (2009) for Maloney's Eye draining the Steenkoppies Basin (Sects. 3.1, 4, 5.1 and 7.1 in Chapter "Physical Hydrogeology") (Fig. 7).

An inspection of the whole record cumulative annual discharge of the Grootvlei Spruit (station A2H033) and the Skeerpoort River (station A2H034) indicates a year-on-year difference in median annual discharge between these stations of 6.2 $Mm^3$ (~197 L/s) for $n = 50$ values (Fig. 9). This approaches the August 2014 SDM value of ~208 L/s (Sect. 7.4.8 in Chapter "Physical Hydrogeology"). The inter-quartile range is defined by values of ~4.8 $Mm^3$ (~152 L/s) and ~ 8.0 $Mm^3$ (~253 L/s). The >75%ile excursions are unequivocally associated with the periods of surface runoff-driven flow (see also Fig. 7), and the <25%ile excursions with the pre-October 1996 period of lower baseflow values (Fig. 8). The year-on-year difference in median discharge places the long-term median discharge of the ungauged Nash Spring at ~200 L/s (~6.3 $Mm^3$/a).

The agreement already demonstrated between the combined discharge of the karst springs (~9.5 $Mm^3$/a) and the long-term median stream discharge (~10.2 $Mm^3$/a) quantifies the baseflow component of the Skeerpoort River with sufficient rigour. As will be made evident in Sect. 8.7 in Chapter "Chemical Hydrogeology", it is not only the volumetric contribution of this component to Hartbeespoort Dam, but also the generally excellent quality of the karst groundwater, that is important to the severely compromised quality of that received by and impounded in this hypertrophic reservoir from other much larger tributaries (Hobbs 2017c).

## 3 Bloubank Spruit System

The Bloubank Spruit system comprises the following main drainages, described sequentially downstream from the headwater reaches (see Fig. 1 for reference).

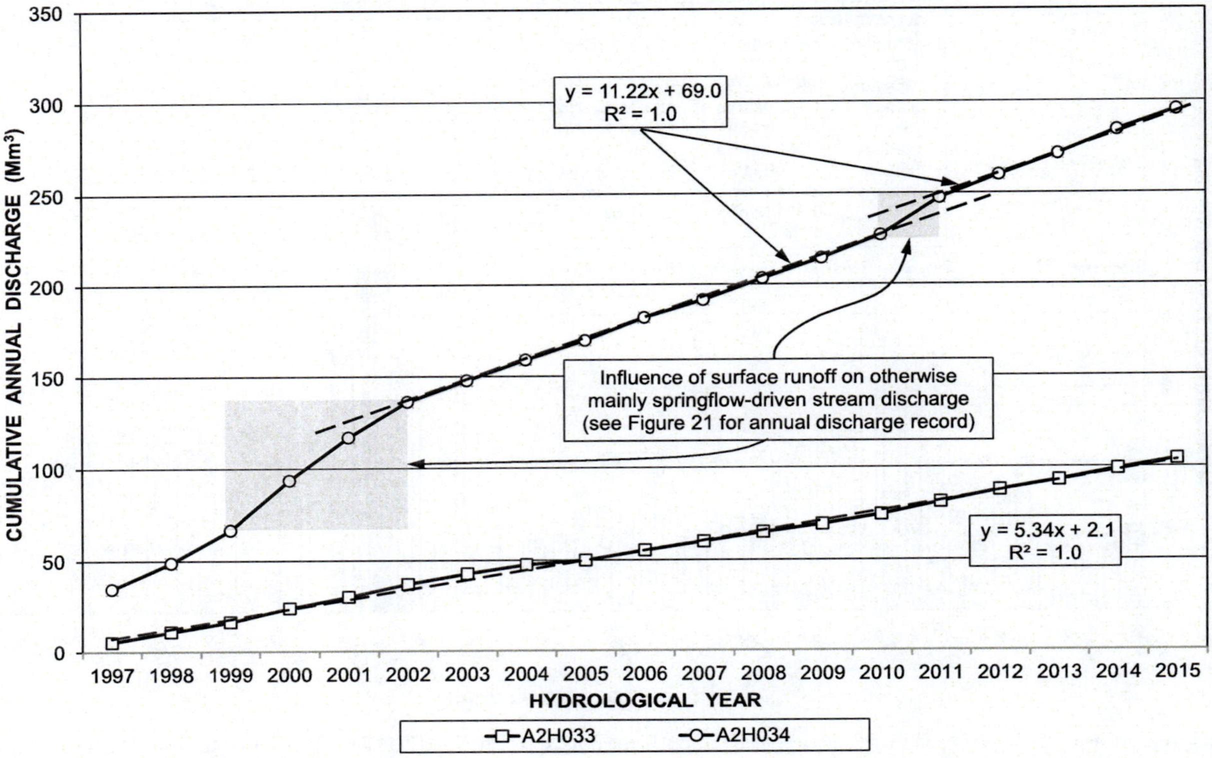


**Fig. 7** Cumulative annual discharge at stations A2H033 and A2H034 illustrating the year-on-year constancy of flow in these springflow-driven drainages

- The Bloubank Spruit rises on the farm Reydal 165IQ north of Tarlton, and flows east-north-eastwards past Sterkfontein Cave and Kromdraai down to its confluence with the Crocodile River in the vicinity of Glenburn Lodge at Zwartkop. Its upper reaches traversing the Zwartkrans Basin (Sect. 7.2.1 in Chapter "Physical Hydrogeology"), are ephemeral.
- The north-westerly draining upper reach of the Riet Spruit, which rises in Riebeeck Lake in Randfontein and flows past the Randfontein WWTW toward Tarlton, where it swings 90° to the east-north-east. Its middle and lower reaches follow the N14 national road down to the confluence with the Bloubank Spruit near the Swartkrans fossil site.
- The Tweelopie Spruit, a tributary of the Riet Spruit, can be traced upstream to Robinson Lake[2] between Randfontein and Krugersdorp, and drains northward through the Krugersdorp Game Reserve, intersecting the N14 before its confluence with the Riet Spruit.
- The Blougat Spruit, another tributary of the Riet Spruit, rises in the Factoria industrial area of Mogale City and first drains in a north-westerly direction through the city as a canal before flowing northward past the Munsieville township and Percy Stewart WWTW, intersecting the N14 before its confluence with the Riet Spruit.

Minor drainages are (a) the Honingklip Spruit,[3] an ephemeral tributary of the Bloubank Spruit that rises on the northern margin of Mogale City in the Rant-en-Dal and Paardekraal areas, and flows northwards to join the Bloubank Spruit at Kromdraai, and (b) the Tweefontein Spruit[4] that rises in the John Nash Nature Reserve, and drains southward across the farm Rietfontein 522JQ to join the Bloubank Spruit near Brookwood Trout Farm and the eastern boundary of the Nedbank Olwazini Estate.

The discharge of the Bloubank Spruit system is gauged at station A2H049 (Fig. 2 in Chapter "Description of the Physical Environment") located ~12 km downstream from Sterkfontein Cave. The ~45-year record yields the monthly

[2] This does not imply that Robinson Lake (an endoreic feature) is the source of the Tweelopie Spruit.

[3] Name assigned to a drainage that for most of its path traverses the farm Honingklip 178IQ, but is unnamed on the 1:50 000 scale topocadastral map 2627BB Roodepoort (RSA, 2002).

[4] Name given to an unnamed drainage on the 1:50 000 scale topocadastral map 2527DD Broederstroom (RSA, 2001).

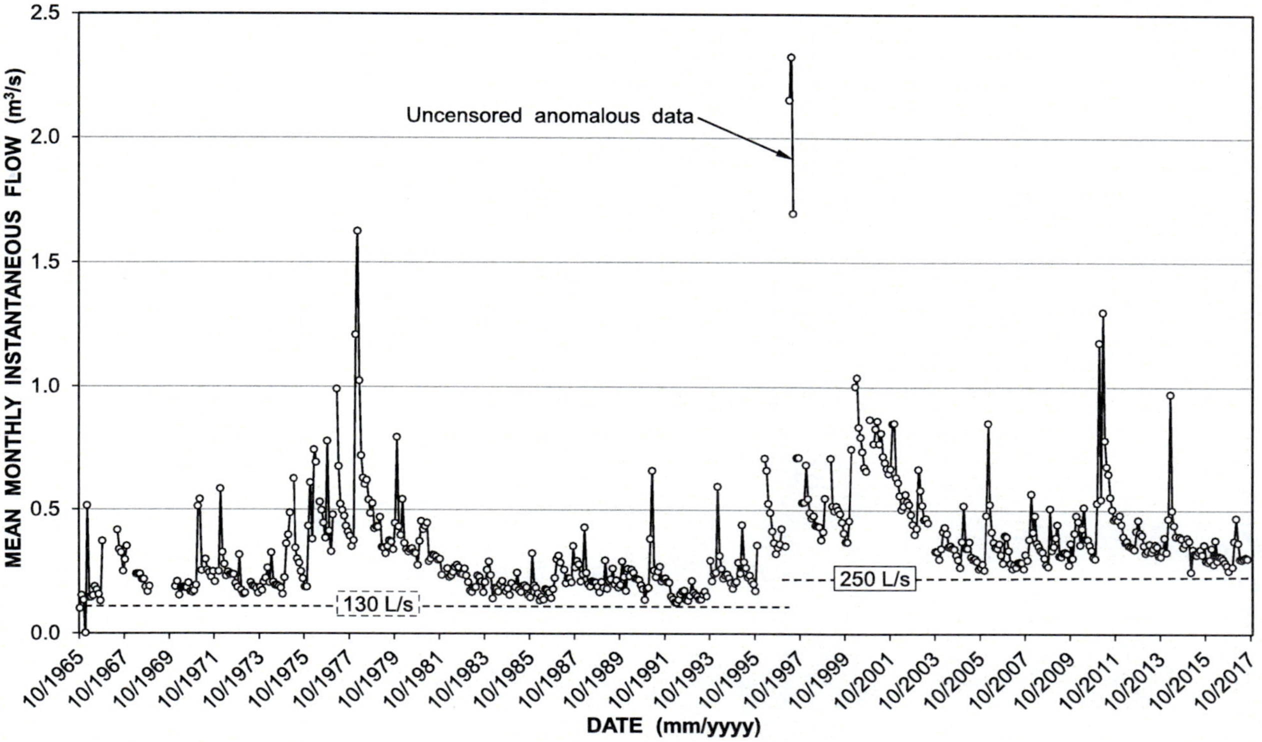


**Fig. 8** Long-term monthly hydrograph of the Skeerpoort River at station A2H034 for the period October 1965 to September 2014

**Plate 2** Damage to infrastructure caused in the upper reaches of the Skeerpoort River valley by a flash flood in October 2009; note flood stage height of ~2 m as indicated by the drift debris on pumphouse roof and in inset picture

discharge statistics presented in Table 4. This shows that the summer month of November has the lowest CoV value of ~44%, September the lowest median discharge of 1.56 $Mm^3$ (~602 L/s ≈ 52.0 ML/d), and October the lowest minimum discharge of 0.68 $Mm^3$ (~254 L/s ≈ 21.9 ML/d).

The discharge per hydrological year is shown in Fig. 10. This indicates that four of the last eight hydrological years experienced the 2nd, 3rd, 4th and 5th highest runoff (59.1, 54.5, 47.0 and 44.9 $Mm^3$ after the 66.9 $Mm^3$ of 1978) in the historical record of this catchment. The significance hereof is evident in the impact on the long-term median discharge of the Bloubank Spruit system, which reflects a median value of ~19.3 $Mm^3$/a for the 37-year period 1973 to 2009, and a ~115% greater value of ~41.4 $Mm^3$ for the last eight years of the record (Fig. 10). These circumstances indicate that an analysis of hydrological data (both quantity and quality) must recognise the influence imposed on the long-term data set by the discharges of the last eight hydrological years. The pre-2010 and post-2009 median annual discharges represent 10 and 22% respectively of the FSC (~190 $Mm^3$) of Hartbeespoort Dam. Equally significant is the substantially smaller CoV value for the post-2009 record (23%) compared to that of the pre-2010 set (46%). This is irrefutable evidence of the perennial impact of the mine water discharge contribution on the flow regime of the Bloubank Spruit in the recent past.

The instantaneous monthly flow pattern at station A2H049 for the complete record is shown in Fig. 11. This reveals a comparatively constant lowest value of ~260 L/s

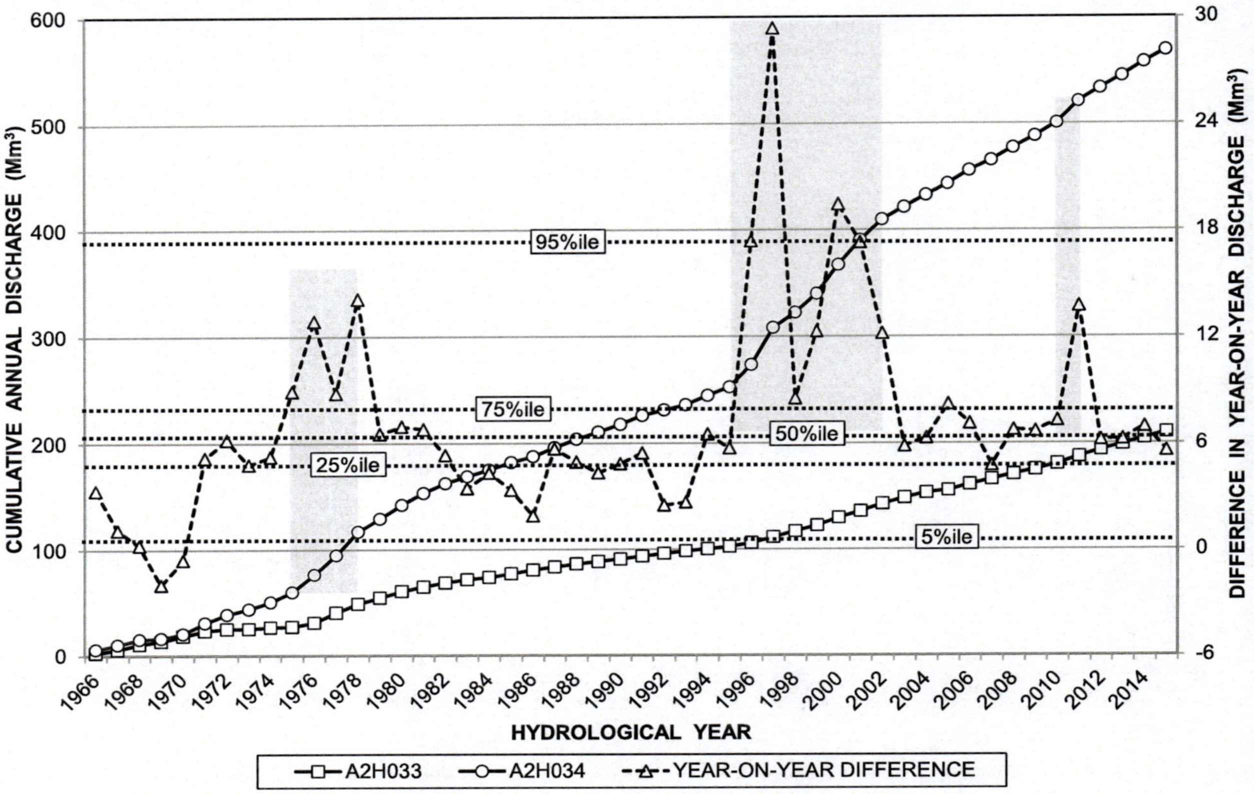


**Fig. 9** Difference in long-term year-on-year discharge between stations A2H033 and A2H034

**Table 4** Statistical analysis of Bloubank Spruit mean monthly discharge gauged at station A2H049 in the period October 1972 to September 2017

| Variable | Month | | | | | | | | | | | |
|---|---|---|---|---|---|---|---|---|---|---|---|---|
| | Oct | Nov | Dec | Jan | Feb | Mar | Apr | May | Jun | Jul | Aug | Sep |
| Count (n) | 43 | 43 | 44 | 44 | 45 | 45 | 45 | 44 | 45 | 45 | 44 | 43 |
| Minimum | 0.682 | 0.815 | 0.711 | 0.721 | 0.706 | 0.828 | 0.886 | 0.847 | 0.894 | 0.939 | 0.890 | 0.770 |
| 5%ile | 0.789 | 0.860 | 1.043 | 1.097 | 0.901 | 1.066 | 1.187 | 0.998 | 0.964 | 0.961 | 0.921 | 0.802 |
| Mean | 1.874 | 1.886 | 2.276 | 2.745 | 2.702 | 3.031 | 2.436 | 2.297 | 2.103 | 2.086 | 1.961 | 1.804 |
| Median | 1.676 | 1.743 | 2.066 | 2.471 | 2.222 | 2.534 | 1.987 | 1.925 | 1.797 | 1.695 | 1.676 | 1.561 |
| 95%ile | 3.824 | 2.952 | 4.501 | 5.355 | 6.328 | 7.863 | 5.355 | 4.882 | 4.115 | 4.058 | 3.658 | 3.509 |
| Maximum | 4.211 | 4.577 | 5.900 | 12.079 | 10.619 | 11.351 | 6.081 | 5.373 | 5.166 | 4.754 | 4.055 | 4.342 |
| SD | 0.921 | 0.830 | 1.094 | 1.931 | 1.932 | 2.208 | 1.301 | 1.197 | 0.976 | 0.944 | 0.876 | 0.883 |
| CoV (%) | 49.1 | 44.0 | 48.1 | 70.4 | 71.5 | 72.8 | 53.4 | 52.1 | 46.4 | 45.3 | 44.7 | 49.0 |

All units are $Mm^3$ unless otherwise indicated
Analysis excludes months with missing and station rating exceedance data, but includes unaudited (recent) and estimated data

up until early-2007. Again evident in the hydrograph are distinct recession curves following peak discharge events, for example ca. January 1978, March 1997 and February 2000. Station A2H049, however, is not only located a substantial distance (5–10 km) downstream of its primary perennial natural sources, the Zwartkrans and Kromdraai springs, but also receives the discharge of other 'smaller' springs (e.g. the Plover's Lake and Aquamine springs) and ephemeral tributaries such as the Honingklip Spruit. These circumstances again negate a correlation between spring discharge and rainfall. The rising trend in lowest monthly mean instantaneous flow observed since October 2007 (Fig. 11) reflects the impact of mine water discharge mentioned earlier. A detailed interrogation of the discharge

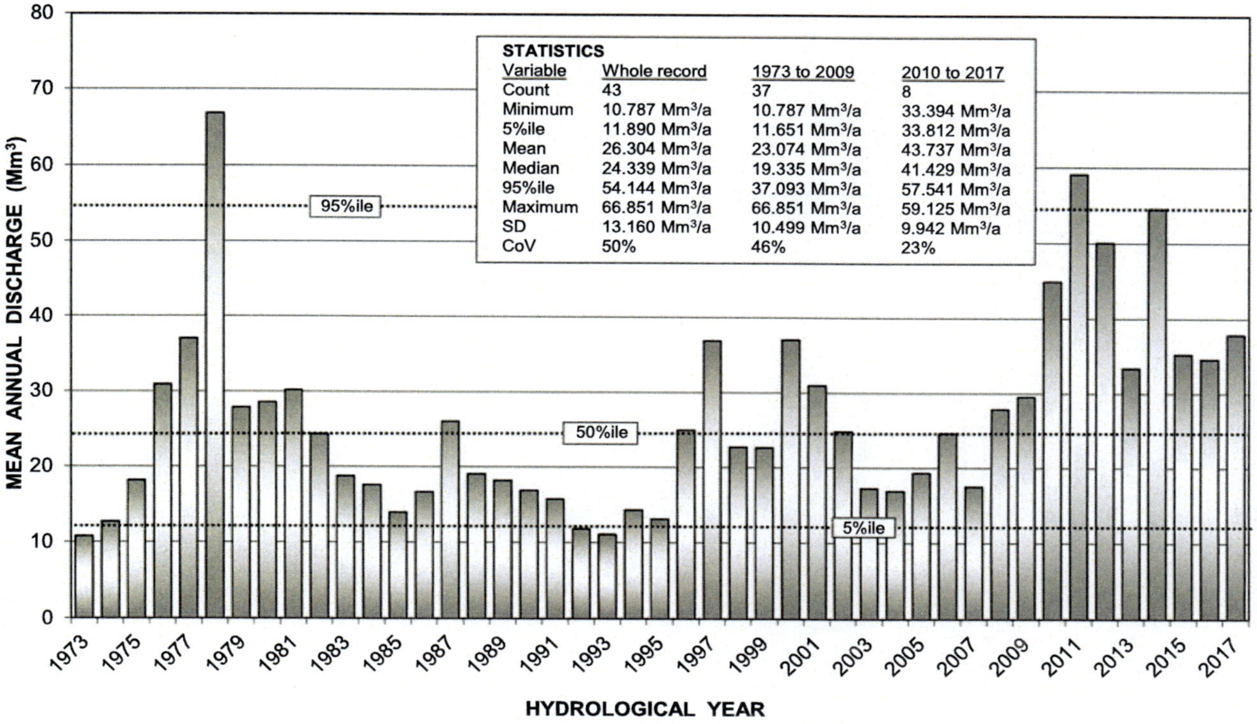


**Fig. 10** Pattern and trend of Bloubank Spruit mean annual ($a_h$) discharge gauged at station A2H049 in the period October 1972 to September 2017

regime in the upper reaches of this system better informs the hydrological and hydrochemical response observed at station A2H049.

A discussion of the two most significant tributary drainages in the Bloubank Spruit system, namely the Tweelopie Spruit and the Blougat Spruit, is presented in the following sections. These two drainages attract attention for (a) the unnatural (anthropogenic) characteristics that dominate their flow regimes in terms of both water quantity and quality, and (b) the availability of historical flow and water quality data associated with the allogenic sources, in particular the mine water driven Tweelopie Spruit.

## 3.1 Tweelopie Spruit

The decant of mine water in the upper reaches of the Tweelopie Spruit commenced in August 2002 (JFA 2006). In 2004, Sibanye-Stillwater (then Harmony Gold) constructed an HDPE-lined dam (Plate 3) at the Black Reef Incline (BRI) shaft. The dam serves two main functions, namely (a) the collection, control and temporary storage of raw mine water from the various point and diffuse sources in the area, and (b) the prevention of raw mine water release into the downstream receiving environment. The raw mine water (RMW) is pumped from here to the high density sludge (HDS) mine water treatment plant on the continental divide to the south for basic treatment (neutralisation).

The earliest record of routinely frequent mine water discharge and quality measurements is traced back to spreadsheeted data commencing in January 2005. These early records reveal several periods of zero discharge (notably June 2005 to January 2006) separating periods of active discharge. The historical record of RMW and treated/neutralised mine water (TMW) discharge as gauged at the end-of-pipe (EoP) (Plate 4) is shown in Fig. 12. The TMW Parshall-type flume gauging structure was installed in the period prior to 2008 when very little RMW left the property, partly because the gradually increasing rate of decant remained within the early capacity (~15 ML/d) of the treatment plant. Fig. 12 indicates that these circumstances changed in early-2008, when discharges >15 ML/d became the norm, and the contribution of RMW to the total mine water discharge became significant. Fig. 12 suggests that total mine water discharges >20 ML/d were irregular and occasional. Nevertheless, these circumstances resulted in the installation of a second flume (Fig. 15) adjacent to the EoP flume to also gauge the RMW component leaving the mine property. Later proving inadequate (in the 2011 wet season) despite being extended (Fig. 16), a third RMW flume was

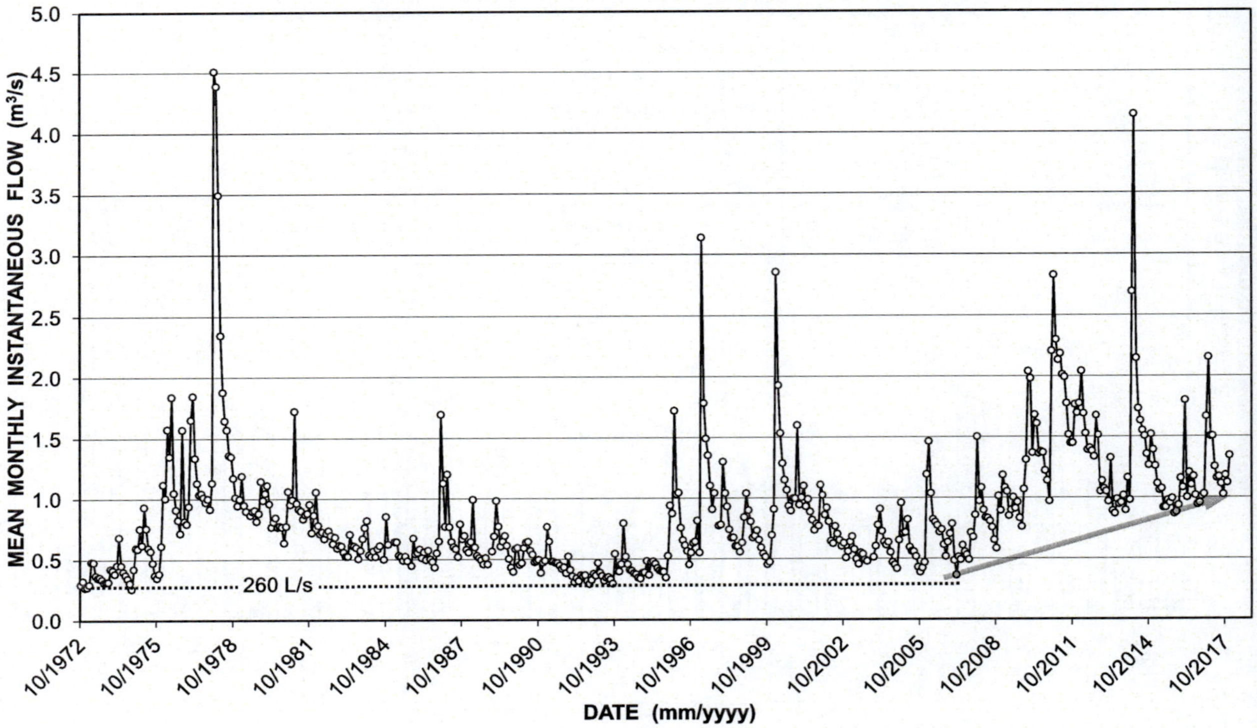


**Fig. 11** Long-term monthly hydrograph of the Bloubank Spruit at station A2H049 for the period October 1972 to September 2017

**Plate 3** Aerial view (looking north) of the HDPE-lined off-channel storage dam shortly after construction at the Black Reef Incline (circular structure at centre right) to contain mine water decanting from this structure; also visible are the pump station (centre left), the gravity-fed pipeline (45° diagonal from centre bottom) draining the upstream in-channel Portuguese Dam, and the course of the Tweelopie Spruit exiting picture at centre top (photo G Krige, 05/10/2004)

installed (Fig. 17). The accuracy of flow measurements associated with all of the flumes remains questionable, especially for high flows.

The circumstances illustrated in Fig. 12 describe the 'tipping point' for the Bloubank Spruit system, when RMW discharges started to consistently exceed 10 ML/d in February 2010. These exceedances are evident in Fig. 13, indicating their duration for extended periods (months). Together with their considerable magnitude (often $\gg$ 30 ML/d), these circumstances manifested a marked impact on surface water chemistry (Sect. 6.1.2 in Chapter "Chemical Hydrology") observed ~28 km downstream from the mine area at station A2H049 (Table 9 in Chapter "Chemical Hydrology"). Fig. 14 shows that for the first time since the advent of mine water treatment, the RMW component regularly equalled or exceeded the TMW component in the total discharge released into the Tweelopie Spruit and downstream receiving environment. The trigger of this 'tipping point', dated to 30/01/2010, rests with the chemistry associated with the contributing RMW and TMW sources, and the impact thereof on the hydrochemical composition of the combined discharge to the Tweelopie Spruit. This influence is interrogated in Sects. 6.1.2 and 6.2.2 in Chapter "Chemical Hydrology".

The very high discharges in four of the last eight hydrological years (Fig. 10) reflect an abnormal mine water discharge regime which the immediate and short-term mine water interventions have sought to normalise. A closer inspection of the discharge statistics for station A2H049

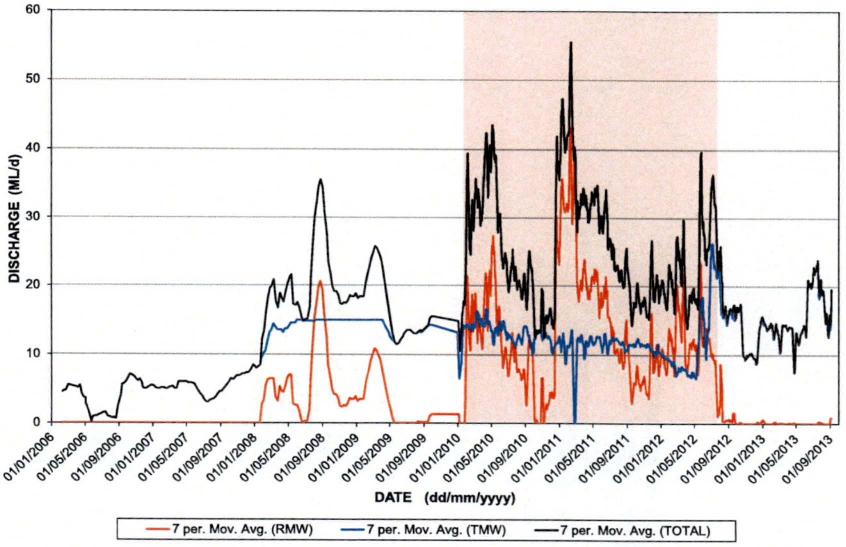


**Fig. 12** Pattern and trend of mine water discharge to the Tweelopie Spruit

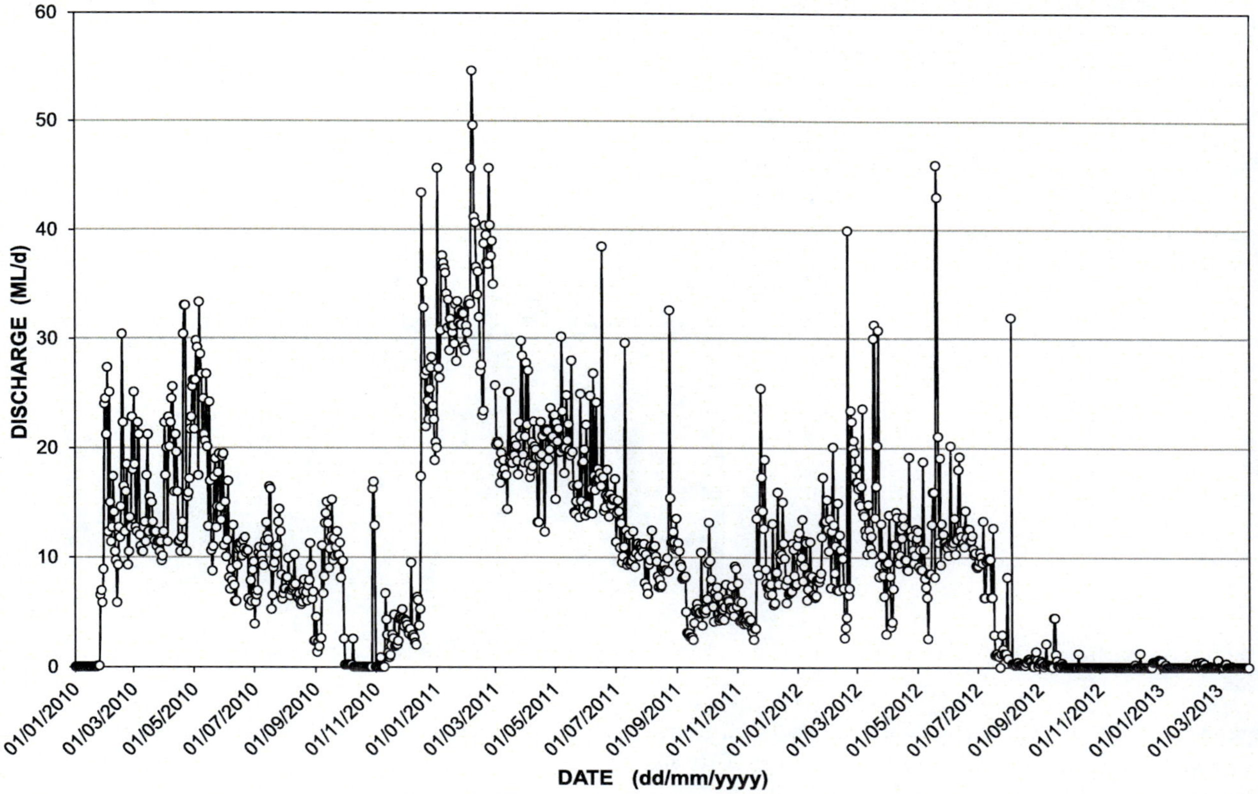


**Fig. 13** Pattern and trend of raw mine water discharge to the Tweelopie Spruit

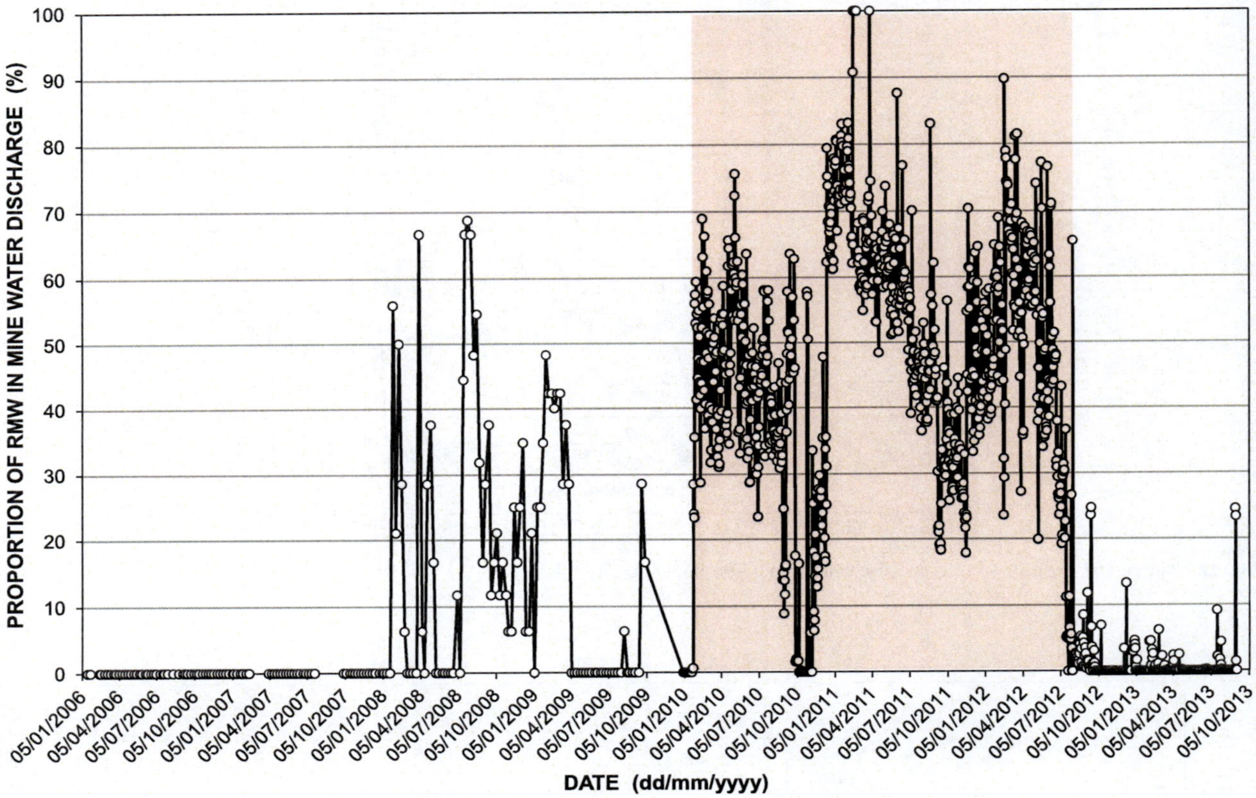


**Fig. 14** Pattern and trend of RMW proportion in total mine water discharge

**Plate 4** Treated mine water EoP flow gauging weir

presented in Fig. 10 reveals a difference in median discharge of ~22.1 $Mm^3/a$ (~61 ML/d) between the pre-2010 and the post-2009 periods of record. The mine water treatment plant capacity of 34 ML/d since mid-2013 (Text Box 1) explains ~56% of the much greater median annual discharge of the post-2009 period. The remaining ~9.9 $Mm^3/a$ (~27 ML/d) comprises raw mine water discharges via the Tweelopie Spruit, municipal wastewater effluent via the Blougat Spruit (Sect. 3.2), and natural surface water runoff.

Alternative measures to mitigate the impact of adverse outcomes likely to be generated by the immediate and short-term mine water intervention measures include distributing the treated/neutralised mine water discharge between two or more drainages. Candidate drainages are the

**Plate 5** Raw mine water flow gauging weir (original)

**Plate 6** Raw mine water flow gauging weir showing inadequacy at high flows despite modification of original; note extended superstructure relative to position of lower front 'lifting eye' (circled) in common with Plate 10

**Plate 7** Additional flow measurementstructure (at right) installed to better gauge uncontrolled high flows of raw mine water; compare with Plate 6 setup

Blougat Spruit to the east of the Tweelopie Spruit, and the upper Riet Spruit to the west (Fig. 1). The latter drains into the Steenkoppies Basin which supports large-scale and intensive irrigation, and discharges via Maloney's Eye into the Magalies River. In both cases, the target drainages receive treated municipal wastewater effluent, in the case of the Blougat Spruit from the Percy Stewart WWTW, and in that of the upper Riet Spruit from the Randfontein WWTW.

**Text Box 1** Mine water control and management in the Western Basin

As was the case for all of the gold mines on the Witwatersrand in the first century of mining, mine water control and management in the West Rand Goldfield (aka the Western Basin) focussed on the removal of water 'made' in the underground workings to surface. This water was pumped out to keep the workings safe and productive (Handley 2004). The volumes involved were generally manageable, and it was only when mining descended beneath the productive karst aquifers of the overlying Malmani Subgroup dolomite that considerable and sometimes catastrophic volumes were encountered.

In the Western Basin, the pumping of mine water resulted in the drying up of a number of springs in the Krugersdorp Game Reserve. The water was released into the Tweelopie Spruit, from where it contributed to the discharge of the Bloubank Spruit and Crocodile River. The quality of this water was considered suitable for discharge without prior treatment beyond settling of suspended material and fines in a natural holding facility such as Robinson Lake. Nevertheless, the impact of this discharge on groundwater quality in the receiving karst environment of the Zwartkrans Basin was already evident in the data generated by the DWS emergency water supply study of the mid-1980s (Bredenkamp et al. 1986) (see Section 8.2.2 in Chapter "Chemical Hydrogeology" and Fig. 20 in Chapter "Physical Hydrogeology").

In the late-1990s, geologists and environmental scientists in the employ of Randfontein Estates Gold Mine drew attention to the probability of acid mine drainage impacting on surface water resources following the cessation of mining. Calculations based on the observed rewatering rate and knowledge of the mine geometry predicted the 'when' and 'where' with remarkable accuracy. These predictions and cautions were ignored by the mining company and authorities.

The recent effort to control and manage mine water decant in the Witwatersrand Basin is based on recommendations by Coetzee et al. (2010) to the Inter-Ministerial Committee on Acid Mine Drainage. Amongst other factors, the use of 1000 ML/d of high quality water from the Lesotho Highlands Water Scheme to maintain a TDS level of 600 mg/L in the Vaal River System drives the need for such control and management. The DWS tasked the Trans-Caledon Tunnel Authority (TCTA) with the implementation of those recommendations required immediately and in the short-term, such as demanded by the active decant in the Western Basin. The scope and magnitude of these interventions in the Western Basin comprise the following activities (after TCTA 2012).

- *The immediate upgrading and retrofitting of the existing high density sludge (HDS) treatment plant to a capacity of ~24 ML/d. This was achieved and the plant commissioned in June 2012.*
- *The pumping of mine water from #8 Shaft to reduce the potentiometric head in the flooded underground mine workings and so eliminate active decant from mine structures such as #17 Winze, #18 Winze and the Black Reef Incline (BRI) Shaft in the locus of decant.*
- *The further upgrade of the HDS plant to a capacity of ~34 ML/d in the short-term. This was achieved in June 2013.*

The immediate and short-term intervention measures have been amended (DEA 2014b) to encompass pumping an average of 53 ML/d (peak of 60 ML/d) from #9 Shaft in order to reduce the potentiometric head in the flooded mine void to an elevation of ~1550 m amsl (~165 m below surface) from its current ~1670 m amsl. This will meet the obligation of maintaining an environmental critical level (ECL) at an elevation where the mine water potentiometric head has a negligible impact on the water resources environment. The treatment capacity in late-2017, however, was only some 40 ML/d.

The DWS embarked (mid-2017) on the implementation of a long-term solution (LTS) formulated in a multi-disciplinary study (DWA 2013a) completed in mid-2013 for all three basins in the Witwatersrand Goldfield. Strydom et al. (2016) provide a synopsis of the complex inter-relationships that describe the Witwatersrand AMD conundrum.

The potential benefits of mixing mine water with municipal wastewater effluent are discussed in Sect. 8.11 in Chapter "Chemical Hydrogeology". An obvious advantage for the Steenkoppies Basin is the enhanced recharge of a karst aquifer that already suffers from over-abstraction for irrigated agriculture. The magnitude of this impact has resulted in a drastic reduction in the discharge of Maloney's Eye (Holland et al. 2009), as well as the implementation of regulatory measures by the DWS to control abstraction (RSA 2008).

## 3.2 Blougat Spruit

The Blougat Spruit receives the effluent discharge from the Percy Stewart WWTW. This has turned an ephemeral stream into a perennial drainage since the commissioning of the plant in the 1950s. Designed to treat domestic sewage at a rate of 25 ML/d based on a design norm chemical oxygen demand (COD) of 600 mg/L, the inclusion of low-cost housing and industrial wastewater increased the COD loading to 1000 mg/L, reducing the plant capacity to 15 ML/d. Other factors such as poor maintenance and failure of structures (the facility straddles the regional Rietfontein Fault) reduced this even further to 5 ML/d. Following refurbishment completed in 2010, the treatment capacity is currently 24 ML/d (DWA 2011). This compares favourably with the recent inflow of 16 to 18 ML/d, i.e. 67–75% of capacity.

The record of effluent discharge (Fig. 15) indicates a median value of ~13 ML/d (~4.7 $Mm^3$/a) and a 95%ile value of ~17.4 ML/d (~6.4 $Mm^3$/a) in the period July 2007 to June 2009.[5] The median discharge equates to ~21% of the long-term median annual discharge of the Bloubank Spruit system at station A2H049 (Fig. 10). The 2008 summer discharge of 16 to 18 ML/d reflects the rainfall of 883 mm gauged at the WWTW in this season (Fig. 3 in Chapter "Description of the Physical Environment"), compared to the 584 mm gauged in the 2009 summer and the average MAP of 767 mm for the 14 hydrological years since 2001 (Sect. 3.2 in Chapter "Description of the Physical Environment").

Fig. 18 also shows the pattern of median daily inflow received by the facility. A comparison of this volume with that discharged to the Blougat Spruit indicates that the latter represented 44% (on average) of the inflow in the period up to September 2008, and 77% (on average) in the subsequent period of lesser inflow. The Percy Stewart WWTW also discharges a portion of its treated effluent to the adjacent Bergland property for the irrigation of cultivated lawn (unnumbered Plate, p 10). This amounted to ~1.3 ML/d (on average) in the period July 2007 to June 2009, i.e. the equivalent of 4.5% of the inflow into the plant, and 10% of the median discharge to the Blougat Spruit.

The results of opportunistic SDMs carried out at the causeway across the Blougat Spruit downstream of the Percy Stewart WWTW are presented in Table 5. The 2012 results indicate good agreement with the July 2007 to June 2008 record (Fig. 18). A significant increase is evident in the 2013 results, although the December 2013 result is probably biased by heavy rainfall experienced in the region earlier in the week of 09/12/2013. The receiving drainage, the Blougat Spruit, drains much of the central Mogale City business district and north-western suburbs, including the Munsieville township and Delporton industrial area. The 'return' in 2017 to pre-2013 levels is not readily explained, but confirms the most probable minimum discharge of the plant.

The provision of treated effluent to the KGR for game watering and the flood irrigation of kikuyu grass in the northern portion of the reserve amounted to ~0.97 ML/d (on average). This is equivalent of 3.3% of the inflow received, and 7% of the discharge released to the Blougat Spruit. This practice was curtailed on 08/08/2008 (Brink 2008). The primary reason for the curtailment of this transfer and use was concern expressed by the West Rand District Municipality (WRDM) for the bacteriological quality of groundwater in the dolomitic aquifer that underlies this portion of the study area. The primary driver for this concern was a complaint by the landowner of Ptn 2/7/Rem of the farm Sterkfontein 173IQ regarding the poor quality of groundwater produced by a water supply borehole on the property. These circumstances were investigated and reported on by Hobbs (2008b), who concluded that the deterioration in groundwater quality was attributable to the excessive discharge of treated and untreated mine water, exacerbated by impoundment of this water in the stream channel promoting allogenic recharge (Sect. 8.2.3.3 in Chapter "Chemical Hydrogeology" and Fig. 10 in Chapter "Chemical Hydrogeology"). The absence of any indication of bacteriological contamination that might reasonably be attributed to the municipal wastewater impact confounds the decision to curtail the supply to the KGR on these grounds. It also begs the question why the similar practice on the adjoining Bergland property continues to be sanctioned, despite evidence that the ambient groundwater quality in Month 200 already reflected elevated nitrate concentrations (Sect. 8.2.3.2 in Chapter "Chemical Hydrogeology" and Fig. 10 in Chapter "Chemical Hydrogeology").

In light of the above, it is lamentable that the apparent impact of the Percy Stewart WWTW as a significant contributor of poor quality water in the study area, remains poorly defined. Fortunately, routine monitoring at the Nedbank Olwazini Estate (NOE) further downstream in the Bloubank Spruit system provides data on bacterial levels (Sect. 6.1.2.6 in Chapter "Chemical Hydrology") and nutrient concentrations (Sect. 6.3.2 in Chapter "Chemical Hydrology") that bridge this deficiency to some extent. The veracity of this monitoring is, of course, subject to the possible opportunistic high volume release of poorly treated effluent at unconventional times such as at night.

[5] The Mogale City Local Municipality has not made any monitoring data available since the provision of that for the period July 2007 to June 2009. The contribution of this facility to the flow and quality of the Bloubank Spruit in the last decade is therefore largely inscrutable.

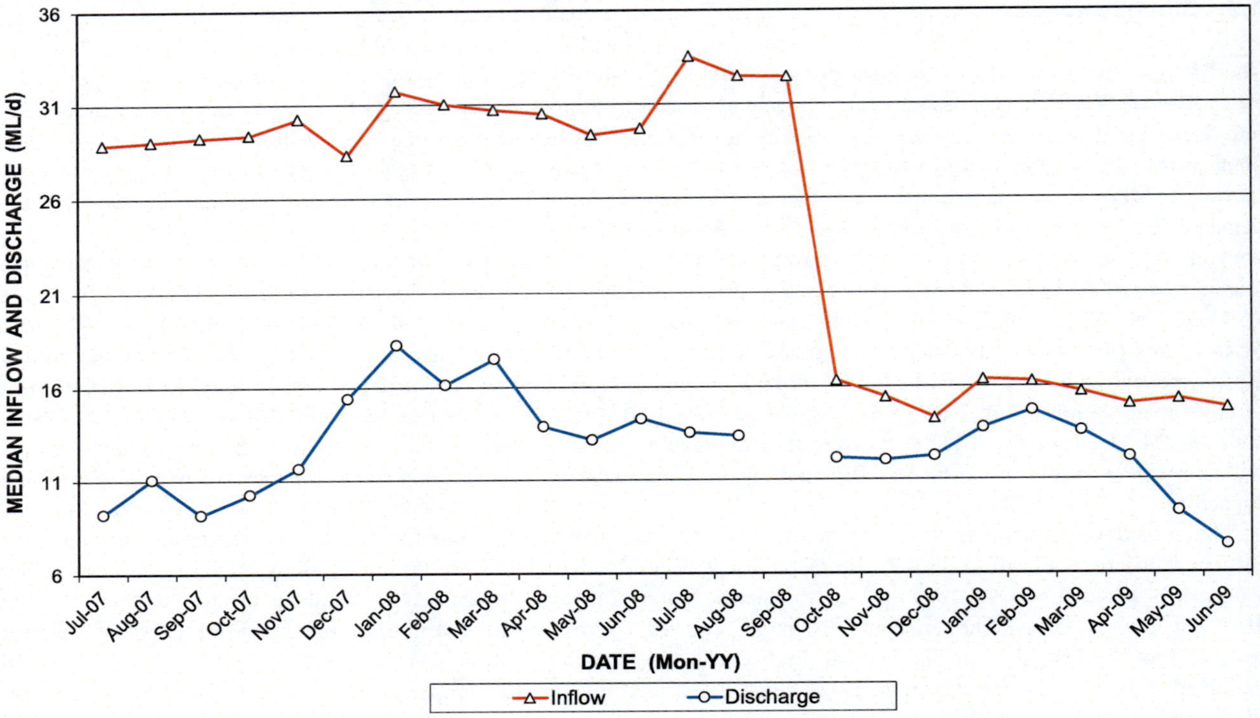


**Fig. 15** Pattern and trend of inflow to and effluent discharge from the Percy Stewart WWTW

**Table 5** Results of opportunistic flow measurements downstream of the Percy Stewart WWTW

| Date | 26/09/2012 | 24/10/2012 | 15/08/2013 | 12/12/2013 | 04/12/2017 |
|---|---|---|---|---|---|
| SDM Result (ML/d) | 15.9 ± 1.6 | 16.3 ± 1.6 | 24.0 ± 2.4 | 39.4 ± 5.9 | 14.8 ± 1.5 |

## 3.3 Crocodile River

The Crocodile River drains the north-western quadrant of the Johannesburg Metropole in Quaternary catchment A21E. This drainage flows northwards into Hartbeespoort Dam. Although it does not enter the COH (Fig. 2 in Chapter "Introduction and Background"), it is included in the study area because it traverses the COH buffer zone and receives the treated wastewater effluent discharged from the Driefontein WWTW operated and managed by the City of Johannesburg Metropolitan Municipality. The Driefontein WWTW collects and treats sewage from the northern areas of Roodepoort and Mogale City, has a treatment capacity of 35 ML/d (DWA 2011), and currently treats ~31 ML/d, or 88% of its capacity (DWA 2011). Gauging station A2H050 is located downstream of this facility (Fig. 2 in Chapter "Description of the Physical Environment") before the confluence with the Bloubank Spruit.

The DWS flow record for station A2H050 () spans a period of 44 years. It provides the monthly discharge statistics presented in Table 6. This shows that the month of August reflects the lowest CoV value of ~76%, September the lowest median discharge value of 0.518 $Mm^3$ (200 L/s), and October the lowest minimum discharge of 0.003 $Mm^3$ (1.2 L/s).

The long-term (whole record) median annual discharge of ~13.5 $Mm^3/a$ is ~56% that of the Bloubank Spruit (Fig. 10), and represents ~7% of the Hartbeespoort Dam FSC. A more robust comparison with the Bloubank Spruit discharge record must, however, consider separating the A2H050 hydrograph into two similar periods, namely a pre-2010 and a post-2009 period. The substantial difference in discharge statistics between these two periods and the whole record analysis is clearly evident (text box Fig. 16). The pre-2010 median contribution to Hartbeespoort Dam of 5% FSC more than trebles to ~17% FSC in the post-2009 period. Any number of reasons might account for an increase of this magnitude, but the expanding densely urbanised nature of the catchment is considered the major contributor.

The instantaneous monthly flow pattern at station A2H050 (Fig. 17) reveals a comparatively constant minimum value of <0.1 $m^3/s$ for the period up to at least October

**Table 6** Statistical analysis of upper Crocodile River mean monthly discharge gauged at station A2H050 in the period October 1973 to September 2017

| Variable | Month | | | | | | | | | | | |
|---|---|---|---|---|---|---|---|---|---|---|---|---|
| | Oct | Nov | Dec | Jan | Feb | Mar | Apr | May | Jun | Jul | Aug | Sep |
| n | 43 | 44 | 44 | 44 | 42 | 43 | 44 | 43 | 44 | 43 | 44 | 43 |
| Minimum | 0.003 | 0.107 | 0.196 | 0.262 | 0.151 | 0.157 | 0.131 | 0.141 | 0.153 | 0.130 | 0.081 | 0.039 |
| 5%ile | 0.104 | 0.188 | 0.329 | 0.339 | 0.224 | 0.205 | 0.191 | 0.172 | 0.181 | 0.183 | 0.164 | 0.137 |
| Mean | 0.860 | 1.073 | 1.527 | 2.304 | 2.440 | 2.152 | 1.527 | 1.240 | 1.018 | 0.871 | 0.762 | 0.709 |
| Median | 0.620 | 0.846 | 1.001 | 1.605 | 1.546 | 1.752 | 1.093 | 0.962 | 0.725 | 0.670 | 0.618 | 0.518 |
| 95%ile | 2.198 | 2.458 | 4.104 | 5.061 | 7.427 | 5.442 | 4.238 | 3.640 | 2.452 | 2.069 | 1.823 | 1.498 |
| Maximum | 2.612 | 2.696 | 6.715 | 10.084 | 9.583 | 7.350 | 5.097 | 4.203 | 3.574 | 2.478 | 2.290 | 3.200 |
| SD | 0.714 | 0.796 | 1.358 | 2.176 | 2.403 | 1.813 | 1.316 | 1.125 | 0.861 | 0.666 | 0.583 | 0.641 |
| CoV (%) | 83.1 | 74.2 | 88.9 | 94.4 | 98.5 | 84.2 | 86.2 | 90.7 | 84.6 | 76.5 | 76.4 | 90.4 |

All units are $Mm^3$ unless otherwise indicated. Analysis excludes months with missing and station rating exceedance data, but includes unaudited (recent) and estimated data

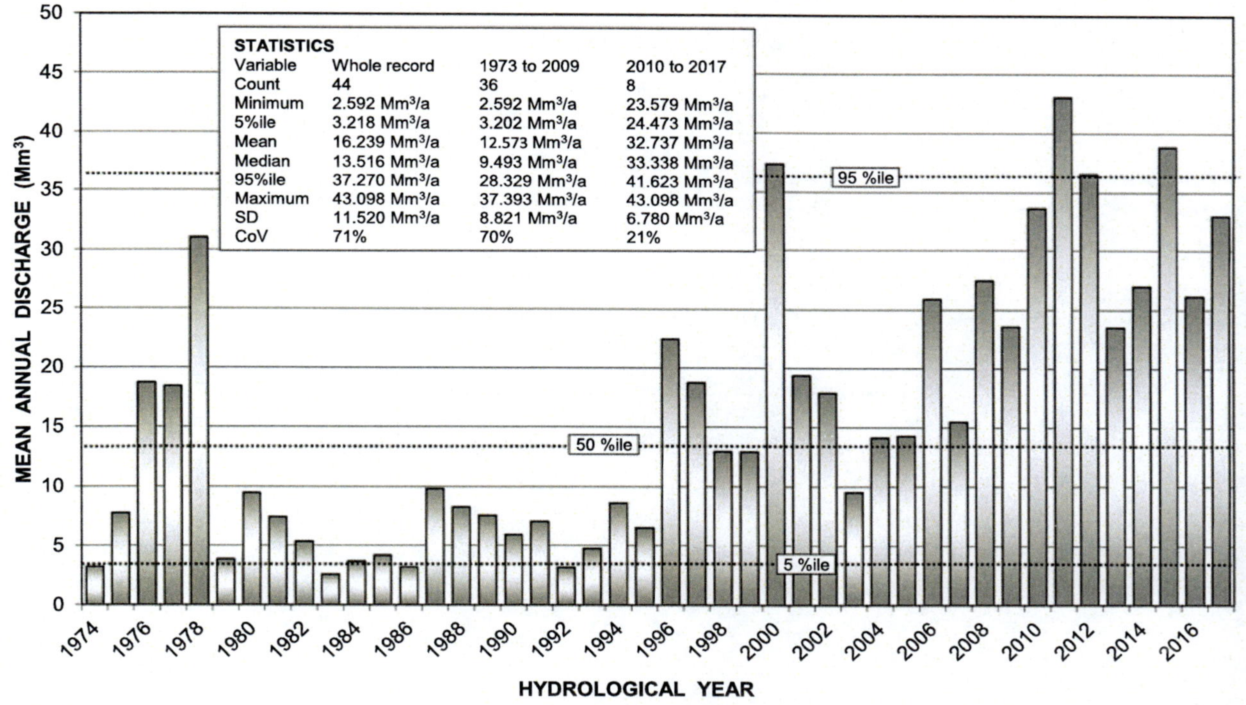


**Fig. 16** Pattern and trend of (upper) Crocodile River annual ($a_h$) discharge at station A2H050 in the period October 1973 to September 2017

1995, followed by a gradual increase since October 2002 to a value of ~0.5 $m^3/s$ (15.8 $Mm^3/a$) by August 2017. This pattern is attributed to discharge from the Driefontein WWTW where the treatment capacity of Unit 2, originally 15 ML/d when commissioned in 1988, was increased to 25 ML/d in 2002. The 'base' discharge of 15.8 $Mm^3/a$ is roughly half the volume currently treated by the plant.

## 4 Regional Context and Synthesis

The significance of the Bloubank Spruit system in the catchment of the regionally important Hartbeespoort Dam (Sect. 1) is evident in Table 7 and Fig. 18. Although the long-term normalised median discharge of the Bloubank Spruit, expressed as $Mm^3/km^2$ (mm) ranks only 5th out of

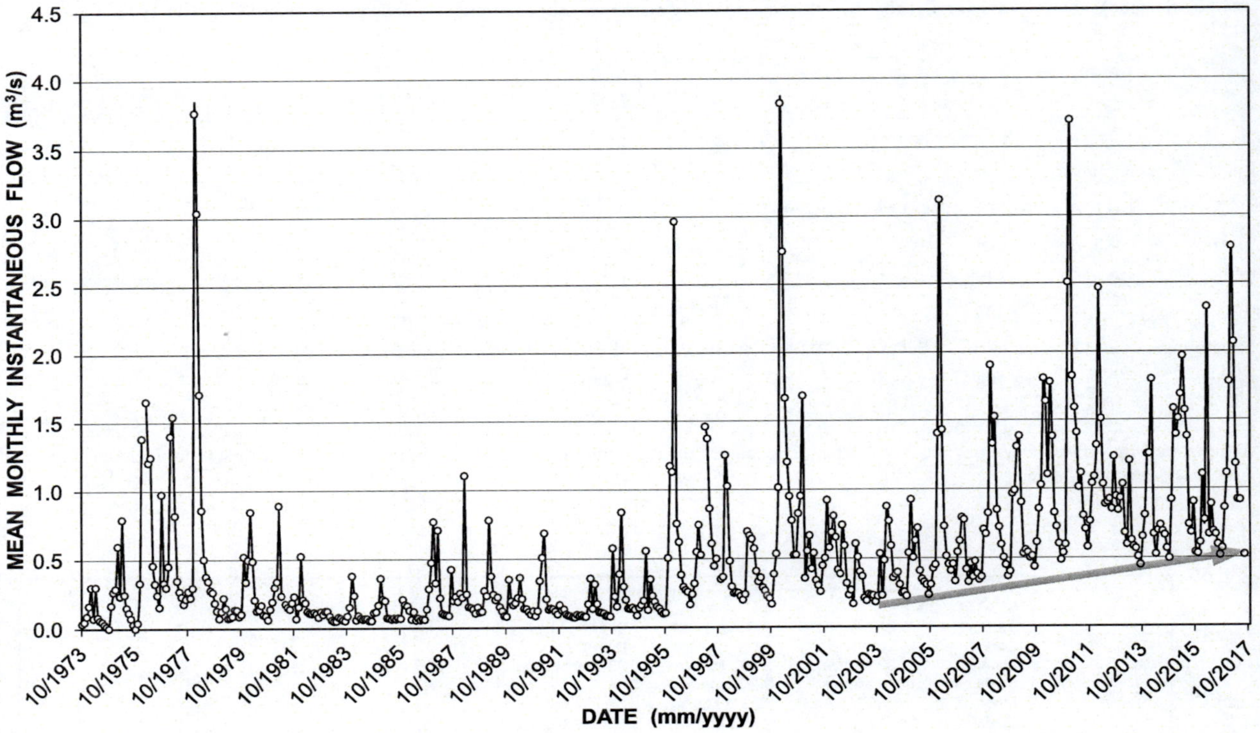


**Fig. 17** Long-term monthly hydrograph of the upper reach of the Crocodile River at station A2H050 for the period October 1973 to September 2017

**Table 7** Statistical analysis of the long-term annual ($a_h$) discharge contributions by the main rivers draining the Hartbeespoort Dam catchment

| Statistical parameter | Gauging station # and drainage | | | | | |
|---|---|---|---|---|---|---|
| | A2H013 magalies R | A2H014 hennops R | A2H034 skeerpoort R | A2H044 jukskei R | A2H049 bloubank Sp | A2H050 crocodile R |
| Area (km$^2$) | 1171 | 1007 | 150 | 798 | 371 | 148 |
| Minimum | 0.799 | 11.422 | 4.858 | 66.027 | 11.051 | 2.592 |
| 5%ile | 1.179 | 16.259 | 5.478 | 70.729 | 12.688 | 3.215 |
| Mean | 30.607 | 70.373 | 12.497 | 155.710 | 26.674 | 15.604 |
| Median | 19.642 | 68.281 | 11.040 | 129.844 | 24.485 | 12.905 |
| 95%ile | 100.273 | 155.707 | 23.410 | 309.312 | 54.371 | 37.352 |
| Maximum | 127.094 | 187.304 | 34.745 | 326.546 | 66.851 | 43.098 |
| SD | 32.504 | 47.353 | 6.408 | 76.858 | 13.092 | 11.380 |
| CoV (%) | 106 | 67 | 51 | 49 | 49 | 73 |
| 50%ile MAR Ranking | 17 mm 6th | 68 mm 4th | 74 mm 3rd | 163 mm 1st | 66 mm 5th | 87 mm 2$^{nd}$ |

All units are Mm$^3$ unless otherwise indicated. Analysis excludes months with missing and station rating exceedance data, but includes unaudited (recent) and estimated data

The analysis period spans the 42 hydrological years 1974–2015 common to all six stations; $A_h$ 2016 and 2017 not included because of periods of no record at the highest ranking station A2H044

The Crocodile River data exclude the ungauged discharge generated downstream of gauging stations A2H014, A2H044, A2H049 and A2H050

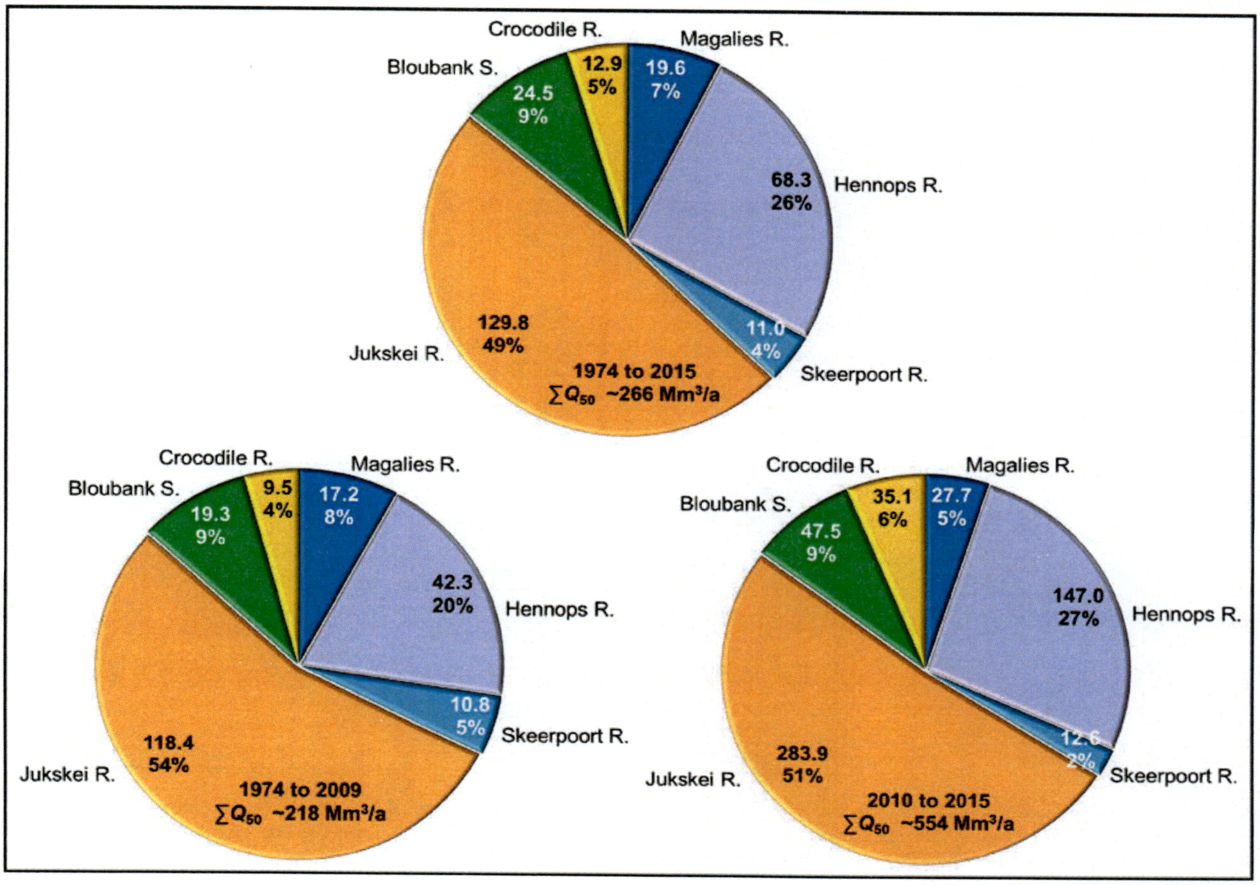


**Fig. 18** Comparison of whole record (top), pre-2010 (left) and post-2009 (right) median annual (**a**$_h$) discharge contributions as Mm$^3$ and % $\Sigma Q$ of the main rivers draining to Hartbeespoort Dam

six (Table 7), this system contributes the third highest discharge (9%) to the dam after the Jukskei River (48%) and Hennops River (26%) in the long-term (Fig. 18).

The long-term annual discharge contributions are also compared to those for the more recent 2010 to 2015 period[6] (Fig. 18). This reveals the similar proportional contributions despite the ~126% greater aggregate discharge in the much shorter more recent period. The lowest CoV values reported in Table 7 are associated with those drainages that host sources of perennial discharge either in the form of an artificial contribution such as the Northern WWTW to the Jukskei River, or a natural contribution such as the karst springs of Quaternary basins A21D (Bloubank Spruit system) and A21G (Skeerpoort River). These 'moderating' contributions are reflected in CoV values of ~50% (Skeerpoort River, Jukskei River and Bloubank Spruit). The poorest CoV value of 106% is associated with the Magalies River (Quaternary basin A21F), and reflects the two-fold impact of excessive groundwater abstraction in the Steenkoppies Basin on the discharge of Maloney's Eye (the source of the river), and agricultural use in the intensively developed fertile valley. Proposed opencast gold mining in the Blaauwbank Spruit[7] valley upstream (west) of the town of Magaliesburg (Fig. 2 in Chapter "Description of the Physical Environment") poses a further threat to the water resources of the Magalies River.

Further inspection of the discharge data for the different drainages in the period 1974 to 2015 indicates that the ~190 Mm$^3$ FSC of Hartbeespoort Dam was equalled or exceeded in 30 of the 42 hydrological years on record (Fig. 19). The years of 'deficit' ($\Sigma Q$ < MAR) all precede 1994. The sustained period of 'below normal' runoff

[6] Hydrological years 2016 and 2017 excluded because of periods of no record at station A2H044.

[7] Not to be confused with the Bloubank Spruit of the COH WHS.

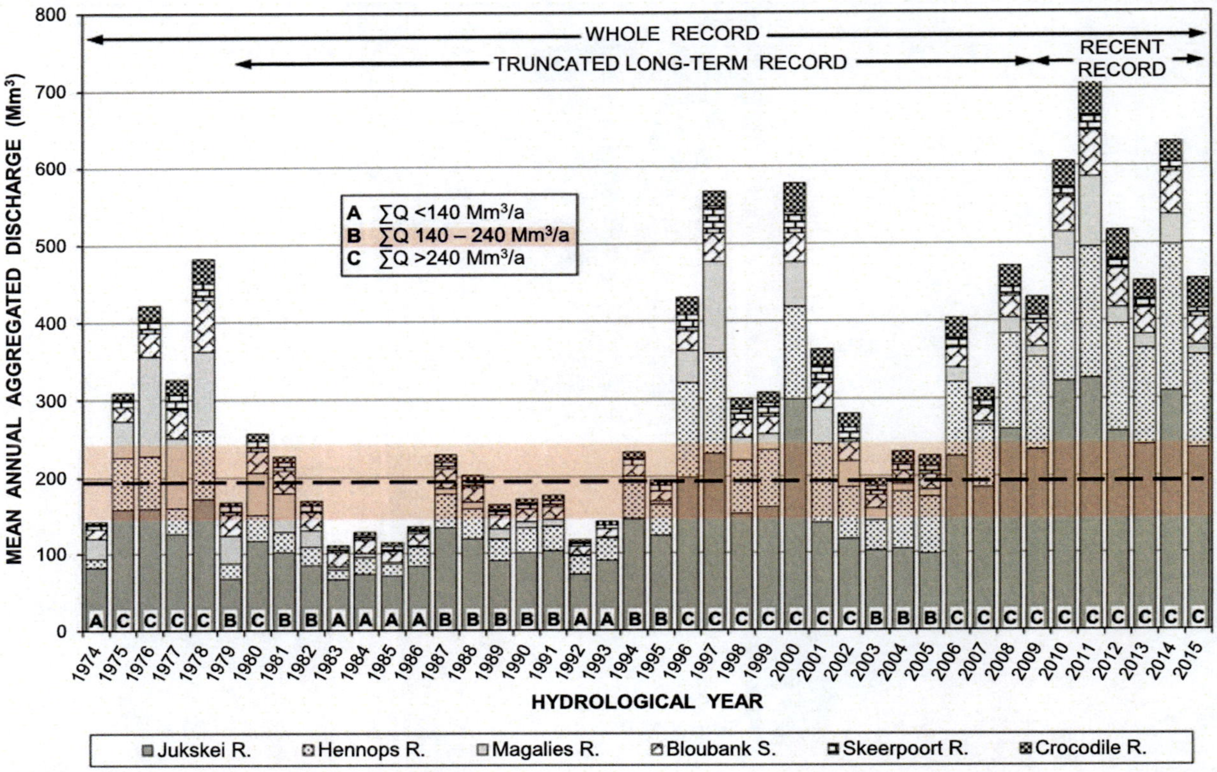


**Fig. 19** Pattern and trend of combined annual discharge ($\Sigma Q$) by main drainages in the Hartbeespoort Dam catchment in the period of common record, compared to the ~190 Mm³ FSC of the dam (horizontal pecked line) that approximates the mean annual runoff to the dam; left to right legend order is stacked from bottom to top in bar graph

commencing in 1979, terminated with > MAR conditions in 1986. The relative constancy of the springflow-driven Bloubank Spruit and Skeerpoort River in the 'lean' runoff years is evident. Equally evident is the sustained period of above normal runoff commencing in 2006, and including the two highest runoff years (2011 and 2014) in the 42-year record.

As shown in Fig. 20, the proportional contribution of the karst basins to discharge in the Hartbeespoort Dam catchment is greater during periods when the aggregate runoff ($\Sigma Q$/a) is < FSC of the dam. An analysis of the $\Sigma Q < 140$ $Mm^3$/a (FSC – 25%) threshold returns a median contribution of 16.3%, and that for the $\Sigma Q^{>}$ 240 $Mm^3$/a (FSC + 25%) threshold a median contribution of 11.8%. The $\Sigma Q$/a = FSC ± 25% (140–240 $Mm^3$) data bracket defines a median contribution of 14.1%. Expressed as a percentage of the FSC of the dam for the 36-year truncated long-term record, the two karst-dominated basins A21D and A21G together delivered a median annual contribution of ~16% (~32 $Mm^3$) in the range 8% (2008) to 24% (1979).

## 5 Surface Water Gains/Losses

### 5.1 Historical Information

In a regional DWS groundwater study in the mid-1980s (Bredenkamp et al. 1986), flow gaugings made at six locations in the Blougat Spruit, the Riet Spruit and the Bloubank Spruit (Table 8) sought to quantify surface water losses to the karst aquifer. The results for the first four localities representing three water loss transects, are graphed in Fig. 21.

The results suggest that flow losses occur more or less uniformly at a rate of ~23 L/s/km along the reach of measurement. Immediately downstream of the Zwartkrans Spring (locality #5, Table 8), the gaugings reflect a surface flow of 258 L/s (22.3 ML/d) in the Bloubank Spruit. This value is replicated in the 2004 water balance for the Zwartkrans Basin presented by van Biljon (2006). Subtracting the 13 L/s of locality #4 from this value suggests a discharge of 245 L/s for the Zwartkrans Spring. The flow

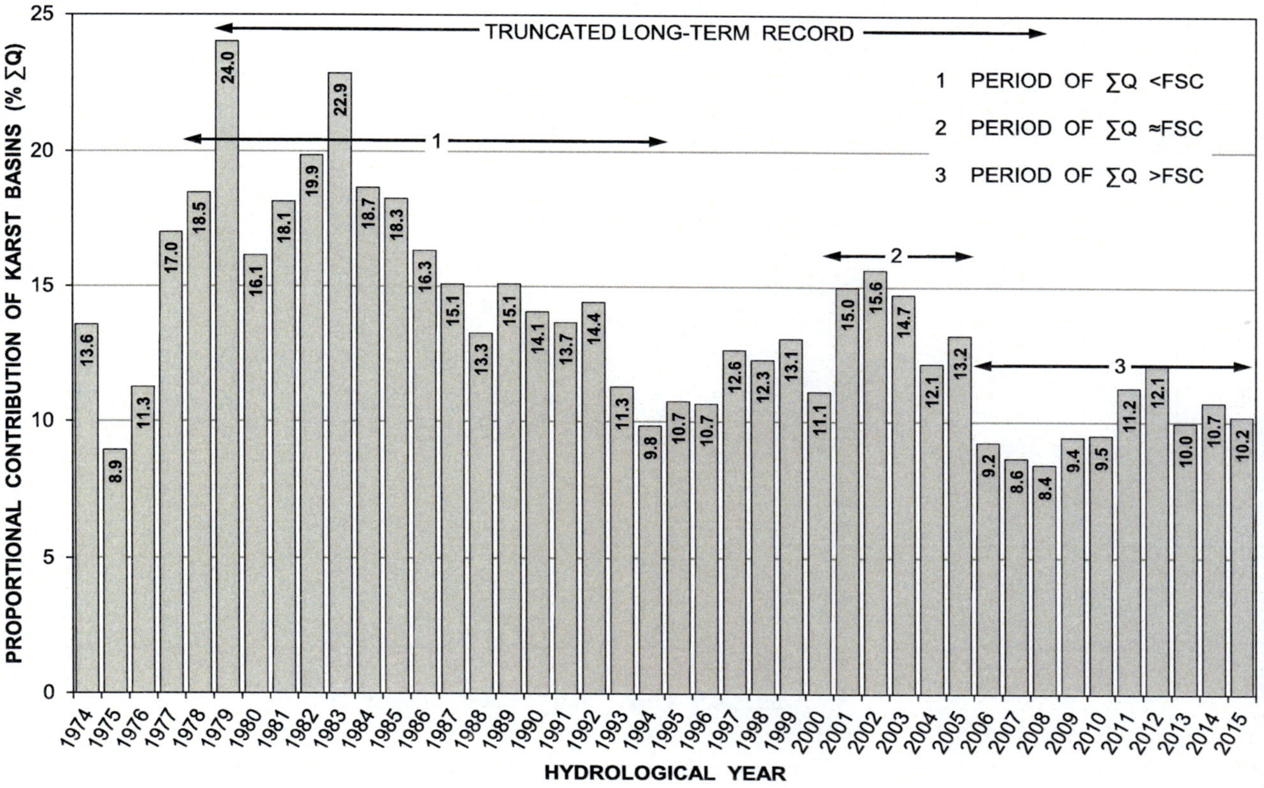


**Fig. 20** Pattern and trend of proportional contribution of karst basins to the total annual discharge ($\Sigma Q$) by main drainages in the Hartbeespoort Dam catchment in the period of common record

**Table 8** Flow gauging results ca. 1985 in the middle reaches of the Bloubank Spruit system (from Fig. 7.1 in Chapter "Physical Hydrogeology" of Bredenkamp et al. 1986)

| Locality # | Description | Distance (km) | Flow | | Gain/Loss Rate (L/s/km) | |
|---|---|---|---|---|---|---|
| | | | L/s | ML/d | | |
| 1 | Blougat Spruit downstream of Percy Stewart WWTW | 0 | 200 | 17.3 | –22.2 | – |
| 2 | Riet Spruit upstream of Oaktree agricultural holdings | 4.5 | 100 | 8.6 | | –40.7 |
| 3 | Riet Spruit upstream of Sterkfontein Cave | 6.0 | 39 | 3.4 | –21.7 | |
| 4 | Riet Spruit opposite Sterkfontein Cave | 7.2 | 13 | 1.1 | | +123 |
| 5 | Riet Spruit downstream of Zwartkrans Spring | 9.2 | 258 | 22.3 | +23.3 | |
| 6 | Bloubank Spruit downstream of Plover's Lake | 14.0 | 370 | 32.0 | | – |

then increases by 112 L/s (43%) between gauging localities #5 and #6, indicating a significant contribution from a source (or sources) not identified by Bredenkamp et al. (1986). The Kromdraai Spring, attributed a discharge of 30 L/s by Bredenkamp et al. (1986), is certainly a contributor, but leaves the balance of 82 L/s unaccounted for. An answer in this regard is provided in Sect. 7.4.3 in Chapter "Physical Hydrogeology".

The results suggest that flow losses occur more or less uniformly at a rate of ~23 L/s/km along the reach of measurement. Immediately downstream of the Zwartkrans Spring (locality #5, Table 8), the gaugings reflect a surface flow of 258 L/s (22.3 ML/d) in the Bloubank Spruit. This value is replicated in the 2004 water balance for the Zwartkrans Basin presented by van Biljon (2006). Subtracting the 13 L/s of locality #4 from this value suggests a discharge of 245 L/s for the Zwartkrans Spring. The flow then increases by 112 L/s (43%) between gauging localities #5 and #6, indicating a significant contribution from a source (or sources) not identified by Bredenkamp et al. (1986).

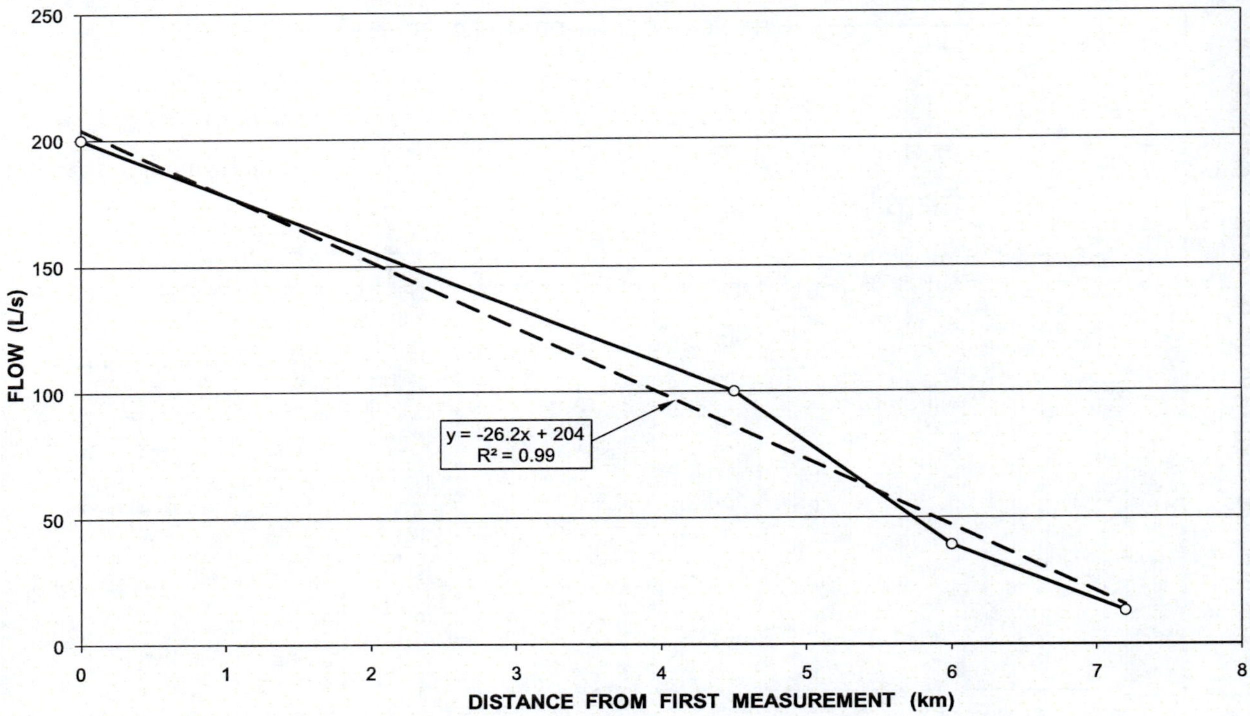


**Fig. 21** Flow reduction with distance along the middle reaches of the Bloubank Spruit system upstream of Sterkfontein Cave (data from Table 8, localities 1 to 4)

The Kromdraai Spring, attributed a discharge of 30 L/s by Bredenkamp et al. (1986), is certainly a contributor, but leaves the balance of 82 L/s unaccounted for. An answer in this regard is provided in Sect. 7.4.3 in Chapter "Physical Hydrogeology".

A more detailed 'final' water balance reported by JFA (2006) reflects a discharge of 208 L/s (18 ML/d) in the Bloubank Spruit (presumably in the vicinity of the Zwartkrans Spring) This comprised a 172 L/s component described as 'Canal leaving stream at Danielsrust' and a 36 L/s component described as 'River flow leaving compartment'. The first (larger) component represents the A-furrow (Sect. 6.2) immediately downstream of the Zwartkrans Spring. Although the gain from locality #5 to locality #6 is also reported as a L/s/km value, it is more realistic to regard much (if not all) of this as concentrated point source input from the Plover's Lake springs (Sect. 7.4.2 in Chapter "Physical Hydrogeology") and the Kromdraai Spring (Sect. 7.4.3 in Chapter "Physical Hydrogeology"). Flow measurements generated in this study (Sect. 7.2.2.2 in Chapter "Physical Hydrogeology") indicate no losses from the Bloubank Spruit in the reach between the Zwartkrans and Kromdraai springs.

Under normal flow conditions, the middle reach of the Riet Spruit between Tarlton and its confluence with the Tweelopie Spruit is dry, having lost the flow from its upper reaches (sustained mainly by the treated effluent discharge from the Randfontein WWTW[8]) to the westerly Steenkoppies Basin by the time it reaches Tarlton (Barnard 1996; Holland et al. 2009). Whilst Hobbs and Cobbing (2007) verified only the former of these circumstances, Holland et al. (2009) present groundwater quality information that shows a WWTW effluent discharge signature in two samples, one sourced from a borehole located south of Tarlton in the Steenkoppies Basin, and the other north-east of Tarlton in the Zwartkrans Basin. The latter instance is provisionally considered to reflect the ingress of surface water bearing a WWTW effluent signature into the Zwartkrans Basin under abnormally high flow conditions. It is under such circumstances that the upper Riet Spruit carries water past Tarlton in a north-easterly direction across the hydrogeologic boundary between the Steenkoppies and Zwartkrans basins, after which it is lost through evaporation, evapotranspiration and infiltration in the river channel within a distance of ~3.4 km. Such circumstances were observed in early-2009 at the time the Holland et al. (2009) sample was

[8] Data presented by Holland et al. (2009) indicate that the Randfontein WWTW discharged on average some 3.1 ML/d (36 L/s) to the upper Riet Spruit in the 5-year period 2004 to 2008.

collected in January 2009, and again in the first six months of 2010.

Flow rates measured in the 2005 winter at various locations on the Tweelopie Spruit as part of the then Harmony Gold (now Sibanye-Stillwater) Environmental Impact Assessment (EIA) (JFA 2006) represent a more recent attempt to quantify surface discharge in this portion of the study area. These measurements record an increase in discharge from 17 L/s (1.47 ML/d) designated as '....... originating from the mine void .........' (i.e. in the mine area), to 29 L/s (2.51 ML/d) where the 4 × 4 trail crosses the Tweelopie Spruit at the Oukraal Lapa in the KGR, to 50 L/s (4.32 ML/d) where the tar road crosses the stream near the entrance to the lion enclosure in the KGR, to 59 L/s (5.10 ML/d) at the Aviary Dam outlet in the KGR, before declining to only 9 L/s (0.78 ML/d) immediately upstream of where the N14 national road crosses the stream. Discounting the possibility of surface water abstraction, these circumstances suggest that 50 L/s (4.32 ML/d) of stream flow was lost to the karst aquifer over a reach length of ~1.5 km. Although the loss rate of ~33 L/s/km is not too dissimilar to the ~23 L/s/km obtained from the Bredenkamp et al. (1986) data, it remains to be established whether groundwater resources (primarily springs) in the KGR contribute as much as 42 L/s (3.63 ML/d) to the flow in the Tweelopie Spruit upstream of the reach in question. Especially the almost doubling of flow between the 4 × 4 crossing and the entrance to the predator sanctuary (a distance of only ~730 m) seems extraordinary in the absence of a known source or sources. Admittedly outdated, especially against the background of known changes in the mine water discharge regime since 2005, these results are reported for the context they provide for similar measurements generated in this study.

Krige (2009) further illuminates the JFA (2006) set of historical flow measurements for the Blougat Spruit, from the point of discharge of Percy Stewart WWTW effluent to a position 5.86 km downstream on the Bloubank Spruit near the Sterkfontein Cave, reporting a loss rate of 26.6 L/s/km in late-winter 2005. This is again similar to the ~23 L/s/km obtained from the Bredenkamp et al. (1986) data, although the Percy Stewart WWTW contribution is represented by a mean annual discharge value of 19.3 ML/d rather than by a contemporaneous SDM. Nevertheless, this result is again reported for the context it provides to this study.

A re-assessment of the Krige (2009) reported loss rate of ~26.6 L/s/km suggests that this is an underestimate for the following reason. The distance of 5.86 km along which the loss rate is calculated, traverses 1.46 km of quartzitic bedrock followed by 4.4 km of dolomitic substrate, the reach ending in the vicinity of Sterkfontein Quarry. The loss rate should therefore be calculated only for the karst reach of 4.4 km, which yields a value of ~35 L/s/km. Even then, the greater portion (2.84 km) of the 4.4 km karst reach traverses chert-poor dolomite of the Oaktree Formation over which losses to the aquifer might reasonably be expected to be minimal. If only the last 1.56 km of river reach traversing chert-rich Monte Christo Formation dolomite is considered, then the loss rate amounts to ~100 L/s/km. Of course, this loss rate is readily accomplished via a single point source infiltration feature such as an instream swallet.

The critical review of the JFA (2006) and Krige (2009) flow data is equally relevant to the Bredenkamp et al. (1986) data reported in Table 8. This is especially true for the flow loss of 100 L/s observed between localities 1 and 2 in Table 8. The last ~1 km of the 4.5 km reach between these localities traverses chert-rich dolomite of the Monte Christo Formation. The loss rate of ~100 L/s/km is similar to that re-calculated from the JFA (2006) and Krige (2009) observations for an overlapping reach.

It is also important to recognise that the results obtained by both Bredenkamp et al. (1986) and Krige (2009) pertain to the hydrodynamic conditions that existed in the relevant river/stream reaches ca. 1985 and 2005, respectively. To what extent these conditions have changed over time has not been investigated, although the similar results obtained in 1985 and 2005 suggest that any changes were negligible and probably absorbed by the unreported error margin(s) associated with the original flow measurements. Despite these qualifications, the historical discharge measurement and flow loss rate data provide an important reference for the similar assessment presented in Sect. 5.2.

## 5.2 Recent/Current Information

### 5.2.1 Tweelopie Spruit and Riet Spruit

Flow gauging at the inlet to the KGR and streamflow measurements at the Aviary Dam outlet in the KGR, as reported by DD Science Laboratory cc. at Western Basin Technical Working Group[9] meetings, unfortunately reflect equivocal results. This is attributed to the lower confidence afforded the accuracy of gauged flow measurements at the downstream station (D Dorling, personal communication). The exceptionally wet 2010 and 2011 summer rainfall seasons again manifested excessive discharge in the Tweelopie Spruit. This was driven mainly by the resumption in late-January 2010 of uncontrolled mine water discharge in the mine area upstream of the KGR. The resulting discharge, comprising a mixture

[9] The Western Basin Technical Working Group (WBTWG) was a forum convened by the Department of Water Affairs to consider the issue of AMD in the West Rand Goldfield. Its bi-annual meetings were attended by a wide range of stakeholders representing environmental lobby groups, mining houses, provincial and national government departments, local and district municipalities and research councils.

of treated/neutralised mine water from the HDS Plant and raw mine water subjected to emergency in-stream liming[10] at the point of release, again peaked at ~50 ML/d as it did in early-2008.

Field (2006) and Walter et al. (2012) recognise estimations of leakage rate through a streambed by any means other than synoptic discharge measurements (Plate 8) as a poor substitute for in-stream flow gain–loss measurements between upstream and downstream gauging stations. Further, that the error will be unquantifiable and will range from insignificant to severe in all instances. The SDM data were obtained using a current meter[11] applied in two basic scenarios, namely (a) the classical cross-sectional flow section (width x depth x velocity) method (Plate 8) and (b) the cross-sectional pipe section (chord length[12] x velocity) method. Computation of the data was facilitated through spreadsheet-based algorithms.

The results of SDMs carried out on two occasions at a number of localities (Fig. 22) in the downstream receiving drainages are presented in Tables 9 and 10. These indicate a general loss of water down to locality #6, followed sequentially by a water gain to locality #7 (Plate 8) and a substantial loss to locality #8 (Table 9). Whereas localities #1, #5, #6 and #7 reflect flow in the Tweelopie Spruit, locality #8 reflects flow in the lower reaches of the Riet Spruit immediately before its confluence with the Blougat Spruit tributary. SDMs carried out at localities #6 and #7 on 11 occasions returned equivocal results (Table 11). The comparatively small differences seldom exceed the error margin (±10%) of the measurements. Nevertheless, the results indicate a gain on six occasions, and a loss on five occasions. The gains fall in the range 9 to 46 L/s/km over the ~1.5 km distance between the sites, and the losses a similar range of 9 to 35 L/s/km. These findings contextualise reports by JFA (2006), Holland (2007) and Krige (2009) that the Tweelopie Spruit loses water[13] to the karst via the Rietfontein Fault (Fig. 22 and Fig. 14 in Chapter "Physical Hydrogeology") that intersects the stream between these two sites (Plate 9).

**Plate 8** View of synoptic discharge measurement with current meter in progress at site F11S12, showing both the relatively 'clean' cross-sectional area of flow (from bottom left to right foot of observer) and the laminar nature of flow over the crest of the weir

The stream flow discharge and loss data are illustrated in Fig. 23. This shows that prior to the 2010 summer, site MRd witnessed surface flow only under exceptional discharge conditions,[14] when under 'normal' circumstances all of the discharge entering the Riet Spruit at Glen Almond via the Tweelopie Spruit was lost primarily to recharge of the karst aquifer before reaching this station. This is exemplified in the measurements recorded on 09 and 22/09/2009 respectively (Table 12 and Fig. 23). These indicate an absorptive capacity defined by an ingress value of ~14 ML/d (~41 L/s/km). A similar situation is described by Katz et al. (1998) and Katz et al. (2004) for sinkhole lakes in the Suwannee and northern Leon counties, respectively, in northern Florida, USA. These lakes overflow when inflow exceeds ~200 L/s (~17 ML/d). Sasowsky and White (1993) describe similar circumstances for the East Fork of the Obey River in north-central Tennessee (USA), reporting that at discharges of <4.5 $m^3/s$ the entire flow of the river disappears into the subsurface. Bailly-Comte et al. (2009) relate the 'swallow capacity' of a cave in a karst watershed in southern France with a losing stream.

[10] Carried out in the period mid-March to mid-May 2010 (Sect. 6.5.1 in Chapter "Chemical Hydrology" and Footnote 59).

[11] Flow velocity determined with an OTT C20 current meter and Z400 signal counter set using impellor # 1–239627 (diameter = 125 mm, pitch = 0.25 m) mounted on a 20 mm diameter rod.

[12] Height of water flow through circular pipe section of known diameter.

[13] JFA (2006) report a single measurement (see Table 4 in Chapter "Description of the Physical Environment", p 46) of a water loss of 4.3 ML/d in late-winter 2005, which translates to a loss of ~40 L/s/km over the 1.25 km distance between the JFA localities.

[14] Caused by excessive and uncontrolled AMD overflow from the mine area together with excess surface runoff associated with very high rainfall events.

**Table 9** Change in flow with distance downstream of the mine area on 05/02/2010

| Locality # and description | | Distance (m) | | Flow | | |
|---|---|---|---|---|---|---|
| | | segmental | cumulative | ML/d | ΔQ (+ gain; – loss) | |
| 1 | KGR inflow | 0 | 0 | 40 | | –8 ML/d [–19 L/s/km] |
| 6 | Aviary dam inlet in KGR | 4905 | 4905 | 32 | + 3 ML/d [+23 L/s/km] | |
| 7 | KBW dam outlet at N14 (F11S12) | 1540 | 6445 | 35 | | –28 ML/d [–83 L/s/km] |
| 8 | Riet Spruit at Malmani road (MRd) | 3900 | 10 345 | 7 | | |

**Table 10** Change in flow with distance downstream of the mine area on 01/04/2010

| Locality # and description | | Distance (m) | | Flow | | |
|---|---|---|---|---|---|---|
| | | Segmental | cumulative | ML/d | ΔQ (+ gain; – loss) | |
| 5 | 4 × 4 Track at Oukraal Lapa in KGR | 2815 | 2815 | 52 | | –16 ML/d [–88 L/s/km] |
| 6 | Aviary Dam inlet in KGR | 2090 | 4905 | 34 | + 6 ML/d [+46 L/s/km] | |
| 7 | KBW Dam (F11S12) | 1540 | 6445 | 40 | | –30 ML/d [–89 L/s/km] |
| 8 | Riet Spruit at Malmani Road (MRd) | 3900 | 10,345 | 10 | | |

**Table 11** Quantification of stream flow gains/losses in the lower reach of the Tweelopie Spruit

| Date | Flow @ aviary dam inlet (ML/d) | Flow @ F11S12 (ML/d) | Flow difference (ML/d) | Flow gain/Loss rate[a] (L/s/km) |
|---|---|---|---|---|
| 22/09/2009 | 19.4 ± 1.9 | 14.9 ± 1.5 | **–4.5** | –35 |
| 05/02/2010 | 32.1 ± 3.2 | 35.2 ± 3.5 | +3.1 | +24 |
| 01/04/2010 | 34.4 ± 3.4 | 40.4 ± 4.0 | **+6.0** | +46 |
| 19/08/2010 | 22.0 ± 2.2 | 25.8 ± 2.6 | **+3.8** | +44 |
| 19/11/2010 | 21.0 ± 2.1 | 22.2 ± 2.2 | +1.2 | +9 |
| 27/07/2011 | 30.2 ± 3.0 | 31.9 ± 3.2 | +1.7 | +13 |
| 14/08/2012 | 23.7 ± 2.4 | 22.5 ± 2.3 | –1.2 | –9 |
| 06/03/2013 | 22.7 ± 2.3 | 20.7 ± 2.1 | –2.0 | –15 |
| 15/08/2013 | 31.9 ± 3.2 | 30.1 ± 3.0 | –1.8 | –14 |
| 15/10/2013 | 31.4 ± 3.1 | 29.6 ± 3.0 | –1.8 | –14 |
| 12/12/2013 | 18.3 ± 1.8 | 22.2 ± 2.2 | **+3.9** | +30 |

[a] Based on a distance of ~1.5 km between localities
Bold text denotes result exceeds the error margin of ±10%

Flow measurements made on 31 occasions at stations F11S12 and MRd (Table 12) further quantify and elucidate the magnitude of surface water loss to the karst aquifer. It is notable that eight of the 14 measured discharges at the downstream station MRd in Period 3 substantially exceed the previous highest measured value of 11.7 ML/d recorded on 06/05/2010 (Table 12). Further, that five of the 15 Period 3 surface flow losses between stations F11S12 and MRd represent the lowest in the record of measurements since flow at station MRd was first recorded. These circumstances suggest that the absorptive capacity of the karst aquifer underlying the losing ~3.9 km reach of the Riet Spruit reached a new equilibrium condition in the 2011 wet season that continued through to the end of $a_h$ 2015.

It is recognised that not all of the water loss is to the karst aquifer of the Zwartkrans Subcompartment. A part thereof is lost through evaporation and riparian evapotranspiration (Sect. 7.4.1 in Chapter "Physical Hydrogeology"). Nevertheless, the measured flow losses reflect mean and median values of ~54 L/s/km (18.1 ML/d) and 51 L/s/km (17.3 ML/d), respectively (Table 12). Although the minimum value of ~20 L/s/km (Period 3) compares favourably with the typical leakage rate of ~25 L/s/km reported in earlier studies (Sect. 5.1), the maximum value of 95 L/s/km (period 2) approaches the re-calculated upper limit of ~ 100 L/s (Sect. 5.1). Fig. 23 suggests that a threshold flow value exists at station F11S12 below which all surface flow is lost to the karst aquifer, leaving the downstream

**Plate 9** Original SDM flow gauging station MRd in separated pipe (below left, 05/02/2010) and one of two identical replacement culverts (below right, 05/10/2010) beneath the reconstructed gravel Malmani Road across the Riet Spruit at this location

station MRd dry. Conversely, flow that exceeds the threshold value at F11S12 results in surface water flow past station MRd. This observation is explored in Fig. 24. The linear regression equation for Period 2 yields a cut-off y-value of ~15 ML/d. The similar equation for Period 3 returns a cut-off value of ~14 ML/d. These flows define the 'threshold' value range.

The existence of two different regression equations associated with different time periods of a single data set supports the conclusion regarding a change in the hydraulic response that characterises the surface and subsurface hydrologic interaction along the subject stream reach. The set of 'before-and-after' historical Google Earth® images presented in Fig. 25 provide a possible explanation. These images (and Plate 10) show the formation of ferrous hydroxide [$Fe(OH)_2$] deposits in the stream channel that might reduce streambed permeability. In any event, it is concluded that the 'absorptive capacity' (and therefore also the 'transmissive capacity') of the epikarst along the ~3.9 km reach of the Riet Spruit between stations F11S12 and MRd, functions with 100% efficiency at surface discharges of up to ~15 ML/d (175 L/s). Above this threshold, the 'absorptive capacity' is exceeded, the per cent infiltration of surface flow reduces, and the volume of surface water discharge passing station MRd is greater.

**Text Box 2** The permanence (insolubility) of iron (oxy)hydroxide precipitates

A key environmental concern regarding the impact of AMD on receiving watercourses is the permanence of the iron oxyhydroxide efflorescence (precipitate) that coats/encrusts stream channels. This depends on a number of factors including pH, temperature, ionic strength and organic ligands, e.g. humic acid (Liu and Millero 1999), and microbial activity and specific surface area (Bonneville et al. 2009). These factors influence the solubility of the precipitate, which ranges from fairly soluble poorly crystalline forms such as freshly precipitated amorphous ferrihydrite and hydrous ferric oxide, to insoluble crystalline forms such as goethite and hematite. The former present as a slimy/gelatinous ferric hydroxide [$Fe(OH)_3$] coating that is quite easily rubbed off the surface on which it has formed. The latter form the hard ferrous hydroxide [$Fe(OH)_2$] crust that 'cements' the streambed and encapsulates stones and vegetation.

The solubility of a substance in a given solvent is measured as the Saturation concentration, where adding more solute does not increase the concentration of the solution. Solubility is commonly expressed as a concentration, e.g. mg/kg, mg/L, molarity, molartity mole, function or similar. The maximum equilibrium amount of solute that can dissolve per amount of solvent is its solubility in that solvent under prevailing specific conditions, e.g. temperature and pressure. Solubility products of ferric oxyhydroxides reported in the literature span more than three orders of magnitude (Bonneville et al. 2009). The fairly soluble poorly crystalline forms pose a lesser risk in regard to permanency than do the insoluble crystalline forms. The solubility of the former are $\sim 10^{-14}$, and of the latter $\sim 10^{-36}$ (Gayer and Woontner 1956). Fox (1988) derived a solubility equation for aqueous ferric

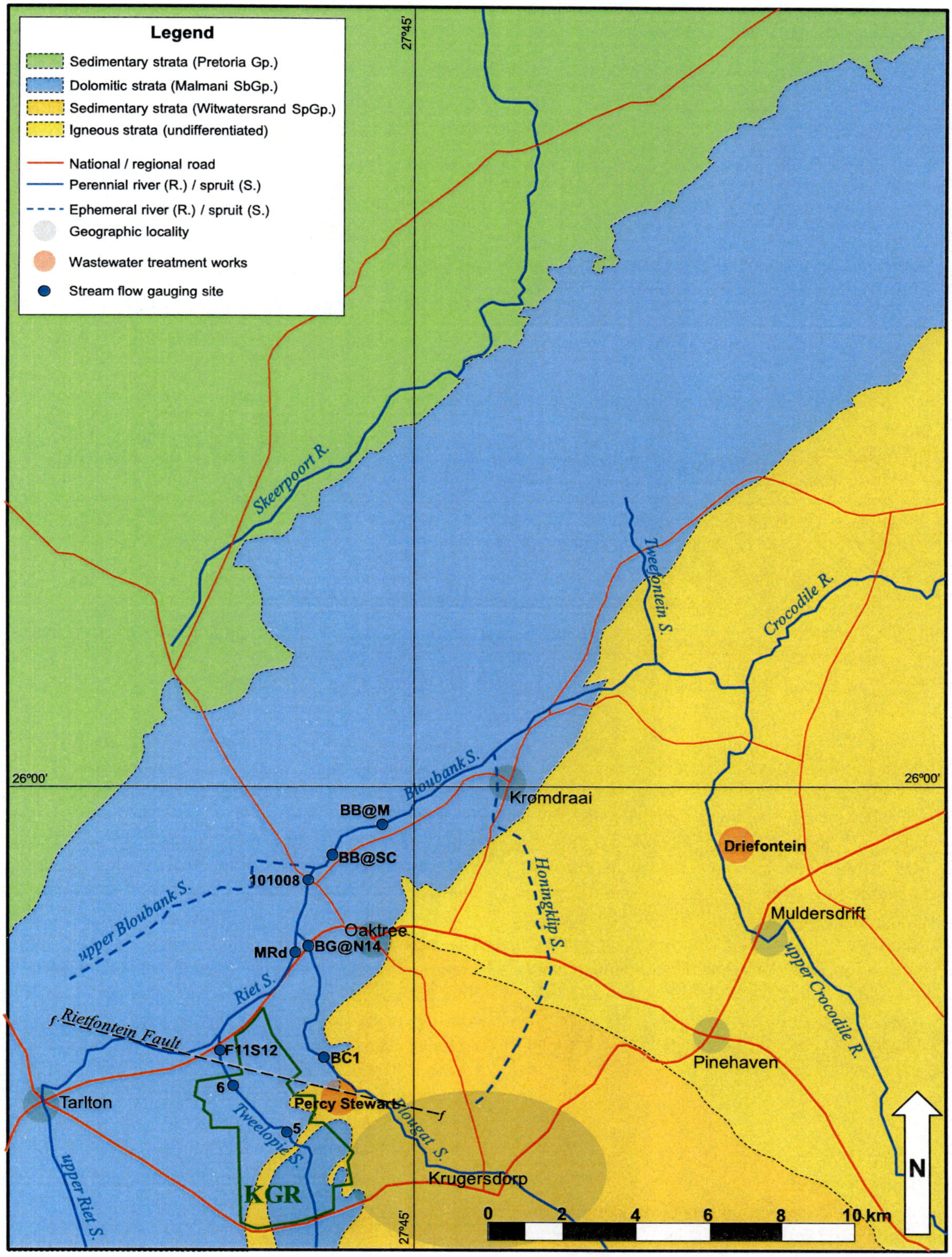


**Fig. 22** Manually gauged stream flow measurement sites employed in this study

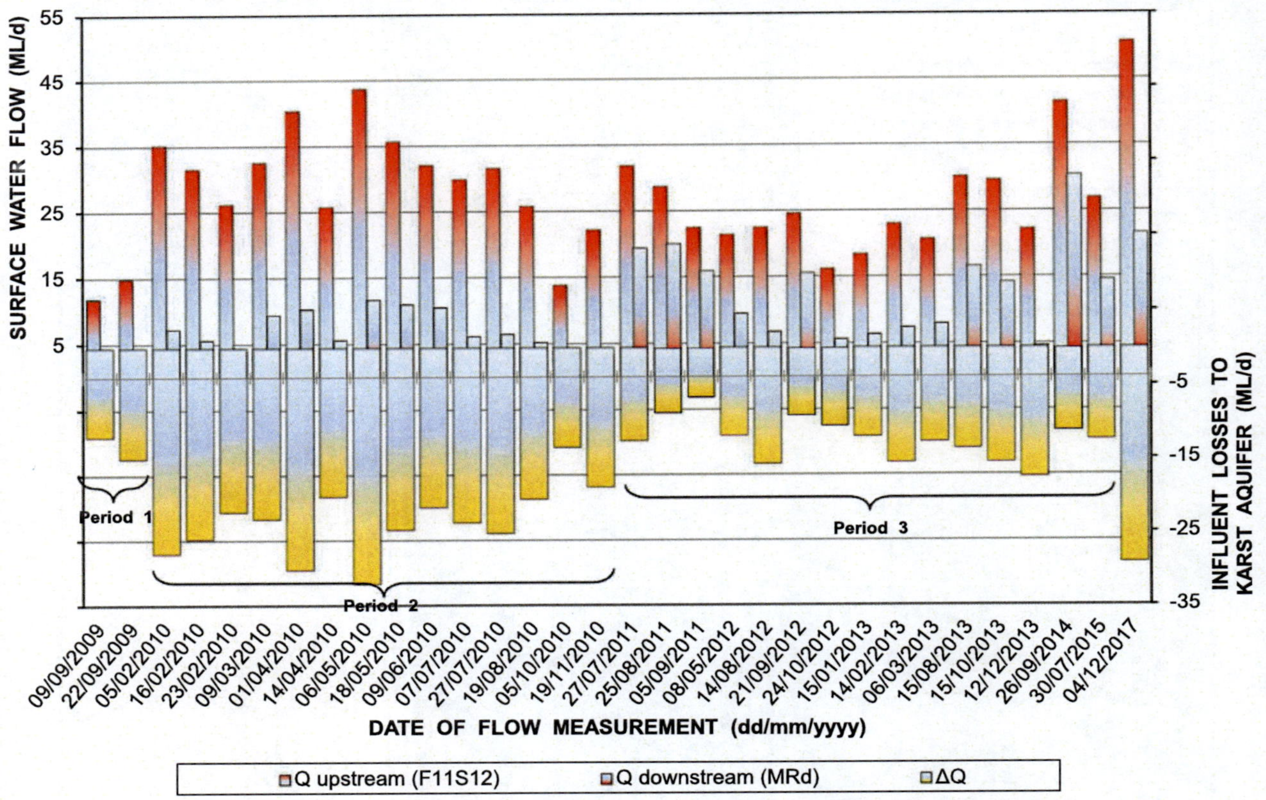


**Fig. 23** Pattern and trend of discharge and flow losses in the lower reach of the Riet Spruit (data from Table 12)

hydroxide colloids that returned a solubility value of $10^{-31.7}$. Appelo and Postma (2009) report solubility product (K) values of common Fe-oxyhydroxides in the range $10^{-37}$ to $10^{-43}$.

As with all solid material exposed to the environment, the insoluble crystalline forms are susceptible to weathering and decomposition. In this instance, this takes the form of erosion through in-stream mechanical abrasion by continuous flow and suspended matter that, with time, will remove the precipitate and transport it downstream. It will eventually erode to fine particulate matter visually indistinguishable from natural stream sediments.

### 5.2.2 Blougat Spruit

Flow measurements carried out at two stations on the Blougat Spruit on 18/05/2010 (Table 13) corroborate the figures reported by Bredenkamp et al. (1986) for flow losses downstream of the Percy Stewart WWTW. The upstream station BC1 (Fig. 26 and Plate 11) is a causeway across this stream located ~1.1 km downstream of the WWTW effluent discharge point (EoP) into this drainage. The downstream station BG@N14 (Fig. 26 and Plate 12) is located at the intersection with the N14 national road. The apparent loss rate of ~23 L/s/km (Table 13) is similar to the ~22 L/s/km reported by Bredenkamp et al. (1986) between localities 1 and 2 (Table 8).

The result, however, is subject to a similar re-calculation as was applied to the Bredenkamp et al. (1986) data, returning a slightly higher loss rate of ~26 L/s/km across the 3.1 km reach that crosses over dolomite. Restricting the losing reach to <1 km over chert-rich dolomite of the Monte Christo Formation increases the loss rate to at least ~81 L/s/km. This again approaches the ~100 L/s/km obtained for the re-calculated historical data (Sect. 5.1). Nevertheless, the observations confirm the Blougat Spruit as a source of allogenic recharge to the karst system in the amount of ~7 ML/d.

### 5.2.3 Bloubank Spruit

This drainage rises on dolomite in the western portion of the Zwartkrans Basin, and traverses only dolomite in its easterly path to the confluence with the Riet Spruit at Oaktree.

**Table 12** Quantification of stream flow losses in the lower reach of the Riet Spruit

| Date | Flow @ F11S12 (ML/d ± 10%) | | Flow @ MRd (ML/d ± 5%) | | Flow loss (ML/d) | | Flow loss rate (L/s/km)[a] | |
|---|---|---|---|---|---|---|---|---|
| 09/09/2009 | 11.9 ± 1.2 | Period 1 | 0 | Period 1 | 11.9 | Period 1 | 35 | Period 1 |
| 22/09/2009 | 14.9 ± 1.5 | | 0 | | 14.9 | | 44 | |
| 05/02/2010 | 35.2 ± 3.5 | Period 2 | 7.3 ± 0.4 | Period 2 | 27.9 | Period 2 | 83 | Period 2 |
| 16/02/2010 | 31.6 ± 3.2 | | 5.7 ± 0.3 | | 25.9 | | 77 | |
| 23/02/2010 | 26.2 ± 2.6 | | 4.0 ± 0.2 | | 22.2 | | 66 | |
| 09/03/2010 | 32.6 ± 3.3 | | 9.4 ± 0.5 | | 23.2 | | 69 | |
| 01/04/2010 | 40.4 ± 4.0 | | 10.3 ± 0.5 | | 30.1 | | 89 | |
| 14/04/2010 | 25.8 ± 2.6 | | 5.7 ± 0.3 | | 20.1 | | 60 | |
| 06/05/2010 | 43.7 ± 4.4 | | 11.7 ± 0.6 | | 32.0 | | 95 | |
| 18/05/2010 | 35.7 ± 3.6 | | 11.0 ± 0.6 | | 24.7 | | 73 | |
| 09/06/2010 | 32.1 ± 3.2 | | 10.5 ± 0.5 | | 21.6 | | 64 | |
| 07/07/2010 | 29.9 ± 3.0 | | 6.2 ± 0.3 | | 23.7 | | 70 | |
| 27/07/2010 | 31.6 ± 3.2 | | 6.5 ± 0.3 | | 25.1 | | 74 | |
| 19/08/2010 | 25.8 ± 2.6 | | 5.3 ± 0.3 | | 20.5 | | 61 | |
| 05/10/2010 | 13.8 ± 1.4 | | 0.4 | | 13.4 | | 40 | |
| 19/11/2010 | 22.2 ± 2.2 | | 3.4 ± 0.2 | | 18.8 | | 56 | |
| 27/07/2011 | Period 3 | 31.9 ± 3.2 | Period 3 | 19.4 ± 1.0 | Period 3 | 12.5 | Period 3 | 37 |
| 25/08/2011 | | 28.7 ± 2.9 | | 20.0 ± 1.0 | | 8.7 | | 26 |
| 05/09/2011 | | 22.5 ± 2.3 | | 15.9 ± 0.8 | | 6.6 | | 20 |
| 08/05/2012 | | 21.4 ± 2.1 | | 9.6 ± 0.5 | | 11.9 | | 35 |
| 14/08/2012 | | 22.5 ± 2.3 | | 6.8 ± 0.3 | | 15.7 | | 47 |
| 21/09/2012 | | 24.6 ± 2.5 | | 15.5 ± 0.8 | | 9.1 | | 27 |
| 24/10/2012 | | 16.2 ± 1.6 | | 5.7 ± 0.3 | | 10.5 | | 31 |
| 15/01/2013 | | 18.4 ± 1.8 | | 6.4 ± 0.3 | | 12.0 | | 36 |
| 14/02/2013 | | 23.0 ± 2.3 | | 7.5 ± 0.4 | | 15.5 | | 46 |
| 06/03/2013 | | 20.7 ± 2.1 | | 8.0 ± 0.4 | | 12.7 | | 38 |
| 15/08/2013 | | 30.1 ± 3.0 | | 16.5 ± 0.8 | | 13.6 | | 40 |
| 15/10/2013 | | 29.6 ± 3.0 | | 14.1 ± 0.7 | | 15.5 | | 46 |
| 12/12/2013 | | 22.2 ± 2.2 | | 4.7 ± 0.2 | | 17.5 | | 52 |
| 26/09/2014 | | 41.5 ± 4.2 | | 30.3 ± 1.5 | | 11.2 | | 33 |
| 30/07/2015 | | 26.9 ± 2.7 | | 14.5 ± 0.7 | | 12.4 | | 37 |
| 04/12/2017 | 50.6 ± 5.1 | | 21.5 ± 1.1 | | 29.1 | | 86 | |
| n | 32 | | 32 | | 32 | | 14[b] | 15[c] |
| Minimum | 11.9 | | 0.0 | | 6.6 | | 39.8 | 19.6 |
| Mean | 27.6 | | 9.8 | | 17.5 | | 70.9 | 36.7 |
| Median | 26.6 | | 7.7 | | 15.5 | | 70.3 | 36.8 |
| Maximum | 50.6 | | 30.3 | | 32.0 | | 95.0 | 51.9 |

[a] Based on a distance of ~3.9 km between localities
[b] Period 2 statistics
[c] Period 3 statistics

Although it is an ephemeral drainage upstream of this junction, these circumstances identify it as a source of autogenic recharge to the Zwartkrans Basin when it carries runoff.

Of greater importance are the Tweelopie and Riet spruit tributaries that deliver mine water from the Western Basin in the south-western portion of the study area. A 17.7 km long stream profile spanning an elevation difference of 150 m is

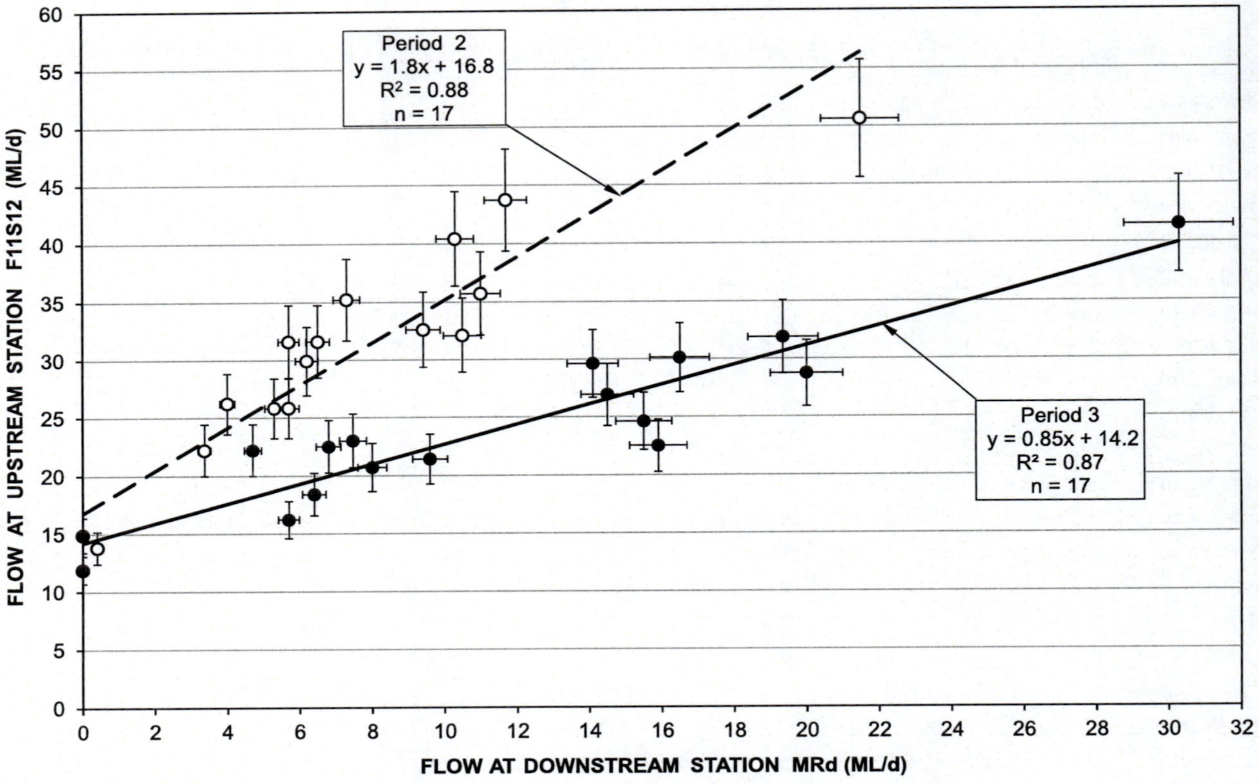


**Fig. 24** Correlation of SDMs at stations F11S12 and MRd in the Riet Spruit valley; error bars denote ±10% at F11S12 and ±5% at MRd (data from Table 12)

presented in Fig. 26. The profile follows the Tweelopie Spruit from the vicinity of Kemp's Cave near the predator sanctuary entrance in the KGR, down the Riet Spruit and the Bloubank Spruit, ending on the south-eastern margin of the carbonate strata near the Nedbank Olwazini Estate (NOE) downstream of Kromdraai.

Traversing the karst terrane from end to end, the profile reveals two distinct gradients. As might be expected, the steeper gradient is associated with the Tweelopie Spruit in the upper 3 km section of the profile. This section cuts orthogonally across the regional strike of the karst strata (Fig. 22) with an average gradient of 0.0164 (1:61). The slope associated with the Riet and Bloubank spruits is ~60% shallower at 0.0068 (1:147) on average. This section of the profile, which is subparallel to the regional SW–NE strike of the strata, shows no marked deviation in slope at the locus of either the Zwartkrans or the Kromdraai springs. The relevance of the 'step' evident in the Bloubank Spruit opposite Sterkfontein Cave (Fig. 26) is discussed later in this section.

Flow measurements carried out in the Bloubank Spruit at localities bracketing (i.e. located upstream and downstream) the Sterkfontein Cave indicate an increase in surface water discharge rather than a loss as observed further upstream. The outcome of two sets of measurements reflecting this are presented in Table 14. The September 2009 value of 260 L/s for the downstream station closely approximates the 258 L/s reported by Bredenkamp et al. (1986) for locality 5 (Table 8) located downstream of the Zwartkrans Spring. Given that the downstream station BB@M is located ~460 m upstream of the Zwartkrans Spring, these circumstances indicate that recent late-winter flow in the Bloubank Spruit upstream of the Zwartkrans Spring is considerably greater than the historical observation suggests. This is also evident in Fig. 11 The discharge of 258 L/s reported by Bredenkamp et al. (1986) is all the more impressive considering that the mid-1980s experienced one of the longer and more severe drought periods in the region, and which was broken only in 1987 (Sect. 4.2.1 in Chapter "Literature Review").

A set of flow measurements made on 16/05/2012 at the localities BB@M and BB@L bracketing the Zwartkrans Spring, returned values of ~306 and ~631 L/s respectively. The difference of ~325 L/s must represent a groundwater contribution that includes the Zwartkrans Spring discharge. The discharge of the spring has been set at ~136 L/s. If it is assumed that the yield of the spring is comparatively constant, then it would appear that ~190 L/s (~16.4 ML/d) represents

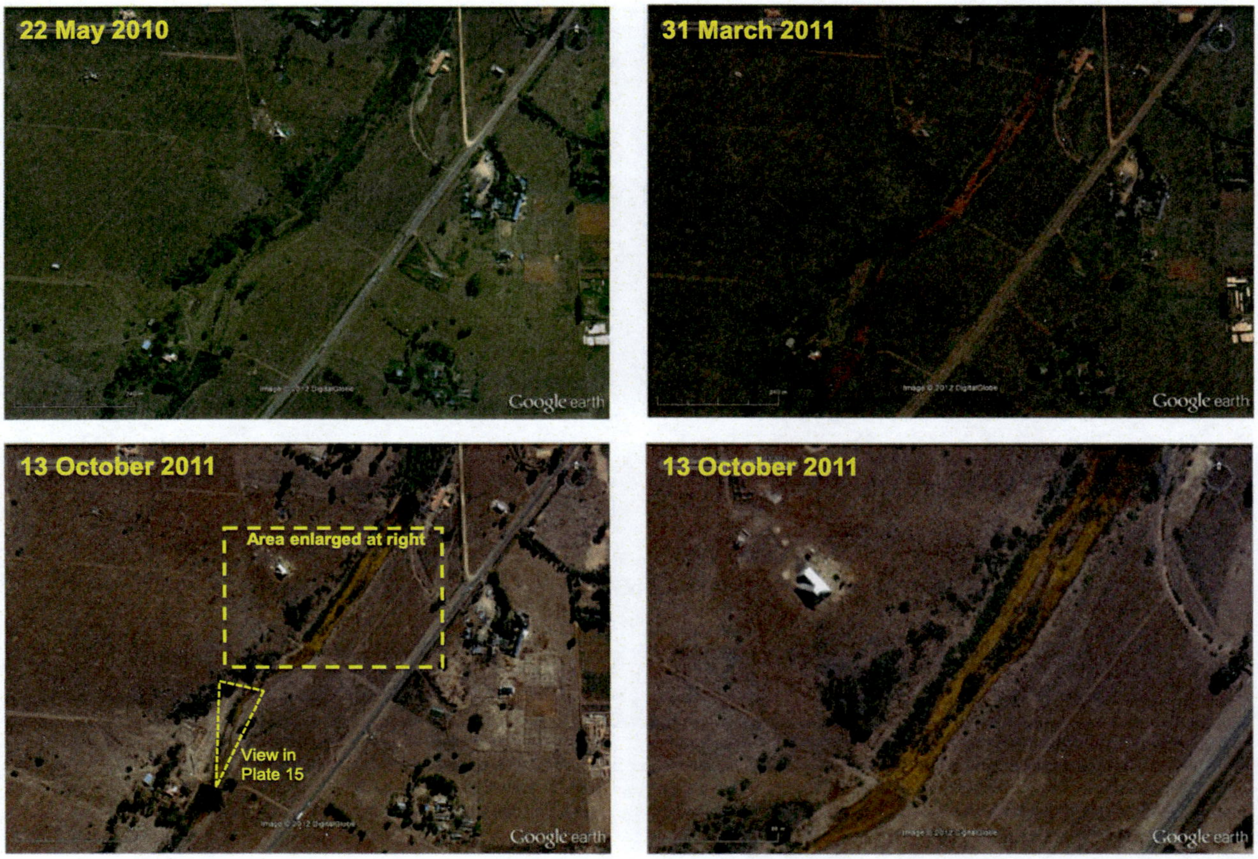


**Fig. 25** Historical Google Earth® images showing the development of ferrous hydroxide deposits in the channel of the Riet Spruit sometime between 22/05/2010 (top left) and 31/03/2011 (top right); stream section located between stations F11S12 and MRd as shown in Fig. 22

groundwater entering the river channel in the ~460 m reach between station BB@M and the spring. This conclusion finds support in the similar electrical conductivity values of 92 and 99 mS/m recorded for the springwater and the river water at station BB@M, respectively, on 16/05/2012. The surface water 'gain' of ~16.4 ML/d is similar to the ~17 ML/d measured on 18/05/2010 for a 2.1 km reach upstream of station BB@M (Table 14). The flow and specific electrical conductivity (SEC) values measured on 16/05/2012 at stations BB@M, BB@L and Zwartkrans Spring provide data for a mass balance calculation (see Sect. 7.4.1 in Chapter "Physical Hydrogeology") that supports the observed total groundwater contribution of ~325 L/s.

Synoptic discharge measurements carried out on 24/10/2012 at the three stations MRd, BG@N14 and BB@M returned flows of ~5.7, ~16.2 and ~29.7 ML/d respectively. The stations MRd and BG@N14 contribute the only visible surface flow to the Bloubank Spruit. As reported earlier, the station BB@M is located ~460 m upstream of the Zwartkrans Spring. The flow at this station should therefore represent the sum of that at stations MRd and BG@N14. The reported SDM values, however, indicate that the flow exceeds the sum of the upstream stations by ~7.8 ML/d. Further analysis of this value (~90 L/s) indicates that it represents a resurgence rate of ~56 L/s/km over the ~1600 m reach extending upstream from this station to opposite Sterkfontein Cave. This is in reasonable agreement with the mean and median loss rates (~70 L/s/km) calculated for the ~3.9 km reach of the Riet Spruit between stations F11S12 and MRd in Period 2 (Table 12). The only possible source of the additional flow is groundwater resurgence in the channel of the Bloubank Spruit.

Flow measurements carried out at station BB@M and ~20 m upstream of Zwartkrans Spring on 15/01/2013 provide additional data that further elucidate the postulated resurgence of groundwater in the stream channel upstream of the spring. The data presented in Table 15 describe the results obtained.

**Plate 10** View looking downstream of relic ferrous hydroxide deposits in the stream channel of the Riet Spruit (see Fig. 25 for photo location)

**Plate 11** Causeway across the Blougat Spruit on the Bergland Instant Lawn property downstream of the Percy Stewart WWTW, and site of station BC1 (photo P Hobbs; date 18/02/2010)

The veracity of the flow measurements presented in Table 15 is again interrogated on the basis of a mass balance calculation as follows:

$$\begin{aligned} Q_Q &= [(Q_D C_D) - (Q_U C_U)] \div C_0 \\ &= [301656 - 215600] \div 665 = 129\,\mathrm{L/s} \end{aligned}$$

The derived groundwater resurgence value of ~125 L/s ($Q_G$, Table 15) is similar to the mass balance calculated value of 129 L/s. These results indicate a surface water gain of ~125 L/s over a stream reach of ~460 m, which equates to ~270 L/s/km. This result offers further insight into the stream gain values reported in Table 14. It is likely that these circumstances are similar to the Pescara River 'linear' or 'streambed' springs described by Salvati (2002) as draining, together with the Giardino Spring, the Morrone Mount Ridge carbonate massif in the Central Apennines of west-central Italy.

The stream gain value of 77 L/s/km reported in Table 14 for the reach immediately upstream of station BB@M clearly falls well short of the ~270 L/s/km reported above for the ~460 m stream reach between station BB@M and the Zwartkrans Spring. This 'discrepancy' is explained by recognising the improbability of a constant 'gain rate' under hydraulic circumstances that describe a varying potentiometric head along a stream reach. These circumstances posit that the potentiometric head that drives groundwater resurgence in the stream channel is greatest at the spring, and decreases with distance upstream to extinction at the point where the potentiometric surface no longer intersects the stream channel. This position is postulated to lie to the north opposite Sterkfontein Cave, and could be marked by the 'step' evident in the channel gradient shown in Figs. 26 and 27.

**Table 13** Results of stream flow measurements in the Blougat Spruit on 18/05/2010

| Station | Description | Flow[a] | | Loss Rate[b] (L/s/km) |
|---|---|---|---|---|
| | | L/s | ML/d | |
| BC1 (upstream) | Causeway on the Bergland property downstream of the Percy Stewart WWTW | 655 | 56.6 | 23.1 |
| BG@N14 (downstream) | Intersection of the Blougat Spruit with the N14 national road | 574 | 49.6 | |

[a] Flow measurement based on the cross-sectional flow method using the average of six measured surface velocities of styrofoam flotsam over known distances. The relatively shallow flow depth (80–150 mm) negates reduction of the calculated flow values by 25% to account for a mean velocity that is less than the surface velocity (Brassington 1998)

[b] Based on a distance of ~3.5 km between the stations

**Plate 12** Bridge culvert at the N14 national road crossing of the Blougat Spruit, and site of station BG@N14

In light of the above, it is more realistic to increment groundwater resurgence in the stream channel over shorter reach lengths of say 100 m rather than 1 km. In the case of the Bloubank Spruit upstream of Zwartkrans Spring, this might amount to unit resurgences of up to 30 L/s per 100 m of stream reach. Assigning a uniform channel width of 5 m to each such reach gives a streambed area of 500 $m^2$ delivering a unit resurgence of up to 0.06 L/s/$m^2$ (6 L/s per 100 $m^2$). This could be expected to gradually decrease in a headward (upstream) direction. Closer to Sterkfontein Cave, the unit resurgence might reduce to only ~0.01 L/s/$m^2$ (1 L/s per 100 $m^2$) of stream channel area before reaching extinction.

On the basis of groundwater level and streambed elevation information, it is postulated that the water table and land surfaces intersect at an elevation of ~1439 to 1440 m amsl in the Sterkfontein Valley (Bloubank Spruit) opposite the caves. These circumstances drive the resurgence of groundwater in the stream at the downstream end of the Zwartkrans Basin between Sterkfontein Cave and the Zwartkrans Spring, and partly explain the observed decrease in SEC values and increase in pH values observed at station BB@M since late-2011 (Fig. 23 in Chapter "Chemical Hydrology", Sect. 6.1.2.6 in Chapter "Chemical Hydrology"). The surface water gain/loss data and associated stream reaches in the Zwartkrans Basin are illustrated schematically in Fig. 27.

An attempt to quantify the 'not measured' loss component (C in Fig. 27) in the upper portion of the Zwartkrans Subcompartment is based on the following simple water balances[15]:

[Minimum loss B] + [Loss C] = [Minimum gain D] + [Zwartkrans Spring]

$$7 + C = 3 + 11.7 ML/d, \text{ so that } C = 7.7 ML/d = 89 L/s$$

[Maximum loss B] + [Loss C] = [Maximum gain D] + [Zwartkrans Spring]

$$32 + C = 23 + 11.7 ML/d, \text{ so that } C = 2.7 ML/d = 31 L/s$$

The above results show the following encouraging agreement:

- the loss of 7.7 ML/d effectively balances the measured groundwater resurgence of 7.8 ML/d reported earlier in this section for component D (Fig. 27);
- the unit loss rates across the ~3.8 km reach of component C amount to 23 and 8 L/s/km, in keeping with values determined for other stream reaches in the study area (e.g. Sections 5.2.1 and 5.2.2); and
- the highest component C loss occurs when the component B loss (across a reach length of ~3.9 km) is least, and vice versa.

[15] The discharge of the Zwartkrans Spring is assigned a 'constant' value of ~11.7 ML/d (~136 L/s) derived from an analysis of SDM data and TDS load calculations (**Sect. 7.4.1**).

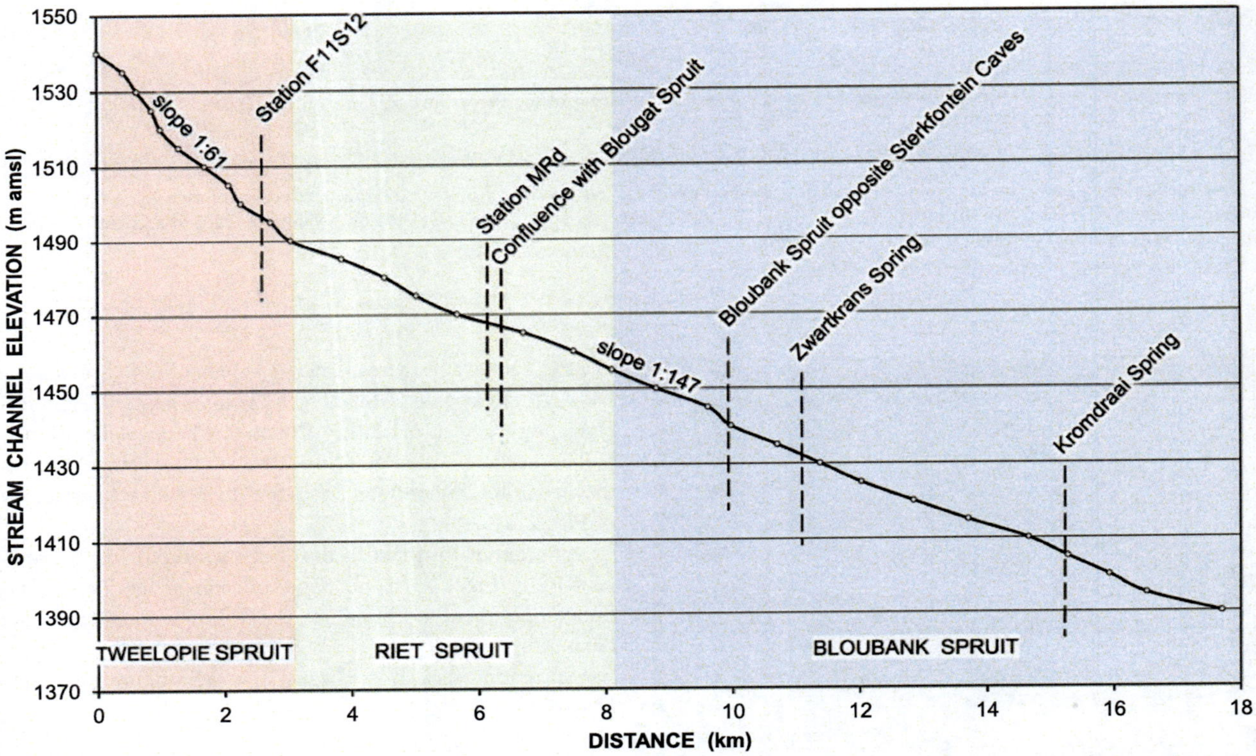


**Fig. 26** Longitudinal profile along the course of the Tweelopie, Riet and Bloubank spruits from Kemp's Cave in the KGR to the south-eastern margin of the carbonate strata at the Nedbank Olwazini Estate

## 6 Surface Water Use

### 6.1 WARMS Data

From an assessment of water use information sourced from the DWS Water Authorisation and Registration Management System (WARMS), Holland and Cobbing (2008) report a total surface water use of 3.5 $Mm^3/a$ as being registered for mainly agricultural purposes in the Zwartkrans Basin. Bredenkamp et al. (1986) and van Biljon (2006) reported values of 5.52 and 6.62 $Mm^3/a$, respectively.

### 6.2 Canals

The Skeerpoort River feeds a canal that starts immediately upstream of where the R400 regional road crosses this drainage on the farm Hartebeesthoek 498JQ. This canal, however, does not influence the flow regime of the Skeerpoort River in the study area, and is therefore not considered further. By comparison, the network of canals/furrows[16] that distribute water to riparian users in the Bloubank Spruit system represent an important component of surface water use in the study area. The following canal systems are identified and described.

- The defunct canal that starts at Glen Almond on the farm Sterkfontein 173IQ and follows the right bank[17] of the drainage for a distance of ~1.7 km, ending on Ptn 8/2 of Sterkfontein 173IQ.
- The so-called A-furrow (canal) managed by the Kromdraai Irrigation Board, as described by Mr M Gomes[18] (personal communication). This canal starts immediately downstream of the Zwartkrans Spring on the farm Zwartkrans 172IQ, and follows the right bank for a distance of ~3.6 km, ending in the Honingklip Spruit

[16] A canal is a properly designed, engineered and constructed water transfer structure over a substantial distance traversing numerous properties, and a furrow a simple (typically hand-dug) trench serving a similar local purpose over a considerably shorter distance typically limited to an individual property.

[17] By convention, the bank to the right of an observer facing downstream.

[18] Current Chairman of the Kromdraai Irrigation Board.

**Table 14** Results of stream flow measurements in the Bloubank Spruit

| Date | Station | Description | Distance (km) | Flow[a] | | Gain Rate |
|---|---|---|---|---|---|---|
| | | | | L/s | ML/d | |
| 09/09/2009 | 101,008 (upstream) | R563 regional road crossing | 2.95 | 225[b] | 19.5[b] | + 3.0 ML/d + 12 L/s/km |
| | BB@M (downstream) | Crossing of Danielsrust road downstream of Sterkfontein Cave | | 260[c] | 22.5[c] | |
| 18/05/2010 | BB@SC (upstream) | Crossing of Swartkrans fossil site road upstream of Sterkfontein Cave | 2.1 | 311[c] | 23.9[c] | +16.9 ML/d +77 L/s/km |
| | BB@M (downstream) | Crossing of Danielsrust road downstream of Sterkfontein Cave | | 472[c] | 40.8[c] | |

[a] Flow measurements carried out with the current meter described in Footnote 28 (Sect. 5.2.1)
[b] Flow measurement based on the cross-sectional flow method (width x depth x velocity)
[c] Flow measurements based on the cross-sectional pipe method (chord length x velocity)
*Note* Although damage to the outlets of two of the three culverts at the downstream locality (BB@M) influence the accuracy of the 18/05/2010 SDMs, the margin of error is not considered greater than the observed difference in flow. This damage had not yet occurred on 09/09/2010, generating greater confidence in these SDMs at station BB@M

**Table 15** Calculation of groundwater resurgence in the Bloubank Spruit upstream of Zwartkrans Spring on 15/01/2013 using measured flows and SEC-based TDS load values

| Flow location | Field SEC (mS/m) | TDS [*C*] (mg/L) | Discharge [*Q*] | | TDS Load | |
|---|---|---|---|---|---|---|
| | | | (L/s) | (ML/d) | mg/s [=*QXCX*] | t/d |
| Downstream[a] | 105 | 735[c] [$=C_D$] | ~ 410[d] [$=Q_D$] | 35.5 | 301,656 | 26.1 |
| Upstream[b] | 108 | 756[c] [$=C_U$] | ~285[d] [$=Q_U$] | 24.6 | 215,600 | 18.6 |
| Calculated difference in TDS load | | | | | 86,056 | 7.5 |
| Groundwater resurgence | 95[e] | 665[c] [$=C_G$] | ~ 125[f] [$=Q_G$] | 10.8 | 83,125 | 7.2 |

[a] Located ~20 m upstream of the Zwartkrans Spring
[b] Located at station BB@M
[c] SEC × 7.0 used as a proxy to derive a theoretical TDS value
[d] Synoptic discharge measurement (SDM) value
[e] Assumed similar to that of the Zwartkrans Spring (Table 13 in Chapter "Physical Hydrogeology")
[f] Derived value from the difference of the SDM values

immediately downstream of the R540 road crossing at the Kromdraai T-junction. This canal (Plate 13) serves 18 properties. The allocation of water from the canal is made on a rotational basis that at any one time provides three properties with 70 L/s each for two days per week,[19] and two half-days every other week. Theoretically, therefore, it carries an absolute minimum permanent discharge of 210 L/s ostensibly fed by the Zwartkrans Spring.[20] The discharge measured at Ptn 8 of Kromdraai 520JQ on 27/07/2010 was ~154 L/s. A second measurement carried out further upstream at Ptn 5 on 13/08/2010 returned a slightly higher value of ~176 L/s. Both measurements are in reasonable agreement with the 172 L/s reported by JFA (2006) (Sect. 5.1). The very high flow conditions in mid-December 2010 not only caused significant damage to this furrow, but also silted up the sluice gate at its entrance, cutting off flow in the furrow. These circumstances continued to prevail as at December 2017, and will remain so for as long as the entrance to the furrow remains obstructed.

- The so-called B-furrow, also managed by the Kromdraai Irrigation Board, which starts at the position where the A-furrow enters the Honingklip Spruit. A weir constructed in this spruit at this position provides the head that feeds water into the B-furrow, which follows the topographic contour along the right bank of the Bloubank Spruit for a distance of ~3.7 km before entering the Bloubank Spruit on the Nedbank Olwazini Estate property. Since the ephemeral Honingklip Spruit delivers an excellent quality surface water into the Bloubank Spruit system (Table 21 in Chapter "Chemical Hydrology"), the quality of water entering the B-furrow in the summer months must similarly reflect a better quality than that

[19] Based on a 6-day week, respecting Sunday as the Christian Sabbath, day of rest.
[20] See Sect. 7.4.1 in Chapter "Physical Hydrogeology" for a discussion of groundwater discharge from the Zwartkrans Compartment.

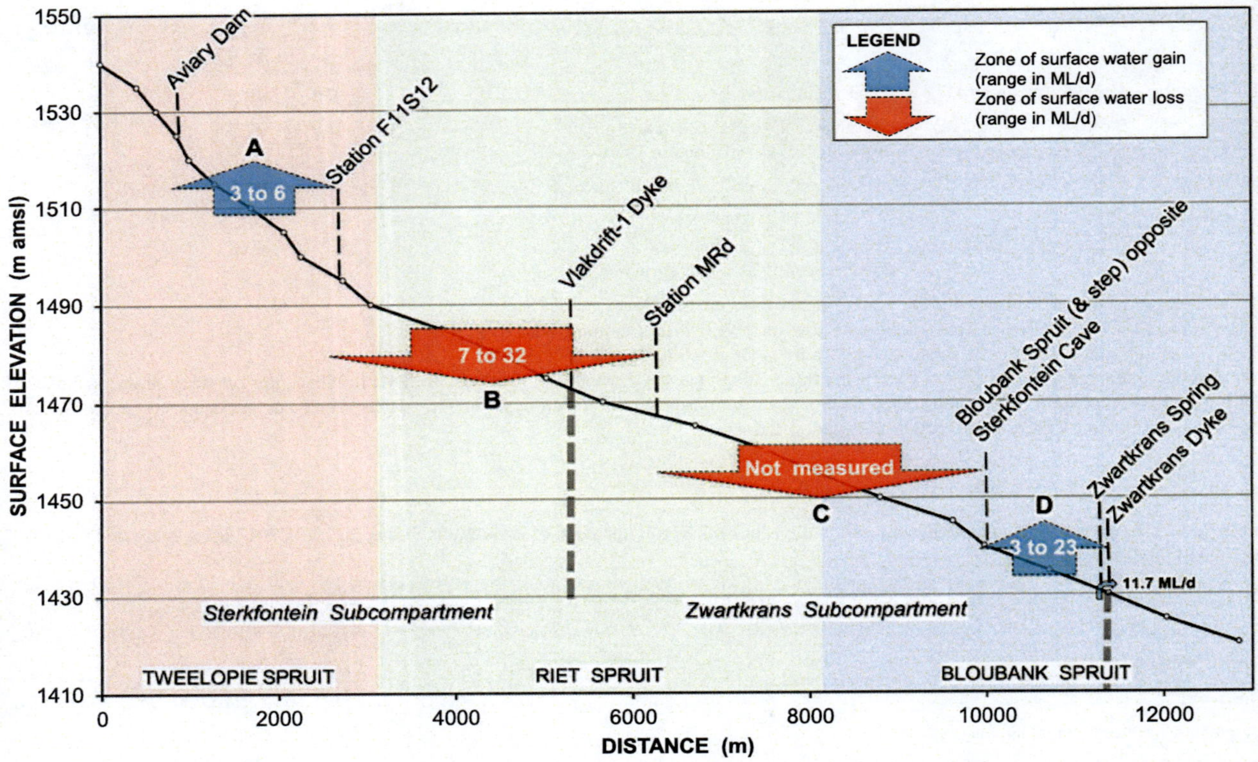


**Fig. 27** Schematic profile illustrating surface water gain/loss zones (A to D) and associated rates along the Tweelopie, Riet and Bloubank spruits from the Aviary Dam in the KGR to the Zwartkrans Spring

delivered by the A-furrow at this location, representing as it does a blend of Bloubank Spruit and Honingklip Spruit water. In the winter months, flow in the B-furrow is sustained solely from the A-furrow. A site inspection of the B-furrow on 27/03/2012 in the presence of others[21] confirmed that it was blocked where the furrow passes beneath the R540 road at the Nirox property. The blockage was caused by damage from trenching carried out in the road reserve in mid-2011, severing the culvert and cutting off flow to the downstream properties (including Kenjara Lodge) that used the water for landscape watering and the irrigation of fodder. The blockage remained in place as at December 2014.

- The furrow that draws water from the Bloubank Spruit immediately upstream of the Nedbank Olwazini Estate (NOE), and follows the right bank of this drainage for a distance of ~3.9 km before re-entering the spruit on the Kloofzicht Lodge Estate. The 'management' of this canal does not fall within the ambit of the Kromdraai Irrigation Board. The NOE draws water from this canal for on-site treatment in a purification plant to potable quality for own use in the average amount of 40 $m^3$/d (14,600 $m^3$/a) (H Carpenter, personal communication). As this canal/furrow also passes through a number of dams, many of which are stocked with trout, it is evident that this water supply function probably represents the main purpose of this structure. The estate also operates its own wastewater treatment plant, using the treated wastewater effluent for landscape irrigation purposes. A water use licence application to the DWS for return of the treated wastewater to the Bloubank Spruit has been unsuccessful (H Carpenter, personal communication) on the basis of unacceptably high nitrate and phosphate levels produced by the treatment process.

## 6.3 Conclusion

Although surface water use on the property is poorly quantified, the ephemeral nature of the good quality streams and the poor water quality of the perennial streams are limiting factors in this regard (Plate 13).

[21] Peter Mills of the COH WHS MA, and local residents/landowners Hannes Fourie and Belinda Cooper.

**Plate 13** View of the ~100-year old A-furrow at Ptn 8 of Kromdraai 520JQ, looking upstream; flow measured at ~154 L/s on date of photo

**Text Box 3** The veracity of SDM-based surface water loss calculations

The synoptic discharge measurements (SDMs) carried out at stations F11S12 and MRd are typically made within 30 minutes of each other. The flow velocity at the upstream station (F11S12) falls in the range 0.3–0.4 m/s (1.1–1.4 km/h), which translates into a travel time of ~5.6 to 4.4 hours between stations. This suggests that the equivalent representative flow at the downstream station will only be manifested ~5 hours after a measurement at the upstream station. As this does not include the retention time of water impounded in dams (e.g. that on Ptn 8/2 of Sterkfontein 173IQ) along the intervening stream reach, nor the lower velocities in wider, shallower and/or more vegetated sections of the river channel, a longer time interval might be considered more applicable. It must also be recognised that flow losses introduced by other agents such as evaporation and evapotranspiration are not factored into the SDM-based inter-station water loss calculations. It is apparent, therefore, that the interval of ~0.5 hours between sequential SDMs does not accommodate changes in discharge that are mutually observable only over periods ≥5 hours.

The p uced by neglect of the above-mentioned factors must be gauged against the understanding that significant variations in discharge along the stream reach in question result from event-driven high runoff associated with rainfall or abnormal mine water discharges. Further, that the SDMs themselves incorporate an inaccuracy estimated to be in the ranges ±5 and ±10%. The lower value is associated with station MRd where the smaller volumes passing through culvert(s) provide for a more rigorous flow measurement than the weir at F11S12, which is assigned an error margin of ±10% also because of the higher flows. Against this background, the short period between measurements is not considered to introduce a significant gauging error when carried out under 'normal' flow conditions. The calculated losses therefore represent the worst case scenario in regard to the possible volume of poor quality water lost to the karst aquifer.

# Chemical Hydrology

Harrison Pienaar, Philip J. Hobbs, Sebinasi Dzikiti, and Ranya Amer

## 1 Surface Water Chemistry

### 1.1 Skeerpoort River

Water quality in this drainage is monitored by the DWS at station A2H034 (Fig. 4 in "Introduction and Background") shortly before its confluence with the Magalies River. Monthly water quality monitoring by the DWS at station A2H035Q01 located ~6 km upstream on the farm Hartebeesthoek 498JQ in the middle reaches of the Skeerpoort River, was discontinued in July 1982 after only ~2.5 years. This was followed by four opportunistic analyses, one each in late 1986, early 1996, late 1998 and early 1999. These data have not been evaluated in this study. The water quality delivered by the Nouklip Spring located ~7 km upstream in the John Nash Nature Reserve, is monitored at station A2H033 on the Grootvlei Spruit tributary of the Skeerpoort River. Representing mainly groundwater, these data are evaluated in Sect. 7.7 in Chapter "Chemical Hydrogeology". An evaluation of the A2H034 water chemistry record, which represents the integrated Skeerpoort River quality, yields the information presented in Table 1.

It is evident from Table 1 that the SANS (2015a) recommended limit for Class 1 drinking water is easily met in respect of the 95%ile value of each analyte reported. A high degree of confidence is afforded the analytical results on the basis that the median and mean electrical balance (EB) values fall well within the ± 5% limit (i.e. 95–105%) that defines an acceptable error margin for freshwater (Hem, 1985). The mean and median TDS:SEC ratio value of 8.1 is a useful factor for calculating the TDS of this water from an SEC measurement. It is also evident that sulfate ($SO_4$) typically comprises only 3% of the TDS associated with this water. The small variability in the chemical composition of Skeerpoort River water evident in the narrow 2$\sigma$ (5%ile to 95%ile) range shown for each analyte, with the possible exception of bicarbonate ($HCO_3$), in Fig. 1, is attributed to the karstic Tweefontein, Nouklip and Nash springs (Table 8 in Chapter Physical Hydrogeology) as main perennial source. The aggregate discharge of ~300 L/s (~9.5 $Mm^3$/a) approximates the long-term median discharge of the Skeerpoort River at station A2H034 (Sect. 2 in Chapter "Physical Hydrology"), which also explains the dominant MgCa–$HCO_3$ composition of the surface water at this station.

### 1.2 Bloubank Spruit System

#### 1.2.1 Locus of Mine Water Discharge (Decant)

The monitoring of surface water chemistry (quality) in the study area is heavily biased in favour of the headwaters of the Bloubank Spruit system. Especially the Tweelopie Spruit, which encompasses the locus of mine water discharge/decant (LoD) in its upper reaches, is intensively monitored by SS. Water quality monitoring in the mine area originally focussed on two sources. One of these was the composite chemistry represented by the raw mine water

P. J. Hobbs: Deceased

H. Pienaar (✉) · P. J. Hobbs
Smart Places, Council for Scientific and Industrial Research, Pretoria, South Africa
e-mail: hpienaar@csir.co.za

H. Pienaar
Hebei University of Engineering, Handan, China

S. Dzikiti
Department of Horticultural Science, Stellenbosch University, Cape Town, South Africa
e-mail: sdzikiti@sun.ac.za

R. Amer
Genetic Engineering and Biotechnology Research Institute (GEBRI), City of Scientific Research and Technological Applications (SRTA-City), New Borg El Arab City, Alexandria, Egypt
e-mail: ramer@srtacity.sci.eg



P. J. Hobbs et al. (eds.), *Anatomy of a South African Karst Hydrosystem*, Cave and Karst Systems of the World,
https://doi.org/10.1007/978-3-030-95829-9_6

**Table 1** Statistical analysis of Skeerpoort River water chemistry at station A2H034 in the period January 1976 to March 2017

| Variable/Analyte | Statistical parameter | | | | | | | SANS (2015a)[1] |
|---|---|---|---|---|---|---|---|---|
| | n | 5%ile | Mean | Median | 95%ile | SD | CoV (%) | |
| pH ($-\log_{10}\alpha_{H+}$) | 1631 | 7.5 | – | 8.2 | 8.6 | 0.4 | 4 | 5.0 – 9.7 |
| SEC (mS/m) | 1718 | 28.5 | 32.2 | 32.5 | 35.2 | 2.6 | 8 | $\leq 170$ |
| TDS (mg/L) | 1407 | 229 | 262 | 265 | 284 | 20.3 | 8 | $\leq 1200$ |
| Ca (mg/L) | 1512 | 26.1 | 31.6 | 32.1 | 35.0 | 3.2 | 10 | n.s |
| Mg (mg/L) | 1503 | 18.4 | 21.3 | 21.4 | 23.7 | 2.3 | 11 | n.s |
| Na (mg/L) | 1472 | 1.0 | 2.9 | 2.8 | 4.5 | 2.1 | 72 | $\leq 200$ |
| K (mg/L) | 1480 | 0.15 | 0.72 | 0.49 | 0.72 | 0.93 | 129 | n.s |
| Cl (mg/L) | 1513 | 1.5 | 4.2 | 4.0 | 8.6 | 2.8 | 66 | $\leq 300$ |
| $SO_4$ (mg/L) | 1514 | 2.0 | 6.7 | 6.1 | 13.7 | 4.2 | 62 | $\leq 500$ |
| $HCO_3$ (mg/L) | 1522 | 156 | 191 | 194 | 212 | 18.3 | 10 | n.s |
| $NO_3 + NO_2$ (mg N/L) | 1577 | 0.09 | 0.65 | 0.60 | 1.27 | 0.53 | 81 | $\leq 11$ |
| F (mg/L) | 1474 | 0.05 | 0.17 | 0.14 | 0.38 | 0.17 | 99 | 1.5 |
| Si (mg/L) | 1568 | 5.0 | 6.2 | 6.1 | 7.3 | 1.3 | 21 | n.s |
| EB (%) | – | 4.0 | 1.1 | 1.0 | –1.4 | – | – | ±5 |
| TDS:SEC | – | 8.0 | 8.1 | 8.2 | 8.1 | – | – | n.s |
| $SO_4$:TDS (%) | – | 1 | 3 | 3 | 5 | – | – | n.s |

[1]Standard health-related limit for consumption of 2 L/d over 70 years by a 60 kg person

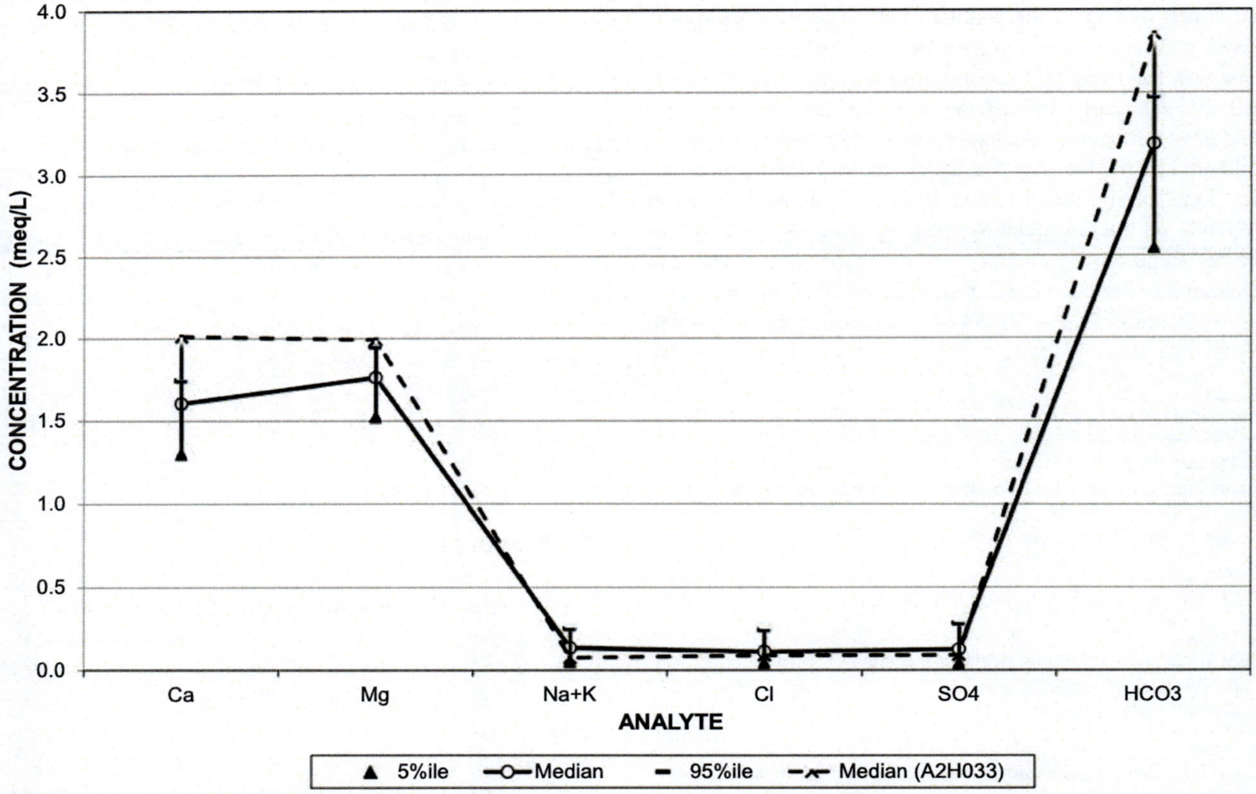


**Fig. 1** Variability of Skeerpoort River water major ion chemistry at station A2H034 in the period January 1976 to March 2017 (data from Table 1)

(RMW) collected in the HDPE-lined dam (Plate 3 in Chapter "Physical Hydrology") at the Black Reef Incline (BRI) from various point and diffuse sources. The other was at the point of discharge (end-of-pipe or EoP) (Plate 4 in Chapter "Physical Hydrology") of treated/neutralised mine water (TMW) into the stream immediately upstream of the R24 regional road and the Krugersdorp Game Reserve. The main point sources of mine water discharge are the Black Reef Incline shaft, #17 Winze and #18 Winze. Individual monitoring of the RMW chemistry from these mine structures commenced in early to mid-2009.

Mine water discharge to the Tweelopie Spruit ideally comprises only a TMW component from the HDS treatment plant (Fig. 12 in Chapter "Physical Hydrology"). The RMW collected in the HDPE-lined dam at the BRI shaft (Plate 3 in Chapter "Description of the Physical Environment" and Plate 3 in Chapter "Physical Hydrology") overflows from this dam in times of excessive discharge (Sect. 3.1 in Chapter "Physical Hydrology"), and leaves the mine property untreated. The chemistry of the two mine water components is summarised in Tables 2 and 3 respectively, and illustrated in Fig. 2 for the analytes pH, SEC, $SO_4$ and Fe. An important aspect of the data presented in Tables 2 and 3 is the recognition of detection limits (DLs) (see Table 6 for temporal variation of DLs) and the influence of non-detect values on the statistical results. For example, the metal analytes Al, Cr, V and Pb reflect a particularly high proportion of non-detect values (conversely few > DL or 'actual' values), which contextualises the veracity of the associated statistical results. Also indicated is the extent to which the reported analytes exceed the SANS (2015a) standard health-related limit for drinking water.

Also shown in Tables 2 and 3 is the TDS:SEC ratio that for natural surface water in the region is typically 7.5 (Text Box 1 in Chapter "Study Framework"), but for RMW has a value of 12.9 (Table 2), and for TMW a value of 10.9 (Table 3). Sulfate typically comprises ~63% of the TDS (Tables 2 and 3) irrespective of whether it is RMW or TMW.

The pattern and trend associated with the RMW produced by the two most productive point sources of AMD, viz. the BRI and #18 Winze (#18 W), is illustrated in Fig. 2 for pH, SEC, $SO_4$ and Fe. The linear regression analyses applied to each of the variables reflect a general increase in pH and decrease in SEC, $SO_4$ and Fe levels. A further notable characteristic is the greater variability in Fe concentration in the first half of the BRI record (Fig. 2), compared to the later half of the record. This is associated with the apparent 'levelling out' of values evident in the latter portion of each record (to the right of each of the vertical chevron lines in Fig. 2), an explanation for which remains obscure. Perhaps a clue in this regard is the timing of the start of the 'levelling out' trend, which is similar for the interdependent variables

**Table 2** Statistical analysis of raw mine water chemistry delivered by the BRI in the period January 2005 to September 2017

| Variable/Analyte | Statistical parameter | | | | | | | SANS (2015a)[2] |
|---|---|---|---|---|---|---|---|---|
| | $n_T$ ($n_A$)[1] | 5%ile | Mean | Median | 95%ile | SD | CoV (%) | |
| pH ($-\log_{10}\alpha_{H+}$) | 501 | **3.2** | – | **4.3** | 6.2 | 1.0 | 21 | 5.0–9.7 |
| SEC (mS/m) | 500 | **361** | **471** | **476** | **571** | 74 | 16 | ≤170 |
| TDS (mg/L) | 263[3] | **4434** | **5316** | **5240** | **6309** | 618 | 12 | ≤1200 |
| $SO_4$ (mg/L) | 501 | **2860** | **3773** | **3700** | **4790** | 725 | 19 | ≤500 |
| Fe (mg/L) | 499 | **283** | **712** | **642** | **1271** | 345 | 49 | ≤2 |
| Mn (mg/L) | 272 (271) | **40** | **86** | **60** | **217** | 84 | 97 | ≤0.5 |
| Al (mg/L) | 262 (104) | 0.005 | **1.042** | 0.100 | **4.965** | 3.1 | 302 | ≤0.3 |
| Cr (mg/L) | 174 (120) | 0.001 | 0.031 | 0.001 | 0.013 | 0.276 | 892 | ≤0.05 |
| V (mg/L) | 174 (17) | 0.001 | 0.003 | 0.001 | 0.012 | 0.007 | 230 | ≤0.2 |
| Pb (mg/L) | 215 (129) | 0.001 | 0.008 | 0.004 | **0.030** | 0.011 | 147 | ≤0.01 |
| Co (mg/L) | 174 (173) | 0.07 | **1.27** | **1.05** | **2.80** | 0.84 | 66 | ≤0.5 |
| Ni (mg/L) | 262 (262) | **0.20** | **1.70** | **0.80** | **7.28** | 2.12 | 125 | ≤0.07 |
| $U_T$ (μg/L) | 265 (244) | 7 | **61** | **55** | **135** | 46 | 76 | ≤15[4] |
| TDS:SEC | 263 | 10.8 | 12.9 | 12.9 | 15.3 | 1.4 | 11 | n.s |
| $SO_4$:TDS (%) | 263 | 54 | 62 | 62 | 71 | 6 | 10 | n.s |

[1]$n_T$ denotes total number of analyses; $n_A$ denotes number of analyses exceeding detection limit
[2]Standard health-related limit for consumption of 2 L/d over 70 years by a 60 kg person
[3]Monitoring of this variable commenced in August 2009
[4]The SANS (2015a) only recognises U; it uses the WHO (2011) *Guidelines for Drinking-water Quality* as reference, which sets a provisional value for $U_T$ of 30 μg/L (Table 9.2, footnote c, p 211)
Bold text denotes value exceeds standard health-related limit as described in note 1

**Table 3** Statistical analysis of treated/neutralised mine water chemistry discharge to the Tweelopie Spruit in the period April 2005 to September 2017

| Variable/Analyte | Statistical parameter | | | | | | | SANS(2015a)[1] |
|---|---|---|---|---|---|---|---|---|
| | $n_T$ ($n_A$)[(2)] | 5%ile | Mean | Median | 95%ile | SD | CoV (%) | |
| pH ($-\log_{10}\alpha_{H+}$) | 458 | **3.0** | – | 7.0 | 9.9 | 1.8 | 26 | 5.0–9.7 |
| SEC (mS/m) | 458 | **305** | **382** | **381** | **459** | 56.5 | 15 | ≤170 |
| TDS (mg/L) | 458 | **3207** | **4168** | **4200** | **4982** | 582 | 14 | ≤1200 |
| $SO_4$ (mg/L) | 458 | **2051** | **2647** | **2680** | **3102** | 379 | 14 | ≤500 |
| Fe (mg/L) | 457 (401) | 0.006 | **21.0** | 0.8 | **100.3** | 73.5 | 351 | ≤2 |
| Mn (mg/L) | 264 (260) | **0.9** | **15.6** | **9.5** | **60.1** | 19.5 | 125 | ≤0.5 |
| Al (mg/L) | 261 (87) | 0.002 | 0.154 | 0.010 | 0.100 | 0.871 | 567 | ≤0.3 |
| Cr (mg/L) | 174 (27) | 0.001 | **0.120** | 0.001 | **0.400** | 0.507 | 423 | ≤0.05 |
| V (mg/L) | 174 (6) | 0.001 | 0.001 | 0.001 | 0.002 | 0.000 | 44 | ≤0.2 |
| Pb (mg/L) | 262 (42) | 0.001 | 0.009 | 0.002 | **0.030** | 0.015 | 171 | ≤0.01 |
| Co (mg/L) | 174 (171) | 0.020 | 0.221 | 0.050 | **1.480** | 0.515 | 234 | ≤0.5 |
| Ni (mg/L) | 262 (256) | 0.022 | **0.348** | **0.100** | **1.625** | 0.899 | 259 | ≤0.07 |
| $U_T$ (μg/L) | 174 (119) | 5.9 | **75.9** | **16.0** | **672.0** | 191.1 | 252 | ≤15 |
| $U_D$ (μg/L) | 204 (88) | 5.1 | **15.1** | 9.9 | **42.3** | 18.4 | 122 | ≤15[(3)] |
| TDS:SEC | 458 | 9.0 | 11.0 | 10.9 | 13.3 | 1.4 | 12 | n.s |
| $SO_4$:TDS (%) | 458 | 55 | 64 | 64 | 74 | 7.3 | 12 | n.s |

[1]Standard health-related limit for consumption of 2 L/d over 70 years by a 60 kg person

[2]$n_T$ denotes total number of analyses; $n_A$ denotes number of analyses exceeding detection limit

[3]The SANS (2015a) only recognises U; it uses the WHO (2011) *Guidelines for Drinking-water Quality* as reference, which sets a provisional value for $U_T$ of 30 μg/L (Table 9.2, footnote c, p. 211)

Bold text denotes value exceeds standard health-related limit as described in note 1.

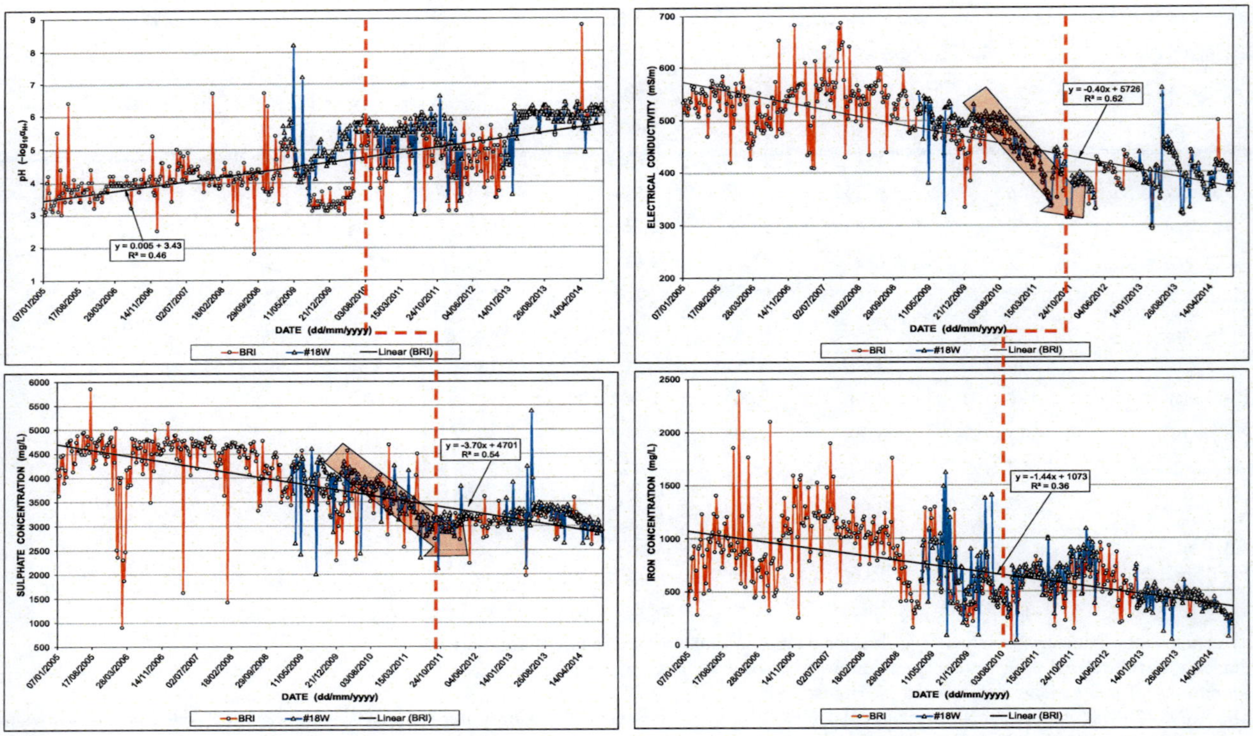


**Fig. 2** Long-term pattern and trend of pH (top left), EC (top right), $SO_4$ (bottom left) and Fe (bottom right) associated with raw mine water produced by the BRI and #18 Winze point sources of AMD; significance of broken linking chevron lines and arrows explained in text

pH and Fe ca. mid-2010, and later (ca. mid-2011) for the interdependent variables EC and $SO_4$. Neither of the ca. mid-2010 or mid-2011 timelines have any relevance in the timeframe of mine water management intervention measures[1] (Text Box 1 in Chapter "Physical Hydrology"). The mid-2010 timeline, however, occurs ~6 months after the onset of uncontrollable RMW decant (Sect. 3.1 in Chapter "Physical Hydrology") precipitated by the very high rainfall in December 2009 and January 2010 (Sect. 2 in Chapter "Description of the Physical Environment").

It is postulated within the framework of the above observations, that the pH and opposite concomitant Fe response patterns reflect the influence of 'fresh' and oxygenated recharge to the flooded mine void. The increase in pH might be driven by redox reactions in the mine water body, leading to the subaerial precipitation of iron hydroxide in the flooded mine workings. The reactions driving this response took a longer time to manifest steeply over a period of ~1 year (arrowed in Fig. 2) on the SEC (and less markedly on the $SO_4$) of the RMW discharge. Understanding the hydrogeochemistry behind these observed responses will advance the understanding of AMD evolution in the Western Basin and, by extrapolation, possibly elsewhere in the region and globally.

The variability associated with the RMW produced not only by the BRI and #18 Winze, but also that associated with #17 Winze and the mixture drawn from the BRI Dam for treatment in the HDS Plant, is illustrated in Fig. 3. The data set of $n = 91$ values spans the period of common record December 2012 to September 2014. The results present a striking reflection of the complexity that informs the hydrochemistry of AMD at source in the Western Basin revealing, amongst others, the following characteristics:

- BRI consistently produces the least variable RMW chemistry;
- BRI and #18 Winze consistently produce the most similar RMW chemistry;
- #17 Winze consistently produces RMW with the most variable pH and SEC; and
- the low median pH and Fe values associated with the RMW mixture compared to each of the sources are consistent with the production of free hydrogen ions and precipitation of iron associated with hydrolysis (Sect. 5.1) in the oxygenated BRI Dam.

Turning attention to the TMW produced by the HDS mine water treatment plant and discharged to the Tweelopie Spruit at the end-of-pipe, an inspection of the pH, SEC, $SO_4$, Fe and Mn associated with this water is presented in Fig. 4. The patterns and trends are readily explained by various factors such as the onset of uncontrollable RMW discharge (Sect. 3.1 in Chapter "Physical Hydrology") in February 2010, which explain the period of consistently low pH values (<4) and elevated Fe concentrations in the second half of the record. The marked difference in mean and median Fe values in this water (reflected in the CoV value of 356%, Table 3), is attributed to the bias introduced by the ca. May 2006, February 2010 to July 2012 and March to July 2014 concentrations illustrated in Figs. 4 and 8. The efficacy of the mine water treatment process especially since the commissioning of the immediate and short-term intervention measures in mid-2012 and expanded in mid-2013 (Text Box 4 in Chapter "Physical Hydrology"), is clearly reflected in the more acceptable pH, Fe and $SO_4$ levels during periods of controlled mine water discharge.

### 1.2.2 Tweelopie Spruit

The chemistry of Tweelopie Spruit water is monitored by SG at five localities from where it leaves the mine property down to the confluence with the Riet Spruit north of the KGR, a distance of ~6.4 km. These localities are identified as (a) the inlet to the KGR, (b) the Hippo Dam, (c) the Charles Fourie Dam, (d) the Aviary Dam and (e) the (Krugersdorp) Brickworks Dam (station F11S12). The monitoring therefore provides a record of the mine water quality discharge that eventually enters the karst terrane of the COH.

The weekly monitoring of the variables pH, EC and $SO_4$ dates back to May 2004. The results for the upstream Hippo Dam and the downstream F11S12 (Brickworks Dam) are presented in Fig. 5 (pH), Fig. 6 (SEC) and Fig. 7 ($SO_4$). The KGR inlet data are excluded due to their close proximity to the Hippo Dam, and consideration of the fact that the residence time of this water in the Hippo Dam renders the data for the latter location more representative of the mine water entering the Tweelopie Spruit. The Charles Fourie Dam and Aviary Dam data are excluded because of their congruence as intermediate compositions between those observed at the Hippo and Brickworks dams (Volume 2).

The patterns revealed in the graphs indicate the trend and variation in the respective variable concentrations through the KGR over time. De Villiers and Mkwelo (2009) have demonstrated the use of $SO_4$ as a measure of surface water deterioration in the Olifants River catchment similarly influenced by mine water impacts, using the long-term record of this analyte to assess the efficacy of the DWSs surface water quality monitoring.

It is clear from Fig. 5, and to a lesser extent from Figs. 6 and 7, that the severest and most sustained mine water impact on the receiving surface water environment of the Tweelopie Spruit commenced ca. February 2010. This is unequivocally shown in the somewhat shorter record of Fe (Fig. 8) and

[1] Which have no influence on the chemistry of the raw mine water produced by any AMD source.

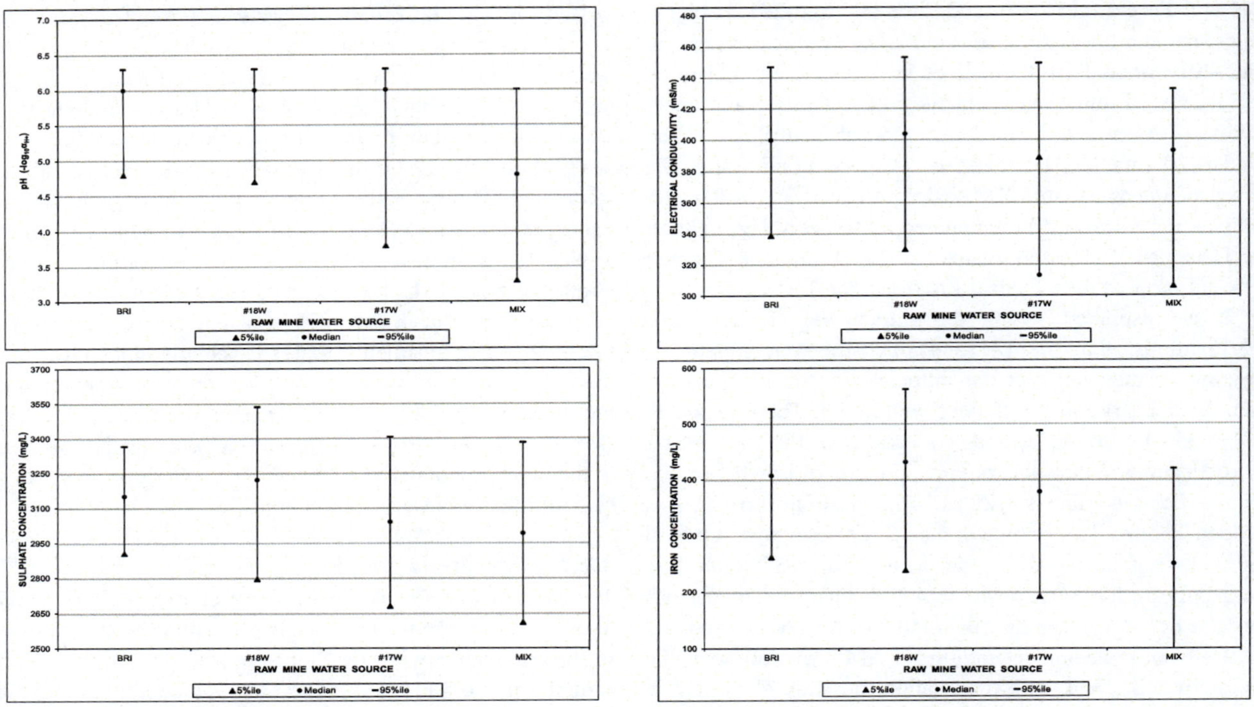


**Fig. 3** Variability of pH (top left), EC (top right), $SO_4$ (bottom left) and Fe (bottom right) associated with raw mine water produced by the BRI, #18 Winze and #17 Winze point sources of AMD, and the RMW mixture (MIX) drawn from the BRI Dam for treatment in the HDS Plant

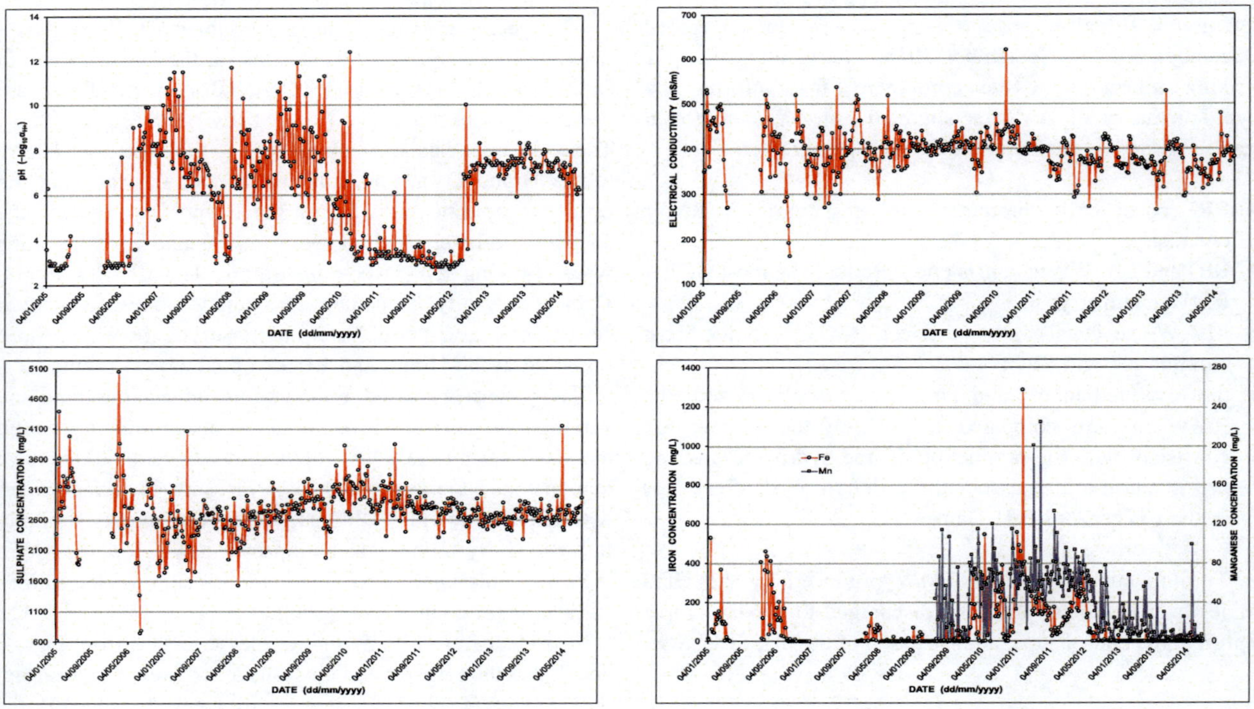


**Fig. 4** Long-term pattern and trend of pH (top left), EC (top right), $SO_4$ (bottom left) and Fe and Mn (bottom right) associated with treated/neutralised mine water produced by the HDS mine water treatment plant and discharged into the Tweelopie Spruit at the end-of-pipe

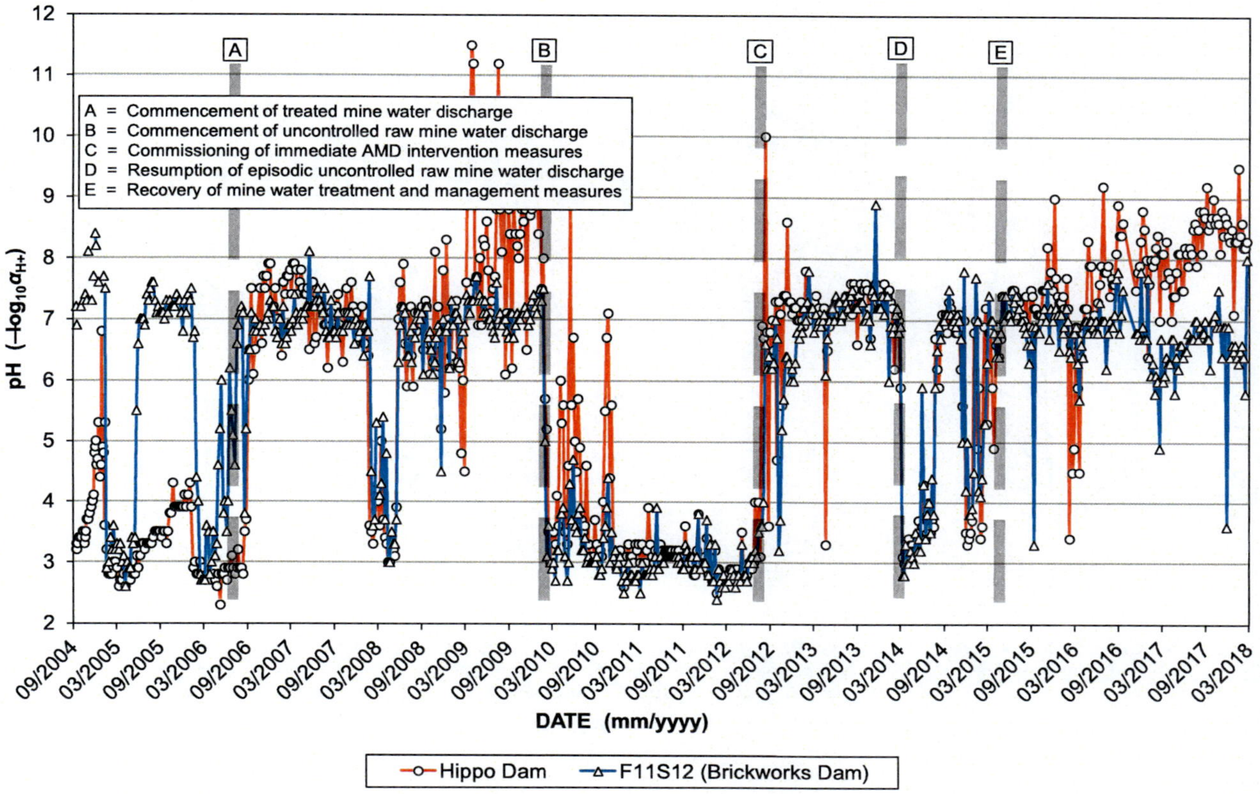


**Fig. 5** Pattern and trend of pH of Tweelopie Spruit water from May 2004 to September 2014

especially the Mn values (Fig. 9). Prior to this, the gradual 'increase' in impact since 2004 is evidenced by the increasing SEC and $SO_4$ trends up to the persistent elevated levels in the second half of the record period (Figs. 6 and 7). The discharge regimes associated with these observations are discussed in Sect. 3.1 in Chapter "Physical Hydrology".

A scrutiny of the differences between the three periods of record defined by the divisions recognised in Figs. 5, 6 and 7 returns the information presented in Table 4 and illustrated in Fig. 10. The graphs not only illustrate the differences, most notably the 'poorer' values in the B–C period of severest mine water impact, but also reveal other salient aspects such as the following.

- The generally greater variability in analyte/variable concentrations at the upstream Hippo Dam station compared to the F11S12 station.
- The typically lower analyte/variable concentrations (except pH) at the F11S12 station compared to the upstream Hippo Dam station.

The statistics for total uranium ($U_T$) indicate the need for caution when considering U concentrations under circumstances where significant differences between the 'normal' A–B and C–D periods and the 'abnormal' B–C period are evident. The mean and median values at both stations in the B–C period exceed the SANS (2015a) limit of 0.015 mg/L associated with a chronic health risk attributable to ingestion over an extended period. In contrast, the mean and median values for the other two periods meet the limit. A factor that is not considered, however, is the potential under-evaluation of U concentrations associated with a 'biased' sample collection regimen. For example, Winde et al. (2004) report differences in U concentrations between daytime and night-time samples, the latter typically returning higher concentrations. Reasons put forward for this include preferential groundwater exfiltration and lower pH values at night when biological decalcification and the associated U immobilisation is also at a minimum.

The association of mine water with the presence of other metals such as aluminium (Al), cadmium (Cd), copper (Cu), mercury (Hg), nickel (Ni), lead (Pb) and zinc (Zn) raises concern for the levels of these elements in the water discharged into the environment. The weekly surface water quality monitoring programme carried out by SG, and which forms the basis for the analysis presented in Sect. 1.2, has since January 2005 included the metal analytes listed in Table 5 for the periods indicated (Table 6).Interrogation of

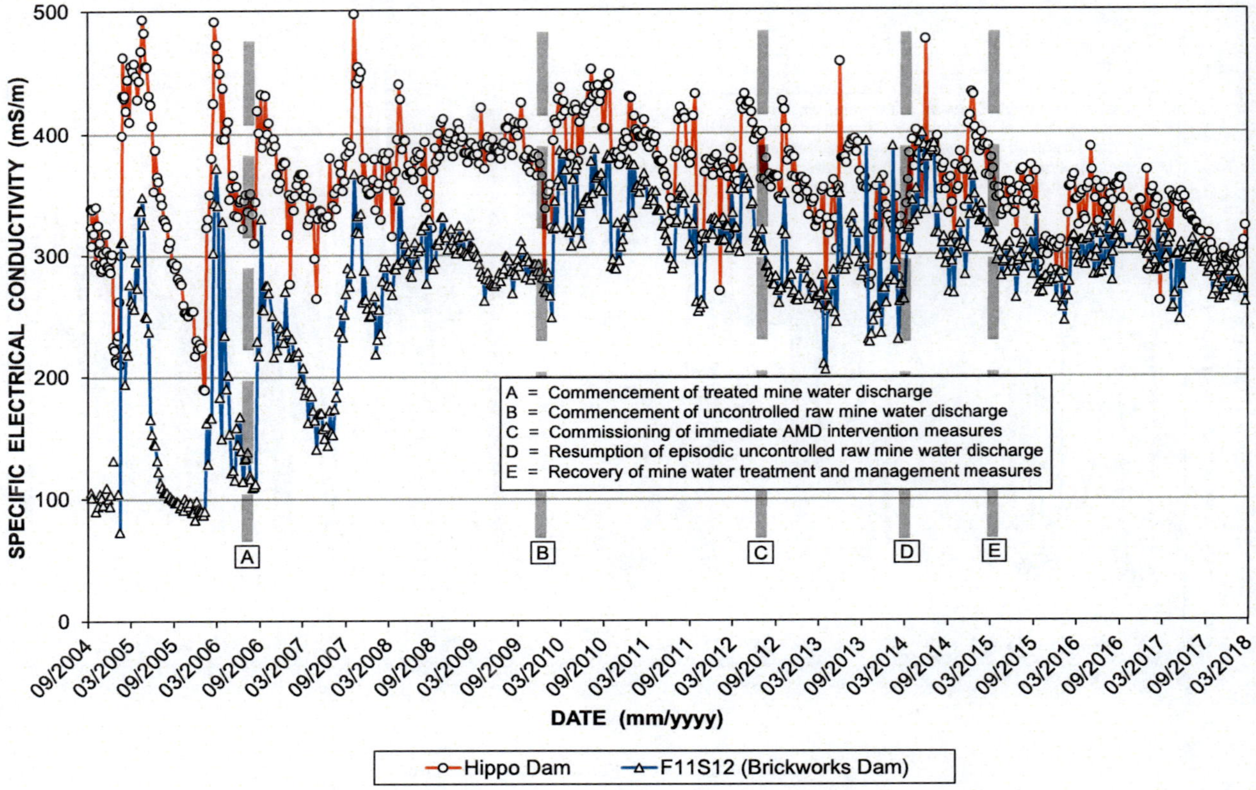


**Fig. 6** Pattern and trend of SEC of Tweelopie Spruit water from September 2004 to September 2014

the data set presented in Table 5 returns the statistical results given in Table 6. The results reflect concern for the analytes Mn, Ni and Cd at the 50%ile level.

A slightly different picture emerges when an interrogation of the data set represented in Table 5 is carried out for the first three periods of distinctly different discharge chemistry recognised in Figs. 5, 6, 7 and 8 and 9 and Table 4. This is presented in Table 7, and despite the differences with Table 6, the results again reflect concern for the analytes Mn, Ni and Cd at the 50%ile level.

Although Heath et al. (2009) report a poor correlation between electrical conductivity and sulfate in the Lower Vet River Catchment, the congruence that is apparent between the SEC values and the $SO_4$ concentrations illustrated in Figs. 6 and 7 prompts an inspection of the correlation between these hydrochemical variables for the study area. In a study of AMD produced by the copper and sulfur mines at Avoca, South-East Ireland, Gray (1996) reports the use of EC to predict the $SO_4$ concentration in both AMD and contaminated surface waters using regression analysis. The correlation between these variables at three sites in the Tweelopie Spruit and at station A2H049 at the end of the Bloubank Spruit system, computed on the basis of the $SO_4$:SEC ratio, is summarised in Table 8. Graphs of the correlations are presented in Volume 2. The similarity of the linear regression expressions for the Tweelopie Spruit stations and that of Gray (1996) is apparent. Also evident is the improvement in the regression coefficient value ($R^2$) from 0.63 at the upstream Hippo Dam station to 0.92 at station A2H049.[2] This result is at odds with the finding by Gray (1996) that the strength of the correlation improves with increasing AMD contamination which, in this instance, would hold for the upstream Hippo Dam and Charles Fourie Dam stations.

Unlike other ions, $SO_4$ is unaffected by variations in pH and is not significantly removed by sorption or precipitation processes (Gray 1996). A provisional explanation for the improved correlation is therefore attributed to an initially less stable water chemistry which approaches chemical equilibrium with distance along the flow path. This might be associated with the continued precipitation of dissolved ferrous iron as ferric iron (oxy)hydroxide (FeOOH).

Whatever the reason for the improved correlation along the flow path, it is evident that the reasonable correlation

[2] It is shown in **Annexure B** that the A2H049 data support either an exponential trend or a bi-linear trend with an inflection point at 60 mS/m, where values <60 mS/m reflect the mitigating influence of the karst springs, and values ≥60 mS/m the militating influence of AMD.

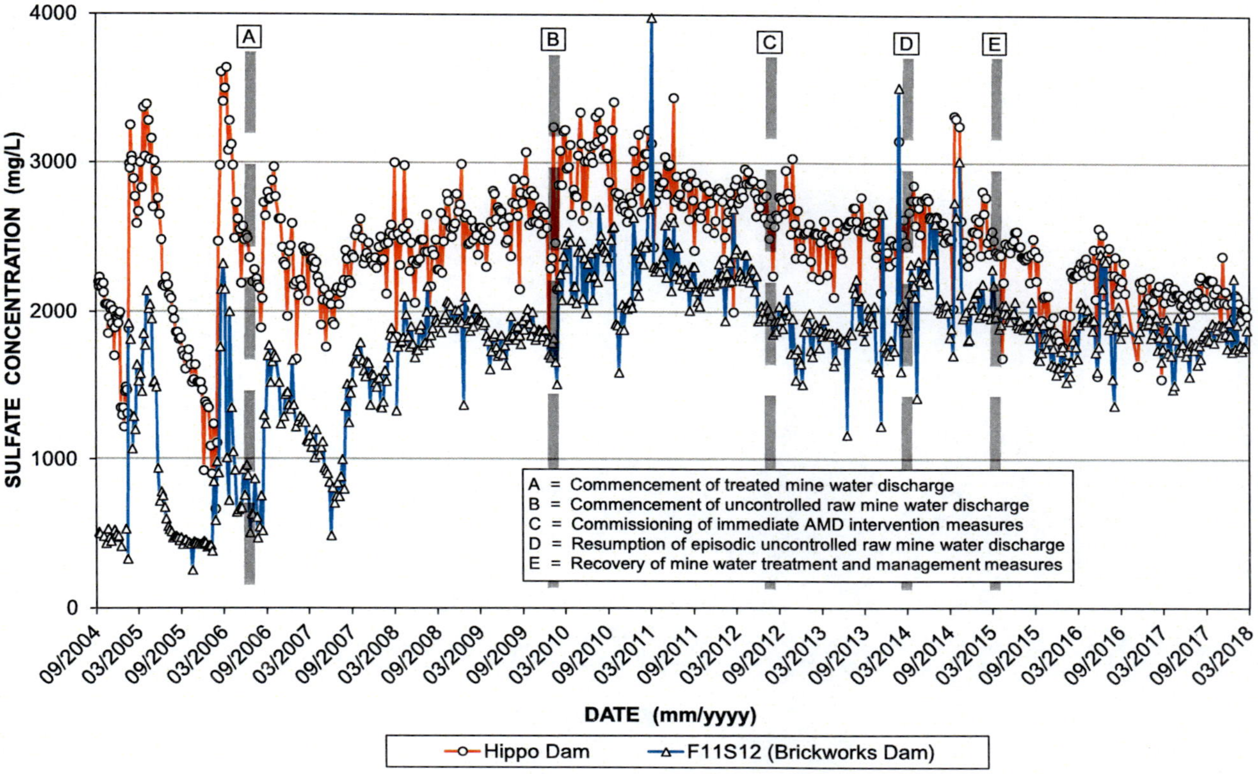


**Fig. 7** Pattern and trend of $SO_4$ levels in Tweelopie Spruit water from September 2004 to September 2014

allows the readily determined SEC value to be used to calculate an approximate $SO_4$ concentration for AMD-impacted surface water in the study area. This is especially useful under circumstances where $SO_4$ analysis in the field is difficult due to a lack of ion-specific electrodes. As noted by Gray (1996), it also has application where the optimal measure of dilution of a water sample prior to analysis of $SO_4$ in a laboratory (e.g. by ICP) must be determined.

The data presented in Table 9 summarise the results of the DWS water quality monitoring[3] at station F11S12. This period precedes the 2010, 2011 and 2014 hydrological years when the chemistry of water in the drainages receiving mine water experienced a much greater raw mine water influence (see discussion earlier in this section). Nevertheless, it is evident from Table 9 that the historical median SEC, TDS, $SO_4$ and Mn values exceed the SANS (2015a) limits for these analytes, and that this exceedance extends to Fe and Al at the 95%ile level. All these six variables/analytes show non-compliance in regard to their mean values. Despite these circumstances, the mean and median electrical balance values remain acceptable. Sulfate typically comprises ~65% of the TDS, very similar to the ~64% that characterises RMW (Table 2) and TMW (Table 3). Together with the Ca–$SO_4$ composition (Fig. 11), the mine water dominance in the surface water at this location is again unequivocal, as is the significant variance associated with Ca and especially $SO_4$.

### 1.2.3 Riet Spruit

Although the Riet Spruit rises in Riebeeck Lake in Randfontein (Sect. 3 in Chapter "Physical Hydrology"), it is only from its confluence with the Tweelopie Spruit tributary at Glen Almond north of the Krugersdorp Game Reserve that it gains relevance for the COH, and for the following reasons:

- the Riet Spruit receives the mine water discharge from the Western Basin via the Tweelopie Spruit;
- the confluence is located just outside the south-western boundary of the COH, and surface water drains into the WHS via both the surface and subsurface from this position; and
- the receiving reach of the Riet Spruit is a losing drainage contributing allogenic mine water recharge to the karst aquifer of the Zwartkrans Basin (Sect. 5.2.1 in Chapter "Physical Hydrology").

[3] Data sourced from the DWS in November 2013 indicates that the most recent analysis on record for station F11S12 is dated 29/09/2008.

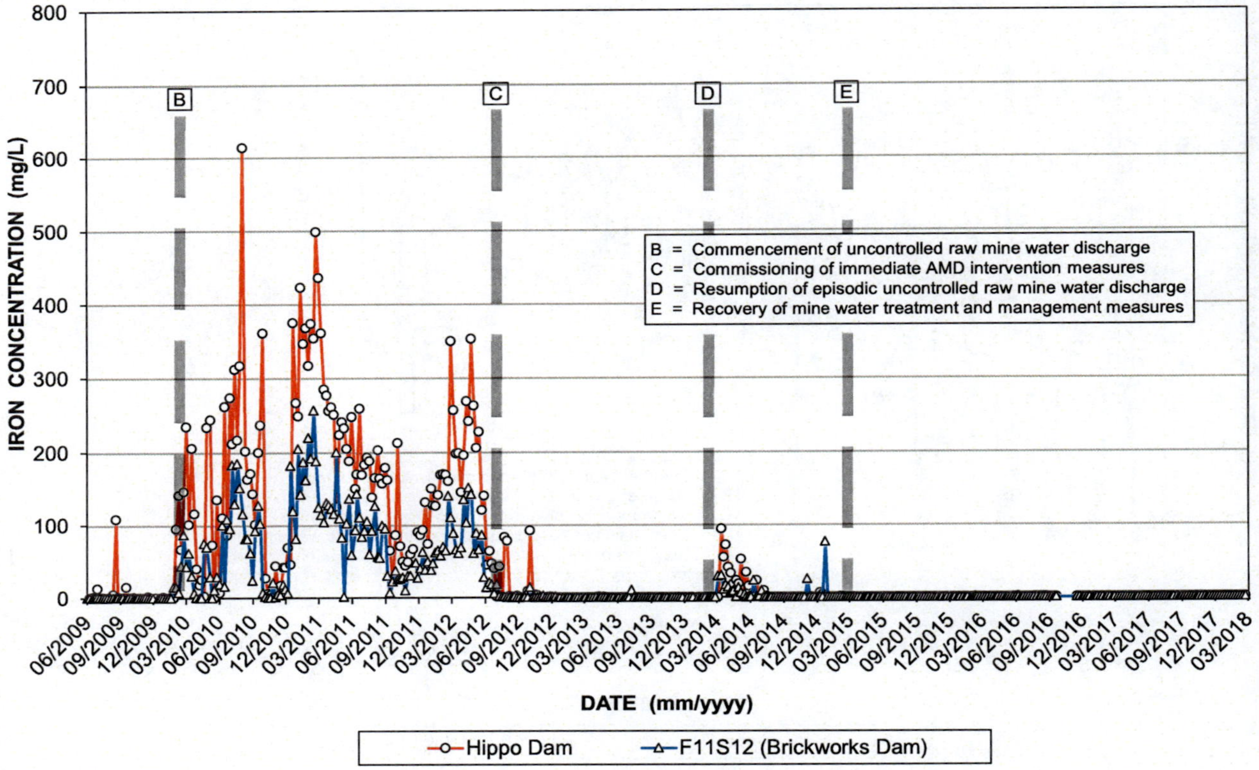


**Fig. 8** Pattern and trend of Fe levels in Tweelopie Spruit water from June 2009 to September 2014

The temporal behaviour of SEC and pH at stations F11S12 and MRd over a period of four years is illustrated in Fig. 12. Based on the data presented in Table 10, the record starts after the resumption of uncontrolled raw mine water decant on 30/01/2010. These circumstances have been discussed in Sect. 3.1 in Chapter "Physical Hydrology" and illustrated in Fig. 13 in Chapter "Physical Hydrology" in regard to discharge rates, and earlier in this section in regard to hydrochemistry.

The record in Fig. 12 reflects already acidic conditions (pH 3.2–4.4) that worsen from August 2010 (pH < 3) until an improvement is manifested in late 2012. A similar response is observed in regard to the SEC values, which typically exceed 350 mS/m until late 2012, after which values reduce to <300 mS/m. It is also evident from Fig. 12 that the pH value at station MRd is generally lower than at station F11S12, especially in the latter part of the record. This response is readily explained as the result of hydrolosis (Sect. 5).

Although spanning a distance of only ~3.9 km, the chemistry of surface water passing stations F11S12 and MRd is revealing in itself. It reflects both the hydrochemistry of allogenic recharge to the karst aquifer and the evolution of water quality in this section of river reach. The close similarity of the major ion composition at the sampling stations is reflected in Fig. 13. The concentrations of selected trace metals (Fig. 14), on the other hand, indicate increases in Mn and Al and a decrease in Fe concentration at the downstream station. The lower Fe concentration is readily explained on the basis of precipitation of this metal out of solution in the presence of oxygen. The low pH, in turn, might also explain the increase in Al and Mn levels (Fig. 14) because of mobilisation of these metals from streambed sediments (Beltrán et al., 2010). This aspect is discussed in Sect. 5.4. The reader is reminded of the significance of the sampling date in the timeframe of the mine water discharge regime (Sect. 3.1 in Chapter "Physical Hydrology").

### 1.2.4 Blougat Spruit

The next tributary to the east of the Tweelopie Spruit, the Blougat Spruit is monitored by the DWS at station 188048 (Fig. 4) downstream of the Percy Stewart WWTW treated sewage effluent discharge point. This position is ~3.5 km above the confluence with the Riet Spruit. The data presented in Table 11 summarise the results of this monitoring.[4] The median and mean electrical balance values only

[4] Data obtained from the DWS in November 2013 for station 188048 indicated a most recent sample date of 30/09/2008.

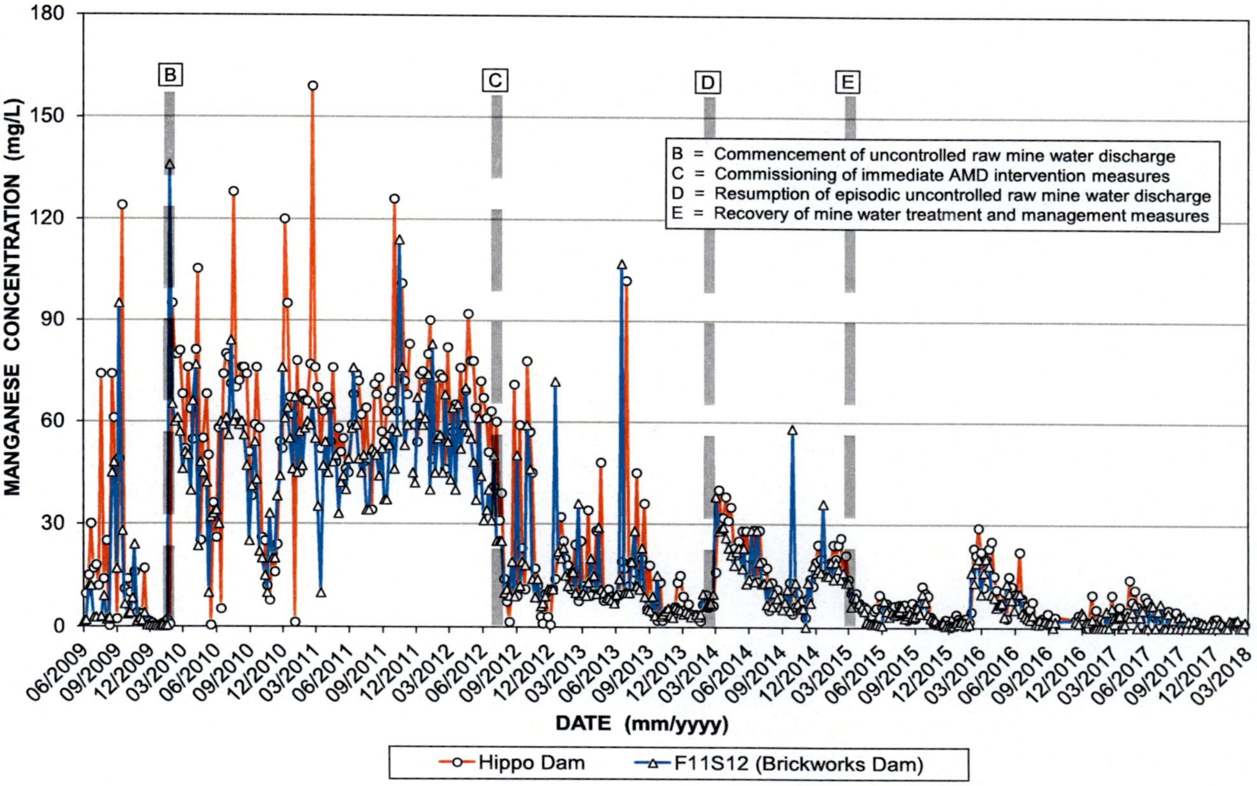


**Fig. 9** Pattern and trend of Mn levels in Tweelopie Spruit water from June 2009 to September 2014

marginally exceed the acceptable error margin of ±5% and, considering the source of this water, render the veracity of the analytical results acceptable. The median values of $NO_3 + NO_2$ and Mn, and the 95%ile values of SEC, $SO_4$, $NO_3 + NO_2$, Fe and Mn exceed the standard health related limits (SANS, 2015a). The treated wastewater origin of this discharge readily explains the $SO_4$ and $NO_3 + NO_2$ exceedances, as well as the high measure of variance in the $SO_4$ concentration (Fig. 15). Sulfate typically comprises ~33% of the TDS of this water (Table 11 and Text Box 1), whilst TDS typically represents 6.3 times the SEC of this water.

Awofolu et al. (2007) report the results (Table 12) of the trace metal analyses Cd, Pb, Mn, Zn, Ni and Cu in water samples collected on four occasions in late 2006 from three sites located downstream of the Percy Stewart WWTW. The sites coincide with the stations identified in this study as BG@N14 (Blougat Spruit at the N14), BB@M (Bloubank Spruit at Makiti) and BB@PL (Bloubank Spruit at Plover's Lake) (Fig. 24). The results are in agreement regarding an elevated Mn concentration, which raises concern for the presence of this metal in the municipal wastewater discharge. It is postulated that this result reflects the industrial wastewater component received by the WWTW, which was originally designed to receive only domestic wastewater.

The data presented in Table 13 summarise the chemistry of the treated wastewater effluent discharged by the Percy Stewart WWTW as monitored and reported by the MCLM. The analysis reveals compliance of all the inorganic and organic analytes (including $NO_3 + NO_2$) with the SANS (2015a) limits. This observation suggests that the $NO_3 + NO_2$ exceedances noted in Table 11 for station 188048 either pre-date July 2007 or derive from a source located upstream of the Percy Stewart WWTW. This disparity warrants interrogation in a separate study. The significant exceedances shown by the bacteriological variables in Table 13 are cause for considerable concern, as are the coliform counts reported in the text box in Fig. 15. Similarly, the mean and median $PO_4$ values of >4 mg P/L and the elevated chemical oxygen demand (COD) values generate concern for potential eutrophication in downstream receiving water bodies. A notable exclusion from the MCLM analytical suite is $SO_4$.

The data presented in Table 14 summarise the chemistry of the treated wastewater effluent discharged by the Percy Stewart WWTW as reported to the DWS for station PSFE.

Table 14 data record goes back to late 2002, compared to the shorter MCLM record (Table 13). The data support the observed exceedances associated with the bacteriological

**Table 4** Summary statistics of period-specific surface water chemistry changes in the Tweelopie Spruit associated with four distinct mine water discharge regimes

| Variable /Analyte | Statistical parameter | Hippo Dam | | | | F11S12/Brickworks Dam | | | |
|---|---|---|---|---|---|---|---|---|---|
| | | A–B[1] | B–C[2] | C–D[3] | D–[4] | A–B[1] | B–C[2] | C–D[3] | D–[4] |
| pH ($-\log_{10}\alpha_{H+}$) | n | 176 | 129 | 83 | 20 | 173 | 128 | 83 | 20 |
| | 5%ile | 3.6 | 2.8 | 5.9 | 3.0 | 3.9 | 2.7 | 5.3 | 2.8 |
| | Mean | – | – | – | – | – | – | – | – |
| | Median | 7.2 | 3.2 | 7.2 | 3.4 | 6.9 | 3.0 | 7.0 | 3.4 |
| | 95%ile | 9.3 | 5.7 | 7.6 | 4.0 | 7.4 | 3.9 | 7.4 | 4.5 |
| | SD | 1.5 | 1.0 | 0.8 | 0.3 | 0.9 | 0.4 | 0.9 | 0.7 |
| | CoV | 22 | 30 | 11 | 10 | 14 | 14 | 13 | 20 |
| SEC (mS/m) | n | 175 | 129 | 83 | 20 | 172 | 128 | 83 | 20 |
| | 5%ile | 324 | 320 | 285 | 342 | 167 | 288 | 230 | 305 |
| | Mean | 374 | 391 | 350 | 382 | 268 | 332 | 281 | 356 |
| | Median | 379 | 393 | 354 | 383 | 283 | 330 | 276 | 349 |
| | 95%ile | 426 | 438 | 395 | 406 | 329 | 378 | 350 | 400 |
| | SD | 32 | 33 | 34 | 28 | 48 | 29 | 34 | 34 |
| | CoV | 9 | 8 | 10 | 7 | 18 | 9 | 12 | 10 |
| $SO_4$ (mg/L) | n | 176 | 128 | 82 | 20 | 171 | 128 | 83 | 20 |
| | 5%ile | 2017 | 2511 | 2221 | 2381 | 893 | 1947 | 1600 | 1990 |
| | Mean | 2445 | 2846 | 2520 | 2638 | 1636 | 2264 | 1879 | 2260 |
| | Median | 2460 | 2815 | 2525 | 2655 | 1760 | 2240 | 1870 | 2250 |
| | 95%ile | 2810 | 3220 | 2770 | 2774 | 2015 | 2593 | 2148 | 2640 |
| | SD | 259 | 226 | 193 | 153 | 349 | 245 | 268 | 273 |
| | CoV | 11 | 8 | 8 | 6 | 21 | 11 | 14 | 12 |
| Fe (mg/L) | n | 33 | 129 | 83 | 20 | 33 | 128 | 82 | 20 |
| | 5%ile | 0.1 | 6.5 | 0.004 | 0.2 | 0.1 | 1.2 | 0.006 | 0.1 |
| | Mean | 4.7 | 168.4 | 2.490 | 24.6 | 0.3 | 72.9 | 0.466 | 8.8 |
| | Median | 0.4 | 163.0 | 0.030 | 18.0 | 0.2 | 64.0 | 0.075 | 6.7 |
| | 95%ile | 13.8 | 365.2 | 3.090 | 73.0 | 0.8 | 186.3 | 1.000 | 29.1 |
| | SD | 18.8 | 116.2 | 13.150 | 26.7 | 0.3 | 57.7 | 1.896 | 9.7 |
| | CoV | 399 | 69 | 528 | 109 | 94 | 79 | 407 | 110 |
| Mn (mg/L) | n | 34 | 129 | 83 | 20 | 33 | 128 | 83 | 20 |
| | 5%ile | 0.2 | 22.2 | 1.9 | 13.9 | 0.1 | 20.7 | 3.3 | 12.6 |
| | Mean | 18.1 | 62.7 | 16.5 | 25.5 | 10.3 | 50.3 | 14.4 | 21.9 |
| | Median | 9.8 | 65.0 | 11.0 | 26.5 | 2.7 | 50.0 | 10.0 | 22.5 |
| | 95%ile | 74.0 | 95.0 | 56.1 | 38.1 | 46.2 | 76.0 | 45.0 | 29.5 |
| | SD | 27.6 | 23.5 | 18.0 | 7.7 | 19.4 | 17.6 | 15.8 | 7.0 |
| | CoV | 153 | 38 | 109 | 30 | 188 | 35 | 110 | 32 |
| $U_T$ (mg/L) | n | 56 | 129 | 61 | – | 56 | 128 | 61 | – |
| | 5%ile | 0.003[5] | 0.003[5] | 0.003[5] | – | 0.003[5] | 0.003[5] | 0.003[5] | – |
| | Mean | 0.013 | 0.049 | 0.009 | – | 0.008 | 0.035 | 0.004 | – |
| | Median | 0.010 | 0.043 | 0.006 | – | 0.003[5] | 0.030 | 0.003[5] | – |
| | 95%ile | 0.030 | 0.109 | 0.025 | – | 0.026 | 0.076 | 0.011 | – |
| | SD | 0.018 | 0.042 | 0.008 | – | 0.011 | 0.032 | 0.003 | – |
| | CoV | 135 | 85 | 98 | – | 147 | 91 | 95 | – |

[1]September 2006–January 2010
[2]February 2010–July 2012
[3]August 2012–February 2014
[4]March 2014–September 2017
[5]Value biased by the detection limit (DL) of 0.005 mg/L, analysed as 50% DL (0.0025 mg/L) if < DL

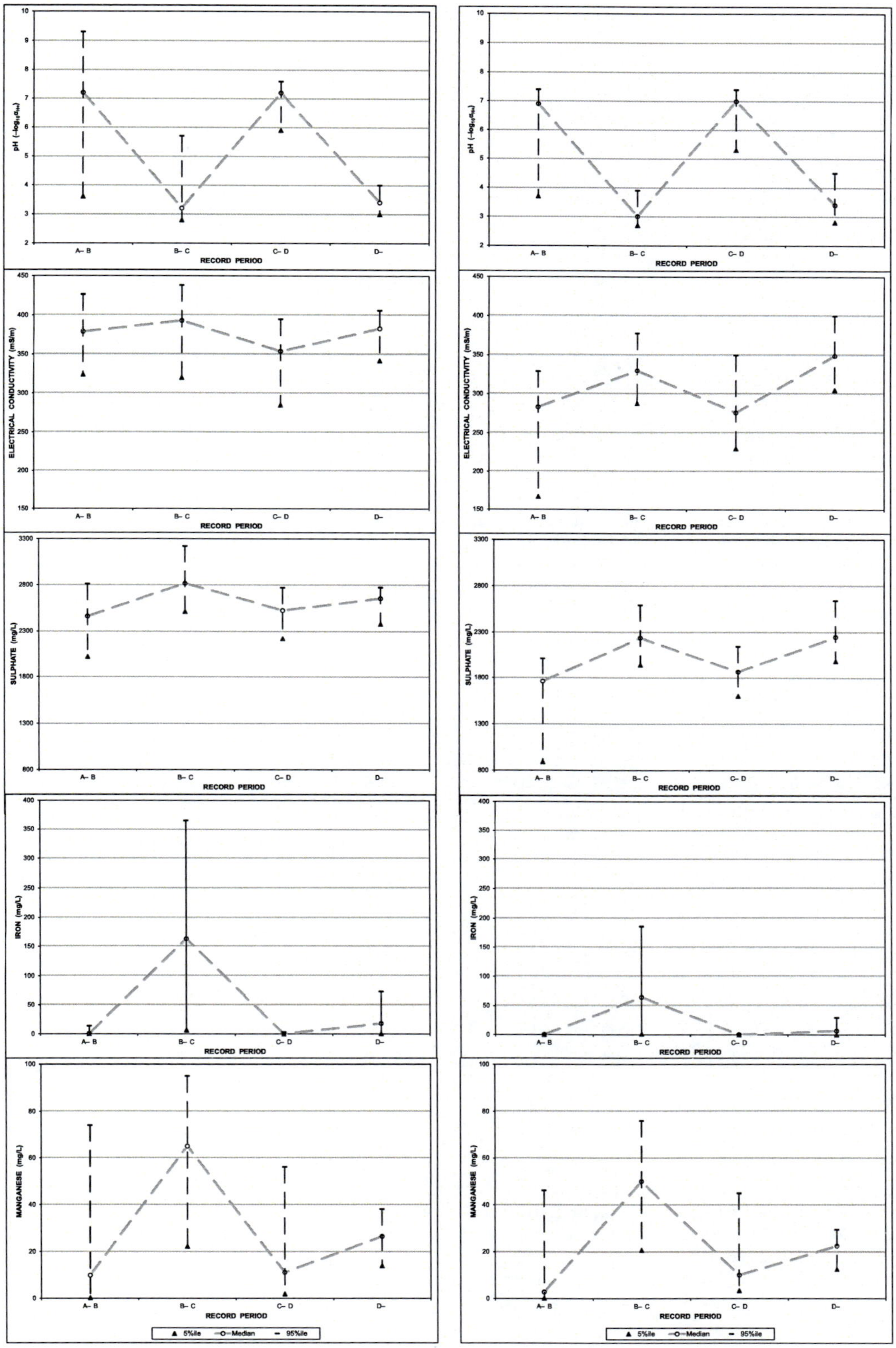


**Fig. 10** Period-specific surface water chemistry changes in the Tweelopie Spruit of the variables/analytes (from top to bottom) pH, SEC, $SO_4$, Fe and Mn at the Hippo Dam (left) and F11S12 (right) (data from Table 4)

**Table 5** List of metal analytes monitored in treated/neutralised mine water discharged to the Tweelopie Spruit from the mine area per distinctive monitoring period

| Analyte | Monitoring period | | | | SANS (2015a)[1] |
|---|---|---|---|---|---|
| | 01/2005–07/2009 | 08/2009–09/2010 | 10/2010–11/2012 | 12/2012–05/2014 | |
| Fe (mg/L) | **0.1** | **0.1** | **0.1** | **0.001** | ≤2 |
| Al (mg/L) | | **1.0** | **1.0** | **0.001** | ≤0.3 |
| Mn (mg/L) | | **0.1** | **0.1** | **0.1** | ≤0.5 |
| $U_T$ (µg/L) | | **5.0** | **5.0** | | ≤15 |
| Cr (mg/L) | | **0.001** | **0.001** | | ≤0.05 |
| V (mg/L) | | **0.001** | **0.001** | | ≤0.2 |
| Pb (mg/L) | | **0.001** | **0.001** | **0.001** | ≤0.01 |
| Co (mg/L) | | **0.001** | **0.001** | | ≤0.5 |
| Ni (mg/L) | | **0.001** | **0.001** | **0.001** | ≤0.07 |
| $U_D$ (µg/L) | | | **5.0** | **5.0** | ≤15 |
| Cd (mg/L) | | | | **0.001** | ≤0.003 |
| Cu (mg/L) | | | | **0.001** | ≤2 |
| Zn (mg/L) | | | | **0.001** | ≤5 |

[1]Standard health-related limit for consumption of 2 L/d over 70 years by a 60 kg person
$U_T$ denotes total uranium, and $U_D$ denotes dissolved uranium
Bold cells denote analyte monitoring period and associated detection limit
Blank cells denote absence/discontinuance of analyte monitoring

**Table 6** Statistical analysis of metal analyte concentrations in treated/neutralised mine water discharged to the Tweelopie Spruit from the mine area per distinctive monitoring period

| Analyte | Monitoring period | | | | | | | | | | | |
|---|---|---|---|---|---|---|---|---|---|---|---|---|
| | 08/2009–09/2010 | | | | 10/2010–11/2012 | | | | 12/2012–05/2014[2] | | | |
| | $\Sigma n$ | >DL | 50%ile | 95%ile | $\Sigma n$ | >DL | 50%ile | 95%ile | $\Sigma n$ | >DL | 50%ile | 95%ile |
| Fe[1] | 254 | 246 | 0.7 | **162** | 113 | 108 | 1.0 | **31.5** | 77 | 77 | 0.025 | 0.336 |
| Al | 61 | 0 | – | – | 113 | 0 | – | – | 74 | 74 | 0.009 | 0.061 |
| Mn | 61 | 58 | **8.65** | **66.35** | 113 | 112 | **11.00** | **60.90** | 77 | 77 | **8.600** | **50.40** |
| $U_T$ | 61 | 30 | 14.00 | **73.25** | 113 | 89 | **16.00** | **700.20** | | | | |
| Cr | 61 | 9 | 0.001 | 0.019 | 113 | 18 | 0.001 | **0.866** | | | | |
| V | 61 | 4 | 0.001 | 0.002 | 113 | 2 | 0.001 | 0.001 | | | | |
| Pb | 61 | 16 | 0.002 | **0.033** | 113 | 22 | 0.005 | **0.030** | 75 | 2 | 0.002 | 0.002 |
| Co | 61 | 60 | 0.070 | **2.005** | 113 | 111 | 0.050 | 0.350 | | | | |
| Ni | 61 | 56 | **0.150** | **3.725** | 113 | 113 | **0.170** | **0.940** | 75 | 75 | **0.100** | **0.118** |
| $U_D$[2] | | | | | 113 | 51 | 9.9 | 35.0 | 77 | 32 | 7.850 | **26.35** |
| Cd | | | | | | | | | 75 | 2 | **0.007** | **0.010** |
| Cu | | | | | | | | | 75 | 55 | 0.002 | 0.010 |
| Zn | | | | | | | | | 75 | 75 | 0.017 | 0.100 |

[1]Values for period 08/2009–09/2010 calculated from 01/2005 as per Table 5
[2]Detection limit for $U_D$ raised from 5 µg/L to 20 µg/L as from June 2014
Analyte units (and symbols) as per Table 5
>DL denotes number of analyte values that exceed the detection limit as per Table 5
Bold text denotes value exceeds standard limit as described in Table 5 (note 1)

**Table 7** Statistical analysis of metal analyte concentrations in treated/neutralised mine water discharged to the Tweelopie Spruit from the mine area per distinctive discharge period

| Analyte | Discharge period | | | | | | | | | | | |
|---|---|---|---|---|---|---|---|---|---|---|---|---|
| | A–B[1] | | | | B–C[2] | | | | C–D[3] | | | |
| | n | >DL | 50%ile | 95%ile | n | >DL | 50%ile | 95%ile | n | >DL | 50%ile | 95%ile |
| Fe | 116 | 103 | 0.5 | **77.4** | 130 | 121 | 1.200 | **29.0** | 82 | 80 | 0.033 | **4.07** |
| Al | 26 | 0 | – | – | 130 | 0 | – | – | 79 | 61 | 0.009 | 0.100 |
| Mn | 26 | 24 | **6.5** | **72.7** | 130 | 128 | **10.5** | **46.95** | 82 | 82 | **9.450** | **59.85** |
| $U_T$ | 26 | 13 | 12.0 | **66.2** | 130 | 90 | **16.5** | **700** | 18 | 16 | **15.5** | **79.0** |
| Cr | 26 | 3 | 0.010 | 0.024 | 130 | 19 | 0.001 | **0.764** | 18 | 5 | 0.001 | 0.003 |
| V | 26 | 2 | 0.002 | 0.002 | 130 | 4 | 0.001 | 0.001 | 18 | 0 | – | – |
| Pb | 26 | 6 | 0.002 | **0.023** | 130 | 27 | 0.005 | **0.037** | 80 | 7 | 0.002 | 0.003 |
| Co | 26 | 25 | 0.039 | **2.900** | 130 | 130 | 0.050 | **1.063** | 18 | 16 | 0.023 | **0.523** |
| Ni | 26 | 22 | 0.069 | **7.995** | 130 | 129 | **0.180** | **1.820** | 80 | 80 | **0.100** | **0.210** |
| $U_D$ | | | | | 95 | 43 | 9.8 | **22.9** | 82 | 35 | 7.550 | **40.50** |
| Cd | | | | | | | | | 62 | 2 | **0.007** | **0.010** |
| Cu | | | | | | | | | 62 | 46 | 0.002 | 0.010 |
| Zn | | | | | | | | | 62 | 62 | 0.020 | 0.100 |

[1]September 2006–January 2010
[2]February 2010–July 2012
[3]August 2012–February 2014
Analyte units (and symbols) as per Table 5
>DL denotes number of analyte values that exceed the detection limit as per Table 5
Bold text denotes value exceeds standard limit as described in Table 5 (note 1)

**Table 8** Summary of $SO_4$: SEC ratio statistics for surface water in the Tweelopie Spruit and the Bloubank Spruit for the period September 2004 to September 2014

| Statistical parameter | Drainage and monitoring station | | | |
|---|---|---|---|---|
| | Tweelopie Spruit | | | Bloubank Spruit |
| | Hippo Dam | Charles Fourie Dam | F11S12/Brickworks Dam | A2H049 |
| n | 535 | 476 | 509 | 159 |
| 5%ile | 5.64 | 5.11 | 4.60 | 1.20 |
| Mean | 6.80 | 6.39 | 6.19 | 1.79 |
| Median | 6.82 | 6.41 | 6.33 | 1.48 |
| 95%ile | 7.89 | 7.70 | 7.46 | 3.52 |
| SD | 0.92 | 0.88 | 1.02 | 0.71 |
| CoV (%) | 13.5 | 13.8 | 16.5 | 39.4 |
| Regression Eq. (1) | $y = 6.89x - 17$ | $y = 7.31x - 243$ | $y = 7.19x - 223$ | $y = 6.39x - 315$[2] |
| | $y = 7.7x - 69.5$[3] | | | |
| $R^2$ | 0.63 | 0.85 | 0.88 | 0.92 |
| Distance from LoD (m) | 935 | 2820 | 6390 | 28 160 |

[1]y = $SO_4$ as mg/Lx = SEC as mS/m
[2]For SEC $\geq$ 60 mS/m
[3]Relationship reported by Gray (1996) for AMD-impacted surface waters

variables. Similarly, the even higher mean and median $PO_4$ values of > 5 mg P/L again generate concern for potential eutrophication in downstream receiving impoundments. Encouragingly, the longer term mean and median COD values are significantly lower (~50%) than reported in Table 13. The water quality situation in regard to the Blougat

**Table 9** Statistical analysis of Tweelopie Spruit water chemistry data associated with station F11S12 for the period November 2003 to September 2008 sourced from the DWS in November 2013

| Variable/Analyte | Statistical parameter | | | | | | | SANS (2015a)[1] |
|---|---|---|---|---|---|---|---|---|
| | n | 5%ile | Mean | Median | 95%ile | SD | CoV (%) | |
| pH ($-\log_{10}\alpha_{H+}$) | 54 | 3.3 | – | 7.2 | 8.0 | 1.4 | 21 | 5.0–9.7 |
| SEC (mS/m) | 52 | 86 | **172** | **158** | **296** | 76.3 | 44 | ≤170 |
| TDS (mg/L) | 44 | 583 | **1262** | **1150** | **2300** | 602.8 | 48 | ≤1200 |
| Ca (mg/L) | 53 | 108.0 | 243.2 | 210.7 | 482.1 | 123.3 | 51 | n.s |
| Mg (mg/L) | 52 | 32.1 | 60.7 | 55.9 | 109.2 | 22.8 | 38 | n.s |
| Na (mg/L) | 53 | 21.4 | 70.0 | 44.0 | 141.7 | 45.1 | 64 | ≤200 |
| K (mg/L) | 52 | 1.3 | 4.1 | 3.8 | 7.1 | 2.3 | 56 | n.s |
| Cl (mg/L) | 53 | 13.7 | 23.2 | 20.8 | 36.1 | 7.9 | 34 | ≤300 |
| $SO_4$ (mg/L) | 52 | 295 | **916** | **811** | **1848** | 527 | 58 | ≤500 |
| $HCO_3$ (mg/L) | 47 | 11.8 | 39.2 | 36.5 | 77.2 | 19.7 | 50 | n.s |
| $NO_3 + NO_2$ (mg N/L) | 47 | 0.60 | 1.15 | 0.98 | 2.04 | 0.6 | 50 | ≤11 |
| Si (mg/L) | 48 | 3.32 | 4.99 | 4.88 | 7.07 | 1.2 | 24 | n.s |
| Fe (mg/L) | 52 | 0.01 | **2.68** | 0.01 | **10.11** | 11.8 | 439 | ≤2 |
| Mn (mg/L) | 52 | 0.01 | **14.30** | **7.99** | **60.37** | 19.3 | 135 | ≤0.5 |
| Al (mg/L) | 53 | 0.01 | **1.38** | 0.09 | **8.57** | 3.0 | 217 | ≤0.3 |
| EB (%) | 46 | −8.2 | 1.1 | 0.8 | 12.8 | 8.5 | 737 | ±5 |
| TDS:SEC | 44 | 6.7 | 7.8 | 7.7 | 9.3 | 0.9 | 11 | n.s |
| $SO_4$:TDS (%) | 42 | 53 | 64 | 66 | 69 | 1 | 9 | n.s |

[1]Standard health-related limit for consumption of 2 L/d over 70 years by a 60 kg person
Bold text denotes value exceeds standard limit as described in note 1

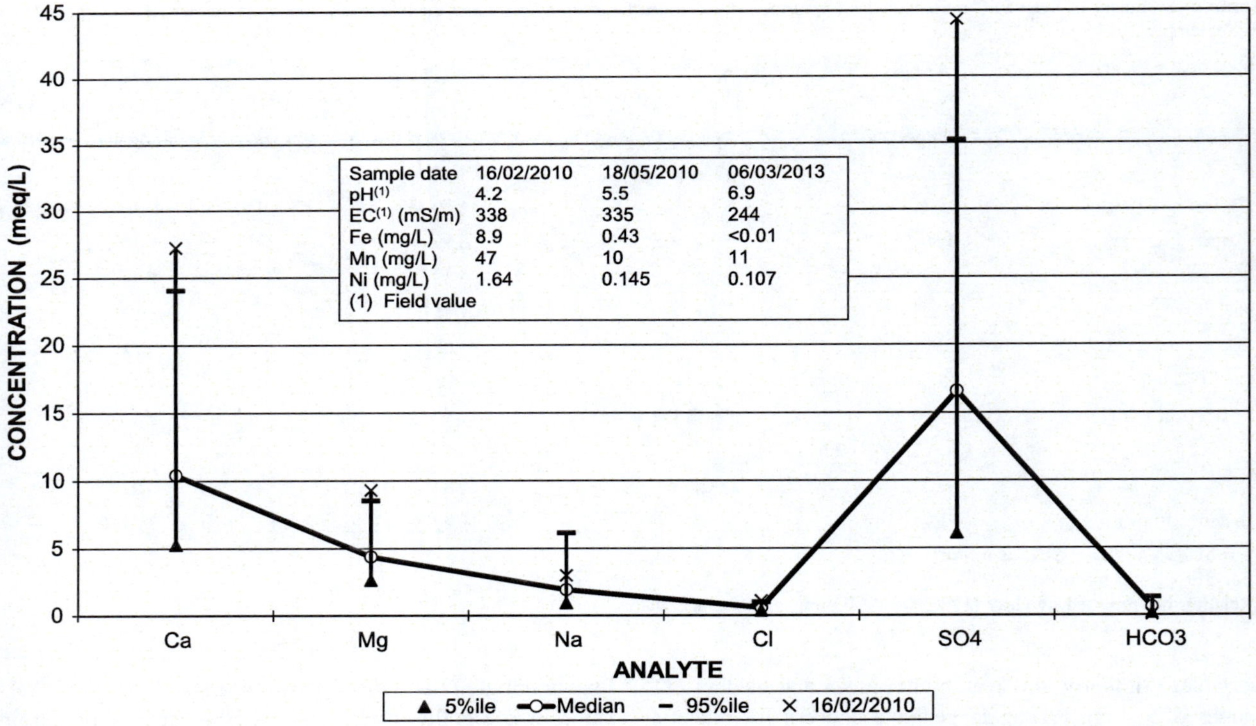


**Fig. 11** Variability of Tweelopie Spruit water major ion chemistry at station F11S12 in the period November 2003 to September 2008, also showing mid-February 2010 values (X symbol) for comparison with more impacted conditions, and subsequent improvement in specific analytes/variables (text box)

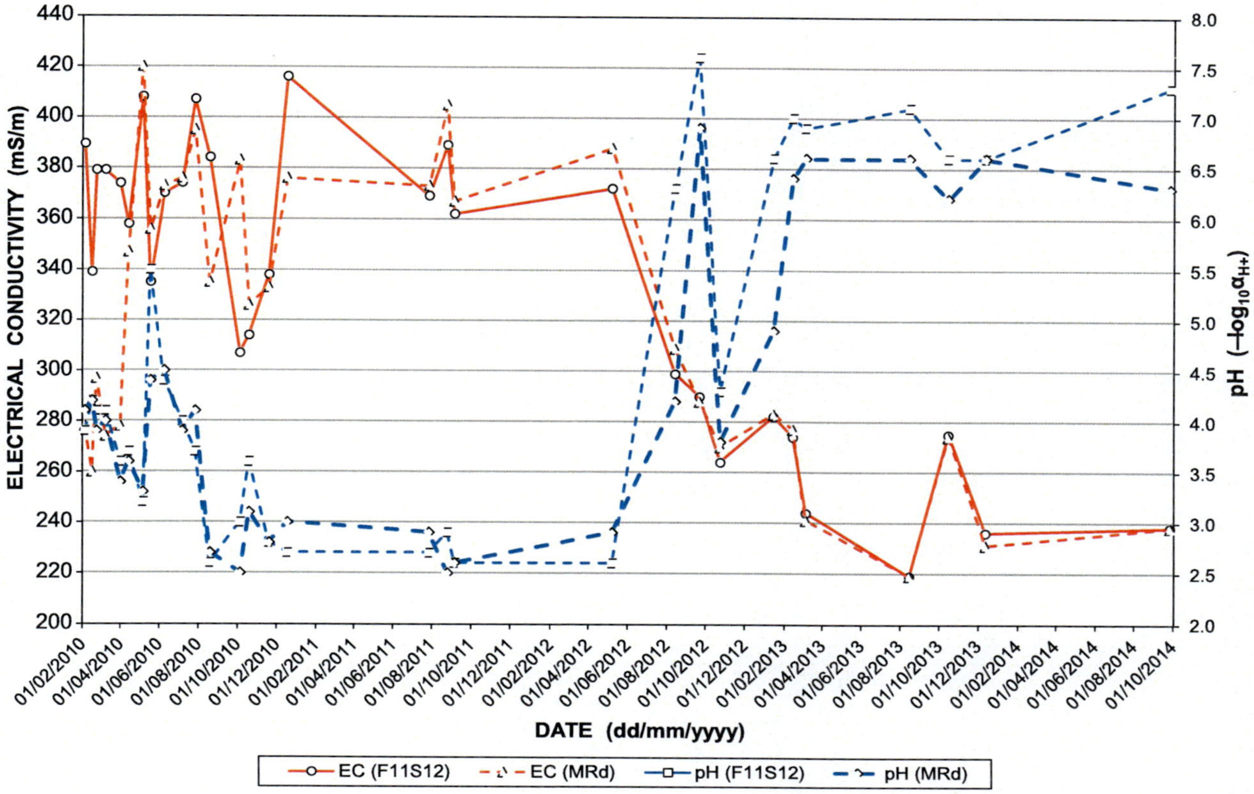


**Fig. 12** Pattern and trend of SEC and pH at stations F11S12 and MRd in the Tweelopie/Riet Spruit system in the period February 2010 to September 2014 (data from Table 10)

Spruit is further informed by the DWS monitoring records for stations located upstream (station BGS2) and downstream (BGS1) of the Percy Stewart WWTW. These data are presented in Tables 15 and 16, respectively.

The results for SEC and the inorganic analytes Cl, $SO_4$ and $HCO_3$ are shown in Fig. 16, and for the organic analytes $NO_3$ + $NO_2$, $NH_4$ and $PO_4$ in Fig. 17. The bacteriological variables faecal coliforms and *E. coli* are compared in Fig. 18. The comparison of the upstream and downstream water quality data indicates that the Percy Stewart WWTW discharge leads to a doubling of the inorganic analyte concentration values in the downstream water (Fig. 16). Further, that the $NH_4$ and $PO_4$ nutrient concentrations similarly reflect substantial increases (Fig. 17). Most notable, however, are the orders of magnitude increase in the bacteriological variables (Fig. 18).

The lower $NO_3$ + $NO_2$ concentration at the downstream station BGS1 (Fig. 17) is the single exception to the trend shown by the other variables. A possible explanation lies in a combination of factors such as the relatively large difference in discharge at the two stations and a comparatively small difference in concentration to begin with. The WWTW discharges at least 16 ML/d (5.8 $Mm^3$/a) to the Blougat Spruit, although SDMs downstream of the plant suggest values closer to the current 24 ML/d capacity of the facility (Sect. 3.2 in Chapter "Physical Hydrology"). The headwater contribution is unknown, but is almost certainly much less under 'normal' flow conditions.

An inspection of the surface water quality results associated with monitoring station BC1 provides an indication of the chemical composition of the surface water discharged from the WWTW. Although the results are related to opportunistic sampling activities rather than to regular and routine monitoring activities, they nevertheless provide a rigorous snapshot of the surface water quality in the Blougat Spruit downstream of the WWTW at the time of sampling. The analytical results of a water sample collected at station BC1 on 06/03/2013 are presented in Table 17. The results show similar values for ammonia ($NH_3$–N) and ortho-phosphate ($PO_4$–P) as are reflected for station BGS1 in Fig. 17. The dissolved organic carbon (DOC) concentration of 10 mg/L matches the standard limit for total organic

**Table 10** Record of SEC and pH measurements at stations F11S12 and MRd on occasion of flow gauging measurements (SDMs), also showing derived $SO_4$ and TDS concentrations

| Date | Station F11S12 | | | | Station MRd | | | |
|---|---|---|---|---|---|---|---|---|
| | SEC (mS/m) | $SO_4$[1] (mg/L) | TDS[2] (mg/L) | pH ($-\log_{10}\alpha_{H^+}$) | SEC (mS/m) | $SO_4$[(1)] (mg/L) | TDS[(2)] (mg/L) | pH ($-\log_{10}\alpha_{H^+}$) |
| 22/09/2009 | 322 | 2092 | 2479 | 6.7 | – | – | – | – |
| 05/02/2010 | 389 | 2574 | 2997 | 3.9 | 358 | 2351 | 2759 | 4.1 |
| 16/02/2010 | 339 | 2214 | 2610 | 4.2 | 335 | 2186 | 2581 | 4.2 |
| 23/02/2010 | 379 | 2502 | 2918 | 4.1 | 383 | 2531 | 2948 | 3.9 |
| 09/03/2010 | 379 | 2502 | 2918 | 4.1 | 353 | 2315 | 2720 | 4.0 |
| 01/04/2010 | 374 | 2466 | 2878 | 3.6 | 358 | 2351 | 2759 | 3.4 |
| 14/04/2010 | 358 | 2351 | 2757 | 3.7 | 347 | 2272 | 2672 | 3.6 |
| 06/05/2010 | 408 | 2711 | 3142 | 3.2 | 420 | 2797 | 3234 | 3.3 |
| 18/05/2010 | 335 | 2186 | 2580 | 5.5 | 356 | 2337 | 2741 | 4.4 |
| 09/06/2010 | 370 | 2437 | 2 849 | 4.4 | 373 | 2459 | 2872 | 4.5 |
| 07/07/2010 | 374 | 2466 | 2880 | 4.0 | 376 | 2480 | 2895 | 3.9 |
| 27/07/2010 | 407 | 2703 | 3134 | 3.7 | 395 | 2617 | 3042 | 4.1 |
| 19/08/2010 | 384 | 2538 | 2957 | 2.6 | 335 | 2186 | 2580 | 2.7 |
| 05/10/2010 | 307 | 1984 | 2364 | 3.0 | 383 | 2531 | 2949 | 2.5 |
| 19/10/2010 | 314 | 2035 | 2418 | 3.6 | 326 | 2121 | 2510 | 3.1 |
| 19/11/2010 | 338 | 2207 | 2603 | 2.8 | 333 | 2171 | 2564 | 2.8 |
| 18/12/2010 | 416 | 2768 | 3203 | 2.7 | 376 | 2480 | 2895 | 3.0 |
| 27/07/2011 | 369 | 2430 | 2841 | 2.7 | 373 | 2459 | 2872 | 2.9 |
| 25/08/2011 | 389 | 2574 | 2995 | 2.9 | 405 | 2689 | 3119 | 2.5 |
| 05/09/2011 | 362 | 2380 | 2787 | 2.6 | 367 | 2416 | 2826 | 2.6 |
| 08/05/2012 | 372 | 2452 | 2864 | 2.6 | 388 | 2567 | 2988 | 2.9 |
| 14/08/2012 | 299 | 1927 | 2302 | 6.3 | 309 | 1999 | 2379 | 4.2 |
| 21/09/2012 | 290 | 1862 | 2233 | 7.6 | 288 | 1848 | 2218 | 6.9 |
| 24/10/2012 | 264 | 1675 | 2033 | 4.3 | 270 | 1718 | 2079 | 3.8 |
| 15/01/2013 | 282 | 1805 | 2171 | 6.6 | 283 | 1812 | 2179 | 4.9 |
| 14/02/2013 | 274 | 1747 | 2110 | 7.0 | 277 | 1769 | 2133 | 6.4 |
| 06/03/2013 | 244 | 1531 | 1879 | 6.9 | 241 | 1510 | 1856 | 6.6 |
| 15/08/2013 | 219 | 1352 | 1686 | 7.1 | 219 | 1352 | 1686 | 6.6 |
| 15/10/2013 | 275 | 1754 | 2118 | 6.6 | 274 | 1747 | 2110 | 6.2 |
| 12/12/2013 | 236 | 1474 | 1817 | 6.6 | 231 | 1438 | 1779 | 6.6 |
| 26/09/2014 | 238 | 1488 | 1833 | 7.3 | 238 | 1488 | 1833 | 6.3 |
| 30/07/2015 | | | | | | | | |
| 04/12/2017 | | | | | | | | |
| n | 31 | 31 | 31 | 31 | 30 | 30 | 30 | 30 |
| Minimum | 219 | 1352 | 1686 | 2.6 | 219 | 1352 | 1686 | 2.5 |
| Mean | 333 | 2167 | 2563 | – | 332 | 2166 | 26,559 | – |
| Median | 349 | 2214 | 2683 | 4.1 | 350 | 2294 | 2696 | 4.0 |
| Maximum | 416 | 2768 | 3203 | 7.6 | 420 | 2797 | 3234 | 6.9 |
| SD | 58 | 411 | 447 | 1.7 | 56 | 404 | 433 | 1.4 |
| CoV (%) | 17 | 19 | 17 | 38 | 17 | 19 | 17 | 34 |

[1]SO4 = 7.19 × SEC − 223 ($R^2 = 0.88$) from Table 8 for station F11S12

[2]SEC × 7.7 to derive a theoretical representative TDS value (from Text Box 1)

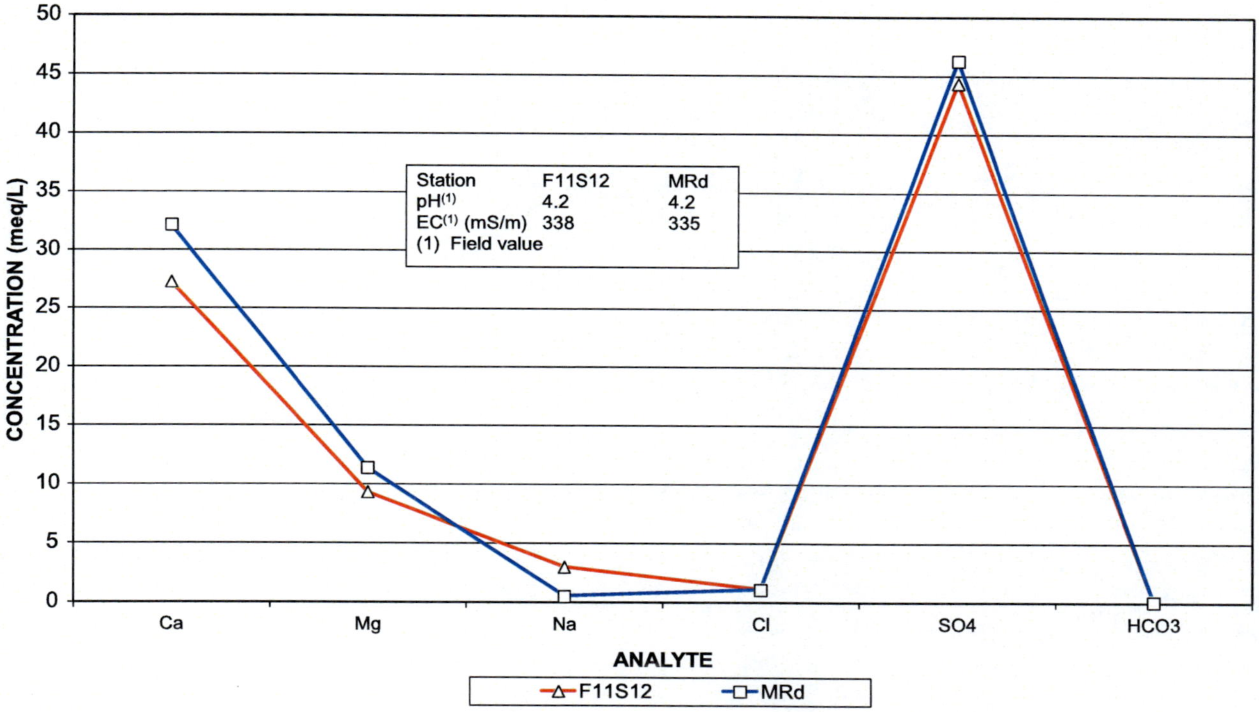


**Fig. 13** Comparison of major ion chemistry at stations F11S12 and MRd in the Tweelopie/Riet Spruit system in mid-February 2010

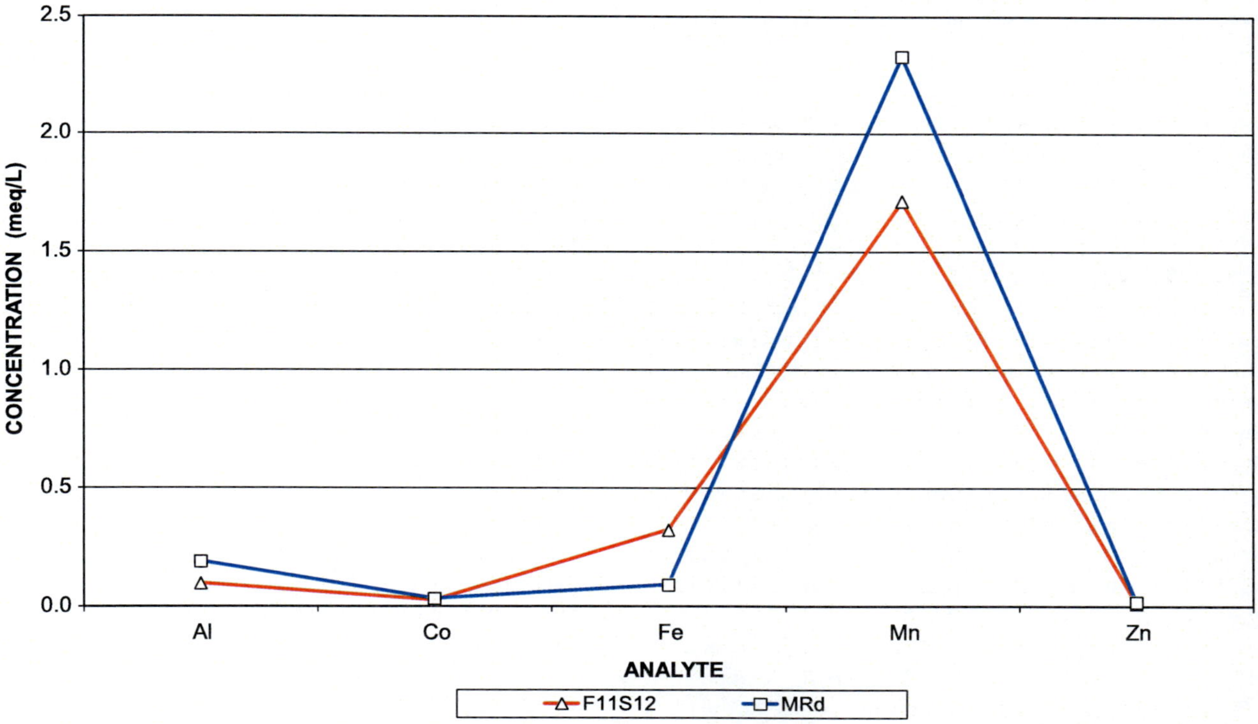


**Fig. 14** Comparison of some trace metals concentration at stations F11S12 and MRd in the Tweelopie/Riet Spruit system in mid-February 2010

**Table 11** Statistical analysis of Blougat Spruit water chemistry at station 188048 for the period June 2004 to September 2008 (raw data sourced from the DWS in November 2013)

| Variable/Analyte | Statistical parameter | | | | | | | SANS (2015a)[1] |
|---|---|---|---|---|---|---|---|---|
| | n | 5%ile | Mean | Median | 95%ile | SD | CoV (%) | |
| pH ($-\log_{10}\alpha_{H+}$) | 48 | 4.2 | – | 6.7 | 7.8 | 1.0 | 16 | 5.0–9.7 |
| SEC (mS/m) | 48 | 58 | 89 | 82 | 156 | 27.3 | 31 | ≤ 170 |
| TDS (mg/L) | 43 | 384 | 551 | 502 | 927 | 172.3 | 31 | ≤ 1200 |
| Ca (mg/L) | 46 | 34.8 | 56.5 | 54.0 | 77.1 | 15.6 | 28 | n.s |
| Mg (mg/L) | 46 | 6.6 | 9.4 | 9.5 | 12.6 | 1.9 | 21 | n.s |
| Na (mg/L) | 46 | 52.1 | 86.0 | 85.2 | 119.4 | 21.1 | 25 | ≤ 200 |
| K (mg/L) | 45 | 8.4 | 13.2 | 13.1 | 19.8 | 3.7 | 28 | n.s |
| Cl (mg/L) | 48 | 45.3 | 65.3 | 64.3 | 79.5 | 12.9 | 20 | ≤ 300 |
| $SO_4$ (mg/L) | 48 | 71 | 187 | 151 | 430 | 120 | 64 | ≤ 500 |
| $HCO_3$ (mg/L) | 45 | 12 | 59 | 48 | 158 | 47 | 79 | n.s |
| $NO_3 + NO_2$ (mg N/L) | 45 | 0.27 | **12.63** | 10.63 | **30.52** | 10.0 | 79 | ≤ 11 |
| $PO_4$ (mg P/L) | 47 | 0.72 | 3.79 | 2.68 | 11.19 | 3.2 | 85 | n.s |
| Si (mg/L) | 47 | 3.59 | 6.41 | 6.32 | 9.37 | 2.3 | 37 | n.s |
| Fe (mg/L) | 48 | 0.006 | 0.678 | 0.068 | **4.116** | 2.0 | 290 | ≤ 2 |
| Mn (mg/L) | 48 | 0.004 | **0.903** | 0.267 | **2.011** | 3.5 | 392 | ≤ 0.5 |
| Al (mg/L) | 46 | 0.008 | 0.156 | 0.035 | 0.280 | 0.7 | 428 | ≤ 0.3 |
| EB (%) | 46 | **–13.8** | **5.7** | **5.3** | **20.7** | 10.1 | 178 | ± 5 |
| TDS:SEC | 43 | 5.0 | 6.2 | 6.3 | 7.3 | 0.7 | 12 | n.s |
| $SO_4$:TDS (%) | 43 | 16 | 32 | 33 | 51 | 11 | 33 | n.s |

[1]Standard health-related limit for consumption of 2 L/d over 70 years by a 60 kg person
Bold text denotes value exceeds standard limit as described in note 1

carbon (TOC) specified in SANS 241-1 (SANS, 2015a). These analytes represent nutrient contaminants that hold severe repercussions for the eutrophic status of the receiving surface water resources.

### 1.2.5 Tweefontein Spruit

Still further downstream, the Tweefontein Spruit draining southwards from the ~7300 ha (H Visser, personal communication) John Nash Nature Reserve area is monitored for water quality at DWS station F14S15 (Fig. 4). Although rising on dolomite, this drainage for much of its flow path follows the contact between older Witwatersrand Supergroup and Ventersdorp Supergroup strata that outcrop along the south-eastern margin of the study area. The Witwatersrand Supergoup strata are represented primarily by quartzite of the Hospital Hill Subgroup, and the Ventersdorp Supergroup by shaly sandstone (Table 5).

The generally good quality water (Table 18) delivered by this drainage exhibits a distinct CaMg–$HCO_3$ composition (Fig. 19). This suggests that the principal source feeding this drainage is the karst system that forms the headwaters of this catchment. The slightly elevated 95%ile Fe and Mn values are attributed to the ferruginous nature of strata such as the Hospital Hill Subgroup quartzite (Table 5) that underlie part of this catchment.

The acceptable mean and median electrical balance values again testify to the veracity of the analytical results. It is a little surprising, however, that the $SO_4$:TDS ratio is as high as 18%. Although this is similar to the 19% of its main stem the Bloubank Spruit at station A2H049 (Text Box 1), it is an order of magnitude greater than the 2% associated with pristine karst springwater.

### 1.2.6 Bloubank Spruit

The significant loss of poor quality water[5] from the Riet Spruit into the karst aquifer upstream of Sterkfontein Cave and the Oaktree area has been discussed in Sect. 5.2.1 in Chapter "Physical Hydrology" and illustrated in Fig. 23. In the middle reaches of the Bloubank Spruit downstream of the confluence[6] with the Riet Spruit in the vicinity of Sterkfontein Cave and down to the Zwartkrans Spring, groundwater

[5] Mainly water that originates in the mine area located in the upper reaches of the Tweelopie Spruit.
[6] Located immediately downstream of the R540 road crossing over the Riet Spruit near the Swartkrans fossil site.

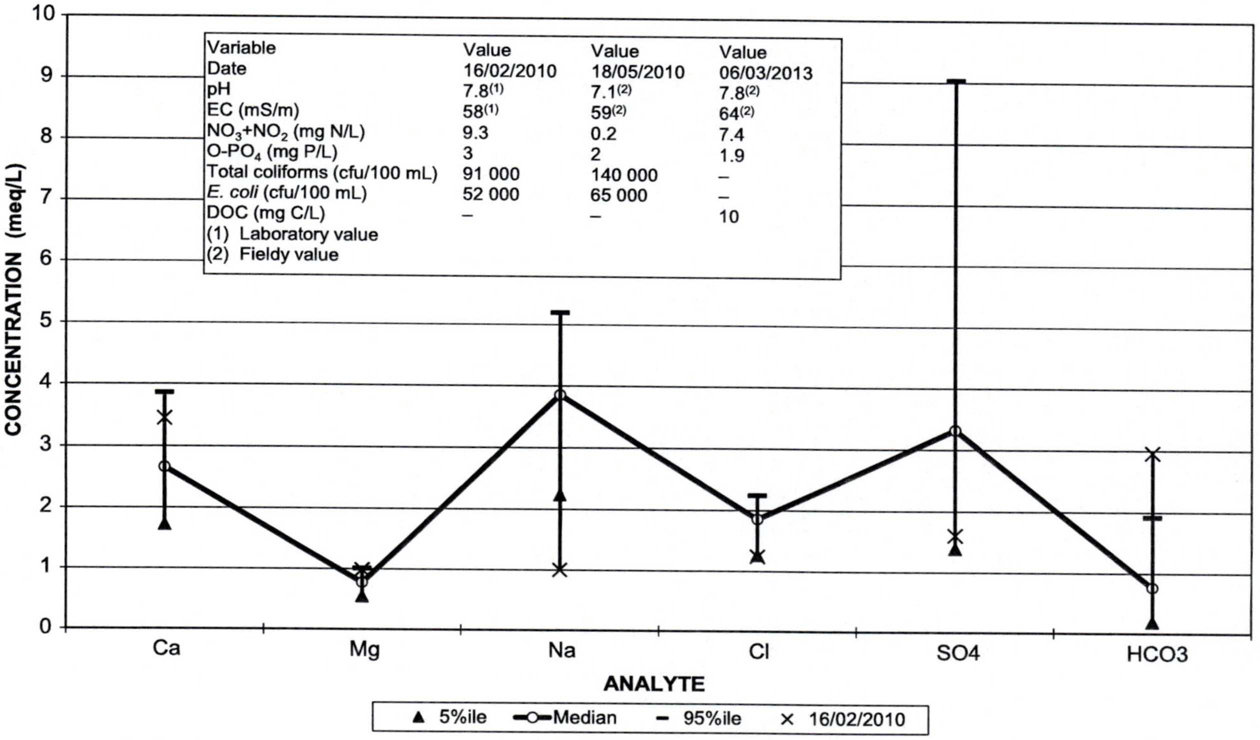


**Fig. 15** Variability of Blougat Spruit water major ion chemistry at station 188048 in the period June 2004 to September 2008, also showing more recent selected variable/analyte values (text box) for comparison

**Table 12** Concentration range of selected trace metals in Blougat Spruit water at three sites downstream of the Percy Stewart WWTW (from Awofolu et al., 2007)

| Trace metal | Concentration range | | SANS (2015a)[1] |
|---|---|---|---|
| | Minimum (mg/L) | Maximum (mg/L) | |
| Cd | Trace (<0.002) | **0.05** | ≤0.003 |
| Pb | Trace (<0.004) | **0.13** ± 0.03 | ≤0.01 |
| Mn | **4.35** ± 0.004 | **942** ± 7.26 | ≤0.5 |
| Zn | 0.10 ± 0.002 | 0.41 ± 0.001 | ≤5.0 |
| Ni | **0.08** ± 0.004 | **0.88** ± 0.002 | ≤0.07 |
| Cu | 0.15 ± 0.003 | 0.742 ± 0.001 | ≤2.0 |

[1]Standard health-related limit for consumption of 2 L/d over 70 years by a 60 kg person
Bold text denotes value exceeds standard limit as described in note 1

resurgence in the streambed partly explains the observed decrease in SEC values and increase in pH values observed at station BB@M since late 2011 (Fig. 20). Average SEC and pH values of 107 mS/m and 7.5, respectively, are similar to the contemporary Zwartkrans Spring values of 95 mS/m and 7.5 (Sect. 7.1 in Chapter "Chemical Hydrogeology").

The elevated SEC (mean of 159 mS/m) and depressed pH values in the period after mid-December 2010 (Fig. 20) are attributed to the increasing contribution of mine water[7] entering the system as surface flow via the Riet Spruit (Sect. 5.2.1 in Chapter "Physical Hydrology"). This also applies to the impacts associated with the very high flow conditions experienced in mid-December 2010 (Plate 5). As is evident in Fig. 20, the mid-December 2010 flood event[8] precipitated the second highest SEC and second lowest pH

[7] The fate of mine water in the receiving drainages is discussed in greater detail in Sect. 5.

[8] The DWS flow gauging station A2H049 at the lower end of the Bloubank Spruit recorded a second highest daily average flow rate of 18.6 $m^3$/s on 16/12/2010. The maximum (highest) historical daily average flow rate gauged at this station is 34.3 $m^3$/s recorded on 28/01/1978.

**Table 13** Statistical analysis of Percy Stewart WWTW discharge chemistry data (from MCLM) for the period July 2007 to June 2009

| Variable/Analyte | Statistical parameter | | | | | | | SANS (2015a)[1] |
|---|---|---|---|---|---|---|---|---|
| | n | 5%ile | Mean | Median | 95%ile | SD | CoV (%) | |
| pH ($-\log_{10}\alpha_{H+}$) | 24 | 7.5 | – | 7.8 | 8.2 | 0.3 | 4 | 5.0–9.7 |
| SEC (mS/m) | 24 | 83 | 100 | 101 | 120 | 11.7 | 40 | ≤170 |
| TDS[(2)] (mg/L) | 16 | 699 | 783 | 781 | 883 | 72 | 9 | ≤1200 |
| Na (mg/L) | 19 | 62.1 | 81.6 | 84.5 | 102.4 | 12.9 | 16 | ≤200 |
| Cl (mg/L) | 24 | 63.3 | 72.3 | 73.9 | 80.2 | 5.7 | 8 | ≤300 |
| $HCO_3$ (mg/L) | 16 | 251 | 312 | 315 | 367 | 34.0 | 13 | n.s |
| $NO_3 + NO_2$ (mg N/L) | 24 | 0.8 | 2.4 | 1.4 | 7.4 | 2.3 | 94 | ≤11 |
| $NH_3$ (mg N/L) | 24 | 14.2 | 30.7 | 31.5 | 42.0 | 8.1 | 26 | ≤1.5 |
| $PO_4$ (mg P/L) | 24 | 2.05 | 4.55 | 4.10 | 7.42 | 1.85 | 41 | n.s |
| COD (mg/L) | 24 | 112 | 202 | 197 | 288 | 58.3 | 29 | n.s |
| Faecal coliforms (cfu/100 mL) | 22 | **69 150** | **359 075** | **244 834** | **801 883** | **371 560** | 104 | ≤10 in 1% of samples |
| *E. coli* (cfu/100 mL) | 6 | **151 417** | **264 333** | **241 167** | **408 000** | **109 408** | 41 | ≤1 in 1% of samples |
| TDS:SEC | 16 | 6.3 | 7.4 | 7.5 | 8.1 | 0.6 | 8 | n.s |

[1]Standard health-related limit for consumption of 2 L/d over 70 years by a 60 kg person
[2]TDS calculated as the sum of the reported TDS value (determined @ 180 °C) and $HCO_3$
Bold text denotes value exceeds standard limit as described in note 1

**Table 14** Statistical analysis of Percy Stewart WWTW effluent chemistry data associated with station PSFE for the period November 2002 to February 2010 sourced from the DWS November 2013

| Variable/Analyte | Statistical parameter | | | | | | | SANS (2015a)[1] |
|---|---|---|---|---|---|---|---|---|
| | n | 5%ile | Mean | Median | 95%ile | SD | CoV (%) | |
| pH ($-\log_{10}\alpha_{H+}$) | 31 | 6.2 | – | 7.2 | 7.9 | 0.6 | 8 | 5.0–9.7 |
| SEC (mS/m) | 30 | 46 | 99 | 103 | 136 | 29.1 | 29 | ≤170 |
| Cl (mg/L) | 19 | 61.6 | 81.5 | 82.0 | 102.9 | 20.6 | 25 | ≤300 |
| $HCO_3$ (mg/L) | 20 | 23 | 194 | 180 | 401 | 158.1 | 81 | n.s |
| $SO_4$ (mg/L) | 24 | 103 | 235 | 199 | **579** | 151.6 | 64 | ≤500 |
| $NO_3 + NO_2$ (mg N/L) | 28 | 0.05 | 6.20 | 3.00 | **18.50** | 8.21 | 132 | ≤11 |
| $NH_3$ (mg N/L) | 29 | 0.01 | 0.55 | 0.17 | **2.83** | 0.99 | 180 | ≤1.5 |
| $NH_4$ (mg N/L) | 29 | **5.38** | **26.23** | **21.90** | **55.08** | 16.85 | 64 | n.s |
| $PO_4$ (mg P/L) | 31 | 1.30 | 5.01 | 3.70 | 10.00 | 3.06 | 61 | n.s |
| COD (mg/L) | 30 | 5 | 118 | 68 | 336 | 149.3 | 126 | n.s |
| Total coliforms (count/100 mL) | 11 | **6885** | **814 750** | **155 000** | **4 140 000** | 1 330 961 | 163 | n.s |
| Faecal coliforms (cfu/100 mL) | 33 | **218** | **763 983** | **131 000** | **4 500 000** | 1 676 161 | 219 | ≤10 in 1% of samples |
| *E. coli* (cfu/100 mL) | 22 | 830 | 231 705 | 30 000 | 1 150 000 | 480 840 | 208 | ≤1 in 1% of samples |

[1]Standard health-related limit for consumption of 2 L/d over 70 years by a 60 kg person
Bold text denotes value exceeds standard limit as described in note 1

values recorded at station BB@M in the course of ththe monitoring record.

The elevated SEC is also manifested in the lower reaches of the Bloubank Spruit as revealed by the DWS surface water quality record for station A2H049. Both the long-term mean and median SEC values of ~60 mS/m and the 95%ile value of 70 mS/m (Table 19) have consistently been exceeded since mid-2010 (Volume 2). A historical

**Table 15** Statistical analysis of Blougat Spruit water chemistry at station BGS2 for the period November 2002 to July 2009 (raw data sourced from the DWS in November 2013)

| Variable/Analyte | Statistical parameter | | | | | | | SANS (2015a)[1] |
|---|---|---|---|---|---|---|---|---|
| | n | 5%ile | Mean | Median | 95%ile | SD | CoV (%) | |
| pH ($-\log_{10}\alpha_{H+}$) | 27 | 6.3 | – | 7.5 | 8.5 | 0.8 | 10 | 5.0–9.7 |
| SEC (mS/m) | 27 | 25 | 53 | 32 | 124 | 36.9 | 70 | ≤ 170 |
| Cl (mg/L) | 22 | 19.3 | 44.6 | 31.5 | 84.9 | 24.1 | 54 | ≤ 300 |
| $HCO_3$ (mg/L) | 22 | 26.9 | 72.4 | 73.2 | 115.8 | 32.4 | 45 | n.s |
| $SO_4$ (mg/L) | 26 | 24 | 120 | 36 | 379 | 149.5 | 124 | ≤ 500 |
| $NO_3 + NO_2$ (mg N/L) | 26 | 0.43 | 5.10 | 4.15 | **13.20** | 4.00 | 79 | ≤ 11 |
| $NH_3$ (mg N/L) | 26 | 0.000 | 0.068 | 0.041 | 0.191 | 0.09 | 136 | ≤ 1.5 |
| $NH_4$ (mg N/L) | 26 | 0.05 | **8.47** | **1.50** | **39.25** | 14.61 | 172 | n.s |
| $PO_4$ (mg P/L) | 27 | 0.05 | 2.04 | 0.05 | 9.04 | 3.43 | 169 | n.s |
| COD (mg/L) | 27 | 5 | 38 | 17 | 143 | 48.7 | 128 | n.s |
| Total coliforms (count/100 mL) | 11 | 310 | 48 029 | 19 000 | 165 000 | 65 369 | 136 | n.s |
| Faecal coliforms (cfu/100 mL) | 25 | 1 | **6984** | **980** | **29 200** | 11 851 | 170 | ≤ 10 in 1% of samples |
| *E. coli* (cfu/100 mL) | 14 | 1 | **1610** | **245** | **8100** | 3388 | 211 | ≤ 1 in 1% of samples |

[1]Standard health-related limit for consumption of 2 L/d over 70 years by a 60 kg person
Bold text denotes value exceeds standard limit as described in note 1

**Table 16** Statistical analysis of Blougat Spruit water chemistry at station BGS1 for the period November 2002 to February 2010 sourced from the DWS in November 2013

| Variable/Analyte | Statistical parameter | | | | | | | SANS (2015a)[1] |
|---|---|---|---|---|---|---|---|---|
| | n | 5%ile | Mean | Median | 95%ile | SD | CoV (%) | |
| pH ($-\log_{10}\alpha_{H+}$) | 33 | 6.9 | – | 7.4 | 8.2 | 0.4 | 6 | 5.0–9.7 |
| SEC (mS/m) | 33 | 24 | 68 | 78 | 105 | 27.1 | 40 | ≤ 170 |
| Cl (mg/L) | 22 | 28.1 | 56.0 | 61.5 | 80.8 | 22.0 | 39 | ≤ 300 |
| $HCO_3$ (mg/L) | 22 | 37 | 165 | 138 | 314 | 111.5 | 67 | n.s |
| $SO_4$ (mg/L) | 26 | 21 | 109 | 110 | 199 | 64.0 | 59 | ≤ 500 |
| $NO_3 + NO_2$ (mg N/L) | 32 | 0.35 | 5.26 | 2.65 | **17.58** | 6.23 | 116 | ≤ 11 |
| $NH_3$ (mg N/L) | 32 | 0.002 | 0.432 | 0.105 | **2.049** | 0.77 | 179 | ≤ 1.5 |
| $NH_4$ (mg N/L) | 32 | 0.05 | **13.40** | **9.25** | **34.68** | 12.38 | 92 | n.s |
| $PO_4$ (mg P/L) | 33 | 0.05 | 1.87 | 1.90 | 4.56 | 1.55 | 83 | n.s |
| COD (mg/L) | 33 | 5 | 59 | 50 | 143 | 44.8 | 75 | n.s |
| Total coliforms (count/100 mL) | 11 | 25 | 8 652 | 740 | 41 000 | 20 659 | 239 | n.s |
| Faecal coliforms (cfu/100 mL) | 32 | **52** | **169,396** | **18,000** | **989,000** | 354,833 | 210 | ≤ 10 in 1% of samples |
| *E. coli* (cfu/100 mL) | 21 | **1 900** | **179,433** | **76,000** | **700,000** | 258,409 | 144 | ≤ 1 in 1% of samples |

[1]Standard health-related limit for consumption of 2 L/d over 70 years by a 60 kg person
Bold text denotes value exceeds standard limit as described in note 1

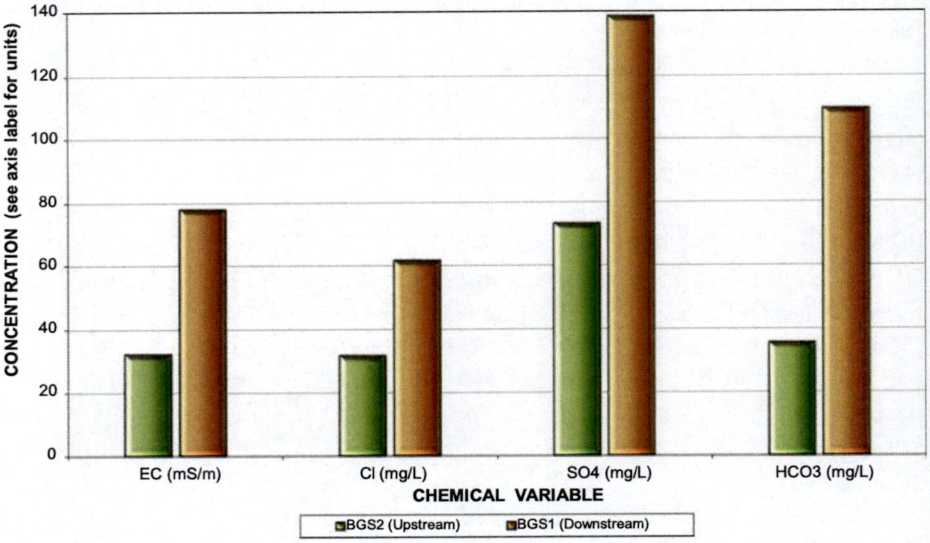


**Fig. 16** Comparison of inorganic quality of Blougat Spruit water upstream and downstream of the Percy Stewart WWTW (data from Tables 15 and 16)

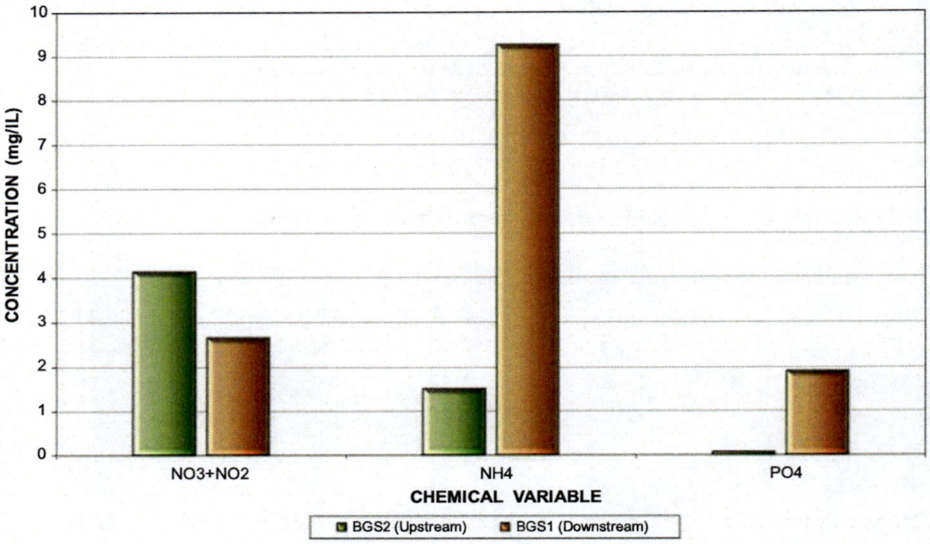


**Fig. 17** Comparison of organic quality of Blougat Spruit water upstream and downstream of the Percy Stewart WWTW (data from Tables 15 and 16)

maximum value of ~122 mS/m was recorded as recently as May 2014. The latter part of the record shown in Fig. 20 indicates the recovery to pre-2010 conditions in regard to pH values, with SEC values that are lower but still almost double those of the pre-2010 period. This 'reversal' occurred in the 2012 winter following the commissioning of the refurbished and upgraded HDS mine water treatment facility (Text Box 4 in Chapter "Physical Hydrology").

The contribution of surface water losses in the Blougat Spruit (Sect. 5.2.2 in Chapter "Physical Hydrology") adds another dimension to the quantity and quality of the groundwater resources in this portion of the Zwartkrans Basin. A particular concern relates to the performance of the Percy Stewart WWTW, in particular the unacceptable bacteriological quality of the treated effluent discharged to the Blougat Spruit by this facility (Sect. 1.2.4). These circumstances have resulted in bacterially contaminated surface water entering the karst aquifer along the lower reaches of the Blougat Spruit at a rate of up to ~ 7 ML/d (Sect. 5.2.2 in Chapter "Physical Hydrology"). The presence of this contamination in surface water is reflected in *E. coli* concentrations of >50,000 cfu/100 mL (Fig. 18).

Whilst the introduction of bacterial contamination into a groundwater resource is normally regarded an environmental tragedy, it is also possible that such contamination might in fact provide an energy (carbon) source that promotes the natural chemical process of bacterial sulfate reduction (BSR). As discussed in Sect. 11 in Chapter "Chemical

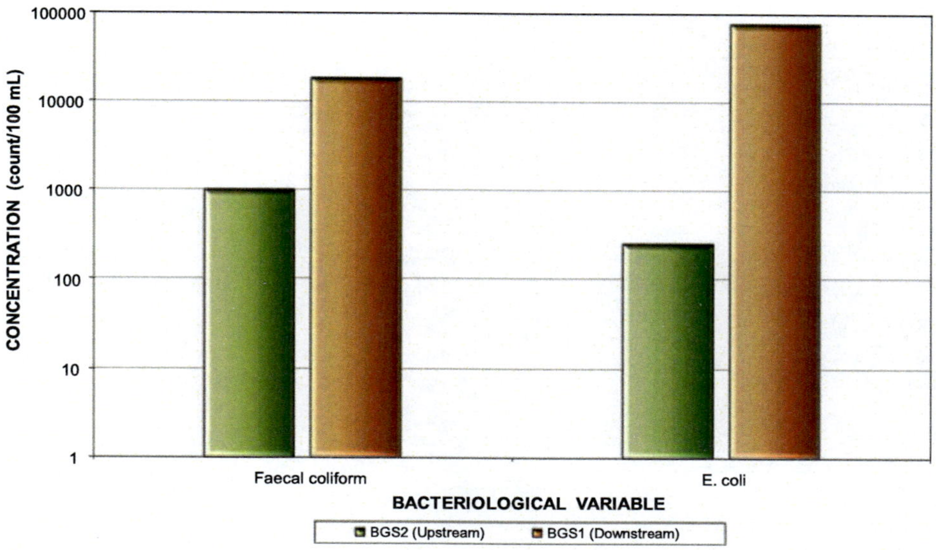


**Fig. 18** Comparison of bacteriological quality of Blougat Spruit water upstream and downstream of the Percy Stewart WWTW (data from Tables 15 and 16)

**Table 17** Water chemistry results for station BC1 on 06/03/2013

| Variable/Analyte | Value | SANS (2011a)[1] | Variable/Analyte | Value | SANS (2015a)[1] |
|---|---|---|---|---|---|
| pH ($-\log_{10}\alpha_{H+}$) | 7.6 | 5.0–9.7 | Sr (mg/L) | 0.113 | n.s |
| SEC (mS/m) | 70 | ≤170 | Se (mg/L) | **0.011** | ≤0.01 |
| Ca (mg/L) | 39 | n.s | Zn (mg/L) | 0.015 | ≤5 |
| Mg (mg/L) | 9.7 | n.s | Al (mg/L) | <0.010 | ≤0.3 |
| Na (mg/L) | 70 | ≤200 | As (mg/L) | <0.010 | ≤0.01 |
| K (mg/L) | 11 | n.s | Be (mg/L) | <0.010 | n.s |
| Cl (mg/L) | 57 | ≤300 | Cd (mg/L) | <0.010 | ≤0.003 |
| $SO_4$ (mg/L) | 84 | ≤500 | Co (mg/L) | <0.010 | ≤0.5 |
| $HCO_3$ (mg/L) | 171 | n.s | Cr (mg/L) | <0.010 | ≤0.05 |
| $NO_3 + NO_2$ (mg N/L) | 7.4 | ≤11 | Cu (mg/L) | <0.010 | ≤2.0 |
| $NH_3$ (mg N/L) | **11** | ≤1.5 | Fe (mg/L) | <0.010 | ≤2.0 |
| F (mg/L) | 0.2 | ≤1.5 | Hg (mg/L) | <0.001 | ≤0.006 |
| $PO_4$ (mg P/L) | 1.9 | n.s | Mn (mg/L) | <0.010 | ≤0.5 |
| DOC (mg C/L) | 10 | n.s | Pb (mg/L) | <0.010 | ≤0.01 |
| Ba (mg/L) | 0.022 | n.s | Sb (mg/L) | <0.010 | ≤0.02 |
| B (mg/L) | 0.109 | n.s | Ti (mg/L) | <0.010 | n.s |
| Mo (mg/L) | 0.079 | n.s | U (mg/L) | <0.001 | ≤0.015 |
| Ni (mg/L) | 0.044 | ≤0.07 | V (mg/L) | <0.010 | ≤0.2 |

[1]Standard health-related limit for consumption of 2 L/d over 70 years by a 60 kg person
Bold text denotes value exceeds standard limit as described in note 1

Hydrogeology", this process has relevance for the potentially ameliorative influence it might have on the mine water impact in the karst aquifer in the zone of wastewater convergence.

At the bottom end of the Bloubank Spruit system, surface water quality is monitored at flow gauging station A2H049 (Fig. 4 in Chapter 2) located ~700 m before the confluence with the Crocodile River. A summary of the statistics that characterise this water quality record is presented in Table 19. Again, the median (and mean) electrical balance values afford the analytical results a high degree of confidence. The 95%ile EB value of 9.5% suggests the increasing inaccuracy of analyses at higher $SO_4$ concentrations. None of the analytes recorded in Table 19 exceed the respective SANS (2015a) health-related limit where specified. The distinct CaMg–$HCO_3$ composition of the water (Fig. 21)

**Table 18** Statistical analysis of Tweefontein Spruit water chemistry at station F14S15 for the period June 2004 to September 2008 sourced from the DWS in November 2013

| Variable/Analyte | Statistical parameter | | | | | | | SANS (2015a)[1] |
|---|---|---|---|---|---|---|---|---|
| | n | 5%ile | Mean | Median | 95%ile | SD | CoV (%) | |
| pH ($-\log_{10}\alpha_{H+}$) | 46 | 7.6 | – | 8.2 | 8.5 | 0.7 | 8 | 5.0–9.7 |
| SEC (mS/m) | 46 | 51 | 59 | 60 | 66 | 4.8 | 8 | ≤ 170 |
| TDS (mg/L) | 42 | 383 | 439 | 448 | 485 | 34.8 | 8 | ≤ 1200 |
| Ca (mg/L) | 45 | 43.9 | 50.7 | 51.3 | 58.3 | 4.9 | 10 | n.s |
| Mg (mg/L) | 45 | 25.8 | 30.6 | 30.9 | 34.2 | 2.6 | 8 | n.s |
| Na (mg/L) | 45 | 11.1 | 24.2 | 26.3 | 32.4 | 7.4 | 30 | ≤ 200 |
| K (mg/L) | 43 | 0.3 | 0.3 | 0.3 | 0.3 | 0.0 | 0 | n.s |
| Cl (mg/L) | 46 | 19.5 | 33.6 | 35.8 | 42.5 | 7.5 | 22 | ≤ 300 |
| $SO_4$ (mg/L) | 46 | 39.9 | 77.9 | 81.6 | 107.9 | 22.1 | 29 | ≤ 500 |
| $HCO_3$ (mg/L) | 44 | 167 | 203 | 199 | 246 | 24.5 | 12 | n.s |
| $NO_3 + NO_2$ (mg N/L) | 44 | 0.97 | 4.22 | 3.97 | 6.91 | 2.5 | 59 | ≤ 11 |
| $PO_4$ (mg/L) | 45 | 0.02 | 0.18 | 0.13 | 0.48 | 0.2 | 91 | n.s |
| Si (mg/L) | 45 | 5.26 | 6.22 | 6.11 | 7.33 | 0.7 | 12 | n.s |
| Fe (mg/L) | 49 | 0.006 | 0.066 | 0.014 | 0.370 | 0.2 | 254 | ≤ 2 |
| Mn (mg/L) | 49 | 0.001 | 0.031 | 0.006 | 0.172 | 0.1 | 182 | ≤ 0.5 |
| Al (mg/L) | 47 | 0.006 | 0.049 | 0.026 | 0.091 | 0.1 | 286 | ≤ 0.3 |
| EB (%) | 42 | −2.1 | 1.7 | 2.5 | **5.2** | 2.8 | 162 | ± 5 |
| TDS:SEC | 42 | 7.0 | 7.4 | 7.4 | 8.0 | 0.3 | 5 | n.s |
| $SO_4$:TDS (%) | 42 | 10 | 17 | 18 | 23 | 4 | 25 | n.s |

[1]Standard health-related limit for consumption of 2 L/d over 70 years by a 60 kg person
Bold text denotes value exceeds standard limit as described in note 1

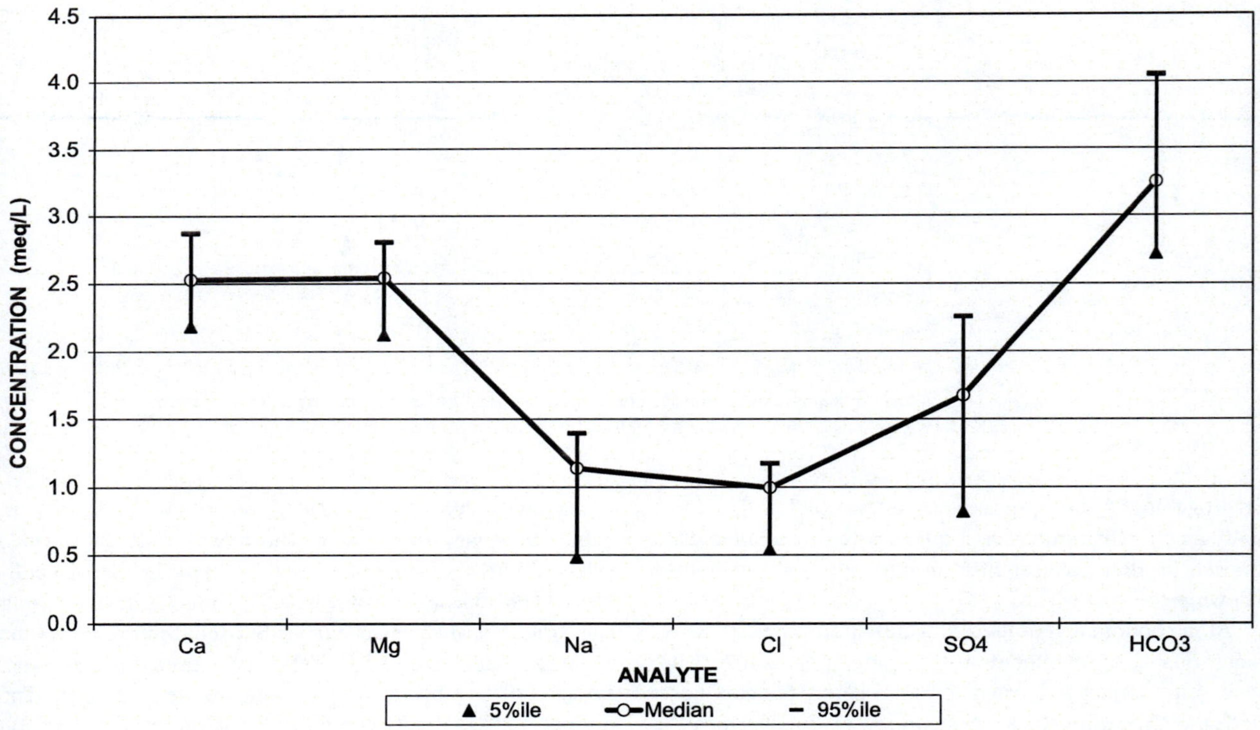


**Fig. 19** Variability of Tweefontein Spruit water major ion chemistry at station F14S15 in the period June 2004 to September 2008

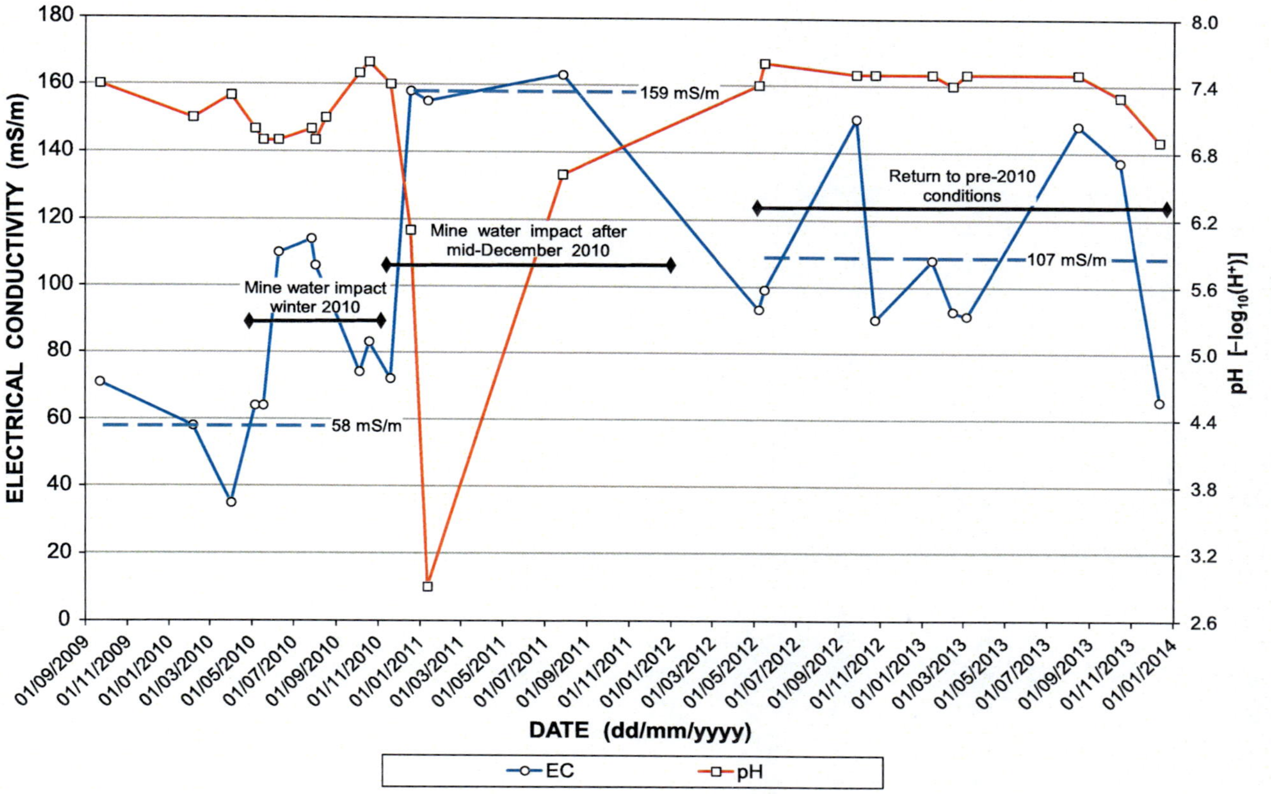


**Fig. 20** Recent electrical conductivity and pH pattern and trend in Bloubank Spruit water at station BB@M between Sterkfontein Cave and Zwartkrans Spring

again reflects the significant contribution of dolomitic groundwater discharged from the karst aquifer in this catchment (Sect. 4.16 in Chapter "Physical Hydrogeology").

Further inspection of the long-term water quality record is premised on four periods of observation, namely May 1979 to August 2002, September 2002 to January 2010, February 2010 to July 2012 and August 2012 to February 2014. These timeframes bracket the pre-decant period, the period of 'normal' mine water discharge up to the commencement of abnormally high decant volumes, the period of abnormally high mine water discharge and, finally, the period following the commissioning of the immediate and short-term mine water control and management measures (Text Box 1 in "Physical Hydrology"). The latter resulted in some recovery of the situation before the resumption of high mine water discharge in the period March to July 2014.

The data reveal little difference in the CaMg–$HCO_3$ chemical composition between the pre-decant period and first period of 'normal' decant. The slight but notable increases in the median Na and Cl concentrations in the decant period are considered to reflect the municipal wastewater influence on surface water quality. Evident in Fig. 21, however, is the change to a Ca–$SO_4HCO_3$ chemical composition in the subsequent two periods. The change is further characterised by a significantly greater variability in the concentrations of Ca, Mg, $SO_4$ and $HCO_3$ as reflected in the range of 5–95%ile values. In the case of Ca, Mg and $SO_4$, the variability is biased in favour of the 95%ile values, and in the case of $HCO_3$ in favour of the 5%ile value. These circumstances together reflect the impact of mine water on the chemical composition of the water discharged by the Bloubank Spruit system since February 2010. A more insightful understanding of the hydrochemical 'evolution' of surface water at station A2H049 over three decades is presented in Volume 2. The results are synthesised as follows.

The recorded SEC and $SO_4$ pattern and trend clearly reflect the manifestation of a strong mine water component in the recent discharge hydrochemistry of the Bloubank Spruit system. This finds support in the equivalent Ca:$SO_4$ ratios that reflect a declining trend indicative of a greater proportion of $SO_4$ relative to Ca in the latter part of the record. Both Na and Cl exhibit an increasing trend over the historical record. Attributing the increasing Cl trend to a municipal wastewater influence finds support in the $PO_4$-P (phosphorus) pattern and trend. A similar trend is not as evident in the $NO_3$–N (nitrate) values, possibly for reasons

**Table 19** Statistical analysis of Bloubank Spruit water chemistry at station A2H049 for the period May 1979 to September 2017

| Variable/Analyte | Statistical parameter | | | | | | | SANS (2015a)[1] |
|---|---|---|---|---|---|---|---|---|
| | n | 5%ile | Mean | Median | 95%ile | SD | CoV (%) | |
| pH ($-\log_{10}\alpha_{H+}$) | 966 | 7.4 | – | 8.2 | 8.5 | 0.3 | 4 | 5.0–9.7 |
| SEC (mS/m) | 1071 | 51 | 60 | 61 | 70 | 8.4 | 14 | ≤ 170 |
| TDS (mg/L) | 1071 | 356 | 438 | 444 | 496 | 62 | 14 | ≤ 1200 |
| Ca (mg/L) | 888 | 42.9 | 54.2 | 53.6 | 63.5 | 11.5 | 21 | n.s |
| Mg (mg/L) | 886 | 25.1 | 32.5 | 32.4 | 37.9 | 5.0 | 15 | n.s |
| Na (mg/L) | 882 | 10.0 | 21.9 | 21.8 | 33.4 | 7.1 | 32 | ≤ 200 |
| K (mg/L) | 880 | 0.7 | 1.9 | 1.8 | 3.5 | 0.9 | 47 | n.s |
| Cl (mg/L) | 892 | 20.0 | 31.8 | 32.1 | 40.8 | 6.0 | 19 | ≤ 300 |
| $SO_4$ (mg/L) | 889 | 65.1 | 91.3 | 83.6 | 134.1 | 43.9 | 48 | ≤ 500 |
| $HCO_3$ (mg/L) | 883 | 146 | 192 | 197 | 219 | 24.6 | 13 | n.s |
| $NO_3 + NO_2$ (mg N/L) | 923 | 3.00 | 4.55 | 4.36 | 6.41 | 1.74 | 38 | ≤ 11 |
| $PO_4$ (mg P/L) | 962 | 0.005 | 0.093 | 0.054 | 0.317 | 0.106 | 115 | n.s |
| Si (mg/L) | 962 | 5.08 | 5.98 | 5.98 | 6.82 | 0.81 | 14 | n.s |
| Fe (mg/L) | 116 | 0.004 | 0.029 | 0.014 | 0.118 | 0.097 | 338 | ≤ 2 |
| Mn (mg/L) | 116 | 0.001 | 0.113 | 0.002 | 0.145 | 0.645 | 571 | ≤ 0.5 |
| Al (mg/L) | 111 | 0.003 | 0.044 | 0.011 | 0.091 | 0.218 | 493 | ≤ 0.3 |
| EB (%) | 840 | −1.2 | 3.6 | 3.6 | **9.5** | 3.9 | 109 | ± 5 |
| TDS:SEC | 1070 | 6.7 | 7.3 | 7.2 | 8.1 | 0.5 | 7 | n.s |
| $SO_4$:TDS (%) | 889 | 16 | 20 | 19 | 27 | 7 | 36 | n.s |

[1]Standard health-related limit for consumption of 2 L/d over 70 years by a 60 kg person
Bold text denotes value exceeds standard limit as described in note 1

such as denitrification and uptake of this nutrient by aquatic vegetation and sequestration in eutrophic systems along the ~23-km reach between the Percy Stewart WWTW and station A2H049.

Perhaps the clearest indication of a mine water impact on the chemistry of water discharged by the Bloubank Spruit at station A2H049 is provided by the Piper diagram presented in Fig. 22. The shift towards the apex of the central diamond field that characterises the Ca–$SO_4$ composition of the February 2011 chemistry is reversed in the return to a more 'normal' pre-2009–'10 chemistry in January 2012.

A suite of surface water samples collected on 18/05/2010 at seven stations in the Bloubank Spruit system provide a 'snapshot in time' of the water chemistry in this system. A graphical comparison of the major ion results presented in Table 20 is made in Fig. 23. The sampling positions are shown in Fig. 24.

The graphical comparison (Fig. 23) reveals the extremely dominant Ca–$SO_4$ composition of the water at the Riet Spruit station MRd influenced by the mine water received via the Tweelopie Spruit. It also reveals the similar Na–$HCO_3$ composition at the Blougat Spruit stations BC1 and BG@N14, and further downstream at BB@M on the Bloubank Spruit. The manifestation of Na as the dominant cation matches the long-term record at station 188048 (Fig. 15). The ranking of $HCO_3$ as the dominant anion in the May 2010 analysis (compared to $SO_4$ in the long-term record) is attributed to the greater contribution of rainwater runoff compared to effluent discharge from the Percy Stewart WWTW in early to mid-2010.

The Ca–$HCO_3$ composition of the water at station BB@PL further downstream on the Bloubank Spruit at Kromdraai reflects the contribution of karst groundwater upstream of this station. The principal sources of this water are the Kromdraai, Zwartkrans and Plover's Lake springs (Sect. 4 in Chapter "Physical Hydrogeology"). The Mg–$HCO_3$ composition of the Honingklip Spruit water (HS1) at Kromdraai contrasts sharply with that of the Bloubank Spruit at this location, and reflects the different geology of the contributing catchment.

With the exception of the ephemeral Honingklip Spruit tributary, it is apparent from Table 20 that Fe, Mn and Ni are persistent trace metals in the other Bloubank Spruit system surface waters. This is also apparent for the persistence of poor bacteriological quality as reflected in the *E. coli* values (see also Fig. 15). Encouragingly, the Honingklip Spruit

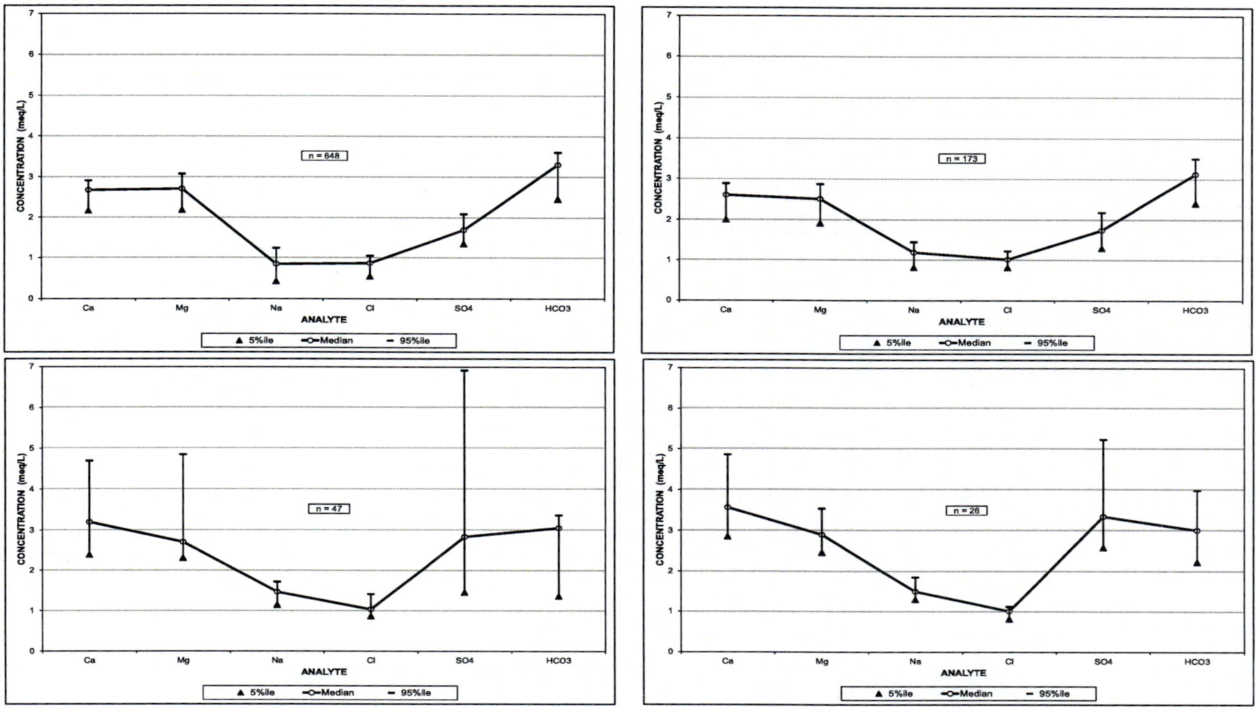


**Fig. 21** Variability of Bloubank Spruit water major ion chemistry at station A2H049 for the period May 1979 to August 2002 (top left), the period September 2002 to January 2010 (top right), the period February 2010 to July 2012 (bottom left), and the period August 2012 to September 2017 (bottom right)

**Fig. 22** Piper diagram of pre-2010 (long-term as per Fig. 29) and more recent Bloubank Spruit water chemistry at station A2H049

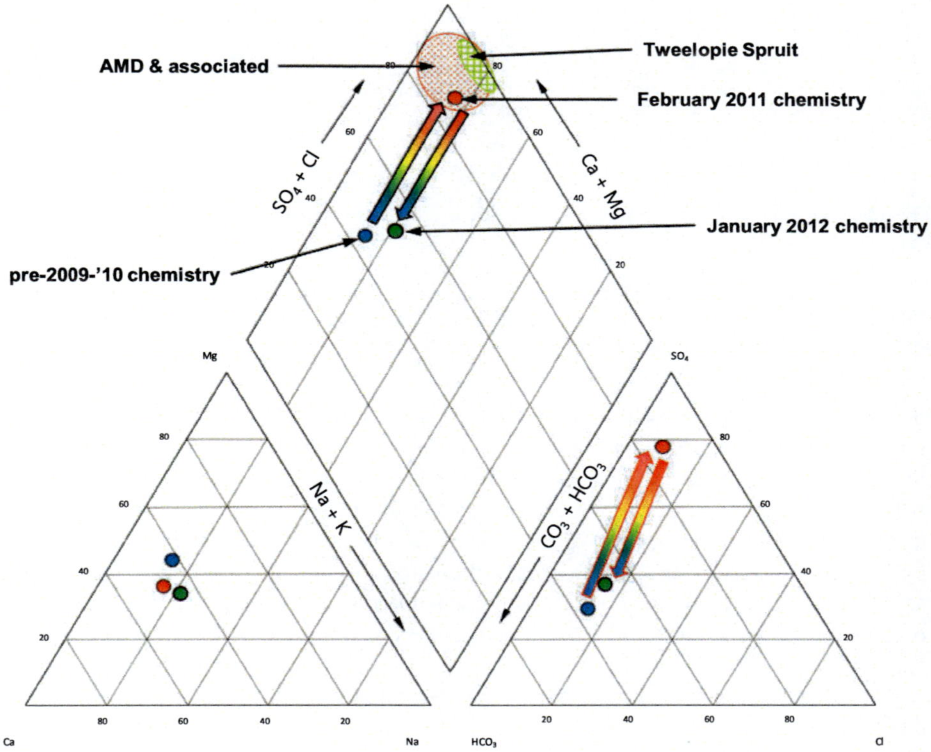

**Table 20** Comparison of mid-2010 surface water chemistry at various positions in the Bloubank Spruit system

| Variable/Analyte | Sampling station | | | | | | | SANS (2015a)[1] |
|---|---|---|---|---|---|---|---|---|
| | F11S12 | MRd | BC1 | BG@N14 | BB@M | HS1 | BB@PL | |
| Date | 18/05/2010 | | | | | | | |
| pH ($-\log_{10}\alpha_{H+}$) (field) | 5.5 | **4.4** | 7.1 | 6.9 | 6.9 | 6.9 | 7.0 | 5.0–9.7 |
| SEC (mS/m) (field) | **355** | **356** | 59 | 51 | 64 | 22 | 60 | ≤ 170 |
| TDS (mg/L) | **2700** | **2751** | 384 | 338 | 400 | 164 | 409 | ≤ 1200 |
| Ca (mg/L) | 662 | 689 | 41 | 38 | 45 | 15 | 54 | n.s |
| Mg (mg/L) | 70 | 81 | 10 | 11 | 15 | 18 | 29 | n.s |
| Na (mg/L) | 115 | 103 | 55 | 48 | 52 | 5 | 34 | ≤ 200 |
| K (mg/L) | 10.4 | 9.4 | 8.6 | 8.0 | 8.1 | 1.1 | 3.0 | n.s |
| Cl (mg/L) | 43 | 44 | 52 | 45 | 50 | 12 | 40 | ≤ 300 |
| $SO_4$ (mg/L) | **1794** | **1819** | 56 | 42 | 74 | 15 | 78 | ≤ 500 |
| $HCO_3$ (mg/L) | <6 | <6 | 161 | 146 | 156 | 98 | 171 | n.s |
| $NO_3 + NO_2$ (mg N/L) | 0.6 | 0.6 | 0.2 | 0.5 | 3.2 | 1.9 | 6.0 | ≤ 11 |
| F (mg/L) | <0.2 | <0.2 | 0.2 | 0.2 | <0.2 | <0.2 | <0.2 | ≤ 1.5 |
| $PO_4$ (mg P/L) | <0.2 | <0.2 | 2.0 | 2.5 | 2.4 | <0.2 | 0.5 | n.s |
| Total coliforms (cfu/100 mL) | n.d | n.d | 140,000 | 34,000 | 3500 | 0 | 720 | n.s |
| *E. coli* (cfu/100 mL) | n.d | n.d | **65,000** | **26,000** | **3000** | 0 | **470** | ≤ 1 in 1% of samples |
| Si (mg/L) | 0.8 | 1.7 | 5.2 | 4.9 | 5.4 | 5.3 | 5.3 | n.s |
| Fe (mg/L) | 0.431 | 0.934 | 0.528 | 0.682 | 0.906 | 0.073 | 0.358 | ≤ 2 |
| Mn (mg/L) | **10** | **18** | 0.279 | 0.470 | **0.665** | 0.087 | 0.293 | ≤ 0.5 |
| Al (mg/L) | < 0.1 | 0.121 | 0.118 | < 0.1 | 0.134 | < 0.1 | 0.137 | ≤ 0.3 |
| Hg (mg/L) | < 0.001 | < 0.001 | < 0.001 | < 0.001 | < 0.001 | < 0.001 | < 0.001 | ≤ 0.001 |
| U (mg/L) | < 0.001 | < 0.001 | < 0.001 | < 0.001 | < 0.001 | < 0.001 | < 0.001 | ≤ 0.015 |
| Ni (mg/L) | **0.145** | **0.117** | 0.049 | 0.035 | 0.029 | < 0.025 | < 0.025 | ≤ 0.07 |
| Sn (mg/L) | 0.097 | < 0.025 | 0.153 | 0.138 | 0.167 | < 0.025 | 0.091 | n.s |
| Sr (mg/L) | 0.500 | 0.395 | 0.123 | 0.117 | 0.096 | 0.036 | 0.054 | n.s |
| Zn (mg/L) | 0.048 | 0.379 | 0.043 | 0.025 | < 0.025 | < 0.025 | < 0.025 | ≤ 5.0 |
| EB (%) | **6.6** | **7.8** | 1.9 | **5.6** | 3.4 | 4.1 | **8.1** | ± 5 |
| TDS:SEC | 7.6 | 7.7 | 6.5 | 6.6 | 6.3 | 7.5 | 6.8 | n.s |
| $SO_4$:TDS (%) | 66 | 66 | 15 | 12 | 19 | 9 | 19 | n.s |

[1]Standard health-related limit for consumption of 2 L/d over 70 years by a 60 kg person
Bold text denotes value exceeds standard limit as described in note 1

displayed no bacteriological contamination on the date sampled. Perhaps even more encouraging is the observation that the *E. coli* value at the most downstream station BB@PL is two orders of magnitude less than at the most upstream station BC1 on the day of sampling. This situation by no means suggests that a value of 470 cfu/100 mL is acceptable, merely that a measure of natural amelioration of bacteriological contamination in the drainage system is apparent.

The results of water chemistry monitoring carried out at the NOE (Nedbank Olwazini Estate) property downstream of station BB@PL substantiate the above observations. The NOE monitoring programme provides a valuable additional measure of water quality in the lower reaches of the Bloubank Spruit. This is particularly relevant under circumstances where discharge quality data since mid-2009 for the Percy Stewart WWTW and the DWSs water quality data for upstream monitoring stations are unavailable. The monitoring has further significance for its location in proximity to where the Bloubank Spruit drainage leaves the karst environment and traverses older strata down to its confluence with the Crocodile River. Unlike the A2H049 water chemistry record, therefore, the NOE record represents almost exclusively the water which drains the karst portion

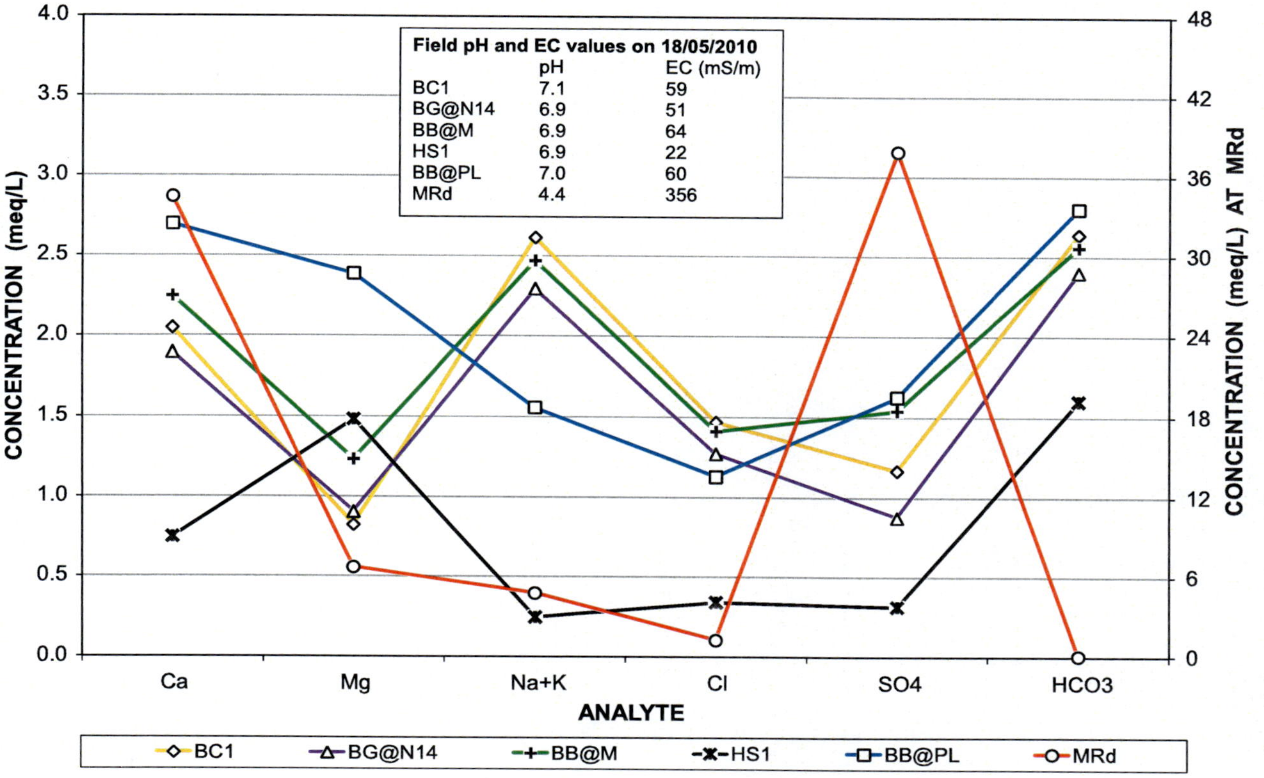


**Fig. 23** Graphical comparison of surface water chemistry at various localities in the Bloubank Spruit system on 18/05/2010

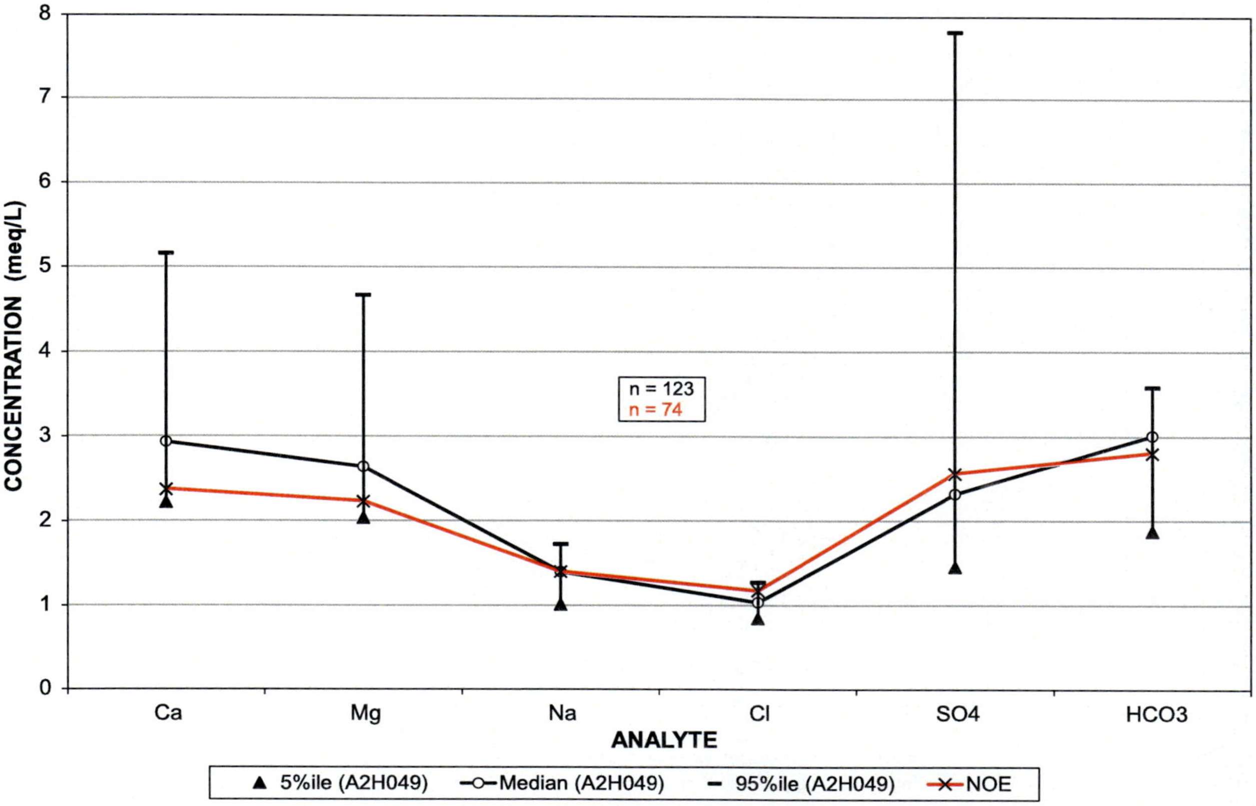


**Fig. 24** Comparison of the Bloubank Spruit water chemistry at the NOE property and station A2H049 in the period of common record November 2007 to September 2014

**Table 21** Chemical characterisation of surface water drawn from the Bloubank Spruit at the NOE property in the period November 2007 to September 2014

| Variable/Analyte | Statistical parameter | | | | | | | SANS (2015a)[1] |
|---|---|---|---|---|---|---|---|---|
| | n | 5%ile | Mean | Median | 95%ile | SD | CoV (%) | |
| pH ($-\log_{10}\alpha_{H+}$) | 74 | 7.0 | – | 7.7 | 8.3 | 0.4 | 5 | 5.0–9.7 |
| SEC (mS/m) | 74 | 52 | 73 | 68 | 117 | 20 | 27 | ≤ 170 |
| TDS (mg/L) | 74 | 325 | 490 | 427 | 866 | 176 | 36 | ≤ 1200 |
| Ca (mg/L) | 74 | 26.0 | 48.1 | 47.5 | 77.9 | 15.4 | 32 | n.s |
| Mg (mg/L) | 72 | 13.4 | 29.2 | 27.1 | 44.4 | 15.5 | 53 | n.s |
| Na (mg/L) | 72 | 16.7 | 33.0 | 32.2 | 44.9 | 11.9 | 36 | ≤ 200 |
| K (mg/L) | 72 | 2.3 | 3.4 | 3.0 | 4.5 | 2.4 | 72 | n.s |
| Cl (mg/L) | 74 | 34.0 | 44.8 | 41.6 | 63.6 | 9.7 | 21 | ≤ 300 |
| $SO_4$ (mg/L) | 73 | 22.4 | 164.6 | 123.0 | 357.0 | 116.3 | 71 | ≤ 500 |
| $HCO_3$ (mg/L) | 74 | 65.6 | 161.8 | 171.3 | 212.9 | 41.5 | 26 | n.s |
| $NO_3$ (mg N/L) | 70 | 2.6 | 6.5 | 6.4 | 10.2 | 3.3 | 51 | ≤ 11 |
| Fe (mg/L) | 41 | 0.00 | 0.05 | 0.03 | 0.12 | 0.08 | 150 | ≤ 2 |
| Mn (mg/L) | 54 | 0.00 | **0.49** | 0.06 | **3.04** | 1.01 | 206 | ≤ 0.5 |
| Al (mg/L) | 35 | 0.015 | 0.061 | 0.050 | 0.146 | 0.044 | 73 | ≤ 0.3 |
| EB (%) | 74 | **–49.3** | **–8.3** | **–5.2** | **22.4** | 20.7 | –250 | ± 5 |
| TDS:SEC | 74 | 5.2 | 7.7 | 6.3 | 9.6 | 8.3 | 108 | n.s |
| $SO_4$:TDS (%) | 73 | 6 | 33 | 31 | 59 | 19 | 59 | n.s |

[1]Standard health-related limit for consumption of 2 L/d over 70 years by a 60 kg person

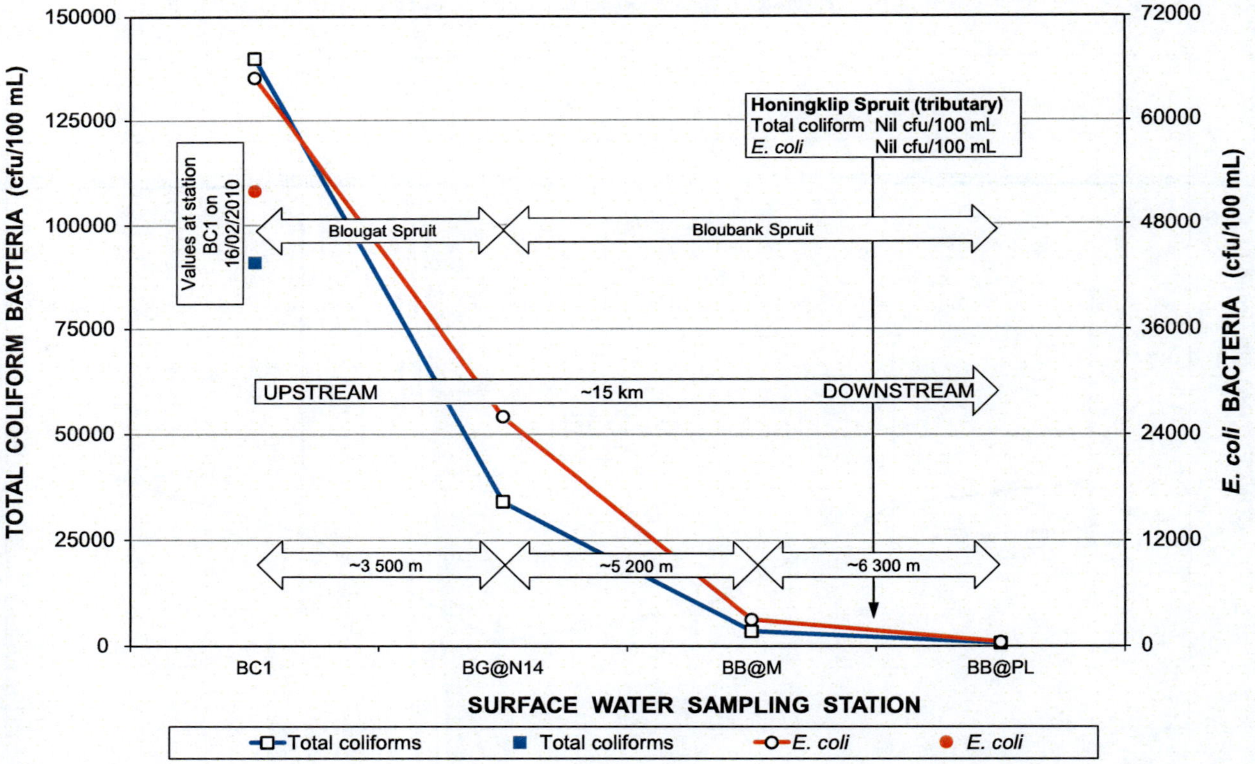


**Fig. 25** Comparison of bacterial concentrations in surface water at various locations in the Bloubank Spruit system on 18/05/2010; for comparison, note the earlier values at station BC1 as well as the ~2× difference in scale between the left and right vertical axes (data from Table 20)

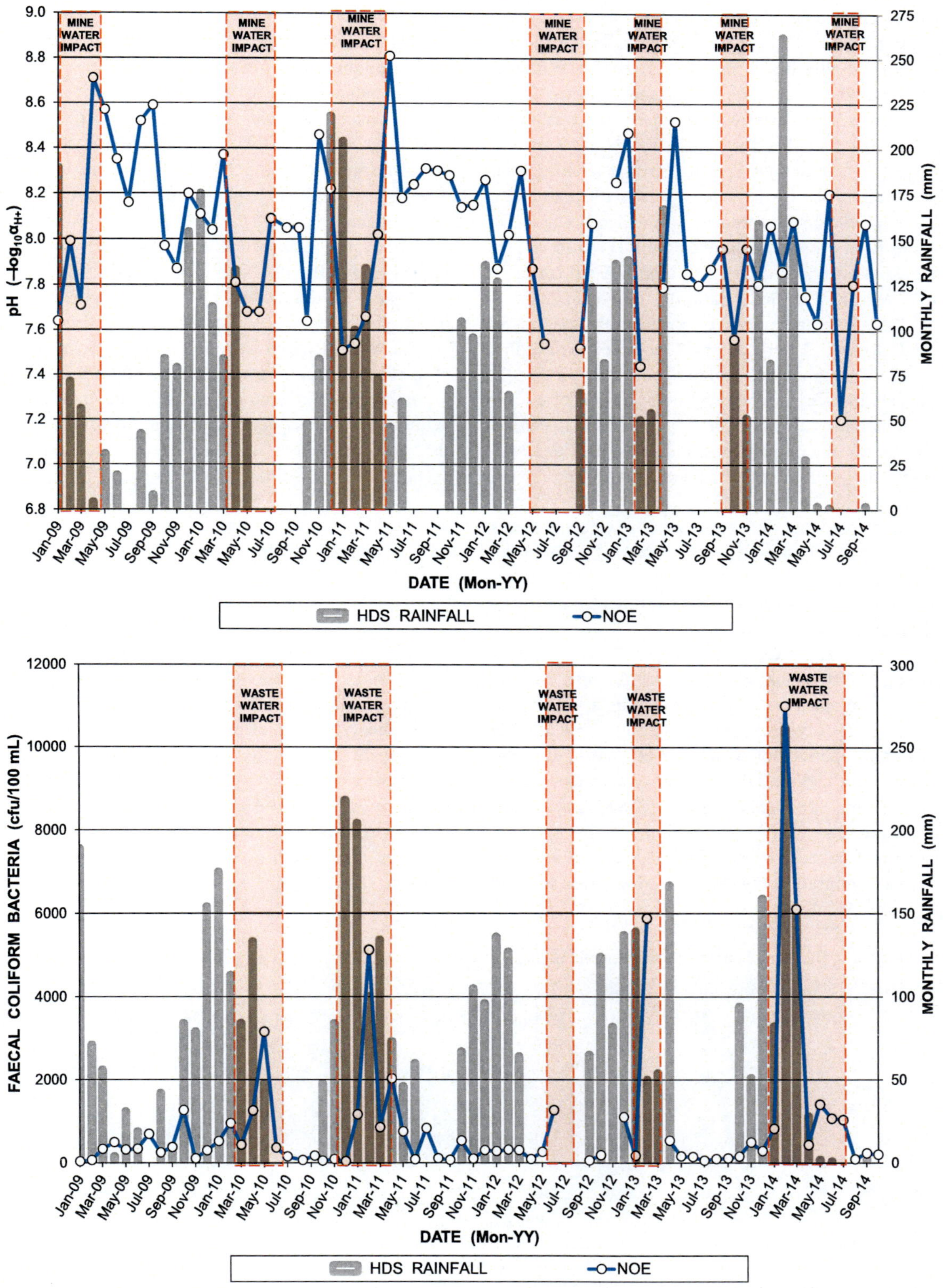


**Fig. 26** Correlation of pH (top) and faecal coliform bacteria (bottom) levels in the Bloubank Spruit at the NOE property with monthly rainfall in the headwaters of the catchment

of the catchment. These circumstances are explored in the following text.

The NOE complex draws its water from the Bloubank Spruit, subjects the raw water to on-site purification for potable and general domestic use, and treats the wastewater generated by the complex for garden irrigation and landscaping use. These circumstances inform the sophistication of the purification plant, which must deal with both bacteriological and trace metal impacts on the feed water. The complex has been granted authority by the DWS to dispose of its treated wastewater effluent to land via irrigation (H Carpenter, personal communication). The chemistry of Bloubank Spruit water prior to purification for on-site potable and 'domestic' use at the NOE complex is described in Table 21. The graphical comparison made in Fig. 24 further reveals the similar chemical composition of surface water at NOE and A2H049. The congruence between the respective chemical compositions indicates that the NOE water quality data record serves as a suitable proxy for that generated at A2H049.

The trend in bacterial concentrations in surface water at various locations in the Bloubank Spruit system on 18/05/2010 is shown in Fig. 25. This reveals a marked decrease in concentrations with distance downstream of the principal source.

The pH values and $NO_3$–N, $PO_4$–P, COD and bacterial concentrations in Bloubank Spruit water at the upstream end of the NOE complex in the period January 2009 to September 2014 ($\sim$6 years) are presented in Table 22. This record covers the period for which the MCLMs Percy Stewart WWTW data and the DWSs water quality data for the upstream stations 188048, PSFE, BGS1 and BGS2 (Sect. 1.2.4) are unavailable.

The temporal patterns of the pH and bacterial variables are compared to rainfall in Fig. 26. The congruence between lower pH values (associated with mine water impacts) and elevated faecal coliform levels (associated with municipal wastewater impacts) is evident. Congruence of the variable response patterns with the monthly rainfall is more obscure.

It is concerning that the faecal coliform count as far downstream as the NOE property reflects elevated levels ($\sim 4\times$ the limit) at the 1%ile level. The mean and median values of 882 and 315 cfu/100 mL respectively (Table 22), are compared to August 2014 *E. coli* levels of > 2419 MPN/100 mL obtained at the upstream stations BC1, BB@M and BB@PL. The lower NOE values are commensurate with the decreasing pattern associated with 'distance from source' that is illustrated in Fig. 26. Nevertheless, the results indicate a severe non-conformance of faecal coliforms (and therefore almost certainly of *E. coli*) in regard to potable, animal and recreational use in the Bloubank Spruit system. There is no doubt that the principal source of this contamination is the Percy Stewart WWTW (Sect. 1.2.4).

## 1.3 Crocodile River

Water quality in this drainage is monitored at the DWS station A2H050 (Fig. 4 in Chapter "Introduction and Background") shortly before its confluence with the Bloubank Spruit. The data (Table 23) show that the SANS (2015a) drinking water limit is not exceeded in regard to any variable/analyte. The electrical balance mean and median values again indicate the comparative reliability and accuracy of the chemical analyses in the data record. The water exhibits a distinct Na–$HCO_3$ composition (Fig. 27) that is representative of the predominantly granitic nature of this catchment. The alkali-rich granitic rocks comprise sodium feldspars ($NaAlSi_3O_8$), e.g. the sodium-plagioclase albite

**Table 22** Analysis of Bloubank Spruit water nutrient and bacterial content at the NOE property in the period January 2009 to September 2014

| Variable/Analyte | Statistical parameter | | | | | | | | SANS (2015a)[1] | TWQR[2] TWQR[3] |
|---|---|---|---|---|---|---|---|---|---|---|
| | n | 1%ile | 5%ile | Mean | Median | 95%ile | SD | CoV (%) | | |
| pH ($-\log_{10}\alpha_{H+}$) | 66 | – | 7.5 | – | 8.0 | 8.6 | 0.3 | 4 | 5.0–9.7 | – |
| $NO_3$ (mg N/L) | 57 | – | 2.7 | 7.5 | 7.1 | **14.9** | 3.6 | 48 | < 11 | – |
| O-$PO_4$ (mg P/L) | 66 | – | 0.02 | 0.4 | 0.3 | 1.1 | 0.4 | 94 | n.s | – |
| COD (mg/L) | 48 | – | 6.4 | 54.6 | 43.0 | 158.3 | 51.3 | 93 | n.s | – |
| N:P (ratio)[(4)] | 54 | – | 6.8 | 37.1 | 28.8 | 116.5 | 35.5 | 96 | – | – |
| Faecal coliforms (cfu/100 mL) | 66 | **36** | **57** | **882** | **315** | **4627** | 1757 | 199 | $\leq$10 in 1% of samples | $\leq 200^2$ $\leq 130^3$ |

1Standard health-related limit for consumption of 2 L/d over 70 years by a 60 kg person
[2]Target Water Quality Range for livestock watering as per DWAF (1996b)
[3]Target Water Quality Range for recreational water use as per DWAF (1996d)
[4]Derived from $NO_3$–N and $PO_4$–P values
Bold text denotes value exceeds standard limit as described in note 1

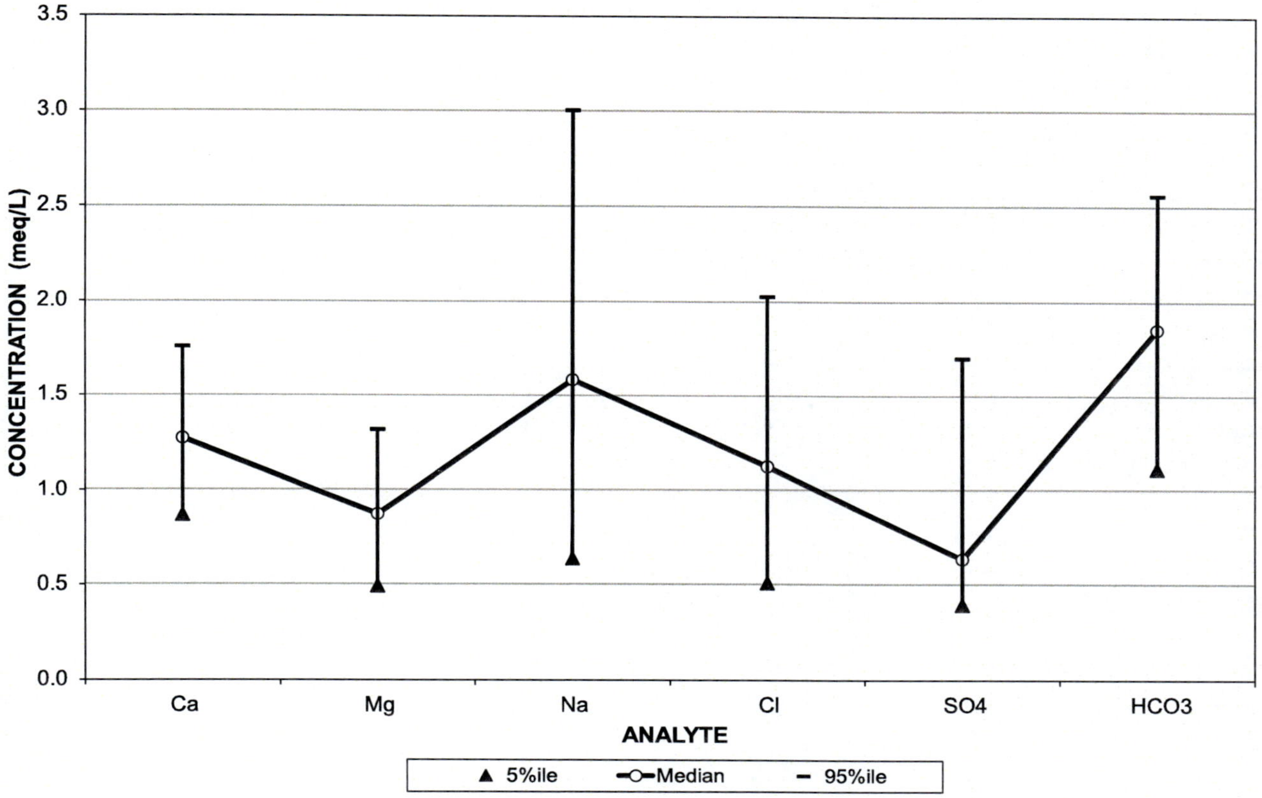


**Fig. 27** Variability of Crocodile River water major ion chemistry at station A2H050 in the period May 1979 to September 2014 (data from Table 3)

that contributes Na as a silicate weathering product. The higher Cl concentration (also the second most abundant anion) compared to that of the Skeerpoort River and Bloubank Spruit water most likely reflects the influence of municipal wastewater discharge from the Driefontein WWTW, which also explains the greater variability evident in the analytes Na and Cl.

The data presented in Table 24 summarise the chemistry of the treated wastewater effluent discharged by the Driefontein WWTW as reported for station DFE by the DWS. Although this record only goes back to late 2002 compared to the much longer record of station A2H050 (Table 23), it provides an assessment of the effluent chemistry more recently discharged from this facility.

The statistics indicate that the Driefontein WWTW meets the SANS (2015a) limit in regard to all of the inorganic and organic variables adjudicated with the exception of ammonium ($NH_4$) at the mean and 95%ile levels. The exceedances in regard to the bacteriological variables, whilst a concern, are orders of magnitude smaller than those associated with the Percy Stewart WWTW (Table 14). Similarly, the mean and median $PO_4$ values are an order of magnitude smaller than those for the Percy Stewart WWTW.

## 1.4 Synthesis of Surface Water Chemistry

The Piper and Stiff diagrams presented in Fig. 28, and the Schoeller diagrams in Fig. 28, characterise the hydrochemistry of the respective sources. They are based on the median long-term major ion chemistry of surface water monitored in the Skeerpoort River, the Bloubank Spruit system and the upper Crocodile River. A comparison of the Piper diagram characterisation with that of groundwater sources in the south-western portion of the study area (Fig. 2 in Chap. "Chemical Hydrogeology") shows a marked similarity reflecting the large measure of hydraulic continuity between surface water and groundwater resources.

- The Skeerpoort River water (A2H034) with its MgCa–$HCO_3$ composition characterises pristine dolomitic (karst) groundwater. As observed by Huizenga (2004), this drainage does not show any sign of pollution. Its water quality can therefore serve as a benchmark against which to gauge the level of degradation experienced by other karst-fed rivers in the study area.
- The Bloubank Spruit system water chemistry varies with the main source as follows:

**Table 23** Statistical analysis of Crocodile River water chemistry at station A2H050 for the period May 1979 to September 2014

| Variable/Analyte | Statistical parameter | | | | | | | SANS (2015a)[1] |
|---|---|---|---|---|---|---|---|---|
| | n | 5%ile | Mean | Median | 95%ile | SD | CoV (%) | |
| pH ($-\log_{10}\alpha_{H+}$) | 1007 | 7.0 | – | 7.9 | 8.2 | 0.4 | 5 | 5.0–9.7 |
| SEC (mS/m) | 1113 | 27 | 45 | 44 | 63 | 17.6 | 39 | ≤ 170 |
| TDS (mg/L) | 872 | 177 | 289 | 286 | 412 | 71.3 | 25 | ≤ 1200 |
| Ca (mg/L) | 933 | 17.54 | 25.9 | 25.5 | 35.2 | 5.3 | 21 | n.s |
| Mg (mg/L) | 934 | 6.0 | 10.8 | 10.6 | 16.0 | 3.2 | 30 | n.s |
| Na (mg/L) | 915 | 14.9 | 38.0 | 36.5 | 68.9 | 16.8 | 44 | ≤ 200 |
| K (mg/L) | 922 | 4.0 | 7.7 | 7.4 | 13.1 | 2.8 | 37 | n.s |
| Cl (mg/L) | 934 | 18.2 | 42.1 | 39.8 | 71.6 | 16.6 | 39 | ≤ 300 |
| $SO_4$ (mg/L) | 933 | 19.1 | 36.7 | 30.6 | 81.1 | 20.5 | 56 | ≤ 500 |
| $HCO_3$ (mg/L) | 927 | 68 | 112 | 113 | 156 | 26.9 | 24 | n.s |
| $NO_3 + NO_2$ (mg N/L) | 977 | 0.584 | 3.067 | 2.710 | 6.569 | 2.1 | 68 | ≤ 11 |
| $PO_4$ (mg P/L) | 1006 | 0.026 | 0.531 | 0.145 | 2.909 | 1.251 | 236 | n.s |
| Si (mg/L) | 1006 | 4.7 | 6.6 | 6.6 | 8.6 | 1.2 | 18 | n.s |
| F (mg/L) | 896 | 0.13 | 0.24 | 0.21 | 0.43 | 0.10 | 44 | ≤ 1.5 |
| Fe (mg/L) | 116 | 0.004 | 0.066 | 0.022 | 0.300 | 0.114 | 172 | ≤ 2 |
| Mn (mg/L) | 116 | 0.001 | 0.029 | 0.004 | 0.150 | 0.063 | 219 | ≤ 0.5 |
| Al (mg/L) | 111 | 0.002 | 0.025 | 0.012 | 0.091 | 0.038 | 154 | ≤ 0.3 |
| EB (%) | 804 | **–5.1** | 2.5 | 2.5 | **10.1** | 5.0 | 199 | ± 5 |
| TDS:SEC | 772 | 6.0 | 6.7 | 6.7 | 7.6 | 0.5 | 7 | n.s |
| $SO_4$:TDS (%) | 773 | 8 | 13 | 11 | 22 | 5 | 40 | n.s |

[1]Standard health-related limit for consumption of 2 L/d over 70 years by a 60 kg person
Bold text denotes value exceeds standard limit as described in note 1

- the Ca–$SO_4$ composition of the Tweelopie Spruit (F11S12) and Riet Spruit waters characterises both raw and treated mine water;
- the Blougat Spruit water (188048) reflects a Na–$SO_4$ composition when treated municipal wastewater contributes the bulk of flow in this drainage, and a Na–$HCO_3$ composition when rainfall runoff provides a significant contribution;
- the Honingklip Spruit[9] delivers a good quality Mg–$HCO_3$ type water; and
- the Bloubank Spruit downstream of its tributaries (A2H049) exhibits a CaMg–$HCO_3$ composition that reflects the significant contribution of karst springwater draining from the carbonate strata that dominate this catchment; the elevated $SO_4$ and Cl concentrations from the influence of mine water (F11S12) and municipal wastewater (188048) sources respectively, explain the shift in the Piper diagram plotting position relative to the Skeerpoort River water (A2H034).

[9]Not shown in Fig. 28 because of the lack of a long-term water chemistry record for this drainage.

- The Crocodile River water (A2H050) exhibits a distinctly different Na–$HCO_3$ composition representative of the predominantly granitic nature of this catchment. This water also shows an elevated Cl concentration compared to that of the Skeerpoort River and Bloubank Spruit water that most likely reflects the municipal wastewater influence on the chemical composition of this water.

## 2 Salt Load Assessment

### 2.1 Catchment Scale

A second aspect of water quality is the salt load associated with the water chemistry of the main drainages in the study area. Calculated as the product of concentration [*C*] and discharge [*Q*], yielding a load value typically expressed as tons per day (t/d), the calculation might target individual analytes such as $SO_4$ or $NO_3 + NO_2$, or it may target a composite variable such as TDS. The latter approach has been taken to compare the annual TDS load that is discharged at each of the sampling stations A2H034, A2H049

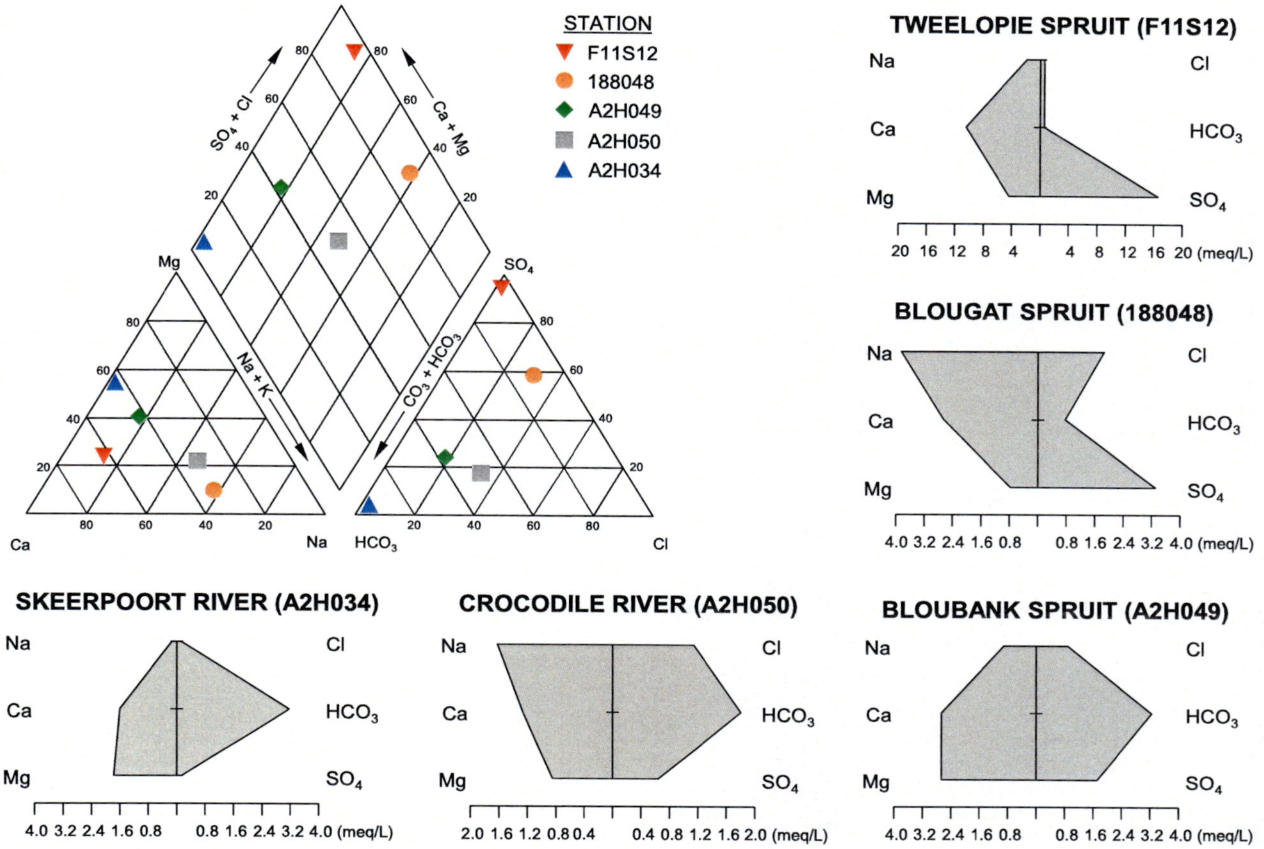


**Fig. 28** Piper diagram (top left) and Stiff diagrams (bottom and right) characterising pre-2010 surface water chemistry in the study area; note the different Stiff diagram horizontal scales

**Table 24** Statistical analysis of Driefontein WWTW discharge water chemistry (at station DFE) for the period November 2002 to February 2010 sourced from the DWS in November 2013

| Variable/Analyte | Statistical parameter | | | | | | | SANS (2015a)[1] |
|---|---|---|---|---|---|---|---|---|
| | n | 5%ile | Mean | Median | 95%ile | SD | CoV (%) | |
| pH ($-\log_{10}\alpha_{H+}$) | 31 | 6.9 | – | 7.3 | 7.9 | 0.5 | 7 | 5.0–9.7 |
| SEC (mS/m) | 31 | 50 | 56 | 56 | 67 | 7.7 | 14 | $\leq 170$ |
| Cl (mg/L) | 21 | 55.0 | 74.5 | 73.0 | 125.0 | 26.0 | 35 | $\leq 300$ |
| $HCO_3$ (mg/L) | 21 | 89 | 141 | 132 | 198 | 60.0 | 43 | n.s |
| $SO_4$ (mg/L) | 25 | 34.4 | 49.2 | 46.0 | 58.8 | 24.5 | 50 | $\leq 500$ |
| $NO_3 + NO_2$ (mg N/L) | 30 | 0.35 | 3.92 | 3.50 | 8.00 | 3.54 | 90 | $\leq 11$ |
| $NH_3$ (mg N/L) | 30 | 0.000 | 0.020 | 0.005 | 0.102 | 0.04 | 206 | $\leq 1$ |
| $NH_4$ (mg N/L) | 30 | 0.05 | **1.10** | 0.30 | **4.20** | 2.51 | 229 | $\leq 1$ |
| $PO_4$ (mg P/L) | 31 | 0.05 | 1.31 | 0.20 | 1.60 | 5.56 | 425 | n.s |
| COD (mg/L) | 31 | 5.0 | 25.7 | 25.0 | 45.5 | 15.5 | 60 | n.s |
| Total coliforms (cfu/100 mL) | 10 | 16 | 578 | 59 | 2880 | 1 556 | 269 | n.s |
| Faecal coliforms (cfu/100 mL) | 30 | 2 | **212** | **25** | **1400** | 439 | 208 | $\leq 10$ in 1% of samples |
| *E. coli* (cfu/100 mL) | 20 | **4** | **469** | **46** | **2160** | 1206 | 257 | $\leq 1$ in 1% of samples |

[1]Standard health-related limit for consumption of 2 L/d over 70 years by a 60 kg person.
Bold text denotes value exceeds standard limit as described in note 1

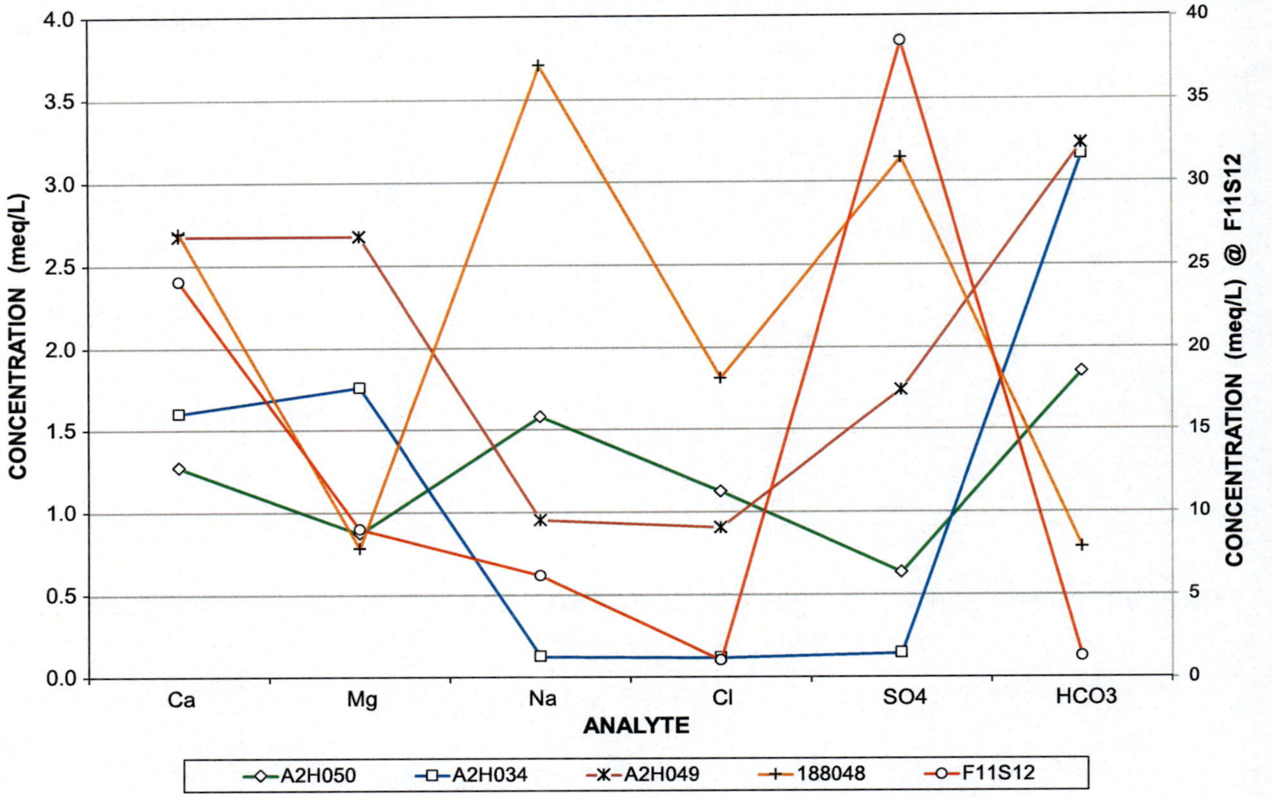


**Fig. 29** Schoeller diagrams characterising pre-2010 surface water chemistry in the study area; note the 10× difference in scale between the left and right vertical axes

and A2H050 (Table 25). The comparison is made on the basis of the median long-term discharge and TDS values previously reported for each station. It would appear that the upper reach of the Crocodile River contributes only a slightly greater TDS load than does the pristine Skeerpoort River.

A more detailed comparison is made on the basis of the median long-term TDS concentration calculated for each month of mutual flow and water quality record (Fig. 30). This shows that the Skeerpoort and (upper) Crocodile rivers carry similar TDS loads that are typically 3–4 times less than that carried by the Bloubank Spruit system.

**Text Box 1** TDS: Sec. and SO4:TDS ratios for various water sources in the study area

The ratio of total dissolved solids (TDS) to electrical conductivity (SEC) provides a useful factor to estimate the TDS (in mg/L) from an easily obtained SEC measurement. The value of this ratio for natural water typically falls in the range 5.5–7.5 (Hem, 1985) for SEC in mS/m. An average value of 6.5 is generally taken as the most applicable empirical factor. More acidic, poorly buffered and/or more saline water will differ from this empirical norm, with higher values generally being associated with water containing higher $SO_4$ concentrations (Hem 1985). The dominance of $SO_4$ in the chemical composition of mine water facilitates the recognition of a mine water impact in other water sources. In this regard, the $SO_4$:TDS ratio provides a useful measure to identify sources impacted by AMD. The appropriate data sourced by this study facilitates exploration of the above relationships in greater detail for the various water sources in the study area. The following analysis is based on median values reported in the referenced sections.

The long-term monthly TDS load data are interrogated further in Figs. 31, 32, 33, 34 and 35 to determine the variance associated with this variable at stations A2H034, A2H049 (for three periods of record) and A2H050.

All the graphs reveal the significant variance in TDS load associated with the summer months (especially February and March). Also evident in all five figures, although muted in

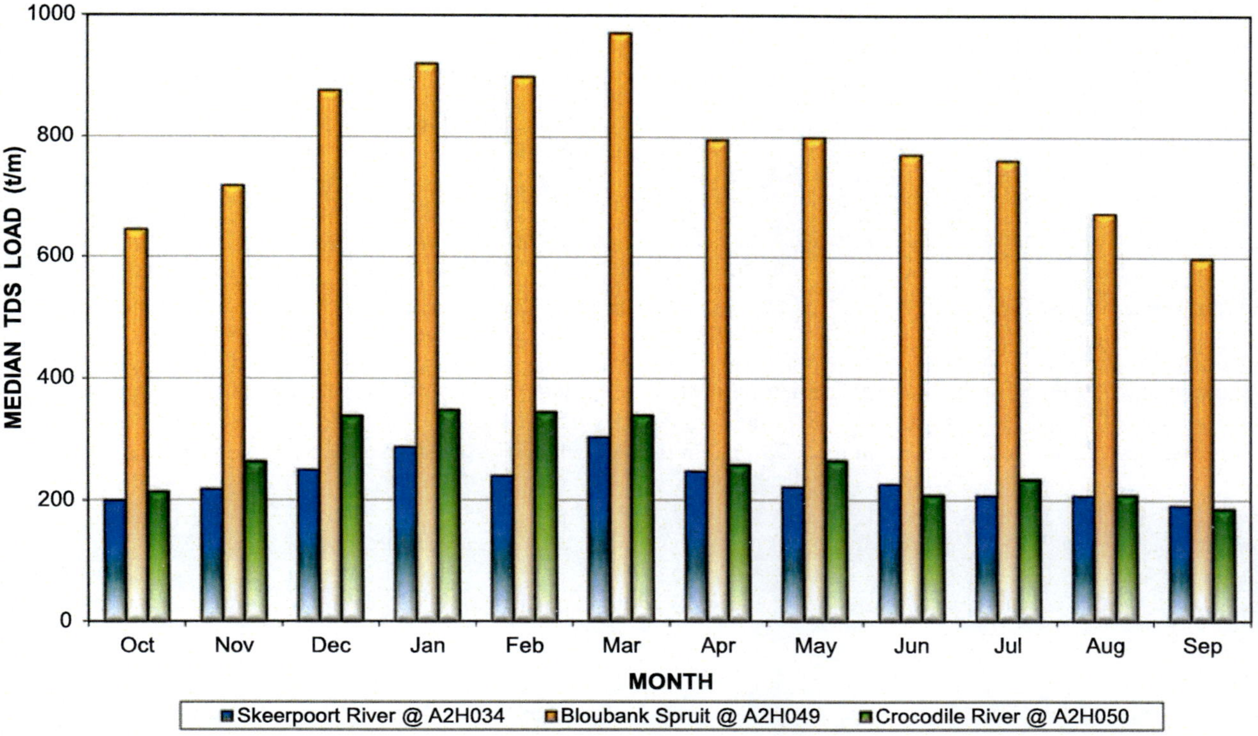


**Fig. 30** Comparison of the long-term (October 1976 to September 2014) median monthly TDS loads carried by the main drainages in the study area

Fig. 33, is the bimodal nature of discharge in the respective drainages. The peaks associated with largely natural discharge regimes occur in January and March (Figs. 31, 33 and 35). Those associated with an anthropogenically modified discharge regime occur in February and May (Fig. 34). The pattern reflected in Fig. 34 reveals the extent to which mine water discharges in the last five hydrological years have altered the flow regime of the Bloubank Spruit system.

The modification of the Bloubank Spruit system flow regime is also reflected in the hydrochemistry. The median TDS load pattern (as a measure of water quality) associated with a largely natural water quality regime exhibits a unimodal shape peaking in summer (Figs. 31, 33 and 35). By comparison, that associated with an anthropogenically modified water quality regime exhibits peaks in February, May and August (Fig. 34). Further, the bias of variance at the 95%ile level evident in January, March and June (Fig. 34) again reflects the recent influence of mine water discharge in the Bloubank Spruit system. This is substantiated by comparison with the graph for the pre-2009–'10 period of the A2H049 record (Fig. 33).

The long-term constancy of the monthly median TDS load in the Skeerpoort River (Fig. 31) again reflects the dominant contribution of the perennial karst springs to the water quality in this drainage. The discharge from the Driefontein WWTW on the Crocodile River upstream of station A2H050 explains the similar circumstances describing the monthly median TDS load in this drainage (Fig. 35).

The long-term monthly pattern and trend in the TDS loads delivered by the Bloubank Spruit and upper reach of the Crocodile River are shown in Figs. 36 and 37, respectively. Inspection of the graphs shows that both stations reflect an increasing TDS load (as indicated by the visually inserted arrows) since the early to mid-2000s. In the case of station A2H050, the commissioning of Unit 2 of the Driefontein WWTW might explain this observation (refer Sect. 3.3 in Chapter "Physical Hydrology").

The text box in Fig. 36 lists the median and 95%ile values associated with different periods of the record. The post-January 2010 period reveals the greatest difference with the long-term record. This is readily attributable to the contemporary very high mine water discharges (Sect. 3.1 in Chapter "Physical Hydrology"). Less obvious are the following comparisons:

- the similar median and 95%ile values of the pre-August 2002 and post-July 2002 periods—this indicates little change in the water chemistry following the commencement of decant from the Western Basin in August 2002; and

**Table 25** Annual long-term TDS load delivered by the main drainages in the study area

| Drainage | Station | Median TDS (mg/L) | Median discharge ($Mm^3/a$) | TDS Load | | |
|---|---|---|---|---|---|---|
| | | | | t/d | t/a | % of total |
| Skeerpoort River | A2H034 | 265 (from Table 22) | 10.21 (from Fig. 6 in Chapter "Physical Hydrology") | 7.4 | 2703 | 16.1 |
| Bloubank Spruit | A2H049 | 444 (from Table 19) | 23.52 (from Fig. 11 in Chapter "Physical Hydrology") | 28.6 | 10 443 | 62.0 |
| Crocodile River | A2H050 | 286 (from Table 23) | 12.89 (from Fig. 22 in Chapter "Physical Hydrology") | 10.1 | 3687 | 21.9 |
| Total | | | | 46.1 | 16 833 | 100 |

- the similar 95%ile values but doubled median value of the post-July 2012 period compared to the whole record period—this indicates an increase in the TDS load in the most recent past without manifesting excursions at the 95%ile level. Figure 36 suggests that this increase amounts to ~100% (from 750 to 1500 t/m).

| Surface water (Station) | Section | TDS (mg/L) | SEC (mS/m) | TDS:SEC | $SO_4$:TDS (%) |
|---|---|---|---|---|---|
| Skeerpoort River (A2H034) | 1.1 | | | | |
| Tweelopie Spruit (F11S12)[1] | 1.2.2 | 2890 | 283 | 10.2 | 63 |
| Tweelopie Spruit (F11S12)[2] | 1.2.2 | 3500 | 330 | 10.5 | 64 |
| Tweelopie Spruit (F11S12)[3] | 1.2.2 | 3030 | 295 | 10.3 | 63 |
| Blougat Spruit (188,048) | 1.2.4 | 502 | 82 | 6.1 | 33 |
| Tweefontein Spruit (F14S15) | 1.2.5 | 448 | 60 | 7.5 | 18 |
| Bloubank Spruit (A2H049) | 1.2.6 | 451 | 61 | 7.4 | 19 |
| Honingklip Spruit | 1.2.6 | 164 | 22 | 7.5 | 9 |
| Crocodile River (A2H050) | 1.3 | 290 | 44 | 6.6 | 11 |
| *Groundwater* | | | | | |
| Pristine karst springwater | 7 in Chapter "Chemical Hydrogeology" | 310 | 36 | 8.6 | 2 |
| Impacted karst groundwater | 8.4 in chapter "Chemical Hydrogeology" | 1699 | 192 | 8.5 | 62 |
| Quartzitic strata[4] | — | 40 | 9 | 4.6 | 11 |
| *Wastewater* | | | | | |
| Raw mine water[5] | 1.2.1 | | | | |
| Treated mine water | 1.2.1 | | | | |
| *Municipal sewage effluent* | | | | | |
| Percy Stewart WWTW | 1.2.4 | 649[(6)] | 103 | (6.3) | 31[(6)] |
| Driefontein WWTW | 1.3 | 358[(6)] | 56 | (6.4) | 13[(6)] |

[1]Associated with the pre-February 2010 period of record
[2]Associated with the February 2010 to July 2012 period of record
[3]Associated with the August 2012 to September 2017 2014 period of record
[4]Derived from data presented by Hobbs and Cobbing (2007) and Hobbs (2011a)
[5]Associated with the Black Reef Incline discharge
[6]Value based on the TDS:SEC ratio shown in brackets

*Notes*

(a) The similar $SO_4$:TDS values reported for the Blougat Spruit / Percy Stewart WWTW (33%/31%), the Crocodile River/Driefontein WWTW (11%/13%) and the Tweelopie Spruit/mine water (66%/64%) sources indicate the veracity of the assessments for these sources

(b) The similar $SO_4$:TDS values reported for Honingklip Spruit and Crocodile River surface water on the one hand, and groundwater sourced from quartzitic strata on the other, reflects the similar geologic environments common to the provenance areas of these water sources

(c) The similar $SO_4$:TDS values reported for Tweelopie Spruit surface water, raw and treated mine water and impacted karst groundwater similarly reflects the common influence of AMD

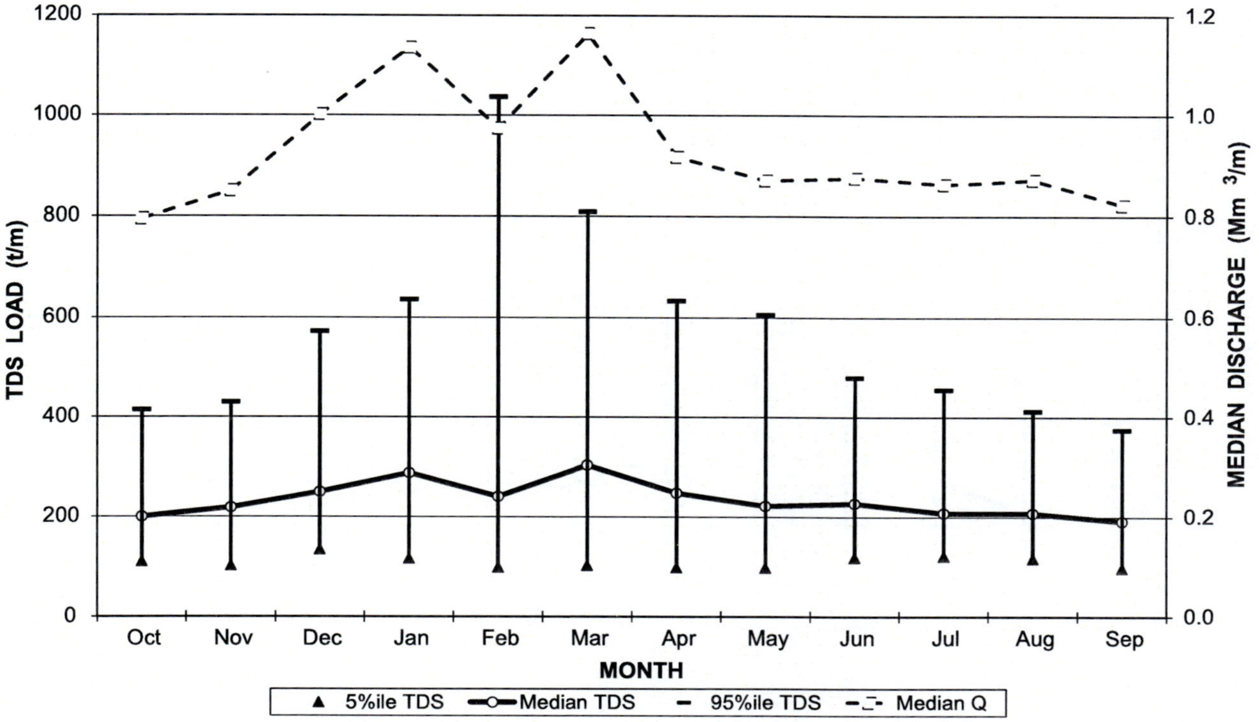


**Fig. 31** Variance in the long-term (October 1976 to September 2014) monthly Skeerpoort River TDS load at station A2H034

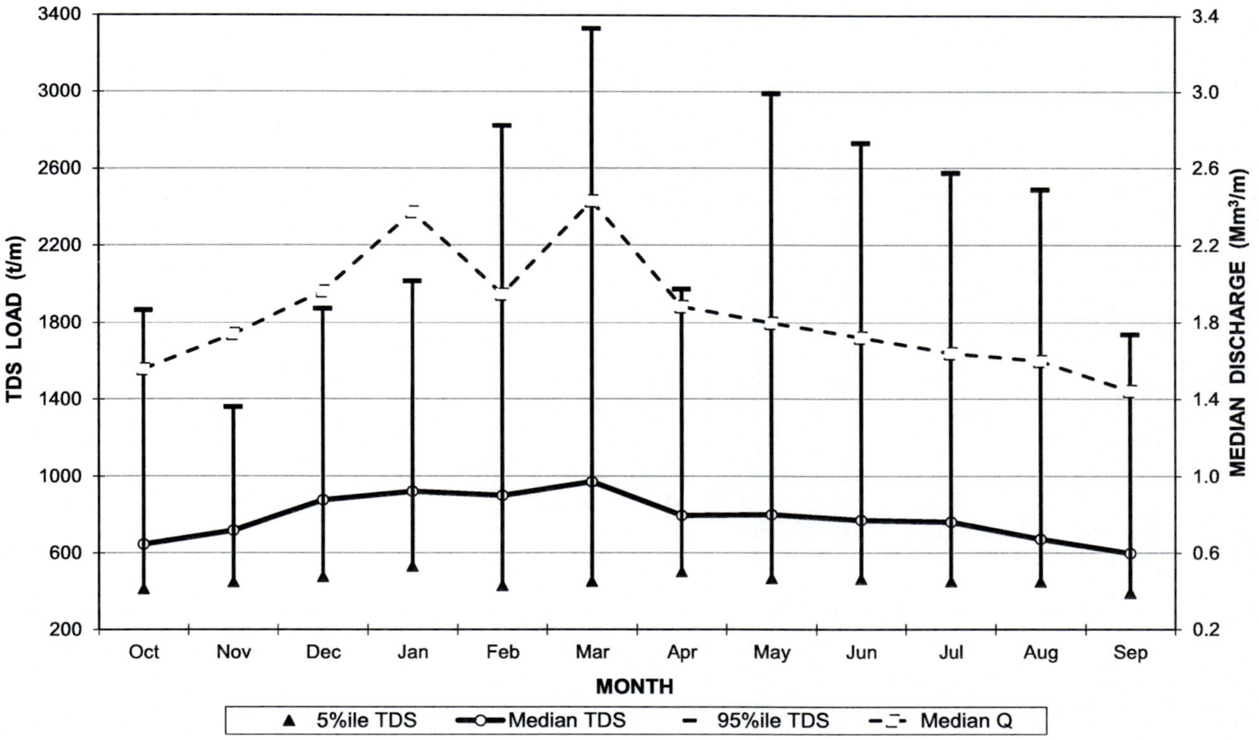


**Fig. 32** Variance in the long-term (October 1979 to September 2014) monthly Bloubank Spruit TDS load pattern at station A2H049

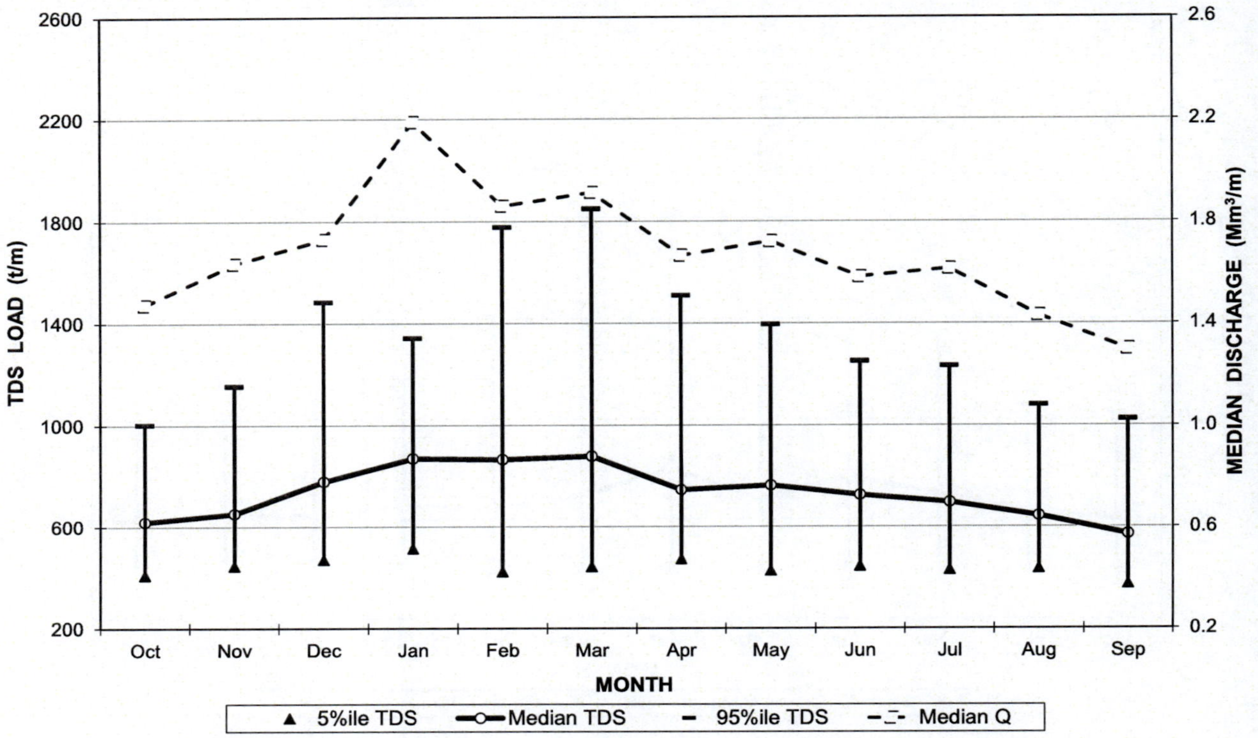


**Fig. 33** Variance in the pre-2010 monthly Bloubank Spruit TDS load pattern at station A2H049

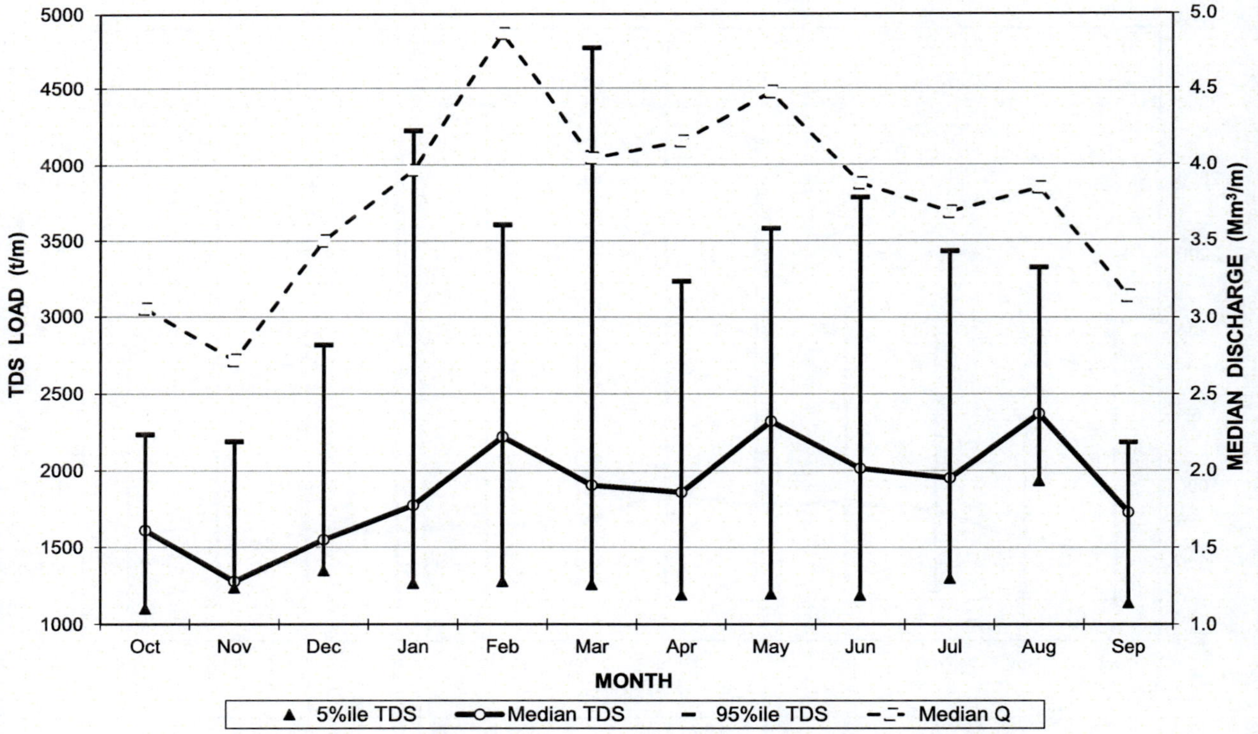


**Fig. 34** Variance in the recent (2010 to 2014) monthly Bloubank Spruit TDS load pattern at station A2H049

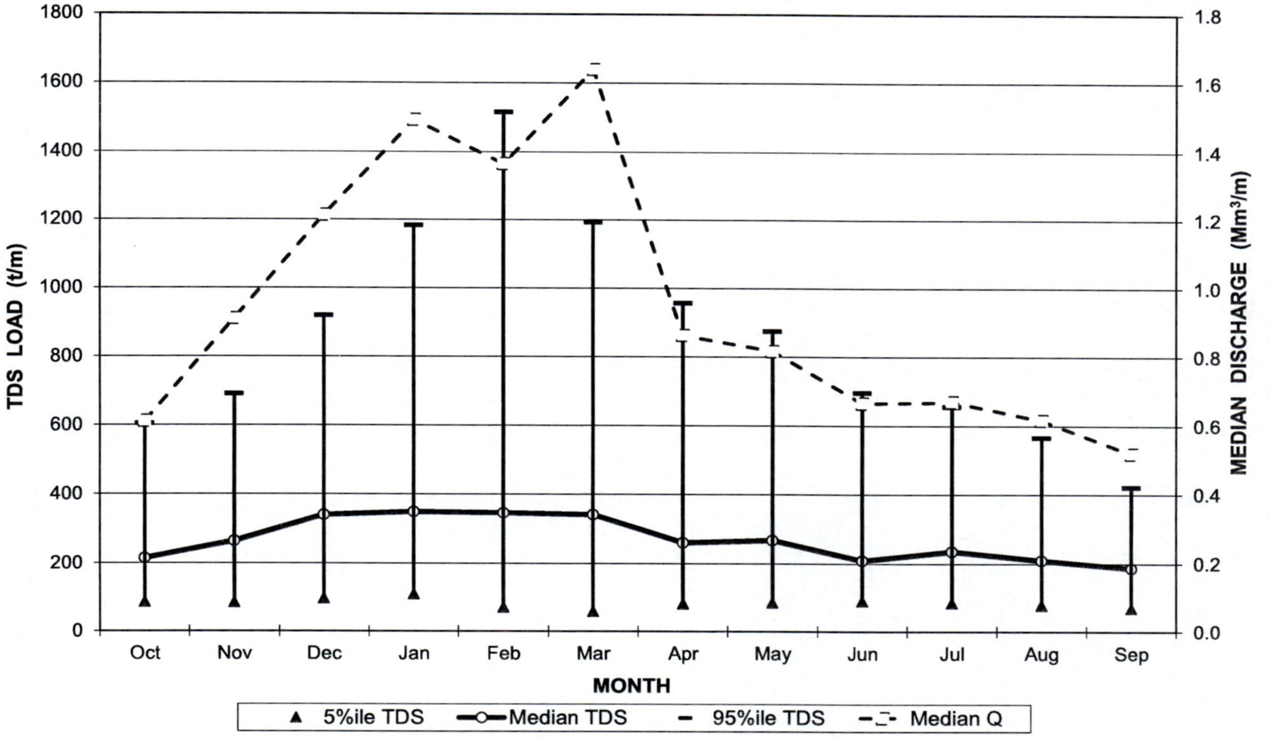


**Fig. 35** Variance in the long-term (October 1979 to September 2014) monthly Crocodile River TDS load pattern at station A2H050

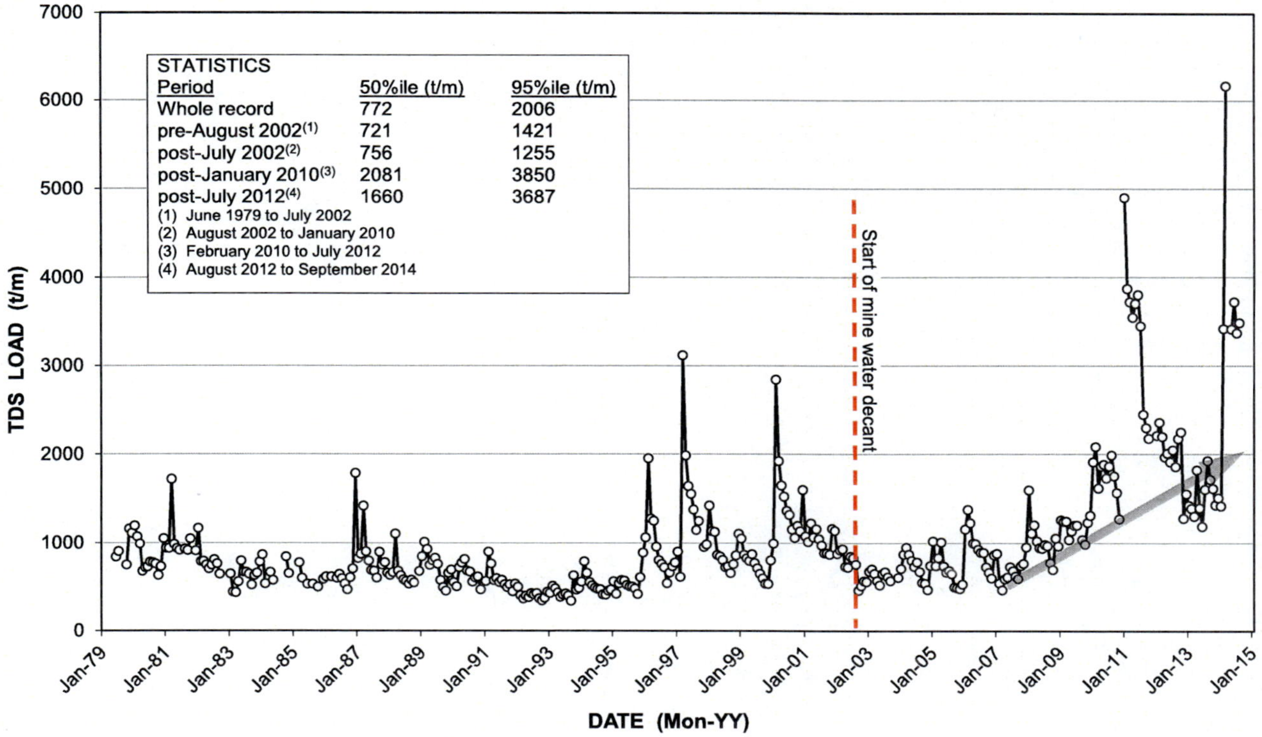


**Fig. 36** Long-term (June 1979 to September 2014) monthly TDS load pattern and trend in the Bloubank Spruit at station A2H049

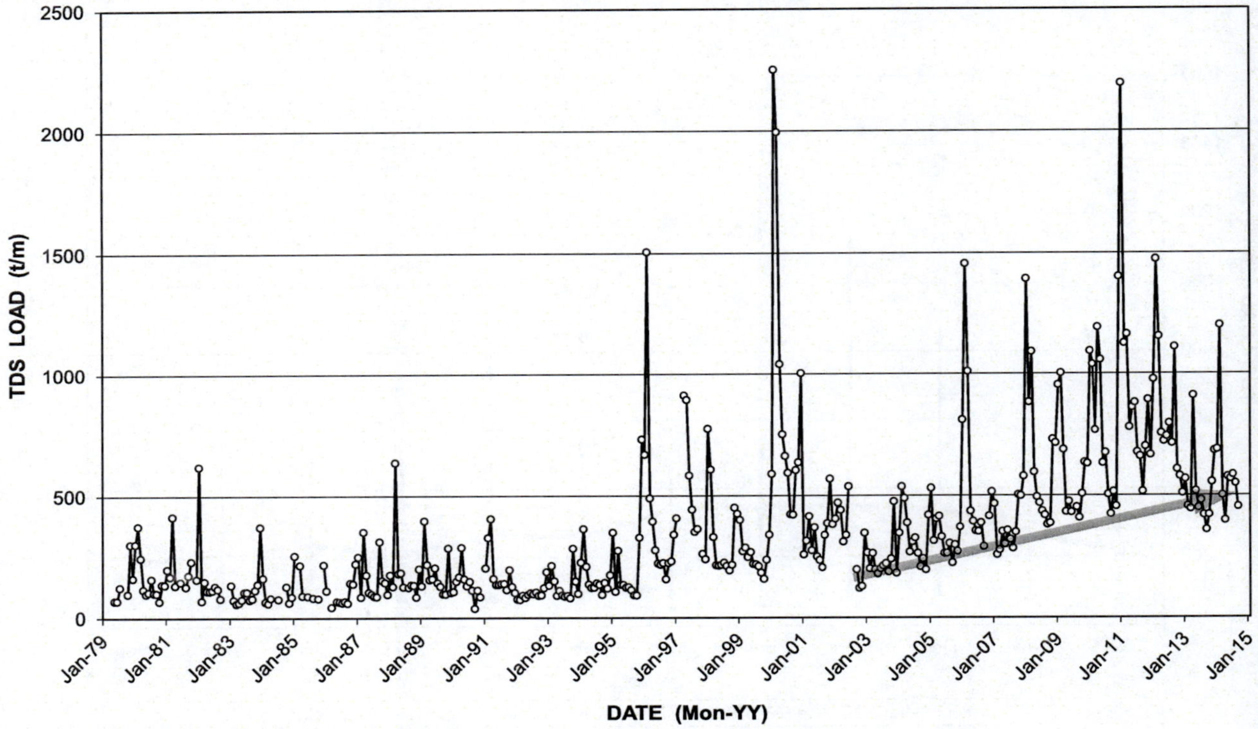


**Fig. 37** Long-term (May 1979 to September 2014) monthly TDS load pattern and trend in the upper reach of the Crocodile River at station A2H050

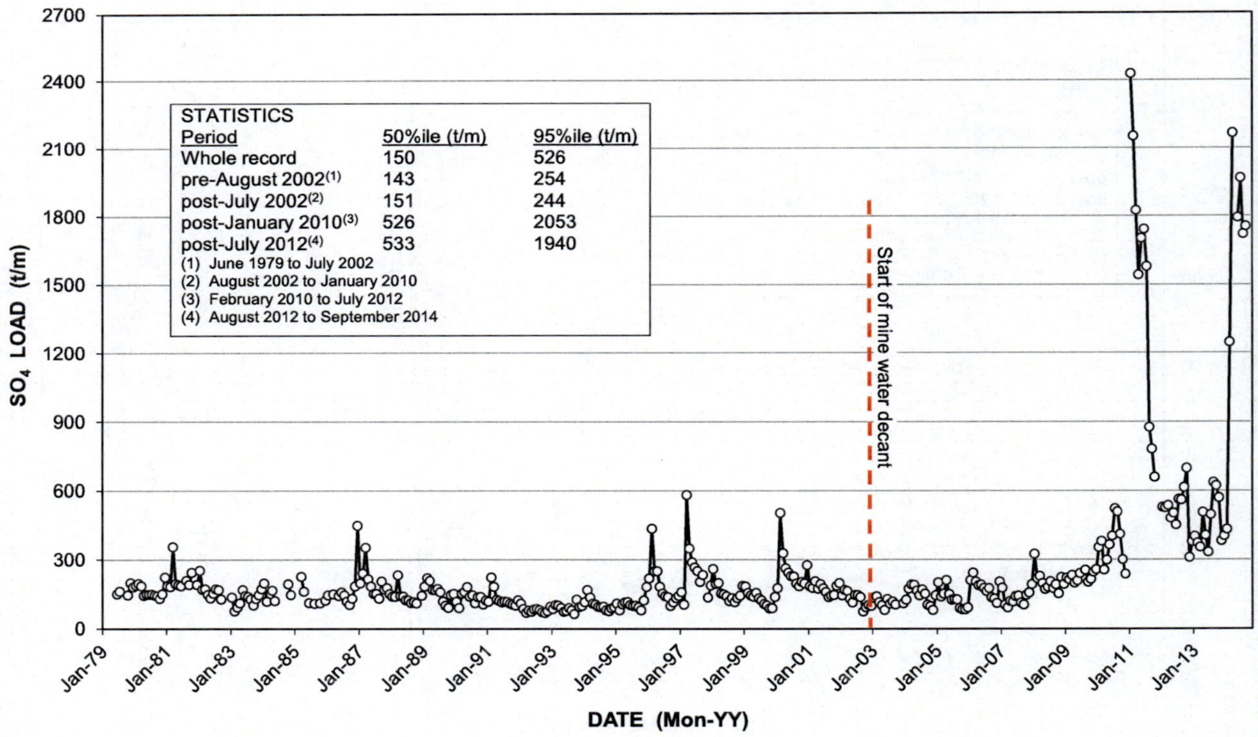


**Fig. 38** Long-term (October 1979 to September 2014) monthly $SO_4$ load pattern and trend in the Bloubank Spruit at station A2H049

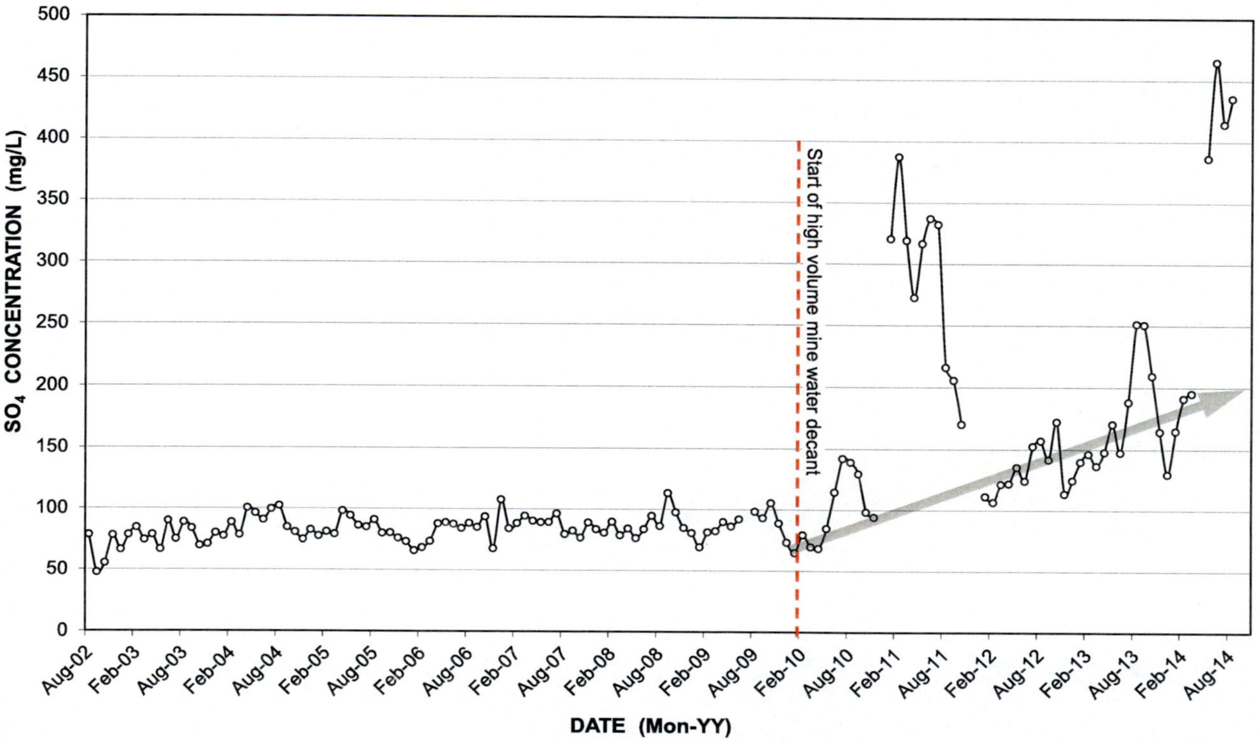


**Fig. 39** Pattern and trend in the $SO_4$ concentration at station A2H049 since the start of mine water decant in the Western Basin

The long-term monthly trend in the $SO_4$ load delivered by the Bloubank Spruit is shown in Fig. 38, and since the start of decant in Fig. 39. Unsurprising under circumstances where $SO_4$ comprises ~63% of the major ion concentration in mine water (Text Box 1), the most recent period of record mimics the TDS load pattern and trend (Fig. 36 and 38).

The $SO_4$:TDS ratio provides a further measure of the mine water impact on the chemistry of surface water at the downstream end of the Bloubank Spruit system. This is illustrated in Fig. 40 for the long-term record, and in Fig. 41 for the period since the start of decant.

The rising trends (arrowed) indicate that $SO_4$ contributes an increasing proportion of the TDS concentration in the more recent past. Quantification of this ratio for various sources of water in the study area is presented in Text Box 1. Clearly the 19% value reported in Text Box 1 for the Bloubank Spruit at A2H049 more recently approaches 30 to 35% (ignoring the greater short-term exceedances in mid-2011 and mid-2014). This trend is in contrast to the pre-decant trend revealed in Fig. 40, which reflects a distinct decline in the $SO_4$:TDS ratio in the period 1986 to 2001.

A possible explanation for the 1986 to 2001 trend is the greater contribution of karst groundwater, typically with a very low $SO_4$ concentration, draining from the Zwartkrans and Krombank compartments following the breaking of the drought that characterised the early 1980s (Sect. 2.1 in Chapter 4). The preceding (1979 to 1985) rising $SO_4$ trend (Fig. 40) possibly reflects the increasing impact of contemporary mine water releases from still active mining operations in the Western Basin, exacerbated by a concurrent drought-related decrease in spring discharges. These observations are considered to provide the clearest indication that mine water originating in the Western Basin has had both an historical and a recent impact on the quality of surface water (and groundwater) resources in the south-western portion of the study area. Further discussion of this impact is presented in Sect. 2.3.3 in Chapter "Chemical Hydrogeology".

The magnitude of the large mine water discharges in 2011 and 2014 is also illustrated in Fig. 42, which depicts the monthly TDS load in each year of record. The high loads observed in 1997 and 2000 (concentrated in the months February, March and April in both instances) are surpassed by the 2011 and 2014 loads.

The data presented in Fig. 42 further indicate that the month of March has produced the highest monthly TDS load on seven occasions (1981, 1985, 1987, 1988, 1997, 2004 and 2014) in the 36-year record period. The most recent is also the highest monthly load on record. March has also produced the 2nd highest load on a further seven occasions (1989, 1991, 1994, 1996, 2000, 2006 and 2008). These

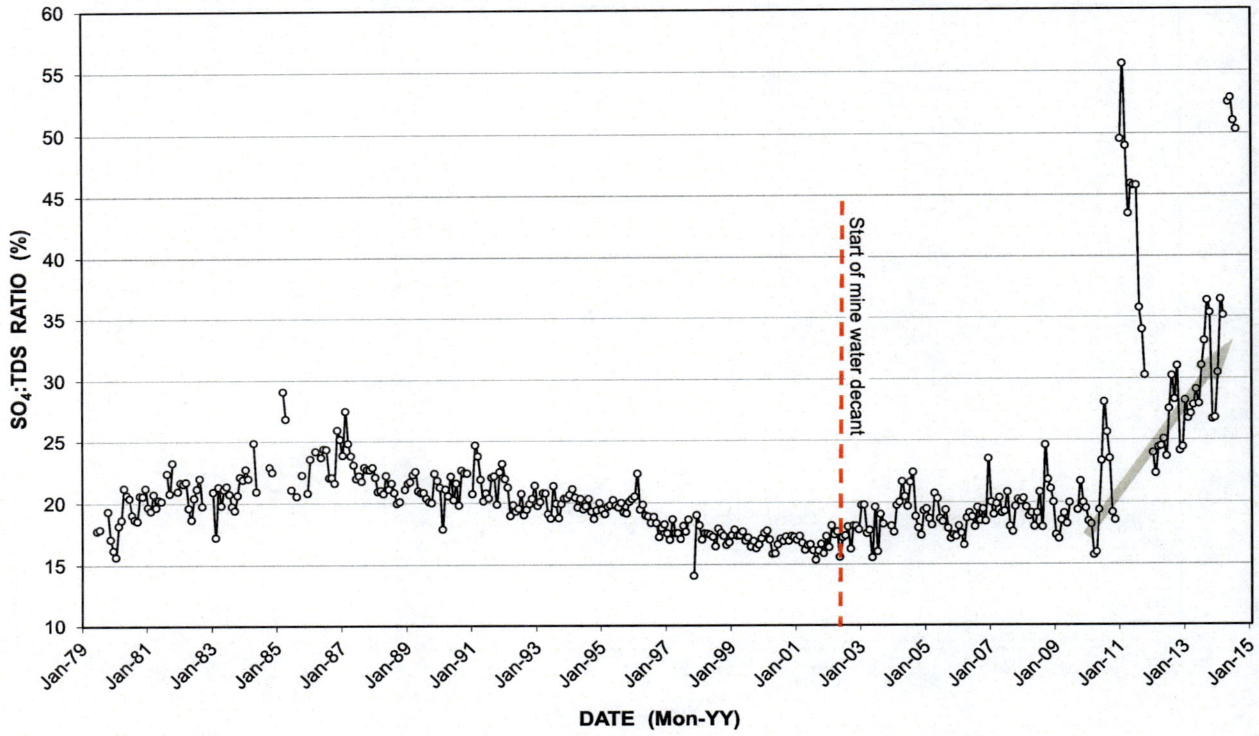


**Fig. 40** Long-term (June 1979 to September 2014) trend in the $SO_4$:TDS ratio at station A2H049

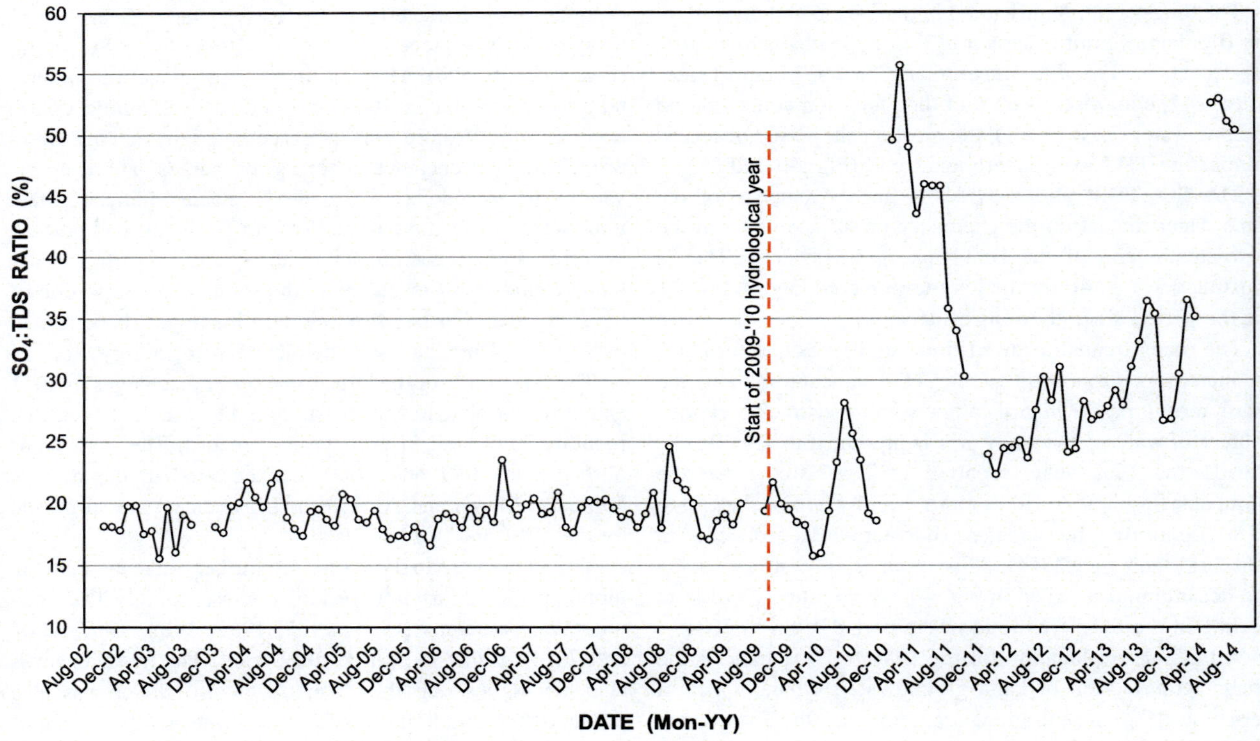


**Fig. 41** Pattern and trend in the $SO_4$:TDS ratio at station A2H049 since the start of mine water decant in the Western Basin

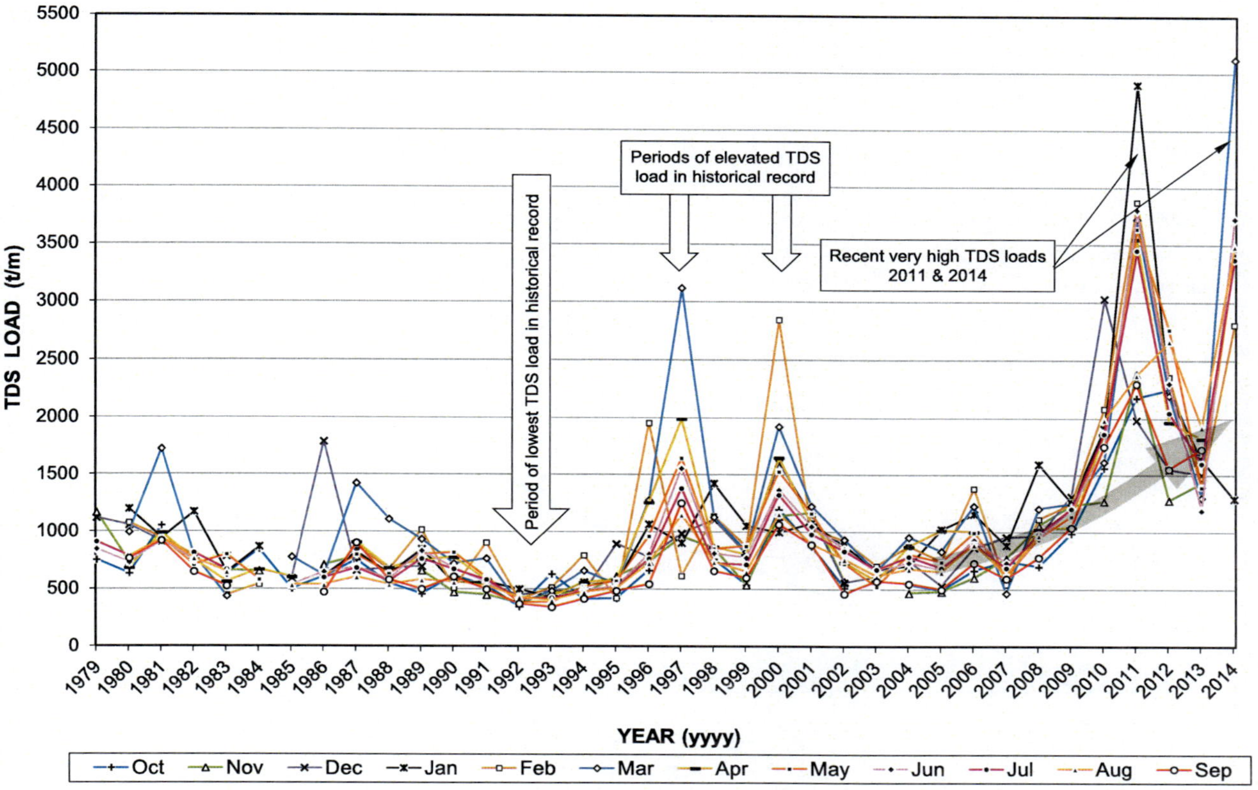


**Fig. 42** Monthly TDS load pattern and trend at station A2H049 in each year of historical record

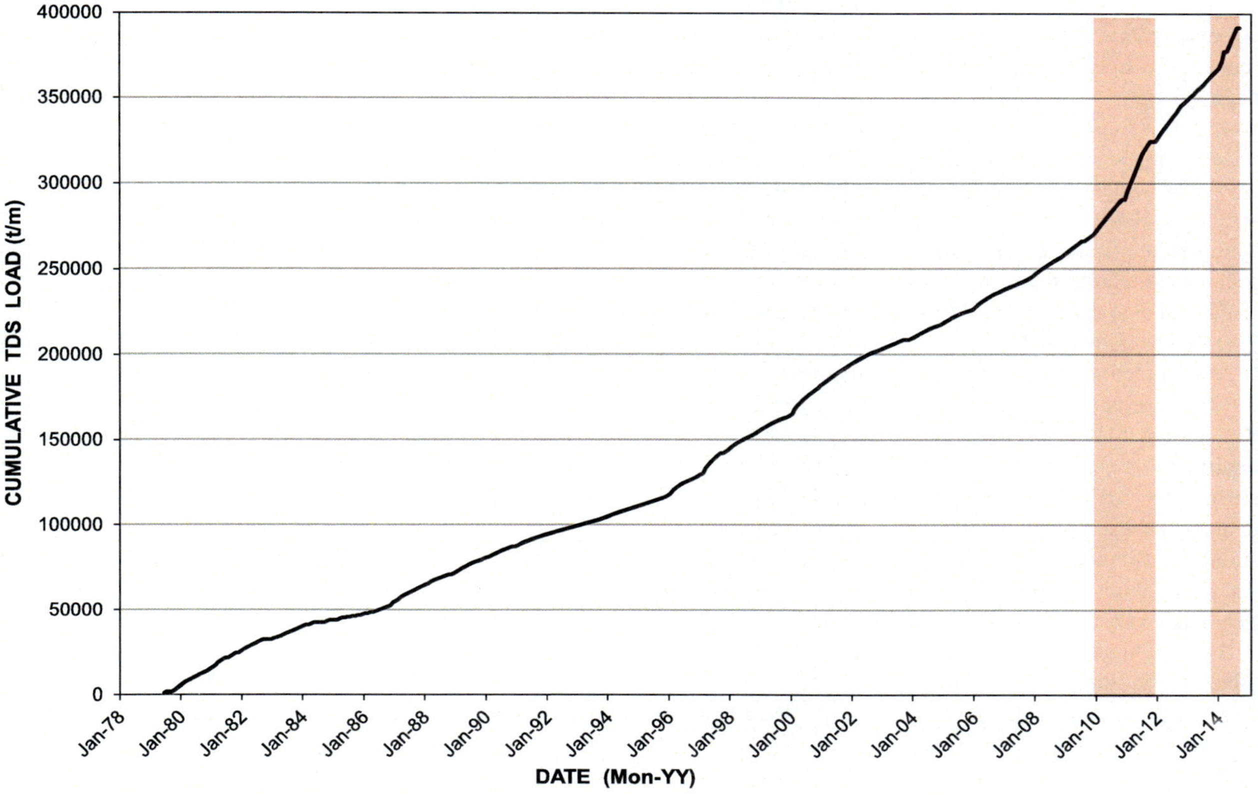


**Fig. 43** Pattern and trend of cumulative monthly TDS load delivered by the Bloubank Spruit system in the long-term

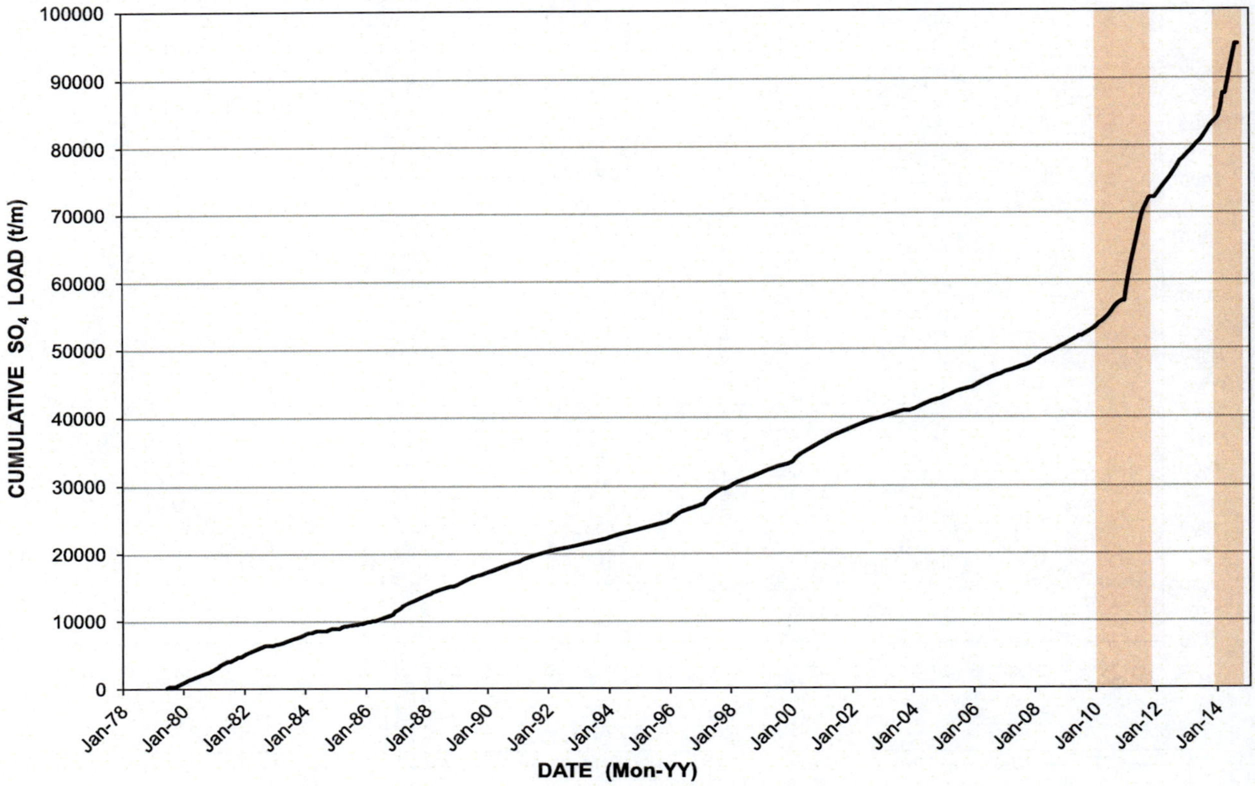


**Fig. 44** Pattern and trend of cumulative monthly $SO_4$ load delivered by the Bloubank Spruit system in the long-term

occurrences together represent 39% of the record period, suggesting that March is the most 'prolific' load-bearing month of a hydrological year in the Bloubank Spruit drainage system.

It is since mid-2010 that the large mine water discharges resulted in a significant change in the chemical composition of surface water leaving the Bloubank Spruit system at station A2H049. This serves as confirmation that a significantly greater mine water component, and in particular raw mine water, characterised the surface water chemistry in the middle and lower reaches of the Bloubank Spruit system than before mid-2010.

The reduction in monthly TDS loads through 2012 into 2013 was reversed by the excessive mine water discharge in the period March to July 2014. It is therefore worth exploring the effects of the large mine water discharges during the 2010, 2011 and 2014 rainfall seasons in the context of the long-term TDS and $SO_4$ loads delivered by the Bloubank Spruit system. This is attempted in Figs. 43 and 44, which show the respective cumulative monthly TDS and $SO_4$ loads at station A2H049 in the long-term.

The cumulative TDS and $SO_4$ loads in the pre-January 2010 period suggest a slight change in slope ca. January 1996. This change, although interesting, is not considered significant enough to warrant more detailed inspection in this study. More significant is the sharper change in slope commencing January 2010 and expressed most dramatically (especially in the case of $SO_4$) ca. January 2011. Similar circumstances are again evident in early 2014. These are unequivocal manifestations of the impact of mine water on the surface water chemistry delivered by the Bloubank Spruit system.

The cumulative long-term median contribution to the TDS load delivered by each of the Bloubank Spruit and the (upper) Crocodile River is shown in Fig. 45. The analysis, based on the hydrological years 1980 to 2014, reflects the ~3 times greater contribution of the Bloubank Spruit system.

A similar inspection of the TDS load data for only the Bloubank Spruit system over the period of record 1980 to 2014 is presented in Fig. 46. The similarity of the pre-2010 and whole record data contrasts markedly with the pattern and trend associated with the period 2010 to 2014. The analysis reveals the ~140% greater load discharged since the 2009 hydrological year, compared to the preceding ~30 years. These circumstances further illustrate the impact of mine water discharges on the recent water chemistry of the Bloubank Spruit system. This analysis also provides the

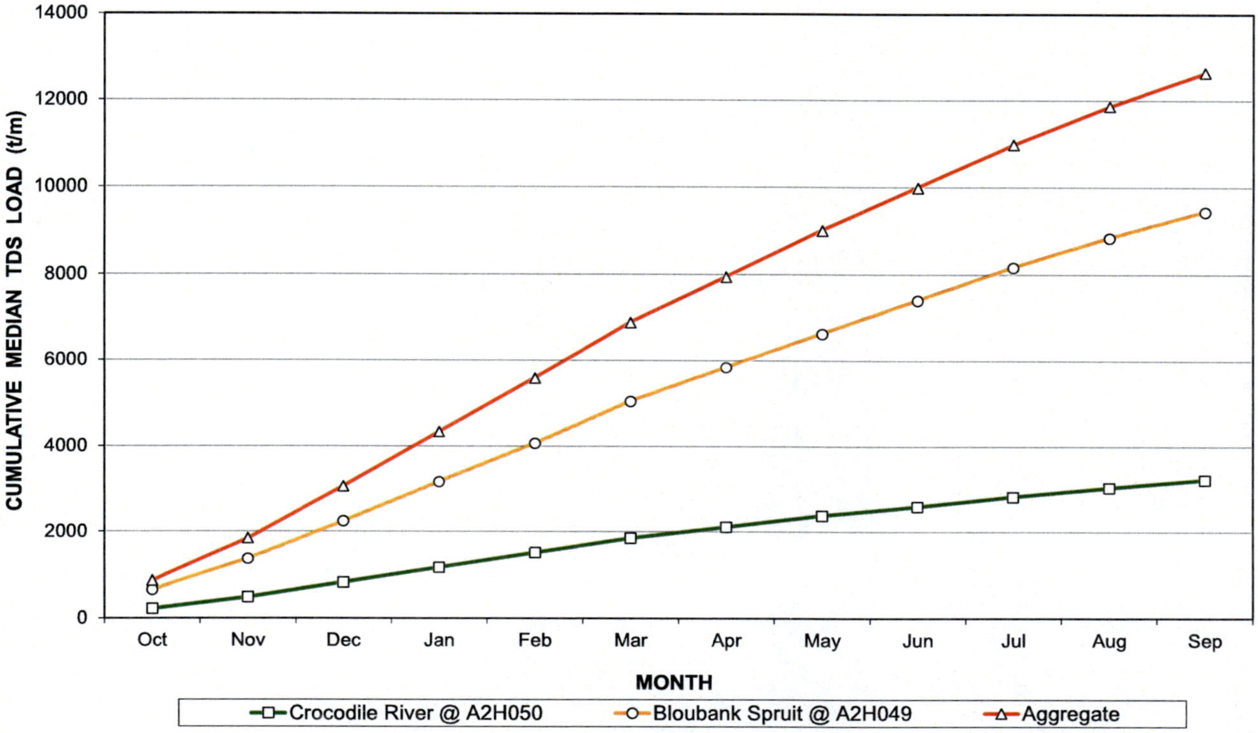


**Fig. 45** Cumulative long-term median monthly TDS loads discharged by the Bloubank Spruit and the upper reach of the Crocodile River

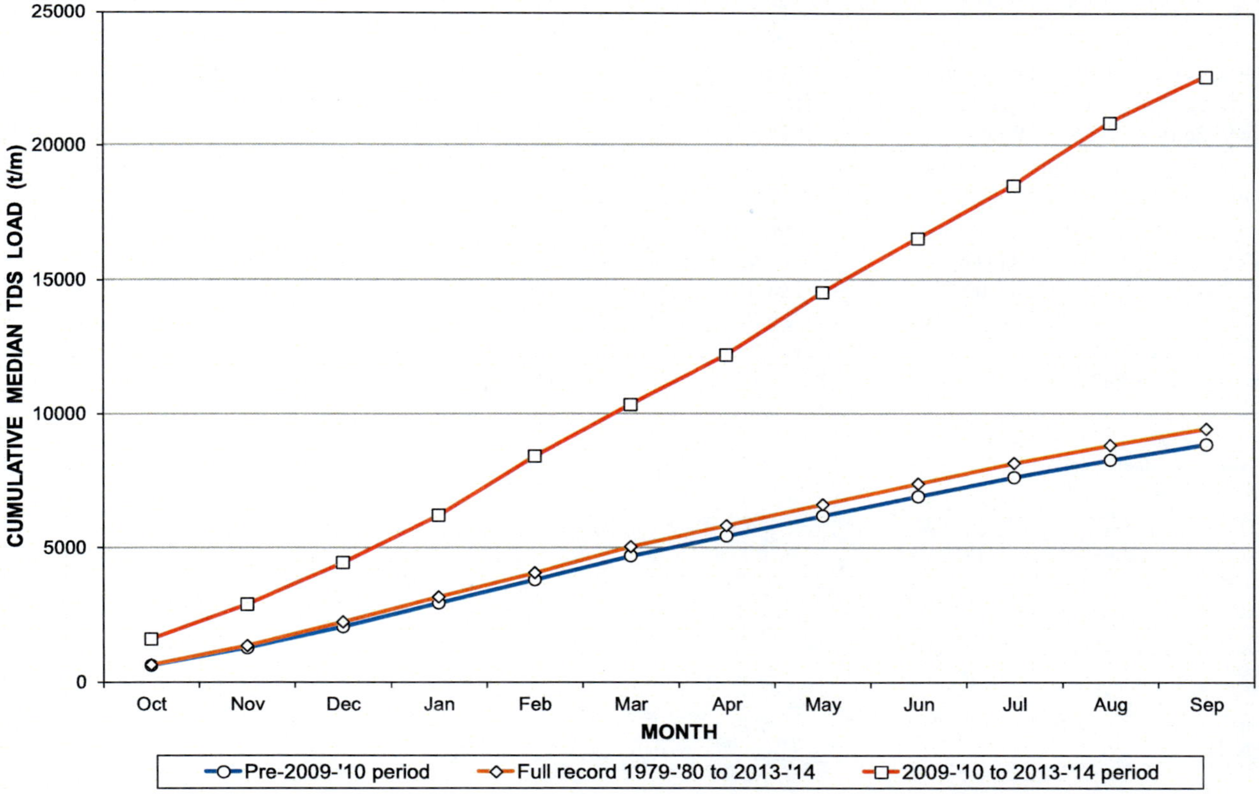


**Fig. 46** Comparison of cumulative median monthly Bloubank Spruit system TDS loads for three periods of record

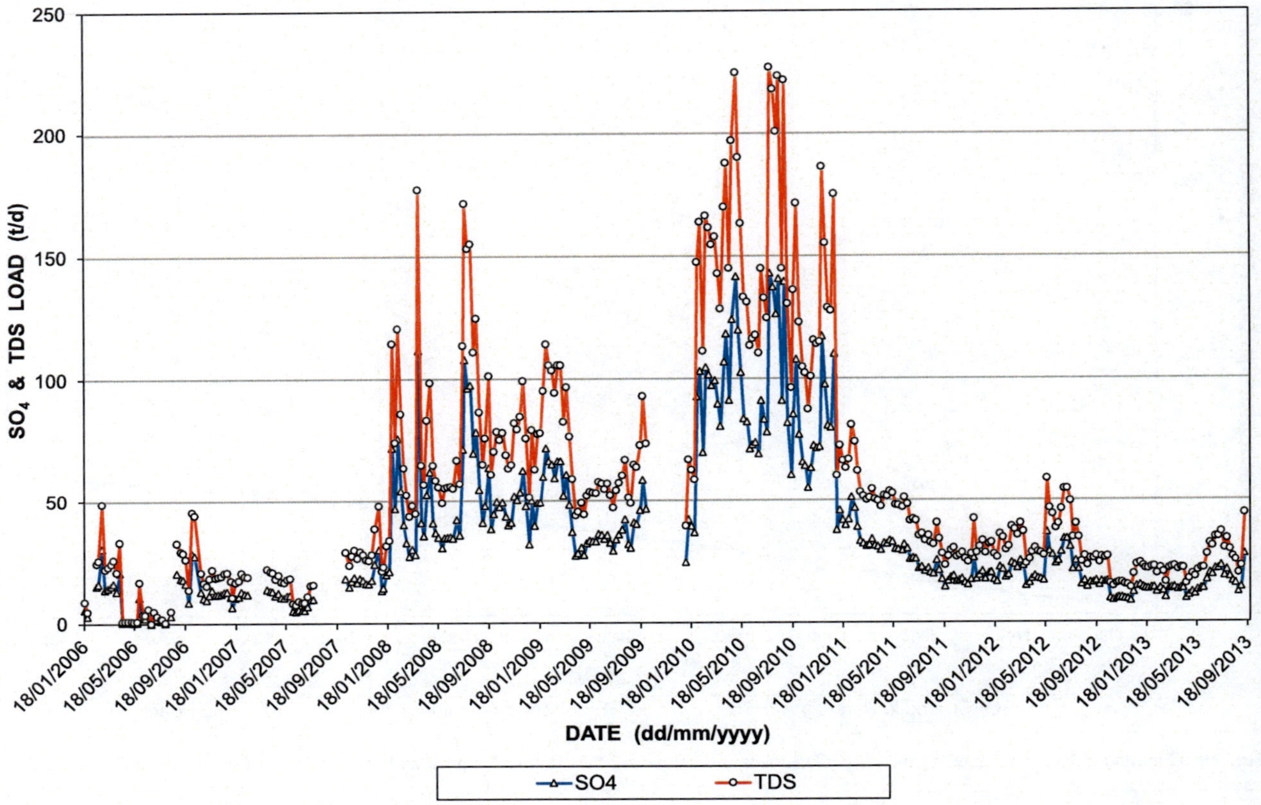


**Fig. 47** Pattern and trend of daily $SO_4$ and TDS loads in mine water delivered to the Tweelopie Spruit

background for the interrogation of the mine water impacts in a regional context as set out in Sect. 2.3. The relevance of specific timeframes in such an exercise is evident.

Perhaps of greater consequence for the downstream environment that includes the Hartbeespoort Dam, is the nutrient load delivered by the Bloubank Spruit and the (upper) Crocodile River catchments. The evaluation of nutrients such as $NO_3$–N and $PO_4$–P as indicators of pollution is presented in greater detail in Sect. 3.2 in Chapter 6.

The Percy Stewart WWTW on the Blougat Spruit in the Bloubank Spruit system is reported to have delivered a median orthophosphate ($PO_4$ as mg P/L) load of ~19.3 t/a to this catchment in the period July 2007 to June 2009 (Hobbs, 2011a). This is calculated from a median $PO_4$ concentration of 4.1 mg P/L (Table 13) and a median plant discharge of ~4.7 $Mm^3$/a (Sect. 3.2 in Chapter "Physical Hydrology"). An inspection of the average monthly concentrations and flows recorded at stations A2H049 (0.25 mg P/L at 27 $Mm^3$/a) and A2H050 (0.18 mg P/L at 25.3 $Mm^3$/a) in a similar period (mid-2007 to mid-2008) indicates loads of ~6.7 and ~4.6 t/a passing these stations, respectively.

**Table 26** Median annual nutrient loads (as $NO_3$–N and $PO_4$–P) delivered by the Crocodile River system to Hartbeespoort Dam in the period 1979–'80 to 2013–'14

| Drainage (station #) | Median discharge ($Mm^3$/a) | Median $NO_3$–N (mg/L) | Median $PO_4$–P (mg/L) | Median Load | | % of Total | |
|---|---|---|---|---|---|---|---|
| | | | | $NO_3$–N (t/a) | $PO_4$–P (t/a) | $NO_3$–N | $PO_4$–P |
| Hennnops River (A2H014) | 67.570 | 4.72 | 0.617 | 319 | 41.2 | 22.4 | 36.6 |
| Jukskei River (A2H044) | 122.845 | 7.88 | 0.557 | 968 | 68.4 | 68.2 | 60.7 |
| Bloubank Spruit (A2H049) | 22.701 | 4.36 | 0.054 | 99 | 1.2 | 7.0 | 1.1 |
| Crocodile River (A2H050) | 12.892 | 2.66 | 0.139 | 34 | 1.8 | 2.4 | 1.6 |
| Total | | | | 1 420 | 112.6 | 100 | 100 |

The median phosphorus load of ~19.3 t/a delivered by the Percy Stewart WWTW contrasts markedly with the much smaller median load of ~6.7 t/a passing station A2H049 located further downstream. This discordance remains valid even if the comparative coarseness of the calculations is considered, and prompts the following discussion.

It is shown in Table 20 that $PO_4$ concentrations of ~2.4 mg P/L were observed at stations BB@N14 and BB@M on 18/05/2010, whilst at station BB@PL further downstream it amounted to only 0.5 mg P/L on the same day. The reduction is readily attributed to the dilution provided by the Kromdraai Spring, and might account for the smaller downstream $PO_4$-P load. It is also possible, however, that immobilisation processes involving phosphorus and heavy metals such as aluminium and iron (see for example Johnson and Younger, 2006; Omoike and Vanloon, 1999; Strosnider and Nairn, 2010) might account for this. This possibility is explored in Sect. 11 in Chapter "Chemical Hydrogeology". It is worth noting that even the most severely impacted karst groundwater source (Sect. 4 in Chapter "Chemical Hydrogeology") returned an orthophosphate concentration of <0.2 mg P/L, i.e. below the analytical detection limit.

It is shown in Table 26 that the Bloubank Spruit and Crocodile River drainages together contributed median $NO_3$–N and $PO_4$–P loads of 133 and 3 t/a, respectively, to Hartbeespoort Dam in the period 1980 to 2014. By comparison, an appraisal of similar DWS data for this period shows that the Jukskei and Hennops river catchments contributed combined median $NO_3$–N and $PO_4$–P loads of 1287 and ~110 t/a, respectively.

Seen in context, therefore, the southern and south-western catchments represented by the (upper) Crocodile River and Bloubank Spruit systems respectively, contribute <10% to the median $NO_3$–N load entering Hartbeespoort Dam. This contribution is substantially exceeded by those of the Jukskei River (~68%) and the Hennops River (~22%). The situation in regard to the $PO_4$-P load is even more weighted towards the Jukskei and Hennops rivers, with the Crocodile River and Bloubank Spruit systems delivering < 3% of the $PO_4$–P load (Table 26). These results find support in an evaluation presented by Roux (2010) and Oelofse et al. (2012). They are also in agreement with an earlier finding by Harding et al. (2004), who report a total phosphorus (TP) load of ~166 t/a (within a range of 80 to 300 t/a) entering the dam, >99% of which is via the Crocodile River including the contributions of all its tributaries.

## 2.2 Subcatchment Scale

The information in Table 27 provides an estimate of the TDS load contributed by various sources in the Bloubank Spruit system compared to that leaving this system at station A2H049. The data are applicable to the prevailing conditions preceding the 2010 hydrological year since when, as has been shown, mine water discharges have substantially modified the water chemistry delivered by this drainage.

Table 27 shows that the historical TDS load at station A2H049 (8560 t/a) is within 3% of the sum (8827 t/a) of the contributions from the upstream sources represented by the Zwartkrans, Kromdraai and Plover's Lake springs, the Blougat Spruit and groundwater resurgence in the Bloubank Spruit upstream of the Zwartkrans Spring. It has been shown that the Tweelopie and Blougat spruits represent the two most impacted contributing sources within the Bloubank Spruit system. The contribution of the Tweelopie Spruit is not explicitly accounted for in this estimate, because a significant proportion of the discharge in this drainage into the Riet Spruit is lost as allogenic recharge to the karst aquifer. This contribution is reflected in the groundwater resurgence upstream of the Zwartkrans Spring. In summary, the Blougat Spruit reflects a ~27% contribution to the load at station A2H049 compared to the ~53% of the combined springwater sources. The contribution of groundwater resurgence accounts for the balance of 20%. As discussed in Sect. 1.2.6, however, the very wet 2010, 2011 and 2014 summer rainfall seasons (Sect. 2 in Chapter "Description of the Physical Environment") precipitated a significant change in these

**Table 27** Estimated TDS load contributed by various sources in the Bloubank Spruit system prior to the 2010 hydrological year

| Source (station #) | Description | TDS (mg/L) | Discharge ($Mm^3/a$) | | TDS load (t/a) | |
|---|---|---|---|---|---|---|
| Blougat Spruit (188,048) | Treated sewage effluent | 502 | 4.75 | 23.00 | 2385 | 8 827 |
| Groundwater resurgence[1] | Surface water | 665 | 2.63 | | 1749 | |
| Zwartkrans Sp. | Groundwater | 244 | 4.27 | | 1042 | |
| Plover's Lake Sp. | Groundwater | 110 | 1.89 | | 208 | |
| Kromdraai Sp. | Groundwater | 364 | 9.46 | | 3443 | |
| Bloubank Spruit (A2H049) | Surface water | 444 | 22.70 | | 8560 | |

[1]Refer Sect. 5.2.3 in Chapter "Physical Hydrology" for a comprehensive discussion of this component

**Table 28** Statistical analysis of $SO_4$ and TDS loads in composite mine water entering the Tweelopie Spruit in the period January 2006 to September 2013

| Load variable /Analyte | Statistical parameter | | | | | | |
|---|---|---|---|---|---|---|---|
| | n | 5%ile | Mean | Median | 95%ile | SD | CoV (%) |
| Sulfate (t/d) | 365 | 5.36 | 35.23 | 25.47 | 102.52 | 34.19 | 97 |
| TDS (t/d) | | 8.51 | 55.92 | 40.43 | 162.73 | 49.59 | 89 |

circumstances. Drainage-specific TDS loads that reflect the latter are discussed in more detail in the following subsections.

### 2.2.1 Tweelopie Spruit

A TDS load assessment for this drainage must be based on the blended raw mine water and treated/neutralised mine water components leaving the mine property (Fig. 13 in Chapter "Physical Hydrology"). The outflow from the Hippo Dam provides a measure of this variable with due cognisance of the following. The Hippo Dam data represents a poorer condition than that of station F11S12 located much closer to where the allogenic recharge of the karst aquifer occurs (Sect. 5.2.1 in Chapter "Physical Hydrology").

Both the Hippo Dam and F11S12 stations suffer the drawback that their water quality record is not matched by a similarly abundant set of discharge data. The Hippo Dam station, however, benefits from its proximity to the end-of-pipe (EoP) stations where the mine water discharge (both raw and treated/neutralised) is gauged (Fig. 13 in Chapter "Physical Hydrology"). Accepting that evaporation and evapotranspiration losses do not significantly reduce the outflow from the dam compared to that entering the dam from the mine area, it is possible to derive a $SO_4$ load for the Tweelopie Spruit in its upper reaches below the mine area.

Although TDS is not a monitoring variable downstream of the mine area, a theoretical TDS load can be derived

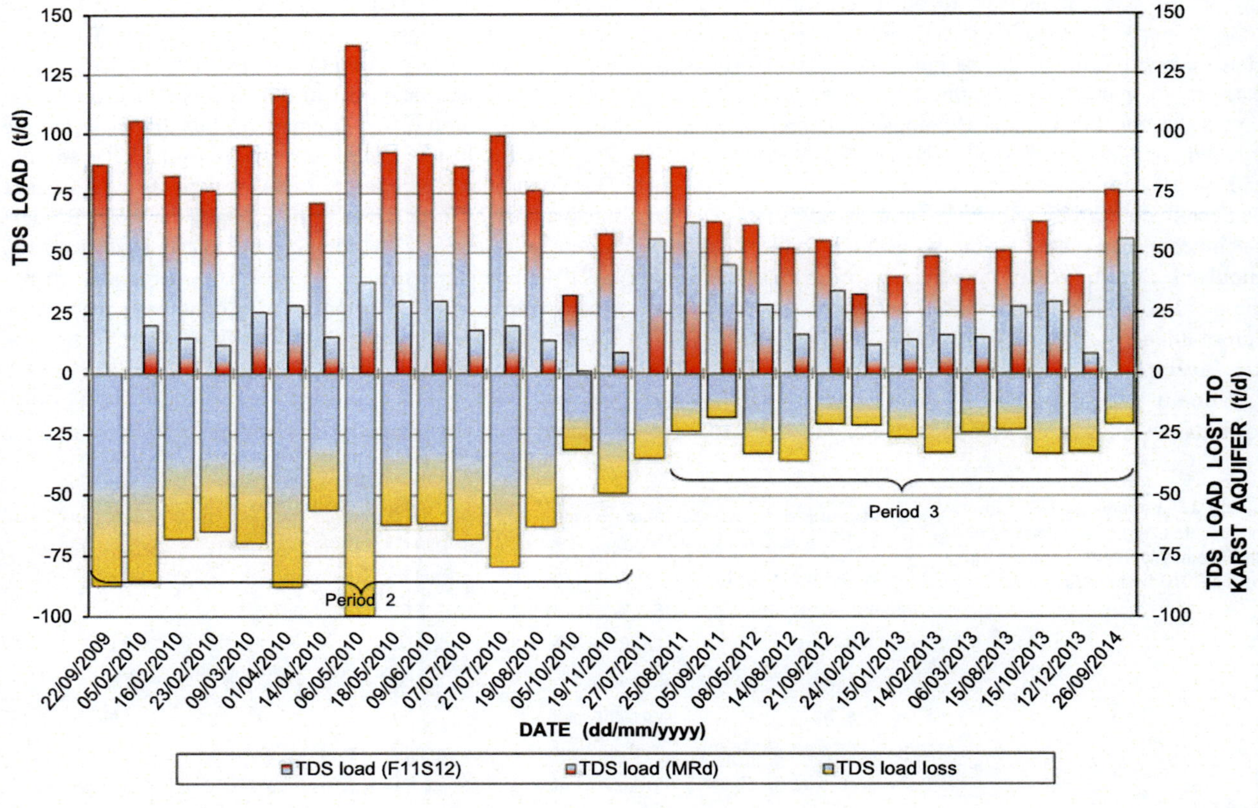


**Fig. 48** Pattern and trend of TDS load losses in the lower reach of the Riet Spruit

from the empirical observation that $SO_4$ typically represents ~63% of the TDS concentration in mine water (Text Box 1). This ratio also applies at station F11S12. Salient statistics that describe this information are presented in Table 28, and the relationships used to compile the record of $SO_4$ and TDS loads associated with composite mine water entering the Tweelopie Spruit illustrated in Fig. 47.

**Table 29** Assessment of TDS load losses to the karst aquifer between stations F11S12 and MRd

| Date | Station F11S12 | | | | Station MRd | | | | Δ TDS Load (t/d) |
|---|---|---|---|---|---|---|---|---|---|
| | SEC[1] (mS/m) | TDS[(1)] mg/L | Flow (ML/d) | Load (t/d) | SEC[1] (mS/m) | TDS[1] (mg/L) | Flow (ML/d) | Load (t/d) | |
| 22/09/2009 | 322 | 2479 | 14.9 | 36.9 | – | – | 0.0 | – | –36.9 |
| 05/02/2010 | 389 | 2997 | 35.2 | 105.5 | 358 | 2759 | 7.3 | 20.1 | –85.4 |
| 16/02/2010 | 339 | 2610 | 31.6 | 82.5 | 335 | 2581 | 5.7 | 14.7 | –67.8 |
| 23/02/2010 | 379 | 2918 | 26.2 | 76.5 | 383 | 2948 | 4.0 | 11.8 | –64.7 |
| 09/03/2010 | 379 | 2918 | 32.6 | 95.1 | 353 | 2720 | 9.4 | 25.6 | –69.5 |
| 01/04/2010 | 374 | 2878 | 40.4 | 116.3 | 358 | 2759 | 10.3 | 28.4 | –87.9 |
| 14/04/2010 | 358 | 2757 | 25.8 | 71.1 | 347 | 2672 | 5.7 | 15.2 | –55.9 |
| 06/05/2010 | 408 | 3142 | 43.7 | 137.3 | 420 | 3234 | 11.7 | 37.8 | –99.5 |
| 18/05/2010 | 335 | 2580 | 35.7 | 92.1 | 356 | 2741 | 11.0 | 30.2 | –61.9 |
| 09/06/2010 | 370 | 2849 | 32.1 | 91.5 | 373 | 2872 | 10.5 | 30.2 | –61.3 |
| 07/07/2010 | 374 | 2880 | 29.9 | 86.1 | 376 | 2895 | 6.2 | 17.9 | –68.2 |
| 27/07/2010 | 407 | 3134 | 31.6 | 99.0 | 395 | 3042 | 6.5 | 19.8 | –79.3 |
| 19/08/2010 | 384 | 2957 | 25.8 | 76.3 | 335 | 2580 | 5.3 | 13.7 | –62.6 |
| 05/10/2010 | 307 | 2364 | 13.8 | 32.6 | 383 | 2949 | 0.4 | 1.2 | –31.4 |
| 19/11/2010 | 338 | 2603 | 22.2 | 57.8 | 333 | 2564 | 3.4 | 8.7 | –49.1 |
| 27/07/2011 | 369 | 2841 | 31.9 | 90.6 | 373 | 2872 | 19.4 | 55.7 | –35.0 |
| 25/08/2011 | 389 | 2995 | 28.7 | 86.0 | 405 | 3119 | 20.0 | 62.4 | –23.6 |
| 05/09/2011 | 362 | 2787 | 22.5 | 62.7 | 367 | 2826 | 15.9 | 44.9 | –17.8 |
| 08/05/2012 | 372 | 2864 | 21.4 | 61.3 | 388 | 2988 | 9.6 | 28.7 | –32.9 |
| 14/08/2012 | 299 | 2302 | 22.5 | 51.8 | 309 | 2379 | 6.8 | 16.2 | –35.6 |
| 21/09/2012 | 290 | 2233 | 24.6 | 54.9 | 288 | 2218 | 15.5 | 34.4 | –20.6 |
| 24/10/2012 | 264 | 2033 | 16.2 | 32.9 | 270 | 2079 | 5.7 | 11.9 | –21.1 |
| 15/01/2013 | 282 | 2171 | 18.4 | 39.9 | 283 | 2179 | 6.4 | 13.9 | –26.0 |
| 14/02/2013 | 274 | 2110 | 23.0 | 48.5 | 277 | 2133 | 7.5 | 16.0 | –32.5 |
| 06/03/2013 | 244 | 1879 | 20.7 | 38.9 | 241 | 1856 | 8.0 | 14.8 | –24.0 |
| 15/08/2013 | 219 | 1686 | 30.1 | 50.7 | 219 | 1686 | 16.5 | 27.8 | –22.9 |
| 15/10/2013 | 275 | 2118 | 29.6 | 62.7 | 274 | 2110 | 14.1 | 29.8 | –32.9 |
| 12/12/2013 | 236 | 1817 | 22.2 | 40.3 | 231 | 1779 | 4.7 | 8.4 | –32.0 |
| 26/09/2014 | 238 | 1833 | 41.5 | 76.1 | 238 | 1833 | 30.3 | 55.5 | –20.5 |
| 30/07/2015 | | | | | | | | | |
| 04/12/2017 | | | | | | | | | |
| n | 29 | 29 | 29 | 29 | 28 | 28 | 29 | 28 | 29 |
| Minimum | 219 | 1686 | 13.8 | 32.6 | 219 | 1686 | 0.0 | 1.2 | –17.8 |
| Mean | 330 | 2543 | 27.4 | 70.8 | 331 | 2549 | 9.6 | 24.8 | –46.8 |
| Median | 339 | 2610 | 26.2 | 71.1 | 350 | 2696 | 7.5 | 20.0 | –35.6 |
| Maximum | 408 | 3142 | 43.7 | 137.3 | 420 | 3234 | 30.3 | 62.4 | –99.5 |
| SD | 57 | 438 | 7.7 | 26.6 | 58 | 444 | 6.5 | 15.3 | 23.8 |
| CoV (%) | 17 | 17 | 28 | 38 | 17 | 17 | 67 | 61 | –51 |

[1]From Table 10

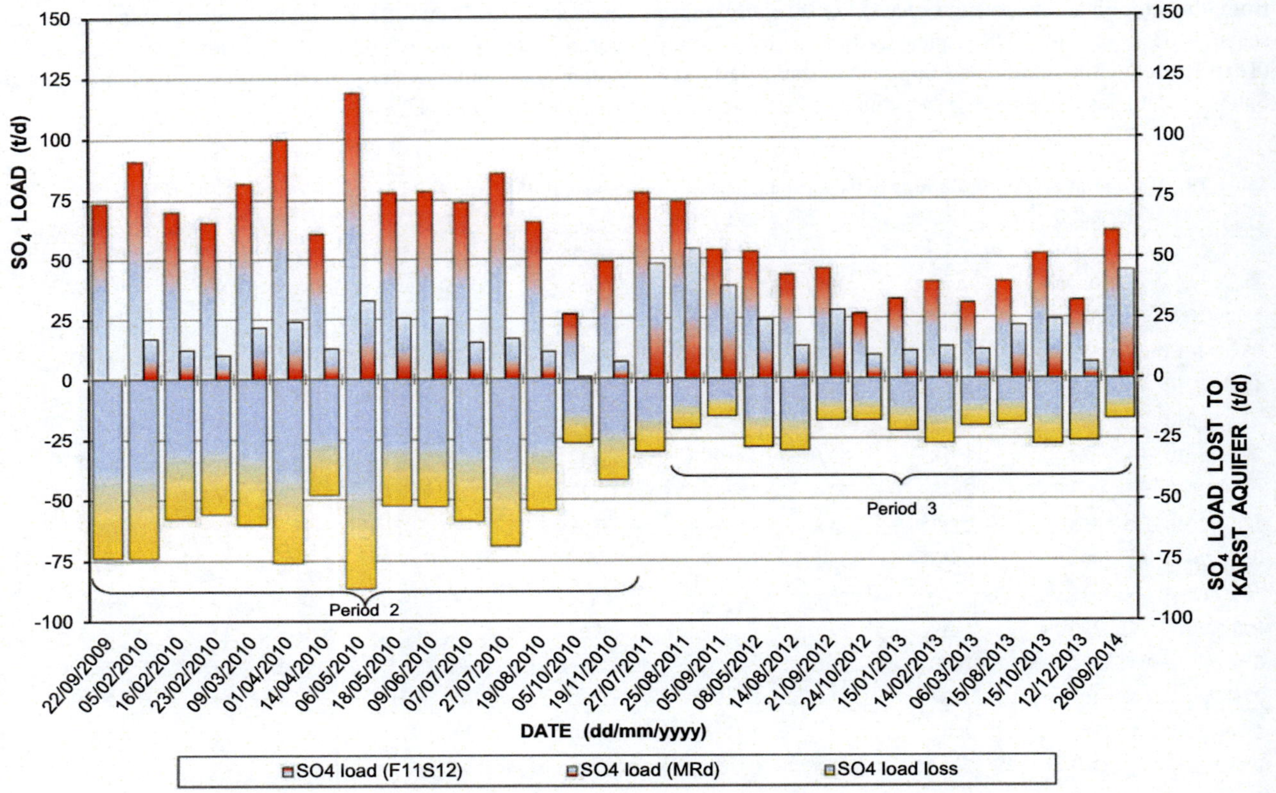


**Fig. 49** Pattern and trend of $SO_4$ load losses in the lower reach of the Riet Spruit

The information presented in Fig. 47 clearly reveals the exceptionally high loads delivered to the Bloubank Spruit system from the Western Basin throughout 2010. It also reveals that elevated loads prior to 2010 were common since early 2008. It is evident from Table 28 that median $SO_4$ and TDS loads of ~ 5 and ~40 t/d, respectively, have entered the Tweelopie Spruit from the mine area since January 2006. The 95%ile values of ~103 t $SO_4$/d and ~163 t TDS/d provide an indication of the upper limits associated with these loads.

The $SO_4$ and TDS loads since mid-2011 are noticeably lower than the long-term median values (Fig. 47). Against the historic record, this reduction is a positive development for the receiving downstream environment, and once again demonstrates the value of the immediate and short-term mine water control and management interventions (Text Box 4in Chapter "Physical Hydrology").

### 2.2.2 Riet Spruit

The record of surface water losses between stations F11S12 and MRd documented in Table 18, together with the contemporary SEC measurements documented in Table 10, provide material for an assessment of the TDS load lost to the karst aquifer between these two stations. This assessment is set out in Table 29 and illustrated in Fig. 48. A similar assessment for $SO_4$, also based on data presented in Table 10, is set out in Table 30 and illustrated in Fig. 49.

The concern previously expressed for the quality of the allogenic recharge entering the karst aquifer (Sect. 1.2.6), is echoed in the median TDS and $SO_4$ load losses of ~36 and ~30 t/d, respectively (Table 29 and Table 30). In recognition of the two distinct periods[10] identified in each of Figs. 48 and 49, however, it is warranted to explore the respective loads (and load losses) associated with each of these periods. The respective load losses are summarised in Table 31, the information reflecting the smaller losses associated with Period 3 compared to Period 2. These circumstances mimic the change in flow regime shown in Fig. 31 in Chapter "Physical Hydrology", and also span the period of positive impact following the start of the immediate mine water intervention measures in mid-2012 (Text Box 4).

[10] The periods are numbered '2' and '3' in accordance with those labeled in Fig. 35.

**Table 30** Assessment of $SO_4$ load losses to the karst aquifer between stations F11S12 and MRd

| Date | Station F11S12 | | | | Station MRd | | | | Δ $SO_4$ Load (t/d) |
|---|---|---|---|---|---|---|---|---|---|
| | SEC[1] (mS/m) | $SO_4$ (mg/L) | Flow (ML/d) | Load (t/d) | SEC[1] (mS/m) | $SO_4$[1] (mg/L) | Flow (ML/d) | Load (t/d) | |
| 22/09/2009 | 322 | 2092 | 14.9 | 73.6 | – | – | 0.0 | – | –73.7 |
| 05/02/2010 | 389 | 2574 | 35.2 | 90.6 | 358 | 2351 | 7.3 | 17.2 | –73.9 |
| 16/02/2010 | 339 | 2214 | 31.6 | 70.0 | 335 | 2186 | 5.7 | 12.5 | –57.9 |
| 23/02/2010 | 379 | 2502 | 26.2 | 65.6 | 383 | 2531 | 4.0 | 10.1 | –55.7 |
| 09/03/2010 | 379 | 2502 | 32.6 | 81.6 | 353 | 2315 | 9.4 | 21.8 | –60.0 |
| 01/04/2010 | 374 | 2466 | 40.4 | 99.6 | 358 | 2351 | 10.3 | 24.2 | –75.7 |
| 14/04/2010 | 358 | 2351 | 25.8 | 60.7 | 347 | 2272 | 5.7 | 13.0 | –47.9 |
| 06/05/2010 | 408 | 2711 | 43.7 | 118.4 | 420 | 2797 | 11.7 | 32.7 | –86.2 |
| 18/05/2010 | 335 | 2186 | 35.7 | 78.0 | 356 | 2337 | 11.0 | 25.7 | –52.4 |
| 09/06/2010 | 370 | 2437 | 32.1 | 78.2 | 373 | 2459 | 10.5 | 25.8 | –52.6 |
| 07/07/2010 | 374 | 2466 | 29.9 | 73.7 | 376 | 2480 | 6.2 | 15.4 | –58.6 |
| 27/07/2010 | 407 | 2703 | 31.6 | 85.4 | 395 | 2617 | 6.5 | 17.0 | –68.8 |
| 19/08/2010 | 384 | 2538 | 25.8 | 65.5 | 335 | 2186 | 5.3 | 11.6 | –54.2 |
| 05/10/2010 | 307 | 1984 | 13.8 | 27.4 | 383 | 2531 | 0.4 | 1.0 | –26.4 |
| 19/11/2010 | 338 | 2207 | 22.2 | 49.0 | 333 | 2171 | 3.4 | 7.4 | –41.8 |
| 27/07/2011 | 369 | 2430 | 31.9 | 77.5 | 373 | 2459 | 19.4 | 47.7 | –30.0 |
| 25/08/2011 | 389 | 2574 | 28.7 | 73.9 | 405 | 2689 | 20.0 | 53.8 | –20.1 |
| 05/09/2011 | 362 | 2380 | 22.5 | 53.5 | 367 | 2416 | 15.9 | 38.4 | –15.2 |
| 08/05/2012 | 372 | 2452 | 21.4 | 52.5 | 388 | 2567 | 9.6 | 24.6 | –28.1 |
| 14/08/2012 | 299 | 1927 | 22.5 | 43.4 | 309 | 1999 | 6.8 | 13.6 | –29.7 |
| 21/09/2012 | 290 | 1862 | 24.6 | 45.8 | 288 | 1848 | 15.5 | 28.6 | –17.1 |
| 24/10/2012 | 264 | 1675 | 16.2 | 27.1 | 270 | 1718 | 5.7 | 9.8 | –17.3 |
| 15/01/2013 | 282 | 1805 | 18.4 | 33.2 | 283 | 1812 | 6.4 | 11.6 | –21.5 |
| 14/02/2013 | 274 | 1747 | 23.0 | 40.2 | 277 | 1769 | 7.5 | 13.3 | –26.8 |
| 06/03/2013 | 244 | 1531 | 20.7 | 31.7 | 241 | 1510 | 8.0 | 12.1 | –19.5 |
| 15/08/2013 | 219 | 1352 | 30.1 | 40.7 | 219 | 1352 | 16.5 | 22.3 | –18.2 |
| 15/10/2013 | 275 | 1754 | 29.6 | 51.9 | 274 | 1747 | 14.1 | 24.6 | –27.2 |
| 12/12/2013 | 236 | 1474 | 22.2 | 32.7 | 231 | 1438 | 4.7 | 6.8 | –25.8 |
| 26/09/2014 | 238 | 1488 | 41.5 | 61.8 | 238 | 1488 | 30.3 | 45.1 | –16.5 |
| 30/07/2015 | | | | | | | | | |
| 04/12/2017 | | | | | | | | | |
| n | 29 | 29 | 29 | 29 | 28 | 28 | 29 | 28 | 29 |
| Minimum | 219 | 1352 | 13.8 | 27.1 | 219 | 1352 | 0.0 | 1.0 | –15.1 |
| Mean | 330 | 2151 | 27.4 | 60.0 | 331 | 2157 | 9.6 | 21.0 | –39.8 |
| Median | 339 | 2214 | 26.2 | 60.7 | 350 | 2294 | 7.5 | 17.1 | –29.8 |
| Maximum | 408 | 2711 | 43.7 | 118.4 | 420 | 2797 | 30.3 | 53.8 | –85.7 |
| SD | 57 | 409 | 7.7 | 23.4 | 58 | 414 | 6.5 | 13.0 | –20.8 |
| CoV (%) | 17 | 19 | 28 | 39 | 17 | 19 | 67 | 62 | –52 |

[1]From Table 10.

### 2.2.3 Blougat Spruit

The pattern of TDS load associated with treated effluent released into the Blougat Spruit from the Percy Stewart WWTW (Fig. 50) resembles the discharge pattern (Fig. 17 in Chapter "Physical Hydrology", Sect. 3.2 in Chapter "Physical Hydrology"). The median value of 9.4 t/d is significantly less than the ~ 61 t/d delivered by the Tweelopie Spruit in the same period (Fig. 47). Although the

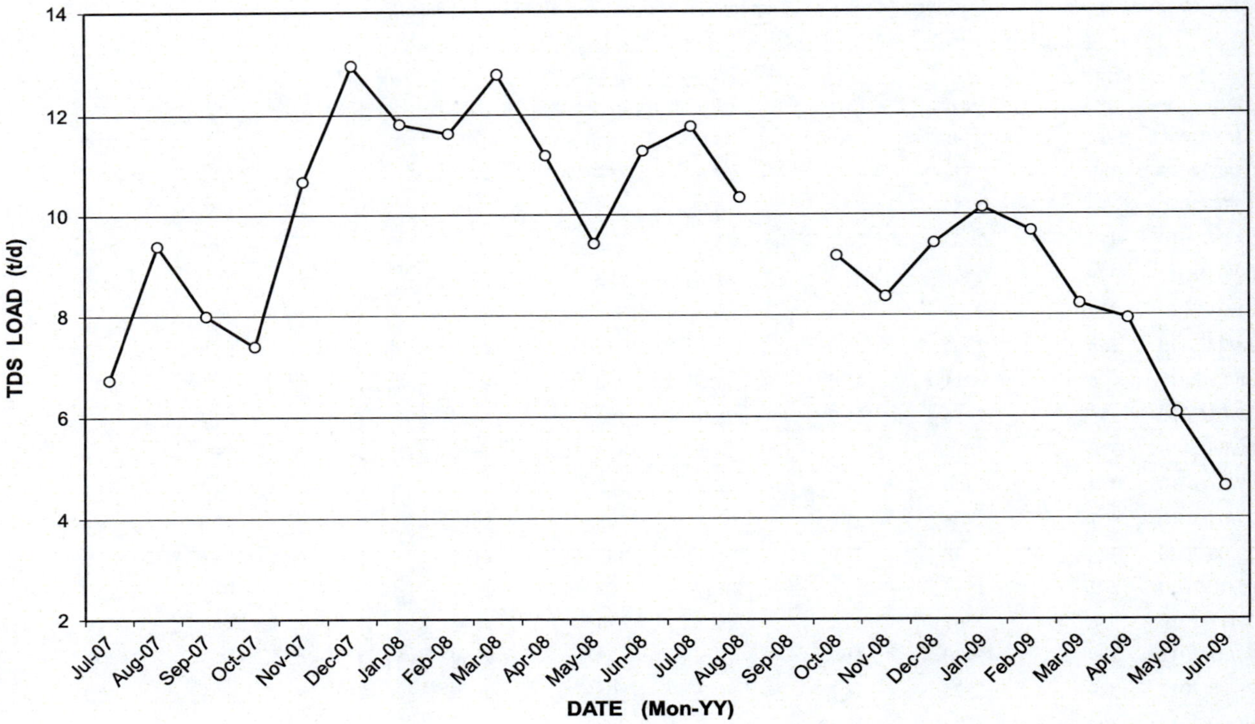


**Fig. 50** Pattern and trend of recent TDS load associated with treated municipal effluent discharge to the Blougat Spruit

**Table 31** Statistical analysis of TDS and $SO_4$ load losses in the lower reach of the Riet Spruit

| Statistical parameter | TDS load loss (t/d) | | $SO_4$ load loss (t/d) | | % Loss | |
|---|---|---|---|---|---|---|
| | Period 2 | Period 3 | Period 2 | Period 3 | Period 2 | Period 3 |
| n | 15 | 14 | 15 | 14 | 15 | 14 |
| Minimum | 31.4 | 17.8 | 26.4 | 15.2 | 67 | 27 |
| Mean | 68.8 | 27.0 | 59.0 | 22.4 | 80 | 51 |
| Median | 67.8 | 25.0 | 57.6 | 20.8 | 80 | 53 |
| Maximum | 99.5 | 35.6 | 86.2 | 30.0 | 100 | 79 |
| SD | 17.2 | 6.3 | 15.0 | 5.3 | 9.2 | 17.3 |
| CoV | 25 | 23 | 25 | 23 | 11 | 34 |

DWS reports the effluent chemistry at station PSFE dating back to November 2002, the exclusion of TDS from this data set (Table 14) limits a longer term load assessment for this variable.

## 2.3 Regional Context and Synthesis

Given the regional importance of Hartbeespoort Dam as a storage reservoir (Sect. 1 in Chapter "Physical Hydrology"), an assessment of the TDS load must include that delivered by all of the contributing main drainages (Table 32). The assessment spans the complete hydrological and hydrochemical record common to all six drainages. It provides a regional context for the TDS load delivered by the Bloubank Spruit system.

The assessment shows that although the Bloubank Spruit contributed the third highest TDS load to Hartbeespoort Dam (after the combined ~79% of the Jukskei and Hennops rivers) in the period 1980 to 2014, this represented only ~ 10% of the total load. It is shown in Sect. 3 in Chapter "Physical Hydrology", however, that the quality of

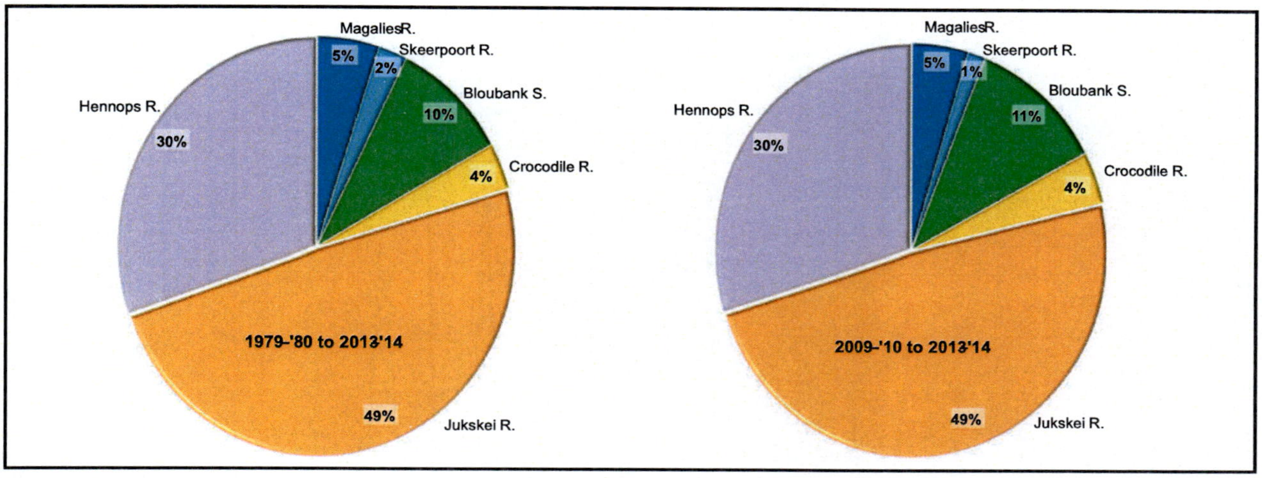


**Fig. 51** Comparison of annual TDS load contributions of the main drainages in the Hartbeespoort Dam catchment in the long-term (left) (ΣTDS = 104 943 t/a) and the recent past (right) (ΣTDS = 240 089 t/a) (data from Tables 32 and 33)

**Table 32** Median annual TDS load delivered by main drainages in the Hartbeespoort Dam catchment in the period of common hydrological and hydrochemical record (1980 to 2014)

| Drainage | Station | Median TDS (mg/L) | Median Discharge[(1)] ($Mm^3/a$) | Median TDS Load | | |
|---|---|---|---|---|---|---|
| | | | | t/d | t/a | % of Total |
| Magalies River | A2H013 | 336 | 15.689 | 14.4 | 5272 | 5.0 |
| Hennnops River | A2H014 | 470 | 67.570 | 87.0 | 31 758 | 30.3 |
| Skeerpoort River | A2H034 | 218 | 11.083 | 6.6 | 2416 | 2.3 |
| Jukskei River | A2H044 | 420 | 122.845 | 141.4 | 51 595 | 49.2 |
| Bloubank Spruit | A2H049 | 446 | 22.701 | 27.7 | 10 125 | 9.6 |
| Crocodile River | A2H050 | 293 | 12.892 | 10.3 | 3777 | 3.6 |
| TOTAL | | | | 287.4 | 104 943 | 100 |

**Table 33** Median annual TDS load delivered by main drainages in the Hartbeespoort Dam catchment in the hydrological and hydrochemical record period 2010 to 2014

| Drainage | Station | Median TDS (mg/L) | Median Discharge ($Mm^3/a$) | Median TDS Load | | |
|---|---|---|---|---|---|---|
| | | | | *t/d* | *t/a* | % of Total |
| Magalies River | A2H013 | 323 | 34.026 | 30.1 | 10 990 | 4.6 |
| Hennops River | A2H014 | 458 | 156.672 | 196.6 | 71 756 | 29.9 |
| Skeerpoort River | A2H034 | 274 | 12.626 | 9.5 | 3460 | 1.4 |
| Jukskei River | A2H044 | 378 | 309.817 | 320.9 | 117 111 | 48.8 |
| Bloubank Spruit | A2H049 | 539 | 50.054 | 73.9 | 26 979 | 11.2 |
| Crocodile River | A2H050 | 291 | 33.655 | 26.8 | 9794 | 4.1 |
| TOTAL | | | | 657.8 | 240 089 | 100 |

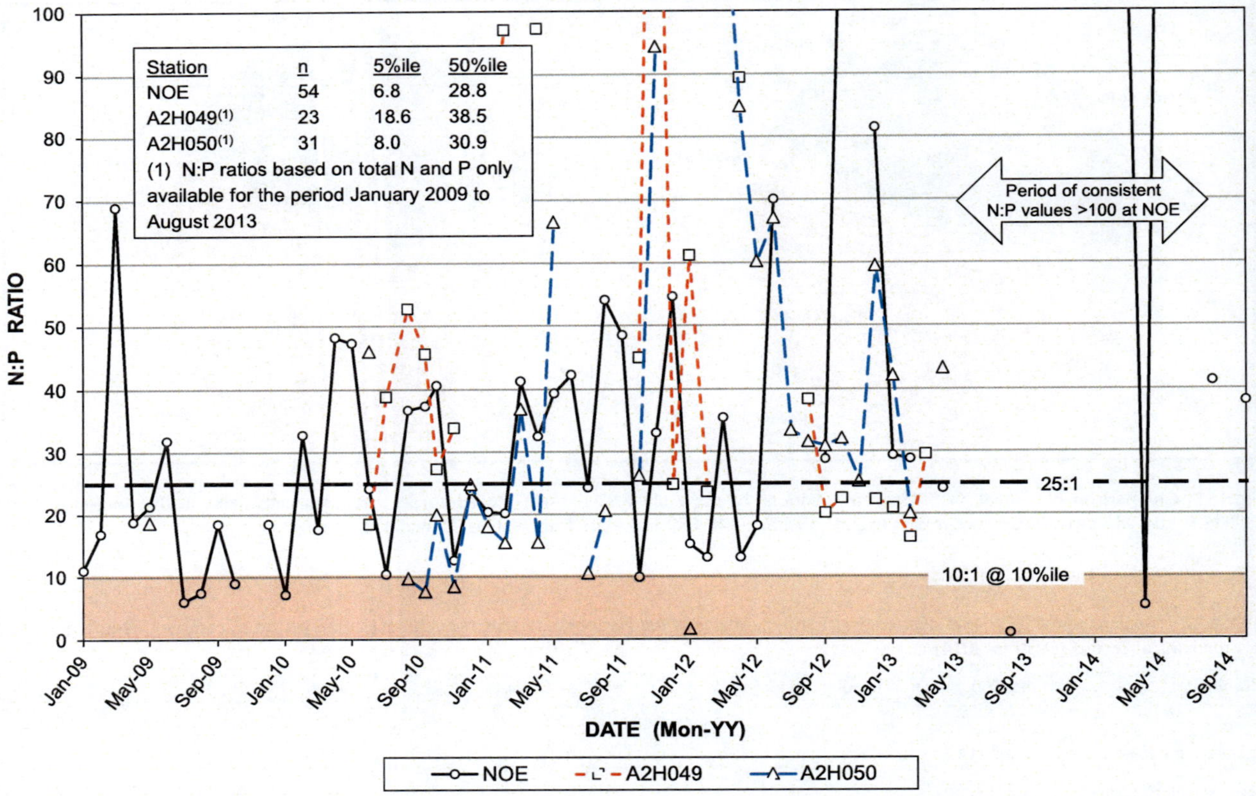


**Fig. 52** Pattern and trend of the N:P ratio for Bloubank Spruit water at the NOE and station A2H049 and for Crocodile River water at station A2H050 in the period January 2009 to September 2014 (note cut-off at 100:1 to magnify graphs at lower ratio values)

**Table 34** Comparison of the long-term (1980 to 2014) median TDS concentrations with those for the recent (2010 to 2014) hydrological years for the regional drainages

| Drainage | Station | Long-term Median TDS (mg/L) | Recent Median TDS (mg/L) | Change in TDS | |
|---|---|---|---|---|---|
| | | | | mg/L | % Change |
| Magalies River | A2H013 | 336 | 323 | –13 | –3.9 |
| Hennnops River | A2H014 | 470 | 458 | –12 | –2.6 |
| Skeerpoort River | A2H034 | 218 | 274 | 56 | 25.7 |
| Jukskei River | A2H044 | 420 | 378 | –42 | –10.0 |
| Bloubank Spruit | A2H049 | 446 | 539 | 93 | 20.9 |
| Crocodile River | A2H050 | 293 | 291 | –2 | –0.7 |

surface water discharged by the Bloubank Spruit deteriorated since the 2008–'09 hydrological year, and especially so in the 2010, 2011 and 2014 hydrological years (Fig. 39). To examine the relative impact of this change in both a regional and historical context, a similar assessment as above but spanning only the 2010 to 2014 period is presented in Table 33.

This assessment shows an overall increase in total TDS load of ~129% in the more recent period. The contribution of the Bloubank Spruit system to this increase is negligible, increasing from ~ 10% to only ~11% at the cost of the Skeerpoort River. As in the long-term, the Jukskei and Hennops rivers again share the bulk (~79%) of the recent total TDS load also in similar proportions, viz. ~3:2, as in the long-term. These circumstances are illustrated and compared in Fig. 51.

Closer examination of the data presented in Tables 32 and 33 indicates that the changes in TDS concentration between the two periods of analysis are comparatively small (Table 34). In four instances (the Magalies, Hennops,

**Table 35** Comparison of the long-term (1980 to 2014) median discharges with those for the recent (2010 to 2014) hydrological years for the regional drainages

| Drainage | Station | Long-term median discharge ($Mm^3$) | Recent median discharge ($Mm^3$) | Change in discharge | |
|---|---|---|---|---|---|
| | | | | $Mm^3$ | % Change |
| Magalies River | A2H013 | 15.689 | 34.026 | 18.337 | 117 |
| Hennnops River | A2H014 | 67.570 | 156.672 | 89.102 | 132 |
| Skeerpoort River | A2H034 | 11.083 | 12.626 | 1.543 | 14 |
| Jukskei River | A2H044 | 122.845 | 309.817 | 186.972 | 152 |
| Bloubank Spruit | A2H049 | 22.701 | 50.054 | 27.353 | 120 |
| Crocodile River | A2H050 | 12.892 | 33.655 | 20.763 | 161 |

Jukskei and Crocodile rivers), the TDS levels reduced. This is attributed to dilution by greater volumes of 'fresh' surface water runoff. Surprisingly, the pristine Skeerpoort River reflects a greater increase (~26%) than the Bloubank Spruit (~21%). This is partly attributed to an increase in $CaCO_3$ delivered by the perennial karst springs that feed the Skeerpoort River[11] in the median amount of ~9.5 $Mm^3/a$ (Sect. 2 in Chapter "Physical Hydrology").

Significantly greater changes, without exception increases, are reflected in the comparison of discharges (Table 35). The Crocodile River leads in this regard with 161%. The Bloubank Spruit reflects the third lowest increase (120%).

Further analysis of the TDS load delivered to Hartbeespoort Dam in the 2010 to 2014 period compared to the long-term median load, is summarised in Table 36. This indicates that the Bloubank Spruit delivered the second lowest percentage increase (69%) after the Skeerpoort River (44%). The increase in the other drainages exceeded 100%, headed by the Crocodile River with 160%.

The Jukskei and Hennops rivers share the 2nd greatest percentage increase of ~127%. The present mean discharge of 380 ML/d (~139 $Mm^3/a$) of the 450 ML/d (~164 $Mm^3/a$) capacity Northern WWTW represents ~45% of the median annual discharge (~310 $Mm^3$) of the Jukskei River (Table 33). A contribution of this magnitude provides a quite plausible explanation for the relative 'insensitivity' of the Jukskei River water quality to the higher natural discharges, compared to that experienced by the other drainages (with the exception of the Skeerpoort River) in the Hartbeespoort Dam catchment in the 2010 to 2014 period. As might be expected, the spring flow dominated Skeerpoort River also exhibits an 'insensitivity' to the contemporary higher discharges, the increase in TDS load amounting to 44% (Table 36).

[11] The median total alkalinity concentration at stations A2H033 and A2H034 increased from 190 to 202 mg $CaCO_3$/L, and from 158 to 167 mg $CaCO_3$/L respectively, in the two periods of analysis.

## 3 Pollution Indicators and Pollutants

### 3.1 Sulfate-to-Chloride Ratio

A further perspective on surface water quality is provided by the $SO_4$:Cl ratio. This ratio (based on milliequivalent values) is an indicator of potential mining and/or industrial impact on water quality. Its median value for the near-pristine Skeerpoort River water is ~ 1.1:1, with a 95%ile value of 3.5:1 (Table 37). Against this local 'natural' benchmark, the median ratio value of ~45–48:1 associated with raw mine water and slightly 'better' value of ~ 28:1 associated with surface water discharge in the Tweelopie Spruit at the downstream station F11S12, provides further unequivocal proof of a mine water impact. Encouragingly, the median values further downstream and elsewhere in the Bloubank Spruit system are only marginally greater than that for the Skeerpoort River. This reflects the comparatively unimpacted inorganic chemistry of the surface water in these drainages.

The Crocodile River reflects a median $SO_4$:Cl value of 0.6:1, indicating the dominance of Cl over $SO_4$ in this surface water. This is most reasonably attributable to the influence of treated effluent discharge from the Driefontein WWTW (Sect. 1.3).

### 3.2 Nitrate-to-Phosphorus Ratio

Any assessment of the impact of inorganic nitrogen concentrations should be coupled to an evaluation of the inorganic nitrogen to inorganic phosphorus ratio (DWAF 1996a). This also applies to inorganic phosphorus concentrations. The N:P ratio for unimpacted (oligotrophic) systems typically is >25:1, reducing to <10:1 for impacted (eutrophic or hypertrophic) systems. The median N:P ratio values for selected drainages in the study area for various timeframes are presented in Table 38. The information

**Table 36** Comparison of the long-term (1980 to 2014) median TDS load delivered to Hartbeespoort Dam with the median TDS load delivered in the 2010 to 2014 hydrological years

| Drainage | Station | Long-term median TDS Load (t/d) | Recent median TDS Load (t/d) | Increase in TDS load | |
|---|---|---|---|---|---|
| | | | | t/d | % Increase |
| Magalies River | A2H013 | 14.4 | 30.1 | 15.7 | 109 |
| Hennnops River | A2H014 | 87.0 | 196.6 | 109.6 | 126 |
| Skeerpoort River | A2H034 | 6.6 | 9.5 | 2.9 | 44 |
| Jukskei River | A2H044 | 141.4 | 320.9 | 179.5 | 127 |
| Bloubank Spruit | A2H049 | 27.7 | 74.6 | 46.9 | 69 |
| Crocodile River | A2H050 | 10.3 | 26.8 | 16.5 | 160 |
| COMPOSITE ANALYSIS | | 287.4 | 658.5 | 371.1 | 129 |

**Table 37** Statistical characterisation of the $SO_4$:Cl ratio for drainages in the study area for the full available record up to September 2014 and the common record period since June 2004

| Drainage | Station | Statistical Parameter | | | | | | |
|---|---|---|---|---|---|---|---|---|
| | | n | 5%ile | Mean | Median | 95%ile | SD | CoV (%) |
| Skeerpoort River | A2H034 | 1447(1) | 0.3 | 1.5 | 1.1 | 3.5 | 1.9 | 129 |
| | | 247 | 0.4 | 1.8 | 1.2 | 3.8 | 1.9 | 105 |
| Tweelopie Spruit | BRI(2) | 88 | 19.2 | 47.1 | 47.4 | 78.3 | 17.5 | 37 |
| | #17 Winze(2) | 88 | 17.5 | 45.0 | 44.5 | 71.2 | 17.0 | 38 |
| | #18 Winze(2) | 88 | 20.5 | 50.8 | 48.2 | 87.3 | 19.6 | 39 |
| | F11S12(3) | 138 | 14.2 | 29.0 | 28.3 | 48.0 | 13.5 | 47 |
| Blougat Spruit | 188,048(4) | 42 | 1.1 | 2.1 | 1.9 | 4.4 | 1.1 | 53 |
| Tweefontein Spruit | F14S15(4) | 42 | 1.4 | 1.7 | 1.7 | 2.3 | 0.3 | 19 |
| Bloubank Spruit | A2H049 | 889(5) | 1.5 | 2.1 | 2.0 | 3.0 | 0.8 | 40 |
| | | 209 | 1.4 | 2.5 | 1.8 | 6.6 | 1.6 | 64 |
| Crocodile River | A2H050 | 833(5) | 0.3 | 0.7 | 0.6 | 1.5 | 0.4 | 51 |
| | | 238 | 0.4 | 0.6 | 0.6 | 0.8 | 0.1 | 22 |

[1]Since January 1976
[2]Raw mine water since December 2012
[3]Since November 2003
[4]Record period terminates in September 2008 as per Table 11 (188048) and Table 18 (F14S15)
[5]Since May 1979

presented indicates the measure of impact associated with the Percy Stewart WWTW effluent discharge into the Blougat Spruit compared to the Bloubank Spruit and Crocodile River.

Given that Zohary et al. (1988) already reported an N:P ratio of $\leq 7{:}1$ for Hartbeespoort Dam more than two decades ago, the longer records of stations A2H049 (Bloubank Spruit) and A2H050 (Crocodile River) prompt an inspection of the variability of this parameter over time. This shows a smaller variability when only the last ten years of record is considered, whilst still reflecting a largely unimpacted character. Accepting that much of the nutrient loads in the Bloubank Spruit and Crocodile River derive from the Percy Stewart and Driefontein WWTWs, respectively, then closer scrutiny of the data in Table 38 reveals the following circumstances:

- the smaller mean and median $NO_3$–N:$PO_4$–P ratio values for the last ten years compared to the long-term values suggest that the performance of the Percy Stewart WWTW has deteriorated; and
- the greater mean and median $NO_3$–N:$PO_4$–P ratio values for the last ten years compared to the long-term values suggest that the performance of the Driefontein WWTW has improved.

**Table 38** Statistical values of the N:P ratio for drainages in the study area for various periods

| Drainage (station) | Period | Statistical Parameter | | | | | | |
|---|---|---|---|---|---|---|---|---|
| | | n | 5%ile | Mean | Median | 95%ile | SD | CoV (%) |
| Blougat Spruit (Percy Stewart WWTW) | 07/2007–06/2009 | 24 | **3.8** | **9.0** | **7.5** | 18.4 | 5.2 | 58 |
| Blougat Spruit (188048) | 06/2004–03/2008 | 41 | **0.03** | **6.3** | **3.3** | 26.0 | 9.6 | 152 |
| Bloubank Spruit (A2H049) | 06/2004–03/2008 | 99 | 11.1 | 32.6 | 28.7 | 68.2 | 19.1 | 59 |
| | | *46* | *11.1* | *24.5* | *22.1* | *45.7* | *11.12* | *45* |
| | 05/1979–09/2014 | 920 | 16.0 | 219.7 | 83.5 | 820.7 | 477.6 | 227 |
| | | *76* | *11.6* | *39.1* | *25.0* | *97.1* | *49.9* | *128* |
| | 10/2003–09/2014 | 199 | 11.6 | 140.4 | 32.7 | 876.7 | 325.4 | 232 |
| | | *75* | *11.6* | *37.6* | *24.8* | *91.6* | *48.6* | *129* |
| Crocodile River (A2H050) | 06/2004–03/2008 | 99 | **3.7** | 27.3 | 25.9 | 57.8 | 18.2 | 67 |
| | | *47* | ***4.1*** | *17.3* | *17.4* | *31.9* | *9.7* | *56* |
| | 05/1979–09/2014 | 795 | **0.3** | 42.3 | 18.3 | 162.2 | 92.7 | 219 |
| | | *89* | ***4.1*** | *25.0* | *20.1* | *63.9* | *20.5* | *82* |
| | 10/2003–09/2014 | 219 | **5.3** | 83.3 | 38.1 | 335.1 | 148.9 | 179 |
| | | *89* | ***4.1*** | *25.0* | *20.1* | *63.9* | *20.5* | *82* |

*Notes* Bold text indicates ratio values < 10:1 indicative of eutrophic or hypertrophic systems
Italicised text (highlighted rows) indicates N:P ratio derived from total N and P values, compared to $NO_3$–N and $PO_4$–P derived values reported in preceding row

The above circumstances provide further substance to the results reported in the 'Green Drop' report (DWA, 2011) in regard to the Percy Stewart WWTW (Sect. 1.2.4). By comparison, the Driefontein WWTW recorded a 100% score for the wastewater quality compliance performance area (DWA 2011).

Further indictment of the poor performance of the Percy Stewart WWTW is provided by the following observations. Station A2H049 is located much further (~22 km) downstream from the Percy Stewart WWTW than station A2H050 is (~2 km) from the Driefontein WWTW. In addition, the quality at station A2H049 benefits from the significant volumes of good quality karst groundwater contributed to the river (Sect. 4.16 in Chapter 7), which benefit is absent in the case of A2H050. Finally, the discharge of the Percy Stewart WWTW is significantly less than that of the Driefontein facility. These circumstances suggest that the water quality at station A2H049 should be considerably less sensitive to poor quality municipal wastewater discharges than that at station A2H050.

A concern in regard to $PO_4$–P is the concentrations observed at several surface water sampling stations in the Bloubank Spruit system (see for example Table 20). The values of 0.5 mg P/L and 6 mg N/L observed at station BB@PL (Table 20) return a N:P ratio of 12:1. This compares unfavourably with the median value of 41:1 observed for the shorter record at station A2H049 (Table 38). It also explains the degree of eutrophication observed in the defunct trout dams at the Nirox Estate at Kromdraai.

The routine water quality monitoring carried out at the Nedbank Olwazini Estate (NOE) provides an additional source of $NO_3$–N and $PO_4$–P data. This data set has already been interrogated in regard to its nutrient and bacterial content (Table 22), which returns a median N:P ratio value of ~29:1 and a 5%ile value of ~ 7:1 in the period January 2009 to September 2014. The pattern and trend described by the N:P ratio at the NOE and stations A2H049 and A2H050 in this period is compared in Fig. 52. This reveals a generally acceptable N:P ratio, where the limit of 10:1 below which eutrophic or hypertrophic systems are identified is defined by the 10%ile value of the NOE data set, i.e. is exceeded in 47 of the 54 values (Table 22 and text box Fig. 52).

## 3.3 Trace/Heavy Metals and Metalloids

The presence of trace/heavy metal(loid)s provides a further indication of especially mine water pollution. These analytes therefore represent a concern for their impact on the quality of surface water in the Bloubank Spruit system. The metals iron (Fe) and manganese (Mn) arguably receive most attention,[12] the former for its very high concentrations in raw mine water and the latter for its persistence in solution at higher pH levels. In addition to these analytes, this section

[12] Refer to Tables 2, 3, 4, 7, 9, 11, 12, 18, 19, 21 and 23.

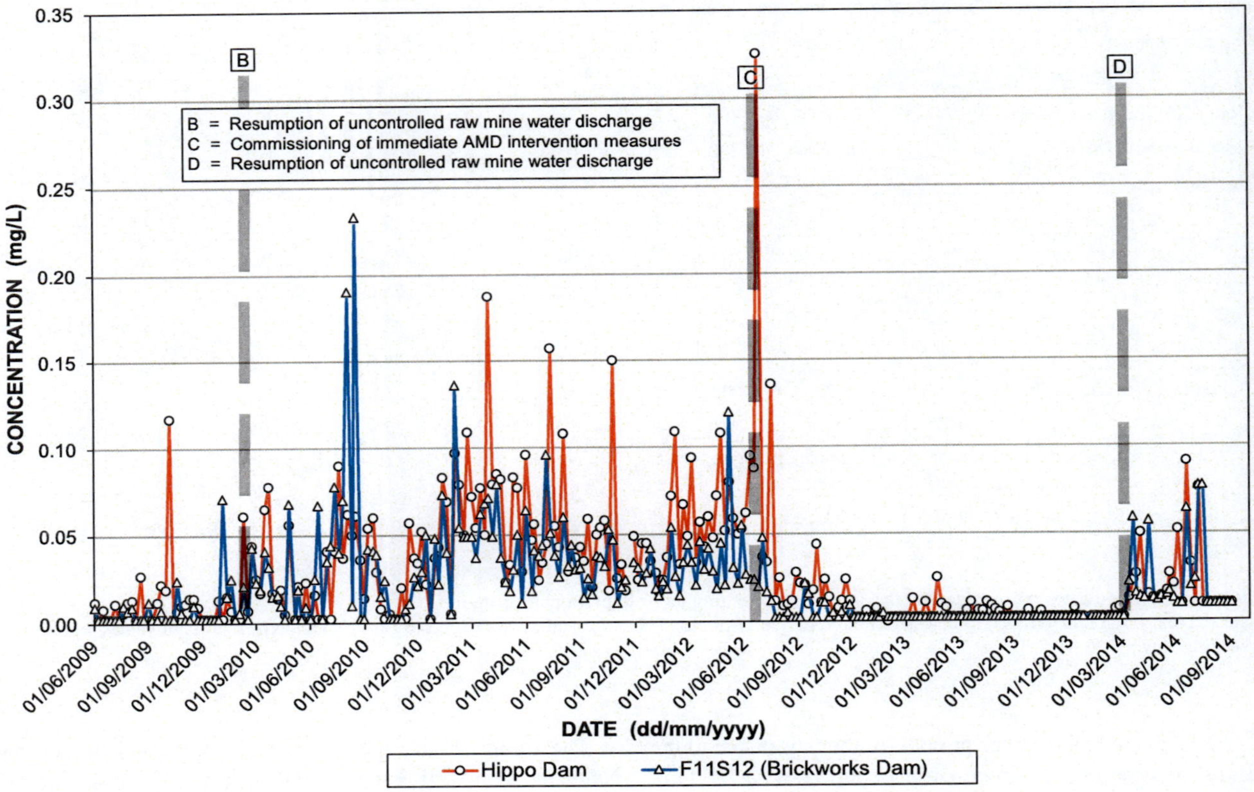


**Fig. 53** Pattern and trend of $U_T$ in Tweelopie Spruit surface water

reports on a much broader suite of analytes that are not often afforded attention under circumstances such as occur in the study area.

### 3.3.1 Iron (Fe) and Manganese (Mn)

A surface water sample collected in May 2010 near Sterkfontein Cave returned Fe and Mn values of 0.91 and 0.67 mg/L, respectively (station BB@M, Table 20). A sample collected at Plover's Lake further downstream on the same day (station BB@PL, Table 20) returned Fe and Mn values of 0.36 and 0.29 mg/L, respectively. Although both sets of values significantly exceed the long-term median values of 0.014 and 0.003 mg/L determined for these analytes at station A2H049 (Table 19), the significant differences between the 'upstream' and 'downstream' samples caution against a direct comparison with the long-term values at A2H049. A more appropriate comparison is considered to be with the values reported for mine water discharged from the mine area (Table 7) for which median values of 1.2 mg Fe/L and 10.5 mg Mn/L characterise these analytes in the period B–C (February 2010 to July 2012).

It has been shown that the period February 2010 to July 2012 experienced the most serious impact from mine water discharges (Sects. 3.1 in Chapter "Physical Hydrology" and 1.2 in Chapter 6). It is therefore appropriate to interrogate the contemporary A2H049 water chemistry record for at least Fe and Mn for comparison with these analyte values in the composite [13]mine water discharge in this timeframe. This comparison is presented in Table 39. The reduction in concentrations at A2H049 even in the period of most likely greatest impact is self-evident, questioning reports (Allison et al., 2011) that the AMD impact on the quality of surface water entering Hartbeespoort Dam has compromised the quality of water impounded in this ~190 $Mm^3$ reservoir.

### 3.3.2 Mercury (Hg)

Mercury (Hg), a heavy metal known for its propensity for bioaccumulation in the food chain (particularly predatory fish species) and impact on human health (Holmes et al. 2009), reveals equivocal levels in mine water in the study area, and generally very low levels (<0.002 mg/L) in surface water and groundwater sources. SANS (2015a) sets a

[13] It is shown in Sect. 1.2.2 that the raw mine water component greatly exceeded the treated/neutralised mine water component for much of this period (Fig. 16 in "Physical Hydrology").

**Table 39** Statistical comparison of Fe and Mn concentrations in composite mine water discharge with Bloubank Spruit surface water at A2H049 in the period February 2010 to July 2012

| Statistical parameter | Analyte and source | | | |
|---|---|---|---|---|
| | Fe | | Mn | |
| | Hippo Dam | A2H049 | Hippo Dam | A2H049 |
| n | 129 | 33 | 129 | 33 |
| 5%ile (mg/L) | **6.5** | 0.001 | **22.2** | 0.001 |
| Mean (mg/L) | **168.4** | 0.020 | **62.7** | 0.292 |
| Median (mg/L) | **163.0** | 0.013 | **65.0** | 0.003 |
| 95%ile (mg/L) | **365.2** | 0.086 | **95.0** | **1.692** |
| SD (mg/L) | 116.2 | 0.025 | 23.5 | 1.136 |
| CoV (%) | 69 | 1 | 38 | 4 |
| SANS (2015a)[1] | $\leq 2.0$ | | $\leq 0.5$ | |

[1]Standard health-related limit for consumption of 2 L/d over 70 years by a 60 kg person
Bold text denotes value exceeds standard limit as described in note 1

standard health-related limit of $\leq 0.006$ mg/L for this element, and the USAs Safe Drinking Water Act (SDWA) a maximum contaminant level (MCL) of 0.002 mg/L.

Although a single mine water sample reported by Hobbs and Cobbing (2007) returned a value below a detection limit of 0.0001 mg/L (Table 40), a larger population ($n$ = 6) reported by Hobbs et al. (2010) returned values in the range 0.398 mg/L (for treated/neutralised mine water) to 3.61 mg/L (for raw mine water sourced from mine shafts). Actively decanting shafts such as the Black Reef Incline and #17 and #18 winzes exhibited the highest levels (2.07 to 3.61 mg/L), compared to #8 Shaft mine water (0.634 mg/L). By comparison, Hg levels in surface water typically do not exceed the strictest analytical detection limit of 0.001 mg/L (Table 40), and are therefore not considered a threat in this resource.

The higher levels in raw mine water will only become a concern if circumstances similar to those experienced in the period February 2010 to July 2012 are manifested (Sect. 3.1 in Chapter "Physical Hydrology", Figs. 13 and 16). More analyses, and especially of the composite well-mixed raw and treated/neutralised mine water discharge, would be required to better inform such concern.

### 3.3.3 Aluminium (Al)

Aluminium (Al) is set a standard health-related limit of $\leq 0.3$ mg/L (SANS 2015a). Its threat to human health especially via the consumption of water remains largely unknown, as is borne out by its SANS (2015a) association with an operational risk rather than a chronic health risk. The level in treated/neutralised mine water since November 2012[14] (to September 2013) exhibits a median value of 0.009 mg/L and a 95%ile value of 0.12 mg/L (Table 7). These values are similar to the 0.007 and 0.105 mg/L ($n$ = 40) for the mixed raw and treated/neutralised mine water leaving the Hippo Dam. The level in raw mine water since November 2012 reflects a median value of 0.1 mg/L and a 95%ile value of 0.88 mg/L ($n$ = 41). It has also been recorded on two occasions, however, that the Al concentration in raw mine water returned conflicting values of 3.9 mg/L (in May 2012) and <0.01 mg/L (in March 2013).

By comparison, the level in surface water at the lower end of the Tweelopie Spruit (Brickworks Dam/F11S12) since November 2012 reflects median and 95%ile values of 0.007 and 0.03 mg/L, respectively. The activity of aluminium in water is strongly dependent on pH (Appelo and Postma, 2009). It is least soluble at a pH in the range 6 to 7, and increasingly soluble at pH <4.5 and >7. The pH of the surface water at the Brickworks Dam/F11S12 since November 2012 ranged from 6.2 (5%ile value) to 7.4 (95%ile value) with a median of 7.2, i.e. favouring aluminium insolubility circumstances. Unfortunately a similar evaluation for the pre-December 2012 period, and in particular the period February 2010 to July 2012 when pH values ranged from 2.7 (5%ile) to 3.9 (95%ile) at the Brickworks Dam/F11S12 (Table 4), is thwarted by the paucity of data due to non-exceedance of the coarse detection limit of 1 mg/L.

Nevertheless, the higher levels in raw mine water might only become a concern if circumstances similar to those experienced in the period February 2010 to July 2012 are manifested (Sect. 3.1 in Chapter "Physical Hydrology", Figs. 13 and 16). More analyses, and especially of the composite well-mixed raw and treated/neutralised mine water discharge, would be required to better inform such concern.

### 3.3.4 Arsenic (As)

The metalloid arsenic (As), another element of concern for human health, is set a standard health-related limit of

[14] The detection limit for Al was lowered from 1 mg/L to 0.001 mg/L in December 2012 (Table 5).

**Table 40** Summary of Hg values for mine and surface water sources in the study area

| Detection limit (DL) | Mine Water[1] | | Surface water | | Σ*n* | Data source |
|---|---|---|---|---|---|---|
| | n | >DL | *n* | >DL | | |
| 0.0001 mg/L | 1 | 0 | 1 | 0 | 2 | Hobbs and Cobbing (2007) |
| Value(s) > DL | – | | – | | | |
| 0.002 mg/L | 6 | 6 | 2 | 0 | 8 | Hobbs et al. (2010) |
| Value(s) > DL | 0.398 to 3.61 mg/L | | – | | | |
| 0.001 mg/L | 3 | 0 | 17 | 0 | 20 | This study |
| Value(s) > DL | – | | – | | | |

[1]Comprises both raw and treated/neutralised mine water

≤ 0.01 mg/L (SANS, 2015a). Hobbs and Cobbing (2007) report values of 0.159 and 0.214 mg/L for the actively decanting Black Reef Incline and #18 Winze, respectively. In all but one of the seven surface water sources sampled (including a treated/neutralised mine water sample), the detection limit of 0.002 mg/L was not exceeded. This pattern has subsequently been replicated on four further sampling campaigns targeting surface water in the Tweelopie, Riet, Blougat and Bloubank spruits.

The higher levels in raw mine water will only become a concern if circumstances similar to those experienced in the period February 2010 to July 2012 are manifested (Sect. 3.1 in Chapter "Physical Hydrology", Figs. 13 and 16). More analyses, and especially of the composite well-mixed raw and treated/neutralised mine water discharge, would be required to better inform such concern.

### 3.3.5 Other Metals

The activity of other metals/metalloids in Tweelopie Spruit surface water is reported in Table 41. Of the listed analytes, only nickel (Ni) poses a threat on the basis of its concentration even at the 5%ile level. The threat from exposure to Ni or Ni compounds to human health is primarily via dermal (skin) and respiratory (inhalation) exposure (ATSDR 2011; RAIS 1995).

The results for cadmium (Cd) and lead (Pb) are not afforded much significance because of the very small number of analyses that exceed the detection limit(s), militating against meaningful interpretation. What is considered significant is that >90% of the analyses do not exceed the detection limit(s).

The routine water quality monitoring carried out at NOE also provides an additional source of data for the trace/heavy

**Table 41** Summary of trace/heavy metal(loid) concentrations in Tweelopie Spruit water at the Hippo Dam and Brickworks Dam stations in the period December 2012 to September 2014

| Statistical parameter | Analyte and source | | | | | | | | | | | |
|---|---|---|---|---|---|---|---|---|---|---|---|---|
| | B | | Cd | | Cu | | Ni | | Pb | | Zn | |
| | HD[1] | BD[2] | HD[1] | BD[2] | HD[1] | BD[2] | HD[1] | BD[2] | HD[1] | BD[2] | HD[1] | BD[2] |
| n | 88 | 88 | 88 | 88 | 88 | 88 | 88 | 4188 | 88 | 88 | 88 | 88 |
| *n* > DL[3] | 88 | 88 | 6 | 5 | 76 | 75 | 88 | 87 | 24 | 26 | 86 | 88 |
| 5%ile (mg/L) | 0.20 | 0.20 | 0.001 | 0.001 | 0.001 | 0.001 | 0.04 | **0.10** | 0.001 | 0.001 | 0.01 | 0.02 |
| Mean (mg/L) | 0.56 | 0.49 | **0.031** | 0.002 | 0.006 | 0.005 | **0.17** | **0.16** | **0.011** | 0.006 | 0.07 | 0.09 |
| Median (mg/L) | 0.40 | 0.30 | 0.001 | 0.001 | 0.003 | 0.002 | **0.10** | **0.10** | 0.008 | 0.005 | 0.03 | 0.10 |
| 95%ile (mg/L) | 1.38 | 1.31 | **0.130** | **0.006** | 0.019 | 0.021 | **0.40** | **0.34** | **0.046** | 0.010 | 0.17 | 0.18 |
| SD (mg/L) | 0.59 | 0.52 | 0.068 | 0.003 | 0.007 | 0.008 | 0.13 | 0.11 | 0.014 | 0.003 | 0.07 | 0.05 |
| CoV (%) | 107 | 107 | 222 | 143 | 125 | 138 | 75 | 73 | 134 | 60 | 104 | 58 |
| SANS (2015a)[4] | n.s | | ≤ 0.003 | | ≤ 2.0 | | ≤ 0.07 | | ≤ 0.01 | | ≤ 5.0 | |

[1]Hippo Dam
[2]Brickworks Dam (station F11S12)
[3]See Table 5 for analyte detection limits
[4]Standard health-related limit for consumption of 2 L/d over 70 years by a 60 kg person
Bold text denotes value exceeds standard limit as described in note 4

**Table 42** Summary of trace/heavy metal concentrations in Bloubank Spruit water at the NOE in the period May 2011 to December 2014

| Statistical parameter | Analyte | | | | | | | | |
|---|---|---|---|---|---|---|---|---|---|
| | Cd | Cr | Co | Cu | Fe | Pb | Mn | Ni | Zn |
| N | 11 | 11 | 12 | 12 | 12 | 11 | 12 | 12 | 12 |
| 5%ile (mg/L) | 0.000 | 0.000 | 0.000 | 0.000 | 0.000 | 0.000 | 0.000 | 0.000 | 0.000 |
| Mean (mg/L) | 0.000 | 0.000 | 0.003 | 0.006 | 0.023 | 0.000 | 0.345 | 0.009 | 0.023 |
| Median (mg/L) | 0.000 | 0.000 | 0.000 | 0.000 | 0.010 | 0.000 | 0.008 | 0.000 | 0.020 |
| 95%ile (mg/L) | 0.000 | 0.000 | 0.019 | 0.033 | 0.089 | 0.000 | **1.832** | 0.044 | 0.054 |
| SD (mg/L) | 0.000 | 0.000 | 0.009 | 0.017 | 0.042 | 0.000 | 1.139 | 0.018 | 0.021 |
| CoV (%) | – | – | 266 | 297 | 188 | – | 330 | 200 | 93 |
| SANS (2015a)(1) | ≤0.003 | ≤0.05 | ≤0.5 | ≤2.0 | ≤2.0 | ≤0.01 | ≤0.5 | ≤0.07 | ≤5.0 |

[1]Standard health-related limit for consumption of 2 L/d over 70 years by a 60 kg person
Bold text denotes value exceeds standard limit as described in note 1

metals listed in Table 42. The analyses reflect a nominal sampling frequency of three months.

The information presented in Table 42 largely corroborates the analysis presented in Table 39 for Fe and Mn (the latter again at the 95%ile level). The results similarly reflect non-detect activities at 50%ile levels for all but Fe, Mn and Zn, and generally do not raise concern for the listed analytes at this locality.

Reference to Table 12 suggests that a greater concern should be given to the metal concentrations in municipal wastewater discharged from the Percy Stewart WWTW. This is at least relevant for Cd, Mn, Pb and Ni.

## 3.4 Radionuclides

### 3.4.1 Uranium ($^{238}U$ and $^{234}U$)

Both SANS (2015a) and the WHO (2011) set a drinking water limit of 15 µg/L for uranium (U). The USAs SDWA proposes a MCL value of 30 µg/L. The comparatively recent inclusion of U (both total U ($U_T$) and dissolved U ($U_D$) as per Table 5) as an analyte in the weekly surface water monitoring programme carried out by SG in the Tweelopie Spruit, provides a more detailed record of the pattern and trend of this element in the receiving aquatic environment. Table 7 indicates that median $U_T$[15] and $U_D$ levels of 16.5 and 9.8 µg/L, respectively, were observed in treated/neutralised mine water discharged to the Tweelopie Spruit in the period B–C (February 2010 to July 2012) of maximum mine water impact. Raw mine water in the same period carried a median $U_T$ concentration of 61 µg/L (n = 131), quadruple both the treated/neutralised mine water median $U_T$ value and the SANS (2015a) limit, and a 95%ile value of twice that (122 µg/L).

The pattern and trend of $U_T$ at the upper and lower ends of the Tweelopie Spruit is illustrated in Fig. 53. Importantly, the water represents a mixure of raw and treated/neutralised mine water. Salient statistics presented in Table 4 (Sect. 1.2.2) more clearly reveal the differences between the two stations in each distinctive period of mine water discharge.

The statistics show that in the period of maximum mine water impact (B–C), the $U_T$ level at the distal Brickworks Dam (F11S12) station exhibited median and 95%ile values of 30 and 76 µg/L, respectively. The 95%ile values of 26 and 11 µg/L in the preceding and subsequent periods at this station (Table 4) do not represent levels of undue concern if it is considered that Roychoudury and Starke (2006) report a median $^{238}U$[16] value of 26 µg/L in surface water of the East Rand Basin.

The results in Table 4 indicate a 70% reduction in $U_T$ concentration between the two stations located ~5.4 km apart under 'normal' discharge conditions. Uranium in its most soluble oxidation state of U(VI), takes the form of the uranyl ion ($UO_2^{2+}$) which, in oxidising water/systems, is both stable and strongly adsorbed onto Fe oxides (Wanty and Nordstrom, 1993 and references therein). Adsorption is stronger at circum-neutral pH, and weaker at higher and lower values. This behaviour readily explains the observed reduction in $U_T$.

Although the Hobbs et al. (2010) study targeted $^{222}Rn$ as a hydrological tracer in natural and polluted environments, other readionuclides such as $^{238}U$, $^{234}U$ and $^{226}Ra$ were also investigated. The Western Basin (amongst other areas) was identified as a suitable candidate for testing this hypothesis,

[15] SANS (2015a) requires that the method used to analyse metal analytes be capable of determining the acid soluble (pH < 2) metal present, i.e. the total concentration and not the dissolved fraction only.

[16] Elemental U comprises ~99.3% $^{238}U$, rendering $^{238}U$ concentrations practically equivalent to $U_T$.

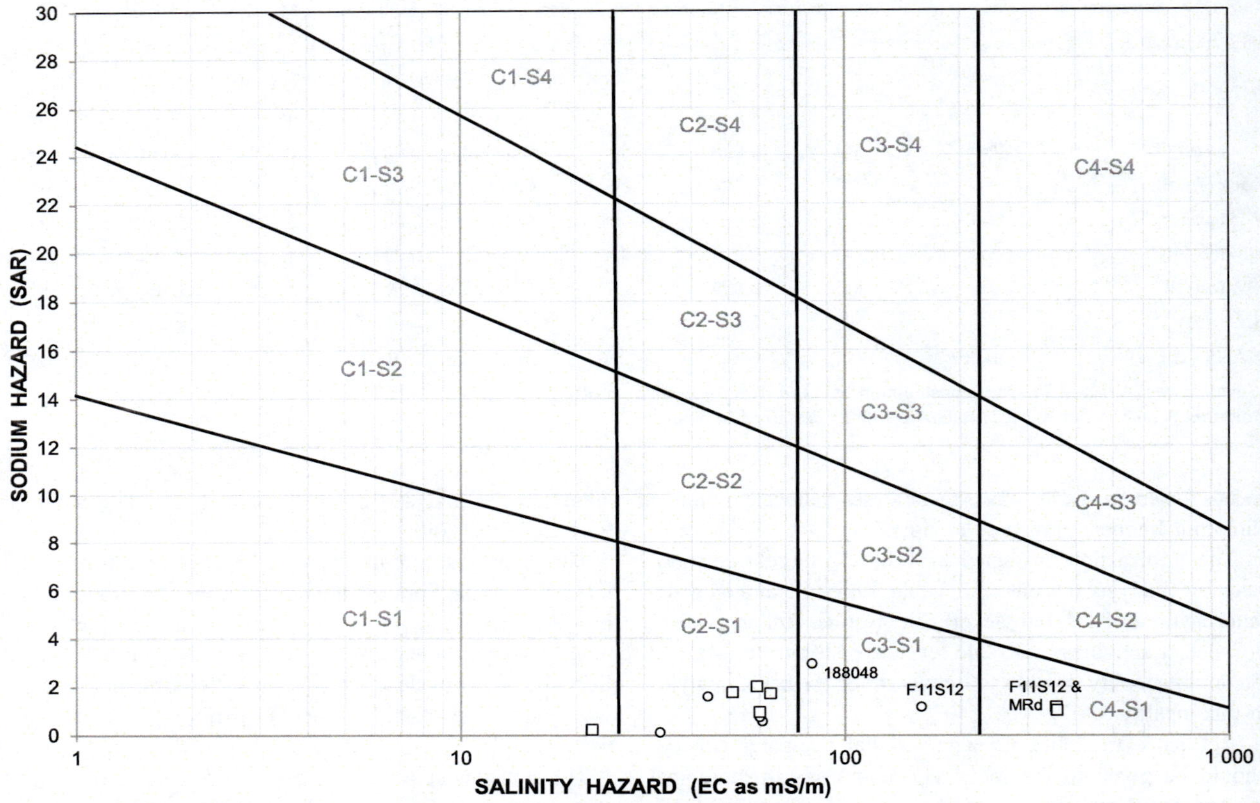


**Fig. 54** Wilcox diagram illustrating the classification of surface water chemistry for irrigation purposes (data from Table 46); circle represents long-term median data and square single values

and various sources of water were targeted for radionuclide sampling. A treated/neutralised mine water sample returned a similar value of 0.02 ± 0.005 Bq/L for both $^{238}U$ and $^{234}U$. The equivalent $U_T$ level of 1 µg/L compares favourably with the results reported for the periods A–B and C– in Table 7.

Of potentially greater concern is the level of $^{238}U$ activity in raw mine water in the study area. Five mine water samples analysed by Hobbs et al. (2010) returned levels in the range 0.82 to 2.47 Bq/l (147 to 200 µg/L). These are an order of magnitude greater than the $U_T$ limit of 15 µg/L set by SANS (2011a) for drinking water, and 5 to 6 times greater than the median $U_T$ level observed in surface water at the lower end of the Tweelopie Spruit.

### 3.4.2 Radon ($^{222}Rn$)

The USAs SDWA proposes a MCL of 300 pCi/L (11.1 Bq/L ≡ $1.953 \cdot 10^{-12}$ mg/L) for radon ($^{222}Rn$) in drinking water. Results reported by Hobbs (2010) indicate $^{222}Rn$ activity levels in the range 0.06 ± 0.16 Bq/L (MDA[17] = 0.54 Bq/L) to 0.15 ± 0.12 Bq/L (MDA = 0.39 Bq/L) for surface water in the Tweelopie Spruit. It is evident that the very low $^{222}Rn$ concentrations do not exceed the minimum detectable activity (MDA) levels associated with the analytical accuracy. These circumstances reflect both the comparatively short half-life (3.82 d) of this radionuclide, and the ready degassing in the turbulent flow regime that characterises the Tweelopie Spruit drainage through the KGR.

In light of the above, the radionuclide $^{222}Rn$ is not considered to present a threat to the surface water resources in the study area. Concerns for its airborne threat from mine shafts and residues (FSE, 2013) through inhalation in a confined environment would be justified for activity levels > 100 Bq/m³ (2.7 pCi/L), the WHO (2009) recommended reference level.

### 3.4.3 Radium ($^{226}Ra$ and $^{228}Ra$)

The USAs SDWA proposes a MCL of 5 pCi/L (0.185 Bq/L) for the aggregate of $^{226}Ra$ and $^{228}Ra$ in drinking water. Values for $^{226}Ra$ reported by Hobbs et al. (2010) provide a

[17] Minimum detectable activity associated with NECSA liquid scintillation counting (LSC) analysis method.

**Table 43** Net acidity and net alkalinity of various mine and surface waters in the study area

| Water source | Water type[1] | Acidity calculation analytes | | | | | | | |
|---|---|---|---|---|---|---|---|---|---|
| | | Total alkalinity (mg $CaCO_3$/L) | pH[2] | Fe (mg/L) | Mn (mg/L) | Al (mg/L) | Total acidity | Net alkalinity | Net acidity |
| | | | | | | | (mg $CaCO_3$/L) | | |
| BRI | 1 | <5 | 4.0 | 1153 | 69 | n.a | 2197 | −2192 | 2192 |
| #18 Winze | 1 | <5 | 4.2 | 1306 | 47 | n.a | 2429 | −2424 | 2424 |
| #8 Shaft | 1 | <5 | 3.6 | 725 | 148 | 25 | 1743 | −1738 | 1738 |
| CSIR26 | 1 | <5 | 2.6 | 190 | 477 | n.a | 1335 | −1330 | 1330 |
| End-of-pipe | 2 | 36 | 8.5 | 82 | 51 | n.a | 153 | −117 | 117 |
| F11S12[3] | 3 | <5 | 4.2 | 8.94 | 47 | 0.86 | 109.56 | −105 | 105 |
| F11S12[4] | 3 | <5 | 5.5 | 0.431 | 10 | 0.1 | 19.70 | −15 | 15 |
| F11S12[6] | 3 | <5 | 3.0 | 33 | 29 | n.a | 163 | −158 | 158 |
| MRd[3] | 3 | <5 | 4.2 | 2.49 | 64 | 1.7 | 133.64 | −129 | 129 |
| MRd[4] | 3 | <5 | 4.4 | 0.934 | 18 | 0.121 | 37.12 | −32 | 32 |
| MRd[6] | 3 | <5 | 3.0 | 18 | 28 | n.a | 134 | −129 | 129 |
| BB@M[4] | 4 | 128 | 6.9 | 0.906 | 0.665 | 0.134 | 3.6 | 124 | −124 |
| BB@M[5] | 3 | 5 | 2.9 | 44 | 16 | 0.78 | 175 | −170 | 170 |
| BB@M[6] | 4 | 66 | 7.1 | 8.1 | 11 | n.a | 36 | 30 | −30 |

[1] 1 = raw mine water; 2 = treated mine water; 3 = impacted surface water; 4 = slightly impacted surface water
[2] $-\log_{10}\alpha_{H+}$
[3] Sample date 16/02/2010
[4] Sample date 18/05/2010
[5] Sample date 12/01/2011
[6] Sample date 16/05/2014

measure of its presence in various sources of mine water and groundwater. The two values available for surface water indicated a $^{226}Ra$ level of 0.008 ± 0.002 Bq/L for treated/neutralised mine water, and 0.002 ± 0.001 Bq/L for water from the toe of a slimes dam. These levels compare favourably with those for five mine water sources which ranged from 0.12 to 1.38 Bq/L with a mean of 0.65 Bq/L. The results support the following observations by de Jesus (1985):

- that mine waters obtained at source in various underground mines sporadically showed $^{226}Ra$ values > 0.37 Bq/L (10 pCi/L), and
- that tailings dams do not represent a significant threat as a source of $^{226}Ra$ in the receiving surface and groundwater environments.

The extreme sparseness of $^{226}Ra$ values for surface water in the study area cautions against conclusive observations regarding the threat of this radionuclide in rivers and streams downstream of the mine area. Nevertheless, the very low level of $^{226}Ra$ in treated/neutralised mine water suggests that this source does not pose a threat in this regard. The higher levels in raw mine water will only become a concern if circumstances similar to those experienced in the period February 2010 to July 2012 are manifested (Sect. 3.1 in Chapter "Physical Hydrology", Figs. 13 and 16). More analyses, and especially of the composite well-mixed raw and treated/neutralised mine water discharge, would be required to better inform such concern.

## 3.5 Net Alkalinity and Net Acidity

It has been shown by Wirries and McDonnell (1983) that pH is a relatively poor measure of the acidity of AMD. This strengthens the need to understand the chemical characteristic of mine water defined by its net alkalinity, or more importantly the converse, net acidity. The latter variable presents as an indicator of AMD in a water resource, and an evaluation of this characteristic further informs an assessment of mine water impact in the receiving hydrologic environment.

The concepts of net alkalinity and net acidity are discussed at length by Kirby and Cravotta (2005a; 2005b). The term net alkalinity describes a water that is suitable for discharge into a stream after treatment (typically neutralisation by the addition of alkaline material) to remove metals to concentrations low enough to avoid negative impact on aquatic life or human health. As alkalinity is equal to zero at

**Table 44** Free acidity and metal acidity associated with the results presented in Table 43

| Water source | Water type[1] | Free acidity (mg $CaCO_3$/L) | Metal acidity (mg $CaCO_3$/L) | Total (calculated) Acidity (mg $CaCO_3$/L) | Free: total Acidity ratio (%) |
|---|---|---|---|---|---|
| BRI | 1 | 5 | 2192 | 2197 | 0.23 |
| #18 Winze | 1 | 3 | 2426 | 2429 | 0.12 |
| #8 Shaft | 1 | 13 | 1730 | 1743 | 0.75 |
| CSIR26 | 1 | 126 | 1209 | 1335 | 9.4 |
| End-of-pipe | 2 | 0 | 153 | 153 | 0 |
| F11S12 | 3 | 3 | 106 | 109 | 2.75 |
| | | 0.2 | 19.5 | 19.7 | 1.02 |
| | | 50 | 113 | 163 | 30.7 |
| MRd | 3 | 3.2 | 130.5 | 133.7 | 2.39 |
| | | 2 | 35 | 37 | 5.41 |
| | | 50 | 84 | 134 | 37.2 |
| BB@M | 64 | 0.01 | 3.58 | 3.59 | 0.2 |
| BB@M | 3 | 63 | 112 | 175 | 36.0 |
| BB@M | 4 | 0 | 36 | 36 | 0 |

[1]1 = raw mine water; 2 = treated mine water; 3 = impacted surface water; 4 = slightly impacted surface water

a pH of 6.4, a net alkaline water ideally has a pH $\geq$ 6.4. This also applies to complex metal-rich solutions that have been allowed to react to reach equilibrium. Water that is net acidic would have a pH < 6.4 after hydrolysis and precipitation of metals.

The (total) acidity is calculated from the pH and dissolved metals concentrations ($C_{Metal}$ in mg/L) as follows (from Kirby and Cravotta, 2005a):

$$\text{Acidity}_{\text{calc}}(\text{mg CaCO}_3/\text{L}) = 50\left(10^{(3-\text{pH})} + 2\text{CFe}/55.8 + 2\text{CMn}/54.9 + 3\text{CA}/27.0\right) \quad (1)$$

where the term $50\left(10^{(3-\text{pH})}\right)$ represents the free acidity contributed by $HSO_4^-$ and hydrogen ion ($H^+$) activity (pH), and the term $50\left(2\text{CFe}^2/55.8 + 2C_{\text{Mnn}}/54.9 + 3C_{A/}/27.0\right)$ represents the metal acidity (acidity contributed by metals). The difference between the Total Alkalinity and the Acidity$_{\text{calc}}$ values gives the Net Alkalinity value (all as mg $CaCO_3$/L). Net Acidity is simply the negative of the Net Alkalinity, i.e.

Net Alkalinity (mg $CaCO_3$/L) = Alkalinity (mgCaCO/L) – Aciditysals = – Net Acidity or

Net Acidity (mg $CaCO_3$/L) = Acidity$_{\text{calc}}$ – Alkalinity ($\text{mgCaCO}_3/\text{L}$) = – Net Alkalinity

Applying these concepts to raw mine water, treated mine water and various surface waters in the study area gives an indication of the hydrochemical suitability (or otherwise) of these waters in the environment. The outcome is presented in Table 43. A comprehensive discussion of net alkalinity and net acidity in regard to groundwater is provided in Sect. 13.2 in Chapter "Chemical Hydrogeology".

The results reveal the extremely high net acidity ($\sim$1700 to 2400 mg/L) associated with the raw mine water. The single treated mine water sample similarly reflects a positive net acidity (negative net alkalinity), albeit typically $\leq$5% of that of raw mine water. The marginally positive net acidity in the range 15 to 158 mg $CaCO_3$/L (Table 43) associated with the surface water collected at stations F11S12 and MRd confirms the very strong mine water presence in this water on the sampling dates. This contrasts with the −124 mg/L net acidity value of the surface water at station BB@M on 18/05/2010 attributed to the municipal wastewater effluent contribution. The latter contribution is negated by mine water at station BB@M on 12/01/2011 (net acidity = 170 $\text{mgCaCO}_3/L$), a characteristic that violates the positive net alkalinity requirement for surface water suitability. It also helps explain the role of mine water in precipitating the Koelenhof Farm fish mortality event (Hobbs and Mills 2011).

The free ($H^+$ activity) and metal (mineral) acidity components associated with the various waters listed in Table 43 yields the information presented in Table 44. This shows that the free acidity contribution to total acidity ranges from 0 to ~37%. Without exception, the greater contributions (30–37%) are associated with surface water at stations located >6 km downstream of the mine area. These circumstances support the influence of hydrolysis (Sects. 3.5 in Chapter 6 and 5.1 in Chapter 6) on the chemistry of mine water impacted surface water in its passage downstream from the mine area.

The net acidity of Tweelopie Spruit water at station F11S12 (Brickworks Dam) is of particular interest for the reason that this station is located shortly before the losing reach of the Riet Spruit that straddles the south-western boundary of the COH. The magnitude of influent losses to the karst aquifer has previously been quantified in Sect. 5.2.1 in Chapter "Physical Hydrology". Coupling this loss with the net acidity of the influent surface water provides an indication of the acid load that enters the karst aquifer as a component of the allogenic recharge. A coupling of the surface water losses in the period mid-December 2012 to late July 2014 with the two periods of distinctively different net acidity concentrations, namely:

- mid-December 2012 to early March 2014 (coinciding with the period C–D in Sect. 1.2.2); and
- early March 2014 to late July 2014 (coinciding with the period D– in Sect. 1.2.2),

yields the acid loads presented in Table 45. These are calculated as the product of the 5%ile, median and 95%ile flow losses and the 5%ile, median and 95%ile net acidity concentrations (as mg $CaCO_3$/L).

It is shown that the aquifer received a net alkalinity load equivalent to 36.5 t $CaCO_3$/a at the 5%ile level in the mid-December 2012 to early March 2014 period only. An acid load equivalent to $\geq$15.7 t $CaCO_3$/a otherwise characterises this addition to the karst aquifer in both periods, attaining a value of 1042 t $CaCO_3$/a at the 95%ile level in the period spanning both a 'moderate' and a 'severe' mine water impact. The impact of the acid load on the receiving karst groundwater resources is discussed in Sect. 13.2 in Chapter "Chemical Hydrogeology".

## 4 Surface Water Fitness

### 4.1 Potable Use

The fitness of water for human consumption is gauged against the SANS (2015a) standard for drinking water quality. The extent to which the water quality analytes associated with the various surface water sources in the study area meet the limits of the SANS (2015a) standard in the long-term is already reflected in various tables (e.g. Tables 2, 9, 11, 18, 19, 22 and 23). It is evident that drainages such as the Skeerpoort River (Table 22) and the Tweefontein Spruit (Table 18) experience an excellent quality water in all respects. It is equally evident, however, that the drainages receiving mine water (Tweelopie and Riet spruits) and treated municipal wastewater (Blougat Spruit) experience exceedances in regard to SEC, Ca, $SO_4$, Fe, Mn and Al.

The information in Tables 13 and 14 indicates that the Blougat Spruit additionally experiences severe exceedances in regard to the bacteriological quality as represented by the faecal coliform and *E. coli* counts. These circumstances seriously compromise the use of the surface water in these drainages for human consumption. However, Table 20 and Fig. 26 indicate that the situation improves significantly in the main stem, the Bloubank Spruit. Nevertheless, the persistent presence of bacteriological contamination in surface water of the Bloubank Spruit system cautions strongly against its unqualified use for human consumption without treatment.

**Table 45** Summary of the acid load addition to the karst aquifer of the Zwartkrans Subcompartment in the period December 2012 to July 2014

| Period | Variable/Analyte | Statistical Parameter Value | | |
|---|---|---|---|---|
| | | 5%ile | Median | 95%ile |
| Mid-December 2012 to early March 2014 | n | 63 | 63 | 63 |
| | pH | 6.6 | 7.1 | 7.5 |
| | Discharge (ML/d) | 8.0 | 12.3 | 16.3 |
| | Acid load (t $CaCO_3$/a) | –36.5 | 15.7 | 224.9 |
| Early March 2014 to late July 2014 | n | 20 | 20 | 20 |
| | pH | 2.8 | 3.4 | 4.5 |
| | Discharge (ML/d) | 8.0 | 12.3 | 16.3 |
| | Acid load (t $CaCO_3$/a) | 61.3 | 357.8 | 1042.4 |

## 4.2 Agricultural Use

### 4.2.1 Livestock Watering

Guidelines in this regard are provided in the DWAF (1996b) publication. The target water quality range (TWQR) limits for the major inorganic analytes are typically less strict than for human consumption. For example, the TWQR for $SO_4$ (an analyte that, together with TDS and Cl, effects the palatability of water also for livestock) is 1000 mg/L compared to the 400 mg/L for humans. This is also true for trace/heavy metals, e.g. Mn and Fe with a TWQR of $\leq 10$ mg/L for livestock compared to $\leq 0.1$ mg Mn/L and $\leq 0.2$ mg Fe/L for humans. Even the observed presence of bacteriological contamination in surface water of the Bloubank Spruit system is rendered more acceptable for livestock by the faecal coliforms TWQR of $\leq 1000$ c/100 mL in 20% of samples, compared to the $\leq 10$ c/100 mL in 1% of samples for humans.

The chemical analysis results for surface water in the study area cautions against the unqualified use thereof for livestock watering. Water of the Tweelopie/Riet Spruit subsystem is characterised by an elevated salt load reflecting a dominant Ca-$SO_4$ composition that accounts for 84% of the TDS load in this water (Sect. 8.4 in Chapter "Chemical Hydrogeology"). This water is not, however, compromised in regard to its trace/heavy metals or bacteriological composition for livestock watering. Water of the Blougat Spruit, on the other hand, exhibits a suitable inorganic and trace/heavy metals composition, but is compromised by unacceptable levels of bacteriological contamination. Although these circumstances improve markedly in the Bloubank Spruit downstream of Sterkfontein Cave (Sect. 1.2.6 and Fig. 25), caution is still advised against its unqualified use for livestock watering.

### 4.2.2 Irrigation

The fitness of water for irrigation use is assessed on the basis of the sodium adsorption ratio (SAR) and salinity (expressed as electrical conductivity) of the water. The SAR is calculated as the ratio between the Na concentration and the combined Ca and Mg concentrations using the formula

$$\mathrm{SAR} = \mathrm{Na} \div [(\mathrm{Ca} + \mathrm{Mg}) \div 2]^{0.5}\ \text{(concentrations as meq/L)} \tag{2}$$

to derive the sodium hazard (designated S) associated with the water. This value is graphed against the electrical conductivity value, which represents the salinity hazard (designated C), to derive an alpha-numeric classification (e.g. C#-S#) defined by the Wilcox diagram (Fig. 54), and where the numeric component ranks as follows: 1 = low, 2 = medium, 3 = high and 4 = very high.

The range of SAR values that represent the various surface water sources in the study area is presented in Table 46. The results (also shown graphically in Fig. 54) indicate that the sodium hazard in no instance exceeds a low (S1) classification. However, the salinity hazard associated with the drainage impacted by mine water (stations F11S12 and MRd) reflects a very high (C4) ranking. This limits the use of this water to the irrigation of crops with a very good salt tolerance on well-drained permeable soils.

Other aspects of the water chemistry that have relevance to the irrigation use of the water are the N:P ratio and the corrosion tendency ratio (CTR) values. The N:P ratio is

**Plate 1** Corrosion of large-diameter pipe draining the Aviary Dam on the Tweelopie Spruit in the KGR; photo at left shows remnant (at bottom centre) tenuously attached to flange, and photo at right shows close-up view from opposite side

**Table 46** Range of representative SAR values for various surface water sources in the study area

| Station | Range of representative SAR values | | | | | Classification per median[1] |
|---|---|---|---|---|---|---|
| | Source | 5%ile | Mean | Median | 95%ile | |
| A2H034 | Table 22 | 0.07 | 0.10 | 0.10 | 0.15 | C2–S1 |
| A2H049 | Table 19 | 0.30 | 0.59 | 0.58 | 0.83 | C2–S1 |
| A2H050 | Table 23 | 0.77 | 1.58 | 1.53 | 2.42 | C2–S1 |
| F11S12 | Table 9 | 0.46 | 1.04 | 0.70 | 1.52 | C3–S1 |
| | Table 20 | 1.14[(2)] | | | | C4–S1 |
| MRd | Table 20 | 0.99[(2)] | | | | C4–S1 |
| 188,048 | Table 11 | 2.12 | 2.79 | 2.81 | 3.32 | C3–S1 |
| BC1 | Table 20 | 2.00[(2)] | | | | C2–S1 |
| BG@N14 | Table 20 | 1.76[(2)] | | | | C2–S1 |
| F14S15 | Table 18 | 0.33 | 0.66 | 0.72 | 0.83 | C1–S1 |
| BB@M | Table 20 | 1.71[(2)] | | | | C2–S1 |
| BB@PL | Table 20 | 0.93[(2)] | | | | C2–S1 |
| HS1 | Table 20 | 0.21[(2)] | | | | C1–S1 |

[1]Where applicable (refer source tables for sample population sizes)
[2]Single value as per source table

discussed in greater detail in Sect. 6.3.2, where concern is expressed in regard to N:P ratio values as low as 12 recently observed at several stations in the Bloubank Spruit system.

The CTR is calculated as the ratio between the sum of the Cl and $SO_4$ concentrations and the Total Alkalinity ($CaCO_3$) concentration using the formula

$$CTR = [Cl + SO_4] \div TAlk \quad (concentrations as meg/L) \tag{3}$$

It is used to estimate the corrosive tendency of Cl and $SO_4$ ions. A ratio of $\leq 0.1$ indicates general freedom from corrosion in neutral to slightly alkaline (pH 7–8) oxygenated waters (DWAF 1996c), and increasingly higher ratios a tendency towards greater corrosivity.

The CTR values associated with the surface water samples collected at seven stations in the Bloubank Spruit system in May 2010 are presented in Table 47, together with other analytes for which the DWAF (1996c) provides a target water quality range (TWQR) for irrigation use. The information presented in regard to stations F11S12 and MRd, and to a lesser extent stations BB@M and BB@PL, reflects the severest impact of mine water on the Bloubank Spruit system as a result of the discharge conditions discussed in Sect. 5.3.1.

The corrosivity of Tweelopie Spruit surface water is evident in Plate 1, although the severity of corrosive attack in this instance is most probably exacerbated by the high degree of aeration and the high flow velocity of the water through the pipe.

A chemical analysis of surface water collected at station BB@M in mid-January 2011 yielded a CTR value of 338 which, together with a pH of ~3, indicated the water to be extremely corrosive. This is commensurate with the manifestation of a greater mine water component in the middle to lower reaches of the Bloubank Spruit system especially during the 2011 rainfall season (Figs. 13 and 15). However, corrosion of pipelines and fittings is only likely to manifest as a problem through continued medium- to long-term use of water with this chemical characteristic. The likelihood of this occurring depends on the success (or otherwise) of mine water treatment interventions in the Western Basin proposed by the Inter-Ministerial Committee (IMC) on AMD (Coetzee et al. 2010).

## 4.3 Recreational Use

Guidelines in this regard are provided in the DWAF (1996d) publication. The need to assess the fitness of surface water for recreational use is indicated by businesses that provide adventure activities in the area. Of concern is that the target group for such activities is mainly schoolchildren, and that some of the activities also involve full-body water contact activities. This is especially problematic where the surface water environment is strongly impacted by acid mine water and/or bacteriological contamination.

An inspection of the TWQR for recreational use (DWAF 1996d) reveals limits that are much less rigorously defined than for other uses. In many instances, the criteria are

**Table 47** Evaluation of Bloubank Spruit system surface water quality for irrigation use

| Variable/Analyte | Sampling station | | | | | | | DWAF TWQR[1] |
|---|---|---|---|---|---|---|---|---|
| | F11S12 | MRd | BC1 | BG@N14 | BB@M | HS1 | BB@PL | |
| Date | 18/05/2010 | | | | | | | |
| pH ($-\log_{10}\alpha_{H+}$) (field) | **5.5** | **4.4** | 7.1 | 6.9 | 6.9 | 6.9 | 7.0 | 6.5–8.4 |
| SEC (mS/m) (field) | **355** | **356** | **59** | **51** | **64** | 22 | **60** | ≤40 |
| Na (mg/L) | **115** | **103** | 55 | 48 | 52 | 5 | 34 | ≤70 |
| Cl (mg/L) | 43 | 44 | 52 | 45 | 50 | 12 | 40 | ≤100 |
| $NO_3 + NO_2$ (mg N/L) | **0.6** | **0.6** | 0.2 | 0.5 | **3.2** | **1.9** | **6.0** | ≤5.0[(2)]<br>≤0.5[(3)] |
| F (mg/L) | < 0.2 | < 0.2 | 0.2 | 0.2 | < 0.2 | < 0.2 | < 0.2 | ≤2.0 |
| *E. coli* (count/100 mL) | n.a | n.a | **65 000** | **26 000** | **3000** | 0 | **470** | ≤1 in 1% of samples |
| Fe (mg/L) | **0.431** | **0.934** | **0.528** | **0.682** | **0.906** | 0.073 | **0.358** | ≤5.0[(2)]<br>≤0.2[(3)] |
| Mn (mg/L) | **10** | **18** | **0.279** | **0.470** | **0.665** | **0.087** | **0.293** | ≤0.02[(2)]<br>≤0.2[(3)] |
| Al (mg/L) | < 0.1 | 0.121 | 0.118 | < 0.1 | 0.134 | < 0.1 | 0.137 | ≤5.0 |
| U (mg/L) | < 0.001 | < 0.001 | < 0.001 | < 0.001 | < 0.001 | < 0.001 | < 0.001 | ≤0.01 |
| Ni (mg/L) | 0.145 | 0.117 | 0.049 | 0.035 | 0.029 | < 0.025 | < 0.025 | ≤0.2 |
| Zn (mg/L) | 0.048 | 0.379 | 0.043 | 0.025 | < 0.025 | < 0.025 | < 0.025 | ≤1.0 |
| N:P value | 6 | 6 | 0.1 | 0.2 | 1.3 | 19 | 12 | n.s |
| SAR value | 1.14 | 0.99 | 2.00 | 1.76 | 1.71 | 0.21 | 0.93 | ≤2.0 |
| CTR value | 772 | 783 | 1.0 | 0.9 | 1.2 | 0.4 | 1.0 | n.s |

[1]Recommended target water quality range (TWQR) limit for irrigation use (from DWAF, 1996c)
[2]Effect on crop yield and quality
[3]Effect on irrigation equipment (mainly drip irrigation systems)
Bold text denotes values that exceed the TWQR recommended limit for irrigation use

**Plate 2** View of station F11S12 showing concrete weir (dark linear shape below right hand of person) and object (circled) on causeway for common reference with Plate 3 (photo W Basson)

**Table 48** Change in surface water chemistry with distance downstream of the mine water source

| Locality | | Distance (m) | | SEC (mS/m) | pH[(1)] | Flow (ML/d) |
|---|---|---|---|---|---|---|
| | | Segmental | Cumulative | | | |
| 1 | KGR Inflow (raw mine water) | 0 | 0 | 354 | 4.5 | 21.0 |
| 2 | KGR Inflow (treated mine water) | | | 306 | 6.3 | 19.0 |
| 3 | Hippo Dam outlet in KGR | 950 | 950 | 332 | 5.6 | n.m |
| 5 | Poplar Grove in KGR | 850 | 1800 | 324 | 4.3 | n.m |
| 6 | Aviary Dam inlet in KGR | 3105 | 4905 | 302 | 4.0 | 32.1 |
| 7 | KBW Dam outlet at N14 (F11S12) | 1540 | 6445 | 302 | 3.9 | 35.2 |
| 8 | Riet Spruit at Malmani Road (MRd) | 3900 | 10 345 | 278 | 4.1 | 7.3 |
| 9 | Bloubank Spruit downstream of Makiti (BB@M) | 5050 | 15 395 | 45 | 7.1 | 43.2 |
| 10 | Bloubank Spruit at A2H049 | 12 000 | 27 255 | 44 | 7.2 | n.m |
| 11 | Blougat Spruit tributary of the Riet Spruit at N14 (BB@N14) | n/a | 10 835[(2)] | 51 | 6.9 | ~ 36.0[(3)] |
| 12 | Honingklip Spruit tributary of the Bloubank Spruit at Kromdraai (HS1) | n/a | 18 905[(2)] | 19 | 7.2 | n.m |

[1]($-\log_{10}\alpha_{H+}$)
[2]Distance of confluence with main stem downstream from the mine water source
[3]Difference between flow measured at localities 8 and 9
n.m. denotes not measured

**Table 49** Change in surface water chemistry with distance downstream of the mine water source

| Locality | | Distance (m) | | SEC (mS/m) | pH ($-\log_{10}\alpha_{H+}$) | Flow (ML/d) |
|---|---|---|---|---|---|---|
| | | Segmental | Cumulative | | | |
| 3 | Hippo Dam outlet in KGR | 950 | 950 | 406 | 5.5 | n.m |
| 4 | Poplar Grove in KGR | 850 | 1800 | 322 | 3.6 | n.m |
| 5 | 4 × 4 Track at Oukraal Lapa in KGR | 1015 | 2815 | 302 | 3.5 | 52.3 |
| 6 | Aviary Dam inlet in KGR | 2090 | 4905 | 293 | 3.4 | 34.4 |
| 7 | Brickworks Dam outlet / F11S12 | 1540 | 6445 | 290 | 3.6 | 40.4 |
| 8 | Riet Spruit at Malmani Road (MRd) | 3900 | 10,345 | 278 | 3.4 | 10.3 |
| 9 | Bloubank Spruit downstream of Makiti (BB@M) | 5050 | 15,395 | 27 | 7.3 | 128.7 |
| 11 | Blougat Spruit tributary of the Riet Spruit at N14 (BB@N14) | n/a | 10,835[1] | 28 | 7.5 | ~ 118[2] |
| 12 | Honingklip Spruit tributary of the Bloubank Spruit at Kromdraai (HS1) | n/a | 18 905[1] | 14 | 7.4 | n.m |

[1]Distance of confluence with main stem downstream from the mine water source
[2]Difference between flow measured at localities 8 and 9
n.m. denotes not measured

qualitative and adherence to general principles is advocated. The greater part of the guidelines pertains to indicator organisms, e.g. faecal coliforms, *E. coli*, faecal streptococci, coliphages, enteric viruses and schistosoma/bilharzia. The limits for these are necessarily less stringent than the TWQR limits for drinking water (Sect. 6.4.1), e.g. faecal coliforms $\leq 30$ c/100 mL for full-contact recreation (swimming and diving) and $\leq 1000$ c/100 mL for intermediate contact recreation (waterskiing, canoeing, angling, paddling and wading) versus $\leq 10$ c/100 mL for drinking water.

## 5 Mine Water Chemistry in the Receiving Drainages

### 5.1 Ad Hoc Observations

The ad hoc measurement of field pH and SEC values triggered by the advent of excessive mine water decant in late January 2010 provided the opportunity to gather valuable information on the chemistry of surface water in the receiving environment. Comprising a mixture of treated/neutralised mine water from the HDS Plant and raw mine water, the pH and SEC of this water was measured at various localities in the downstream receiving drainages. The data presented in Tables 48 and 49 reflect the degree of change recorded on 05/02/2010 and 01/04/2010, respectively, in the analytes pH and SEC with distance downstream from where the mine water leaves the mine property. The latter data set occurs in the period mid-March to mid-May 2010 when the raw mine water discharge was subjected to in-stream liming (as a DWS-funded emergency treatment measure) at the point of concentrated release into the environment (refer Footnote 59).

The measurements reveal the following hydrochemical responses.

- The decrease in the pH of the water with distance downstream of the mine water source area, from a value of ~5.6 at the Hippo Dam outlet to a value of ≤4.1 at the Brickworks Dam/F11S12 outlet on the Tweelopie Spruit and at the Malmani Road crossing on the Riet Spruit. This response is readily accounted for by hydrolysis reactions (Hem, 1985), in this instance the precipitation of iron out of the water producing free hydrogen ions, with the increase in hydrogen ion activity lowering the pH as illustrated in the following reaction (amongst many similar such reactions)

$$Fe^{3+}(aq) + 3H_2O_{(aq)} \rightarrow Fe(OH)_{3(s)} + 3H^+(aq) \quad (4)$$

Hydrolysis of the ferric iron ($Fe^{3+}$) by water forms an amorphous ferric hydroxide $[Fe(OH)_3]$ efflorescence, a residue also called 'ochre' or 'Yellow boy'. The production of free hydrogen ions generates acidity which maintains the solubility of the Fe(III) (Banks, 2004). The rapidity of this process is illustrated in Plate 2 and 3, which show the appearance of the concrete weir at the outlet of station F11S12 on 28/01/2010 and 08/02/2010, a period of only 10 days apart. At this stage, the ferrihydrite efflorescence visible in Plate 3 merely formed a thin (<1 mm thick) slick film coating the concrete and that could be rubbed off with a finger. Through aggregation and ageing of the ferrihydrite (Webb and Sasowsky, 1994), it had formed a hard goethite (FeO · OH) crust ~1–2 cm thick by ca. mid-2012. This crust gradually eroded away following the improvement in mine water quality discharge in mid-2012.
The situation illustrated in Plate 2 and 3 marks the transition from a surface water chemistry characterised by the discharge of mainly treated/neutralised mine water in which there is very little iron, to a situation where raw mine water comprises ≥50% of the mine water waste stream discharged from the mine area (Sect. 3.1 in Chapter "Physical Hydrology" and Fig. 16). The very

**Plate 3** View of station F11S12 showing concrete weir coated with freshly precipitated hydrous ferric oxide or ferrihydrite ($5Fe_2O_3 \cdot 9H_2O$) and object (circled) on causeway for common reference with Plate 2

**Plate 4** Discharge at station F11S12 (Tweelopie Spruit) on 18/12/2010. Compare with Plate 13, Plates 2, 3 or Plate 41 to note the complete submergence of the causeway

**Plate 5** View looking north of the high discharge at BB@M (Bloubank Spruit) on 18/12/2010

high concentrations of iron (5% ile = 322 mg/L, 50% ile = 687 mg/L) dissolved in the raw mine water (Table 2) lead to the rapid precipitation of ferrihydrite much further downstream (as observed at station F11S12, Plate 3) than was observed previously (Plate 2). Figure 16 indicates that 30/01/2010 marks the setting of the surface water chemical regime in the drainage system to a lower or 'poorer' state, which monitoring data indicates returned to 'normal' ca. mid-2012.

- The significant improvement in surface water quality downstream of the Riet Spruit/Blougat Spruit confluence, as indicated by the SEC values of 45 and 27 mS/m and the pH values of 7.1 and 7.3 measured in the Bloubank Spruit downstream of the Makiti Wedding and Conference Centre, compared to the 278 mS/m and pH of $\leq$4.1 measured in the Riet Spruit at the Malmani Road crossing (station MRd). This is attributed to the mitigating influence of the significant discharges (estimated at $\sim$36 and $\sim$118 ML/d, respectively) contributed by the better inorganic quality of Blougat Spruit water characterised by an SEC in the range 30–50 mS/m and a pH in the range 6.9–7.3. It is evident from the information presented and discussed in Sect. 1.2.4, however, that the microbiological quality of this water is a grave concern.

## 5.2 Opportunistic Observations

Peak flow (flood) conditions were experienced in the Tweelopie Spruit, the lower Riet Spruit and the Bloubank Spruit following the high rainfall experienced in mid-December 2010 (Sect. 3.2). Various localities in the area reported precipitation in the order of 130 to 140 mm in a period of some 24 h ca. 16/12/2010. The rainfall gauging station operated by SG at the Black Reef Incline in the LoD in the upper reaches of the Tweelopie Spruit recorded 90 mm (B van der Walt, personal communication) between 07h00 on 16 and 17/12/2010. The DWS flow gauging station A2H049 at the lower end of the Bloubank Spruit recorded a second highest recorded daily average flow rate of 18.6 $m^3/s$ on 16/12/2010, followed by rates of 6.9 and 2.9 $m^3/s$ on the following two days. The maximum (highest) historical daily average flow rate gauged at this station is 34.3 $m^3/s$ recorded on 28/01/1978.

The aftermath of the mid-December 2010 flood event was still visible on 18/12/2010 (Plates 3 and 4) when a series of opportunistic surface water field chemistry measurements were made at the 'key' locations along the effected drainages shown in Fig. 55. The results, obtained roughly a month before the fish mortality event that occurred in the study area in mid-January 2011 (Hobbs and Mills, 2011), are presented in Table 50.

Perhaps the most significant impact of the flood was the disruption of flow into the A-furrow (Sect. 5.6.2). Flow into this furrow was only restored, albeit partially, on 10/01/2011 (M Gomes, personal communication), i.e. a day before the fish mortality event. At this time, Mr. Gomes (Chairman of the Kromdraai Irrigation Board) expressed concern for the yellow appearance of the furrow water and its suitability for various uses including stock watering and irrigation. Unaware of the fish mortality event, this communication precipitated another round of field water quality monitoring on 12/01/2011, the results of which are also reported in Table 50.

**Plate 6** Examples of Fe-(orange) and Mn-oxide (black) precipitation as concentric scaling in pipe (left) and as 'Yellow boy' in ditch (right) (photo at left J Davies; photo at right W Basson).

A closer inspection of the 12/01/2011 data (Table 50) reveals the following hydrochemical characteristics:

- The very low pH values and elevated SEC values of ~ 400 mS/m at the uppermost two stations F11S12 and MRd indicating a strong mine water signature at these localities.
- The generally good inorganic quality associated with the Percy Stewart WWTW discharge via the Blougat Spruit (station BG@N14) into the Bloubank Spruit system as shown by the high pH value and low SEC value; this by no means implies an acceptable bacteriological quality.
- The very low pH and elevated SEC of the surface water at station BB@M, again indicative of a mine water impact on the quality of the surface water at this locality. Whereas the pH value represents the most extreme value observed at this station in the project period, the SEC value is only exceeded by that observed on 18/12/2010 (Table 50). Significantly, the ORP value of 221 mV (indicative of an oxidising environment) approximates those observed at the upstream stations F11S12 and MRd.
- The improvement in quality at the downstream station BB@PL, as reflected in a higher pH and lower SEC compared to that at station BB@M, is attributed to the mitigatory influence contributed by the ~307 L/s Kromdraai and ~60 L/s Plover's Lake springs.
- The still quite 'acceptable' pH and SEC values at station A2H049 at the bottom end of the Bloubank Spruit system at Glenburn Lodge near Zwartkop, although the pH value approaches the long-term 5%ile value of 7.4, and the SEC value of 87 mS/m exceed the long-term 95%ile value of 67 mS/m (Table 19) recorded at this DWS gauging station. Even so, the observed SEC value does not exceed the maximum value of ~122 mS/m recorded in the 36-year period May 1979 to September 2014 that represents the historical record of water chemistry data available for this station from the DWS.

The SEC values of 158 and 155 mS/m (Table 50) recorded at station BB@M approach the highest recorded at this station in the course of the study, and the pH value of 2.9 similarly the lowest observed. A surface water sample collected for more complete laboratory chemical analysis[18] returned the results presented in Table 51. These circumstances reflect the dominant contribution of raw mine water discharge to the chemistry of surface water at this locality on this occasion. Of specific relevance to the fish mortality event is the very low field pH value of 2.9 observed on 12/01/2011 (Plate 5).

**Plate 7** Excessive sludge deposition in the upper reaches of the Hippo Dam in the Krugersdorp Game Reserve as a result of emergency liming of raw mine water to raise its pH (photo S du Toit)

The author was notified on 13/01/2011 of fish mortalities that had occurred on Koelenhof Farm upstream of the Kromdraai T-junction (M Liefferink, personal communication). The mortalities occurred in an off-channel irrigation water storage dam fed by the A-furrow, and not in the natural stream course of the Bloubank Spruit. It is therefore probable that the restored flow of A-furrow water into this dam resulted in oxygen depletion associated with the precipitation of iron hydroxides in this low energy impounding environment. It is also probable that the slow rate of iron precipitation resulted in the accumulation of this efflorescence in the gills of fish, compounding the suffocating effect of a depleted oxygen supply and contributing to the observed fish mortalities in this impoundment.

The above circumstances are also likely to have similar consequences for crustaceans and macroinvertebrates resident in these dams. Enquiries further established that above average fish mortalities had also been experienced at the Brookwood Trout Farm located downstream of station BB@PL (Table 50) (H Carpenter, personal communication). The Koelenhof Farm fish mortality event precipitated further field studies on 14/01/2011 by the author and P Mills of the MA (Hobbs and Mills, 2011). The results are summarised in Table 52.

The nature of the event that precipitated the 14/01/2011 studies necessitated the inclusion of dissolved oxygen (DO) as a field water chemistry variable. An inspection of the data presented in Table 52 reveals the following hydrochemical characteristics:

- In regard to the station BB@M;
  - the low pH value indicative of, amongst other sources, a combination of very low pH mine water and low pH

[18] The oxidation of dissolved $Fe^{2+}$ in the water sample resulted in the precipitation of an iron oxide in the sample bottle during transport and laboratory storage. This was not seen in the acidified sample bottle.

**Table 50** Surface water field chemistry variable values

| Station | Field chemical variable | | | | | | | |
|---|---|---|---|---|---|---|---|---|
| | pH ($-\log_{10}\alpha_{H+}$) | | SEC (mS/m) | | ORP (mV) | | Temp. (°C) | |
| | 18/12/2010 | 12/01/2011 | 18/12/2010 | 12/01/2011 | 18/12/2010 | 12/01/2011 | 18/12/2010 | 12/01/2011 |
| F11S12 | 2.7 | 2.5 | 416 | 397 | 230 | 243 | 21.3 | 21.1 |
| MRd | 3.0 | 2.3 | 276 | 410 | 217 | 254 | 24.1 | 25.0 |
| BG@N14 | n.m | 8.2 | n.m | 53 | n.m | –65 | n.m | 26.1 |
| BB@M | 6.1 | 2.9 | 158 | 155 | 51 | 221 | 24.8 | 23.9 |
| BB@PL | n.m | 6.8 | n.m | 94 | n.m | 12 | n.m | 21.3 |
| A2H049 | n.m | 7.7 | n.m | 87 | n.m | –37 | n.m | 22.3 |

**Table 51** Chemical analysis laboratory report for surface water sample collected at station BB@M

| Variable/Analyte | Value | Source Indicator Analyte | | SANS (2011a)[1] |
|---|---|---|---|---|
| | | Mine water | Sewage wastewater | |
| Laboratory | CSIR (Pretoria) | n/a | | n.a |
| Laboratory report # | 9156 | n/a | | n.a |
| Laboratory sample # | 82,235 | n/a | | n.a |
| Report date (dd/mm/yyyy) | 19/01/2011 | n/a | | n.a |
| Sample date (dd/mm/yyyy) | 12/01/2011 | n/a | | n.a |
| pH ($-\log_{10}\alpha_{H+}$) (field) | **2.9** | Yes | No | 5.0–9.5 |
| pH ($-\log_{10}\alpha_{H+}$) (lab value) | **3.2** | Yes | No | 5.0–9.5 |
| SEC (mS/m) (field) | **155** | Yes | No | $\leq 170$ |
| SEC (mS/m) (lab) | 122 | Yes | No | $\leq 170$ |
| Ca (mg/L) | 149 | Yes | No | $\leq 150$ |
| Mg (mg/L) | 49 | Dual | | $\leq 70$ |
| Na (mg/L) | 50 | Dual | | $\leq 200$ |
| K (mg/L) | 9 | Dual | | $\leq 50$ |
| Cl (mg/L) | 40 | No | Yes | $\leq 200$ |
| $SO_4$ (mg/L) | **758** | Yes | No | $\leq 400$ |
| $HCO_3$ (mg/L) | < 6 | Dual | | n/s |
| $NO_2 + NO_3$ (mg N/L) | 4.8 | No | Yes | $\leq 11$ |
| $NH_3 + NH_4$ (mg N/L) | 2 | No | Yes | n/s |
| $PO_4$ (mg P/L) | <0.2 | No | Yes | n/s |
| Fe (mg/L) | **44** | Yes | Subordinate | $\leq$ 0.2 |
| Mn (mg/L) | **16** | Yes | Subordinate | $\leq$ 0.1 |
| Al (mg/L) | **0.78** | Yes | Yes | $\leq$ 0.3 |
| Ni (mg/L) | **0.35** | Yes | Yes | $\leq$ 0.15 |
| CN (mg/L) | <0.01 | | | n/s |
| EB (%) | –3.0[2] | n/a | | ± 5 |
| Chemical character | $CaSO_4$ | Yes | No | |

[1]Standard health-related limit for consumption of 2 L/d over 70 years by a 60 kg person
[2]Includes Fe and Mn in the calculation; value deteriorates to –10% without these cations.
Bold text denotes value exceeds standard limit as described in note 1

rainwater runoff in the discharge at this position in the Bloubank Spruit;
  - the very low SEC value of ~22 mS/m compared to the value of 155 mS/m measured two days earlier on 12/01/2011 (Table 50), which similarly reflects the significant contribution of fresh water in the river on 14/01/2011; and
  - the high level of oxygen saturation reflected by the DO value of 106%, which is readily explained by the degree of turbulence exhibited in the strong-flowing river and natural diffusion of gaseous oxygen ($O_2$) from the atmosphere into the water (DWAF, 1996a); the supersaturated oxygen state is possibly indicative of eutrophication associated with a high nutrient load attributable to the presence of sewage wastewater effluent
- The similar analyte values observed at the station A-furrow @ Lotz as were observed at station BB@M, which establishes the direct hydraulic link between the Bloubank Spruit surface water and that flowing in the A-furrow.
- The very low pH and DO[19] values, and the elevated SEC and temperature[20] values of the Koelenhof Farm dam water, which likely represent an artefact of the water quality conditions that gave rise to the fish mortality event.
- The saturated oxygen content of the furrow water downstream of the Koelenhof Farm dam compared to the 60% DO concentration of the dam water.
- The similar analyte values (except for temperature) associated with the Zwartkrans Spring and Sterkfontein Cave Lake water, which establishes the hydraulic connection between these two groundwater sources. [Note: The cave water level response is discussed in greater detail in Sect. 7.3.2] Whereas the Zwartkrans springwater field SEC of ~74 mS/m is similar to its more recent 'historical' value, the cave water field SEC of ~56 mS/m is slightly lower than the typical 59–62 mS/m range that characterises this water quality variable. The fresher nature of the cave water is attributed to the influence of considerably fresher rainwater directly infiltrating the cave environment from above. For comparison, the DWS reports an SEC value[21] of 74 mS/m in April 2001.

[19] The Target Water Quality Range (TWQR) for the DO level in aquatic ecosystems brackets the range 80–120% of saturation, and a DO level below the Minimum Allowable Value (MAV) defined by a 7-day mean minimum value of >60% (sub-lethal) in combination with a 1-day minimum value of >40% (lethal), is likely to cause acute toxic effects on aquatic biota (DWAF, 1996).

[20] High water temperatures combined with low DO levels can compound stress effects on aquatic organisms (DWAF, 1996).

[21] It is not indicated whether this is a field- or laboratory-based value.

## 5.3 Visual Observations

Perhaps the most striking behaviour of mine water in the receiving drainages is the visual impact in the form of efflorescence. These take a variety of forms ranging from pipe scaling (Plate 6) to sludge deposits (Plate 7) and surface encrustation (Plate 8).

A chemical analysis of the pipe scaling precipitate shown in Plate 6 is presented in Table 53. The excessive sludge build-up in the Hippo Dam (Plate 7) was the result of in-stream liming[22] of the raw mine water where it leaves the mine property. The dam is located ~700 m downstream of this position.

Plate 8 shows the dual precipitation of iron hydroxides and gypsum on the floodplain of the Tweelopie Spruit at its confluence with the Riet Spruit on the Glen Almond property. This position is located a distance of ~6.3 km downstream (north) of the Hippo Dam. The permanence of these precipitates is a function of the solubility of Fe(III) oxyhydroxides which, as is evident from the discussion in Text Box 5, depends on numerous factors.

## 5.4 Sediment Chemistry

Although sensu stricto not a mine water chemistry issue, the uptake/immobilisation of trace metals and metalloids in streambed sediments and the possible release/remobilisation of these through acidification, is a topical subject. This is informed by the outcome of a streambed sediment chemistry investigation carried out in the south-western portion of the study area originally reported by Venter et al. (2010) and augmented by Hobbs and Venter (2010). The study included an assessment of remobilisation (solubilisation) of metals with exposure of the sediment through laboratory batch leach (BL), toxicity characterising leaching procedure (TCLP) and acid rain (AR) leaching tests.

### 5.4.1 Chemical Composition

The sediment chemistry results determined from XRF analysis are presented in Venter et al. (2010). The elements are ranked in order of decreasing maximum detected value in Table 54 for each of the sites and samples. Site 1 (the Hippo Dam) supports the maximum concentration in 18 of the 32 analytes listed,[23] sharing arsenic with site 2. This site is located closest to the mine area. Site 2 (the Lion Camp dam) supports a further six of the elements. Site 6 (the Riet Spruit

[22] In-stream liming funded by the DWS at a cost of R6.9 million was employed as an emergency water treatment measure in the period mid-March to mid-May 2010.

[23] Counting only analytes where the detection limit is exceeded in at least one instance.

**Table 52** Surface water and groundwater field chemistry variables on 14/01/2011

| Station | Field Chemical Variable | | | | |
|---|---|---|---|---|---|
| | pH ($-\log_{10}[H^+]$) | SEC (mS/m) | ORP (mV) | DO (%) | Temp (°C) |
| Bloubank Spruit at Makiti (BB@M) | 6.0 | 21.9 | 48 | 106 | 21.1 |
| Zwartkrans Spring (ZSp) | 7.3 | 73.6 | –19 | 66 | 19.1 |
| A-furrow @ Lotz (upstream of KFD) | 6.2 | 24.0 | 44 | 101 | 21.1 |
| Koelenhof Farm dam (KFD) | 3.6 | 128.7 | 178 | 60 | 24.0 |
| Koelenhof Farm furrow downstream of KFD | 5.6 | 126.8 | 72 | 103 | 21.5 |
| Sterkfontein Cave (Main Lake) | 7.8 | 55.8 | –49 | 66 | 16.7 |

immediately before its confluence with the Bloubank Spruit) supports five of the remaining six elements, notably including Fe and Mn. Site 3 supports the highest hafnium (Hf) concentration of the eight sites.

The high Fe and Mn levels at site 6 are not readily explained on the basis of proximity to the mine area, given the distance of ~12.5 km from the latter. It is possible that these circumstances reflect the historical impact of Percy Stewart WWTW effluent, which also contains elevated Fe and Mn concentrations (Table 11), discharged via the Blougat Spruit. Whether this also explains the elevated levels of Ba, Rb and Mo observed at this locality is unclear.

The maximum values range from ~66 000 mg Al/kg (66 kg Al/t) to 1.6 mg Ge/kg (0.002 kg Ge/t) of sediment. A similar perspective on elements such as Cr, U, As and Pb yields values of 0.6 kg Cr/t, 0.1 kg U/t, 0.08 kg As/t and 0.05 kg Pb/t. The USACE (reported in O'Bourke et al., 2008) indicates an average in situ density for mud (clay) and silt in the range 1200 to 1400 g/L. Taking the higher value, then a cubic metre of this material covering an area of 4 $m^2$ to a uniform depth of 0.25 m contains up to ~800 g Cr, 140 g U, ~110 g As and 70 g Pb. These values reduce to ~330 g Cr, 36 g U, 34 g As and 25 g Pb for mean element values. Each 100 m of stream reach 2 m wide would therefore contain 16.5 kg Cr, 1.8 kg U, 1.7 kg As and 1.25 kg Pb in elemental terms. The measure to which complexation and other geochemical (e.g. redox) and biogeochemical (e.g. BSR) reactions will modify these values has not been considered.

Table 12 indicates that Awofolu et al. (2007) recorded maximum Zn, Ni and Cu values at site S2, the same locality as site 7 in Table 54. A comparison of the S2/site7 results reveals significant differences, primarily the order of magnitude lower values reported for all analytes by Awofolu et al. (2007). It is possible that a similar difference might apply to the other results. An explanation for the observed differences has not been sought.

**Plate 8** View of the Tweelopie Spruit at its confluence with the Riet Spruit at Glen Almond showing the iron hydroxide and gypsum efflorescence on the flood plain at this juncture; the taller vegetation in the middle distance marks the stream channel

### 5.4.2 Solubility and Remobilisation

The leach tests sought to quantify the extractable (mobile) fraction of the total concentration of analytes (primarily metals/metalloids) by ICP-MS (inductively coupled plasma —mass spectrometry). The results are discussed by Venter et al. (2010) and Hobbs and Venter (2010). Their significance warrants a little further interrogation in this work. Except in the case of aluminium, the TCLP method generally mobilised the greatest fraction, followed by the AR and the BL methods. The BL test manifested an equal or greater Al solubilisation result.

The trace elements that showed the greatest propensity for solubilisation/remobilisation (mobility) are Al, Ba, Co, Cu, Mn, Ni, Sr and Zn (in simply alphabetic order). The magnitude of solubilisation of these analytes at each of the eight sampling sites is shown in Fig. 55. The apparently random pattern of the localities where elements are mobilised in greatest concentrations (Fig. 55) is in keeping with observations by Sahuquillo et al. (2002) that:

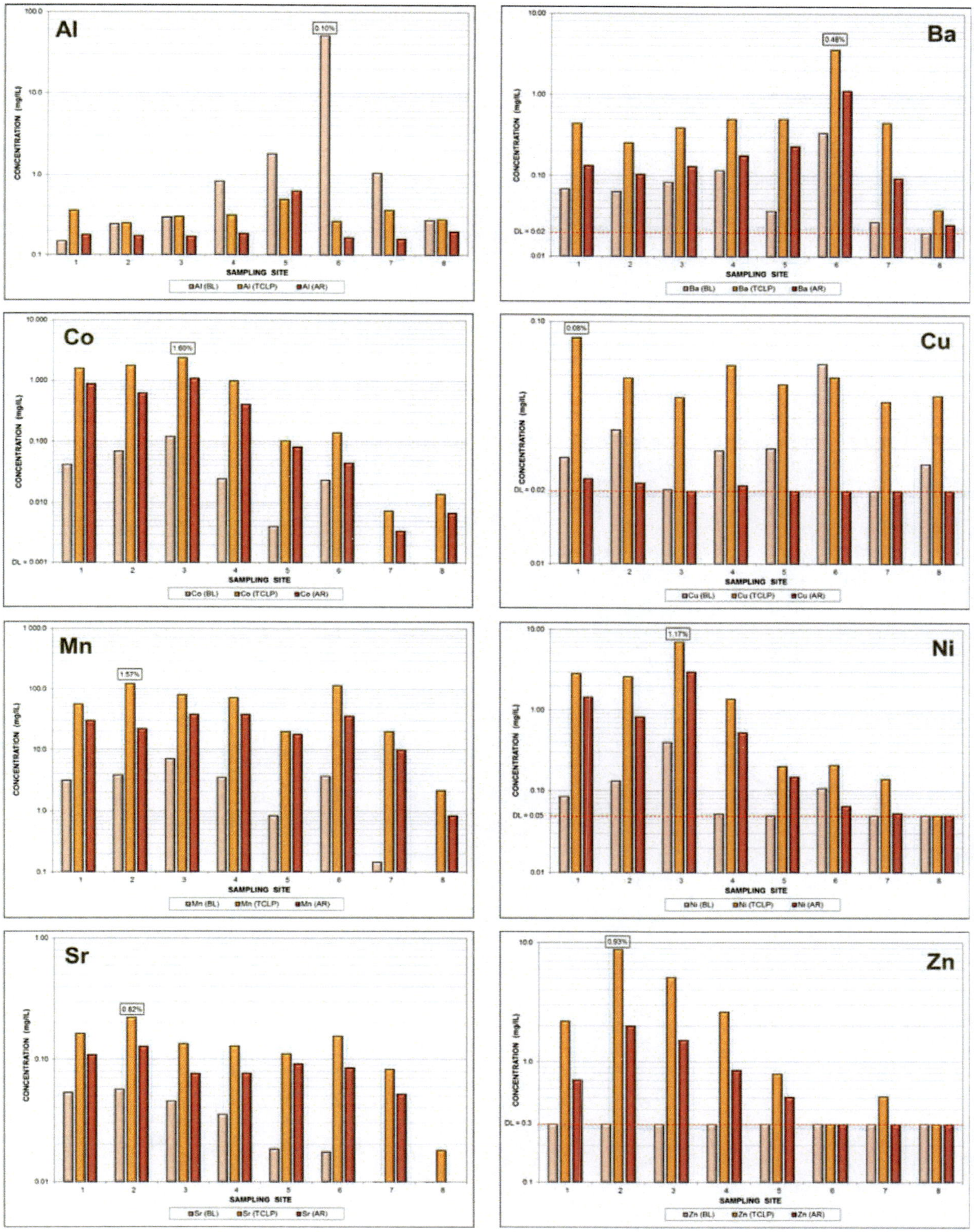


**Fig. 55** Magnitude of solubilisation of trace elements from streambed sediments in the Bloubank Spruit system

**Table 53** Chemical analysis of pipe scaling precipitate as seen in Plate 1 in Chapter "Physical Hydrology" (data courtesy of Helmholtz Centre for Environmental Research—UFZ, 2007)

| Analyte | Value | Analyte | Value | Analyte | Value |
|---|---|---|---|---|---|
| MgO | 0.50% | Cr | 200 mg/L | Br | <2 mg/L |
| $Al_2O_3$ | 0.13% | $P_2O_5$ | <200 mg/L | Rb | <2 mg/L |
| $SiO_2$ | 0.55% | Sr | 79 mg/L | Pb | <2 mg/L |
| $K_2O$ | <0.01% | Co | 59 mg/L | U | <2 mg/L |
| CaO | 29.00% | Zn | 37 mg/L | Th | <2 mg/L |
| $TiO_2$ | 0.005% | V | <10 mg/L | Mo | <2 mg/L |
| $Fe_2O_3$ | 2.30% | Y | 6 mg/L | Ag | <2 mg/L |
| S | 160 000 mg/L | As | 4 mg/L | Cd | <2 mg/L |
| Mn | 1540 mg/L | Cu | 3 mg/L | Hg | <2 mg/L |
| Ni | 213 mg/L | Zr | 3 mg/L | Sb | <2 mg/L |

**Table 54** Ranking of analyte concentrations in sediment by maximum value (data from Venter et al., 2010)

| Analyte | Site and sample # | | | | | | | | | | Max value (mg/kg) |
|---|---|---|---|---|---|---|---|---|---|---|---|
| | 1 | | | 2 | 3 | 4 | 5 | 6 | 7 | 8 | |
| | 001 | 002 | 003 | 005 | 007 | 008 | 010 | 012 | 014 | 016 | |
| Al[1] | 48 314 | **66,014** | 38,001 | 56,290 | 5897 | 41,198 | 13,037 | 50,322 | 17,144 | 5897 | 66,014 |
| Fe[2] | 50 579 | 9658 | 53,900 | 60,900 | 4276 | 36,934 | 18,783 | **62,916** | 34,821 | 4276 | 62,916 |
| Mn[3] | 4565 | 213 | 77 | 7850 | 120 | 4349 | 1057 | **8747** | 8498 | 120 | 8747 |
| Zn | 216 | 72 | 460 | **943** | 430 | 261 | 53 | 56 | 81 | 12 | 943 |
| Ba | 531 | 120 | 122 | 173 | 549 | 307 | 101 | **751** | 446 | 21 | 751 |
| Ni | 313 | 134 | **720** | 497 | 607 | 196 | 38 | 111 | 68 | 6.7 | 720 |
| Cr | **603** | 425 | 290 | 401 | 367 | 339 | 223 | 413 | 214 | 95 | 603 |
| Co | 119 | 17 | **202** | 154 | 149 | 72 | 11 | 28 | 25 | 1.7 | 227 |
| Zr | **227** | 213 | 143 | 197 | 154 | 164 | 61 | 177 | 50 | 21 | 227 |
| Cu | 105 | 60 | 135 | **174** | 49 | 57 | 15 | 46 | 35 | 4.2 | 174 |
| V | 126 | **137** | 100 | 116 | 99 | 84 | 30 | 97 | 51 | 9.6 | 137 |
| Ce | 100 | 126 | **130** | 83 | 100 | 48 | <10 | 48 | 31 | <10 | 130 |
| U | 41 | 35 | **107** | 50 | 6.3 | 9.9 | <2 | 3.6 | <2 | <2 | 107 |
| As | **83** | 9.3 | 65 | **83** | 18 | 14 | 6.4 | 15 | 7.8 | <4 | 83 |
| Nd | 30 | 61 | **73** | 61 | 46 | 29 | 10 | 27 | 10 | <10 | 73 |
| La | 37 | 57 | 62 | **66** | 64 | 37 | <10 | 33 | 18 | <10 | 66 |
| Pb | **52** | 25 | 31 | 32 | 29 | 37 | 9.4 | 13 | 17 | 2.9 | 52 |
| Sr | 30 | 11 | 39 | 27 | 38 | 16 | 14 | 17 | 19 | 2.1 | 49 |
| Rb | 28 | 35 | 23 | 43 | 37 | 42 | 9.4 | **47** | 17 | 7.5 | 47 |
| Y | 18 | 31 | 36 | **41** | 26 | 18 | 4.8 | 18 | 7.6 | 1.9 | 41 |
| Sc | 16 | **19** | 17 | 17 | 16 | 14 | 5.1 | 16 | 7.4 | <3 | 19 |
| Ga | 13 | **16** | 11 | 14 | 13 | 11 | 3.7 | 12 | 4.9 | 2 | 16 |
| Sm | <10 | 12 | 15 | <10 | <10 | <10 | <10 | <10 | <10 | <10 | 15 |
| Nb | 10 | **13** | 8.9 | 10 | 9.4 | 9 | 3.6 | 9.6 | 4 | 2.5 | 13 |
| Th | **10** | 8.6 | 8.7 | 8.4 | 7 | 6.3 | <3 | 7.2 | 3.7 | <3 | 10 |
| Br | <2 | 3.4 | 8.4 | 11 | 4.3 | 3.7 | <2 | 4.8 | <2 | <2 | 8.4 |

(continued)

**Table 54** (continued)

| Analyte | Site and sample # | | | | | | | | | | Max value (mg/kg) |
|---|---|---|---|---|---|---|---|---|---|---|---|
| | 1 | | | 2 | 3 | 4 | 5 | 6 | 7 | 8 | |
| | 001 | 002 | 003 | 005 | 007 | 008 | 010 | 012 | 014 | 016 | |
| Cs | 6.3 | <5 | **8.2** | <5 | <5 | 5.5 | <5 | <5 | <5 | <5 | 8.2 |
| Hf | 5.4 | 5.4 | 4 | 4.1 | **6.2** | 4.2 | <3 | 5.6 | <3 | <3 | 6.2 |
| Yb | <3 | 3.8 | **5.0** | 4.3 | <3 | <3 | <3 | <3 | <3 | <3 | 5.0 |
| Mo | 2 | <2 | <2 | <2 | <2 | <2 | <2 | **2.3** | <2 | <2 | 2.3 |
| Se | 1.2 | 1.2 | 2 | **1.8** | <1 | <1 | <1 | <1 | <1 | <1 | 1.8 |
| Ge | <1 | **1.6** | <1 | 1 | 1.5 | 1 | <1 | 1.5 | <1 | 1.3 | 1.6 |
| Bi | <3 | <3 | <3 | <3 | <3 | <3 | <3 | <3 | <3 | <3 | – |
| Tl | <3 | <3 | <3 | <3 | <3 | <3 | <3 | <3 | <3 | <3 | – |
| W | <3 | <3 | <3 | <3 | <3 | <3 | <3 | <3 | <3 | <3 | – |
| Ta | <2 | <2 | <2 | <2 | <2 | <2 | <2 | <2 | <2 | <2 | – |
| Maxima | 5 | 6 | 7 | 6 | 1 | 0 | 0 | 5 | 0 | 0 | |

[1]Calculated from weight% concentration of $Al_2O_3$
[2]Calculated from weight% concentration of $Fe_2O_3$
[3]Calculated from weight% concentration of MnO
All values are as mg/kg
Bold text denotes location of analyte maximum value

**Table 55** Estimate of soluble mass of individual elements per length of stream reach

| Analyte | Mean concentration in sediment (mg/kg) | Mass per Unit volume of sediment (kg/m$^3$) | Mass per unit length of stream reach (kg/m) | Mean solubility (%) | Soluble Mass per length of stream reach | |
|---|---|---|---|---|---|---|
| | | | | | (kg/m) | (kg/km) |
| Al | 29 762 | 41.667 | 20.83 | 0.02 | 0.4167 | 417 |
| Mn | 4413 | 6.179 | 3.090 | 1.39 | 4.2944 | 4294 |
| Zn | 257 | 0.359 | 0.180 | 1.01 | 0.1813 | 181 |
| Ba | 360 | 0.504 | 0.252 | 0.22 | 0.0554 | 55.4 |
| Ni | 230 | 0.321 | 0.161 | 0.80 | 0.1284 | 128 |
| Cr | 332 | 0.465 | 0.233 | 0.01 | 0.0023 | 2.33 |
| Co | 70 | 0.098 | 0.049 | 1.26 | 0.0617 | 61.7 |
| Cu | 61 | 0.085 | 0.043 | 0.10 | 0.0043 | 4.25 |
| U | 22 | 0.031 | 0.016 | 0.12 | 0.0019 | 1.86 |
| Sr | 20 | 0.029 | 0.014 | 0.63 | 0.0090 | 8.99 |

- extractability is not proportional to the total metal content; and
- samples with similar metal concentrations can exhibit very different metal extractability.

The magnitude of resolubility evident in Fig. 55 varies according to total concentration and element. For example, Mn typically produces > 10 mg/L; Ni and Zn > 1 mg/L; Al, Ba and Co < 1 mg/L; and Cu and Sr < 0.1 mg/L. The results provide context for concerns in regard to the remobilisation of trace/heavy metals by acidic mine water discharges in the receiving drainages, and the resulting impact on the aquatic ecosystem(s).

The information presented in Table 55 attempts to quantify (a) the individual element mass per cubic metre of sediment, (b) the mass per metre length of stream reach and (c) mass of element potentially solubilised from streambed sediments per unit stream reach. The element mass per unit

Sediment sampling in the Tweelopie Spruit

length of stream reach is based on the product of the mean concentration (mg/kg) of the seven sampling sites and a sediment density of 1400 kg/m$^3$ over a streambed area of 4 m$^2$ (2 × 2 m) to a uniform depth of 0.25 m.

## 5.5 Historical Observations

It is worth noting that the $SO_4$ concentrations in DWS monitoring boreholes[24] (Sect. 7.1) in the area were substantially higher in 1986 than in 2006. This is discussed in greater detail in Sect. 8.2.2. These circumstances, however, illustrate the fact that a mine water impact on the water resources of the area is not a recent or new phenomenon. It also illustrates the fact that under suitable conditions, a karst aquifer is a resilient and self-cleansing system in the medium to long term.

[24] These boreholes were originally established as potential emergency water supply production boreholes (Bredenkamp et al., 1986).

# Physical Hydrogeology

Philip J. Hobbs, Harrison Pienaar, and Sebinasi Dzikiti

## 1 Introduction

An early reference to the hydrogeology of the study area is found in Cooke (1969) who states simply that "The water table is regional and the underground lake is 186 feet below the crest of the hill with the ape-man quarry." The ape-man quarry is the location of the *Australopithecine* find 'Mrs Ples' at Sterkfontein Cave. The depth of 186 feet (56.7 m) places the 'water table' at an elevation of ~ 1430 m amsl for a crest elevation of ~ 1487 m amsl as interpolated from the 1:10,000 scale orthophoto map 2627BA5 Sterkfontein (2nd ed., 1987). Wilkinson places the top of the "hillock" at 1485 m amsl, and Martini et al. (2003) the top of the 'small hill' at 1491 m amsl. Recent information places the current (since ca. January 2011) elevation of the free water surface in Sterkfontein Cave at ~ 1439 m amsl (Sect. 3.2.1 and Fig. 10). The interpolated elevation of the Zwartkrans Spring located ~ 1300 m down-gradient/downstream (north-east) of the caves is ~ 1433 m amsl.

The most comprehensive regional groundwater study is considered to be that carried out in the mid-1980s by the then DWAF (Bredenkamp et al. 1986), primarily for the reason that it saw the execution of a regional gravimetric survey and the establishment of 38 large-diameter boreholes directed at evaluating the groundwater supply potential of portions of the dolomitic aquifers represented by the Steenkoppies and Zwartkrans compartments. A test pumping programme found 18 of the exploration boreholes to be high-yielding ($\geq$ 25 L/s), with a maximum recommended long-term yield of 100 L/s in three instances. The boreholes were sited primarily on the basis of gravimetric survey data (Sect. 4 in Chapter "Description of the Physical Environment"). From Table 1 it is evident that the Bredenkamp et al. (1986) study enumerated 326 private water supply boreholes that produced 131 groundwater level measurements (a conversion ratio of sites enumerated to water level measurements of only ~ 40%), and 29 groundwater chemistry analyses. The poor conversion ratio of the latter (~ 9% of the number of boreholes enumerated) is attributed to the volumetric focus of this and similar contemporary studies, with groundwater quality attracting only secondary consideration.

Bredenkamp et al. (1986) state that "Ground-water in the Zwartkrans compartment drains north-east to the Danielspruit and Kromdraai springs." It is presumed that reference to the Danielspruit Spring should, in fact, read Danielsrust Spring. Although not mentioned, it must also be presumed that the much stronger Zwartkrans Spring reportedly delivering 258 L/s (Bredenkamp et al. 1986), also drains this compartment under circumstances where the Danielsrust and Kromdraai springs[1] reportedly delivered only 3 and 28 L/s respectively (Bredenkamp et al. 1986). A later

[1] Field data and information sourced in this study indicates that the Kromdraai Spring(s) referred to by Bredenkamp et al. (1986) and Holland (2009) are in fact the Plover's Lake springs referred to in this report. This study identifies the Kromdraai Spring as a single groundwater source quite separate from the Plover's Lake springs; the plural in regard to the latter denotes a grouping of springs and seeps, as opposed to a single source. The Kromdraai Spring discharge of ~ 307 L/s is considerably greater than the ~ 60 L/s of the Plover's Lake springs. It is presumed, therefore, that the Kromdraai Springs referred to by Bredenkamp et al. (1986) and more recently by Holland (2008), Holland and Cobbing (2008) and Holland et al. (2009), in fact represent the Plover's Lake springs. Both Bredenkamp et al. (1986) and Holland and Cobbing (2008) show the position of the Kromdraai Spring(s) to be coincident with the Plover's Lake springs location. Further discussion of this 'discrepancy' is presented in Text Box 1.

P. J. Hobbs: Deceased

P. J. Hobbs · H. Pienaar (✉)
Smart Places, Council for Scientific and Industrial Research, Pretoria, South Africa
e-mail: hpienaar@csir.co.za

H. Pienaar
Hebei University of Engineering, Handan, China

S. Dzikiti
Department of Horticultural Science, Stellenbosch University, Stellenbosch, South Africa
e-mail: sdzikiti@sun.ac.za



P. J. Hobbs et al. (eds.), *Anatomy of a South African Karst Hydrosystem*, Cave and Karst Systems of the World,
https://doi.org/10.1007/978-3-030-95829-9_7

**Table 1** Groundwater data sources relevant to the COH obtained from the Bredenkamp et al. (1986) study

| Topocadastral farm | # Of private boreholes enumerated[a] | # Of water levels determined | # Of water quality determinations | # Of state boreholes drilled |
|---|---|---|---|---|
| Vlakplaats 160IQ | 141 | 75 | 5 | 12 |
| Vlakdrift 163IQ | 86 | 21 | 1 | 3 |
| Sterkfontein 173IQ | 49 | 22 | 4 | 15 |
| Zwartkrans 172IQ | 7 | 3 | 1 | 0 |
| Reydal 165IQ | 41 | 10 | 0 | 3 |
| Eljeesee agicultural holdings | 2 | 0 | 0 | 0 |
| Danielsrust 518JQ | 0 | 0 | 3 | 0 |
| Elandsvlei 249IQ | 0 | 0 | 3 | 0 |
| Kromdraai 520JQ | 0 | 0 | 8 | 0 |
| Rietfontein 522JQ | 0 | 0 | 4 | 0 |
| TOTAL | 326 | 131 | 29 | 33 |

[a]Data pertain only to the Zwartkrans Basin

study by Holland et al. (2009) reports discharges of 260, 3.2 and 371 L/s for the Zwartkrans, Danielsrust and Kromdraai springs, respectively. The order of magnitude discrepancy in regard to the Kromdraai Spring discharge is a confirmed typing error (K Witthüser, personal communication). In an earlier issue paper compiled for the IUCN Karst Working Group, Holland et al. (2005) discuss the geology, surface water and groundwater aspects of the COH in very broad and qualitative terms. A revision of this paper subsequently appeared in print in the WRC/IUCN/SAKWG (2010) report.

The portion of the Riet Spruit upstream of the Oaktree area, including its tributary the Tweelopie Spruit, has recently been the subject of hydrogeological studies (Hobbs and Cobbing 2007; Hobbs 2008b) that have helped define the impact of acid mine drainage on the groundwater resources in the LoD in the upper reaches of this drainage, as well as downstream of its confluence with the Riet Spruit at Glen Almond. The study by Hobbs and Cobbing (2007) enumerated 54 geosites (6 springs, 7 mine shafts and 41 boreholes), the latter two providing 48 potentiometric measurements. These data facilitated the construction of a potentiometric contour map for the south-western portion of the study area (Fig. 1). The data also informed a critique of the Bredenkamp et al. (1986) description of compartments (and subcompartments) and groundwater drainage patterns (Hobbs and Cobbing 2007).

An inspection of aquifer transmissivity (T) data generated by the Bredenkamp et al. (1986) study from pumping tests performed on exploration boreholes in the Zwartkrans Basin indicates a zone of higher transmissivity as defined in Fig. 3. The zone is demarcated by 13 exploration boreholes that returned T-values greater than ~ 1000 $m^2$/d (Table 2). The geometry of this zone (Fig. 3), notably its width, slightly more northerly position and SW–NE orientation, suggests that it coincides with the position of the chert-rich Monte Christo Formation and the generally greater water-bearing potential associated with this formation (Sect. 4 in Chapter "Description of the Physical Environment"). The Bredenkamp et al. (1986) study, however, also reports T-values as low as 1 $m^2$/d. This corroborates the substantial heterogeneity and anisotropy that characterises the karst aquifers of the region as found by numerous contemporary studies (e.g. Kafri et al. 1986; Leskiewicz 1986; Hobbs 1988a; Kuhn 1989) and synthesised by Hobbs (1988b). Transmissivity values ranging through five orders of magnitude (< $10^1$ to > $10^4$ $m^2$/d) are common to these aquifers.

A study by Leyland et al. (2008) applied the concept of vulnerability mapping to the COH. This study built on various internationally recognised methods and approaches to develop a South African version that takes account of local conditions. An improvement on this approach, including the results of its application to the COH, is presented by Leyland (2010). An investigation by Holland (2007) applied the concept of groundwater resource directed measures (GRDM) to the COH.

The Holland (op. cit.) study reports estimated average annual groundwater recharge volumes of 16.8 $mm^3$ (from Bredenkamp et al. 1986) and 14.3 $mm^3$ (from Krige 2006) for the Zwartkrans Basin. The Holland (op. cit.) study puts forward its own estimates of 16.6 $mm^3$/a for the Zwartkrans Basin and 13.4 $mm^3$/a for the Tweefontein Basin. The veracity of these estimates is explored in the discussion on groundwater recharge in Sect. 2.12. In addition to the 26 geosites provided by the Hobbs and Cobbing (2007) study, Holland (2007) enumerated a further 41 geosites (boreholes) in the COH (including 16 DWS boreholes) for which potentiometric measurements are given.

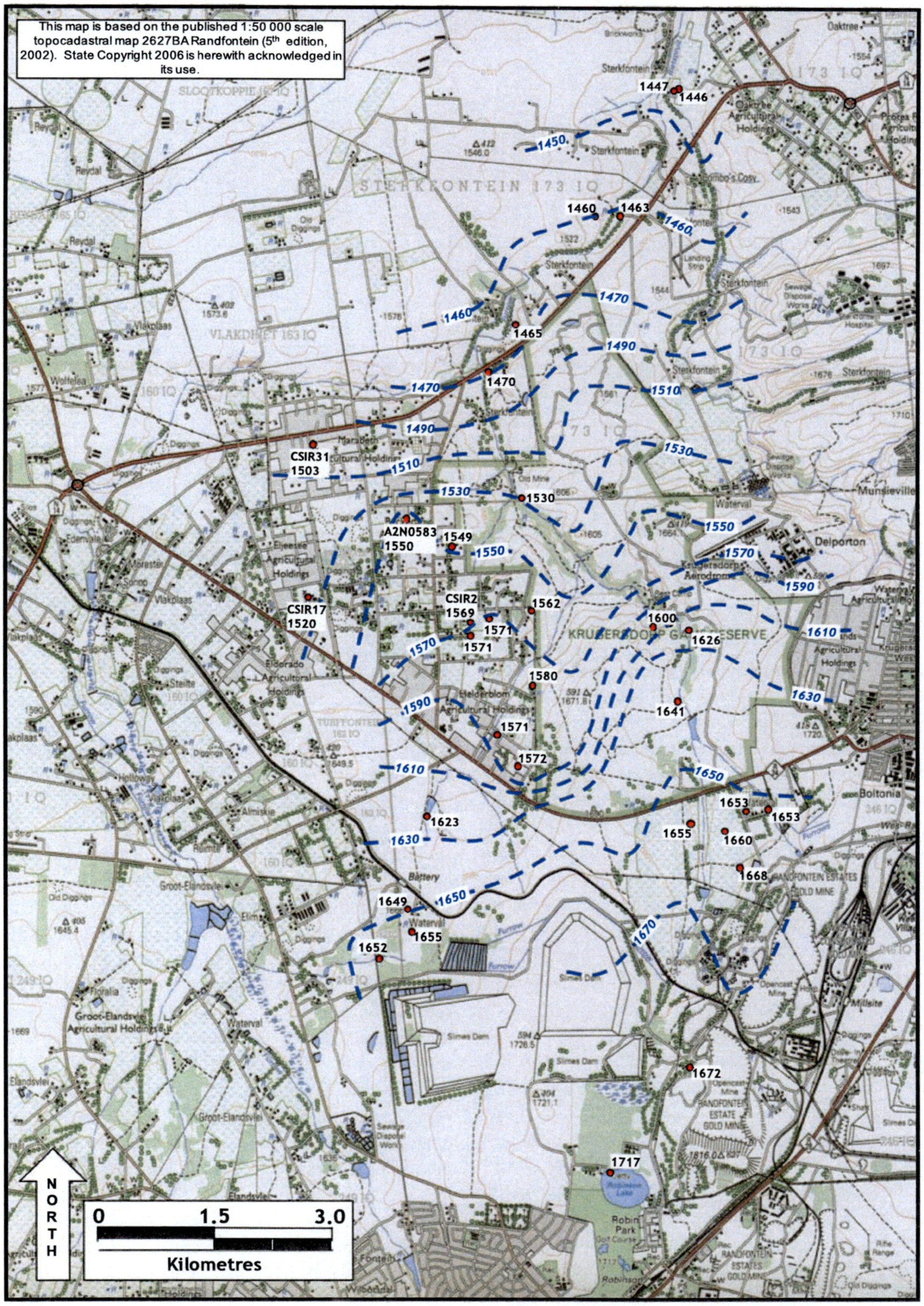


**Fig. 1** Potentiometric map of the south-western portion of the study area (modified after Hobbs and Cobbing 2007)

**Table 2** DWS exploration boreholes with transmissivity values greater than ~ 1000 m²/d (from Bredenkamp et al. 1986)

| DWS exploration borehole # | DWS Geosite # | Coordinates | | Transmissivity (m²/d) |
|---|---|---|---|---|
| | | Latitude (dd. ddddd°S) | Longitude (dd. ddddd°E) | |
| G36320 | A2N0589 | *26.01632* | *27.70823* | 3000 |
| G36321 | A2N0593 | *26.02319* | *27.68882* | 13,900 |
| G36323 | A2N0590 | 26.02441 | 27.71527 | 20,500 |
| G36324 | A2N0592 | *26.03134* | *27.69753* | 11,200 |
| G36326 | A2N0588 | *26.03766* | *27.70903* | 3700 |
| G36327 | A2N0587 | 26.03668 | 27.71534 | 900 |
| G36328 | A2N0595 | *26.04684* | *27.67811* | 13,600 |
| G36329 | A2N0596 | *26.05139* | *27.67215* | 2400 |
| G36330 | A2N0597 | *26.05403* | *27.68415* | 5400 |
| G36331 | A2N0586 | 26.04761 | 27.70903 | 2100 |
| G36332 | A2N0598 | 26.06581 | 27.66622 | 4300 |
| G36335 | A2N0599 | *26.06585* | *27.66623* | 1500 |
| G36336 | A2N0581 | *26.07942* | *27.66628* | 1200 |

*Note* Italicised coordinate values denote DWS coordinates not verified in this study

## 2 Basin (Compartment) Definition

The phenomenon of compartmentalisation in the South African Neo-Proterozoic (2.64–2.50 Ga) telogenetic karst aquifers developed in diagenetically mature, well-compacted carbonate rocks, is well known (Martini and Kavalieris 1976). Osborne (2003) reports a similar phenomenon for the Palaeozoic (545–250 Ma) karst of eastern Australia. The phenomenon transforms a heterogeneous and anisotropic aquifer of considerable regional extent into smaller portions, allowing for more meaningful subregional analysis and understanding of the hydrogeologic environment. White (1993) uses the term 'karstic ground-water basins', recognition of which in the COH reveals a degree of compartmentalisation that contributes significantly to a more informed understanding of the groundwater environment.

The GRDM focus of the Holland (2007) study (Sect. 1) necessarily included the recognition of groundwater resource units (GRUs). These were identified as the Zwartkrans and Tweefontein compartments (Fig. 2). Both Holland (2007) and Holland and Cobbing (2008) accepted the subdivision of the Zwartkrans Basin into subunits defined by dykes originally put forward by Bredenkamp et al. (1986), and later adopted by JFA (2006), van Biljon (2006) and Krige (2009). Holland (2007) also presents a potentiometric map for the Zwartkrans Basin, which map identifies dykes (including sections of dykes) as variously representing actual or perceived impermeable or permeable hydrogeologic boundaries.

A later evaluation of the "West Rand Dolomite Compartments" by Holland (2009) for the DWS presents a more elaborate recognition of compartmentalisation of the karst formations in the region (Fig. 2). The revised rendition still identifies three compartments represented by the Steenkoppies Basin in the south-west, the Zwartkrans Basin in the centre and the Tweefontein Basin in the north-east, each of these being defined as groundwater management areas (GMAs). The GMAs are subdivided into groundwater management units (GMUs), each of which is further subdivided into GRUs. The scope of this subdivision, which is based simply on the geographic distribution of mapped geological structures (i.e. not considering groundwater potentiometry, spring locations and discharges, groundwater chemistry, etc. to any significant degree), is reflected in Table 3.

The tenuous and arguably over-complicated (consider the 51 karst GRUs as per Table 3) definition of the karst environment in the study area in terms of compartmentalisation (both compartments and subcompartments) by Holland (2009), and replicated by Meyer (2014), precipitated closer inspection. Another factor prompting such review is the seemingly different interpretation of the term compartment by Holland (2009). The review focusses on the karst environment that supports the COH property, and therefore excludes the Steenkoppies Basin. The latter has been the subject of a study by Holland et al. (2009) which has adequately elucidated the karst environment of this compartment. The outcome in regard to the karst environment of the less studied and more poorly understood WHS property, presented in Fig. 3, recognises a total of eleven compartments,[2] two of which comprise subcompartments. The revised physical hydrogeologic layout is informed by a

[2] Distributed between the Zwartkrans and Tweefontein compartments of Holland (2009), and excluding the Steenkoppies Compartment.

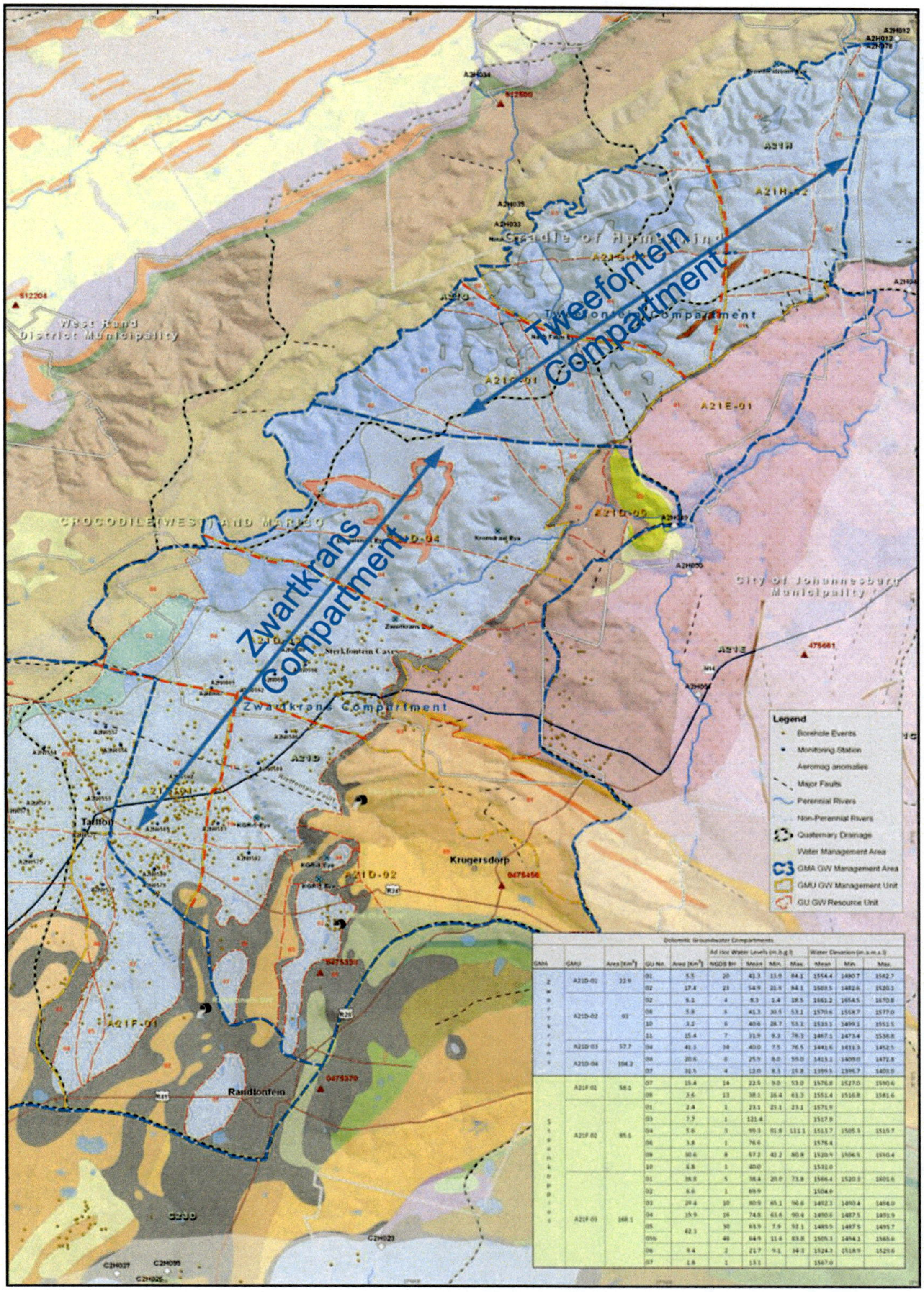


| GMA | GMU | Area [Km²] | GU No. | Area [Km²] | NGDB BH | Ad Hoc Water Levels (m.b.g.l) Mean | Min. | Max. | Water Elevation (m.a.m.s.l) Mean | Min. | Max. |
|---|---|---|---|---|---|---|---|---|---|---|---|
| Zwartkrans | A21D-01 | 22.9 | 01 | 5.5 | 20 | 41.3 | 13.9 | 84.1 | 1554.4 | 1490.7 | 1582.7 |
| | | | 02 | 17.4 | 23 | 54.9 | 21.9 | 94.1 | 1503.5 | 1482.6 | 1520.1 |
| | A21D-02 | 93 | 02 | 6.1 | 4 | 8.3 | 1.4 | 18.5 | 1661.2 | 1654.5 | 1670.8 |
| | | | 08 | 5.8 | 5 | 41.3 | 30.5 | 53.1 | 1570.6 | 1558.7 | 1577.0 |
| | | | 10 | 3.2 | 6 | 40.6 | 28.7 | 53.1 | 1535.1 | 1499.1 | 1551.5 |
| | | | 11 | 15.4 | 7 | 31.9 | 8.3 | 78.3 | 1467.1 | 1473.4 | 1538.8 |
| | A21D-03 | 57.7 | 04 | 41.1 | 34 | 40.0 | 7.5 | 76.5 | 1441.6 | 1411.3 | 1452.5 |
| | A21D-04 | 104.2 | 06 | 20.6 | 8 | 25.9 | 8.0 | 59.0 | 1413.1 | 1409.0 | 1472.8 |
| | | | 07 | 10.5 | 4 | 13.0 | 8.1 | 15.8 | 1399.5 | 1395.7 | 1403.0 |
| Steenkoppies | A21F-01 | 58.1 | 07 | 15.4 | 14 | 22.5 | 9.0 | 53.0 | 1576.8 | 1527.0 | 1590.6 |
| | | | 08 | 3.6 | 13 | 38.1 | 16.4 | 61.3 | 1551.4 | 1516.8 | 1581.6 |
| | A21F-02 | 85.1 | 01 | 2.4 | 1 | 23.1 | 23.1 | 23.1 | 1571.9 | | |
| | | | 03 | 7.7 | 1 | 121.4 | | | 1517.8 | | |
| | | | 04 | 5.6 | 3 | 99.3 | 91.8 | 111.1 | 1513.7 | 1505.3 | 1515.7 |
| | | | 06 | 3.8 | 1 | 76.6 | | | 1578.4 | | |
| | | | 09 | 30.6 | 8 | 57.2 | 42.2 | 80.8 | 1520.9 | 1506.9 | 1550.4 |
| | | | 10 | 6.8 | 1 | 40.0 | | | 1532.0 | | |
| | A21F-03 | 168.1 | 01 | 38.8 | 5 | 38.4 | 20.0 | 73.8 | 1566.4 | 1520.1 | 1601.6 |
| | | | 02 | 6.6 | 1 | 69.9 | | | 1504.0 | | |
| | | | 03 | 29.4 | 10 | 80.9 | 65.1 | 96.6 | 1492.3 | 1490.4 | 1494.0 |
| | | | 04 | 15.9 | 16 | 74.8 | 61.6 | 90.4 | 1490.6 | 1487.5 | 1491.9 |
| | | | 05 | 62.3 | 30 | 63.9 | 7.9 | 92.1 | 1489.5 | 1487.5 | 1495.7 |
| | | | 05b | | 40 | 64.9 | 11.6 | 83.8 | 1505.1 | 1494.1 | 1565.6 |
| | | | 06 | 9.4 | 2 | 21.7 | 9.1 | 34.3 | 1524.3 | 1518.9 | 1529.6 |
| | | | 07 | 1.8 | 1 | 13.1 | | | 1567.0 | | |

**Fig. 2** Previous definition of dolomitic basins/subcompartments as proposed by Holland (2007), and later modified by Holland (2009) and replicated by Meyer (2014)

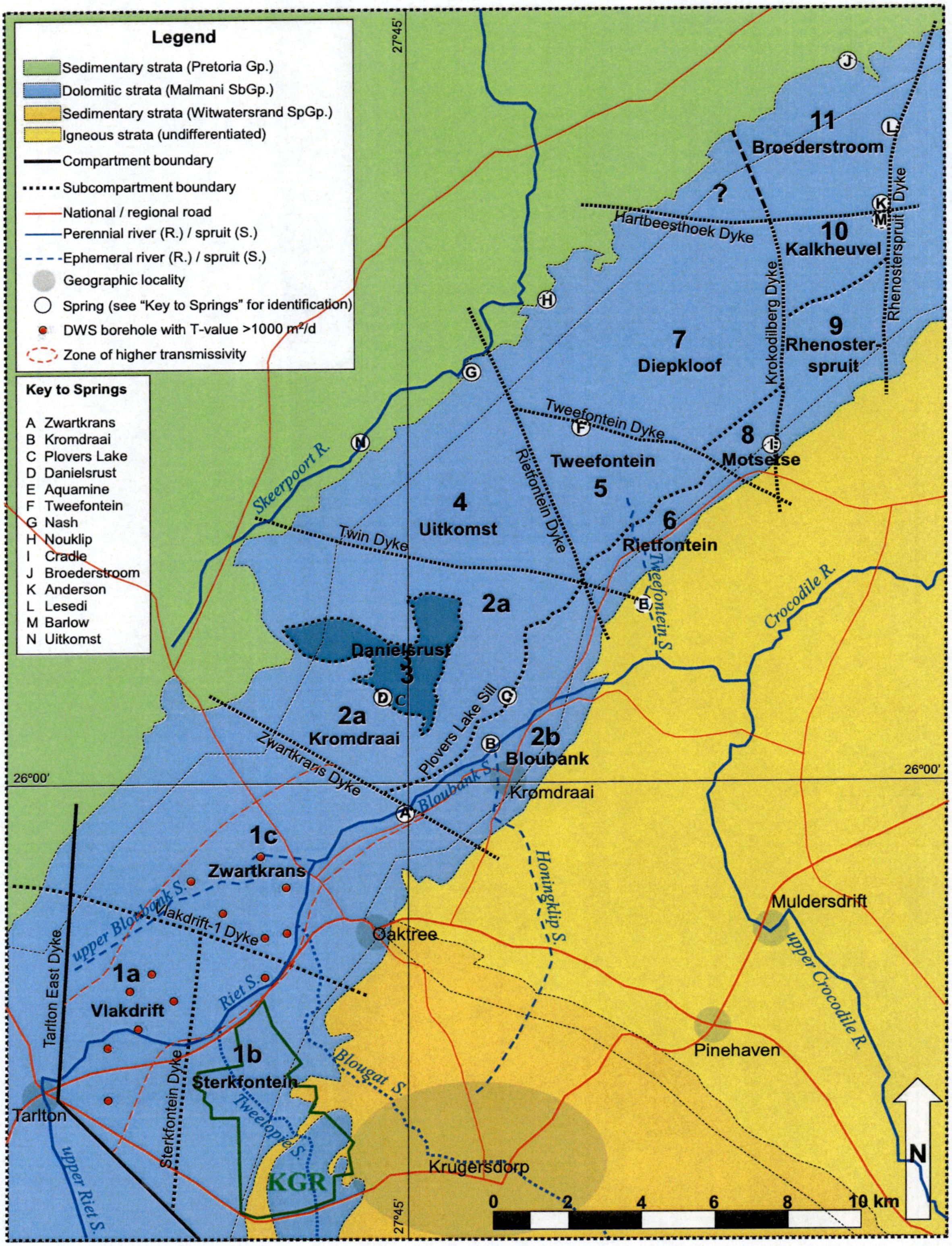


**Fig. 3** Current understanding of dolomitic compartment/subcompartment distribution showing relationship to the spring outlets draining the hydrogeologic units

**Table 3** Enumeration of compartment and subcompartment definition from Fig. 2

| Groundwater management area | Number of subdivisions | | |
|---|---|---|---|
| | Groundwater management unit (undifferentiated[a]) | Groundwater resource unit | |
| | | Karst | Non-karst |
| Steenkoppies Basin | 3 | 20 | 7 |
| Zwartkrans Basin | 5 | 15 | 11 |
| Tweefontein Basin | 4 | 16 | 1 |
| Total | 12 | 51 | 15 |

[a] Undifferentiated in regard to 'karst' or 'non-karst' environment implies a 'mixed' hydrogeologic regime, i.e. management units straddle lithologic boundaries to encompass different geologic strata

combination of groundwater level data, spring locations and discharges, groundwater chemistry and geological structures. It is described and discussed in greater detail in the following sections, with the schematic layout (Fig. 3) providing a visual reference for this discussion.

## 2.1 Zwartkrans Basin

The Zwartkrans Basin is drained by the Zwartkrans Spring located ~ 1.4 km north-east (downstream) of Sterkfontein Cave on the farm Zwartkrans 172IQ. The compartment spans an area of ~ 9800 ha bounded in the south-west by the Tarlton Dyke system, in the north-east by the Zwartkrans Dyke, and along the south-eastern boundary by the geological contact between the Malmani Subgroup and Witwatersrand Supergroup quartzitic strata. To the north-west, the contact between the Malmani Subgroup dolomite and the overlying Pretoria Group sedimentary strata represents a pseudo-boundary under circumstances where the dolomite continues at depth beneath the latter.

The Zwartkrans Basin was subdivided by Bredenkamp et al. (1986) into nine subcompartments. The subdivision was based mainly on sharp transitions in water level over short distances, with partial verification of bounding dyke structures using ground and airborne geophysical information. This reasoning rests on the premise of very weak/flat hydraulic gradients in each subcompartment. These were also put forward by Bredenkamp et al. (1986) as the reason for not compiling a water level contour (potentiometric) map. It is notable, however, that the difference in 'representative water level' between five contiguous subcompartments is ~ 70 m over a distance of only ~ 8 km. This is considered to be an artefact of the comparatively sparse set of water level data used to define the subcompartments, and which raises doubt over the recognition of an overly disrupted groundwater flow pattern informed by perceived and unverified barrier boundaries (Hobbs and Cobbing 2007). For example, one of the Bredenkamp et al. (1986) subcompartments is recognised in this study as a separate compartment, namely the Krombank Basin (Sect. 2.2).

A simplified subdivision of the Zwartkrans Basin is proposed. This comprises only three subcompartments as described in the following sections.

### 2.1.1 Vlakdrift Subcompartment

The Vlakdrift Subcompartment (~ 2360 ha) forms the upper reaches of the Zwartkrans Basin. Its western margin is the Tarlton East Dyke that marks the boundary with the Steenkoppies Basin to the west (Holland et al. 2009). The eastern boundary is represented by the roughly N–S striking Sterkfontein Dyke, the northern boundary by the E–W striking Vlakdrift-1 Dyke, and the southern margin by the lithological contact with the older Witwatersrand Supergroup. Both the Sterkfontein Dyke[3] and the Vlakdrift-1 Dyke are recognised by Holland et al. (2009) as hydrogeological boundaries. An upward step in water level elevation of 45–50 m from west to east across the Sterkfontein Dyke characterises the influence of this boundary on the potentiometric surface in the southern portion of the Vlakdrift Subcompartment (1a) and the adjoining Sterkfontein Subcompartment (1b). This is reflected in the groundwater elevations in boreholes CSIR17 (1520 m amsl) and CSIR2 (1569 m amsl), and CSIR31 (1503 m amsl) and A2N0583 (1550 m amsl) as shown in Fig. 1.

The main recharge to this subcompartment is from rainfall, although the anthropogenic (and allogenic) contribution from excess surface flow via the upper Riet Spruit past Tarlton, as is described in Sect. 5.2.1 in Chapter "Physical Hydrology", represents a significant secondary input. The outflow from this subcompartment northwards into the Zwartkrans Subcompartment is a postulated subsurface drainage that coincides with a zone of higher transmissivity as described in Sect. 1 and illustrated in Fig. 3, as well as

[3] Note that Holland et al. (2009) assigned the name Beckedan Barrier to the Vlakdrift-1 Dyke, whereas the names as applied in this study (Fig. 63 in Chapter "Chemical Hydrology") conform to those assigned by Bredenkamp et al. (1986). It also seems probable that the western extension of the Vlakdrift-1 Dyke might form part of the Maloney's Eye Dyke recognised by Bredenkamp et al. (1986).

significant grike features that intersect the Vlakdrift-1 Dyke (Sect. 6 and Fig. 18).

### 2.1.2 Sterkfontein Subcompartment

The Sterkfontein Subcompartment (~ 2450 ha) forms the south-central portion of the Zwartkrans Basin. It receives primarily allogenic recharge in the form of mine water delivered via the Tweelopie Spruit. The boundaries of this subcompartment are defined by the Sterkfontein Dyke in the west, the Vlakdrift-1 Dyke to the north, and the lithologic contact with the older Witwatersrand Supergroup strata to the south-east. The discontinuity in the potentiometric surface (water table) in the Oaktree area created by the Vlakdrift-1 Dyke is revealed by the following:

- potentiometric elevation differences of 15–20 m between boreholes AM1 and ME1 (at 1467 m amsl) to the south, and boreholes PD1 and CSIR9 (at 1450 m amsl) to the north; and
- the apparent northerly displacement of the mine water impact on karst groundwater quality (as revealed by elevated SEC and $SO_4$ levels) in the Oaktree area.

The outflow of this subcompartment over/through the Vlakdrift-1 Dyke into the Zwartkrans Subcompartment is similar to that of the adjoining Vlakdrift Subcompartment, namely subsurface flow via a zone of higher transmissivity that intersects the structure (Fig. 3). The principal groundwater flow vector at this position is north-northeast, and is considered to be partly driven by an artificial groundwater 'mounding' effect created by a combination of (a) the accumulation of infiltrating mine water lost from the Riet Spruit as described in Sect. 5.2.1 in Chapter "Physical Hydrology" and (b) this water damming up behind the Vlakdrift-1 Dyke (Fig. 4). The latter forms the boundary between the Sterkfontein and Zwartkrans subcompartments (see also Fig. 3 and Fig. 18), and influences the elevation of the water table either side of this feature as is shown for the years 2007 and 2011. The ~ 12 m step in water level from west to east across the Zwartkrans Dyke is also illustrated.

The profile further shows the intersection of the potentiometric surface with the land surface (represented by the riverbed of the Bloubank Spruit) at the downstream end of the Zwartkrans Subcompartment. This supports the observation (Sect. 2.1.3) that the Zwartkrans Subcompartment is drained not only by the Zwartkrans Spring, but also by groundwater losses to the effluent section of the Bloubank Spruit stream channel for some distance upstream of the spring. These circumstances also explain the difficulty in accurately gauging the discharge of the Zwartkrans Spring, and therefore also the wide range of values obtained for this feature (Sect. 4.1).

### 2.1.3 Zwartkrans Subcompartment

The Zwartkrans Subcompartment covers ~ 51% (~ 4990 ha) of the Zwartkrans Basin. Its boundaries are defined by the Vlakdrift-1 Dyke to the south, the lithologic contact with the younger Pretoria Group strata to the north-west, the lithologic contact with the older Witwatersrand Supergroup strata to the south-east, and the Zwartkrans Dyke to the north-east. The latter gives rise to the Zwartkrans Spring. It also gives rise to diffuse (and difficult to quantify) seepage in an area on the opposite (left) bank that is collected and channelled into trout dams on the Boland Farm property on Danielsrust 518JQ.

Groundwater rest levels measured in three boreholes that straddle the dyke structure reveal that it marks a step down in groundwater level, from west to east across the structure, of ~ 13 m. The potentiometric surface in the Zwartkrans Basin immediately upstream of the dyke occupies an elevation of ~ 1433 m amsl (the approximate elevation of the Zwartkrans Spring), and an elevation of ~ 1420 m amsl in the Krombank Basin (Sect. 2.2) immediately downstream of the dyke. It is postulated that the zone of higher transmissivity (Fig. 3) might extend in a north-easterly direction to intersect the Zwartkrans Dyke, thus also providing a subsurface outflow component into the Krombank Basin.

An equally over-looked discharge component is represented by the intersection of the water table with the streambed of the Bloubank Spruit for some distance (~ 2 km) upstream of the Zwartkrans Spring. Flow measurements described in Sects. 5.2.1 and 5.2.3 in Chapter "Physical Hydrology" indicate a surface water 'gain' of as much as 23 ML/d (266 L/s) in this reach of the Bloubank Spruit. This 'gain' is considered to be provided by diffuse in-channel (streambed) groundwater resurgence, in contrast to the 'point-source' off-channel Zwartkrans Spring. Note in this regard that monitoring borehole GP00314 sunk within 50 m of the Bloubank Spruit ~ 700 m upstream of the Zwartkrans Spring revealed unconsolidated sedimentary material down to 14 m bs, with a water table at ~ 2 m bs, i.e. a saturated thickness of ~ 12 m. Together with the ~ 136 L/s discharge of the Zwartkrans Spring (Sect. 4.1 and Table 13), the aggregate groundwater discharge from the Zwartkrans Subcompartment exceeds 300 L/s, and on occasion possibly even 400 L/s. Given the hydraulic continuity with the 'upstream' Sterkfontein and Vlakdrift subcompartments, this discharge represents that of the Zwartkrans Basin as a whole. For this reason, it necessarily excludes the potential contribution 'lost' to groundwater abstraction, e.g. for agriculture, in this compartment. Estimates of this use range from ~ 9 to 26 $mm^3/a$ (Sect. 9), which equate to values of ~ 285 to 824 L/s.

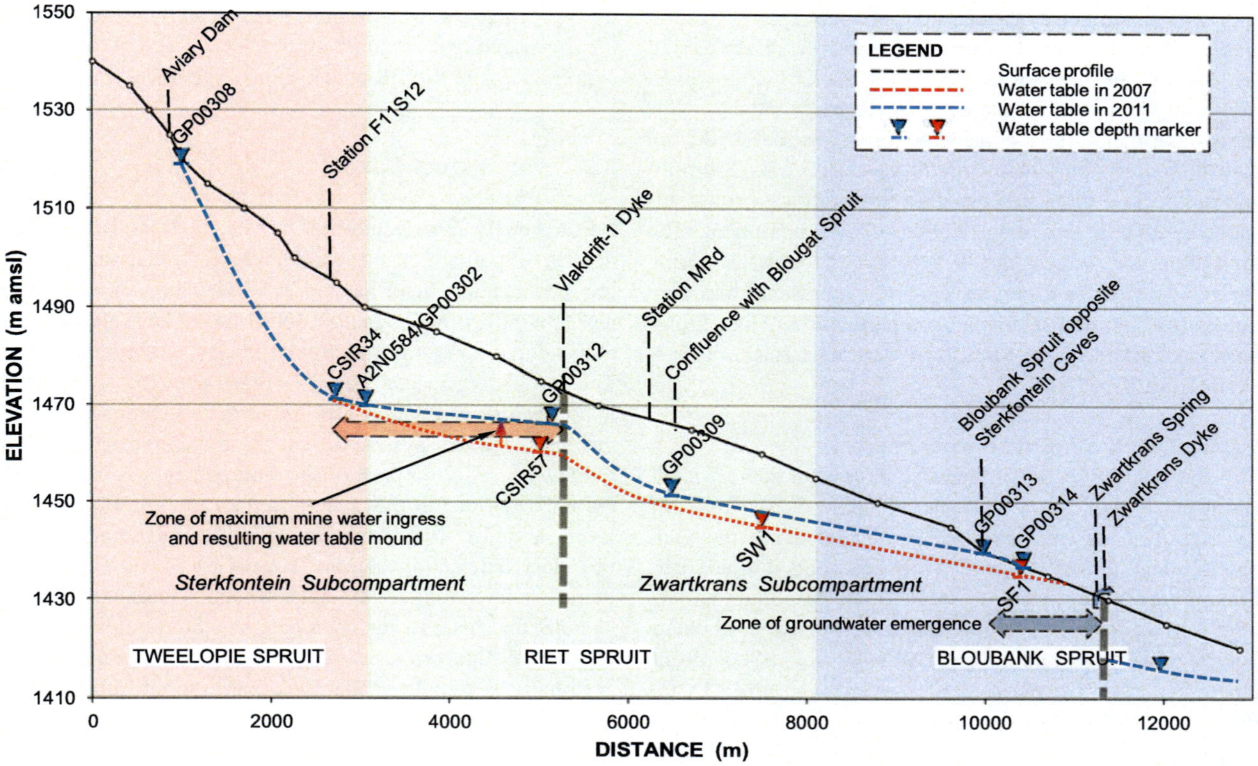


**Fig. 4** Longitudinal surface and water table profile along the course of the Tweelopie, Riet and Bloubank Spruits from Kemp's Cave in the Krugersorp Game Reserve to beyond the Zwartkrans Spring

Although the Zwartkrans Subcompartment hosts the Sterkfontein Cave fossil site, the latter is located along the south-eastern margin of this subcompartment. It is postulated that this places the Sterkfontein fossil site to the south and outside the main groundwater flow vector that collects the subsurface discharge of both the Vlakdrift and Sterkfontein subcompartments, and directs these towards the Zwartkrans Spring along the trace of the 'dry' Bloubank Spruit upstream of its confluence with the Riet Spruit. These circumstances recognise the possible existence of a subsurface 'thalweg'[4] associated with the relict surface expression of the ephemeral upper reaches of the Bloubank Spruit. The most visual discharge feature of this subcompartment and its contributing upstream subcompartments remains the Zwartkrans Spring (Sect. 4.1).

[4] See GLOSSARY. Used here to describe a preferential subsurface flowpath such as a large channel (or network of channels) that might develop and occur in a karst aquifer and act as a "subsurface groundwater stream" (from Martini, 2006, p 664) or a "master drain" (after White, 1993).

## 2.2 Krombank Basin

This compartment (total area ~ 5080 ha) comprises two subcompartments identified as the Kromdraai Subcompartment drained by the Plover's Lake springs and the Kromdraai Spring located on the farm Kromdraai 520 JQ, and the Bloubank Subcompartment that straddles the Bloubank Spruit. The name 'Krombank' (a combination of the names Kromdraai and Bloubank) used to describe this compartment denotes the degree of hydraulic continuity considered to exist between these subcompartments.

### 2.2.1 Kromdraai Subcompartment

The Kromdraai Subcompartment covers an area of ~ 3470 ha (excluding the footprint of the separate superimposed barré associated with the Danielsrust Basin) bounded to the south-west by the Zwartkrans Dyke, to the north-west by the geological contact between the Malmani Subgroup dolomite and the overlying Pretoria Group sedimentary strata, to the north-east by the Twin Dyke structure, and along the south-eastern boundary by the Plover's Lake Sill, a north-westerly dipping syenite intrusion in the dolomite. This sill marks the common boundary with the Bloubank Subcompartment.

The potentiometric surface intersects the land surface at an elevation of ~ 1419 m amsl at the Plover's Lake Spring (s). The discharge of these springs (~ 60 L/s) is, however, too small to account for the postulated surface extent of this subcompartment. It is considered probable that structural discontinuities along the trace of the Plover's Lake Sill form additional flow paths for groundwater discharge from this subcompartment into the Bloubank Subcompartment. The fault zone on which the ~ 307 L/s Kromdraai Spring (Sect. 4.3) is located, almost certainly represents one such outlet. It is likely that this subcompartment receives a proportion of its recharge from the Zwartkrans Basin. This is explored in Sect. 2.12

### 2.2.2 Bloubank Subcompartment

The Bloubank Subcompartment straddles the course of the Bloubank Spruit for a distance of some 8 km from the Zwartkrans Dyke to where this drainage leaves the karst environment in the vicinity of Brookwood Trout Farm. Encompassing an area of ~ 1610 ha, its south-eastern boundary is formed by the contact between the Malmani Subgroup dolomite and the older Witwatersrand Supergroup quartzitic strata. The north-western margin is formed by the Plover's Lake Sill intrusion that follows a trace which runs subparallel to that of the south-eastern boundary. This subcompartment receives:

- the overflow from the Zwartkrans Basin via the Zwartkrans Spring and groundwater resurgence in the Bloubank Spruit upstream of the Zwartkrans Spring (Sect. 5.2.3 in Chapter "Physical Hydrology", Sect. 7.4.1, and Fig. 32 in Chapter "Physical Hydrology");
- discharge from the Kromdraai Subcompartment along its north-western margin via structural discontinuities in the bounding Plover's Lake Sill (Sect. 2.2.1); and
- infiltrating river water from losing sections of the Bloubank Spruit between the Zwartkrans and Kromdraai springs.

As shown in Table 4, the combined discharge of the Kromdraai and Plover's Lake springs (339 L/s excluding the Danielsrust Spring contribution), returns an unrealistically high equivalent recharge of ~ 30% of MAP as representing the 'recharge input' on the Krombank Basin footprint of 5080 ha. Circumstances that explain this anomaly are discussed in Sect. 2.12.

Anomalous groundwater rest level elevations in the extreme north-eastern corner of the subcompartment (e.g. boreholes VW1 and PvW1) are attributed to further sub-compartmentalisation formed by the NW–SE striking diabase Rietfontein Dyke at this location (Fig. 3). The small area (~ 30 ha) encompassed by this subcompartment, however, negates recognition thereof as an important part of the hydrogeologic regime in the COH, and it has been aggregated with the Bloubank Subcompartment.

## 2.3 Danielsrust Basin

Located on the farm Danielsrust 518JQ, the Danielsrust Basin defines a perched karst aquifer (barré[5]) that supports a groundwater rest level located at an elevation of > 1490 m amsl, i.e. 50–70 m above the 1420–1440 m amsl elevation in the regional karst aquifer defined by the Kromdraai Basin. The cross-section in Fig. 17 illustrates the postulated hydrogeological relationship that describes the position of this groundwater system in the regional hydrogeological framework. The footprint of this compartment is defined by the dolerite outcrop (not recognised by Obbes 2001) which forms a quasi 'ring' structure (Fig. 3), and straddles the boundary between the farms Danielsrust 518JQ (west) and Sterkfontein 519JQ (east). This structure, measured around its outer perimeter to accommodate its entire footprint on the Kromdraai Subcompartment, encompasses an area of ~ 740 ha. It is postulated that the dolerite forms an impermeable basin that serves to contain groundwater which decants naturally at the Danielsrust Spring with an elevation of ~ 1490 m amsl. According to the landowner, the long-term discharge of this spring is 100,000 L/h (28 L/s) (P van der Merwe, personal communication). The establishment of three water supply boreholes within this compartment has added an artificial loss component to the water balance of this aquifer.

In order for the ~ 740 ha extent of the Danielsrust Basin to support a spring discharge of 28 L/s, it would be required that rainfall recharge amounts to ~ 17% of a mean annual precipitation (MAP) of 710 mm. This value compares favourably with the results of four different calculation methods used in the earlier study by Bredenkamp et al. (1986). These yielded recharge values in the range 15–21% of an MAP of 630 mm. Further support for recognition of the Danielsrust Basin as a distinctly separate hydrosystem is found in water chemistry data. Firstly, the SEC field values of 23–26 mS/m associated with the Danielsrust Spring water is notably lower than the 40–46 mS/m associated with the dolomitic groundwater in the wider area. Secondly, the stable isotope results (Holland 2007 and this study as reported in Sect. 5 in Chapter "Physical Hydrology") compare favourably with those of 'pristine' karst groundwater in the study area.

The above circumstances place all of the responsibility for the protection of this groundwater resource on those landowners of property located within the Danielsrust Basin. The greatest threat to this resource is considered to be that

[5] See GLOSSARY.

**Table 4** Summary information of groundwater compartments in the COH

| Basin | Sub-compartment | Extent (ha) | | Outflow feature | Discharge | | | Recharge | |
|---|---|---|---|---|---|---|---|---|---|
| | | | | | (L/s) | | (ML/d) | mm | %MAP[a] |
| Zwartkrans | | | 9800 | Zwartkrans Sp. | | 136 | 11.75 | 130 | ~ 18 |
| | | | | Resurgence[b] | | 270 | 23.33 | | |
| Krombank | Kromdraai | 3470 | 5080 | Plover's Lake Sp. | 60 | 339[c] | 29.23[c] | 210 | ~ 30 |
| | Bloubank | 1610 | | Kromdraai Sp. | 279 | | | | |
| Danielsrust | | | 740 | Danielsrust Sp. | | 28 | 2.42 | 119 | ~ 17 |
| Uitkomst | | | 2860 | Nash Sp. | | 130 | 11.23 | 143 | ~ 20 |
| Tweefontein | | | 1160 | Tweefontein Sp. | | 30 | 2.59 | 81 | ~ 11 |
| Rietfontein | | | 540 | Aquamine Sp. | | < 2 | < 0.17 | 11 | ~ 2 |
| Diepkloof | | | 3800 | Nouklip Sp. | | 143[d] | 12.36[d] | 119 | ~ 17 |
| Motsetse | | | 270 | Cradle Sp. | | 2 | 0.17 | 23 | ~ 3 |
| Rhenosterspruit | | | 850 | Not established | | – | – | – | – |
| Kalkheuvel | | | 720 | Barlow Sp. | | 2 | 0.17 | 48 | ~ 1 |
| Broederstroom | | | 1810 | Broederstroom Sp. | | 21 | 1.81 | 49 | ~ 7 |
| | | | | Lesedi Sp. | | 2 | 0.17 | | |
| | | | | Anderson Sp. | | 5 | 0.43 | | |
| (see Fig. 3) | | | 450 | Not established | | – | – | – | – |
| TOTAL AVERAGE | | 28,080 | | | 1110 | | 95.83 | 93 | 12.6 |

[a] Mean annual precipitation = 710 mm
[b] See discussion in Sect. 4.1 regarding groundwater resurgence upstream of the spring
[c] The Plover's Lake and Kromdraai springs are combined to represent the theoretical groundwater discharge of the Krombank Basin excluding the autogenic recharge of 28 L/s contributed by the Danielsrust Spring
[d] Excludes the autogenic recharge of 30 L/s contributed by the Tweefontein Spring

associated with on-site sanitation facilities, particularly those which carry a potentially elevated load. The greatest part of the compartment supports game farming, which affords most of this resource an excellent measure of protection from pollution. However, the adjacent commercial tourist lodge and camp facility on the farm Sterkfontein 519JQ represents a concern that needs to be considered.

The location of the lodge near the 'downstream' end of the compartment means that much of this resource (i.e. the area located upstream of the camp) is 'immune' to contamination from this source. This does not apply to the spring located only ~ 500 m down-gradient of the lodge. Furthermore, abstraction from the lodge water supply borehole located immediately upstream of the spring exacerbates this threat by drawing groundwater closer. Although the spring does not currently serve a water supply function for any purpose, protection of its quality is mandated by Sect. 19 of the NWA (Act No 36, 1998), "Prevention and remedying effects of pollution.", which requires of a landowner, person in control of land or person occupying or using land to take all reasonable measures to prevent pollution of a water resource from occurring, continuing or recurring. These circumstances identify the Danielsrust Basin as a candidate groundwater management unit (GMU).

Although recognition of the Danielsrust Basin as a 'stand-alone' groundwater resource unit is fully justified, it could also be aggregated with the Krombank Basin in regard to both extent and yield/recharge. This is possible because the full discharge of the Danielsrust Spring drains into the Kromdraai Subcompartment of the Krombank Basin, manifesting an autogenic recharge component that contributes to the discharge of this karst basin. The Krombank and Danielsrust compartments are listed as separate basins in Table 4.

## 2.4 Uitkomst Basin

This compartment encompasses the southern portion of the ~ 7300 ha John Nash Nature Reserve. The Nash Spring (JNNR3), located at an elevation of ~ 1360 m amsl, represents the most logical outlet of this compartment. The ~ 130 L/s discharge of this spring suggests a substantial catchment area. A provisional assessment recognises the triangular footprint defined in Fig. 3 as a suitable description of this compartment spanning ~ 2860 ha (Table 4). It is named the Uitkomst Basin, as ~ 55% of its postulated extent is located on the farm Uitkomst 499 JQ. The spring discharge of ~ 130 L/s represents 21% of an MAP of 710 mm. The compartment boundaries are defined by the Twin Dyke structure in the south, the geological contact between the Malmani Subgroup dolomite and the overlying Pretoria Group sedimentary strata along the north-western margin, and the NW–SE striking Rietfontein Dyke to the north-east.

The Nash Spring elevation of ~ 1360 m amsl (Table 9) ties in with the groundwater elevation of 1365 m amsl reported for the borehole MVR1 on the Moon Valley Ranch property (Ptn 38 of Rietfontein 522JQ) in the upper reaches of this compartment. Although these geosites are located at distance of ~ 3.8 km apart, they are both located in the Skeerpoort River catchment (A21G). Data sourced from station MVR1 and station RLGR5 on the Rhino and Lion Game Reserve property in the Kromdraai Subcompartment to the south, reveals a difference in groundwater level elevation of ~ 55 m across the Twin Dyke structure that separates these two stations and compartments (Figs. 3 and 17). This is an abnormally large difference between contiguous karst compartments, and indicates as complete a hydraulic separation as might reasonably be conceived in practical terms.

## 2.5 Tweefontein Basin

The enumeration of the Tweefontein and Nash springs in the John Nash Nature Reserve (Sects. 4.6 and 4.8) and the Cradle Spring in the Motsetse Nature Reserve (Sect. 4.10) provide a better definition of groundwater elevations (and associated compartmentalisation) in the central north-eastern portion of the COH. This area is characterised by a paucity of hydrogeological information, the scarcity of boreholes in this largely pristine natural landscape contributing to the poor definition of groundwater rest level elevations in this portion of the study area. This limitation is countered by the lack of groundwater abstraction via boreholes, so that the spring discharges approximate the full delivery of such karst systems.

It is postulated that the Tweefontein Basin encompasses a much smaller area than suggested by previous studies. For example, Holland (2007) and Holland and Witthüser (2009) indicate this compartment to extend from the Twin Dyke system north-eastwards all the way to almost 27.9°E, a longitude that coincides with the eastern extremity of Hartbeespoort Dam north of the village of Broederstroom, and encompassing an area of ~ 12,750 ha. New information such as quantified spring discharges and spring surface elevations (that serve as a proxy for the lowest potentiometric surface elevation of the contributing compartment), together with springwater chemistry (including isotope data) and likely recharge rates, support a redefinition of this compartment as shown in Fig. 3.

The Tweefontein Spring elevation of ~ 1450 m amsl is some 30 m above the potentiometric surface in the Krombank Basin to the south-west. It is hypothesised on the basis of the relatively small discharge (~ 30 L/s) of the Tweefontein Spring (Sect. 4.6), that it drains a shallow aquifer of limited surface extent described by the ~ 1160 ha triangular footprint shown in Fig. 3. Under these circumstances, the spring discharge of ~ 30 L/s represents 12% of an MAP of 710 mm. It is proposed, therefore, that the boundaries of this compartment are defined by the NW–SE striking Rietfontein Dyke to the west, the Tweefontein Dyke to the north-east, and the northern extension of the syenitic Plover's Lake Sill along the south-eastern margin. The tritium concentration of ~ 3 ± 0.3 TU in the springwater (Sect. 7.5 in "Chapter Chemical Hydrogeology") suggests that this compartment is also characterised by comparatively recent and rapid rainfall recharge and a 'shallow' subsurface circulation regime (see Footnote 8 in Chapter "Chemical Hydrogeology").

In light of the above, it is postulated that the Tweefontein Basin located to the north-east of the Krombank Basin constitutes a shallow aquifer of comparatively limited extent. It is similar to the Danielsrust Basin in that its discharge via the Tweefontein Spring also provides autogenic recharge, in this instance northwards into the Diepkloof Basin (Sect. 2.7) drained by the Nouklip Spring.

## 2.6 Rietfontein Basin

The Rietfontein Basin (~ 540 ha) comprises a rectangular footprint adjoining the south-eastern margin of the Tweefontein Basin. It therefore shares the northern extension of the Plover's Lake Sill as the common boundary with the Tweefontein Basin to the north-west. Although the Aquamine Spring represents the most logical discharge (outflow) of groundwater from this compartment, the springflow of < 2 L/s does not adequately reflect the extent of the contributing karst aquifer. Not considered to be of consequence to the broader aims and objectives of this study, these circumstances nevertheless might attract further investigation in future hydrogeologic studies.

## 2.7 Diepkloof Basin

This compartment represents the comparatively poorly defined karst environment in the north-eastern portion of the COH. The north-westerly groundwater flow pattern in the Diepkloof Basin is based mainly on the surface drainage pattern that is characterised by deeply incised streams such as the Snake Stream and the Grootvlei Spruit. The elevation of the Nouklip Spring (~ 1330 m amsl) in the lower reaches of the Grootvlei Spruit supports this hypothesis. The substantial yield of this spring (Sect. 4.7) again reflects a highly transmissive aquifer of substantial surface extent.

A provisional assessment recognises the rectangular footprint defined in Fig. 3 as a suitable geometric description of this compartment spanning ~ 3800 ha. This recognises the Tweefontein Dyke as the southern boundary, the E–W striking Hartbeesthoek Dyke as the northern boundary, the N–S striking Krokodilberg Dyke as the eastern boundary, and the geological contact between the Malmani Subgroup dolomite and the overlying Pretoria Group sedimentary strata the western margin. The mean spring discharge of 143 L/s[6] (Sect. 4.7) represents 17% of an MAP of 710 mm.

## 2.8 Motsetse Basin

The Motsetse Basin (name coined in this study) forms a small catchment (Fig. 3) located in the Motsetse Nature Reserve. It encompasses only ~ 270 ha, and is drained by the Cradle Spring at a rate of ~ 2 L/s. This discharge represents ~ 3% of an MAP of 710 mm. Circumstances that might explain this anomalously low recharge value are discussed in Sect. 4.10. As this compartment is largely enclosed within a proclaimed nature reserve, its preservation in a natural state is reasonably well ensured.

## 2.9 Rhenostersprui Basin

This compartment of ~ 850 ha is located in the north-eastern portion of the COH (Fig. 3) where the karst environment is poorly defined in terms of its hydrogeology. This is attributable to the more populated nature of this landscape, where numerous properties support residences mainly reliant on their own borehole water supplies for general domestic use. Abstraction for this use impacts negatively on the yield of springs that would drain this karst basin naturally. Not considered to be of consequence to the broader aims and objectives of this study, these circumstances nevertheless might attract further investigation in future hydrogeologic studies.

The north-western boundary of this compartment is associated with a set of SW–NE striking diabase sill intrusions, its western boundary with the Krokodilberg Dyke, its eastern boundary with the Rhenosterspruit Dyke, and its south-eastern boundary with the contact between the Malmani Subgroup dolomite and the older undifferentiated igneous strata.

[6] Mean flow of 163 L/s minus the 14% mean contribution of the Tweefontein Spring (Sect. 4.6).

## 2.10 Kalkheuvel Basin

Encompassing and area of ~ 720 ha, this compartment is drained by the Barlow Spring (~ 2 L/s) located in the north-eastern corner of this compartment (Fig. 3). The E–W striking Hartbeesthoek Dyke that separates the Barlow Spring from the Anderson Spring to the north, forms the northern boundary of this compartment. Its western boundary is formed by the Krokodilberg Dyke, its eastern boundary by the Rhenosterspruit Dyke, and its south-eastern boundary by the set of SW–NE striking diabase sill intrusions that forms the north-western boundary of the Rhenosterspruit Basin (Sect. 2.9).

The compartment hosts numerous residential properties (mainly associated with the Kalkheuwel West Estate) reliant to a large extent on own borehole water supplies for potable and general domestic water use. The impact of this use on natural spring discharge is seen in the observation that the flow of the Barlow Spring falls significantly short of the natural yield (~ 20–35 L/s) that might be expected for a recharge of 17 ± 5% of 710 mm MAP on the basin footprint area.

## 2.11 Broederstroom Basin

This compartment is located in the extreme north-eastern portion of the COH (Fig. 3) where the karst environment is comparatively poorly defined in terms of its hydrogeology. It encompasses an area of ~ 1810 ha. The ~ 21 L/s Broederstroom Spring (Sect. 4.14) on the Glen Afric Country Lodge property south of Broederstroom represents one of the more prominent natural features draining this compartment. Others include the ~ 5 L/s Anderson Spring and the ~ 2 L/s Lesedi Spring. Further, numerous boreholes abstract groundwater for potable and general domestic use. This use, when added to the aggregate ~ 28 L/s of the enumerated spring discharges, might well double the groundwater loss from this compartment and approach a natural rainfall recharge contribution of 17 ± 5%, compared to the ~ 7% represented by the currently known natural discharges (Table 9).

## 2.12 Discussion of Basin Definition and Groundwater Recharge

The information presented in Table 4 summarises the definition of groundwater compartments in the COH as has been established from available information and data. The estimated size of each compartment and subcompartment is derived from planimetric measurements of its footprint. Where possible, the veracity of these values is gauged against the natural discharge (as represented by measured spring flows and other quantified groundwater discharge), converted to an equivalent depth of rainfall. This approach encapsulates the groundwater basin concept described, amongst others, by White (1993; 2007a).

It was reported in Sect. 1 that the Holland (2007) study reports estimated average annual groundwater recharge volumes of 16.8 $mm^3$ (from Bredenkamp et al. 1986) and 14.3 $mm^3$ (from Krige 2006) for the Zwartkrans Basin based on water budget calculations. The Holland (op. cit.) study puts forward its own estimates of 16.6 $mm^3/a$ for the Zwartkrans Basin and 13.4 $mm^3/a$ for the Tweefontein Basin based on chloride mass balance (CMB) calculations. In regard to the Zwartkrans Basin with a surface extent of ~ 9800 ha, the recharge estimates represent ~ 24% (~ 170 mm) of an MAP of 710 mm. The Holland (op. cit.) estimate for the 12,230 ha Tweefontein Basin represents 15.4% (~ 110 mm) of 710 mm MAP, and which would increase marginally to 16.4% for a slightly lower MAP of 668 mm (from Table 1 in Chapter "Description of the Physical Environment") as characterises the north-eastern portions of the property.

The 'basinal' rainfall recharge values reported in Table 4 range from ~ 1 to ~ 30% of MAP, but if the anomalously high and low values are excluded, more typically fall in the range ~ 11 to ~ 20%. This is in reasonable agreement with the range of ~ 9 to ~ 18% of MAP (excluding the extremes of ~ 3 to ~ 23% that defines the full range) reported by Kok (1992) for 12 dolomitic springs in South Africa. It is also in reasonable agreement with the range 15–21% of an MAP of 630 mm cited by Bredenkamp et al. (1986).

It is posited that recharge in the range 17 ± 5% of MAP, where MAP ranges from ~ 670 to ~ 710 mm (Table 1 in Chapter "Description of the Physical Environment"), represents a typical value for the karst terrane of the COH. Gondwe (2010) reports a similar range (17 ± 3%) for the karst system of the Sian Ka'an Biosphere Reserve, Yucatan Peninsula, Mexico.

The approach described above fails in those areas where groundwater abstraction from boreholes is significant, and reduces the natural discharge via springs. This is particularly relevant in the north-eastern portion of the property that encompasses the Broederstroom, Kalkheuwel West and Rhenosterspruit rural communities. This failing is evidenced in Table 4, which reflects significantly lower annual recharge percentages than the 17 ± 5% of MAP posited above. Although the water demand on an individual borehole in such a setting is low (because of the lack of high-demand uses such as irrigated agriculture and the low-density estate character of the land development), annual groundwater use will be significant.

An indication of this significance is provided by the results of a hydrocensus carried out by Maidment (2015), who enumerated 43 boreholes and three springs distributed over 38 properties in the Kalkheuwel West and Broederstroom basins. Water use data for 14 boreholes returned a mean rate of ~ 1450 L/d (~ 0.5 ML/a) per installation. Aggregated for 44 equipped boreholes, this amounts to a groundwater use of ~ 0.022 $mm^3/a$. The Broederstroom Basin also supports the Lesedi African Lodge and Cultural Village, a facility with a restaurant and overnight visitor accommodation, staffed by ~ 70 permanent residents. The borehole serving this facility produces ~ 12 $m^3/d$ (~ 0.004 $mm^3/a$).

As shown in Table 4, the combined discharge of the Kromdraai and Plover's Lake springs (339 L/s excluding the Danielsrust Spring autogenic contribution), returns an unrealistically high equivalent recharge of ~ 30% of MAP as representing the 'recharge input' on the Krombank Basin footprint of 5080 ha. An explanation for the anomalously high recharge associated with the Krombank Basin is sought in the following possibilities:

- that surface water is lost from an influent Bloubank Spruit downstream of the Zwartkrans Dyke; and
- that groundwater is gained from the Zwartkrans Basin via a subsurface route over/through the Zwartkrans Dyke.

A surface water contribution of 140 L/s would reduce the groundwater component of discharge from the Krombank Basin to 200 L/s, equivalent to a recharge of 17.5%. SDMs at localities BB@Lotz and ~ 450 m upstream of the Kromdraai Spring on 13/08/2010, however, returned similar values of 459 and 466 L/s respectively. This result eliminates a significant surface water contribution from a losing Bloubank Spruit as possible explanation. The SEC of 56 mS/m associated with the Kromdraai Spring water (Table 14), compared to the 38 mS/m of the Plover's Lake springwater (Fig. 32 "Chapter Chemical Hydrogeology"), rather suggests some measure of allogenic impact. The autogenic recharge contributed by the ~ 28 L/s Danielsrust Spring introduces a further complication that needs to be accounted for in this regard. It is appropriate to again employ a mass balance approach in an attempt to resolve this conundrum.

Let

$$(Q_{DS}C_{DS}) + (Q_G C_G) + (Q_R C_R) = (Q_{KS}C_{KS}) + (Q_{PLSP}C_{PLSP})$$

then

$$(28 \times 182) + (140 \times C_G + (137 \times 35) = (307 \times 392) + (60 \times 266)$$

And

$$C_G = 126413 \div 140 = 903\, mg \frac{TDS}{L} = 129 \frac{mS}{m} Sec$$

where

$Q_{DSp}$ is the discharge (L/s) of the Danielsrust Spring;
$C_{DSp}$ is the SEC (mS/m) of the Danielsrust Spring water converted to TDS by the multiplier factor 7;
$Q_G$ is the subsurface input (L/s) of groundwater from the Zwartkrans Basin;
$C_G$ is the SEC of the Zwartkrans Basin groundwater water converted to TDS as above;
$Q_R$ is the recharge (L/s) to the Krombank Basin at ~ 17% of an MAP of 710 mm;
$C_R$ is the SEC of the rainwater recharge converted to TDS as above;
$Q_{KS}$ is the discharge (L/s) of the Kromdraai Spring;
$C_{KS}$ is the SEC of the Kromdraai Spring water converted to TDS as above;
$Q_{PLS}$ is the discharge (L/s) of the Plover's Lake Spring(s); and
$C_{PLS}$ is the SEC of the Plover's Lake Spring(s) water converted to TDS as above.

The mass balance result suggests that ~ 140 L/s of groundwater with an SEC of 129 mS/m would explain and account for the 'missing' water budget component of the Krombank Basin. It is encouraging, therefore, that the SEC of groundwater sampled from monitoring borehole GP00313 (Fig. 21) on eight occasions between July 2012 and September 2014 return an average value of ~ 129 mS/m in the range 120–136 mS/m.

The slightly elevated recharge value reported for the Uitkomst Basin is attributed to the more rugged nature of the terrain that characterises this compartment. These circumstances promote runoff in the incised drainages which, owing to their geologic setting, are controlled by geologic structures (fracture and fault zones), and typically represent losing streams contributing autogenic recharge to the karst aquifer. This has been observed in the Grootvlei Spruit draining the Diepkloof Basin.

Whilst it is acknowledged that an assessment such as provided above is hypothetical and necessarily coarse, it does provide plausible order of magnitude volumes that offer a 'reality check'. A second 'reality check' of specifically the subcompartment definition in the south-western portion of the study area, based on a statistical analysis of groundwater rest level data, is presented in Table 5. Salient results of this analysis are summarised as follows:

- The difference in groundwater rest level depth and elevation between subcompartments 1a and 1b. Especially the difference of 35–40 m in terms of the mean and median elevation (absolute) values supports the observations documented in Sect. 2.1.1 regarding the difference in groundwater rest level between these two subcompartments. Note that it is shown in Sect. 6 and Fig. 18 that this difference reduces at the contiguous northern ends of these two subcompartments.
- The difference in groundwater rest level depth and elevation between subcompartments 2a and 2b. Although the difference in mean and median elevation values is comparatively small (~ 3 m), the much shallower values that characterise subcompartment 2b reflect the association of this subcompartment with the Bloubank Spruit valley between the Zwartkrans Spring and Plover's Lake.

## 3 Groundwater Level Behaviour

### 3.1 Zwartkrans Basin

The behaviour of groundwater levels (the hydrostatic response) associated with the karst aquifer is reflected in the long-term water level records for the 15 oldest DWS monitoring boreholes in the study area. An assessment of these data returns the statistics presented in Table 6. A graphical representation of the information is shown in Fig. 5. An analysis of the %ile $\Delta h$ data yields a 25%ile value of 3.86 m, a mean value of 5.35 m, a median value of 4.93 m, and a 75%ile value of 6.98 m. The information presented in Table 6 and Fig. 5 indicates that there is little correlation between the depth to groundwater rest level and the magnitude of variation evident in this hydrologic variable.

A selection of the hydrographs is compared in Fig. 6, which indicates two distinct groupings of hydrograph, namely Group A occupying an elevation of > 1530 m amsl, and Group B occupying an elevation < 1490 m amsl. The elevation difference of > 40 m reflects the location of these groupings in two different basins/subcompartments. It is evident from Table 6 that the maximum difference in groundwater rest level amounts to ~ 12.2 m (station A2N0598), with a mean of ~ 6.2 m and a median of ~ 5.6 m. The slightly smaller differences between the 5%ile and 95%ile limits are characterised by a maximum value of ~ 9.8 m (station A2N0598), and mean and median values of ~ 5.2 and ~ 4.6 m respectively. These results are in agreement with similar results obtained for the karst aquifer elsewhere in Gauteng Province, e.g. for the karst aquifer extending from south of Pretoria to Kempton Park, and from Rietvlei Dam in a south-easterly direction to Bapsfontein (Hobbs 2004). Previous analyses of these data (Hobbs and Cobbing 2007) indicate natural fluctuations in the order of

**Table 5** Definition of groundwater rest level variables for the Bloubank Spruit system

| Basin/subcompartment | | Groundwater rest level variable | Statistical parameter | | | | | |
|---|---|---|---|---|---|---|---|---|
| | | | n | 25% ile | Mean | Median | 75% ile | Inter-quartile Range (m) |
| Zwartkrans | Vlakdrift (1a) | Depth (m bs)<br>Elevation (m amsl) | 11 | 48<br>1461 | 63<br>1502 | 62<br>1503 | 75<br>1535 | 27<br>74 |
| | Sterkfontein (1b) | Depth (m bs)<br>Elevation (m amsl) | 16 | 20<br>1471 | 31<br>1536 | 28<br>1540 | 44<br>1570 | 24<br>99 |
| | Zwartkrans (1c) | Depth (m bs)<br>Elevation (m amsl) | 25 | 17<br>1439 | 28<br>1446 | 22<br>1444 | 30<br>1450 | 13<br>11 |
| Krombank | Kromdraai (2a) | Depth (m bs)<br>Elevation (m amsl) | 12 | 44<br>1418 | 60<br>1422 | 59<br>1420 | 74<br>1424 | 30<br>6 |
| | Bloubank (2b) | Depth (m bs)<br>Elevation (m amsl) | 14 | 6<br>1410 | 12<br>1419 | 11<br>1417 | 18<br>1420 | 12<br>10 |
| | TOTAL | | 78 | – | – | – | – | – |

**Table 6** Salient statistics for long-term DWS groundwater level monitoring data

| Station | Groundwater rest level (m bc) | | | | | Variability | | Record period[a] |
|---|---|---|---|---|---|---|---|---|
| | n | 5%ile | Mean | Median | 95%ile | Max Δh[b] | %ile Δh[c] | |
| A2N0580 | 274 | 51.01 | 54.55 | 54.32 | 59.95 | 11.13 | 8.94 | 05/1985–09/2014 |
| A2N0582 | 210 | 35.77 | 40.11 | 40.15 | 42.91 | 8.53 | 7.14 | 05/1985–12/2010 |
| A2N0583 | 234 | 44.40 | 44.95 | 44.90 | 45.54 | 1.84 | 1.14 | 05/1985–10/2014 |
| A2N0584 | 268 | 21.82 | 25.52 | 26.26 | 28.09 | 8.19 | 6.81 | 05/1985–09/2014 |
| A2N0586 | 289 | 21.34 | 26.23 | 27.14 | 28.67 | 8.54 | 7.33 | 05/1985–10/2014 |
| A2N0589 | 169 | 27.92 | 28.89 | 28.97 | 29.90 | 3.85 | 1.98 | 05/1985–06/2010 |
| A2N0590 | 188 | 31.50 | 34.62 | 35.14 | 36.46 | 6.11 | 4.96 | 05/1985–10/2014 |
| A2N0592 | 278 | 73.62 | 76.92 | 77.33 | 78.53 | 6.33 | 4.91 | 06/1985–10/2014 |
| A2N0594 | 183 | 70.86 | 72.79 | 72.80 | 74.41 | 4.91 | 3.55 | 01/1985–09/2008 |
| A2N0598 | 89 | 53.53 | 58.76 | 58.84 | 63.32 | 12.17 | 9.79 | 07/1985–05/2010 |
| A2N0600 | 208 | 21.22 | 23.92 | 24.39 | 25.38 | 5.07 | 4.16 | 04/1989–10/2014 |
| A2N0602 | 233 | 51.02 | 54.30 | 54.85 | 55.95 | 6.44 | 4.93 | 06/1987–10/2014 |
| A2N0605 | 202 | 60.18 | 62.49 | 62.74 | 63.65 | 4.16 | 3.47 | 04/1989–09/2014 |
| A2N0606 | 72 | 64.83 | 67.07 | 67.06 | 69.51 | 5.11 | 4.68 | 08/1989–10/2014 |
| A2N0607 | 169 | 64.20 | 67.00 | 66.97 | 70.68 | 7.95 | 6.48 | 10/1993–10/2014 |

[a] From month of first measurement to month of most recent available measurement as at November 2014 update from DWS; shaded rows denote stations no longer in service
[b] Difference between minimum and maximum values (not shown in this table)
c Difference between the 5%ile and 95%ile values

5–8 m manifested over a period of 2–5 years, i.e. not from season to season. In regard to the Sterkfontein Cave system, Martini et al. (2003) report a gradual water level fluctuation in response to the preceding precipitation pattern in the range ~ 2 m. The CGS (undated) also reports that the water level in the cave in 1983 was 2–3 m higher than its current (Presumably 2004) level.

The groupings shown in Fig. 6 are produced separately in Fig. 7 (Group A) and Fig. 8 (Group B). The large measure of similarity in the hydrostatic response of the Group B stations is evident. By comparison, the Group A stations exhibit a poor correlation that is particularly evident in station A2N0583. An inspection of the original borehole drilling and test pumping records reported by Bredenkamp et al. (1986) provides a possible explanation for the observed response patterns.

The data in Table 7 identify station A2N0580 as intersecting an unproductive aquifer (also reflected in the very low transmissivity value of 1 $m^2$/d), which explains the more 'ragged' trace of this graph compared to the 'smoother'

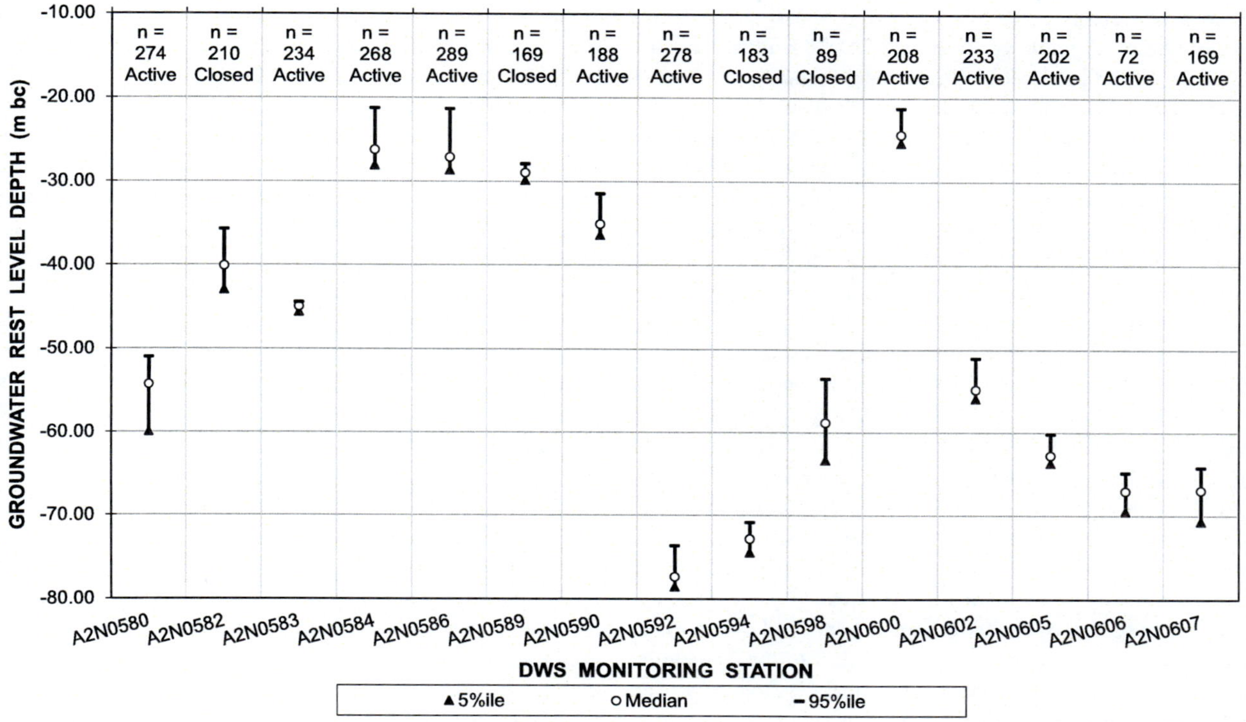


**Fig. 5** Graphic comparison of the statistical hydrographic response observed in DWS groundwater level monitoring stations in the period 1985–2014 (data from Table 6)

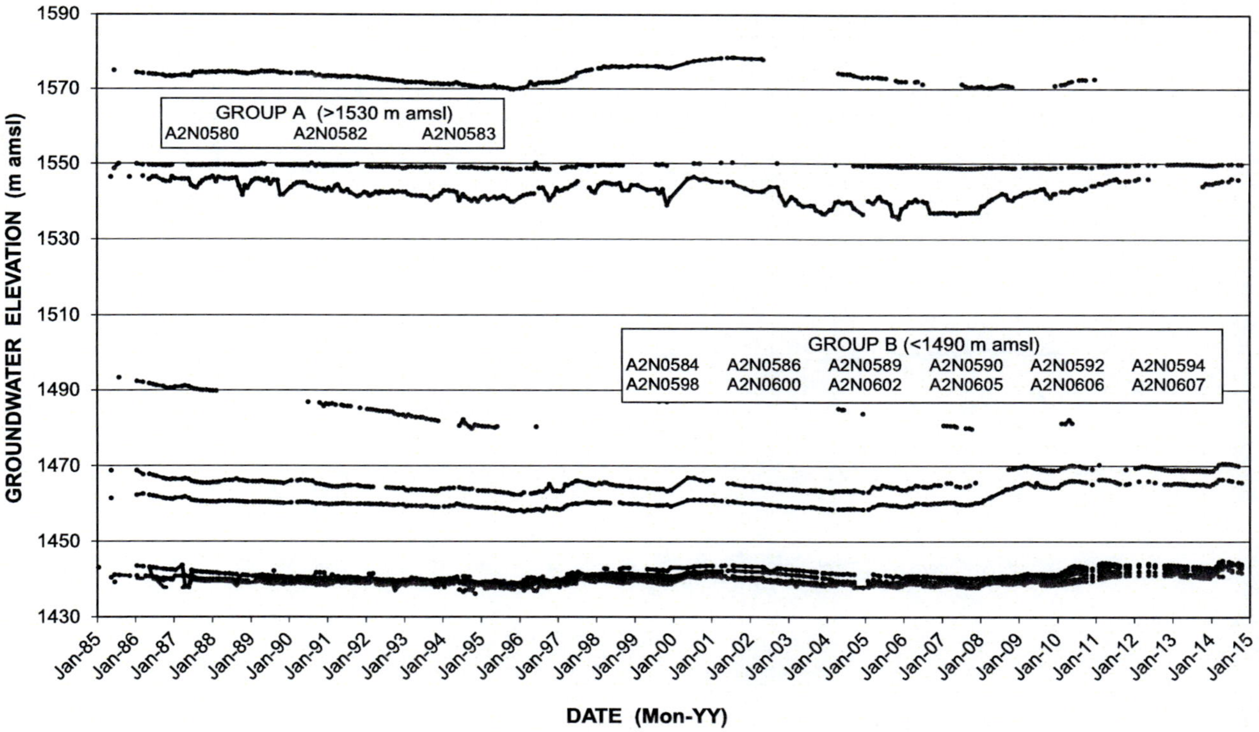


**Fig. 6** Long-term groundwater level response pattern in DWS monitoring boreholes

graphs of stations A2N0582 and A2N0583 (Fig. 7). Especially the flat graph of station A2N0583 illustrates the capacity of the intersected groundwater resource to 'absorb' and 'cushion' the impact of recharge- or abstraction-induced stresses in the aquifer. This capacity derives from high transmissivity and/or storativity characteristics (Table 7).

The very productive aquifer(s) that support stations A2N0584 and A2N0586 (Fig. 8) similarly explain the relative 'smoothness' of these hydrographs. It is probable that the other stations reflected in Fig. 8 also penetrate productive groundwater systems.

The unprecedented rise in the groundwater level observed in stations A2N0584 and A2N0586 since late-2007 (Fig. 8) reflects the impact of exceptional recharge associated with raw and/or treated mine water being lost from the lower reaches of the Tweelopie Spruit and its receiving main stem, the Riet Spruit (Sect. 5.2.1 in Chapter "Physical Hydrology" and Sect. 2.1.2). Both these stations are located in proximity to the Riet Spruit. These circumstances were precipitated by the abnormally wet summers experienced in the region starting with the 2008 hydrological year, and resulting in treated mine water discharges > 25 ML/d to the Tweelopie Spruit (see Fig. 15 in Chapter "Physical Hydrology"). The additional contribution of raw mine water to this discharge in the much wetter 2010, 2011 and 2014 rainy seasons has, on occasion, increased the flow in this drainage to > 60 ML/d (Sect. 5.2 in Chapter "Physical Hydrology").

A comparison of ca. 2006 and ca. 2010 water level data for the same sites is provided in Table 8. This reveals an unequivocal rise in water level. The greatest rises (> 3 m) associated with boreholes located in the Riet Spruit valley upstream of Oaktree (CSIR57 and GB1) are again readily attributed to the ingress of surface water contributed primarily by the Tweelopie Spruit from the mining area. Further away from the Riet Spruit, the rise amounts to 0.5–1.5 m. Although the rise of ~ 0.6 m in the Danielsrust Basin (sites DRGF1 and DRGF2) is similar to the 0.4–0.8 m observed nearby in the main karst aquifer (sites BolandB1, SF1 and MB1), these compartments are not in hydraulic continuity with each other.

The resumption of uncontrolled mine water discharge in late-January 2010, combined with the exceptionally high rainfall experienced in the 2010, 2011 and 2014 summers, precipitated exceptional recharge conditions in the receiving Zwartkrans Basin. These conditions manifest as a rise in groundwater level of ~ 1 m within the space of only a few months in boreholes located next to the Riet Spruit, e.g. ≥ 0.86 m and ≥ 1.27 m in boreholes CSIR57 and

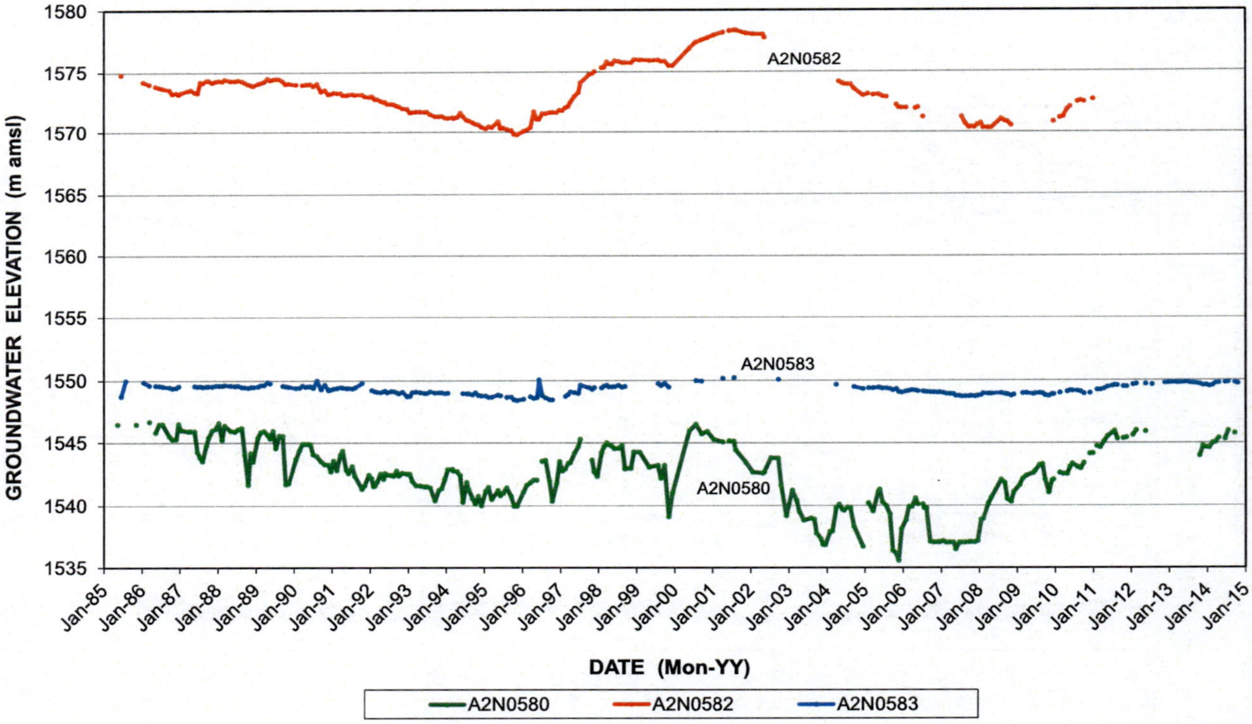


**Fig. 7** Long-term groundwater level response pattern in Group A boreholes from Fig. 6

**Table 7** Salient hydrogeologic attributes associated with selected DWS monitoring stations (from Bredenkamp et al. 1986)

| Station | Blowout yield (L/s) | Maximum tested yield (L/s) | Maximum drawdown (m) | Transmissivity ($m^2$/d) | Comment |
|---|---|---|---|---|---|
| A2N0580 | 0 | 0.6 | 38.5 | 1 | Unproductive aquifer |
| A2N0582 | 15 | 25.2 | 4.52 | 1600 | Highly productive aquifer |
| A2N0583 | > 40 | 94.7 | 4.15 | 750 | Extremely productive aquifer |
| A2N0584 | > 40 | 18 | 13.17 | 34 | Moderately productive aquifer |
| A2N0586 | > 40 | 106.9 | 3.18 | 2100 | Extremely productive aquifer |

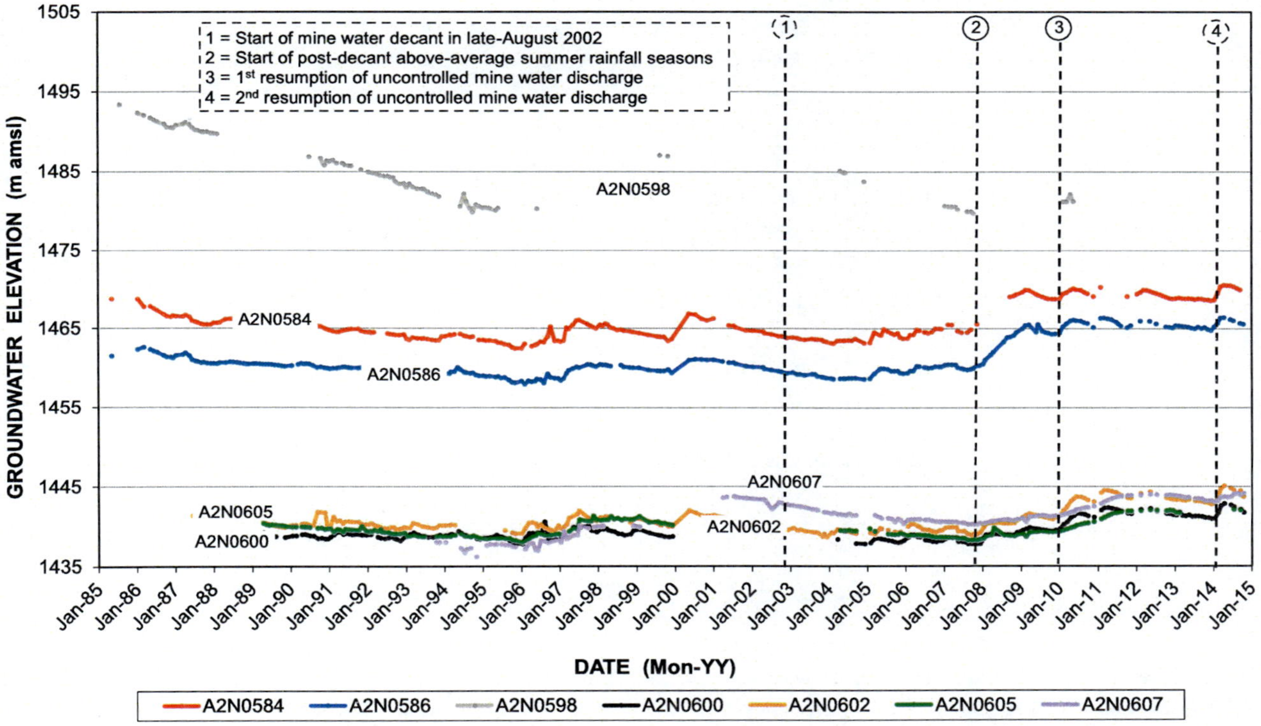


**Fig. 8** Long-term groundwater level response pattern in Group B boreholes from Fig. 6

GB1, respectively. Groundwater elevations are currently the highest (e.g. stations A2N0584, A2N0586 and A2N0600) in the ~ 30-year record of measurements.

## 3.2 Sterkfontein Cave System Water Level

### 3.2.1 General Discussion

It is common cause that the water level in the Sterkfontein Cave system has been the subject of considerable debate and at least some confusion. The water level in question is that associated with what Martini et al. (2003) refer to as the 'Main Lake', which is the most readily accessible water body in the cave system. Wilkinson (1973) refers to this water body simply as the 'Lake', a convention followed in this manuscript. A recent substantial rise (~ 3 m) in the Lake water level has caused Maropeng āAfrika (the authority responsible for managing the tourist component of the site) to reroute the tourist path through the caves to successively higher elevations on three occasions since early-2010. The circumstances warrant separate and specific discussion.

**Table 8** Comparison of historical and more recent groundwater levels in the study area

| Station | Historic water level | | | More recent water level | | | Difference (m bc) |
|---|---|---|---|---|---|---|---|
| | Date | Depth (m bc) | Elevation (m amsl) | Date | Depth (m bc) | Elevation (m amsl) | |
| SW1 | 09/12/2005 | 18.00 | 1441 | 11/02/2010 | 11.36 | 1448 | + 6.64 |
| A2N0594 | 25/01/2006 | 74.00 | 1439 | 08/02/2010 | 73.43 | 1440 | + 0.57 |
| DRP15 | 30/07/2006 | 59.00 | 1416 | 02/12/2009 | 58.63 | 1417 | + 0.37 |
| ZW1 | 30/07/2006 | 29.64 | 1438 | 17/02/2010 | 27.64 | 1440 | + 2.00 |
| A2N0600 | 30/07/2006 | 25.06 | 1438 | 17/02/2010 | 23.03 | 1440 | + 2.03 |
| DRGF1 | 30/07/2006 | 11.00 | 1486 | 11/02/2010 | 10.45 | 1487 | + 0.55 |
| DRGF2 | 30/07/2006 | 26.93 | 1486 | 11/02/2010 | 26.30 | 1487 | + 0.63 |
| BolandB1 | 30/07/2006 | 22.15 | 1438 | 17/02/2010 | 21.77 | 1438 | + 0.38 |
| SF1 | 01/10/2007 | 17.43 | 1437 | 17/02/2010 | 16.13 | 1438 | + 1.30 |
| | | | | 09/06/2010 | 15.54 | 1438 | + 1.89 |
| GB1 | 30/07/2006 | 20.36 | 1444 | 11/02/2010 | 17.64 | 1447 | + 2.72 |
| MB1 | 30/07/2006 | 15.07 | 1435 | 13/05/2010 | 13.79 | 1436 | + 1.28 |
| VW1 | 30/07/2006 | 13.94 | 1445 | 06/05/2010 | 9.71 | 1449 | + 4.23 |
| HW1 | 30/07/2006 | 8.00 | 1410 | 13/05/2010 | 6.09 | 1412 | + 2.09 |
| SWBH1 | 30/07/2006 | 28.00 | 1417 | 13/05/2010 | 27.30 | 1418 | + 0.70 |
| SBH1 | 30/07/2006 | 22.26 | 1415 | 13/05/2010 | 21.18 | 1416 | + 1.08 |
| A2N0584 | 08/02/2007 | 25.71 | 1466 | 16/02/2010 | 21.78 | 1469 | + 3.93 |
| CSIR34 | 13/02/2007 | 35.10 | 1470 | 14/04/2010 | 33.99 | 1471 | + 1.11 |
| CSIR8 | 13/02/2007 | 28.42 | 1447 | 14/04/2010 | 24.28 | 1451 | + 4.14 |
| A2N0586 | 08/03/2007 | 26.96 | 1460 | 16/02/2010 | 21.98 | 1465 | + 4.98 |
| CSIR57 | 08/03/2007 | 12.32 | 1461 | 16/02/2010 | 7.06 | 1466 | + 5.26 |
| A2N0598 | 16/05/2007 | 63.22 | 1480 | 09/06/2010 | 61.14 | 1482 | + 2.08 |
| Statistical analysis (excluding stations DRGF1 and DRGF2) | | | | | n | | 19 |
| | | | | | Minimum value | | 0.37 |
| | | | | | 5%ile value | | 0.38 |
| | | | | | Mean value | | 2.42 |
| | | | | | Median value | | 2.03 |
| | | | | | 95%ile value | | 5.40 |
| | | | | | Maximum value | | 6.64 |

The elevation of the Lake water level was reported as 1450.88 m amsl by JFA (2006). The discrepancy with the groundwater level elevation of 1437.5 m amsl in the nearby borehole SF1 (op. cit.) was attributed to perched aquifer conditions associated with the cave environment. A Western Basin Void Technical Group (WBVTG) meeting held on 28/02/2007 at the Rand Uranium Office Complex in Randfontein was informed that the Zwartkrans Spring was dry, and that any flow in this vicinity represented surface runoff. This was in response to a query regarding the potentiometric level of water in Sterkfontein Cave (reportedly 1436 m amsl) versus the elevation of the spring (reportedly 1439 m amsl). These circumstances were quite logically put forward as evidence that the water level in cave could not rise more than 3 m, i.e. up to the level at which the dolomitic Basin would overflow via the spring. This information clearly contradicted the JFA (2006) Lake water level elevation. Besides, it was determined that the Zwartkrans Spring had not dried up in the ~ 31 years that the owner of the property had resided there (H Roos, personal communication).

A re-survey on 12/12/2007 of the benchmark in the cave from which the Lake water level is derived, placed it an elevation of 1437.94 m amsl compared to its previous elevation of 1452.37 as reported by JFA (2006). The difference of 14.43 m, when applied to the cave water level of 1450.88 reported by JFA (2006), returns a cave water level of 1436.5 m amsl, i.e. in agreement with that reported to the WBVTG. This compares favourably with both the reported groundwater level elevation of 1437.5 m amsl (JFA 2006) in the nearby monitoring borehole SF1, and the elevation of the

Zwartkrans Spring which is placed at ~ 1433 m amsl (this study).

A rest water level measurement in borehole SF1 on 01 October 2007 returned a value of 17.4 m bc which, for an interpolated surface elevation of 1454 m amsl, gives a potentiometric level of 1436.6 m amsl. This closely resembled that of the cave water level, suggesting that the groundwater rest level in SF1 and the caves represents a single potentiometric surface. This observation found support from Krige (2009, 2010) following a revision of an earlier report by Krige and van Biljon (2006). In practical terms, this indicates that the caves share the same aquifer as the nearby borehole SF1, and which is drained by the Zwartkrans Spring. Martini et al. (2003) report earlier similar discrepancies in elevation (up to 9 m) between various water bodies (up to 30 reported 'static' pools) inside the caves, but conclude that more recent measurements suggest elevation differences in the order of decimetres (tens of centimetres) rather than metres.

The above circumstances bring into question the relationship with the channel of the Bloubank Spruit north of (opposite) the cave system, which is placed at ~ 1440 m amsl (this study) compared to the 1445 m amsl attributed it by JFA (2006). The elevation difference of ~ 3 m between the surface drainage and the October 2007 groundwater level is not uncommon in the study area. It has been shown by Hobbs and Cobbing (2007) that the potentiometric surface in the Riet Spruit upstream of the Oaktree area (and therefore the caves) was separated from the overlying surface drainage by 12–30 m (Fig. 4). This difference has subsequently decreased significantly as a consequence of surface water inflow into the karst aquifer (see Sect. 5.2 in Chapter "Physical Hydrology" and Fig. 4). For example, by 09/06/2010 the water table in borehole SF1 had risen ~ 1.9 m from its October 2007 depth of ~ 17.4 m bc (Table 8). This places the water table elevation in June 2010 at ~ 1438 m amsl, which is only some 2 m below the ~ 1440 m amsl of the Bloubank Spruit channel. The difference can be expected to reduce even more as the streambed elevation approaches that of the Zwartkrans Spring, these surfaces coinciding in proximity to the spring. Under these circumstances, it is probable that groundwater also 'resurfaces' in the channel of the Bloubank Spruit upstream of the Zwartkrans Spring. Observations that support such resurgence are presented and discussed in Sect. 5.2.3 in Chapter "Physical Hydrology".

Further informative aspects of the cave water level reported by Martini et al. (2003) are the following.

- The indication that the cave has dewatered from 20 to 25 m above the present water level, in the past 3.3 Ma ago. The water level decline was irregular, being punctuated by temporary rises as indicated by re-solution of calcified deposits ~ 12 m above the current water level, and speleothems corroded up to 6 m above this level. The dewatering interval extends to an upper bounding elevation of 1456–1461 m amsl which, in terms of the present landscape, implies that the valley to the north of the caves would have been under water to a depth of 15–20 m (see Sect. 12 "Chapter Chemical Hydrogeology" for a perspective on this likelihood).
- The observation that the ca. 2003 cave lake water level fluctuation was within a range of ~ 2 m, and varied gradually in response to a prior rainfall pattern. Further, that speleothems (flowstone and stalactites) were corroded in the interval of roughly 3–6 m above the current (ca. 2003) water level, and that no speleothems occur below 2–3 m above this level in the vicinity of the lake. The latter observation indicates that this interval is the most aggressive in terms of carbonate re-solution, at least in the more recent speleogenesis of the cave system. This observation also accords with the evaluation of the ambient potentiometric response pattern (**Sect.** 3.2.2).

### 3.2.2 Potentiometric Response Pattern

The installation of a continuous water level monitoring device in the cave lake by the DWS in May 2005 yields information on the cave water level response pattern to July 2007. The recorded pattern is shown in Fig. 9. It reveals three periods of steady decline at a rate of between 0.08 and 0.06 m/month, the latter occurring after an upward 'adjustment' of ~ 0.5 m in the 2007 summer.

Also shown in Fig. 9 is the SF1 groundwater level elevation on 30/07/2006. This derives from a value of 17.00 m bc reported by Holland (2007), and which indicates a < 1 m discrepancy with that of the contemporary Lake water level elevation. The datum for the latter is a surveyed elevation (Krige 2009), and must therefore be regarded as accurate to < 0.1 m. By comparison, the SF1 datum is a surface elevation interpolated from the 1:10,000 scale orthophoto map 2627BA5 Sterkfontein (RSA 1987). With a contour interval of 5 m, an interpolation accuracy of < 2 m is feasible with knowledge of the landscape.

In sympathy with the observed rise in water levels in the study area in the 2010 and 2011 hydrological years (Sect. 3 and Table 8), a similar response is observed in the cave. In mid-May 2010, cave guide K Mangole (personal communication) estimated a rise of ~ 1–2 feet (0.3–0.6 m) since late-2009. This is in good agreement with the ~ 0.6 m rise observed in the nearby borehole SF1 between February 2010 and June 2010 (Table 8), and the ~ 0.4 m rise in borehole MB1 between February 2010 and May 2010. A water level

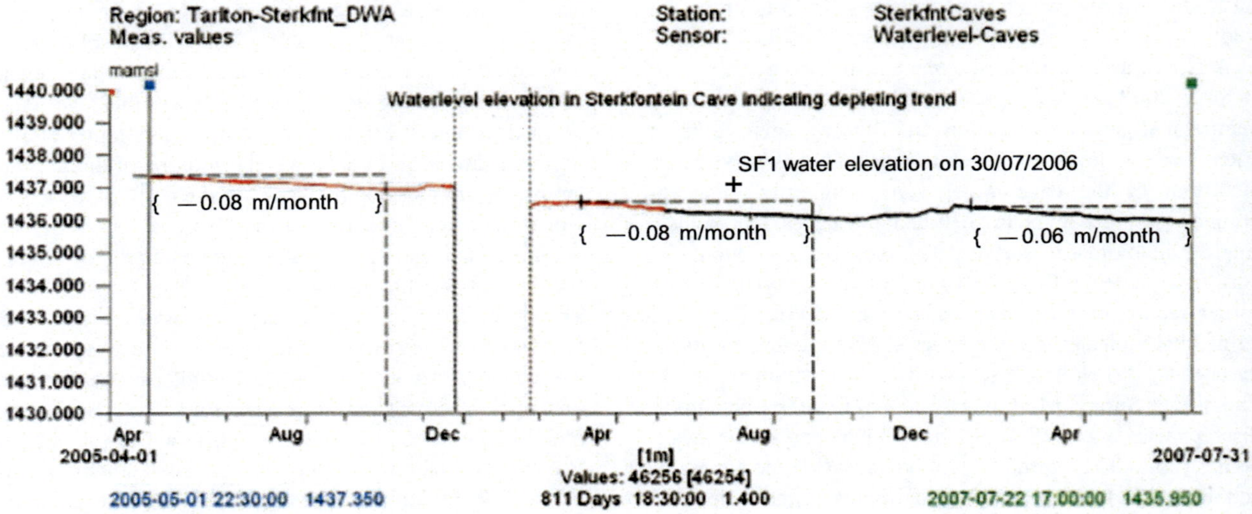


**Fig. 9** Continuous groundwater level response pattern in Sterkfontein Cave over a period of 27 months (use of image courtesy of DWS)

measurement in borehole SF1 on 14/01/2011 indicated a further rise of ~ 0.7 m since June 2010, for a total rise of ~ 2.6 m between October 2007 and January 2011, increasing to ~ 2.8 m with a further rise of ~ 0.2 m being manifested in the 5-month period January 2011 to June 2011. The rates of rise in the three periods February 2010 to June 2011, June 2010 to January 2011, and January 2011 to June 2011 amount to ~ 0.15, ~ 0.09 and ~ 0.04 m/month respectively. The latest measurements reflect a decline in the Lake water level.

The cave water level trend is shown in Fig. 10, and suggests that the water level might have started rising in mid-2009. This followed a decline (the early part of the SF1 record shown in Fig. 10) at a rate of ~ 0.03 m/month, i.e. roughly half the rate of 0.06–0.08 m/month reflected in the three earlier periods marked in Fig. 9.

The 'maximum' elevation of ~ 1439.5 m amsl (Fig. 10) approaches the ~ 1440 m amsl assigned to the Bloubank Spruit channel to the north of the caves. This suggests that the cave water level reaches equilibrium at an elevation of ~ 1440 m amsl when the karst water table intersects the stream channel of the Bloubank Spruit opposite the caves. If so, then the maximum possible rise of ~ 3 m agrees well with the zone of perceived most aggressive carbonate re-solution that defines the more recent speleogenetic evolution of the cave system as observed by Martini et al. (2003) (Sect. 3.2.1). Groundwater level measurements ca. mid- to late-2010 in boreholes indicate that the hydraulic gradient in the valley of the Bloubank Spruit to the north of Sterkfontein Cave amounted to 0.003 in an ENE direction. This is equal to a barely discernible fall of 0.3 m per 100 m in an ENE direction, which provides the context for the 'apparently static' nature of pools in the cave system observed by Martini et al. (2003).

The decline in water level in borehole SF1 in the period May 2012 to November 2013 indicates a rate of decline of ~ 0.03 m/month (Fig. 10). This is similar to the ~ 0.03 m/month reported for the earlier (pre-rise) portion of the SF1 record shown in Fig. 10. It is postulated that the Lake will maintain a high water level into the future because of sustained above-normal discharge in the upper tributaries of the Bloubank Spruit, provided that abnormal precipitation does not trigger excessive recharge resulting in a rise in groundwater levels in the Zwartkrans Basin. As is evident in Fig. 10, the high rainfall in February and March 2014 precipitated a rapid rise in the contemporary cave water level. It is highly unlikely, however, that the cave water level will rise above the elevation of ~ 1440 m amsl.

Martini et al. (2003) also report the observation of cave divers that the depth of the Lake does not exceed 4 m. Unfortunately, the authors do not provide a datum for this measurement. Wilkinson (1973), however, reports a similar depth for the Lake, and a depth of 6 m for water body W5 in Lincoln's Cave. Although there is some doubt over the Wilkinson (1973) datum, the reported depth of 48 m below the top of the 'hillock' places the Lake water level at an elevation of ~ 1437 m amsl. This is in agreement with recent measurements, and places the deepest point of the Lake at ~ 1433 m amsl, similar to that of the Zwartkrans Spring (Table 9). The 6 m depth reported for water body W5 therefore occupies a base elevation of ~ 1431 m amsl. These circumstances prompt an assessment of the conduit flow depth associated with the karst aquifer as attempted in Sect. 8.

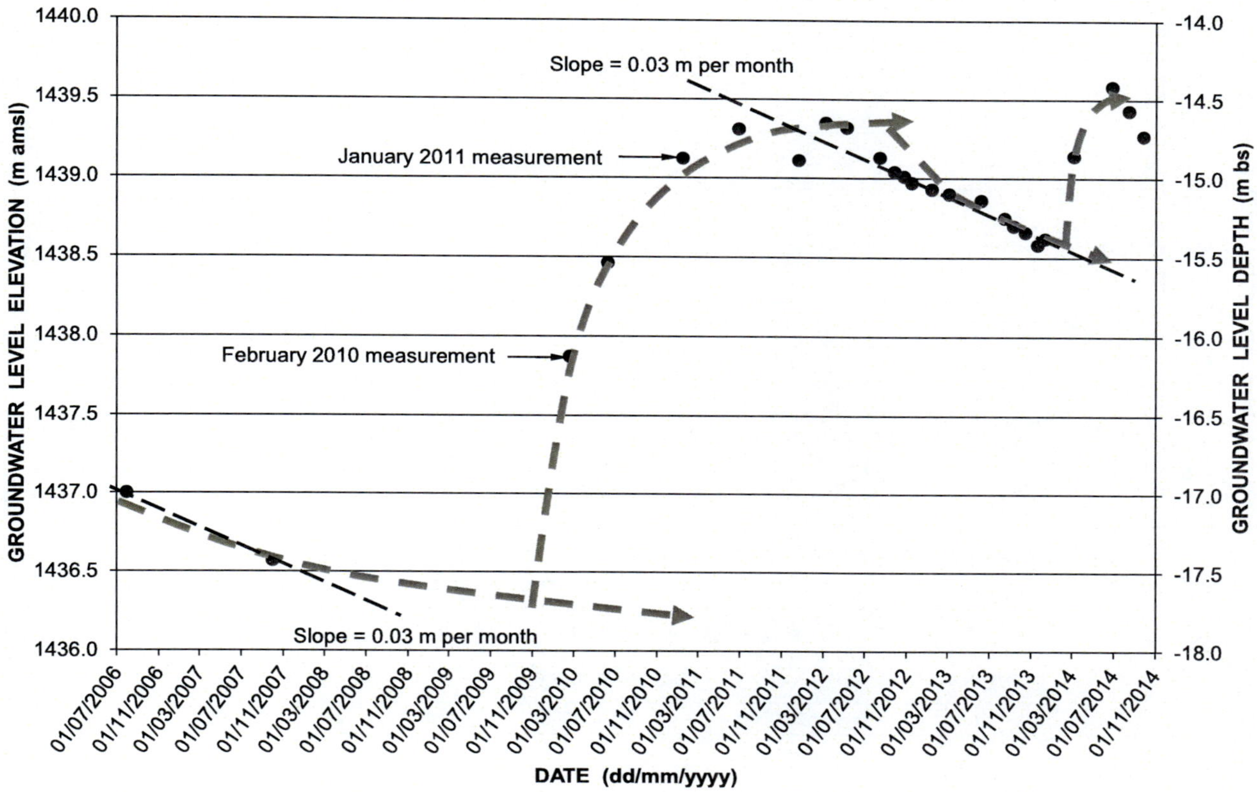


**Fig. 10** Groundwater level response pattern and trend in borehole SF1 that serves as a proxy for the Lake water level in Sterkfontein Cave

Wilkinson (1973) suggested that resurgence occurred as underflow on the buried contact between the alluvial valley fill deposit and the underlying dolomitic bedrock located 12 m below the streambed (op cit., Fig. 7.5). Further, that the resurgence was driven by a vector directed normal to the stream channel rather than by an upstream migration driven by a rising water table. The hydraulic gradient of 0.1 (6°) reported by Wilkinson (1973) contrasts markedly with the much shallower slope of 0.003–0.005 (0.2–0.3°) established from recent information, and which is more typical of highly transmissive karst strata of the South African interior.

Exploration borehole information suggests that the main flowpath (thalweg) of groundwater discharge through the Zwartkrans Basin toward the spring lies ~ 600–700 m to the north-west of the caves coincident with the Bloubank Spruit drainage. It is postulated, therefore, that the location of the cave system on the south-eastern periphery of the main flowpath offers an explanation for the muted mine water impact on the cave water chemistry. These circumstances are discussed in greater detail in Sect. 8.4 "Chapter Chemical Hydrogeology". It is also worth noting that the caves straddle the contact between the chert-poor Oaktree Formation that forms the south-eastern (and oldest) lithostratigraphic margin of the Malmani Subgroup, and the overlying chert-rich strata of the Monte Christo Formation. The latter is generally associated with a more 'productive' aquifer, in contrast to the typically more 'barren' character of the Oaktree Formation. Nevertheless, the potentiometric response exhibits a spatially more uniform behaviour (Fig. 11) as might be expected in a highly transmissive karst aquifer.

## 4 Springs

Springs (or 'eyes') are hydrogeologic features that typically mark the intersection of the potentiometric surface with the land surface.[7] For this reason, they represent the lowest point in a groundwater basin to which flow converges (Quinlan and Ewers 1986), and their flow behaviour, turbidity and

[7] The position of springs in the landscape is determined by numerous geologic factors including structural features such as intrusive dykes and sills and lithological contact zones between strata. These factors are used to describe each of the enumerated springs as either (1) a barrier dyke, (2) a barrier sill or (3) a lithologic contact type, respectively (see Table 8). Ford and Williams (2007) simply refer to such springs by the collective term 'dammed springs'.

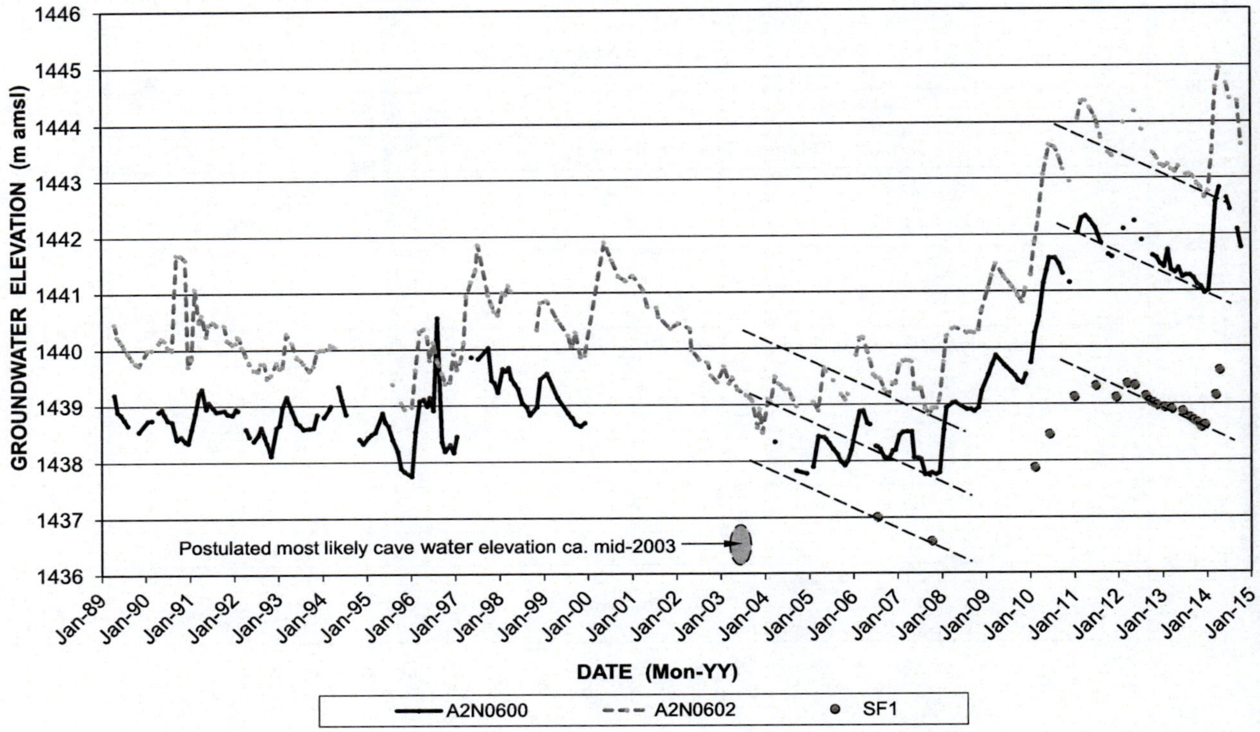


**Fig. 11** Long-term potentiometric response in monitoring boreholes A2N0600 and A2N0602 compared to the response in borehole SF1 serving as a proxy for the cave water level

chemistry reflect a composite of everything that has happened upstream (White 2002). It is for this reason that Quinlan and Ewers (1985) stress that reliable monitoring in karst must involve springs, and not just boreholes. Bakalowicz (2005) goes as far as to state "The spring is the only place where one can obtain information on the functioning of the whole system, ...." and "Consequently, if the main spring cannot be monitored, .... the system cannot be simulated nor managed properly". These circumstances define the recognition of springs as the most appropriate gauging, sampling and monitoring points also, and probably more particularly so, in a karst environment (Williams 2008).

Vegter (1995) lists 57 cold springs in South Africa yielding > 1000 $m^3/d$ (~ 11.6 L/s). An inspection of this list and its accompanying map suggests that only one of these springs, namely #13 on Kromdraai 520JO (25.969°S–27.788°E), might form part of those enumerated in the COH. It must be noted, however, that the Kromdraai Spring(s) reported by Bredenkamp et al. (1986), and echoed by Holland (2008) and Holland and Cobbing (2008) as representing #13 of Vegter (1995), cannot be associated with the Kromdraai Spring enumerated in this study for the reasons set out in Text Box 1.

Salient information pertaining to the 14 springs (Fig. 2) enumerated in the COH, is presented in Table 9. Some of these represent groups of springs (and seeps) located in close proximity to one another.

## 4.1 Zwartkrans Spring

Located ~ 1.4 km north-east (downstream) of Sterkfontein Cave, this spring drains the Zwartkrans Basin spanning the south-western dolomitic portion of Quaternary basin A21D. It represents a typical barrier contact spring formed by the Zwartkrans Dyke at this position, which also explains a seepage area located on the opposite (north) bank of the Bloubank Spruit on the farm Boland. The spring occupies a position immediately adjacent to the Bloubank Spruit at an elevation that is similar to that of the streambed (Plate 1). Discharging directly into the stream, these features share a common elevation of ~ 1433 m amsl at this location.

Flow measurements carried out on 27/07/2010 at locations ~ 460 m upstream and ~ 2.5 km downstream of the Zwartkrans Spring provided an early estimate of the spring discharge. The upstream measurement yielded a discharge of ~ 426 L/s. The downstream discharge comprised the

**Table 9** Salient information pertaining to enumerated springs in the COH

| Spring | Basin | Coordinates (dd.dddd)[a] | Elevation (m amsl) | Discharge[b] | | Type and Source Aquifer |
|---|---|---|---|---|---|---|
| | | | | (L/s) | (ML/d) | |
| Zwartkrans | Zwartkrans | 26.01°S 27.75°E | ~ 1433 | ~ 136 | ~ 11.7 | Barrier dyke contact: Dolomite (Malmani Sbgp.) |
| Plover's Lake | Krombank | 25.99°S 27.78°E | ~ 1419 | ~ 60 | ~ 5.2 | Barrier sill contact: Dolomite (Malmani Sbgp.) |
| Kromdraai | | 25.98°S 27.78°E | ~ 1406 | ~ 279[c] | ~ 24.1 | Structural lineament: Dolomite (Malmani Sbgp.) |
| Danielsrust | Danielsrust | 25.98°S 27.74°E | ~ 1486 | ~ 28 | ~ 2.4 | Barrier sill contact: Dolomite (Malmani Sbgp.) |
| Aquamine | Rietfontein | 25.95°S 27.81°E | ~ 1436 | < 2 | < 0.2 | Lithologic contact: Shaly sandstone (Ventersdorp Spgp.) |
| Tweefontein | Tweefontein | 25.91°S 27.80°E | ~ 1450 | ~ 30 | ~ 2.6 | Barrier dyke contact: Dolomite (Malmani Sbgp.) |
| Nouklip | Diepkloof | 25.88°S 27.79°E | ~ 1330 | ~ 143[d] | ~ 12.4 | Lithologic contact: Dolomite (Malmani Sbgp.) |
| Nash | Uitkomst | 25.89°S 27.77°E | ~ 1360 | ~ 130 | ~ 11.2 | Lithologic contact: Dolomite (Malmani Sbgp.) |
| Barlow | Kalkheuvel | 25.85°S 27.88°E | ~ 1363 | ~ 2 | ~ 0.2 | Lithologic contact: Dolomite (Malmani Sbgp.) |
| Uitkomst | (non-karst) | 25.91°S 27.74°E | ~ 1410 | ~ 2 | ~ 0.2 | Lithologic contact: Shale (Timeball Hill Fm.) |
| Cradle | Motsetse | 25.91°S 27.85°E | ~ 1405 | ~ 2 | ~ 0.2 | Barrier sill contact: Dolomite (Malmani Sbgp.) |
| Broederstroom | Broederstroom | 25.82°S 27.87°E | ~ 1281 | ~ 21 | ~ 1.8 | Lithologic contact: Dolomite (Malmani Sbgp.) |
| Anderson | | 25.85°S 27.88°E | ~ 1365 | ~ 5 | ~ 0.4 | Lithologic contact: Dolomite (Malmani Sbgp.) |
| Lesedi | | 25.84°S 27.88°E | ~ 1337 | ~ 2 | ~ 0.2 | Lithologic contact: Dolomite (Malmani Sbgp.) |
| | | | TOTAL | 842 | 72.7 | |

[a] Coordinates truncated to two decimals to prevent unwelcome visitation of these features
[b] Measured by the author using an OTT C20 current meter as described in Footnote 28
[c] Excludes the autogenic recharge produced by the Danielsrust Spring
[d] Excludes the autogenic recharge produced by the Tweefontein Spring

**Plate 1** View of the Zwartkrans Spring rising in basin at front right of picture, discharging into the Bloubank Spruit

sum of the A-furrow discharge (~ 154 L/s) and the Bloubank Spruit discharge (~ 477 L/s) for a total of ~ 631 L/s. The difference of ~ 205 L/s (17.7 ML/d) might represent the discharge of the Zwartkrans Spring. However, the flow conditions and setting of the upstream and downstream SDM stations indicated that the calculated spring discharge value should be regarded as a relatively coarse approximation for the following reasons (amongst others):

- the distance (~ 2450 m) of the river measurement station downstream of the spring;
- the shallow and turbulent flow conditions that characterised the culvert settings of the downstream river measurement station;
- the possibility of unaccounted-for abstraction from the A-furrow upstream of the canal measurement station; and

- the unaccounted flow through streambed vegetation at the upstream gauging site.

As a consequence, a second set of measurements was carried out closer to the spring on 13/08/2010. These returned a discharge of ~ 535 L/s (46.2 ML/d) at the upstream station, and a total downstream discharge of ~ 635 L/s (54.8 ML/d) comprising the sum of the A-furrow discharge (~ 176 L/s) and the Bloubank Spruit discharge (~ 459 L/s). It is encouraging that the downstream river and A-furrow discharges on the two measurement dates are similar, viz. ~ 477 versus ~ 459 L/s and ~ 154 versus ~ 176 L/s, respectively. The total downstream discharges of ~ 631 and ~ 635 L/s on the two measurement dates differs by < 1%. The 13/08/2010 set of measurements return a significantly lower discharge of ~ 100 L/s (8.64 ML/d) for the spring. These circumstances indicated the need to more accurately determine the discharge of the spring, perhaps focussing especially on the upstream measurement. The veracity of both sets of flow measurements is interrogated in Tables 10 and 11 on the basis of mass balance calculations as follows:

$$Q_S = [(Q_D C_D) - (Q_U C_U)] \div C_S = [416\,171 - 339\,948] \div 595 = 128\,L/s$$
$$Q_S = [(Q_D C_D) - (Q_U C_U)] \div C_S = [440\,055 - 396\,970] \div 595 = 72\,L/s$$

The mass balance results are in better agreement with each other than the SDM-based results. Further, the average of the mass balance results (~ 100 L/s) is similar to the SDM-based result of 13/08/2010. This finding suggests that the discharge of the Zwartkrans Spring on the gauging occasions is more likely in the order of 100 L/s than 200 L/s or greater.

The SDMs carried out at stations BB@M and BB@L on 16/05/2012 (Sect. 5.2.3 in Chapter "Physical Hydrology") provided a third opportunity to establish the flow of the Zwartkrans Spring. The data presented in Table 12 describe the results obtained. The veracity of the flow measurements given in Table 12 is again interrogated on the basis of a mass balance calculation as follows:

$$\begin{aligned} Q_S &= [(Q_D C_D) - (Q_U C_U)] \div C_S \\ &= [415\,198 - 212\,058] \div 644 = 315\,L/s \end{aligned}$$

The derived spring discharge value of 325 L/s ($Q_S$, Table 12) is encouragingly close to the mass balance calculated value of ~ 315 L/s. The difference of 10 L/s is readily accounted for by the negligible flow in the A-furrow ($Q_F$) which, if added to the downstream flow ($Q_D$) and assuming an SEC value ($C_F$) of 94 mS/m, returns a mass balance calculated value of ~ 327 L/s for the Zwartkrans Spring. The result, however, is clearly significantly greater than obtained in either of the previous attempts (Tables 10 and 11).

As postulated in Sect. 5.2.3 in Chapter "Physical Hydrology", it is questionable whether the Zwartkrans Spring represents the only outlet of the Zwartkrans Basin, i.e. that it is the sole source of groundwater discharge. There is sufficient evidence to indicate that a significant proportion of the observed groundwater contribution to surface flow derives from effluent conditions in the stream channel upstream of the spring. It is quite plausible that the magnitude of this contribution (~ 90 L/s/km of stream reach) exceeds that of the spring itself. Given a stream reach length of ~ 460 m between the upstream BB@M SDM position and the spring, groundwater resurgence could amount to at least ~ 40 L/s. Subtracting this from the spring discharge of 325 L/s yields a value of 285 L/s. Although this is in reasonable agreement with historical reported spring flow values (Bredenkamp et al. 1986), it still far exceeds the results reported in Tables 10 and 11. Unfortunately these circumstances do not resolve the ambiguity that still exists in regard to more accurate quantification of the Zwartkrans Spring discharge.

A fourth set of SDMs carried out on 15/01/2013 provided an opportunity to resolve the existing ambiguity regarding the Zwartkrans Spring discharge. The difference on this occasion was the much tighter bracketing of the Zwartkrans Spring by the in-stream SDM positions. The upstream position was located ~ 20 m 'above' the spring, and the downstream position ~ 15 m 'below' the spring. Additional rigour was secured by having the irrigation water supply pumps installed at the spring switched off prior to the SDMs, a measure that was not implemented with any of the previous SDMs. The data presented in Table 13 describe the results obtained.

The veracity of the flow measurements given in Table 13 is again interrogated on the basis of a mass balance calculation as follows:

$$\begin{aligned} Q_S &= [(Q_D C_D) - (Q_U C_U)] \div C_S \\ &= [393\,713 - 301\,656] \div 665 = 138\,L/s \end{aligned}$$

The derived spring discharge value of ~ 136 L/s ($Q_S$, Table 13) is very close to the mass balance calculated value of ~ 138 L/s. These results are in reasonable agreement with those obtained from the 13/08/2010 (Table 11) SDMs, as well as the mass balance derived discharges associated with the 27/07/2010 and 13/08/2010 measurements, which yielded a spring discharge in the range 72–128 L/s. These findings indicate that a discharge of ~ 136 L/s in mid-January 2013 may confidently be assigned to the Zwartkrans Spring.

A groundwater loss factor that is not accounted for in the preceding discussion, and which is additional to the groundwater resurgence, is that of riparian evapotranspiration (RET) (Sect. 2 in Chapter "Description of the Physical

**Table 10** SDM and load-based calculation of the Zwartkrans Spring discharge on 27/07/2010

| Flow location | | Field SEC (mS/m) | TDS [$C$] (mg/L) | Discharge [$Q$] | | TDS load | |
|---|---|---|---|---|---|---|---|
| | | | | (L/s) | (ML/d) | mg/s [$=Q_XC_X$] | t/d |
| Downstream | Stream | 93 | 651[a] [$=C_D$] | ~ 477[b] [$=Q_D$] | 41.2 | 310,527 | 26.8 |
| | A-furrow | 98 | 686[a] [$=C_F$] | ~ 154[b] [$=Q_F$] | 13.3 | 105,644 | 9.1 |
| Upstream | | 114 | 798[a] [$=C_U$] | ~ 426[b] [$=Q_U$] | 36.8 | 339,948 | 29.4 |
| Calculated difference in salt load | | | | | | 76,223 | 6.5 |
| Zwartkrans Spring | | 85 | 595[a] [$=C_S$] | ~ 205[c] [$=Q_S$] | 17.7 | 121,975 | 10.5 |

[a] SEC · 7 used as a proxy to derive a theoretical TDS value
[b] Synoptic discharge measurement (SDM) value
[c] Derived value from the difference of the SDM values

**Table 11** SDM and load-based calculation of the Zwartkrans Spring discharge on 13/08/2010

| Flow location | | Field SEC (mS/m) | TDS [$C$] (mg/L) | Discharge [$Q$] | | TDS load | |
|---|---|---|---|---|---|---|---|
| | | | | (L/s) | (ML/d) | mg/s [$=Q_XC_X$] | t/d |
| Downstream | Stream | 99 | 693[a] [$=C_D$] | ~ 459[b] [$=Q_D$] | 39.7 | 318,087 | 27.5 |
| | A-furrow | 99 | 693[a] [$=C_F$] | ~ 176[b] [$=Q_F$] | 15.2 | 121,968 | 10.5 |
| Upstream | | 106 | 742[a] [$=C_U$] | ~ 535[b] [$=Q_U$] | 46.2 | 396,970 | 34.3 |
| Calculated difference in salt load | | | | | | 43,085 | 3.7 |
| Zwartkrans Spring | | 85 | 595[a] [$=C_S$] | ~ 100[c] [$=Q_S$] | 8.6 | 59,500 | 5.1 |

[a] SEC · 7 used as a proxy to derive a theoretical TDS value
[b] Synoptic discharge measurement (SDM) value
[c] Derived value from the difference of the SDM values

**Table 12** SDM and load-based calculation of the Zwartkrans Spring discharge on 16/05/2012

| Flow location | | Field SEC (mS/m) | TDS [$C$] (mg/L) | Discharge [$Q$] | | TDS load | |
|---|---|---|---|---|---|---|---|
| | | | | (L/s) | (ML/d) | mg/s [$=Q_XC_X$] | t/d |
| Downstream | Stream | 94 | 658[a] [$=C_D$] | ~ 631[b] [$=Q_D$] | 54.5 | 415,198 | 35.9 |
| | A-furrow | 94 | 658[a] [$=C_F$] | Negligible [$=Q_F$] | | | |
| Upstream | | 99 | 693[a] [$=C_U$] | ~ 306[b] [$=Q_U$] | 26.4 | 212,058 | 18.3 |
| Calculated difference in salt load | | | | | | 203,140 | 17.6 |
| Zwartkrans Spring | | 92 | 644[a] [$=C_S$] | ~ 325[c] [$=Q_S$] | 28.1 | 209,300 | 18.1 |

[a] SEC · 7 used as a proxy to derive a theoretical TDS value
[b] Synoptic discharge measurement (SDM) value
[c] Derived value from the difference of the SDM values

Environment"). The magnitude of such loss has been estimated at 25–50 mm/a in the Valley and Ridge, Blue Ridge and Piedmont physiographic provinces in the eastern USA (Rutledge and Mesko 1996; Swain et al. 2004). These values equate to losses of 1250–2500 $m^3$/a/km (0.04–0.08 L/s/km) for a 50-m wide zone[8] spanning the Bloubank Spruit. It is evident that even if the width of this zone was an order of magnitude greater, groundwater losses attributable to RET in the study area would still not approach those of groundwater resurgence along stream reaches where conditions favoured both these processes.

In conclusion, it is worth noting that the combined discharge of at least ~ 260 L/s represented by the Zwartkrans Spring and groundwater resurgence as set out in Table 13 and Table 5.15 respectively, agrees remarkably well with the 258 L/s reported by Bredenkamp et al. (1986) (Sect. 1) for the Zwartkrans Spring alone. To this must be added the

[8] Bredenkamp and van Staden (2009) report that the riparian tree zone is typically 10–20 m wide and the vegetation is dense and almost forest-like, dominated by indigenous trees and very dense forest-like bush on both banks of rivers, streams and creeks (Sect. 3.2).

**Table 13** SDM and load-based calculation of the Zwartkrans Spring discharge on 15/01/2013

| Flow location | Field SEC (mS/m) | TDS [$C$] (mg/L) | Discharge [$Q$] | | TDS load | |
|---|---|---|---|---|---|---|
| | | | (L/s) | (ML/d) | mg/s [=$Q_X C_X$] | t/d |
| Downstream | 103 | 721[a] [=$C_D$] | ~ 546[b] [=$Q_D$] | 47.2 | 393,713 | 34.0 |
| Upstream | 105 | 735[a] [=$C_U$] | ~ 410[b] [=$Q_U$] | 35.5 | 301,656 | 26.1 |
| Calculated difference in salt load | | | | | 92,057 | 7.9 |
| Zwartkrans Spring | 95 | 665[a] [=$C_S$] | ~ 136[c] [=$Q_S$] | 11.7 | 90,206 | 7.8 |

[a] SEC · 7 used as a proxy to derive a theoretical TDS value
[b] Synoptic discharge measurement (SDM) value
[c] Derived value from the difference of the SDM values

groundwater resurgence that occurs upstream of station BB@M, which has been measured at between 90 and 190 L/s (Sect. 5.2.3 in Chapter "Physical Hydrology"), for a total groundwater discharge from the Zwartkrans Basin in the range 350–450 L/s. These discharges equate to recharge values of ~ 16 and ~ 20%, respectively, for the Zwartkrans Basin.

## 4.2 Plover's Lake Springs

Located on the Crab Farm and Plover's Lake estates in the Kromdraai area of Quaternary A21D, the features that comprise the group of springs and seeps collectively referred to as the Plover's Lake springs[9] produced a combined yield of ~ 60 L/s (5.2 ML/d) in both September 2009 and May 2010, and ~ 64 L/s (5.5 ML/d) in August 2014. The excellent quality karst groundwater (Sect. 7.3 "Chapter Chemical Hydrogeology") is channelled into a landscaped central drainage feature with canals and furrows on either side that deliver water to various downstream properties on the Plover's Lake and Crab Farm estates before discharging into the Bloubank Spruit. These springs also provide a basis for the extrapolated elevation of the potentiometric surface associated with the Plover's Lake fossil site (Fig. 9 and Table 6 in Chapter "Description of the Physical Environment") located ~ 200 m upgradient (west) of these features.

## 4.3 Kromdraai Spring

Also located in Quaternary catchment A21D, the Kromdraai Spring discharges directly into the Bloubank Spruit on the left bank of this drainage on Ptn 35 (Ekuthuleni Estate) of the farm Kromdraai 520 JQ. Its submerged discharge is distinguishable from that of the passing flow in the Bloubank Spruit on the basis of the following observations:

- displacement towards the opposite bank of floating particulate matter from upstream;
- a distinctive body of much clearer water in the river, together with a perceptibly perturbed flow environment directed at right-angles to the passing surface flow;
- a typical groundwater temperature of 19.4 °C at the discharge point in late winter,[10] compared to the 13.3 °C of upstream river water (difference of 6.1 °C) at this time; and
- an SEC value of ~ 55 mS/m (Fig. 33 "Chapter Chemical Hydrogeology") at the discharge point into the river, compared to an SEC of 84 mS/m for the river water upstream of the spring.

An assessment of the spring discharge obtained from the difference between flow measurements made immediately upstream and downstream of the site (Table 14) on 13/08/2010 returned a value of ~ 307 L/s (26.5 ML/d). This is by far the highest discharge of the eleven springs enumerated in the study area (Table 9). Understandably, this observation raised concern for the veracity of the SDMs. As applied in the Zwartkrans Spring discharge assessment (Sect. 4.1), a mass balance comparison of the upstream and downstream springwater TDS loads offers a means to verify the result. Using the values in Table 14, the mass balance calculation yields the following result.

$$Q_S = [(Q_D C_D) - (Q_U C_U)] \div C_S$$
$$= [400\,414 - 274\,008] \div 392 = 322\,L/s$$

Accepting the margin of error associated with the flow measurements (possibly as much as ± 15%), the result indicates that the Kromdraai Spring TDS load closely approximates the difference between the upstream and downstream SDM values as illustrated in Fig. 12. A spring discharge of 322 L/s would account exactly for the calculated difference in TDS load of 10.9 t/d (Table 14). Although this result breeds confidence in the spring discharge value derived from the SDMs, it does not explain the anomalously large recharge estimation of ~ 30% which this result

[9] Referred to as the Kromdraai Springs in previous studies, e.g. Bredenkamp et al. (1986) (see Text Box 1).

[10] The measurements were made on 13/08/2010.

**Table 14** Verification of the Kromdraai Spring discharge using TDS load calculations

| Flow location | Temp. (°C) | SEC (mS/m) | TDS [*C*] (mg/L) | Discharge [*Q*] | | TDS load | |
|---|---|---|---|---|---|---|---|
| | | | | (L/s) | (ML/d) | mg/s [$=Q_XC_X$] | t/d |
| Downstream | 14.8 | 74 | 518[a] [$=C_D$] | ~ 773[b] [$=Q_D$] | 66.8 | 400,414 | 34.6 |
| Upstream | 13.3 | 84 | 588[a] [$=C_U$] | ~ 466[b] [$=Q_U$] | 40.3 | 274,008 | 23.7 |
| Calculated difference in salt load | | | | | | 126,406 | 10.9 |
| Kromdraai Spring | 19.4 | 56 | 392[a] [$=C_S$] | ~ 307[c] [$=Q_S$] | 26.5 | 120,344 | 10.4 |

[a] SEC · 7 used as a proxy to derive a theoretical TDS value
[b] Synoptic discharge measurement (SDM) value
[c] Derived value from the difference of the SDM values

attributes to the basin footprint area (Table 4). This circumstance is discussed in Sect. 2.12.

The discharge values suggest that the Kromdraai Spring contributed ~ 40% of the flow in the Bloubank Spruit at this location on the day of measurement. The downstream discharge of 66.8 ML/d equates to 24.4 mm$^3$/a, which is slightly greater than the long-term median MAR value of 22.6 mm$^3$/a recorded at gauging station A2H049 (Fig. 10 in Chapter "Physical Hydrology"). Accepting the extreme coarseness of such a comparison, these circumstances reflect the abnormally high flow conditions experienced in the inter-connected surface water and groundwater components of the Bloubank Spruit system during the 2010 and 2011 winters. More significantly, however, the Kromdraai Spring discharge of 26.5 ML/d (9.7 mm$^3$/a) represents ~ 43% of the long-term median MAR at station A2H049. This contribution from karst groundwater increases to ~ 53% with the inclusion of the Plover's Lake springs discharge of ~ 60 L/s (5.2 ML/d).

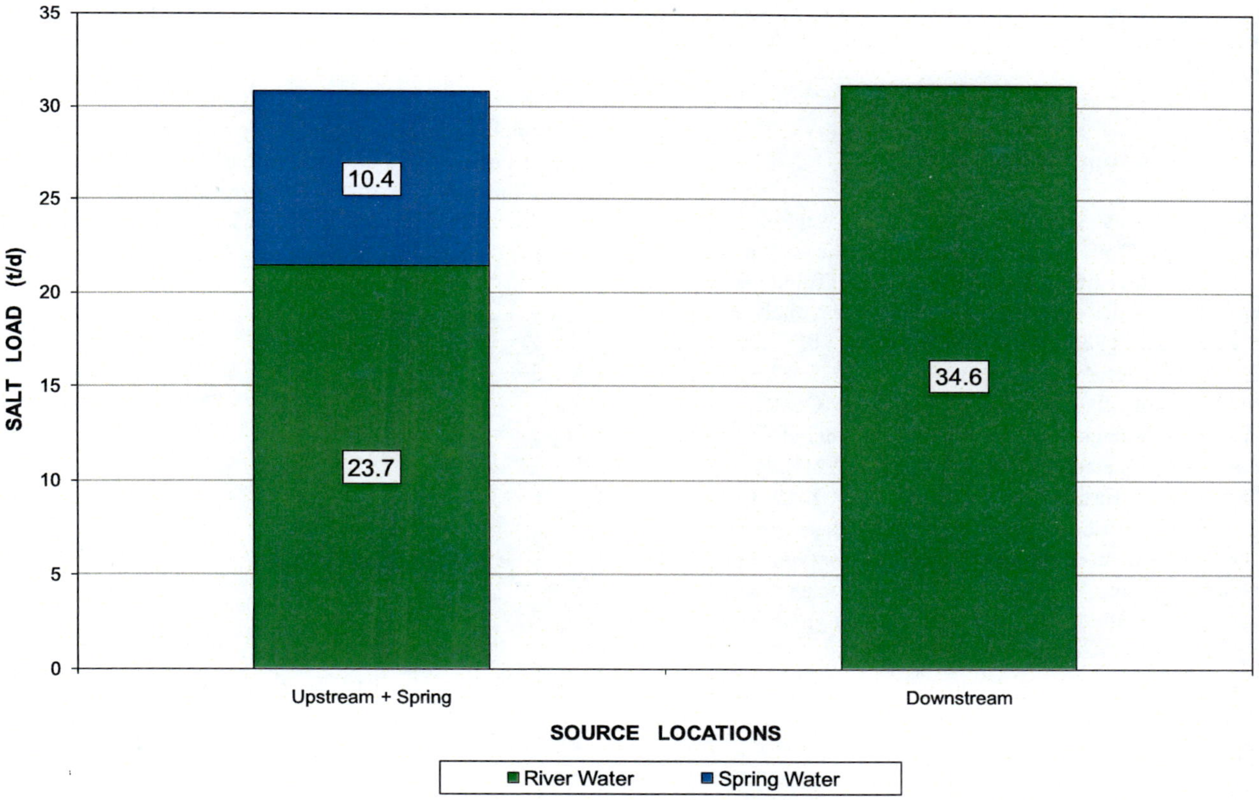


**Fig. 12** Verification of the Kromdraai Spring discharge based on mass balance calculations

## 4.4 Danielsrust Spring

The Danielsrust Spring drains the Danielsrust Basin, a 'perched' karst aquifer ~ 740 ha in extent located within the Krombank Basin, and which clearly exhibits hydraulic separation from the latter (Sect. 2.3). The long-term discharge of 28 L/s (2.4 ML/d) is not used at present, and is lost to a combination of evaporation, evapotranspiration and infiltration into the Krombank Basin. These circumstances simultaneously describe autogenic discharge of the Danielsrust Basin and autogenic recharge of the Krombank Basin, both located in Quaternary catchment A21D.

## 4.5 Aquamine Spring

The Aquamine Spring rises in the upper reaches of the Tweefontein Spruit on the contact between Ventersdorp Supergroup and Dominium Group strata, but is provisionally considered to include the ~ 725 ha Rietfontein Basin in its catchment. The yield of the spring (< 2 L/s), however, is notably less than that of the karst springs draining karst basins of similar size, e.g. the 28 L/s Danielsrust Spring draining the ~ 740 ha Danielsrust Basin. These circumstances might attract further investigation in future hydrogeologic studies.

## 4.6 Tweefontein Spring

Of the 11 karst springs surveyed, the Tweefontein Spring (aka JNNR1) in the JNNR occupies the second highest elevation after the Danielsrust Spring (Table 9). It also exhibits a similar flow to that of the latter, namely ~ 30 L/s (SDM on 19/05/2010). SDMs on 12/08/2014 and 27/02/2018 returned flows of ~ 26 and ~ 12 L/s respectively. Although the spring is located in Quaternary basin A21G, its catchment of ~ 1160 ha includes dolomitic portions of Quaternary basins A21D and A21E. Rising in the Tweefontein Basin at an elevation of ~ 1450 m amsl, its discharge contributes autogenic recharge to the Diepkloof Basin to the north drained by the Grootvlei Spruit and Nouklip Spring. Contemporary SDM values indicate an average contribution of 14% to the discharge of the Nouklip Spring.

## 4.7 Nouklip Spring

The Nouklip Spring (aka JNNR2) rises in the lower reach of the Grootvlei Spruit and is impounded in the so-called 'swimming pool' dam in the north-western portion of the JNNR. This impoundment also collects the discharges (< 2 L/s each) of at least 11 other seepages in the vicinity (T Abiye, personal communication), and releases the aggregate flow (measured at ~ 173 L/s on 19/05/2010, ~ 211 L/s on 12/08/2014 and ~ 12 L/s on 27/02/2018) via the Grootvlei Spruit into the Skeerpoort River (Plate 2). Rising at an elevation of ~ 1330 m amsl at the contact between the Rooihoogte Formation and the overlying ferruginous shales of the Timeball Hill Formation (Pretoria Group), it drains the ~ 4125 ha Diepkloof Basin. This basin includes part of the dolomitic portions of Quaternary basins A21G and A21H.

The SDM flows of ~ 173, ~ 211 and ~ 105 L/s translate into discharges of ~ 5.5, ~ 6.7 and ~ 3.3 $mm^3/a$. The mean discharge of 163 L/s (5.1 $mm^3/a$) is in fair agreement with the median discharge per hydrological year of ~ 5.4 $mm^3$ reported for gauging station A2H033 (Fig. 2 in Chapter "Physical Hydrology") in the period 1997–2015. Subtracting the autogenic recharge delivered by the Tweefontein Spring returns 'stand-alone' flows for the Nouklip Spring in the range ~ 93 to ~ 183 L/s. On 19/05/2010, the losing reach downstream of the Tweefontein Spring extended for ~ 700 m. On 19/01/2012 it had increased to ~ 1450 m, and on 12/08/2014 to ~ 1550 m. By 27/02/2018, however, the losing reach had shrunk to < 600 m. This is considered to reflect the coeval lower SDM yield of ~ 12 L/s for the Tweefontein Spring (Sect. 4.6).

A closer inspection of the A2H033 hydrograph for the last seven hydrological years returns the information presented in Fig. 13. The median discharge of 170 L/s (~ 5.4 L/s) is similar to that for the longer period going back to 1997. The relative accuracy of the May 2010 and August

**Plate 2** Discharge of the Grootvlei Spruit (centre right) into the Skeerpoort River (at left, looking downstream); flow in the Skeerpoort River before this confluence (in foreground) derives mainly from the ~ 130 L/s discharge of the Nash Spring

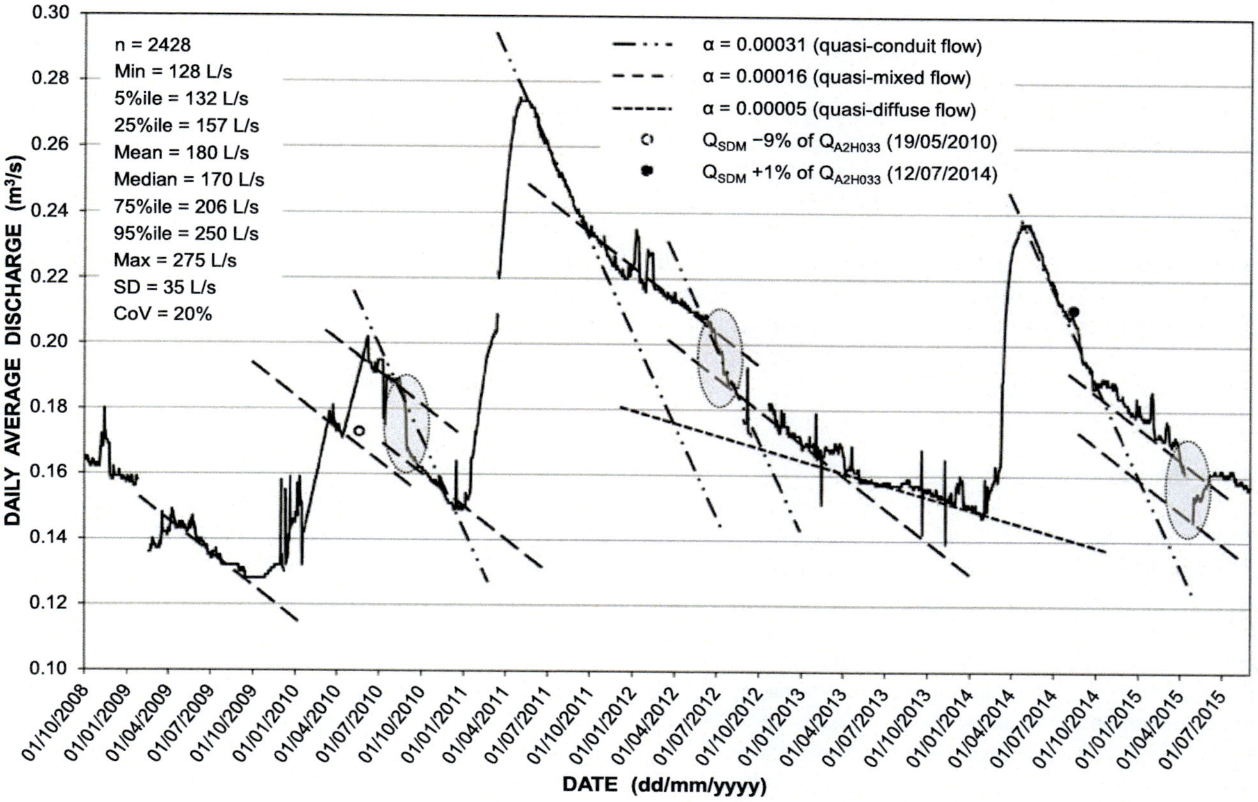


**Fig. 13** Hydrograph of the Nouklip Spring as gauged at station A2H033 on the Grootvlei Spruit ~ 525 m downstream of the spring

2014 SDM values to the gauged values are shown to be ± 10%. The hydrograph reveals a complex recession comprising three distinct limbs defined by α–values (slopes) of 0.00031, 0.00016 and 0.00005 respectively. Taylor and Greene (2008) show that slopes of this magnitude describe the diffuse flow or slow flow components of karst drainage systems. This is somewhat misleading, as the steepest recession segments typically represent the conduit flow component, and the intermediate segments the mixed flow component of such systems.

A further characteristic of the hydrograph is the periods (shaded oval areas) of quasi-conduit ($\alpha_{q\text{-}c}$) flow that 'interrupt' the periods of quasi-mixed ($\alpha_{q-m}$) flow. These suggest that the discharge characteristics of the Diepsloot karst basin reflect a multi-layered karst system comprising a hydraulic continuum between three sets of hydrogeologic units where the porosity and permeability of $\alpha_{q-c} > \alpha_{q-m} > \alpha_{q\text{-}d}$ (quasi-diffuse flow) all within the domain of classic diffuse flow. The volume of water in storage in the saturated zone above the elevation of the spring, known as the dynamic volume $V$[11] (Ford and Williams 2007), is calculated at 530 mm$^3$, being the sum of the component volumes $V_{R0}$ (328 mm$^3$), $V_{R1}$ (124 mm$^3$) and $V_{R2}$ (78 mm$^3$). The sum of $V_{R1}$ and $V_{R2}$ (202 mm$^3$) represents the flood-water volume. The $V_{RT}$ of 530 mm$^3$ is ~ 280% greater than the nominal FSC (190 mm$^3$) of Hartbeespoort Dam, equivalent to a uniform water depth of ~ 14 m over the 3800 ha extent of the Diepkloof Basin, or else a uniform depth of ~ 70 m for a storativity of 20%. Readers conversant in French are referred to the articles by A. Mangin, as reported in Ford and Williams (2007), for further literature on this topic.

It is postulated that the discharge characteristics of the Nouklip Spring are representative of the other high-yielding springs in the COH and, for that matter, in the wider distribution of Malmani Subgroup strata. The relative constancy and low variability (~twofold difference between $Q_{min}$ and $Q_{max}$) of these features supports the Martini (2006) observation regarding the 'very regular discharges at resurgences' that characterise karst springs of the South African interior.

---

[11] $V = cQ_0/\alpha$, where the constant $c$ = 86 400 for baseflow $Q_0$ in m$^3$/s and slope α in days.

## 4.8 Nash Spring

The location of the Nash Spring (aka JNNR3) close to the contact between the dolomitic (Malmani Subgroup) and sedimentary (Timeball Hill Formation) strata in the north-western portion of the JNNR, together with the substantial discharge (~ 130 L/s on 19/05/2010, ~ 208 L/s on 12/08/2014 and > 100 L/s on 27/02/2018), indicates that this feature represents the main outflow of the ~ 2750 ha Uitkomst Basin north-westwards into the Skeerpoort River catchment (Quaternary basin A21G). The catchment of this spring also includes a small dolomitic portion of Quaternary catchment A21D.

## 4.9 Uitkomst Spring

This spring (aka JNNR4) issues from shale of the Timeball Hill Formation (Pretoria Group) in close proximity to the contact with the underlying older dolomite of the Eccles Formation, or possibly the Rooihoogte Formation, in the north-western portion of the JNNR. The discharge of ~ 2 L/s is similar to that of the Aquamine Spring which also issues from non-dolomitic strata (Sect. 4.5). These circumstances again reflect the poorer water-bearing potential of the sedimentary rocks of the younger (overlying) Pretoria Group in Quaternary basin A21G compared to the dolomitic formations.

## 4.10 Cradle Spring

Located in the Motsetse Nature Reserve (MNR), the Cradle Spring is another example of a barrier sill contact spring in the COH. It drains the comparatively small (270 ha) Motsetse Basin in Quaternary basin A21E at an elevation of ~ 1405 m amsl. The groundwater level elevation of ~ 1432 m amsl determined for geosite MNR1, a borehole located ~ 1100 m upgradient of the spring, indicates a hydraulic gradient of 0.0245 between these two sources. This is 'normal' for a karst environment, and it is therefore surprising that the Cradle Spring delivers only ~ 2 L/s. This flow equates to a natural recharge of only ~ 3% of an MAP of 710 mm on the Motsetse Basin (Table 4). This presumes, of course, that the spring is the only outlet for this compartment. Other water 'loss' components that might factor into this equation, in order of decreasing likelihood, are:

- groundwater abstraction for water supply to the Cradle Boutique Hotel and Cradle Restaurant on the property;
- unaccounted for subsurface losses through the barrier sill; and
- evapotranspiration under circumstances of a comparatively shallow depth to groundwater level (~ 12 m bs) as observed in borehole MNR1 located upstream of the spring.

These components together need only amount to ~ 5 L/s to produce a natural recharge that is equivalent to ~ 12% of an MAP of 710 mm for this compartment. It is also possible, of course, that the spring drains mainly chert-poor strata of the Lyttelton Formation.

## 4.11 Anderson Spring

The Anderson Spring occupies an elevation of ~ 1365 m amsl in an easterly draining valley on the Kalkheuwel West Estate. It is located in the south-eastern corner of the Broederstroom Basin defined by the E–W striking Hartbeesthoek Dyke and the N–S striking Rhenosterspruit Dyke (Fig. 3). Its yield was measured at ~ 5 L/s on 27/11/2014.

## 4.12 Barlow Spring

The Barlow Spring is located ~ 1600 m south-southeast of the Anderson Spring, which places it in the north-eastern corner of the Kalkheuvel Basin at a similar elevation (~ 1365 m amsl) as the Anderson Spring. It comprises a number of seeps that individually are difficult to quantify, but which drain from the receiving dam at a rate of < 2 L/s.

## 4.13 Lesedi Spring

The Lesedi Spring is located at an elevation of ~ 1337 m amsl on the property of the Lesedi African Lodge and Cultural Village. This locates it in the Broederstoom Basin. It is a protected feature encased in a caisson-like structure which drains via a buried 100 mm uPVC pipe to a collection chamber. The collection chamber is fitted with a centrifugal pump to distribute water to residences located on the high ground (> 1450 m amsl) in the western portion of the Kalkheuwel West Estate.

## 4.14 Broederstroom Spring

Located on the Glen Afric Country Lodge property south of Broederstroom, this spring occupies an elevation of ~ 1281 m amsl in a northerly draining valley in Quaternary basin A21H. It is another example of a lithologic contact spring in the COH. Its measured yield of ~ 21 L/s compares favourably with a reported maximum yield of 25 L/s (J Brooker, personal communication). Nevertheless, this equates to a natural recharge of only ~ 5% of an MAP of 710 mm on the 1810 ha Broederstroom Basin (Table 4), which supports the finding that other 'drains' (e.g. the Anderson Spring, Lesedi Spring and borehole abstractions) impact on the Broederstroom Basin groundwater store.

### 4.15 Krugersdorp Game Reserve Springs

The Krugersdorp Game Reserve hosts at least eight springs. As shown in Table 15, seven of these are associated with dolomite and one with quartzitic strata. The spring SP3 represents a group of distributed seeps rather than a single recognisable 'eye'. In this instance, the discharge represents the aggregate flow of the contributing seeps as measured at a position downstream of the topographically lowest seep in the group.

Five of the seven karst springs drain the karst outlier that extends from the mine area located to the south, into the southern portion of the KGR (Fig. 14). The other two karst springs are located in the Zwartkrans Basin to the north of the quartzite ridge that strikes NE–SW diagonally through the centre of the KGR.

The springs typically support discharges < 5 L/s, although at least three (SP2, SP3 and Buffalo Sp.) yield > 5 L/s. The yield of Spring 2 (CSIR30) has not been measured, but JFA (2006) reported a yield of 12 L/s. SP3 delivers ~ 10 L/s on average, its catchment defined by the Tweelopie Spruit on the east and the Sterkfontein Dyke on the west (Fig. 3). It is doubtful whether the combined yield of all these features exceeds 30 L/s (2.6 ML/d).

### 4.16 Spring Discharge Context and Significance

The total number of springs in the study area is probably an order of magnitude greater than the eleven enumerated in this study (Table 9). Substantially greater investigative effort relying largely on the contribution of local residents is required to properly evaluate this aspect of the hydrogeologic environment. Nevertheless, the springs enumerated in the study area form two groups, namely those draining the Skeerpoort River catchment (Quaternary basin A21G) and those draining the Bloubank Spruit system (A21D).

The A21G springs together deliver ~ 305 L/s (9.62 $mm^3/a$), the equivalent of ~ 5% of the MAR of Hartbeespoort Dam. The A21D springs together deliver ~ 504 L/s (15.89 $mm^3/a$), to which must be added groundwater resurgence in the stream channel of at least ~ 130 L/s, delivering the equivalent of ~ 10.5% of the Hartbeespoort Dam MAR. Although certainly not all of the spring discharge reaches the dam, the combined flow of 29.6 $mm^3/a$ still represents ~ 16% of the full supply capacity of the reservoir. More significantly, the collective groundwater discharge in the A21D catchment represents ~ 88% of the long-term median annual discharge (22.7 $mm^3$) of this basin. The value of this generally good to excellent quality groundwater in mitigating the negative impact of the mine and municipal wastewater cannot be over-emphasised.

It is evident, therefore, that the COH is not only one of the most productive palaeo-anthropological terranes in southern Africa, but also one of the most productive karst terranes from a groundwater yield perspective. Calculating the financial value of the 'goods and services' delivered by this karst terrane[12] in mitigating the negative impact of the mine and municipal wastewater components on downstream water resources presents a further challenging exercise that is beyond the scope of this study.

## 5 Water in the Mining Environment

It is common cause that the underground void created by gold mining in the West Rand Goldfield is flooded and actively decanting acid mine drainage (AMD). The import hereof for the study area is woven into the fabric of the water resource environment (Sect. 5 in Chapter "Description of the Physical Environment"). The conceptual model of the mine water relationship to the natural hydrogeologic setting is illustrated in Fig. 15 using the geologic profile presented in Fig. 7 in Chapter "Description of the Physical Environment" as basis. The hydraulic link between the subsurface Main Reef mine workings and the LoD via the Black Reef Incline (BRI), #17 and #18 winzes is evident. Also shown in Fig. 15 is the position of monitoring boreholes RG1 and MGP2 that aid in elucidating the occurrence of mine water in the hydrogeologic environment of the Western Basin. The need to lower the hydrostatic head in the flooded mine void if active decant is to be curtailed, is also evident from Fig. 15. This might be achieved with a reduction in head of as little as ~ 25 m, i.e. from ~ 1675 to ~ 1650 m amsl.

It has also been recognised, however, that sufficient freeboard needs to be created in the mine void to serve as buffer for excessive recharge during wetter than normal summer rainfall seasons. It is partly for this reason that a reduction in head of ~ 150 m (equivalent to an elevation of ~ 1525 m amsl) has been proposed (Coetzee et al. 2010). This has, however, already been revised upward

[12] The reader is referred to a discussion piece by Blackeagle (2014) on the distinction between 'terrane' and 'terrain', on the basis of which use of the term 'terrane' in this context is explicit.

**Table 15** Description of springs enumerated in the Krugersdorp Game Reserve

| Spring | Alternative ID | Latitude[a] (dd.ddddd°S) | Longitude[a] (dd.ddddd°E) | Elevation (m amsl) | Source Aquifer | |
|---|---|---|---|---|---|---|
| CSIR30 | Spring 2 @ cemetery | 26.09808 | 27.71892 | 1641 | Karst outlier | Also referred to as the cemetery group of springs |
| CSIR63 | Spring 1 @ cemetery | 26.09814 | 27.71958 | 1638 | | |
| SP1 | F6S7 (Cemetery) | 26.09694 | 27.71898 | 1637 | | |
| SP2 | F8S9 (Poplar Grove) | 26.09092 | 27.72011 | 1626 | Karst outlier | |
| Buffalo Sp | – | 26.09073 | 27.72230 | 1627 | Karst outlier 'toe' | |
| Lodge Sp | CSIR20 | 26.09044 | 27.71625 | 1600 | Witwatersrand quartzite | |
| Aviary Sp | CSIR55 | 26.07753 | 27.70092 | 1530 | Zwartkrans Basin | |
| SP3 | Flip-se-Gat stream | 26.08149[b] | 27.69589[b] | 1553 | Zwartkrans Basin | |

[a] For a discussion of spring coordinates and disparity resolution, see Sect. 3.5 in Hobbs (2011)
[b] Coordinates of the most westerly (highest) seepage contributing to the aggregate discharge measured and sampled at coordinates 26.07692°S–27.69928°E

to ~ 1550 m amsl (BKS 2011; DEA 2014), and might even be relaxed further to ~ 1575 m amsl if circumstances allow (DWA 2013a).

**Text Box 1** The Kromdraai Spring/Plover's Lake Spring(s) conundrum

*The spring listed as #13 by Vegter (1995), and located on the farm Kromdraai 520JO[13] located at 25.969°S–27.788°E[14], has historically been regarded as the Kromdraai Spring. For example, Bredenkamp et al. (1986) refer to the Kromdraai Spring(s), as do Holland (2008) and Holland and Cobbing (2008). The Kromdraai Spring enumerated in this study, however, is located at coordinates 25.990°S–27.776°E (**Table 85** and **Fig.** 5.27). This equates to a linear geometric disparity of ~ 2600 m compared to the Vegter (1995) coordinates, and cautions against a direct correlation based simply on the spring and topocadastral farm names.*

*Under circumstances where none of Bredenkamp et al. (1986), Holland (2008) and Holland and Cobbing (2008) report coordinates for the Kromdraai Spring, it is difficult to unravel the geographic uncertainty between this feature and the Plover's Lake Spring(s). Further to the discussion presented in **Footnote 62**, an inspection of Fig. 7.1 in Bredenkamp et al. (1986), Figs. 2.1 and 2.2in Holland and Cobbing (2008) and Fig.100 (from Holland, 2009) clearly show the position of the Kromdraai Spring at the head of a short (~ 1000 m) left bank tributary of the Bloubank Spruit. As described in* Sect. 4.3 *, the Kromdraai Spring enumerated in this study discharges directly into the Bloubank Spruit.*

*The coordinates of the Kromdraai "Eye" taken off Fig. 7.1 (scale 1:50,000) of Bredenkamp et al. (1986) are X 2,874,550 m and Y –77,870 m (Cape Datum), which translate to X 2,874,850 m and Y –77,835 m (Hartebeesthoek94 Datum). The latter coordinates are within 205 m of one of the group of springs that comprise the Plover's Lake Spring(s) enumerated in this study. It is therefore reasonable to presume that the Kromdraai Spring referred to in earlier studies (e.g. Bredenkamp et al. 1986; Holland 2008; Holland and Cobbing 2008) are the Plover's Lake Spring(s) of this study.*

*The preceding discussion does not, however, resolve the disparity in reported yield values for the Plover's Lake Spring(s). This is given as ~ 30 L/s (0.946 mm$^3$/a) by Bredenkamp et al. (1986) and ~ 60 L/s by this study* (Sect. 4.2). *Likely explanations for the disparity are that (a) the Plover's Lake springs comprise two groups of springs and seeps delivering ~ 45 and ~ 15 L/s respectively, and (b) the mid-1980s was characterised by a drought period when spring yields in the region might reasonably be expected to have been lower than at present.*

The reduction in mine water head pertains to the so-called environmental critical level (ECL) that defines the elevation above which AMD will not impact on the environment (Coetzee et al. 2010). In most instances, securing and/or maintaining the ECL requires pumping of the mine water at

[13] *The topocadastral property is identified as Kromdraai 520JQ on the 1:50 000 scale topocadastral map sheet 2527DD Broederstroom (RSA, 2001).*

[14] *Converted from the Cape Datum to the Hartebeesthoek94 Datum; the accuracy of the conversion is limited by the coarse resolution of the Vegter (1995) coordinates reported to only three decimals.*

significant cost. The latter varies according to the pumping volume (typically 10's of ML/d) and the head (typically > 100 m) against which this must be achieved, which is a direct function of the specified ECL. It is therefore imperative that an ECL be selected which accommodates both the environmental and economic concerns. The feasibility study commissioned by the DWS to investigate the long-term solution to control and manage acid mine drainage in the Witwatersrand underground mining basins, has sought to refine aspects of this imperative also in regard to the Western Basin. These have been discussed in Text Box 1 in Chapter "Physical Hydrology". This study, referred to as the AMD FSLTS study, introduced the additional terms socio-economic critical level (SECL) and target operating level (TOL) (DWA 2013b) into the discourse on mine water management.

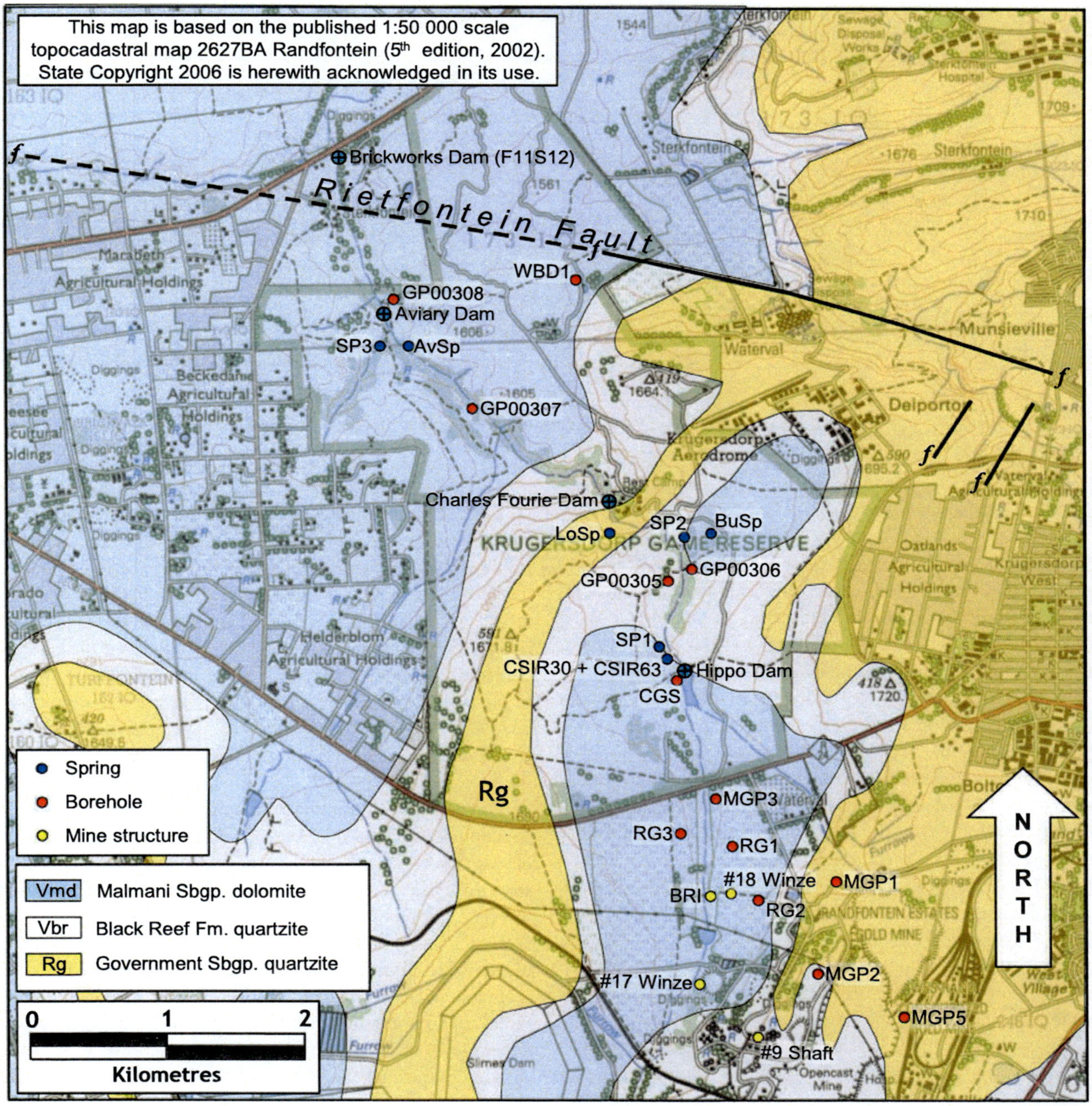


**Fig. 14** Locality map of geosites in the mine area and Krugersdorp Game Reserve

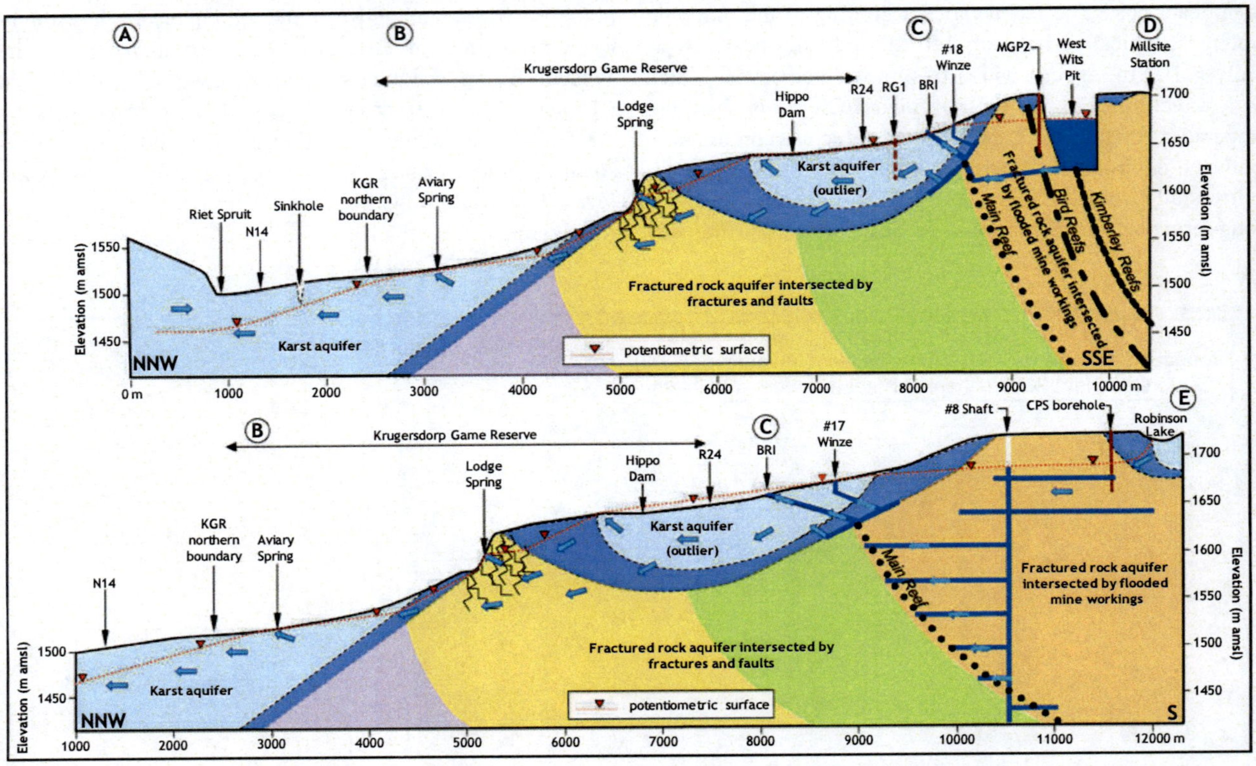


**Fig. 15** Schematic profile of hydrogeologic and mining-related features in the West Rand Goldfield (see Fig. 10 for profile positions)

## 6 Groundwater Drainage Pattern

The groundwater drainage pattern in the karst of the WHS property has undoubtedly been one of the poorer understood hydrogeologic aspects of the study area. Hobbs and Cobbing (2007) criticise the Bredenkamp et al. (1986) subdivision of the Zwartkrans Basin into nine subunits that, together with a comparatively sparse set of water level data, raises doubt over the recognition of an overly disrupted groundwater flow pattern due to barrier boundaries and associated sub-compartmentalisation. Hobbs and Cobbing (2007) employed 48 potentiometric measurements and six spring elevations to develop a potentiometric map for the south-western portion of the study area (Fig. 1). Holland (2007) and Holland and Cobbing (2008) provide groundwater drainage maps for the Zwartkrans Basin that suffer a lack of rigorous discussion of their derivation. Nevertheless, groundwater drainage patterns in the south-western portion of the property have historically enjoyed a much better understanding than in the north-eastern portion.

Data sourced for the situation assessment study (Hobbs 2011a) forms the basis for an improved understanding of groundwater drainage patterns throughout most of the study area. The drainage patterns rendered in Fig. 16 are based on potentiometric measurements associated with 160 geosites (mine shafts, boreholes, wells and springs) distributed across the WHS property. The information presented in Table 5 reflects the distribution of these sources in the Zwartkrans and Krombank basins, the total of 78 reflecting the relative abundance of geosites in this part of the property compared to the more remote and pristine north-eastern part dominated by the John Nash Nature Reserve. The relative density of enumerated geosites amounts to 0.006 per ha, or ~ 1.2 per 2 km$^2$.

The groundwater drainage patterns are represented by the conceptual principal flow vectors associated with each of the basins/subcompartments, and recognise the 'drain' function served by the enumerated springs. This is further illustrated in Fig. 17, which presents a hydrogeologic profile along the section line A–B–C–D in Fig. 16. The surface profile is based on the 20-m contour interval of the published 1:50,000 scale topocadastral maps RSA (1996), RSA (2001) and RSA (2002). The potentiometric surfaces derive from groundwater rest level data sourced in boreholes or representing spring surface elevations as listed in Table 9. The horizontal

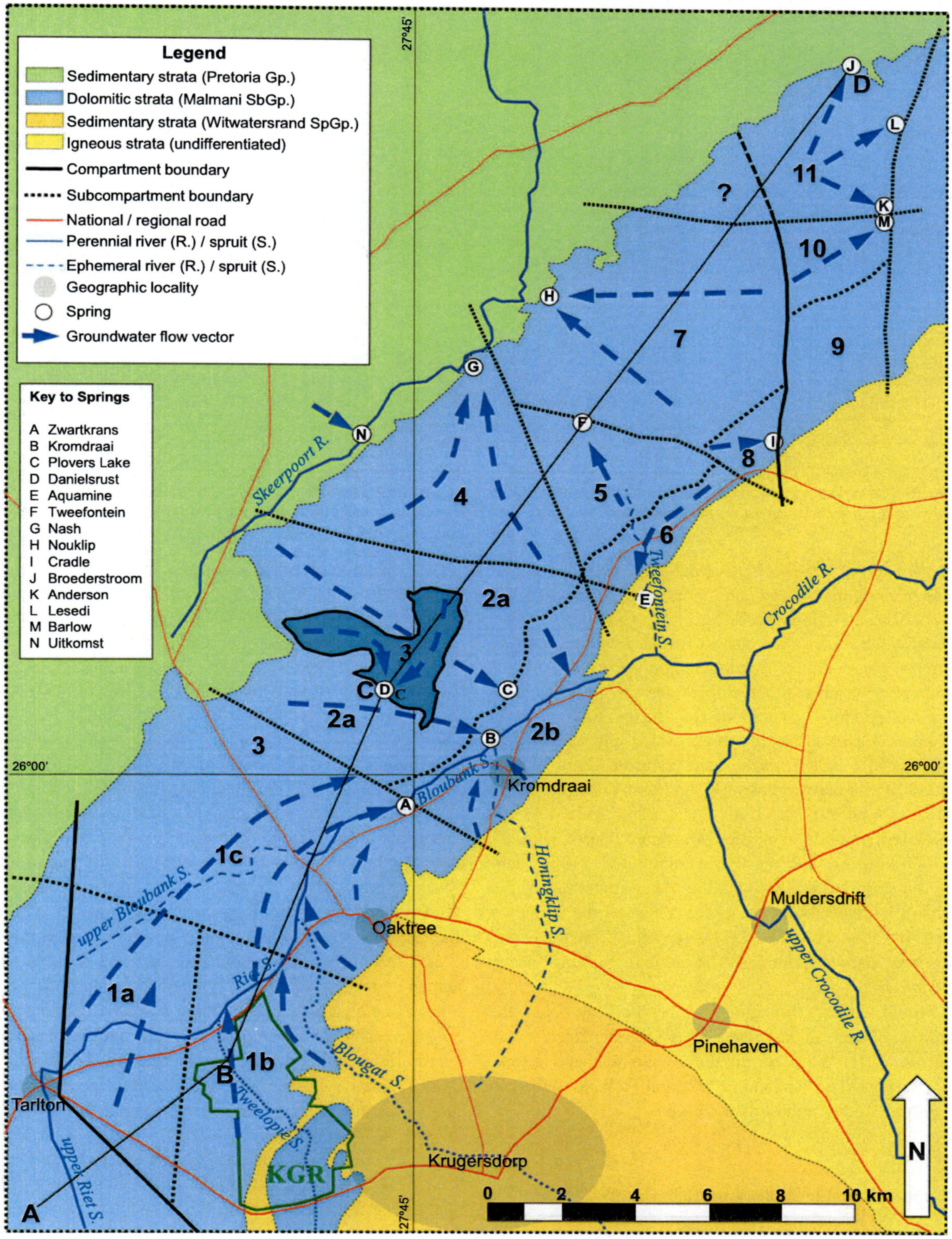


**Fig. 16** Groundwater drainage map illustrating the conceptual flow patterns associated with the basins/sub-compartments that subdivide the karst aquifer in the study area; see Table 5 for statistical analysis of groundwater level data in the Bloubank Spruit system

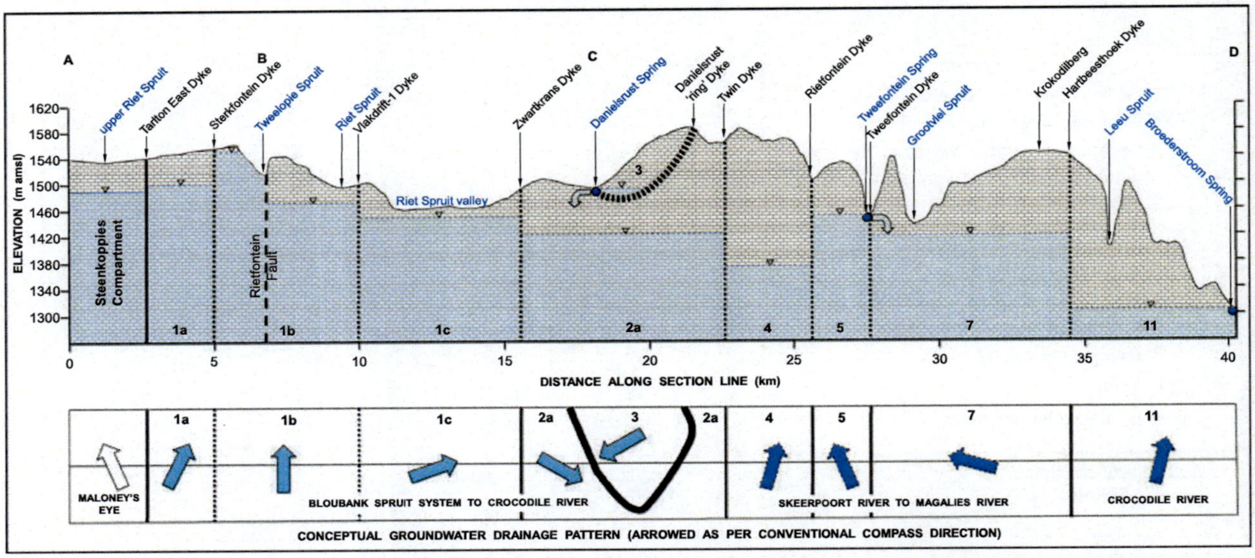


**Fig. 17** Hydrogeological cross-section (top) through the karst aquifer from south-west to north-east (see Fig. 16) illustrating the differences in groundwater rest level between basins/subcompartments, and plan view (bottom) illustrating the corresponding conceptual groundwater drainage pattern as per conventional compass direction; note the autogenic recharge associated with the Danielsrust and Tweefontein springs

attitude of these surfaces is accounted for mainly by their coincidence with equipotential drainage contours, i.e. orthogonal to the drainage vectors.

Perhaps the most salient aspect revealed by the drainage pattern is the complete hydrogeologic separation of the groundwater systems either side of the Twin Dyke. This contradicts previous contentions (e.g. Krige 2010) that the Zwartkrans Basin drains north-eastwards into the Tweefontein Basin[15] and towards Broederstroom. In a revised rendition of compartments by Holland and Cobbing (2008) and Holland (2009), the Twin Dyke is recognised as the north-eastern boundary of the Zwartkrans Basin (Fig. 2). This is in keeping with the Bredenkamp et al. (1986) definition (in Fig. 7.1 op. cit.) of the Zwartkrans Basin. No indication is given, however, of the nature of the boundary formed by this structure, i.e. whether it constitutes an impervious barrier boundary or a semi-pervious leaky boundary.

It would appear that groundwater drainage from the Zwartkrans Basin as identified in the Bredenkamp et al. (1986) study, is focussed on the Zwartkrans and Kromdraai[16] springs. The Holland (2007) and Holland and Cobbing (2008) studies echo this definition.

The movement of groundwater northwards from the Sterkfontein Subcompartment into the Zwartkrans Subcompartment through/over/across the Vlakdrift-1 Dyke is a particularly vexing question given the observed ~ 20 m step in groundwater rest level elevation that exists from south (~ 1465 m amsl) to north (~ 1445 m amsl) across this 'barrier' (Fig. 18). Field observations at the four localities identified as 'a', 'b', 'c' and 'd' in this part of the study area (Fig. 18) reveal the existence of grikes[17] with a distinct NW–SE orientation. These features are accessible to humans (a measure of their size), and features 'c' and 'd' extend many tens of metres into the subsurface. Some of these also demonstrate a significant visible linear length amounting to 10's of metres, most notably the feature identified as 'b', which extends ~ 60 m north-westwards from the valley of the Riet Spruit near its confluence with the Blougat Spruit.

It is reasonable to presume that the grike features represent an expression of similar sympathetic features that occur at depth, i.e. that intersect the water table of the karst aquifer. For this reason, they represent adequate conduits to establish a subsurface hydraulic connection through the Vlakdrift-1 Dyke 'barrier', linking the Sterkfontein and Zwartkrans subcompartments. These circumstances, together with those set out in Sect. 5.2.1 in Chapter "Physical Hydrology", inform the observed escalation of a mine water impact on the karst groundwater in this portion of the COH (Hobbs 2013a, b, c; 2014a, b).

---

[15] Term used in earlier studies (e.g. Holland, 2007; Holland and Cobbing, 2008; Holland et al., 2010) for the north-eastern expanse of karst strata extending from the Twin Dyke to Broederstroom at the north-eastern end of the COH (Fig. 2); also referred to as the North Compartment by Krige (2009; 2010).

[16] The (mis)naming of this spring is discussed in Text Box 1.

[17] See GLOSSARY.

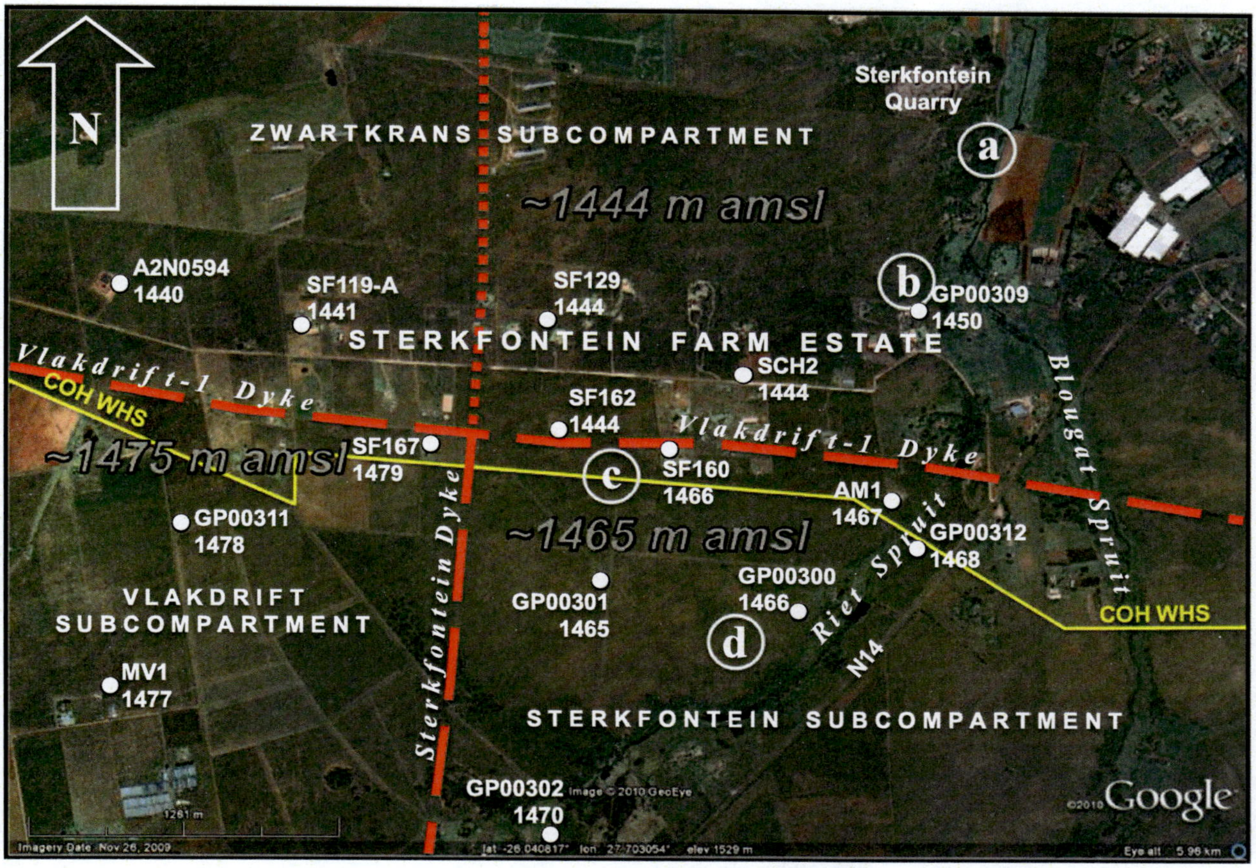


**Fig. 18** Elucidation of hydrological, geological and hydrogeological aspects along the southern boundary of the COH south-west of Sterkfontein Quarry; circled localities labelled 'a' to 'd' mark the position of significant grike features, and italicised values indicate the recent ambient groundwater elevation in each subcompartment as derived from individual borehole measurements (base image courtesy of Google Earth®)

The rendition of the groundwater drainage pattern in the extreme north-eastern portion of the property is obscured by the extremely rugged nature of the terrain. Centred on the high ground defined by Krokodilberg (Fig. 17) at an elevation of 1526 m amsl in the Broederstroom Basin, this terrain is drained to the west by the Witwatersrand Spruit, to the north-west by the Leeu Spruit (Fig. 17), and to the east by the Jukskei Spruit. The Broederstroom, Lesedi and Anderson springs (Fig. 16) represent the few natural outlets enumerated in this basin. It is probable that numerous other groundwater emergences occur in the valleys and ravines that dissect this terrain, and which contribute to the groundwater loss component of a basinal water budget. It is further apparent from Fig. 16 that the western boundary of the Broederstroom Basin is poorly understood in regard to the following aspects:

- the hydraulic characteristics of the possible northerly extension of the Krokodilberg Dyke indicated by the pecked line in Fig. 16; and
- the existence of natural groundwater emergences in the valley of the westerly draining Witwatersrand Spruit located in the triangular compartment/subcompartment designated '?' in Fig. 16.

| Key to basins/subcompartments | | | | | |
|---|---|---|---|---|---|
| 1a | Vlakdrift Subcompartment | 1b | Sterkfontein Subcompartment | 1c | Zwartkrans Subcompartment |
| 2a | Kromdraai Subcompartment | 3 | Danielsrust Basin | 4 | Uitkomst Basin |
| 5 | Tweefontein Basin | 7 | Diepkloof Basin | 11 | Broederstroom Basin |

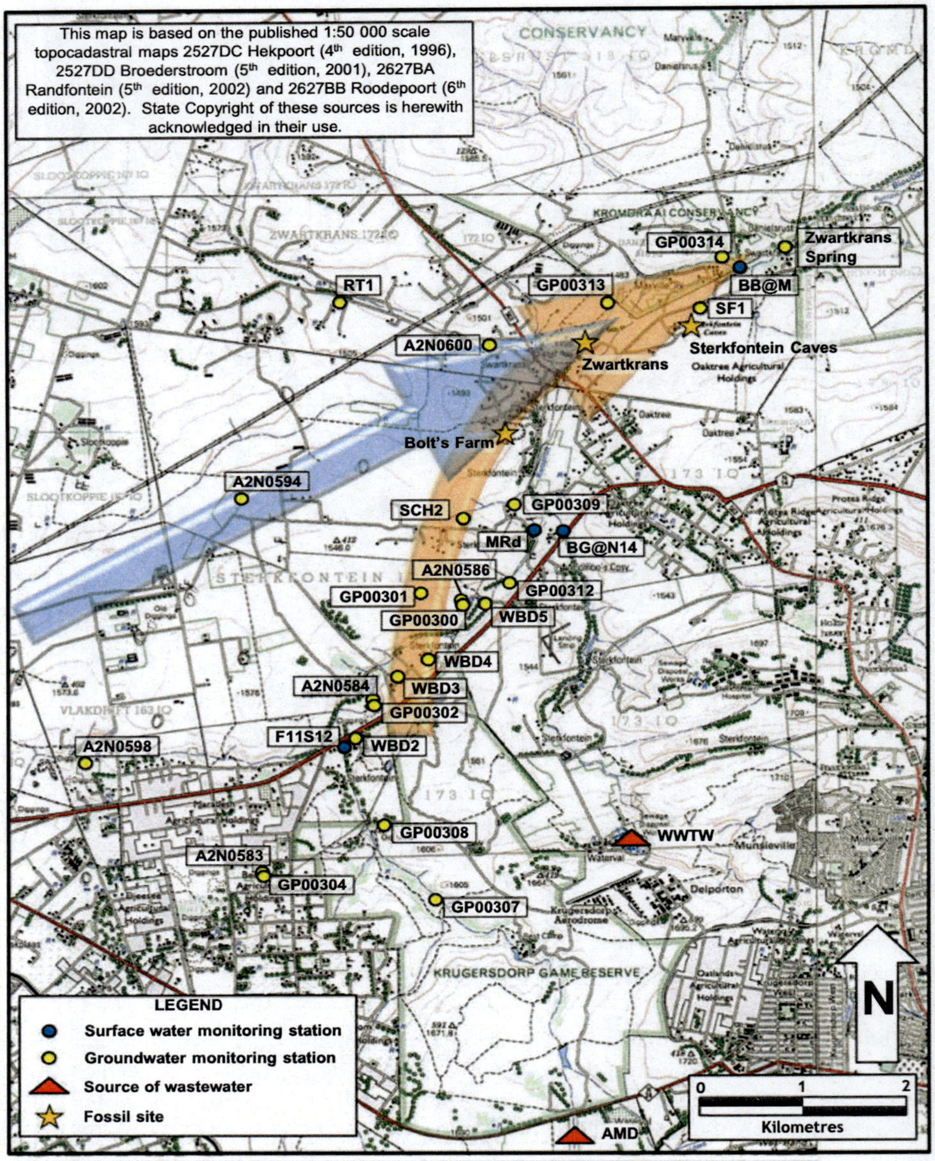


**Fig. 19** Distribution of selected surface water and groundwater monitoring stations in the Zwartkrans Basin, also showing the groundwater flow paths followed by allogenic recharge (brown) and autogenic recharge (blue)

## 7 Groundwater Fluxes

The surface water fluxes (gains and losses) discussed in Sect. 5 in Chapter "Physical Hydrology" necessarily impact on the groundwater environment. This is manifested as groundwater resurgence at springs and in stream channels in the discharge reaches of groundwater basins. It is most pertinent in regard to the Zwartkrans Basin, where allogenic recharge associated with the mine water and wastewater sources caused an unprecedented volumetric addition to the quantity of water in this aquifer. The fluxes associated with the spring and groundwater resurgence in the stream channel upstream thereof are discussed in Sect. 4.1.

The magnitude of allogenic recharge of mine water to the karst aquifer along the Riet Spruit between stations F11S12 and MRd amounted to a median value of 17.3 ML/d in the last four years (Table 12 in Chapter "Physical Hydrology"). The path followed by this recharge is shown in Fig. 19 as encompassing monitoring stations A2N0584 and A2N0586 at the influent (upstream) end, and station A2N0600 further north midway to its discharge at the Zwartkrans Spring. The coarse dimensions of the flowpath from the upstream end to station A2N0600 are 4500 m (length) by 2200 m (width), with a potentiometric difference of ~ 30 m. These values define a hydraulic gradient (i) of 0.0067. Applying Darcy's law and assuming an aquifer transmissivity (T) value of ~ 1000 $m^2/d$ (roughly 50% of the T-value determined for the aquifer intersected in borehole G36331/A2N0586 as per Table 2), yields the following calculated flow estimate:

$$Q = T_{i,w} = 1000 \times 0.0067 \times 2200$$
$$= 14,740\frac{m^3}{d}(\sim 14.7\frac{ML}{d})$$

This value is similar to the minimum of 14–15 ML/d associated with allogenic mine water recharge as per Fig. 24 in Chapter "5" (Sect. 5.2.1 in Chapter "Physical Hydrology"). It also finds support in semi-quantified fluxes such as those of the Zwartkrans Spring and related groundwater resurgence in the Bloubank Spruit at the discharge end of the Zwartkrans Basin. These have been estimated at between 11.6 and 25.6 ML/d (Sect. 5.2.1 in Chapter "Physical Hydrology"), with an average of 16.4 ML/d. The values are not in conflict with one another, and are considered to represent the pre-2010 hydrodynamic situation. A different picture emerges, however, from a mass balance calculation that considers the volume and concentration of the allogenic recharge at the influent end of the flow path, compared to that at station A2N0600. Using the relationship:

$$Q_U C_U = Q_D C_D$$

previously employed in Sect. 4.1, and where

| | |
|---|---|
| $Q_U$ = 17,300 $m^3/d$ | [median influent flow volume from Table 12 in Chapter "Physical Hydrology"] |
| $C_U$ = 2170 mg $SO_4$/L | [$SO_4$ concentration at influent end (station GP00312) [18]] |
| $Q_D$ = $m^3/d$ | [flow volume] |
| $C_D$ = 720 mg $SO_4$/L | [$SO_4$ concentration at downstream end (station A2N0600)] |

yields a $Q_D$ value of ~ 52.1 ML/d. This represents an additional contribution of ~ 34.8 ML/d, of which 3–7 ML/d derives from municipal wastewater recharge (Sect. 5.2.2 in Chapter "Physical Hydrology"). The most probable source of the balance (~ 31.8–27.8 ML/d) is autogenic recharge (Fig. 19 and Fig. 20) from rainfall (Table 4). These volumes represent a water depth of 118–104 mm (16.6–14.6% of an MAP of 710 mm), well within the recharge rate of 17 ± 5% of MAP that generally applies in the COH (Sect. 2.12). The substantial agreement and congruence evident between the various figures derived, generates cautious confidence in the groundwater fluxes that broadly characterise the Zwartkrans Basin since January 2010, when the mine water impact has been greatest. The caution derives from the use of $SO_4$ as an anionic tracer because of its propensity to participate in any number of bio (geo)chemical reactions, e.g. bacterial sulfate reduction as discussed in Sect. 11 "Chapter Chemical Hydrogeology". In their discussion on the application of 'conservative' anionic tracers, Korom and Seaman (2012) elaborate on the properties of Br and Cl, and make no mention of $SO_4$ in this regard. Further caution is indicated by non-recognition of factors such as longitudinal dispersion in the aquifer and 3-component mixing.

[18] See also borehole WBD5 in Fig. 7 in Chapter "Chemical Hydrogeology"

The groundwater flux values presented in Fig. 19 provide a means to derive an approximate groundwater balance for the Zwartkrans Basin. Input and output values for such a calculation are known to varying degrees of accuracy. For example, readily quantifiable parameters such as recharge and spring discharge present more accurate values than groundwater use. The basic input values to a groundwater balance calculation for the Zwartkrans Basin are autogenic recharge of ~ 30 ML/d and allogenic recharge of ~ 22 ML/d (as derived earlier in this section) for a total of ~ 52 ML/d. Basic output values are groundwater discharge of ~ 30 ML/d (as determined in Sect. 4.1) and groundwater use of ~ 22 ML/d to balance the inputs. The reported values are considered to be in error by < 10%. The veracity of the calculation is demonstrated by the plausibility of the groundwater use value, which compares favourably with the values reported in Sect. 9.

Bredenkamp et al. (1986) report a flux of ~ 11.7 $mm^3/a$ leaving the karst environment of the Bloubank Spruit system downstream of Kromdraai. Being representative of discharge in the mid-1980s, this provides a historical perspective on this aspect. An inspection of the A2H049 hydrograph (Fig. 5.12) indicates that mean discharge in the 1985 and 1986 hydrological years amounted to ~ 13.9 and ~ 16.6 $mm^3/a_h$ respectively. The discharge of ~ 11.7 $mm^3/a$ approximates the 5%ile value of the long-term annual flow at A2H049. These observations support the comparatively low value reported by Bredenkamp et al. (1986). Under circumstances where the mid-1980s were a dry period, it is presumed that the flow of ~ 11.7 $mm^3/a$ must represent mainly the groundwater contribution from the karst aquifer to discharge in the Bloubank Spruit system. This flow necessarily included the ~ 0.9 $mm^3/a$ (~ 30 L/s) contribution of the Kromdraai 'Eye' as identified by Bredenkamp et al. (1986).

Further interrogation of the Bredenkamp et al. (1986) flow data indicates an upstream contribution of ~ 8.1 $mm^3/a$ (~ 257 L/s) immediately below the Zwartkrans Spring. The difference of 3.6 $mm^3/a$ (~ 114 L/s) must therefore represent the combined contribution of the Plover's Lake and Kromdraai springs identified and enumerated in

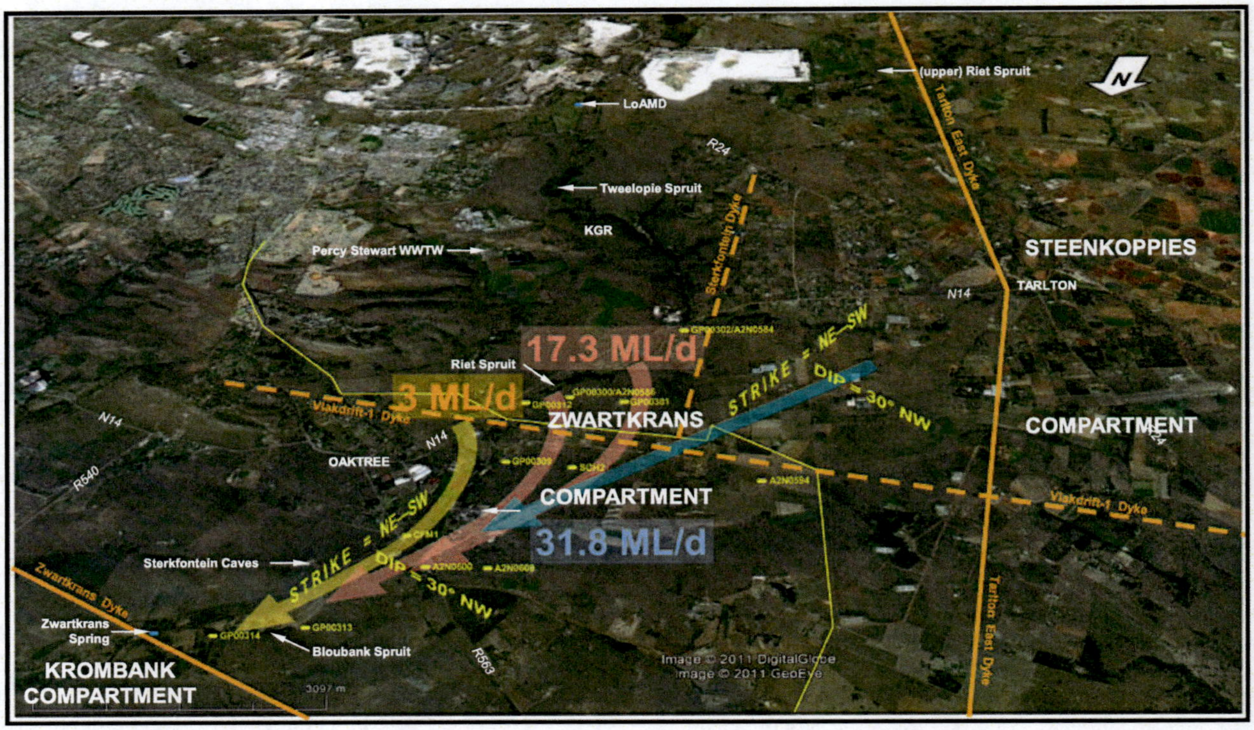


**Fig. 20** Oblique Google Earth® view of the Zwartkrans Basin illustrating the nexus of allogenic (municipal and mine wastewater) and autogenic groundwater fluxes

this study. Excluding the ~ 0.9 mm$^3$/a (~ 30 L/s) contribution of the Kromdraai 'Eye' leaves ~ 2.7 mm$^3$/a (~ 84 L/s) as the possible contribution of the 'real' Kromdraai Spring (refer Text Box 8) at this time.

## 8 Conduit Flow Depth and Path

Mundane to the subject of speleogenesis is the question of whether cave formation occurs at, above or below the water table (Ford 2003; Worthington 2004). The relationship to the water table is described by the conduit flow depth and the flow path. In regard to the latter, Worthington (2004) recognises the following basic geometries each incorporating flow from a recharge to a discharge area described by Ford (2003) as 'from sink to spring':

- flow is in the same direction as the stratal dip; or
- flow is along strike, generally following a bedding plane; with
- various combinations of the above two geometries also possible.

This topic is explored in regard to the Zwartkrans Basin[19] which represents a combination of the two geometries. The flow path of allogenic recharge[20] in the Zwartkrans Basin initially follows a northerly direction which bisects the NW stratal dip and the NE–SW strike, before diverting along strike to its outlet in the locus of the Zwartkrans Spring (Figs. 19 and 20). The path followed by natural groundwater conforms to the NE–SW strike sub-paralleling the upper Bloubank Spruit (Fig. 19). Some of the other basins almost certainly represent a singular relationship, for instance the stratal dip geometry in the Tweefontein and Nouklip basins (Fig. 3).

Worthington (2001; 2004) developed a set of empirical relationships to describe the depth of circulation (D) in an unconfined carbonate aquifer. These relationships incorporate variables that describe the flow path (L), the stratal dip (θ), and the elevation difference (E) between the highest point of recharge and the lowest spring draining the aquifer.

[19] The assessment focusses on this compartment for various reasons, including its relatively straight forward geometry and its comparatively rich hydrologic and hydrogeologic data record.

[20] Associated with the ingress of mine water contaminated surface water from the Riet Spruit.

The main relationships developed by Worthington (2004) are defined by the following equations:

$$D = 0.18(L \times \theta)^{0.81} \quad [R^2 = 0.79] \tag{1}$$

$$D = 0.061 \times L^{0.91} \times \theta^{0.72} \quad [R^2 = 0.90] \tag{2}$$

$$D = 0.047 \times L^{0.85} \times \theta^{0.64} \times E^{0.11} \quad [R^2 = 0.90] \tag{3}$$

where

- $D$ mean conduit depth (m) below the water table;
- $L$ flow path length (m) from farthest input to spring;
- $\theta$ dimensionless stratal dip (equal to the sine of the dip in degrees); and
- $E$ elevation difference (m) between the lowest spring and the highest point of recharge.

The two variables of stratal dip ($\theta$) and flow path length ($L$) explain most of the variability in conduit flow depth. More recently, Boudinet (2012) has produced similar results using statistical modelling in an analysis of karstic flow conduits. In the case of Sterkfontein Cave with a stratal dip of ~ 30° (Martini et al. 2003), the farthest input to the Zwartkrans Spring falls in the range 5000–9000 m. These limits define the distance along strike[21] to the Vlakfontein-1 Dyke and the Tarlton East Dyke, respectively.

The application of the above values in Eqs. 1 and 2 places the mean conduit depth at between 118 and 164 m bwl (Eq. 1), and between 86 and 147 m bwl (Eq. 2). These values appear unlikely in proximity to the caves, especially when gauged against the drilling information provided by DWS monitoring borehole GP00313. Located ~ 800 m to the north-west across the valley of the Bloubank Spruit from the cave (Fig. 19), this borehole intersected very weathered to decomposed and cavernous dolomite down to its completion depth of 37 m bs (~ 1423 m amsl), i.e. 17 m below the water table depth of ~ 20 m bs (~ 1440 m amsl). The most substantial cavity was encountered in the interval 17–32 m bs, i.e. 12 m bwl. Applied in Eq. 1, values for D in the range 12–17 m return L-values of only 180–280 m, which appear equally unlikely. It is possible that borehole GP00313 was terminated above the deepest conduit in this area.

It is informative to note that applying the distance of ~ 1400 m between the caves and the Zwartkrans Spring in Eqs. 1 and 2 returns D-values of 36 and 27 m bwl, respectively. It is also instructive to interrogate the Bredenkamp et al. (1986) drilling records within the context of Eqs. 1, 2 and 3.

The results for the 13 high-yielding boreholes listed in Table 2 are presented in Table 16. The outcome indicates that Eq. 1 returns the greatest values which in each instance far exceed the empirical value derived from the drill record. Equation 3 returns the smallest values, which therefore also show the least difference with the empirical values.

## 9 Groundwater Use

From an assessment of water use information sourced from the DWS Water Authorisation and Registration Management System (WARMS), Holland and Cobbing (2008) report a total groundwater use of 8.7 $mm^3/a$ as being registered for mainly agricultural use in the Zwartkrans Basin. This figure contrasts sharply with those estimated by Schoeman and Associates (2006) of 14.1 $mm^3/a$ based on an interpretation of satellite imagery, and of 25.7 $mm^3/a$ estimated by JFA (2006) from a water budget calculation. The latter quantity is accepted by Krige (2009). A further measure of the variance in groundwater uses figures for the Zwartkrans Basin is provided by those of 15 and 18 $mm^3/a$ reported earlier by Bredenkamp et al. (1986) and van Biljon (2006), respectively.

The Rhino and Lion Game Reserve maintains an excellent record of the abstraction from borehole RLGR4 which serves the chalets and Wonder Cave. An inspection of this record indicates that an average daily abstraction of 17 $m^3/d$ (~ 6200 $m^3/a$) is made from this borehole. Further to the north-east, developments such as the Kalkheuwel West Estate support numerous properties that are dependent on own borehole water supplies for potable and general domestic water use.

The occasional business opportunity where groundwater is bottled for wholesale distribution as 'mineral water' is also manifested in the study area. The number of such businesses and their combined annual abstraction is difficult to quantify. It is unlikely, however, that their combined groundwater use is sufficiently great to significantly impact on the groundwater resource supply potential of the karst aquifer(s) in the study area.

## 10 Closure of DWS Monitoring Stations

The comparatively recent closure of the DWS monitoring stations A2N0589, A2N0594, A2N0598 and A2N0606 is reported in Table 6. This circumstance was informed by the factors listed in Table 17.

It is envisaged by the DWS (E Bertram, personal communication) that in those instances where the stations have

[21] According to Worthington (2001), this relationship is applicable irrespective of whether flow is along, oblique or normal to the strike of the strata.

**Table 16** Interrogation of drill records of high-yielding DWS exploration boreholes located in the Zwartkrans Basin (from Bredenkamp et al. 1986) for mean conduit depth

| DWS Geosite # | Deepest Karst Horizon (m bs)[a] | Water level depth (m bs) | Depth of Karst Horizon (m bwl)[b] | Elevation difference (m)[c] | Mean conduit depth (m bwl) | | | Distance to ZSp (m) |
|---|---|---|---|---|---|---|---|---|
| | | | | | 7–1 | 7–2 | 7–3 | |
| A2N0589 | 80 | 28 | 52 | 35 | 80 | 66 | 49 | 3742 |
| A2N0593 | 103 | 74 | 29 | 81 | 114 | 98 | 77 | 5773 |
| A2N0590 | 87 | 34 | 53 | 42 | 76 | 62 | 47 | 3496 |
| A2N0592 | 85 | 76 | 9 | 83 | 108 | 92 | 73 | 5406 |
| A2N0588 | 50 | 48 | 2 | 59 | 100 | 85 | 65 | 4915 |
| A2N0587 | 89 | 19 | 70 | 34 | 93 | 78 | 56 | 4463 |
| A2N0595 | 150 | 58 | 92 | 110 | 147 | 130 | 104 | 7881 |
| A2N0596 | 110 | 55 | 55 | 113 | 159 | 142 | 113 | 8681 |
| A2N0597 | 150 | 80 | 70 | 113 | 148 | 131 | 104 | 7911 |
| A2N0586 | 130 | 24 | 106 | 55 | 114 | 98 | 74 | 5748 |
| A2N0598 | 120 | 73 | 47 | 131 | 181 | 164 | 132 | 10,189 |
| A2N0599 | 130 | 51 | 79 | 110 | 180 | 163 | 128 | 10,124 |
| A2N0581 | 95 | 56 | 39 | 141 | 195 | 178 | 143 | 11,149 |

[a] Intersected by drilling; in some instances onerous drilling conditions prevented further advance
[b] Empirical value derived from the drill record
[c] Difference between geosite surface elevation and Zwartkrans Spring elevation of ~ 1433 m amsl

**Table 17** Assessment framework for the closure/recommissioning of DWS water level monitoring stations

| Station | Record length (months) | # Of records | Record frequency (per month) | Reason for closure[a] | Additional information (provided by author) |
|---|---|---|---|---|---|
| A2N0589 | 285 | 170 | 0.60 | Borehole collapsed | Redundant and expendable due to proximity to A2N0600 with record frequency of 0.66 |
| A2N0594 | 288 | 187 | 0.65 | Borehole equipped | Should be reopened due to strategic position in upper reaches of the Zwartkrans Basin |
| A2N0598 | 283 | 93 | 0.33 | Borehole equipped | Should be reopened due to strategic position adjacent to the losing middle reaches of the Riet Spruit; record frequency is easily improved |
| A2N0606 | 243 | 36 | 0.15 | Borehole collapsed | Redundant and expendable due to proximity to A2N0594 with record frequency 0.65 |

[a] E Bertram (personal communication)

been equipped with pumps and serve an active water supply function, the landowners will be required to provide alternative monitoring stations. This implies that the landowner (s) will establish a unique monitoring station that replaces the function of the original station. In both instances where this applies (stations A2N0594 and A2N0598, Table 17), such an approach is considered to be unrealistic and therefore not workable. Further, if deemed necessary for reasons such as long record or strategic location, the collapsed boreholes stations A2N0589 and A2N0606 (Table 17) will be rehabilitated and returned to duty (E Bertram, personal communication).

A review of the situation within the framework of the additional information presented in Table 17, suggested that the closure of only two of the four stations was warranted and justifiable. A recommendation to the DWS that the decision in regard to stations A2N0594 and A2N0598 be rescinded has been acknowledged by the return of these stations to service.

# Chemical Hydrogeology

Harrison Pienaar, Philip J. Hobbs, and Sebinasi Dzikiti

## 1 Regional Groundwater Chemistry

The main lithostratigraphic units in the study area are identified as the younger units of the Witwatersrand Supergroup and the older units of the overlying Transvaal Supergroup (Table 5, Sect. 4 in Chapter "Description of the Physical Environment"). Barnard (2000) provides information regarding the chemistry of groundwater regionally (and typically) associated with each of the relevant rock units.[1] This information is summarized in Table 1 and illustrated in Fig. 1.

Figure 1 reflects the similarity in chemical composition (MgCa–$HCO_3$) of groundwater associated with the Timeball Hill Formation, Black Reef Formation, and, to a lesser extent, Central Rand Group strata. This is attributed to the similar rock type (mainly quartzite) that builds these formations. Although the volcanic rocks (lava) of the Hekpoort Formation also exhibit a CaMg–$HCO_3$ composition, they can be distinguished on the basis of generally higher analyte concentrations, especially $SO_4$ and Na, as well as higher SEC and TDS values (Table 1). The Malmani Subgroup dolomitic strata exhibit the most dominant MgCa–$HCO_3$ composition of all the formations represented. The higher Cl, $SO_4$, and N concentrations (Table 1) in the karst groundwater are, however, indicative of contamination (Barnard 2000) which is uncharacteristic of pristine karst groundwater that still occurs in much of the study area. The Piper diagram presented in Fig. 2 provides further and possibly more distinct characterization of the water chemistry associated with the various surface and subsurface water resource environments.

## 2 Subregional Groundwater Chemistry

### 2.1 Background

Hobbs and Cobbing (2007) report groundwater (including raw and treated mine water) chemistry information for 42 sources enumerated in the southwestern portion of the study area. Surface water chemistry was determined for a further seven sites mostly located on the Tweelopie Spruit, but including one Blougat Spruit sample. This information has enabled the characterization of the hydrochemistry associated with the various water sources shown in Fig. 2. A similar characterization was also reported by Holland and Witthüser (2009).

#### 2.1.1 West Rand Group (Witwatersrand Supergroup)

Groundwater from these strata is typically very weakly mineralized, as reflected in SEC values <20 mS/m and TDS values <120 mg/L, and exhibits a low pH (in the range 5.4–6.0). The latter is attributed to the very low alkalinity of this groundwater, with total alkalinity typically <10 mg $CaCO_3$/L (Hobbs and Cobbing 2007). This water is therefore also corrosive with a typical corrosion tendency ratio (CTR) of »1.[2] An example from this study is the water from station

P. J. Hobbs: Deceased

[1] No groundwater chemistry data are available for the Rooihoogte Formation sandwiched between the Timeball Hill Formation and the Malmani Subgroup strata (Table 5, Sect. 4 in Chapter "Description of the Physical Environment").

[2] For pH values in the range 7–8 and in the presence of dissolved oxygen, CTR values $\leq 0.1$ indicate general freedom from corrosion, whereas higher values are indicative of more corrosive waters.

H. Pienaar (✉) · P. J. Hobbs
Council for Scientific and Industrial, Research, Smart Places, Water Centre, Pretoria, South Africa
e-mail: hpienaar@csir.co.za

H. Pienaar
Hebei University of Engineering, Handan, China

S. Dzikiti
Department of Horticultural Science, Stellenbosch University, Stellenbosch, South Africa
e-mail: sdzikiti@sun.ac.za



P. J. Hobbs et al. (eds.), *Anatomy of a South African Karst Hydrosystem*, Cave and Karst Systems of the World,
https://doi.org/10.1007/978-3-030-95829-9_8

**Table 1** Characterization of groundwater hydrochemistry per lithostratigraphic unit represented in the study area (data from Barnard 2000)

| Variable/analyte | Lithostratigraphic unit | | | | | SANS (2011a)[a] |
|---|---|---|---|---|---|---|
| | Hekpoort Fm. | Timeball Hill Fm. | Malmani Sbgp. | Black Reef Fm. | Central Rand Gp. | |
| pH ($-\log_{10}\alpha_{H+}$) | 7.5 | 7.2 | 7.6 | 7.0 | 7.3 | 5.0–9.7 |
| SEC (mS/m) | 52 | 34 | 63 | 34 | 29 | ≤ 170 |
| TDS (mg/L) | 398 | 278 | 444 | 238 | 207 | ≤ 1200 |
| Ca (mg/L) | 44.0 | 29.0 | 52.7 | 28.0 | 17.6 | n.s. |
| Mg (mg/L) | 26.0 | 19.0 | 35.4 | 18.0 | 13.7 | n.s. |
| Na (mg/L) | 30.0 | 15.0 | 24.1 | 14.0 | 20.0 | ≤ 200 |
| K (mg/L) | 2.0 | 1.6 | 2.3 | 1.7 | 2.6 | n.s. |
| Cl (mg/L) | 23.5 | 15.0 | 37.7 | 15.0 | 17.9 | ≤ 300 |
| $SO_4$ (mg/L) | 79.0 | 26.0 | 70.5 | 36.0 | 33.5 | ≤ 500 |
| $HCO_3$ (mg/L) | 190 | 152 | 216 | 120 | 104 | n.s. |
| $NO_3$ (mg N/L) | 3.5 | 2.9 | 5.6 | 2.8 | 2.0 | ≤ 11 |
| F (mg/L) | 0.3 | 0.3 | 0.3 | 0.2 | 0.3 | ≤ 1.5 |
| TDS:SEC | 7.7 | 8.2 | 7.1 | 6.9 | 7.1 | n.s. |
| $SO_4$:TDS (%) | 20 | 9 | 16 | 15 | 16 | n.s. |
| Chemical character | CaMg–$HCO_3$ | MgCa–$HCO_3$ | MgCa–$HCO_3$ | MgCa–$HCO_3$ | MgCa–$HCO_3$ | – |
| *n* | 41 | 81 | 223 | 52 | 18 | |

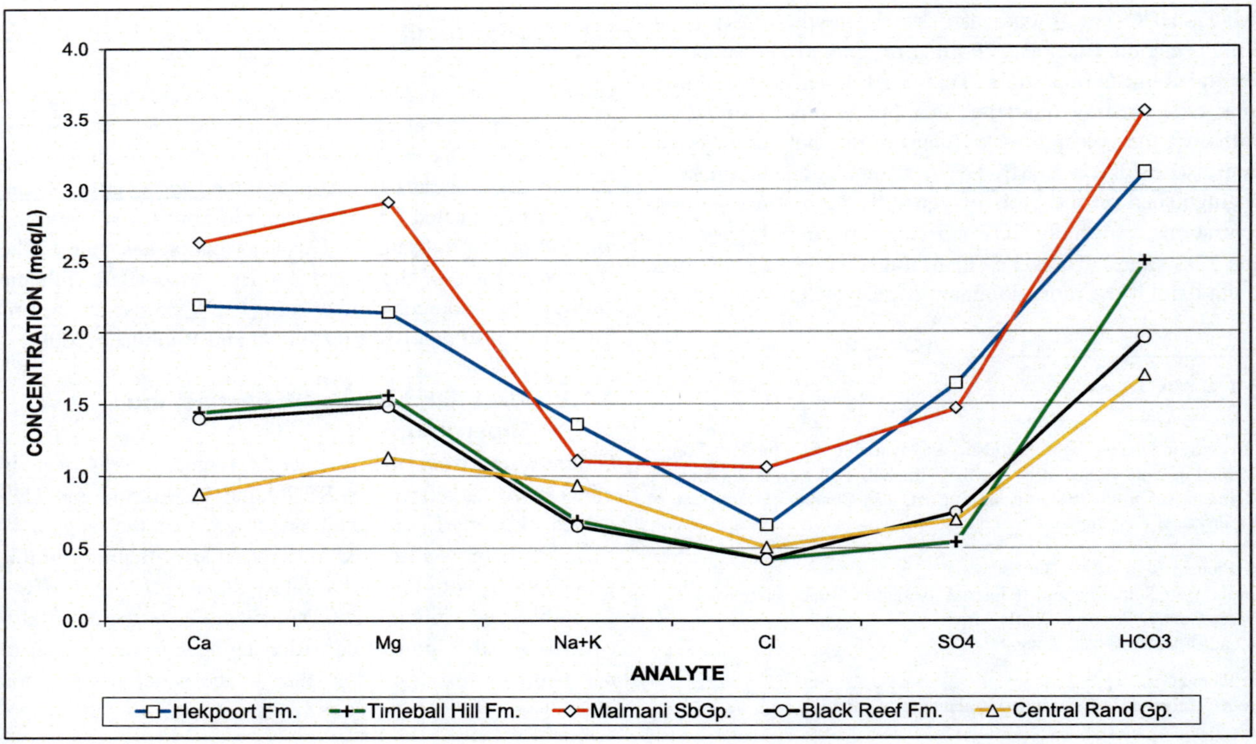


**Fig. 1** Schoeller graphical comparison of natural hydrochemistry associated with the respective lithostratigraphic units represented in the study area (data from Table 1)

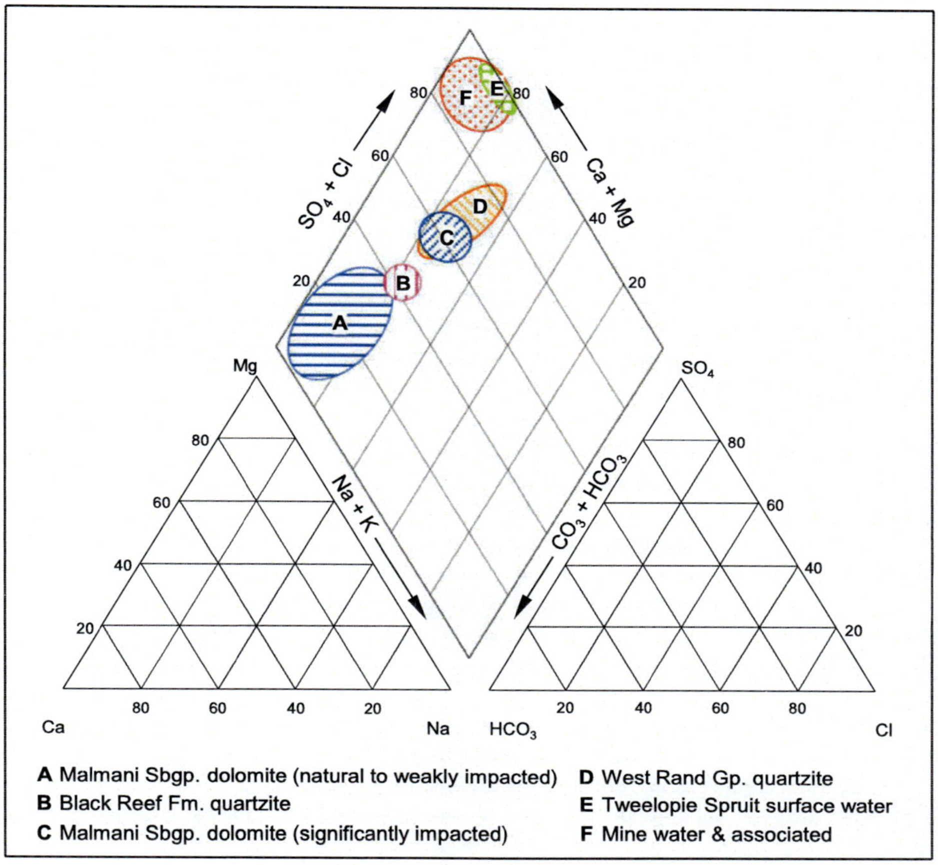


**Fig. 2** Piper diagram characterization of hydrochemistry associated with various surface water and groundwater sources in the study area (modified after Hobbs and Cobbing 2007)

PS3, a borehole located on the quartzite ridge to the east of the Blougat Spruit, which reflects an Mg–$HCO_3$ composition and the following variable-specific values:

$$\mathrm{pH} = 6.0\,(\mathrm{field});\quad \mathrm{SEC} = 4\,\frac{\mathrm{mS}}{\mathrm{m}}\,(\mathrm{field});$$
$$\text{Total alkalinity} = \frac{8\,\mathrm{mg\,CaCO_3}}{\mathrm{L}}\,(\mathrm{lab})$$

### 2.1.2 Malmani Subgroup (Chuniespoort Group)

Dolomitic groundwater is characterized by a CaMg–$HCO_3$ composition which, in its pristine state, also exhibits very low concentrations of Cl and $SO_4$ (generally <5 mg/L), a low SEC (in the range 20–40 mS/m), and a typically alkaline (basic) pH of 7.5–8.0. The water delivered by the Danielsrust, Nouklip, and Nash springs exemplify this observation. Anthropogenic impacts on this water quality are recognized on the basis of elevated Cl and $SO_4$ concentrations, higher SEC values (>60 mS/m), and lower pH values (<7.2).

### 2.1.3 Pretoria Group (Transvaal Supergroup)

The composition of groundwater associated with these strata is similar to that described for the Witwatersrand Supergroup strata (Sect. 2.1.1). An example from this study is the analysis for the Uitkomst Spring (Fig. 16 in Chapter "Physical Hydrogeology") draining interlayered shales and quartzites of the Timeball Hill Formation in the Skeerpoort River Valley, which reflects an Mg–$HCO_3$ composition together with the following variable-specific values:

$$\mathrm{pH} = 7.0\,(\mathrm{field});\quad \mathrm{SEC} = 1.0\,\frac{\mathrm{mS}}{\mathrm{m}}\,(\mathrm{field});$$
$$\text{Total alkalinity} = \frac{16\,\mathrm{mg\,CaCO_3}}{\mathrm{L}}\,(\mathrm{lab})$$

This groundwater is generally distinguishable from that of the West Rand Group strata (Sect. 2.1.1) on the basis of a more neutral pH, a slightly greater salinity, and detectable Fe and Mn concentrations which, in the case of the Uitkomst Spring, amounted to ~0.3 mg Fe/L and ~0.5 mg Mn/L in May 2010 (Table 5).

## 2.2 Temporal Groundwater Chemistry Assessment

The geographic extent of a mine water impact on the karst groundwater resources of the Zwartkrans Basin is postulated in the EIA for Harmony Gold Mine (JFA 2006). The EIA report also, however, records $SO_4$ levels in DWS monitoring boreholes in the area that were substantially higher, in some instances by a factor of 4–6, in 1986 than in 2006. These circumstances are illustrated in Fig. 3, revealing an $SO_4$ reduction in the range of 30–83% (average 57%) from 1986 to 2006. The mine water impact on the groundwater resources of the area is therefore not a recent or new phenomenon. The improvement between 1986 and 2006 illustrates the resilience and self-cleansing nature of karst groundwater systems in the long-term, given conditions that are conducive to this.

Further illustration of the long-term pattern and trend observed in the groundwater chemistry recorded at the DWS groundwater quality monitoring stations A2N0583, A2N0584, A2N0586, and A2N0600. The position of these stations is shown in Fig. 19 in Chapter "Physical Hydrogeology". The information shown in Fig. 4 compares the 'oldest' (first) and the latest groundwater chemical compositions at the longest-operating DWS monitoring stations. Apart from the discussion of SEC and $SO_4$ patterns and trends accompanying Fig. 46 in Sect. 8.4, the Volume 2 and Fig. 4 results are summarized as follows.

- Station A2N0583 reveals a gradual but persistent increase in the $SO_4$ concentration since June 2004 to produce a MgCa–$SO_4$ composition compared to the historical (ca. 1985) CaMg–$HCO_3$ composition. This is both surprising and a concern given the position of this borehole to the west of the Krugersdorp Game Reserve ostensibly outside the path of mine water impact.
- Station A2N0584 reflects a sharp increase in Ca and $SO_4$ concentrations since 2008, imparting a Ca–$SO_4$ character to this groundwater. It is also evident, however, that this chemistry is not too dissimilar to the historical chemistry as observed in July 1985 and March 1989, when $SO_4$ concentrations of 368 and 420 mg/L, respectively were recorded. Both the historical and current compositions indicate mine water infiltration into the karst aquifer at this locality.
- Station A2N0586 reflects a sharp increase in Ca and $SO_4$ concentrations since 2008, again imparting a Ca–$SO_4$ character to this groundwater. These circumstances are again indicative of mine water infiltration into the karst aquifer at this locality.
- Station A2N0600 reflects a less sharp increase in Ca, Mg, and $SO_4$ concentrations that is manifested over a longer period of time since 2008. The current Ca–$SO_4$ composition of the groundwater is much more pronounced than the similar historical composition. The 'prominence' of $SO_4$ throughout the observational record is out of character for karst groundwater, and not readily explained on the basis of available data. It may be postulated, however, that the earlier presence of notable but muted $SO_4$ levels ($\sim$170 mg/L) in the groundwater at this position again reflects a historical mine water impact.

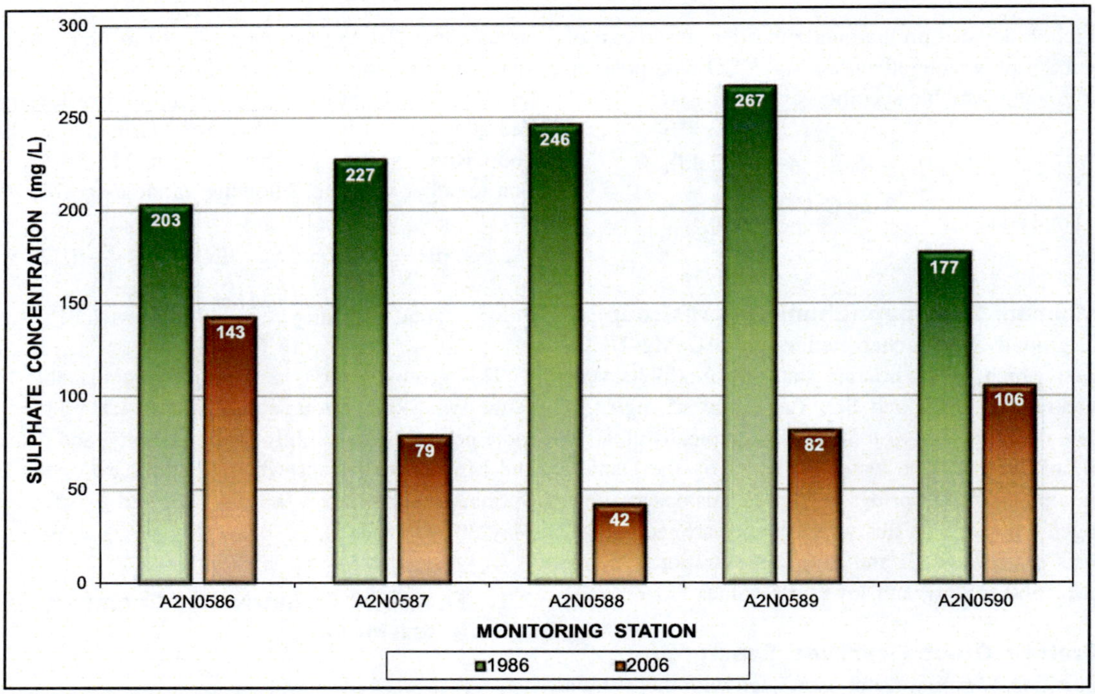


**Fig. 3** Comparison of historical and more recent $SO_4$ levels in groundwater from DWS monitoring boreholes in the Zwartkrans Basin

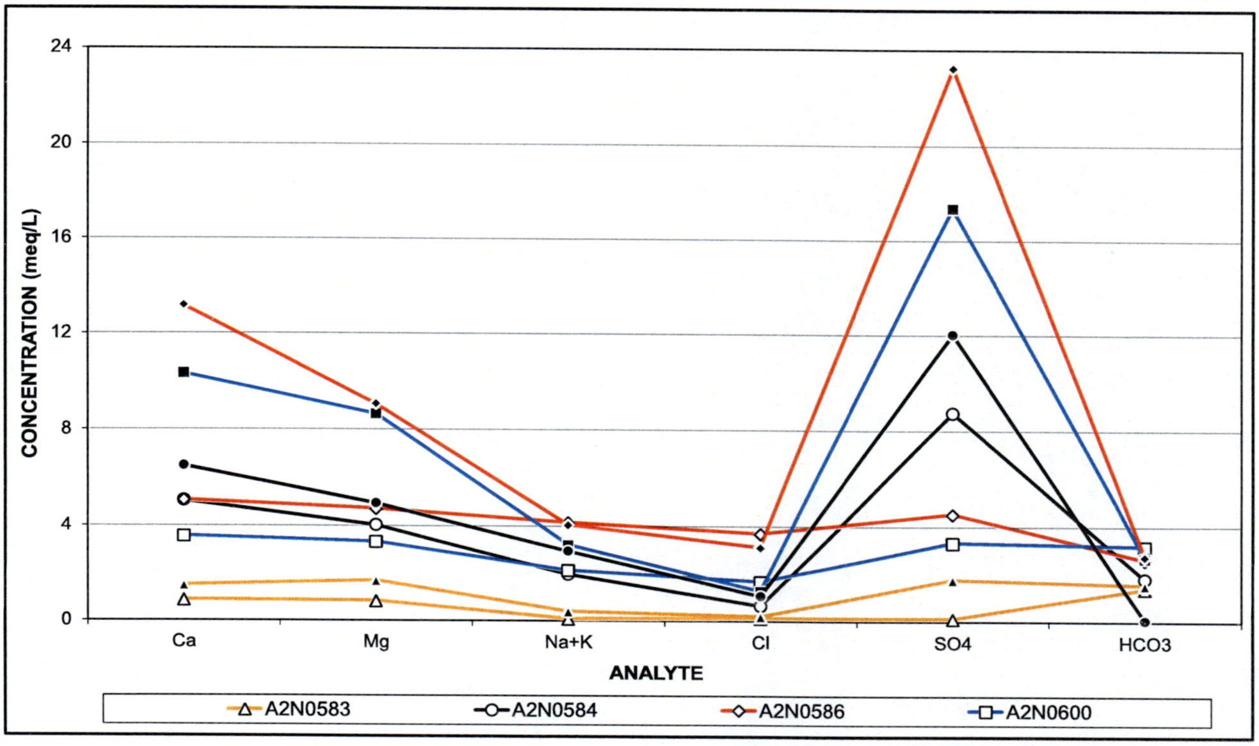


**Fig. 4** Comparison of the 'oldest' (open symbol ca. 1989) and most recent (solid symbol December 2014) groundwater chemical compositions at the longest-operating DWS monitoring stations

## 2.3 Source-Specific Temporal Assessment

### 2.3.1 Mine Area (Locus of Decant)

The groundwater quality monitoring programs carried out by the mining houses (SG and MG/MSA) and the DWS historically focused on the area of mine water decant. The footprint of this monitoring gradually expanded to include areas adjacent to and further downstream of the mine area (S. du Toit, personal communication) as described in Sect. 2.3.3.

The groundwater chemistry monitoring in the area of mine water decant provides insight into the geochemical characterization of AMD in its host receiving environment that includes dolomite.[3] Eight boreholes and three mine structures in this area (Fig. 5) provide information of varying relevance in this regard, as shown in the following discussion. Visualization of the geologic and hydrogeologic regimes that elucidate this discussion is provided in Figs. 7 in Chapter "Description of the Physical Environment" and 15 in Chapter "Physical Hydrogeology", respectively.

Key to an understanding of the mine water relationship with the karst outlier is recognition of the situation that the Black Reef Incline (BRI), together with #17 and #18 Winzes (Fig. 5), allowed for mining of the Black Reef at comparatively shallow depth (<100 m) below surface in this area (Fig. 15 in Chapter "Physical Hydrogeology"). The Main Reef was accessed at greater depths via vertical shafts such as #8 and #9 shafts and inclines driven from the base of opencast mine workings such as the West Wits Pit. Although the opencast workings exploited the auriferous package comprising the Kimberley Reefs, inclines driven from the base of these pits targeted the Main Reef at shallower depth than the vertical shafts. This geometry explains the hydraulic continuity between the Black Reef Incline and #17 and #18 winzes, and the deeper-seated mine workings on the Main Reef.

The BRI and #17 and #18 winzes represent the most prolific point sources of mine water discharge (Plate 1). Whereas the BRI and #17 Winze were the more productive sources of mine water discharge in the earlier period of decant because of their lower surface elevations, #18 Winze gained prominence in this role (Plate 1) ca. early-2010 as the other two sources, which had gradually become blocked

[3] Most, if not all, of the locus of mine water decant area is underlain by Malmani Subgroup dolomite forming an elongated basin-like outlier contained within Black Reef Formation and West Rand Group (Witwatersrand Supergroup) strata. This is illustrated in Figs. 7 in Chapter "Description of the Physical Environment" and 15 in Chapter "Physical Hydrogeology".

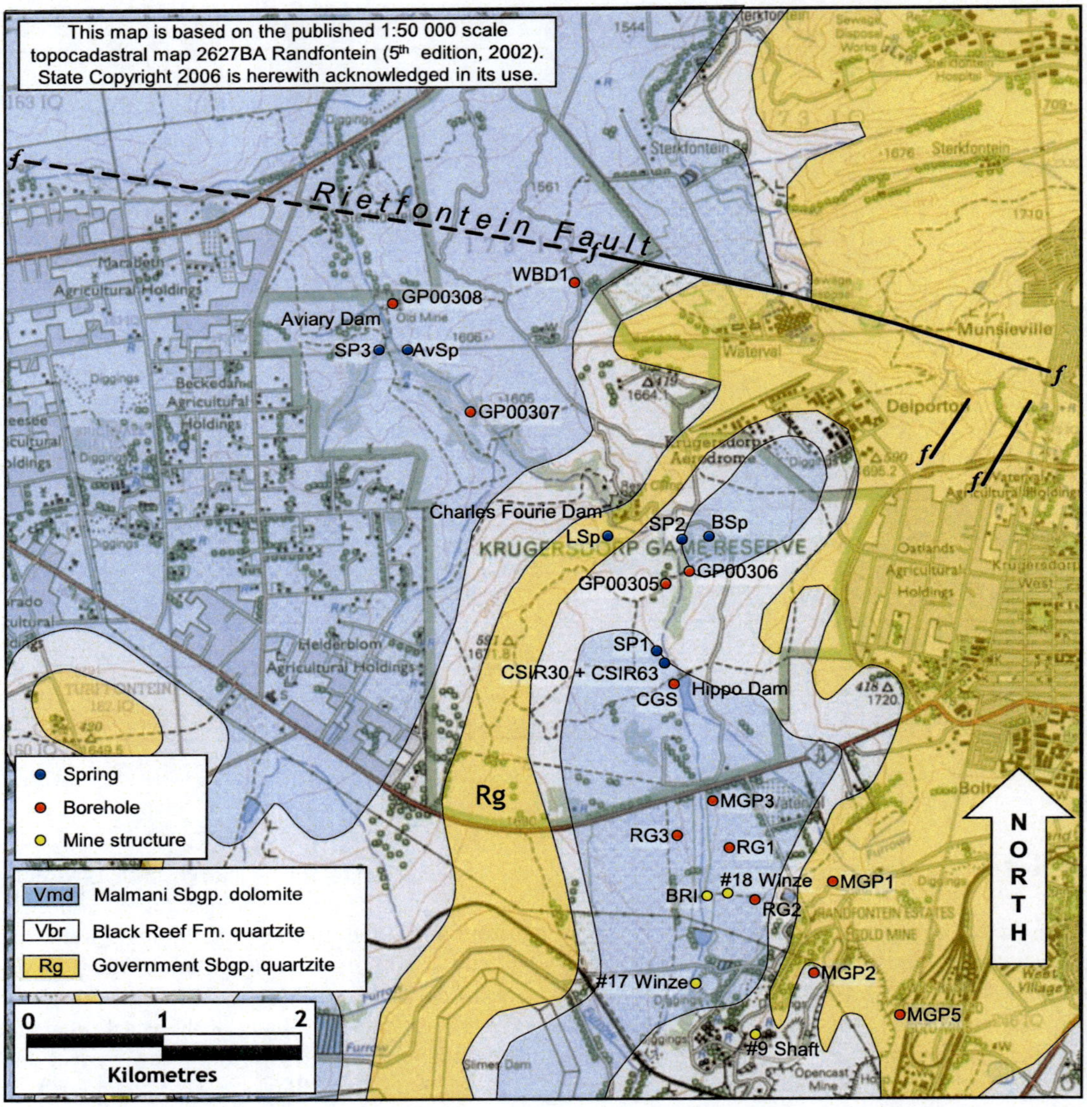


**Fig. 5** Locality map of mine water and groundwater monitoring stations in the locus of decant and the Krugersdorp Game Reserve; borehole MGP4 located off southern edge of map below MGP5

with gypsiferous precipitate, could not cope with the substantial increase in discharge (Sect. 3.1 in Chapter "Physical Hydrology") driven by the increasing potentiometric head associated with the flooded mine void. The presence of numerous surface seeps in the area is a further manifestation of the potentiometric head in the mine void.

The chemistry of the mine water produced by the BRI is characterized in Table 2 in Chapter "Chemical Hydrology" and discussed in detail in Sect. 1.2.1 in Chapter "Chemical Hydrology". Familiarity with this material will promote a better grasp of the following discussion.

In addition to routine monitoring of the point source mine water discharges, the mines also serve a number of monitoring boreholes. The SG boreholes RG1, RG2, and RG3, and the MG/MSA monitoring borehole MGP3, intersect the karst outlier in the mine area (Fig. 5). The MG/MSA monitoring boreholes MGP1, MGP2, MGP4, and MGP5 surround the West Wits Pit located on the water divide in the mine area. Borehole MGP3 lies at the northern end of the mine area. The results of monthly water chemistry monitoring associated with these boreholes yield the information presented in Table 2. The results for boreholes MGP1, MGP2, and MGP3 are graphed in Fig. 6. Although the monitoring results indicate that borehole MGP1 reflects SEC and pH values that represent natural quartzitic groundwater (Sect. 2.1.1), profiles of the water column in this borehole

**Plate 1** View looking south-west from #18 Winze over the locus of decant with Dump 39 (see Fig. 5 in Chapter "Description of the Physical Environment") on the horizon (photo W. Basson)

reveal a very different situation (Sect. 10.3.6). The median SEC and pH values of 9 mS/m and 5.4, respectively (Table 2), mask values of 100 mS/m and 4.0 for these water chemistry variables at a depth >15 m below water level in this borehole (Fig. 57). This phenomenon also applies to borehole MGP2 (Fig. 58) and MGP3 (Fig. 59), although it is much less pronounced in borehole MGP2.

### 2.3.2 Krugersdorp Game Reserve

The establishment of four monitoring boreholes in the KGR by the DWS in late-2010 has provided hydrogeological information for an area that previously suffered from a dearth of such information. The boreholes GP00305, GP00306, GP00307, and GP00308 (Fig. 5) each target a specific aspect of the groundwater environment in the KGR. Borehole GP00305 targets the faulted, fractured, and jointed quartzitic strata that form the escarpment which separates the KGR into an elevated southern plateau and more incised lower northern portion. Borehole GP00306 targets the northern margin of the karst outlier in close proximity to the Tweelopie Spruit. Borehole GP00307 targets the karst aquifer of the Zwartkrans Basin at a position next to the Tweelopie Spruit very near the Kemp's Cave fossil site at the entrance to the predator sanctuary. Borehole GP00308 targets the

**Table 2** Median chemistry variable/analyte values for groundwater associated with MG/MSA monitoring boreholes in the mine area in the period January 2011–April 2012

| Variable/analyte | Detection limit | Monitoring borehole # and median value | | | | | | | | | | SANS (2011a)[a] |
|---|---|---|---|---|---|---|---|---|---|---|---|---|
| | | MGP1 | | MGP2 | | MGP3 | | MGP4 | | MGP5 | | |
| *n* | | | 16 | | 15 | | 16 | | 16 | | 16 | |
| pH ($-\log_{10}\alpha_{H+}$) | | 5.4 | 16 | 5.5 | 15 | 8.8 | 16 | 6.1 | 16 | 4.9 | 16 | 5.0–9.7 |
| SEC (mS/m) | | 9.0 | 16 | 6.1 | 15 | 183 | 16 | 8.5 | 16 | 25.0 | 16 | ≤170 |
| Ca (mg/L) | | 6.2 | 16 | 3.8 | 14 | 234 | 16 | 7.0 | 16 | 8.5 | 16 | n.s. |
| Mg (mg/L) | | 4.2 | 16 | 2.0 | 15 | 105 | 16 | 1.6 | 16 | 5.2 | 16 | n.s. |
| Na (mg/L) | 10 | 7.2 | | 8.2 | | 73.0 | | 8.2 | | 28.0 | | ≤200 |
| Cl (mg/L) | 5.0 | 13.5 | 12 | 10.0 | 9 | 42.0 | 16 | 10.5 | 12 | 36.5 | 16 | ≤300 |
| $SO_4$ (mg/L) | 50 | <50 | 0 | <50 | 0 | 1105 | 16 | <50 | 0 | 58.0 | 1 | ≤500 |
| $HCO_3$ (mg/L) | 2.4 | 4.3 | 14 | 6.7 | 14 | 9.8 | 14 | 9.8 | 16 | 6.1 | 12 | n.s. |
| Fe (mg/L) | 0.1 | 0.1 | 3 | 0.2 | 4 | 0.2 | 4 | 1.5 | 14 | 0.1 | 7 | ≤2 |
| Mn (mg/L) | 0.1 | 0.3 | 15 | 0.2 | 6 | 0.2 | 7 | 0.5 | 15 | 0.2 | 13 | ≤0.5 |
| U (μg/L) | 5.0 | 7.0 | 7 | 12.5 | 6 | 7.1 | 8 | 8.6 | 8 | 10.5 | 12 | ≤15 |

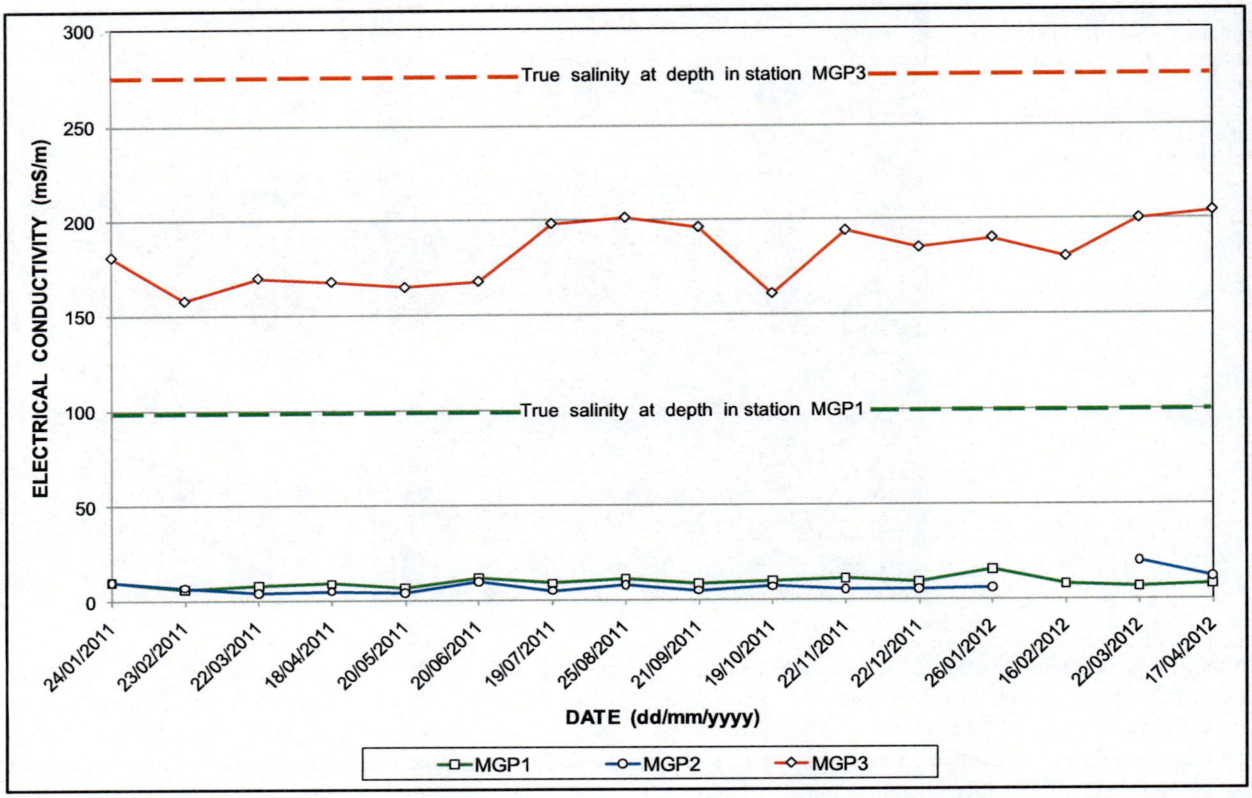


**Fig. 6** Pattern and trend of SEC monitored by MG/MSA in the mine area, compared to the true SEC values at depth in boreholes MGP1 and MGP3

chert-poor dolomite of the Oaktree Formation near the Aviary Dam close to the northern boundary of the KGR.

Together with six springs and an earlier borehole (WBD1), the four recently established monitoring boreholes provide a deeper perspective on the hydrogeologic environment downstream of the mine area. This perspective includes the following hydrogeologic observations:

- a subsurface mine water impact manifested in the karst outlier;
- a surface water driven mine water impact in the south-eastern portion of the Zwartkrans Basin;
- a municipal wastewater impact in the south-eastern portion of the Zwartkrans Basin;
- a natural radiogenic impact associated with quartzitic groundwater; and
- the natural discharge of nearly pristine karst groundwater along the north-western margin of the KGR.

### 2.3.3 Lower Riet Spruit

This discussion considers those groundwater monitoring activities carried out to the north of the KGR, a distance of >6000 m from the area of mine water decant and its immediate downstream receiving environment. This is informed by the need to better understand the impact of mine water discharge on groundwater quality in closer proximity to the COH, as the upstream impacts already receive considerable attention (Sect. 2.3.1). Routine sampling by SG of the four stations WBD2, WBD3, WBD4, and WBD5 (Fig. 19 in Chapter "Physical Hydrogeology") reflect the $SO_4$ trends (and recent pH and SEC values) shown in Fig. 7. These are described as follows.

WBD2: The most upstream station, located on Ptn 106 of Sterkfontein 173IQ (the old Krugersdorp Brickworks property) where the water table occupies a depth of ~25 m below that of the nearby Tweelopie Spruit streambed, displays a gradual but comparatively linear increase through to February 2010.[4] The $SO_4$ concentration doubled in this 3-year period when the borehole was in frequent production, thereby inducing flow from the Tweelopie Spruit located <50 m to the west of the borehole. After a hiatus of ~2½

[4] Access to this property and station was curtailed in March 2010 when the business closed down and borehole use terminated. Access resumed in November 2012 following discussions with the new owner.

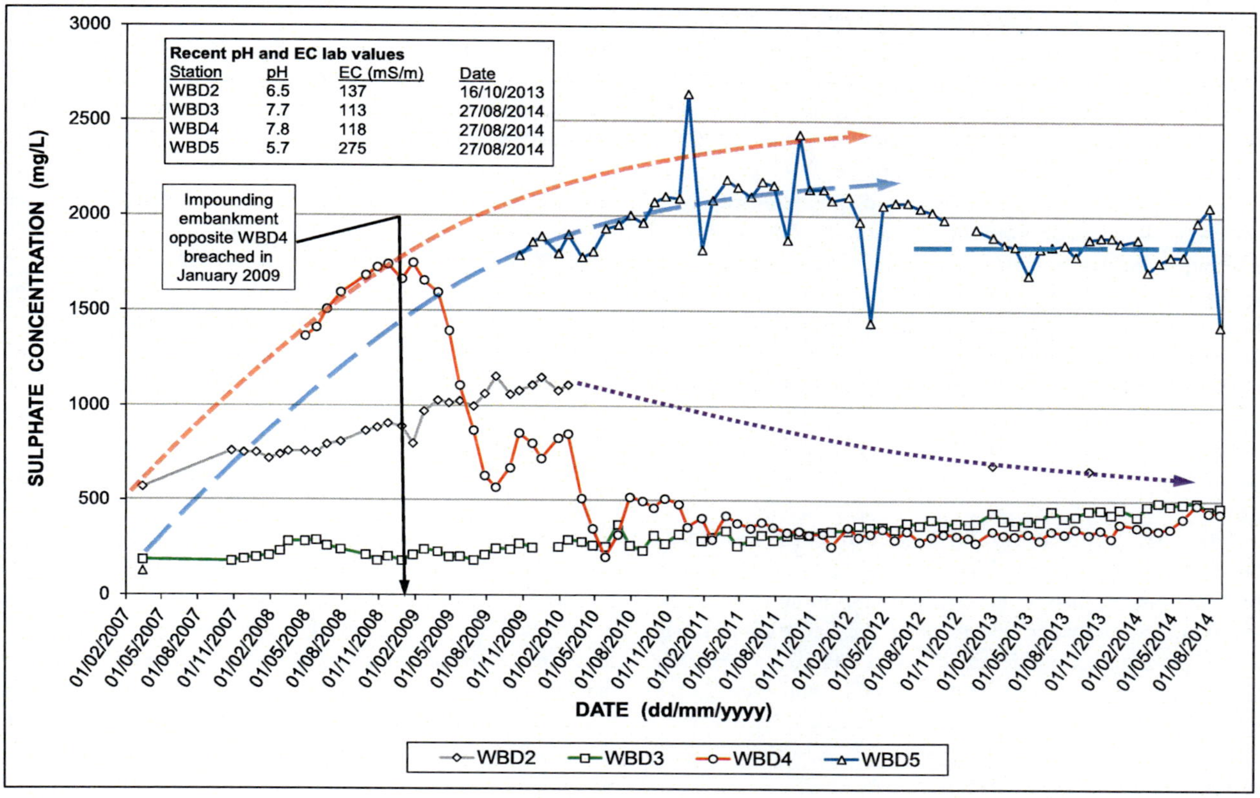


**Fig. 7** Observed $SO_4$ trend at four groundwater quality monitoring stations along the losing reach of the Riet Spruit upstream of Oaktree

years (Footnote 4), the borehole was re-commissioned by the new owner in late-2012. The $SO_4$ concentrations of 689 and 661 mg/L for samples collected in January and October 2013, respectively are shown in Fig. 7. These indicate a significant decrease from the 1110 mg/L of the next earliest 2010 analysis, the October 2013 concentration of 661 mg $SO_4$/L comparing favorably with the earliest value of 568 mg $SO_4$/L.

WBD3: The next station downstream, a distance of ~880 m on the Ptn 96/7 of Sterkfontein 173IQ (the Môrester property), displays a gradually increasing $SO_4$ concentration from ~200 to ~450 mg/L in the period of record. This is a somewhat anomalous trend compared to that observed at the other stations. A water level measurement in 2007 placed the groundwater rest level in this borehole at a depth of ~21 m below that of the nearby streambed (Hobbs and Cobbing 2007). The potentiometric response observed in the DWS monitoring borehole A2N0584 (Fig. 8 in Chapter "Physical Hydrogeology") indicates that the water level in WBD3 has almost certainly also risen since the 2007 measurement was made.

WBD4: Located ~550 m downstream of WBD3 on the Remainder Ptn 7/2 of Sterkfontein 173IQ (the Royal Engineering property), a mid-2008 water level measurement in this borehole placed the groundwater rest level at a depth of ~12 m below that of the nearby streambed (Hobbs 2008b). The groundwater chemistry shows an increase in $SO_4$ to a maximum of ~1750 mg/L in February 2009, followed by a rapid decline to a level of 322 mg/L in September 2013. The initial response is attributed to the damming up of the Riet Spruit on this property in early-2008, resulting in the enhanced infiltration of surface water comprising mainly treated/neutralized mine water into the karst aquifer at this location. Breaching of the impoundment in January 2009, following recommendations by Hobbs (2008b) to mitigate the adverse groundwater quality impact, produced the decrease in $SO_4$ concentrations observed since February 2009.

WBD5: This borehole is located ~950 m downstream of WBD4 on Ptn 8/2 of Sterkfontein 173IQ (the Chrisuél property) on the left bank of the Riet Spruit adjacent to a dam. Having risen by ~5 m to an elevation of ~1466 m amsl by February 2010 (Table 8 in Chapter "Physical Hydrogeology", station CSIR57), the water level in this borehole remains ~5 m[5] below that of the Riet Spruit streambed at this

[5] A groundwater rest level measurement on 01/09/2014 returned a value of 6.69 m bs.

position. The station displays an elevated and rising $SO_4$ concentration leveling out at ~2200 mg/L in late-2011. The extrapolation of the upward trend from the early-2007 value mimics that of the WBD4 data up to January 2009 (Fig. 7). A decrease in $SO_4$ concentration from ~2200 to ~1800 mg/L from late-2011 to September 2013 is also evident. The measure of water quality deterioration as represented by the increase from the earliest $SO_4$ concentration of 125 mg/L in March 2007 to ~2200 mg/L in late-2011, is again attributed to the damming up of surface water at this location. It is likely that a similar intervention as was applied in the case of WBD4, namely breaching of the impoundment to establish the free flow of surface water past this station, will most probably produce a similar result as observed at WBD4. The landowner, however, is not amenable to such an intervention.

The establishment in 2010 of additional monitoring boreholes in the lower Riet Spruit Valley by the DWS provides greater insight into the impact of mine water on the receiving groundwater resources of the Zwartkrans Basin. These stations are identified as GP00300, GP00301, GP00302, GP00309, GP00312, GP00313 and GP00314 in Fig. 19 in Chapter "Physical Hydrogeology".

### 2.3.4 Sterkfontein Cave System

Holland et al. (2009) report $SO_4$ and Cl concentrations of 154 and 55 mg/L, respectively for groundwater sourced from a borehole (presumably SF1) near the Sterkfontein Cave. This is put forward as "… undoubtedly indicating anthropogenic impacts." These values agree with the averages of 147 mg $SO_4$/L and 66 mg Cl/L for three boreholes (CSIR7, CSIR8, and CSIR9) in the upstream Oaktree area reported by Hobbs and Cobbing (2007), and raise concern for the quality of the cave water.

A more complete comparison of earlier cave water chemistry with the present is provided by the analyses of April 2001 (from SG records), March 2005 (from DWS records), February 2006 (from Harmony Gold records), May 2010, January 2011, and May 2014 (this study), and August 2012 from Maropeng āAfrika. The data are presented in Table 3, and the graphical comparison is made in Fig. 8, which also shows the October 2013 and May 2014 results for Zwartkrans Spring water. The Piper diagram (Fig. 9) provides a synoptic picture of the respective groundwater chemistries. The similar chemical composition of the pre-2014 cave waters is readily apparent in Fig. 8. Also notable and significant is the increase in $SO_4$ in the latest cave water sample, and the very different Ca–$SO_4$ chemical composition of the Zwartkrans Spring water, which reflects an unequivocal mine water impact.

A comparison of the Sterkfontein Cave water chemistry with that of 'pristine' karst springs in the COH is shown in Fig. 9. The Piper diagram indicates that the difference is primarily associated with the anionic composition (Cl and $SO_4$) of the cave water.

The installation of a continuous EC monitoring device in the Lake by the DWS in May 2005 yields data on the cave water salinity response pattern in the 27 months to July 2007. This pattern (Fig. 10) reveals an initial SEC value of ~47 mS/m, followed after a hiatus of ~32 weeks by a notably higher value of ~60 mS/m. An analysis of the cave water sampled on 29/04/2005 (sourced from the DWS records), i.e., within a month of installation of the monitoring device, returned an SEC value of 59 mS/m (Fig. 8). As shown in Fig. 10, this suggests that the early salinity record might be in error by an 'under-reading' of ~12 mS/m. Later field SEC values of 62, 56, 63, and 69 mS/m (in May 2010, January 2011, May 2012, and May 2014, respectively) for the cave water confirm the constancy reflected in the continuous DWS record (Fig. 10). The observation that the SEC of the cave water has not changed much in the decade since mid-2005 must be welcomed under the circumstances described by acid mine drainage. This observation finds support in the PHREEQC-derived calcite ($SI_C$) and dolomite ($SI_D$) saturation index values (Table 3), which show a saturated to supersaturated state.

The similar chemical composition of Sterkfontein Cave water reflected in Figs. 8 and 9, despite the difference of some ten years between analyses, mimics the situation sketched by the near-continuous SEC record (Fig. 10). This is encouraging against the background of recent groundwater quality trends observed elsewhere in the Zwartkrans Basin (Sects. 2.3.3 and 4). Equally encouraging are the results of stable isotope analyses carried out on cave water in 2009 and 2010. The comparative data tabulated below indicate very little difference in the span of one year between analyses.

| Date | $\delta^2H$ (‰) | $\delta^{18}O$ (‰) | $^3H$ |
|---|---|---|---|
| 28/05/2009 | −19.8 | −3.53 | 1.3 ± 0.3 TU |
| 13/05/2010 | −20.3 | −3.55 | n.a. |

A comparison of the May 2009 cave water tritium value with that obtained in July 2009 for the Zwartkrans Spring (Sect. 7.1) reflects a significant difference of 1.1 ± 0.3 TU. The higher value associated with the springwater indicates that this feature delivers a younger groundwater than that which is represented by the cave water. The most plausible explanation for this situation is that the cave system does not lie in the main flow path of groundwater discharge toward the Zwartkrans Spring. Similarly, Groenewald (2010b) refers

**Table 3** Historical and recent chemistry of Sterkfontein Cave groundwater

| Variable/analyte | Sample date | | | | | | | SANS (2011a)[a] |
|---|---|---|---|---|---|---|---|---|
| | 04/2001 | 04/2005 | 02/2006 | 05/2010 | 01/2011 | 08/2012 | 05/2014 | |
| pH $(-\log_{10}\alpha_{H+})$ | 7.5 | 8.1 | 7.9 | 7.9 | 7.6 | 8.1 | 7.5 | 5.0–9.7 |
| SEC (mS/m) | 74 | 59 | 57 | 59 | 46 | 66 | 69 | ≤170 |
| Ca (mg/L) | 47 | 44 | 46 | 47 | 48 | 48 | 61 | n.s. |
| Mg (mg/L) | 27 | 28 | 37 | 26 | 27 | 34 | 42 | n.s. |
| Na (mg/L) | 14 | 23 | 20 | 20 | 22 | 20 | 26 | ≤200 |
| K (mg/L) | 0.2 | 1.0 | <0.1 | 1.1 | 1.0 | 1.0 | 1.5 | n.s. |
| Cl (mg/L) | 38 | 31 | 46 | 27 | 37 | 44 | 39 | ≤300 |
| $SO_4$ (mg/L) | 46 | 58 | 61 | 58 | 71 | 78 | 126 | ≤500 |
| $HCO_3$ (mg/L) | 193 | 184 | 196 | 185 | 188 | 198 | 206 | n.s. |
| $NO_3$ (mg N/L) | 8.4 | 7.0 | 7.5 | 8.2 | 9.3 | 1.0 | 7.5 | ≤11 |
| $SI_C$ | −0.10 | 0.44 | 0.28 | 0.27 | −0.01 | 0.49 | −0.01 | n.s. |
| $SI_D$ | −0.17 | 0.94 | 0.73 | 0.56 | −0.01 | 1.10 | 0.09 | n.s. |
| EB (%) | −1.3 | 3.4 | 2.5 | 2.3 | −1.3 | −0.3 | 3.0 | n.s. |

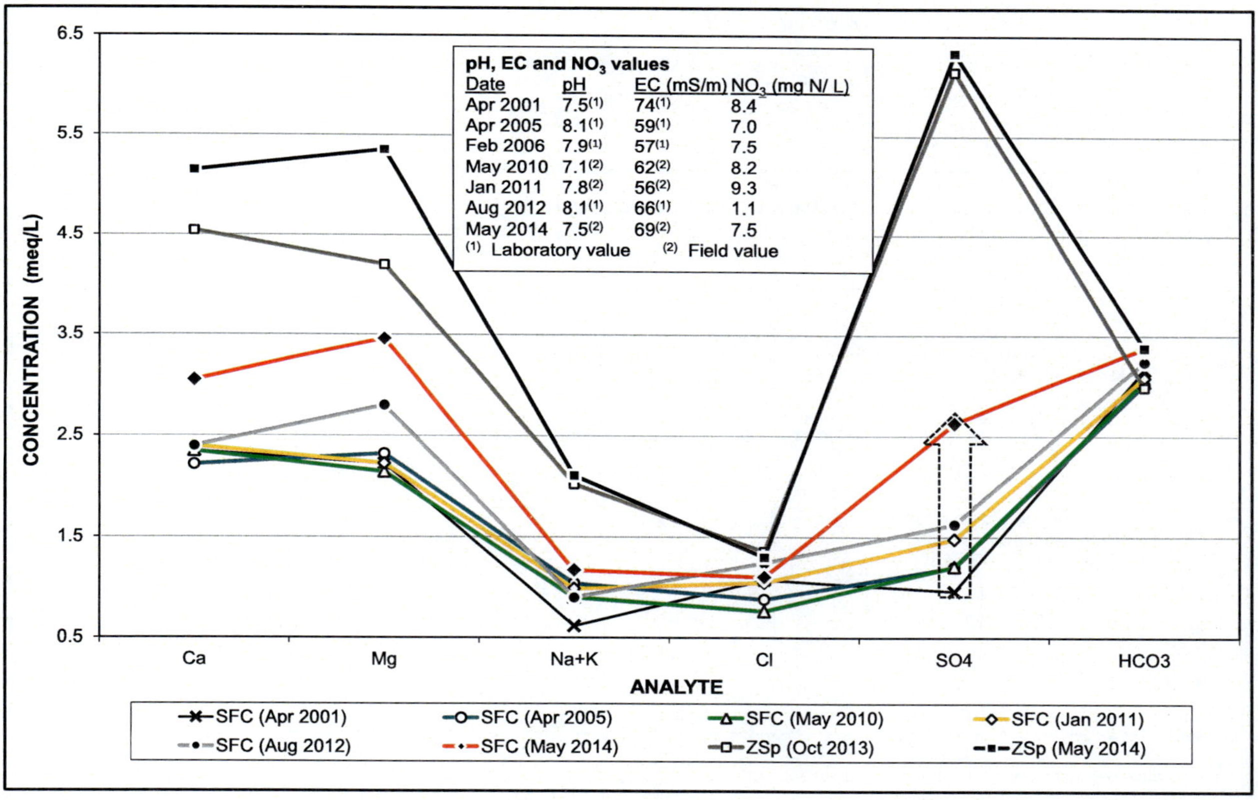


**Fig. 8** Graphic comparison of historical and current Sterkfontein Cave groundwater chemistry

to the cave as occupying a low energy groundwater system, and Martini et al. (2003) refer to the "… apparently static …" pools in the cave system. These circumstances are discussed in greater detail in Sect. 4.

Attention must also be drawn to the $NO_3$ concentrations of 8.2 and 9.3 mg N/L recorded for the 13/05/2010 and 14/01/2011 samples of cave water (Table 3 and text box Fig. 8). This level of $NO_3$ is uncharacteristic of mine water

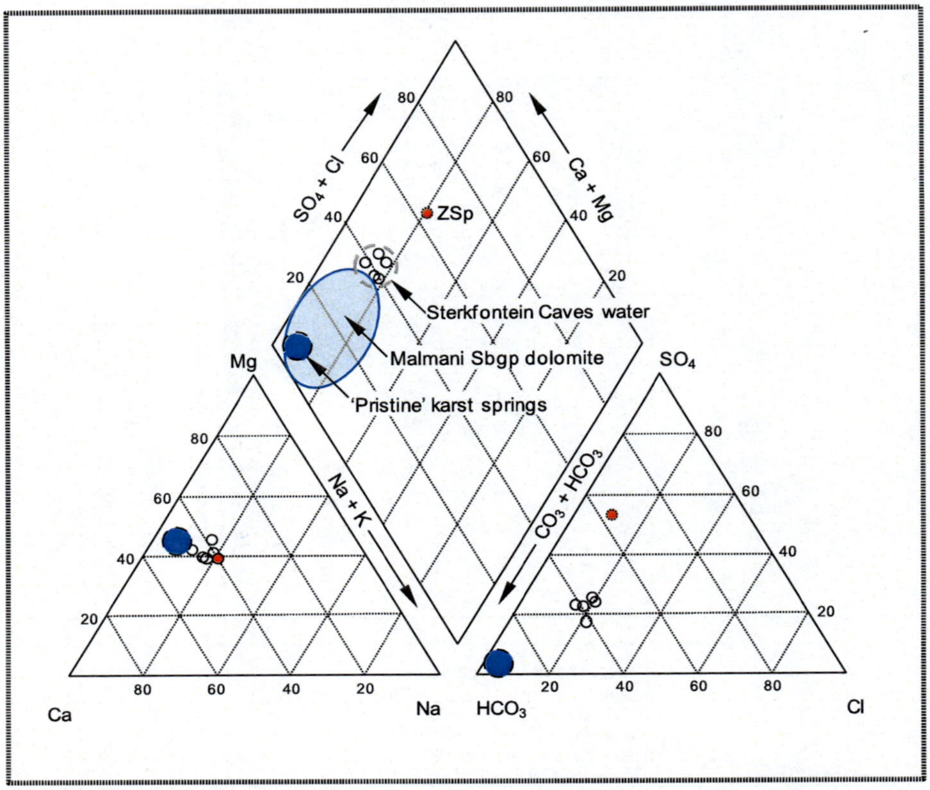


**Fig. 9** Piper diagram of historical and recent Sterkfontein Cave water chemistry compared to other karst groundwater and October 2012 and May 2014 Zwartkrans Spring (ZSp) water

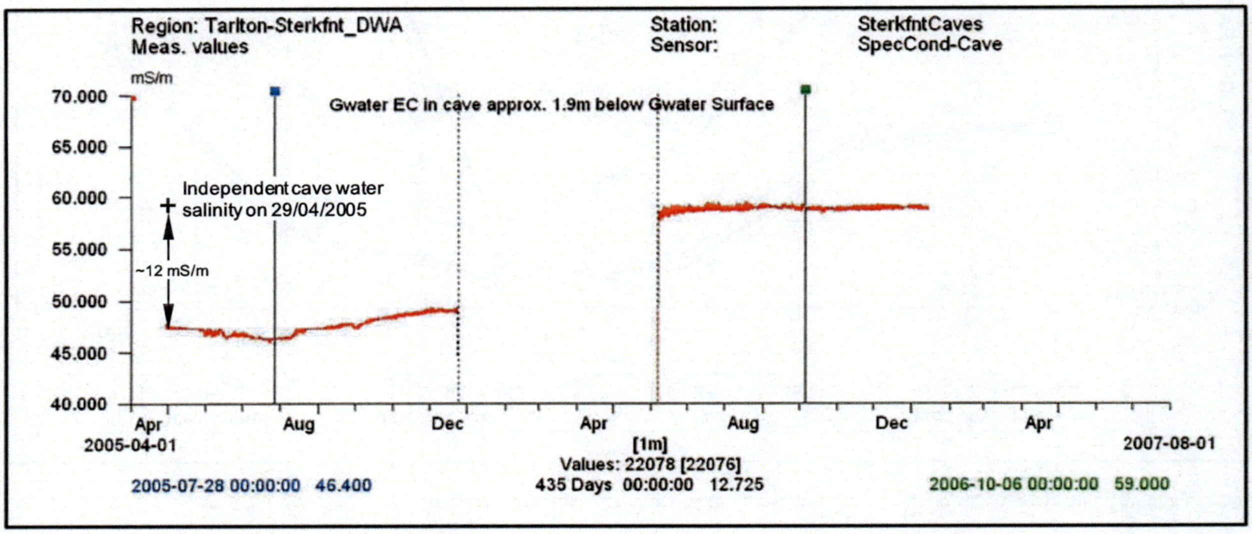


**Fig. 10** Continuous electrical conductivity response pattern in Sterkfontein Cave water over a period of 27 months (use of image courtesy of DWS)

and indicates another source (or sources) of water contamination possibly reflecting a historical legacy. Nevertheless, the nitrate concentration suggests a measure of contamination by nutrient-rich water. Possible sources of such water include the following:

- upstream agricultural land-use practices—the Oaktree area supports extensive irrigated agriculture that includes maize, vegetables, and tunnel farming (of cut-flowers) using groundwater drawn locally from the karst aquifer;
- municipal wastewater effluent—the discharge from Percy Stewart WWTW imposes a median $SO_4$ concentration of ~160 mg/L in the receiving Blougat Spruit (Hobbs 2011a), an upper tributary of the Bloubank Spruit;
- on-site sanitation facilities on the numerous smallholdings in the Oaktree area—this area is not served by the Mogale City Local Municipality sewerage system;
- a piggery[6] that used to be operated ~1150 m upstream (south-west) of Sterkfontein Cave on the right bank of the Bloubank Spruit—anecdotal evidence suggests that this facility routinely washed its surface accumulation of piggery waste into the adjacent stream as a means of disposal; and
- the sanitation facility serving the Sterkfontein Cave Visitors Center which, until mid-2011, comprised a septic tank system that was replaced with an environment-riendly self-contained bioreactor system (Sect. 4).

The source of the $NO_3$ contamination, i.e., whether from agriculture or municipal wastewater effluent, has not been established because of a paucity of historical groundwater chemistry information for this portion of the study area. In addition to the possibilities raised above, it might be speculated that the circumstances which precipitated the study by Barnard (1996) possibly could account for the elevated nitrate levels in the cave water. The possible impact of on-site sanitation facilities serving the visitor center at the caves also cannot be disregarded, especially in light of the following observations.

The septic tank and wetland sanitation system that previously served the visitor complex (Plate 2) was replaced with an environment-friendly self-contained bioreactor system in mid-2011 (I. Wright, personal communication). The 20 $m^3$/d design capacity of the SAFF (submerged aeration fixed film) technology biozone system (Plate 3) is indicative of the sanitation load generated by the visitor center. This center typically receives 110,000 people per annum (300 per day on average), with a peak of ~400 per day experienced during the 2010 soccer World Cup (T. Rubin, personal communication).

Replacement of the septic tank sanitation system has eliminated this as a possible/probable source of nutrient contamination and provides the most reasonable explanation for the 1 mg N/L concentration (i.e., an order of magnitude reduction) observed in the August 2012 cave water analysis (Table 3). The low $PO_4$ concentration (<0.2 mg P/L) and the absence of sunlight suggest that the trophic status of the Sterkfontein Cave water system is unlikely to change provided the current hydrochemical conditions prevail. Assurance in this regard can be obtained from a regular biomonitoring program (as described by Culver and Sket 2002) that targets stygobitic fauna.

Closer inspection of the data presented in Table 3 indicates that the $SO_4$ concentration in the cave water has increased by ~174% from 46 mg/L in April 2001 to 126 mg/L in May 2014 (Fig. 11). It has been shown (Sect. 1.4 in Chapter "Chemical Hydrology" and Text Box 1 in Chapter "Chemical Hydrology") that $SO_4$ comprises ~63% of the TDS concentration associated with Western Basin mine water, ~19% of the TDS typical of surface water, and ~2% in the case of pristine karst groundwater. The $SO_4$:TDS ratio value, therefore, serves as an indicator of mine water in receiving water resources. The $SO_4$:TDS ratio associated with the cave water chemistry over time is shown in Fig. 11. The latest value (~25%) represents an increase of ~92% over the ca. 2001 value (~13%), and of ~56% over the ca. 2010 value (16%). This provides a clear indication of a surface water influence on the cave water chemistry. Given the previously demonstrated evidence of a mine water impact on the surface water chemistry of the Bloubank Spruit system, the corollary transposes a muted mine water impact also on the cave water environment. The reasons for a muted mine water impact as opposed to a more prominent impact are explored in Sect. 8.4 and illustrated in Fig. 48.

## 3 Groundwater Chemistry Assessment by GRU

Chemical analyses representing a combination of inorganic, organic, bacteriological, and stable isotope results for groundwater sourced from boreholes and springs provide information on the spatial distribution of groundwater chemistry (quality) in the COH. A representative sample was ensured by obtaining water only from equipped and active borehole installations, and as close to the installed headworks as possible.

[6] Located in the north-eastern quadrant of the Kromdraai Road/R563 junction opposite the Cradle Food Market, this property now hosts the Sterkfontein Public Storage Facility.

**Plate 2** View from the Sterkfontein Cave Visitor Center looking north over the septic tank system soak-away area (circled) and the Bloubank Spruit valley (middle foreground) at the highest point in the COH WHS (arrowed)

**Plate 3** View of the biozone sewage treatment plant installed under the Sterkfontein Cave Visitor Center in mid-2011

The results of 48 analyses of water sourced from 39 boreholes and nine springs (excluding the KGR springs, which are discussed separately in Sect. 7.15) are grouped geographically according to the basins/subcompartments identified in Sect. 2 in Chapter "Physical Hydrogeology", and which are recognized as groundwater resource units[7] (GRUs). The grouping of analyses on this basis, and

[7] See Glossary.

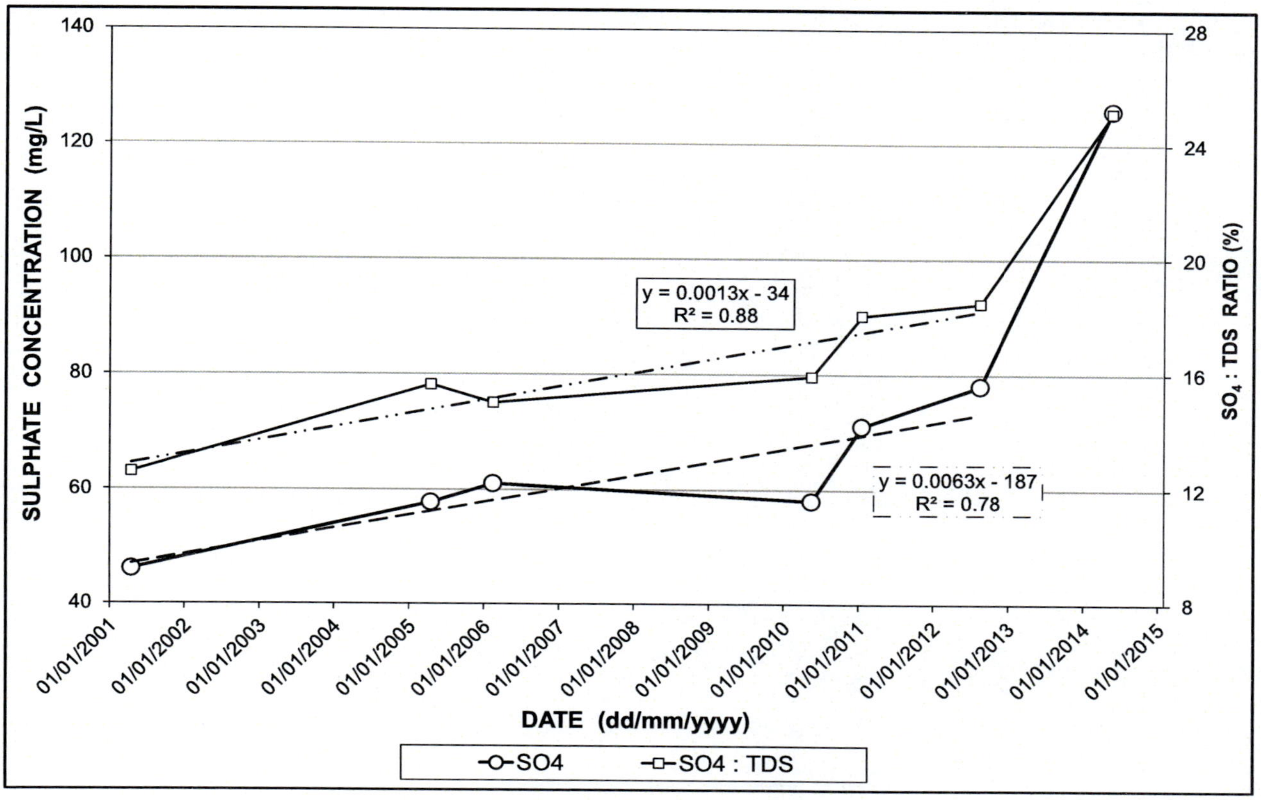


**Fig. 11** Pattern and trend of the $SO_4$ concentration and $SO_4$:TDS ratio in Sterkfontein Cave Lake water since 2001

considering the mean concentration of analytes per group, provides a measure of the current groundwater chemistry associated with these hydrogeologic units. This information is presented in Figs. 12, 13 and 14 and discussed in the following sections.

## 3.1 Zwartkrans Basin

### 3.1.1 Vlakdrift Subcompartment

It is evident from Figs. 12 and 13 that the groundwater chemistry associated with subcompartment GRU1a approaches that of a 'pristine' karst aquifer. The only concern is the indication of bacteriological contamination in the upper reaches of the subcompartment where it receives allogenic recharge via the upper Riet Spruit (Sects. 5.1 in Chapter "Physical Hydrology" and 6).

### 3.1.2 Sterkfontein Subcompartment

Figure 12 and 13 show the extent to which the groundwater chemistry in this subcompartment (GRU1b) differs from that in the other GRUs. The Ca–$SO_4$ groundwater character and mean SEC value of 133 mS/m reflect the impact of mine water on the karst groundwater associated with this GRU.

### 3.1.3 Zwartkrans Subcompartment

Albeit less pronounced compared to that of GRU1b, the Ca–$SO_4$ composition of the GRU1c groundwater (Figs. 12 and 13) still reflects the impact of mine water on the karst groundwater. Figure 13 also shows that this subcompartment supports the highest mean Cl concentration, which reflects the additional impact associated with the ingress of municipal wastewater effluent discharged from the Percy Stewart WWTW via the Blougat Spruit.

## 3.2 Krombank Basin

### 3.2.1 Kromdraai Subcompartment

Figure 14 indicates that the groundwater chemistry in this subcompartment (GRU2a) is similar to that in the other 'pristine' dolomite compartments (e.g., GRUs 3, 4, 5, and 7). However, station RLGR5 located at the Kiosk/Camp Site in the Rhino and Lion Game Reserve (RLGR) in this

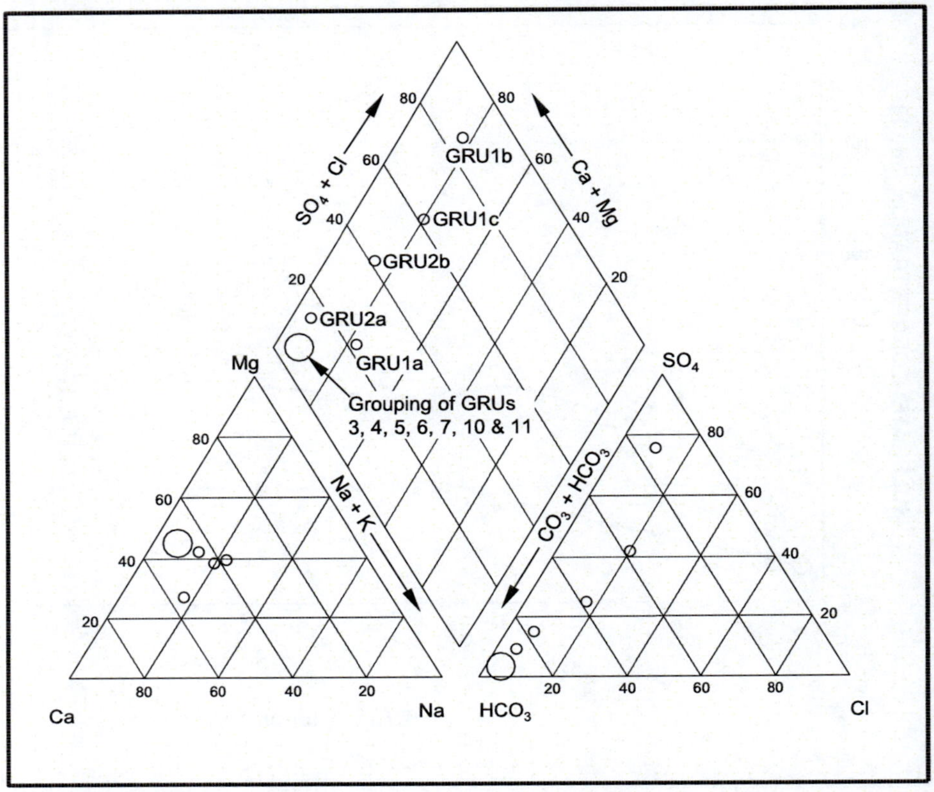


**Fig. 12** Piper diagram of GRU water chemistry

subcompartment indicates slight bacteriological contamination (Sect. 6 and Fig. 17). The *E. coli* concentration of 4 cfu/100 mL is cause for concern in the medium- to long-term, although routine monitoring of groundwater quality by the RLGR provides an appropriate management measure in this regard.

### 3.2.2 Bloubank Subcompartment

The similarity in mean groundwater chemical composition between GRU2b and GRU1c is evident in Figs. 12 and 13. The more muted composition associated with GRU2b groundwater reflects the situation where this unit is the main receiving groundwater resource for the discharge from GRU1c. It is therefore also not surprising that Fig. 14 indicates that groundwater quality in the Bloubank Subcompartment (GRU2b) is the 'poorest' of the 'pristine' dolomite compartments. This is also reflected in the mean SEC value of 49 mS/m that characterizes this groundwater.

## 3.3 Danielsrust Basin

Figure 14 shows that the groundwater chemistry in this compartment (GRU3) reflects a composition that is typical of a 'pristine' dolomite aquifer. Again the only concern is the indication of bacteriological contamination, albeit slight, in the water produced by the Danielsrust Spring (Sect. 6 and Fig. 17). A more complete discussion on the springwater chemistry is presented in Sect. 7.2.

## 3.4 Uitkomst Basin

The Uitkomst Basin (GRU4) supports a pristine karst groundwater environment that is reflected in the chemical composition of the groundwater discharged from this compartment via the Nash Spring. The quality of this groundwater is described and discussed in Sect. 7.8.

## 3.5 Tweefontein Basin

The groundwater chemistry in GRU5 conforms to that associated with the more 'pristine' karst environments. Figure 14 indicates that the composition of this groundwater is virtually indistinguishable from that associated with GRU7 (the Diepkloof Basin drained by the Nouklip Spring). These circumstances support the postulated

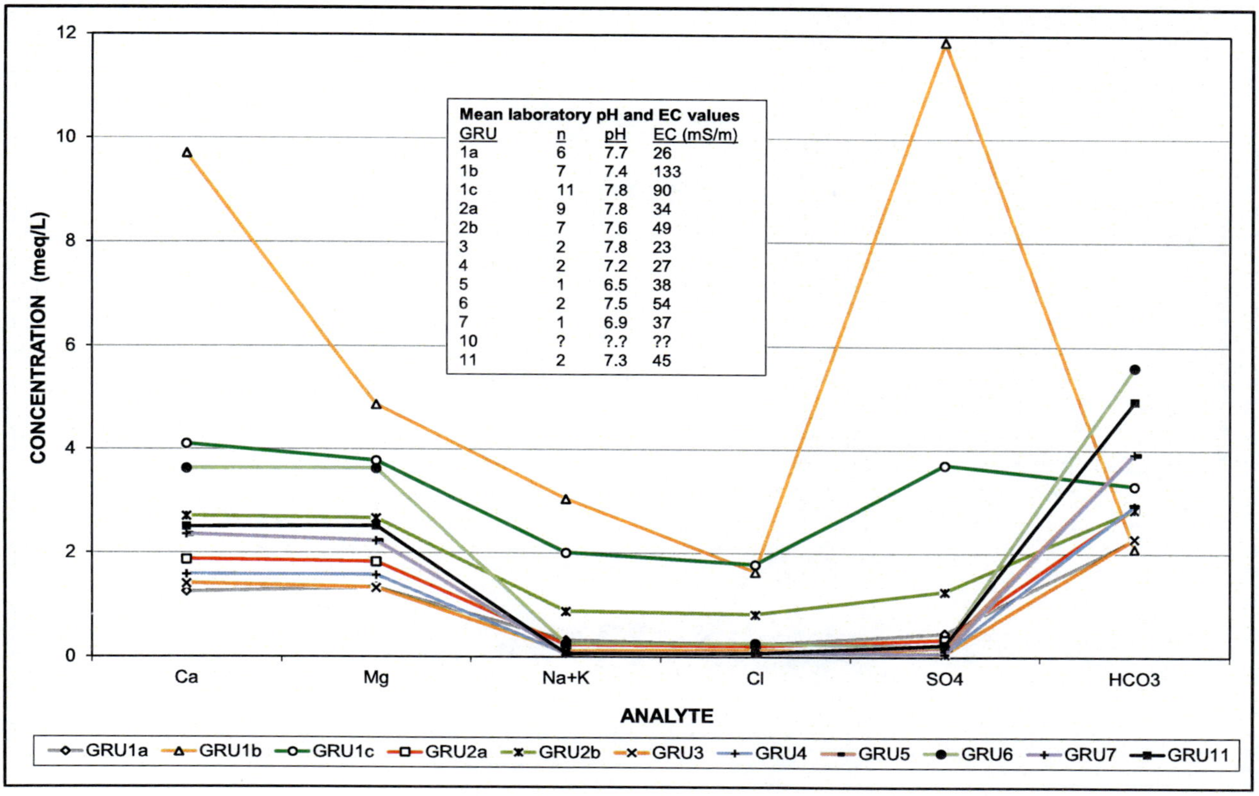


**Fig. 13** Graphic comparison of groundwater chemistry associated with GRUs

autogenic recharge of GRU7 via the Tweefontein Spring discharge. Further information on the chemistry of this groundwater is provided in Sect. 7.5.

## 3.6 Rietfontein Basin

The Rietfontein Basin (GRU6) produces a good quality groundwater (Sect. 7.6) which reflects the different rock types (dolomite and undifferentiated igneous strata) that comprise this hydrogeologic unit. The influence of the non-karst groundwater on the mixture of karst and non-karst groundwater produced by GRU6 is evident in detectable concentrations of metals such as Fe (~2.6 mg/L) and Mn (~0.4 mg/L) (Table 5).

## 3.7 Diepkloof Basin

The Diepkloof Basin (GRU7) reflects a similar groundwater chemistry to that of GRU5, the Tweefontein Basin (Sect. 3.5). For this reason, it reflects a 'pristine' karst environment that produces good quality groundwater (see Sect. 7.7).

## 3.8 Motsetse Basin

The groundwater chemistry associated with GRU8 is represented by the field variables listed in Sect. 7.10. The SEC value of 64 mS/m is indicative of a slightly impacted karst groundwater chemistry.

## 3.9 Rhenosterspruit Basin

There are no groundwater chemistry data available for this hydrogeologic unit (GRU9).

## 3.10 Kalkheuvel Basin

Represented by the Barlow Spring and GF1 borehole groundwater (GRU10).

## 3.11 Broederstroom Basin

The groundwater chemistry of this compartment (GRU11) is represented by that of the Broederstroom, Anderson, and

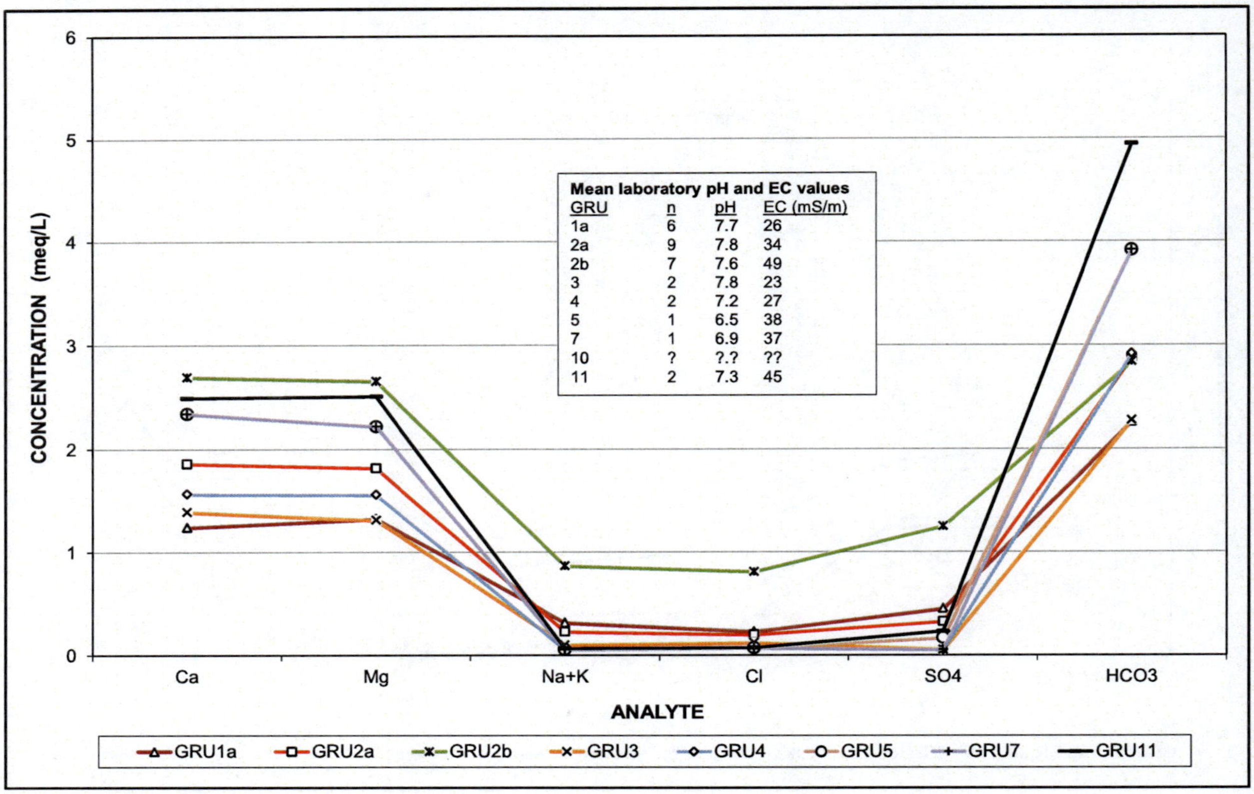


**Fig. 14** Graphic comparison of groundwater chemistry associated with 'pristine' GRUs

Lesedi springs. The composition of the Anderson and Broederstroom springwaters (Sects. 7.12 and 7.14, respectively) again reflect that which characterizes the more 'pristine' karst environments (Fig. 14). Concern regarding the Lesedi springwater quality is discussed in Sect. 7.13.

## 4 Local Groundwater Chemistry Assessment

A real and present concern is the result returned by a chemical analysis carried out on the groundwater produced by geosite WBD1, the water supply borehole in the KGR established in 2009 to mitigate the termination of the wastewater supply from the Percy Stewart WWTW (see hereunder). This analysis yielded a nitrate concentration of 51 mg N/L, a Cl value of 103 mg/L, an $SO_4$ value of 110 mg/L, a Na concentration of 82 mg/L, and a laboratory SEC value of 103 mS/m. All of these concentrations are anomalous for karst groundwater and point to a significant impact on the karst groundwater delivered by this borehole. The most likely source of this impact is the irrigation of ~65 ha of instant lawn with treated wastewater effluent obtained from the Percy Stewart WWTW as discussed in Sect. 3.2 in Chapter "Physical Hydrology" and shown in the unnumbered Plate on p. 10.

These circumstances are not new, as a similar situation precipitated the study by Barnard (1996). The Barnard (1996) study found that the practice of flood irrigation in the KGR using wastewater effluent from the Percy Stewart WWTW was the most likely cause of elevated $NO_3$, $SO_4$, and Cl concentrations in the groundwater produced by the borehole WBD4 (Sect. 2.3.3), as well as the source of bacteriological contamination of this groundwater.

The provision of wastewater to the KGR was terminated in September 2008 (Sect. 3.2 in Chapter "Physical Hydrology") on the presumption that wastewater irrigation of kikuyu in the KGR was again the cause for the observed deterioration in mid-2008 of the groundwater quality produced by borehole WBD4 (Fig. 7). The more probable cause, however, was the ingress of a mixture of raw and treated mine water into the karst aquifer from the Riet Spruit (Sect. 2.3.3) as discussed by Hobbs (2008b). Other factors that inform this probability are the following:

- the borehole WBD4 is located ~180 m from the losing Riet Spruit drainage;

- the depth to groundwater rest level in the northern portion of the KGR where irrigation was practiced, exceeds 40 m from surface;
- the method of irrigation practiced in the KGR followed best practice for irrigation on dolomitic terrains such as ensuring no ponding of water (S. du Toit, personal communication) in dolines and other surface depressions; and most convincingly;
- the chemistry of the groundwater produced by borehole WBD4 in early- to mid-2008.

It remains a travesty, therefore, that the same practice that precipitated termination of the provision of ~0.97 ML/d wastewater to the KGR was conveniently overlooked in regard to the similar supply of ~1.3 ML/d to irrigate the 65 ha of instant lawn on the adjoining property to the east. Some of the wastewater provided to the KGR was also used for game watering.

## 5 Isotope Chemistry

The use of isotope chemistry is well established in both surface water and groundwater investigations (Cecil and Green 2000; Osmond and Cowart 2000; Wanty and Nordstrom 1993). Ford and Williams (2007), however, caution that the use of stable isotopes in hydrogeology merely provides information on the general source of waters, and does not identify the exact source. This is seen as a significant limitation. The principal isotopes are those of hydrogen (protium $^{1}$H and deuterium $^{2}$H) and oxygen ($^{18}$O and $^{16}$O). These are not subjected to nuclear transformation, and therefore also known as stable isotopes. Radioactive isotopes, e.g., tritium ($^{3}$H), are subject to decay over time, which explains their use for dating.

It is standard practice to report stable isotope results as the ratio of the least abundant (heaviest) isotope to the most abundant (lightest) isotope (Kendall and Caldwell 1998). For oxygen, the ratio is $^{18}$O/$^{16}$O, and for hydrogen $^{2}$H/$^{1}$H. As variations in the isotopic ratios are small, the δ (delta) notation is used to express the deviation of the isotopic ratio in the sample with respect to the ratio in the standard (e.g., $\delta^{18}$O) in units of parts per thousand (‰) relative to a standard of known composition represented by that of Standard Mean Ocean Water (SMOW).

### 5.1 Previous Studies

The use of isotopes in the study of water chemistry in the project area (primarily the area of mine water decant) has been documented, among others, by CGS (undated), Holland (2007), Hobbs, and Cobbing (2007), and Hobbs et al. (2010). The CGS (undated) employed a variety of isotopes ($^{2}$H/$^{1}$H, $^{18}$O/$^{16}$O, $^{87}$Sr/$^{86}$Sr, $^{34}$S/$^{32}$S, $^{206}$Pb/$^{207}$Pb, and $^{208}$Pb/$^{207}$Pb) to investigate the chemical (isotopic) relationship between various water sources in the area of mine water decant and immediate downstream environment. In regard to the $\delta^{34}$S data, this study reported interpretation difficulties associated with the available information, due mainly to a lack of background reference values. Nevertheless, the observed $\delta^{34}$S values are ascribed to natural oxidative weathering of pyritic Witwatersrand strata interacting with and leached by the water.

The $^{87}$Sr/$^{86}$Sr data, viewed in conjunction with the $\delta^{34}$S data, suggest the gradual mixing between waters of probably dolomitic provenance and mine effluents. The Pb isotope results, on the other hand, clearly reflect the difference between raw mine water and surface water and groundwater sources that exhibit a mine water impact (Fig. 15).

Both Holland (2007) and Hobbs and Cobbing (2007) used the stable isotope ratios $\delta^{2}$H and $\delta^{18}$O to distinguish between various natural and impacted water sources in the area (see Sect. 5.2). In addition, Holland (2007) reports the results of 21 tritium ($^{3}$H) analyses for groundwater and springwater samples in the study area. Hobbs et al. (2010) report on the additional application of the radio-isotopes $^{222}$Rn (radon), $^{226}$Ra (radium), and $^{238}$U as possible hydrological tracers in a natural and polluted environment (Sect. 5.2.2). In a much wider geographic

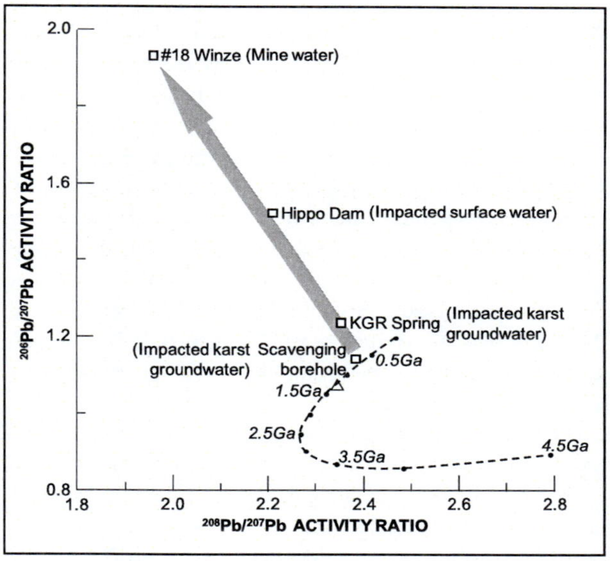


**Fig. 15** Relationship of Pb isotopic ratios of water in the West Rand area to the Stacey–Kramers curve (dotted line); South African leaded petrol plots at Δ on this curve; different stages of crustal evolution shown by ages in Ga; arrow shows the direction of deviation of the Pb isotope signature associated with mine water impacts (modified from Coetzee and Rademeyer 2006)

**Table 4** Comparison of ca. mid-2009 isotope data for cave and springwater

| Variable/analyte | Sterkfontein Cave | Zwartkrans Spring | Plover's Lake springs |
|---|---|---|---|
| Date | 28/05/2009 | 17/07/2009 | 17/07/2009 |
| $\delta^2H$ (‰ SMOW) | −19.8 | −16.8 | −28.9 |
| $\delta^{18}O$ (‰ SMOW) | −3.53 | −3.17 | −5.01 |
| $^3H$ (TU ± 1σ) | 1.3 ± 0.3 | 2.4 ± 0.3 | 1.6 ± 0.3 |
| $^{222}Rn$ (Bq/L) | n.a. | 8.8 | 10.5 |

context, Bredenkamp et al. (2007) report on the use of $^{14}C$, $^3H$, and CFCs in providing new insights into recharge and circulation processes in the karst aquifers of the South African interior. These insights extend to a derivation of mean residence (turnover) periods for groundwater in these hydrogeologic systems. For the Pretoria Springs, identified by Bredenkamp et al. (2007) as the Erasmus, Grootfontein, Pretoria Upper, Pretoria Lower, and Sterkfontein springs, these authors report residence periods in the range of 12.7–20.3 years.

## 5.2 Isotopic Information

### 5.2.1 Stable Isotopes

Isotope analyses performed on Sterkfontein Cave and Zwartkrans and Plover's Lake springs groundwaters in mid-2009 are presented in Table 4. The more similar $\delta^2H$ and $\delta^{18}O$ values of the Sterkfontein Cave and Zwartkrans Spring groundwaters support the water level (Sect. 3.2.1 in Chapter "Physical Hydrogeology") and inorganic chemistry (Sect. 7.1) data that indicates that the cave system penetrates and shares the karst aquifer drained by the spring (Hobbs and De Meillon 2017). This observation also supports the use of these isotopes to distinguish between various water sources (and mixture of sources) in the study area.

Perhaps more significantly, however, is the difference in $^3H$ levels between the Sterkfontein Cave and Zwartkrans Spring groundwater. These values indicate that the springwater represents a younger water than that of the cave water. This observation supports the hypothesis (Sect. 2.1.3 in Chapter "Physical Hydrogeology") that the Sterkfontein Cave does not lie in the main flow path of groundwater discharging to the Zwartkrans Spring.

The tritium value of the cave water is closer to that of the pristine Plover's Lake springs groundwater. The value of 1.3 ± 0.3 TU is representative of groundwater that is several decades old,[8] compared to the ostensibly 'younger' Zwartkrans Spring water with a tritium value of 2.4 ± 0.3 TU. The circumstances that give rise to these differences are important in evaluating the risk posed to the Sterkfontein Cave system by poor quality groundwater in the Zwartkrans Basin that originates from anthropogenic impacts. For example, Martini et al. (2003) report the development of a dolerite sill immediately below Sterkfontein, but which has not been observed in the cave. It is possible that the resistivity surveys reported by Chirenje et al. (2010) might have recognized this structure.

Although the chemistry of springwater (including environmental isotope data) is discussed in more detail in Sect. 7, it is appropriate to note here that the relative age of the groundwater as indicated by the environmental tracers principally reflects the much slower flow paths through the matrix and narrow fractures in the rock (Worthington and Gunn 2009). These circumstances contribute to the apparently large difference in relative groundwater age (from <10 years to several decades) between different karst springs in the study area.

The stable isotope results generated by this study are illustrated in Fig. 16. The comparatively close grouping of six of the springwater samples in the 'pristine' karst groundwater field is apparent. Further up the theoretical 'local' evaporation line is the Sterkfontein Cave water sample, and beyond that the impacted Zwartkrans Spring groundwater. The elongated shape of the marginally impacted karst groundwater field reflects the wider mixture of isotopic composition associated with this groundwater, ranging from nearly pristine at the lower-left ('heavier') end to moderately impacted at the upper right ('lighter') end as the evaporative water signature increases. Covering, among others, the same area investigated by Hobbs and Cobbing (2007), a Water Research Commission (WRC) study (Hobbs et al. 2010) that included $\delta^{18}O$ and $\delta^2H$ analyses associated with groundwater, mine water, and surface water sources, further established the use of these analytes to distinguish between the various water sources as reflected in Fig. 16.

## 5.3 Radioactive Isotopes

The study by Hobbs et al. (2010) investigated the use of $^{222}Rn$ as a hydrological tracer in natural and polluted environments and added radioactive isotope data to the water

[8] Tritium values <2 TU are generally associated with groundwater residence times in the order of several decades, whereas higher values are indicative of more recent recharge and faster circulation (see for e.g., Hershey et al. 2010).

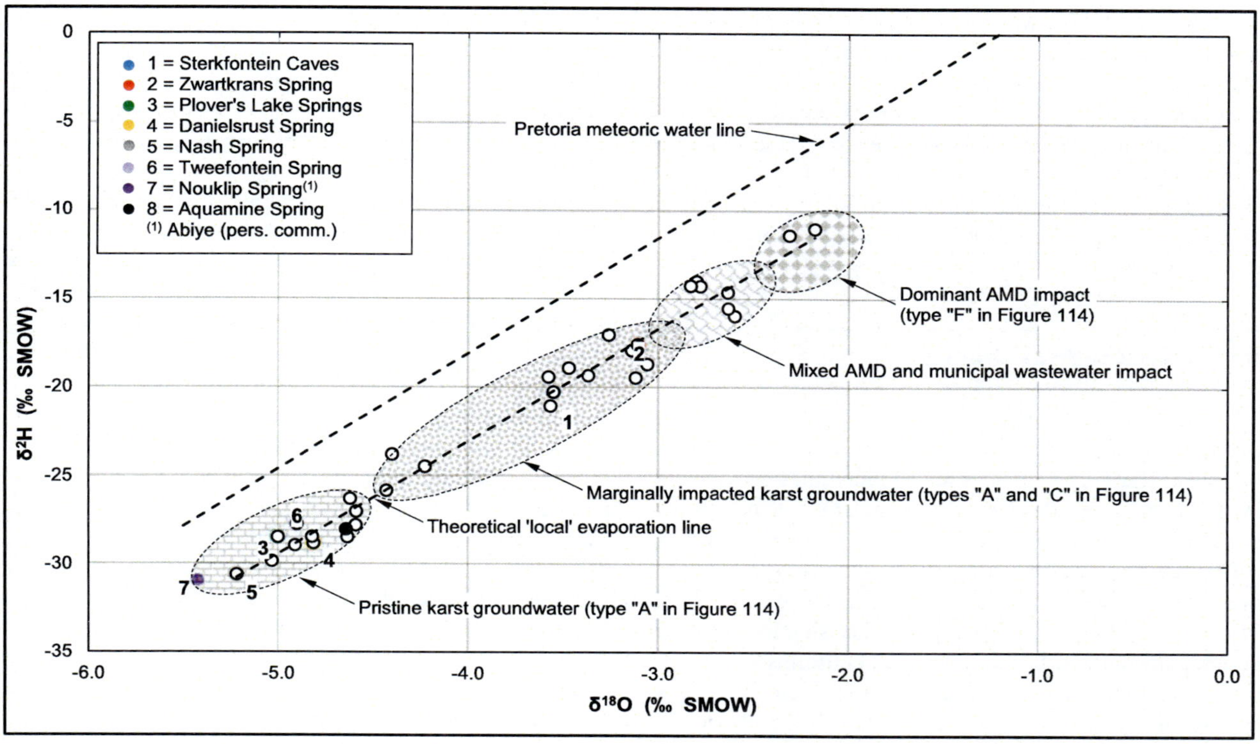


**Fig. 16** Stable isotope composition of groundwater in the study area, also showing the characteristic fields associated with various sources (modified from Hobbs et al. 2010)

quality data set generated by the Hobbs and Cobbing (2007) study. The data included $^{222}$Rn (radon), $^{226}$Ra (radium), and $^{238}$U (uranium) analyses associated with groundwater, mine water, and surface water sources collected on two occasions, once in early-2007 and again in early-2008. The radio-isotope data ($^{222}$Rn, $^{226}$Ra, and $^{238}$U) proved less successful as hydrological tracers than $\delta^{18}$O and $\delta^2$H. For example, $^{222}$Rn activity levels (concentrations) as high as 65 Bq/L were found in otherwise pristine groundwater issuing from quartzitic strata of the Witwatersrand Supergroup, whereas the median $^{222}$Rn concentration associated with mine water sources was ~27 Bq/L. 'Pristine' karst groundwater yielded a median $^{222}$Rn concentration of ~7 Bq/L, in good agreement with the values reported in Table 4 for the Zwartkrans and Plover's Lake springs.

The $^{226}$Ra concentrations in mine water sources ranged from 0.12 to 1.38 Bq/L, the mean of 0.65 Bq/L exceeding the DWS (1999) target water quality range for domestic water use of 0.42 Bq/L. The comparatively low $^{226}$Ra values of 0.029 and 0.002 Bq/L associated with a tailings dam source support findings by de Jesus (1985) that tailings dams do not represent a significant threat as a source of this radionuclide in the receiving surface and groundwater environments.

It should be considered that radium shows an affinity for manganese oxide, for example as represented by MnO and $MnO_2$ in manganese-impregnated fiber (Moore and Reid 1976). In this regard, the substantial presence of wad in the Malmani Subgroup dolomite is especially significant. Wad (manganiferous earth) is a weathering/decomposition product of dolomite. It comprises of manganese–and iron–oxides that exhibit insoluble and highly compressible properties (Brink 1996). The chemical composition of wad associated with the Oaktree Formation reflects an $MnO_2$ content of ~42% (and an $Fe_2O_3$ content of ~20%). Its association with more leached dolomite and therefore preferential water-bearing horizons suggests that this material may selectively scavenge radium from any mine water that may recharge the karst aquifer. The median $^{238}$U concentration of 1.82 Bq/L (0.15 mg/L) associated with mine water is an order of magnitude greater than the SANS (2011a) limit of 0.015 mg/L for this analyte.

The chemical analyses carried out as part of the current study in most instances included the determination of uranium (U) concentrations. Values in excess of the 0.001 mg/L detection limit were only observed in two surface water samples associated with mine water discharge in the Tweelopie/Riet Spruit system in mid-February 2010 when 0.026 mg U/L was recorded at stations F11S12 and MRd. Strachan et al. (2008) report that the Klip River Radioactivity Monitoring Program found a good linear correlation between the U concentration in water and the total radiation dose from all radionuclides. This prompted these authors to suggest that the U concentration might serve as an indirect measure of the total radiation dose from all radionuclides for screening and routine monitoring purposes in a catchment.

## 6 Bacteriological Quality

The bacteriological quality of groundwater in terms of the microbiological variables total coliform bacteria and *E. coli* has been determined for 16 sources (boreholes and springs) in the study area. Total coliforms and *E. coli* have traditionally been considered as the best fecal indicator bacteria (FIB) tracers of surface water and human fecal contamination in groundwater, respectively (Kozuskanich et al. 2010). Payment and Locus (2010) suggest that simple 'presence-absence' methods detecting both total coliforms and *E. coli* provide a reasonable approach when combined with an appropriate sampling frequency, to establish the measure of risk from enteric pathogens in water. The presence of total coliforms in groundwater indicates that microorganisms in surface water have entered the aquifer, and a more rigorous monitoring should begin for pathogens that might also reach the aquifer (Payment and Locus 2010).

Groundwater samples were collected from sources where the likelihood of bacteriological contamination was considered significant. The results are presented in Fig. 17, and indicate the presence of bacteriological contamination at station A2N0598 in the upper reaches of the Zwartkrans Basin, in the Oaktree area (stations CFM1, CSIR8, CSIR9, and NR1) and further downstream (stations MB1, BolandA, BolandC, ZSp, and OB1). Especially the presence of *E. coli* in some of the samples, notably the Zwartkrans Spring (ZSp) and station OB1, is a concern.

Station OB1, located the furthest downstream at Kromdraai, exhibits a surprising result given the following circumstances:

- the distance of ~870 m from the Bloubank Spruit with its elevated bacterial load (Table 20 in Chapter "Chemical Hydrology"); and
- the groundwater rest elevation of ~1420 m amsl compared to the ~1410 m amsl of the Bloubank Spruit, which indicates a hydraulic gradient toward the river at this locality.

Although the above circumstances suggest the Bloubank Spruit is an unlikely source of this contamination, the use of the borehole for large-scale agricultural water supply might cause a reversal of the hydraulic gradient and draw in water from this source. In common with stations MB1 and ZSp each producing groundwater with an $NO_3$ concentration of 12 mg N/L, station OB1 similarly yields groundwater with an elevated nitrate level of 8 mg N/L. Another possible source of both the bacteriological and the nitrate 'contamination' in the aquifer supporting borehole OB1 is the presence of livestock (cattle) in the area. A cattle feedlot located ~600 m to the northwest of the borehole was decommissioned ca. 2002, although the surface infrastructure was only completely demolished sometime between July 2008 and November 2009.[9]

Also anomalous is the bacteriological impact is seen in the Danielsrust Spring (DSp) water, which otherwise reflects a pristine dolomitic groundwater quality. The exposed and unprotected setting of the spring likely attracts a greater risk of bacteriological contamination from game such as Eland (*Taurotragus oryx*) that roam freely in this setting. Such instances of bacteriological contamination might attract further investigation in future hydrogeologic studies.

## 7 Springwater Chemistry

The importance of springs in the study area warrants a comprehensive discussion of the chemical composition of the water produced by each of the enumerated features (Sect. 4 in Chapter "Physical Hydrogeology"). The discussion is informed by the data in Table 5, which also form the basis for the graphed stable isotope data (Fig. 16), the Schoeller graphs (Fig. 18), and the Piper diagram (Fig. 19) representation. It is also important to read this discussion on the springwater chemistry in conjunction with the description of the hydrophysical environment that defines each of the spring features (Sect. 4 in Chapter "Physical Hydrogeology"). Both Figs. 18 and 19 reveal the deviation (for reasons discussed in Sect. 7.1) of the dolomitic Zwartkrans Spring water chemistry from that of the other dolomitic sources. The deviation of the Uitkomst Spring (JNNR4) water chemistry with its Mg–$HCO_3$ composition and extremely low SEC is readily attributed to the non-dolomitic source, namely shale of the Pretoria Group, drained by this spring (Sect. 7.9). The circumstances that inform the exaggerated Ca–$HCO_3$ character of the Aquamine Spring (AqSp) water are discussed in Sect. 7.6.

The calcite ($SI_C$) and dolomite ($SI_D$) saturation index values reported in Table 5 reveal the typically undersaturated nature of the pristine springwaters in regard to these

[9] Based on an inspection of historical Google Earth® imagery that brackets this time interval.

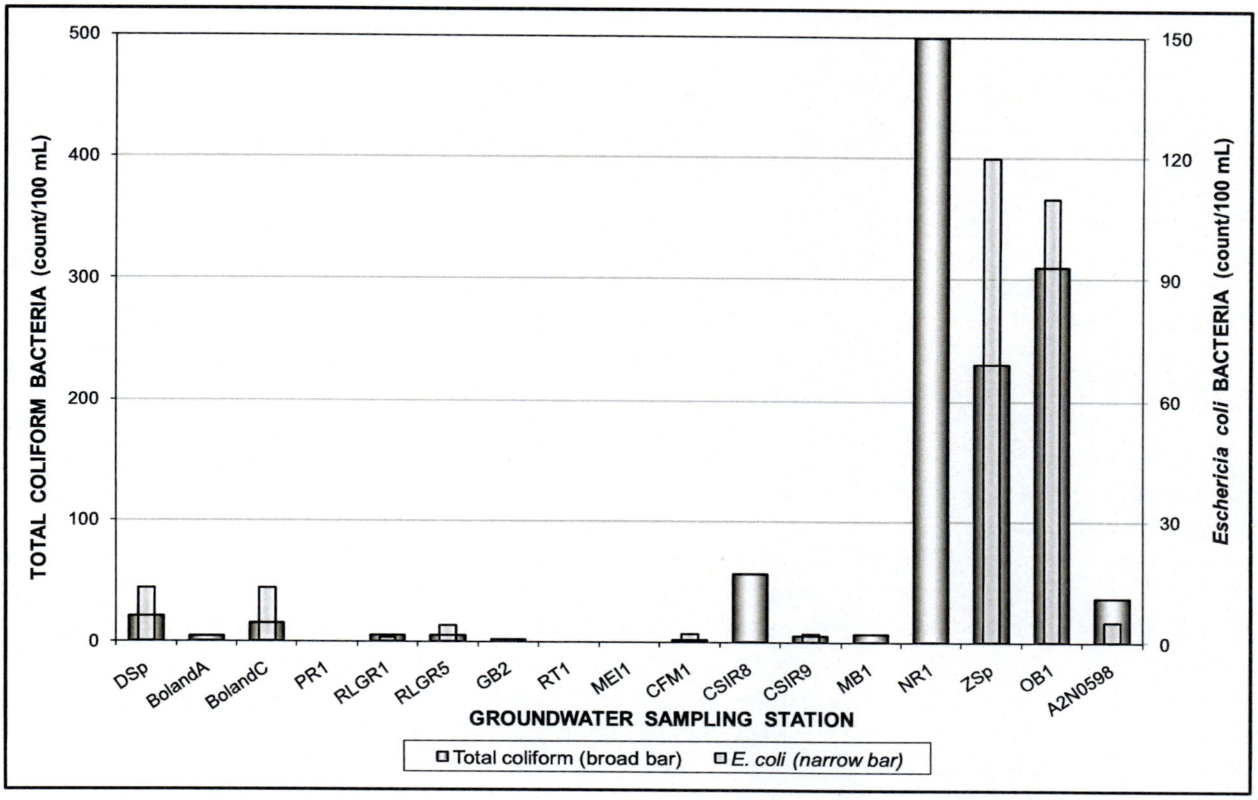


**Fig. 17** Graph of groundwater bacteriological quality in the study area; the ~3× difference in scale between the vertical axes explains those *E. coli* values that exceed total coliform counts

minerals. The Zwartkrans (ZSp) and Broederstroom (BSp) springs reflect saturated[10] (equilibrium) conditions for these minerals.

More recent (ca. 2014) springwater chemistry data (Table 6) provides a measure of the change in chemical composition of most of the springwaters represented in Table 5 and Fig. 18. The changes are discussed and illustrated individually in the following sections, and the grouped data are presented in Figs. 19 and 20 for comparative purposes.

The trend described by the $SI_C$ and $SI_D$ values associated with the various springwaters sampled in 2010 (Table 5) and 2014 (Table 6) is illustrated in Fig. 21. The comparison reveals that some of the springwaters (e.g., ZSp, PLSp, and DSp) reflect a change toward greater undersaturation, whereas others (e.g., JNNR1, JNNR2, and JNNR3) reflect a change toward lesser undersaturation.

[10] Adopting the Langmuir (1971) procedure where $-0.1 \leq SI \leq 0.1$ (SI expressed in logarithmic form) denotes saturated (equilibrium) conditions.

## 7.1 Zwartkrans Spring

Bredenkamp et al. (1986) report an SEC value of 94 mS/m for the Zwartkrans Spring water. Figure 22 indicates that the chemical composition of the springwater has changed significantly since May 2006. The principal change is the increase in $SO_4$ concentration from 123 to 304 mg/L in May 2014. Compared to that of the other karst springs, the Ca–$SO_4$ composition of this water reflects the impact of poorer water quality sources (mainly mine water) on the karst groundwater draining the Zwartkrans Basin (Sect. 3.1). The springwater chemistry represents the composition of the groundwater aggregated upstream thereof. The $SI_C$ and $SI_D$ values (text box in Fig. 22) indicate equilibrium to slightly undersaturated conditions in regard to calcite and dolomite, respectively. It has previously been shown (Sects. 3 in Chapter "Physical Hydrology" and 1.2 in Chapter "Chemical Hydrology") that the contributing sources include Ca–$SO_4$ dominated mine water, and Na–$SO_4$ dominated treated municipal wastewater effluent with its elevated Cl levels, all of which are evident in this springwater. The stable isotope chemistry of this springwater is reflected in analyses dated July 2009 (Table 4) and May

**Table 5** Springwater chemistry ca. mid-2010

| Variable/analyte | Spring | | | | | | | | |
|---|---|---|---|---|---|---|---|---|---|
| | ZSp | PLSp | DSp | JNNR1 | JNNR2 | JNNR3 | JNNR4 | AqSp | BSp |
| Date (dd/mm)[a] | 13/05 | 13/05 | 10/02 | 19/05 | 19/05 | 19/05 | 19/05 | 10/02 | 21/12 |
| pH ($-\log_{10}\alpha_{H+}$) (field) | 7.5 | 7.7 | 6.9 | 6.5 | 6.9 | 7.0 | 7.0 | 6.8 | 7.3 |
| SEC (mS/m) (field) | 84 | 38 | 26 | 39 | 39 | 25 | 1 | 63 | 49 |
| Ca (mg/L) | 66 | 51 | 33 | 47 | 47 | 30 | <2 | 87 | 54 |
| Mg (mg/L) | 37 | 29 | 18 | 27 | 27 | 18 | 3 | 48 | 34 |
| Na (mg/L) | 36 | 2 | <2 | <2 | <2 | <2 | 2 | 4 | <2 |
| K (mg/L) | 1.5 | <1 | <1 | <1 | <1 | <1 | <1 | <1 | <1 |
| Cl (mg/L) | 57 | <5 | <5 | <5 | <5 | <5 | <5 | 10 | <5 |
| $SO_4$ (mg/L) | 154 | 5 | <5 | 8 | <5 | <5 | <5 | 7 | 11 |
| $HCO_3$ (mg/L) | 195 | 234 | 156 | 239 | 239 | 161 | 20 | 390 | 332 |
| $NO_3$ (mg N/L) | 12 | 0.9 | 0.7 | 0.8 | 0.6 | 0.5 | 0.4 | <0.2 | 0.4 |
| Si (mg/L) | 6.0 | 5.9 | 5.9 | 5.0 | 5.2 | 5.5 | 4.1 | 7.5 | 6.5 |
| Fe (mg/L) | <0.025 | <0.025 | <0.025 | <0.025 | <0.025 | <0.025 | 0.315 | 2.57 | <0.02 |
| Mn (mg/L) | <0.025 | <0.025 | <0.025 | <0.025 | <0.025 | <0.025 | 0.466 | 0.38 | <0.005 |
| Al (mg/L) | 0.116 | <0.1 | <0.1 | <0.1 | <0.1 | <0.1 | <0.1 | 0.128 | <0.04 |
| Sr (mg/L) | 0.036 | <0.025 | <0.025 | <0.025 | <0.025 | <0.025 | <0.025 | 0.043 | 0.01 |
| V (mg/L) | 0.034 | 0.039 | 0.029 | <0.025 | <0.025 | <0.025 | <0.025 | 0.035 | <0.005 |
| Hg (mg/L) | <0.001 | <0.001 | <0.001 | n.a. | n.a. | n.a. | n.a. | 0.001 | n.a. |
| U (mg/L) | <0.001 | <0.001 | <0.001 | n.a. | n.a. | n.a. | n.a. | <0.001 | n.a. |
| $\delta^2H$ (‰) | −17.6 | −28.5 | −28.8 | −27.7 | −31.0[b] | −30.6 | n.a. | −28.0 | n.a. |
| $\delta^{18}O$ (‰) | −3.11 | −5.00 | −4.81 | −4.90 | −5.42[b] | −5.22 | n.a. | −4.64 | n.a. |
| $^3H$ (TU) | 2.1 ± 0.3 | 1.4 ± 0.3 | 1.3 ± 0.2 | 3.1 ± 0.3 | 0.6 ± 0.2[b] | 0.1 ± 0.2 | n.a. | 3.4 ± 0.3 | n.a. |
| $SI_C$ | −0.01 | 0.24 | −0.88 | −0.98 | −0.58 | −0.81 | −3.09 | −0.26 | 0.00 |
| $SI_D$ | 0.00 | 0.50 | −1.76 | −1.93 | −1.12 | −1.57 | −5.43 | −0.50 | 0.06 |
| EB (%) | −1.6 | 11.1 | 8.4 | 5.2 | 6.6 | 4.6 | −11.9 | 10.8 | −1.6 |

[a]All dates are dd/mm/2010
[b]ca. 2009 data from Abiye (see Footnote 17)
See text box in Fig. 18 for association of spring identifiers with spring names

2010 (Table 5). The results are compared below and reflect ostensibly small differences.

| Date | $\delta^2H$ (‰) | $\delta^{18}O$ (‰) | $^3H$ |
|---|---|---|---|
| 17/07/2009 | −16.8 | −3.17 | 2.4 ± 0.3 TU |
| 13/05/2010 | −17.6 | −3.11 | 2.1 ± 0.3 TU |

A data logger installed in a stilling well at the spring in mid-2013 was set to record SEC and temperature at 30-min intervals. The SEC data captured by the logger are presented in Fig. 23. The near-continuous SEC record reveals a gradual and steady increase at a rate of 0.8 mS/m per month from ca. August 2013 to ca. March 2014, followed by a decrease of similar proportions (0.7 mS/m per month) to mid-2014, before reverting to an increase (again at 0.8 mS/m per month) to the end of record in September 2014. The peak reached ca. March 2014 is interpreted as the passage through the spring of mine water impacted groundwater with a maximum SEC of ~103 mS/m, triggered by the ingress of poorer quality mine water in the period February 2010–June 2012. Hobbs (2013c) forecast that the worst of this water would pass through the spring by the end of 2013. The subsequent declining trend indicates a gradual and consistent (albeit slight) improvement in the quality of the springwater. The reversal of this trend in mid-2014 is interpreted to reflect the impact of poor quality mine water input in the period March–July 2014 into the karst system following the resumption of uncontrolled mine water discharge into the environment at end-February 2014 (Sect. 2 in Chapter "Description of the Physical Environment"). This triggered a second flush of more saline water through the aquifer toward the spring. These circumstances find support in the

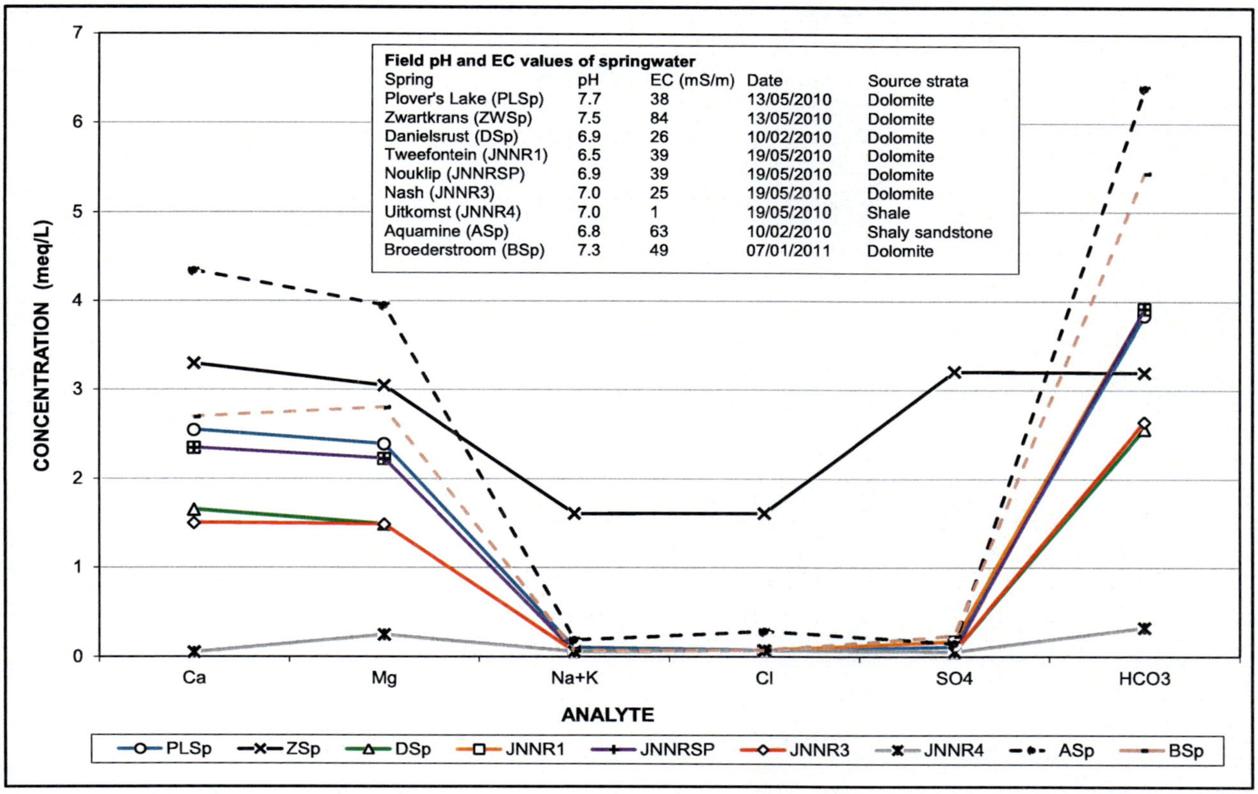


**Fig. 18** Graphic comparison of ca. 2010 springwater chemistry (data from Table 5)

**Fig. 19** Piper diagram of ca. 2010 springwater chemistry (data from Table 5) compared to typical Malmani Subgroup karst groundwater

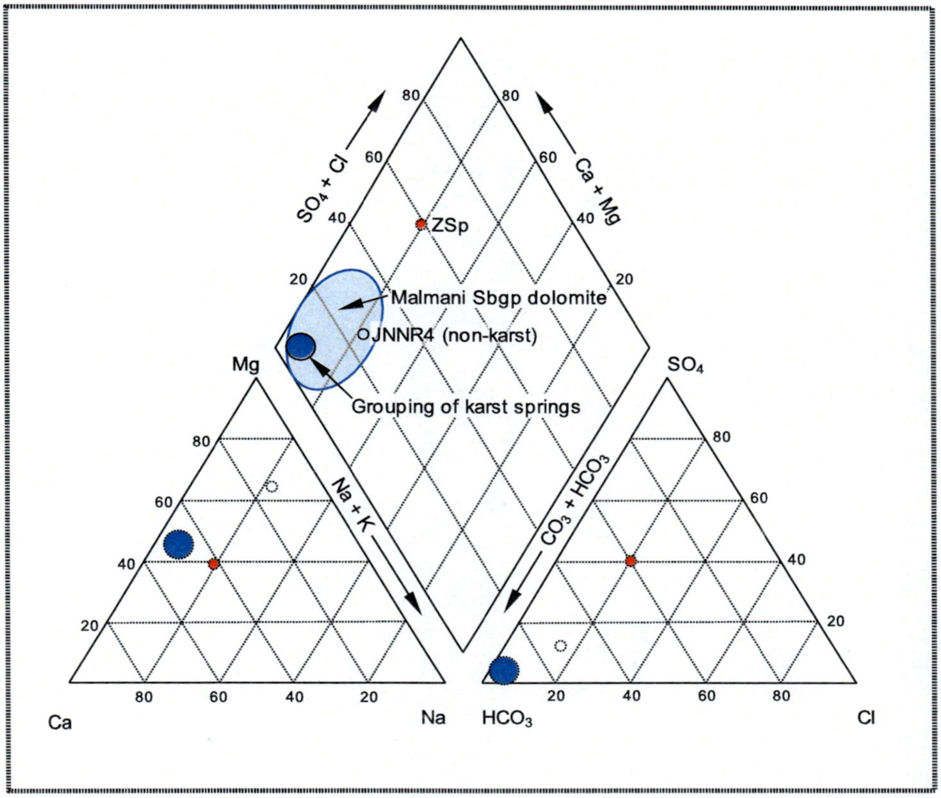

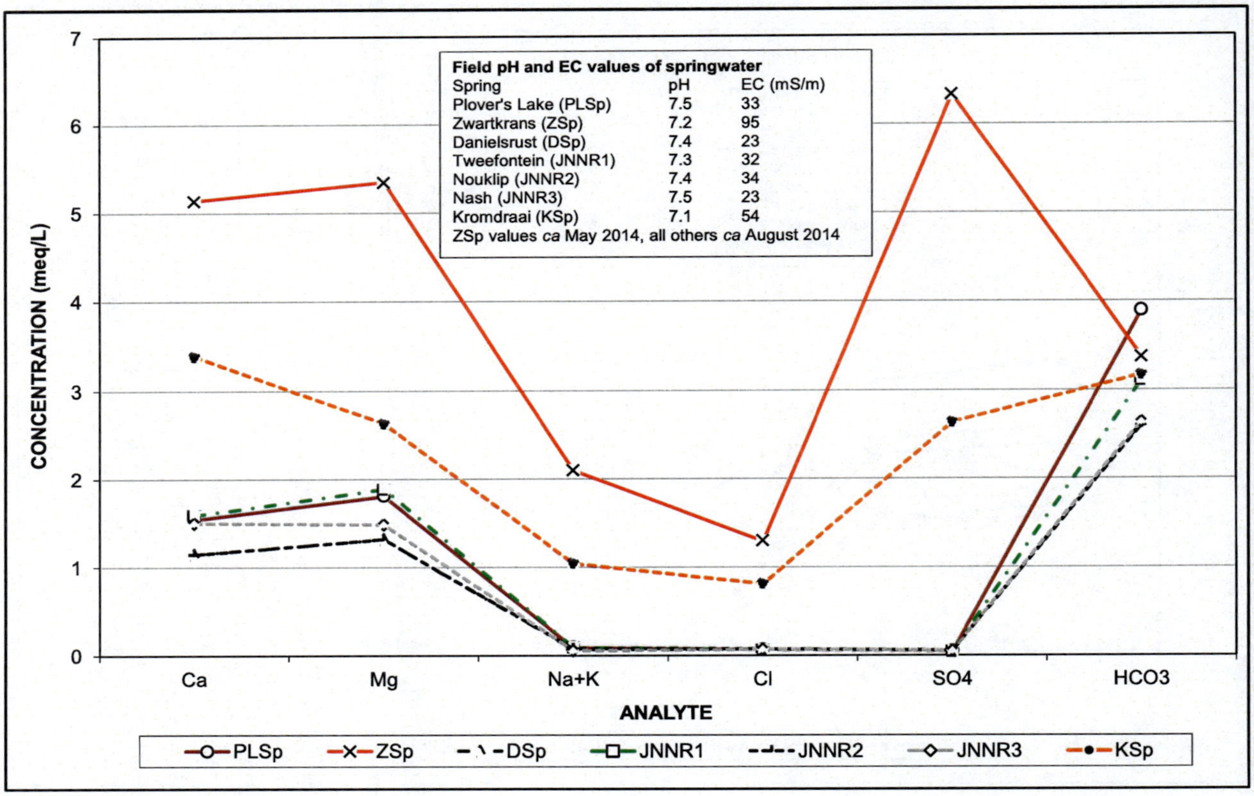


**Fig. 20** Graphic comparison of ca. 2014 springwater chemistry (data from Table 6); compositions of DSp and JNNR2 are virtually identical

**Table 6** Springwater chemistry ca. mid-2014

| Variable/analyte | Spring | | | | | | |
|---|---|---|---|---|---|---|---|
| | ZSp | PLSp | DSp | JNNR1 | JNNR2 | JNNR3 | KSp |
| Date (dd/mm)[a] | 16/05 | 14/08 | 14/08 | 14/08 | 14/08 | 14/08 | 20/08 |
| pH ($-\log_{10}\alpha_{H+}$) (field) | 7.2 | 7.5 | 7.4 | 7.3 | 7.4 | 7.5 | 7.1 |
| SEC (mS/m) (field) | 95 | 33 | 23 | 32 | 34 | 23 | 54 |
| Ca (mg/L) | 103 | 31 | 23 | 32 | 43 | 23 | 68 |
| Mg (mg/L) | 65 | 22 | 16 | 23 | 24 | 16 | 32 |
| Na (mg/L) | 47 | 1.9 | 1.3 | 1.6 | 1.4 | 1.5 | 23 |
| K (mg/L) | 2.2 | <1 | <1 | 1.2 | <1 | <1 | 1.9 |
| Cl (mg/L) | 46 | <5 | <5 | <5 | <5 | <5 | 29 |
| $SO_4$ (mg/L) | 304 | <5 | <5 | <5 | 7.8 | <5 | 127 |
| $HCO_3$ (mg/L) | 206 | 238 | 160 | 190 | 240 | 158 | 194 |
| $NO_3$ (mg N/L) | 9.1 | 0.3 | 0.3 | 7.4 | 5.6 | 0.4 | 3.9 |
| Si (mg/L) | 6.2 | 5.8 | 5.2 | 54.8 | 4.9 | 5.0 | 5.7 |
| Fe (mg/L) | 0.04 | <0.02 | <0.02 | <0.02 | 0.022 | <0.02 | 0.049 |
| Mn (mg/L) | <0.005 | <0.005 | <0.005 | 0.006 | <0.005 | <0.005 | 0.16 |
| Hg (mg/L) | <0.001 | | n.a. | n.a. | n.a. | n.a. | n.a. |
| $SI_C$ | −0.25 | −0.15 | −1.02 | −0.42 | −0.11 | −0.42 | −0.46 |
| $SI_D$ | −0.44 | −0.17 | −1.92 | −0.72 | −0.21 | −0.74 | −0.98 |
| EB (%) | 6.1 | −7.6 | −4.0 | 3.3 | −0.8 | −3.5 | 2.7 |

[a]All dates are dd/mm/2014

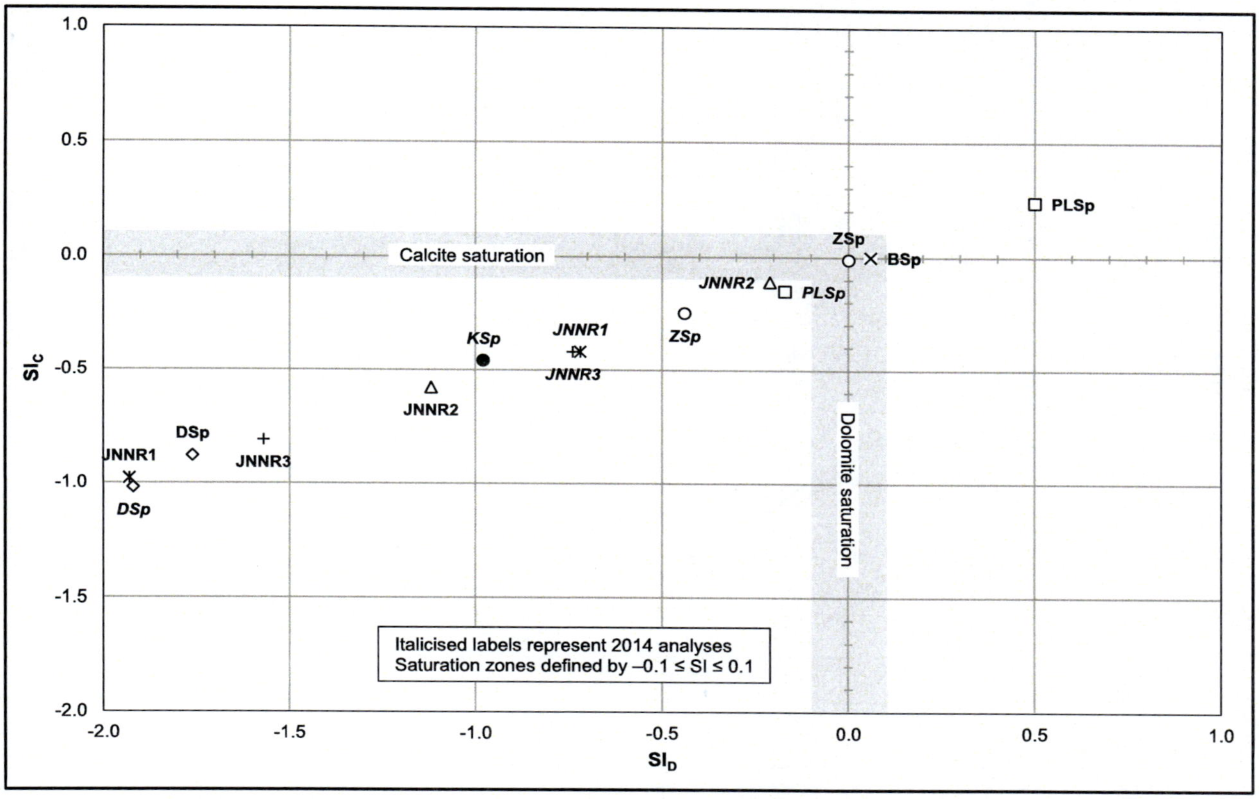


**Fig. 21** Saturation state of springwaters with respect to calcite ($SI_C$) and dolomite ($SI_D$) ca. 2010 and ca. 2014 (data from Tables 5 and 6)

excursion of the $SI_C$ and $SI_D$ values from equilibrium to slightly undersaturated conditions (text box in Fig. 22) in 2012 and 2014.

The anomalies evident in the record are associated with two flood events (on 31/01/2014 and 03/02/2014, respectively), on which occasions the spring was completely submerged under fresher surface floodwater to a depth approaching 2 m. These are quite common occurrences given the location of the spring next to the Bloubank Spruit (Plate 1 in Chapter "Physical Hydrogeology"), and undoubtedly reflect the circumstances that prevailed during the 16/12/2010 flood event discussed in Sect. 5.1 in Chapter "Chemical Hydrology". If nothing else, then the digital record captures and reflects the rapid recovery to 'normal' conditions following such flood events.

A slightly longer-term picture of the SEC trend of the Zwartkrans Spring water is shown in Fig. 24. Based on field measurements dating back to mid-2010, the gradually increasing trend supports the observation provided by Fig. 23. The rate of increase of 0.9 mS/m per month of the longer Fig. 24 record, is similar to the 0.8 mS/m per month of the much shorter rising limb segments of Fig. 23 record.

### 7.2 Danielsrust Spring

Together with that of the Nash Spring (JNNR3), the inorganic quality of the water delivered by this spring (DSp) is the best of any karst groundwater sourced by this study. The spring drains the Danielsrust Basin (Sect. 2.3 in Chapter "Physical Hydrogeology"). The water composition (Fig. 25) is virtually identical to that of the Nash Spring (JNNR3) (Fig. 18). The isotope hydrochemistry of the water (see comparison hereunder) similarly reflects a ca. 2010 composition that differs only marginally from an earlier analysis. The observed difference of +0.38‰ in $\delta^{18}O$ (ignoring possible analytical error margins) reflects a shift toward a marginally heavier isotopic composition. This suggests either a slightly greater evaporative component in this groundwater or a greater component of rainwater. Although the wetter than average past two summer rainfall seasons favor the rainfall influence, a possible anthropogenic influence from immediate upstream land-use practices in the form of a commercial tourist lodge facility (Sect. 2.3 in Chapter "Physical Hydrogeology") cannot be ruled out. The $SI_C$ and $SI_D$ values (text box in Fig. 25) indicate

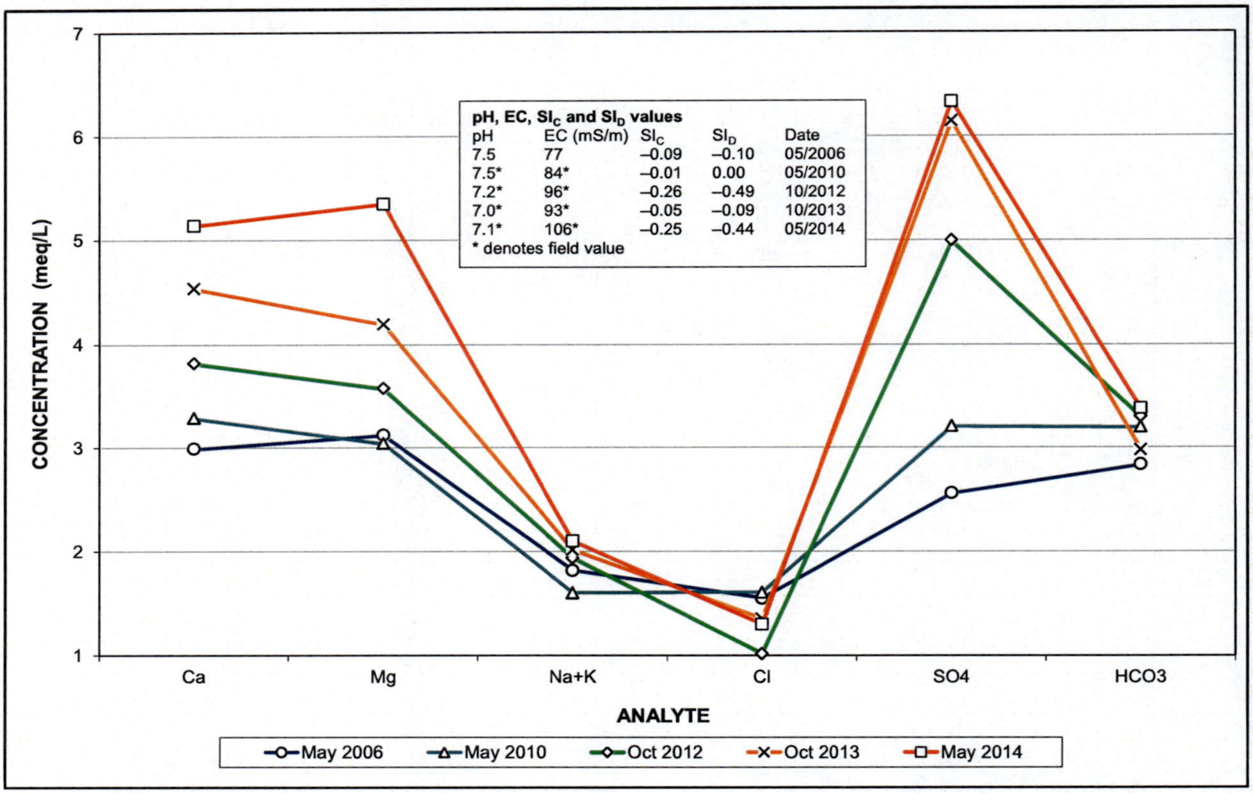


**Fig. 22** Graphic comparison of Zwartkrans Spring water chemistry since 2006

undersaturation of the water with respect to calcite and dolomite, respectively, indicating the continued dissolution of these minerals in the Danielsrust Basin.

| Date | $\delta^2$H (‰) | $\delta^{18}$O (‰) | $^3$H |
|---|---|---|---|
| ca. 2006[11] | −28.4 | −5.19 | 1.2[12] TU |
| 10/02/2010 | −28.9 | −4.81 | 1.3 ± 0.2 TU |

## 7.3 Plover's Lake Springs

Figure 26 shows that the chemical composition of the springwater has changed very little since May 2006. The chemical composition also shows a close resemblance (virtually superimposed in Fig. 18) to that produced by the Tweefontein (JNNR1) and Nouklip (JNNR2) springs. All of these springs are representative of natural dolomitic groundwater with its typical CaMg–$HCO_3$ composition.

The Plover's Lake springs (PLSp) drain the Kromdraai Subcompartment of the Krombank Basin (Sect. 2.2 in Chapter "Physical Hydrogeology"). The stable isotope values of this springwater (compared hereunder) show very little difference between the two more recent sampling dates, but a significant difference compared to the earliest date. The heavier isotopic values of the two later analyses suggest that a greater component of comparatively recent recharge is manifested in these samples compared to the earlier sample. The $SI_C$ and $SI_D$ values (text box in Fig. 26) indicate near-equilibrium of the water with respect to calcite and dolomite, respectively.

| Date | $\delta^2$H (‰) | $\delta^{18}$O (‰) | $^3$H |
|---|---|---|---|
| ca. 2006[13] | −30.3 | −5.62 | 0.7[14] TU |
| 17/07/2009 | −28.9 | −5.01 | 1.6 ± 0.3 TU |
| 13/05/2010 | −28.5 | −5.00 | 1.4 ± 0.3 TU |

[11] Data from Holland (2007); no exact date of sampling readily discernible from source.

[12] Data from Holland (2007); no error margin reported.

[13] See Footnote 11.

[14] See Footnote 12.

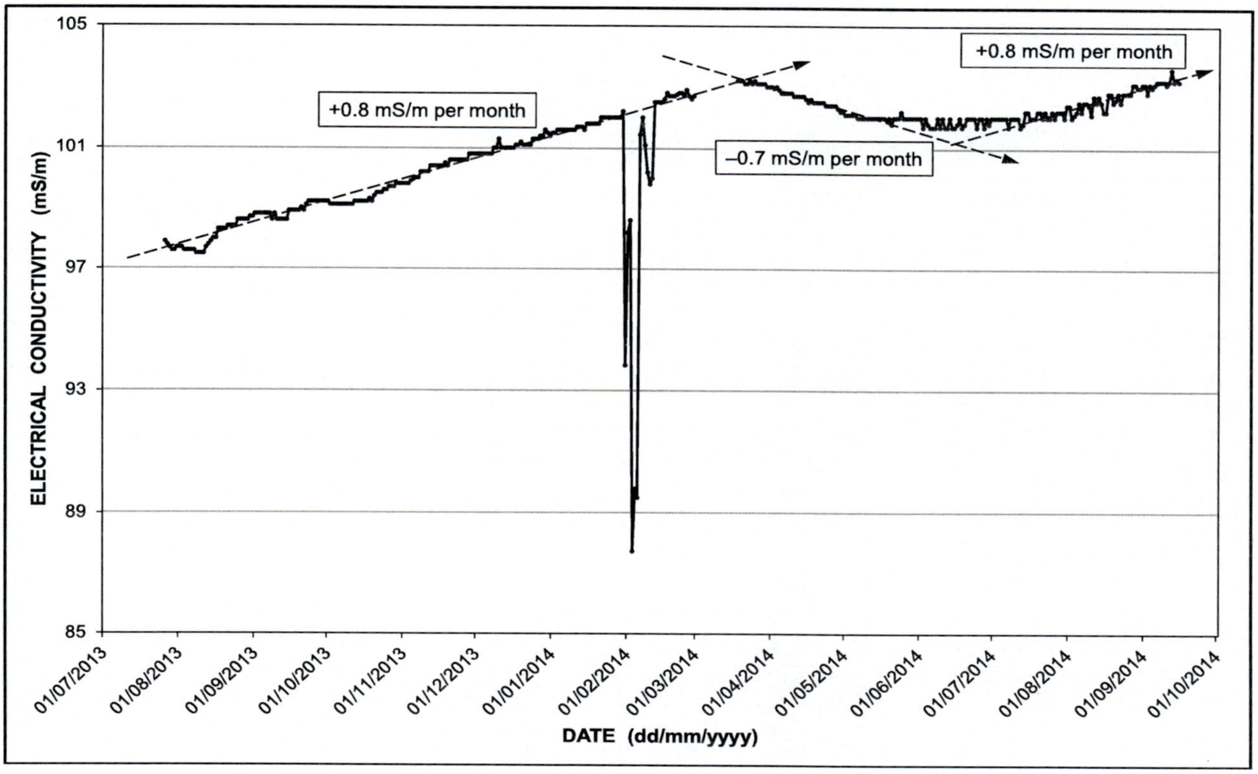


**Fig. 23** Pattern and trend of Zwartkrans Spring water salinity since mid-2013

## 7.4 Kromdraai Spring

The location of this source complicates the collection of a representative sample. In situ measurements carried out from the riverbank at the locus of discharge gave an SEC value of 56 mS/m, a pH value of 7.2, and a temperature of 19.4 °C on 13/08/2010. These values are compared with those of river water upstream and downstream of the spring in Table 14 in Chapter "Physical Hydrogeology". A water sample collected on 20/08/2014 returned the more definitive chemical composition reflected in Fig. 27. The Schoeller graph indicates a distinctive Ca–$HCO_3$ composition which is at variance with a typical pristine dolomitic water composition in regard to the following:

- the slight imbalance in the Ca:Mg ratio with a value of ~1.3 compared to the ~1 of pristine dolomitic groundwater; and
- the significantly elevated $SO_4$ level and elevated Na and Cl levels, all of which are minor constituents in pristine dolomitic groundwater.

It is evident that the Kromdraai Spring catchment receives water of a compromised quality. The $SO_4$:TDS ratio of ~27% points to mine water as the most likely source. The circumstances sketched in Fig. 28 provide a conceptual model of the pathway taken by this contamination. Representing allogenic recharge of the Kromdraai Subcompartment, these circumstances also explain the anomalously high recharge rate of ~30% of MAP (Table 80) attributed to this karst basin. The import of these circumstances for the elucidation of water budget calculations is explored in Sect. 2.12 in Chapter "Physical Hydrogeology". Nevertheless, $SI_C$ and $SI_D$ values of −0.46 and −0.98 (Fig. 27) associated with this water indicate undersaturation in regard to calcite and dolomite, respectively.

## 7.5 Tweefontein Spring

This spring (JNNR1) drains the Tweefontein Basin (Sect. 2.5 in Chapter "Physical Hydrogeology"). Its water chemistry is virtually identical in inorganic composition to that of the Nouklip Spring (JNNR2) (Fig. 18), and very similar to that produced by the Plover's Lake springs. The chemical composition of the springwater has remained largely unchanged since May 2006 (Fig. 29). The stable isotope composition of this water, as sampled on various dates reported hereunder, reflects a shift from a lighter to a heavier $\delta^2H$ and $\delta^{18}O$ isotopic ratio. The shift in $^3H$ values supports

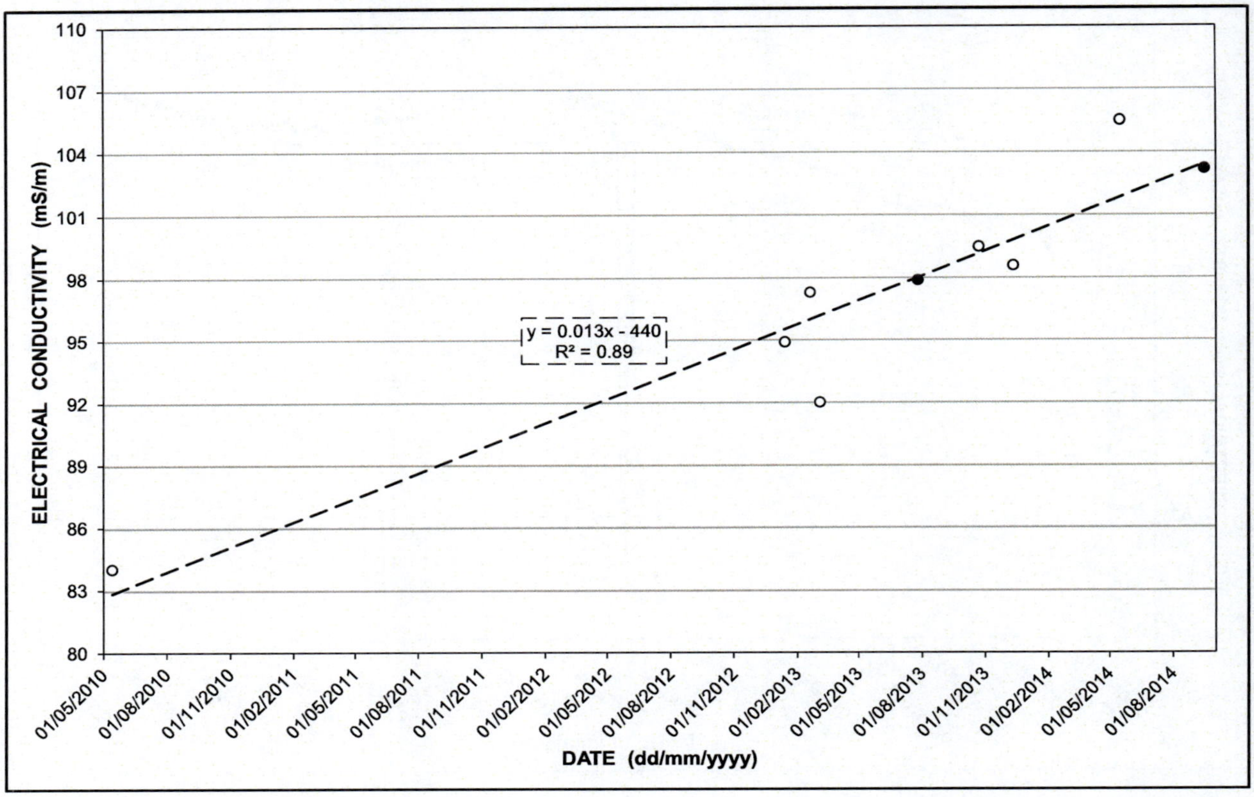


**Fig. 24** Pattern and trend of Zwartkrans Spring water electrical conductivity since mid-2010; the solid symbols mark the start and end values of Fig. 23 data record

an interpretation that attributes this to a greater proportion of younger rainwater recharge in the 2010 spring discharge. The $SI_C$ and $SI_D$ values of (text box in Fig. 29) indicate undersaturation of the water with respect to calcite and dolomite, respectively.

| Date | $\delta^2H$ (‰) | $\delta^{18}O$ (‰) | $^3H$ |
|---|---|---|---|
| ca. 2006[15] | −29.2 | −5.52 | 2.1[16] TU |
| ca. 2009[17] | −29.0 | −5.17 | 2.9 ± 0.3 TU |
| 19/05/2010 | −27.7 | −4.90 | 3.1 ± 0.3 TU |

## 7.6 Aquamine Spring

The chemistry of the Aquamine Spring (AqSp) water, although also reflecting a Ca–$HCO_3$ character, is clearly different from that of the karst groundwater. It is postulated that the upper reaches of the spring catchment comprise the dolomitic strata of the Rietfontein Basin (Sect. 2.6 in Chapter "Physical Hydrogeology"). The difference in chemistry is also evident in the detectable trace metal concentrations in this water (Table 5), compared to the generally undetectable levels in many respects in a number of the other springwaters. A further difference is the elevated $^3H$ level in the Aquamine Spring water (Table 5) compared to that of some of the karst springwaters. This indicates the comparatively 'young' age, possibly coupled with a shallow circulation pattern, associated with this groundwater system. The $SI_C$ and $SI_D$ values of −0.26 and −0.50 (Table 5) indicate undersaturation of the water with respect to calcite and dolomite, respectively.

## 7.7 Nouklip Spring

This water is of excellent quality in respect of all the elements included in the analytical suite. The Nouklip Spring (JNNR2) drains the Diepkloof Basin (Sect. 2.7 in Chapter "Physical Hydrogeology") and, as shown in Fig. 30, produces water that is virtually identical in composition to

[15] See Footnote 11.
[16] See Footnote 12.
[17] Data from T. Abiye (personal communication).

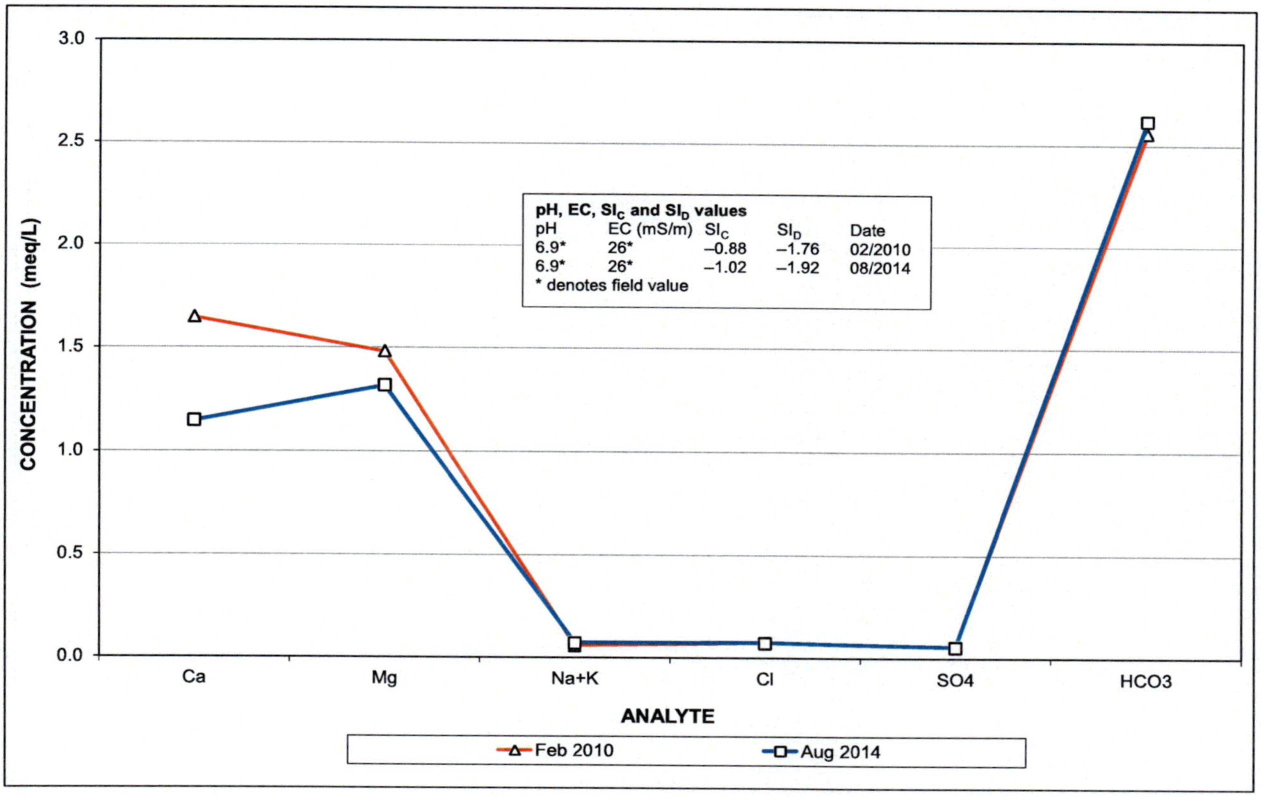


**Fig. 25** Graphic comparison of Danielsrust Spring water chemistry between 2010 and 2014

that produced by the Tweefontein Spring (Fig. 18). It is, therefore, reasonable to presume that the May 2006 composition of the Tweefontein Spring water (Fig. 29) also reflects that of the Nouklip Spring at the time. The similarity, however, is not reflected in the isotope results as shown hereunder. These results, and in particular the $^{3}H$ values, indicate a large proportion of substantially older groundwater in the discharge of the Nouklip Spring compared to the Tweefontein Spring. The $SI_C$ and $SI_D$ values (text box in Fig. 30) indicate undersaturation of the water with respect to calcite and dolomite, respectively.

| Date | $\delta^{2}H$ (‰) | $\delta^{18}O$ (‰) | $^{3}H$ |
|---|---|---|---|
| ca. 2006[18] | −32.1 | −5.78 | 0.6[19] TU |
| ca. 2009[20] | −31.0 | −5.42 | 0.6 ± 0.2 TU |

A notable aspect associated with the Nouklip Spring is the formation of tufa deposits in the lower reach of the Grootvlei Spruit near its confluence with the Skeerpoort River. The deposits have formed in three environments identified (after Ford and Pedley 1996) as follows:

- a massive hummock type deposit in the high energy (turbulent) environment on the downstream side of a gauging weir where degassing of mainly $CO_2$ occurs (Plate 4);
- as arcuate terraces (barrages) spanning the 5–8 m wide stream channel further downstream of the gauging weir (Plate 5); and
- as a waterfall deposit where the stream cascades ~4 m into the Skeerpoort River (Plate 6).

## 7.8 Nash Spring

The quality of the typically CaMg–$HCO_3$ karst groundwater produced by this spring (JNNR3) reflects a pristine condition. This is unlikely to deteriorate for as long as the largely 'wilderness' character of the Uitkomst Basin (Sect. 2.4 in Chapter "Physical Hydrogeology") is maintained. The isotope results presented hereunder reflect similar circumstances to those associated with the Nouklip Spring

[18] See Footnote 11.
[19] See Footnote 12.
[20] See Footnote 17.

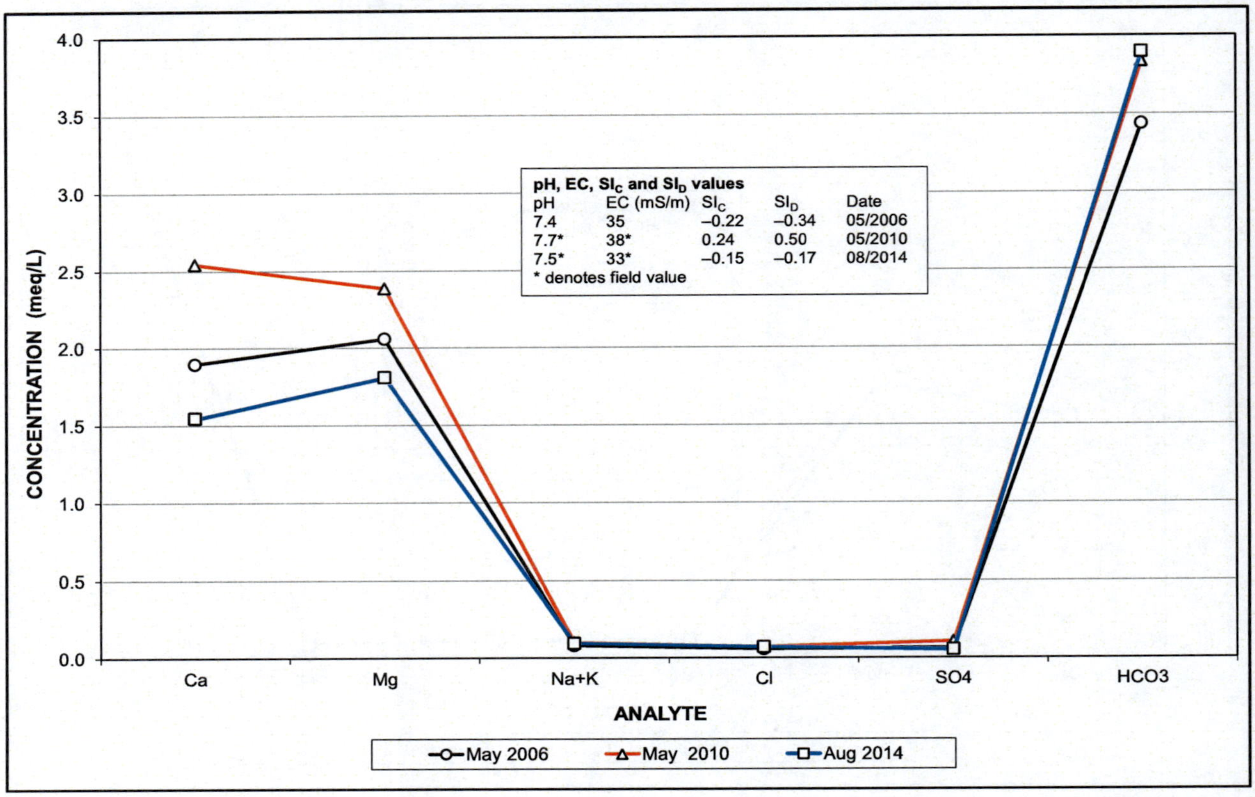


**Fig. 26** Graphic comparison of Plover's Lake springs water chemistry between 2006 and 2014

(Sect. 7.7). The $SI_C$ and $SI_D$ values (text box in Fig. 31) indicate undersaturation of the water with respect to calcite and dolomite, respectively.

| Date | $\delta^2H$ (‰) | $\delta^{18}O$ (‰) | $^3H$ |
|---|---|---|---|
| ca. 2009[21] | −32.4 | −5.64 | 0.2 ± 0.2 TU |
| 19/05/2010 | −30.6 | −5.22 | 0.1 ± 0.2 TU |

The inorganic chemical composition reflected in Fig. 31 is virtually identical to that of the Danielsrust Spring water (Fig. 18) is not, however, matched by the respective $^3H$ values (Table 5). These reveal a distinct difference in the relative age of the two springwaters. The smaller $^3H$ concentration of the Nash Spring water suggests a comparatively older water that accords with a larger and almost certainly much deeper groundwater circulation pattern in the Uitkomst Basin compared to that of the Danielsrust Basin.

[21] See Footnote 17.

## 7.9 Uitkomst Spring

The Uitkomst Spring (JNNR4) water chemistry reflects an extremely fresh character with a field SEC of 1 mS/m. Together with the Mg–$HCO_3$ composition of this water, these characteristics help to distinguish groundwater draining the Pretoria Group sedimentary strata from that draining the Malmani Subgroup carbonate strata (Fig. 19). A further distinguishing hydrochemical characteristic of this groundwater is the elevated Fe and Mn concentrations (Table 5). The $SI_C$ and $SI_D$ values of −3.09 and −5.43 (Table 5) indicate significant undersaturation of the water with respect to calcite and dolomite, respectively. This is expected of groundwater draining from non-carbonate strata.

## 7.10 Cradle Spring

The Cradle Spring (CSp) drains the Motsetse Basin (Sect. 2.8 in Chapter "Physical Hydrogeology"). Its chemistry is characterized by the field variable values listed hereunder. The SEC value of 64 mS/m is indicative of a slightly impacted karst groundwater chemistry.

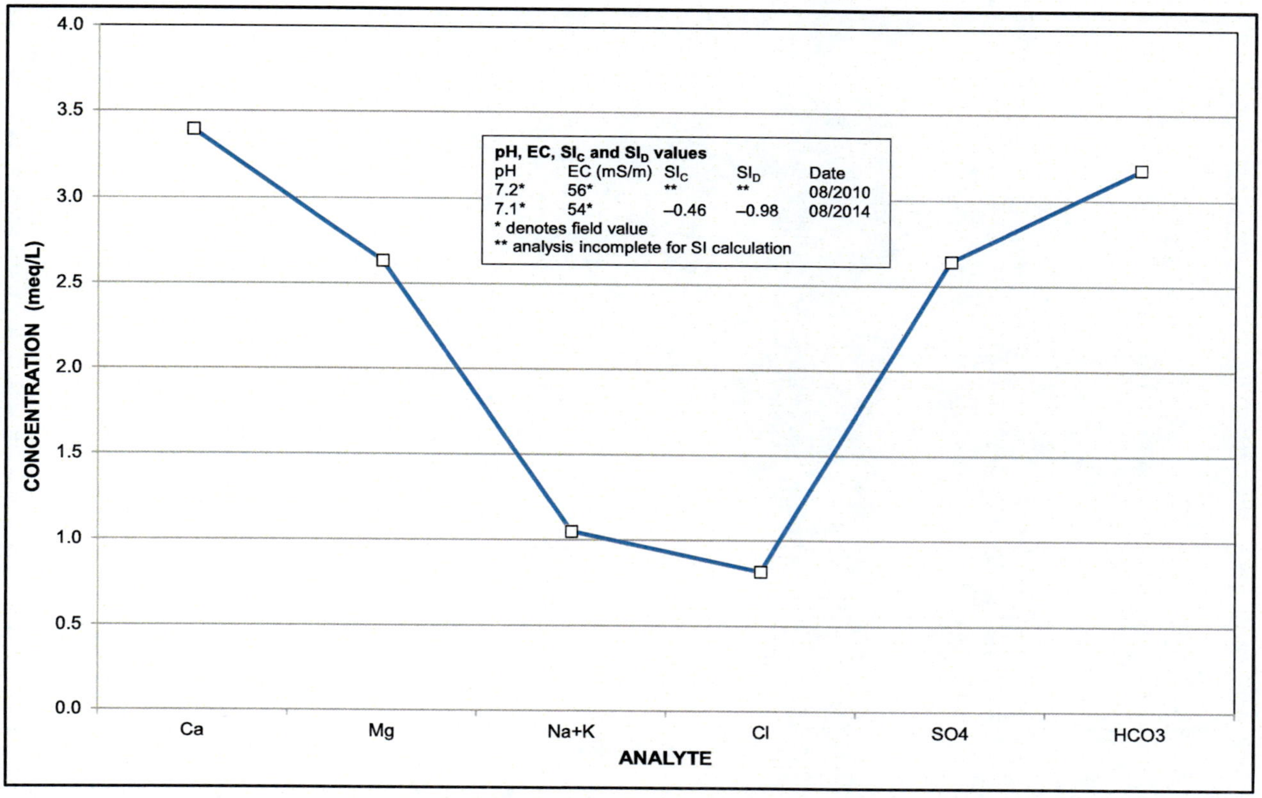


**Fig. 27** Graphic of Kromdraai Spring water chemistry in 2014

| Date | pH | SEC | ORP | Temperature |
|---|---|---|---|---|
| 21/12/2010 | 7.7 | 64 mS/m | −35 mV | 21.0 °C |

## 7.11 Barlow Spring

The chemistry of the Barlow Spring (BaSp) water is characterized by the field values listed hereunder. Notably, the pH value is lower than that of the nearby Anderson Spring (Sect. 7.12).

| Date | pH | SEC | ORP | Temperature |
|---|---|---|---|---|
| 27/11/2014 | 7.0 | 42 mS/m | −16 mV | 20.4 °C |

## 7.12 Anderson Spring

The Anderson Spring (AnSp) water chemistry reflects a typical $CaMg–HCO_3$ composition.

| Date | pH | SEC | ORP | Temperature |
|---|---|---|---|---|
| 27/11/2014 | 7.3 | 40 mS/m | −6 mV | 20.0 °C |

## 7.13 Lesedi Spring

The chemistry of the Lesedi Spring (LeSp) water reflects a significantly higher SEC of 66 mS/m compared to that of the Anderson Spring measured on the same day. These springs are provisionally recognized as draining the same compartment. The more complete analysis of the LeSp springwater reveals further differences in significantly higher Ca, $SO_4$, and $HCO_3$ concentrations and a $MgCa–HCO_3$ composition.

| Date | pH | SEC | ORP | Temperature |
|---|---|---|---|---|
| 27/11/2014 | 6.9 | 66 mS/m | −13 mV | 19.7 °C |

The Lesedi property has 70 permanent residents who serve the cultural visitor center, restaurant, and overnight accommodation facilities. All of these are served by an

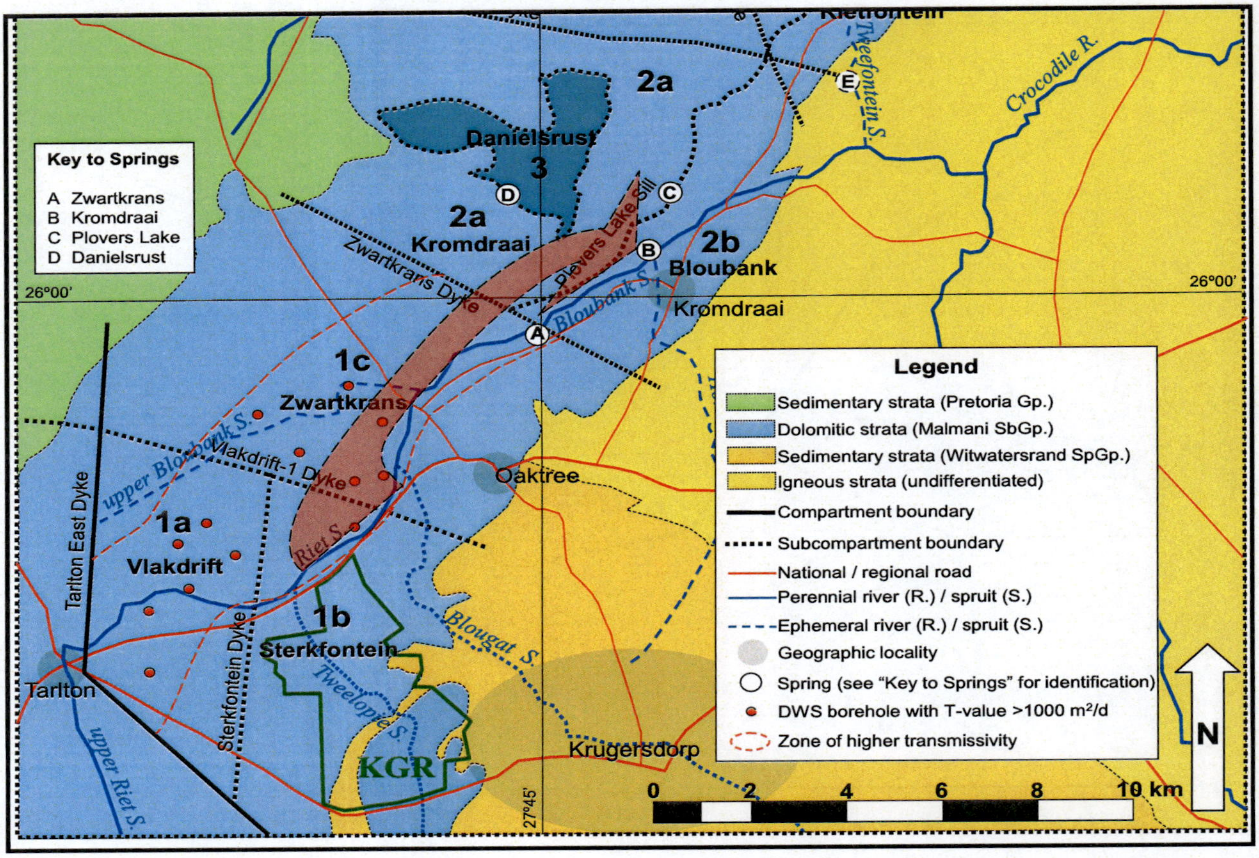


**Fig. 28** Schematic illustration of the hypothetical flow path (arrowed) taken by groundwater bearing a mine water impact into the Kromdraai Subcompartment toward the Kromdraai Spring

on-site sanitation system comprising a septic tank/French drain located?

## 7.14 Broederstroom Spring

The Broederstroom Spring (BSp) water chemistry is characterized by the following field variable values.

| Date | pH | SEC | ORP | Temperature |
|---|---|---|---|---|
| 21/12/2010 | 7.3 | 49 mS/m | −13 mV | 22.8 °C |

Holland (2007) reports the following values for the listed isotope analytes. The tritium value of 5.2 TU is the highest of any springwater in the COH for which a $^{3}$H value is known. It indicates a relatively young groundwater that is probably associated with a shallow groundwater system. This observation finds support in the higher than normal ambient springwater temperature of ~23 °C as indicated above. The $SI_C$ and $SI_D$ values of 0.00 and 0.06 (Table 5) indicate equilibrium conditions in regard to calcite and dolomite, respectively.

| Date | $\delta^2$H | $\delta^{18}$O | $^3$H |
|---|---|---|---|
| ca. 2006[22] | −29.2‰ | −5.37‰ | 5.2 TU |

## 7.15 Krugersdorp Game Reserve Springs

A description of the springs enumerated in the KGR is given in Sect. 4.15 in Chapter "Physical Hydrogeology" and Table 15 in Chapter "Physical Hydrogeology" (see Fig. 5 for locations). Some of these springs are important for the reason that they reflect the chemistry of groundwater in the receiving groundwater environment located immediately downstream of the mine area. This was recognized as early as mid-2003 when the DWS commenced water quality

[22] See Footnote 11.

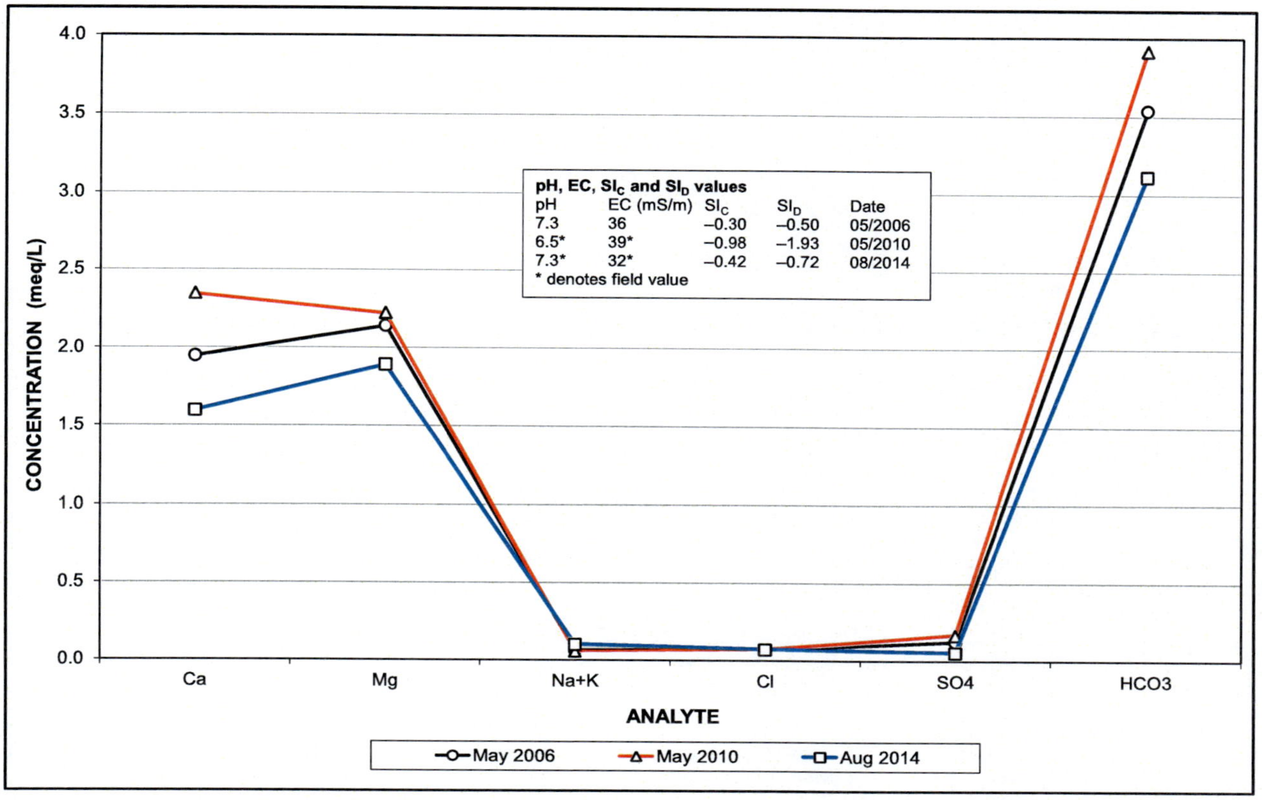


**Fig. 29** Graphic comparison of Tweefontein Spring water chemistry between 2006 and 2014

monitoring at these sources. The results of field variable measurements are reported in Table 7.

It is evident that elevated SEC values distinguish the springs SP1, SP2, CSIR30, CSIR63, and AvSp that deliver a compromised groundwater quality indicative of an impacted resource, from the springs LSp, SP3, and BuSp that drain an unimpacted resource. The groundwater produced by the spring sources SP1 and SP2 is subject to routine chemical analysis by various parties. The last DWS directive issued to RU also includes the spring SP3 in this program. The results of the SG monitoring summarized in Table 8, are reflected in Fig. 32. This shows the median inorganic chemical composition of the groundwater produced by these sources in the 8-year period to September 2014.

A similar but more muted impact to that observed in regard to SP2 is reflected in the Ca–$SO_4$ composition of the SP1 springwater (Fig. 32). This, together with the median pH of 6.5 and comparatively low median total alkalinity of 34 mg $CaCO_3$/L (Table 8), is evidence of a mine water impact. Spring source SP3, on the other hand, produces typically fresh CaMg–$HCO_3$ karst groundwater draining from the area of smallholdings located to the west. The excellent quality of this water indicates that extensive modification of the natural landscape mainly by agricultural land-use practices has not yet manifested a discernible impact on the ambient karst groundwater resource. It should be noted, however, that nitrate as an indicator analyte of agricultural impact has only been incorporated in the analytical suite since December 2012. The median value of 2.6 mg N/L for source SP3 is bracketed by those of the sources SP1 and SP2 (text box in Fig. 32). The $SI_C$ and $SI_D$ values for the springwaters (Table 8) reflect their typically undersaturated calcite and dolomite states, raising concern for the structural stability of the carbonate strata in the karst outlier, especially in the case of spring SP2.

The SEC and $SO_4$ trends associated with the water produced by the three monitored springs in the KGR are shown in Figs. 33 and 34, respectively. Whereas spring source SP1 shows a hardly discernible trend in these analytes, the source SP2 reflects an apparent rising trend until early-2011, followed by a falling trend until mid-2013 before again showing an increase at the end of the record. The $SO_4$ trend is more clearly evident than that of SEC.

The SEC and $SO_4$ trends are not readily explained because of the variability exhibited, with the SEC values bracketing an interval of ~100–450 mS/m (Fig. 33), and the $SO_4$ values an interval of 500–3200 mg/L (Fig. 34). The

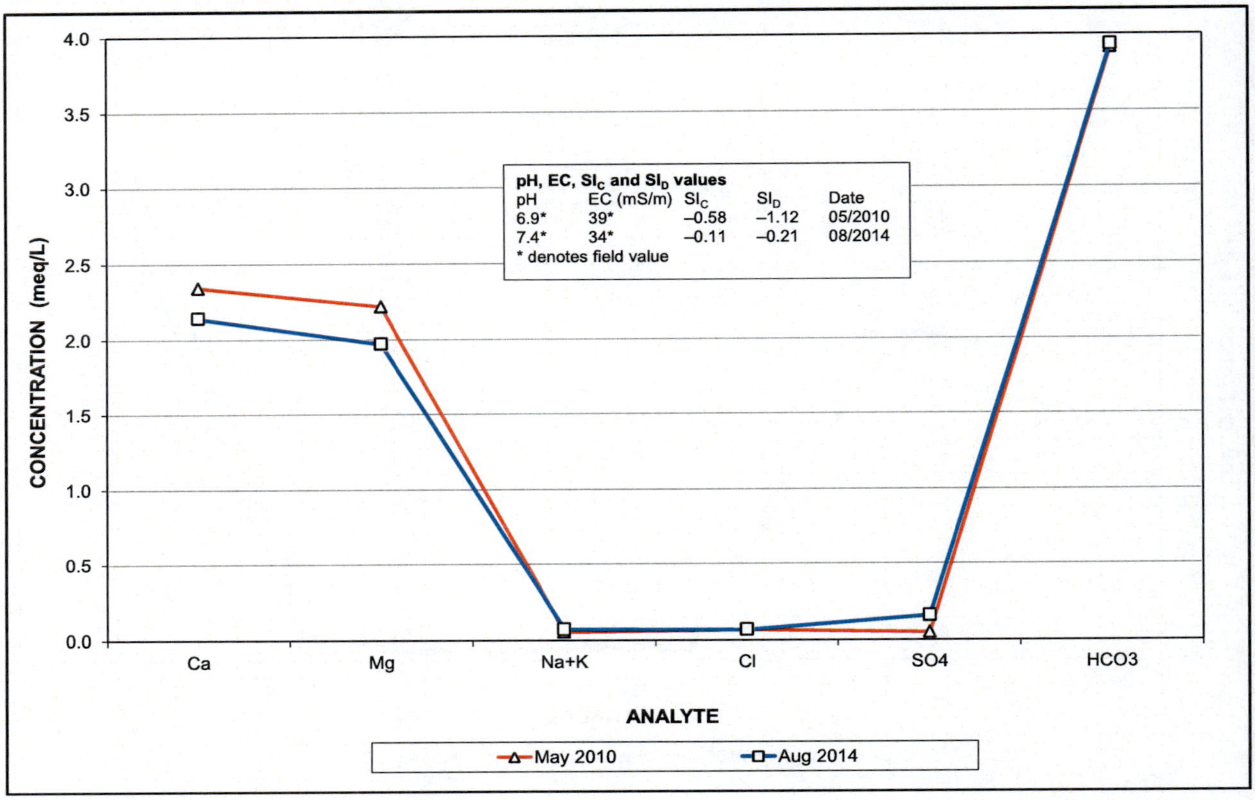


**Fig. 30** Graphic comparison of Nouklip Spring water chemistry in 2010 and 2014

**Plate 4** Massive hummock tufa deposit (under cascading water) on the downstream side of gauging weir A2H033 ~525 m downstream of the spring

inclusion in Fig. 34 of the contemporary $SO_4$ values observed for the Hippo Dam explores the possibility of a correlation with the SP2 sulfate trend. The 3rd order polynomial trendlines suggest that a correlation does exist. The excellent correlation of pH values in Fig. 35 dispels any doubts in this regard.

**Plate 5** Uppermost tufa terrace (barrage) extending across the channel of the Grootvlei Spruit ~600 m downstream of the spring

**Plate 6** Waterfall tufa deposit at left of picture where the Grootvlei Spruit cascades ~4 m into the Skeerpoort River gorge ~650 m downstream of the spring

It is evident from Fig. 35 that the pH of SP2 springwater exhibits a sharp drop from 6.5 in January 2010 to 3.3 in February 2010, after which a consistently declining trend to a value of ~3.0 in mid-2012 is demonstrated. The pH drop coincides with the substantial increase in raw mine water discharge from the mine area (Sect. 3.1 in Chapter "Physical Hydrology"). Similarly, the equally rapid increase to values >6 in the period mid-2012 to early-2014 coincides with the commissioning of the immediate mine water control, management and treatment measures in June 2012 (Text Box 1 in Chapter "Physical Hydrology"). The subsequent period of lower pH values ending in mid-2014 again reflects the uncontrolled discharge of raw mine water from the mine area.

The hydrochemical character of the springwaters is compared with pristine karst groundwater in Fig. 36.

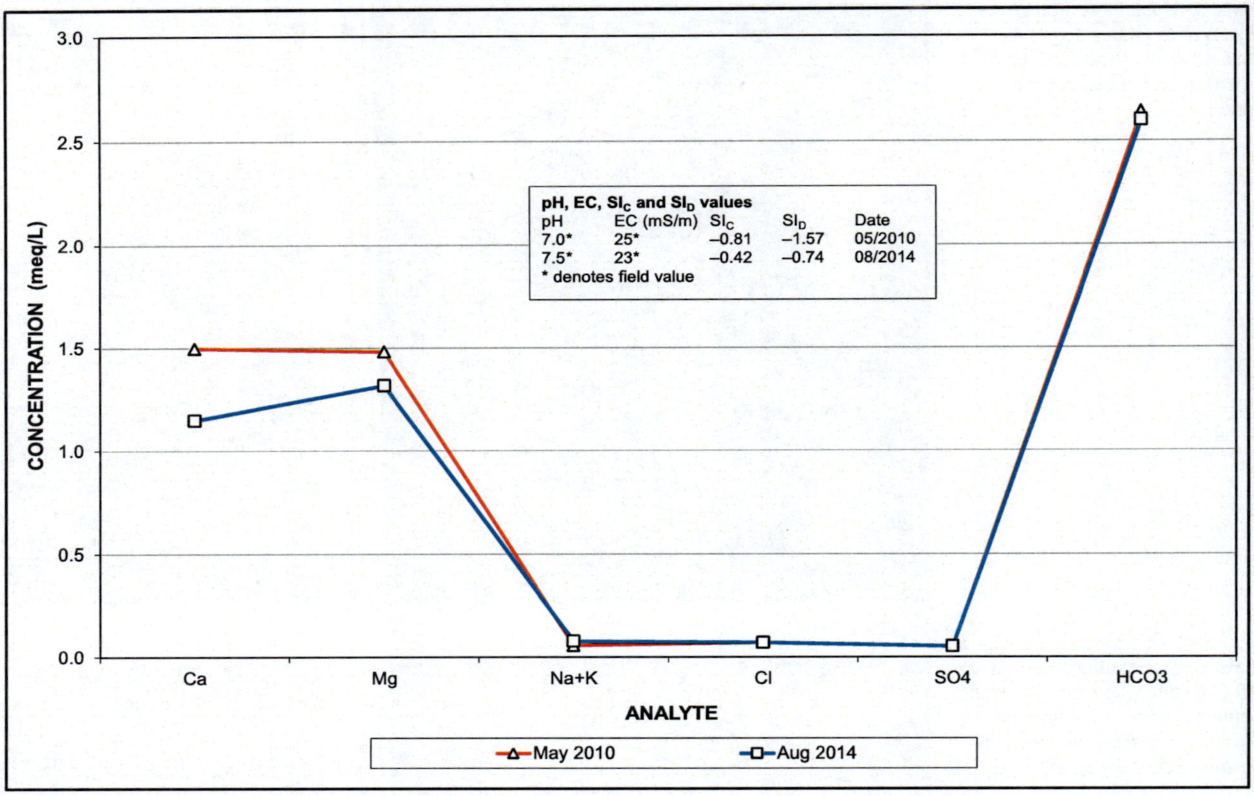


**Fig. 31** Graphic comparison of Nash Spring water chemistry in 2010 and 2014

## 8 Selected Impacts

### 8.1 Agriculture

The impact of agriculture on the quality of groundwater resources is assessed on the basis of analytical results for pesticide residues in water. Groundwater samples collected from two boreholes (CSIR8 and CFM1) in the Oaktree area were subjected to multi-residue pesticide analysis by the South African Bureau of Standards (SABS) Chromatographic Services. The sources were selected for their position in proximity to and downgradient of intensive agricultural activity including the tunnel farming of flowers for export (Fig. 37). These show that no residues were detected in either of the samples. Similarly, Groenewald (2010b) reports that no knowledge of pesticide contamination is currently available. Although these circumstances suggest that no impact is present, future groundwater quality studies will further inform this aspect.

### 8.2 Historical Mining

The COH and surrounding area is witness to numerous forms of mining activity at various scales. These vary from artisanal gold mine workings represented by stopes in the Black Reef Formation such as are seen in the KGR, to opencast workings for limestone, for example, the Aviary Quarry in the KGR, and dolomite, e.g., the large Sterkfontein Quarry at Bolt's Farm near Oaktree. The (often destructive) exploitation of numerous caves (including Sterkfontein) in the study area for limestone at the end of the nineteenth and beginning of the twentieth centuries is also well documented (e.g., McCarthy and Rubidge 2005).

The COH hosts the earliest proclaimed gold mine on the Witwatersrand. The Kromdraai Gold Mine is located on Ptn 43 (Ibis Ridge Farm) of the farm Kromdraai 520JQ commenced operations in 1881 and closed down in 1914 (G. Whatley, personal communication). In an article titled 'The Kromdraai Gold Field', The Star newspaper of 07 October 1893 reports on a substantial quantity of water encountered in the mine workings. As was common to many of the early Witwatersrand gold mines, the Kromdraai Gold Mine exploited gold-bearing horizons in the Black Reef Formation located at the base of the Malmani Subgroup dolomite succession. Only one of the original three adits to the Kromdraai Gold Mine still allows risk-free access to the underground workings, which comprise at least three levels linked by winzes. The water level in the mine fluctuates in the interval defined by the 3rd (deepest known) and the 2nd

**Table 7** Field chemical variable values of springwater in the KGR

| Spring[a] | Date (dd/mm/yyyy) | SEC (mS/m) | pH ($-\log_{10}\alpha_{H+}$) | Eh (mV) | Temp (° C) | Source aquifer |
|---|---|---|---|---|---|---|
| SP1/Spring 1/F6S7 | 24/11/2010 | 79.4 | 6.1 | 48.4 | 18.2 | Main karst outlier |
| | 14/08/2012 | 101.6 | 6.0 | 37.2 | 17.9 | |
| | 15/10/2013 | 96.0 | 6.0 | 23.4 | 18.3 | |
| CSIR30/Spring 2 | 24/11/2010 | 90.2 | 6.0 | 55.0 | 17.9 | Main karst outlier |
| | 14/08/2012 | 108.8 | 5.9 | 42.3 | 17.9 | |
| | 15/10/2013 | 101.5 | 6.0 | 23.1 | 17.8 | |
| CSIR63/Spring 3 | 24/11/2010 | 101.2 | 5.9 | 58.3 | 17.7 | Main karst outlier |
| | 14/08/2012 | 115.4 | 5.9 | 38.3 | 17.8 | |
| | 15/10/2013 | 113.7 | 5.9 | 26.8 | 17.5 | |
| SP2/CSIR37/Poplar Sp./F8S9 | 24/11/2010 | 90.9 | 5.4 | 85.2 | 17.6 | Main karst outlier |
| | 14/08/2012 | Spring inundated by Tweelopie Spruit | | | | |
| Lodge Sp. (LoSp) | 24/11/2010 | 12.6 | 5.3 | 92.2 | 19.2 | Wits quartzite |
| | 14/08/2012 | 21.1 | 5.3 | 77.0 | 19.1 | |
| | 15/10/2013 | 19.3 | 5.3 | 60.5 | 19.0 | |
| Aviary Sp. (AvSp) | 24/11/2010 | 201 | 7.3 | −16.5 | 17.2 | Zwartkrans Basin |
| | 14/08/2012 | 236 | 7.3 | −36.0 | 16.5 | |
| | 15/10/2013 | 209 | 7.1 | −42.9 | 16.6 | |
| SP3/Flip-se-Gat | 24/11/2010 | 21.5 | 8.2 | −61.3 | 19.1 | Zwartkrans Basin |
| | 14/08/2012 | 25.6 | 8.2 | −84.3 | 16.6 | |
| | 15/10/2013 | 23.6 | 8.0 | −84.0 | 18.3 | |
| Buffalo Sp. (BuSp) | 15/10/2013 | 7.7 | 5.7 | 37.8 | 18.3 | Toe of karst outlier |

[a]Cross-reference Table 15 in Chapter "Physical Hydrogeology"

Levels and reached a maximum elevation just above the 2nd Level (G. Whatley, personal communication) in the year 2000, notably an extremely wet year.

The current water level occupies an elevation of ~1420 m amsl, as inferred from the depth to groundwater rest level of ~12 m below surface measured in a water supply borehole (geosite KGM1) on the property in May 2010. The quality of the groundwater produced by this borehole (SEC = 18 mS/m, pH = 5.6) does not reflect an acidic mine water signature. The low SEC and pH values are characteristic of groundwater associated with quartzitic strata of the Witwatersrand Supergroup (Sect. 2.1.1). The absence of an acidic mine water signature is supported by 'in-mine water' EC, pH, and temperature measurements that reveal even lower SEC values (in the range 10–13 mS/m) but slightly higher pH values (in the range 6.1–6.4), at two localities within the partially flooded mine workings.

The differences with the borehole KGM1 groundwater might be attributed to the depth from which the latter is drawn; the pump has installed a depth of ~55 m below surface (G. Whatley, personal communication). This would indicate that the borehole penetrates some distance into the underlying Witwaterstrand Supergroup quartzites, which host a naturally acidic groundwater.

Figure 38 provides a comparison of the mine water and groundwater in the vicinity of the mine with that of surface water in the nearby Honingklip Spruit. Despite the ostensible similarity of Mg–$HCO_3$ chemical composition reflected in the graphs, a distinctive grouping of 'mine' water (samples KGM1 and KGM-MW) on the one hand, and groundwater (OB1) and surface water (HS1) on the other, is also apparent. Nevertheless, the absence of acid mine water development in the mine workings as well as the absence of visible decant after more than a century of 'flooding' is notable.

The groundwater elevation of ~1420 m amsl is slightly lower than the streambed elevation of the nearby Honingklip Spruit, which is interpolated at ~1423 m amsl. Circumstances, where the water level in the mine workings is partly influenced by discharge in the Honingklip Spruit, can therefore not be discounted. The mine water level would need to rise more than ~3 m above its current level before a reversal of the hydraulic gradient between the mine and the Honingklip Spruit is established.

Although two of the adits that provide access to the mine workings (one risk-free and the other hazardous) occupy a

**Table 8** Statistical analysis of groundwater chemistry delivered by springs SP1, SP2, and SP3 in the KGR in the period September 2009–September 2014

| Spring | Chemical analyte | Statistical parameter | | | | | | | SANS (2011a)[a] |
|---|---|---|---|---|---|---|---|---|---|
| | | *n* | 5%ile | Mean | Median | 95%ile | SD | CoV (%) | |
| SP1 | pH[b] @ 25 °C | 59 | 6.2 | – | 6.5 | 7.1 | 0.4 | 6 | 5.0–9.7 |
| | SEC (mS/m) | 59 | 91 | 99 | 99 | 106 | 4.9 | 5 | ≤ 170 |
| | TDS (mg/L) | 59 | 744 | 830 | 824 | 1010 | 123 | 15 | ≤ 1200 |
| | Ca (mg/L) | 59 | 58.6 | 109.2 | 113.0 | 143.1 | 25.7 | 24 | n.s. |
| | Mg (mg/L) | 59 | 44.8 | 57.8 | 55.0 | 79.6 | 11.0 | 19 | n.s. |
| | Na (mg/L) | 58 | 22.6 | 29.5 | 28.0 | 37.9 | 10.8 | 37 | ≤ 200 |
| | Cl (mg/L) | 59 | 16.0 | 26.0 | 21.0 | 58.8 | 13.9 | 53 | ≤ 300 |
| | $SO_4$ (mg/L) | 59 | 413 | 470 | 469 | **528** | 37.5 | 8 | ≤ 500 |
| | $HCO_3$ (mg/L) | 59 | 29.1 | 48.2 | 42.7 | 86.8 | 17.1 | 35 | n.s. |
| | Fe (mg/L) | 57 | 0.002 | 0.425 | 0.050 | 1.740 | 2.029 | 477 | ≤ 2.0 |
| | Mn (mg/L) | 58 | 0.008 | 0.091 | 0.050 | 0.415 | 0.148 | 163 | ≤ 0.5 |
| | Al (mg/L)[c] | 57 | 0.002 | **0.276** | **0.500** | **0.500** | 0.242 | 89 | ≤ 0.3 |
| | U (mg/L) | 50 | 0.003 | 0.106 | 0.003 | 0.017 | 0.721 | 678 | ≤ 0.015 |
| | $SI_C$ | – | −2.24 | −1.50 | −1.55 | 0.56 | – | – | n.s. |
| | $SI_D$ | – | −4.34 | −3.02 | −3.16 | −1.12 | – | – | n.s. |
| | EB (%) | 58 | −8.9 | 0.4 | 0.2 | 10.8 | 6.3 | 1475 | n.s. |
| SP2 | pH[b] @ 25 °C | 58 | **2.9** | – | **3.7** | 7.0 | 1.7 | 37 | 5.0–9.7 |
| | SEC (mS/m) | 58 | 139 | 252 | 260 | 351 | 68.3 | 27 | ≤ 170 |
| | TDS (mg/L) | 58 | **1094** | **2561** | **2500** | **3970** | 832 | 33 | ≤ 1200 |
| | Ca (mg/L) | 58 | 53.0 | 361.0 | 373.0 | 609.1 | 155.6 | 43 | n.s. |
| | Mg (mg/L) | 58 | 50.7 | 105.8 | 100.0 | 151.2 | 45.0 | 43 | n.s. |
| | Na (mg/L) | 58 | 38.4 | 70.2 | 70.0 | 104.9 | 21.0 | 30 | ≤ 200 |
| | Cl (mg/L) | 58 | 22.0 | 46.7 | 36.5 | 78.0 | 47.0 | 101 | ≤ 300 |
| | $SO_4$ (mg/L) | 58 | **642** | **1589** | **1560** | **2508** | 543 | 34 | ≤ 500 |
| | $HCO_3$ (mg/L) | 58 | 1.2 | 12.1 | 1.2 | 47.3 | 17.5 | 145 | n.s. |
| | Fe (mg/L) | 58 | 0.01 | **28.9** | **1.3** | **109.8** | 58.8 | 203 | ≤ 2.0 |
| | Mn (mg/L) | 58 | **0.99** | **22.2** | **17.0** | **57.2** | 20.4 | 92 | ≤ 0.5 |
| | Al (mg/L)[c] | 57 | 0.001 | **0.333** | **0.500** | **0.616** | 0.313 | 94 | ≤ 0.3 |
| | U (mg/L) | 58 | 0.003 | **0.021** | 0.011 | **0.074** | 0.033 | 157 | ≤ 0.015 |
| | $SI_C$ | – | −5.07 | −4.13 | −4.28 | −0.66 | – | – | n.s. |
| | $SI_D$ | – | −9.90 | −8.54 | −8.87 | −1.67 | – | – | n.s. |
| | EB (%) | 58 | −26.6 | −7.6 | −5.4 | 6.4 | 11.2 | −148 | n.s. |
| SP3 | pH[b] @ 25 °C | 60 | 6.5 | – | 7.8 | 8.2 | 0.5 | 7 | 5.0–9.7 |
| | SEC (mS/m) | 60 | 24.0 | 26.2 | 26.0 | 30.1 | 2.5 | 10 | ≤ 170 |
| | TDS (mg/L) | 60 | 127 | 180 | 162 | 210 | 127 | 70 | ≤ 1200 |
| | Ca (mg/L) | 60 | 19.0 | 26.0 | 24.0 | 35.2 | 10.0 | 39 | n.s. |
| | Mg (mg/L) | 60 | 11.0 | 14.4 | 14.0 | 18.1 | 3.7 | 26 | n.s. |
| | Na (mg/L) | 59 | 1.8 | 6.3 | 4.7 | 14.4 | 7.4 | 118 | ≤ 200 |
| | Cl (mg/L) | 59 | 2.5 | 10.0 | 7.0 | 25.1 | 10.0 | 100 | ≤ 300 |
| | $SO_4$ (mg/L)[d] | 60 | 20.0 | 28.2 | 25.0 | 46.0 | 8.4 | 30 | 25.0 |
| | $HCO_3$ (mg/L) | 60 | 81 | 95 | 96 | 115 | 14.6 | 15 | n.s. |
| | Fe (mg/L) | 58 | 0.001 | 0.143 | 0.050 | 0.815 | 0.244 | 170 | ≤ 2.0 |
| | Mn (mg/L) | 59 | 0.003 | 0.141 | 0.050 | 0.520 | 0.388 | 274 | ≤ 0.5 |
| | Al (mg/L)[c] | 59 | 0.010 | **0.332** | **0.500** | **0.500** | 0.228 | 69 | ≤ 0.3 |

(continued)

**Table 8** (continued)

| Spring | Chemical analyte | Statistical parameter | | | | | | | SANS (2011a)[a] |
|---|---|---|---|---|---|---|---|---|---|
| | | *n* | 5%ile | Mean | Median | 95%ile | SD | CoV (%) | |
| | U (mg/L) | | 0.003 | 0.005 | 0.003 | 0.012 | | 94 | ≤0.015 |
| | $SI_C$ | – | −1.78 | −0.32 | −0.35 | 0.25 | – | – | n.s. |
| | $SI_D$ | – | −3.53 | −0.63 | −0.67 | 0.47 | – | – | n.s. |
| | EB (%) | 59 | −8.2 | 6.6 | 6.3 | 17.4 | 9.8 | 150 | n.s. |

[a]Standard health-related limit for consumption of 2 L/d over 70 years by a 60 kg person
[b]Laboratory value ($-\log_{10}\alpha_{H+}$)
[c]Range includes default values equal to 50% of a 1 mg Al/L detection limit
[d]Range includes default values equal to 50% of a 50 mg $SO_4$/L detection limit
Bold text denotes value exceeding standard limit as described in note a

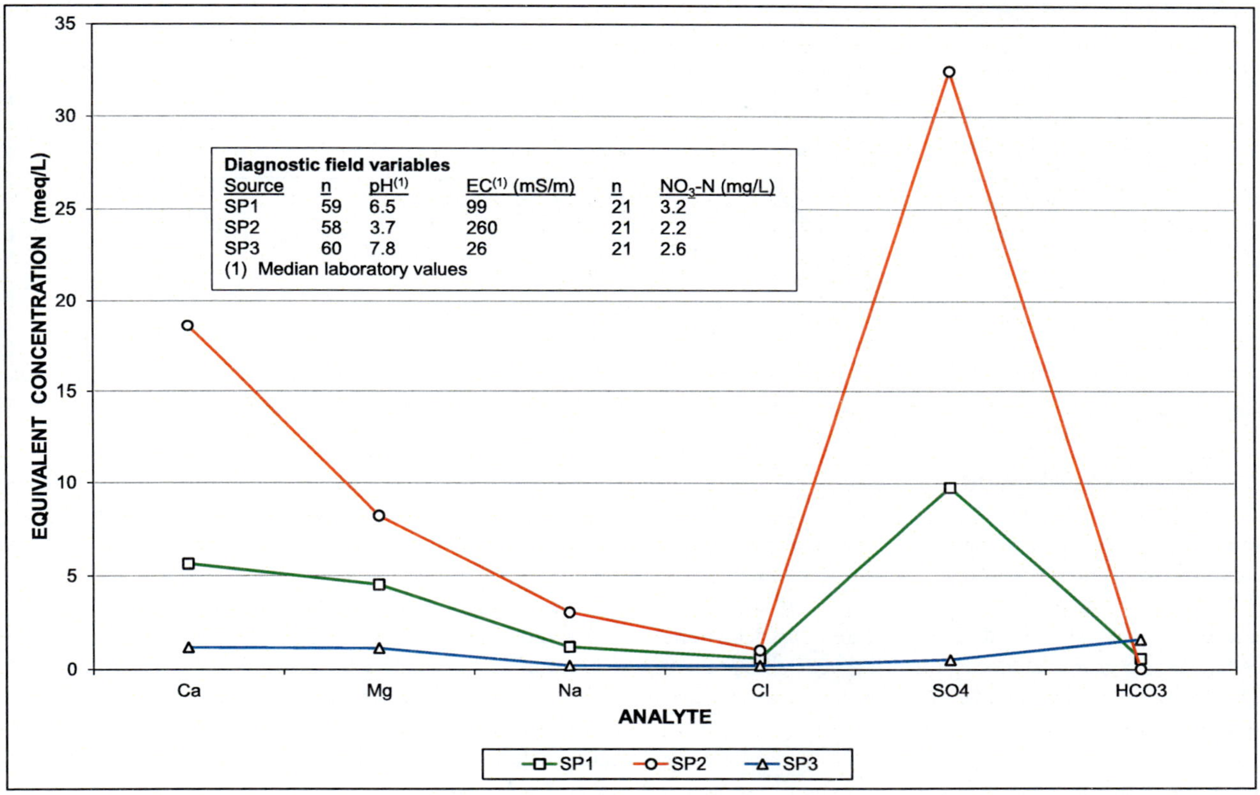


**Fig. 32** Graphic comparison of median water chemistry associated with selected springs in the KGR in the period September 2009–September 2014 (data from Table 8)

similar elevation of ~1435 m amsl, the 3rd so-called 'lost adit' (G. Whatley, personal communication) was uncovered during earthworks along the regional road (R540) that passes the property. This adit occupies a lower surface elevation of ~1425 m amsl, which is only ~2 m above the level of the nearby Honingklip Spruit, and ~5 m above the mine water level elevation. It is probable that the highest recorded mine water level observed in 2000 (G. Whatley, personal communication) approached this elevation. If the Kromdraai Gold Mine is therefore to decant, it will most likely be via this adit. Based on the present quality of the mine water, however, such decant is unlikely to contribute AMD into the receiving surface and groundwater environments.

It bears mention that the historical Kromdraai Gold Mine shares the ecological significance associated with many of the natural cave features in the study area, namely that of providing a habitat and breeding ground for bats (G. Whatley, personal communication).

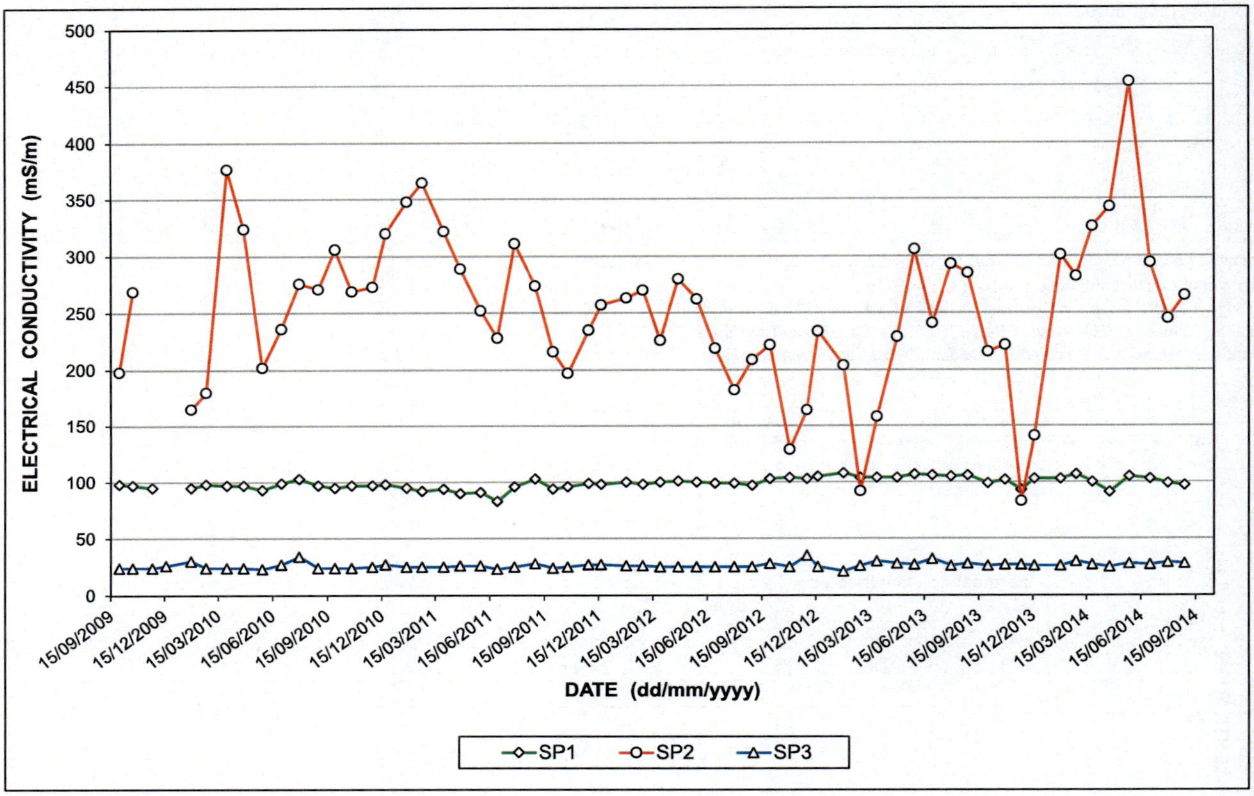


**Fig. 33** Salinity trend for water produced by springs SP1, SP2, and SP3 in the KGR in the period September 2009–September 2014

### 8.3 Human Settlement

The issue of informal housing developments in the COH assumed disturbing proportions when such a community on Ptn 26 at Kromdraai resorted to civil unrest in order to air grievances they felt toward the West Rand District Municipality as responsible authority for the area (Damons 2010). The geo-environmental factors that inform the situation are summarized as follows:

- location of the settlement on 'untested'[23] dolomitic ground;
- location of the settlement on undermined ground associated with the historical Kromdraai Gold Mine workings, the entrance to which is located on the adjoining property; and
- a comparatively shallow (<20 m) depth to groundwater rest level.

[23] The development of dolomitic terrain is subject to prior geotechnical investigation and assessment in accordance with guidelines put forward in the CGS/SAIEEG (2003) publication.

The combination of these factors creates understandably significant concern for a variety of reasons that include compromising human health and safety and water resource quality. It is therefore imperative that provincial government and district/local authorities are equally aware of the sensitivity and value of the COH environment.

A further concern associated with both informal and formal low-cost residential areas is the proliferation of refuse disposal sites associated with these areas. Examples are the refuse sites centered on the coordinates 26.07385° S–27.65700° E and 26.07073° S–27.65820° E to the north-east of the informal residential settlement located on Ptn 6 of Vlakplaats 160IQ in the south-eastern quadrant of the N14/R24 junction at Tarlton. Both sites are readily identifiable on the Google Earth® image of the area, which also reveals their location in the middle reaches of the Riet Spruit valley. This segment of the drainage only carries surface water into the Zwartkrans Basin during periods of excessive rainfall and runoff from the Randfontein area. This runoff is impounded in several dams, notably one located at 26.06515° S–27.66104° E. However, no surface water flow passes the downstream (eastern) boundary, located at 26.06436° S–27.66960° E, of the Jomajoco Farms property (J. van den Bosch, personal communication). Although the

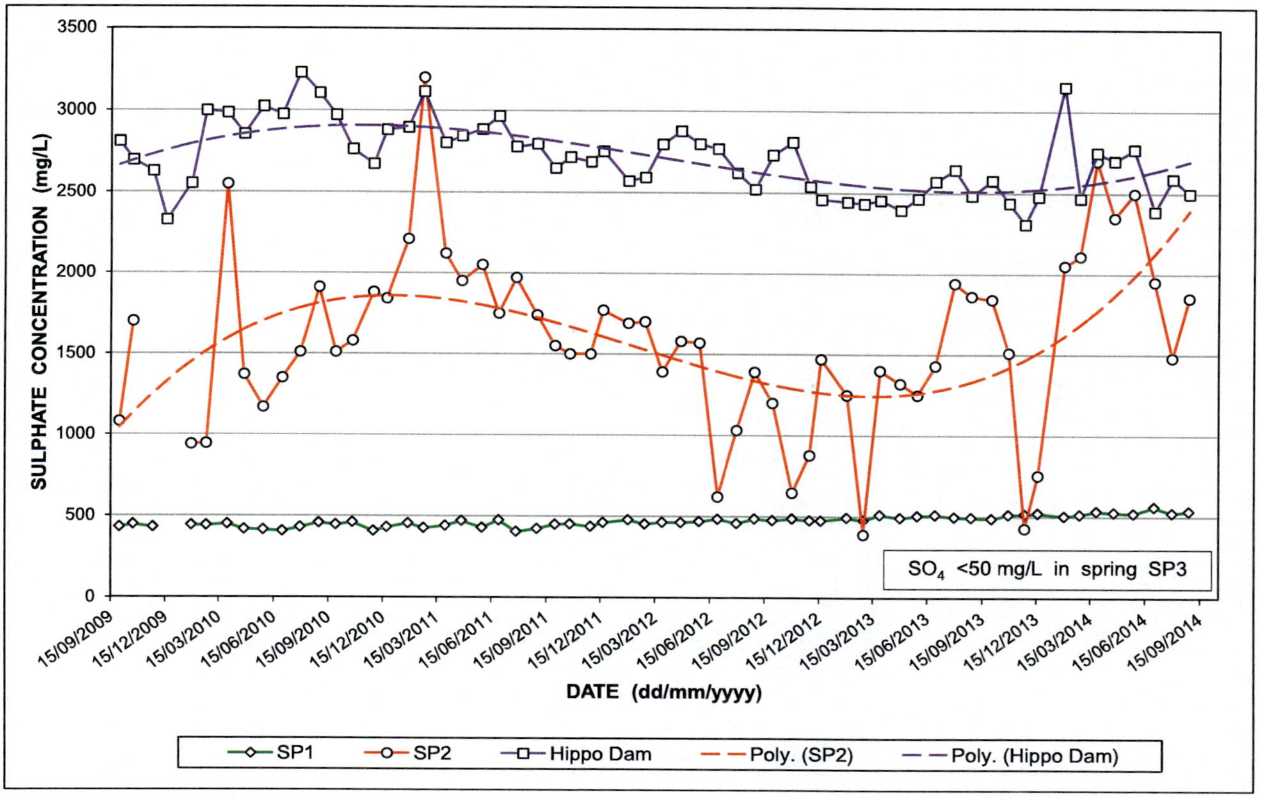


**Fig. 34** Sulfate trend for water produced by springs SP1, SP2, and SP3 in the KGR in the period September 2009–September 2014; the Hippo Dam record provides a reference of contemporary surface water discharge quality

depth to groundwater rest level along this segment of the Riet Spruit channel is ~60 m bs (station A2N0598, Table 8 in Chapter "Physical Hydrogeology"), this otherwise substantial separation with the surface is short-circuited by swallow holes in the river bed that allow poorer quality surface water more rapid access to the aquifer. This might explain the total coliform and *E. coli* values of 36 and 5 c/100 mL, respectively, detected in a groundwater sample from station A2N0598 collected on 09/06/2010.

The provision of suitably adequate refuse/waste removal facilities for the low-cost and/or informal residential areas that have been developed on dolomitic ground in the study area is no less important than the provision of water and sanitation facilities to these communities. These represent opportunities for integrated land-use development between local (MCLM) and district (WRDM) municipal authorities. The illegal dumping of refuse in the COH is, however, not limited to areas of low-cost and/or informal housing. The image in Plate 7 shows an illegal dumpsite located immediately downstream of where the Malmani Road which serves the Sterkfontein Country Estates crosses the Riet Spruit. The refuse covered the entire drainage channel, including the active flow section. Exercising control over such practices is extremely difficult, and must rely as much on pride in a 'sense of place' as it does on community 'policing'. The situation has since been remedied through the intervention of local residents and landowners. The placement of a new cemetery in the upper reaches of the Zwartkrans Basin at 26.05667° S–27.65645° E is considered a responsible development as it recognizes the ~70 m depth to groundwater level in this part of the karst environment.

## 8.4 Mine Water

The groundwater chemistry data generated by the DWS monitoring program in the Zwartkrans Basin provides an indication of the extent and magnitude of the mine water impact on the karst aquifer in this portion of the COH. The subsurface impact of mine water is defined by the pseudo-plume and principal flow path illustrated in Fig. 39. This path extends northwards from the Sterkfontein Subcompartment into the contiguous portion of the Zwartkrans Subcompartment and is characterized by SEC values of up to 300 mS/m and $SO_4$ concentrations of up to 1800 mg/L,

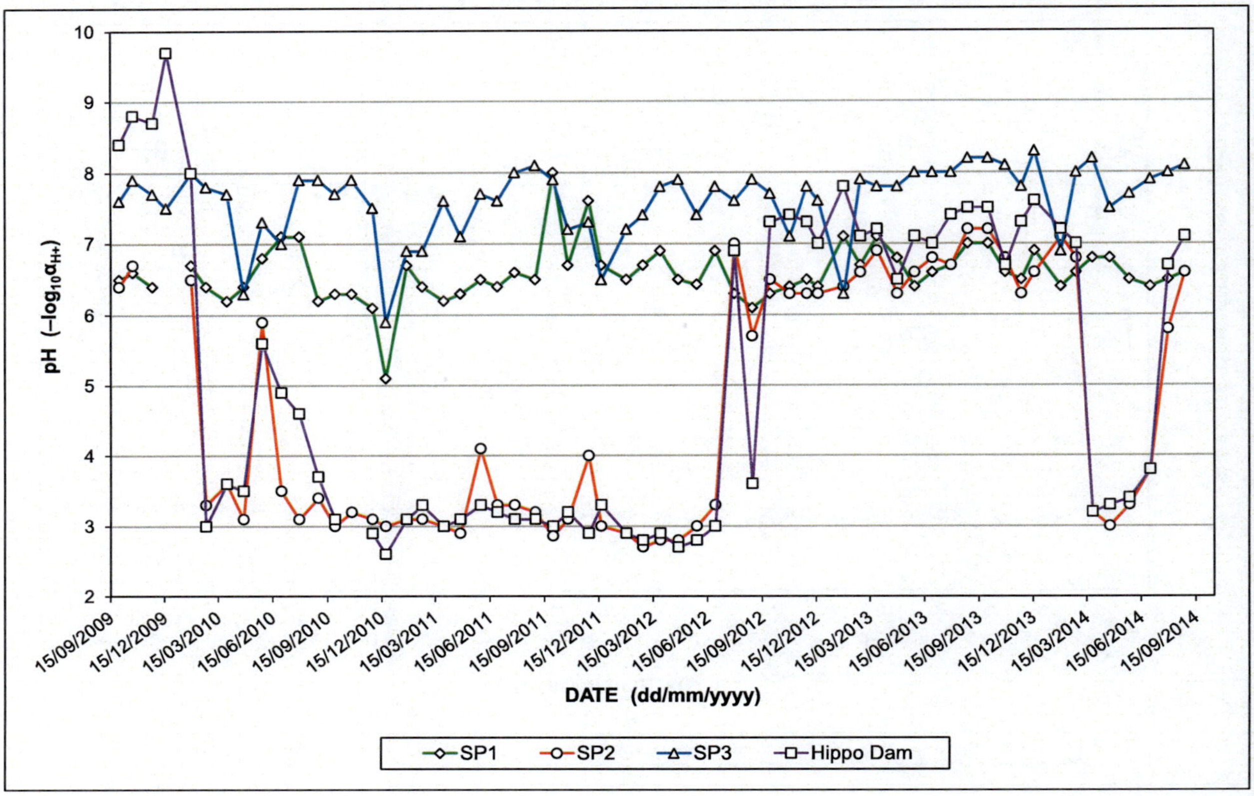


**Fig. 35** Trend in pH for water produced by springs SP1, SP2, and SP3 in the KGR in the period September 2009–September 2014

although pH values >6.4 indicate that net alkaline conditions still prevail (see Sect. 13.2) as at late-2014. This is illustrated in Figs. 40, 41 and 42 with the aid of bar graphs for the chemical variables pH, EC, and $SO_4$, respectively.

The SEC and pH values measured on the occasion of each SDM reported for stations F11S12 and MRd in Table 10 in Chapter "Chemical Hydrology", are graphed in Fig. 43 (for SEC) and Fig. 44 (for pH). These data reflect the elevated SEC values (>350 mS/m) and depressed pH values (<3) that characterized the surface water lost to the karst aquifer in the period mid-2010 to mid-2012 (Sect. 5.2.1 in Chapter "Physical Hydrology"). It is this allogenic recharge that has manifested the mine water imprint on the karst groundwater of the Zwartkrans Basin observed in Figs. 40, 41 and 42.

The footprint of the pseudoplume shown in Fig. 39 encompasses ~1030 ha. This represents ~11% of the Zwartkrans Basin, and only ~4% of the entire COH karst footprint. These observations provide context for the spatial extent of the mine water impact on the karst water resources of the COH.

The positive influence of the immediate mine water control and management intervention measures since mid-2012, as evidenced by the lower SEC values (<300 mS/m as per Fig. 43) and higher pH values (>4 as per Fig. 44), signifies an improvement in the situation. This improvement is associated in part with the containment of uncontrolled raw mine water discharge into the environment in mid-2012, and except for a week or two in December 2012, was maintained until February 2014 (Hobbs 2014b). The subsequent 'excursion', described by Tancott (2014) as a "big problem", was brought under control in July 2014. Hobbs (2013a, b, 2014a, b) has demonstrated that an observed improvement continues to manifest as decreasing SEC and $SO_4$ values in the monitoring boreholes located in each of the segments as of December 2014. Nevertheless, the undersaturated state of the karst groundwater in regard to calcite and dolomite reflects conditions of dissolution of these minerals, especially in proximity to the losing reach of the Riet Spruit. These circumstances are illustrated in Fig. 45, which also reveals the saturated state of the Sterkfontein Cave (SC) groundwater.

The magnitude of the mine water impact on the receiving karst groundwater resources is explored further in Sect. 13.2 in regard to the net alkalinity and net acidity of the groundwater. The demonstrated presence of karst

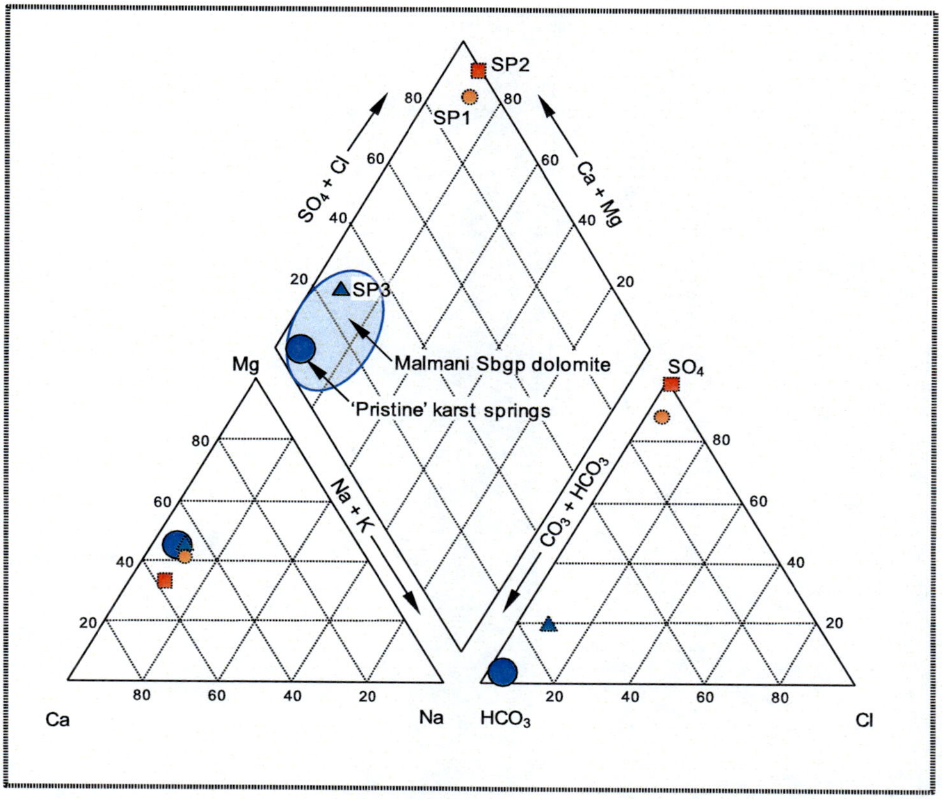


**Fig. 36** Piper diagram of KGR median springwater chemistry (data from Table 8) compared to pristine karst springwater chemistry

groundwater exhibiting a net positive acidity, albeit marginal and generally associated with metal (mineral) acidity, is another concern.

The $SI_C$ values of the more recent allogenic recharge from the Riet Spruit (Sect. 5.2.1 in Chapter "Physical Hydrology") fall in the range −4.0 to −3.2, reflecting undersaturation of the water in regard to this mineral. These circumstances equate to the conditions identified by Palmer (2003) and Worthington and Ford (2009) where water that is undersaturated with regard to calcite (typical of allogenic recharge) enters a karst aquifer via point recharge from a sinking stream carrying a high discharge. The information presented in Fig. 45 indicates that these circumstances extend to the karst groundwater, at least in proximity to the losing stream reach. Of concern in this regard is that Worthington and Ford (2009) indicate that these circumstances typically result in the concentrated dissolution of carbonate strata in the receiving karst aquifer. Together with the net acidity associated with this water (Sect. 13.2), the aggressive character of the allogenic recharge represents a concern that requires concerted assessment and evaluation.

The historical pattern and trend of groundwater SEC and $SO_4$ concentrations in proximity to the losing reach of the Riet Spruit between stations F11S12 and MRd is reflected in the long-term monitoring data associated with the paired stations A2N0584–GP00302 and A2N0586–GP00300, and station A2N0600 (see Fig. 39 for station positions). These are presented in Fig. 46, and again reveal the comparatively recent increase in SEC and $SO_4$ concentration levels. The positions of these stations in relation to the principal direction of groundwater flow are shown in Figs. 39, 40 and 41.

As might be expected given their respective position along the principal flow path, the $SO_4$ concentration pattern shown in Fig. 46 indicates a later response at station A2N0600 than at station A2N0584. A concentration of ~300 mg $SO_4$/L was only manifested at the former ~600 days after this value was measured at station A2N0584. Given the distance of ~4500 m between these two stations, the travel time equates to a hydraulic conductivity (*K*) of 7.5 m/d. This value compares favorably with the range of *K* values (~1–27 m/d) obtained by Holland (2007) from slug tests carried out on four boreholes in the area. It compares particularly well with the *K* values of ~4–7 m/d obtained for borehole SCH2 (identified as SBH-2 in Fig. 4.21 and Table 4.9 of Holland 2007) located in the flow path (Fig. 19 in Chapter "Physical Hydrogeology").

Accepting the transmissivity value of ~1000 $m^2$/d previously assigned to this aquifer, a flow depth (*d*) of ~130 m is obtained from the relationship

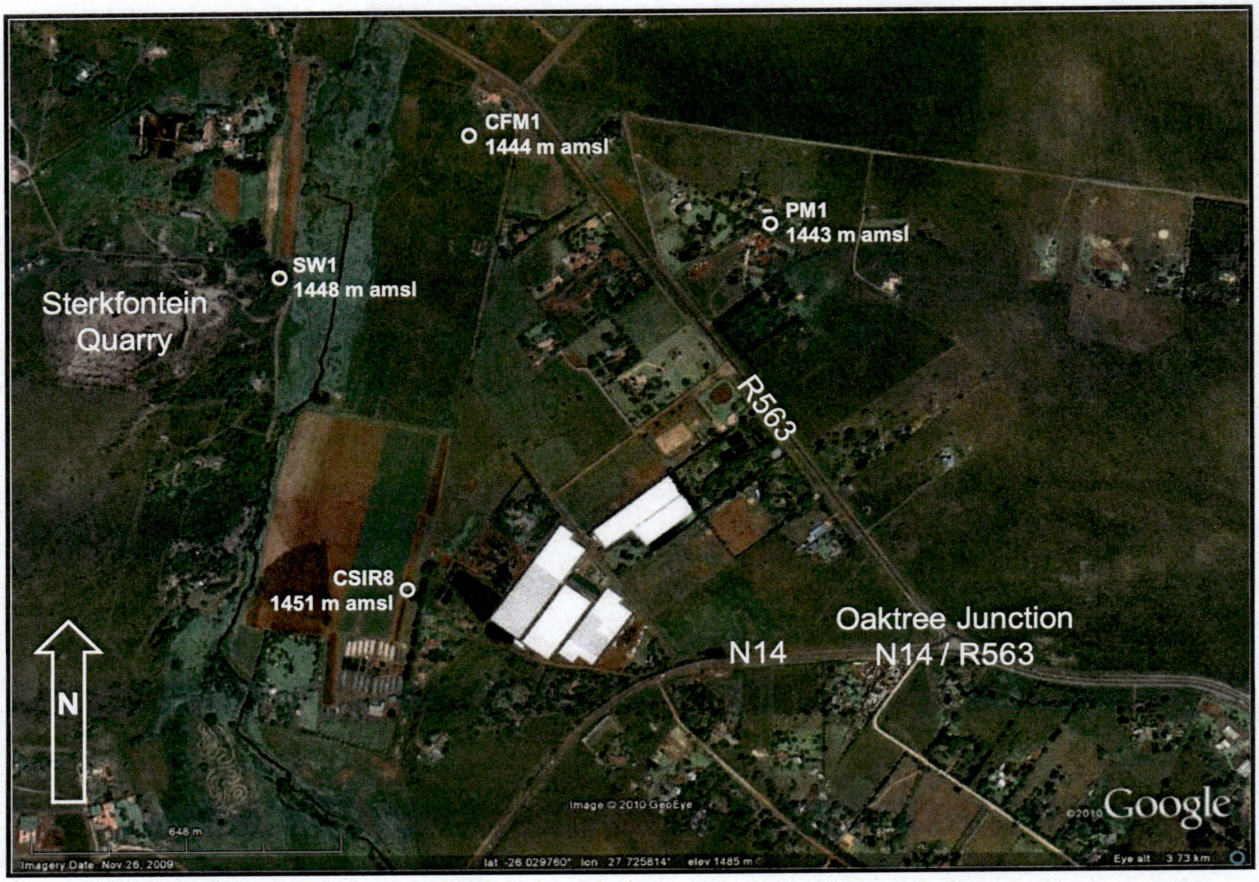


**Fig. 37** Locality map of pesticide residue sample sources CSIR8 and CFM1

$$T = K \times d$$

The veracity of this value can be tested against the geological log (record) for station A2N0584 (originally assigned the number G36331) as reported by Bredenkamp et al. (1986). This record indicates that two horizons of leached and fractured chert-rich dolomite were intersected in the depth intervals of 20–40 m and 120–132 m below surface. The deeper aquifer produced an airlift yield of 50 L/s. It is the shallower karst aquifer at this location that has been targeted for monitoring by the recently established borehole GP00302, and which yields the slightly poorer quality groundwater evident in Fig. 41. The same drilling record, however, reports that a ~70-m thick chert-poor horizon separates the two aquifer horizons in borehole A2N0584. These circumstances illustrate the extremely heterogeneous and anisotropic nature that is typical of the so-called 'Transvaal' karst aquifers, and which favor the development of a pseudoplume (Ewers et al. 2012). This derives from the typically convergent nature of flow in the conduit system that militates against the hydrodynamic dispersion of contaminants in a karst aquifer.

Further chemical characterization of the groundwater associated with the mine water impact in the Zwartkrans Basin is shown in the Piper diagram in Fig. 47. The diagram represents all the stations sampled in late-2012, excluding the five stations enumerated on an ad hoc basis in December 2012.

It is evident from Fig. 39 that the pseudoplume of mine water contaminated groundwater discharges via the Zwartkrans Spring. The circumstances that describe this phenomenon are discussed in Sect. 7.1. Reference by Groenewald (2010) to the caves as occupying a low energy groundwater system, and by Martini et al. (2003) to the "… apparently static …" pools in the cave system, support a distal location on the south-eastern margin of a postulated subsurface thalweg. This hypothesis finds support in the "flood induced pseudoplume" reported by Ewers et al. (2012), which describes the intrusion of contaminated groundwater from a flooded major conduit into peripheral porous karst strata during storm/flood events. Although the intruded water returns to the conduit as groundwater levels recede following passage of the flood, it leaves behind contaminants in various settings in the intruded portion of the karst aquifer. These circumstances are more likely to

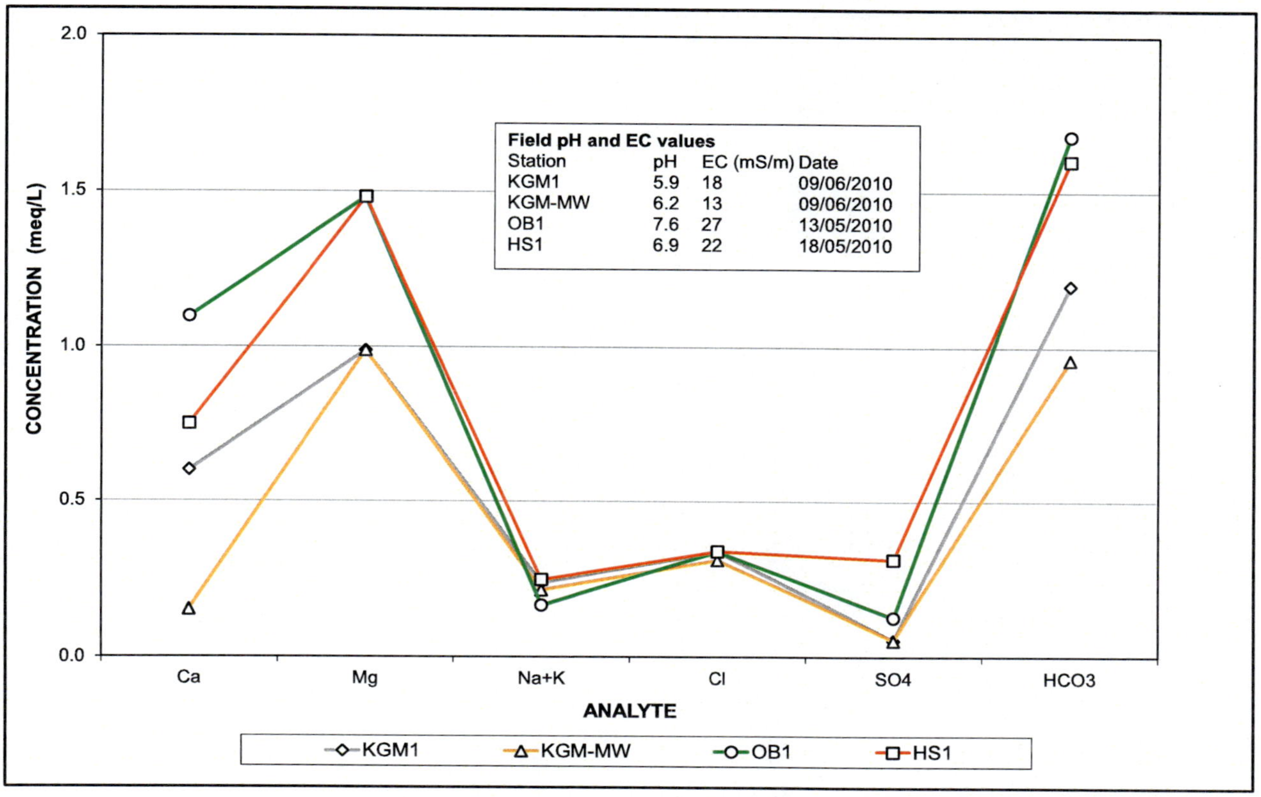


**Fig. 38** Comparison of surface water and groundwater chemistry in the vicinity of the historic Kromdraai Gold Mine

**Plate 7** View of an 'illegal' refuse disposal site in the channel of the Riet Spruit near its confluence with the Blougat Spruit

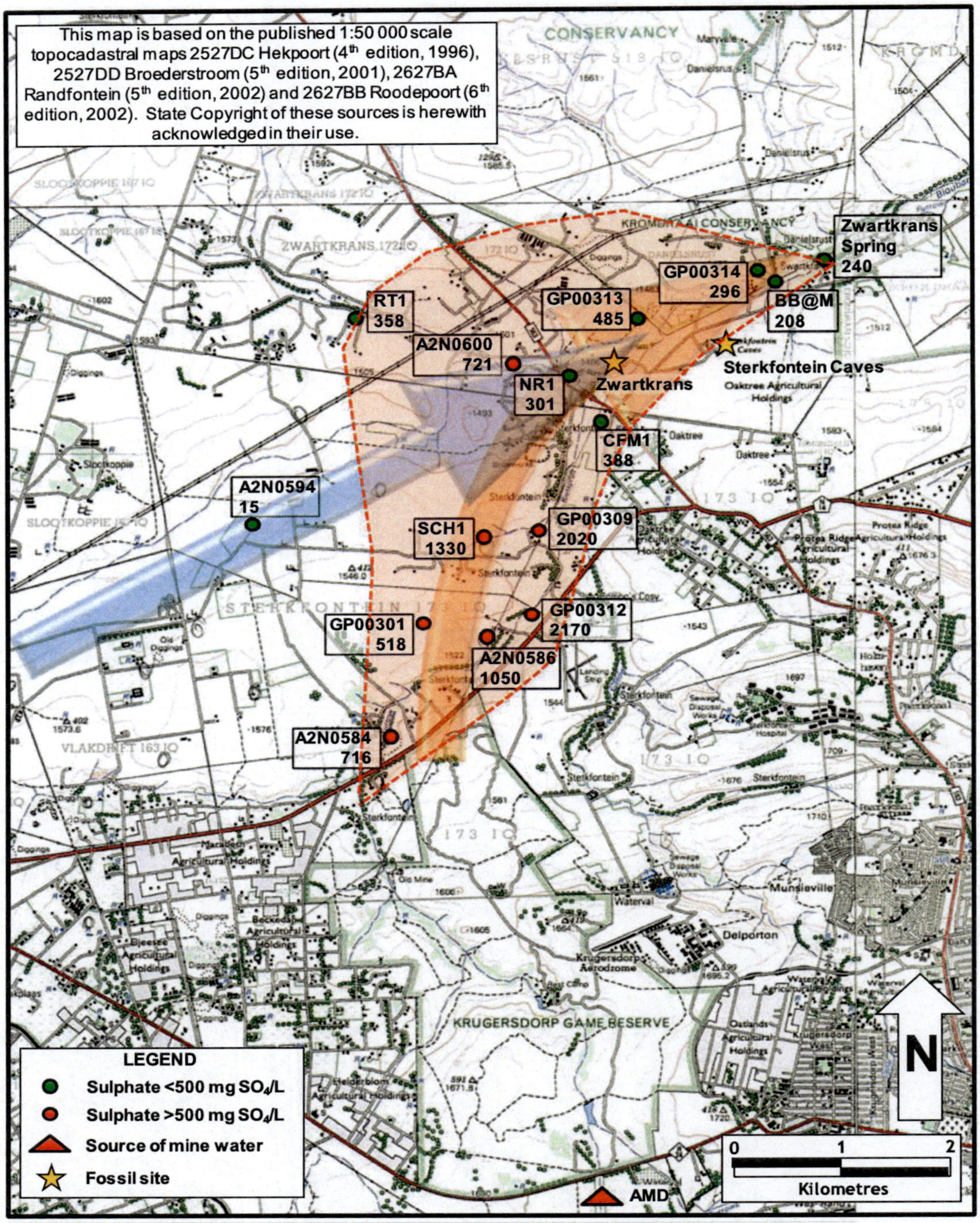


**Fig. 39** Definition of the principal flow path describing the passage of allogenic recharge through the Zwartkrans Basin to the Zwartkrans Spring based on late-2014 sulfate concentrations in groundwater

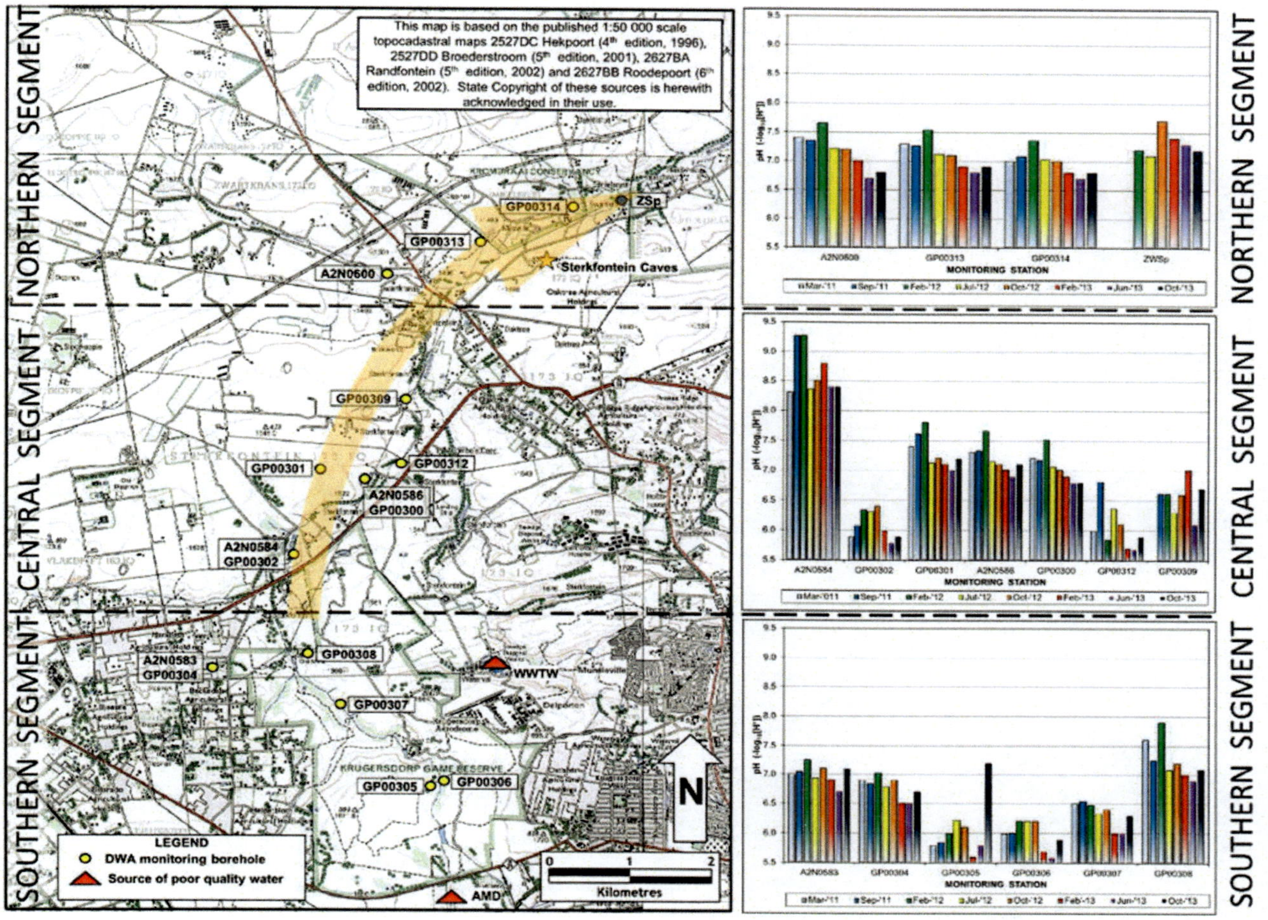


**Fig. 40** Distribution of DWS monitoring boreholes with pH pattern and trend as bar graphs; arrow denotes principal direction of groundwater flow

prevail in the lower (downstream) reaches of a karst basin where conduits are convergent (Ewers et al. 2012), a hydrogeologic setting that suitably describes the location of the caves in the landscape. These circumstances underscore the distal location of the caves relative to a postulated subsurface thalweg as illustrated in Fig. 48. This reflects a combination of the hydrophysical and hydrochemical dynamics that describe the hydrogeology of the cave system within its broader hydro-environment along two transects.

The occurrence of mercury in mine and surface water sources in the study area has been discussed in Sect. 3.3.2 in Chapter "Chemical Hydrology". Concern for its occurrence in groundwater sources precipitates the comparison made in Table 9, where the mine and surface water values derive from Table 40 in Chapter "Chemical Hydrology" (Sect. 3.3.2 in Chapter "Chemical Hydrology"). The comparison reveals the typically low concentrations of mercury in groundwater as evidenced by the singular exceedance of the various detection limits, under circumstances where SANS (2011a) sets a standard chronic health limit of $\leq 0.006$ mg/L for this metal in drinking water. The 'poorest' (highest) detection limit in Table 9 is 3× more stringent than this limit, which is exceeded in only two out of 66 groundwater samples.

## 9 Groundwater Fitness

### 9.1 Potable Use

The project has defined the quality of groundwater resources in far the greater portion of the study area as being suitable for human consumption in terms of the SANS 241 (2006) standard for drinking water quality. This does not, of course, provide solace to those groundwater users who have experienced a compromised quality of groundwater attributable to extraneous impacts associated with mine water and municipal wastewater effluent. The extent to which

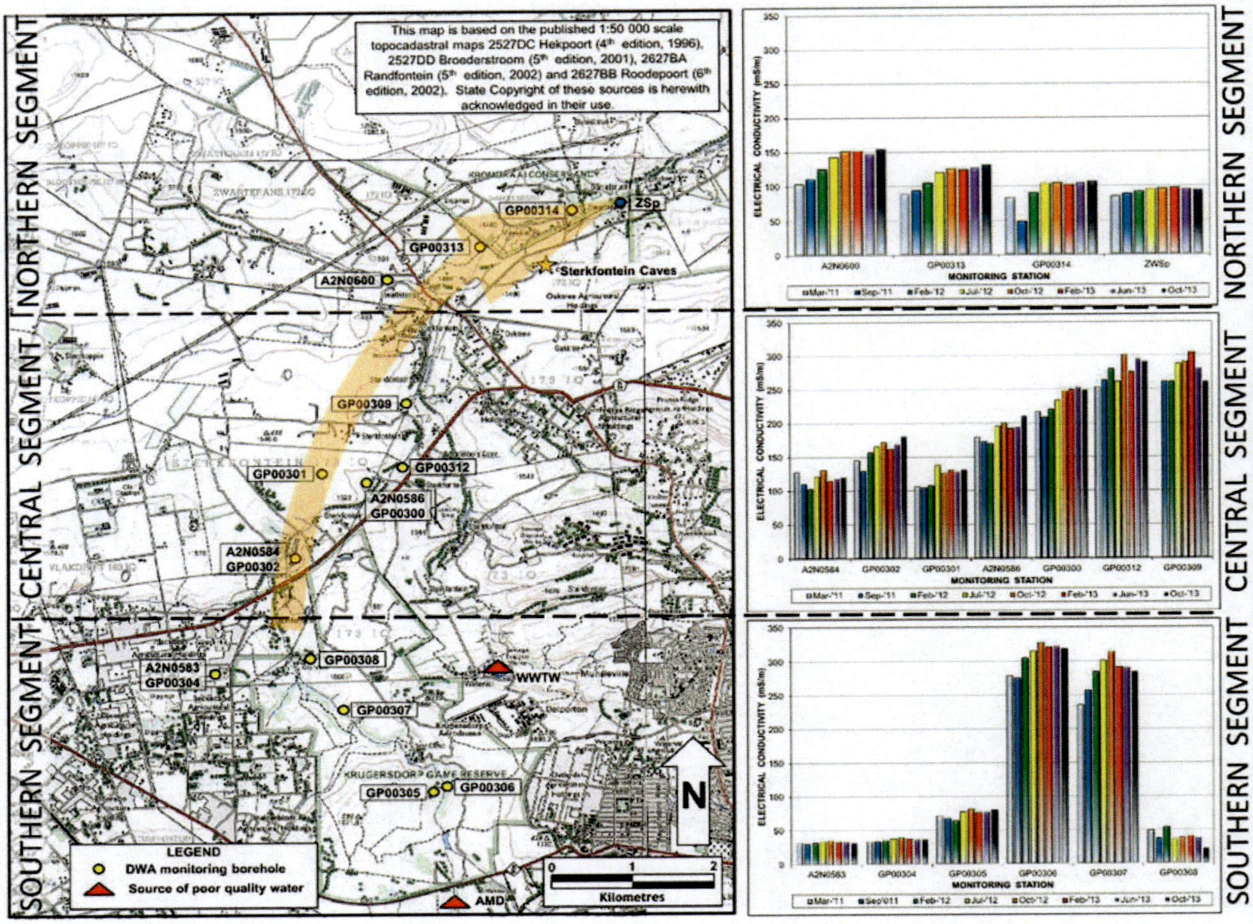


Fig. 41 Distribution of DWS monitoring boreholes with SEC pattern and trend as bar graphs; arrow denotes principal direction of groundwater flow

these impacts extend into the broader hydrogeologic environment is described by the water resources monitoring program implemented by the COH WHS MA and the DWS. The mine water impact has been discussed in Sect. 8.4.

The presence of heavy metals in both the mine wastewater and the municipal sewage wastewater also raises a concern for the potability of groundwater resources in the study area. As shown in Sect. 11, the presence of heavy metals such as Al, Co, Mn, Ni, and Zn could not be detected (for the detection limits reported in Sect. 9.3) in the groundwater from sources otherwise reflecting the most impacted quality in terms of SEC and $SO_4$. The association of radionuclides with mine water raises a further concern. However, it is shown in Sect. 5.2.2 that none of the groundwater samples exhibited a U concentration greater than the detection limit of 0.001 mg/L (1 μg/L).

The above observations suggest that the fitness of groundwater in the study area for potable use is not unduly compromised by inorganic elements, trace/heavy metals, or radionuclides. However, the observed presence of bacteriological contamination in some sources (refer to Sect. 6) cautions against the unqualified use of groundwater for drinking water purposes.

## 9.2 Agricultural Use

### 9.2.1 Livestock Watering

Guidelines in this regard are provided in the DWAF (1996b) publication. The target water quality range (TWQR) limits for the major inorganic analytes are typically less strict than for human consumption. For example, the TWQR for $SO_4$ (a variable that affects the palatability of water also for livestock) is 1000 mg/L compared to the 400 mg/L for humans. This is also true for trace/heavy metals, e.g., Mn and Fe $\leq$ 10 mg/L for livestock compared to $\leq 0.1$ and $\leq 0.2$ mg/L, respectively, for humans.

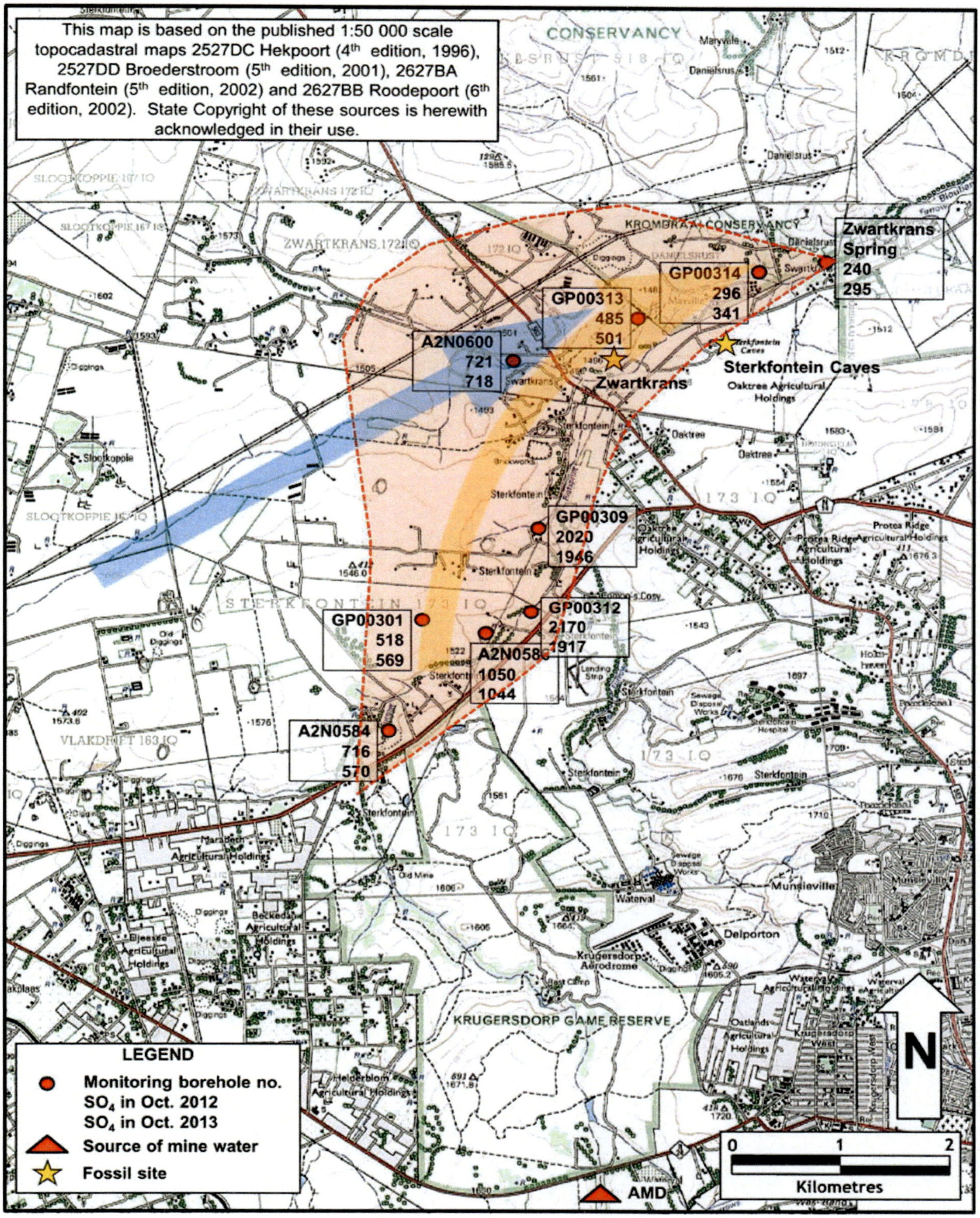


**Fig. 42** Distribution of DWS monitoring boreholes with $SO_4$ pattern and trend as bar graphs; arrow denotes principal direction of groundwater flow

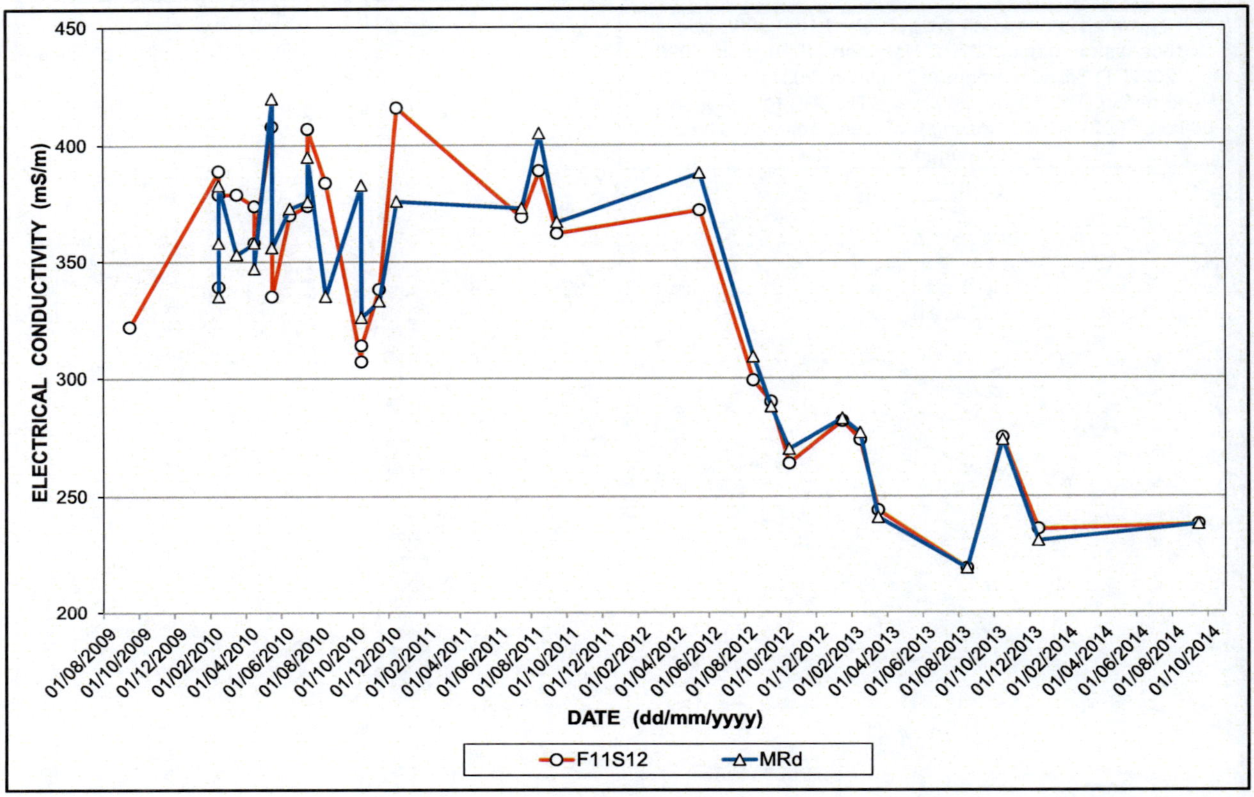


**Fig. 43** Pattern and trend of SEC at flow gauging stations F11S12 and MRd on occasion of each synoptic discharge measurement (data from Table 10 in Chapter "Chemical Hydrology")

Similarly, the observed presence of bacteriological contamination in some groundwater sources (refer to Sect. 6) is mitigated by the fecal coliforms TWQR of $\leq 1000$ c/100 mL in 20% of samples, compared to the $\leq 10$ c/100 mL in 1% of samples for humans. Under circumstances where the highest total coliform count (which includes fecal coliforms) recorded in this study for groundwater amounted to 500 c/100 mL, the fitness of groundwater in the study area for livestock watering is not unduly compromised by inorganic elements, trace/heavy metals, or bacteriological considerations.

### 9.2.2 Irrigation

An analysis of the 51 groundwater chemistry analyses generated by this study returned mean and median SAR values of 0.52 and 0.27, respectively. Grouped by compartment/subcompartment, this analysis yields the information in Table 10. Wilcox diagrams reflecting the SAR classification are presented in Figs. 49 and 50.

It is evident from Table 10 that the mean value is matched or exceeded in only three basins/subcompartments. Significantly, these are the compartments that receive allogenic recharge from the poorest quality surface water sources (Sects. 5.2 in Chapter "Physical Hydrology" and 1.2 in Chapter "Chemical Hydrology"). Nevertheless, the classification of the groundwater for irrigation use (refer to Sect. 4.2.2 in Chapter "Chemical Hydrology") is no worse than a C3–S1 quality class rating as illustrated in Fig. 50. More typically, however, it exhibits a C2–S1 rating (Table 10), which allows its use for the irrigation of crops with a moderate salt tolerance on soils where a moderate amount of leaching occurs.

## 9.3 Cave Ecosystems

The fitness of groundwater quality in regard to cave ecosystems is important where the latter intersects the water table and provide refugia for stygobitic (groundwater) fauna (Tasaki 2006), also referred to as stygobionts or stygofauna (aquatic species) by Culver and Sket (2000, 2002). As far as could be established, such biota (typically blind amphipods) have been found in Sterkfontein Cave and two other caves (Koelenhof and Yom Tov) in the study area (Tasaki 2006). Durand and Peinke (2010) indicate that certain amphipod species (notably *Sternophysinx transvaalensis*) are also

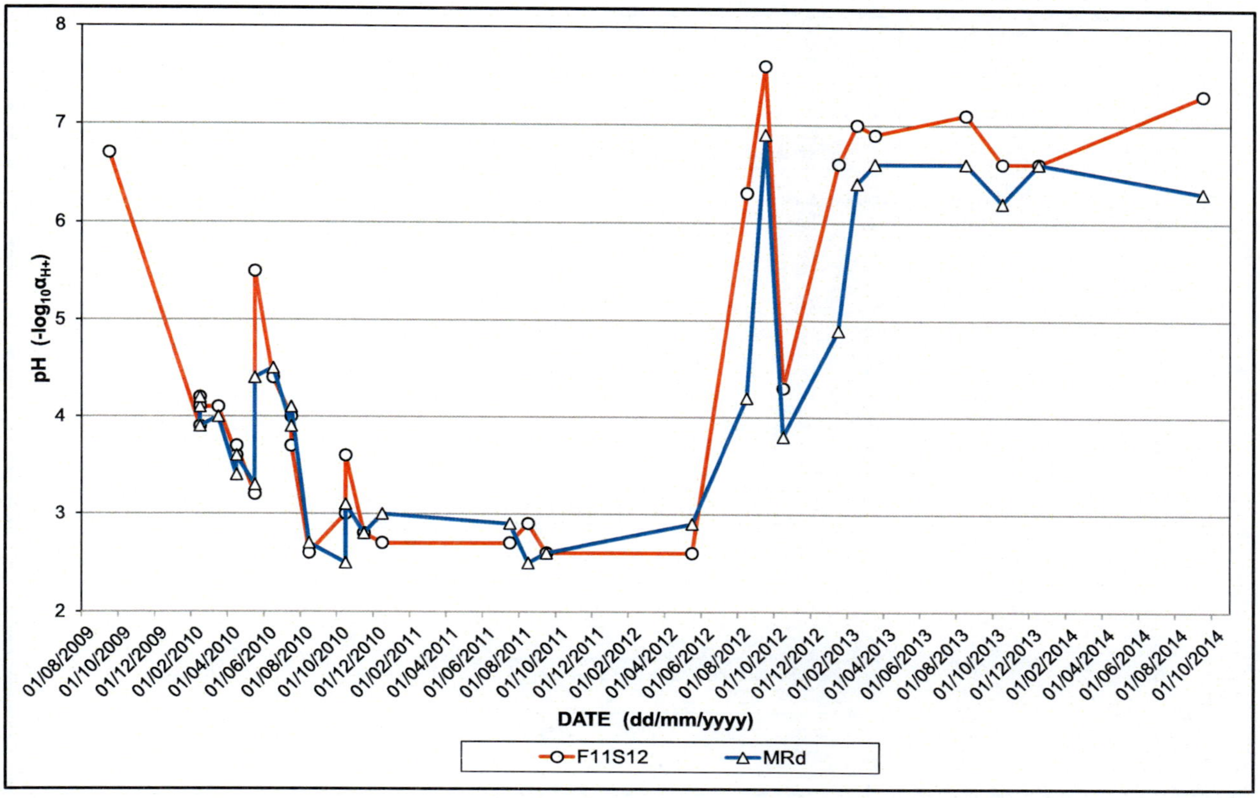


**Fig. 44** Pattern and trend of pH at flow gauging stations F11S12 and MRd on occasion of each synoptic discharge measurement (data from Table 10 in Chapter "Chemical Hydrology")

associated with springs in the John Nash Nature Reserve. As this report is concerned with water resources, the consideration of subterranean (cave-dwelling) terrestrial species (troglobitic fauna or troglobionts) is not considered. However, Durand and Peinke (2010) indicate that the humidity levels associated with 'wet' caves might be an important environmental factor for resident bat colonies.

The impact of pollution effects on cave systems is related to the same factors that inform the contamination of surface water and/or groundwater resources. These have previously been recognized as elevated nutrients (nitrate and phosphorus) associated with municipal wastewater effluent and agricultural return flows, and trace metals associated with wastewater effluent and mine water. The treated wastewater and/or agriculturally derived nutrients promote eutrophication in surface water systems through photosynthesis. However, the absence of sunlight in cave ecosystems precludes the excessive growth of aquatic vegetation in the subterranean environment (Graening and Brown 2000; Boulton et al. 2003). These authors recognize the negative impacts of organic pollution on cave ecosystems as being related to the following:

- alteration of the community assemblage;
- impoverishment of biodiversity; and
- increased risk of predation from surface fauna.

Culver and Sket (2002) identify the usual sequence of eutrophic impacts in caves as an initial increased population size of stygobites, followed by increased population sizes of non-specialized species usually accompanied by a decline in population sizes of stygobites, and culminating in the extirpation of the latter. As population increases in the food-poor environment of caves usually means an increase in available food (typically associated with eutrophication), observations of such a trend through routine biomonitoring activities offer an early warning of potentially catastrophic longer-term impacts on stygobitic fauna (Culver and Sket 2002).

Graening and Brown (2000) recognize that increased concentrations of Pb and Zn in cave water are manifested in the accumulation of these trace metals in cave sediments and the tissues of cave organisms rather than in their biomagnification in the food chain. This is identified as a concern under circumstances where heavy metals might be present in concentrations that are acutely toxic to aquatic organisms.

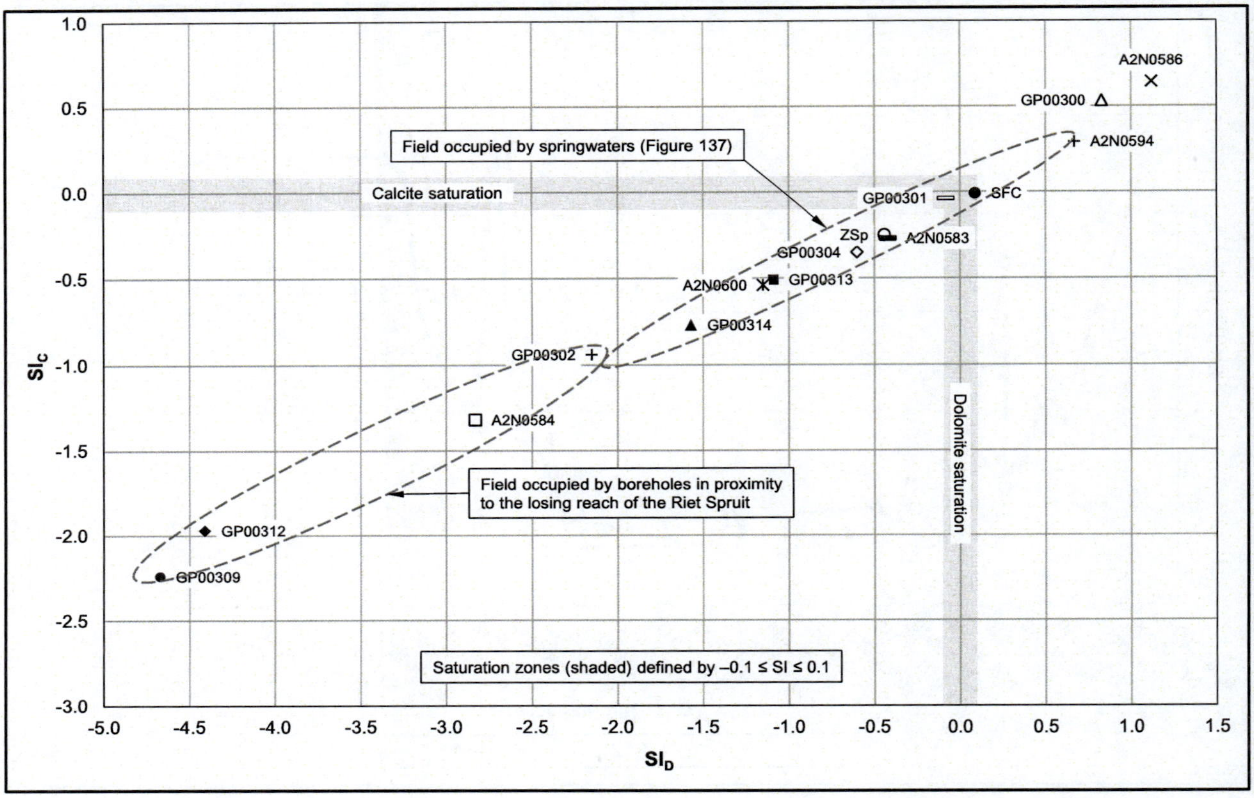


**Fig. 45** Saturation state of karst groundwaters in the Zwartkrans Basin with respect to calcite ($SI_C$) and dolomite ($SI_D$)

The chemical analysis of a Sterkfontein Cave water sample collected on 13/05/2010 returned a $NO_3$–N concentration of 8.2 mg/L and a $PO_4$–P concentration of <0.2 mg/L but revealed no concentrations of the following trace metals above the specified detection limits (DLs).

| DL ( mg/L) | Analyte |
|---|---|
| ≥0.100 | Al |
| ≥0.025 | Ag, B, Ba, Be, Bi, Co, Cr, Cu, Fe, Li, Mn, Mo, Ni, Sn, Sr, Ti, V, W, Zn, Zr |
| ≥0.020 | Pb, Se |
| ≥0.010 | As, Sb |
| ≥0.005 | Cd |

Whether the detection limits mask the possible presence of trace metals that at lower concentrations might still constitute a threat to the continued existence of stygobionts in the Sterkfontein Cave system (and other similar refugia in the study area) is beyond the scope of this study. Further, it is important to note that single measurements of inorganic nitrogen ($NH_3$, $NH_4$, $NO_2$, and $NO_3$) are a poor basis for assessment (DWAF 1996a). The likely biological consequences (at least in surface water systems) are best estimated from average summer inorganic nitrogen concentrations.

Nevertheless, the concurrently low $PO_4$ concentration and the absence of sunlight suggest that the trophic status of the Sterkfontein Cave water system is unlikely to change provided the current hydrochemical conditions prevail. Assurance in this regard can be obtained from regular biomonitoring programs (as described by Culver and Sket 2002) that target stygobitic fauna. Schneider and Culver (2004) have shown that focusing on the largest caves improves sampling efficiency for both terrestrial and aquatic underground species.

In support of Kenyon and Ellis (2010), it is also recommended that any cave biomonitoring program that is implemented in the COH includes the participation of recognized caving associations and their members. The contribution of such I&Aps, especially if these include knowledgeable and experienced local 'cavers', will extend the benefit of biomonitoring activities to less well-known and frequented caves.

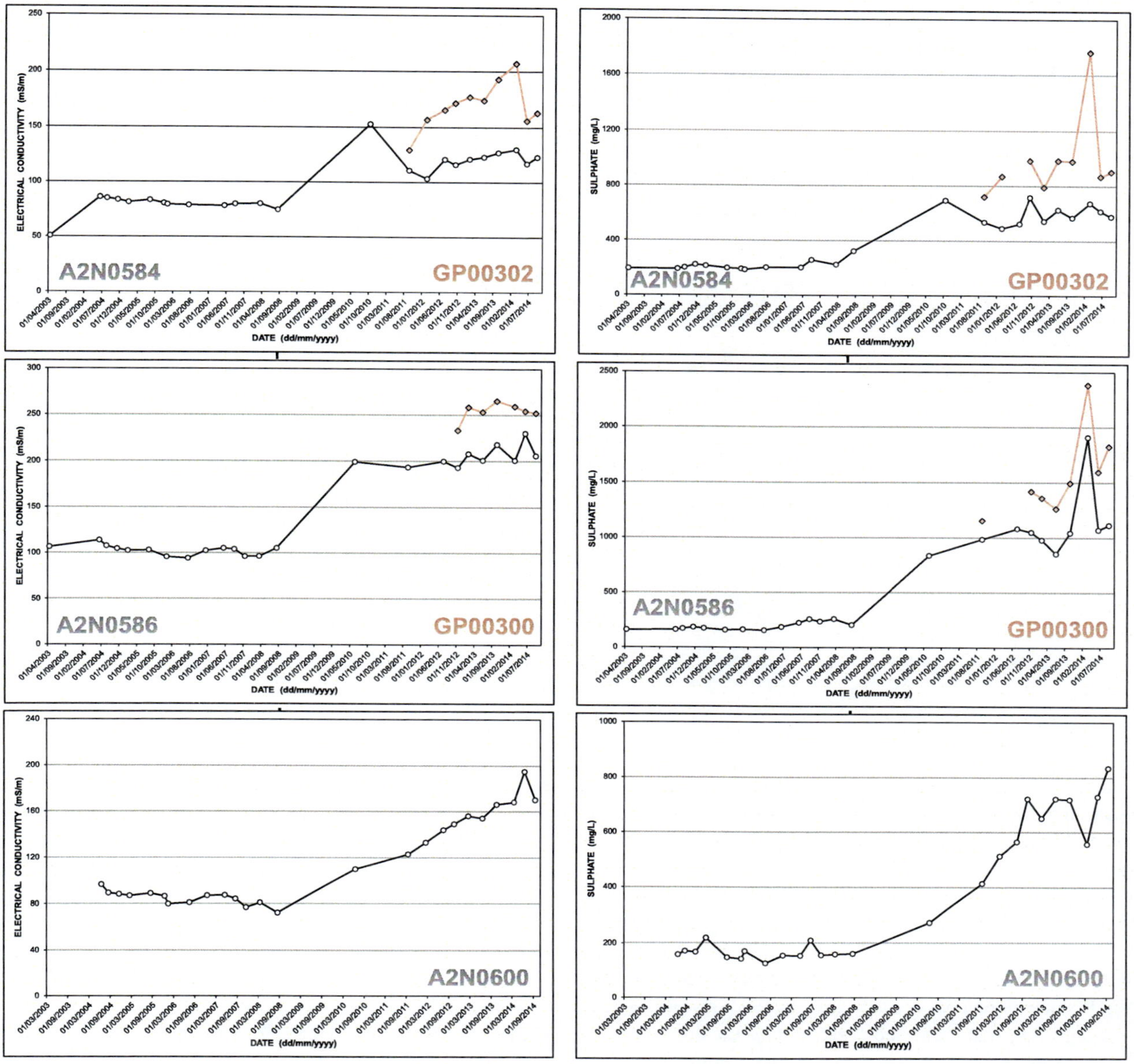


**Fig. 46** Long-term pattern and trend of SEC (at left) and $SO_4$ (at right) in karst groundwater from DWS monitoring stations A2N0584/GP00302, A2N0586/GP00300, and A2N0600; note common time scales and postulated start of rise in concentrations (long vertical pecked line)

In conclusion, it is acknowledged that cave ecosystem monitoring extends well beyond the water resource environment alone. In the pursuit of 'integrated monitoring' across a broad spectrum of disciplines including geomicrobiology, geobiology, biogeochemistry, and biogeography (Engel et al. 2008), the author wholly supports the monitoring requirements put forward by Durand and Peinke (2010) in developing a holistic understanding of the karst ecology of the COH within the paradigm advanced by Williams (2008).

## 10 Vertical Hydrochemical Variation

### 10.1 Introduction

Although stratification is a common occurrence in many mines, the published literature seldom provides an explanation for this phenomenon. Wolkersdorfer (2008) discusses 13 case studies in which depth profiles for a range of variables (temperature, EC, TDS, pH, redox, resistivity, flow

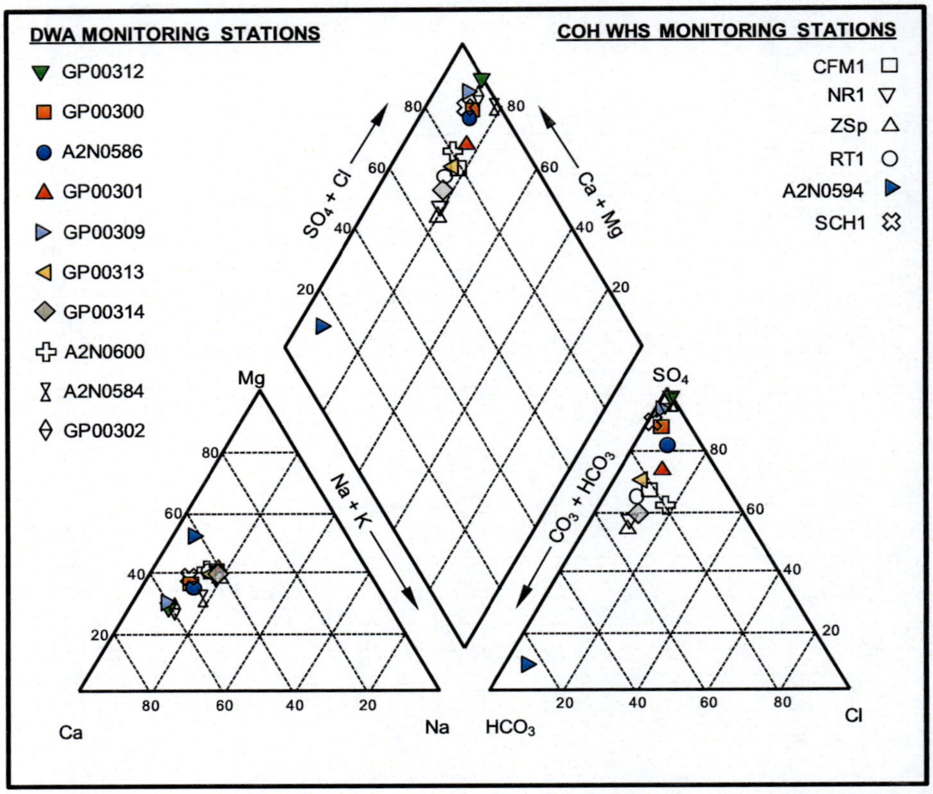


**Fig. 47** Piper diagram of recent groundwater chemistry defining the mine water impact on the karst groundwater of the Zwartkrans Basin

velocity, and density) are presented. The case studies represent three coal mines, five base metal mines, and five other mines (fluorspar, uranium, and salt). Their distribution ranges between North America (2 #), Europe (9 #), and the United Kingdom (2 #). The results indicate the association of various patterns with the following broad circumstances:

- 'sharp' changes in a profile are generally associated with the shaft interception of drives, tunnels, ore passes, etc. suggesting that these represent significant flow conduits (Erickson et al. 1982 cited in Wolkersdorfer 2008);
- 'smoother' profiles indicate a well-mixed flow regime or flow rates that are slow enough to allow chemical equilibrium;
- profiles exhibiting a gradually increasing (or decreasing) trend reflect diffusion-driven flow, whereas 'constant-value' profiles reflect convection-driven flow mechanisms; and
- in general, large-diameter shafts exhibit 'smaller' transition zones than 'slim' boreholes.

The manifestation of hydrochemical stratification in a flooded mine void is a particular aspect of mine water that attracts attention for the numerous implications it has for the management and treatment of AMD. These include the following:

- the presence of fresher and typically less acidic water at the top of the water column (Wolkersdorfer 2008 and references therein), suggesting a lesser impact on the receiving environment from uncontrolled and unmanaged decant;
- the association of less saline and circum-neutral pH water at shallower depth with younger recharge and shorter residence times than might be the case for 'deeper' mine water; and
- the abstraction of 'better' quality mine water from a shallow depth in a pump-and-treat management scenario to reduce the cost of treatment.

A fundamental question that arises from a consideration of stratification in a mine water body [Wolkersdorfer (2008) refers to mine water lakes], and especially under circumstances where hydrodynamic and hydrogeochemical factors inform the stratification, is whether or not a mine water lake exhibits a meromictic[24] character. Moreira et al. (2011) have shown that stratification models using electrical conductivity for density are not suited for simulating density created by

[24] See Glossary.

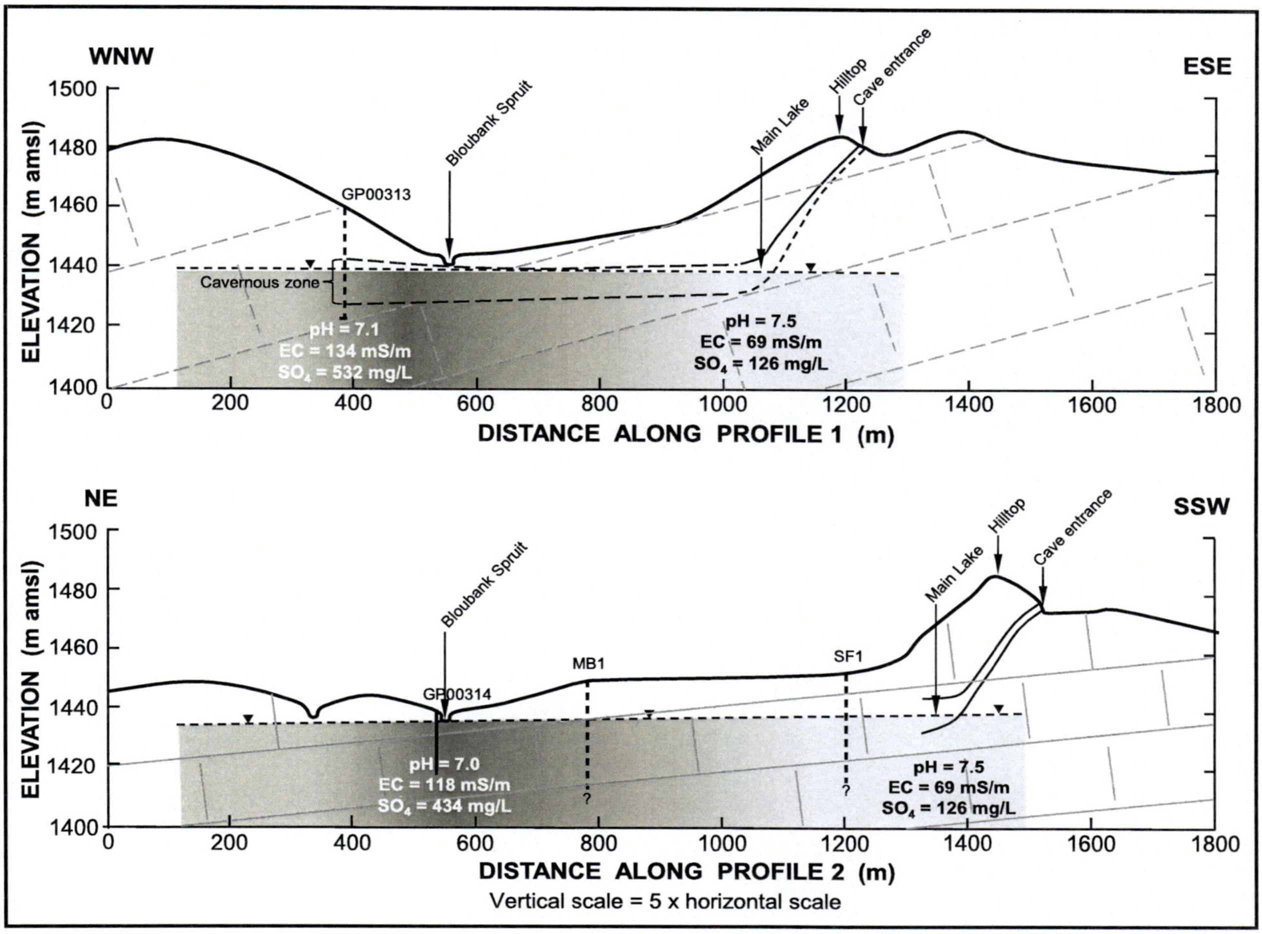


**Fig. 48** Schematic profiles through the Bloubank Spruit valley and Sterkfontein Cave illustrating the relationship between the water table and lake elevations and the stream channel; lateral gradational shading reflects relative intensity of mine water impact on karst groundwater as shown by recent pH, SEC, and $SO_4$ values; see Fig. 39 for location of the transects

**Table 9** Summary of mercury values for various water sources in the study area

| Detection limit (DL) | Mine water[a] | | Surface water | | Groundwater | | Σ*n* | Data source |
|---|---|---|---|---|---|---|---|---|
| | *n* | >DL | *n* | >DL | *n* | >DL | | |
| 0.0001 mg/L | 1 | 0 | 1 | 0 | 23 | 1 | 25 | Hobbs and Cobbing (2007) |
| Value(s) >DL | – | | – | | 0.0056 mg/L | | | |
| 0.002 mg/L | 6 | 6 | 2 | 0 | 10 | 1 | 18 | Hobbs et al. (2010) |
| Value(s) >DL | 0.398–3.61 mg/L | | – | | 0.008 mg/L | | | |
| 0.001 mg/L | 3 | 0 | 17 | 0 | 33 | 1 | 53 | This study |
| Value(s) >DL | – | | – | | 0.01 mg/L | | | |

[a]Comprises both raw and treated/neutralized mine water

lake-internal geochemical transformation of solutes. The application of the UNESCO equation for deriving density from salinity and temperature data (Fofonoff and Millard 1983) as demonstrated by Wolkersdorfer (2008), assumes non-reactive solutes and was shown by Moreira et al. (2011) to be inappropriate for modeling density-based stratification associated with reactive mineralized solutes, including iron and sulfate.

**Table 10** Range of SAR values for basins/subcompartments

| Basin/subcompartment | Range of SAR values | | | | Classification per mean[a] |
|---|---|---|---|---|---|
| | *n* | Minimum | Mean | Maximum | |
| 1a | 6 | 0.07 | 0.27 | 0.65 | C2–S1 |
| 1b | 7 | 0.78 | 1.10 | 1.78 | C3–S1 |
| 1c | 11 | 0.40 | 0.99 | 1.99 | C3–S1 |
| 2a | 9 | 0.04 | 0.16 | 0.27 | C2–S1 |
| 2b | 7 | 0.11 | 0.50 | 0.91 | C2–S1 |
| 3 | 2 | 0.03 | 0.07 | 0.12 | C1–S1 |
| 4 | 2 | 0.03 | 0.03 | 0.04 | C2–S1 |
| 5 | 1 | | 0.03 | | C2–S1 |
| 6 | 2 | 0.09 | 0.10 | 0.12 | C2–S1 |
| 7 | 1 | | 0.03 | | C2–S1 |

[a]Where applicable

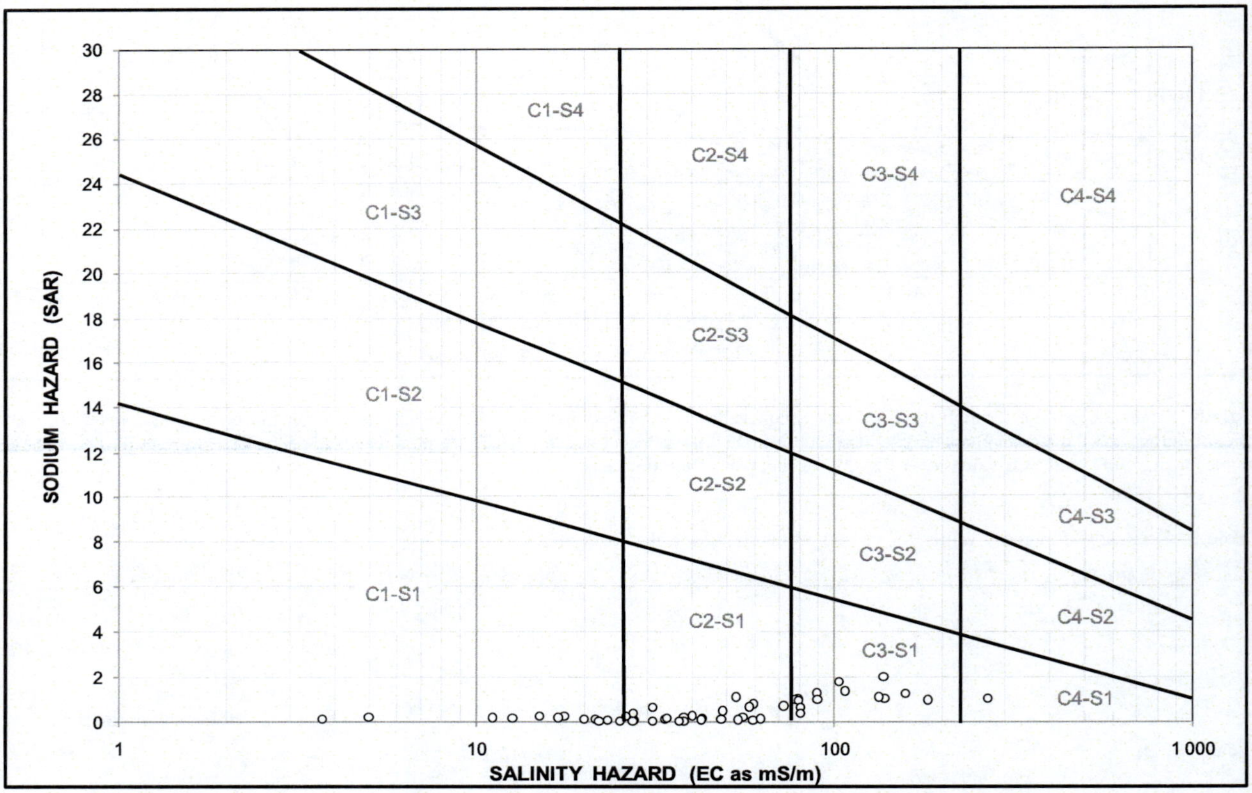


**Fig. 49** Wilcox diagram illustrating the classification of groundwater chemistry in the study area for irrigation purposes

While the preceding discussion focuses on mine water lakes as intersected by shafts, the following discussion draws primarily on data generated from vertical physicochemical profiling carried out in monitoring boreholes. These facilities demonstrate varying degrees of mine water contamination as manifested in natural aquifers, in particular the karst groundwater resource associated with the Malmani Subgroup dolomite that forms the Zwartkrans Basin in the downstream hydro-environment.

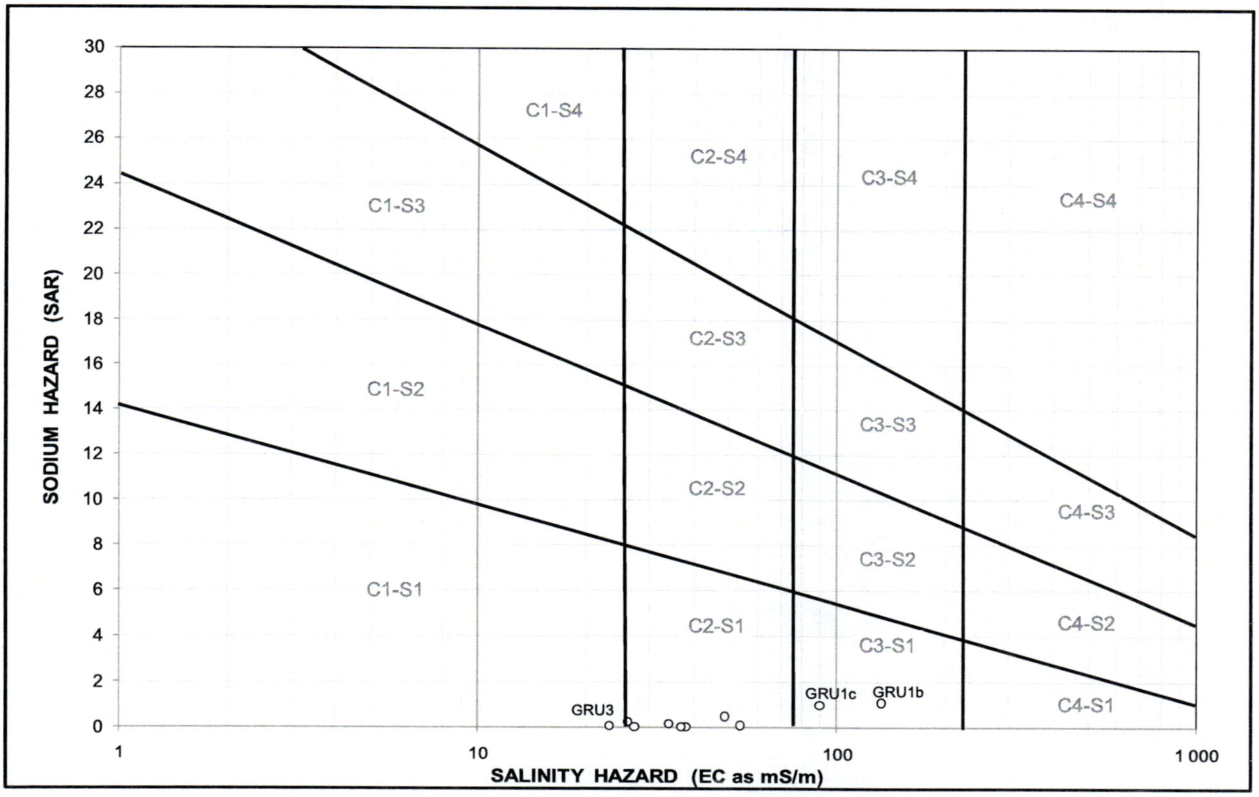


**Fig. 50** Wilcox diagram illustrating the classification of groundwater chemistry per GRU for irrigation purposes

## 10.2 Methodology and Approach

The presence of vertical hydrochemical variation in a mine water or groundwater column was explored on the basis of profiles of the field variables EC, pH, ORP (oxidation–reduction potential or redox), and temperature measured on a number of occasions. The initial profiling was carried out by the DWS in late-2011 (October/November/December) in 22 monitoring boreholes (Table 11) using a YSI 6-series multiparameter sonde (model 600XLMD-BRC-DEEP, serial # 06K002D). The sonde was set to record a measurement every 5 s and lowered on a mechanically-driven winch running at ~0.05 m/s, giving a measurement every ~0.25 m. The instrument has a maximum operating head of 90 m. This was reached in only one borehole, namely GP00301 (Sect. 10.3.17).

The profiled boreholes are distributed in the southern part of the study area where stratification associated with poorer quality water ingress into the karst aquifer is more probable. The pattern and trend of mine water discharge into the karst environment since the start of gauged monitoring is shown in Fig. 51. The graph distinguishes between raw mine water (RMW) and treated/neutralised mine water (TMW), the aggregate of which represents the total mine water discharge. The period during which the RMW contribution to total mine water discharge regularly exceeded 10 ML/d is clearly evident in Fig. 51. The DWS boreholes in the set (Table 11) had been sampled in early September 2011 following standard procedures that included purge pumping prior to sample capture only after stabilization of the field variables EC, pH, and temperature. Although these data provide a useful reference against which to evaluate the vertical profiling data as is discussed in the following sections, a different multiparameter sonde in the form of a CyberScan PC 10 pH/Conductivity meter was used (T. Moolman, personal communication). This limits a direct comparison of the respective results.

A second profiling run was carried out by the DWS in late-February 2013 with the same instrument used during the late-2011 profiling exercise. The 12 monitoring boreholes sampled on this occasion (Table 12) had been profiled following a routine sampling exercise carried out in mid-February 2013 during which the CyberScan PC10 multiparameter sonde was used for pH, conductivity, and temperature measurement.

Later profiling runs were carried out in June 2013 (Table 13), September and October 2013 (Table 14), and March 2014 (Table 15). The September and October 2013

**Table 11** Summary of salient hydrophysical and hydrochemical data for monitoring boreholes subjected to vertical hydrochemical profiling in late-2011

| Station# | General location and station owner | Date profiled | Total depth[a] (m bs) | Rest water level depth (m bc) | Date sampled | Field variable value[b] | | |
|---|---|---|---|---|---|---|---|---|
| | | | | | | SEC (mS/m) | pH | Temp (°C) |
| RG1 | Locus of decant; SG | 20/12/2011 | ~27 | 4.72 | Not sampled | Not available | | |
| RG2 | | | ~31 | 8.20 | | | | |
| RG3 | | | ~19 | 6.82 | | | | |
| MGP1 | Mine area; MG/MSA | | ~48 | 24.11 | | | | |
| MGP2 | | | ~59 | 34.66 | | | | |
| MGP3 | | | ~56 | 3.82 | | | | |
| GP00300 | Zwartkrans Basin; DWS | 19/10/2011 | ~36 | 21.70 | 05/09/2011 | 208 | 7.2 | 20.1 |
| GP00301 | | 21/10/2011 | ~148 | 58.77 | 05/09/2011 | 104 | 7.6 | 20.0 |
| GP00302 | | 01/11/2011 | ~33 | 21.98 | 06/09/2011 | 129 | 6.1 | 18.1 |
| GP00304 | | 04/11/2011 | ~53 | 44.74 | 06/09/2011 | 34 | 6.8 | 20.4 |
| GP00305 | KGR; DWS | 02/11/2011 | ~41 | 3.49 | 07/09/2011 | 67 | 5.9 | 18.7 |
| GP00306 | | 02/11/2011 | ~13 | 3.09 | 07/09/2011 | 276 | 6.0 | 17.5 |
| GP00307 | | 02/11/2011 | ~11 | 2.03 | 07/09/2011 | 257 | 6.5 | 15.8 |
| GP00308 | | 02/11/2011 | ~81 | 5.59 | 07/09/2011 | 38 | 7.3 | 19.6 |
| GP00309 | Zwartkrans Basin; DWS | 11/11/2011 | ~35 | 16.68 | 05/09/2011 | 261[c] | 6.6[c] | 18.8[c] |
| GP00311 | | 21/10/2011 | ~119 | 68.38 | 05/09/2011 | 36 | 7.6 | 20.2 |
| GP00312 | | 21/10/2011 | ~23 | 7.14 | 05/09/2011 | 263 | 6.8 | 19.5 |
| GP00313 | | 11/11/2011 | ~36 | 20.34 | 07/09/2011 | 98 | 7.3 | 20.1 |
| GP00314 | | 01/11/2011 | ~13 | 3.46 | 06/09/2011 | 49 | 7.1 | 20.1 |
| A2N0583 | | 04/11/2011 | ~95 | 44.70 | 06/09/2011 | 32 | 7.0 | 20.6 |
| A2N0584 | | 01/11/2011 | ~55 | 21.87 | 06/09/2011 | 111 | 9.3 | 18.1 |
| A2N0586 | | 19/10/2011 | ~55 | 21.62 | 05/09/2011 | 172 | 7.3 | 20.5 |

[a]Depth to which sonde was lowered on date profiled before encountering a terminating obstruction
[b]After stabilization of variable value during purge pumping on date sampled
[c]Measured with a YSI 6-series model 600XLMD-BRC-DEEP multiparameter sonde

runs explored the influence of sampling activity on the post-sampling profile in the five DWS monitoring boreholes A2N0586, GP00300, GP00305, GP00309, and GP00313. The pre-sampling and post-sampling profiles carried out on 11/09/2013 and 22/10/2013, respectively, bracket the 01–03/10/2013 routine sampling exercise period. These results are discussed in the opening paragraphs of each of the relevant sections.

## 10.3 Results

The discussion of the individual profiling results follows the 'south-to-north' sequence described by the geographic location of the boreholes (Fig. 51).

The interrogation and discussion of the ORP measurements that form part of the profile data are subject to the caveat expressed in Text Box 1. The data are considered sufficiently valuable to present for their general interest and completeness rather than for their definitive substance. Such a caution is also expressed by Montgomery et al. (2002) who state "… measurements of Eh in natural systems must be cautiously evaluated and not used strictly for calculations of chemical equilibria."

Wolkersdorfer (2008) recognizes that the small diameter-to-length ratio of monitoring boreholes (compared to that of mine shafts) limits the influence of free convective flow on the water profile in such especially deep (»100 m) installations. The implication is that diffusion informs the hydrochemical profile in the standing water column in boreholes. It should be recognized, however, that the monitoring boreholes in this instance are mostly comparatively shallow (<50 m), and intersect natural aquifers rather than a mine void.

Finally, the evaluation of the profile data must also recognize the possible 'disturbing' influence on the water column by prior routine sampling. This is further complicated by the different 'rest' periods between sampling and profiling, viz. 6–8 weeks in late-2011 (Table 11), and ~2 weeks in February 2013 (Table 12). As already mentioned, the September and October 2013 runs explore this influence.

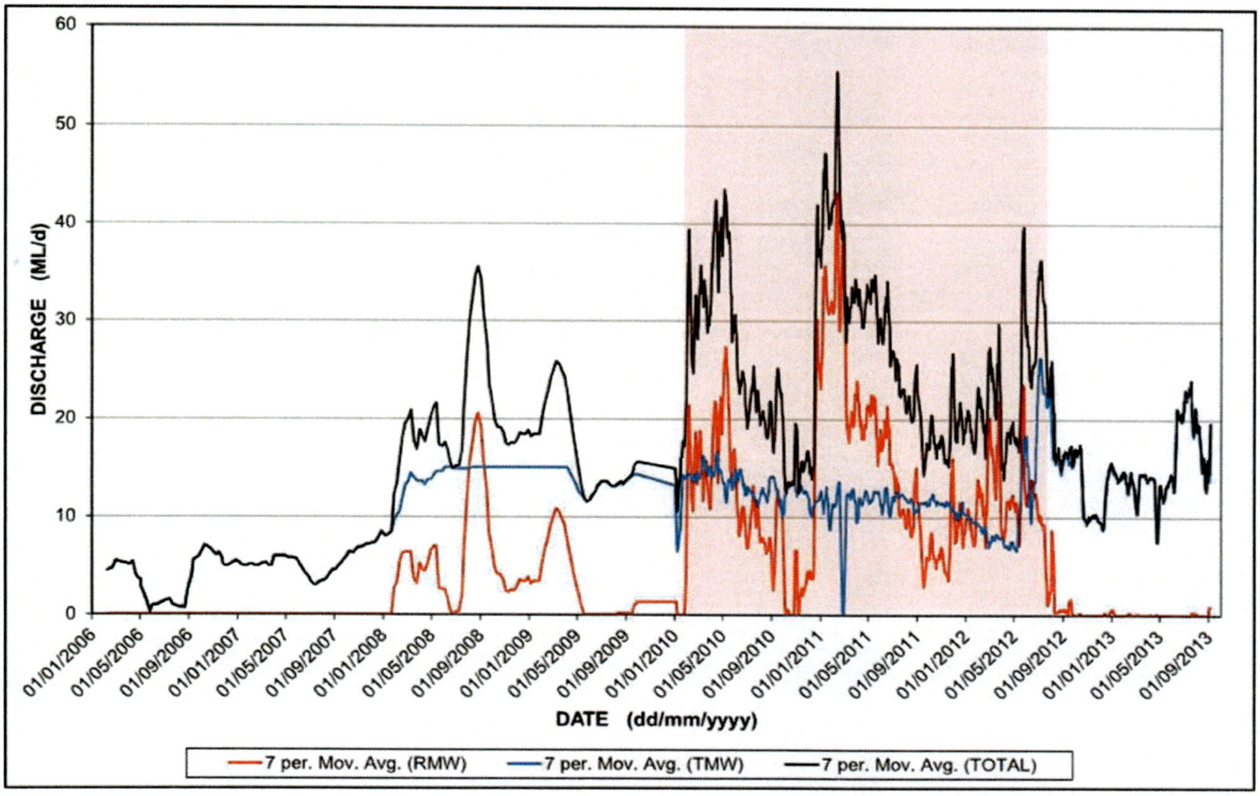


**Fig. 51** Pattern and trend of mine water discharge into the karst environment (RMW = raw mine wate; TMW = treated mine water)

**Table 12** Summary of salient hydrophysical and hydrochemical data for monitoring boreholes subjected to vertical hydrochemical profiling in February 2013

| Station # | General location and station owner | Date profiled | Total depth[a] (m bs) | Rest water level depth (m bc) | Date sampled | Field variable value[b] | | |
|---|---|---|---|---|---|---|---|---|
| | | | | | | SEC (mS/m) | pH | Temp (°C) |
| CPS borehole | Mine area; SG | 26/02/2013 | ~60 | 49.56 | Not sampled | Not available | | |
| RG1 | | 26/02/2013 | ~27 | 4.52 | | | | |
| MGP1 | Mine area; MG/MSA | 28/02/2013 | ~48 | 25.04 | | | | |
| MGP3 | | | ~56 | 4.10 | | | | |
| MGP4 | | | ~80 | 43.35 | | | | |
| GP00300 | Zwartkrans Basin; DWS | 28/02/2013 | ~36 | 21.41 | 21/02/2013 | 249 | 6.9 | 19.6 |
| GP00302 | | 26/02/2013 | ~33 | 22.00 | 19/02/2013 | 162 | 6.0 | 18.4 |
| GP00305 | KGR; DWS | 26/02/2013 | ~41 | 3.38 | 20/02/2013 | 76 | 5.6 | 19.3 |
| GP00309 | Zwartkrans Basin; DWS | 28/02/2013 | ~35 | 16.91 | 20/02/2013 | 304 | 7.0 | 17.3 |
| GP00313 | | 28/02/2013 | ~36 | 20.45 | 20/02/2013 | 125 | 6.9 | 20.2 |
| A2N0584 | | 26/02/2013 | ~55 | 21.94 | 19/02/2013 | 116 | 8.8 | 19.1 |
| A2N0586 | | 28/02/2013 | ~55 | 21.84 | 21/02/2013 | 192 | 6.9 | 19.9 |

[a]Depth to which sonde was lowered on date profiled before encountering a terminating obstruction
[b]After stabilization of variable value during purge pumping on date sampledable

**Table 13** Summary of salient hydrophysical and hydrochemical data for monitoring boreholes subjected to vertical hydrochemical profiling in June 2013

| Station # | General location and station owner | Date profiled | Total depth[a] (m bs) | Rest water level depth (m bc) | Date sampled | Field variable value[b] | | |
|---|---|---|---|---|---|---|---|---|
| | | | | | | SEC (mS/m) | pH | Temp (°C) |
| GP00300 | Zwartkrans Basin; DWS | 06/06/2013 | ~36 | 21.74 | 27/06/2013 | 253 | 6.8 | 19.0 |
| GP00302 | | | ~33 | 22.20 | 25/06/2013 | 169 | 5.8 | 18.1 |
| GP00309 | | | ~36 | 16.86 | 25/06/2013 | 281 | 6.1 | 17.5 |
| GP00313 | | | ~37 | 20.60 | 26/06/2013 | 127 | 6.8 | 19.7 |
| A2N0584 | | | ~56 | 22.15 | 25/06/2013 | 118 | 8.4 | 18.4 |
| A2N0586 | | | ~77 | 22.19 | 27/06/2013 | 195 | 6.9 | 19.3 |

[a]Depth to which sonde was lowered on date profiled before encountering a terminating obstruction
[b]After stabilization of variable value during purge pumping on date sampled

**Table 14** Summary of salient hydrophysical and hydrochemical data for monitoring boreholes subjected to vertical hydrochemical profiling in September and October 2013

| Station # | General location and station owner | Date profiled | Total depth[a] (m bs) | Rest water level depth (m bc) | Date sampled | Field variable value[b] | | |
|---|---|---|---|---|---|---|---|---|
| | | | | | | SEC (mS/m) | pH | Temp (°C) |
| GP00305 | KGR; DWS | 11/09/2013 | ~42 | 3.29 | 02/10/2013 | GP00305 | | |
| GP00300 | Zwartkrans Basin; DWS | | ~36 | 21.73 | | 80 | 7.2 | 18.3 |
| GP00309 | | | ~36 | 17.08 | 03/10/2013 | GP00300 | | |
| GP00313 | | | ~37 | 20.68 | | 249 | 6.8 | 19.2 |
| A2N0586 | | | ~90 | 22.16 | 02/10/2013 | GP00309 | | |
| GP00305 | KGR; DWS | 22/10/2013 | ~41 | 3.38 | | 261 | 6.7 | 20.4 |
| GP00300 | Zwartkrans Basin; DWS | | ~33 | 21.72 | 01/10/2013 | GP00313 | | |
| GP00309 | | | ~36 | 17.46 | | 131 | 6.9 | 19.6 |
| GP00313 | | | ~37 | 20.78 | 03/10/2013 | A2N0586 | | |
| A2N0586 | | | ~121 | 22.16 | | 210 | 7.1 | 19.5 |
| GP00302 | | | ~33 | 22.13 | 01/10/2013 | 249 | 6.8 | 19.2 |
| A2N0584 | | | ~55 | 22.09 | | 120 | 8.4 | 18.2 |

[a]Depth to which sonde was lowered on date profiled before encountering a terminating obstruction
[b]After stabilization of variable value during purge pumping on date sampled

**Table 15** Summary of salient hydrophysical and hydrochemical data for monitoring boreholes subjected to vertical hydrochemical profiling in March 2014

| Station # | General location and station owner | Date profiled | Total depth[a] (m bs) | Rest water level depth (m bc) | Date sampled | Field variable value[b] | | |
|---|---|---|---|---|---|---|---|---|
| | | | | | | SEC (mS/m) | pH | Temp (°C) |
| GP00305 | KGR; DWS | 06/03/2014 | ~41 | 2.23 | 11/03/2014 | 76 | 6.1 | 18.6 |
| GP00300 | Zwartkrans Basin; DWS | | ~35 | 20.90 | 13/03/2014 | 249 | 7.2 | 19.6 |
| GP00302 | | | ~33 | 20.70 | 11/03/2014 | 194 | 6.5 | 18.1 |
| GP00309 | | | ~36 | 15.28 | 13/03/2014 | 301 | 7.8 | 19.5 |
| GP00313 | | | ~36 | 20.23 | 13/03/2014 | 132 | 7.2 | 20.1 |
| A2N0584 | | | ~55 | 21.24 | 11/03/2014 | 125 | 8.8 | 18.1 |
| A2N0586 | | | ~121 | 21.34 | 13/03/2014 | 183 | 7.3 | 20.1 |

[a]Depth to which sonde was lowered on date profiled before encountering a terminating obstruction
[b]After stabilization of variable value during purge pumping on date sampled

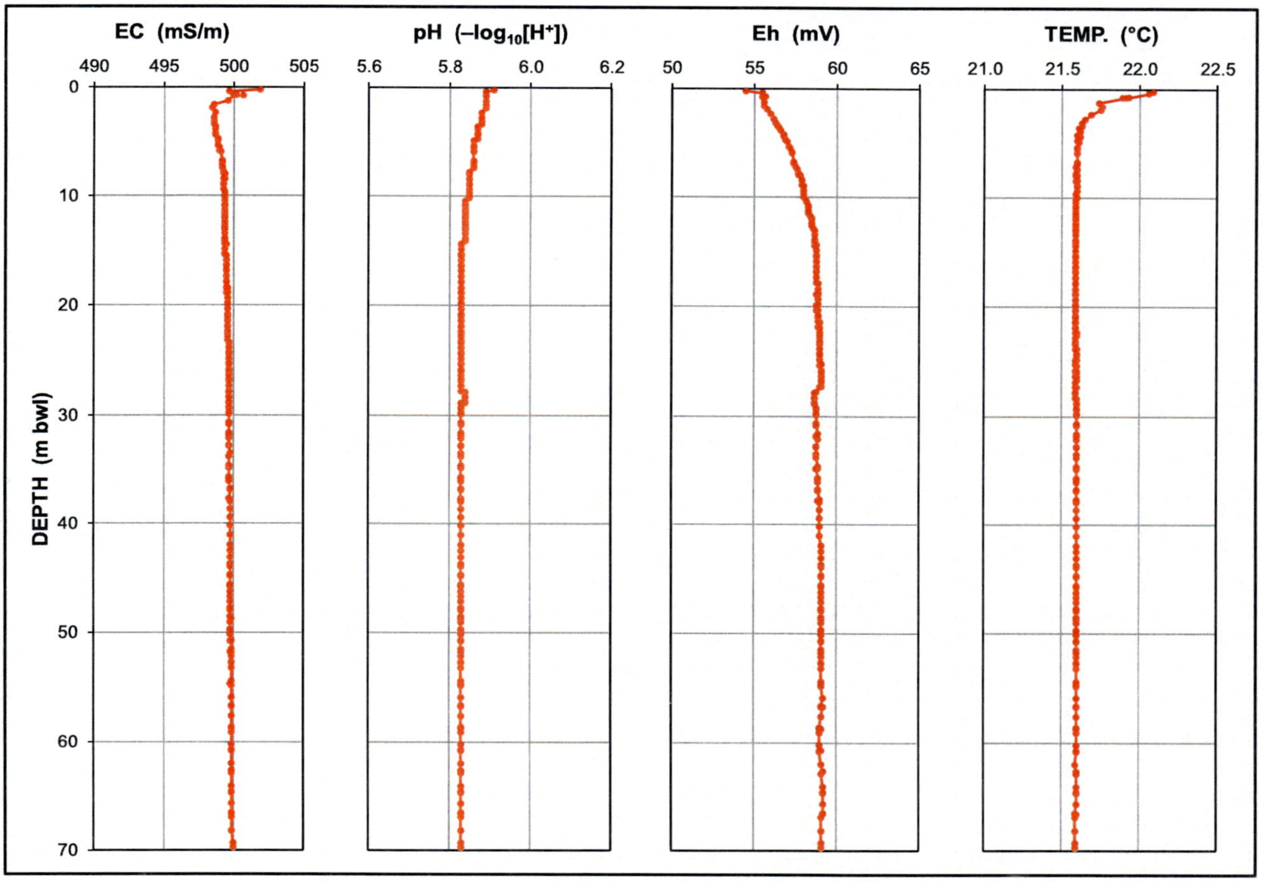


**Fig. 52** Vertical profiles of field water chemistry variables in #18 Winze in October 2010

### 10.3.1 #18 Winze

Together with the BRI and #17 Winze, #18 Winze is one of three actively decanting shaft-like mine structures in the Western Basin (Plate 1). It is the most readily accessible of these structures for vertical profiling purposes. The results provide an example of the patterns associated with the profiled variables in this structure. It would be unreasonable, however, to view the profiling results as typical of all such structures in the Western Basin. It is also indefensible to extrapolate the profiling results to mine shafts in which the mine water rest level occurs tens of meters below surface, and which will never manifest active decant. The depth of ~70 m reached by the probe indicates that this structure extends at least this vertical distance into the subsurface. It is also reasonable to expect that the structure is lined with brickwork to at least this depth. The ~2 × 4 m (~8 m$^2$) rectangular shape of the structure decants at a rate of 10–25 ML/d (Plate 1), which equates to an upward flow velocity of ~0.01–0.04 m/s. These circumstances define the structure-specific hydrodynamic environment in which the profiling was carried out.

The profiles presented in Fig. 52 reflect a remarkably constant pattern over the entire water column profiled. The SEC value of 500 mS/m represents an unmistaken mine water signature which is accompanied by a pH value of little more than 5.8. The pH value is surprising under circumstances where raw mine water is typically characterized as highly acidic, i.e., pH <5.0. It is reasonable to presume that the pH value of 5.8 characterizes that of the raw mine water rising from depth in the mine void. Its rapid drop to a pH of <4.0 and even <3.0 within tens of meters of the decant position exemplifies the rapidity of the hydrolysis reaction.

### 10.3.2 CPS Borehole

The CPS (Central Power Station) borehole is located ~250 m south of the overflow of the CPS pit and 1400 m south of #8 Shaft. The borehole is an old mine rescue facility, and would therefore have intersected underground mine workings at depth (B. van der Walt, personal communication). The interest in this borehole is its proximity to the CPS pit that is used as a settling facility for the

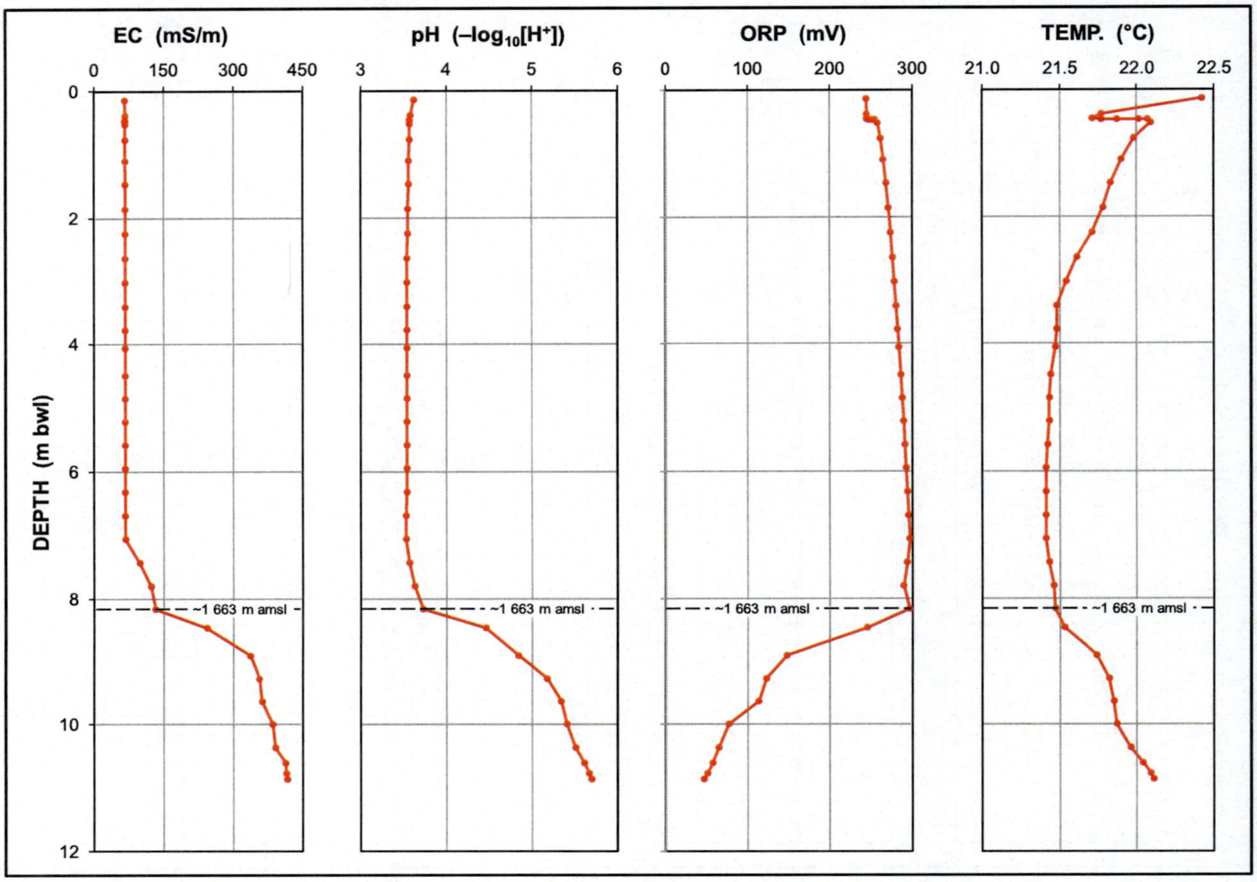


**Fig. 53** Vertical profiles of field water chemistry variables in the CPS borehole in February 2013

sludge produced by the mine water treatment plant. A video log of the borehole revealed its construction to comprise plain steel casing to a depth of ~57 m bs (~1663 m amsl) (N. de Meillon, personal communication). The rest water level depth of 49.1 m bs equates to an elevation of ~1671 m amsl that correlates with the similar contemporary mine water elevation observed in the nearby #8 Shaft (Fig. 51).

The profiles presented in Fig. 53 indicate the borehole to be accessible to a depth of only ~60 m bs (~11 m bwl). Each profile also indicates a shift in trend at a depth of ~8 m bwl (~1663 m amsl). This shift is characterized by a dramatic increase in SEC (by ~350 mS/m) and pH (by ~2 pH units), and a substantial drop in ORP value (by ~250 mV). If the higher SEC and pH and lower ORP at depth is attributed to a mine water influence derived from the flooded mine workings, which observation finds support in the #18 Winze profile results for SEC, pH, and ORP (Fig. 52), then the overlying less saline, more acidic and more oxic conditions characterize the stagnant water column confined to the cased section of the borehole.

### 10.3.3 Borehole RG1

The December 2011 vertical profiles (Fig. 54) reflect a distinct change at a depth of ~17 m bwl (~22 m bs; ~1646 m amsl). This depth does not coincide with the top of the piezometer perforated interval, which extends from a depth of 22 m bwl (27 m bs; ~1641 m amsl) to the original end-of-hole at 31 m bs (36 m bs; ~1632 m amsl). The profile depth data indicate that the borehole is currently only accessible to a depth of ~23 m bwl (~28 m bs; ~1640 m amsl), having lost 8 m possibly due to caving or siltation. The SEC value increases from a constant ~45 to ~330 mS/m over a depth interval of only ~6 m. The temperature profile increases by ~1.0 °C in this interval. In the lower portion of this interval (21–23 m bwl; <1642 m amsl), the pH and ORP profiles reflect a sharp change characterized by a decrease in pH values by 3.5 pH units and an increase in ORP values by ~430 mV.

The February 2013 profiles for SEC, pH and ORP reveal a significant change from the December 2011 profiles. This change marks a migration of the fresh/saline water interface and the neutral/acidic water interface up the water column by

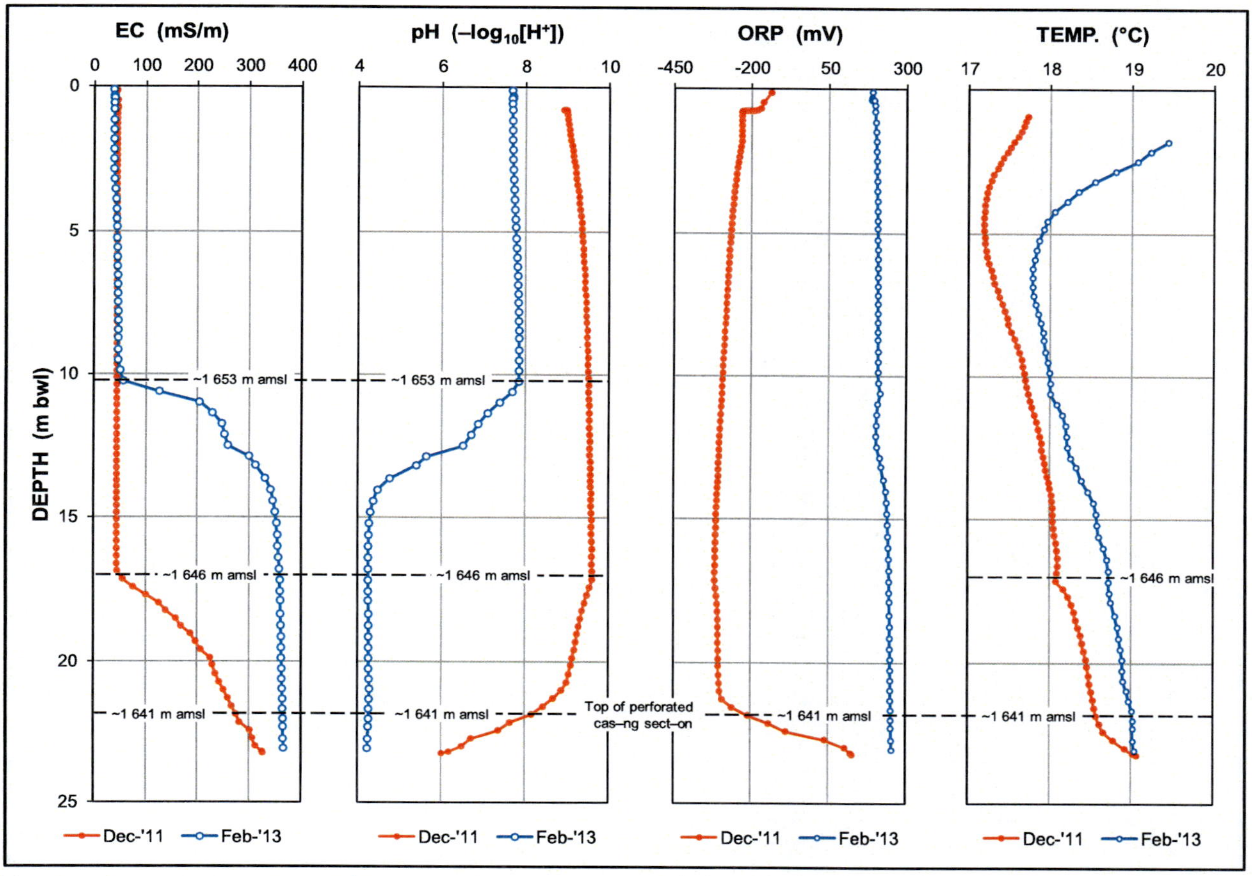


**Fig. 54** Vertical profiles of field water chemistry variables in borehole RG1

~8 m. The change in ORP marks a shift toward a uniform more oxic condition across the entire water column. The temperature profile for the most part reflects a uniform positive shift of ~0.5 °C, suggestive of exothermic reactions driving a shift to oxic (aerobic) conditions generating a more acidic groundwater. The common mid-summer period of profiling negates a seasonal cause for the temperature shift.

The RG1 profiles unequivocally identify a zone (or at least the upper portion thereof) of impacted preferential groundwater flow at this location. The upward 'growth' in SEC and pH of this zone together with the ORP profile transition from strongly reducing (anoxic) to moderately oxidizing (oxic) conditions in a period of 14 months, seriously challenge an understanding of the geochemical (biogeochemical?) processes that occur in the subsurface of a karst environment impacted by acid mine water.

The constant and low SEC in the upper 10 m of the water column, compared to the steeply increasing SEC below this depth, highlights the potential misrepresentation of the groundwater salinity (and probably also water chemistry) at this location based on a grab sample taken from <10 m bwl. This is exemplified by the results of routine chemical analyses reported by SG for the groundwater obtained from this borehole (Sect. 2.3.1).

### 10.3.4 Borehole RG2

The 'unremarkable' profiles (Fig. 55) are readily explained by the very weak yield (<0.2 L/s) as determined in 2007 when this borehole was subjected to purge pumping for water sample collection purposes. The borehole is therefore assessed as being in poor hydraulic connection with the contaminated karst aquifer of the dolomitic outlier. As a consequence, any water quality data obtained from this borehole must be viewed with caution because of its 'isolated and non-representative' nature. Nevertheless, the comparatively low pH of 4.8 is not readily explained under circumstances where the SEC of ~37 mS/m does not reflect a mine water influence at this location.

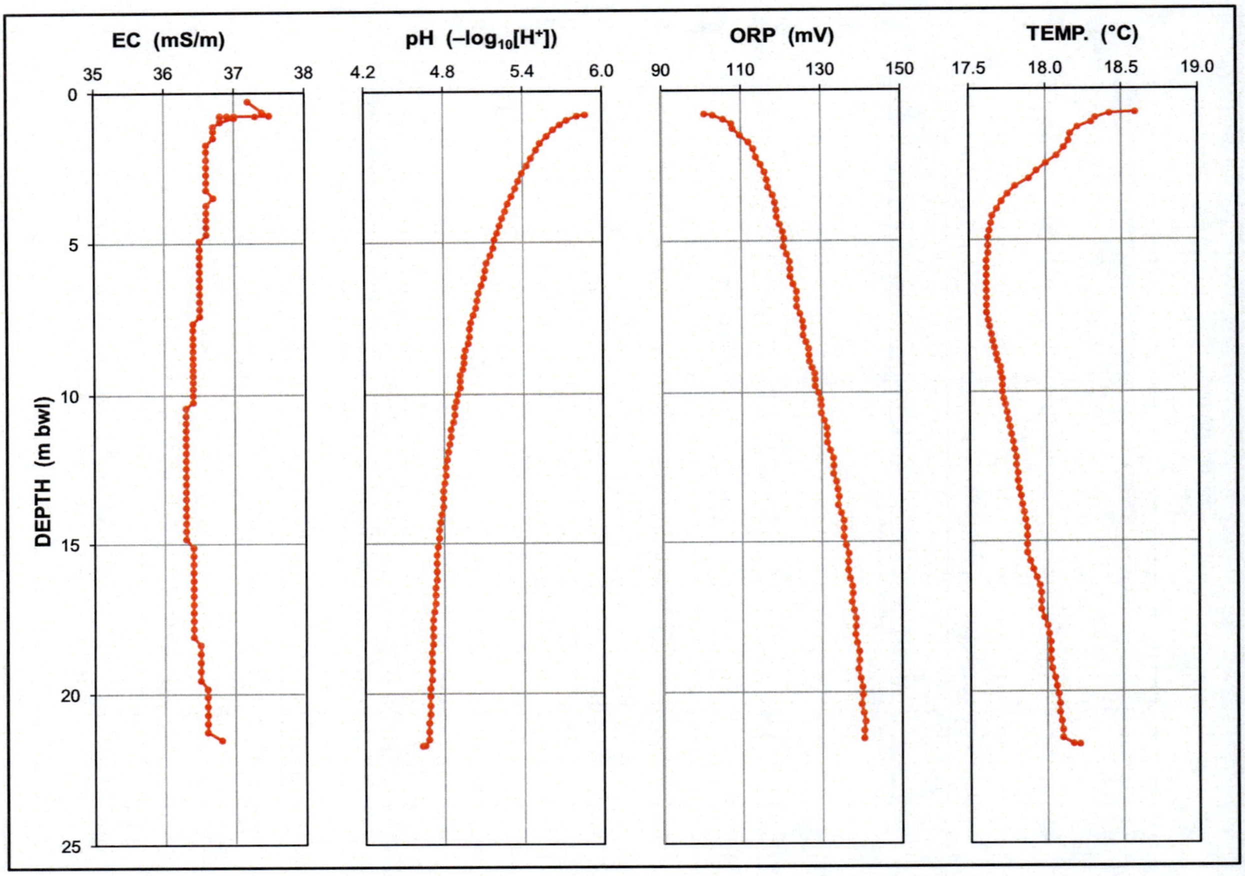


**Fig. 55** Vertical profiles of field water chemistry variables in borehole RG2 in December 2011

### 10.3.5 Borehole RG3

The profiles obtained for this borehole are as 'unremarkable' as for borehole RG2, except for the very low (~9 mS/m) SEC values (Fig. 56). The latter suggests that there is no influence of mine water at this location ~150 m west of the Tweelopie Spruit, despite the groundwater level elevation of ~1657 m amsl in the borehole equaling the streambed elevation. Together with the slightly acidic pH values (~6.5), these results rather indicate an association with natural quartzitic groundwater most likely related to the West Rand Subgroup strata. The latter form the enclosing older strata along the western margin of the dolomite outlier. These strata are earmarked for future extraction, by the opencast mining method, of the gold and uranium reserves associated with the Government Reef in this area.

### 10.3.6 Borehole MGP1

The location of this borehole to the north of the West Wits Pit places it in the subsurface flow path of water that might drain from this opencast mining feature along the NE–SW striking (Fig. 6 in Chapter "Description of the Physical Environment") and steeply SSE-dipping (Fig. 7 in Chapter "Description of the Physical Environment") Kimberley Reefs exploited by the West Wits Pit workings. The depth of ~49 m (Table 11) places the base of the borehole at an elevation of ~1651 m amsl, i.e., below the Tweelopie Spruit valley bottom of ~1657 m amsl to the west. With a groundwater level elevation at ~1676 m amsl, the profiles (Fig. 57) indicate changes in EC, pH, and ORP below an elevation of ~1666 m amsl. These changes are not evident in the temperature profile.

The profiles indicate a shift in EC and pH values between December 2011 and February 2013. The EC and pH shifts are again to more saline and more acidic conditions, respectively. Further, the step in EC values also reflect a vertical upward shift as was observed in boreholes RG1 (Fig. 54) and GP00305 (Fig. 62). The ORP shift is toward a more uniform vertical distribution of redox values that occupy a narrower range (240–280 mV) than that of the earlier December 2011 profile (220–320 mV). The temperature profiles are virtually identical over 80% of the water column.

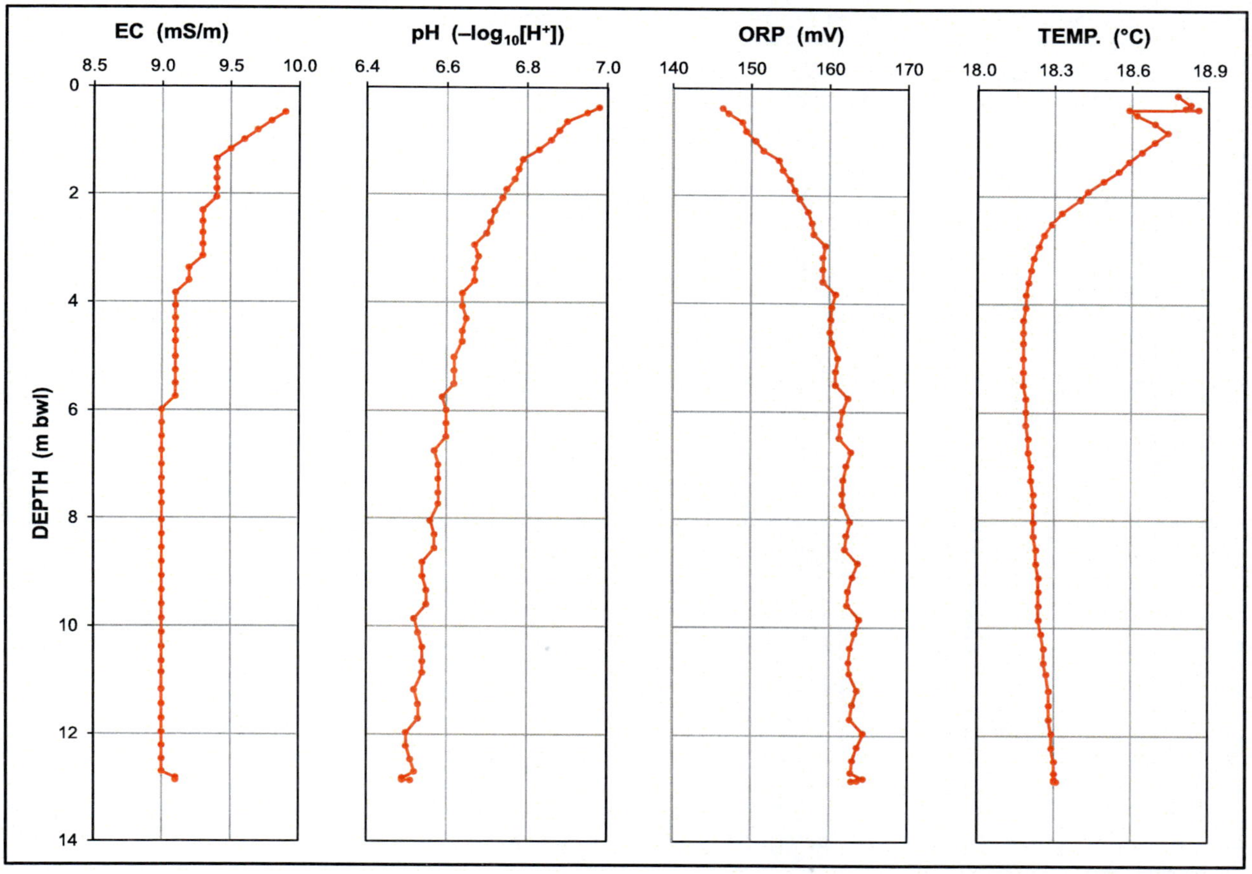


**Fig. 56** Vertical profiles of field water chemistry variables in borehole RG3 in December 2011

The elevated SEC and low pH values associated with the deepest horizon suggest the influence of an acid mine water presence. The relatively constant SEC values of <50 mS/m in the upper 10 m of the profile, compared to the >100 mS/m in the lowest horizon, again highlights the potential misrepresentation of the groundwater salinity and chemistry at this location based on a grab sample taken from a shallow depth (<10 m) below the water level. It is shown in later sections (e.g., Sect. 10.3.16) that even a sample obtained after purge pumping might not provide a representative value for the variables EC, pH, ORP, and temperature (among others). The use of non-representative hydrochemical data presents another (and very basic) challenge to an understanding of the hydrogeochemical processes that occur in the subsurface.

### 10.3.7 Borehole MGP2

Located adjacent to the north-western rim of the West Wits Pit, this borehole also targets any possible mine water migration from the opencast facility. The profiling results (Fig. 58) show a hydrochemical profile characterized by an acidic (pH ~ 4.5) fresh groundwater (SEC <15 mS/m) in a strongly oxidizing environment (ORP >400 mV). Except for the lower SEC, this is similar to that of borehole MGP1 (Fig. 57). The most plausible explanation for the low SEC and pH values is the possible association of this groundwater with the quartzitic strata of the Central Rand Group in the Witwatersrand Supergroup at this location. It should also be noted that the borehole is located on the north-western flank of the West Wits Pit workings, which places it in an unfavorable position relative to the hydrogeologic regime of the West Wits Pit as described for borehole MGP1 (Sect. 10.3.6). Together with the relatively shallow depth of 59 m (Table 11), these circumstances militate against the manifestation of an acid mine water presence at this location.

### 10.3.8 Borehole MGP3

The location of this borehole in the dolomite outlier provides a second opportunity (with borehole RG1) to evaluate the dynamic of the hydrogeochemical environment impacted by mine water. The profiles of this borehole (Fig. 59) indicate that SEC reaches and then maintains a maximum value of

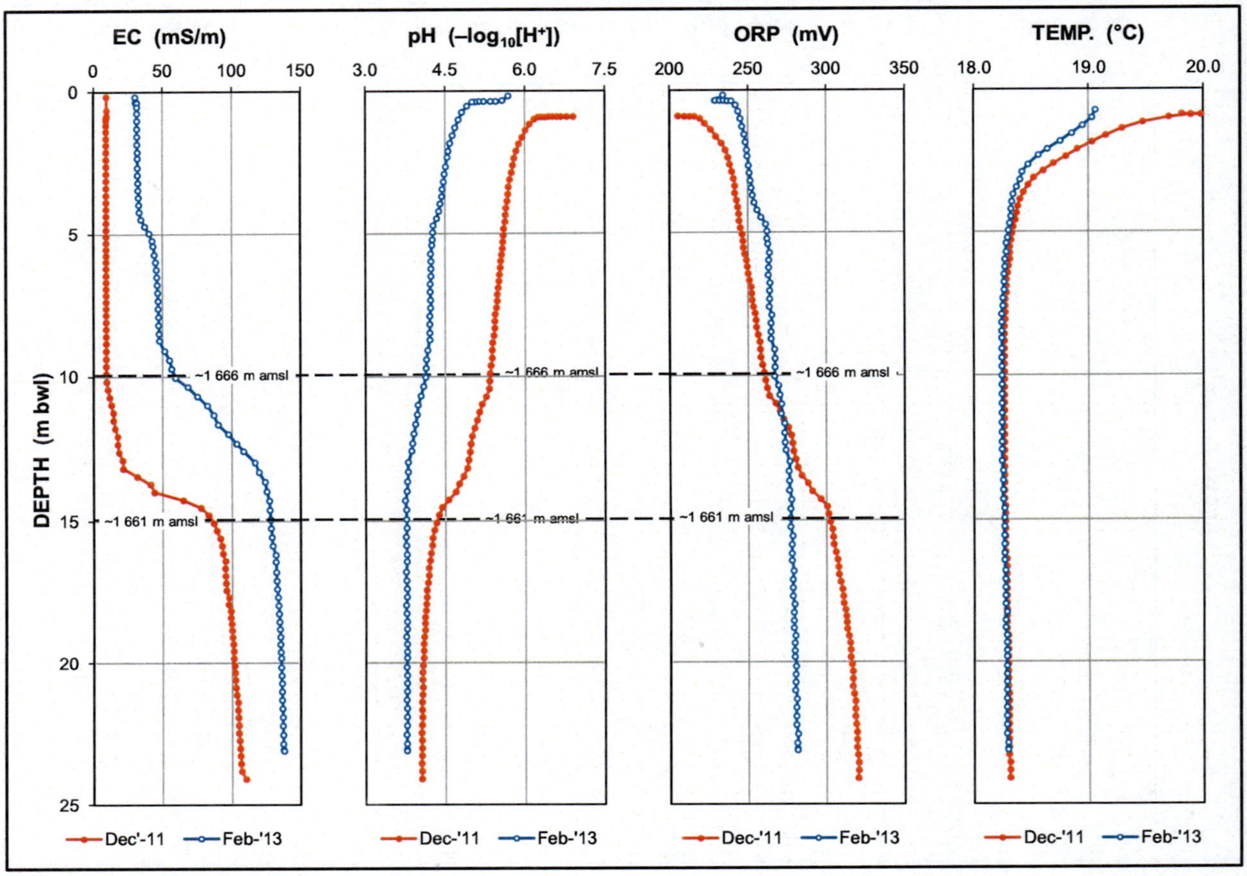


**Fig. 57** Vertical profiles of field water chemistry variables in borehole MGP1

270–300 mS/m at a depth of ~20 m bwl (~24 m bs, ~1632 m amsl) which corresponds with the depth of the first water strike (Mothoa and Lombaard 2008).

The pH profiles on the two profiling occasions are similar, the February 2013 profile reflecting a shift to a slightly more acidic condition below ~1639 m amsl. This shift is accompanied by a prominent ORP shift to more oxic conditions. An anomaly in the February 2013 ORP profile at an elevation of ~1639 m amsl is identified, and a plausible adjustment is applied. The temperature profiles are practically identical over >80% of the water column.

The measure of congruence between the MGP3 and RG1 borehole profiles is explored in Fig. 60. The data sets have been normalized to absolute elevations on the vertical axis. The horizontal axes represent common scales. The groundwater rest level elevations of ~1662 and ~1651 m amsl in boreholes RG1 and MGP3, respectively, are evident in the profiles. The water table elevation in borehole MGP3 approximates the Tweelopie Spruit streambed elevation at this location. The difference of ~11 m between the water table elevations (the tops of each profile) in the two boreholes reflects the hydraulic gradient of 0.022 that exists over the ~500 m distance between the two boreholes.

The congruence shown by the MGP3 profiles is discussed earlier in this section. Similarly, the geometry and interpretation of the RG1 profiles have been discussed in section. The purpose of Fig. 60 is to establish whether any congruence exists between boreholes RG1 and MGP3 given their locations within the dolomite outlier that hosts the locus of decant.

The assessment is somewhat constrained by the comparatively shallow depth (~25 m) of borehole RG1. Nevertheless, Fig. 60 suggests that the shallower profiles of RG1 represent an exaggerated form of the MGP3 profiles. For example, the step in SEC values from ~40 to ~350 mS/m at an elevation of ~1640 to ~1650 m amsl in RG1 is an exaggeration of the step from ~200 to ~300 mS/m at an elevation of ~1635 m amsl in MGP3. Similarly, the steps in pH and ORP are of greater magnitude in RG1 than in MGP3. Finally, the increase in temperature with depth is greater in RG1 than it is in MGP3. The similar pattern and trend shown by the profiled variables is therefore evident.

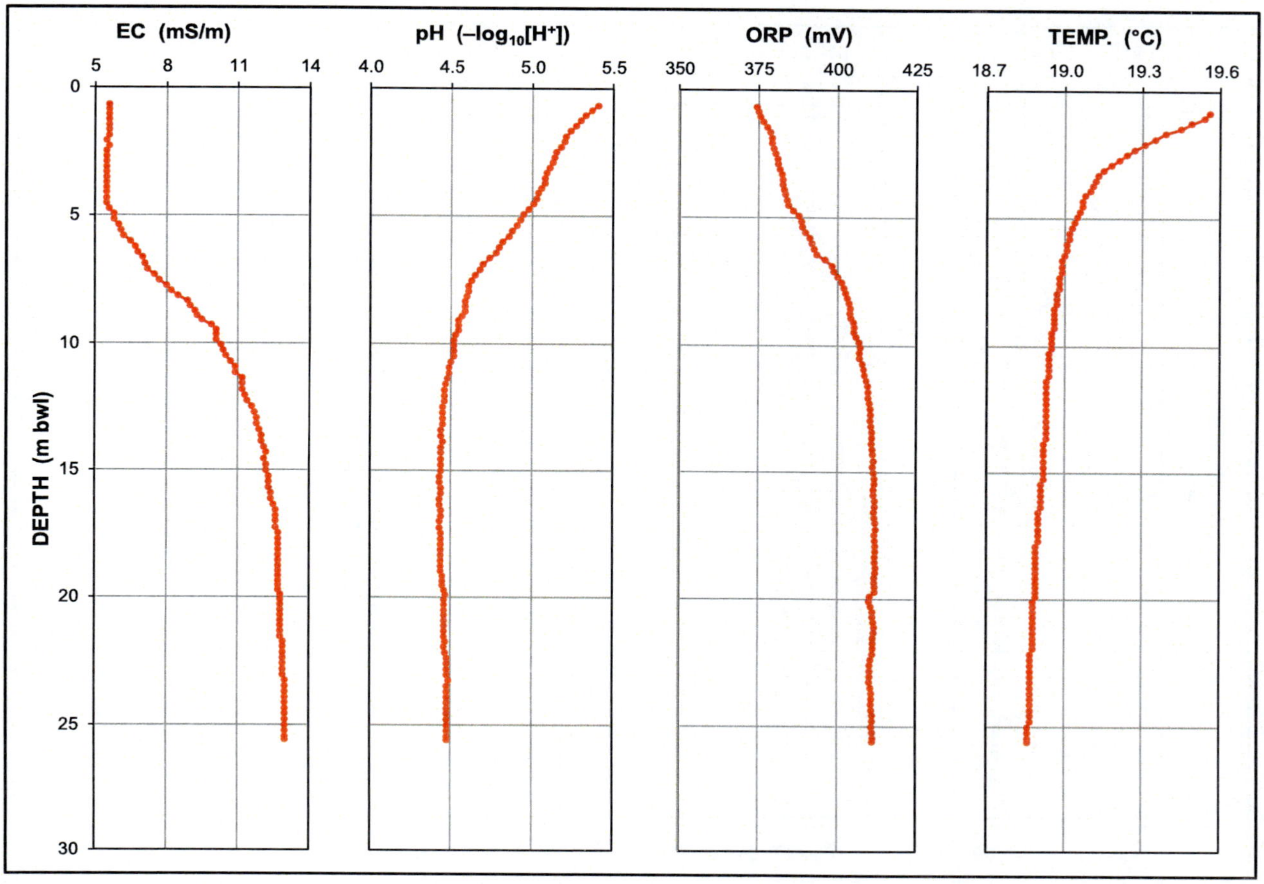


**Fig. 58** Vertical profiles of field water chemistry variables in borehole MGP2 in December 2011

The profiles indicate that some measure of isotropy and homogeneity exists in the otherwise heterogeneous and anisotropic karst environment of the dolomitic outlier. The influence of factors such as borehole construction, e.g., depth, type, placement of casing, etc. on the profiles has not been investigated.

### 10.3.9 Borehole MGP4

This borehole is located on the south-eastern flank of the West Wits Pit and was established to meet the DWS directive that informs the disposal of high-density sludge generated by the refurbished and upgraded SG mine water treatment plant into this excavation. As in the case of borehole MGP2 (Sect. 10.3.7), its location places it in an unfavorable position relative to the hydrogeologic regime of the West Wits Pit as described for borehole MGP1 (Sect. 10.3.6). Despite its substantial depth of 80 m (Table 12), these circumstances militate against the manifestation of an acid mine water presence at this location, at least at the time of profiling in early-2013. This finds support in the fresh (SEC 15–21 mS/m) and slightly acidic (pH 4.5–5.5) groundwater (Fig. 61) that is characteristic of the very weakly buffered quartzitic groundwater encountered in the West Rand Group strata of the Witwatersrand Supergroup (Sect. 2.1.1).

### 10.3.10 Borehole GP00305

Located in the southern portion of the KGR, this borehole penetrates the quartzite of the West Rand Group that forms the SW–NE striking ridge through the reserve. Water strikes at depths of ~25 and ~39 m bs (1608 and 1594 m amsl) inform a better understanding of the hydrochemical profiles.

The influence of purge pumping for routine sampling purposes on the pre- and post-sampling profile is explored in Fig. 62. The results suggest that purge pumping exerts little influence on the pH and temperature profiles, but causes significant shifts in SEC and ORP. The SEC profile reflects an upward migration by ~10 m of the SEC increase from ~40 mS/m in the upper section of the water column, to ~90 mS/m in the lower section of the water column.

The coincidence of the pH and ORP inflection points at ~22 m bwl (~1608 m amsl) with the first water strike is

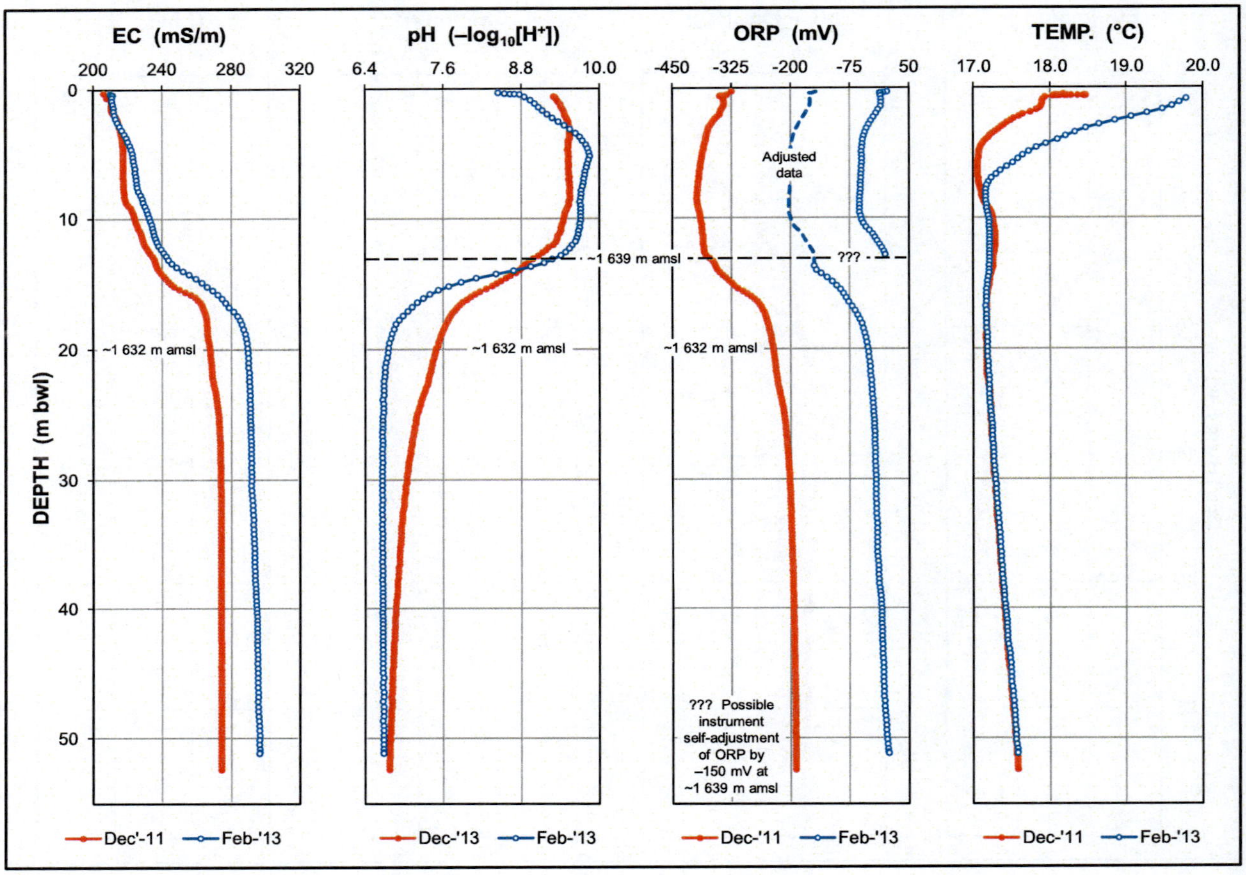


**Fig. 59** Vertical profiles of field water chemistry variables in borehole MGP3

noteworthy. It is below this level in the borehole that the pH trend becomes negative and the ORP trend becomes positive. The association of these circumstances with an elevated SEC value of ~90 mS/m suggests a mine water influence rather than the naturally acidic but extremely fresh groundwater that characterizes the quartzitic strata in the study area (Sect. 2.1.1). The $SO_4$ concentrations of ~300–350 mg/L and Fe concentration of ~20 mg/L returned by recent chemical analyses of groundwater from this borehole provide substantiation in this regard.

The results of five profiling exercises are shown in Fig. 63. Only the temperature profiles below ~5 m bwl would appear to be consistent over time. The SEC, pH and ORP profiles exhibit a temporal variance that is difficult to explain. A noteworthy aspect of the SEC profile is the congruence exhibited by the earliest (November 2011) and the latest (March 2014) profiles, compared to the stepped nature of the other SEC profiles. The development and progression of the stepped character of the intermediate SEC profiles, however, is not readily explained, nor are the ostensibly random patterns that characterize the pH and ORP profiles. For example, the negative ORP excursions at depths of ~5, ~15, and ~23 m bwl do not follow a temporal pattern, being manifested first at ~15 m bwl in February 2013, then at ~23 m bwl in September 2013, and most recently at ~5 m bwl in March 2014.

### 10.3.11 Borehole GP00306

Borehole GP00306 located on the east (right) bank of the Tweelopie Spruit in the southern portion of the KGR, targets the northern end of the dolomitic outlier. The very saline character (SEC ~328 mS/m) of the groundwater across the entire water column in this shallow (13 m) borehole indicates the well-mixed character of the groundwater in the saturated profile of the aquifer at this location. This is also true for the pH and the ORP profiles (Fig. 64), which reflect acidic and reducing conditions, respectively. The low pH value is of particular concern, as it indicates the lack of alkalinity in the karst system to maintain neutralization (buffering) of the acidic mine water entering this aquifer. This 'deficiency' might be attributed to the highly leached nature of the dolomite intersected in this profile.

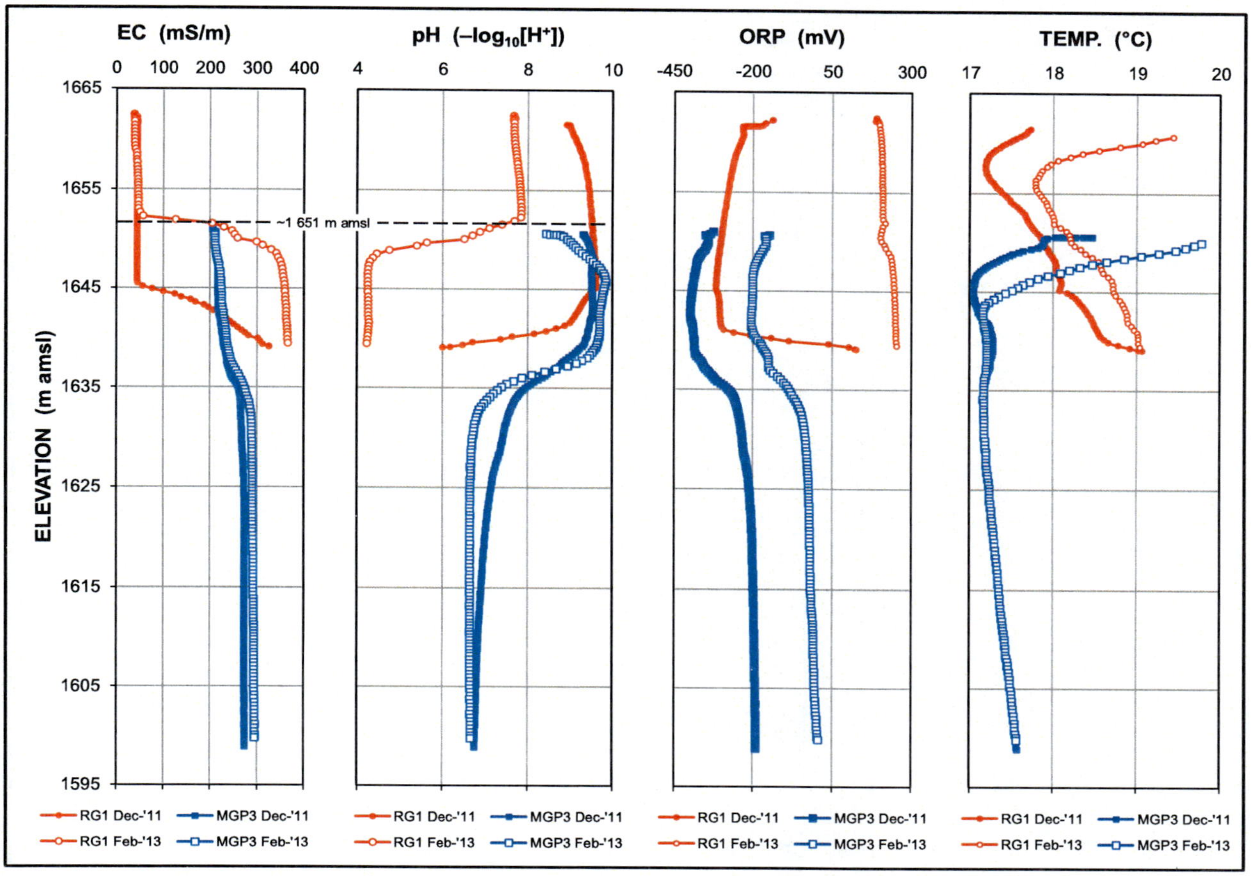


**Fig. 60** Comparison of contemporary field water chemistry variable profiles in boreholes RG1 and MGP3 on two occasions

A comparison of the profile values with those of the September 2011 sampling exercise (Table 11) shows a slight discrepancy. In regard to SEC, this is represented by a profile value of 328 mS/m compared to 276 mS/m of the sample. A more pronounced difference in temperature of <16.5 °C (profile) compared to 17.5 °C (sample) is observed. The pH values, however, are similar at 6.1 (profile) and 6.0 (sample). An explanation of these differences is not readily apparent unless it is associated with the use of two different instruments as described in Sect. 10.2. This possibility, however, is not afforded much credence under circumstances where different instruments are more likely to return disparate pH values rather than either SEC or temperature values.

### 10.3.12 Borehole GP00307

Located on the west (left) bank of the Tweelopie Spruit in the KGR, this borehole explores the karst aquifer of the Zwartkrans Basin at the confluence of this stream with a 'blind' drainage that enters the KGR from the south-west. The SEC profile (Fig. 65) clearly reflects a mine water influence with values of ~335 mS/m. The pH of 6.7 at depth, however, suggests the presence of buffering that might be attributed to the better quality karst groundwater contributed by the 'blind' drainage.

The temperature profile shows a marked decrease from ~17 °C near the water table to ~14.5 °C at >6 m bwl, a rate of −0.4 °C/m. Neither the pH nor ORP profiles indicate extraordinary trends that might suggest a subsurface environment favoring endothermic reactions that could drive a decrease in temperature of this magnitude.

### 10.3.13 Borehole GP00308

Borehole GP00308 is located at the northern end of the KGR on the right bank of the Tweelopie Spruit ~50 m from the stream. This borehole yielded little more than seepage water encountered at a depth of ~54 m bs (~49 m bwl;

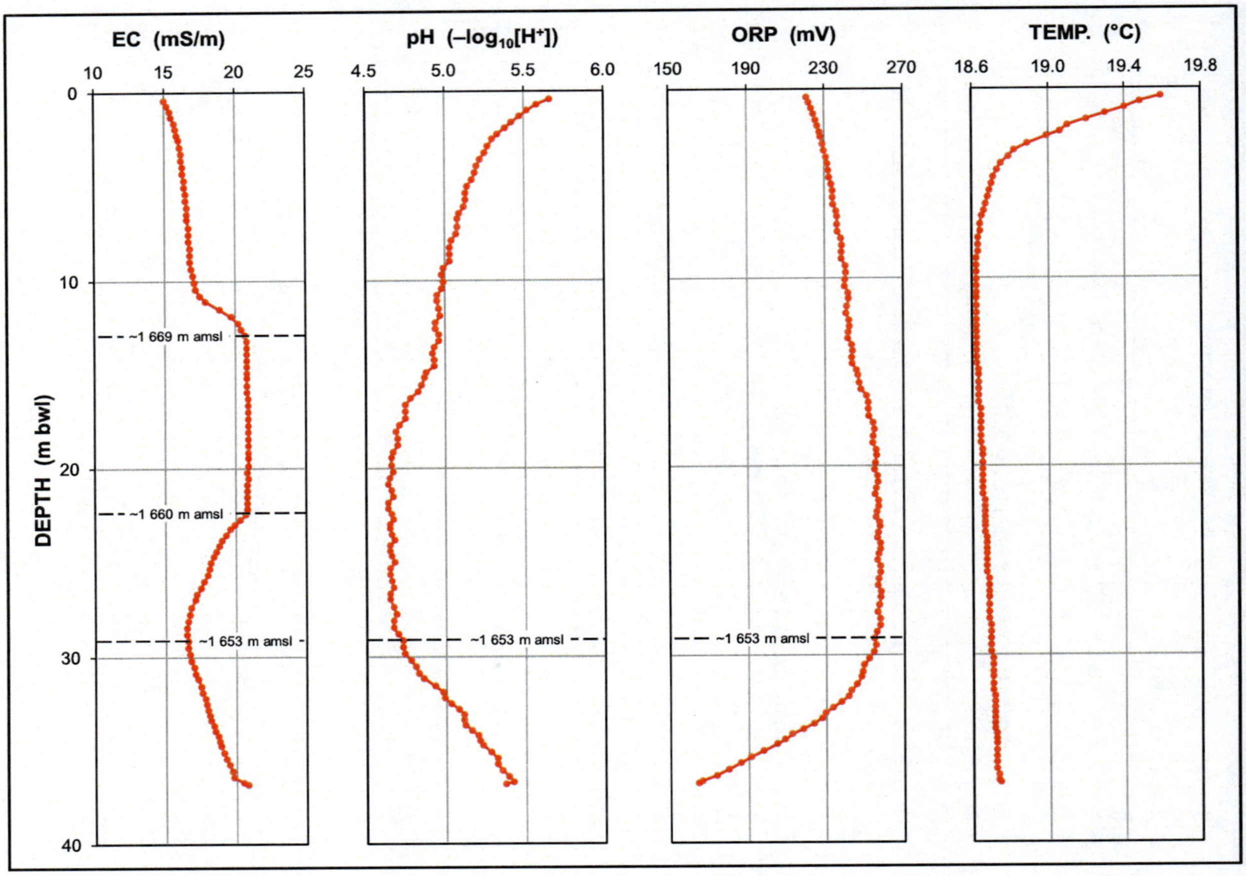


**Fig. 61** Vertical profiles of field water chemistry variables in borehole MGP4 in February 2013

~1471 m amsl). These circumstances and the competence of the fresh chert-poor dolomitic strata informed the completion of the bore as an open hole except for 19 m of plain steel casing at the top.

The profiles, in particular those of SEC, pH, and ORP (Fig. 66) exhibit distinct 'anomalies' in the interval 50–57 m bwl (55–62 m bs, ~1470–1463 m amsl) located immediately below the seepage water strike zone. The pattern of the ORP anomaly in this interval is similar to that observed in two other boreholes, namely A2N0583 (Fig. 67) and GP00311 (Fig. 78). The increase in SEC from ~40 to ~65 mS/m at depth in this borehole is not considered to reflect a mine water influence on the basis of the mutual increase in pH. A further anomalous pattern is reflected in the pH and ORP profiles which are similar and, except at ~1466 ± 3 m amsl, unlike the mirror-like images observed in most of the other profiles. This pattern might be associated with the static nature of the groundwater environment intersected by the borehole, the insignificant yield indicating a very low transmissivity and poor hydraulic connectivity.

### 10.3.14 Boreholes A2N0583 and GP00304

Located to the west of the KGR in the area of the Beckedan Smallholdings, this set of boreholes represents the hydrochemical profile where ostensibly no mine water impact[25] is manifested at the time (November 2011) of profiling. The similar surface elevations (~1594 m amsl) and depths to groundwater rest level (~44 m bc) of the boreholes again negate the need to normalize the data sets to a common datum.

The profiles (Fig. 67) reflect some measure of congruence and values that are indicative of a natural karst groundwater. This supports information generated for this portion of the study area that is exemplified in the chemistry of water discharged from the seeps that feed the Flip-se-Gat drainage in the KGR. These seeps deliver a combined perennial yield of 8–12 L/s characterized by an SEC of ~25 mS/m, a pH of ~8.0, and an Eh of −84 mV (Table 7). The deeper

[25] It is shown in Sect. 2.2 and Volume 2 that a gradual but persistent mine water impact is manifested at station A2N0583 as reflected in a doubling of the $SO_4$ concentration since 2004.

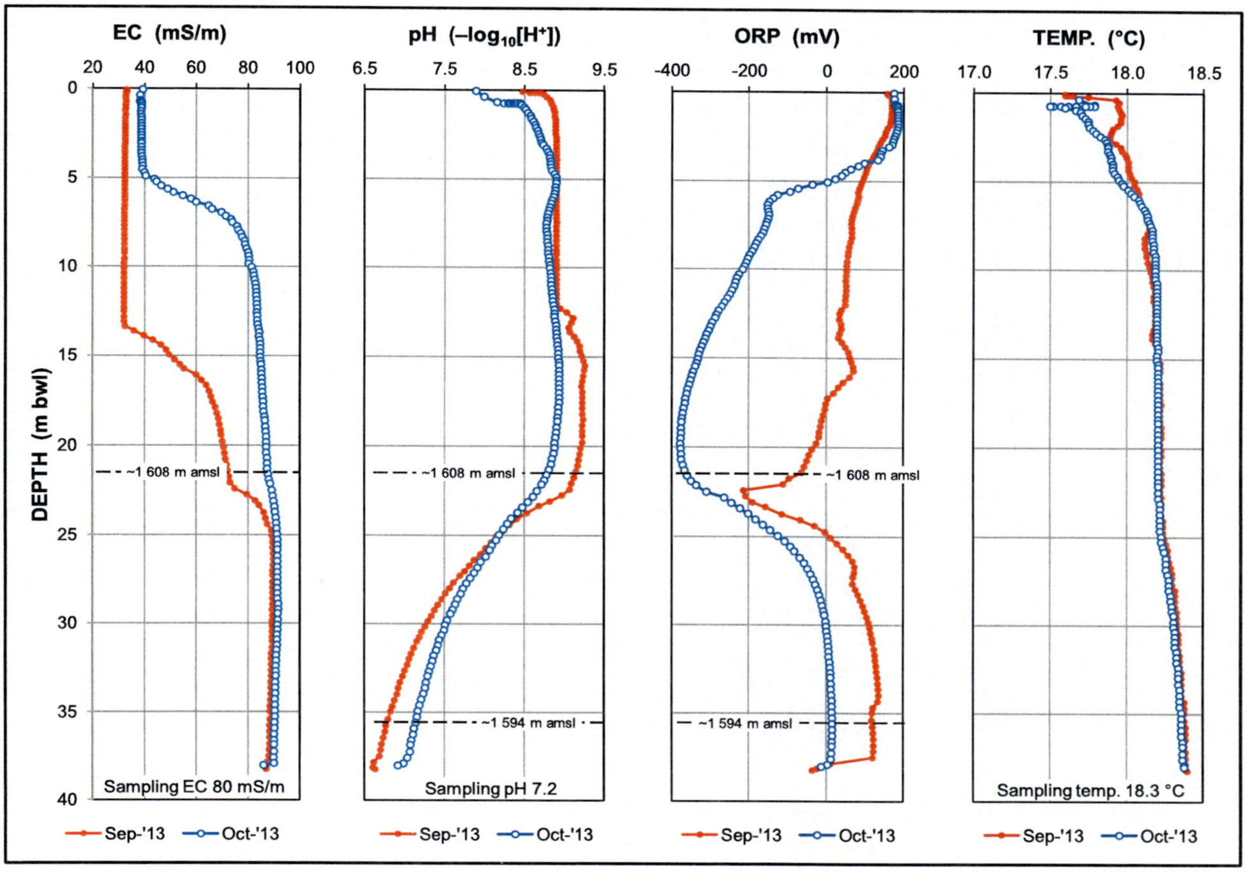


**Fig. 62** Pre- and post-sampling vertical profiles of field water chemistry variables in borehole GP00305

A2N0583 profiles reveal an inflection point at a depth of ~12 m bwl (~1538 m amsl) that is evident for each profiled variable. A reason for this quite sharply defined horizon has not been established.

An anomalous aspect of the profiles is the sympathetic decrease in pH and ORP trend in the uppermost ~12 m of the water column. Only below this depth do these variables illustrate the typical mirror-like image that in most other instances in this study characterizes their relationship. Changes in the SEC and temperature profiles are also discernible below the inflection point in the water column.

### 10.3.15 Boreholes A2N0584 and GP00302

These boreholes are located within 20 m of each other at the confluence of the Tweelopie Spruit with its main stem, the Riet Spruit, on the farm Glen Almond. The ingress of surface water with a very strong mine water composition into the karst aquifer along this stream reach is documented in Sect. 5.2.1 in Chapter "Physical Hydrology". The similar surface elevations (~1491 m amsl) and depths to groundwater rest level (~22 m bc) shared by the boreholes again negate the need to normalize the data sets to a common datum. An assessment of the profile data must also recognize the fact that borehole A2N0584 extends to a depth of only ~57 m, and therefore intersects the same aquifer horizon as the shallower (~35 m) borehole GP00302. Onerous drilling conditions associated with highly leached and cavernous dolomite were encountered in GP00302 and contributed to the shallow final depth of this borehole. Unsurprisingly, Bredenkamp et al. (1986) reported an airlift yield of >40 L/s for the nearby borehole A2N0584.

The March 2014 profiles measured in these boreholes are compared in Fig. 68 and indicate largely congruent SEC and temperature results (similarly shaped profiles) but incongruent pH and ORP results. The SEC profile in borehole A2N0584 reveals a marked increase of ~70 mS/m over the ~2 m interval from elevation ~1461 m amsl to elevation ~1459 m amsl. Although both boreholes reveal a consistent elevated SEC at depth, the difference of ~60 mS/m is not readily explained. The higher SEC in the shallower borehole GP00302 is, however, consistent with the slightly higher temperature of this groundwater column. It is the difference

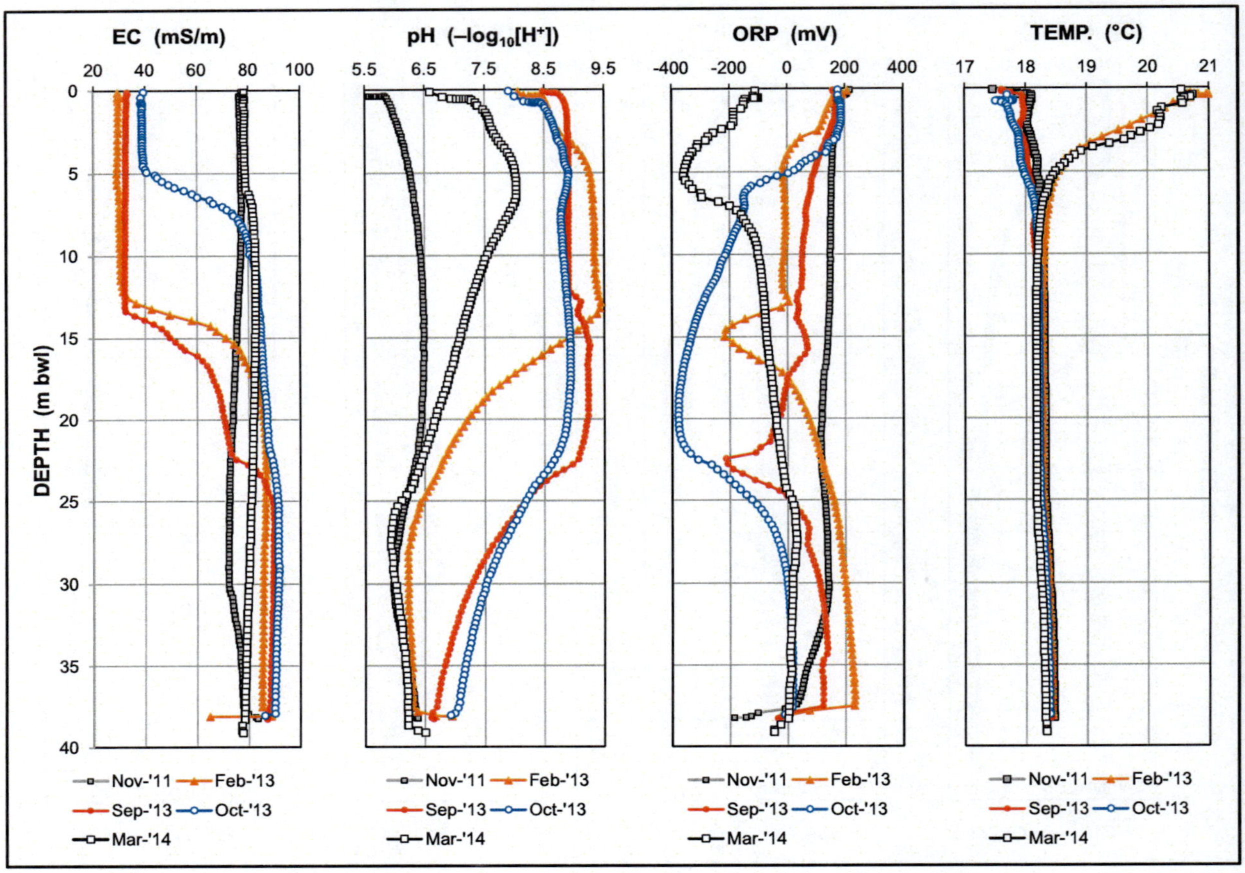


**Fig. 63** Temporal variation in vertical profiles of field water chemistry variables in borehole GP00305

of ~0.4 °C between the two groundwater columns in the first ~12 m below the water table which is intriguing, especially if the difference is maintained at depth as logical extrapolation would suggest. Further, the increasing temperature gradient of ~0.4 °C per 5 m below an elevation of ~1451 m amsl in A2N0584 is also not readily explained, although its coincidence with the most reducing section of the groundwater column characterized by ORP values in the range −225 to −350 mV cannot go unnoticed. The most plausible explanation associates this trend with exothermic biogeochemical reactions in a strongly reducing (anoxic) environment at depth in the subsurface. The perforated casing in A2N0584 extends from ~5 m bwl to the end-of-hole at ~35 m bwl, and therefore extends to above the zone of maximum change in the SEC and pH profiles evident in Fig. 68.

Perhaps most surprising, however, is the considerable difference reflected in the pH profiles and, to a lesser extent, the ORP profiles. Not only are the pH profiles in the first ~12 m below the water table divergent, but the pH of ~9.2 at depth in borehole A2N0584 is three full pH units (1000-fold) more alkaline than in borehole GP00302. The high pH is also in severe conflict with a mine water impact as is suggested by the SEC value of ~130 mS/m.

The ~3 m interval from elevation ~1464 m amsl to elevation ~1461 m amsl in borehole GP00302 brackets a step in each of the SEC ($\Delta = +45$ mS/m), pH ($\Delta = -1.1$) and ORP ($\Delta = +150$ mV) profiles. This horizon is not observed in borehole A2N0584.

It is apparent that the hydrochemical environment intersected in boreholes A2N0584 and GP00302 severely challenges an understanding of this environment at this location. The temporal change in profile variables in each borehole is illustrated in Fig. 69 (A2N0584) and Fig. 70 (GP00302), which provides an opportunity to determine the veracity of the profiling results and ascertain any site-specific differences.

Figure 69 indicates that the shape of the March 2014 profiles in borehole A2N0584 is replicated in each of the profiling events. The measure of temporal consistency is greater for SEC and temperature than for pH and ORP. The latter two variables reveal a greater variance over time in

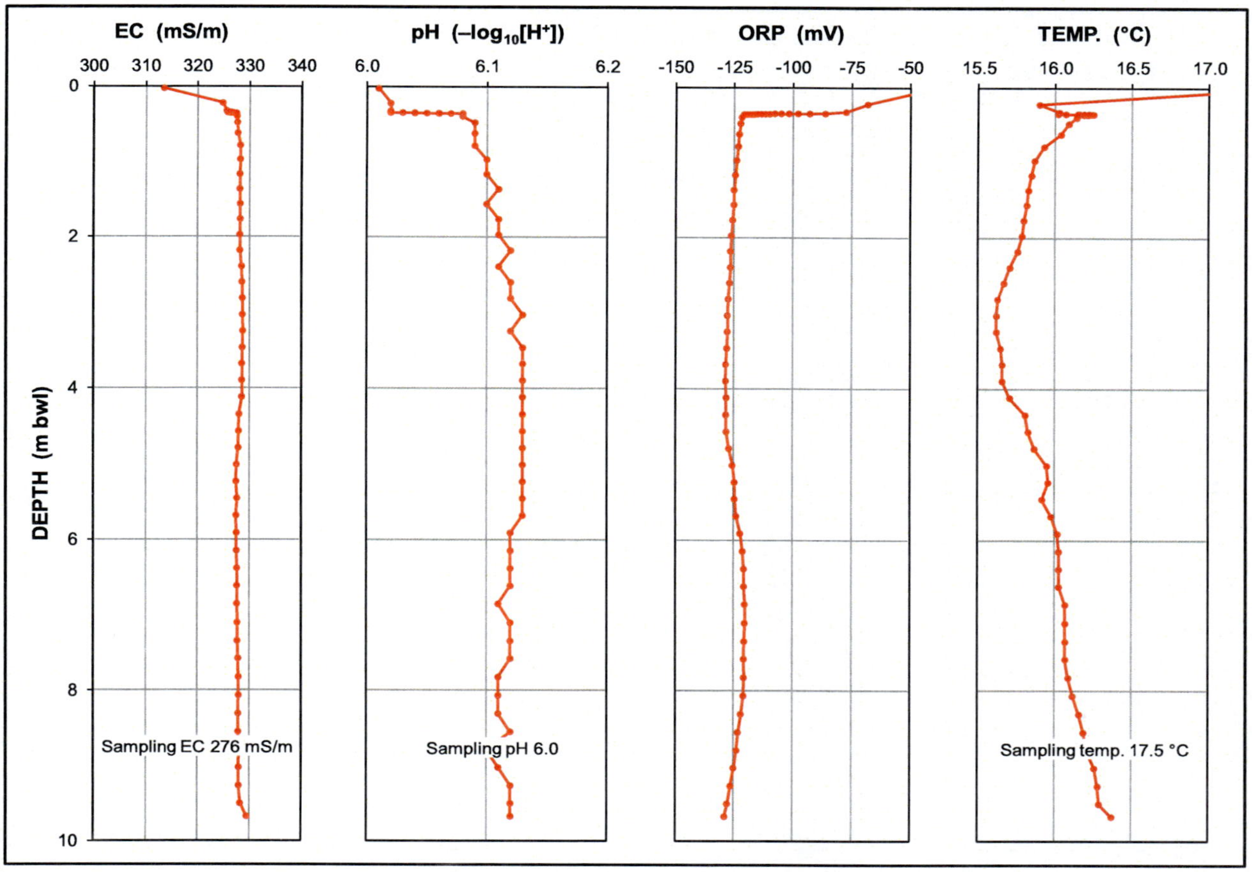


**Fig. 64** Vertical profiles of field water chemistry variables in borehole GP00306 in November 2011

the upper 10 m of the groundwater column. The variance reduces considerably in the groundwater column below an elevation of ~1459 m amsl (>10 m bwl). The observed temporal consistency exhibited by each variable below ~1459 m amsl suggests that hydrochemical equilibrium exists in the deeper portion of the karst aquifer intersected by this borehole. The shallower portion of the karst aquifer experiences hydrochemical fluxes that are most likely driven by allogenic recharge associated with mine water discharges from the Western Basin. Nevertheless, it must be concluded that the profiling events present unequivocal results that do not reflect significant site-specific differences over time.

The groundwater column in borehole GP00302 represents the same hydrostratigraphic horizon that forms the upper 10 m of the karst aquifer intersected in borehole A2N0584. Figure 70 similarly indicates that the shape of the March 2014 profiles in this borehole is replicated in each of the profiling events. The measure of temporal consistency is again greater for SEC and temperature than for pH and ORP. It must again be concluded that the profiling events present unequivocal results that do not reflect significant site-specific differences over time.

In light of the preceding discussion, the significantly different hydrochemical signatures that characterize the common hydrostratigraphic horizon in the groundwater columns of the two boreholes remain inexplicable. These circumstances are not readily explained on the basis of the current understanding of the hydrophysical and hydrochemical subsurface environments at this location.

It is possible that the observed profile patterns are informed by the 'as built' construction characteristics of the two boreholes. Borehole GP00302 was constructed with a single 140 mm diameter perforated uPVC casing 'string' installed to a depth of 33 m, and extending to surface with a perforated interval from ? to ? m bs. Borehole A2N0584 was constructed with two sets of perforated steel casing as follows (from Bredenkamp et al. 1986):

- an outer 330 mm diameter 'string' installed to a depth of ~40 m, extending to surface and perforated in the interval 27–40 m bs; and

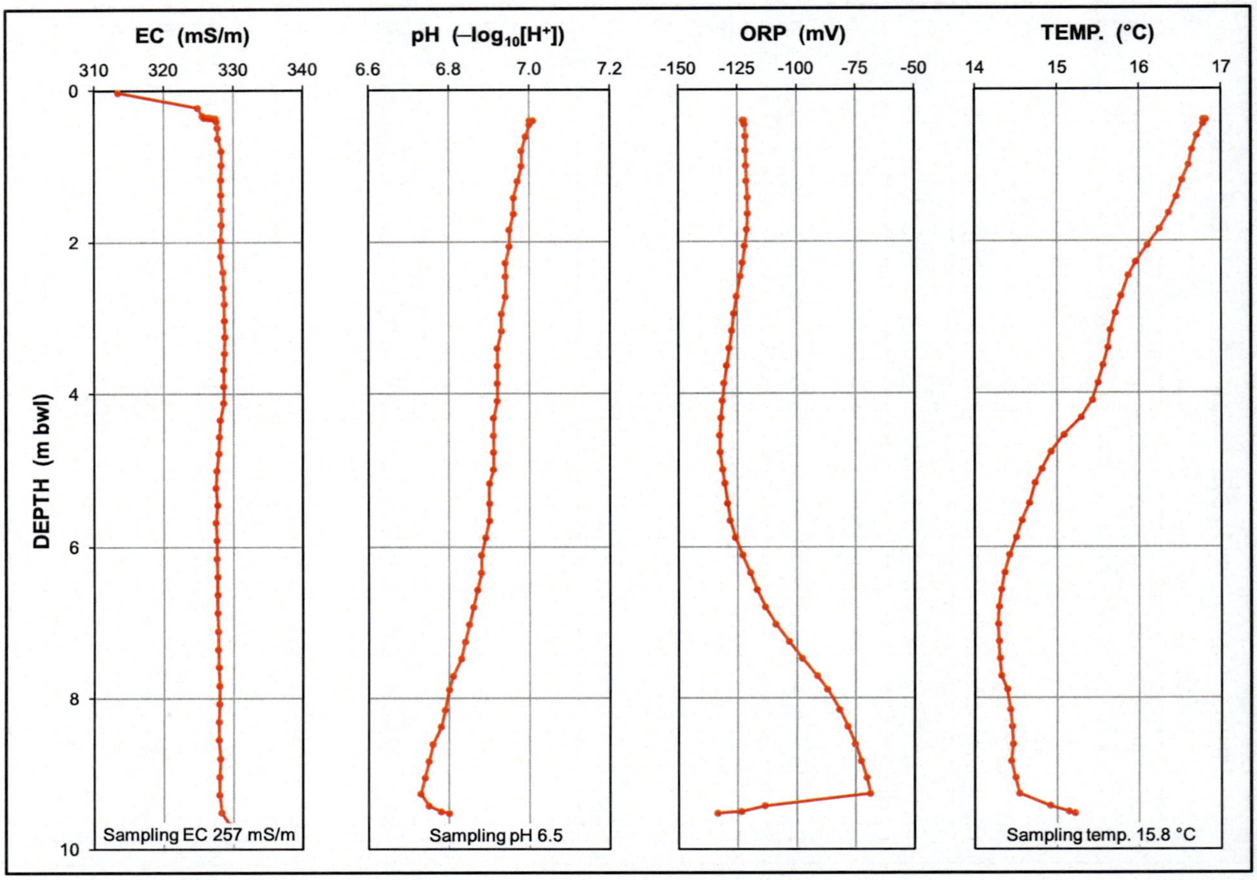


**Fig. 65** Vertical profiles of field water chemistry variables in borehole GP00307 in November 2011

- an inner 250 mm diameter 'string' installed to a depth of ~57 m, extending to surface and perforated in the interval 26.5–57 m bs.

The greater complexity associated with the A2N0584 construction might account for the 'anomalous' pH and ORP profiles obtained for this borehole. It is not known to what extent the different casing materials used in the two constructions might account for the observed differences. A more likely cause for the differences must, however, be sought in the poorly understood hydro(geo)chemical domain and the complex interaction between mine water and dolomitic groundwater in a karst aquifer. Part of this complexity almost certainly relates to processes such as sulfidogenesis (sulfate reduction), $Fe^{2+}$ oxidation, $Fe^{3+}$ reduction, and production of carbonate alkalinity.

### 10.3.16 Boreholes A2N0586 and GP00300

This grouping of boreholes is located in the Sterkfontein Subcompartment ~270 m north-west of the Riet Spruit midway along the stream reach where surface water with a very strong mine water composition enters the karst aquifer (Sect. 5.2.1 in Chapter "Physical Hydrology"). The rate of ingress varies from ~14 to >30 ML/d. The principal groundwater flow vector in this portion of the study area is shown in Fig. 19 in Chapter "Physical Hydrogeology", indicating the boreholes to be located immediately down-gradient of the losing Riet Spruit. The impact of this ingress on ambient groundwater chemistry has been discussed at length in Sect. 2.3.3, especially in regard to borehole WBD5. An assessment of the profile data must recognize the fact that borehole A2N0586 (G36331 in Bredenkamp et al. 1986) intersects two horizons of leached and fractured chert-rich dolomite in the depth intervals 20–40 and 120–132 m bs, respectively. The deeper aquifer produced an airlift yield of 50 L/s (Bredenkamp et al. 1986). A ~80-m thick layer of chert-poor dolomite separates the two aquifer horizons. Borehole GP00300 intersects only the shallow aquifer.

The influence of purge pumping for routine sampling purposes on the pre- and post-sampling profiles in boreholes A2N0586 and GP00300 is explored in Figs. 71 and 72,

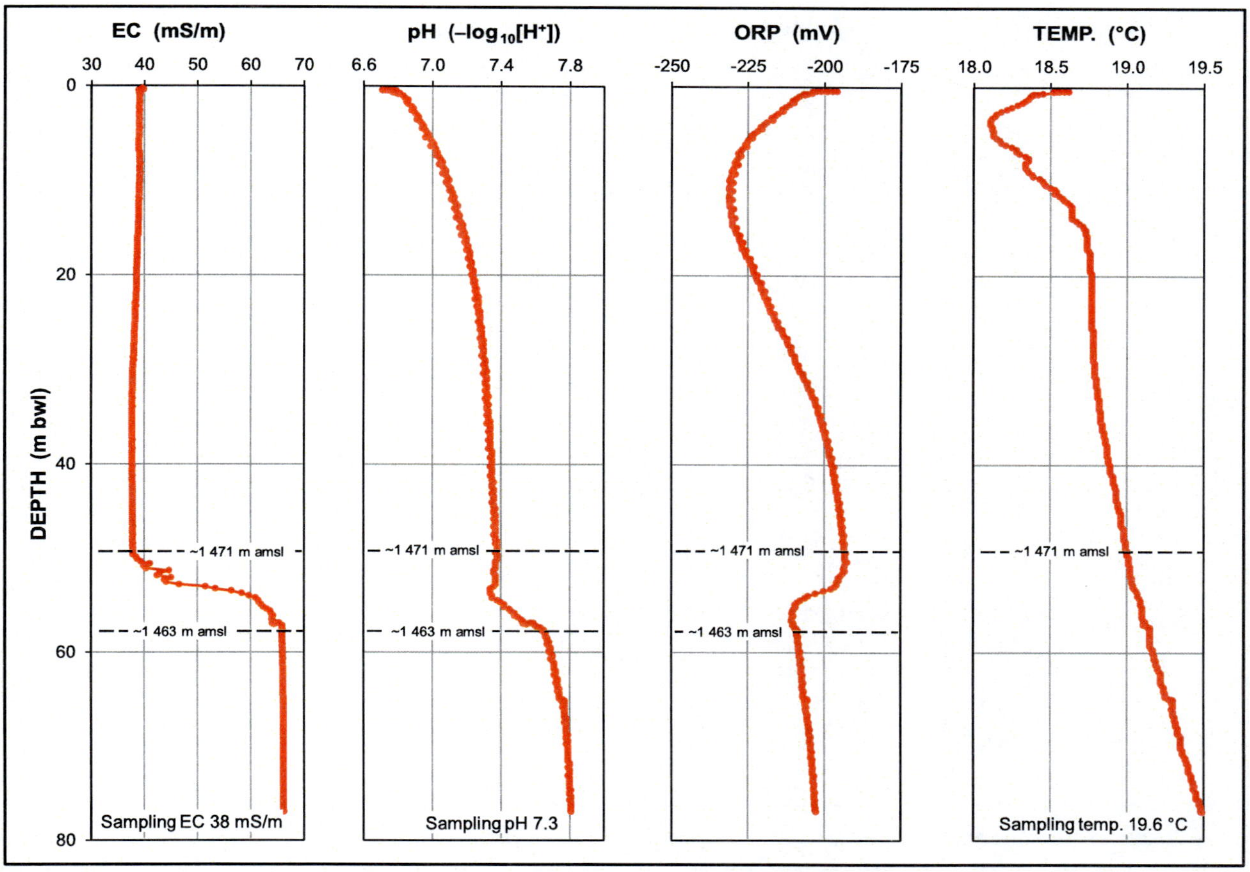


**Fig. 66** Vertical profiles of field water chemistry variables in borehole GP00308 in November 2011

respectively. The A2N0586 results suggest that purge pumping exerts little influence on any of the variable profiles, causing only minor negative and positive lateral shifts in pH and ORP, respectively. A transition zone at ~10 m bwl is marked by an SEC, pH, and ORP anomaly that occurs in association with the center of the upper karst horizon. The reasonable measure of congruence[26] between the two sets of profiles is self-evident. The greatest difference between the older and more recent data sets is evident in the pH profile of borehole GP00300, the latest profile reflecting an accentuated replica of the earlier profile, but shifted by 3 full pH units (a factor of 1000) from an acidic to a basic character. The step in SEC and pH between elevations of ~1458 and ~1454 m amsl in borehole A2N0586 are similarly accentuated and is also manifested in the latest ORP profile. The latter further indicates the development of a redox anomaly at an elevation of ~1440 m amsl (~26 m bwl) that is only hinted at in the earlier profile.

The influence of purge pumping for routine sampling purposes on the pre- and post-sampling profiles in borehole GP00300 (Fig. 72) is ostensibly more significant than in borehole A2N0586. The GP00300 results suggest that purge pumping exerts a 'normalizing' influence resulting in 'smoother' SEC, pH, and ORP profiles. Equally intriguing is the ~1 °C negative shift in temperature. These observations are not readily explained on the basis of current knowledge and understanding of the complex hydrophysical and hydrochemical interactions that characterize the hydrogeologic system. The established presence of a shallower and a deeper aquifer horizon at this location, however, suggests the delivery of different groundwater chemistries from the boreholes.

The deeper borehole A2N0586 which intersects both aquifers delivers a mixture of the shallow more saline and the deeper fresher groundwater. This is exemplified in the values obtained during the routine sampling of the boreholes

[26] The similar surface elevations (~1487 m amsl) and depths to groundwater rest level (~22 m bc) shared by the boreholes negate the need to normalise the data sets to a common vertical datum.

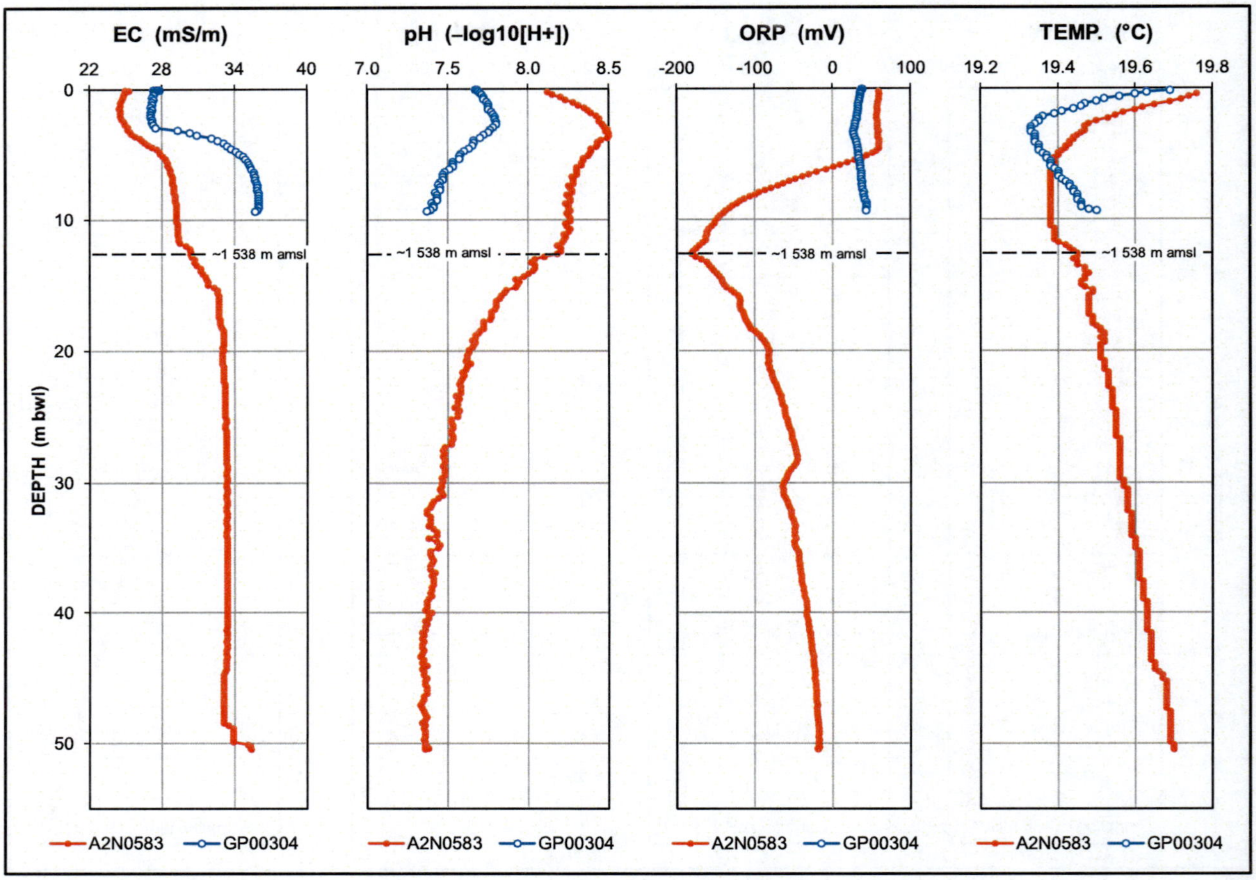


**Fig. 67** Vertical profiles of field water chemistry variables in boreholes A2N0583 and GP00304 in November 2011

(e.g., Tables 11 and 12). The comparison, however, also indicates that none of the stabilized[27] SEC, pH, and temperature values is similar to that of any profile value. For example, the stabilized SEC values of 172 and 208 mS/m (for boreholes A2N0586 and GP00300, respectively) do not correlate with the 'final' values of ~210 and ~250 mS/m indicated by the SEC profiles for these boreholes (Figs. 73 and 74, respectively).

The above circumstances indicate that a different mixture of 'shallow' and 'deep' groundwater is produced by each borehole during pumping. This again poses the question of when might a 'representative' sample be considered 'truly representative'. Of course, discrepancies introduced by factors such as the use of different instruments and the lapse of time between sampling and profiling also need to be included in an analysis of these circumstances.

[27] The field chemistry variables are monitored in a flow through cell during purge pumping of the borehole being sampled, and a sample collected only once the variable values have stabilised.

### 10.3.17 Borehole GP00301

Located ~600 m north-west of the Riet Spruit on the postulated western margin of the mine water impacted zone in the Sterkfontein Subcompartment, this borehole reports a depth to groundwater rest level of ~59 m bs. Notable changes in SEC, pH and ORP occur at an elevation of ~12 m bwl (~71 m bs, ~1453 m amsl), which coincides with the first water strike in the borehole.

The SEC value of ~125 mS/m in the lower portion of the profile (>35 m bwl; >94 m bs; <1430 m amsl) (Fig. 75) indicates the influence of poorer quality water on the typically much lower SEC associated with unimpacted natural karst groundwater. This influence is not yet evident in the pH values of ~7.4, possibly the result of effective neutralization (buffering) by abundant alkalinity in the karst hydro system. The moderately reducing conditions are characterized by ORP values in the range −200 to −150 mV. A slight freshening of the groundwater at the base of the water column (>80 m bwl, <1385 m amsl) is also evident.

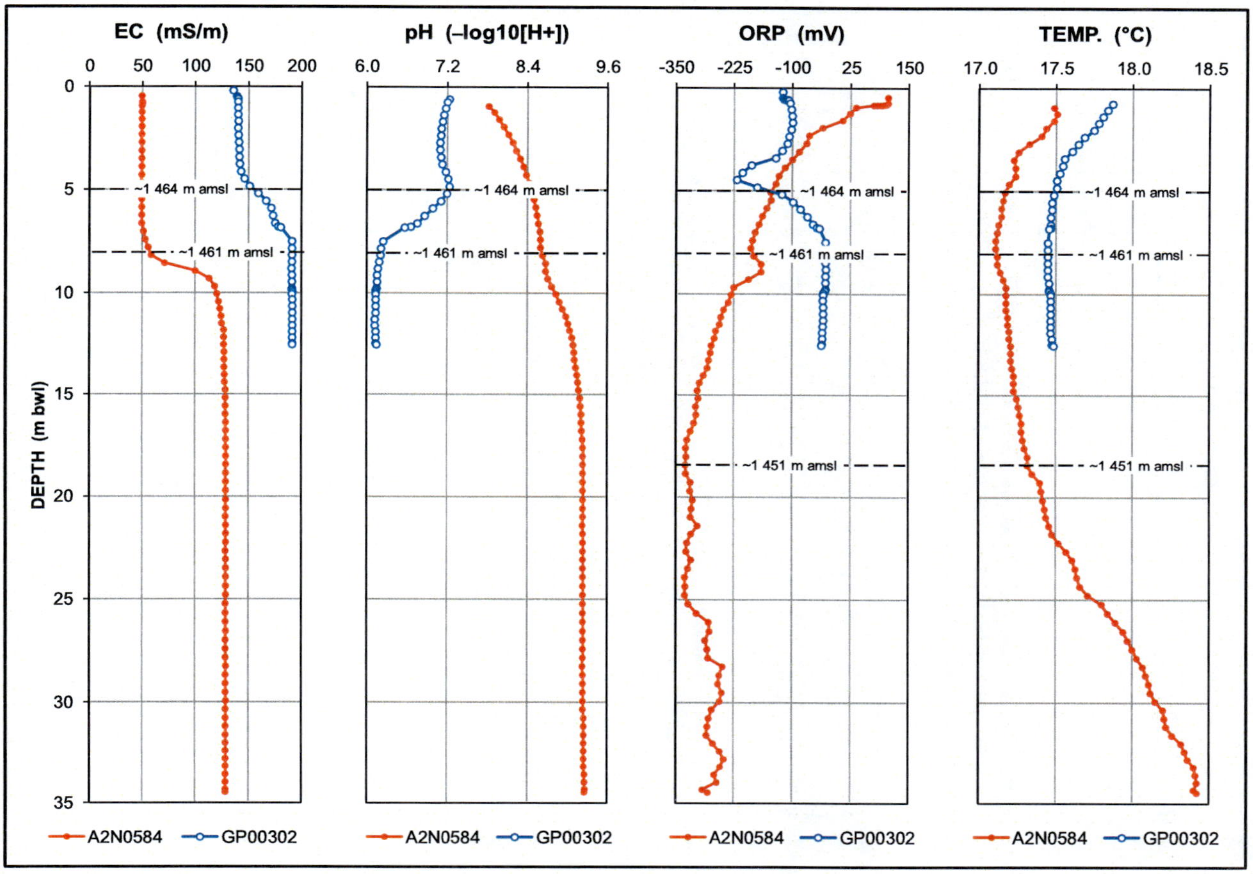


**Fig. 68** Vertical profile of field water chemistry variables in boreholes A2N0584 and GP00302 in March 2014

### 10.3.18 Borehole GP00309

This borehole occupies a strategic position at the confluence of the Riet Spruit and the Blougat Spruit upstream of Oaktree. This location marks the subsurface confluence of infiltrated mine water (from the Riet Spruit) and infiltrated municipal wastewater (from the Blougat Spruit). The groundwater is drawn from the karst aquifer at this location, therefore, representing a mixture of these two allogenic sources of recharge.

The influence of purge pumping for routine sampling purposes on the pre- and post-sampling profile is explored in Fig. 76. The results suggest that purge pumping exerts little influence on the SEC profile, causes slight shifts in pH and ORP toward more alkaline and reducing conditions, respectively, and causes a significant negative shift in temperature. The pH and ORP changes are attributed to the drawing of a greater proportion of better quality karst groundwater into the radius of hydrogeochemical influence around the borehole.

The large measure of congruence between the respective SEC, pH, and ORP profiles generates confidence in the veracity of the data sets. The significant difference between the sampled SEC and temperature values compared to the profiled values is attributed to the use of two different multiparameter probes as discussed in Sect. 10.2.

The results of six profiling exercises are shown in Fig. 77. The SEC increase from <200 to ~290 mS/m at ~4 m bwl is attributed to a fresher water horizon overlying the mine water impacted karst groundwater below this depth. The transition zone of ~3 m is also characterized by significant reversals in pH and ORP values. In all instances, the pH profiles trend toward more acidic conditions (by almost one full pH unit) with depth, and the ORP profiles toward more oxic conditions.

The low temperature of <14 °C that characterizes the bottom section of the November 2011 profile extends much further up the water column in the February 2013 profile. The four later profiles demonstrate first a positive migration to a value of almost 19 °C (through June–September 2013), followed by a reversal to a value of ~14.5 °C (through October 2013–March 2014). The latter is at least 4 °C lower than that which characterizes natural karst groundwater in

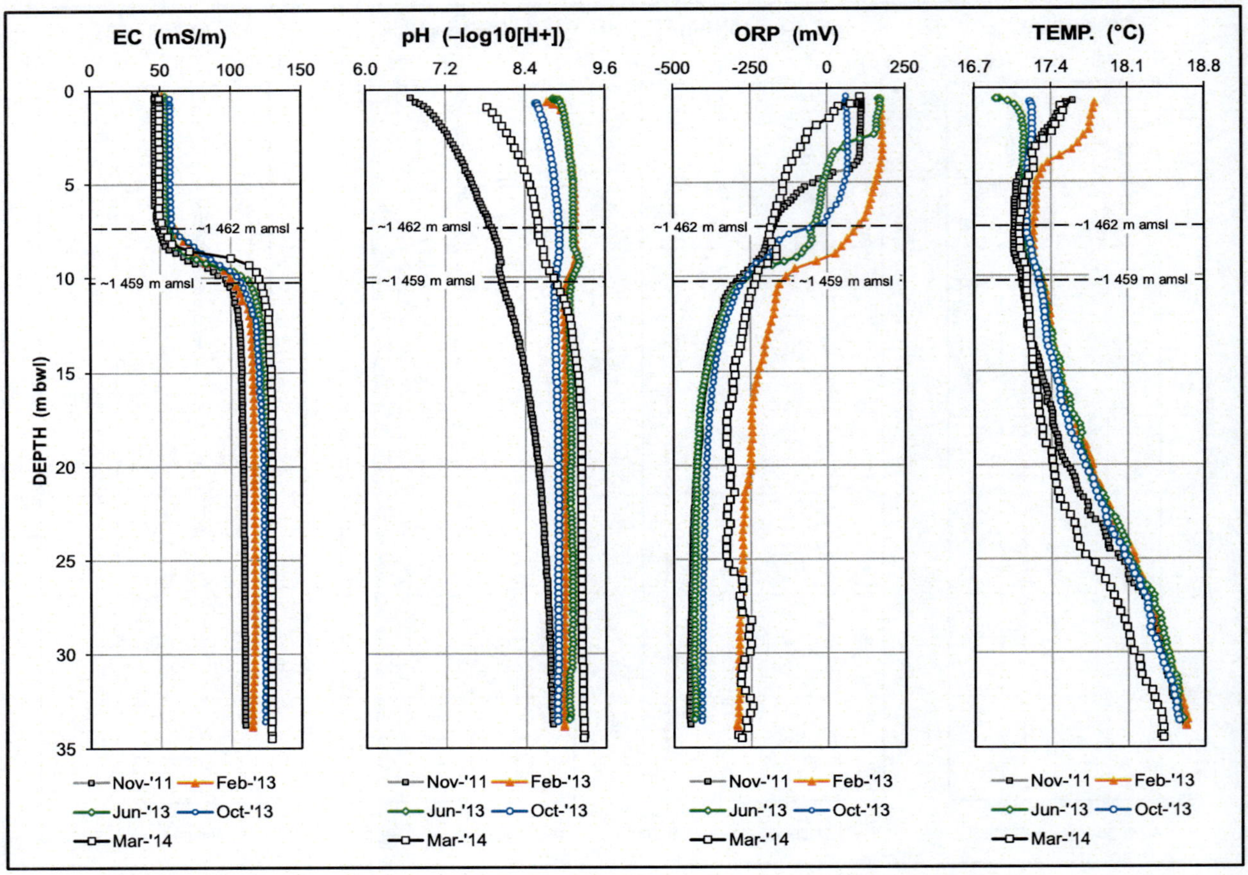


**Fig. 69** Temporal variation in vertical profiles of field water chemistry variables in borehole A2N0584

the study area. An explanation for these circumstances has not yet been established, although endothermic redox reactions that accompany a shift from reducing (anaerobic) to oxic (aerobic) and more acidic conditions provide the most plausible cause. Support in this regard might be found in the thermodynamic complexity of redox processes that involve terminal electron-accepting processes (TEAPs) as discussed by Hoehler et al. (1998) and Jakobsen et al. (1998).

### 10.3.19 Borehole GP00311

The position of this borehole in the upper reaches of the Zwartkrans Basin characterizes the hydrochemistry of the dolomitic groundwater outside the influence of the mine water impact. This is supported by the SEC value of ~38 mS/m and the pH value of ~7.5 (Fig. 78). The step-wise increase in SEC at an elevation of ~1473 m amsl, and the synchronous but opposite pH and ORP patterns at this elevation, are not readily explained within the current knowledge framework. The redox profile indicates a step at a depth of ~39 m bwl (~107 m bs; ~1441 m amsl) that does not manifest as a change in any of the other three variable profiles at this depth. The reason for a shift toward reducing conditions in a comparatively narrow horizon (<3 m in extent) at this depth is unclear. Nevertheless, the profile values compare favorably with the stabilized SEC value of 36 mS/m, pH of 7.6, and temperature of 20.2 °C obtained during the sampling exercise in September 2011 (Table 11).

### 10.3.20 Borehole GP00312

The October 2011 profiles (Fig. 79) reveal the very strong presence of mine water in the karst aquifer at this location on the right bank of the Riet Spruit within 50 m of the stream channel. The SEC profile shows two step-wise increases, the first occurring within the upper 5 m of the profile, and the second below an elevation of ~1462 m amsl (>5 m bwl). As the water table occupies an elevation of ~1467 m amsl, which is ~4 m below the streambed elevation of ~1471 m amsl, the most saline horizon (~330 mS/m) is encountered >10 m below the streambed, in the lower section of the borehole. The pH profile indicates a uniform increase by ~0.3 units in this section. Nevertheless, of concern is the

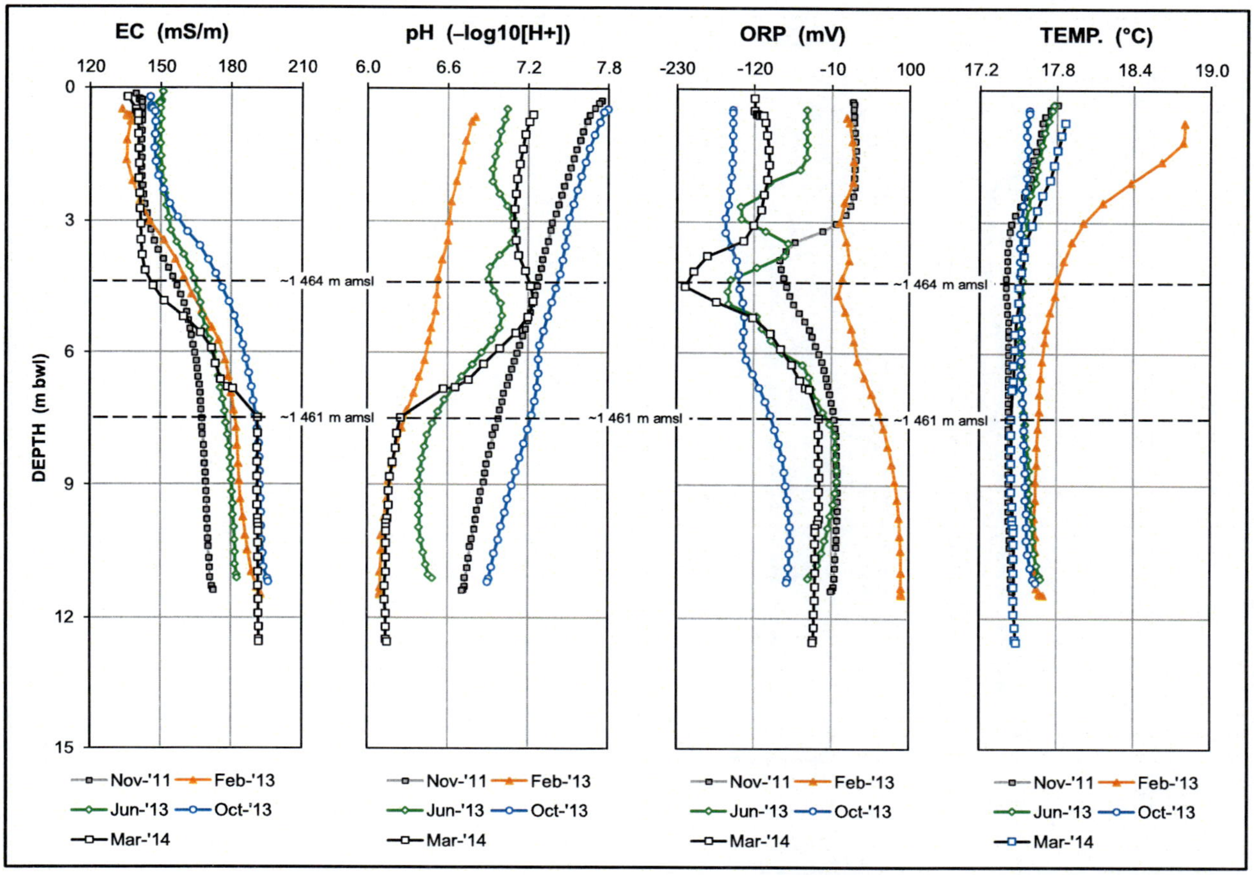


**Fig. 70** Temporal variation in vertical profiles of field water chemistry variables in borehole GP00302

pH value of <6.0, which is below the value of 6.4 that describes the limit at which net alkalinity is zero (Sect. 13.2). These circumstances suggest that the dolomite at this location no longer provides a neutralizing function. The ~0.5 °C decrease in temperature below ~1454 m amsl is not readily explained.

The stabilized values obtained during the September 2011 sampling exercise again show little correlation with the respective profile values. Neither the pH nor temperature values of 6.8 and 19.5 °C (Table 11), coincide with the profiles. Especially the pH value provides a false 'positive' indication in regard to this variable. Similarly, the SEC value of 263 mS/m (Table 11) under-reports the 'final' SEC profile values by ~70 mS/m.

### 10.3.21 Borehole GP00313

This borehole occupies an important position ~750 m to the northwest of Sterkfontein Cave, where it reflects the chemistry of karst groundwater draining from the Zwartkrans Basin toward the Zwartkrans Spring. The influence of purge pumping for routine sampling purposes on the pre- and post-sampling profile is explored in Fig. 80.

The results suggest that purge pumping exerts little influence on the SEC and temperature profiles, and causes slight shifts in pH and ORP toward more alkaline and reducing conditions, respectively. The latter changes are attributed to the drawing in of a greater proportion of better quality karst groundwater into the radius of hydrogeochemical influence around the borehole. The significant difference between the sampled SEC, pH, and temperature values compared to the profiled values is attributed to the use of two different multiparameter probes as discussed in Sect. 10.2.

The results of six profiling exercises are shown in Fig. 81, together with the absolute elevation that brackets the Zwartkrans Spring. The 'spikes' on the November 2011 SEC profile coincide with the joint interval of 2.3 m between slotted screen sections. These are not evident in the other profiles. The SEC increase from ~105 mS/m in November 2011 to ~135 mS/m in the later profiles are attributed to the

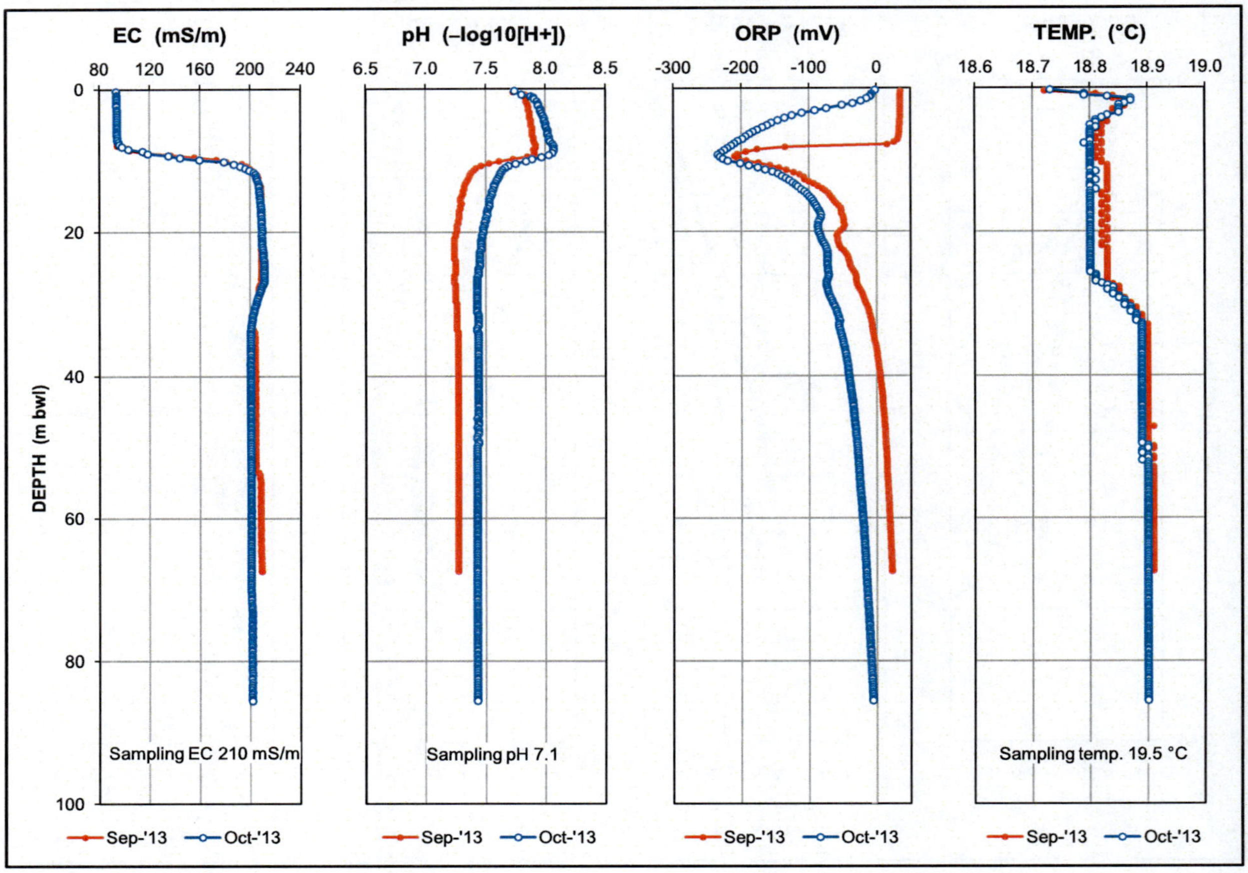


**Fig. 71** Pre- and post-sampling vertical profiles of field water chemistry variables in borehole A2N0586

mine water influence on the karst water chemistry at this location. This surprisingly does not find support in the more alkaline pH reflected in the post-November 2011 profiles, all of which exhibit a remarkably similar trend. The initial significant shift from reducing to oxidizing conditions shown in the November 2011 and February 2013 ORP profiles is again attributed to a mine water influence. The subsequent incomplete return to reducing conditions in the three most recent profiles is attributed to a partial re-establishment of the natural karst hydrogeochemical system. The position of this borehole in the lower reaches of the Zwartkrans Basin imposes a complexity (including a municipal wastewater influence) that cautions against an unequivocal conclusion in this regard.

The relevance of the water-bearing horizon intersected by this borehole within the broader hydrogeologic framework has been discussed in Sect. 3.2.2 in Chapter "Physical Hydrogeology". The base of the cavernous zone intersected by the borehole approximates the elevation of ~1429 m amsl (~11.5 m bwl; ~31 m bs) shown in Fig. 81. Both the pH and the ORP profiles show a slight 'change' (most noticeable in the November 2011 profiles) in pattern below this elevation, the change representing a trend reversal.

It is worth repeating that this borehole intersected a cavity between 17 and 32 m bs. The elevation of ~1429 m amsl is also located ~10 m below the current Lake water level of ~1439 m amsl in Sterkfontein Cave (Sect. 3.2.2 in Chapter "Physical Hydrogeology"). The temperature data >4 m bwl show little difference between the profiling events.

### 10.3.22 Borehole GP00314

This borehole is located between borehole GP00313 (Sect. 10.3.21) and the Zwartkrans Spring on the left (north) bank of the Bloubank Spruit within 10 m of the stream channel. Only 14 m deep, the borehole is in direct hydraulic continuity with the stream, an observation that finds support in the constancy of the profile values shown in Fig. 82. This constancy also indicates the well-mixed character of the groundwater in the near-surface saturated profile of the aquifer comprising highly leached to decomposed dolomite at this downstream location.

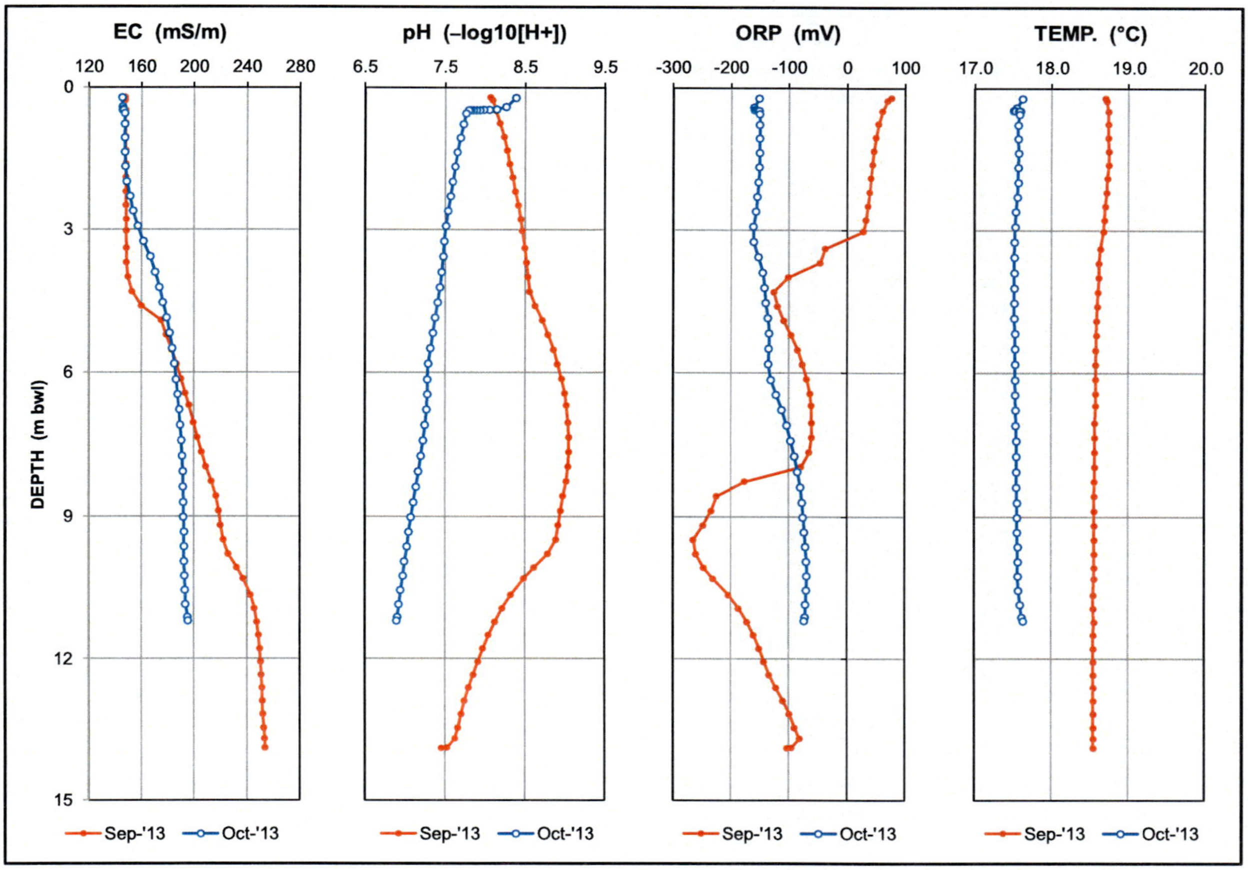


**Fig. 72** Pre- and post-sampling vertical profiles of field water chemistry variables in borehole GP00300

The discrepancy between the profile SEC value of ~98 mS/m and the sampling SEC value of 49 mS/m is not readily explained. The profile value, however, is similar to the 84 mS/m of the Zwartkrans Spring water in May 2010, at which time the SEC of groundwater in the lower reaches of the Zwartkrans Basin had already started increasing (Fig. 24). The borehole is located ~700 m upstream of the spring, and its shallow depth suggests that it shares the shallow karst horizon draining toward the spring.

## 10.4 Discussion

The vertical hydrochemical profiling results reflect a substantially more complex subsurface hydrogeochemical environment than is apparent from the more mundane data and information collected in the course of routine hydrogeologic investigations. This complexity poses fundamental questions regarding the measure of understanding of the hydrogeochemical and biogeochemical environments and their interaction in the karst system of the study area.

The added complexity associated with the allogenic sources of recharge, primarily mine water and municipal wastewater effluent, challenge this understanding even further. Recognition of this complexity precipitates an admission of the inadequacy of the current understanding of the interaction between impacting water resources and the natural receiving karst groundwater environment. This limitation, however, must not stand in the way of the proactive implementation of water resource management measures required to mitigate any negative impacts within the framework of the current understanding of these inter-relationships.

Figure 83 shows that the Eh values that characterize the mine water and groundwater environments span a comparatively narrow range compared to that of pH. The limits are defined by values in the range ~200 mV at pH ~2.5, to −150 MV at pH ~9.3. These circumstances describe an Eh–pH condition which falls mainly in the stability field of ferrous iron at values more representative of mine water where $Fe^{2+}$ is the dominant species, and in the amorphous Fe-oxyhydroxide [$Fe(OH)_3$] stability field at values most

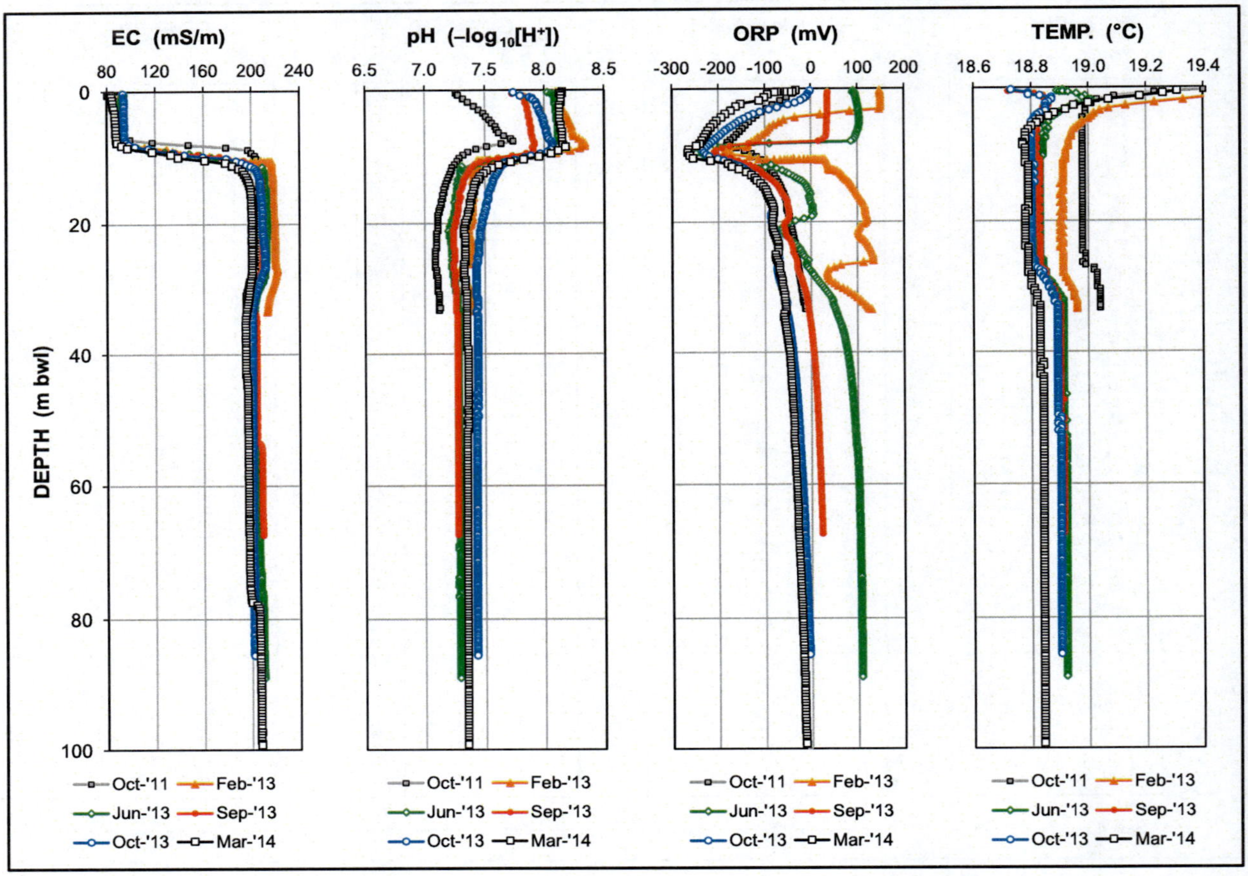


**Fig. 73** Temporal variation in vertical profiles of field water chemistry variables in borehole A2N0586

representative of mine water impacted karst groundwater where ferric hydroxide is the dominant species.

The Eh–pH condition that characterizes the mine water and groundwater environments correlates well with that which defines environments that are isolated from the atmosphere. It is therefore surprising that the raw mine water (RMW) Eh–pH reference condition which forms the one end-member of the Eh–pH range, differs significantly from the condition that characterizes acidic mine drainage as defined by Baas-Becking et al. (1960) and Langmuir (1996) reported in USDA (2012). This is defined by the Eh–pH condition at A in Fig. 83 representing an acidic mine drainage environment in contact with the atmosphere, circumstances which closely describe the RMW sampling environment.

The other reference Eh–pH condition shown in Fig. 83 is that for the Zwartkrans Spring (ZSp). This source produces groundwater that exhibits a slight to moderate impact of contamination by mine water, as is reflected in the SEC value of ~98 mS/m and a pH of ~7.0 that is notably lower than that of pristine karst groundwater. The Eh–pH condition of the ZSp springwater is appropriate to these circumstances, and points to the Eh–pH condition of those springs in the COH delivering pristine karst groundwater as defining the other end-member of the Eh–pH range opposite the RMW environment in Fig. 83.

The association of some of the impacted karst groundwater with the stability field of amorphous iron oxyhydroxide [$Fe(OH)_3$], the most unstable (soluble) but also 'freshest' mineral form of Fe (Plate 3 in Chapter "Chemical Hydrology"), supports the postulated presence of dolomite armoring by these precipitates (Sect. 12.2.3).

**Text Box 1** The veracity of ORP measurements

Oxidation reduction potential (ORP or redox) measurements provide an indication of the electron activity [$e^-$] that informs the potential for electron transfer in a medium (usually water) as sensed by an inert metal electrode immersed in the medium, the measurement being compared to a reference electrode of known potential immersed in the same medium, and corrected

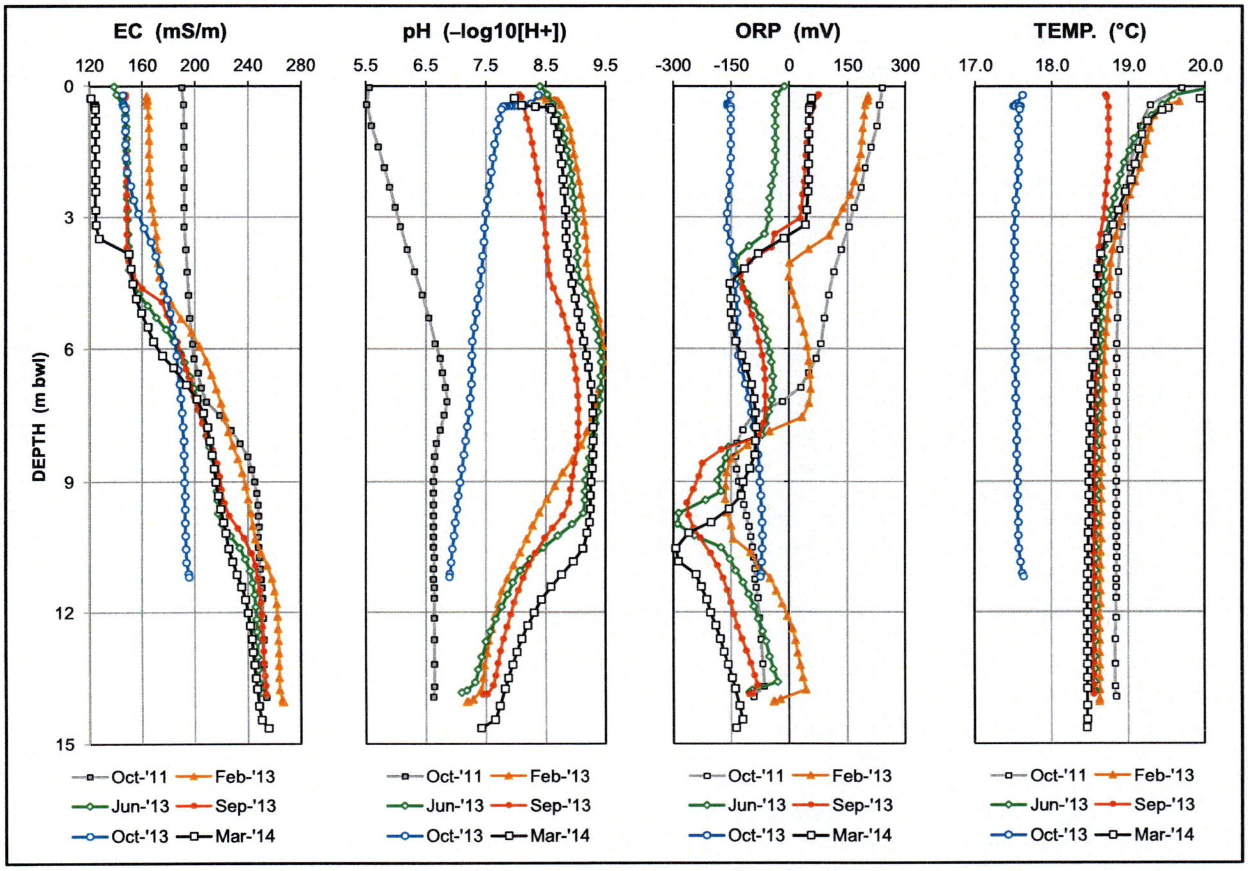


**Fig. 74** Temporal variation in vertical profiles of field water chemistry variables in borehole GP00300

accordingly (Appelo and Postma 2009). Electron transfer potential is characterised as either the ability to accept electrons (indicating an oxidising environment recognised by positive ORP values) or to furnish electrons (indicating a reducing environment recognised by negative ORP values). ORP is a generic (less specific) term for *Eh* (*pε*) and defines measurements made relative to any reference electrode. *Eh*, however, defines the ORP value measured specifically against a standard hydrogen electrode (SHE) which has a reference point of 0 mV, and where $p\varepsilon = -\log_{10}[e^-]$. The inert metal electrode on the ORP sensor model 6032/6565 of the YSI 6-series multiparameter sonde is made of platinum (Pt). The reference electrode is that of the pH sensor, a Ag/AgCl (silver:silver chloride) electrode fitted to the sonde. The sensor reading reported is a voltage (accuracy ±20 mV) relative to the reference electrode, e.g. +250 or −400 mV relative to Ag/AgCl. The difference in value between an $E_h$ reading and an ORP reading reflects the voltage offset between the different reference electrodes used. The ORP value can therefore be converted to an equivalent Eh value if the voltage offset is known, by adding/subtracting the offset value to/from the ORP value. This has not been determined for the ORP measurements reported in this document, which explains why the term Eh is not used in this context. A cursory review of the literature on redox measurements, in particular non-SHE reference electrode readings, reveals repeated cautions issued against the representativeness of these measurements (Stumm and Morgan 1981; Hem 1985; Appelo and Postma 2009; Weaver et al. 2007). The reasons for caution range from lack of equilibrium between different redox couples in the same medium, to "poisoning" of the Pt-electrode, e.g., the development of an iron oxyhydroxide (FeOOH) coating on the electrode when immersed in anoxic $Fe^{2+}$-rich water because of $O_2$ adsorbed on the electrode surface (Appelo and Postma 2009). Nevertheless, ORP measurements are

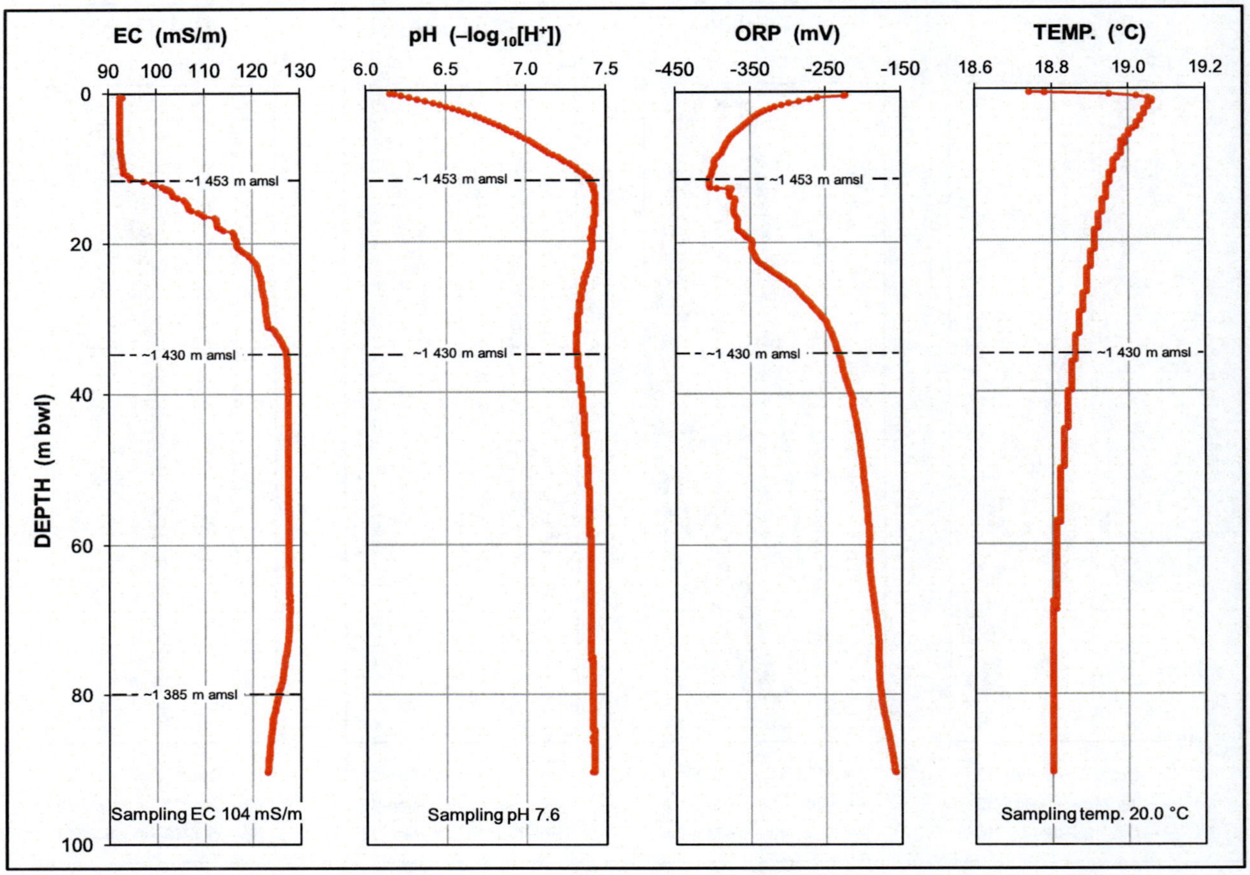


**Fig. 75** Vertical profiles of field water chemistry variables in borehole GP00301 in October 2011

considered particularly useful in water containing a comparatively high concentration of a redox-active species, examples being acid mine water containing metals ($Fe^{2+}$, $Fe^{3+}$) and municipal wastewater containing strong oxidising (chlorine) and reducing (sulfite) agents. From this it is apparent that ORP measurements only gain usefulness for media where a predominant redox-active species is known to be present, or specific information about the host environment is known.

Recognition of the complexities that inform the determination of ORP values as described above, as well as lack of rigorous control over what is really being measured (Appelo and Postma 2009), militates against an over-interpretation of the ORP (and for that matter also the Eh) data presented in this study. Montgomery et al. (2002) express a similar caveat in regard to the use of Eh and ORP measurements. Hem (1985) reports that the numerical value of Eh is an indicator of redox **intensity** with little value as an indicator of redox **capacity**. These limitations do not, however, detract from the value of qualitative inference provided by ORP (Eh) patterns and trends in regard to redox zones and conditions in the subsurface.

## 11 Bacterial Sulfate Reduction

The possibility that bacterial sulfate reduction (BSR) might be occurring in that portion of the karst aquifer where mine water entering the groundwater environment encounters infiltrating wastewater effluent must be considered. The organic matter associated with the latter might serve as the driving reductant, and positive indications in this regard are the comparatively low $SO_4$ and trace metal concentrations in groundwater in this portion of the Zwartkrans Basin. The process reduces $SO_4$ to $H_2S$ (hydrogen sulfide), which in

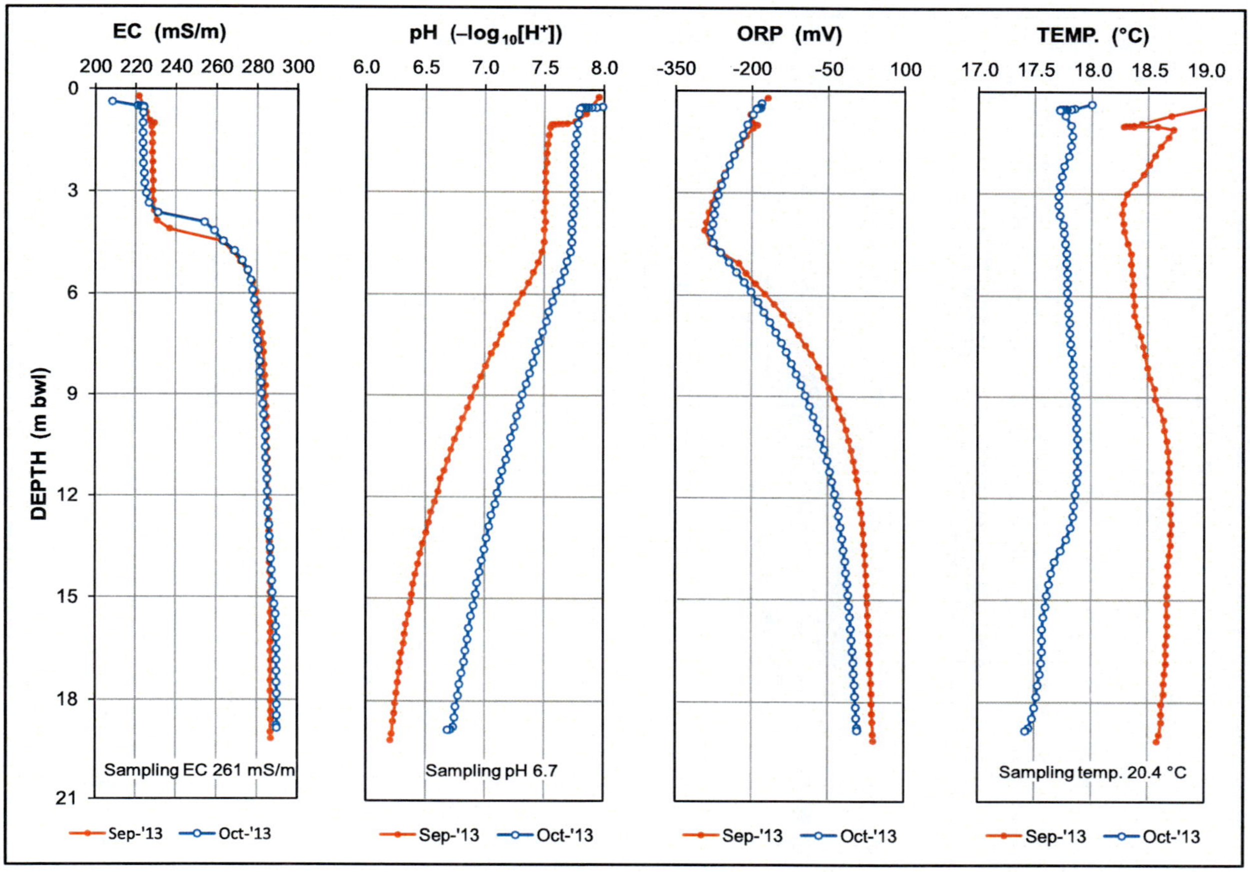


**Fig. 76** Pre- and post-sampling vertical profiles of field water chemistry variables in borehole GP00309

turn reacts with metal ions such as Pb, Cu, Hg, and As to form metallic sulfides. The availability of excess Fe-oxide may consume all $H_2S$ present in the groundwater (Appelo and Postma 2009), removing the characteristic 'rotten egg' odor from the water.

The precipitation of metal sulfides out of solution in a reducing (anaerobic) environment leads to a reduction in the concentrations of these metals in the groundwater. Further, Klaine et al. (2008) indicate that aquatic colloids comprise macromolecular organic materials that include the typical colloidal inorganic species of hydrous iron and manganese oxides. Their small nanoparticle size range (1 nm–1 μm) and large surface area per unit mass make them important binding phases for both organic and inorganic contaminants. Vesper et al. (2003) report the absorption of metals onto clays and other clastic particulates, onto organic material in the water, and onto iron or manganese oxides. Especially manganese oxides, recognized as a scavenger of radium (Sect. 5.2.2), are also recognized as extremely efficient scavengers of heavy metals. This possibility is demonstrated in Fig. 84.

Figure 84 shows the larger number of trace metals that exhibit elevated concentrations in the Riet Spruit surface water compared to those in the Blougat Spruit. Significant in the latter regard are Fe, Mn, and Ni. A further observation is the ubiquitous presence of Sr and, to a lesser extent Fe and V, in both surface water and groundwater sources.

Perhaps most significantly, however, is the undetectable presence (for the specified detection limits) of metals such as Al, Co, Mn, Ni, U, and Zn in the groundwater. This observation lends support for the possible occurrence of BSR in the karst aquifer. A study by Strosnider and Nairn (2010) demonstrated that concentrations of various trace metals including Al, As, Cd, Cr, Fe, Pb, V, and Zn reduced significantly when incubated in a mixture of acid mine water and municipal wastewater (see also Omoike and Vanloon 1999; Johnson and Younger 2006; McCullough 2008; McCullough et al. 2008; Strosnider et al. 2011; Hughes and Gray 2012). In essence, the mixing of these two environmental 'liabilities' is likely to generate beneficial reactions such as the following:

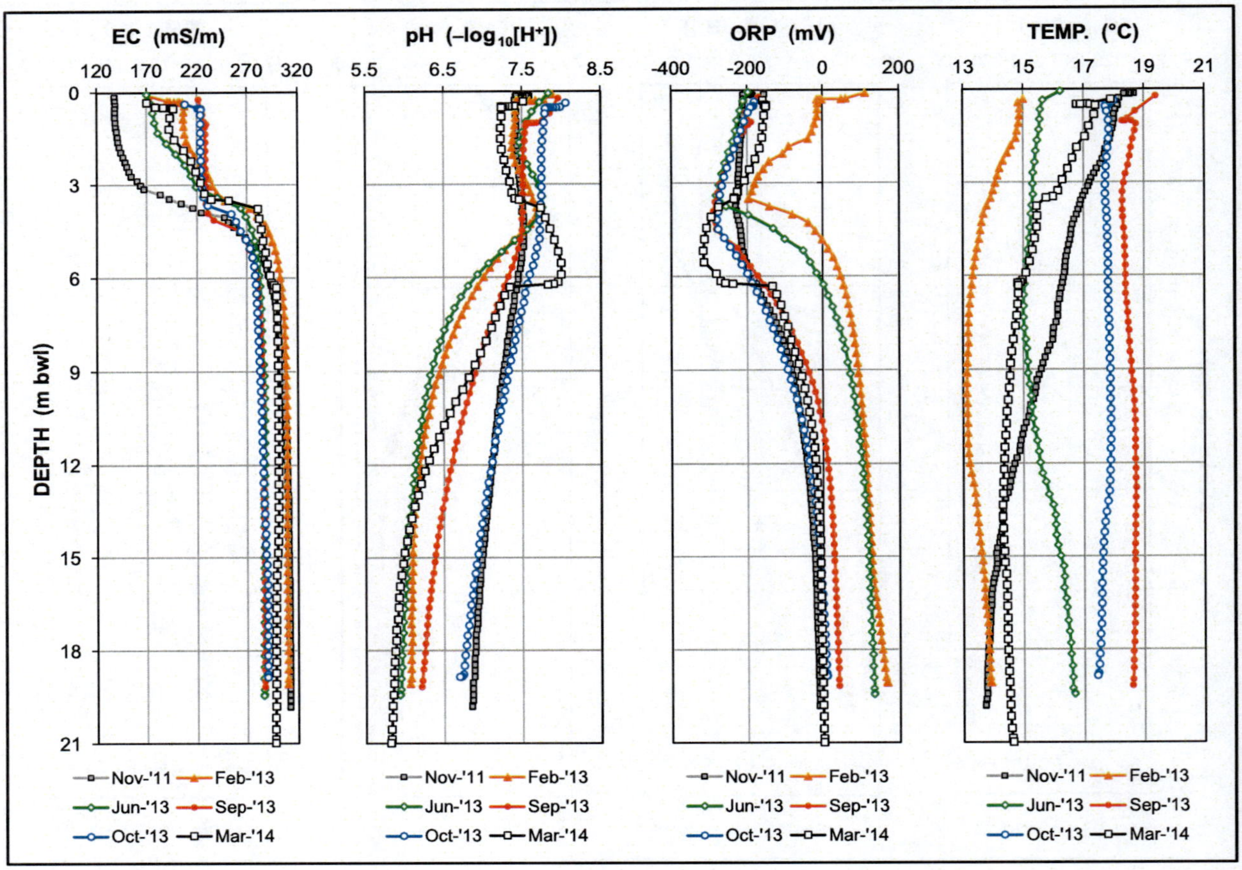


**Fig. 77** Temporal variation in vertical profiles of field water chemistry variables in borehole GP00309

- dilution of hydrogen ion concentrations in mine water leading to reduced solubility of many metals at higher pH;
- reaction of phosphorus with dissolved Al and Fe in mine water resulting in precipitation;
- binding of many metals to organic ligands present in the municipal wastewater; and
- organic matter in the municipal wastewater provides a source of carbon that sulfate-reducing bacteria utilize to generate alkalinity and precipitate metals as sulfides.

It has similarly been demonstrated by Winfrey et al. (2010) that fecal indicator bacteria (FIB) counts reduced significantly when subjected to co-treatment with municipal wastewater and acid mine drainage in an engineered laboratory-scale passive system. Tutu et al. (2008) also attribute the observed increase in pH of AMD-related water in wetlands in the Central Rand Goldfield to desulfurization by bacteria.

Further evidence for sulfate reduction (sulfidogenesis) in the karst aquifer is provided by the metallic sulfide coating (Plate 8) on the data loggers in boreholes A2N0584, GP00302, and GP00305. The black deposit suggests the conversion of $H_2S$ gas in the water, which might explain the absence of a 'rotten egg' odor (Cullimore 1999). The vertical hydrochemical profiles of these boreholes (Figs. 63, 69 and 70, respectively) do not reveal a common pattern or trend in the recorded variables, with only A2N0584 (Fig. 69) exhibiting strongly reducing conditions in the deeper section of the water column which favor the presence of sulfate-reducing bacteria (SRB). Similarly, an inspection of the groundwater chemistry associated with these boreholes indicates little congruence. It is postulated that anaerobic activities cause the increase in pH via (a) reduction of $Fe^{3+}$ hydroxides as per Reaction ?, (b) conversion of strongly acidic sulfate to weakly acidic sulfide as per Reaction ?, and (c) production of carbonate alkalinity as per Reactions ? and ?.

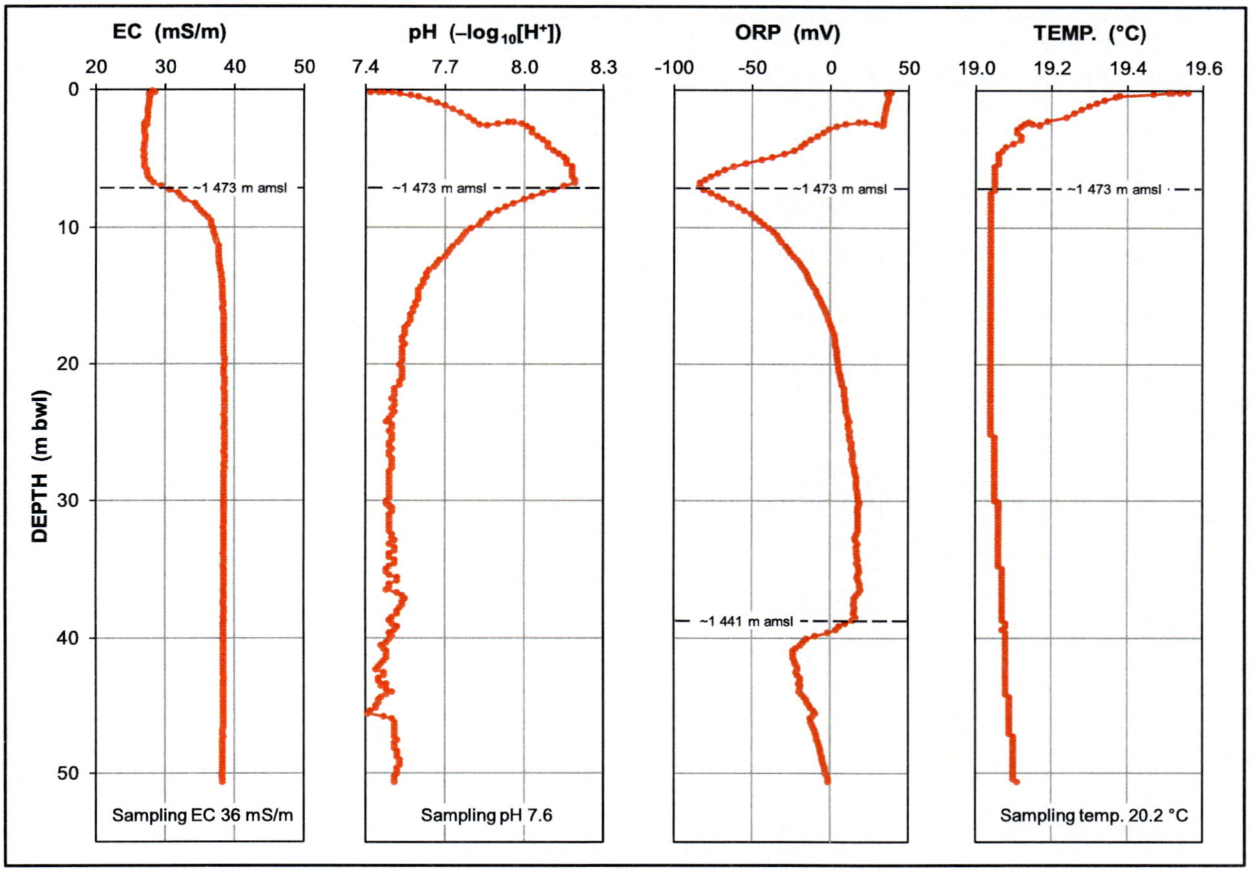


**Fig. 78** Vertical profiles of field water chemistry variables in borehole GP00311 in October 2011

## 12 Dissolution of Carbonate Strata

The subject of karst denudation is afforded a 25-page chapter in the authoritative publication *Karst Hydrogeology and Geomorphology* by Ford and Williams (2007), and discussed extensively by Gabrovšek (2009). An earlier comprehensive review of carbonate rock dissolution by Morse and Arvidson (2002) reports the limited understanding of the kinetics of dolomite dissolution. In a study of the role of karstic dissolution in the global carbon cycle, Gombert (2002) mainly considered calcite ($CaCO_3$) in this role. Locally, Dirks et al. (2010) report long-term erosion rates of 3–5 m/Ma (3–5 mm/Ka), derived from cosmogenic $^{10}$Be analysis of a quartz sample, for the plateau representing the African Erosion Surface south of the Malapa fossil site in the COH. More recently, Dirks and Berger (2013) describe the use of geologic and geomorphic evidence to explain landscape dynamics in the COH and propose an erosion rate of 5–6 m/Ma (5–6 mm/Ka) for dolomite along the plateau south of Malapa. The marginally greater erosion rate accommodates differential erosion between more resistant chert and less resistant dolomite. Although centered on an iconic South African karst environment, these studies afford the process of solutional karst denudation little attention as an agent in the development of this landscape.

Morse and Arvidson (2002) point out the complexities revealed by two approaches to the study of karst denudation, namely solution-orientated studies (those that consider the chemical composition of solutions) and surface complexation studies (those that consider the effect of reaction processes on mineral surfaces). The first of these approaches is explored on the basis of newly acquired data and an improved understanding that informs the hydrophysical and hydrochemical environments in the COH. The results obtained from formula-derived, empirically-derived, and theoretically-derived approaches, are compared with observationally-derived values for this landscape sourced from the literature. The results are applied to various fossil sites on the property to demonstrate their relevance in reconstructing the probable prehistoric landscape at these localities.

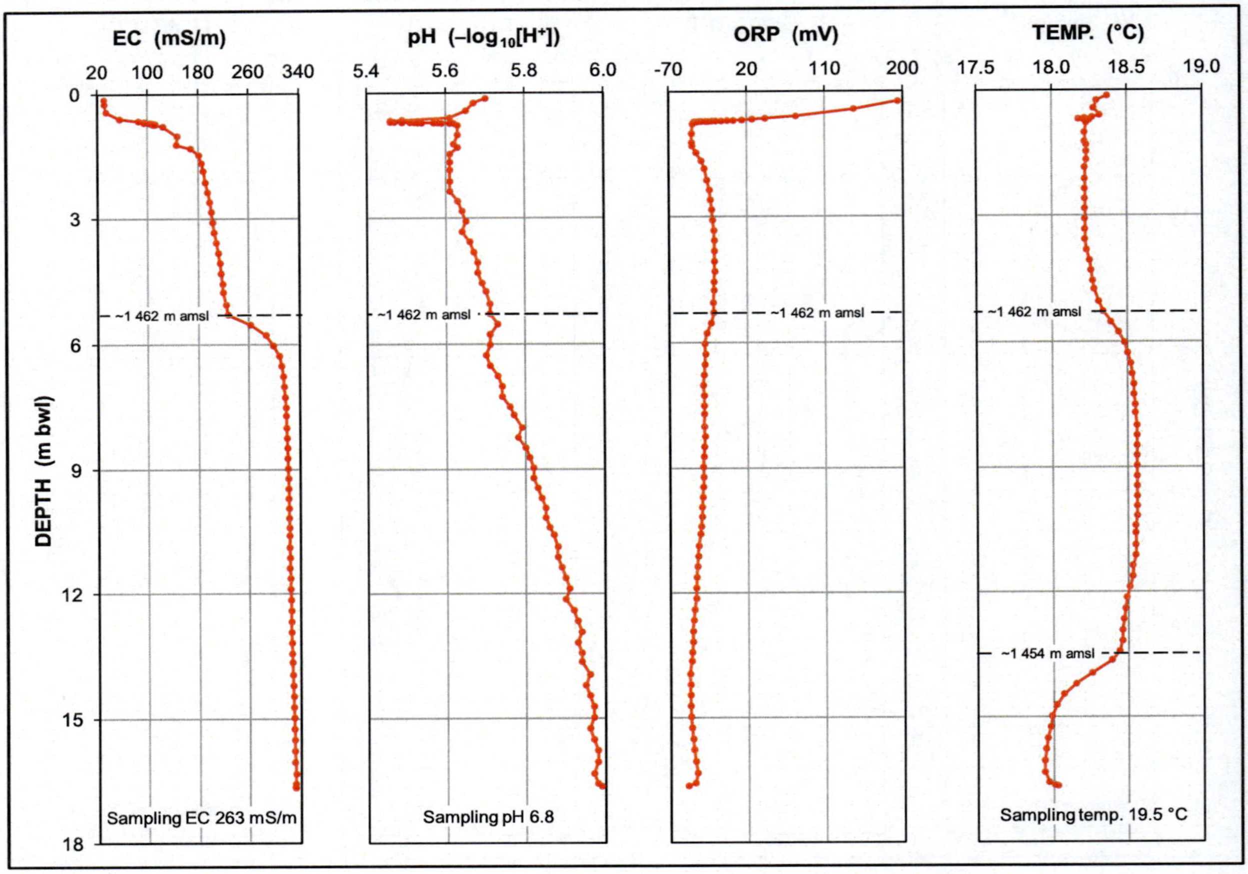


**Fig. 79** Vertical profiles of field water chemistry variables in borehole GP00312 in October 2011

## 12.1 Natural Solutional Denudation

Kaufmann (2003) recognizes three principal agents in the evolution of karst landscapes as being those associated with hillslope, fluvial and chemical processes, respectively. Embedded in karst development, these processes act in tandem, the hillslope processes comprising weathering, slope wash, and soil creep representing short-range agents, and the fluvial (erosion and sedimentation) and chemical (karst denudation) processes representing long-range agents, i.e., controlling landscape evolution over long distances. Chemical dissolution of dolomite is a natural phenomenon described by the reaction

$$\begin{aligned} &\mathrm{CaMg(CO_3)_{2(s)}} + 2\mathrm{CO_{2(g)}} + 2\mathrm{H_2O_{(l)}} \\ &\leftrightarrow \mathrm{Ca^{2+}_{(aq)}} + \mathrm{Mg^{2+}_{(aq)}} + 4\,\mathrm{HCO^-_{3\,(aq)}} \end{aligned} \quad (1)$$

The rate of karst denudation by chemical dissolution is controlled mainly by the partial $CO_2$ pressure ($P_{CO2}$) (White 1984; Kaufmann 2003), with temperature of lesser importance. The denudation rate increases by ~30% with a decrease in water temperature from 25 to 5 °C (White 1984). Under circumstances where soil $CO_2$ is the primary driver of dissolution in karst environments, Brook et al. (1983) derived an empirical relationship between soil $\log[P_{CO2}]$ values and mean annual actual evapotranspiration (AET) in mm described by the equation

$$\log[P_{CO2}] = -3.47 + 2.09\left(1 - e^{(-0.00172\,\mathrm{AET})}\right) \quad (2)$$

Applying a study area AET value of 1540 mm (Sect. 2 in Chapter "Description of the Physical Environment") in Eq. 2 gives a $\log[P_{CO2}]$ value of −1.53, i.e., $P_{CO2} = 10^{-1.53}$ (0.03% or 300 mg $CO_2$/L air).

In addition to the dissolution of carbonate strata, Bakalowicz (2005) recognizes the transport and removal of the dissolved material as a necessary condition for karst development. This is typically provided by groundwater flow, a condition met in the COH by spring discharges.

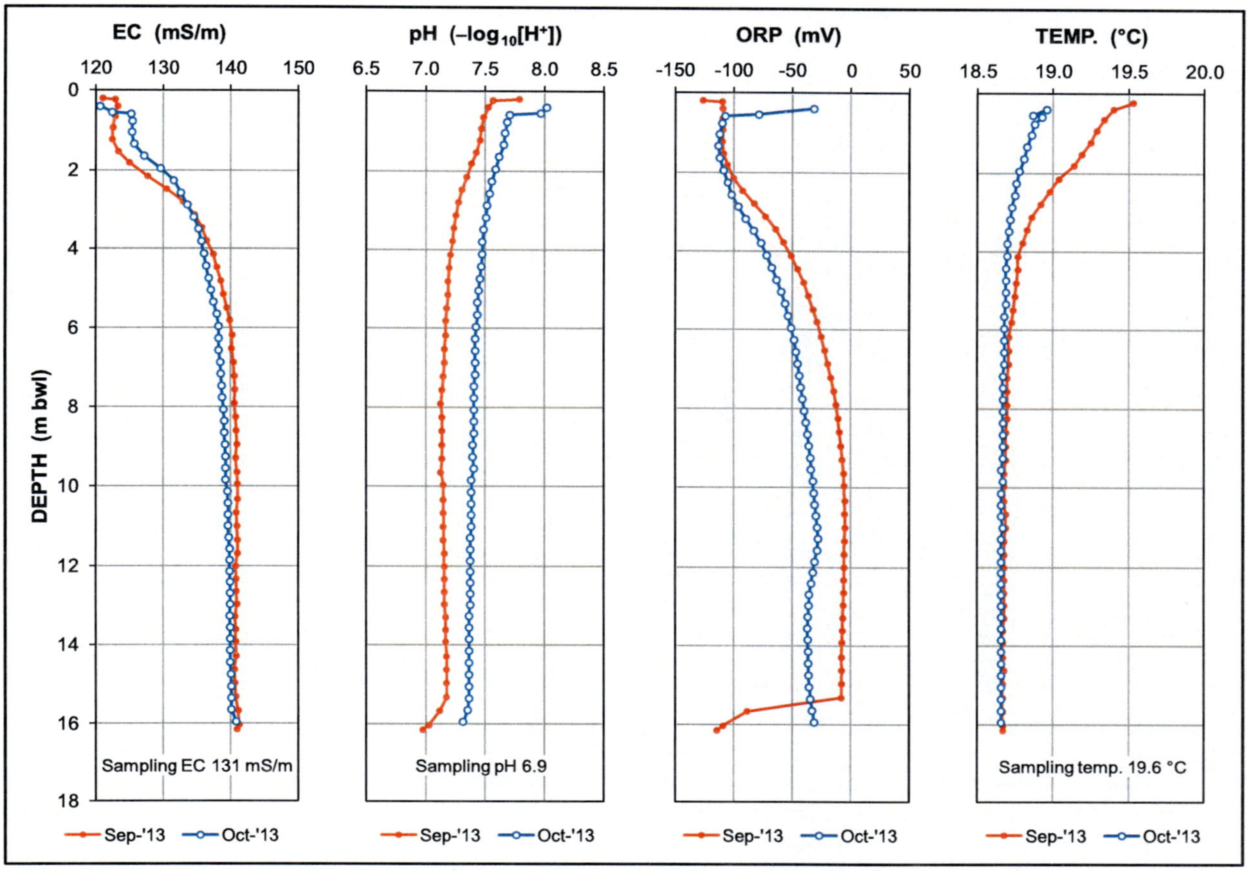


**Fig. 80** Pre- and post-sampling vertical profiles of field water chemistry variables in borehole GP00313

### 12.1.1 Formula Approach

The Corbel (1959) formula is the simplest and most generic equation for calculating a first-order solutional denudation rate of a karst environment (Ford and Williams 2007; Gabrovšek 2009). It calculates limestone solution ($D$ as mm/a) according to the relationship

$$D = 4 \cdot 1 \cdot T \div 100 \tag{3}$$

where $I$ is runoff (dm) and $T$ is the mean total alkalinity concentration (mg $CaCO_3$/L) of the water draining the karst environment, and the fraction 4/100 (≡ 100/2500) incorporates both the unit conversion factor of runoff to mm/a and the density of limestone (2500 kg/m$^3$). The application of the formula to the seven dolomitic compartments drained at rates >20 L/s returns the solutional denudation rates in the range ~4–16 mm/Ka reported in Table 16 for spring discharge and springwater chemistry ca. May 2010. The karst basins encompass 25,250 ha, which represents ~94% of the 26,860 ha footprint of dolomitic strata that underlie the COH.

A second set of spring discharge and springwater chemistry data for a smaller set of springs ca. August 2014, returns the solutional denudation rates reported in Table 17. Considering only those springs enumerated on both occasions, then the greater average runoff value of 1.62 dm/a (vs. 1.35 dm/a) and lower mean $CaCO_3$ value of 155 mg/L (vs. 169 mg/L) yield a denudation rate in the range ~4–13 mm/Ka for a mean of 10 mm/Ka compared to the 9.1 mm/Ka of the earlier data (Table 16).

The results might be questioned on the basis that the formula applies to limestone with a density of ~2500 kg/m$^3$, whereas that of dolomite is ~2850 kg/m$^3$ (Ford and Williams 2007). Although the formula also incorporates surface denudation, the results reflect primarily in situ dissolution as the $T$ values represent the $CaCO_3$ concentration in the primarily autogenic springwater. Gabrovšek (2007) postulates that of the many assumptions that underpin estimation of the denudation rate, at least one of the following two, namely that (a) most of the dissolution occurs close to surface in the epikarst, or (b) dissolution in the

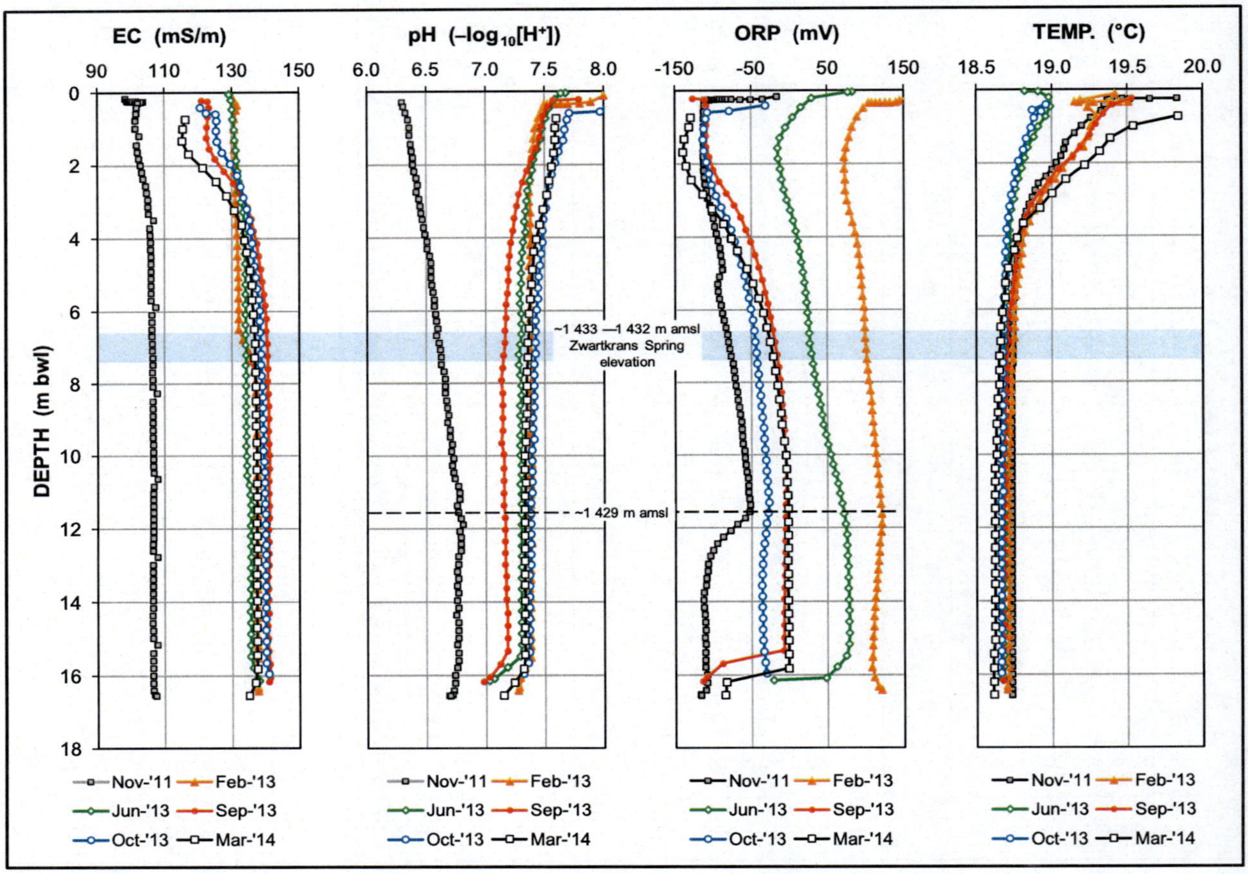


**Fig. 81** Temporal variation in vertical profiles of field water chemistry variables in borehole GP00313

long-term at depth is integrated into a surface denudation, must be valid. The karst basins generate surface water runoff only intermittently following episodic rainfall events.

Gabrovšek (2009) reports the following formula that considers the density of rock

$$D = I \times c_{eq} \div \rho \quad (4)$$

where $I$ is the difference between precipitation and evaporation (mm/a), $c_{eq}$ is mineral solubility under given temperature and $P_{CO2}$ conditions, and $\rho$ is the rock density (kg/m$^3$). The similarity to Eq. 3 is evident. Taking $P_{CO2} = 1^{-1.53}$ (0.03% or 300 mg $CO_2$/L air) as per Eq. 2 at a temperature of 20 °C, and equating $I$ to the groundwater recharge (Table 4 in Chapter "Physical Hydrogeology"), then Eq. 4 returns the dissolution rates presented in Table 18. Although generally slightly less than the values returned by Eq. 3 (Table 16), mainly attributable to the greater density of dolomite compared to limestone, the results returned by Eq. 4 are nevertheless in reasonable agreement.

Gabrovšek (2009) introduces a third equation for calculating the maximal solutional denudation as follows:

$$D = [M \times (P - E) \div 1000\,\rho] \times c_{eq}(P_{CO2}, T) \quad (5)$$

where $M$ is the molar mass of calcite (100 g/mol), $P–E$ is infiltration $I$ as in Eq. 4, $c_{eq}$ is the equilibrium concentration of calcite (mol/m$^3$) in an open system for prevailing $P_{CO2}$ and temperature, and $\rho$ is the rock density (g/cm$^3$). As might be expected, this equation returns the same results as Eq. 4 for $P_{CO2} = 1^{-1.53}$ (0.03% or 300 mg $CO_2$/L air) and 20 °C.

### 12.1.2 Empirical Approach

AL du Toit (in Brink 1996) calculated that the Gerhardminnebron Eye draining the Gerhardminnebron Basin in the Lower Wonderfontein Spruit at a rate of 57 ML/d (~660 L/s), removed >10 tons of dolomite in solution daily. This equates to a dissolution-to-flow rate ratio of 0.2 t/ML. For dolomite, this translates into the creation of a void volume at a rate of ~3.5 m$^3$/d (~1280 m$^3$/a). Similarly, Swart et al. (2003) report analyses of springwater

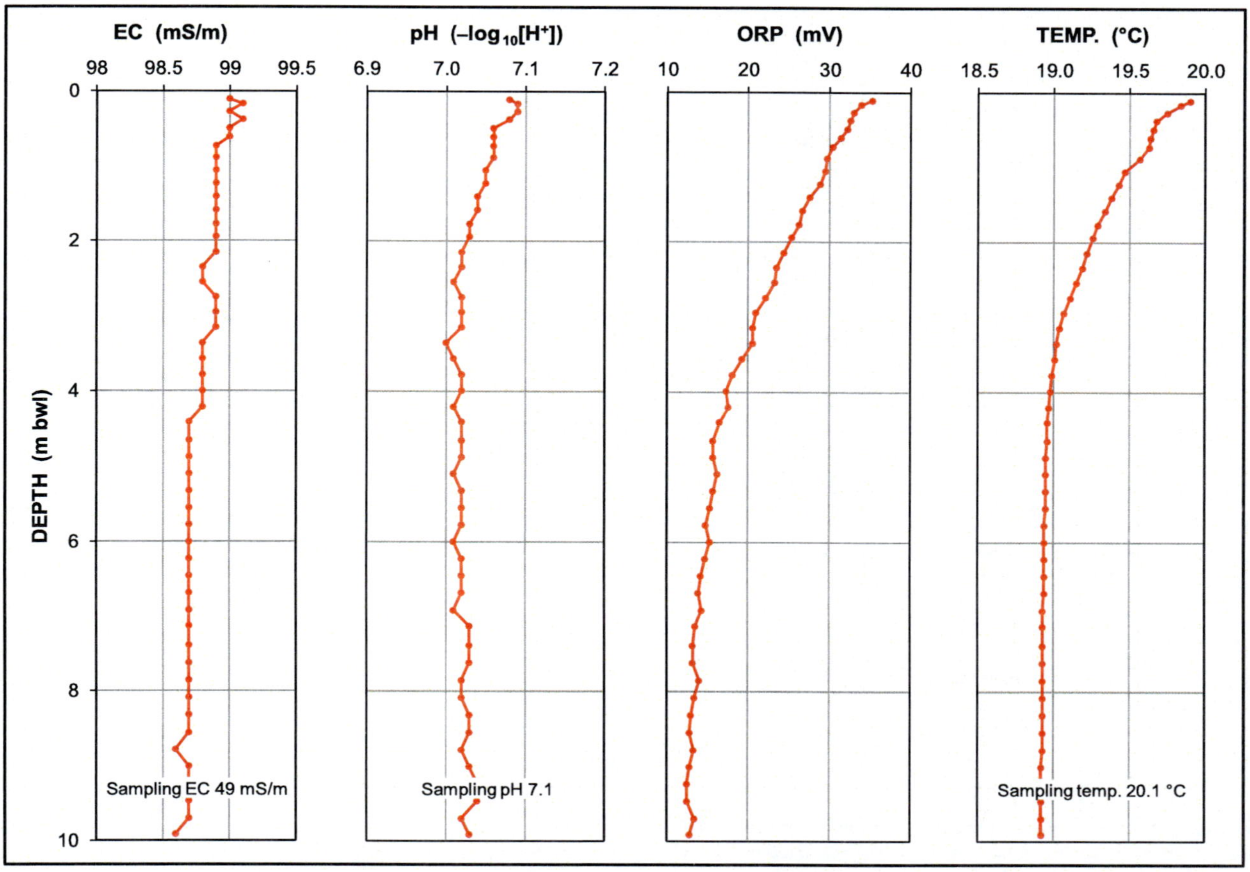


**Fig. 82** Vertical profiles of field water chemistry variables in borehole GP00314 in November 2011

draining the Wonderfontein Valley as containing ~200 kg of dissolved limestone per million liters (≡ 200 mg $CaCO_3$/L) in 1964. For a discharge of 200 ML/d, this implies the removal of >40 tons of limestone daily from this karst system. At a rate of ~14 $m^3$/d, this represents the development of a void volume of ~5110 $m^3$/a. A comparison with du Toit's ca. 1954 calculation reveals a similar dissolution-to-flow rate ratio of ~0.2 t/ML indicative of, among other factors, similar groundwater chemistry, and host groundwater-bearing strata.

Application of the dissolution-to-flow rate ratio of ~0.2 t/ML to the seven karst basins drained at rates >20 L/s, returns the dissolution rates reported in Table 19. The conversion from mass to volume is based on a rock density of 2850 kg/$m^3$. The conversion of the dissolution rates to representative void development rates provides a simplistic volumetric (3-dimensional) perspective for comparison with the above-mentioned spatial (2-dimensional) denudation rate values.

The similar dissolution (denudation) rates returned by Eq. 3 (Tables 16 and 17) and Eq. 4 (Table 18), and the empirically-derived values (Table 19) are evident in the respective mean values of 7.8, 7.2 and 7.9 mm/Ka. The correlation of the formula-derived values with the corresponding empirically-derived value is explored in Fig. 85. This suggests that the empirically-derived values exhibit a slightly better correlation with the Eq. 3 results ($R^2 = 0.79$) than with the Eq. 4 results ($R^2 = 0.73$). This observation is a little surprising seeing that Eq. 4 considers the density of dolomite rather than of limestone as in the case of Eq. 3 and should, therefore, return the better correlation.

### 12.1.3 Theoretical Approach

White (1984) and Kaufmann (2003) developed theoretical relationships between the solutional denudation rate and discharge/runoff (precipitation less evaporation) for different $P_{CO2}$ values and water temperatures. The equations incorporate terms that represent the strata and equilibria factors as well as rainfall, evapotranspiration, and temperature. Solutional denudation varies with the cube root of $P_{CO2}$ (Gabrovšek 2009) reflecting the dominance of the partial $CO_2$ pressure in the relation. This is illustrated in Fig. 86 for $P_{CO2}$

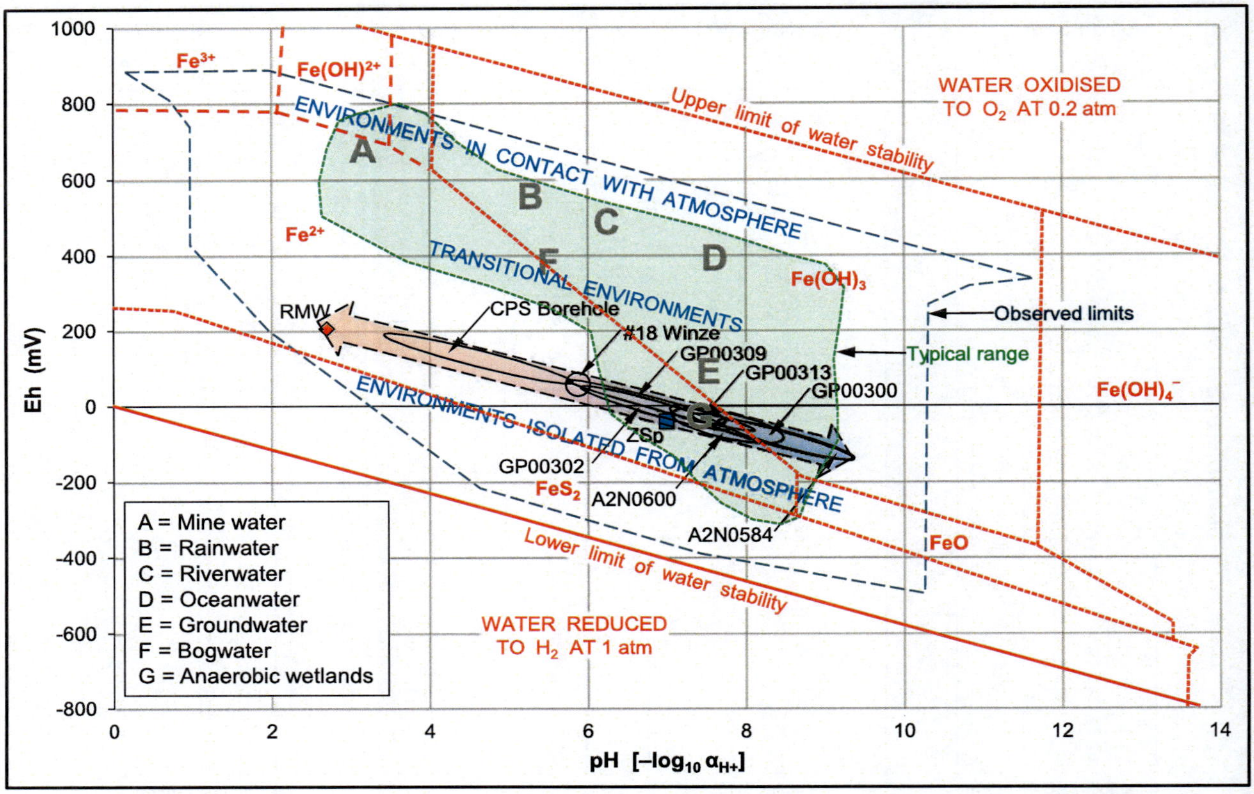


**Fig. 83** Pourbaix diagram showing the distribution of Eh–pH values in mine water and groundwater environments in the subregion in relation to limits in the natural environment (redrawn in green and blue from USDA 2012) as defined by Baas-Becking et al. (1960), Garrels and Christ (1965) and Langmuir (1996), and stability fields for solid and dissolved forms of Fe at 25 °C and 1 atm (redrawn in red from Hem 1985)

values of $10^{-1.5}$, $10^{-2.5}$, and $10^{-3.5}$ atm at a water temperature of 25 °C.

Figure 86 also translates the range defined by the runoff rates listed in Tables 16, 18 and 19, to the equivalent denudation rate ranges of White (1984) and Kaufmann (2003). Reverse extrapolation of the mean value of 7.6 mm/Ka (from Table 19) to the Kaufmann (2003) graph (Fig. 86) suggests that an average (groundwater) discharge (runoff) value of ~118 mm/a (~1.2 dm/a) characterizes the karst environment in the COH. This equates to the discharge of 955 L/s from a karst area of 25,250 ha reported in Table 16.

### 12.1.4 Application

The observation by Martini et al. (2003) that the Sterkfontein Cave system hosted a water level 20–25 m above the present water level ~3.3 Ma ago, offers an opportunity to apply and test the karst dissolution rates reported above. Hobbs (2011) reports the association of this dewatering interval with elevations in the range 1456–1461 m amsl. In terms of the present landscape, the Blaauwbank Spruit valley to the northwest of the caves would have been underwater to a depth of 15–20 m. It is more likely that the ~3.3 Ma BP landscape occupied a higher elevation, the prehistoric karst environment even then supporting a subterranean cave lake system as at present, with a water table located 20–25 m above the present water level.

Applying the Zwartkrans Basin denudation rate 'extremes' of 5.5 and 7.3 mm/Ka (Table 20) to the dewatering timeframe of ~3.3 Ma indicates a lowering of the land surface by ~18 m and ~24 m, respectively. Cognisant of the observation by White (2007b) that caves are the only features of the karst landscape for which "the age is locked in" and, as such, represent fixed elevation markers, then the values account quite reasonably for the magnitude of cave dewatering observed by Martini et al. (2003), lending support for the derived solutional denudation rates.

The above results also apply to the site of the Rising Star Expedition (Howley 2013) located in the Bolt's Farm complex of fossil sites. Located at a depth of ~30 m below present surface, the fossil 'deposit' would have occupied a depth of 42–44 m below surface some 2 Ma ago.

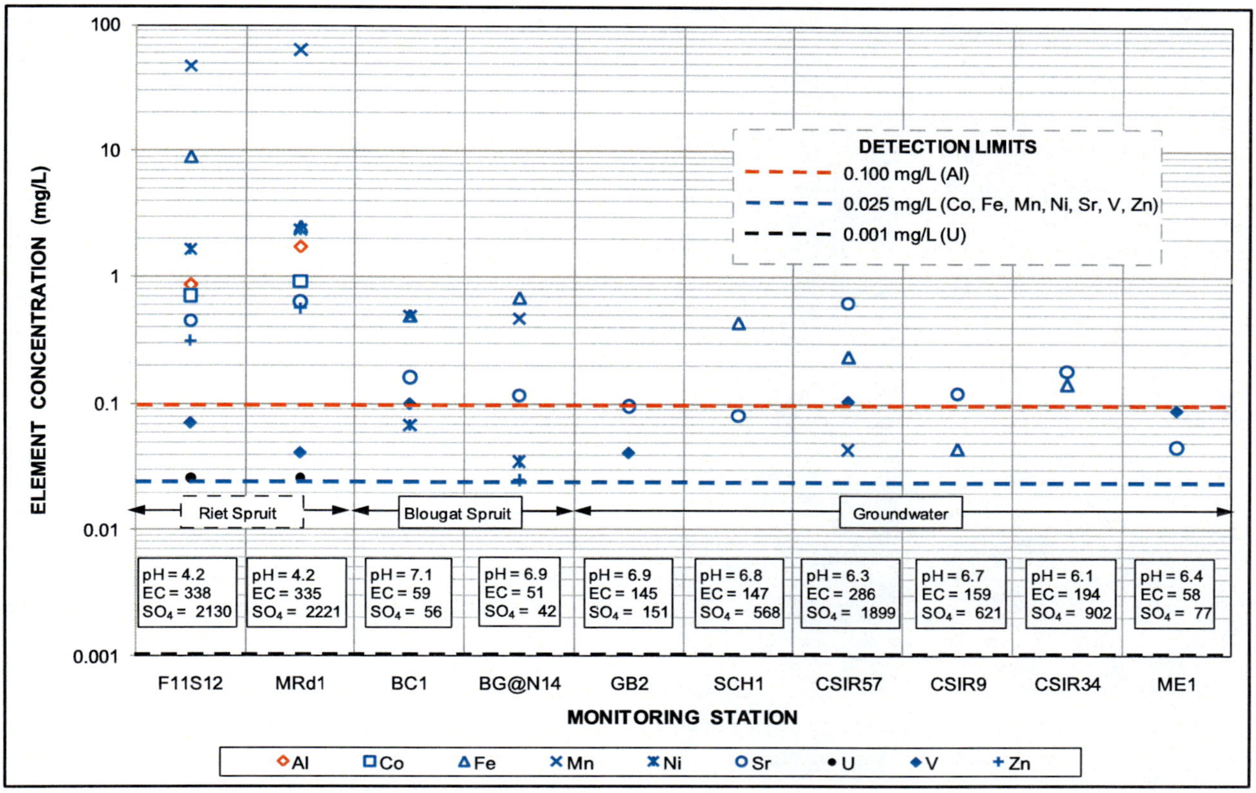


**Fig. 84** Comparison of trace metal and other variable concentrations in surface water and proximate groundwater sources (SEC as mS/m; $SO_4$ as mg/L)

A similar analysis for the Malapa site, employing the denudation rates of 9.3 and 8.2 m/Ma, suggests that this site was located ~16–19 m below surface at ~2 Ma BP. This is considerably less than the ~50 m suggested by Dirks and Berger (2013) using a relational computation that gives estimated denudation rates of ~28 and ~24 m/Ma. The difference might be accounted for by exaggerated erosion rates associated with hillslope and fluvial processes (Kaufmann 2003) in the aptly-named more rugged Diepkloof Basin in which the Malapa site is located.

The relative constancy of spring discharge in the COH has been discussed in Sects. 2 and 4 in Chapter "Physical Hydrology", and the similar constancy of springwater chemistry in Sects. 2.1 in Chapter "Chemical Hydrology" and 4 in Chapter "Physical Hydrogeology".

### 12.1.5 Discussion

The average values of 8.0, 7.4 and 8.1 mm/Ka reported in Tables 16, 18 and 19, respectively, are bracketed by the theoretically-based White (1984) and Kaufmann (2003) values of ~9 and ~7.5 mm/Ka (Fig. 86, main diagram). The study area soil $P_{CO2}$ of $10^{-1.53}$ and mean water temperature of 20.9 °C (Table 16) approach the values of $P_{CO2} = 10^{-1.5}$ atm (0.032%) and 25 °C employed in the theoretical relationships described in Fig. 86. Under these circumstances, the theoretical denudation rate of Kaufmann (2003) is in better agreement with the average values than that of White (1984).

The formula-derived solutional denudation rate of the seven karst basins falls in the range of 4.7–16.2 mm/Ka, compared to the 3.4–14.8 mm/Ka range of the empirically-derived rate (Table 20). Both these ranges of denudation rate accommodate the 5–6 mm/Ka reported by Dirks and Berger (2013) for the dolomite plateau south of the Malapa fossil site. It must be recognized that the development of the karst landscape in the COH has also involved the hillslope and fluvial processes recognized by Kaufmann (2003) as completing the trilogy of principal agents in landscape evolution. The erosion associated with fluvial processes is given context by Dirks et al. (2010), who report a rate of vertical incision of the Skeerpoort River below the Gladysvale fossil site of 53 ± 9 mm/Ka based on a cosmogenic $^{10}$Be study of valley bedrock strata. Dirks and Berger (2013) simplify this to ~55 mm/Ka.

Gladysvale cave is located in the Uitkomst Basin, which adjoins the Diepkloof Basin in which the Malapa site is located. These compartments occupy a similar

**Plate 8** Data logger (length ~20 cm) from monitoring borehole GP00305 completely coated with black iron sulfide (FeS) precipitate on stainless steel instrument (photo N. de Meillon)

hydrostratigraphic position along the north-western margin of the COH and exhibit similar mean denudation rates of 8.6 and 8.1 mm/Ka, respectively (Table 20). Compared to the ~85% greater fluvial down-cutting rate, the difference demonstrates the greater efficacy of the latter as an agent in landscape development in the presence of conducive physiographic conditions. These conditions also underpin the caution by Gabrovšek (2009) that methods that employ hydrochemistry (in this instance springwater chemistry) return a "batch result" of regional denudation that is subject to variation in climate and catchment over time.

The denudation rate values reported in Table 20 are similar to the range of 13.1–17.9 mm/Ka determined for the Cumberland Plateau, USA, from a geochemical mas balance (Florea 2015). Denudation rate values obtained by different methods in different environments listed by Gabrovšek (2009) and Florea (2015) as sourced from international literature provide for a more comprehensive comparison with the COH results reported in Table 20.

Values returned by the solute load method for four locations in Europe and one in New Zealand fall in the range of 48–95 mm/Ka. The order of magnitude greater results than are reported for the COH can mainly be attributed to much wetter climates. The local MAP of ~700 mm is >50% drier than the range 1450–3000 mm/a that characterizes the international locations. Much more comparable values in the range 6.0–14.6 mm/Ka are reported for limestone tablet exposure studies in tropical humid, temperate mountainous and Alpine environments with precipitation in the range 1300–3000 mm/a. Micro-erosion meter (MEM) and cosmogenic nuclide methods return higher values (15–150 mm/Ka) in similar MAP regimes. The lowest value of 1 mm/Ka was obtained by the cosmogenic nuclide method applied in an arid environment in Australia.

A further aspect of relevance to the historical landscape and cave water levels is the position of the 'Little Foot' fossil site in Sterkfontein Cave. The site is located ~25 m below surface and for a (maximum) present site surface elevation of 1487 m amsl places 'Little Foot' at an elevation of ~1462 m amsl (Bruxelles et al. 2014). This is at least ~22 m above the current maximum attainable cave water level of ~1440 m amsl. If the age of 'Little Foot' is now

**Table 16** Estimate of solutional denudation rate based on the Corbel formula (Eq. 3) for spring discharges and springwater chemistry in May 2010

| Basin | Area (ha) | Spring | Discharge (L/s) | Runoff (dm/a) | $CaCO_3$ (mg/L) | Temp (°C) | Denudation rate ($m^3/km^2/a \equiv mm/Ka$) |
|---|---|---|---|---|---|---|---|
| Zwartkrans | 9800 | Zwartkrans | 265 | 0.85 | 160 | 18.8 | 5.5 |
| Krombank | 5080 | Plover's Lake | 60 | 2.10 | 192 | 19.8 | 16.2 |
| | | Kromdraai | 279 | | | | |
| Danielsrust | 740 | Danielsrust | 28 | 1.19 | 128 | 20.4 | 6.1 |
| Tweefontein | 1160 | Tweefontein | 30 | 0.82 | 196 | 20.6 | 6.4 |
| Diepkloof | 3800 | Nouklip | 143 | 1.19 | 196 | 22.6 | 9.3 |
| Uitkomst | 2860 | Nash | 130 | 1.43 | 132 | 21.4 | 7.6 |
| Broederstroom | 1810 | Broederstroom | 20 | 0.35 | 272 | 22.8 | 3.8 |
| Total average | 25,250 | | 955 | 1.13 | 182 | 20.9 | 7.8 |

**Table 17** Estimate of solutional denudation rate based on the Corbel formula for spring discharges and springwater chemistry in August 2014

| Basin | Area (ha) | Spring | Discharge (L/s) | Runoff (dm/a) | $CaCO_3$ (mg/L) | Temp (°C) | Denudation rate ($m^3/km^2/a \equiv mm/Ka$) |
|---|---|---|---|---|---|---|---|
| Krombank | 5080 | Plover's Lake | 65 | 2.14 | 159 | 20.1 | 13.6 |
| | | Kromdraai | 279 | | | 19.4 | |
| Danielsrust | 740 | Danielsrust | 28 | 1.19 | 131 | 19.6 | 6.3 |
| Tweefontein | 1160 | Tweefontein | 26 | 0.71 | 156 | 20.5 | 4.4 |
| Diepkloof | 3800 | Nouklip | 211 | 1.75 | 197 | 21.4 | 13.8 |
| Uitkomst | 2860 | Nash | 208 | 2.29 | 130 | 21.3 | 11.9 |
| Total average | 13,640 | | 817 | 1.62 | 155 | 20.4 | 10.0 |

**Table 18** Estimate of solutional denudation rate based on Eq. 4

| Spring(s) | Recharge | | $c_{eq}$ (mg $CaCO_3$/L) | Denudation rate ($m^3/km^2/a \equiv mm/Ka$) |
|---|---|---|---|---|
| | % MAP[a] | mm/a | | |
| Zwartkrans | 18 | 130 | 160 | 7.3 |
| Plover's Lake + Kromdraai | 30 | 210 | 192 | 14.1 |
| Danielsrust | 17 | 119 | 128 | 5.3 |
| Tweefontein | 11 | 81 | 196 | 5.6 |
| Nouklip | 17 | 119 | 196 | 8.2 |
| Nash | 20 | 143 | 132 | 6.6 |
| Broederstroom + Anderson + Lesedi | 7 | 49 | 272 | 4.7 |
| Average | 17.1 | 122 | 182 | 7.4 |

[a]For an MAP of 710 mm

**Table 19** Empirically-derived estimate of dissolution rate based on the mass of $CaCO_3$ removed

| Spring | Discharge (ML/a) | Runoff (mm/a) | Mass of $CaCO_3$ removed (t/a) | Dissolution rate (mm/Ka) | Void development rate ($m^3/a$) |
|---|---|---|---|---|---|
| Zwartkrans | 8357 | 85 | 1671 | 6.0 | 586 |
| Plover's Lake + Kromdraai | 10,691 | 210 | 2138 | 14.8 | 750 |
| Danielsrust | 883 | 119 | 177 | 8.4 | 62 |
| Tweefontein | 946 | 82 | 189 | 5.7 | 66 |
| Nouklip | 4510 | 119 | 902 | 8.3 | 316 |
| Nash | 4100 | 143 | 820 | 10.1 | 288 |
| Broederstroom | 883 | 49 | 132 | 3.4 | 62 |
| Total average | 30,369 | 115 | 6030<br>861 | 8.1 | 2131 |

thought to be ~3.7 Ma (Gardner 2015), then it is conceivable that this individual met its end close to the contemporary cave water level at an elevation in the range of ~20–27 m above the current water table. Further, it might well have fallen further (as much as ~50 m) than the ~20 m reported in the Balter (2014) article if this is what led to its demise.

### 12.1.6 Conclusion

An assessment of karst solutional denudation rates in the COH based on formula-derived, empirically-derived, and theoretically-derived approaches yields results that are in agreement with recently reported values and with the observed dewatering horizon developed over 3.3 Ma in Sterkfontein Cave. Cognisant of the variation in

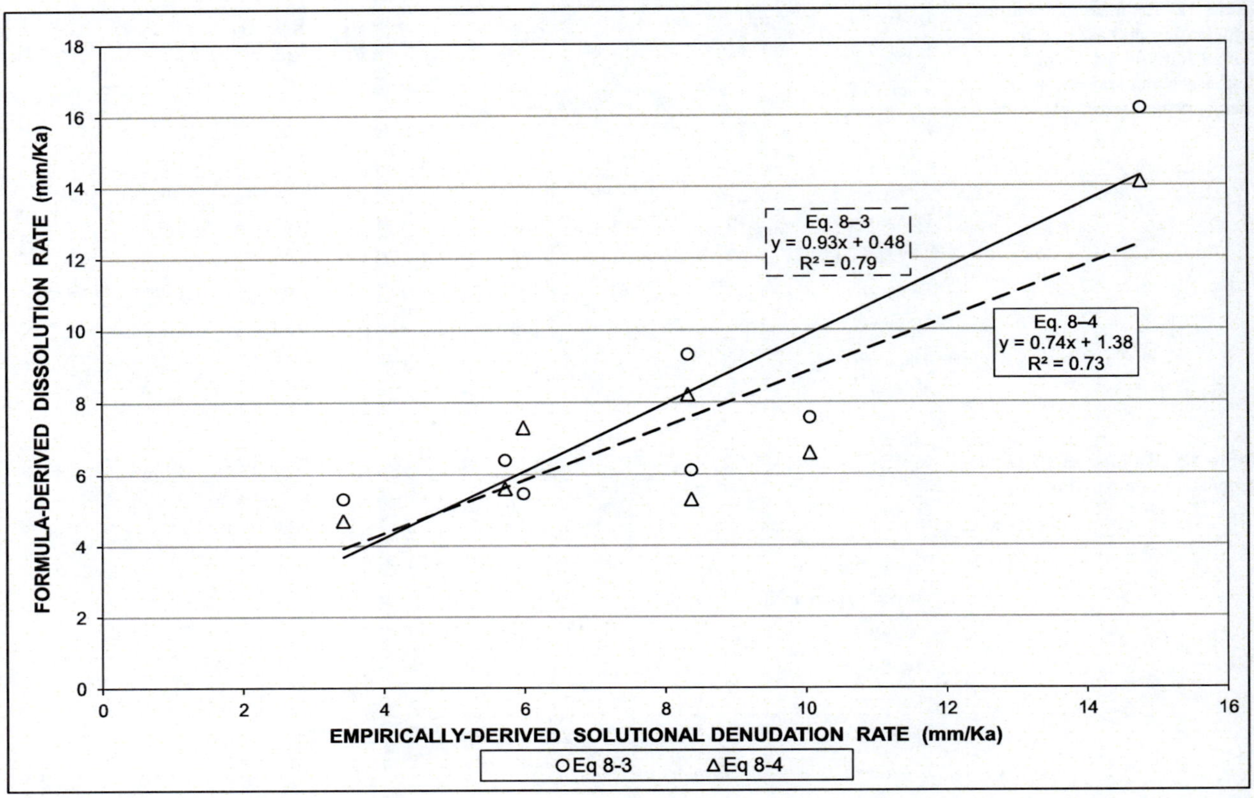


**Fig. 85** Correlation of empirically-derived and formula-derived karst denudation/dissolution rates

climatological and physiographical factors that influence landscape evolution in the long-term (White 2007b), the mean solutional denudation rate in the COH is placed at 7.8 mm/Ka in the range 5–15 mm/Ka.

Studies that consider the effect of reaction processes on mineral surfaces (the surface complexation studies defined by Morse and Arvidson 2002) in the evaluation of karst denudation remain unexplored in a South African context. The manifestation of a combined mine water and municipal wastewater impact on a portion of the COH karst environment provides an opportunity to explore this aspect against a comparison with the pristine portions of this environment.

## 12.2 Interaction with Mine Water

Sherlock et al. (1995) provide a synthesis of the neutralization of acid rock drainage (ARD) by carbonate and silicate minerals in the context of predicting and managing site-specific mining and processing wastes. Studies by Sasowsky and White (1993), Webb and Sasowsky (1994), and Sasowsky et al. (1995) provide valuable information on various aspects of the interaction between AMD and karst aquifers in the Obey River Basin in north-central Tennesee (USA). These studies illuminate the conjecture that characterizes the geochemical interaction between acidic mine water and the karst aquifer in another location where such interaction occurs, namely the Zwartkrans Basin in the COH.

The two main hypotheses that inform this conjecture are summarized as either dissolution or armoring of dolomite (carbonate strata). It is apparent that these mechanisms relate to the two approaches discussed by Morse and Arvidson (2002), namely studies that consider the chemical composition of solutions, and those that consider the effect of reaction processes on mineral surfaces. These are explored in the following sections.

### 12.2.1 Carbonate Dissolution

Neutralization of $SO_4$-rich AMD by calcite precipitates gypsum according to the reaction

$$CaCO_{3(s)} + H_2SO_{4(aq)} + 2H_2O_{(l)} \rightarrow CaSO_4 \cdot 2H_2O_{(s)} + 2H_2CO_{3(aq)} \quad (6)$$

and neutralization by dolomite precipitates epsomite according to the reaction (Lottermoser 2010)

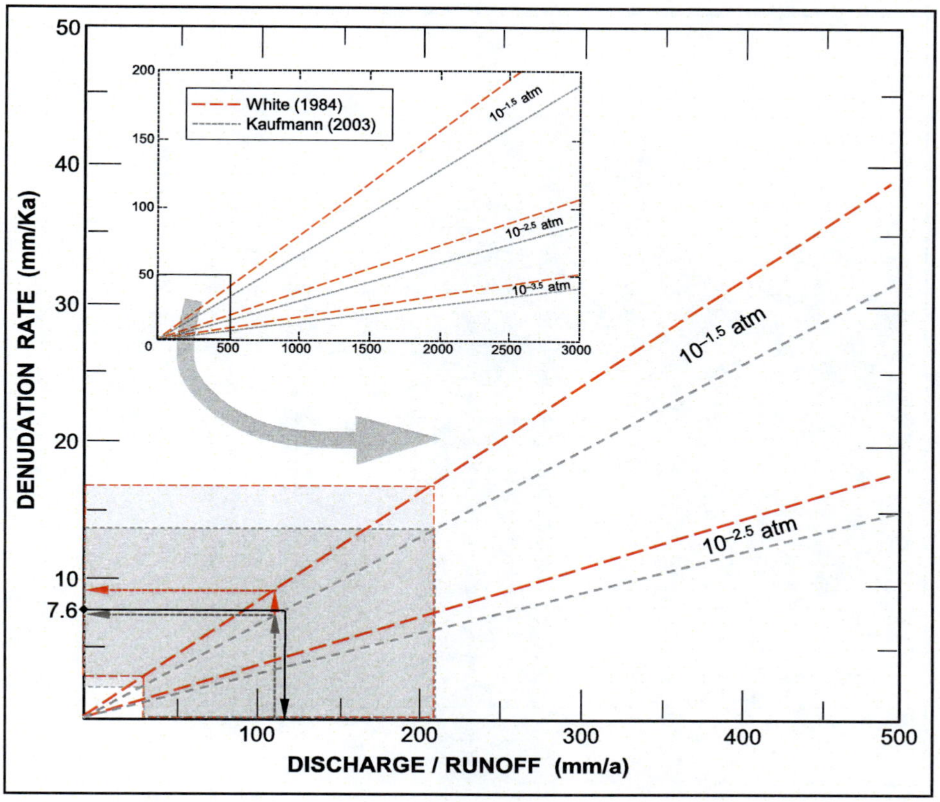


**Fig. 86** Denudation rate as a function of discharge for different $P_{CO2}$ values (insert diagram) at a temperature of 25 °C [redrawn from Ford and Williams (2007) and Kaufmann (2003)]; the shaded areas (main diagram) define the identical runoff ranges listed in Tables 16, 18 and 19, and the pecked arrows the mean discharge/runoff rate (113 mm/a) and associated theoretical denudation rates for a $P_{CO2}$ of $10^{-1.5}$ atm; see text for discussion of reverse (anti-clockwise) arrow

**Table 20** Comparison of formula-derived and empirically-derived solutional denudation rates per karst basin and associated fossil site(s)

| Karst Basin | Solutional denudation rate ($m^3/km^2/a$ ≡ mm/Ka ≡ m/Ma) | | | | Fossil site |
|---|---|---|---|---|---|
| | Formula-derived | | Empirically-derived (Table 19) | Average | |
| | Equation 3 (Table 16) | Equation 3 (Table 18) | | | |
| Zwartkrans | 5.5 | 7.3 | 6.0 | 6.3 | Sterkfontein<br>Bolt's Farm<br>Swartkrans<br>Minnaar's Cave<br>Cooper's Cave<br>Kromdraai |
| Krombank | 16.2 | 14.1 | 14.8 | 15.0 | Plover's Lake<br>Wonder Cave |
| Danielsrust | 6.1 | 5.3 | 8.4 | 6.6 | Drimolen |
| Tweefontein | 6.4 | 5.6 | 5.7 | 5.9 | – |
| Diepkloof | 9.3 | 8.2 | 8.3 | 8.6 | Malapa |
| Uitkomst | 7.6 | 6.6 | 10.1 | 8.1 | Gladysvale |
| Broederstroom | 5.3 | 4.7 | 3.4 | 4.5 | Gondolin |
| Average | 8.0 | 7.4 | 8.1 | 7.8 | |

**Plate 9** Examples of **a** hard ferrihydrite crust effectively sealing the bottom of a stream channel, and **b**, **c** entombing/entombed in-stream vegetation. (Photo **b**, B Genthe)

$$CaMg(CO_3)_{2(s)} + 2H_2SO_{4(aq)} + 9H_2O_{(l)} \rightarrow MgSO_4 \cdot 7H_2O_{(s)} + CaSO_4 \cdot 2H_2O_{(s)} + 2H_2CO_{3(aq)} \quad (7)$$

The dissolution of dolomite in the presence of free protons is also more simply described by the reaction

$$CaMg(CO_3)_{2(s)} + 4H^+_{(aq)} \rightarrow Ca^{2+}_{(aq)} + Mg^{2+}_{(aq)} + 2H_2CO_{3(aq)} \quad (8)$$

which consumes four moles of $H^+$ for each mole of Ca and Mg released into solution and produces carbonic acid ($H_2CO_3$) which adds alkalinity to the groundwater system through the release of bicarbonate ($HCO_3$) according to the reaction

$$H_2CO_{3(aq)} \rightarrow H^+_{(aq)} + HCO^-_{3(aq)} \quad (9)$$

Further dissociation of the $H_2CO_3$ results in the exsolution (degassing or release) of carbon dioxide ($CO_2$) according to the reaction

$$H_2CO_{3(aq)} \rightarrow H_2O_{(aq)} + CO_{2(g)} \quad (10)$$

whereas Reaction 8 can proceed at a pH of up to slightly >8 where solid $CaMg(CO_3)_2$ is stable, i.e., the solution is in equilibrium with the atmosphere where $P_{CO2} \approx 10^{-3.5}$ atm, Reactions 9 and 10 proceed only below a pH of ~6.4 (Barnes and Romberger 1968).

Scott (1995) reports the 'consumption' (dissolution) of dolomite in the East Rand Basin as ranging from ~0.9 t/d (321 t/a) to ~18 t/d (~6400 t/a) for 'flow' rates of 1 ML/d (most realistic) and 20 ML/d (worst case scenario), respectively, for 'modeled' ambient water chemistry. These values represent void volumes of ~115 to ~2250 $m^3$ being created annually, i.e., at a rate of ~0.3 to ~6 $m^3$/d. They also represent a dissolution-to-flow rate ratio of ~0.9 t/ML, which is nearly five times the value of ~0.2 t/ML associated with 'pristine' karst groundwater (Sect. 12.1.2).

Apart from the dissolution of carbonate rocks, however, a second main source of positive alkalinity in mine drainage is $SO_4$ reduction (Kirby and Cravotta 2005a). This releases bicarbonate ($HCO_3$) to solution according to the reaction

**Table 21** Net acidity and net alkalinity of various groundwaters in the COH environment

| Water source | Water type[a] | Acidity calculation analytes | | | | | | | |
|---|---|---|---|---|---|---|---|---|---|
| | | Total alkalinity (mg $CaCO_3$/L) | pH[b] | Fe (mg/L) | Mn (mg/L) | Al (mg/L) | Total acidity | Net alkalinity | Net acidity |
| | | | | | | | (mg $CaCO_3$/L) | | |
| CSIR57[c] | 1 | 28 | 6.3 | 0.237 | 0.044 | <0.1 | 0.53 | 27 | −27 |
| CSIR57[d] | | 108 | 6.8 | <0.025 | <0.025 | n.a. | 0.10 | 108 | −108 |
| CSIR57[e] | | 4.1 | 5.7 | 0.003 | 14 | 0.2 | 27 | −23 | 23 |
| CSIR8 | 2 | 140 | 6.9 | 0.036 | n.a. | <0.1 | 0.63 | 139 | −139 |
| CSIR9 | 2 | 128 | 6.7 | 0.045 | n.a. | <0.1 | 0.65 | 127 | −127 |
| CSIR34 | 2 | 56 | 6.1 | 0.145 | 0.025 | <0.1 | 0.90 | 55 | −55 |
| NR1 | 2 | 224 | 7.1 | 1.62 | 0.061 | <0.1 | 3.6 | 220 | −220 |
| DC1 | 2 | 148 | 7.5 | 0.037 | n.a. | <0.1 | 0.62 | 147 | −147 |
| GB2 | 2 | 252 | 6.9 | 0.025 | 0.025 | <0.1 | 0.65 | 251 | −251 |
| SCH1 | 2 | 160 | 6.8 | 0.436 | 0.025 | <0.1 | 1.4 | 159 | −159 |
| ZSp[e] | 2 | 160 | 7.5 | <0.025 | <0.025 | 0.116 | 0.74 | 159 | −159 |
| ZSp[f] | | 169 | 7.1 | 0.040 | 0.005 | 0.20 | 1.20 | 168 | −168 |
| SFC[f] | 2 | 169 | 7.6 | 0.032 | 0.011 | 0.20 | 1.19 | 168 | −168 |
| SFC[g] | | 152 | 7.9 | <0.025 | <0.025 | 0.20 | 1.20 | 151 | −151 |
| SFC[h] | | 154 | 7.6 | <0.02 | <0.005 | 0.20 | 1.16 | 153 | −153 |

[a]1 = impacted karst groundwater; 2 = slightly impacted karst groundwater
[b]$-\log_{10}\alpha_{H+}$
[c]Sample date 16/02/2010
[d]Sample date 18/03/2007
[e]Sample date 27/08/2014
[f]Sample date 16/05/2014
[g]Sample date 13/05/2010
[h]Sample date 14/01/2011

**Table 22** Free acidity and metal acidity associated with the results presented in Table 21

| Water source | Water type | Free acidity (mg $CaCO_3$/L) | Metal acidity (mg $CaCO_3$/L) | Total (calculated) acidity (mg $CaCO_3$/L) | Free: total acidity ratio (%) |
|---|---|---|---|---|---|
| CSIR57 | 1 | 0.01 | 0.09 | 0.10 | 10.0 |
| | | 0.03 | 1.06 | 1.09 | 2.75 |
| | | 0.1 | 12.6 | 26.7 | 0.37 |
| CSIR8 | 2 | 0.01 | 0.62 | 0.63 | 1.59 |
| CSIR9 | 2 | 0.01 | 0.64 | 0.65 | 1.54 |
| CSIR34 | 2 | 0.04 | 0.86 | 0.90 | 4.44 |
| NR1 | 2 | 0 | 3.57 | 3.57 | 0.1 |
| DC1 | 2 | 0 | 0.62 | 0.62 | 0.3 |
| GB2 | 2 | 0.01 | 0.65 | 0.66 | 1.52 |
| SCH1 | 2 | 0.01 | 1.38 | 1.39 | 0.72 |
| ZSp | 2 | 0 | 0.73 | 0.74 | 0.2 |
| | | 0 | 1.19 | 1.20 | 0.3 |
| SFC | 2 | 0.0006 | 1.2 | 1.2 | 0.05 |
| | | 0.0013 | 1.2 | 1.2 | 0.11 |
| | | 0.0013 | 1.2 | 1.2 | 0.11 |

$$2CH_2O_{(s)} + SO^{2-}_{4(aq)} \rightarrow H_2S_{(g)} + 2HCO^{-}_{3(aq)} \quad (11)$$

where $2CH_2O$ represents organic matter.

Kirby and Cravotta (2005a) also report that reductive dissolution of metal hydroxides may contribute to alkalinity according to the reaction

$$CH_2O_{(s)} + 4Fe(OH)_{3(s)} + 7H^{+}_{(aq)} \rightarrow 4Fe^{2+}_{(aq)} + HCO_{3(aq)} + 10H_2O_{(l)} \quad (12)$$

It has previously been reported (Sect. 1.2.4 in Chapter "Chemical Hydrology") that the treated effluent discharged from the Percy Stewart WWTW carries an elevated bacteriological and nutrient (N, P, and COD) load (see Tables 11, 13 and 14 in Chapter "Chemical Hydrology"). Further, this water also enters the karst aquifer at a rate of ~7 ML/d (Table 19). This water mixes in the subsurface with the allogenic recharge of mine water (Sect. 5.2.1 in Chapter "Physical Hydrology"), the mixing nexus coinciding with the area around the confluence of the Riet and Blougat spruits targeted for monitoring by borehole GP00309. These circumstances reflect the addition of an organic matter component that potentially promotes in situ 'natural' bioremediation associated with the generation of alkalinity through bacteriological sulfate reduction (Knoeller et al. 2004; McCullough 2008; McCullough et al. 2008) in this portion of the karst aquifer.

Finally, it is worth noting the findings of Brown and Glynn (2003) that the in situ field dissolution rates of calcite and dolomite experimentally exposed to active acid mine drainage were 1000 times less than the rate calculated from laboratory-derived equations. Further, the dolomite in situ dissolution rate was typically an order of magnitude smaller than that of calcite. This is in keeping with the observation by Busenberg and Plummer (1982) that the $CaCO_3$ component of dolomite dissolves preferentially relative to $MgCO_3$ on exposure to acidic conditions. The import of these findings for the impact of AMD on the karst environment in the study area cannot be ignored.

### 12.2.2 Dedolomitization

The possibility that the ingress of $SO_4$-rich mine water into the dolomitic aquifer of the Zwartkrans Basin may cause dedolomitization similarly cannot be ignored. Although this phenomenon is comparatively rare in natural karst environments (Ford and Williams 2007), Plummer et al. (1990) describe its manifestation as calcite precipitation and dolomite dissolution driven by anhydrite (anhydrous $CaSO_4$) dissolution in the carbonate Madison aquifer, USA. The total mass transfer is given by the reaction (Appelo and Postma 2009)

$$1.8\,CaSO_4 + 0.8\,CaMg(CO_3)_2 \rightarrow 1.6\,CaCO_3 + Ca^{2+} + 0.8\,Mg^{2+} + 1.8\,SO_4^{2-} \quad (13)$$

where for each mole of dolomite that dissolves, two moles of calcite are precipitated and the $Ca^{2+}$, $Mg^{2+}$ and $SO_4{}^{2-}$ concentrations increase. The ingress of $SO_4$-rich mine water as allogenic recharge into the dolomitic aquifer of the Zwartkrans Basin provides the driver for potential dedolomitization. This possibility adds another layer of complexity to the hydrogeochemical framework that informs the mine water impacted karst water resource.

### 12.2.3 Dolomite Armoring

Armoring (or protection) of dolomite results from the precipitation of insoluble oxyhydroxides [mainly ferric hydroxide or ferrihydrite, $Fe(OH)_3$] on the surface of the dolomite within fractures and karst features according to the reaction

$$4Fe^{2+}_{(aq)} + 10\,H_2O_{(l)} + O_{2(g)} \rightarrow 4\,Fe(OH)_{3(s)} + 8H^{+}_{(aq)} \quad (14)$$

Mineral surface passivation (Lindsay et al. 2015) by the efflorescent coating inhibits any further dissolution and reaction(s) that mediate neutralizing acidity in the water through dissolution of the dolomite. It is this coating/crust that also protects structures such as the concrete weir visible in Plate 3 in Chapter "Chemical Hydrology" from attack by the acidic surface water. Unfortunately, the precipitate is fatal to aquatic systems because the crust seals channel bottoms and entombs in-stream vegetation (Plate 9). It is also the armoring effect that inhibits the neutralizing capacity of crushed limestone when used for this purpose in passive AMD treatment systems such as open limestone channels (OLCs), limestone leach beds (LLBs), and anoxic limestone drains (ALDs). Ziemkiewicz et al. (1997), however, established from field studies that armored limestone in OLCs reduced acidity by between 10 and 60% of the rate of unarmored limestone. The greatest reductions were associated with steeper channel gradients (45–60%) and for mine water with a net acidity of 500–2600 mg $CaCO_3$/L.

Barnes and Romberger (1968) recognized >40 years ago the possible treatment option of introducing AMD via boreholes into carbonate strata. Clearly, environmental considerations at the time were considerably less emotive than today. The principal disadvantage is that neutralization occurs at the expense of added hardness. Further, dolomite as a source of carbonate produces undesirable quantities of magnesium salts ($MgSO_4$).

Sasowsky et al. (1995) posit the possibility that the availability of large amounts of $CO_2$ produced by Reaction 10 might also promote the precipitation of siderite ($FeCO_3$) according to the reaction

$$Fe^{2+}_{(aq)} + CO_{2(g)} + H_2O_{(l)} \rightarrow FeCO_{3(s)} + 2H^{+}_{(aq)} \quad (15)$$

Webb and Sasowsky (1994) further report the experimental results of Bradley and Herman (1989) that indicate

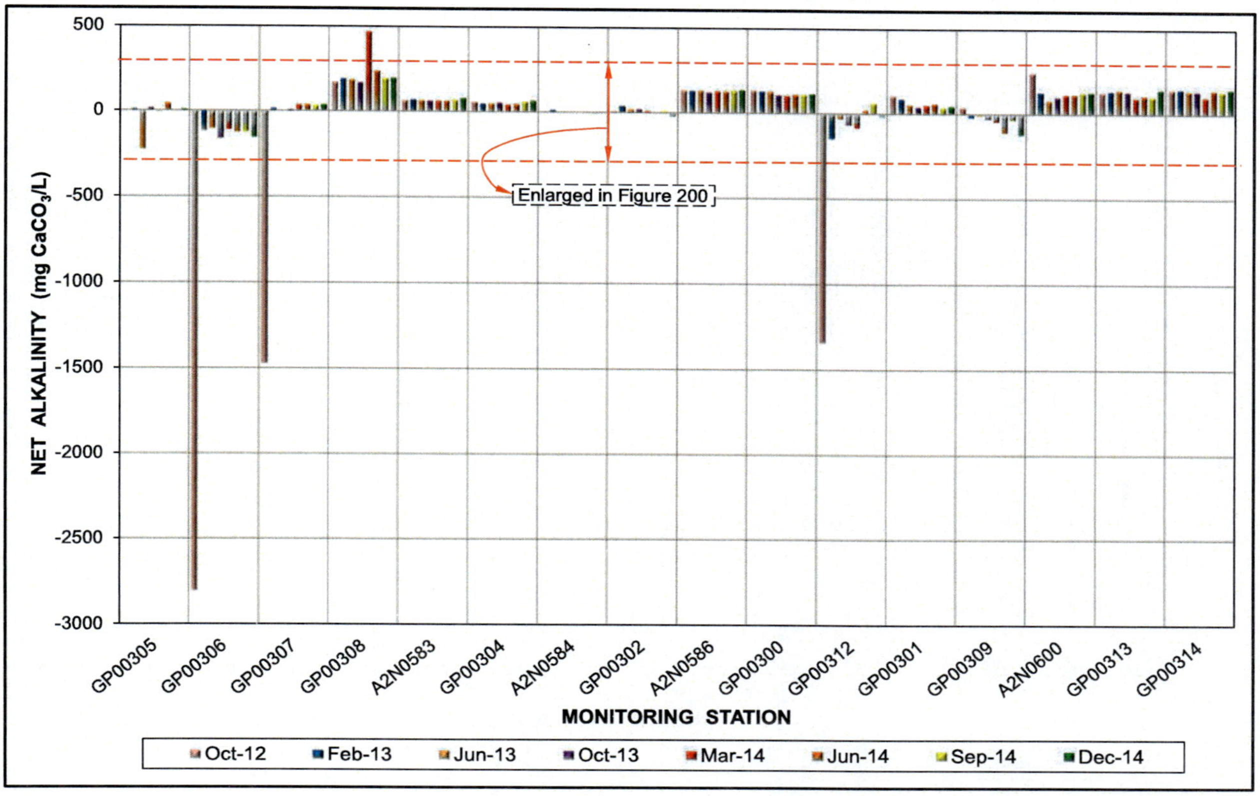


**Fig. 87** Pattern and trend of net alkalinity in groundwater from monitoring boreholes

the adsorption of sulfate onto ferrihydrite at low pH values. Together with bacterial sulfate reduction (Sect. 11), this process might also account for the comparatively low $SO_4$ concentrations observed in the AMD pseudoplume in the Zwartkrans Basin (Sect. 8.4).

## 13 Hydrochemistry of Mine Water

The importance of acid mine drainage as a hydrologic agent in the study area warrants an understanding of salient aspects of the hydrochemistry associated with mine water. The analytes of primary interest in this regard are iron and alkalinity/acidity, and elements of their hydrochemistry are discussed in the following sections.

### 13.1 Iron Chemistry

Ferrous iron ($Fe^{2+}$) is present in much higher concentrations in raw mine water than other metals such as Mn, Cu, Pb, Zn, Co, Ni, and U. It is formed when iron sulfide ($FeS_2$) comes into contact with water, oxygen and, often catalyzed by iron-oxidizing bacteria (e.g., *Gallionella* and *Sphaerotilus*) or sulfur-oxidizing bacteria (e.g., *Thiobacillus ferro-oxidans, Thiobacillus thio-oxidans and Ferrobacillus ferro-oxidans*), forms ferrous sulfate ($FeSO_4$) and sulfuric acid ($H_2SO_4$) resulting from oxidation by $O_2$ according to the reactions

$$2FeS_{2(s)} + 2H_2O_{(l)} + 7O_{2(g)} \rightarrow 2FeSO_{4(s)} + 2H_2SO_{4(aq)} \quad (16)$$

with limited available oxygen as in a closed system evidenced by reducing conditions (very low Eh), and

$$2FeS_{2(s)} + H_2O_{(l)} + 7\,1/2\,O_{2(g)} \rightarrow Fe_2(SO_4)_{3(s)} + H_2SO_{4(aq)} \quad (17)$$

with excess available oxygen as in an open system evidenced by oxidizing conditions (very high Eh).

Oxidation by $Fe^{3+}$ proceeds according to the reaction

$$FeS_{2(s)} + 14Fe^{3+}_{(aq)} + 8H_2O_{(l)} \rightarrow 15Fe^{2+}_{(aq)} + 2SO^{2-}_{4\,(aq)} + 16H^{+}_{(aq)} \quad (18)$$

The $Fe^{2+}$ produced by Reaction 18 oxidizes to $Fe^{3+}$ at pH <3 according to the reaction

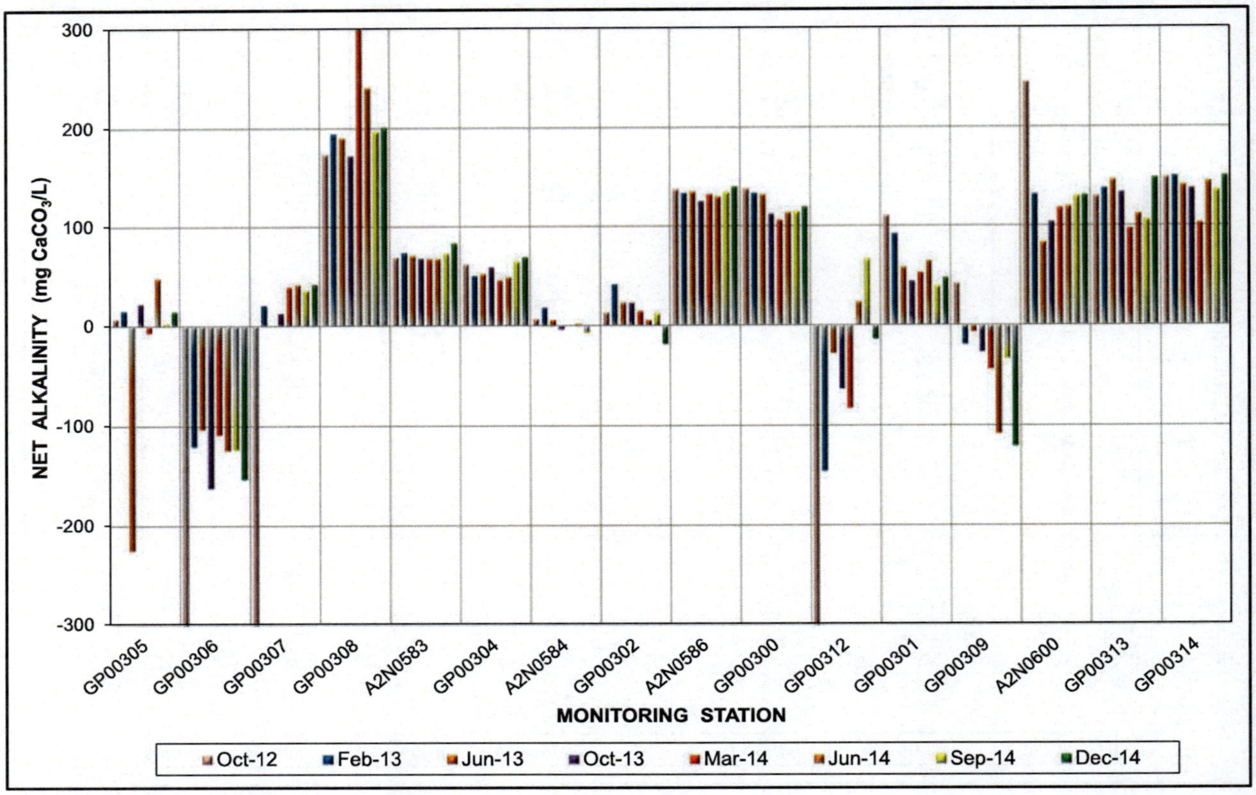


**Fig. 88** Enlarged view of pattern and trend of net alkalinity shown in Fig. 87

$$Fe^{2+}_{(aq)} + 1/4\,O_{2(g)} + H^{+}_{(aq)} \rightarrow Fe^{3+}_{(aq)} + 1/2\,H_2O_{(l)} \quad (19)$$

and at pH >3 according to the reaction

$$2\,Fe^{2+}_{(aq)} + 1/2\,O_{2(g)} + 5\,H_2O_{(l)} \rightarrow 2\,Fe(OH)_{3(s)} + 4\,H^{+}(aq) \quad (20)$$

which produces the ferric hydroxide precipitate known as 'Yellow boy'.

## 13.2 Net Alkalinity and Net Acidity

A discussion of these characteristics in regard to mine and surface waters in the study area is presented in Sect. 3.5 n Chapter "Chemical Hydrology". Their evaluation in regard to groundwaters further informs an assessment of water impacts on the karst water resource. Natural groundwater with a pH <6.4, such as the groundwater draining the quartzitic strata of the Witwatersrand Supergroup (Sect. 2.1.1), has no acid neutralizing (or buffering) capacity. This also applies to impacted karst groundwater that otherwise (and typically) would have a pH in the range of 7.5–8.0 and a total alkalinity of >150 mg $CaCO_3$/L in its unimpacted state (Sect. 2.1.2). Calculations of the net alkalinity and acidity associated with various groundwater sources are presented in Table 21 and of the free and metal acidity components associated with these waters in Table 22. The results indicate that the karst groundwater, impacted to various degrees, maintains a positive net alkalinity, although instances of this reducing over time are observed adjacent to the losing reach of the Riet Spruit, e.g., station CSIR57 (aka WBD5). Further, free acidity contributes «5% to the total acidity in most instances, with >95% being contributed by metals. Gammons and Tucci (2013) similarly report that dissolved metals are a much greater source of acidity in the Berkely pit-lake, Butte, Montana, pit-lake waters than $H^+$ or $HSO_4^-$.

Calculation of the net alkalinity of groundwater sampled from the DWS monitoring boreholes downgradient of the mine area on eight occasions in the period October 2012 to December 2014 returns the results presented in Figs. 87 and 88. The results reflect the impact of the acid load contributed by influent surface water carrying a mine water signature as allogenic recharge to the karst hydro system traversed by the lower Riet Spruit in the Zwartkrans Basin. This load has been discussed in Sect. 3.5 in Chapter "Chemical Hydrology".

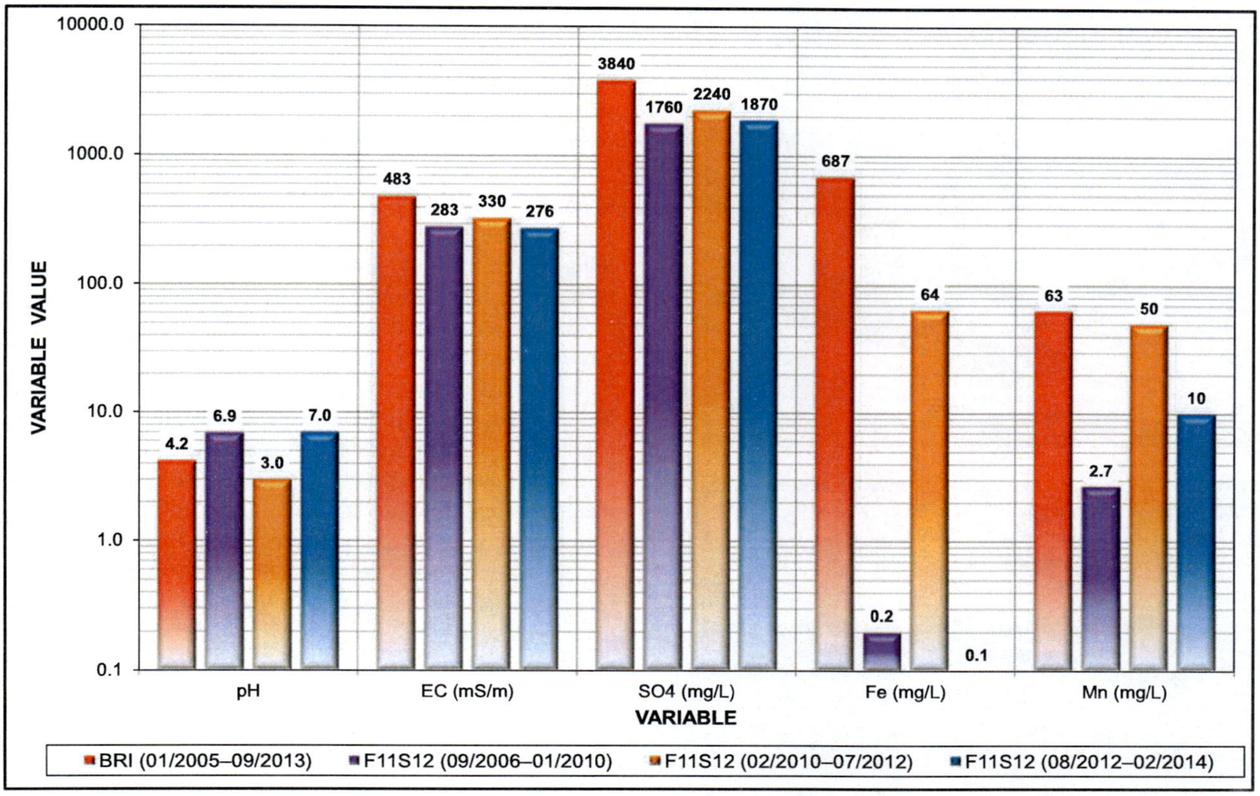


**Fig. 89** Comparison of 'diagnostic' chemical variable values for raw mine water (BRI) and downstream surface water (F11S12) (data from Table 23)

**Plate 10** Waterfall over the quartzite ridge south of the Kruger Kloof Lodge in the KGR (photo H Weiss, date 13/05/2008)

**Plate 11** Outlet of the Charles Fourie Dam in the KGR

The analysis reveals the generally positive net alkalinity of the groundwater, the exceptions being associated with boreholes located in very close proximity to the losing surface water drainages carrying mine water discharge. These boreholes are identified as GP00306 adjoining the Tweelopie Spruit at the northern end of the karst outlier in the Krugersdorp Game Reserve, and GP00309 and GP00312 adjoining the Riet Spruit in the Zwartkrans Basin. The very low (and even negative) net alkalinity associated with borehole GP00305 is attributed mainly to the West Rand

**Plate 12** Overflow of the Aviary Dam (impounded by the earth embankment at rear of photo) at the northern end of the KGR; compare condition of outlet pipe at left with that of same pipe in Plate 1 in Chapter "Chemical Hydrology"

**Plate 13** Cascade and aeration of water over the causeway at station F11S12 (see Plate 8 in Chapter "Physical Hydrology" for scale of monitoring station)

Group quartzitic strata intersected by this borehole and that typically host a weakly mineralized and acidic groundwater (Sect. 1), although a more recent mine water impact must also be considered.

A closer inspection of the histograms suggests that in some instances the net alkalinity values indicate a declining trend, e.g., GP00300, GP00301, GP00302, and GP00309. This observation suggests a generally increasing mine water influence on the karst groundwater chemistry at these locations. Especially the persistent trend from a positive to an increasingly negative net alkalinity associated with the karst groundwater at station GP00309 is concerning and needs to be monitored closely. By comparison, the positive net alkalinity associated with the groundwater in Sterkfontein Cave (Table 22) is another encouraging sign of a limited impact from mine water on the karst groundwater in this environment. A similar observation in regard to the Zwartkrans springwater is testimony to the capacity of the karst hydro system to mitigate a mine water impact. However. The longevity and threshold of this capacity still need to be established.

## 13.3 Local Studies

Krige (2009) reports on studies carried out by DD Science Laboratory CC., using the acid–base titration method employed in standard acid–base accounting (ABA) determinations, to establish the neutralization potential of dolomite iro raw mine water from the decant area. The outcome suggested that 1 ML of mine water would dissolve 900 kg of $CaCO_3$, creating a theoretical void of ~0.32 $m^3$.

By extrapolation of the volume of mine water that had entered the karst aquifer since the commencement of decant in August 2002 up until the 'capture-and-control' of this discharge ca. February 2005, Krige (2009) reports a total void volume of ~4500 $m^3$ that potentially had developed in

**Table 23** Water chemistry data used in Fig. 89 to compare raw mine water from station BRI with Tweelopie Spruit surface water at station F11S12

| Station (period) | 'Diagnostic' chemical variable median value and count ($n$) | | | | |
|---|---|---|---|---|---|
| | pH ($-\log_{10}\alpha_{H+}$) | SEC (mS/m) | $SO_4$ (mg/L) | Fe (mg/L) | Mn (mg/L) |
| Black Reef Incline (01/2005–09/2013) | 4.2 | 483 | 3840 | 687 | 63 |
| | $n$ = 453 | $n$ = 452 | $n$ = 453 | $n$ = 452 | $n$ = 224 |
| F11S12 (09/2006–01/2010) | 6.9 | 283 | 1760 | 0.2 | 2.7 |
| | $n$ = 173 | $n$ = 172 | $n$ = 171 | $n$ = 33 | $n$ = 33 |
| F11S12 (02/2010–07/2012) | 3.0 | 330 | 2240 | 64.0 | 50.0 |
| | $n$ = 128 | $n$ = 128 | $n$ = 128 | $n$ = 128 | $n$ = 128 |
| F11S12 (08/2012–02/2014) | 7.0 | 276 | 1870 | 0.075 | 10.0 |
| | $n$ = 83 | $n$ = 83 | $n$ = 83 | $n$ = 82 | $n$ = 83 |

the karst aquifer in this time. Further, it was suggested by Krige (2009) that this volume could be doubled (i.e., to ~9000 $m^3$) on the basis of the additional carbonic acid ($H_2CO_3$) generated by the reaction between mine water and dolomite creating an equivalent volume of dissolution. These figures, and a value of 16,700 $m^3$ quoted by Liefferink (2010), have been placed in the public domain. Establishing the veracity of these figures is crucial to an informed assessment of this impact.

The procedure followed in the assessment described by Krige (2009) used raw mine water in the calculation. The karst aquifer that receives this water is, however, located north of the KGR a distance of ~4.9 km from where the mine water enters the environment (Tables 48 and 49 in Chapter "Chemical Hydrology"). In its path down the Tweelopie Spruit, this water cascades over a number of natural and man-made features, e.g., the waterfall over the quartzite ridge to the south of the Kruger Kloof Lodge (formerly the Ngonyama Lion Lodge) in the KGR (Plate 10), the spillway of the Charles Fourie Dam at the lodge itself (Plate 11), the outflow of the Aviary Dam (Plate 12) where this water leaves the KGR, and the causeway at station F11S12 (Plate 13). The cascading action serves to thoroughly aerate the water (Benedini and Tsakiris 2013; Nakasone 1987), the oxidization causing any ferrous iron ($Fe^{2+}$) in solution to precipitate out of solution as seen in the orange-brown staining and encrustation on surfaces at each of these locations. As a consequence, the chemical composition of the mine water has changed significantly by the time it reaches the Riet Spruit where it traverses the karst aquifer of the Zwartkrans Basin. The nature of this change is shown in the comparison of the long-term median concentration of elements monitored at the BRI (representative of raw mine water chemistry), with that of these elements monitored at station F11S12 for both the entire record and the early record. This is shown in Fig. 89 (see Table 23 for data), which reflects the very different $SO_4$ and Fe concentrations in the water at the two stations.

Another contentious aspect of the procedure described by Krige (2009) is the use of hydrochloric acid (HCl) to determine the mass of dolomite involved in a reaction between the mine void water and the acid according to the reaction

$$2HCl_{(aq)} + CaCO_{3(s)} \rightarrow CaCl_{2(aq)} + CO_{2(g)} + H_2O_{(l)} \quad (21)$$

It is understood that HCl was used to speed up the reaction because of budget and time constraints (D. Dorling, personal communication). It is contended that the use of raw mine water and HCl in the acid–base titration exercise does not replicate the 'real-world' situation. This exercise would need to be repeated with $H_2SO_4$ and surface water having a composition similar to that passing station F11S12 (Table 23) in order to have greater relevance and application to the situation in the study area.

Studies carried out by the Council for Geoscience (CGS) to further inform the topic of dolomite dissolution in regard to the potential impact of AMD on the karst environment of the COH have entailed a comparative petrographic study (Roelofse 2010) as well as laboratory-based column leaching tests (Coetzee et al. 2010).

In the petrographic study, portions of two polished thin sections of a sample of dolomite were photographed under magnification before and after being exposed to acid mine water. Any changes associated with the exposure to the mine water were studied under a microscope using both plane and cross polarized light. A comparison of the 'before' and 'after' photomicrographs allowed for a qualitative assessment of observed changes/differences. Based on the following observations, it appears that partial dissolution of the dolomite occurred on exposure to acid mine water.

- A lower apparent birefringence after exposure compared to before exposure. This is most likely attributable to a loss of thin section thickness during exposure.
- Greater definition (under magnification) of carbonate grain boundaries after exposure.
- Preferential exploitation of cleavages within carbonate grains, resulting in greater definition of cleavage patterns in the post-exposure image.
- In some instances, a decrease in carbonate grain size following exposure.

The kinetic (column leaching) experiment was devised by the CGS using synthetic AMD prepared in the laboratory as a leachant to closely mimic the bulk composition of Western Basin AMD while maintaining the iron in a dissolved state. The pH of the leachant was adjusted to a value of 3.6 using $H_2SO_4$. The leachant was passed through five columns each packed with dolomite of a different size fraction, i.e., <2, 2–4, 4–8, 8–16, and >16 mm, at a rate of 0.1 L/kg/d to simulate that of slow groundwater flow through a dolomitic aquifer. The kinetic tests revealed the following processes that occurred during the interaction between dolomite and synthetic AMD solutions:

- dissolution of the dolomite as indicated by the neutralization of the AMD at sufficiently high solid to liquid ratios; whether this will lead to the development of new cavities (or the enlargement of existing cavities) within the affected dolomitic areas cannot be determined, as this will be influenced by a combination of the flow rate, the surface area exposed and the rate at which mineral precipitation armors the reactive surface of the dolomite within the aquifer;

- precipitation of new minerals as identified at both macro- and micro-scales using optical and scanning electron microscopy, respectively; the minerals that precipitate will depend on the chemistry of the input solutions as well as the Eh–pH conditions under which precipitation occurs;
- armoring of the dolomite surface with reaction products, reducing the quantity of reactive material exposed;
- an increase in the pH of the solution, reducing the solubility of a number of its components; iron and aluminum are almost completely attenuated, manganese is partially attenuated with an apparent strong dependence on the available reactive surface area, while sodium and sulfate are not attenuated significantly; and
- increased concentrations of calcium and magnesium as a result of the dissolution of dolomite, with an apparent preferential dissolution of calcium over magnesium, in the final solutions produced by the tests.

### 13.4 Conclusions

The likely effects of AMD ingress into the karst aquifer(s) of the COH will be a rise in the pH of the influent mine water (as a result of the dissolution of dolomite and associated neutralization) to a value comparable to that of the ambient karst groundwater. This might lead to a structural weakening of the subsurface formation if the process continued over a long period of time without the dolomite surfaces becoming armored. The rise in pH would result in the reduction of the solubility of some of the contaminant species (e.g., the metals) in the mine water, significantly reducing Fe and Al in solution, partially removing Mn, and not having a significant effect on the Na or $SO_4$ concentration.

Influent mine water would therefore tend to react at the point of first contact with the dolomite resulting in neutralization and the precipitation of metals. The metal precipitates, which would include gypsum ($CaSO_4$), goethite [FeO (OH)], and crystalline and amorphous (botryoidal) aluminum oxyhydroxide [AlO(OH)] species, are likely to form on the reactive surfaces of the dolomite, preventing further neutralization from taking place. Influent mine water impacted surface water would therefore migrate further and further into the karst aquifer, most likely taking the form of a plume comprising a migrating reaction front ahead of various zones each representing a reaction subdomain controlled by factors such as oxidation–reduction state, microbial degradation, and mineral precipitation-dissolution reactions (e.g., Bjerg et al. 1995; Sherlock et al. 1995; Hunter et al. 1998; Elsner et al. 2004), and leaving behind an aquifer hosting an altered biogeochemical environment. How and to what extent the altered aquifer environment differs from the original is unknown, as is the ability of this aquifer to regenerate its original hydrogeochemical character following a return to largely autogenic natural recharge.

In conclusion, natural attenuation of AMD by allowing it into dolomitic aquifer systems is not regarded as an appropriate or sustainable management strategy. However, a better understanding of the hydrochemical interaction between surface water of compromised quality and karst groundwater, and of the hydrogeochemical and biogeochemical processes that control the composition of groundwater, will assist authorities in making informed decisions for protection of the valuable water resources environment in the COH.

## 14 Carbon Flux in Springwater

### 14.1 Introduction

The role of karst aquifers in the transport and storage of carbon has recently received attention in publications by Veni (2013) and White (2013), among others. Veni (2013) estimates the global volume of organic carbon stored in karst aquifers at 200 $km^3$. This does not include the amount of organic carbon stored in paleokarst, which is indicated to be substantially greater. Falkowski et al. (2000) indicate that sedimentary carbonate strata in the lithosphere represent a carbon reservoir of >60 petatons (1 Pt = $10^{15}$ t). This is ~800 times more than the reservoir represented by the oceans. Pokrovsky et al. (2009) demonstrated the efficiency of $CO_2$ sequestration in Mg-bearing carbonate strata.

Although karst environments undoubtedly represent a major sink for atmospheric carbon and $CO_2$ inputs from losing (effluent) streams and the epikarst, they also constitute a source of $CO_2$ from release by speleothem deposition, tufa or travertine deposition, and degassing from spring discharges and spring runs (White 2013). The data acquired in this study, together with the improved understanding of the hydrophysical and hydrochemical environments in the COH, provides material that facilitates a provisional and rudimentary assessment of $CO_2$ evasion via the karst springs on the property.

### 14.2 Approach

Given that the total inorganic carbon (TIC) in solution is the sum of all dissolved carbon species

$$[TIC] = [CO_2(aq)] + [H_2CO_3] + [HCO_3^-] + [CO_3^{2-}] \quad (22)$$

all as mol/L concentrations, and that the combined [$CO_2$ (aq)] and [$H_2CO_3$] can be calculated from alkalinity and pH as follows:

**Table 24** Summary of carbon flux and export rates associated with COH spring discharges ca. May 2010

| Spring | Karst basin area ($km^2$) | Discharge ($m^3/s$) | $HCO_3$ (mmol/L) | $\gamma_{HCO3}$ | $F_c$ (kg/d) | C (mg/L) | C-export rate ($kg/km^2/a$) |
|---|---|---|---|---|---|---|---|
| Zwartkrans | 98.0 | 0.136 | 3.20 | 0.89 | 451 | 38 | 1681 |
| Plover's Lake Kromdraai | 50.8 | 0.339 | 3.84 | 0.92 | 1350 | 46 | 9701 |
| Danielsrust | 7.4 | 0.028 | 2.56 | 0.93 | 74 | 31 | 3669 |
| Tweefontein | 11.6 | 0.030 | 3.92 | 0.92 | 122 | 47 | 3847 |
| Nouklip | 38.0 | 0.143 | 3.92 | 0.92 | 582 | 47 | 5590 |
| Nash | 28.6 | 0.130 | 2.64 | 0.93 | 356 | 32 | 4545 |
| Broederstroom | 18.1 | 0.028 | 5.44 | 0.91 | 158 | 65 | 2814 |
| Total average | 252.5 | 0.834 | 3.65 | 0.92 | 3094 442 | 44 | 4550 |

**Table 25** Summary of carbon flux and export rates associated with COH spring discharges ca. August 2014

| Spring | Karst basin area ($km^2$) | Discharge ($m^3/s$) | $HCO_3$ (mmol/L) | $\gamma_{HCO3}$ | $F_c$ (kg/d) | C (mg/L) | C-export rate ($kg/km^2/a$) |
|---|---|---|---|---|---|---|---|
| Plover's Lake Kromdraai | 50.8 | 0.343 | 3.18 | 0.91 | 1133 | 38 | 8138 |
| Danielsrust | 7.4 | 0.028 | 2.62 | 0.94 | 76 | 32 | 3760 |
| Tweefontein | 11.6 | 0.026 | 3.11 | 0.93 | 84 | 37 | 2645 |
| Nouklip | 38.0 | 0.185 | 3.93 | 0.92 | 756 | 47 | 7258 |
| Nash | 28.6 | 0.208 | 2.59 | 0.94 | 559 | 31 | 7136 |
| Total average | 252.5 | 0.790 | 3.09 | 0.93 | 2608 522 | 37 | 5787 |

$$[H_2CO_3] = \frac{[HCO_3]\gamma_{HCO_3} 10^{-pH}}{K_1} \quad (23)$$

where $[H_2CO_3]$ is the carbonic acid concentration as mol/L, $[HCO_3^-]$ is the molar concentration of bicarbonate, $\gamma_{HCO3}$ is the activity coefficient of bicarbonate, and $K_1$ is the ionization constant of carbonic acid, then the carbon flux can be calculated from known spring discharge rate and springwater pH and bicarbonate concentration values as follows (White 2013):

$$F_C = 17.012\, C_{HCO_3} \left\{ 1 + \frac{[HCO_3]\gamma_{HCO_3} 10^{-pH}}{K_1} \right\} Q \quad (24)$$

where $F_C$ is the total carbon flux as kg/d, $C_{HCO3}$ is the bicarbonate concentration as mg/L, $Q_S$ is the spring discharge as $m^3/s$, and all other terms are as for Eq. 23.

## 14.3 Results

Application of Eq. 25 to the seven karst basins (and eight springs) discussed in Sect. 12 returns the $F_C$ values reported in Table 24.

The average total carbon loading in the karst spring waters is found to be 44 mg C/L in the range of 31–65 mg/L. The average total carbon flux carried is found to be ~442 kg/d (~161 t/a), with the eight springs delivering a combined total C-flux of 3094 kg/d (1129 t/a). The average C-export rate is calculated at 4550 $kg/km^2/a$ in the range 2814–9701 $kg/km^2/a$ for the seven karst basins.

A second set of SDMs and chemical analyses carried out on five of the springs listed in Table 24 in mid-August 2014, returned the results reported in Table 25. The average total carbon loading in the karst spring waters is found to be 37 mg C/L in the range 31–47 mg/L, compared to the 41 mg C/L and range 31–47 mg/L of the May 2010 data. The average total carbon flux carried is found to be ~522 kg/d (~190 t/a), with the five springs delivering a combined total C-flux of 2608 kg/d (952 t/a), compared to the ~497 kg/d and 2484 kg/d of the respective May 2010 results. The average C-export rate is calculated at 5787 $kg/km^2/a$ in the range 2645–8138 $kg/km^2/a$ for the six karst basins, compared to the 5470 $kg/km^2/a$ and range 3669–9701 $kg/km^2/a$ of the respective May 2010 results.

The results are not only in reasonable agreement with each other (Table 26), but also with values reported in the

**Table 26** Comparison of carbon flux and export rates associated with COH spring discharges ca. May 2010 and ca. August 2014

| Spring | $F_c$ (kg/d) | | C (mg/L) | | C-export rate (kg/km$^2$/a) | |
|---|---|---|---|---|---|---|
| | May 2010 | August 2014 | May 2010 | August 2014 | May 2010 | August 2014 |
| Plover's Lake Kromdraai | 1350 | 1133 | 46 | 38 | 9701 | 8138 |
| Danielsrust | 74 | 76 | 31 | 32 | 3669 | 3760 |
| Tweefontein | 122 | 84 | 47 | 37 | 3847 | 2645 |
| Nouklip | 582 | 756 | 47 | 47 | 5590 | 7258 |
| Nash | 356 | 559 | 32 | 31 | 4545 | 7136 |
| Total average | 2484 497 | 2608 522 | 41 | 37 | 5470 | 5787 |

international literature. For example, Florea (2015) reports the conveyance of 25.8–62.4 Gg/a (~6900–16,700 kg/km$^2$/a) of $CO_2$ (equivalent to ~1900–4600 kg C/km$^2$/a), based on dissolved inorganic carbon concentrations, from the atmosphere to the aqueous system through the dissolution of carbonate strata spanning an outcrop area of 3730 km$^2$ in the Cumberland River watershed of south-east Kentucky, USA.

## 14.4 Discussion

It is common cause that TIC dominates the total carbon (TC) concentration in springwater, the role is reversed in surface water where total organic carbon (TOC) dominates. The TIC derives from 'dead' carbon (weathering of carbonate strata), and the TOC from 'live' carbon (e.g., decaying plant and animal matter). It has been shown by White (2013) that whereas the carbon loading (mg/L) in springwater responds only weakly to increased discharge, carbon fluxes (kg/d) reflect a much more prominent response. In other words, a significant (>tenfold) increase in spring discharge manifests only as a slight (<twofold) decrease in carbon loading attributable to dilution. Under similar circumstances, the carbon flux might well reflect a sympathetic response by increasing tenfold.

The total alkalinity (mg $CaCO_3$/L) concentrations associated with the bicarbonate ($HCO_3$) values listed in Table 5 allow an estimate of the total mass of calcium carbonate exported via the karst springs. This amounts to ~6000 t/a as shown in Table 19 (Sect. 12.1), and broadly represents a rate of total void development of ~2100 m$^3$/a, or 8.2 m$^3$/km$^2$/a over a 25,490 ha extent of dolomite outcrop in the COH. This is equivalent to a denudation rate of 8.2 mm/Ka. A more detailed discussion in Sect. 12.1 provides greater elucidation of solutional denudation and void development rates in the COH.

# Conclusions

Harrison Pienaar and Philip J. Hobbs

The understanding of the surface water and groundwater environments in the Cradle of humankind (COH), also in regard to their inter-relationship, is considerably expanded by this study. This understanding extends as much to the water chemistry aspect as it does to the water quantity aspect. Outcomes of the study that are considered especially significant are summarised as follows:

- the quantification of surface water flow losses in the lower Riet Spruit valley;
- the quantification of spring discharges;
- the definition of basins/subcompartments and corresponding groundwater resource units associated with the karst formations in the study area;
- the evaluation of the geochemical interaction between acidic mine water and the dolomitic strata that form the receiving karst aquifer in the study area;
- the development of semi-quantitative resource water quality objectives (RWQOs) to inform groundwater resource directed measures for the karst portions of the study area;
- the derivation of a fossil site risk assessment that informs the vulnerability of each declared fossil site and its associated cave system in the context of its hydrogeologic setting; and
- the development of a monitoring programme for the COH.

The platform built from historical data integrated with a wide range of rigorous and defensible newly generated and interpreted hydrologic and hydrogeologic data and information, underpins the situation assessment of the surface water and groundwater environments. This, in turn, has provided the means to objectively gauge the impact of varied and numerous threats on the water resources in the study area. The salient conclusions drawn from the study are summarised in the following sections.

P. J. Hobbs: Deceased

H. Pienaar (✉) · P. J. Hobbs
Smart Places, Council for Scientific and Industrial Research, Pretoria, South Africa
e-mail: hpienaar@csir.co.za

H. Pienaar
Hebei University of Engineering, Handan, China

## 1 Surface Water Resources

### 1.1 Quantity

The Skeerpoort River system fed by dolomitic springs remains in a near pristine condition. The Bloubank Spruit system, on the other hand, has experienced above average discharges in its upper reaches via the Tweelopie Spruit and the Riet Spruit since the resumption of mine water decant in late-January 2010. Together with the treated sewage effluent discharged from the Percy Stewart WWTW via the Blougat Spruit, these circumstances have resulted in an unprecedented volume and duration of surface water flow in the Bloubank Spruit system through the 2010 winter. Further, the combined discharge of raw and treated mine water witnessed surface water losses of between 20 and 32 ML/d to the dolomitic aquifer in the lower Riet Spruit, equating to an infiltration rate of as much as ~ 90 L/s/km.

### 1.2 Quality

The Skeerpoort River system continues to reflect a karst-dominated water composition of excellent quality. Up until mid-2010, the impact of the poor quality associated with the abnormal combined discharge of treated and raw mine water in the upper reaches of the Bloubank Spruit system was mitigated by (a) the contribution of treated wastewater effluent discharged by the Percy Stewart WWTW and (b) the above average surface water runoff associated with the extremely wet 2009–10 summer. Since mid-2010, the increase in SEC of Bloubank Spruit water from ~ 50 to > 100 mS/m, together with a decrease in pH



P. J. Hobbs et al. (eds.), *Anatomy of a South African Karst Hydrosystem*, Cave and Karst Systems of the World,
https://doi.org/10.1007/978-3-030-95829-9_9

from 7.2 to 6.9 at the downstream boundary of the Zwartkrans Basin, is considered to reflect an increasing contribution of mine water to the middle reaches of this drainage. This impact is less evident in the Zwartkrans Spring water, which in August 2010 still reflected a field SEC of ~ 77 mS/m. Similarly, the Lake water in Sterkfontein Cave continues to reflect an SEC of ~ 60 mS/m as it did in June 2006, a year after the start of monitoring of this variable by the DWS.

The quality of surface water resources in the Bloubank Spruit system is further compromised by bacterial contamination derived from wastewater effluent. These circumstances also make it difficult to assess agricultural impacts on the surface water resources, as these impacts are similarly associated with nutrient inputs.

## 2 Groundwater Resources

### 2.1 Quantity

The very wet 2010 and 2011 summers precipitated an exceptional recharge of groundwater resources in the study area. A rise in groundwater rest levels by at least ~ 1 m is testimony to these circumstances. Greater water level recoveries of as much as ~ 5 m are attributed to allogenic recharge associated with the infiltration of surface water contributed from extraneous sources including mining and municipal wastewater effluent. This infiltration has amounted to as much as 32 ML/d in the case of mine water, and a more modest 7 ML/d in the case of municipal wastewater effluent. It is concluded, therefore, that the aspect of groundwater quantity as represented by current 2014 groundwater rest levels compared to historical levels, reflects an extremely positive outlook.

### 2.2 Quality

As might be expected, the chemistry of groundwater in the study area reflects the greatest spatial variation depending on the position in the physical hydrogeologic environment. For instance, the subcompartments receiving water of compromised quality in terms of either trace/heavy metals and elevated salt loads associated with mine water, and/or elevated bacterial loads associated with municipal wastewater (both representing allogenic recharge), reflect the poorest groundwater chemical composition. In contrast, compartments (and subcompartments) receiving only autogenic recharge remain largely unaffected in terms of groundwater chemistry (and quality).

Despite the magnitude, severity and duration of allogenic recharge associated especially with the mine water, the receiving karst environment continues to reflect a resilience to the negative impact from this source that is both surprising and comforting. This observation, however, raises the question as to whether the ostensible resilience has a threshold beyond which the biophysical condition of this resource degrades markedly to the detriment of karst ecosystem(s) hosted in this environment.

## 3 Mine Water Impact on Dolomite

A comparative petrographic study together with laboratory-based column leaching tests carried out by the CGS further elucidated the potential impact on the karst environment of the COH dolomite by infiltrating mine water.

The likely effects of AMD ingress into the karst aquifer(s) will be a rise in the pH of the influent mine water (as a result of the dissolution of dolomite) to a value comparable to that of the ambient karst groundwater. The rise in pH would result in the reduction of the solubility of some of the contaminant species in the mine water, significantly removing iron and aluminium from solution, partially removing manganese and not having a significant effect on the sodium or sulphate concentration. The metal precipitates, which would include gypsum, goethite and crystalline and amorphous (botryoidal) aluminium oxyhydroxide species, are likely to form on the reactive surfaces of the dolomite, preventing further neutralisation reactions from taking place. Influent mine water would therefore tend to react at the point of first contact with the dolomite, resulting in both neutralisation of the mine water and the prevention of further neutralisation from taking place. Over time the mine water would migrate further and further into the karst aquifer, and the buffering capacity of the aquifer would reduce progressively.

It is concluded, however, that natural attenuation of AMD by allowing it to enter dolomitic (carbonate) aquifer systems is not regarded as an appropriate or sustainable management strategy. This can only hasten the manifestation of a threshold to the observed resilience alluded to in Sect. 2.2, if such exists.

## 4 Resource Water Quality Objectives

The setting of resource water quality objectives (RWQOs) for groundwater quantity and quality in the COH expands the impact of the study.

### 4.1 Quantity

The statistical assessment of long-term groundwater rest level trends provides a means to quantify the setting of

RQOs for groundwater quantity in a part of the study area. The proposed RQOs recognise 'permissible' changes in groundwater rest level that vary across a compartment. Although the proposed RQOs do not find a priori support in the DWS library of existing RQOs in this regard, they are nevertheless put forward for consideration as advancing the knowledge base in regard to this GRDM component.

Long-term groundwater level monitoring data are only available for the Zwartkrans Basin. The proposed RQO for the upper reaches represented by the Vlakdrift Subcompartment, assigned a *'modified' class C* SoE condition, is a change in groundwater rest level of $\leq$ 6.1 m. The middle reaches represented by the Sterkfontein Subcompartment, assigned a *'largely modified' class D* SoE condition, is set a change in groundwater rest level of $\leq$ 3.6 m, and the Zwartkrans Subcompartment in the lower reaches, assigned a *'largely modified' class D* SoE condition, is set a value of $\leq$ 2.3 m.

The remaining nine karst basins, which are either assigned a *'slightly modified' class B/BC* or a *'natural' class AB* SoE condition, represent a natural environment that remains largely unaffected by changes in land use activities and practices. For this reason, the setting of RQOs as above is not an imperative at this stage. Further, it would be required that a better understanding of groundwater rest level behaviour and trends in these groundwater resource units is developed before RQOs can be set with the necessary statistical support to generate confidence.

### 4.2 Quality

The setting of RQOs in regard to groundwater chemistry (quality) is a comparatively much simpler exercise given the relationship of analytical groundwater chemistry data with the discharge(s) of associated groundwater compartments, especially where these are coupled with springwater chemistries. The springs draining the largely pristine Danielsrust, Tweefontein, Uitkomst and Diepkloof basins do not reflect anthropogenic impacts on the chemistry (quality) of the groundwater produced by these hydrogeologic environments, compared to the chemistry of groundwater produced via the Zwartkrans Spring. The latter drains the most severely compromised hydrogeologic regime in terms of groundwater quality in the study area, with groundwater analyses typically exhibiting electrical conductivity values > 100 mS/m, $SO_4$ concentrations > 150 mg/L, and Cl concentrations > 50 mg/L. The geographic extent of this footprint in the host aquifer is still comparatively limited.

## 5 Karst Solutional Denudation

An assessment of karst solutional denudation rates in the COH based on formula-derived, empirically-derived, and theoretically-derived approaches yields results which are in agreement with recently reported values and with the observed dewatering horizon developed over 3.3 Ma in Sterkfontein Cave. Cognisant of the variation in climatological and physiographical factors that influence landscape development in the long term, the mean solutional denudation rate in the COH is placed at 7.6 mm/Ka in the range 3–15 mm/Ka.

Studies that consider the effect of reaction processes on mineral surfaces (the surface complexation studies defined by Morse and Arvidson, 2002) in the evaluation of karst denudation remain unexplored in a South African context. The manifestation of a combined mine water and municipal wastewater impact on a portion of the COH karst environment provides an opportunity to explore this aspect against a comparison with the pristine portions of this environment.

## 6 Fossil Site Hydro-vulnerability

It is concluded that nine of the 14 fossil sites in the COH exhibit a *very low* or low vulnerability. This is attributed to their location (a) in groundwater compartments that are hydrogeologically separate from those where the contaminated water impact is manifested, and (b) at substantial elevations above the ambient groundwater level. Only the Bolt's Farm site exhibits a *very high vulnerability*. This is because of its position immediately down-gradient of the two main sources of poor quality surface water, namely mine water via the Riet Spruit and treated municipal wastewater via the Blougat Spruit. This is compounded by the fact that (a) both these drainages lose water to the karst aquifer, and (b) at least one cave system in this area, Alladin's (aka Quarry) Cave, is known to intersect the water table (P Boshoff, personal communication). The Sterkfontein site is assigned a *high vulnerability*. The fact that this cave system intersects the water table is mitigated by the observed long-term cave water quality record and other hydrochemical data. These reflect circumstances where the cave water does not exhibit the measure of impact observed in the Zwartkrans Subcompartment as is recorded for the Zwartkrans Spring. The Swartkrans, Minnaar's and Plover's Lake sites exhibit a *moderate vulnerability*. This is mainly because of a relatively shallow (5–25 m bs) water table. In conclusion, therefore, it is apparent that the majority of the fossil sites in the COH are not under threat from either changes in surface water or groundwater levels and/or changes in surface water or groundwater chemistry (quality), whether due to mine water or treated sewage effluent ingress, or from agricultural land use practices. The sites that are the most vulnerable have been identified and are earmarked for specific monitoring activities.

# Recommendations

Harrison Pienaar and Philip J. Hobbs

Recommendations that address aspects specific to the water resources and ancillary aspects such as fossil site hydrovulnerability and cave ecosystem integrity have been put forward in the body of the report. They are not repeated in this section. The following recommendations are of a more general nature, and are put forward for consideration by the Cradle of Humankind Word Heritage Site Management Authority (COH WHS MA) and broader scientific community.

- *Hydrovulnerability assessment*: It is advisable to extend the hydrovulnerability assessment to other cave systems in the study area. These will comprise those non-inscribed sites that researchers and cavers are familiar with, and therefore can be subjected to such assessment in cooperation with the caving fraternity.
- *Gravimetric survey*: It is advisable to carry out gravimetric surveys in the lower Riet Spruit valley extending from the confluence of the Tweelopie Spruit and the Riet Spruit down to the confluence of the Blougat Spruit and the Riet Spruit, as well as along the southern boundary of the KGR across the footprint of the dolomite outlier. The results of these surveys will indicate the measure of karst dissolution present in these important corridors that host the N14 national and the R24 regional roads, respectively.
- *Mine water treatment*: The Inter-Ministerial Committee (IMC) on acid mine drainage (AMD), as per the Coetzee et al. (2010) report, recommends (amongst others) that the mine water treatment capacity in the headwaters of the Tweelopie Spruit be increased to accommodate a decant volume of at least 40 ML/d. This recommendation is supported unequivocally. It is imperative that all raw mine water generated in the upper reaches of the Tweelopie Spruit be captured and prevented from entering the environment in an untreated state. It is recognised, however, that the implementation of this recommendation lies within the jurisdiction of national government in the form of the Department of Water and Sanitation.
  It is further recommended that additional mine water treatment facilities be established in the headwaters of the Tweelopie Spruit to further 'polish' the treated mine water that is generated by the expanded mine water treatment capacity and released into the environment. This might include the construction of artificial wetlands in the lower gradient upper reaches, and the placement of limestone in the river channel at selected high-energy cascade features such as waterfalls and dam spillways.
- *Monitoring:* Empower the tourist guides at Sterkfontein Cave to generate hydrogeologic data through simple observations. For example, tape measurements of the cave water level and regular determination of field water quality parameters with a handheld multi-parameter probe can be recorded daily in a log book. The information can then be communicated to tourists as a demonstration of more 'intimate' knowledge of the host hydrogeologic environment, expanding on the historical focus placed on the palaeoanthropological significance of the site.
- *Monitoring committee*: It is recommended that a monitoring committee comprising a core of key stakeholder groupings (e.g. national, provincial and local government, environment and tourism, agriculture, and NGOs) be established. This committee, chaired by the Management Authority, should function in cooperation with the similar committee established within the Team of Experts that advises the IMC on AMD.
- *Research*: Investigate the possibility of bacterial sulphate reduction in the confluence of the Riet Spruit and the Blougat Spruit. This supports the key science question posed by Banner et al. (2007) (in Martin and White 2007)

P. J. Hobbs: Deceased

H. Pienaar (✉) · P. J. Hobbs
Smart Places, Council for Scientific and Industrial Research, Pretoria, South Africa
e-mail: hpienaar@csir.co.za

H. Pienaar
Hebei University of Engineering, Handan, China



P. J. Hobbs et al. (eds.), *Anatomy of a South African Karst Hydrosystem*, Cave and Karst Systems of the World,
https://doi.org/10.1007/978-3-030-95829-9_10

regarding the fate and transport of redox-sensitive elements and their microbial consequences and feedbacks. Research in this regard will benefit from expertise in the field of biogeochemistry.

- *Communication strategy*: It is important that the COH WHS Management Authority communicates the outcome of the water resources situation assessment and on-going monitoring activities to as a wide an audience as possible. This represents the most powerful means to counter the common and popular misperceptions that prevail regarding the AMD risk to the World Heritage status of the site. The COH WHS MA should consider various avenues of communication including public information meetings in the area, the print media and awareness campaigns.
- *International Recognition*: The COH WHS MA should avail itself of every opportunity to disseminate the results of its hydrologic and hydrogeologic studies to a peer audience using appropriate international platforms such as conferences, symposia and peer-reviewed academic journals.

***Instead of fearing asteroids, we should fear ourselves.***

***Yuval Noah Harari***
***(on Homo sapiens and mass extinctions, in "Homo Deus: A Brief History of Tomorrow", Vintage, London, 2016)***

# Acknowledgements

The author is indebted to the many landowners who granted access to geosites (especially boreholes and springs) at and from which invaluable hydrogeologic data were sourced. The study would not have been possible without the cooperation of the following persons, listed in alphabetical order. The author offers his apologies to any individual inadvertently omitted from this list.

| Stakeholder/landowner/contact | Property and/or geosite description and/or organisation |
|---|---|
| Bacquiere, C | Olera Farmers, Kromdraai 520JQ |
| Bailey, P | Owner, The Cradle/Motsetse Nature Reserve |
| Barnard, P | Danielsrust 518JQ |
| Brengosz, R | Ptn 38 (Moon Valley Ranch), Rietfontein 522JQ |
| Brooker, J | Owner, Glen Afric Country Lodge, Broederstroom |
| Carpenter, H | Facilities Supervisor, Nedbank Olwazini Estate LMDC |
| Carstens, M | Plot 3, Beckedan Agricultural Holdings |
| Coleman, D | Ptn 35, Zwartkrans 172IQ |
| Cox, D | Manager, Nedbank Olwazini Estate LMDC |
| de Klerk, P | Ptn 7, Kromdraai 520JQ |
| de Wit, L | Plateau Farm Estate, Kromdraai 520JQ |
| Deysel, J | Maintenance Supervisor, Sterkfontein Cave, Maropeng |
| Dispan, P | Ptn 143, Sterkfontein 173IQ |
| Dorling, D | Director, DD Science Laboratory cc |
| Doyle, G | Supervisior, Crab Farm Estate, Kromdraai 520JQ |

(continued)

| Stakeholder/landowner/contact | Property and/or geosite description and/or organisation |
|---|---|
| du Plessis, K (Ms) | Data Manager, DD Science Laboratory cc |
| du Toit, S | Specialist: Environmental Protection, Mogale City LM |
| du Toit, S | Manager, Plover's Lake Estate, Kromdraai 520JQ |
| Erasmus, M (Mrs) | Plot 173, Oaktree Agricultural Holdings |
| French, G (Ms) | Kalkheuwel West Estate |
| Fouche, M | Maintenance Manager, Maropeng |
| Fourie, L | Plot 34, Waterval Agricultural Holdings |
| Gaylord, J | Greensleeves Medieval Restaurant, Zwartkrans 172IQ |
| Gomes, M | Chairman, Kromdraai Irrigation Board |
| Gomes, S | Long-term local resident |
| Grobler, H | Ptn 8, Kromdraai 520JQ |
| Hearne, E | Owner, Rhino and Lion Game Reserve, Rietfontein 522JQ |
| Hewitt, J | Owner, Kenjara Lodge, Kromdraai 520JQ |
| Jackson, B | Plot 15, Danielsrust 518JQ |
| Jardim, I | Ptn 66/25, Sterkfontein 173IQ |
| Joubert, W | Savannah Game Reserve, Danielsrust 518JQ |
| Keller, S | ex-Environmental Manager, SibanyeGold |
| Kok, J | Royal Thatch, Zwartkrans 172IQ |
| Kotze, J | Dwarsvlei 503JQ |
| Krige, G | Plot 129, Sterkfontein Farm Estate |
| Kruger, L | Ptn 78, Sterkfontein 173IQ |

(continued)


P. Hobbs et al. (eds.), *Anatomy of a South African Karst Hydrosystem*, Cave and Karst Systems of the World,
https://doi.org/10.1007/978-3-030-95829-9

| Stakeholder/landowner/contact | Property and/or geosite description and/or organisation |
|---|---|
| Lotz, R | Ptn 5, Lotz Kontrei, Kromdraai 520JQ |
| Lourens, W | Chief Ranger, Glen Afric Country Lodge, Broederstroom |
| Maidment, S | Kalkheuwel West Estate |
| Malherbe, C (Mrs) | Plot 72, Sterkfontein 173IQ |
| Malherbe, D | Plot 72, Sterkfontein 173IQ |
| Maré, E (Ms) | Plant Manager, Percy Stewart WWTW, MCLM |
| Marshall, L | Manager, Maropeng āAfrika |
| Mashitisho, D | Municipal Manager, Mogale City Local Municipality |
| McClauchlan, P | Oaktree Agricultural Holdings |
| McDonald, A | Plot 150, Sterkfontein Farm Estate |
| Mitchell, E | Plot 167, Sterkfontein Farm Estate |
| Mostert, J | Game Ranger (former), Krugersdorp Game Reserve |
| Nash, J | Owner, John Nash Nature Reserve |
| Norquoy, N | Wild Cave Adventures |
| Oosthuizen, T | Plot 9, Danielsrust 518JQ |
| Pera, K | Rhino and Lion Game Reserve, Rietfontein 522JQ |
| Potgieter, P | Exclusive Mineral Water, Plot 71, Sterkfontein 173IQ |
| Robeni, B | Chief Ranger, Motsetse Nature Reserve / The Cradle |
| Roodt, P | Plot 66, Oaktree Agricultural Holdings |
| Roos, G | Bergland Instant Lawn, Sterkfontein 173IQ |
| Roos, H (Mrs) | Ptn 5, Zwartkrans 172IQ |
| Roxmouth, C | Plot 56, Zwartkrans 172IQ |
| Rubin, A | ex-Manager, Maropeng āAfrika |
| Rykaart, J | General Manager (former), Krugersdorp Game Reserve |
| Saunders, E | Plot 160, Sterkfontein Farm Estate |
| Scheepers, J | Plot 137, Sterkfontein Farm Estate |
| Schilling, G | AquaMine, Rietfontein 522JQ |
| Schutte, P | Ptn 8/2, Chrisuél, Sterkfontein 173IQ |
| Smith, S | Manager, Ptn 35 (Ekutheni Estate), Kromdraai 520JQ |
| Steyn, C | Electrical Foreman, Sibanye-Stillwater |
| Steyn, D | Plot 119, Sterkfontein Farm Estate |
| Steyn, H | Plot 162, Sterkfontein Farm Estate |
| Tarr, R | Oak Tree Farm, Oaktree Agricultural Holdings |
| van den Bosch, J | Jomajoco Farms, Vlakdrift 163IQ |
| van der Merwe, P | Danielsrust Game Farm, Danielsrust 518JQ |
| van der Merwe, D | Danielsrust Game Farm, Danielsrust 518JQ |
| van der Merwe, P | Glen Almond, Sterkfontein 173IQ |
| van der Walt, B | ex-HDS Plant Manager, SibanyeGold |
| van Ekeren, K | Ptn 3, Dwarsvlei 503JQ |
| van Niekerk, J | Plot 37, Waterval Agricultural Holdings |
| van Rooy, F | Boland Farm, Danielsrust 518JQ |
| van Rooy, J | Boland Farm, Danielsrust 518JQ |
| van Wyk, P | Rietfontein 522JQ |
| Vieira, J (Mrs) | Proprietor, Cradle Food Market, Oaktree |
| Viljoen, W | Ptn 106, Sterkfontein 173IQ |
| Visser, H | Manager, John Nash Nature Reserve, Tweefontein 523JQ |
| Wessels, F | Rhino and Lion Game Reserve, Rietfontein 522JQ |
| Whatley, G | Kromdraai Gold Mine, Kromdraai 520JQ |
| Whillier, J | Plot 127, Sterkfontein Farm Estate |
| Zorab, R | ex-Director, SibanyeGold |

The author also extends a sincere word of thanks to the CSIR and DWS colleagues who participated in various aspects of the field work. These individuals are listed, in alphabetical order, as follows.

| Name | Position/Affiliation |
|---|---|
| Bugan, R | Research Scientist (CSIR Natural Resources & Environment) |
| de Meillon, N | Scientific Technician Production (Department of Water & Sanitation) |
| Leyland, R | Research Scientist (CSIR Built Environment) |
| Masinge, H | Technician (CSIR Natural Resources & Environment) |
| May, F | Research Scientist (ex CSIR Natural Resources & Environment) |
| McMillan, P | Technician (CSIR Natural Resources & Environment) |

(continued)

| Name | Position/Affiliation |
| --- | --- |
| Mmabatswa, S | Trainee (Department of Water & Sanitation) |
| Mokoena, M (Ms) | CSIR Intern (National Research Foundation) |
| Moolman, T | Auxilliary Services Officer (Department of Water & Sanitation) |
| Mukwevhu, L | CSIR Intern (National Research Foundation) |
| Musetsho, M | Trainee (Department of Water & Sanitation) |
| Mvandaba, V (Ms) | CSIR Intern (National Research Foundation) |
| Ngcobo, K (Ms) | Trainee (Department of Water & Sanitation) |
| Shadung, J | CSIR Student Intern (University of Johannesburg) |
| Socosa, L (Ms) | CSIR Student Intern (Tshwane University of Technology) |
| Zengeya, T | Research Scientist (ex CSIR Natural Resources & Environment) |
| Pienaar, H | Comptence Area Manager (CSIR Natural Resources & Environment) |

Especially the efforts and support of Mr T (Theo) Moolman in the field-sourcing of the vertical hydrochemical profiling data presented in Sect. 8.10, Chapter "Chemical Hydrogeology" are gratefully acknowledged.

The final and greatest appreciation must be expressed towards the staff of the Management Authority who have demonstrated absolute confidence in the service provided to this authority over the many years that the CSIR served as Professional Service Provider. The following individuals are listed, in alphabetical order, as follows:

| Name | Position |
| --- | --- |
| Mills, P | Deputy Director (retired) |
| Pienaar, H | Deputy Director |
| Pillay, M | Director |
| Sibanyoni, J | ex-Assistant Director |
| Smith, L (Ms) | Deputy Director |

**Unnumbered plates**

Discharge in the Riet Spruit at its intersection with the Malmani Road; note the red colouration of the water due to dissolved ferrous iron ($Fe^{2+}$) indicative of a strong acid mine water presence as revealed by the following field values: pH = 2.3, SEC = 410 mS/m, ORP = 254 mV (photo P Hobbs, date 12/01/2011)

The ~21 L/s Broederstroom Spring that rises under the concrete slab (at feet of seated figure) which forms part of the spring protection measures (photo P Hobbs, date 21/12/2010)

View of the first portion of the newly constructed and commissioned Train 4 (of 4) of the high density sludge (HDS) mine water treatment plant showing $CaCO_3$-dosed raw mine water moving from stirred condition (left) to aerated condition (right); trains 3 to 1 extend to back of view (photo P Hobbs, date 11/05/2012)

Surface expression of the Moketsi Dyke giving rise to the ~ 2 L/s Moketsi Spring at left rear of picture (photo: P Hobbs)

View of the > 130 L/s Nash Spring (geosite JNNR3, also called the 'Bridge' or 'Rondawel' spring) in the John Nash Nature Reserve (photo P Hobbs, date 19/05/2010)

***"If a man and his story are in conflict, it is the man who must change."***

***Adam Johnson***

***(from The Orphan Master's Son)***

Resistant quartzite of the Witwatersrand Supergroup creating a ~15 m high waterfall (figure at bottom right for scale) on the Blougat Spruit downstream of the Percy Stewart Wastewater Treatment Works; the ameliorative effect of the aeration generated in the cascading water on its downstream quality is a fortuitous 'gratis' natural intervention (photo P Hobbs, date 26/09/2012)

*How many years can a mountain exist, before it is washed to the sea? The answer, my friend, is blowing in the wind.*

*Robert Allen Zimmerman aka Bob Dylan*

*(American folk-rock musician 1941–present)*

# Glossary

**Acid mine drainage** The drainage emanating from mining operations or mine residue deposits as a result of the accelerated oxidation of sulphidic minerals, most commonly iron pyrite ($FeS_2$), due to exposure to water and oxygen, and which process is accelerated in the presence of iron- or sulphur-oxidising bacteria. Generally referred to by the acronym AMD.

**Allogenic** Describes a system that derives its water entirely from that running off a neighbouring non-karst catchment area (Ford and Williams 2007).

**Analyte** A chemical substance whose presence and/or concentration in a sample is determined (Lee 1989).

**Anion** An ion (atom or complex of atoms) that has gained one or more electrons to exhibit a negative electric charge, e.g. $Cl^-$, $SO_4^{2-}$.

**Aquifer** A stratum of water-bearing rock that has the ability to both store and transmit groundwater in quantities that are sufficient to sustain the long-term water-yielding properties of a borehole that intersects such stratum, for the water supply purposes required of the borehole. For example, a low-yielding aquifer that can sustain a sufficient groundwater yield to meet a comparatively low-demand domestic supply is as 'good' as a high-yielding aquifer that can sustain a groundwater yield capable of supplying water for large-scale irrigation or municipal water supply purposes.

**Autogenic** Describes a system composed entirely of *karst* strata, and which derives its water only from rainfall on such a system (Ford and Williams 2007). Andreo et al. (2008) recognise autogenic recharge over carbonate strata as including direct recharge, localised or concentrated recharge via swallow holes and indirect recharge from the bed of superficial water courses.

**Aven** A vertical or highly inclined shaft extending upward from a cave passage, generally to the surface. May equally be described as a shaft when seen from above. Commonly related to enlarged vertical joints. (http://network.speleogenesis.info/directory/glossary/term.php?gloss_id=161)

**Cation** An ion (atom or complex of atoms) that has lost one or more electrons to be left with a positive electric charge, e.g. $Na^+$, $Mg^{2+}$.

**Confined aquifer** An *aquifer* in which the *hydrostatic head* (pressure) is greater than that of atmospheric pressure, generally from the load exerted on the aquifer by overlying impervious (confining) strata, and therefore supports a *potentiometric surface* that is located above the 'roof' of the *aquifer*. This is typically represented by a groundwater level that stands at a shallower depth than that at which water was first encountered in a borehole penetrating the confined aquifer, and which phenomenon finds ultimate expression in the form of an artesian or free-flowing borehole.

**Cutter** Term used by North American karst scientists to describe a soil-covered *grike* that tapers with depth (Ford and Williams 2007).

**Doline** An enclosed topographic depression, basin or typically grassy hollow caused by direct solution and subsidence of the dolomite/limestone surface zone (solution doline), or formed by collapse over a cave (collapse doline) (Field, 2002); a collapse doline more readily conforms to the generally deeper funnel or shaft geometry with rocky cliff-bounded sides typical of a *sinkhole*.

**Electrical balance** Also referred to as the ionic or charge balance, it is a measure of the accuracy of a chemical analysis determined by the formula $(\sum\text{Cations} - \sum\text{Anions}) \div (\sum\text{Cations} + \sum\text{Anions}) \times 100$, which expresses the difference between the sum of the major positively charged *cations* and that of the major negatively charged *anions* as a percentage. Whereas a value of 0% denotes a perfect balance, an error margin of ±5% is generally considered acceptable for fresh water (Hem



P. Hobbs et al. (eds.), *Anatomy of a South African Karst Hydrosystem*, Cave and Karst Systems of the World,
https://doi.org/10.1007/978-3-030-95829-9

1985; Appelo and Postma 2009). The calculation of this parameter in this study has considered only analyses which are complete for their reported sampling date, i.e. are not missing any major anion or cation. Where significant, the calculation has also included elements such as $NO_3$, $PO_4$, Fe and Mn which might materially influence the charge balance.

**Epikarst** The zone of highly weathered strata that occupies the top of the *vadose zone* and might or might not be covered with soil, and gradually grades downward into largely unweathered bedrock that forms the main body of the *vadose zone* (Ford and Williams 2007). Also referred to in some *karst* texts as the subcutaneous zone.

**Gaining stream** A river or stream that receives groundwater discharge; also referred to as an effluent drainage (stream).

**Geosite** The term used to describe any feature that represents a site of geological and/or hydrogeological interest in the environment, whether natural as in the case of a spring or artificial such as a borehole, dug well, mine shaft or tunnel. The DWA/NORAD (2004) recognises ten types of geosite described as the following: borehole, dug well, wellpoint, drain, tunnel, mine, seepage pond, spring, sinkhole and lateral collector.

**Grike** A vertical or sub-vertical cleft in a limestone pavement developed by solution along a joint or system of crisscrossing joints (UNESCO 1972). Ford and Williams (2007) identify these features (also called kluftkarren) as the 'master features' in a karst environment, serving as drains intersecting the *epikarst*.

**Groundwater management unit** An area of a catchment that requires consistent management actions to maintain the desired level of use or protection of groundwater; delineation is based on management considerations rather than geohydrological criteria (Parsons and Wentzel 2007).

**Groundwater resource unit** A groundwater body that has been delineated or grouped into a single significant water resource based on one or more characteristics that are similar across that unit (Parsons and Wentzel 2007).

**Hydraulic conductivity** The rate of flow ($m^3/d$) of water through a 1 $m^2$ cross section of *aquifer* under a *hydraulic gradient* of 1 (Driscoll 1986). Typically expressed as 'k' with the unit $m/d \equiv m^3/d/m^2$.

**Hydraulic gradient** The slope of the *potentiometric surface* that also describes the direction of groundwater movement in an *aquifer*.

**Hydrological year** A hydrological year (aka water year) is defined as the 12-month period October for any given year, through to September of the following year. By convention (see https://water.usgs.gov/nwc/explain_data.html), it is designated by the calendar year in which it ends, and which spans 75% of the composite 12 months.

**Hydrophobic** Lacking an affinity for water.

**Hydrostatic head** Describes the water table in a phreatic aquifer, or the potentiometric head in a *confined aquifer*, that represents the elevation of the groundwater rest level which, in turn, determines the direction of groundwater flow. For example, in a *phreatic aquifer*, groundwater drainage is from a higher elevation to a lower elevation, i.e. driven by gravity drainage. In a *confined aquifer*, groundwater drainage is from an area of a greater head (hydrostatic pressure) to an area of lower head, i.e. driven by differences in pressure (from higher to lower).

**Inlier** An area (usually of limited extent) of older strata completely surrounded by younger strata.

**Juvenile acidity** Describes the acidity (in mine water) that is associated with pyrite oxidation within the zone of mine water level fluctuation once complete rebound has been achieved (Younger 1997).

**Karst** The type of landscape that develops on carbonate rocks such as dolomite and limestone, and is formed principally by the dissolving of rock, giving rise to a landscape characterised by sinkholes and ground subsidence, losing surface water drainages, a well-developed subsurface drainage system, strong interaction between the circulation of surface water and groundwater, and caves (Ford and Williams 2007). Often also supports *aquifers* capable of storing and transmitting substantial quantities of groundwater typically draining to springs or 'eyes'.

**Karst barré** An isolated *karst* that is impounded by impermeable strata (Ford and Williams 2007), examples of which in the study area are represented by the Danielsrust Basin and the dolomitic *outlier* that hosts the *locus of* (mine water) *decant* (LoD). Note that the definition makes no reference to the age of the impounded strata relative to the impounding strata, as is the case in the distinction between *outlier* and *inlier*.

**Locus of decant** The area on the Sibanye-Stillwater property that hosts the various point and diffuse sources of acid mine drainage; includes the Black Reef Incline (BRI), #17 Winze and #18 Winze. Referred to in this manuscript by the acronym LoD.

**Losing drainage** A river or stream that loses some or all of its water into the subsurface, where it becomes groundwater; also referred to as an influent drainage (stream); in a karst environment, such loss typically occurs through a stream sink, swallet or swallow hole.

**Median** Statistical parameter that defines the value which is exceeded 50% of the time in a data set, the remaining 50% of values being below this value, and which means that this value is not unduly influenced (skewed) by anomalously high or low values in the data set.

**Meromictic** A water body (typically a lake) in which the layers of water do not mix or mix only partially/incompletely, leaving a non-mixing bottom layer (monimolimnion) of denser water that remains stagnant and is generally anaerobic. Meromictic surface lakes generally have large relative depths compared to their surface area, and are sheltered from the wind which is the primary driving agent for mixing. The density differences caused by the high dissolved solids concentration are greater than those due to temperature.

**Nanoparticle** Material with at least two dimensions between 1 and 100 nm. Nanoparticles have always existed in our environment, from both natural and anthropogenic sources. Nanoparticles in air were traditionally referred to as ultrafine particles, while in soil and water they were described as colloids (Klaine et al. 2008).

**Occurrence survey** The first phase of study of trace elements and *hydrophobic* organic contaminants in streambed sediment and tissues of aquatic organisms. The primary objective is to determine which target constituents are common and important to water quality conditions in each study unit (Shelton and Capel 1994).

**Outlier** An area (usually of limited extent) of younger strata completely surrounded by older strata.

**Phreatic aquifer** An *aquifer* in which the hydrostatic pressure is equal or similar to that of atmospheric pressure, and therefore supports a *potentiometric surface* that is equivalent to a *water table*. Also referred to as an unconfined *aquifer*.

**Phreatic surface** The surface defined by the *potentiometric surface* in a *phreatic aquifer*, and which is equivalent to the *water table*.

**Potentiometric surface** The theoretical surface fitted to the groundwater level elevation, in layman's terms often referred to as the *water table*, and which represents the spatial dimensions of the *hydrostatic* (pressure) *head* in a *confined aquifer* (Freeze and Cherry 1979; Domenico and Schwartz 1998; Ford and Williams 2007).

**Raw mine water** Mine water (typically acidic) that fills the mine void and decants into the environment either as subsurface flow or upon daylighting in the form of point or diffuse sources of discharge without having been treated in any manner other than that effected by natural factors such as oxygenation/aeration and in situ neutralisation by carbonate strata such as dolomite.

**Resource water quality objective** The measure(s) that must be imposed to ensure that the Class that has been assigned to a groundwater resource as part of a Groundwater Resource Directed Measures assessment is met. The measure(s) might for example consider the maximum allowable lowering and/or fluctuation of groundwater rest level, and the maximum allowable deterioration of groundwater quality. In the absence of data, the measure(s) generally take the form of qualitative descriptions, whereas semi-quantitative or even quantitative measures can be derived from long-term data records.

**Salt load** The mass of dissolved salts carried by a water body per unit time (typically expressed as t/d or t/a), calculated as the product of the flow volume (typically expressed as ML/d, where 1 ML/d $\equiv$ 1000 $m^3$/d) and the salt concentration (typically expressed as mg/L, where 1 mg/L $\equiv$ 0.001 kg/$m^3$).

**Sinkhole** An engineering/engineering geology term of North American origin (Ford and Williams 2007) for a collapse *doline* forming a hole on surface (WRC/IUCN/SAKWG 2010); synonymous terms are closed depression, *doline*, ponor, stream sink, sumidero, swallet and swallow hole (Field 2002).

**Spruit** Afrikaans word for stream or creek, essentially describing a surface water drainage feature (often ephemeral) of lesser magnitude than a river.

**Stygobite** A form of aquatic fauna that completes its entire life cycle in a cave environment.

**Thalweg** Hydrological term that describes the deepest part of a watercourse and, therefore, generally the path of fastest flow in a river or stream.

**Total dissolved solids** Evaporating a liquid and measuring the mass of residues left (drying and weighing), returns a 'total dissolved solids' (TDS) value under circumstances where certain salts decompose during drying, a subtle difference to 'total dissolved salts' as a measure of the combined content of inorganic and organic substances contained in a liquid.

**Transmissivity** The rate at which water is transmitted through a vertical section of an *aquifer* 1 m wide and extending the full saturated height (depth 'd') of the aquifer under a *hydraulic gradient* of 1 (Driscoll, 1986). Typically expressed as 'T' with the unusual unit $m^2/d$ [$L^2/T$], derived from its association with the product of *hydraulic conductivity* 'k' with units m/d [L/T] and 'd' (as previously defined) with units m [L], i.e. T = kd.

**Treated mine water** Mine water (typically acidic) that has been neutralised by means of the high density sludge (HDS) process prior to its release into the environment.

**Troglobite** A form of terrestrial fauna that completes its entire life cycle in a cave environment.

**Vadose zone** The unsaturated zone between the land surface and the **water table**.

**Vestigial acidity** Describes the acidity (in mine water) that is associated with the products of pyrite oxidation taken up in solution during the re-watering of mine workings, and is manifested in the 'first flush' when decant commences (Younger 1997).

**Water table** The surface represented by the groundwater level in a *phreatic aquifer*.

**Winze** A steep (not vertical) shaft which joins different levels in an underground mine. In the strictest sense, a winze has its entrance below surface within the mine, and not at surface as is the case with a shaft, and is excavated downwards, usually on a reef (Handley 2004).

# References


Abiye T (2010) Senior lecturer, hydrogeology. School of Geosciences. University of the Witwatersrand. Personal communication, 1 June 2010.

Abiye TA (2014) Mine water footprint in the Johannesburg area, South Africa: Analysis based on existing and measured data. S Afr J Geol 117(1):87–96

Acocks JPH (1988) Veld types of South Africa, 3rd edn. Botanical Research Institute, South Africa, p 146

Aller L, Bennett T, Lehr JH, Petty RJ (1985) DRASTIC—a standardized system for evaluating ground water pollution potential using hydrogeologic settings. US Environmental Protection Agency. Robert S. Kerr Environmental Research Laboratory. Office of Research and Development. EPA/600/2–85/018, 163 pp

Allison C, Coetzee H, de Wit M (2011) Study of geo-contamination related to acid mine drainage in the West Rand, as a result of South Africa's gold mining industry. Inkaba yeAfrica poster presentation. GeoSynthesis 2011, 28 Aug–2 Sept 2011, Cape Town

Andreo B, Vías J, Durán JJ, Jiménez P, López-Geta JA, Carrasco F (2008) Methodology for groundwater recharge assessment in carbonate aquifers: application to pilot sites in southern Spain. Hydrogeol J 16(5):911–925

Appelo CAJ, Postma D (2009) Geochemistry, groundwater and pollution, 2nd edn. 4th corrected reprint. CRC Press, 649 pp

Arimoro FO (2007) Macroinvertebrates functional feeding groups in River Orogodo, a second order stream in southern Nigeria. Niger J Sci Environ 6

ATSDR (2011) Toxic substances portal: Nickel. Agency for Toxic Substances and Disease Registry. Available at http://www.atsdr.cdc.gov/substances/toxsubstance.asp?toxid=44

Awofolu OR, du Plessis R, Rampedi I (2007) Influence of discharged effluent on the quality of surface water utilized for agricultural purposes. Afr J Biotechnol 6(19):213–223

Baas-Becking LGM, Kaplan IR, Moore D (1960) Limits of the natural environment in terms of pH and oxidation-reduction potentials. J Geol 68:243–284

Bailly-Comte V, Jourde H, Pistre S (2009) Conceptualization and classification of groundwater–surface water hydrodynamic interactions in karst watersheds: case of the karst watershed of the Coulazou River (Southern France). J Hydrol 376:456–462. https://doi.org/10.1016/j.jhydrol.2009.07.053

Bailey GN, King GCP (2011) Dynamic landscapes and human dispersal patterns: tectonics, coastlines, and the reconstruction of human habitats. Quat Sci Rev 30:1533–1553. https://doi.org/10.1016/j.quascirev.2010.06.019

Bakalowicz M (2005) Karst groundwater: a challenge for new resources. Hydrogeol J 13:148–160. https://doi.org/10.1007/s10040-004-0402-9

Bakalowicz M (2006) Importance of regional study site conditions in elaborating concepts and approaches in karst science, pp 15–22. In: Harmon RS, Wicks C (eds) Perspectives on karst geomorphology, hydrology and geochemistry—a tribute volume to Derek C. Ford and William B. White. Special Paper 404. Geological Society of America. 366 pp

Balter M (2014) 'Little Foot' fossil could be human ancestor. Published 14/03/2014 on Science/AAAS. Accessed on 26/03/2014 at http://news.sciencemag.org/africa/2014/03/little-foot-fossil-could-be-human-ancestor

Banks D (2004) Geochemical processes controlling minewater pollution. In: Groundwater management in mining areas proceedings of the 2nd IMAGE-TRAIN advanced study course. Pecs. Hungary, pp 17–44

Banner J, Boston P, Colucci L, Cowan B, Frappier A, Gentry C, Harmon RS, Katz B, Long A, Martin JB, Musgrove M-L, Partin J, Rasmussen J, Wong C, White WB (2007) Geochemistry of modern karst systems. Focus group on geochemistry and climate, pp 82–89. In: Martin JB, White WB (eds) Frontiers of karst research. Special publication 13. Workshop proceedings and recommendations. 3-5/05/2007. San Antonio. Texas. Karst Waters Institute. Leesburg. Virginia, 118 pp

Barnard HC (1996) Investigation into the deterioration of groundwater quality in a part of the Zwartkrans Compartment, District Krugersdorp. Report GH3886. Department of Water Affairs & Forestry. Pretoria, 6 pp

Barnard HC (2000) An explanation of the 1:500 000 general hydrogeological map Johannesburg 2526. Department of Water Affairs & Forestry. Pretoria, 84 pp

Barnes HL, Romberger SB (1968) Chemical aspects of acid mine drainage. J (water Pollut Control Fed) 30(3):371–384

Bauer C, Kellerer-Pirklbauer A, Lieb GK (2008) Human impact on karst environments: a case study from central Styria, Austria. Geophysical Research Abstracts, vol 10. EGU2008-A-01339

Beater M, Kilian D (2009) Environmental management framework and management plan for the Cradle of Humankind World Heritage Site, its proposed buffer zone, and the Muldersdrift area. Status Quo Report. 382,239. Prepared for the Management Authority, Department of Economic Development, Gauteng Province. SRK Consulting, Johannesburg, 138 pp

Béga S (2008a) Pollution threatens Cradle of Humankind. Sourced on 06/01/2010 at http://www.environment.co.za/topic.asp?TOPIC_ID=2124

Béga S (2008b) Cradle's heritage status in danger. Sourced on 06/01/2010 at http://www.iol.co.za/index.php?set_id=1&click_id=13&art_id=vn20081115091504526C425147

Béga S (2010) Is river of acid threatening fossil treasures? Study targets mine water at Cradle of Humankind. Saturday Star, 17 Apr 2010




P. Hobbs et al. (eds.), *Anatomy of a South African Karst Hydrosystem*, Cave and Karst Systems of the World,
https://doi.org/10.1007/978-3-030-95829-9

Béga S (2012a) Study lightens toxic water gloom at Cradle fossil sites. Saturday Star, 14 Jan 2012

Béga S (2012b) A toxic delay. Saturday Star, 28 Jan 2012

Béga S (2012c) There's life in this dead river, say scientists. Saturday Star, 26 May 2012

Béga S (2012d) AMD threat to heritage site – study. Saturday Star, 26 May 2012

Béga S (2013) Water taints life in Cradle: toxic mine water chokes country estate. Saturday Star, 28 Sept 2013

Beltrán R, de la Rosa JD, Santos JC, Beltrán M, Gómez-Ariza JL (2010) Heavy metal mobility assessment in sediments from the Odiel River (Iberian Pyritic Belt) using sequential extraction. Environ Earth Sci 61(7):1493–1503

Benedini M, Tsakiris G (2013) Water quality modelling for rivers and streams. Springer, Dordrecht, p 288

Berger LR (2001) Viewpoint: is it time to revise the system of scientific naming? National Geographic News. Sourced on 28/06/2010 at http://news.nationalgeographic.com/news/2001/12/1204_hominin_id.html

Berger LR, Brink J (undated) An atlas of southern African mammalian fossil bearing sites—late Miocene to late Pleistocene. Chapter 6: The sites: dolomitic and other cave deposits. Sourced on 24/05/2010 at http://www.profleeberger.com/files/An_Atlas_of_southern_African_Fossil_Bearing_Sites.pdf

Berger LR, de Ruiter DJ, Churchill SE, Schmid P, Carlson KJ, Dirks PHGM, Kibii JM (2010) Australopithecus sediba: A new species of Homo-like Australopith from South Africa. Science 328:195–204

Berger LR, Hawks J, De Ruiter DJ, Churchill SE, Schmid P, Delezene LK, Kivell TL, Garvin HM, Williams SA, DeSilva JM, Skinner MM, Musiba CM, Cameron N, Holliday TW, Harcourt-Smith W, Ackermann RR, Bastir M, Bogin B, Bolter D, Brophy J, Cofran ZD, Congdon KA, Deane AS, Dembo M, Drapeau M, Elliott MC, Feuerriegel EM, Garcia-Martinez D, Green DJ, Gurtov A, Irish JD, Kruger A, Laird MF, Marchi D, Meyer MR, Nalla S, Negash EW, Orr CM, Radovcic D, Schroeder L, Scott JE, Throckmorton Z, Tocheri MW, van Sickle C, Walker CS, Wei P, Zipfel B (2015) Homo naledi, a new species of the genus Homo from the Dinaledi Chamber, South Africa. eLife 4:e09560, 35 pp

Bertram E (2008) Assistant director: Groundwater information. Directorate: Hydrological Services. Department of Water Affairs & Forestry. Email communication, 18 Mar 2008

Bertram E (2010) Assistant director: Groundwater information. Directorate: Hydrological Services. Department of Water Affairs. Personal communication, 28 June 2010.

Beukes NJ (1978) Die karbonaat gesteentes en ysterformasies van die Ghaap-groep van die Transvaal-supergroep in die Noord-Kaapland. (E. The carbonate rocks and iron formations of the Ghaap Group of the Transvaal Supergroup in the Northern Cape.) PhD thesis (unpublished). Rand Afrikaans University, Johannesburg, 580 pp

Beukes NJ (1987) Facies relationships, depositional environments and diagenesis in a major early Proterozoic stromatolitic carbonate platform to basinal sequence, Campbell Rand Subgroup, Tranvaal Supergroup, South Africa. Sedimentary Geol 54:1–56

Bjerg PL, Rügge K, Pedersen JK, Christensen TH (1995) Distribution of redox-sensitive groundwater quality parameters downgradient of a landfill (Grindsted, Denmark). Environ Sci Technol 29(5):1387–1394

BKS (2011) Witwatersrand gold fields acid mine drainage: Due diligence report. TCTA Contract 08-041, 154 pp

Blackeagle C (2014) Is it a karst terrane, karst terrain, or both? Sourced on 19/12/2012 at http://speleogenesis.info/community/blog/post/?id=Is-it-a-karst-terrane_-karst-terrain_-or-both

Bollmohr S, Thwala M, Jooste S, Havemann A (2008) An assessment of agricultural pesticides in the Upper Olifants River Catchment. Report N/0000/REQ0801. Resource Quality Services. Department of Water Affairs and Forestry. Pretoria, 45 pp

Bonacci O, Pippan T, Culver DC (2009) A framework for karst ecohydrology. Environ Geol 56:891–900. https://doi.org/10.1007/s00254-008-1189-0

Bonneville S, Behrends T, van Cappellen P (2009) Solubility and dissimilatory reduction kinetics of iron(III) oxyhydroxides: a linear free energy relationship. Geochim Cosmochim 73:5273–5282

Boonzaier B (2010) Proprietor: geowater systems. Personal communication, 15 Dec 2010

Boshoff P (2013) Caver. Personal communication, 26 Nov 2013

Boudinet P (2012) A statistical model of karstic flow conduits. Speleogenesis and Evolution of Karst Aquifers (12):9–16. Sourced on 19/12/2012 at http://www.speleogenesis.info/directory/karstbase/pdf/seka_pdf11251.pdf

Boulton AJ, Humphreys WF, Eberhard SM (2003) Imperilled subsurface waters in Australia: biodiversity, threatening processes and conservation. Aquatic Ecosyst Health Manage 6(1):41–54

Bradley MJ, Herman JS (1989) Sulfate adsorption onto iron and aluminium hydroxides. Eos, Trans Am Geophys Union 70:326

Brassington R (1998) Field hydrogeology, 2nd edn. Wiley, London, p 248

Bredenkamp DB, van der Westhuizen C, Wiegmans FE, Kuhn CM (1986) Ground-water supply potential of dolomite compartments west of Krugersdorp. Report GH3440, vols 1, 2. Department of Water Affairs & Forestry, Pretoria, 81 pp

Bredenkamp DB, Vogel JC, Wiegmans FE, Xu Y, Janse van Rensburg H (2007) Use of natural isotopes and groundwater quality for improved estimation of recharge and flow in dolomitic aquifers. Report KV 177/07. Water Research Commission, Pretoria, 68 pp

Bredenkamp G, van Staden S (2009) Environmental management framework and management plan for the Cradle of Humankind World Heritage Site, its proposed buffer zone, and the Muldersdrift area. Terrestrial and aquatic ecology. Specialist report to SRK Consulting, Johannesburg, p 34

Brink ABA (1996) Case History 18: Malmani subgroup—sequence of events relating to mining and the accelerated development of sinkholes and dolines in the Far West Rand. In: Brink ABA (ed) Engineering geology of South Africa, vol 1, pp 240–248

Brink D (2008) Letter to the Chairperson: western basin technical task team titled crisis situation in the Krugersdorp Game Reserve. 01/09/2008. African Bush Adventures, Krugersdorp, 11 pp

Brook GA, Folkoff ME, Box EO (1983) A world model of soil carbon dioxide. Earth Surface Processes Landforms 8(1):79–88. https://doi.org/10.1002/esp.3290080108

Brooker J (2011) Owner. Glen Afric Country Lodge. Personal communication, 7 Jan 2011

Brown L (2009) Environmental management framework and management plan for the Cradle of Humankind World Heritage Site, its proposed buffer zone, and the Muldersdrift area: Surface and groundwater. Specialist report. Prepared for the Management Authority, Department of Economic Development, Gauteng Province. SRK Consulting. Johannesburg, 80 pp

Brown JG, Glynn PD (2003) Kinetic dissolution of carbonates and Mn oxides in acidic water: Measurement of in situ field rates and reactive transport modelling. Appl Geochem 18:1225–1239

Bruxelles L, Clarke RJ, Maire R, Ortega R, Stratford D (2014) Stratigraphic analysis of the Sterkfontein StW 573 Australopithecus skeleton and implications for its age. J Hum Evol. Available online 31 Mar 2014 and accessed on 7 Apr 2014 at http://www.sciencedirect.com/science/article/pii/S004724841400058X

Burke JJ, Moench MH (2000) Groundwater and society: resources, tensions and opportunities. ST/ESA/265. United Nations. New York. 170 pp

Busenberg E, Plummer LN (1982) The kinetics of dissolution of dolomite in $CO_2$-water systems at 1.5° to 65 °C and 0.0 to 1.0 atmosphere $P_{CO2}$. Am J Sci 282:45–78

Carpenter H (2010) Facilities supervisor. Nedbank Leadership and Management Development Centre. Personal communication, 19 Nov 2010

Carpenter H (2011) Facilities supervisor. Nedbank Leadership and Management Development Centre. Personal communication, 13 Jan 2011

Carpenter H (2013) Energy and water manager. Group Property Services. Nedbank Group Ltd. Personal communication, 25 Feb 2013

Cawthorn RG, Eales HV, Walraven F, Uken R, Watkeys MK (2006) The bushveld complex, pp 261–282. In: Johnson MR, Anhaeusser CR, Thomas RJ (eds) The Geology of South Africa. Geological Society of South Africa/Council for Geoscience, Johannesburg/Pretoria, p 691

Cecil LD, Green JR (2000) Radon-222, pp 175–194. In: Cook PG, Herczeg AL (eds) Environmental tracers in subsurface hydrology. Kluwer Academic Publications, Boston/Dordrecht/London, p 529

CGS (undated) Progress report, Western Mining Basin—15 October to 14 December 2004. Draft report. Witwatersrand Water Ingress Project. Council for Geoscience. Pretoria, 89 pp

CGS/SAIEEG 2003. Guideline for engineering-geological characterisation and development of dolomitic land. Council for Geoscience/South African Institute of Engineering and Environmental Geologists Joint Document. Endorsed by the Geotechnical Division of the South African Institute of Civil Engineers. South Africa, 66 pp

Chirenje E, Nyabeze PK, Coetzee H, Hobbs PJ (2010) Geophysical surveys, Sterkfontein Caves. Specialist report to the COH WHS Management Authority. Council for Geoscience/CSIR. Pretoria, 9 pp

Choquette PW, Pray LC (1970) Geologic nomenclature and classification of porosity in sedimentary carbonates. Am Assoc Petrol Geol 54:207–250

Clendenin CW (1989) Tectonic influence on the evolution of the early Proterozoic Transvaal Sea, southern Africa. PhD thesis (unpublished). University of the Witwatersrand, Johannesburg

Coetzee H, Wade P, Ntsume G, Jordaan W (2002) Radioactivity study on sediments in a dam in the Wonderfontein Spruit catchment. DWAF-Report 2002, Pretoria

Coetzee H, Winde F, Wade P (2006) An assessment of sources, pathways, mechanisms and risks of current and potential future pollution of water and sediments in gold-mining areas of the Wonderfonteinspruit catchment. Report 1214/1/06. Water Research Commission, Pretoria, 202 pp

Coetzee H, Chirenje E, Hobbs P, Cole J (2009) Ground and airborne geophysical surveys identify potential subsurface acid mine drainage pathways in the Krugersdorp Game Reserve, Gauteng Province, South Africa. Short paper. 10th SAGA Biennial Technical Meeting and Exhibition, 13–18 Sept 2009, Swaziland

Coetzee H, Hobbs PJ, Burgess JE, Thomas A, Keet M (2010) Mine water management in the Witwatersrand gold fields with special emphasis on acid mine drainage. Report to the Inter-Ministerial Committee on Acid Mine Drainage, Pretoria, p 128

Cole DI (1998) Uranium, pp 642–658. In: Wilson MGC, Anhaeusser CR (eds) The Mineral Resources of South Africa. Handbook 16. Council for Geoscience, 740 pp

Cooke HBS (1969) Preservation of the Sterkfontein ape-man cave site, South Africa. Reprinted from Studies in Speleology, vol 2, Part 1, 34 pp

Corbel J (1959) Erosion en terrain calcaire (vitesse d'érosion et morphologie). Reprinted in Progress Phys Geogr 32(6):684–690. https://doi.org/10.1177/0309133308096756

Cornaton F, Goldscheider N, Jeannin PY, Perrochet P, Pochon A, Sinreich M, Zwahlen F (2003) The VULK analytical transport model and mapping method, pp 155–160. In: Zwahlen F (ed) Vulnerability and risk mapping for the protection of carbonate (karst) aquifers. Final report. COST Action 620. European Commission, Brussels, 297 pp

Cullimore DR (1999) Microbiology of well biofouling. Lewis Publishers/CRC Press, Boca Raton, p 239

Culver DC, Sket B (2000) Hotspots of subterranean biodiversity in caves and wells. J Cave Karst Stud 62(1):11–17

Culver DC, Sket B (2002) Biological monitoring in caves. Acta Carsologica 31/1(4):55–64

Cummins KW, Merritt RW, Andrade PCN (2005) The use of invertebrate functional groups to characterise ecosystem attributes in selected streams and rivers in south Brazil. Stud Neotr Fauna Environ 40(1):69–89

Dallas HF (2007) River Health Programme: South African Scoring System (SASS) data interpretation guidelines (Draft report). Prepared for the Institute of Natural Resources and the Resource Quality Services River Health, Department of Water Affairs and Forestry, Pretoria/Cape Town

Damons A (2010) Squatters claim Cradle of Humankind. Accessed on 08/06/2010 at http://www.news24.com/SouthAfrica/News/Squatters-claim-Cradle-of-Humankind-20100418

Davidson B, Wei YP (2012) Assessing the 'wicked problems' associated with the quality of groundwater used in irrigation: a case study from the North China Plain. Hydrogeol J. Published online 05 January 2012. https://doi.org/10.1007/s10040-011-0811-5

DEA (2014a) State of Conservation report for the fossil hominid sites of South Africa World Heritage Site (the Sterkfontein, Swartkrans, Kromdraai and environs component) (C 915 BIS). Report to the UNESCO World Heritage Committee. Department of Environmental Affairs, Management Authority and CSIR, Dec 2014, 12 pp

DEA (2014b) Amendment of environmental authorisation: the immediate and short term interventions for the treatment of acid mine drainage in the Western, Central and Eastern basins of the Witwatersrand Gold Fields, Gauteng Province. Department of Environmental Affairs correspondence dated 05/03/2014 to the Department of Water Affairs, 3 pp

DEA (2016) State of Conservation report for the fossil hominid sites of South Africa World Heritage Site (the Sterkfontein, Swartkrans, Kromdraai and environs component) (C 915 BIS). Report to the UNESCO World Heritage Committee. Department of Environmental Affairs, Management Authority and CSIR, Dec 2016, 10 pp

De Jesus ASM (1985) On the behaviour of radium in tailings dams and environmental waters in the Witwatersrand (South Africa) gold/uranium mining area. In: Proceedings of the 2nd international mine water congress, Granada, Spain, pp 633–645

De Villiers S, Mkwelo ST (2009) Has monitoring failed the Olifants River, Mpumalanga? Water SA 35(5):671–676

Dickens CWS, Graham PM (2002) The South African scoring system (SASS) Version 5 rapid bioassessment method for rivers. Afr J Aquatic Sci 27(1):1–10

Dirks PHGM, Berger LR (2013) Hominin-bearing caves and landscape dynamics in the Cradle of Humankind, South Africa. J Afr Earth Sci 78:109–131. https://doi.org/10.1016/j.jafrearsci.2012.09.012

Dirks PHGM, Kibii JM, Kuhn BF, Steininger C, Churchill SE, Kramers JD, Pickering R, Farber DL, Mériaux A-M, Herries AIR, King GCP, Berger LR (2010) Geological setting and age of Australopithicus sediba from Southern Africa. Science 328:205–208. https://doi.org/10.1126/science.1184950

Dirks PHGM, Berger LR, Roberts EM, Kramers JD, Hawks J, Randolph-Quinney PS, Elliott M, Musiba CM, Churchill SE, de Ruiter DJ, Schmid P, Backwell LR, Belyanin GA, Boshoff P,

Hunter KL, Feuerriegel EM, Gurtov A, du Harrison GJ, Hunter R, Kruger A, Morris H, Makhubela TV, Peixotto B, Tucker S (2015) Geological and taphonomic evidence for deliberate body disposal by the primitive hominin species Homo naledi from the Dinaledi Chamber, South Africa. Elife 4:e09651

Doerfliger N, Jeannin PY, Zwahlen F (1999) Water vulnerability assessment in karst environments: a new method of defining protection areas using a multi-attribute approach and GIS tools (EPIK method). J Environ Geol 39(2):165–176

Dorling D (2010) Proprietor. DD Environmental Science cc. Communication to the Western Basin Void Sub-group Monitoring Committee Meeting. Mogale City Local Municipality offices, 17 Aug 2010

Dorling D (2011) Proprietor. DD Environmental Science cc. Personal communication, 13 Oct 2011

Doyle G (2010) Resident Manager. Crab Farm Estate. Personal communication, 13 May 2010

Driscoll FG (1986) Groundwater and Wells, 2nd edn. Johnson Filtration Systems Inc. St. Paul. Minnesota, 1089 pp

Duane MJ, Pigozzi G, Harris C (1997) Geochemistry of some deep gold mine waters from the western portion of the Witwatersrand Basin, South Africa. J Afr Earth Sci 24(1/2):105–123

Dublyansky YV (2014) Hypogene speleogenesis—Discussion of definitions, pp 1–3. In: Klimchouk A, Sasowsky ID, Mylroie J, Engel SA, Engel AS (eds) Hypogene cave morphologies. Karst Waters Institute. Special publication 18, 103 pp. ISBN: 978-0-9.789.976-7-0

Dunne S (2003) A Localised European Approach (LEA), pp 161–163. In: Zwahlen F (ed) Vulnerability and risk mapping for the protection of carbonate (karst) aquifers. Final report. COST Action 620. European Commission, Brussels, 297 pp

Durand JF (2012) The impact of gold mining on the Witwatersrand on the rivers and karst system of Gauteng and North West province, South Africa. J Afr Earth Sci 68:24–43

Durand JF, Peinke D (2010) Issue paper 4: The state of karst ecology research in the Cradle of Humankind World Heritage Site. pp 88–101. In WRC/IUCN-SAKWG. The karst system of the Cradle of Humankind World Heritage Site. Report KV 241/10. Water Research Commission, Pretoria, 402 pp

Durand JF, Meeuvis J, Fourie M (2010) The threat of mine effluent to the UNESCO status of the Cradle of Humankind World Heritage Site. J Transdiscip Res S Afr 6(1):73–92

du Toit S (2010) Environmental Manager. Mogale City Local Municipality. Personal communication, 26 May 2010

DWA/NORAD (2004) Standard Descriptors for Geosites. Version 1.1. Department of Water Affairs and Forestry/Norwegian Agency for Cooperation Development, Pretoria, 122 pp

DWA (2006) Groundwater resource assessment 2. Department of Water Affairs, Pretoria

DWA (2009) Green Drop Report 2009. Version 1. South African Waste Water Quality Management Performance. Department of Water Affairs, Pretoria, 124 pp

DWA (2010a) Document WMA03 CROCODILE WEST MARICO RESERVOIR SITES.doc. Sourced on 18/06/2010 at http://www.dwaf.gov.za/hydrology/dwafapp2_wma/

DWA (2010b) West Rand Dolomite Compartments. Final version 1.0. A1 map at scale 1:85 000 available from Directorate: Water Resources Planning Systems. Department of Water Affairs, Pretoria

DWA (2011) Green drop report 2011. Waste Water Service Regulation. Department of Water Affairs, Pretoria, 450 pp

DWA (2013a) Feasibility study for a long-term solution to address the acid mine drainage associated with the East, Central and West Rand underground mining basins: assessment of the water quantity and quality of the Witwatersrand mine voids. Report No. 5.2. Edition 1. Project No. P RSA 000/00/16512/2, Pretoria. South Africa, 270 pp

DWA (2013b) Newsletter: AMD FS LTS. Ed. 2, 13 pp

DWAF (1996a) South African Water Quality Guidelines. Volume 7: Aquatic Ecosystems, 2nd edn. Department of Water Affairs and Forestry, Pretoria, 159 pp

DWAF (1996b) South African Water Quality Guidelines. Volume 5: Agricultural Use: Livestock Watering, 2nd edn. Department of Water Affairs and Forestry, Pretoria, 163 pp

DWAF (1996c) South African Water Quality Guidelines, vol 4: Agricultural Use: Irrigation, 2nd edn. Department of Water Affairs and Forestry, Pretoria, 199 pp

DWAF (1996d) South African Water Quality Guidelines. Volume 2: Recreational Water Use, 2nd edn. Department of Water Affairs and Forestry, Pretoria, 85 pp

Ellis R, Grove A (2010) Issue paper 12: Legal aspects of karst and cave use in the Cradle of Humankind World Heritage Site, pp 286–352. In: WRC/IUCN-SAKWG. The karst system of the Cradle of Humankind World Heritage Site. Report KV 241/10. Water Research Commission, Pretoria, 402 pp

Elsner M, Schwarzenbach RP, Haderlein SB (2004) Reactivity of Fe (II)-bearing minerals toward reductive transformation of organic contaminants. Environ Sci Technol 38(3):799–807

Engel A, Northup D, Gary M, Gonzalez B, Gonzalez J, Hutchens E, Jones D, Macalady J, Spear J, Spilde M (2008) Focus group on caves and karst as model systems in geomicrobiology, pp 90–95. In: Martin JB, White WB (eds) Frontiers of Karst Research. Special publication 13. Karst Waters Institute, Leesburg, Virginia, 118 pp

EPS (1994) Guidance document on collection and preparation of sediments for physicochemical characterisation and biological testing. Report EPS 1/RM/29. Environmental Protection Series. Environment Canada, 129 pp

Erickson PM, Kleinmann RLP, Posluszny ET, Leonard-Mayer PJ (1982) Hydrogeochemistry of a large mine pool. In: Proceeedings of 1st international mine water congress, Budapest, Hungary, A, pp 27–42

Eriksson PG, Reczko BFF (1995) The sedimentary and tectonic setting of the Transvaal Supergroup floor rocks to the Bushveld Complex. J Afr Earth Sci 21:487–504

Eriksson PG, Altermann W, Hartzer FJ (2006) The Transvaal Supergroup and its precursors. pp 237–260. In: Johnson MR, Anhaeusser CR, Thomas RJ (eds) The Geology of South Africa. Geological Society of South Africa/Council for Geoscience, Johannesburg/Pretoria, p 691

Esterhuysen A (2009) Environmental management framework and management plan for the Cradle of Humankind World Heritage Site, its proposed buffer zone, and the Muldersdrift area: Cultural-Heritage. Specialist report to SRK Consulting, Johannesburg, p 26

Ewers RO, White KA, Fuller JF (2012) Contaminant plumes and pseudoplumes in karst aquifers. Carbonates Evaporites 27(2):153–159. https://doi.org/10.1007/s13146-012-0099-0

Falkowski P, Scholes RJ, Boyle E, Canadell J, Canfield D, Elser J, Gruber N, Hibbard K, Högberg P, Linder S, Mackenzie FT, Moore B III, Pedersen T, Rosenthal Y, Seitzinger S, Smetacek V, Steffen W (2000) The global carbon cycle: a test of our knowledge of Earth as a system. Science 290:291–296

Field MS (2002) A lexicon of cave and karst terminology with special reference to environmental karst hydrology. Report EPA/600/R-02/003. United States Environmental Protection Agency, Washington, DC, 214 pp

Field MS (2006) Tracer-test design for losing stream–aquifer systems. Int J Speleol 35(1):25–36

Fleisher JNE (1981) The geohydrology of the dolomite aquifers of the Malmani Subgroup in the South-Western Transvaal. Report Gh3169. Department of Water Affairs, Pretoria, 228 pp

Florea LJ, Vacher HL (2006) Springflow hydrographs: Eogenetic vs. telogenetic karst. Ground Water 44(3):352–361

Florea LJ (2015) Carbon flux and landscape evolution in epigenic karst aquifers modelled from geochemical mass balance. Earth Syst Processes Landforms. https://doi.org/10.1002/esp.3709

Fofonoff NP, Millard RC, Jr (1983) Algorithms for computation of fundamental properties of seawater. UNESCO Technical Papers in Marine Science 44, Paris, 53 pp

Foose RM (1967) Sinkhole formation by groundwater withdrawal: Far West Rand, South Africa. Science 157(3792):1045–1048

Ford DC (2003) Perspectives in karst hydrogeology and cavern genesis. Speleogenesis and Evolution of Karst Aquifers. Issue 1. 12 pp. Sourced on 19/12/2012 at http://www.speleogenesis.info/directory/karstbase/pdf/seka_pdf4470.pdf

Ford DC (2006) Karst geomorphology, caves and cave deposits: a review of North American contributions during the past half century, pp 1–14. In: Harmon RS, Wicks CW (eds) Perspectives on karst geomorphology, hydrology and geochemistry. Special Paper 404. Geological Society of America, 366 pp

Ford TD, Pedley HM (1996) A review of tufa and travertine deposits of the world. Earth-Sci Rev 41:117–175

Ford D, Williams P (2007) Karst Hydrogeology and Geomorphology. Wiley, Chichester, 562 pp. ISBN: 978-0-470–84.997-2

Förstner U, Wittmann GTW (1976) Metal accumulations in acidic waters from gold mines in South Africa. Geoforum 7(1):9. https://doi.org/10.1016/0016-7185(76)90,056-7

Fourie M (2005) A rising acid tide. EarthYEAR J Sustain Dev 1:36–41

Fourie HE, Thirion C, Weldon CW (2014) Do SASS5 scores vary with season in the South African Highveld? A case study on the Skeerpoort River, North West Province, South Africa. Afr J Aquat Sci 39(4):369–376

Fox LE (1988) The solubility of colloidal ferric hydroxide and its relevance to iron concentrations in water. Geochim Cosmochim Acta 52(3):771–777

FSE (2014) Annual Report 2013. Federation for a Sustainable Environment. Available from http://www.fse.org.za/

Gabrovšek F (2007) On denudation rates in karst. Time in Karst, Postojna, pp 7–13. Available at http://www.carsologica.zrc-sazu.si/downloads/361/gabrovsek1.pdf

Gabrovšek F (2009) On concepts and methods for the estimation of dissolutional denudation rates in karst area. Geomorphology 106:9–14. https://doi.org/10.1016/j.geomorph.2008.09.008

Gammons CH, Tucci NJ (2013) The Berkely Pit Lake, Butte, Montana, pp 362–376. In: Geller W, Schultze M, Kleinmann R, Wolkersdorfer C (eds) Acidic pit lakes: the legacy of coal and metal surface mines. Springer, Heidelberg, 525 pp

Gams I (1991) The origin of the term karst in the time of transition of karst (kras) from deforestation to forestation. In: Proceedings of the international conference on environmental changes in Karst Areas (IGU/UIS). Quaderni del Dipartimento di Geografia 13. Universita di Padova, pp 1–8

Gardner EK (2015) New instrument dates old skeleton; 'Little Foot' 3.67 million years old. Purdue University News. Sourced on 04/04/2014 at http://www.purdue.edu/newsroom/releases/2015/Q2/new-instrument-dates-old-skeleton-little-foot-3.67-million-years-old.html

Gauchon C, Ployon E, Delannoy J-J, Hacquard S, Hobléa F, Jaillet S, Perrette Y (2006) The concepts of heritage and heritage resource applied to karsts: protecting the Choranche Caves (Vercors, France). Acta Carsologica 35(2):37–46

Gayer KH, Woontner L (1956) The solubility of ferrous hydroxide and ferric hydroxide in acidic and basic media at 25 °C. J Phys Chem 60(11):1569–1571

Goldscheider N (2003) The PI method, pp 144–154. In: Zwahlen F (ed) Vulnerability and risk mapping for the protection of carbonate (karst) aquifers. Final report. COST Action 620. European Commission, Brussels, 297 pp

Goldscheider N (2012) A holistic approach to groundwater protection and ecosystem services in karst terrains. AQUA mundi, pp 117–124. https://doi.org/10.4409/Am-046-12-0047

Gombert P (2002) Role of karstic dissolution in global carbon cycle. Glob Planet Change 33:177–184

Gomes M (2010) Chairman. Kromdraai Irrigation Board. Personal communication, 05 Oct 2010

Gomes M (2011) Chairman. Kromdraai Irrigation Board. Personal communication, 11 Jan 2011

Gondwe BRN (2010) Exploration, modelling and management of groundwater-dependent ecosystems in karst—the Sian Ka'an case study, Yucatan, Mexico. PhD thesis. Technical University of Denmark, Lyngby, 87 pp + Appendices

Gordon AK, Muller WJ (2010) Developing sediment quality guidelines for South Africa. Phase 1: Identification of international best practice and applications for South Africa to develop a research and implementation framework. Report KV 242/10. Water Research Commission, Pretoria, 49 pp

Govender B (2010) Assistant Director. DWS. Communication to the Western Basin Void Sub-group Monitoring Committee Meeting. Mogale City Local Municipality offices, 13 July 2010

Graening GO, Brown AV (2000) Trophic dynamics and pollution effects in Cave Springs Cave, Arkansas. Publication MSC-285. Final report to the Arkansas Natural Heritage Commission. Department of Biological Sciences. Arkansas Water Resources Center, 44 pp

Gray NF (1996) Field assessment of acid mine drainage contamination in surface and ground water. Environ Geol 27:358–361

Groenewald Y (2010a) Acid mine water pollution a 'ticking time bomb'. Mail&Guardian, 1–8 Apr 2010

Groenewald J (2010b) Issue paper 11: the impacts of agriculture on the water resources and water-based ecosystems of the Cradle of Humankind World Heritage Site, pp 241–285. In: WRC/IUCN-SAKWG 2010. The karst system of the Cradle of Humankind World Heritage Site. Report KV 241/10. Water Research Commission, Pretoria, 402 pp. ISBN: 978-1-77.005–969-6

Gunn J (ed) (2004) Encyclopedia of caves and karst science. Fitzroy Dearborn/Taylor and Francis, New York/London, p 902

Hamilton-Smith E (2006) Thinking about Karst and World Heritage. Helictite 39(2):51–54

Hammer Ø, Harper DAT, Ryan PD (2001) PAST: Paleontological statistics software package for education and data analysis. Palaeontol Electron 4(1):9. http://palaeo-electronica.org/2001_1/past/issue1_01.htm

Handley JRF (2004) Historic overview of the Witwatersrand Goldfields, Howick, 224 pp.

Hardin G (1968) The tragedy of the commons. Science 162:1243–1248

Harding WR, Thornton JA, Steyn G, Panuska J, Morrison IR (2004) Hartbeespoort Dam remediation project (phase 1). Action plan (Volume 1) final report. DH Environmental Consulting, 166 pp

Harlow GE, Orndorff RC, Nelms DL, Weary DJ, Moberg RM (2005) Hydrogeology and ground-water availability in the carbonate aquifer system of Frederick County, Virginia. Scientific investigations report 2005–5161. US Geological Survey, 30 pp

Harril JR, Prudic DE (1998) Aquifer systems in the Great Basin Region of Nevada, Utah, and adjacent states—summary report. Professional paper 1409-A. US Geological Survey, 66 pp

Heath RG, van Zyl HD, Schutte CF, Schoeman JJ (2009) First order assessment of the quantity and quality of non-point sources of pollution associated with industrial, mining and power generation. Report 1627/1/09. Water Research Commission, Pretoria, 167 pp

Hem JD (1985) Study and interpretation of the chemical characteristics of natural water. Water Supply Paper 2254, 3rd edn. US Geological Survey, 263 pp

Henry AG, Ungar PS, Passey BH, Sponheimer M, Rossouw L, Bamford M, Sandberg P, De Ruiter DJ, Berger L (2012) The diet of Australopithecus sediba. Nature. https://doi.org/10.1038/nature11185

Hershey RL, Mizell SA, Earman S (2010) Chemical and physical characteristics of springs discharging from regional flow systems of the carbonate-rock province of the Great Basin, western United States. Hydrogeol J 18:1007–1026

Herr C, Gray NF (1997) Sampling riverine sediments impacted by acid mine drainage: problems and solutions. Environ Geol 29(1/2):37–45

Hill L, McMillan P, Cheng P (2014) An assessment of the biotic response in streams of the Western Basin that receive neutralised acid mine drainage. Report no. CSIR/NRE/WR/IR/2014/0021/B, Pretoria, 26 pp

Hobbs PJ (1988a) Hydrogeology of the Verwoerdburg dolomite aquifer. Report GH3502. Department of Water Affairs and Forestry, Pretoria, 158 pp

Hobbs PJ (1988b) Recent test pumping of the Malmani Subgroup dolomite aquifer in perspective, pp 35–42. In: GWD-GSSA. Workshop on dolomitic ground water of the PWV area, 28 Mar, Ground Water Division of the Geological Society of South Africa, Pretoria, 145 pp

Hobbs PJ (2004) Intermediate groundwater Reserve determination for Quaternary catchments A21A and A21B. Project 2002–316. Department of Water Affairs & Forestry, Pretoria, 179 pp

Hobbs PJ (2008a) Groundwater levels and dolomite—nuisance or necessity, pp 213–223. In: Proceedings of the "Problem Soils in South Africa" conference. SA Institute for Engineering and Environmental Geologists, 3–4 Nov 2008. ESCOM Conference Centre, Midrand, South Africa, 232 pp

Hobbs PJ (2008b) Situation analysis of hydrologic and hydrogeologic factors informing the Royal Engineering groundwater supply on Sterkfontein 173IQ, Krugersdorp. Report CSIR/NRE/WR/ER/2008/0107/C. Council for Scientific & Industrial Research, Pretoria, 14 pp

Hobbs PJ (ed) (2011a) Situation assessment of the surface water and groundwater resource environments in the Cradle of Humankind World Heritage Site. Report prepared for the COH WHS Management Authority. Department of Economic Development, Gauteng Provincial Government, South Africa, 424 pp. Available online at http://www.dwaf.gov.za/ghreport/ as report 2.2(2657)

Hobbs PJ (2011b) Assessment of the water level rise in Sterkfontein Caves, Cradle of Humankind World Heritage Site, Gauteng Province. Report CSIR/NRE/WR/ER/2011/0083/A. Council for Scientific & Industrial Research, Pretoria, 14 pp. Available online at http://www.dwaf.gov.za/ghreport/ as report 2.2(?)

Hobbs PJ (2011c) Gladysvale fossil site: opinion on the hydrosensitivity of the site. Letter report. Council for Scientific & Industrial Research, Pretoria, 4 pp

Hobbs PJ (2012) Pilot implementation of a surface water and groundwater resources monitoring programme for the Cradle of Humankind World Heritage Site: situation assessment and status report for the period April to September 2012. Report CSIR/NRE/WR/ER/2012/0088/B. Prepared for the Management Authority, Gauteng Tourism Authority. Council for Scientific and Industrial Research. Pretoria, 39 pp. Available online at at http://www.dwaf.gov.za/ghreport/ as report 2.2(?)

Hobbs PJ (2013a) Response to the AED interpretation of groundwater quality at Sterkfontein Country estate in December 2013. Letter report dated 04/04/2013. Prepared for the Department of Water Affairs. Council for Scientific and Industrial Research, Pretoria, 10 pp

Hobbs PJ (2013b) Pilot implementation of a surface water and groundwater resources monitoring programme for the Cradle of Humankind World Heritage Site: situation assessment and status report for the period April 2012 to March 2013. Report CSIR/NRE/WR/ER/2013/0023/B. Prepared for the Management Authority, Gauteng Tourism Authority. Council for Scientific and Industrial Research. Pretoria, 48 pp. Available online at http://www.dwaf.gov.za/ghreport/ as report 2.2(2866)

Hobbs PJ (2013c) Surface water and groundwater resources monitoring, Cradle of Humankind World Heritage Site, Gauteng Province, South Africa: Water resources status report for the period April to September 2013. Report CSIR/NRE/WR/ER/2013/0083/A. Prepared for the Management Authority, Gauteng Tourism Authority. Council for Scientific and Industrial Research, Pretoria, 47 pp. Available online at http://www.dwaf.gov.za/ghreport/ as report 2.2(2866)

Hobbs PJ (2014a) Surface water and groundwater resources monitoring, Cradle of Humankind World Heritage Site, Gauteng Province, South Africa: Water resources status report for the period April 2013 to March 2014. Report CSIR/NRE/WR/IR/2014/0049/A. Prepared for the Management Authority, Gauteng Tourism Authority. Council for Scientific & Industrial Research, Pretoria, 34 pp. Available online at http://www.dwaf.gov.za/ghreport/ as report 2.2(2957)

Hobbs PJ (2014b) Surface water and groundwater resources monitoring, Cradle of Humankind World Heritage Site, Gauteng Province, South Africa: Water resources status report for the period April to September 2014. Report CSIR/NRE/WR/ER/2014/0063/A. Prepared for the Management Authority, Gauteng Tourism Authority. Council for Scientific & Industrial Research, Pretoria, 45 pp. Available online at http://www.dwaf.gov.za/ghreport/ as report 2.2(3087)

Hobbs PJ (2015a) Surface water and groundwater resources monitoring, Cradle of Humankind World Heritage Site, Gauteng Province, South Africa: Water resources status report for the period April 2014 to March 2015. Report CSIR/NRE/WR/ER/2015/0026/A. Prepared for the Management Authority, Gauteng Tourism Authority. Council for Scientific & Industrial Research, Pretoria, 43 pp. Available online at http://www.dwaf.gov.za/ghreport/ as report 2.2(4702)

Hobbs PJ (2015b) Surface water and groundwater resources monitoring, Cradle of Humankind World Heritage Site, Gauteng Province, South Africa: Water resources status report for the period April to September 2015. Report CSIR/NRE/WR/ER/2015/0067/A. Prepared for the Management Authority, Gauteng Tourism Authority. Council for Scientific & Industrial Research, Pretoria, 47 pp. Available online at http://www.dwaf.gov.za/ghreport/ as report 2.2(3296)

Hobbs PJ (2015c) Discussion of surface water and groundwater interaction in the upper Crocodile River basin, Johannesburg, South Africa: Environmental isotope approach by Abiye et al. S Afr J Geol 118(4):511–515

Hobbs PJ (2016a) Surface water and groundwater resources monitoring, Cradle of Humankind World Heritage Site, Gauteng Province, South Africa: water resources status report for the period April 2015 to March 2016. Report CSIR/NRE/WR/ER/2016/0058/A. Prepared for the Management Authority, Gauteng Department of Economic Development. Council for Scientific & Industrial Research, Pretoria, 34 pp. Available online at http://www.dwaf.gov.za/ghreport/ as report 2.2(4703)

Hobbs PJ (2016b) Surface water and groundwater resources monitoring, Cradle of Humankind World Heritage Site, Gauteng Province, South Africa: Water resources status report for the period April to September 2016. Report CSIR/NRE/WR/ER/2016/0073/A. Prepared for the Management Authority, Gauteng Department of Economic Development. Council for Scientific & Industrial

Research, Pretoria, 29 pp. Available online at http://www.dwaf.gov.za/ghreport/ as report 2.2(4704)

Hobbs PJ (2017a) Surface water and groundwater resources monitoring, Cradle of Humankind World Heritage Site, Gauteng Province, South Africa: Water resources status report for the period April 2016 to March 2017. Report CSIR/NRE/WR/ER/2017/0008/A. Prepared for the Management Authority, Gauteng Department of Economic Development. Council for Scientific & Industrial Research, Pretoria, 28 pp. Available online at http://www.dwaf.gov.za/ghreport/ as report 2.2(4705)

Hobbs PJ (2017b) Surface water and groundwater resources monitoring, Cradle of Humankind World Heritage Site, Gauteng Province, South Africa: Water resources status report for the period April to September 2017. Report CSIR/NRE/WR/ER/2017/0072/A. Prepared for the Management Authority, Gauteng Department of Economic Development. Council for Scientific & Industrial Research, Pretoria, 26 pp. Available online at http://www.dwaf.gov.za/ghreport/ as report 2.2(4706)

Hobbs PJ, Hill L, McMillan P (2018) Surface water and groundwater resources monitoring, Cradle of Humankind World Heritage Site, Gauteng Province, South Africa: Water resources status report for the period April 2017 to March 2018. Report CSIR/NRE/WR/ER/2018/0010/A. Prepared for the Management Authority, Gauteng Department of Economic Development. Council for Scientific & Industrial Research, Pretoria, 37 pp. Available online at http://www.dwaf.gov.za/ghreport/ as report 2.2(4782)

Hobbs PJ (2017c) TDS load contribution from acid mine drainage to Hartbeespoort Dam, South Africa. Water SA 43(4):12. https://doi.org/10.4314/wsa.v43i4.10. http://www.wrc.org.za/Knowledge%20Hub%20Documents/Water%20SA%20Journals/Manuscripts/2017/04%20October%202,017/3402.pdf

Hobbs PJ, Cobbing J (2007) A hydrogeological assessment of acid mine drainage impacts in the West Rand Basin, Gauteng Province. Report CSIR/NRE/WR/ER/2007/0097/C. Council for Scientific & Industrial Research, Pretoria, 100 pp

Hobbs PJ, de Meillon N (2017) Hydrogeology of the sterkfontein cave system, cradle of humankind, South Africa. S Afr J Geol 120 (3):403–420

Hobbs PJ, Mills PJ (2011) The Koelenhof Farm fish mortality event of mid-January 2011. Report prepared for the Management Authority. Department of Economic Development, Gauteng Province, South Africa, 21 pp

Hobbs PJ, Mills PJ (2012) Managing the threats to the Cradle of Humankind World Heritage Site, South Africa, pp 53–54. In: Yordanova M, Stefanova D, Mikhova AD (eds) Collection of Abstracts. International Scientific-Practical Conference on Protected Karst Territories—Monitoring and Management, 16–20 Sept 2012. Shumen. Bulgaria, 112 pp. Available at http://www.prokarstterra.bas.bg/forum2012/conf-proceeding.html

Hobbs PJ, Venter J (2010) Situation assessment of the surface water and groundwater environments in the Cradle of Humankind World Heritage Site: Sediment chemistry assessment. Report prepared for the COH WHS Management Authority. Department of Economic Development, Gauteng Provincial Government, South Africa, 22 pp

Hobbs P, Lindsay R, Maherry A, Matshaya M, Newman RT, Talha SA (2010) The use of 222Rn as a hydrological tracer in natural and polluted environments. Report 1685/1/10. Water Research Commission, Pretoria, 96 pp

Hoehler TM, Alperin MJ, Albert DB, Martens CS (1998) Thermodynamic control on hydrogen concentrations in anoxic sediments. Geochim Cosmochim Acta 62(10):1745–1756

Holland M (2007) Groundwater resource directed measures in karst terrain with emphasis on aquifer characterization in the Cradle of Humankind near Krugersdorp. MSc thesis (unpublished). University of Pretoria, Pretoria, 112 pp

Holland M, Cobbing J (2008) Desktop geohydrological assessment of the Tarlton dolomites. Project 14/14/5/2. Activity 13. Water Geosciences Consulting for Department of Water Affairs, Pretoria, 61 pp

Holland M, Witthüser KT (2009) Geochemical characterization of karst groundwater in the Cradle of Humankind World Heritage Site. Environ Geol 57:513–524

Holland M, Wiegmans F, Cobbing J, Witthüser KT (2009) Geohydrological assessment of the Steenkoppies dolomite compartment. Project 14/14/5/2. Activity 25. Water Geosciences Consulting for Department of Water Affairs, Pretoria, 70 pp

Holland M, Witthüser KT, Jamison AA (2010) Issue paper 6: Hydrology of the Cradle of Humankind World Heritage Site; geology, surface and groundwater, pp 125–140. In WRC/IUCN-SAKWG. The karst system of the Cradle of Humankind World Heritage Site. Report KV 241/10. Water Research Commission, Pretoria, 402 pp

Holmes P, James KAF, Levy LS (2009) Is low-level environmental mercury exposure of concern to human health? Sci Total Environ 408:171–182

Howley A (2013) Final day of excavations. Sourced on 20/01/2014 at http://newswatch.nationalgeographic.com/2013/11/27/final-day-of-excavations/

Hughes TA, Gray NF (2012) Acute and chronic toxicity of acid mine drainage to the activated sludge process. Mine Water Environ 31:40–52. https://doi.org/10.1007/s10230-011-0168-y

Huizenga J-M (2004) Natural and anthropogenic influences on water quality: an example from rivers draining the Johannesburg Granite Dome. MSc thesis (unpublished). Rand Afrikaans University (now the University of Johannesburg), Johannesburg, 74 pp

Hunter KS, Wang Y, van Cappellen P (1998) Kinetic modeling of microbially-driven redox chemistry of subsurface environments: Coupling transport, microbial metabolism and geochemistry. J Hydrol 29:53–80

Jakobsen R, Albrechtsen H-J, Rasmussen M, Bay H, Bjerg PL, Christensen TH (1998) H2 concentrations in a landfill leachate plume (Grindsted, Denmark): In situ energetics of terminal electron acceptor processes. Environ Sci Technol 32(14):2142–2148

Jamison AA (In preparation) Structural and geological controls of geohydrology and dolomite cave systems in the Cradle of Humankind World Heritage Site. MSc thesis. University of the Witwatersrand

JFA (2006) Environmental impact document—proposed discharge of treated mine water via the Tweelopies receiving water body into the Rietrivier, Mogale City, Gauteng Province. Final draft. Johan Fourie & Associates for Harmony Gold Mining Company Ltd., 145 pp

Johnson KL, Younger PL (2006) The co-treatment of sewage and mine waters in aerobic wetlands. Eng Geol 35:53–61

Jones WK (2013) Physical structure of the epikarst. Acta Carsol 42(2–3):311–314

Jooste S (2011) Specialist Scientist. Department of Water Affairs. Personal communication, 25 Mar 2011

Kafri U, Foster MBJ, de Tremmerie F, Simonis J, Wiegmans FE (1986) The hydrogeology of the dolomite aquifer in the Klip River—Natalspruit Basin. Report GH3408. Department of Water Affairs and Forestry, Pretoria, 96 pp

Kasum J, Pilić M, Jovanović N, Pienaar H (2019) Model for forensic hydrography. Trans Maritime Sci 2:246–252

Katz BG, Catches JS, Bullen TD, Michel RL (1998) Changes in the isotopic and chemical composition of ground water resulting from a recharge pulse from a sinking stream. J Hydrol 211:178–207

Katz BG, Chelette AR, Pratt TR (2004) Use of chemical and isotopic tracers to assess nitrate contamination and ground-water age, Woodville Karst Plain, USA. J Hydrol 289:36–61

Kaufmann G (2003) Karst landscape evolution. Speleogenesis and Evolution of Karst Aquifers. Issue 3, 10 pp. Sourced on 19/12/2012 at http://www.speleogenesis.info/directory/karstbase/pdf/seka_pdf4473.pdf

Kaufmann JE (2013) Populations living on karst, p 60. In: NCKMS 2013. Program with abstracts. 20th National Cave and Karst Management Symposium—a changing climate. National Cave and Karst Research Institute, Carlsbad, 62 pp

Kemmer FN (ed) (1988) The NALCO water handbook, 2nd edn. Nalco Chemical Company. McGraw-Hill, 1019 pp

Kendall C, Caldwell EA (1998) Fundamentals of isotope geochemistry, pp 51–86. In: Kendall C, McDonnell JJ (eds) Isotope tracers in catchment hydrology. Elsevier, Amsterdam, p 839

Kenyon P, Ellis R (2010) Issue paper 7: the uses of caves and karst in the Cradle of Humankind World Heritage Site. pp 141–162. In WRC/IUCN-SAKWG. The karst system of the Cradle of Humankind World Heritage Site. Report KV 241/10. Water Research Commission, Pretoria, 402 pp

King J, Pienaar H (eds) (2011) Sustainable use of South Africa's inland waters: a situation assessment of Resource Directed Measures 12 years after the 1998 National Water Act. Water Research Commission, Pretoria

Kirby CS, Cravotta CA (2005) Net alkalinity and net acidity 1: theoretical considerations. Appl Geochem 20:1920–1940

Kirby CS, Cravotta CA (2005) Net alkalinity and net acidity 2: practical considerations. Appl Geochem 20:1941–1964

Kirkland DW (2014) Role of hydrogen sulfide in the formation of cave and karst phenomena in the Guadalupe Mountains and Western Delaware Basin, New Mexico and Texas. Special Paper 2. National Cave and Karst Research Institute, Carlsbad, New Mewxico, USA, 87 pp

Kiss M, Tanács E, Bárány-Kevei I (2011) Ecosystem services in Hungarian karst areas. Acta Climatol Chorol Tomus 44–45:41–49

Klaine SJ, Alvarez PJJ, Batley GE, Fernandes TF, Handy RD, Lyon DY, Hahendra S, McClaughlin MJ, Lead JR (2008) Nanomaterials in the environment: behaviour, fate, bioavailability, and effects. Environ Toxicol Chem 27(9):1825–1851

Klimchouk A (2004) Towards defining, delimiting and classifying epikarst: Its origin, processes and variants of geomorphic evolution. Speleogenesis and Evolution of Karst Aquifers. Issue 4. 13 pp. Sourced on 19/12/2012 at http://www.speleogenesis.info/directory/karstbase/pdf/seka_pdf4501.pdf

Klimchouk AB (2006) Unconfined versus confined speleogenetic settings: variations of solution porosity. Int J Speleol 35(1):19–24

Klimchouk AB (2011) Hypogene speleogenesis: hydrogeological and morphological perspective. Special paper 1, 2nd edn. National Cave and Karst Research Institute. Carlsbad. NM. 106 pp. ISBN-13: 978-0-9.795.422-0-6. Available at http://www.speleogenesis.info/directory/karstbase/pdf/seka_pdf9378.pdf

Klimchouk A (2014) The methodological strength of the hydrogeological approach to distinguishing hypogene speleogenesis, pp 4–12. In: Klimchouk A, Sasowsky ID, Mylroie J, Engel SA, Engel AS (eds) Hypogene cave morphologies. Karst Waters Institute. Special publication 18. 103 pp. ISBN: 978-0-9.789.976-7-0

Klimchouk AB, Ford D, Palmer A, Dreybrodt W (eds) (2000) Speleogenesis: evolution of karst aquifers. National Speleological Society, Huntsville, p 527

Klimchouk AB, Ford DC (eds) (2009) Hypogene speleogenesis and karst hydrogeology of artesian basins. Special paper 1. Ukrainian Institute of Speleology and Karstology, Simferopol, 208 pp

Knoëller K, Fauville A, Mayer B, Strauch G, Friese K, Veizer J (2004) Sulfur cycling in an acid mining lake and its vicinity in Lusatia, Germany. Chem Geol 204:303–323

Kok TS, Wiegmans FE, Fayazi M (1985) Results of exploration drilling in dolomite on Rietvallei 377, and at Erasmia and Valhalla, Pretoria. Report GH3379. Department of Water Affairs and Forestry, Pretoria, 91 pp

Kok TS (1992) Recharge of springs in South Africa. Report GH3748. Department of Water Affairs, Pretoria, 9 pp

Korom SF, Seaman JC (2012) When "conservative" anionic tracers aren't. Ground Water 50(6):820–824

Kozuskanich J, Novakowski KS, Anderson BC (2010) Fecal indicator bacteria variability in samples pumped from monitoring wells. Ground Water 10. https://doi.org/10.1111/j.1745-6584.2010.00713.x

Kralik M, Keimel T (2003) The time-input method, pp 172–180. In: Zwahlen F (ed) Vulnerability and risk mapping for the protection of carbonate (karst) aquifers. Final report. COST Action 620. European Commission, Brussels, 297 pp

Krige G (2009) Hydrological/chemical aspects of the Tweelopie-/Riet-/Blaauwbankspruit, with specific reference to the impact water, decanting from the Western Basin mine void, has on this system. Revision 2. African Environmental Development, Sterkfontein, 66 pp

Krige WG (2010) Issue paper 10: the impact of urbanisation on the water resources and water-based ecosystems of the Cradle of Humankind World Heritage Site, pp 211–240. In: WRC/IUCN-SAKWG. The karst system of the Cradle of Humankind World Heritage Site. Report KV 241/10. Water Research Commission, Pretoria, 402 pp

Krige G (2011) Local resident and Environmental Consultant. Personal communication, 13 Jan 2011

Krige G, du Toit S (2005) Report on aquatic biomonitoring and toxicological integrity of the Tweelopiespruit and the upper Wonderfonteinspruit in relation to the decant of mine water from the Western Basin mine void—winter cycle 2005. Revision 00. African Environmental Development, Sterkfontein, 32 pp

Krige G, van Biljon M (2006) Hydrological/chemical aspects of the Tweelopie-/Riet-/Blaauwbankspruit, with specific reference to the impact water, decanting from the Western Basin mine void, has on this system. Revision 1. African Environmental Development, Sterkfontein, 66 pp

Krige WG, van Biljon M (2010) Issue paper 9: the impacts of mining on the water resources and water-based ecosystems of the Cradle of Humankind World Heritage Site, pp 177–210. In: WRC/IUCN-SAKWG. The karst system of the Cradle of Humankind World Heritage Site. Report KV 241/10. Water Research Commission, Pretoria, 402 pp

Kuhn CM (1989) Hydrogeology of the Midrand-Kempton Park dolomite. Report GH3501. Department of Water Affairs and Forestry, Pretoria, 125 pp

Langmuir D (1971) The geochemistry of some carbonate ground waters in central Pennsylvania. Geochim Cosmochim Acta 35:1023–1045

Langmuir D (1996) Aqueous environmental geochemistry: Upper Saddle River. Prentice-Hall, NJ, p 600

Larson EB, Mylroie JE (2013) Quaternary glacial cycles: karst processes and the global $CO_2$ budget. Acta Carsologica 42(2–3):197–202

Lee CC (1989) Environmental engineering dictionary. Government Institutes, Inc., 629 pp

Leskiewicz AF (1986) Exploitable potential of ground water resources in storage in the dolomite of the Bapsfontein, Delmas and Springs area for emergency supplies. Report GH3463. Department of Water Affairs and Forestry, Pretoria, 40 pp

Leyland R (2010) Groundwater risk and hazard maps. Report CSIR/BE/IE/IR/2010/0046/C. Council for Scientific & Industrial Research, Pretoria, 37 pp

Leyland R (2013) Research Scientist. Built Environment. Council for Scientific & Industrial Research. Personal communication, 18 Feb 2013

Leyland RC, Witthüser KT, van Rooy JL (2008) Vulnerability mapping in karst terrains, exemplified in the wider Cradle of Humankind World Heritage Site. Report KV 208/08. Water Research Commission, Pretoria, 97 pp

Liefferink M (2011) Chief Executive Officer. Federation for a sustainable environment. Personal communication, 13 Jan 2011

Lindsay MBJ, Moncur MC, Bain JG, Jambor JL, Ptacek CJ, Blowes DW (2015) Geochemical and mineralogical aspects of sulfide mine tailings. Appl Geochem. https://doi.org/10.1016/j.apgeochem.2015.01.009

Liu X, Millero FJ (1999) The solubility of iron hydroxide in sodium chloride solutions. Geochim Cosmochimica Acta 63(19/20):3487–3497

López-Chicano M, Bouamama M, Vallejos A, Pulido-Bosch A (2001) Factors which determine the hydrogeochemical behaviour of karstic springs: a case study from the Betic Cordilleras, Spain. Appl Geochem 16:1179–1192

Lottermoser BG (2010) Mine wastes: characterization, treatment and environmental impacts, 3rd edn. Springer, Berlin, p 400

Lotter L (2012) Water loss management in the City of Tshwane: updated up to end March 2012. Slide presentation. Sourced on 19/05/2013 at http://www.dwa.gov.za/Projects/Vaal/April2012/6.2%20b%20WC%20WDM%20City%20of%20Tshwane.pdf

Louw L (2012) Mogale Gold: Making the most of AMD. Mining Mirror. October 2012, pp 10–15. Sourced on 11/03/2013 at http://www.myvirtualpaper.com/doc/brookepattrick/mining_mirror_october_2012_e-book/2012100301/12.html#18

Malan JA (1990) The stratigraphy and sedimentology of the Bredasdorp Group, southern Cape Province. MSc thesis (unpublished). University of Cape Town, Rondebosch, 197 pp

Mangole K (2010) Tour guide. Sterkfontein Cave. Personal communication, 13 May 2010

Maré E (2010) Plant Manager. Percy Stewart WWTW. Personal communication, 26 May 2010

Maropeng Staff (2012) Various Visitor Centre personnel. Personal communication, 19 Jan 2012

Martin J (2006) The Kuruman Spring. Soc Dyn 32(2):218–227

Martin JB, Brown A, Ezell J (2013) Do carbonate karst terrains affect the global carbon cycle? Acta Carsologica 42(2–3):187–196

Martin JB, White WB (eds) (2007) Frontiers of karst research. Special publication 13. In: Proceedings and recommendations of the workshop held May 3 through 5, San Antonio, Texas. Karst Waters Institute. Leesburg, Virginia, 118 pp

Martini JEJ (2006) Karsts and caves, pp 661–668. In: Johnson MR, Anhaeusser CR, Thomas RJ (eds) The Geology of South Africa. Geological Society of South Africa/Council for Geoscience. Johannesburg/Pretoria, 691 pp. ISBN: 978-1-919.908-77-9

Martini J, Kavalieris I (1976) The karst of the Transvaal (South Africa). Int J Speleol 8(3):229–251. Available at http://scholarcommons.usf.edu/cgi/viewcontent.cgi?article=1504&context=ijs

Martini JEJ, Wilson MGC (1998) Limestone and dolomite, pp 433–440. In: Wilson MGC, Anhaeusser CR (eds) The mineral resources of South Africa, Handbook 16. Council for Geoscience. Pretoria, 740 pp. ISBN: 1-875.061-52-5

Martini J, Wipplinger PE, Moen HFG, Keyser A (2003) Contribution to the speleology of Sterkfontein Cave, Gauteng Province, South Africa. Int J Speleol 32(1/4):43–69. Available at http://scholarcommons.usf.edu/cgi/viewcontent.cgi?article=1223&context=ijs

Masondo M (2010) R7m to clean up toxic water. The Times, Friday 19 Mar

Maupin TP, Agouridis CT, Barton CD, Warner RC, Yu X (2013a) Specific conductivity sensor performance: I. Laboratory evaluation. Int J Min Reclamat Environ 27(5):329–344. https://doi.org/10.1080/17480930.2013.764701

Maupin TP, Agouridis CT, Edwards DR, Barton CD, Warner RC, Sama MP (2013b) Specific conductivity sensor performance: II. Field evaluation. Int J Min Reclamat Environ 27(5):345–365. https://doi.org/10.1080/17480930.2013.764702

McCarthy TS (2006) The Witwatersrand Supergroup, pp 155–186. In: Johnson MR, Anhaeusser CR, Thomas RJ (eds) The Geology of South Africa. Geological Society of South Africa/Council for Geoscience, Johannesburg/Pretoria, p 691

McCarthy TS, Charlesworth EG, Stannistreet IG (1986) Post-Transvaal structural features of the northern portion of the Witwatersrand Basin. Trans Geol Soc S Afr 89:311–323

McCarthy T, Rubidge B (2005) The story of earth & life: a southern African perspective on a 4.6-billion-year journey. Struik Publishers, Cape Town, 333 pp. ISBN: 1-77.007-148-2

McCarthy TS, Venter JS (2006) Increasing pollution levels on the Witwatersrand recorded in the peat deposits of the Klip River wetland. S Afr J Sci 102:27–34

McCullough CD (2008) Approaches to acid mine drainage water in pit lakes. Int J Min Reclamat Environ 22(2):105–119

McCullough CD, Lund MA, May JM (2008) Field-scale demonstration of the potential for sewage to remediate acidic mine waters. J Mine Water Environ 27(2):31–39. https://doi.org/10.1007/s10230-007-0028-y

McMillan PH (1998) An integrated habitat assessment system for the rapid biological assessment of rivers and streams. Internal STEP report no. ENV-P-I 98,088. CSIR, 37 pp

Mellor ET (1917) The geology of the Witwatersrand. Explanation to the geological map. Special Publication 3. Geological Survey of South Africa, 46 pp

Meyer R (2014) Hydrogeology of groundwater region 10: The karst belt. Report TT 553/14. Water Research Commission, Pretoria, 285 pp

Middleton BJ, Bailey AK (2008a) Water Resources of South Africa, 2005 Study (WR 2005). Book of Maps. Vers. 1. Report TT 382/08. Water Research Commission, Pretoria, 85 pp

Middleton BJ, Bailey AK (2008b) Water Resources of South Africa, 2005 Study (WR 2005). User's Guide. Vers. 1. Report TT 381/08. Water Research Commission, Pretoria, 179 pp

Mills P (2010) Environmental Manager, COH WHS Management Authority. Personal communication, 18 June 2010

Mogale Gold (2009) Ground-water monitoring program (OESH Department), 1 Mar 2009, 8 pp

Montgomery DR, Zabowski D, Ugolini FC, Hallberg RO, Spaltenstein H (2002) Soils, watershed processes, and marine sediments, pp 159–194. In: Jacobson MC, Charlson RJ, Rodhe H, Orians GH (eds) Earth system science: from biogeochemical cycles to global change. International Geophysics Series, vol 72. Academic Press. London, 527 pp

Moolman T (2012) Senior Geotechnician, Department of Water Affairs. Personal communication, 12 Jan 2012

Moore WS, Reid DF (1973) Extraction of radium from natural waters using manganese-impregnated acrylic fibres. J Geophys Res 78 (36):8880–8886

Moreira S, Boehrer B, Schultze M, Dietz S, Samper J (2011) Modeling geochemically caused permanent stratification in Lake Waldsee (Germany). Aquatic Geochem 17:265–280

Morse JW, Arvidson RS (2002) The dissolution kinetics of major sedimentary carbonate minerals. Earth-Sci Rev 58:51–84. https://doi.org/10.1016/S0012-8252(01)00,083-6

Mothoa TP, Lombaard AJ (2008) Drilling of three monitoring boreholes at mogale gold plant—randfontein. Draft report by

AquaEarth Consultants for Mogale Gold/Mintails SA, Randburg, p 12
Mucina L, Rutherford MC (eds) (2006) The vegetation of South Africa, Lesotho and Swaziland. Strelitzia 19. South African National Biodiversity Institute, Pretoria, 807 pp
Mulder MP (1988) Dolomitic resource quantification, pp 54–60. In: GWD-GSSA. Workshop on dolomitic ground water of the PWV area, 28 March. Ground Water Division of the Geological Society of South Africa, Pretoria, 145 pp
Nakasone H (1987) Study of aeration at weirs and cascades. J Environ Eng 113(1):64–81
Nel A (2011) Whole effluent toxicity assessment. Report EUJ11-009. Department of Zoology, University of Johannesburg, 10 pp
Nel A, van der Merwe-Botha M, du Toit S (2011) In-stream biological and chemical integrity of the surface streams at Nedbank Olwazini Conference Centre, Muldersdrift. Assessment of macro-invertebrate communities, aquatic toxicology and hydrochemistry—winter cycle 2011. Report 001/08/2011. Water Group, 32 pp
Obbes AM (2001) The structure, stratigraphy and sedimentology of the Black Reef-Malmani-Rooihoogte succession of the Transvaal Supergroup southwest of Pretoria. Bulletin 127. Council for Geoscience, Pretoria, 89 pp. ISBN: 1-875.061-80-0
O'Bourke NC, Hartman GL, Fuglevand P (2008) In situ specific gravity vs grain size: a better method to estimate new work dredging production. Paper presented at the 39th Texas A&M dredging seminar, 9–11 June 2008. St Louis. Missouri, pp 215–224. Available at https://www.westerndredging.org/index.php/information/proceedings-presentations/category/35-session-2b-dredging-studies-and-projects
Oelofse SHH, Roux S, De Lange W, Mahumani BK, Le Roux W, du Preez M, Greben HA, Steyn M (2012) A comparison of the cost associated with pollution prevention measures to that required to treat polluted water resources. Report 1845/1/11. Water Research Commission, Pretoria, 101 pp
Olivier J, van Niekerk HJ, van der Walt IJ (2008) Physical and chemical characteristics of thermal springs in the Waterberg area in Limpopo Province, South Africa. Water SA 34(2):163–174
Omoike AI, Vanloon GW (1999) Removal of phosphorus and organic matter removal by alum during wastewater treatment. Water Res 33 (17):3617–3627
Osborne RAL (2003) Partitions, compartments and portals: cave development in internally impounded karst masses. Speleog Evol Karst Aquifers 1(4):12
Osmond JK, Cowart JB (2000) U–series nuclides as tracers in groundwater hydrology, pp 145–173. In: Cook PG, Herczeg AL (eds) Environmental tracers in subsurface hydrology. Kluwer Academic Publications, Boston/Dordrecht/London, p 529
Palmer AN (2003) Dynamics of cave development by allogenic water. Speleogenesis and Evolution of Karst Aquifers. Issue 1. 11 pp. Sourced on 19/12/2012 at http://www.speleogenesis.info/directory/karstbase/pdf/seka_pdf4483.pdf
Palmer AN (2007) Cave geology. Cave Books, Dayton, p 454
Palmer AN (2011) Distinction between epigenic and hypogenic maze caves. Geomorphology 134:9–22
Parrish JD, Braun DP, Unnasch RS (2003) Are we conserving what we say we are: measuring ecological integrity within protected areas. BioSciences 53(9):851–860
Parshall RL (1950) Measuring water in irrigation channels with Parshall flumes and small weirs. US Soil Conservation Services. Circular 843. US Department of Agriculture, Washington, DC
Parsons R, Wentzel J (2007) Groundwater resource directed measures manual. Report TT 299/07. Water Research Commission, Pretoria, 109 pp
Partridge TC (1973) Geomorphological dating of cave openings at Makapansgat, Sterkfontein, Swartkrans and Taung. Nature 246:75–79
Partridge TC, Maud RR (1987) Geomorphic evolution of southern Africa since the Mesozoic. S Afr J Geol 90(2):179–208
Partridge TC, Granger DE, Caffee MW, Clarke RJ (2003) Lower Pliocene hominid remains from Sterkfontein. Science 300 (5619):607–612
Paykani ARM, Mardan HR (Undated) Methods of controlling ferric hydroxide precipitation results from corrosion in acidizing treatments. Sourced on 20/07/2012 at http://www.pogc.ir/Portals/0/maghalat/890703-6.pdf
Payment P, Locus A (2010) Pathogens in water: value and limits of correlation with microbial indicators. Issue paper. Ground Water, 8 pp. https://doi.org/10.1111/j.1745-6584.2010.00710.x
Pickering R (2010) Comparison between Australopithecus Sediba (MH1) and other hominin taxa, in the context of probabilities of conspecificity. S Afr J Sci 106(7/8):1–2. https://doi.org/10.4102/sajs.v106i7/8.348
Pickering R, Kramers JD (2010) Re-appraisal of the stratigraphy and determination of new U-Pb dates for the Sterkfontein hominin site, South Africa. J Hum Evol 59:70–86
Pickering R, Dirks PHG, Jinnah Z, De Ruiter DJ, Churchill SE, Herries AIR, Woodhead JD, Hellstrom JC, Berger LR (2011) Australopithecus sediba at 1.977 Ma and implications for the origins of the genus Homo. Science 333(6048):1421–1423. https://doi.org/10.1126/science.1203697
Pienaar H, Xu Y, Braune E, Cao J, Dzikiti S, Jovanovic NZ (2021) Implementation of groundwater protection measures, particularly resource directed measures in South Africa. Water Policy
Plummer LN, Busby JF, Lee RW, Hanshaw BB (1990) Geochemical modeling in the Madison aquifer in parts of Montana, Wyoming and South Dakota. Water Resour Res 26:1981–2014
Pokrovsky OS, Golubev SV, Schott J, Castillo A (2009) Calcite, dolomite and magnesite dissolution kinetics in aqueous solutions at acid to circumneutral pH, 25 to 150 °C and 1 to 55 atm $pCO_2$: new constraints on $CO_2$ sequestration in sedimentary basins. Chem Geol 265:20–30. https://doi.org/10.1016/j.chemgeo.2009.01.013
Pretorius DA (1986) The goldfields of the Witwatersrand Basin. In: Anhaeusser CR, Maske S (eds) Mineral deposits of southern Africa, vol I. Geological Society of South Africa, Johannesburg, pp 489–493
Prudic DE, Harrill JR, Burbey TJ (1995) Conceptual evaluation of regional ground-water flow in the carbonate-rock province of the Great Basin, Nevada, Utah, and adjacent states. Professional paper 1409-D. US Geological Survey, 102 pp
Quinlan JF, Ewers RO (1985) Ground water flow in limestone terrains: Strategy, rationale and procedure for reliable, efficient monitoring of ground water quality in karst areas. In: Proceedings of the National symposium and exposition on aquifer restoration and ground water monitoring. National Water Well Association, Worthington, Ohio, pp 197–234
Quinlan JF, Ewers RO (1986) Reliable monitoring in karst terranes: it can be done, but not by an EPA-approved method. Ground Water Monitoring & Remediation. Winter 1986, pp 4–6
RAIS (1995) Formal toxicity summary for Ni and Ni compounds. Risk Assessment Information System available at http://rais.ornl.gov/tox/profiles/nickel_and_nickel_compounds_f_V1.html
RHP (2005) State-of-rivers report: monitoring and managing the ecological state of rivers in the Crocodile (West) Marico Water Management Area. River Health Programme. Department of Environmental Affairs and Tourism, Pretoria, 53 pp
Reimold WU (2006) Impact structures in South Africa, pp 629–649. In: Johnson MR, Anhaeusser CR, Thomas RJ (eds) The Geology of

South Africa. Geological Society of South Africa/Council for Geoscience, Johannesburg/Pretoria, p 691

Reynders AG (1988) Production borehole drilling in the Middle and Lower Klip River Valley. Report GH3562. Department of Water Affairs and Forestry, Pretoria, 132 pp

Rising Star Expedition Blog (2014) Sourced on 03/08/2014 at http://newswatch.nationalgeographic.com/2014/04/01/wrapping-up-round-two/

Robb LJ, Robb VM (1998a) Gold in the Witwatersrand basin, pp 294–349. In: Wilson MGC, Anhaeusser CR (eds) The Mineral Resources of South Africa. Handbook 16. Council for Geoscience, 740 pp

Robb LJ, Robb VM (1998b) Environmental impact of Witwatersrand gold mining, pp 14–16. In: Wilson MGC, Anhaeusser CR (eds) The Mineral Resources of South Africa. Handbook 16. Council for Geoscience, 740 pp

Roberts CPR (1988) An overview of the proposed emergency ground water supply project in the Pretoria-Witwatersrand-Vereeniging (PWV) area, pp 1–4. In: Proceedings of the workshop on dolomitic ground water of the PWV area, 28 Mar 1988. Ground Water Division of the Geological Society of South Africa, Pretoria, 145 pp

Roberts DL, Botha GA, Maud RR, Pether J (2006) Coastal Cenozoic deposits, pp 605–628. In: Johnson MR, Anhaeusser CR, Thomas RJ (eds) The Geology of South Africa. Geological Society of South Africa/Council for Geoscience, Johannesburg/Pretoria, p 691

Roos H (2011) Landowner. Ptn 5 of Zwartkrans 172IQ. Personal communication, 10 Oct 2011

Rösner T, Boer R, Reyneke R, Aucamp P, Vermaak J (2001) A preliminary assessment of pollution contained in the unsaturated and saturated zone beneath reclaimed gold-mine residue deposits. Report 797/1/01. Water Research Commission, Pretoria, 285 pp

Roux S (2010) Water laundering. Water Sewage & Effluent, Nov 2010, pp 35–45

RSA (1973a) 1:50 000 geological series 2527DC Hekpoort. The Government Printer, Pretoria

RSA (1973b) 1:50 000 geological series 2527DD Broederstroom. The Government Printer, Pretoria

RSA (1986) 1:250 000 geological series 2626 West Rand. The Government Printer, Pretoria

RSA (1987) 1:10 000 orthophoto map 2627BA5 Sterkfontein, 2nd edn. The Government Printer, Pretoria

RSA (1996) 1:50 000 topocadastral series 2527DC Hekpoort, 4th edn. The Government Printer, Pretoria

RSA (1998) National Water Act: Act No 36, 1998. Government Gazette No 19182. Vol 398. 26 August 1998. Government Printer, Pretoria, 201 pp

RSA (2001) 1:50 000 topocadastral series 2527DD Broederstroom, 5th edn. The Government Printer, Pretoria

RSA (2002) 1:50 000 topocadastral series 2627BB Roodepoort, 6th edn. The Government Printer, Pretoria

RSA (2008) Restrictions on taking water from Steenkoppies Dolomitic Compartment. National Gazette No 30863, vol 513, 14 Mar 2008. Government Printer, Pretoria, pp 152–167

Rubin T (2012) Managing Director. Maropeng āAfrika. Personal communication, 09 Oct 2012

Rutledge AT, Mesko TO (1996) Estimated hydrologic characteristics of shallow aquifer systems in the Valley and Ridge, the Blue Ridge, and the Piedmont Physiographic Provinces based on analysis of streamflow recession and base flow. Professional paper 1422-B. US Geological Survey, 58 pp

Sahuquillo A, Rigol A, Rauret G (2002) Comparison of leaching tests for the study of trace metals remobilisation in soils and sediments. J Environ Monit 4:1003–1009

Salvati R (2002) Natural hydrogeological laboratories: a new concept in regional hydrogeology studies. A case history from central Italy. Environ Geol 41:960–965

SANS (2015a) South African National Standard (SANS) 241–1. Drinking water. Part 1: Microbiological, physical, aesthetic and chemical determinands. Edition 2. Standards South Africa, Pretoria, 14 pp

SANS (2015b) South African National Standard (SANS) 241–2. Drinking water. Part 2: Application of SANS 241–1. Edition 2. Standards South Africa, Pretoria, 14 pp

Sasowsky ID, White WB (1993) Geochemistry of the Obey River Basin, north-central Tennessee: a case of acid mine water in a karst drainage system. J Hydrol 146:29–48

Sasowsky ID, White WB, Webb JA (1995) Acid mine drainage in karst terranes: geochemical considerations and field observations, pp 241–247. In: Beck BF, Pearson FM (eds) Karst GeoHazards: engineering and environmental problems in Karst Terrane. Proceedings of the fifth multidisciplinary conference on sinkholes and the engineering and environmental impacts of Karst, 2–5 Apr 1995. Gatlinburg, Tennessee, 581 pp

Schaefer DH, Thiros SA, Rosen MR (2005) Ground-water quality in the carbonate-rock aquifer of the Great Basin, Nevada and Utah, 2003. Scientific investigations report 2005–5232. US Geological Survey, 32 pp

Schneider K, Culver DC (2004) Estimating subterranean species richness using intensive sampling and rarefaction curves in a high density cave region in west Virginia. J Cave Karst Stud 66:39–45

Schrader A, Erasmus E, Winde F (2014) Determining hydraulic parameters of a karst aquifer using unique historical data from large-scale dewatering by deep level mining—a case study from South Africa. Water SA 40(3):555–570

Schulze RE, Maharaj M, Lynch SD, Howe BJ, Melvil-Thomson B (1997) South African atlas of agrohydrology and –climatology. Report TT82/96. Water Research Commission, Pretoria, 276 pp. ISBN: 1-86.845-271-9

Scott R (1995) Flooding of Central and East Rand gold mines: an investigation into controls over the inflow rate, water quality and the predicted impacts of flooded mines. Report 486/1/95. Water Research Commission, Pretoria, 275 pp

Seccombe A (2008) Mine water calamity. Mining mx, 5 Nov 2008. Sourced on 04/01/2011 at http://www.miningmx.com/special_reports/green-book/2008/886408.htm

Slabbert L (2007) Toxicity evcaluation of selected surface and mine water from the West Rand mining basin, Krugersdorp. In: Hobbs P, Cobbing J (eds) A hydrogeological assessment of acid mine drainage impacts in the West Rand Basin, Gauteng Province. Report CSIR/NRE/WR/ER/2007/0097/C. Council for Scientific & Industrial Research, Pretoria, 100 pp

Shelton LR, Capel PD (1994) Guidelines for collecting and processing samples of stream bed sediment for analysis of trace elements and organic contaminants for the National Water-Quality Assessment Program. Open-File Report 94–458. USGS. Sacramento, California, 31 pp

Sherlock EJ, Lawrence RW, Poulin R (1995) On the neutralization of acid rock drainage by carbonate and silicate minerals. Environ Geol 25:43–54

Slabbert L (2004) Methods for direct estimation of ecological effect potential (DEEEP). Report no.: 1313/1/04. Water Research Commission, Pretoria, 100 pp

Strachan L, Ndengu S, Mafanya T, Coetzee H, Wade P, Msezane N, Kwata M, Mengistu H (2008) Regional gold mining closure strategy for the Central Rand Goldfield. (Draft). Report produced for Department of Minerals & Energy, Pretoria, 283 pp

Strosnider WH, Nairn RW (2010) Effective passive treatment of high-strength acid mine drainage and raw municipal wastewater in

Potosí, Bolivia using simple mutual incubations and limestone. J Geochem Explor 105:34–42
Strosnider WH, Winfrey BK, Nairn RW (2011) Alkalinity generation in a novel multi-stage high-strength acid mine drainage and municipal wastewater passive co-treatment system. J Mine Water Environ 30:47–53. https://doi.org/10.1007/s10230-010-0124-2
Strydom HA (2009) Chapter 26: Protected areas, pp 951–970. In: Strydom HA, King ND (eds) Environmental management in South Africa, 2nd edn. JUTA Law, Cape Town, p 1142
Strydom HA, King ND (eds) (2009) Environmental management in South Africa, 2nd edn. JUTA Law, Cape Town, p 1142
Strydom W, Funke N, Hobbs P (2016) The Witwatersrand acid mine drainage conundrum contextualised: mapping and understanding different coalitions, positions and interests in the AMD policy subsystem. Council for Scientific and Industrial Research, Pretoria, p 12
Stumm W, Morgan JJ (1981) Aquatic chemistry, 2nd edn. Wiley Interscience, New York, p 583
Stumpf S, Valentine-Darby P, Gwilliam E (2009) NPS inventory and monitoring program. Aquatic Macroinvertebrates—Ecological role. US National Park Service
Supper R, Motschka K, Bauer-Gottwein P, Ahl A, Römer A, Neumann-Gondwe B, Merediz Alonso G, Kinzelbach W (2008) Spatial mapping of karstic cave structures by means of airborne electromagnetics: an emerging technology to support protection of endangered karst systems. Geophys Res Abst 10. EGU2008-A-01126
Swain LA, Mesko TA, Hollyday EF (2004) Summary of the hydrogeology of the Valley and Ridge, Blue Ridge, and Piedmont Physiographic Provinces in the Eastern United States. Professional paper 1422-A. US Geological Survey, 31 pp
Swart CJU, James AR, Kleywegt RJ, Stoch EJ (2003) The future of the dolomitic springs after mine closure on the Far West Rand, Gauteng, RSA. Environ Geol 44:751–770. https://doi.org/10.1007/s00254-003-0820-3
Tancott G (2014) Excess AMD in West Rand a big problem. April 3 sourced on 18/06/2014 at http://www.infrastructurene.ws/2014/04/03/excess-amd-in-west-rand-a-big-problem/
Tasaki S (2006) The presence of stygobitic macroinvertebrates in karstic aquifers: a case study in the Cradle of Humankind World Heritage Site. M.Sc. thesis (unpubl.). University of Johannesburg, Johannesburg, 144 pp
Taylor CJ, Greene EA (2008) Chapter 3: Hydrogeologic characterization and methods used in the investigation of karst hydrology. In: Rosenberry DO, LaBaugh JA (eds) Field techniques for estimating water fluxes between surface water and ground water. Techniques and methods 4-D2. United States Geological Survey, 128 pp
TCTA (2012) AMD Progress Report August 2012. Report to the Inter-Governmental Task Team meeting, Pretoria, 9 pp
Thompson JG (1980) Acid mine waters in South Africa and their amelioration. Water SA 6(3):130–134
Toens PD, Griffiths GH (1964) The geology of the West Rand, pp 283–321. In: Haughton SH (ed) The geology of some ore deposits in Southern Africa, vol I. Geological Society of South Africa, 625 pp
Tucker RF, Viljoen RP (1986) The geology of the West Rand Goldfield, with special reference to the southern limb, pp 649–688. In: Anhaeusser CR, Maske S (eds) Mineral deposits of Southern Africa, vol I. Geological Society of South Africa, Johannesburg, p 1020
Tutu H, McCarthy TS, Cukrowska E (2008) The chemical characteristics of acid mine drainage with particular reference to sources, distribution and remediation: the Witwatersrand Basin, South Africa as a case study. Appl Geochem 23:3666–3684
UNESCO (1972) Glossary and multilingual equivalents of karst terms. United Nations Educational Scientific and Cultural Organization, Paris, p 72
USDA (2012) Groundwater-dependent ecosystems: level II inventory field guide. Forest Service General Technical Report WO-86b, 124 pp
USEPA (2002) Methods for measuring the acute toxicity of effluents and receiving waters to freshwater and marine organisms, 5th edn. Report no.: EPA-821-R-02–012. U.S. Environmental Protection Agency, USA
van Biljon M (2006) Geohydrological review of the potential impact on the Sterkfontein dolomite during increased surface water runoff. Rison Groundwater Consulting for Johan Fourie & Associates on behalf of Harmony Gold Mining (Ltd.), 67 pp
van den Bosch J (2010) Owner. Jomajoco Farms. Personal communication, 09 June 2010
van der Merwe P (2009) Owner. Danielsrust Game Farm. Personal communication, 01 Dec 2009
van der Walt B (2009) HDS Plant Manager. Rand Uranium. Personal communication, 11 Aug 2009
van der Walt B (2011) HDS Plant Manager. Rand Uranium. Personal communication, 20 Jan 2011
van der Walt B (2013) HDS Plant Manager. SibanyeGold. Personal communication, 26 Feb 2013
van Dyk G, Esterhuyse CJ, Goussard F (2007) Groundwater observations and impact management in the mining, urban and industrial environment of the Sishen Kathu aquifer. In: Proceedings of the groundwater 2007 conference. Ground Water Division. Geological Society of South Africa, 5 − 7 Oct 2007, Bloemfontein
van Schalkwyk A (1981) Ontwikkelingspatroon en risiko-evaluasie in dolomietgebiede. In: Proceedings of the seminar on the engineering geology of dolomite areas. Department of Geology. University of Pretoria, 26 − 27 Nov 1981, Pretoria, pp 9–24
van Schie K (2013) Finally some life back in mine water. The Star. Monday 18 Mar 2013
van Wyk E (2010) Estimation of episodic groundwater recharge in semi-arid fractured hard rock aquifers. PhD thesis (unpublished). University of the Free State, Bloemfontein, 273 pp
Vegter JR (1988) Dolomitic ground water for the metropolitan PWV-area: a resuscitation of ideas, pp 7–11. In: Proceedings of the workshop on dolomitic ground water of the PWV area, 28 Mar 1988. Ground Water Division of the Geological Society of South Africa, Pretoria, 145 pp
Vegter JR (1995) An explanation of a set of national groundwater maps. Report TT 74/95. Water Research Commission, Pretoria, 81 pp
Veni G (2013) A framework for assessing the role of karst conduit morphology, hydrology, and evolution in the transport and storage of carbon and associated sediments. Acta Carsologica 42(2–3):203–211
Venter J, Motlakeng T, Kotoane M, Coetzee H, Hobbs P, Wade P (2010) Establishment of a monitoring system for surface water and groundwater in the Cradle of Humankind World Heritage Site: Sediment sampling and analysis results. Internal project report prepared by the CGS for the CSIR/CGS Joint Venture, Pretoria, 24 pp
Vermeulen A (2013) Return of macroinvertebrates in West Rand stream a sign impact of AMD has lessened. Mining Weekly, Friday 26 Apr 2013. Sourced on 04/05/2013 at http://www.miningweekly.com/page/environmental/
Vesper DJ, Loop CM, White WB (2003) Contaminant transport in karst aquifers. Speleogenesis and Evolution of Karst Aquifers. Issue 2. 11 pp. Sourced on 19/12/2012 at http://www.speleogenesis.info/directory/karstbase/pdf/seka_pdf4490.pdf
Vesper D (2008) Karst resources and other applied issues, pp 65–73. In: Martin JB, White WB (eds) Frontiers of Karst Research. Special publication 13. Karst Waters Institute. Leesburg, Virginia, 118 pp

Vías JM, Andreo B, Perles MJ, Carrasco F, Vadillo I, Jiménez P (2003) The COP method, pp 163–171. In: Zwahlen F (ed) Vulnerability and risk mapping for the protection of carbonate (karst) aquifers. Final report. COST Action 620. European Commission, Brussels, 297 pp

Visser DJL (1989) The geology of the Republics of South Africa, Transkei, Bophuthatswana, Venda and Ciskei and the Kingdoms of Lesotho and Swaziland. Explanation: Geological map (1:1 000 000). Geological Survey and Department of Mineral and Energy Affairs, Pretoria, 491 pp

Visser H (2018) Reserve Manager. John Nash Nature Reserve. Personal communication, 6 Dec 2017

Wade PW, Woodbourne S, Morris WM, Vos P, Jarvis NV (2002) Tier 1 risk assessment of radionuclides in selected sediments of the Mooi River. Report 1095/1/02. Water Research Commission, Pretoria, 92 pp

Walter GR, Necsoiu M, McGinnis R (2012) Estimating aquifer channel recharge using optical data interpretation. Ground Water 50(1):68–76. https://doi.org/10.1111/j.1745-6584.2011.00815.x

Waltham T (2008) Fengcong, fenglin, cone karst and tower karst. Cave Karst Sci 35(3):77–88 (also available at www.speleogenesis.info 2011)

Wanty RB, Nordstrom DK (1993) Natural radionuclides, pp 423–441. In: Alley WM (ed) Regional Ground-Water Quality. Van Nostrand Reinhold, New York, p 634

Watson J, Hamilton-Smith E, Gillieson D, Kiernan K (eds) (1997) Guidelines for Cave and Karst Protection. IUCN Gland, Switzerland, p 63

Weaver JMC, Cavé L, Talma AS (2007) Groundwater sampling, 2nd edn. Report TT 303/07. Water Research Commission, Pretoria, 168 pp

Webb JA, Sasowsky ID (1994) The interaction of acid mine drainage with a carbonate terrane: evidence from the Obey River, north-central Tennessee. J Hydrol 161:327–346

Wells JD, van Meurs LH, Rabie GF, Moir F, Russell J (2009) Chapter 15: Terrestrial minerals, pp 513–578. In: Strydom HA, King ND (eds) Environmental management in South Africa, 2nd edn. JUTA Law, Cape Town, p 1142

WHC (2013) State of conservation of world heritage properties inscribed on the world heritage list. Document WHC-13/37. COM/7B.Add, pp 100–104, Paris, 17 May 2013. Available at http://whc.unesco.org/en/documents/123027

Whatley G (2010) Owner. Kromdraai Gold Mine. Personal communication, 26 May 2010

White WB (1984) Rate processes: chemical kinetics and karst landform development, pp 227–248. In: LaFleur RG (ed) Groundwater as a Geomorphic Agent. Allen & Unwin, London, 390 pp. ISBN-13: 978-0.045.510.696

White WB (1993) Analysis of karst aquifers, pp 471–489. In: Alley WM (ed) Regional ground-water quality. Van Nostrand Reinhold, New York, 634 pp. ISBN: 0-442-00.937-2

White WB (2002) Karst hydrology: recent developments and open questions. Eng Geol 65:85–105

White WB (2007) A brief history of karst hydrogeology: contributions of the NSS. J Cave Karst Stud 69(1):13–26

White WB (2007b) Evolution and age relations of karst landscapes. Time in Karst. Postojna, pp 45–52. Available at http://www.carsologica.zrc-sazu.si/downloads/361/white5.pdf

White WB (2013) Carbon fluxes in karst aquifers: sources, sinks, and the effect of storm flow. Acta Carsologica 42(2–3):177–186

White S, White N (2009) Bullita Cave, Gregory Karst, N.T. Australia: a maze cave in Proterozoic dolomite, pp 1010–1013. In: White WB (ed) Proceedings of the 15th international congress of speleology, vol 2. Symposia. Part 2, 19–26 July, Kerrville, Texas, 680 pp

WHO (2009) Handbook on indoor radon: a public health perspective, 110 pp. Available at http://www.who.int/ionizing_radiation/env/radon/en/index1.html

WHO (2011) Guidelines for drinking-water quality, 4th edn. 564 pp. Available at http://www.who.int/water_sanitation_health/publications/2011/dwq_guidelines/en/index.html

Wiegmans F (2009) Consultant. Golder Associates Africa. Personal communication, 31 July 2009

Wilkinson MJ (1973) Sterkfontein Cave system: evolution of a karst form. MA thesis (unpubl.). University of the Witwatersrand, Johannesburg, 96 pp

Williams PW (1983) The role of the subcutaneous zone in karst hydrology. J Hydrol 61:45–67

Williams PW (2008) World heritage caves and karst: a thematic study. No 2 IUCN/WCPA, Switzerland, 57 pp

Winde F, Wade P, van der Walt IJ (2004) Gold tailings as a source of water-borne uranium contamination of streams—the Koekemoerspruit (South Africa) as a case study Part III of III: fluctuations of stream chemistry and their impacts on uranium mobility. Water SA 30(2):233–239

Winfrey BK, Strosnider WH, Nairn RW, Strevett KA (2010) Highly effective reduction of fecal indicator bacteria counts in an ecologically engineered municipal wastewater and acid mine drainage passive co-treatment system. Ecol Eng. https://doi.org/10.1016/j.ecoleng.2010.06.025 (article in press)

Wirries DL, McDonnell AJ (1983) Drainage quality at deep coal sites. Water Resour Bull 19:235–240

Witthüser KT, Leyland RC, Holland M (2010) Vulnerability of dolomite aquifers in South Africa, pp 119–131. In: Xu Y, Braune E (eds) Sustainable groundwater resources in Africa: Water supply and sanitation environment. UNESCO Paris and CRC Press/Balkema. Taylor & Francis Group, 282 pp

Witthüser K (2011) Consultant. Metago Water Geosciences (Pty) Ltd. Personal communication (in the capacity of COH WHS Project BIQ005/2008 Steering Committee Member), 27 Jan 2011

Wolkersdorfer C (2008) Water management at abandoned flooded underground mines. Fundamentals, tracer tests, modelling, water treatment. International Mine Water Association. Springer, Berlin, 465 pp

Worthington SRH (2001) Depth of conduit flow in unconfined carbonate aquifers. Geology 29(4):335–338

Worthington SRH (2005) Hydraulic and geological factors influencing conduit flow depth. Speleogenesis and Evolution of Karst Aquifers. Issue 7, 19 pp. Sourced on 19/12/2012 at http://www.speleogenesis.info/directory/karstbase/pdf/seka_pdf4514.pdf

Worthington SRH, Ford DC (2009) Self-organized permeability in carbonate aquifers. Ground Water 47(3):326–336

Worthington SRH, Gunn J (2009) Hydrogeology of carbonate aquifers: a short history. Ground Water 47(3):462–467

WRC/IUCN/SAKWG (2010) The karst system of the Cradle of Humankind World Heritage Site. Report KV 241/10. Water Research Commission, Pretoria, 402 pp

Wright I (2012) Managing Director. Ozone Services Industry (Pty) Ltd. Personal communication, 09 Oct 2012

WS&E (2012) Responsible use, pp 17–21. Water Sewage & Effluent 32(3):48

Younger PL (1997) The longevity of minewater pollution: a basis for decision-making. Sci Total Environ 194(195):457–466

Zartman RE, Frimmel HE (1999) Rn-generated $^{206}$Pb in hydrothermal sulphide minerals and bitumen from the Ventersdorp Contact Reef, South Africa. Mineral Petrol 66:171–191

Zhang M, Zhang C, Wang K, Yue Y, Qi X, Fan F (2011) Spatiotemporal variation of karst ecosystem service values and its correlation with environmnental factors in Northwest Guangxi.

Environmental Management, China. https://doi.org/10.1007/s00267-011-9735-z

Ziemkiewicz PF, Skousen JG, Brant DL, Sterner PL, Lovett RJ (1997) Acid mine drainage treatment with armored limestone in open limestone channels. J Environ Qual 26(4):1017–1024. https://doi.org/10.2134/jeq1997.00472425002600040013x

Zohary T, Jarvis AC, Chutter FM, Ashton PJ, Robarts RD (1988) The Hartbeespoort Dam ecosystem programme 1980–1988. CSIR, 12 pp

Zwahlen F (ed) (2003) Vulnerability and risk mapping for the protection of carbonate (karst) aquifers. Final report. COST Action 620. European Commission, Brussels, 297 pp

# Bibliography

Andre BJ, Rajaram H (2005) Dissolution of limestone fractures by cooling waters: early development of hypogene karst systems. Water Resour Res. https://doi.org/10.1029/2004WR003331

Bögli A (1964) Mischungskorrosion – ein Beitrag zur Verkarstungsproblem. Erdkunde 18:83–92. https://doi.org/10.3112/erdkunde.1964.02.02

Botrell SH, Gunn J, Lowe DJ (2000) Calcite dissolution by sulfuric acid. In: Klimchouk A, Ford DC, Palmer AN, Dreybrodt W (eds) Speleogensis: evolution of karst aquifers. National Speleological Society, Huntsville, pp 304–308

Dubljansky JV (1990) Regularities of the formation and modelling of hydrothermal karst. Nauka. Novosibirsk, 151 pp (in Russian)

Dubljansky JuV (2000) Dissolution of carbonates by geothermal waters. In: Klimchouk A, Ford DC, Palmer AN, Dreybrodt W (eds) Speleogensis: evolution of karst aquifers. National Speleological Society, Huntsville, pp 158–159

Dublyansky VN, Lomaev AA (1980) Karst caves of Ukraine. Naukova dumka, Kiev, 180 pp (in Russian)

Ford DC, Williams PW (1989) Karst geomorphology and hydrology. Unwin Human, London, p 601

Hill CA (1996) Geology of the Delaware basin, Guadalupe, Apache and Glass Mountains, New Mexico and West Texas. Permian Basin Section-SEPM Publication 96–39. 480 pp

Hill CA (2000) Sulfuric acid hypogene karst in the Guadalupe Mountainsof New Mexico and West Texas. In: Klimchouk A, Ford DC, Palmer AN, Dreybrodt W (eds) Speleogensis: evolution of karst aquifers. National Speleological Society, Huntsville, pp 309–316

Hill CA (2003a) Caves and karst in New Mexico. New Mexico Earth Matters 3(1). New Mexico Bureau of Geology and Mineral Resources, 1–4

Kaveev MC (1963) About enfluence of carbon dioxide, originated during destruction of oil deposits, on development of karst processes. Doklady. AN SSSR 152(3). (in Russian)

Laptev FF (1939) Aggressive action of water on carbonate rocks, gypsum and concrete. Moscow-Leningrad, 120 pp (in Russian)

Lismonde B (2000) Corrosion des cupoles de plafond par les fluctuations de pression de l'air emprisoné. Karstologica 35:39–46

Lowe DJ (1992) The origin of limestone caverns: an inception horizon hypothesis. PhD thesis (unpublished). Manchester Metropolitan Municipality

Malinin SD (1979) Physical chemistry of hydrothermal systems with carbon dioxide. Nauka, Moscow, 111 pp (in Russian)

Palmer AN (1991) Origin and morphology of limestone caves. Geol Soc Am Bull 103:1–21

Slabe T (1995) Cave rocky relief and its speleological significance. Ljubljana, ZRC ZANU, 128 pp

W&SA (2012) Cradle of Humankind water monitoring results. Water Sanitation Africa 7(3):41–47



P. Hobbs et al. (eds.), *Anatomy of a South African Karst Hydrosystem*, Cave and Karst Systems of the World,
https://doi.org/10.1007/978-3-030-95829-9

# Index





P. Hobbs et al. (eds.), *Anatomy of a South African Karst Hydrosystem*, Cave and Karst Systems of the World,
https://doi.org/10.1007/978-3-030-95829-9